Handbook of SDP for Multimedia Session Negotiations

Handbook of SDP for Multimedia Session Negotiations

SIP and WebRTC IP Telephony

Radhika Ranjan Roy

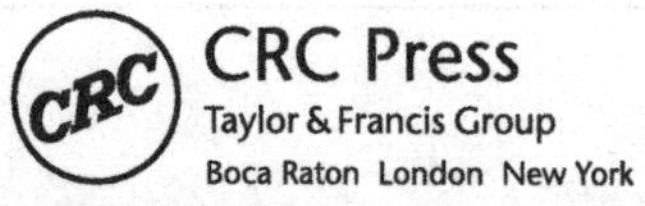

CRC Press is an imprint of the
Taylor & Francis Group, an informa business

CRC Press
Taylor & Francis Group
6000 Broken Sound Parkway NW, Suite 300
Boca Raton, FL 33487-2742

Printed on acid-free paper

International Standard Book Number-13: 978-1-138-48449-8 (Hardback)

Library of Congress Cataloging-in-Publication Data

Names: Roy, Radhika Ranjan, author.
Title: Handbook of SDP for multimedia session negotiations : SIP and WebRTC IP telephony / Radhika Ranjan Roy.
Description: First edition. | Boca Raton, FL : CRC Press/Taylor & Francis Group, 2018. | "A CRC title, part of the Taylor & Francis imprint, a member of the Taylor & Francis Group, the academic division of T&F Informa plc." | Includes bibliographical references and index.
Identifiers: LCCN 2017057107 | ISBN 9781138484498 (hardback : acid-free paper)
Subjects: LCSH: Session Description Protocol (Computer network protocol) | Multimedia communications. | Internet telephony.
Classification: LCC TK5105.15 .R69 2018 | DDC 621.382/12--dc23
LC record available at https://lccn.loc.gov/2017057107

Visit the Taylor & Francis Web site at
http://www.taylorandfrancis.com

and the CRC Press Web site at
http://www.crcpress.com

To our dearest son, Debasri Roy, *Medicinae Doctoris* (M.D.) (January 20, 1988—October 31, 2014) who had brilliant carrier (Summa Cum Laude—undergraduate), had so much more to contribute to this country and the world as a whole in the future, had been so eager for seeing these books get published saying, "Daddy, you are my hero." May God let his soul live in peace in His abode.

To my GrandMa for her causeless love, my parents Rakesh Chandra Roy and Sneholota Roy whose spiritual inspiration remains vividly alive within all of us, my late sisters GitaSree Roy, Anjali Roy, and Aparna Roy and their spouses and my brother Raghunath Roy and his wife Nupur for their inspiration, our daughter Elora and our son-in-law Nick, our son Ajanta, and finally my beloved wife Jharna for their love.

Our special thanks to our son-in-law Nick and our daughter Elora for their hard works and remain engaged with their creative activities how to become financially independent standing on their own feet.

It is also inspiring to see how our son Ajanta who himself is an energetic electrical engineer progressing toward a bright future on his own right keeping his colleagues and all of us amazed.

Contents

List of Figures

List of Tables

Preface

I have been working on networked multimedia communications all along including at my present position at the United States Army Research, Development and Engineering Command (RDECOM), *Communications-Electronics Research, Development, and Engineering Center (CERDEC); Space and Terrestrial Communications Directorate (S&TCD)* Laboratories for large-scale global Session Initiation Protocol (SIP)-based Voice-over-Internet Protocol (VoIP)/ Multimedia networks since 1993 while I had been at AT&T Bell Laboratories. I had been the editor of Multimedia Communications Forum (MMCF) when it was created by many participating companies throughout the world including AT&T for promoting the technical standards for networked multimedia communications to fill an important gap of that time when no standard bodies came forward to do so. Later on, I had the opportunity to participate in International Telecommunication Union—Telecommunication (ITU-T) on behalf of AT&T for standardization of H.323. H.323 has been the first successful technical standard for VoIP/multimedia telephony. However, SIP that has been standardized in the Internet Engineering Task Force (IETF) much later than H.323 and has emulated the simplicity of the protocol architecture of Hypertext Transfer Protocol (HTTP) has started to be popularized for VoIP over the Internet because of its friendliness to be meshed with web services.

After so many years of working experiences on SIP for building the large-scale VoIP networks, I have found that it is an urgent requirement to have a complete book that integrates all SIP-related Request For Comments (RFCs) in a systematic way that network designers, software developers, product manufacturers, implementers, interoperability testers, professionals, professors, and researchers can use as if as a "super-SIP RFC" since the publication of SIP RFC 3261 in the year 2002. To fill this gap, I published the book: *Handbook on SIP: Networked Communications IP Telephony*, CRC Press, March 2016.

One of the most fundamental things of the call control protocols such as SIP is that it requires the help of another protocol known as Session Description Protocol (SDP) for capability negotiations between the call participants. In my earlier book on SIP, SDP has been briefly described. It has almost been an impossible task to provide the full treatment of SDP in that book.

In the meantime, there has been a profound change in the industry to make the web browser itself for initiation, establishment, and maintenance of the real-time multimedia call directly. As a culmination of this, Web Real-Time Control (WebRTC) has emerged as a standardized another call control protocol developed in the IETF in collaboration with World Wide Consortium (W2C). IETF is in charge of the development of WebRTC protocol, whereas W2C is developing WebRTC Application Programming Interfaces (APIs).

The most interesting development is that even WebRTC call control protocol, like SIP, also needs to use another protocol for capability negotiations between the call participants. WebRTC working group (WG) has decided to use SDP as the protocol for capability negotiations. When

WebRTC WG investigated the capabilities of existing SDP standards, they found that WebRTC needs extensions of SDP to accommodate the full spectrum of new capabilities.

Historically, SDP was first developed to be used for Session Announcement Protocol (SAP) (RFC 2974) for distribution of the session information related to audio/video codecs and others to the users who will participate in the session. SDP's main task had been only one-way distribution of the session capability (e.g. audio/video codec type, data applications, and/or others) information to the participants while SIP was developed to two-way conferencing where session capability negotiations among the conference participants on real-time, initial SDP capabilities developed for SAP were not at all suitable. To introduce capability negotiation capabilities, at times it proved so difficult that people became frustrated whether this SDP might have to be discarded.

Finally, the offer/answer model introduced by RFC 3264 (see Section 3.1) has been the watershed that rescues SDP to make a viable candidate for capability negotiations dynamically in real time among the participants. Since then researchers and developers have been extending SDP introducing many more capabilities as time goes by. Recently, WebRTC WG has introduced many other capabilities in SDP.

Although SDP serving quite well for point-to-point multimedia conferencing, its protocol architecture is such that it cannot be extended for distributed multipoint multimedia conference as of today. The only multipoint conferencing that is offered today is the centralized star-like topology with priori known address of the centralized controller. In centralized conference architecture is again capability negotiations are done using SDP in point-to-point fashion only. When we work with SDP, we have to know its limitations how and where we can use SDP for setting up the multimedia conferencing dynamically.

It has created a new urgency to publish a book on SDP providing full treatments due its own merits. Like all other standards, we have to know exactly how many RFCs have been published even related to the base SDP over two decades, and how they are interrelated to each other after so many extensions and enhancements with new features and capabilities, corrections and modifications with latest agreed-upon interpretations based on implementation and interoperability test experiences, and future researches for breaking new ground knowing what already exists in SDP.

Like SIP, this book on SDP is the first of this kind that attempts to put all SIP-related RFCs together with their mandatory and optional texts in a chronological systematic way as if people can use a single "super-SDP RFC" with almost one-to-one integrity from beginning to end to see the big picture of SDP in addition to base SDP functionalities.

It should be noted that the texts of each RFC in the IETF are reviewed by all members of a given WG throughout the world, and rough consensus is made which parts of texts of the draft need to be "mandatory" and "optional" including whether a RFC needs to be "standards track," "informational," or "experimental." The key point is when one tries to put all SIP-related RFCs together making a textbook has serious challenges how to put all texts together because it is not simply to put one RFC after another one chronologically. It takes texts of each RFC needs to be put together for each of these particular functionalities, capabilities, and features keeping integrity. Since this book is planned as if it is like a single-SIP RFC, I have a very limited freedom to change any texts of RFCs other than some editorial texts to make it look like a book. I have used texts, figures, tables, and references from RFCs as much is necessary so that all can use all those as they are found in RFCs. All RFCs along with their authors are provided in references, and all credits of this book go primarily to those authors of these RFCs and many IETF WG members who shaped final RFCs with their invaluable comments and inputs. In this connection, I also extend my sincere thanks to Ms. Alexa, IETF Secretariat, for his kind consent for reproducing texts, figures, and tables with IETF copyright notification. My only credit, as I mentioned earlier, is to put all those RFCs together in a way that looks like one complete SDP RFC.

I have organized this book into eighteen chapters based on their major functionalities, features, and capabilities as follows:

1. Networked Multimedia Protocols (Sections 1–1.6)
   a. Session Initiation Protocol (Sections 1.2–1.2.6)
   b. WebRTC Protocol (Sections 1.3–1.3.12)
2. Basic Session Description Protocol (Sections 2–2.13)
3. Negotiations Model in SDP (Sections 3–3.5)
4. Capability Declaration in SDP (Sections 4–4.3)
5. Media-Level Attribute Support in SDP (Sections 5–5.5)
6. Semantics for Signaling Media Decoding Dependency in SDP (Sections 6–6.4)
7. Media Grouping Support in SDP (Sections 7–7.6)
8. Generic Capability Attribute Negotiations in SDP (Sections 8–8.7)
9. Network Protocol Support in SDP (Sections 9–9.8)
10. Transport Protocol Support in SDP (Sections 10–10.3)
11. RTP Media Loopback Support in SDP (Sections 11–11.3)
12. Quality-of-Service Support in SDP (Sections 12–12.7)
13. Application-Specific Extensions in SDP (Sections 13–13.4)
14. Service-Specific Extension in SDP (Sections 14–14.6)
15. Compression Support in SDP (Sections 15–15.3)
16. Security Capability Negotiation in SDP (Sections 16–16.8)
17. Security Considerations in Using SDP Attributes (Sections 17–17.16)
18. Capability Negotiations Using SDP by SIP and WebRTC (Sections 18–18.5)

Out of this, Chapter 1 deals with SIP and WebRTC call control protocol, whereas Chapter 18 provides how SDP is being used for capability negotiations using SIP and WebRTC. The reason that I have included these two call control protocols is that SDP itself does not say anything how a point-to-point or multipoint multimedia conferencing needs to be configured. It is the use of SDP by call control protocols such as SIP and WebRTC how and where SDP can or cannot be used. In this context, I strongly believe that the readers will appreciate the inclusion of SIP and WebRTC in Chapters 1 and 18. However, this is not the only way that all of SDP RFCs to be grouped or chapters/sections need to be put chronologically. My best intellectual instinct has guided me to arrange these RFCs for the basic SIP functionalities like this although much more complex intelligent capabilities of SDP are yet to be invented in the future. I am looking forward to the readers whether they will validate my judgment. In addition, I am providing a general statement for the IETF copyright information as follows:

> IETF RFCs have texts that are "mandatory" and "optional" including the use of words like "SHALL," "MUST," "MAY," "SHOULD," and "RECOMMENDED." These texts are very critical for providing interoperability for implementation using products from different vendors as well as for inter-carrier communications. The main objective of this book is, as explained earlier, to create as if it is a single integrated SDP RFC, the texts have been reproduced from the IETF RFCs for providing interoperability with permission from IETF in Chapters 1 through 18. The copyright © for the texts

that are being reproduced (with permission) in different chapters and sections of this Part belongs to the IETF. It is recommended that readers should consult original RFCs posted in the IETF website.

I am greatly indebted to many researchers, professionals, software and product developers, network designers, professors, intellectuals, and individual authors and contributors of technical standard documents, drafts, and RFCs throughout the world for learning from their high-quality technical papers and discussions in the group meetings, conferences, and emails in WGs for more than two decades. In addition, I had the privilege to meet many those great souls in person during MMCF, ITU-T, IETF, and other technical standards conferences held in different countries of the world. Their unforgettable personal touch has enriched my heart very deeply as well.

I admire Richard O'Hanley, publisher, ICT Business, and Security, CRC Press, for his appreciative approach in publishing this book. I am thankful to Rebecca Dunn, Project Manager, CodeMantra, Alexandra Andrejevich, Deputy Project Manager, CodeMatra, and reviewers for their sincere proofing of this book and helping in a variety of ways.

Author

Radhika Ranjan Roy is an electronics engineer in United States Army Research, Development and Engineering Command (RDECOM), Communications-Electronics Research, Development, and Engineering Center (CERDEC), Space and Terrestrial Communications Directorate (S&TCD) Laboratories, Aberdeen Proving Ground (APG), Maryland, since 2009. Dr. Roy is leading his research and development efforts in the development of scalable large-scale SIP-based VoIP/Multimedia networks and services, mobile ad hoc networks (MANETs), Peer-to-Peer (P2P) Networks, Cyber Security detecting application software and network vulnerability, Jamming Detection, and supporting array of Army/Department of Defense's (DoD) Nationwide and Worldwide Warfighter Networking Architectures and participating in technical standards development in Multimedia/Real-Time Services Collaboration, IPv6, Radio Communications, Enterprise Services Management, and Information Transfer of DoD Technical Working Groups. He received his PhD in electrical engineering with major in computer communications from the City University of New York, New York, in 1984 and MS in electrical engineering from Northeastern University, Boston, Massachusetts, in 1978. He received his BS in electrical engineering from the Bangladesh University of Engineering and Technology, Dhaka, Bangladesh, in 1967. He was born in a countryside renowned town Derai, Bangladesh.

Prior to joining CERDEC, Dr. Roy worked as the lead system engineer at CACI, Eatontown, New Jersey, from 2007 to 2009 and developed Army Technical Resource Model, Army Enterprise Architecture, DoD Architecture Framework, and Army LandWarNet Capability Sets, and technical standards for Joint Tactical Radio System, Mobile IPv6, MANET, and SIP supporting Army Chief Information Officer (CIO)/G-6. Dr. Roy worked as senior system engineer, SAIC, Abingdon, Maryland, from 2004 to 2007 supporting modeling, simulations, architectures, and system engineering of many Army projects: Warfighter Information Network-Tactical, Future Combat System, and Joint Network Node.

During his career, Dr. Roy worked in AT&T/Bell Laboratories, Middletown, New Jersey, as senior consultant from 1990 to 2004 and led a team of engineers in the designing of AT&T's Worldwide SIP-based VoIP/Multimedia Communications Network Architecture consisting of wired and wireless global carrier network from preparation of Request for Information (RFI), evaluation of vendor RFI responses and interactions with all selected major vendors related to their products. He participated and contributed in the development of VoIP/H.323/SIP multimedia standards in ITU-T, IETF, ATM, and Frame Relay standard organizations.

Dr. Roy worked as senior principal engineer in Computer Sciences Corporation (CSC), Falls Church, Virginia, from 1984 to 1990 and worked in design and performance analysis of the US Treasury nationwide X.25 packet switching network. In addition, he designed many network architectures of many proposed US Government and Commercial Worldwide and Nationwide

Networks: Department of State Telecommunications Network, US Secret Service Satellite Network, Veteran Communications Network, and Ford Company's Dealership Network. Prior to CSC, he worked from 1967 to 1977 as deputy director in the department of design at Power Development Board (PDB)-Carrier Communications, Dhaka, Bangladesh.

Dr. Roy's research interest includes the areas of mobile ad hoc networks, multimedia communications, peer-to-peer networking, and quality of service. He has published more than 50 technical papers and are holding or pending more than 30 patents. He also participates in many IETF WGs. Dr. Roy authored two books: *Handbook on SIP: Networked Communications IP Telephony*, CRC Press, March 2016, and *Handbook of Mobile Ad Hoc Networks for Mobility Models*, Springer, 2010. He lives in a historical district of Howell Township, New Jersey, with his wife Jharna.

Chapter 1

Networked Multimedia Protocols

Abstract

Session Description Protocol (SDP) is used as the mechanism for negotiations of functional, performance, and security parameters of audio, video, and/or data used in multimedia conferencing created by any call control protocol for setting up a session between end users. In order to understand SDP, we need to have a preliminary idea of how it is used in setting up sessions by the call control protocols. Session Initiation Protocol (SIP) is a well-established standardized call control protocol for multimedia conferencing. Recently, the emerging web real-time communication (WebRTC) protocol has also chosen SDP for multimedia session negotiations. The SIP that carries SDP in the message-body of signaling messages for session negotiations between communicating parties is described briefly in this chapter. In addition, the emerging WebRTC that also carries SDP for multimedia session negotiations between browsers is introduced. Both call control protocols have embedded networking capabilities and set up multimedia one-to-one, many-to-many, many-to-one, and one-to-many sessions in communicating environments. Basic information for these call control protocols is provided to show how SDP is used for complicated multimedia session negotiations.

1.1 Introduction

All call control protocols use Session Description Protocol (SDP) that carries the detailed functional, security, and performance description of the session consisting of audio, video, and/or data for negotiations between end parties. The multimedia session that consists of audio, video, and/or data for communications between two or more parties/endpoints needs to be agreed upon by all parties for successful session setup. The capabilities of each kind of media consist of many functional and performance features that need to be negotiated by the communicating parties/endpoints. Each entity should have the same capabilities to ensure successful communication. Multimedia call control signaling protocols, such as Session Initiation Protocol (SIP), have emerged as an important protocol for initiation, modification, and termination of the session over the Internet. SIP, which has embedded networking functions, enables one-to-one, many-to-many, many-to-one, and one-to-many communication over networks.

However, SIP's functional and performance parameters of the multimedia session need to be shared for negotiations between the parties. The SDP representing the functional and performance parameters of the session of each party is carried over by SIP to ensure the negotiation is agreed upon by each party. Recently, another signaling protocol known as web real-time communication (WebRTC), specifically for the initiation of multimedia calls directly by the browser, has emerged. The WebRTC, like SIP, also uses SDP for negotiations of multimedia session parameters between the browsers.

1.2 Session Initiation Protocol

SIP is the call signaling control protocol used to set up, modify, and tear down networked multimedia sessions consisting of audio, video, and/or data applications in accordance with Request for Comments (RFC) 3261. It uses the SDP in its message-body to signal messages for negotiations of session parameters among the communicating parties to support the capabilities needed for communication. In addition, SIP signals the use of the transport parameters of the Real-Time Transport Protocol (RTP)/Real-Time Transport Control Protocol (RTCP) used by audio, video, and/or data. SDP and RTP/RTCP are two protocols needed as a minimum by SIP for multimedia sessions.

However, SIP uses human-understandable text encoding of signaling messages and Hypertext Transfer Protocol (HTTP)/Hypertext Markup Language (HTML)-based web services like protocol architecture to separate the signaling messages into two parts: header and body. This inherent in-built capability of SIP has been used to create an enormous amount of new application services, not only for time- and mission-critical conversational audio and video services but also for the integration of nontime-critical web services defined within the framework of service-oriented architecture primarily as part of application sharing under the same audio conferencing and video conferencing (VC) session.

1.2.1 Session Initiation Protocol

SIP, being an application layer signaling protocol, facilitates the creation of networked multimedia application services such as teleconferencing (TC), video teleconferencing (VTC)/VC, and multimedia application sharing. Multimedia applications may need to support different audio (e.g., International Telegraph Union – Telecommunication (ITU-T) G-series) and video (e.g., Moving Picture Expert Group, Joint Photographic Expert Group, and ITU-T H-series) codecs in multimedia sessions by conference participants. The bit streams of each codec type are transferred using RTP over the User Datagram Protocol (UDP), Stream Control Transmission Protocol (SCTP) over the Internet Protocol (IP) network. However, RTCP is used for the periodic transmission of control packets to all participants in the session, and it provides feedback on the quality of the data (e.g., RTP packets of audio/video) distribution. This is an integral part of the RTP's role as a transport protocol and is related to the flow and congestion control functions of other transport protocols. The data (e.g., text, graphics, and still pictures) application may be transferred using Transmission Control Protocol (TCP). Figure 1.1 shows the relationship between SIP and other protocols.

It is important to note that a huge number of different capabilities are supported by a given multimedia application and that it is possible that not all participants and communicating participants will support or prefer the same capabilities. These functional and performance capabilities need to be negotiated between two communicating entities to ensure they have a common set.

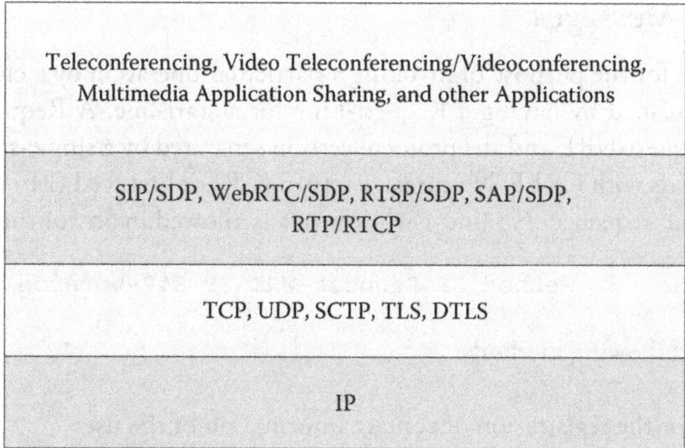

Figure 1.1 SIP and other protocols.

Consequently, a mechanism for session capability negotiations is needed; SDP is used within SIP signaling messages to negotiate capabilities between the communicating parties.

An SIP message consists of two parts: header and message-body. The header is primarily used to route the signaling messages from the caller to the called party and contains a Request-Line composed of the request type, the SIP Uniform Resource Identifier (URI) of the destination or next hop, and the version of SIP being used. The message-body is optional depending on the type of message and where it falls within the establishment process. A blank line separates the header and the message-body part. If SIP invitations used to create sessions carry session descriptions that allow participants to agree on a set of compatible media types and compatible codecs, the message-body part will include this information as described in the SDP.

The domain name system (DNS) and Dynamic Host Configuration Protocol (DHCP) are the integral tools for IP address resolution and allocation respectively for routing of SIP messages between conference participants over the IP network. For example, a host can discover and contact a DHCP server to provide an IP address as well as the addresses of the DNS server and default router that can be used to route SIP messages over the IP network.

1.2.2 SIP Messages

SIP is designed in RFC 3261 that makes RFC 2543 obsolete as a client–server protocol like the HTTP/1.1 syntax, although neither SIP nor its extensions are HTTP. Accordingly, an SIP message is either a request from a client to a server to invoke a particular operation or a response from a server to a client to indicate the status of the request. Both types of messages consist of a start-line, one or more header fields, an empty line indicating the end of the header fields, and an optional message-body.

```
generic-message   =   start-line
                      *message-header
                      CRLF
                      [ message-body ]
start-line        =   Request-Line / Status-Line
```

The request message is also known as method. The start-line, each message-header line, and the empty line must be terminated by a carriage-return line-feed (CRLF) sequence. However, the empty line must be present even if the message-body is not.

1.2.2.1 Request Messages

SIP requests are sent for the purpose of invoking a particular operation by a client to a server. SIP requests are distinguished by having a Request-Line for a start-line. A Request-Line contains a method name, a Request-URI, and the protocol version separated by a single-space (SP) character. The Request-Line ends with CRLF. No carriage return (CR) or line-feed (LF) is allowed except in the end-of-line CRLF sequence. No linear-white-space is allowed in any of the elements.

```
Request-Line  =  Method SP Request-URI SP SIP-Version CRLF
```

SIP has defined the following methods:

- REGISTER for the registration of contact information of the user
- INVITE, ACK, and CANCEL for setting up sessions
- BYE for terminating sessions
- OPTIONS for querying servers about their capabilities
- MESSAGE for chat sessions
- REFER for call transfer
- SUBSCRIBE and NOTIFY for SIP session-related event management
- PUBLISH for the publication of SIP-specific event state
- UPDATE for updating the session parameters
- INFO for mid-session information transfer
- PRACK for the acknowledgment of provisional requests

The most important method in SIP is INVITE, which is used to establish a session between participants. A session is a collection of participants and streams of media between them for the purposes of communication. SIP extensions, documented in standards track RFCs, may define additional methods for accommodating more feature-rich multimedia sessions, especially for multipoint multimedia conferencing services. However, SIP request messages known as methods specify the purpose of SIP messages for taking actions by the SIP user agent (UA) or server. The SIP method (or request message) names are case sensitive and use all uppercase letters to distinguish them from the header fields, which can be a mixture of upper- and lowercase letters. The SIP UAs are required to understand the SIP methods while proxy servers are required to know the relevant header fields for routing of SIP request messages keeping the intermediaries of the SIP network simple thereby, making the SIP-based multimedia communications network more scalable.

1.2.2.2 Response Messages

SIP responses sent from the server to the client indicate the status of the request and are distinguished from requests by having a Status-Line as their start-line. A Status-Line consists of the protocol version followed by a numeric Status-Code and its associated textual phrase, each element separated by an SP character. No CR or LF is allowed except in the final CRLF sequence.

```
Request-Line     =   Method SP Request-URI SP SIP-Version CRLF
```

The Status-Code is a three-digit integer result code that indicates the outcome of an attempt to understand and satisfy a request. The Reason-Phrase is intended to give a short textual description of the Status-Code. The Status-Code is intended for use by automata, whereas the Reason-Phrase is intended

Table 1.1 SIP Response Classes, Descriptions, and Actions

Response Class	*Response Description*	*Action Taken or To Be Taken*
1xx	Provisional	Request received, continuing to process the request
2xx	Success	The action was successfully received, understood, and accepted
3xx	Redirection	Further action needs to be taken in order to complete the request
4xx	Client Error	The request contains bad syntax or cannot be fulfilled at this server
5xx	Server Error	The server failed to fulfill an apparently valid request
6xx	Global Failure	The request cannot be fulfilled at any server

for the human user. SIP response classes are similar to those of HTTP but have been defined in the context of SIP. The first digit of the Status-Code defines the class of response. The last two digits do not have any categorization role. For this reason, any response with a Status-Code between 100 and 199 is referred to as a "1xx response," any response with a Status-Code between 200 and 299 as a "2xx response," and so on. SIP/2.0 allows six values for the first digit, as shown in Table 1.1 (RFC 3261).

1.2.2.2.1 Headers

SIP header fields are similar to HTTP header fields in both syntax and semantics, each carrying its own well-defined information. Each header field is terminated by a CRLF at the end of the header. SIP specifies that multiple header fields of the same field name whose value is a comma-separated list can be combined into one header field whose grammar is of the form:

```
header = "header-name" HCOLON header-value *(COMMA header-value)
```

It allows for combining header fields of the same name into a comma-separated list. The Contact header field allows a comma-separated list unless the header field value is "*."

1.2.2.3 Message-Body

The message-body in SIP messages is an optional component. SIP request messages may contain message-bodies, unless otherwise noted, that can be read, created, processed, modified, or removed as necessary only by the SIP UA. The SIP message-body shall always be opaque to the SIP proxy/redirect/registrar server. Requests, including new requests defined in extensions to this specification, may contain message-bodies unless otherwise noted. The interpretation of the body depends on the request method. For response messages, the request method and the response Status-Code determine the type and interpretation of any message-body. Regardless of the type of body a request contains, certain header fields must be formulated to characterize the contents of the body, such as Allow, Allow-Events, Content-Disposition, Content-Encoding, Content-Language, Content-Length, and Content-Type. All responses may include bodies.

1.2.2.4 Framing SIP Messages

Unlike HTTP, SIP implementations can use the UDP protocol as well. Each such datagram carries one request or response. Implementations processing SIP messages over stream-oriented transports MUST ignore any CRLF appearing before the start-line. The Content-Length header field value is used to locate the end of each SIP message in a stream. It will always be present when SIP messages are sent over stream-oriented transports.

1.2.2.5 SIP Tags

The "tag" parameter is used in the To and From header fields of SIP messages. It serves as a general mechanism to identify a dialog, which is the combination of the Call-ID along with two tags, one from each participant in the dialog. When a UA sends a request outside of a dialog, it contains a From tag only, providing "half" of the dialog identification (ID). The dialog is completed from the response(s), each of which contributes the second half in the To header field.

1.2.2.6 SIP Option Tags

Option tags are unique identifiers used to designate new options (extensions) in SIP. These tags are used in Require, Proxy-Require, Supported, and Unsupported header fields defined in RFC 3261.

1.2.3 SIP Message Elements

The core SIP protocol (RFC 3261) and the SIP network infrastructure that provide end-to-end communications between the SIP functional entities are described. The SIP request messages that have some uniqueness in creating, modifying, and tearing down the sessions are described in detail. The request and response messaging processing, forwarding, and handling of transport errors are explained in great length. The SIP transaction handling by clients and server is articulated. Sending requests and receiving responses of messages by SIP clients are specified. Specifically, SIP message elements contain framing, error handling, common message components, method names, header fields, request and status-line structures and manipulation, message request and response structure and operations, multipart message bodies, parsing of message and message elements, URI types and manipulations, and other related functional features of the core SIP protocol.

1.2.4 SIP Message Structure

SIP is described with some independent processing stages with only a loose coupling between two stages. This protocol is structured in a way to be compliant with a set of rules for operations in different stages of the protocol that provides an appearance of a layered protocol. However, it does not dictate an implementation in any way. Not every element specified by the protocol contains every layer. Furthermore, the elements specified by SIP are logical not physical. A physical realization can choose to act as different logical elements, perhaps even on a transaction-by-transaction basis. An SIP protocol layer can be defined as follows: Syntax and Encoding, Transport, Transaction, and Transaction User. In addition, SIP defines dialog between two UAs. In the syntax and encoding layer, SIP's encoding is specified using the Augmented Backus-Naur Form. The transport layer defines how a client sends requests and receives responses and how a server receives requests and sends responses over the network. All SIP elements contain a transport layer. As explained, TCP, UDP, or SCTP can be used as the transport protocol in SIP, while Transport Layer Security (TLS) can be used as the security transport protocol over TCP.

Request Line	INVITE sip:bob@biloxi.example.com SIP/2.0
Headers	Via: SIP/2.0/TCP
	client.atlanta.example.com:5060;branch=z9hG4bK74bf9
	Max-Forwards: 70
	From: Alice <sip:alice@atlanta.example.com>;tag=9fxced76sl
	To: Bob <sip:bob@biloxi.example.com>
	Call-ID: 3848276298220188511@atlanta.example.com
	CSeq: 1 INVITE
	Contact: <sip:alice@client.atlanta.example.com;transport=tcp>
	Content-Type: application/sdp
	Content-Length: 151
Empty Line	
Message Body	v=0
	o=alice 2890844526 2890844526 IN IP4 client.atlanta.example.com
	s=-
	c=IN IP4 192.0.2.101
	t=0 0
	m=audio 49172 RTP/AVP 0
	a=rtpmap:0 PCMU/8000

Figure 1.2 Example of SIP request message format.

1.2.4.1 Request Message Format

The SIP request message format consists of three important parts: Request-Line, Header, and Message-Body. Figure 1.2 depicts an example of the SIP INVITE method message format. The Request-Line consists of the request type, SIP URI of the destination or next hop, and the SIP version being used.

The SIP header part contains a set of headers, and each head carries its own well-defined information. However, each header is terminated by a CRLF at the end of the header. The SIP message-body of the request method is optional depending on the type of message, as well as where it falls in the call establishment scheme. A blank line defines the boundary between the header part and the Message-Body.

1.2.4.2 Response Message Format

The SIP response message format has three major sections: Status-Line, Header, and Message-Body. Of course, an empty line will separate the header and the Message-Body. Figure 1.3 shows an example of SIP 200 OK response message format. The response message contains the reason header, explaining why this response has been sent, and can be quite large because it has to explain all of the reasons.

The status-line contains the protocol version, the Status-Code, and the Reason-Phrase. The Reason-Phrase makes it easy for human users to understand it, while the protocol version and the Status-Code are processed by the SIP network. For example, the Status-Code 200 is part of the 2xx response class (success), and specifically 200 responses are sent. At this point, the dialog transitions to a confirmed state. When a UA client (UAC) does not want to continue with this dialog, it shall terminate the dialog by sending a BYE request. Again, a header part will contain different headers; each header, like response message, will have its own specific information and

Status Line	SIP/2.0 200 OK
Headers	Via: SIP/2.0/TCP client.atlanta.example.com:5060;branch=z9hG4bK74bf9;received=192.0.2.101
	From: Alice <sip:alice@atlanta.example.com>;tag=9fxced76sl
	To: Bob <sip:bob@biloxi.example.com>;tag=8321234356
	Call-ID: 3848276298220188511@atlanta.example.com
	CSeq: 1 INVITE
	Contact: <sip:bob@client.biloxi.example.com;transport=tcp>
	Content-Type: application/sdp
	Content-Length: 147
Empty Line	
Message Body	v=0
	o=bob 2890844527 2890844527 IN IP4 client.biloxi.example.com
	s=-
	c=IN IP4 192.0.2.201
	t=0 0
	m=audio 3456 RTP/AVP 0
	a=rtpmap:0 PCMU/8000

Figure 1.3 Example of SIP response message format.

be terminated by a CRLF at its end. Like the Request message, the Message-Body is optional and is separated by a blank line from the header part.

1.2.5 SIP Network Functional Elements

The networked multimedia services use SIP to establish, manage, and tear down the multimedia sessions. As a result, the capabilities of SIP need to be used in the context of multimedia service networking context. SIP, being in the session control layer, is also part of the application layer in Open Standard International terminology. SIP has application layer functional entities such as SIP UAs and SIP servers that are described in Sections 1.2.5.1.1–1.2.5.1.4; they communicate among themselves using SIP application/session layer protocol termed the SIP network. Figure 1.4 depicts the logical view of the SIP network and its functional entities.

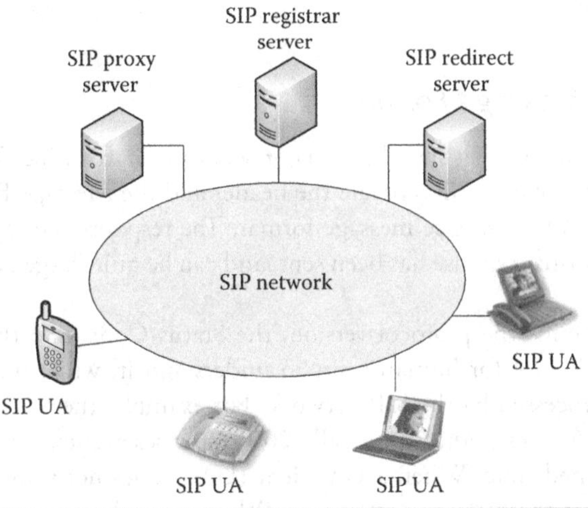

Figure 1.4 Logical view of SIP network and its functional entities.

The functional elements of the SIP network are as follows: SIP UAs and SIP servers. SIP has defined SIP Proxy, SIP Registrar, and SIP Redirect Server. However, location servers and categories of application servers, not shown in Figure 1.4, do not belong to SIP. In this context, SIP; media sessions using RTP/RTCP controlled by SIP; SIP security protocols such as TLS protocol; and SIP transport protocols such as TCP, UDP, and SCTP are carried over the IP network. IP network and SIP application servers such as the TC/VTC/VC server, location server, web conferencing server, media bridging server, chat server, and application sharing server, not shown in the figure, are addressed in Chapters 1–18.

1.2.5.1 SIP UA

SIP has defined some functional entities that can be categorized broadly into SIP UAs and SIP Servers. A UA works in a client–server mode on behalf of the user: UAS and UAC. UAC generates the SIP request and sends it to the UAS directly if the address is known or to the SIP server that routes the request to the UAS. The UAS receives the request, operates on it, and sends the response back to the UAC directly, if the address is known, or via the SIP server that originally sent the responses to the UAC. From a conferencing perspective, a number of different SIP components, such as conference-unaware participant, conference-aware participant, and focus, specified by RFC 4579. Those kinds of SIP UAs are described in Sections 1.2.5.1.1–1.2.5.1.4.

1.2.5.1.1 Focus UA

A focus, as defined in the framework RFC 4579, hosts an SIP conference and maintains an SIP signaling relationship with each participant in the conference. A focus contains a conference-aware UA that supports the conferencing call control conventions as defined in RFC 4579. A focus should support the conference package RFC 4575, behave as a notifier for that package, and indicate its support in the Allow-Events header fields in requests and responses. A focus may include information about the conference in SDP bodies sent as part of normal SIP signaling by populating the Session Information, URI, Email Address, and Phone Number in SDP fields.

1.2.5.1.2 Conference-Unaware UA

The simplest UA can participate in a conference ignoring all SIP conferencing-related information. It is able to dial in to a conference and to be invited to a conference. Any conferencing information is optionally conveyed to/from it using non-SIP means. Such a UA would not usually host a conference (at least, not using SIP explicitly). A conference-unaware UA need only support basic SIP capabilities specified in RFC 3261. Call flows for conference-unaware UAs would be identical to those in the SIP, per the specifications of RFC 3261. Note that the presence of an "isfocus" feature tag in a Contact header field will not cause interoperability issues between a focus and a Conference-Unaware UA since it will be treated as an unknown header parameter and ignored, as per standard SIP behavior.

1.2.5.1.3 Conference-Aware UA

A Conference-Aware UA supports SIP conferencing call control convention, defined in this document as a conference participant, in addition to the support of RFC 3261. It should be able to process SIP redirections such as those described in RFC 3261. It must recognize the "isfocus" feature parameter.

1.2.5.1.4 SIP Back-to-Back UA

A back-to-back UA is a concatenation of a UAC and a UAS functional entity in SIP. That is, it is logical entity that receives a request and processes it as a UAS. It also acts as a UAC and generates requests in order to determine how they should be answered. Unlike a proxy server, it maintains a dialog state and must participate in all requests sent on the dialogs it has established.

1.2.5.2 SIP Servers

An SIP server uses the SIP to manage real-time communication (RTC) among SIP clients. In fact, SIP servers are key to communications among SIP clients; routing SIP messages through the resolution of addresses is the core of the SIP network. SIP servers act on requests sent by SIP clients, process SIP messages, and operate on rules per technical standards defined in the call control protocol SIP. SIP has defined SIP proxy server, SIP registration server, and SIP redirect server.

1.2.5.2.1 Proxy Server

An SIP proxy server receives all SIP request and response messages from UAs or other SIP servers such as proxies. It may use registrars/location servers, DNS, or database servers to route SIP messages for resolving addresses to other UAs or proxies. A proxy is only allowed to forward SIP messages and generate and ACK messages, as described in Section 1.2.5. It should be noted that the proxy might access a database server that will not use SIP. In this case, it is expected that a proxy will use a host of different protocols in its backend servers for address resolution or other purposes. These protocols are outside the scope of SIP. A proxy can be stateful or stateless; a stateless proxy does not keep any state information of the call or transaction, while a stateful proxy keeps all state information of a call or a transaction for the duration of the call or transaction.

1.2.5.2.2 Redirect Server

An SIP Redirect Server receives an SIP request and, unlike a proxy server, responds to it. It usually provides a 3xx (redirection class) response, to a UA or proxy, indicating that the call should be tried at a different location. The main purpose has been to deal with the temporary or permanent location change of a user.

1.2.5.2.3 Registrar Server

An SIP registrar server keeps the contact and other information of UAs sent using a REGISTER message. Registration creates bindings in a location service for a particular domain that associates an address-of-record (AOR) URI with one or more contact addresses. A proxy for that domain receives a request whose Request-URI matches the AOR, and then the proxy forwards the request to the contact addresses registered to that AOR. It is usual to register an AOR at a domain's location service when requests for that AOR would be routed to that domain. In most cases, this means that the domain of the registration will need to match the domain in the URI of the AOR. A registrar may store all of the information including the contact sent via a REGISTER message in a location server. The protocol between the SIP registrar and the location server is outside the scope of SIP.

1.2.5.2.4 Application Server

An SIP application server acting as the SIP UA can send SIP Request messages and receive SIP Response messages. In fact, SIP application servers have emerged as the most important areas for creation and offering of multimedia services using SIP. With the inception of SIP, new feature-rich multimedia services integrated with other services like web services have been enabled; they have opened a new frontier for creating still more real-time multimedia services.

1.2.6 SIP Call and Session Negotiation Using SDP

SIP operation at a high level will now be introduced, although the actual protocol details are provided primarily in Section 18.1 although all other sections refer SIP briefly in the context of session negotiations using SDP. Topics provided will include a review of SIP protocol in a nutshell, how registration is performed, and how a session is created, updated, and terminated. Figure 1.5 illustrates an SIP network with two SIP UAs known as Alice and Bob and two SIP proxies designated as outgoing and incoming.

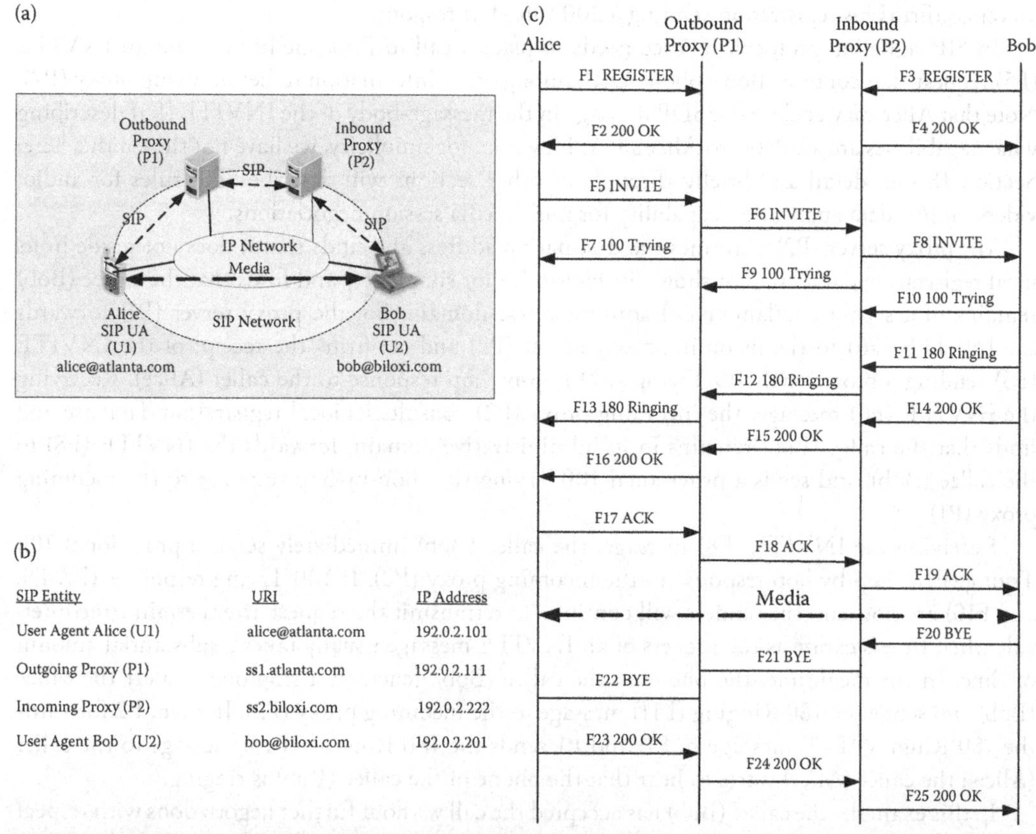

SIP Entity	URI	IP Address
User Agent Alice (U1)	alice@atlanta.com	192.0.2.101
Outgoing Proxy (P1)	ss1.atlanta.com	192.0.2.111
Incoming Proxy (P2)	ss2.biloxi.com	192.0.2.222
User Agent Bob (U2)	bob@biloxi.com	192.0.2.201

Figure 1.5 An SIP network with trapezoid operation with signaling and media: (a) SIP network with two UAs and two Proxies with SIP trapezoid operation; (b) URIs and IP addresses for SIP entities; (c) SIP session establishment and termination.

The SIP signaling between the UAs and the proxies and the media flow directly between the UAs as shown in Figure 1.5a are frequently referred as the SIP trapezoid. Assume that Alice in the atlanta.com domain is trying to establish the call via the proxy (P1) residing in its own administrative domain with Bob, who is residing in the biloxi.com administrative domain. Consequently, the proxy server of the atlanta.com domain is designated as the outgoing proxy (P1) and that of the biloxi.com as the incoming proxy (P2). First, the SIP signaling path is created among Alice's phone, outgoing proxy, incoming proxy, and Bob's phone for the establishment of the session. Then the media is passed between the phones directly, and the human users, Alice and Bob, communicate among themselves using their phones. Assume that the proxy servers will also act as registration servers for their users in the respective domains. Figure 1.5b shows the URIs and IP addresses of the SIP entities over the network.

Figure 1.5c shows an example of SIP call flows for the registration of users along with the establishment and termination of a session between two SIP UAs using an outgoing and an incoming proxy. In the beginning, Alice registers with the proxy of her administrative domain sending a REGISTER (F1) request, and SIP server (P1) confirms the request with the positive response of 200 OK (F2). In this way, the SIP server (P1) knows about the addresses of all phones in its administrative domain through the registration of all users. Similarly, Bob registers his phone with the proxy server (P2) of his administrative domain sending the REGISTER (F3) request, and the server confirms his registration sending a 200 OK (F4) response.

In SIP signaling protocol, if Alice needs to place a call to Bob, she has to send an INVITE (F5) request that contains Bob's phone URI among other information to her outgoing proxy (P2). Note that Alice may enclose the SDP message in the message-body of the INVITE (F5) describing what capabilities are available to Alice's UA. However, for simplicity, we have not shown this here; Section 18.1 in detail and briefly throughout other sections will describe SDP rules for audio, video, and/or data application capability for multimedia session negotiations.

The proxy server (P2) examines the destination address and finds that it does not reside in its local registration database (not shown in Figure 1.5 for simplicity) and finds that the callee (Bob) remains in a separate (atlanta.com) administrative domain. So, the proxy server (P1) forwards the INVITE (F6) to the incoming proxy server (P2) and confirms the receipt of the INVITE (F5) sending a provisional 100 Trying (F7) hop-by-hop response to the caller (Alice). Receiving the INVITE (F6) message, the incoming proxy (P2) consults its local registration database and finds that the callee (Bob) remains in its administrative domain, forwards the INVITE (F8) to the callee (Bob), and sends a provisional 100 Trying (F9) hop-by-hop response to the incoming proxy (P1).

Receiving the INVITE (F8) message, the callee (Bob) immediately sends a provisional 100 Trying (F10) hop-by-hop response to the incoming proxy (P2). If 100 Trying responses (F7, F9, and F10) are not sent, the senders will continue to retransmit the request after certain time intervals when timers expire as the process of an INVITE message usually takes a substantial amount of time. In the meantime, the phone of the callee (Bob) generates a ringtone to alert the callee (Bob) and sends the 180 Ringing (F11) message to the incoming proxy (P2). In turn, P2 forwards the 180 Ringing (F12) message to P1, and P1 sends the 180 Ringing (F13) message to the caller (Alice); the caller (Alice) starts to hear that the phone of the callee (Bob) is ringing.

In this example, the callee (Bob) has accepted the call without further negotiations with respect to audio/video codecs or data applications and their corresponding performances. The callee (Bob) answers the call assuming that Alice's INVITE (F8) message that contains SDP message describing Alice's UA capabilities in its message-body is acceptable to Bob. The callee (Bob) chooses the audio/video codec and/or data application parameter used in the call by the caller (Alice) and sends a final 200 OK (F14) response back to the incoming proxy (P2), and then P2 forwards the 200 OK (F15) response to P1. P1 forwards the 200 OK (F16) response to the caller (Alice).

Again, it should be noted that Bob may enclose the SDP message in the 200 OK (F14) message-body describing Bob's UA audio/video codecs or data applications and their corresponding performance capabilities. Alice accepts Bob's 200 OK response (F16) sent via proxies, assuming that the SDP message enclosed in the response message confirms the fact that Bob has accepted her multimedia session capabilities, and no further negotiations are required.

The caller (Alice) acknowledges this for reliability purposes by sending the ACK (F17) message to P1; P1 forwards the ACK (F18) to P2, and P2 forwards the ACK (19) to the callee (Bob). It should be noted that the ACK message can be sent directly end to end between the UAs without going via the proxies. At this point in time, the session is established and both phones begin to exchange media (audio/video over RTP and/or data applications using respective application protocols).

At the end of the session, either party can decide to terminate the session by sending the BYE request. In this example, the callee (Bob) sends the BYE (F20) request to P2, and P2 forwards the BYE (F21) request to P1; then P1 forwards the BYE (F22) request to the caller (Alice). Like ACK, the BYE message can be sent directly end to end between the UAs without going via the proxies. In turn, the caller (Alice) sends the confirmation with the 200 OK (F23) response to P1, and P1 forwards the 200 OK (F24) response to P2; finally, P2 forwards it to the callee (Bob). Now the session has been terminated, and media no longer flows. The call flows containing SIP signaling message path and media path complete the logical trapezoid path as shown in Figure 1.5a.

The SIP session establishment/termination shown here is a simple one; how a session can be modified and updated, and many other features for each of these audio, video, and/or data applications have not been shown. Security features or quality-of-service issues are not shown either. A call may fail due to issues in the network during call setup or after establishment of the session, and the paths may be shaped differently. Moreover, this is a point-to-point call where only two users are involved. A multipoint conference call with multiple users and multiple media of audio, video, and/or data applications will be much more complex. In the real world, even a point-to-point call is much more complicated; a call may traverse over many more administrative domains with their different security and quality-of-service features, a host of different administrative security policies, and traversal of middleboxes like network address translators (NATs), not to mention different kinds of networks with different call control protocols. Subsequent sections will describe many of those functional features. The detailed use of SDP messages for multimedia session negotiations in SIP is described later in Section 18.1.

1.3 WebRTC Protocol

1.3.1 Introduction

The enormous popularity of the browser platform, not initially designed for interactive RTCs with audio, video, and/or data applications, has recently caused the rise of a new platform for the deployment of services that include interactive RTC: the browser-embedded application or "Web application." It implies that as long as the browser platform has the necessary interfaces, it is possible to deliver almost any kind of service on it.

The WebRTC protocol is designed in the Internet Engineering Task Force (IETF) real-time communication on the web (RTCWeb) working group (WG) to allow the use of interactive audio and video in applications that communicate directly between browsers across the Internet. The resulting WebRTC protocol suite is intended to enable all of the applications that are described as required scenarios in the use case document [1] including access and control through a JavaScript (JS) application programming interface (API) in a browser. The purpose of WebRTC components is to help build a strong RTC platform that works across multiple web browsers and multiple platforms. WebRTC uses the web as the setup and end-point for a call.

In addition to IETF WebRTC protocol efforts, the World Wide Web Consortium (W3C) WebRTC, Web Applications, and Device API WGs, are focusing on making standardized APIs and interfaces available, within or alongside the HTML5 effort for those functions. However, IETF RTCWeb WG is concentrating on specifying the WebRTC protocols and subprotocols needed to specify interactions across the network complementing the W3C efforts.

1.3.2 Protocol Design Principles

The WebRTC protocol is designing to meet the requirements for making the browser suitable for RTC with audio, video, and/or data applications between two or more peers located in the same and/or different geographical locations around the world. It is assumed that user applications are executed on a browser. Whether the capabilities to implement specific browser requirements are implemented by the browser application or are provided to the browser application by the underlying operating system is not a part of this protocol design.

1.3.2.1 Browser Requirements

The browser requirements described here are derived considering the following usecase scenarios [1]:

- Clients can be on IPv4-only.
- Clients can be on IPv6-only.
- Clients can be on dual-stack (IPv4/IPv6).
- Clients can be connected to networks with different throughput capabilities.
- Clients can be on variable-media-quality networks (wireless).
- Clients can be on congested networks.
- Clients can be on firewalled networks with no UDP allowed.
- Clients can be on networks with a NAT or IPv4-IPv6 translation devices using any type of mapping and filtering behavior as described in RFC 4787.

1.3.2.1.1 General Requirements

- The browser must be able to use microphones and cameras as input devices to generate streams.
- The browser must be able to send streams and data to a peer in the presence of NATs.
- Transmitted streams and data must be rate controlled (meaning that the browser must, regardless of application behavior, reduce send rate when there is congestion).
- The browser must be able to receive, process, and render streams and data ("render" does not apply to data) from peers.

- The browser should be able to render good quality audio and video even in the presence of reasonable levels of jitter and packet losses.
- The browser must detect when a stream from a peer is no longer received.
- When there are both incoming and outgoing audio streams, echo cancelation must be made available to avoid disturbing echo during conversation.
- The browser must support synchronization of audio and video.
- The browser should use encoding of streams suitable for the current rendering (e.g., video display size) and should change parameters if the rendering changes during the session.
- The browser must support a baseline audio and video codec.
- It must be possible to protect streams and data from wiretapping in RFCs 2904 and 7258.
- The browser must enable verification, given the right circumstances and by use of other trusted communication, that streams and data received have not been manipulated by any party.
- The browser must encrypt, authenticate, and integrity protect media and data on a per-packet basis and must drop incoming media and data packets that fail the per-packet integrity check. In addition, the browser must support a mechanism for cryptographically binding media and data security keys to the user identity in RFC 5479.
- The browser must make it possible to set up a call between two parties without one party learning the other party's host IP address.
- The browser must be able to collect statistics, related to the transport of audio and video between peers, needed to estimate the quality of experience.

1.3.2.1.2 Requirements Related to Network and Topology

- The browser must support the insertion of reference frames in outgoing media streams when requested by a peer.
- The communication session must survive across a change of the network interface used by the session.
- The browser must be able to send streams and data to a peer in the presence of NATs and firewalls that block UDP traffic.
- The browser must be able to use several simple traversal of UDP around NAT (STUN) and traversal using relays around NAT (TURN) servers described in RFCs 5389 and 5766, respectively.
- The browser must support the use of STUN and TURN servers that are supplied by entities other than the web application (i.e., the network provider).
- The browser must be able to send streams and data to a peer in the presence of firewalls that only allow traffic via an HTTP Proxy, when firewall policy allows WebRTC traffic.
- The browser should be able to take advantage of available capabilities (supplied by network nodes) to differentiate voice, video, and data appropriately.

1.3.2.1.3 Requirements Related to Multiple Peers and Streams

- The browser must be able to transmit streams and data to several peers concurrently.
- The browser must be able to receive streams and data from multiple peers concurrently.
- The browser must be able to render several concurrent audio and video streams.
- The browser must be able to mix several audio streams.

1.3.2.1.4 Requirements Related to Audio Processing

- The browser must be able to apply spatialization effects when playing audio streams.
- The browser must be able to measure the voice activity level in audio streams.
- The browser must be able to change the voice activity level in audio streams.
- The browser must be able to process and mix sound objects (media that is retrieved from a source other than the established media stream(s) with the peer(s) with audio streams.

1.3.2.1.5 Requirements Related to Legacy Interoperability

- The browser must support an audio media format (codec) that is commonly supported by existing telephony services.
- There should be a way to navigate a dual-tone multifrequency (DTMF) signaling based interactive voice response system.
- The browser must be able to initiate and accept a media session where the data needed for establishment can be carried in SIP.

1.3.2.1.6 Miscellaneous Requirements

- The browser must be able to send short latency unreliable datagram traffic to a peer browser specified in RFC 5405.
- The browser must be able to send reliable data traffic to a peer browser.
- The browser must be able to generate streams using the entire user display, a specific area of the user's display, or the information being displayed by a specific application.

1.3.2.2 Relationship between API and Protocol

The total WebRTC effort consists of two pieces:

- A protocol specification, done in the IETF
- A JS API specification, done in the W3C [2,3]

Together, these two specifications aim to provide an environment where JS embedded in any page, viewed in any compatible browser, when suitably authorized by its user, is able to set up communication using audio, video, and auxiliary data, where the browser environment does not constrain the types of application in which this functionality can be used. The protocol specification does not assume that all implementations use this API; it is not intended to be necessary for interoperation to know whether the entity one uses to communicate is a browser or another device implementing this specification.

The goal of cooperation between the protocol specification and the API specification is that for all options and features of the protocol specification, it should be clear which API calls should be made to exercise that option or feature; similarly, for any sequence of API calls, it should be clear which protocol options and features will be invoked, subject to constraints of the implementation, of course. The following terminology will be used to discuss WebRTC.

- A WebRTC browser (also called a WebRTC UA) conforms to both the protocol specification and the JS API defined in a previous section.

- A WebRTC nonbrowser conforms to the protocol specification but does not claim to implement the JS API. This can also be called a "WebRTC device" or "WebRTC native application."
- A WebRTC endpoint is either a WebRTC browser or a WebRTC nonbrowser. It conforms to the protocol specification.
- A WebRTC-compatible endpoint is able to successfully communicate with a WebRTC endpoint but may fail to meet some requirements of a WebRTC endpoint. This may limit where in the network such an endpoint can be attached or may limit the security guarantees it offers to others. It is not constrained by this specification; when it is mentioned at all, it is to note the implications on WebRTC-compatible endpoints of the requirements placed on WebRTC endpoints.
- A WebRTC gateway is a WebRTC-compatible endpoint that mediates media traffic to non-WebRTC entities.

All WebRTC browsers are WebRTC endpoints, so any requirement on a WebRTC endpoint also applies to a WebRTC browser. A WebRTC nonbrowser may be capable of hosting applications in a similar way to that in which a browser can host JS applications, typically by offering APIs in other languages. For instance, it may be implemented as a library that offers a C++ API intended to be loaded into applications. In this case, similar security considerations as for JS may be needed; however, since such APIs are not defined or referenced here, this document cannot give any specific rules for those interfaces. WebRTC gateways are described in a separate document [4].

1.3.2.3 On Interoperability and Innovation

The "Mission Statement of the IETF Described in RFC 3935" states that "The benefit of a standard to the Internet is in interoperability – that multiple products implementing a standard are able to work together in order to deliver valuable functions to the Internet's users." Communication on the Internet frequently occurs in two phases:

- Two parties communicate, through some mechanism, what functionality they both are able to support.
- They use that shared communicative functionality to communicate or, failing to find anything in common, give up on communication.

Often, many choices can be made for communicative functionality; the history of the Internet is rife with the proposal, standardization, implementation, and success or failure of many types of options, in all sorts of protocols. The goal of having a mandatory to implement function set is to prevent negotiation failure, not to preempt or prevent negotiation.

The presence of a mandatory to implement function set serves as a strong changer of the marketplace of deployment—in that it gives a guarantee that, as long as you conform to a specification and the other party is willing to accept communication at the base level of that specification, you can communicate successfully. The alternative, that of having no mandatory to implement, does not mean that you cannot communicate; it merely means that in order to be part of the communication partnership, you have to implement the standard "and then some," that "and then some" usually a profile of some sort. In the version most antithetical to the Internet ethos, that "and then some" consists of having to use a specific vendor's product only.

1.3.2.4 WebRTC Terminology

The following terms are used across the documents specifying the WebRTC suite, and the specific meanings are given here. Not all terms are used in this document. Other terms are used in their commonly used meaning. The list is in alphabetical order.

- Agent: Undefined term. See "SDP Agent" and "Interactive Connectivity Establishment (ICE) UA specified in RFC 5245."
- API: Application Programming Interface—a specification of a set of calls and events, usually tied to a programming language or abstract formal specification such as Web Interface Description Language (WebIDL), with its defined semantics.
- Browser: Used synonymously with "Interactive UA" as defined in the HTML specification [5]. See also "WebRTC UA."
- ICE Agent: An implementation of the Interactive Connectivity Establishment (ICE) specified in RFC 5245 protocol. An ICE Agent may also be an SDP Agent, but some ICE Agents do not use SDP (e.g., those that use Jingle) [6].
- Interactive: Communication between multiple parties, where the expectation is that an action from one party can cause a reaction by another party, and the reaction can be observed by the first party, with the total time required for the action/reaction/observation being on the order of no more than hundreds of milliseconds.
- Media: Audio and video content. Not to be confused with "transmission media" such as wires.
- Media Path: The path that media data follows from one WebRTC endpoint to another.
- Protocol: A specification of a set of data units, their representation, and rules for their transmission, with their defined semantics. A protocol is usually thought of as going between systems.
- Real-time Media: Media where generation and display of content are intended to occur closely together in time (on the order of no more than hundreds of milliseconds). Real-time media can be used to support interactive communication.
- SDP Agent: The protocol implementation involved in the SDP offer/answer exchange, as defined in RFC 3264 (see Section 3.1).
- Signaling: Communication that happens in order to establish, manage, and control media paths.
- Signaling Path: The communication channels used between entities participating in signaling to transfer signaling. There may be more entities in the signaling path than in the media path.

Note: Where common definitions exist for these terms, those definitions should be used to the greatest extent possible.

1.3.3 Architecture and Functionality Groups

The WebRTC model of real-time support for browser-based applications is that the browser will have the functions needed for a Web application, working in conjunction with its backend servers, to implement the functions such as telephone conferencing and/or VC units. Two vital interfaces that are needed are specified (Figure 1.6): the protocols that browsers use to talk to each other, without any intervening servers, and the APIs that are offered for a JS application to take

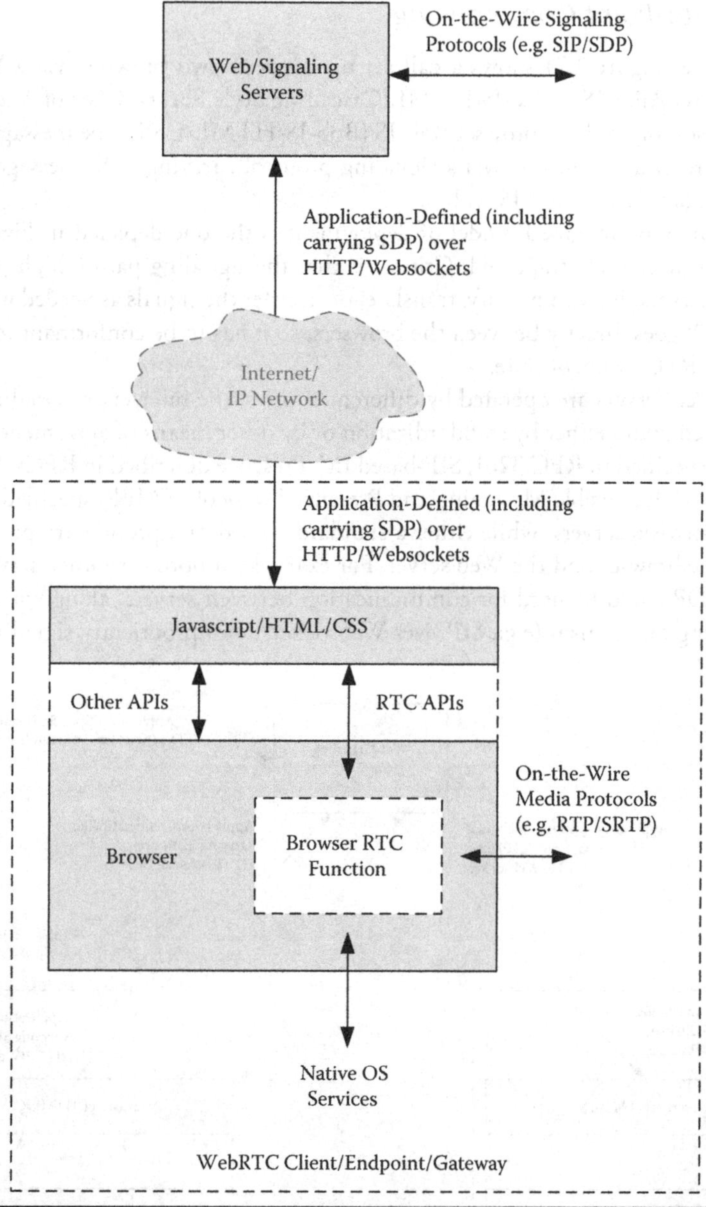

Figure 1.6 Browser model.

advantage of the browser's functionality. Note that the model does not assume that the browser will contain all the functions that are needed for RTC.

The HTTP and Websockets are also offered to the JS application through browser APIs. As for all protocol and API specifications, there is no restriction that the protocols can only be used to talk to another browser; since they are fully specified, any endpoint that implements the protocols faithfully should be able to interoperate with the application running in the browser.

1.3.3.1 Point-to-Point Conferencing

A simple example (Figure 1.7) shows a call set up between two browsers via a Web/Signaling server. The browser Alice-JS (Alice-JS/HTML/Cascading Style Sheets (CSS) of Alice) is initiating the multimedia session to Bob's browser Bob-JS (Bob-JS/HTML/CSS). The messages from Alice's JS to Bob's JS are assumed to flow via a signaling protocol carrying SDP messages to the Web/ Signaling server. Similarly, Bob's JS replies.

A more commonly imagined model of deployment is the one depicted in Figure 1.8 that is termed as the browser RTC trapezoid. Critical is that the signaling path ("high path") goes via Web/Signaling servers that can modify, translate, or transfer the signals as needed while the media path ("low path") goes directly between the browsers, so it has to be conformant to the specifications of the WebRTC protocol suite.

If the two Web servers are operated by different entities, the interserver signaling mechanism needs to be agreed upon, either by standardization or by other means of agreement. Existing protocols (e.g., SIP specified in RFC 3261, SIP-based IM/Presence described in RFCs 3428, 4975-76, and 6914, Jingle, or Extensible Messaging and Presence Protocol (XMPP) specified in RFC 6120) could be used between servers, while either a standards-based or a proprietary protocol could be used between the browser and the Web server. For example, if both operators' servers implement SIP/SDP, SIP/SDP could be used for communication between servers, along with either a standardized signaling mechanism (e.g., SIP over Websockets) or a proprietary signaling mechanism

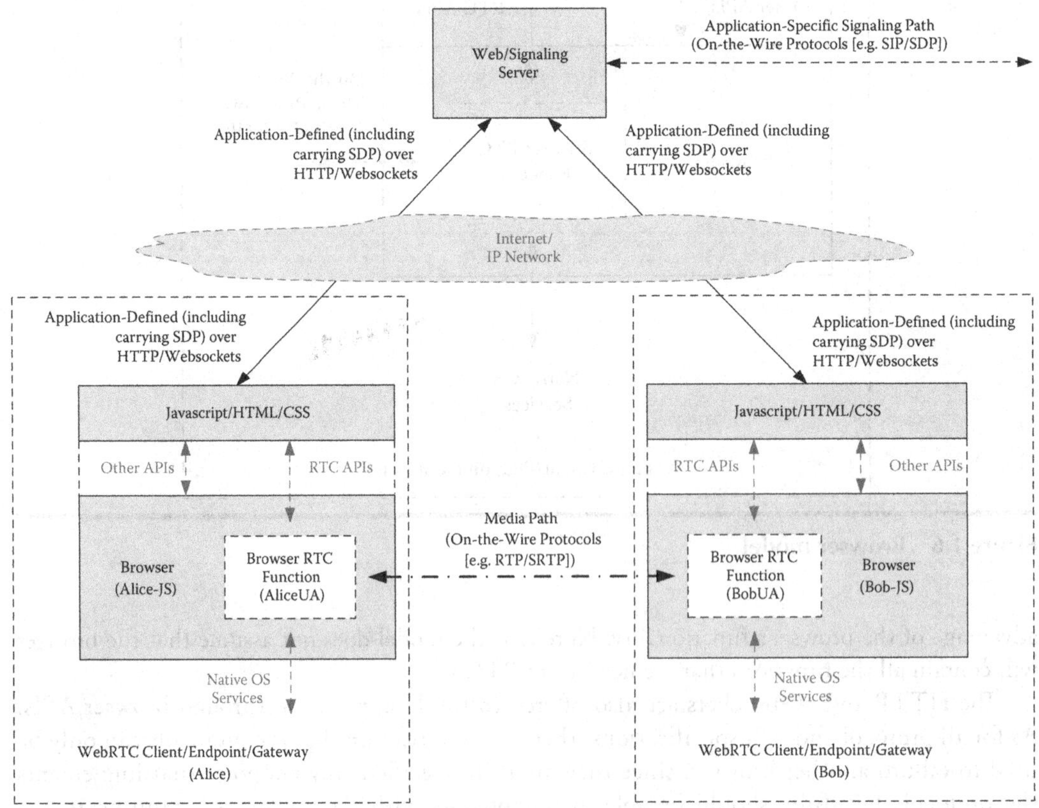

Figure 1.7 Simple example for setting up a call between two browsers.

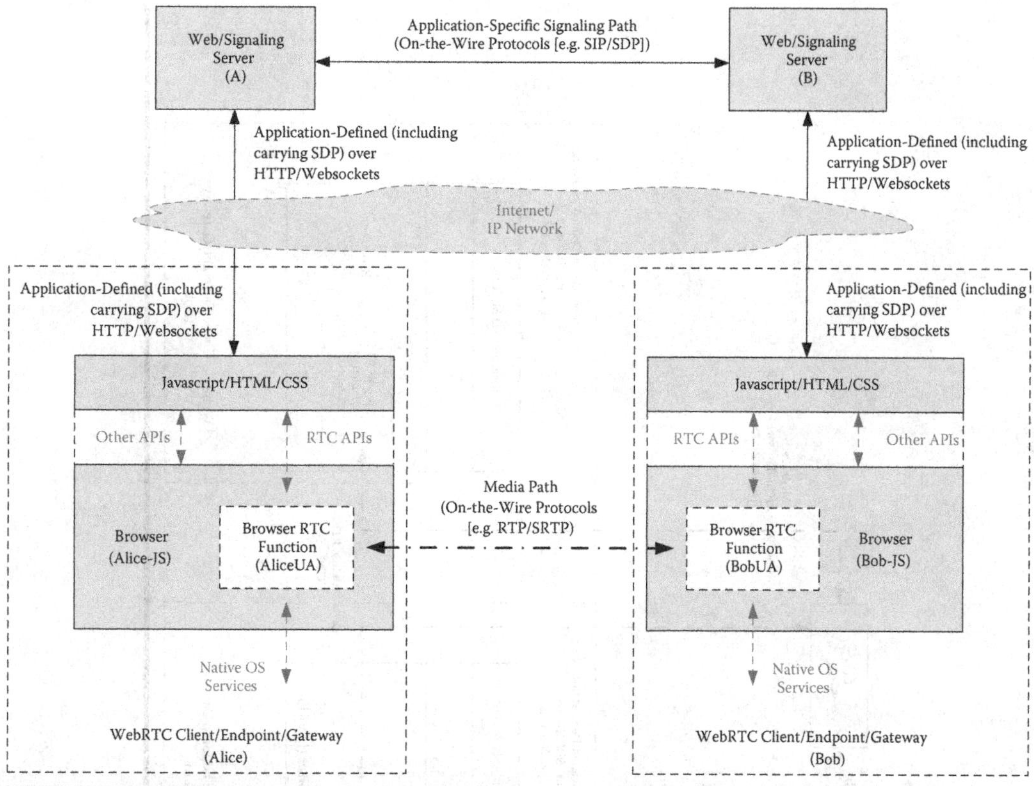

Figure 1.8 Browser RTC trapezoid.

used between the application running in the browser and the Web server. Similarly, if both operators' servers implement SIP/Jingle, or XMPP/SIP-based IM/Presence) could be used for communication between XMPP servers, with either a standardized signaling mechanism (e.g., SIP/Jingle, XMPP/SIP-based IM/Presence over Websockets) or a proprietary signaling mechanism used between the applications running in the browser and the Web server. The choice of protocols, and definition of the translation between them, is outside the scope of the WebRTC protocol suite.

1.3.3.2 Multipoint Conferencing

Figure 1.9 depicts a simple configuration of multipoint conferencing consisting of star topology. The functionality groups needed in the browser can be specified, more or less from the bottom up, as follows:

- **Data transport:** TCP, UDP, and the means to securely set up connections between entities, as well as the functions for deciding when to send data: congestion management, bandwidth estimation, and so on
- **Data framing:** RTP and other data formats that serve as containers and their functions for data confidentiality and integrity
- **Data formats:** Codec specifications, format specifications, and functionality specifications for the data passed between systems. Audio and video codecs, as well as formats for data and

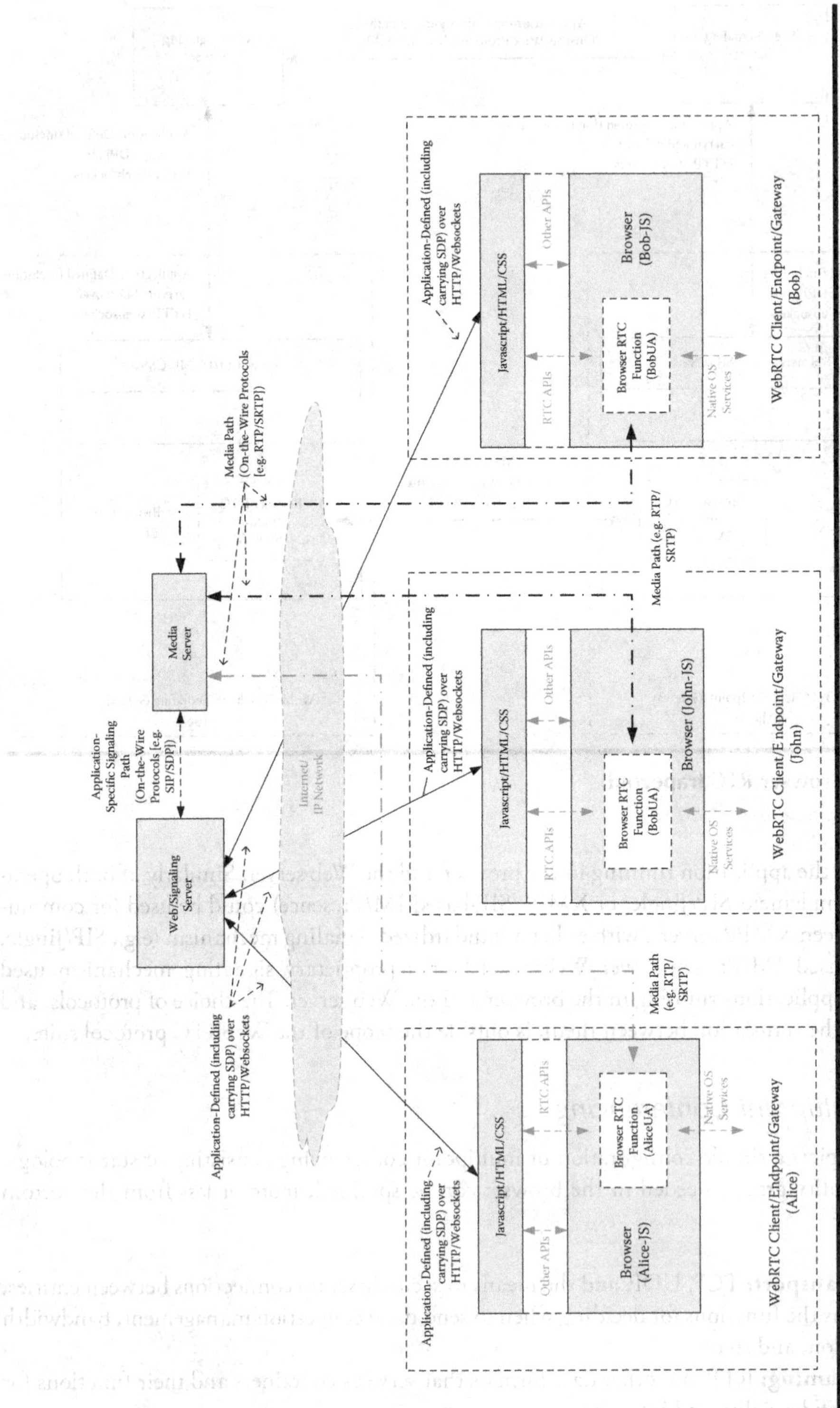

Figure 1.9 Multipoint conferencing.

document sharing, belong to this category. In order to make use of data formats, a way to describe them, a session description, is needed.

- **Connection management:** Setting up connections, agreeing on data formats, changing data formats during the duration of a call; SIP and Jingle/XMPP belong to this category.
- **Presentation and control:** What needs to happen in order to ensure that interactions behave in a nonsurprising manner. This can include floor control, screen layout, voice-activated image switching, and other such functions, where part of the system requires the cooperation between parties. XCON (IETF Multipoint Conferencing WG) that has specified functionalities in RFCs 4597, 5239, and 6503-04 were some attempts at specifying this kind of functionality; many applications have been built without standardized interfaces to these functions.
- **Local system support functions:** These need not be specified uniformly, because each participant may choose to do these in his/her own way, without affecting the bits on the wire in a way that others have to be cognizant of. Examples in this category include echo cancelation (some forms of it), local authentication and authorization mechanisms, OS access control, and the ability to record local conversations.

Within each functionality group, it is important to preserve both freedom to innovate and the ability for global communication. Freedom to innovate is helped by handling specification in terms of interfaces, not implementation; any implementation able to communicate according to the interfaces is a valid implementation. Ability to communicate globally is helped both by having core specifications be unencumbered by intellectual property right issues and by specifying the formats and protocols fully enough to allow for independent implementation.

One can think of the three first groups as forming a "media transport infrastructure" and the three last three as forming a "media service." In many contexts, it makes sense to use a common specification for the media transport infrastructure, which can be embedded in browsers and accessed using standard interfaces, to "let a thousand flowers bloom" in the "media service" layer; to achieve interoperable services, however, at least the first five of the six groups need to be specified.

The sets of APIs [2] that are under development in W3C can be classified into different groups as shown in Table 1.2.

The development of WebRTC protocols under IETF is still in progress, and Table 1.3 provides a list of drafts of the ongoing works.

1.3.4 Data Transport

Data transport refers to the sending and receiving of data over the network interfaces, the choice of network-layer addresses at each end of the communication, and the interaction with any intermediate entities that handle the data but do not modify it (such as TURN relays). It includes necessary functions for congestion control: when not to send data. WebRTC endpoints MUST implement the transport protocols described in the rtcweb-transports draft [7].

1.3.5 Data Framing and Securing

The format for media transport is RTP defined in RFC 3550. Implementation of Secure Real-Time Protocol (SRTP), described in RFC 3711, is REQUIRED for all implementations. The detailed considerations for usage of functions from RTP and SRTP are given in the rtp-usage draft [8]. The security considerations for the WebRTC use case are in the rtcweb-security draft [9], and the resulting

Table 1.2 W3C WebRTC APIs [2] Draft

API Group (API/ Sub-APIs) Classification	*Description*
Peer-to-Peer Connections Group	This API group sets up a connection between the two peer browsers, termed as peer connection, for multimedia (audio, video, and/or data applications) communications. Communications are coordinated via a signaling channel, which is provided by unspecified means but generally by a script in the page via the server (e.g., HTTP/Websockets). However, the media negotiation/renegotiation of the multimedia session for presenting each media on the channel needs to be performed using SDP. The SDP offer/answer model is used for negotiation/renegotiation of each media between the browsers. It should be noted that a request can be made to add or remove a media either locally or remotely, and the browser generates the session object with appropriate description. Similarly, a peer connection can be closed with appropriate application calls.
RTP Media Group	The real-time media such as audio and video is transferred over the network using RTP. The RTP media API group lets a web application send and receive media streams using tracks over a peer-to-peer connection. Tracks, when added to a "RTC Peer Connection," result in signaling; when this signaling is forwarded to a remote peer, it causes corresponding tracks to be created on the remote side. The actual encoding and transmission of "Media Stream Tracks" is managed through objects called "RTC RTP Senders." Similarly, the reception and decoding of "Media Stream Tracks" is managed through objects called "RTC RTP Receivers." Each track to be sent is associated with exactly one "RTC RTP Sender," and each track to be received is associated with exactly one "RTC RTP Receiver." "RTC RTP Senders" are created when the application attaches a "Media Stream Track" to an "RTC Peer Connection," via the "addTrack" scheme. "RTC RTP Receivers," on the other hand, are created when remote signaling indicates new tracks are available, and each new "Media Stream Track" and its associated "RTC RTP Receiver" are surfaced to the application via the "ontrack" event. A RTC Peer Connection object contains a set of "RTC RTP Senders," representing tracks to be sent, and a set of "RTC RTP Receivers," representing tracks that are to be received on this RTC Peer Connection object, and a set of "RTC RTP Transceivers," representing the paired senders and receivers with some shared state. All of these sets are initialized to empty sets when the "RTC Peer Connection" object is created.
Peer-to-Peer Data Group	The Peer-to-Peer Data API group lets a web application send and receive generic application data peer-to-peer. The API for sending and receiving data models the behavior of Websockets.

(Continued)

Table 1.2 (*Continued*) W3C WebRTC APIs [2] Draft

API Group (API/ Sub-APIs) Classification	*Description*
Peer-to-Peer DTMF Group	This describes an interface on RTC RTP Sender to send DTMF (phone keypad) values across an "RTC Peer Connection."
Statistics Model Group	The basic statistics model is that the browser maintains a set of statistics referenced by a "*selector*." The selector may, for example, be a "`Media Stream Track`." For a track to be a valid selector, it *MUST* be a "`Media Stream Track`" sent or received by the "RTP Peer Connection" object on which the stats request was issued. The calling Web application provides the selector to the "getStats" method, and the browser emits (in the JS) a set of statistics it believes is relevant to the selector. The statistics returned are designed in such a way that repeated queries can be linked by the "RTC Statistic Identity (ID)" dictionary member. Thus, a Web application can make measurements over a given time period by requesting measurements at the beginning and end of that period.
Media Stream Extensions for Network Use Group	The "`Media Stream Track`" interface typically represents a stream of data of audio or video. One or more "`Media Stream Tracks`" can be collected in a "`Media Stream`" that may contain zero or more "`Media Stream Track`" objects. A "`Media Stream Track`" may be extended to represent a media flow that either comes from or is sent to a remote peer (and not just the local camera, for instance). The extensions required to enable this capability on the "`Media Stream Track`" object is described. A "`Media Stream Track`" sent to another peer will appear as one and only one "`Media Stream Track`" to the recipient. A peer is defined as a UA that supports this specification. In addition, the sending side application can indicate of what "`Media Stream`" object(s) the "`Media Stream Track`" is a member. The corresponding `Media Stream` object(s) on the receiver side will be created (if not already present) and populated accordingly. As also described earlier in this chapter, the objects "`RTC RTP Sender`" and "`RTC RTP Receiver`" can be used by the application to get more control over the transmission and reception of "`Media Stream Tracks`." Channels are the smallest unit considered in the "`Media Stream`" specification. Channels are intended to be encoded together for transmission as, for instance, an RTP payload type. All of the channels that a codec needs to encode jointly *MUST* be in the same "`Media Stream Track`," and the codecs *SHOULD* be able to encode, or discard, all the channels in the track.

(*Continued*)

Table 1.2 (*Continued*) W3C WebRTC APIs [2] Draft

API Group (API/ Sub-APIs) Classification	*Description*
	The concepts of an input and output to a given `MediaStreamTrack` apply in the case of MediaStreamTrack objects transmitted over the network as well. A "`Media Stream Track`" created by an "`RTC Peer Connection`" object will take as input the data received from a remote peer. Similarly, a "`Media Stream Track`" from a local source, for instance, a camera, will have an output that represents what is transmitted to a remote peer if the object is used with an "`RTC Peer Connection`" object.
Media Stream Extensions for Network Use Group	The concept of duplicating "`Media Stream`" and "`Media Stream Track`" objects is also used. This feature can be used, for instance, in a video-conferencing scenario to display the local video from the user's camera and microphone in a local monitor, while only transmitting the audio to the remote peer (e.g., in response to the user's using a "video mute" feature). Combining different "`Media Stream Track`" objects into new "Media Stream" objects is useful in certain situations.
Identify Group	The "Identity" capability for authenticating the communicating parties based on the recent practices of Web services providers to obtain ID from a trusted third-party "Identify Provider" that is honored by all service providers. As a result, WebRTC offers and answers, hence the channels established by "`RTC Peer Connection`" objects, can be authenticated by using a web-based Identity Provider (IdP). The idea is that the entity sending an offer or answer acts as the Authenticating Party and obtains an identity assertion from the IdP, which it attaches to the session description. The consumer of the session description (i.e., the "`RTC Peer Connection`" on which "Set Remote Description" method is called) acts as the Relying Party and verifies the assertion. The interaction with the IdP is designed to decouple the browser from any particular identity provider; the browser need only know how to load the IdP's JS, the location of which is determined by the IdP's identity, and the generic interface to generating and validating assertions. The IdP provides whatever logic is necessary to bridge the generic protocol to the IdP's specific requirements. Thus, a single browser can support any number of identity protocols.

Table 1.3 IETF WebRTC Protocol Drafts

Title/Draft	*Description*
Web RTC Use cases and Requirements/ draft-ietf-rtcweb-use-cases-and-requirements-16.txt	This document describes web-based RTC use cases. Requirements on the browser functionality are derived from the use cases. This document was developed in an initial phase of the work with rather minor updates at later stages. It has not really served as a tool in deciding features or scope for the WG's efforts so far. It is being published to record the early conclusions of the WG. It will not be used as a set of rigid guidelines that specifications and implementations will be held to in the future.
Local System Support Functions (DSCP) and other packet markings for RTCWeb Quality of Service/ draft-ietf-rtcweb-qos-00.txt	This document defines the recommended DSCP values for browsers to use for various classes of traffic.
JS Session Establishment Protocol/draft-ietf-rtcweb-jsep-14.txt	This document describes the mechanisms for allowing a JS application to control the signaling plane of a multimedia session via the interface specified in the W3C RTCPeerConnection API, and discusses how this relates to existing signaling protocols.
Overview: Real Time Protocols for Browser-based Applications/draft-ietf-rtcweb-overview-15.txt	This document gives an overview and context of a protocol suite intended for use with real-time applications that can be deployed in browsers: "real-time communication on the Web." This draft is an Applicability Statement; it does not itself specify any protocol, but rather specifies which other specifications WebRTC-compliant implementations are supposed to follow.
Security Considerations for WebRTC/draft-ietf-rtcweb-security-08.txt	This document defines the WebRTC threat model and analyzes the security threats of WebRTC in that model. Unlike SIP-based soft phones, WebRTC communications for real-time audio and/or video calls, Web conferencing, and direct data transfer are directly controlled by a Web server between Web browsers, which poses new security challenges. For instance, a Web browser might expose a JS API, which allows a server to place a video call. Unrestricted access to such an API would allow any site a user visited to "bug" a user's computer, capturing any activity that passed in front of the camera.
SIP usage for Trickle ICE/draft-ietf-mmusic-trickle-ice-sip-04.txt	The ICE protocol describes a NAT traversal mechanism for UDP-based multimedia sessions established with the SDP Offer/ Answer model. The ICE extension for Incremental Provisioning of Candidates (Trickle ICE) defines a mechanism that allows ICE agents to shorten session establishment delays by making the candidate gathering and connectivity checking phases of ICE nonblocking and by executing them in parallel. This document defines usage semantics for Trickle ICE with the SIP.

(*Continued*)

Table 1.3 (*Continued*) IETF WebRTC Protocol Drafts

Title/Draft	*Description*
WebRTC: Media Transport and Use of RTP/draft-ietf-rtcweb-rtp-usage-26.txt	This document describes the media transport aspects of the direct interactive rich communication using audio, video, text, collaboration, games, etc., between two peers' web browsers. It specifies how the RTP is used in the WebRTC context, and gives requirements for which RTP features, profiles, and extensions need to be supported.
WebRTC 1.0: RTC between Browsers/ W3C Working Draft 28 January 2016	This document defines a set of ECMAScript APIs in WebIDL to allow media to be sent to and received from another browser or device implementing the appropriate set of real-time protocols. This specification is being developed in conjunction with a protocol specification developed by the IETF RTCWeb group and an API specification to get access to local media devices developed by the Media Capture Task Force.
WebRTC Data Channel Establishment Protocol/draft-ietf-rtcweb-data-protocol-09.txt	The WebRTC framework specifies protocol support for direct interactive rich communication using audio, video, and data between two peers' web browsers. This document specifies a simple protocol for establishing symmetric data channels between the peers. It uses a two-way handshake and allows sending of user data without waiting for the handshake to be complete.
WebRTC Data Channels/draft-ietf-rtcweb-data-channel-13.txt	The WebRTC framework specifies protocol support for direct interactive rich communication using audio, video, and data between two peers' web browsers. This document specifies the nonmedia data transport aspects of the WebRTC framework. It provides an architectural overview of how the SCTP is used in the WebRTC context as a generic transport service allowing Web browsers to exchange generic data from peer to peer.
WebRTC Gateways/ draft-ietf-rtcweb-gateways-02.txt	This document describes interoperability considerations for a class of WebRTC-compatible endpoints called "WebRTC gateways," which interconnect between WebRTC endpoints and devices that are not WebRTC endpoints.
WebRTC Security Architecture/draft-ietf-rtcweb-security-arch-11.txt	The RTCWEB WG is tasked with standardizing protocols for enabling RTCs within UAs using web technologies (commonly called "WebRTC"). This document defines the security architecture for WebRTC.
WebRTC IP Address Handling Recommendations/ draft-ietf-rtcweb-ip-handling-01.txt	This document provides best practices for how IP addresses should be handled by WebRTC applications.

(*Continued*)

Table 1.3 (*Continued*) IETF WebRTC Protocol Drafts

Title/Draft	*Description*
Negotiating Media Multiplexing Using the SDP/draft-ietf-mmusic-sdp-bundle-negotiation-29.txt	This specification defines a new SDP Grouping Framework extension, "BUNDLE." The extension can be used with the SDP Offer/Answer mechanism to negotiate the usage of a single address:port combination (BUNDLE address) for receiving media, referred to as bundled media, specified by multiple SDP media descriptions ("m=" lines). To assist endpoints in negotiating the use of bundle this, which can be used to request that specific media is only used if bundled. There are multiple ways to correlate the bundled RTP packets with the appropriate media descriptions. This specification defines a new RTP source description item and a new RTP header extension that provides an additional way to do this correlation by using them to carry a value that associates the RTP/RTCP packets with a specific media description.
RTP Payload Format for the Opus Speech and Audio Codec/ draft-ietf-payload-rtp-opus-11.txt	This document defines the RTP payload format for packetization of Opus encoded speech and audio data necessary to integrate the codec in the most compatible way. It also provides an applicability statement for the use of Opus over RTP. Further, it describes media-type registrations for the RTP payload format.
XEP-0166: Jingle Version 1.1.1, Standards Track, XMPP Standards Foundation, May 2016	This specification defines an XMPP protocol extension for initiating and managing peer-to-peer media sessions between two XMPP entities in a way that is interoperable with existing Internet standards. The protocol provides a pluggable model that enables the core session management semantics (compatible with SIP) to be used for a wide variety of application types (e.g., voice chat, video chat, file transfer) and with a wide variety of transport methods (e.g., TCP, UDP, ICE, application-specific transports).
Transports for WebRTC/ draft-ietf-rtcweb-transports-17.txt	This document describes the data transport protocols used by WebRTC, including the protocols used for interaction with intermediate boxes such as firewalls, relays, and NAT boxes.

security functions are described in the security-arch draft [10]. Considerations for the transfer of data that is not in RTP format are described in the data-channel draft [11], and a supporting protocol for establishing individual data channels is described in the data-protocol draft [12]. WebRTC endpoints MUST implement these two specifications. WebRTC endpoints MUST implement the rtp-usage draft [13], the rtcweb-security draft [9], the security-arch draft [10], and the requirements they include.

1.3.6 Data Formats

The intent of this specification is to allow each communications event to use the data formats best suited for that particular instance, where a format is supported by both sides of the connection.

However, a minimum standard is greatly helpful in order to ensure that communication can be achieved. This document specifies a minimum baseline that will be supported by all implementations of this specification and leaves further codecs to be included at the will of the implementer. WebRTC endpoints that support audio and/or video MUST implement the codecs and profiles described in the rtcweb-audio draft for audio [14] and specified in RFC 3550 for video.

1.3.7 Connection Management

The methods, mechanisms, and requirements for setting up, negotiating, and tearing down connections is a large subject and one in which it is desirable to have both interoperability and freedom to innovate. The following principles apply:

- The WebRTC media negotiations will be capable of representing the same SDP offer/answer semantics that are used in SIP described in RFC 3264 (see Section 3.1), in such a way that it is possible to build a signaling gateway between SIP and the WebRTC media negotiation.
- It will be possible to connect legacy SIP/SDP devices that support ICE and appropriate RTP mechanisms, codecs, and security mechanisms without using a media gateway. A signaling gateway to convert signaling on the web side to SIP signaling may be needed.
- When a new codec is specified, and the SDP for the new codec is specified in the multiparty multimedia session control (MMUSIC) WG, no other standardization should be required for it to be used in web browsers. The addition of new codecs that might have new SDP parameters should not change the APIs between the browser and the JS application. As soon as browsers support the new codecs, applications written before the codecs were specified should automatically be able to use the new codecs where appropriate with no changes to the JS applications.

The particular choices made for WebRTC, and their implications for the API offered by a browser implementing WebRTC, are described in the rtcweb-jsep draft [15]. WebRTC browsers MUST implement the rtcweb-jsep draft [15].

WebRTC endpoints MUST implement the functions described in that document relating to the network layer (e.g., Bundle, RTCPmux, and Trickle ICE) but do not need to support the API functionality described there.

1.3.8 Presentation and Control

The most important part of control is the user's control over the browser's interaction with input/output devices and communications channels. It is important that the user have some way of figuring out where his/her audio, video, or texting is being sent, for what purported reason, and what guarantees are made by the parties that form part of this control channel. This is largely a local function among the browser, the underlying operating system, and the user interface; this is specified in the peer connection API [2] and the media capture API [3]. WebRTC browsers MUST implement these two specifications.

1.3.9 Local System Support Functions

Local system support functions are characterized by the fact that their quality strongly influences user experience, but the exact algorithm does not need coordination. In some cases (e.g., echo cancelation, as described later in Section 18.2), the overall system definition may need to specify that it needs to have some characteristics for which these facilities are useful, without requiring them to be implemented in a certain way. Local functions include echo cancelation, volume control, camera management including focus, zoom, pan/tilt controls (if available), and more. Certain parts of the system SHOULD conform to certain properties, for instance:

- Echo cancelation should be good enough to achieve the suppression of acoustical feedback loops below a perceptually noticeable level.
- Privacy concerns MUST be satisfied; for instance, if remote control of a camera is offered, the APIs should be available to let the local participant figure out who's controlling the camera and possibly decide to revoke the permission for camera usage.
- Automatic gain control, if present, should normalize a speaking voice into a reasonable dB range.

The requirements on WebRTC systems with regard to audio processing are found in Reference [14]; the proposed API for control of local devices is found in Reference [3].

WebRTC endpoints MUST implement the processing functions in Reference [14]. Together with the requirement in Section 1.3.6, this means that WebRTC endpoints MUST implement the whole document.

1.3.10 Internet Assigned Numbers Authority Considerations

This document makes no request of Internet Assigned Numbers Authority.

1.3.11 Security Considerations

Security of the web-enabled RTC comes in several pieces:

- Security of the components: The browsers and other servers involved. The most target-rich environment here is probably the browser; the aim here should be that the introduction of these components introduces no additional vulnerability.
- Security of the communication channels: It should be easy for a participant to reassure him/herself of the security of the communication, by verifying the crypto parameters of the links participated in and being reassured by other parties to the communication that appropriate measures are taken.
- Security of the partners' identity: verifying that the participants are who they say they are (when positive identification is appropriate) or that their identities cannot be uncovered (when anonymity is a goal of the application). The security analysis, and the requirements derived from that analysis, is contained in Reference [9]. It is also important to read the security sections of References [2,3].

1.3.12 WebRTC Call and Session Negotiation Using SDP

This section shows (Figure 1.10) a very simple example that sets up a minimal audio/video call between two browsers and does not use trickle ICE [16]. The example in the following section provides a more realistic example of what would happen in a normal browser-to-browser connection.

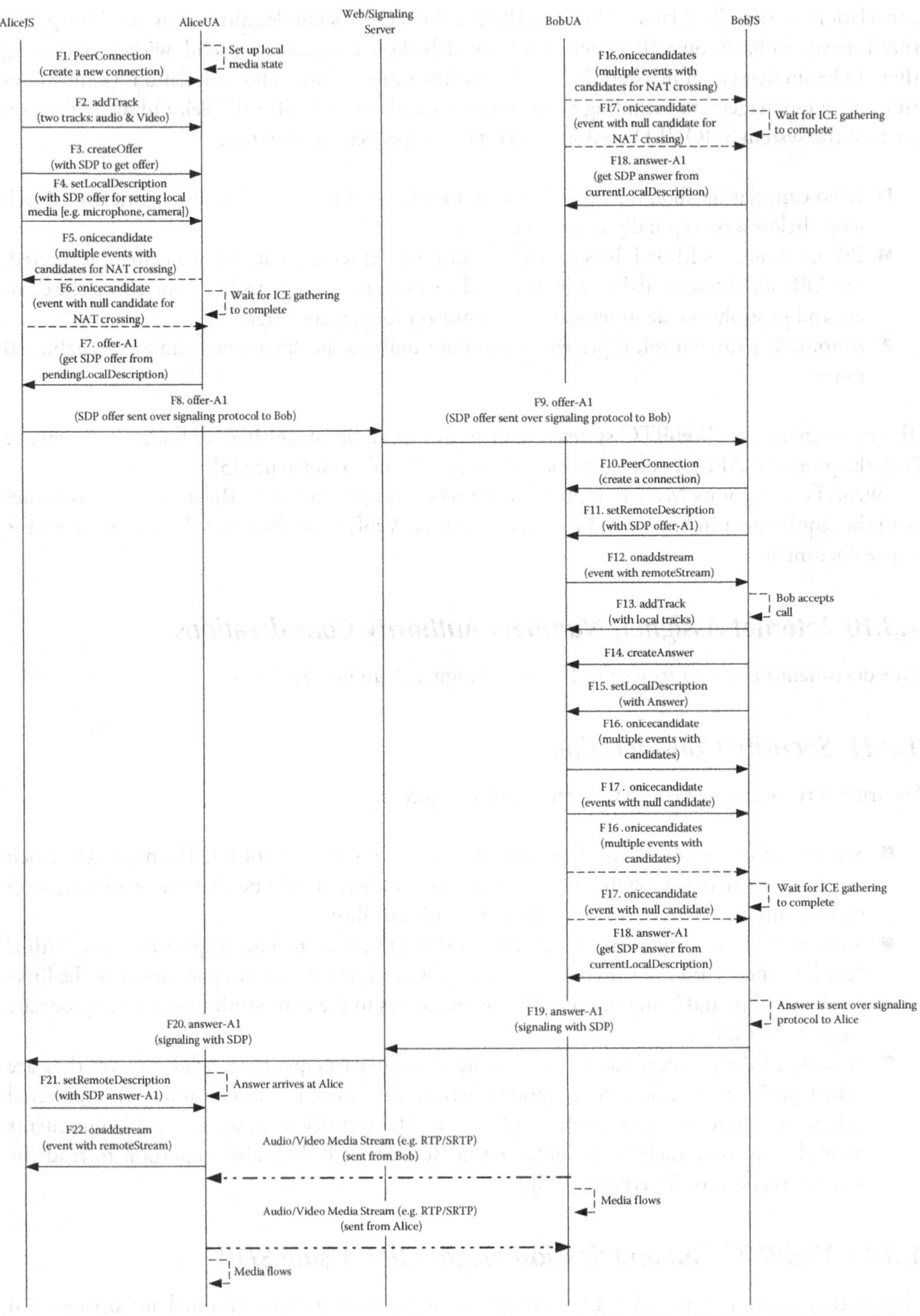

Figure 1.10 WebRTC multimedia session negotiation using SDP (Copyright: IETF).

The flow shows Alice's browser initiating the session to Bob's browser using an SDP offer. Bob's browser responds with an SDP answer for negotiating the parameters for setting up the multimedia session. In this simple call flow, it is assumed that all parameters of the multimedia session are acceptable to both and no further negotiation has been needed.

The messages from Alice's JS to Bob's JS are assumed to flow over some signaling protocol via a Web server. The JS on Alice's side and on Bob's side waits for all candidates before sending the offer or answer, so the offers and answers are complete. Trickle ICE [16] is not used. Both Alice and Bob are using the default policy of signaling message and media transfer described in Section 3. The details of the SDP parameters are described in the next section (that is, Section 2).

1.4 Summary

This chapter has discussed the basic characteristics of two call control protocols, namely SIP and WebRTC, to show why they need SDP for session negotiations. Full treatment of SIP can be found in Reference [17]. A given multimedia application can have many functional, performance, network/transport, and security parameters, and each communicating party may or may not have the capability or willingness to support all of those features equally. SDP is a mechanism to describe of all those session capabilities carried by call control protocols for the purpose of negotiations so that all of the conferencing parties can agree upon some common session parameters. However, the stage has just been set for how SDP is used by call control protocols. The details of SDP functional, performance, network/transport, and security parameters will be addressed in Sections 2–16. The end of this book (that is Section 18) provides detailed examples of session negotiations with all relevant SDP parameters for both SIP and WebRTC.

1.5 Problems

1. Why are both SIP and WebRTC known as networked multimedia call control protocols? How do they differ from a stand-alone nonnetworked protocol?
2. What are the differences between SIP and WebRTC call control protocols? How do they differ in their use of SDP?
3. What are the request and response messages in SIP? Describe each of these SIP messages showing end-to-end call flows.
4. Describe in detail data transport, data framing and securing, data formats, connection management, presentation and control, and local system support functions for WebRTC. Describe each of these WebRTC functions showing end-to-end call flows.
5. What are the WebRTC protocol design principles? How do they differ from SIP design principles?
6. What are the general, network and topology, multiple peers and streams, and audio processing requirements of the browser in supporting WebRTC protocol? How do they differ from those in SIP?
7. Describe the protocol architecture for both SIP and WebRTC including request and response messages, message-body, framing messages, tags, option tags, and request and response format messages, if any.

8. What are the network functional elements that are used for routing of call control messages between communicating parties for both SIP and WebRTC? Describe each of the network functional elements for both call control protocols.
9. Describe the logical call flow diagram for setting up the end-to-end call for point-to-point and multipoint communications using all network functional elements. Provide logical explanations for how SDP is used for session negotiation between the conferencing parties.
10. How does WebRTC provide interoperability with SIP? Describe the interoperability architecture between WebRTC and SIP in detail along with their pros and cons.

References

1. C. Holmberg, S. Hakansson, and G. Eriksson, "Web Real-Time Communication Use-cases and Requirements", draft-ietf-rtcweb-use-cases-and-requirements-16.txt, Informational, January 23, 2015.
2. WebRTC 1.0: Real-Time Communication Between Browsers, W3C Working Draft, June 05, 2017.
3. Media Capture and Streams, W3C Candidate Recommendation, May 19, 2016.
4. H. Alvestrand, "WebRTC Gateways," draft-ietf-rtcweb-gateways-02, Standards Track, January 21, 2016.
5. HTML5: A vocabulary and associated APIs for HTML and XHTML, W3C Working Draft, May 25, 2011.
6. XEP-0166: Jingle Version 1.1.1, Standards Track, XMPP Standards Foundation, May 2016.
7. H. Alvesrand, "Transports for WebRTC," draft-ietf-rtcweb-transports-17, April 29, 2017.
8. C. Perkins, M. Westerlund, and J. Ott, "Web Real-Time Communication (WebRTC): Media Transport and Use of RTP," draft-ietf-rtcweb-rtp-usage-26, Standards Track, March 17, 2016.
9. E. Rescorla, "Security Considerations for WebRTC," draft-ietf-rtcweb-security-08, Standards Track, February 26, 2015.
10. E. Rescorla, "WebRTC Security Architecture," draft-ietf-rtcweb-security-arch-11, Standards Track, March 7, 2015.
11. R. Jesup, S. Loreto, and M. Tuexen, "WebRTC Data Channels," draft-ietf-rtcweb-data-channel-13.txt, Standards Track, January 4, 2015.
12. R. Jesup, S. Loreto, and M. Tuexen, "WebRTC Data Channel Establishment Protocol," draft-ietf-rtcweb-data-protocol-09.txt, Standards Track, January 4, 2015.
13. C. Perkins, M. Westerlund, and J. Ott, "Web Real-Time Communication (WebRTC): Media Transport and Use of RTP," draft-ietf-rtcweb-rtp-usage-26, Standards Track, March 17, 2016.
14. J. Spittka, K. Vos, and J. M. Valin, "RTP Payload Format for the Opus Speech and Audio Codec," draft-ietf-payload-rtp-opus-11, Standards Track, April 14, 2015.
15. J. Uberti, C. Jennings, and E. Rescorla, Ed., "Javascript Session Establishment Protocol," draft-ietf-rtcweb-jsep-14, Standards Track, March 21, 2016.
16. E. Ivov, E. Marocco, and C. Holmberg, "A Session Initiation Protocol (SIP) usage for Trickle ICE," draft-ietf-mmusic-trickle-ice-sip-03, Standards Track, May 17, 2016.
17. R. R. Roy, "*Handbook on Session Initiation Protocol: Networked Multimedia Communications for IP Telephony*," CRC Press, Boca Raton, FL, 2016.

Chapter 2

Basic Session Description Protocol

Abstract

This chapter describes RFC 4566 that obsoletes RFCs 2327 and 3266. The Session Description Protocol (SDP) is designed to negotiate multimedia sessions that may consist of various features/capabilities/functionalities of audio, video, and/or data application between the communicating entities facilitating the common features/capabilities/functionalities of each application that are agreeable to support by the parties. In fact, SDP may not be a termed as a "protocol" per se, as it only provides the description/format of multimedia session parameters. This section defines the basic characteristics of the SDP describing multimedia sessions for the purposes of session announcement, session invitation, and other forms of multimedia session initiation. In addition, all terminologies and their definitions, and Augmented Backus-Naur Form (ABNF) related to SDP have been put together from all Request for Comments (RFCs) in this chapter.

2.1 Introduction

When initiating multimedia teleconferences, voice-over-IP calls, streaming videos, or other sessions, there is a requirement to convey media details, transport addresses, and other session description metadata to the participants. Session Description Protocol (SDP) provides a standard representation for such information, irrespective of how that information is transported. SDP is purely a format for session description: It does not incorporate a transport protocol, and it is intended to use different transport protocols as appropriate, including the Session Announcement Protocol (SAP)(RFC 2974), Session Initiation Protocol (SIP) (RFC 3261, also see Section 1.2, Real-Time Streaming Protocol (RTSP)(RFC 2326), electronic mail using the *Multipurpose Internet Mail Extensions* (MIME) extensions, and the Hypertext Transport Protocol (HTTP). SDP is intended to be general purpose so that it can be used in a wide range of network environments and applications. However, it is not intended to support negotiation of session content or media encodings: This is viewed as outside the scope of session description. The following sections (Sections 2.2–2.11) describe RFC 4566 specification that obsoletes RFC 2327 and RFC 3266.

2.2 Terminologies and Their Definitions

Many terminologies have been defined in SDP over the years since the publication of the original SDP RFCs 2327 and 3266 that are obsoleted by RFC 4566. In addition, many other Request for Comments (RFCs) have enhanced RFC 4566 defining new terminologies. All terminologies in relation to SDP are included in Table 2.1.

Table 2.1 Terminologies and Their Definitions

Terminology	*Definition*
"m=" line	SDP bodies contain one or more media descriptions. Each media description is identified by an SDP "m=" line [1].
5-tuple	A collection of the following values: source address, source port, destination address, destination port, and transport-layer protocol [1].
Actual configuration	An actual configuration specifies which combinations of SDP session parameters and media stream components can be used in the current offer/answer exchange and with what parameters. Use of an actual configuration does not require any further negotiation in the offer/answer exchange (RFC 6871: Standards Track).
Agent	An agent is the protocol implementation involved in the offer/answer exchange. There are two agents involved in an offer/answer exchange (RFC 3264: Standards Track).
Aggressive nomination	The process of picking a valid candidate pair for media traffic by including a flag in every STUN request such that the first one to produce a valid candidate pair is used for media (RFC 5245: Standards Track).
Answer	An SDP message sent by an answerer in response to an offer received from an offerer (RFC 3264: Standards Track).
Answerer	An agent that receives a session description from another agent describing aspects of desired media communication and then responds to that with its own session description (RFC 3264: Standards Track).
Answerer BUNDLE address	Within a given BUNDLE group, an IP address and port combination used by an answerer to receive all media specified by each "m=" line within the BUNDLE group [1].
Answerer BUNDLE-tag	The first identification-tag in a given SDP "group:BUNDLE" attribute identification-tag list in an answer [1].
Base	The base of a server reflexive candidate is the host candidate from which it was derived. A host candidate is also said to have a base, equal to that candidate itself. Similarly, the base of a relayed candidate is that candidate itself (RFC 5245: Standards Track).

(*Continued*)

Table 2.1 (*Continued*) Terminologies and Their Definitions

Terminology	*Definition*
Base attributes	Conventional SDP attributes appearing in the base configuration of a media block (RFC 6871: Standards Track).
Base configuration	The media configuration represented by a media block exclusive of all the capability negotiation attributes defined in this document, the base capability negotiation document (RFC 5939, see Section 8.1) or any other capability negotiation document. In an offer SDP, the base configuration corresponds to the actual configuration as defined in RFC 5939 (RFC 6871: Standards Track).
BUNDLE group	A set of "m=" lines, created using an SDP offer/answer exchange, which uses the same BUNDLE address for receiving media [1].
Bundled "m=" line	An "m=" line, whose identification-tag is placed in an SDP "group:BUNDLE" attribute identification-tag list in an offer or answer. Bundle-only "m=" line: A bundled "m=" line with an associated SDP "bundle-only" attribute [1].
Bundled media	All media specified by a given BUNDLE group. Initial offer: The first offer, within an SDP session (e.g., a SIP dialog when the SIP (RFC 3261, – also see Section 1.2) is used to carry SDP), in which the offerer indicates that it wants to create a given BUNDLE group [1].
Candidate	A transport address that is a potential point of contact for receipt of media. Candidates also have properties: their type (server reflexive, relayed, or host), priority, foundation, and base (RFC 5245: Standards Track).
Candidate pair	A pairing containing a local candidate and a remote candidate (RFC 5245: Standards Track).
Check list	An ordered set of candidate pairs an agent will use to generate checks (RFC 5245: Standards Track).
Check, connectivity check, STUN check	A STUN Binding request transaction for the purposes of verifying connectivity. A check is sent from the local candidate to the remote candidate of a candidate pair (RFC 5245: Standards Track).
Component	A component is a piece of a media stream requiring a single transport address; a media stream may require multiple components, each of which has to work for the media stream as a whole to work. For media streams based on RTP, there are two components per media stream, one for RTP and one for RTCP (RFC 5245: Standards Track).
Conference	A multimedia conference is a set of two or more communicating users along with the software they are using to communicate (RFC 4566: Standards Track).

(*Continued*)

Table 2.1 (*Continued*) Terminologies and Their Definitions

Terminology	*Definition*
Controlled agent	An ICE agent that waits for the controlling agent to select the final choice of candidate pairs
Controlling agent	The ICE agent that is responsible for selecting the final choice of candidate pairs and signaling them through STUN and an updated offer, if needed. In any session, one agent is always controlling. The other is the controlled agent (RFC 5245: Standards Track).
Conventional attribute	Any SDP attribute other than those defined by the series of capability negotiation specifications. Conventional SDP: An SDP record devoid of capability negotiation attributes (RFC 6871: Standards Track).
Core service platform ("core SPF")	Core service platform ("core SPF") is a macro functional block including session routing, interfaces to advanced services, and access control (RFC 6947: Informational).
Data path border element (DBE)	DBE denotes a functional element, located at the boundaries of an IP Telephony Administrative Domain (ITAD), that is responsible for intercepting media/data flows received from user agents (UAs) and relaying them to another DBE (or media servers, e.g., an announcement server or Interactive Voice Response (IVR)). An example of a DBE is a media gateway that intercepts RTP flows. A Signaling Border Element (SBE) may be located at the access segment (i.e., be the service contact point for UAs) or be located at the interconnection with adjacent domains (RFC 6406). (RFC 6947: Informational)
Decoding dependency	The class of relationships media partitions have to each other. At present, this memo defines two decoding dependencies: layered coding and multiple description coding (MDC) (RFC 4566: Standards Track).
Default destination/candidate	The default destination for a component of a media stream is the transport address that would be used by an agent that is not ICE aware. For the RTP component, the default IP address is in the "c=" line of the SDP, and the port is in the "m=" line. For the RTCP component, it is in the rtcp attribute when present, and when not present, the IP address is in the "c=" line and 1 plus the port is in the "m-" line. A default candidate for a component is one whose transport address matches the default destination for that component (RFC 5245: Standards Track).
File receiver	The endpoint that is willing to receive a file from the file sender (RFC 5547: Standards Track).

(*Continued*)

Table 2.1 (*Continued*) Terminologies and Their Definitions

Terminology	*Definition*
File selector	A tuple of file attributes that the SDP offerer includes in the SDP in order to select a file at the SDP answerer. This is described in more detail in Section 13.1.5 (RFC 5547: Standards Track).
File sender	The endpoint that is willing to send a file to the file receiver (RFC 5547: Standards Track).
Foundation	An arbitrary string that is the same for two candidates that have the same type, base IP address, protocol (UDP, TCP, etc.), and STUN or TURN server. If any of these are different, then the foundation will be different. Two candidate pairs with the same foundation pairs are likely to have similar network characteristics. Foundations are used in the frozen algorithm (RFC 5245: Standards Track).
Full	An ICE implementation that performs the complete set of functionality defined by this specification (RFC 5245: Standards Track).
Host candidate	A candidate obtained by binding to a specific port from an IP address on the host. This includes IP addresses on physical interfaces and logical ones, such as those obtained through Virtual Private Networks (VPNs) and Realm Specific IP (RSIP) (RFC 3102) (which live at the operating system level) (RFC 5245: Standards Track).
Identification-tag	A unique token value that is used to identify an "m=" line. The SDP "mid" attribute (RFC 5888), associated with an "m=" line, carries an unique identification-tag. The session-level SDP "group" attribute (RFC 5888) carries a list of identification-tags, identifying the "m=" lines associated with that particular "group" attribute [1].
Latent configuration	A latent configuration indicates which combinations of capabilities could be used in a future negotiation for the session and its associated media stream components. Latent configurations are neither ready for use nor offered for actual or potential use in the current offer/answer exchange. Latent configurations merely inform the other side of possible configurations supported by the entity. Those latent configurations may be used to guide subsequent offer/answer exchanges, but they are not offered for use as part of the current offer/answer exchange (RFC 6871: Standards Track).

(*Continued*)

Table 2.1 (*Continued*) Terminologies and Their Definitions

Terminology	*Definition*
Layered coding dependency	Each media partition is only useful (i.e., can be decoded) when all of the media partitions it depends on are available. The dependencies between media partitions therefore create a directed graph. Note: normally, in layered coding, the more media partitions are employed (following the rule), the better quality is possible (RFC 4566: Standards Track).
Lite	An ICE implementation that omits certain functions, implementing only as much as is necessary for a peer implementation that is full to gain the benefits of ICE. Lite implementations do not maintain any of the state machines and do not generate connectivity checks (RFC 5245: Standards Track).
Local candidate	A candidate that an agent has obtained and included in an offer or answer it sent (RFC 5245: Standards Track).
Media Bitstream	A valid, decodable stream, containing all media partitions generated by the encoder. A media bitstream normally conforms to a media coding standard (RFC 4566: Standards Track).
Media capability	The combined set of capabilities associated with expressing a media format and its relevant parameters (e.g., media format parameters and media specific parameters) (RFC 6871: Standards Track).
Media format capability	A media format, typically a media subtype such as pulse codec modulation (PCM) with μ-law algorithm (PCMU), H263-1998, or T38, expressed in the form of a capability (RFC 6871: Standards Track).
Media format parameter capability	A media format parameter ("a=fmtp" in conventional SDP) expressed in the form of a capability. The media format parameter capability is associated with a media format capability (RFC 6871: Standards Track).
Media partition	A subset of a media bitstream intended for independent transportation. An integer number of media partitions forms a media bitstream. In layered coding, a media partition represents one or more layers that are handled as a unit. In MDC coding, a media partition represents one or more descriptions that are handled as a unit (RFC 4566: Standards Track).

(*Continued*)

Table 2.1 (*Continued*) Terminologies and Their Definitions

Terminology	*Definition*
Media stream	A media stream is a single media instance (e.g., an audio stream or a video stream as well as a single whiteboard or shared application group). In SDP, a media stream is described by an "m=" line and its associated attributes (RFC 3264: Standards Track).
MDC dependency	N of M media partitions are required to form a media bitstream, but there is no hierarchy between these media partitions. Most MDC schemes aim at an increase of reproduced media quality when more media partitions are decoded. Some MDC schemes require more than one media partition to form an operation point (RFC 4566: Standards Track).
Nominated	If a valid candidate pair has its nominated flag set, it means that it may be selected by ICE for sending and receiving media (RFC 5245: Standards Track).
Offer	An SDP message sent by an offerer (RFC 3264: Standards Track)
Offerer	An agent that generates a session description in order to create or modify a session (RFC 3264: Standards Track)
Offerer BUNDLE address	Within a given BUNDLE group, an IP address and port combination used by an offerer to receive all media specified by each "m=" line within the BUNDLE group [1]
Offerer BUNDLE-tag	The first identification-tag in a given SDP "group:BUNDLE" attribute identification-tag list in an offer [1]
Operation point	In layered coding, a subset of a layered media bitstream that includes all media partitions required for reconstruction at a certain point of quality, error resilience, or another property, and that does not include any other media partitions. In MDC coding, a subset of an MDC media bitstream that is compliant with the MDC coding standard in question (RFC 4566: Standards Track)
Ordinary check	A connectivity check generated by an agent as a consequence of a timer that fires periodically, instructing it to send a check (RFC 5245: Standards Track)
Peer	From the perspective of one of the agents in a session, its peer is the other agent. Specifically, from the perspective of the offerer, the peer is the answerer. From the perspective of the answerer, the peer is the offerer (RFC 5245: Standards Track).

(*Continued*)

Table 2.1 (*Continued*) Terminologies and Their Definitions

Terminology	*Definition*
Peer reflexive candidate	A candidate whose IP address and port are a binding allocated by a NAT for an agent when it sent a STUN Binding request through the NAT to its peer (RFC 5245: Standards Track)
Potential configuration	A potential configuration indicates which combinations of capabilities can be used for the session and its associated media stream components. Potential configurations are not ready for use; however, they are offered for potential use in the current offer/answer exchange. They provide an alternative that may be used instead of the actual configuration, subject to negotiation in the current offer/answer exchange. See RFC 5939 for further details (RFC 6871: Standards Track).
Pull operation	A file transfer operation where the SDP offerer takes the role of the file receiver and the SDP answerer takes the role of the file sender (RFC 5547: Standards Track)
Push operation	A file transfer operation where the SDP offerer takes the role of the file sender and the SDP answerer takes the role of the file receiver (RFC 5547: Standards Track)
Regular nomination	The process of picking a valid candidate pair for media traffic by validating the pair with one STUN request and then picking it by sending a second STUN request with a flag indicating its nomination (RFC 5245: Standards Track)
Relayed candidate	A candidate obtained by sending a TURN Allocate request from a host candidate to a TURN server. The relayed candidate is resident on the TURN server, and the TURN server relays packets back toward the agent (RFC 5245: Standards Track)
Remote candidate	A candidate that an agent received in an offer or answer from its peer (RFC 5245: Standards Track)
Selected pair, selected candidate	The candidate pair selected by ICE for sending and receiving media is called the selected pair, and each of its candidates is called the selected candidate (RFC 5245: Standards Track).
Server reflexive candidate	A candidate whose IP address and port are a binding allocated by a NAT for an agent when it sent a packet through the NAT to a server. Server reflexive candidates can be learned by STUN servers using the Binding request or by TURN servers, which provides both a relayed and server reflexive candidate (RFC 5245: Standards Track).

(*Continued*)

Table 2.1 (*Continued*) Terminologies and Their Definitions

Terminology	*Definition*
Session	A multimedia session is a set of multimedia senders and receivers and the data streams flowing from senders to receivers. A multimedia conference is an example of a multimedia session (RFC 4566: Standards Track).
Session description	A well-defined format for conveying sufficient information to discover and participate in a multimedia session (RFC 4566: Standards Track)
Shared address	An IP address and port combination that is associated with multiple "m=" lines within an offer or answer [1]
Signaling path border element (SBE)	SBE denotes a functional element, located at the boundaries of an ITAD (RFC 2871), that is responsible for intercepting signaling flows received from UAs and relaying them to the core service platform. An SBE may be located at the access segment (i.e., be the service contact point for UAs), or be located at the interconnection with adjacent domains (RFC 6406). An SBE controls one or more DBEs. The SBE and DBE may be located in the same device (e.g., the SBC (RFC 5853)) or be separated (RFC 6947: Informational).
Subsequent offer	An offer that contains a BUNDLE group that has been created as part of a previous offer/answer exchange [1]
Transport address	The combination of an IP address and transport protocol (such as UDP or TCP) port (RFC 5245: Standards Track)
Triggered check	A connectivity check generated as a consequence of the receipt of a connectivity check from the peer (RFC 5245: Standards Track)
Unique address	An IP address and port combination that is associated with only one "m=" line in an offer or answer [1]
Valid list	An ordered set of candidate pairs for a media stream that has been validated by a successful STUN transaction (RFC 5245: Standards Track)

2.3 SDP Usage

Conversational multimedia applications are required to carry session descriptions that allow participants to agree on a set of compatible media types for setting up sessions. Multimedia streaming applications require the negotiation of an appropriate set of parameters for media delivery between the client and the server. In all cases, SDP is needed for carrying the multimedia session information between the communicating parties for negotiations. Even simple applications like session announcement require SDP to provide the recommended session description format. We discuss briefly some specific applications that use SDP.

2.3.1 Session Initiation

SIP (RFC 3261, also see Section 1.2) is an application-layer control protocol for creating, modifying, and terminating sessions such as Internet multimedia conferences, Internet telephone calls, and multimedia distribution. The SIP messages used to create sessions carry session descriptions that allow participants to agree on a set of compatible media types. These session descriptions are commonly formatted using SDP. When used with SIP, the offer/answer model (RFC 3264, see Section 3.1) provides a limited framework for negotiation using SDP.

2.3.2 WebRTC Session

Like SIP, the web real-time communication (WebRTC) protocol (see Section 1.3) is an application-layer protocol designed for interactive real-time communication (RTC) with audio, video, and/or data applications directly between browsers across the Internet. The WebRTC messages carry SDP for creation and modification of the conferencing session allowing browser-embedded Web application endpoints to agree on a set of compatible media types. For example, the SDP offer-answer mechanism used with WebRTC can be used for negotiation of the session.

2.3.3 Streaming Media

RTSP, specified in RFC 2326, is an application-level protocol for control over the delivery of data with real-time properties. RTSP provides an extensible framework to enable controlled, on-demand delivery of real-time data, such as audio and video. An RTSP client and server negotiate an appropriate set of parameters for media delivery, partially using SDP syntax to describe those parameters.

2.3.4 Email and the World Wide Web

Alternative means of conveying session descriptions include electronic mail and the World Wide Web (WWW). For both email and WWW distribution, the media type "application/sdp" is used. This enables the automatic launching of applications for participation in the session from the WWW client or mail reader in a standard manner. Note that announcements of multicast sessions made only via email or the WWW do not have the property that the receiver of a session announcement can necessarily receive the session because the multicast sessions may be restricted in scope, and access to the WWW server or reception of email is possible outside this scope.

2.3.5 Multicast Session Announcement

In order to assist the advertisement of multicast multimedia conferences and other multicast sessions, and to communicate the relevant session setup information to prospective participants, a distributed session directory may be used. An instance of such a session directory periodically sends packets containing a description of the session to a well-known multicast group. These advertisements are received by other session directories such that potential remote participants can use the session description to start the tools required to participate in the session. One protocol used to implement such a distributed directory is the SAP (RFC 2974). SDP provides the recommended session description format for such session announcements.

2.4 Requirements and Recommendations

The purpose of SDP is to convey information about media streams in multimedia sessions to allow the recipients of a session description to participate in the session. SDP is primarily intended for use in an internetwork, although it is sufficiently general that it can describe conferences in other network environments. Media streams can be many-to-many. Sessions need not be continually active. Thus far, multicast-based sessions on the Internet have differed from many other forms of conferencing in that anyone receiving the traffic can join the session (unless the session traffic is encrypted). In such an environment, SDP serves two primary purposes. It is a means to communicate the existence of a session, and it is a means to convey sufficient information to enable joining and participating in the session. In a unicast environment, only the latter purpose is likely to be relevant. An SDP session description includes the following:

- Session name and purpose
- Time(s) the session is active
- The media comprising the session
- Information needed to receive those media (addresses, ports, formats, etc.)

As resources necessary to participate in a session may be limited, some additional information may also be desirable:

- Information about the bandwidth to be used by the session
- Contact information for the person responsible for the session

In general, SDP must convey sufficient information to enable applications to join a session (with the possible exception of encryption keys) and to announce the resources to be used to any non-participants that may need to know. (This latter feature is primarily useful when SDP is used with a multicast SAP.)

2.4.1 Media and Transport Information

An SDP session description includes the following media information:

- The type of media (video, audio, etc.)
- The transport protocol (RTP/UDP/IP, H.320, etc.)
- The format of the media (H.261 video, MPEG video, etc.)

In addition to media format and transport protocol, SDP conveys address and port details. For an IP multicast session, these comprise the multicast group address for media.

- The transport port for media: This address and port are the destination address and destination port of the multicast stream, whether being sent, received, or both. For unicast IP sessions, the following are conveyed:
  - The remote address for media
  - The remote transport port for media

The semantics of this address and port depend on the media and transport protocol defined. By default, this SHOULD be the remote address and remote port to which data is sent. Some media types may redefine this behavior, but this is NOT RECOMMENDED since it complicates implementations (including middleboxes that must parse the addresses to open network address translator (NAT) or firewall pinholes).

2.4.2 Timing Information

Sessions may be either bounded or unbounded in time. Regardless of whether they are bounded, they may be only active at specific times. SDP can convey:

- An arbitrary list of start and stop times bounding the session
- For each bound, repeat times such as "every Wednesday at 10 am for one hour"

This timing information is globally consistent, irrespective of local time zone or daylight saving time (see Section 2.5.9).

2.4.3 Private Sessions

It is possible to create both public sessions and private sessions. SDP itself does not distinguish between these; private sessions are typically conveyed by encrypting the session description during distribution. The details of how encryption is performed are dependent on the mechanism used to convey SDP; mechanisms are currently defined for SDP transported using SAP (RFC 2974) and SIP (RFC 3261, also see Section 1.2), and others may be defined in the future. If a session announcement is private, it is possible to use that private announcement to convey encryption keys necessary to decode each of the media in a conference, including enough information to know which encryption scheme is used for each media.

2.4.4 Obtaining Further Information about a Session

A session description should convey enough information to decide whether to participate. SDP may include additional pointers in the form of uniform resource identifiers (URIs) for more information about the session.

2.4.5 Categorization

When many session descriptions are being distributed by SAP, or any other advertisement mechanism, it may be desirable to filter session announcements that are of interest from those that are not. SDP supports a categorization mechanism that is capable of being automated (the "a=cat:" attribute; see Section 2.6).

2.4.6 Internationalization

The SDP specification recommends the use of the International Standard Organization (ISO) 10646 character sets in the Unicode Transformation Format 8 (UTF-8) encoding (RFC 3629) to allow many different languages to be represented. However, to assist in compact representations, SDP also allows other character sets such as ISO 8859-1 to be used when desired. Internationalization only applies to free-text fields (session name and background information) and not to SDP as a whole.

2.5 SDP Specification

In general, the SDP can be grouped into major five components for the session description as shown in Figure 2.1: session metadata, stream, Quality of Service (QOS)/grouping, network, and security.

The session metadata describes the session itself such as SDP protocol version, originator of the session and session identification, and the time a session is active. The stream description contains detail functional parameters of media (e.g., audio, video) streams. The QOS description contains all performance parameters of media streams, and it may be accompanied with information on how multiple media streams can be grouped together for saving bandwidth and other resources. The network description may include what kinds of transport (e.g., TCP, UDP) and network (e.g., IP) protocol may be used for sending and receiving media streams by the conferencing parties. The security description may include security parameters such as encryption key, authentication, authorization, nonrepudiation, and integrity.

An SDP session description is denoted by the media type "application/sdp" (see Section 2.8). An SDP session description is entirely textual using the ISO 10646 character set in UTF-8 encoding. SDP field names and attribute names use only the United States – American Standard Code for Information Interchange (US-ASCII) subset of UTF-8, but textual fields and attribute values MAY use the full ISO 10646 character set. Field and attribute values that use the full UTF-8 character set are never directly compared; hence, there is no requirement for UTF-8 normalization. The textual form, as opposed to a binary encoding such as Abstract Syntax Notation One (ASN.1) or eXternal Data Representation (XDR), was chosen to enhance portability, to enable a variety of transports to be used, and to allow flexible, text-based toolkits to be used to generate and process session descriptions.

However, since SDP may be used in environments where the maximum permissible size of a session description is limited, the encoding is deliberately compact. Also, since announcements may be transported via very unreliable means or damaged by an intermediate caching server, the encoding was designed with strict order and formatting rules so that most errors would result in malformed session announcements that could be detected easily and discarded. This also allows rapid discarding of encrypted session announcements for which a receiver does not have the correct key. An SDP session description consists of a number of lines of text of the form:

```
<type>=<value>
```

where <type> MUST be exactly one case-significant character and <value> is structured text whose format depends on <type>. In general, <value> is either a number of fields delimited by a

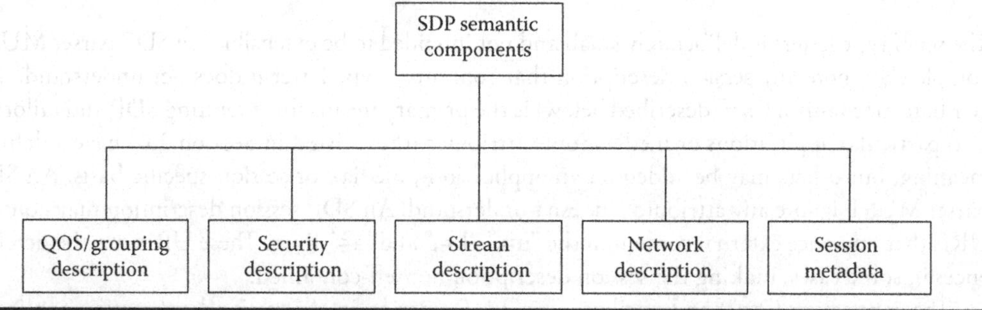

Figure 2.1 SDP semantic components.

single space character or a free format string and is case-significant unless a specific field defines otherwise. Whitespace MUST NOT be used on either side of the "=" sign.

An SDP session description consists of a session-level section followed by zero or more media-level sections. The session-level part starts with a "v=" line and continues to the first media-level section. Each media-level section starts with an "m=" line and continues to the next media-level section or end of the whole session description. In general, session-level values are the default for all media unless overridden by an equivalent media-level value. Some lines in each description are REQUIRED and some are OPTIONAL, but all MUST appear in exactly the order given here (the fixed order greatly enhances error detection and allows for a simple parser). OPTIONAL items are marked with a "*."

Session description

```
v=(protocol version)
o=(originator and session identifier)
s=(session name)
i=* (session information)
u=* (URI of description)
e=* (email address)
p=* (phone number)
c=* (connection information -- not required if included in all media)
b=* (zero or more bandwidth information lines)
One or more time descriptions ("t=" and "r=" lines; see below)
z=* (time zone adjustments)
k=* (encryption key)
a=* (zero or more session attribute lines)
Zero or more media descriptions
```

Time description

```
t= (time the session is active)
r=* (zero or more repeat times)
```

Media description, if present

```
m=(media name and transport address)
i=* (media title)
c=* (connection information -- optional if included at session level)
b=* (zero or more bandwidth information lines)
k=* (encryption key)
a=* (zero or more media attribute lines)
```

The set of type letters is deliberately small and not intended to be extensible; an SDP parser MUST completely ignore any session description that contains a type letter it does not understand. The attribute mechanism ("a=" described below) is the primary means for extending SDP and tailoring it to particular applications or media. Some attributes (those listed in Section 2.6) have a defined meaning, but others may be added on an application-, media-, or session-specific basis. An SDP parser MUST ignore any attribute it doesn't understand. An SDP session description may contain URIs that reference external content in the "u=," "k=," and "a=" lines. These URIs may be dereferences in some cases, making the session description nonself-contained.

The connection ("c=") and attribute ("a=") information in the session-level section applies to all the media of that session unless overridden by connection information or an attribute of the same name in the media description. For instance, in the example here, each media behaves as if it were given a "recvonly" attribute. An example SDP description is:

```
v=0
o=jdoe 2890844526 2890842807 IN IP4 10.47.16.5
s=SDP Seminar
i=A Seminar on the session description protocol
u=http://www.example.com/seminars/sdp.pdf
e=j.doe@example.com (Jane Doe)
c=IN IP4 224.2.17.12/127
t=2873397496 2873404696
a=recvonly
m=audio 49170 RTP/AVP 0
m=video 51372 RTP/AVP 99
a=rtpmap:99 h263-1998/90000
```

Note: Audio-Visual-Profile (AVP)

Text fields such as the session name and information are octet strings that may contain any octet with the exceptions of 0x00 (Null), 0x0a (ASCII newline), and 0x0d (ASCII carriage return). The sequence CRLF (0x0d0a) is used to end a record, although parsers SHOULD be tolerant and accept records terminated with a single newline character. If the "a=charset" attribute is not present, these octet strings MUST be interpreted as containing ISO-10646 characters in UTF-8 encoding (the presence of the "a=charset" attribute may force some fields to be interpreted differently).

A session description can contain domain names in the "o=," "u=," "e=," "c=," and "a=" lines. Any domain name used in SDP MUST comply with RFCs 1034 and 1035. Internationalized domain names (IDNs) MUST be represented using the ASCII Compatible Encoding (ACE) form defined in RFCs 5890 and 5891 that obsolete RFC 3490 and MUST NOT be directly represented in UTF-8 or any other encoding (this requirement is for compatibility with RFC 3513 that obsoletes RFC 2327 and other SDP-related standards, which predate the development of IDNs).

2.5.1 Protocol Version ("v=")

```
v=0
```

The "v=" field gives the version of the SDP. This memo defines version 0. There is no minor version number.

2.5.2 Origin ("o=")

```
o=<username> <sess-id> <sess-version> <nettype> <addrtype>
  <unicast-address>
```

The "o=" field gives the originator of the session (his/her username and the address of the user's host) plus a session identifier and version number:

- <username> is the user's login on the originating host, or it is "-" if the originating host does not support the concept of user identifications (IDs). The <username> MUST NOT contain spaces.

- <sess-id> is a numeric string such that the tuple of <username>, <sess-id>, <nettype>, <addrtype>, and <unicast-address> forms a globally unique identifier for the session. The method of <sess-id> allocation is up to the creating tool, but it has been suggested that a Network Time Protocol (NTP) format timestamp be used to ensure uniqueness defined in RFC 5905 that obsoletes RFC 1305.
- <sess-version> is a version number for this session description. Its usage is up to the creating tool, as long as <sess-version> is increased when a modification is made to the session data. Again, it is RECOMMENDED that an NTP format timestamp be used.
- <nettype> is a text string giving the type of network. Initially "IN" is defined to have the meaning "Internet," but other values MAY be registered in the future (see Section 2.8).
- <addrtype> is a text string giving the type of the address that follows. Initially "IP4" and "IP6" are defined, but other values MAY be registered in the future (see Section 2.8).
- <unicast-address> is the address of the machine from which the session was created. For an address type of IP4, this is either the fully qualified domain name (FQDN) of the machine or the dotted decimal representation of the IP version 4 address of the machine. For an address type of IP6, this is either the fully qualified domain name of the machine or the compressed textual representation of the IP version 6 address of the machine. For both IP4 and IP6, the fully qualified domain name is the form that SHOULD be given unless this is unavailable, in which case the globally unique address MAY be substituted. A local IP address MUST NOT be used in any context where the SDP description might leave the scope in which the address is meaningful (for example, a local address MUST NOT be included in an application-level referral that might leave the scope).

In general, the "o=" field serves as a globally unique identifier for this version of this session description, and the subfields excepting the version taken together identify the session irrespective of any modifications. For privacy reasons, it is sometimes desirable to obfuscate the username and IP address of the session originator. If this is a concern, an arbitrary <username> and private <unicast-address> MAY be chosen to populate the "o=" field, provided these are selected in a manner that does not affect the global uniqueness of the field.

2.5.3 Session Name ("s=")

```
s=<session name>
```

The "s=" field is the textual session name. There MUST be one and only one "s=" field per session description. The "s=" field MUST NOT be empty and SHOULD contain ISO 10646 characters (but see also the "a=charset" attribute). If a session has no meaningful name, the value "s= " SHOULD be used (i.e., a single space as the session name).

2.5.4 Session Information ("i=")

```
i=<session description>
```

The "i=" field provides textual information about the session. There MUST be at most one session-level "i=" field per session description and at most one "i=" field per media. If the "a=charset" attribute is present, it specifies the character set used in the "i=" field. If the "a=charset" attribute is not present, the "i=" field MUST contain ISO 10646 characters in UTF-8 encoding.

A single "i=" field MAY also be used for each media definition. In media definitions, "i=" fields are primarily intended for labeling media streams. As such, they are most likely to be useful when a single session has more than one distinct media stream of the same media type. An example would be two different whiteboards: One for slides, and one for feedback and questions. The "i=" field is intended to provide a free-form human-readable description of the session or the purpose of a media stream. It is not suitable for parsing by automata.

2.5.5 URI ("u=")

```
u=<uri>
```

A URI is a uniform resource identifier as used by WWW clients specified in RFC 3986. The URI should be a pointer to additional information about the session. This field is OPTIONAL, but if it is present it MUST be specified before the first media field. No more than one URI field is allowed per session description.

2.5.6 Email Address and Phone Number ("e=" and "p=")

```
e=<email-address>
p=<phone-number>
```

The "e=" and "p=" lines specify contact information for the person responsible for the conference. This is not necessarily the person that created the conference announcement. Inclusion of an email address or phone number is OPTIONAL. Note that the previous version of SDP specified that either an email field or a phone field MUST be specified, but this was widely ignored. The change brings the specification into line with common usage. If an email address or phone number is present, it MUST be specified before the first media field. More than one email or phone field can be given for a session description.

Phone numbers SHOULD be given in the form of an international public telecommunication number (see ITU-T Recommendation E.164) preceded by a "+." Spaces and hyphens may be used to split up a phone field to aid readability if desired. For example:

```
p=+1 617 555-6011
```

Both email addresses and phone numbers can have an OPTIONAL free text string associated with them, normally giving the name of the person who may be contacted. This MUST be enclosed in parentheses if present. For example:

```
e=j.doe@example.com (Jane Doe)
```

The alternative RFC 2822 (that was made obsolete by RFC 5322) name quoting convention is also allowed for both email addresses and phone numbers. For example:

```
e=Jane Doe <j.doe@example.com>
```

The free text string SHOULD be in the ISO-10646 character set with UTF-8 encoding, or alternatively in ISO-8859-1 or other encodings if the appropriate session-level "a=charset" attribute is set.

2.5.7 Connection Data ("c=")

```
c=<nettype> <addrtype> <connection-address>
```

The "c=" field contains connection data. A session description MUST contain either at least one "c=" field in each media description or a single "c=" field at the session level. It MAY contain a single session-level "c=" field and additional "c=" field(s) per media description, in which case the per-media values override the session-level settings for the respective media. The first subfield ("<nettype>") is the network type, which is a text string giving the type of network. Initially, "IN" is defined to have the meaning "Internet," but other values MAY be registered in the future (see Section 2.8). The second subfield ("<addrtype>") is the address type. This allows SDP to be used for sessions that are not IP based. This memo only defines IP4 and IP6, but other values MAY be registered in the future (see Section 2.8).

The third subfield ("<connection-address>") is the connection address. OPTIONAL subfields MAY be added after the connection address depending on the value of the <addrtype> field. When the <addrtype> is IP4 and IP6, the connection address is defined as follows:

- If the session is multicast, the connection address will be an IP multicast group address. If the session is not multicast, then the connection address contains the unicast IP address of the expected data source or data relay or data sink as determined by additional attribute fields. It is not expected that unicast addresses will be given in a session description that is communicated by a multicast announcement, though this is not prohibited.
- Sessions using an IPv4 multicast connection address MUST also have a time-to-live (TTL) value present in addition to the multicast address. The TTL and the address together define the scope with which multicast packets sent in this conference will be sent. TTL values MUST be in the range 0–255. Although the TTL MUST be specified, its use to scope multicast traffic is deprecated; applications SHOULD use an administratively scoped address instead.

The TTL for the session is appended to the address using a slash as a separator. An example is:

```
c=IN IP4 224.2.36.42/127
```

IPv6 multicast does not use TTL scoping; hence, the TTL value MUST NOT be present for IPv6 multicast. It is expected that IPv6 scoped addresses will be used to limit the scope of conferences. Hierarchical or layered encoding schemes are data streams where the encoding from a single media source is split into a number of layers. The receiver can choose the desired quality (and hence bandwidth) by only subscribing to a subset of these layers. Such layered encodings are normally transmitted in multiple multicast groups to allow multicast pruning. This technique keeps unwanted traffic from sites only requiring certain levels of the hierarchy. For applications requiring multiple multicast groups, we allow the following notation to be used for the connection address:

```
<base multicast address>[/<ttl>]/<number of addresses>
```

If the number of addresses is not given, it is assumed to be one. Multicast addresses so assigned are contiguously allocated above the base address, so that, for example:

```
c=IN IP4 224.2.1.1/127/3
```

would state that addresses 224.2.1.1, 224.2.1.2, and 224.2.1.3 are to be used at a TTL of 127. This is semantically identical to including multiple "c=" lines in a media description:

```
c=IN IP4 224.2.1.1/127
c=IN IP4 224.2.1.2/127
c=IN IP4 224.2.1.3/127
```

Similarly, an IPv6 example would be:

```
c=IN IP6 FF15::101/3
```

which is semantically equivalent to:

```
c=IN IP6 FF15::101
c=IN IP6 FF15::102
c=IN IP6 FF15::103
```

(remembering that the TTL field is not present in IPv6 multicast).

Multiple addresses or "c=" lines MAY be specified on a per-media basis only if they provide multicast addresses for different layers in a hierarchical or layered encoding scheme. They MUST NOT be specified for a session-level "c=" field. The slash notation for multiple addresses just described MUST NOT be used for IP unicast addresses.

2.5.8 Bandwidth ("b=")

```
b=<bwtype>:<bandwidth>
```

This OPTIONAL field denotes the proposed bandwidth to be used by the session or media. The <bwtype> is an alphanumeric modifier giving the meaning of the <bandwidth> figure. Two values are defined in this specification, but other values MAY be registered in the future (see Section 2.8 and RFCs 3556 and 3890):

- CT: If the bandwidth of a session or media in a session is different from the bandwidth implicit from the scope, a "b=CT:..." line SHOULD be supplied for the session giving the proposed upper limit to the bandwidth used (the "conference total" bandwidth). The primary purpose of this is to give an approximate idea as to whether two or more sessions can coexist simultaneously. When using the CT modifier with Real-Time Transport Protocol (RTP), if several RTP sessions are part of the conference, the conference total refers to total bandwidth of all RTP sessions.
- AS: The bandwidth is interpreted to be application specific (it will be the application's concept of maximum bandwidth). Normally, this will coincide with what is set on the application's "maximum bandwidth" control if applicable. For RTP-based applications, AS gives the RTP "session bandwidth" as defined in Section 6.2 of RFC 3550.

Note that CT gives a total bandwidth figure for all media at all sites. AS gives a bandwidth figure for a single media at a single site, although there may be many sites sending simultaneously. A prefix "X-" is defined for <bwtype> names. This is intended for experimental purposes only. For example:

```
b=X-YZ:128
```

Use of the "X-" prefix is NOT RECOMMENDED: Instead, new modifiers SHOULD be registered with Internet Assigned Numbers Authority (IANA) in the standard namespace. SDP parsers MUST ignore bandwidth fields with unknown modifiers. Modifiers MUST be alphanumeric and, although no length limit is given, it is recommended that they be short. The <bandwidth> is interpreted as kilobits per second by default. The definition of a new <bwtype> modifier MAY specify that the bandwidth is to be interpreted in some alternative unit (the "CT" and "AS" modifiers defined in this memo use the default units).

2.5.9 Timing ("t=")

```
t=<start-time> <stop-time>
```

The "t=" lines specify the start and stop times for a session. Multiple "t=" lines MAY be used if a session is active at multiple irregularly spaced times; each additional "t=" line specifies an additional period of time for which the session will be active. If the session is active at regular times, an "r=" line (see below) should be used in addition to, and following, a "t=" line, in which case the "t=" line specifies the start and stop times of the repeat sequence.

The first and second subfields give the start and stop times, respectively, for the session. These values are the decimal representation of NTP time values in seconds since 1900 (RFC 1305). To convert these values to UNIX time, subtract decimal 2208988800. NTP timestamps are elsewhere represented by 64-bit values, which wrap sometime in the year 2036. Since SDP uses an arbitrary length decimal representation, this should not cause an issue (SDP timestamps MUST continue counting seconds since 1900, NTP will use the value modulo the 64-bit limit).

If the <stop-time> is set to zero, then the session is not bounded, though it will not become active until after the <start-time>. If the <start-time> is also zero, the session is regarded as permanent. User interfaces SHOULD strongly discourage the creation of unbounded and permanent sessions as they give no information about when the session is actually going to terminate, and so make scheduling difficult. The general assumption may be made, when displaying unbounded sessions that have not timed out to the user, that an unbounded session will only be active for half an hour from the current time or the session start time, whichever is later. If behavior other than this is required, an end-time SHOULD be given and modified as appropriate when new information becomes available about when the session should really end. Permanent sessions may be shown to the user as never being active unless there are associated repeat times that state precisely when the session will be active.

2.5.10 Repeat Times ("r=")

```
r=<repeat interval> <active duration> <offsets from start-time>
```

"r=" fields specify repeat times for a session. For example, if a session is active at 10 am on Monday and 11 am on Tuesday for one hour each week for three months, then the <start-time> in the corresponding "t=" field would be the NTP representation of 10 am on the first Monday, the <repeat interval> would be one week, the <active duration> would be one hour, and the offsets would be zero and 25 hours. The corresponding "t=" field stop time would be the NTP representation of the end of the last session three months later. By default, all fields are in seconds, so the "r=" and "t=" fields might be the following:

```
t=3034423619 3042462419
r=604800 3600 0 90000
```

To make descriptions more compact, times may also be given in units of days, hours, or minutes. The syntax for these is a number immediately followed by a single case-sensitive character. Fractional units are not allowed; a smaller unit should be used instead. The following unit specification characters are allowed:

```
d - days (86400 seconds)
h - hours (3600 seconds)
m - minutes (60 seconds)
s - seconds (allowed for completeness)
```

Thus, this session announcement could also have been written: r=7d 1h 0 25h. Monthly and yearly repeats cannot be directly specified with a single SDP repeat time; instead, separate "t=" fields should be used to explicitly list the session times.

2.5.11 Time Zones ("z=")

```
z=<adjustment time> <offset> <adjustment time> <offset> ....
```

To schedule a repeated session that spans a change from daylight saving time to standard time or vice versa, it is necessary to specify offsets from the base time. This is required because different time zones change time at different times of day, different countries change to or from daylight saving time on different dates, and some countries do not have daylight saving time at all.

Thus, in order to schedule a session that is at the same time winter and summer, it must be possible to specify unambiguously by whose time zone a session is scheduled. To simplify this task for receivers, we allow the sender to specify the NTP time that a time zone adjustment happens and the offset from the time when the session was first scheduled. The "z=" field allows the sender to specify a list of these adjustment times and offsets from the base time. An example might be the following:

```
z=2882844526 -1h 2898848070 0
```

This specifies that at time 2882844526, the time base by which the session's repeat times are calculated is shifted back by one hour, and that at time 2898848070, the session's original time base is restored. Adjustments are always relative to the specified start time; they are not cumulative. Adjustments apply to all "t=" and "r=" lines in a session description. If a session is likely to last several years, it is expected that the session announcement will be modified periodically rather than transmit several years' worth of adjustments in one session announcement.

2.5.12 Encryption Keys ("k=")

```
k=<method>
k=<method>:<encryption key>
```

If transported over a secure and trusted channel, the SDP MAY be used to convey encryption keys. A simple mechanism for key exchange is provided by the key field ("k="), although this is primarily supported for compatibility with older implementations, and its use is NOT RECOMMENDED. Work is in progress to define new key exchange mechanisms specified in RFCs 4567 and 4568 for use with SDP, and it is expected that new applications will use those mechanisms.

A key field is permitted before the first media entry (in which case it applies to all media in the session), or for each media entry as required. The format of keys and their usage are outside the scope of this document, and the key field provides no way to indicate the encryption algorithm to be used, key type, or other information about the key; this is assumed to be provided by the higher-level protocol using SDP. If there is a need to convey this information within SDP, the extensions mentioned previously SHOULD be used. Many security protocols require two keys: one for confidentiality, another for integrity. This specification does not support the transfer of two keys.

The method indicates the mechanism to be used to obtain a usable key by external means or from the encoded encryption key given. The following methods are defined:

```
k=clear:<encryption key>
```

The encryption key is included untransformed in this key field. This method MUST NOT be used unless it can be guaranteed that the SDP is conveyed over a secure channel. The encryption key is interpreted as text according to the charset attribute; use the "k=base64:" method to convey characters that are otherwise prohibited in SDP.

```
k=base64:<encoded encryption key>
```

The encryption key is included in this key field but has been base64 encoded (RFC 4648 that obsoletes RFC 3548) because it includes characters that are prohibited in SDP. This method MUST NOT be used unless it can be guaranteed that the SDP is conveyed over a secure channel.

```
k=uri:<URI to obtain key>
```

A URI is included in the key field. The URI refers to the data containing the key and may require additional authentication before the key can be returned. When a request is made to the given URI, the reply should specify the encoding for the key. The URI is often an Secure Socket Layer/Transport Layer Security (SSL/TLS)-protected HTTP URI ("https:"), although this is not required.

```
k=prompt
```

No key is included in this SDP description, but the session or media stream referred to by this key field is encrypted. The user should be prompted for the key when attempting to join the session, and this user-supplied key should then be used to decrypt the media streams. The use of user-specified keys is NOT RECOMMENDED, since such keys tend to have weak security properties. The key field MUST NOT be used unless it can be guaranteed that the SDP is conveyed over a secure and trusted channel. An example of such a channel might be SDP embedded inside an S/MIME message or a TLS-protected HTTP session. It is important to ensure that the secure channel is with the party that is authorized to join the session, not an intermediary: if a caching proxy server is used, it is important to ensure that the proxy is either trusted or unable to access the SDP.

2.5.13 Attributes ("a=")

```
a=<attribute>
a=<attribute>:<value>
```

Attributes are the primary means for extending SDP. Attributes may be defined as "session level," "media level," or both. A media description may have any number of attributes ("a=" fields) that are media specific. These are referred to as "media-level" attributes and add information about the media stream. Attribute fields can also be added before the first media field; these "session-level" attributes convey additional information that applies to the conference as a whole rather than to individual media. Attribute fields may be of two forms:

- A property attribute is simply of the form "a=<flag>." These are binary attributes, and the presence of the attribute conveys that it is a property of the session. An example might be "a=recvonly."
- A value attribute is of the form "a=<attribute>:<value>." For example, a whiteboard could have the value attribute "a=orient: landscape."

Attribute interpretation depends on the media tool being invoked. Thus, receivers of session descriptions should be configurable in their interpretation of session descriptions in general and of attributes in particular. Attribute names MUST use the US-ASCII subset of ISO-10646/UTF-8.

Attribute values are octet strings and MAY use any octet value except 0x00 (Null), 0x0A (LF), and 0x0D (CR). By default, attribute values are to be interpreted as in ISO-10646 character set with UTF-8 encoding. Unlike other text fields, attribute values are NOT normally affected by the "charset" attribute as this would make comparisons against known values problematic. However, when an attribute is defined, it can be defined to be charset dependent, in which case its value should be interpreted in the session charset rather than in ISO-10646. Attributes MUST be registered with IANA (see Section 2.8). If an attribute is received that is not understood, it MUST be ignored by the receiver.

2.5.14 Media Descriptions ("m=")

```
m=<media> <port> <proto> <fmt> ...
```

A session description may contain a number of media descriptions. Each media description starts with an "m=" field and is terminated by either the next "m=" field or by the end of the session description. A media field has several subfields:

- <media> is the media type. Currently defined media are "audio," "video," "text," "application," and "message," although this list may be extended in the future (see Section 2.8).
- <port> is the transport port to which the media stream is sent. The meaning of the transport port depends on the network being used as specified in the relevant "c=" field and on the

transport protocol defined in the <proto> subfield of the media field. Other ports used by the media application (such as the RTCP port (RFC 3550) MAY be derived algorithmically from the base media port or MAY be specified in a separate attribute (for example, "a=rtcp:" as defined in RFC 3605). If noncontiguous ports are used or if they don't follow the parity rule of even RTP ports and odd RTCP ports, the "a=rtcp:" attribute MUST be used. Applications that are requested to send media to a <port> that is odd and where the "a=rtcp:" is present MUST NOT subtract 1 from the RTP port: that is, they MUST send the RTP to the port indicated in <port> and send the RTCP to the port indicated in the "a=rtcp" attribute.

For applications where hierarchically encoded streams are being sent to a unicast address, it may be necessary to specify multiple transport ports. This is done using a similar notation to that used for IP multicast addresses in the "c=" field:

```
m=<media> <port>/<number of ports> <proto> <fmt> ...
```

In such a case, the ports used depend on the transport protocol. For RTP, the default is that only the even-numbered ports are used for data with the corresponding one-higher odd ports used for the RTCP belonging to the RTP session and the <number of ports> denoting the number of RTP sessions. For example:

```
m=video 49170/2 RTP/AVP 31
```

would specify that ports 49170 and 49171 form one RTP/RTCP pair and 49172 and 49173 form the second RTP/RTCP pair. RTP/AVP is the transport protocol and 31 is the format (see below). If noncontiguous ports are required, they must be signaled using a separate attribute (for example, "a=rtcp:" as defined in RFC 3605 (see Section 12.3). If multiple addresses are specified in the "c=" field and multiple ports are specified in the "m=" field, a one-to-one mapping from port to the corresponding address is implied. For example:

```
c=IN IP4 224.2.1.1/127/2
m=video 49170/2 RTP/AVP 31
```

would imply that address 224.2.1.1 is used with ports 49170 and 49171, and address 224.2.1.2 is used with ports 49172 and 49173. The semantics of multiple "m=" lines using the same transport address are undefined. This implies that, unlike limited past practice, there is no implicit grouping defined by such means and an explicit grouping framework, for example, RFC 3388 that was made obsolete by RFC 5888, should instead be used to express the intended semantics.

- <proto> is the transport protocol. The meaning of the transport protocol is dependent on the address type field in the relevant "c=" field. Thus a "c=" field of IP4 indicates that the transport protocol runs over IP4. The following transport protocols are defined but may be extended through registration of new protocols with IANA (see Section 2.8):
  - udp: denotes an unspecified protocol running over User Datagram Protocol (UDP).
  - RTP/AVP: denotes RTP (RFC 3550) used under the RTP Profile for Audio and Video Conferences with Minimal Control (RFC 3551) running over UDP. Note: AVP.
  - RTP/SAVP: denotes the Secure Real-time Transport Protocol (SRTP) (RFC 3711) running over UDP. Note: Secure AVP (SAVP).

  The main reason to specify the transport protocol in addition to the media format is that the same standard media formats may be carried over different transport protocols even when the network protocol is the same. A historical example is vat pulse code modulation (PCM) audio and RTP PCM audio; another might be TCP/RTP PCM audio. In addition, relays and monitoring tools that are transport-protocol-specific but format-independent are possible.

- <fmt> is a media format description. The fourth and any subsequent subfields describe the format of the media. The interpretation of the media format depends on the value of the <proto> subfield.

  If the <proto> subfield is "RTP/AVP" or "RTP/SAVP," the <fmt> subfields contain RTP payload type numbers. When a list of payload type numbers is given, this implies that all payload formats MAY be used in the session, but the first of these formats SHOULD be used as the default format for the session. For dynamic payload type assignments the "a=rtpmap:" attribute (see Section 2.6) SHOULD be used to map from an RTP payload type number to a media encoding name that identifies the payload format. The "a=fmtp:" attribute MAY be used to specify format parameters (see Section 2.6).

  If the <proto> subfield is "udp" the <fmt> subfields MUST reference a media type describing the format under the "audio," "video," "text," "application," or "message" top-level media types. The media type registration SHOULD define the packet format for use with UDP transport. For media using other transport protocols, the <fmt> field is protocol specific. Rules for interpretation of the <fmt> subfield MUST be defined when registering new protocols with IANA (see Section 2.8).

  RFC 4566bis that has not yet been standardized in the Internet Engineering Task Force (IETF) has proposed the following text: "Section 3 of RFC 4855 states that the payload format (encoding) names defined in the RTP Profile are commonly shown in upper case, while media subtype names are commonly shown in lower case. It also states that both of these names are case-insensitive in both places, similar to parameter names which are case-insensitive both in media type strings and in the default mapping to the SDP a=fmtp attribute."

2.6 SDP Attributes

The following attributes are defined. Since application writers may add new attributes as required, this list is not exhaustive. Registration procedures with IANA for new attributes are defined in Section 2.8.

```
a=cat:<category>
```

This attribute gives the dot-separated hierarchical category of the session. This is to enable a receiver to filter unwanted sessions by category. There is no central registry of categories. It is a session-level attribute, and it is not dependent on charset.

```
a=keywds:<keywords>
```

Like the cat attribute, this is to assist identifying wanted sessions at the receiver. This allows a receiver to select an interesting session based on keywords describing the purpose of the session; there is no central registry of keywords. It is a session-level attribute. It is a charset-dependent attribute, meaning that its value should be interpreted in the charset specified for the session description if one is specified or by default in ISO 10646/UTF-8.

```
a=tool:<name and version of tool>
```

This gives the name and version number of the tool used to create the session description. It is a session-level attribute, and it is not dependent on charset.

```
a=ptime:<packet time>
```

This gives the length of time in milliseconds represented by the media in a packet. This is probably only meaningful for audio data but may be used with other media types if it makes sense. It should not be necessary to know ptime to decode RTP or vat audio, and it is intended as a recommendation for the encoding/packetization of audio. It is a media-level attribute, and it is not dependent on charset.

```
a=maxptime:<maximum packet time>
```

This gives the maximum amount of media that can be encapsulated in each packet, expressed as time in milliseconds. The time SHALL be calculated as the sum of the time the media present in the packet represents. For frame-based codecs, the time SHOULD be an integer multiple of the frame size. This attribute is probably only meaningful for audio data but may be used with other media types if it makes sense. It is a media-level attribute, and it is not dependent on charset. Note that this attribute was introduced after RFC 2327 (that has been made obsolete by RFC 4566 described in this section), and non-updated implementations will ignore this attribute.

```
a=rtpmap:<payload type> <encoding name>/<clock rate> [/<encoding
   parameters>]
```

This attribute maps from an RTP payload type number (as used in an "m=" line) to an encoding name denoting the payload format to be used. It also provides information on the clock rate and encoding parameters. It is a media-level attribute that is not dependent on charset. Although an RTP profile may make static assignments of payload type numbers to payload formats, it is more common for that assignment to be done dynamically using "a=rtpmap:" attributes.

As an example of a static payload type, consider u-law PCM coded single-channel audio sampled at 8 kHz. This is completely defined in the RTP/AVP as payload type 0, so there is no need for an "a=rtpmap:" attribute, and the media for such a stream sent to UDP port 49232 can be specified as

```
m=audio 49232 RTP/AVP 0
```

An example of a dynamic payload type is 16-bit linear encoded stereo audio sampled at 16 kHz. If we wish to use the dynamic RTP/AVP payload type 98 for this stream, additional information is required to decode it:

```
m=audio 49232 RTP/AVP 98
a=rtpmap:98 L16/16000/2
```

Up to one rtpmap attribute can be defined for each media format specified. Thus, we might have the following:

```
m=audio 49230 RTP/AVP 96 97 98
a=rtpmap:96 L8/8000
a=rtpmap:97 L16/8000
a=rtpmap:98 L16/11025/2
```

RTP profiles that specify the use of dynamic payload types MUST define the set of valid encoding names and/or a means to register encoding names if that profile is to be used with SDP. The "RTP/AVP" and "RTP/SAVP" profiles use media subtypes for encoding names, under the top-level media type denoted in the "m=" line. In the earlier example, the media types are "audio/l8" and "audio/l16."

For audio streams, <encoding parameters> indicates the number of audio channels. This parameter is OPTIONAL and may be omitted if the number of channels is one, provided no additional parameters are needed. For video streams, no encoding parameters are currently specified. Additional encoding parameters MAY be defined in the future, but codec-specific parameters SHOULD NOT be added. Parameters added to an "a=rtpmap:" attribute SHOULD only be those required for a session directory to make the choice of appropriate media to participate in a session. Codec-specific parameters should be added in other attributes (for example, "a=fmtp:").

Note: RTP audio formats typically do not include information about the number of samples per packet. If a nondefault (as defined in the RTP/AVP) packetization is required, the "ptime" attribute is used as given in the earlier example.

```
a=recvonly
```

This specifies that the tools should be started in receive-only mode where applicable. It can be a session- or media-level attribute, and it is not dependent on charset. Note that recvonly applies to the media only, not to any associated control protocol (e.g., an RTP-based system in recvonly mode SHOULD still send RTCP packets).

```
a=sendrecv
```

This specifies that the tools should be started in send and receive mode. This is necessary for interactive conferences with tools that default to receive-only mode. It can be a session- or media-level attribute, and it is not dependent on charset. If none of the attributes "sendonly," "recvonly," "inactive," and "sendrecv" is present, "sendrecv" SHOULD be assumed as the default for sessions that are not of the conference type "broadcast" or "H332" (see here).

```
a=sendonly
```

This specifies that the tools should be started in send-only mode. An example may be where the unicast address to be used for a traffic destination is different from that used for a traffic source. In such a case, two media descriptions may be used, one sendonly and one recvonly. It can be either a session- or media-level attribute but would normally only be used as a media attribute. It is not dependent on charset. Note that sendonly applies only to the media, and any associated control protocol (e.g., RTCP) SHOULD still be received and processed as normal.

```
a=inactive
```

This specifies that the tools should be started in inactive mode. This is necessary for interactive conferences where users can put other users on hold. No media is sent over an inactive media stream. Note that an RTP-based system SHOULD still send RTCP, even if started inactive. It can be a session- or media-level attribute, and it is not dependent on charset.

```
a=orient:<orientation>
```

Normally this is only used for a whiteboard or presentation tool. It specifies the orientation of the workspace on the screen. It is a media-level attribute. Permitted values are "portrait," "landscape," and "seascape" (upside-down landscape). It is not dependent on charset.

```
a=type:<conference type>
```

This specifies the type of the conference. Suggested values are "broadcast," "meeting," "moderated," "test," and "H332." "recvonly" should be the default for "type:broadcast" sessions, "type:meeting" should imply "sendrecv," and "type:moderated" should indicate the use of a floor control tool and that the media tools are started so as to mute new sites joining the conference.

Specifying the attribute "type:H332" indicates that this loosely coupled session is part of an H.332 session as defined in the ITU H.332 specification (ITU Rec. H.332, September 1998). Media tools should be started "recvonly." Specifying the attribute "type:test" is suggested as a hint that, unless explicitly requested otherwise, receivers can safely avoid displaying this session description to users. The type attribute is a session-level attribute, and it is not dependent on charset.

```
a=charset:<character set>
```

This specifies the character set to be used to display the session name and information data. By default, the ISO-10646 character set in UTF-8 encoding is used. If a more compact representation is required, other character sets may be used. For example, the ISO 8859-1 is specified with the following SDP attribute:

```
a=charset:ISO-8859-1
```

This is a session-level attribute and is not dependent on charset. The charset specified MUST be one of those registered with IANA, such as ISO-8859-1. The character set identifier is a US-ASCII string and MUST be compared against the IANA identifiers using a case-insensitive comparison. If the identifier is not recognized or not supported, all strings that are affected by it SHOULD be regarded as octet strings.

Note that a character set specified MUST still prohibit the use of bytes 0x00 (Null), 0x0A (LF), and 0x0d (CR). Character sets requiring the use of these characters MUST define a quoting mechanism that prevents these bytes from appearing within text fields.

```
a=sdplang:<language tag>
```

This can be a session-level attribute or a media-level attribute. As a session-level attribute, it specifies the language for the session description. As a media-level attribute, it specifies the language for any media-level SDP information field associated with that media. Multiple sdplang attributes can be provided either at session or media level if multiple languages provided in the session description or media use multiple languages, in which case the order of the attributes indicates the order of importance of the various languages in the session or media from most important to least important.

In general, sending session descriptions consisting of multiple languages is discouraged. Instead, multiple descriptions SHOULD be sent describing the session, one in each language. However, this is not possible with all transport mechanisms, and so multiple sdplang attributes are allowed although NOT RECOMMENDED. The "sdplang" attribute value must be RFC 3066 (that was made obsolete by RFCs 5646 and 4647) language tag in US-ASCII (RFCs 5646 and 4647). It is not dependent on the charset attribute. An "sdplang" attribute SHOULD be specified when a session is of sufficient scope to cross geographic boundaries where the language of recipients cannot be assumed or where the session is in a different language from the locally assumed norm.

```
a=lang:<language tag>
```

This can be a session-level attribute or a media-level attribute. As a session-level attribute, it specifies the default language for the session being described. As a media-level attribute, it specifies the language for that media, overriding any session-level language specified. Multiple sdplang attributes can be provided either at session or media level if the session description or media use multiple languages, in which case the order of the attributes indicates the order of importance of the various languages in the session or media from most important to least important.

The "lang" attribute value must be RFC 3066 (that was made obsolete by RFCs 5646 and 4647) language tag in US-ASCII (RFCs 5646 and 4647). It is not dependent on the charset attribute. A "lang" attribute SHOULD be specified when a session is of sufficient scope to cross geographic boundaries where the language of recipients cannot be assumed or where the session is in a different language from the locally assumed norm.

```
a=framerate:<frame rate>
```

This gives the maximum video frame rate in frames/sec. It is intended as a recommendation for the encoding of video data. Decimal representations of fractional values using the notation "<integer>.<fraction>" are allowed. It is a media-level attribute, defined only for video media, and it is not dependent on charset.

```
a=quality:<quality>
```

This gives a suggestion for the quality of the encoding as an integer value. The intention of the quality attribute for video is to specify a nondefault trade-off between frame-rate and still-image quality. For video, the value is in the range of 0 to 10, with the following suggested meaning:

10 - the best still-image quality the compression scheme can give.
5 - the default behavior given no quality suggestion.
0 - the worst still-image quality the codec designer thinks is still usable.

It is a media-level attribute, and it is not dependent on charset.

```
a=fmtp:<format> <format specific parameters>
```

This attribute allows parameters that are specific to a particular format to be conveyed in a way that SDP does not have to understand them. The format must be one of the formats specified for the media. Format-specific parameters may be any set of parameters required to be conveyed by SDP and given unchanged to the media tool that will use this format. At most one instance of this attribute is allowed for each format. It is a media-level attribute, and it is not dependent on charset.

2.7 Security Considerations

The security procedures that need to be considered in dealing with the basic SDP messages described in this section (RFC 4566) are provided in Section 17.1.

2.8 IANA Considerations

SDP parameters that are described here have been registered with IANA. We have not included the IANA registration procedures here for the sake of brevity. The procedures for registration of new SDP parameters are described in RFC 4566.

2.9 Augmented Backus-Naur Form for the SDP

Like SIP, SDP uses the Augmented Backus-Naur Form (ABNF) for its messages. However, the syntaxes that are described here contain the SIP messages from the base SIP RFC 3261 and RFCs that extend and update this SIP (RFC 3261). Certain basic rules are in uppercase, such as SP (Space), LWS (Linear Whitespace), HTAB (Horizontal Tab), CRLF (Control Return Line Feed), DIGIT, ALPHA, etc. Angle brackets are used within definitions to clarify the use of rule names. The use of square brackets is redundant syntactically. It is used as a semantic hint that the specific parameter is optional to use.

2.9.1 Basic Rules

The following rules are used throughout this specification to describe basic parsing constructs. The US-ASCII coded character set is defined by American National Standards Institute (ANSI) X3.4-1986.

```
alphanum     =    ALPHA / DIGIT
OCTET        =    <any 8-bit sequence of data>
CHAR         =    <any US-ASCII character (octets 0 - 127)>
UPALPHA      =    <any US-ASCII uppercase letter "A".."Z">
ALPHA        =    UPALPHA | LOALPHA
DIGIT        =    <any US-ASCII digit "0".."9">
CTL          =    <any US-ASCII control character
                  (octets 0 - 31) and DEL (127)>
CR           =    <US-ASCII CR, carriage return (13)>
LF           =    <US-ASCII LF, linefeed (10)>
SP           =    <US-ASCII SP, space (32)>
HT           =    <US-ASCII HT, horizontal-tab (9)>
<">          =    <US-ASCII double-quote mark (34)>
```

Per RFC 3261, several rules are incorporated from RFC 2396 (that was made obsolete by RFC 3986) but are updated to make them compliant with RFC 2234 (that was made obsolete by RFC 4234; again RFC 4234 was made obsolete by RFC 5234). These include:

```
reserved     =    ";" / "/" / "?" / ":" / "@" / "&" / "=" / "+"
                  / "$" / ","
unreserved   =    alphanum / mark
```

```
mark          =     "-" / "ff" / "." / "!" / "~" / "*" / "'"
                    / "(" / ")"
escaped       =     "%" HEXDIG       ; HEXDIG = Hexagonal Digit
```

SIP header field values can be folded onto multiple lines if the continuation line begins with a space or horizontal tab. All linear white space, including folding, has the same semantics as SP. A recipient may replace any linear white space with a single SP before interpreting the field value or forwarding the message downstream. This is intended to behave exactly as HTTP/1.1 as described in RFC 2616. The SWS construct is used when linear white space is optional, generally between tokens and separators.

```
LWS = [*WSP CRLF] 1*WSP ; linear whitespace
SWS = [LWS] ; sep whitespace
```

To separate the header name from the rest of value, a colon is used, which, by this rule, allows whitespace before, but no line break, and whitespace after, including a line-break. The HCOLON defines this construct.

```
HCOLON = *( SP / HTAB ) ":" SWS
```

The UTF-8 (TEXT-UTF-8) rule is only used for descriptive field contents and values that are not intended to be interpreted by the message parser. Words of *TEXT-UTF-8 contain characters from the UTF-8 standard defined in RFC 3629 (that obsoletes RFC 2279). The TEXT-UTF8-TRIM rule is used for descriptive field contents that are not quoted strings, where leading and trailing LWS is not meaningful. In this regard, SIP differs from HTTP, which uses the ISO 8859-1 character set.

```
TEXT-UTF8-TRIM  =    1*TEXT-UTF8char *(*LWS TEXT-UTF8char)
TEXT-UTF8char   =    %x21-7E / UTF8-NONASCII
UTF8-NONASCII   =    %xC0-DF 1UTF8-CONT
                     / %xE0-EF 2UTF8-CONT
                     / %xF0-F7 3UTF8-CONT
                     / %xF8-Fb 4UTF8-CONT
                     / %xFC-FD 5UTF8-CONT
UTF8-CONT       =    %x80-BF
```

A carriage return/line feed (CRLF) is allowed in the definition of TEXT-UTF8-TRIM only as part of a header field continuation. It is expected that the folding LWS will be replaced with a single SP before interpretation of the TEXT-UTF8-TRIM value. Hexadecimal numeric characters are used in several protocol elements. Some elements (authentication) force hex alphas to be lower case.

```
LHEX = DIGIT / %x61-66 ;lowercase a-f
```

Many SIP header field values consist of words separated by LWS or special characters. Unless otherwise stated, tokens are case insensitive. These special characters must be in a quoted string to be used within a parameter value. The word construct is used in Call-ID to allow most separators to be used.

```
token          =     1*(alphanum / "-" / "." / "!" / "%" / "*"
                     / "ff" / "+" / "`" / "'" / "~" )
```

```
separators     =    "(" / ")" / "<" / ">" / "@" /
                    "," / ";" / ":" / "\" / DQUOTE /
                    "/" / "[" / "]" / "?" / "=" /
                    "{" / "}" / SP / HTAB
word           =    1*(alphanum / "-" / "." / "!" / "%" / "*" /
                    "ff" / "+" / "`" / "'" / "~" /
                    "(" / ")" / "<" / ">" /
                    ":" / "\" / DQUOTE /
                    "/" / "[" / "]" / "?" /
                    "{" / "}" )
```

When tokens or separators are used between elements, whitespace is often allowed before or after these characters:

```
STAR      =    SWS "*" SWS ; asterisk
SLASH     =    SWS "/" SWS ; slash
EQUAL     =    SWS "=" SWS ; equal
LPAREN    =    SWS "(" SWS ; left parenthesis
RPAREN    =    SWS ")" SWS ; right parenthesis
RAQUOT    =    ">" SWS ; right angle quote
LAQUOT    =    SWS "<"; left angle quote
COMMA     =    SWS "," SWS ; comma
SEMI      =    SWS ";" SWS ; semicolon
COLON     =    SWS ":" SWS ; colon
LDQUOT    =    SWS DQUOTE; open double quotation mark
RDQUOT    =    DQUOTE SWS ; close double quotation mark
```

Comments can be included in some SIP header fields by surrounding the comment text with parentheses. Comments are only allowed in fields containing "comment" as part of the field value definition. In all other fields, parentheses are considered part of the field value.

```
comment = LPAREN *(ctext / quoted-pair / comment) RPAREN
ctext   = %x21-27 / %x2A-5B / %x5D-7E / UTF8-NONASCII
          / LWS
```

ctext includes all chars except left and right parentheses and the backslash. A string of text is parsed as a single word if it is quoted using double-quote marks. In quoted strings, quotation marks (") and backslashes (\) need to be escaped.

```
quoted-string   =    SWS DQUOTE *(qdtext / quoted-pair ) DQUOTE
qdtext          =    LWS / %x21 / %x23-5B / %x5D-7E
                     / UTF8-NONASCII
```

The backslash character ("\") may be used as a single-character quoting mechanism only within quoted-string and comment constructs. Unlike HTTP/1.1, the characters CR and LF cannot be escaped by this mechanism to avoid conflict with line folding and header separation.

2.9.2 ABNF for SDP Messages

This section provides an ABNF grammar for SDP. Readers are reminded that RFC 5234 obsoletes RFC 4434 (while RFC 4434 again obsoletes RFC 2234) for ABNF standards. We have described ABNF syntaxes and rules from RFC 5234 in Appendix A of this book.

RFC 4566:

```
;SDP Syntax defined in RFC 4566 (that is defined in this section):
    session-description          =     proto-version
                                       origin-field
                                       session-name-field
                                       information-field
                                       uri-field
                                       email-fields
                                       phone-fields
                                       connection-field
                                       bandwidth-fields
                                       time-fields
                                       key-field
                                       attribute-fields
                                       media-descriptions
    proto-version                =     %x76 "=" 1*DIGIT CRLF
                                       ;this memo describes version 0
    origin-field                 =     %x6f "=" username SP sess-id SP sess-
                                       version SP nettype SP addrtype SP
                                       unicast-address CRLF
    session-name-field           =     %x73 "=" text CRLF
    information-field            =     [%x69 "=" text CRLF]
    uri-field                    =     [%x75 "=" uri CRLF]
    email-fields                 =     *(%x65 "=" email-address CRLF)
    phone-fields                 =     *(%x70 "=" phone-number CRLF)
    connection-field             =     [%x63 "=" nettype SP addrtype SP
                                       connection-address CRLF]
                                              ;a connection field must
                                              ;be present
                                              ;in every media
                                              ;description or at the
                                              ;session-level
    bandwidth-fields             =     *(%x62 "=" bwtype ":" bandwidth
                                       CRLF)
    time-fields                  =     1*( %x74 "=" start-time SP
                                       stop-time
                                       *(CRLF repeat-fields) CRLF)
                                       [zone-adjustments CRLF]
    repeat-fields                =     %x72 "=" repeat-interval SP
                                       typed-time
                                       1*(SP typed-time)
    zone-adjustments             =     %x7a "=" time SP ["-"] typed-time
                                       *(SP time SP ["-"] typed-time)
    key-field                    =     [%x6b "=" key-type CRLF]
    attribute-fields             =     *(%x61 "=" attribute CRLF)
    media-descriptions           =     *( media-field
                                       information-field
                                       *connection-field
                                       bandwidth-fields
                                       key-field
                                       attribute-fields )
    media-field                  =     %x6d "=" media SP port ["/"
                                       integer]
```

```
                                 SP proto 1*(SP fmt) CRLF
                                 ; sub-rules of 'o='
username                 =       non-ws-string
                                 ;pretty wide definition, but doesn't
                                 ;include space
sess-id                  =       1*DIGIT
                                 ;should be unique for this
                                 ; username/host
sess-version             =       1*DIGIT
nettype                  =       token
                                        ;typically "IN"
addrtype                 =       token
                                        ;typically "IP4" or "IP6"
                                        ; sub-rules of 'u='
uri                      =       URI-reference
                                        ; see RFC 3986
                                        ; sub-rules of 'e=', see RFC 2822
                                        ; for
                                        ; definitions
email-address            =       address-and-comment / dispname-and-
                                 address / addr-spec
address-and-comment      =       addr-spec 1*SP "(" 1*email-safe ")"
dispname-and-address     =       1*email-safe 1*SP "<" addr-spec ">"
                                        ; sub-rules of 'p='
phone-number             =       phone *SP "(" 1*email-safe ")" /
                                 1*email-safe "<" phone ">" / phone
phone    =                       ["+"] DIGIT 1*(SP / "-" / DIGIT)
                                        ; sub-rules of 'c='
connection-address       =       multicast-address / unicast-address
                                        ; sub-rules of 'b='
bwtype                   =       token
bandwidth                =       1*DIGIT
                                        ; sub-rules of 't='
start-time               =       time / "0"
stop-time                =       time / "0"
time                     =       POS-DIGIT 9*DIGIT
                                 ; Decimal representation of NTP time in
                                 ; seconds since 1900. The representation
                                 ; of NTP time is an unbounded length
                                 ; field
                                 ; containing at least 10 digits. Unlike
                                 ; the
                                 ; 64-bit representation used elsewhere,
                                 ; time
                                 ; in SDP does not wrap in the year 2036.
                                 ; sub-rules of 'r=' and 'z='
repeat-interval          =       POS-DIGIT *DIGIT [fixed-len-time-unit]
typed-time               =       1*DIGIT [fixed-len-time-unit]
fixed-len-time-unit      =       %x64 / %x68 / %x6d / %x73
                                        ; sub-rules of 'k='
key-type                 =       %x70 %x72 %x6f %x6d %x70 %x74 / ;
                                 "prompt"
                                 %x63 %x6c %x65 %x61 %x72 ":" text / ;
                                 "clear:"
```

```
                              %x62 %x61 %x73 %x65 "64:" base64 / ;
                              "base64:"
                          %x75 %x72 %x69 ":" uri ; "uri:"
base64          =         *base64-unit [base64-pad]
base64-unit     =         4base64-char
base64-pad      =         2base64-char "==" / 3base64-char "="
base64-char     =         ALPHA / DIGIT / "+" / "/"
                              ; sub-rules of 'a='
attribute       =         (att-field ":" att-value) / att-field
att-field       =         token
att-value       =         byte-string
                              ; sub-rules of 'm='
Media           =         token
                              ;typically "audio", "video", "text", or
                              ;"application"
fmt             =         token
                          ;typically an RTP payload type for audio
                          ;and video media
proto           =         token *("/" token)
                          ;typically "RTP/AVP" or "udp"
port            =         1*DIGIT
                              ; generic sub-rules: addressing
unicast-address =         IP4-address / IP6-address /Fully Qualified
                          Domain
                          Name (FQDN) / extn-addr
multicast-
  address       =         IP4-multicast / IP6-multicast / FQDN
                          / extn-addr
                          IP4-multicast = m1 3( "." decimal-uchar )
                          "/" ttl [ "/" integer ]
                              ; IPv4 multicast addresses may be in the
                              ; range 224.0.0.0 to 239.255.255.255
m1              =         ("22" ("4"/"5"/"6"/"7"/"8"/"9")) /
                          ("23" DIGIT )
IP6-multicast   =         hexpart [ "/" integer ]
                              ; IPv6 address starting with FF
ttl             =         (POS-DIGIT *2DIGIT) / "0"
FQDN            =         4*(alpha-numeric / "-" / ".")
                              ; fully qualified domain name as
                              ; specified
                              ; in RFC 1035 (and updates)
IP4-address     =         b1 3("." decimal-uchar)
b1              =         decimal-uchar
                              ; less than "224"
                              ; The following is consistent with
                              ; RFC 4291 that obsoletes RFC 3513 while
                              ; RFC 3513 obsoletes RFC 2373
                              .
IP6-address     =         hexpart [ ":" IP4-address ]
hexpart         =         hexseq / hexseq "::" [ hexseq ] /
                          "::" [ hexseq ]
hexseq          =         hex4 *( ":" hex4)
hex4            =         1*4HEXDIG
                              ; Generic for other address families
```

```
extn-addr          =   non-ws-string
                              ; generic sub-rules: datatypes
text               =   byte-string
                              ;default is to interpret this as UTF8
                              ;text.
                              ;ISO 8859-1 requires "a=charset:
                              ; ISO-8859-1"
                              ;session-level attribute to be used
byte-string        =   1*(%x01-09/%x0B-0C/%x0E-FF)
                              ;any byte except NUL, CR, or LF
non-ws-string      =   1*(VCHAR/%x80-FF)
                              ;string of visible characters
token-char         =   %x21 / %x23-27 / %x2A-2B / %x2D-2E / %x30-39
                       / %x41-5A / %x5E-7E
token              =   1*(token-char)
email-safe         =   %x01-09/%x0B-0C/%x0E-27/%x2A-
                       3B/%x3D/%x3F-FF
                              ;any byte except NUL, CR, LF,
                              ; or the quoting
                              ; characters ()<>
integer            =   POS-DIGIT *DIGIT
                              ; generic sub-rules: primitives
alpha-numeric      =   ALPHA / DIGIT
POS-DIGIT          =   %x31-39 ; 1 - 9
decimal-uchar      =   DIGIT
                       / POS-DIGIT DIGIT
                       / ("1" 2*(DIGIT))
                       / ("2" ("0"/"1"/"2"/"3"/"4") DIGIT)
                       / ("2" "5" ("0"/"1"/"2"/"3"/"4"/"5"))
                              ; external references:
                              ; ALPHA, DIGIT, CRLF, SP, VCHAR:
                              ; from RFC 5234 that obsoletes RFC 4234
                              ; URI-reference: from RFC 3986
                              ; addr-spec: from RFC 5322 that obsoletes
                              ; RFC 2822
```

RFC 4566bis:

RFC 4566bis [2] that has not yet been standardized in IETF has suggested modified SDP syntax correcting ABNF especially for IPv6 and IPv4. We have included these suggested changes in Section 2.10.1.

RFC 5245 (Section 3.2):

;This specification (RFC 5245, see Section 3.2) defines seven new SDP attributes: the "candidate," ;"remote-candidates," "ice-lite," "ice-mismatch," "iceufrag," "ice-pwd," and "ice-options" attributes.

; "candidate" attribute (RFC 5245, see Section 3.2)

;The candidate attribute is a media-level attribute only. It contains a transport address for a ;candidate that can be used for connectivity checks. The syntax of this attribute is defined using ;ABNF as defined in RFC 5234:

```
candidate-attribute  = "candidate" ":" foundation SP component-id SP
                       transport SP
                       priority SP
```

```
                              connection-address SP ;from RFC 4566
                              port ;port from RFC 4566
                              SP cand-type
                              [SP rel-addr]
                              [SP rel-port]
                              *(SP extension-att-name SP
                              extension-att-value)
foundation                =   1*32ice-char
component-id              =   1*5DIGIT
transport                 =   "UDP" / transport-extension
transport-extension       =   token ; from RFC 3261
priority                  =   1*10DIGIT
cand-type                 =   "typ" SP candidate-types
candidate-types           =   "host"/ "srflx" / "prflx" / "relay" /
                              token
rel-addr                  =   "raddr" SP connection-address
rel-port                  =   "rport" SP port
extension-att-name        =   byte-string ;from RFC 4566
extension-att-value       =   byte-string
ice-char                  =   ALPHA / DIGIT / "+" / "/"
```

;This grammar encodes the primary information about a candidate: its IP address, port and ;transport protocol, and its properties: the foundation, component ID, priority, type, and related ;transport address:

;<connection-address>: is taken from RFC 4566 described in this section (that is, this Section 2). ;It is the IP address of the candidate, allowing for IPv4 addresses, IPv6 addresses, and FQDNs. ;When parsing this field, an agent can differentiate an IPv4 address and an IPv6 address by ;the presence of a colon in its value the presence of a colon indicates IPv6. An agent MUST ;ignore candidate lines that include candidates with IP address versions that are not supported ;or recognized. An IP address SHOULD be used, but an FQDN MAY be used in place of an IP ;address. In that case, when receiving an offer or answer containing an FQDN in an a=candidate ;attribute, the FQDN is looked up in the domain name servers (DNSs) first using an AAAA ;record (assuming the agent supports IPv6), and if no result is found or the agent only supports ;IPv4, using an A. If the DNS query returns more than one IP address, one is chosen, and then ;used for the remainder of Interactive Connectivity Establishment (ICE) processing.

;<port>: is also taken from RFC 4566 described in this section (that is, this Section 2). It is the ;port of the candidate.

;<transport>: indicates the transport protocol for the candidate. This specification only defines ;UDP. However, extensibility is provided to allow for future transport protocols to be used with ;ICE, such as TCP or the Datagram Congestion Control Protocol (DCCP) (RFC 4340).

;<foundation>: is composed of 1 to 32 <ice-char>s. It is an identifier that is equivalent for two ;candidates that are of the same type, share the same base, and come from the same STUN server. ;The foundation is used to optimize ICE performance in the Frozen algorithm.

;<component-id>: is a positive integer between 1 and 256 that identifies the specific component ;of the media stream for which this is a candidate. It MUST start at 1 and MUST increment by ;1 for each component of a particular candidate. For media streams based on RTP, candidates ;for the actual RTP media MUST have a component ID of 1, and candidates for RTCP MUST ;have a component ID of 2. Other types of media streams that require multiple components ;MUST develop specifications that define the mapping of components to component IDs. See ;Section 3.2.14 for an additional discussion on extending ICE to new media streams.

;<priority>: is a positive integer between 1 and (2**31 - 1).

;<cand-type>: encodes the type of candidate. This specification defines the values "host," "srflx," ;"prflx," and "relay" for host, server reflexive, peer reflexive, and relayed candidates, respectively. ;The set of candidate types is extensible for the future.

;<rel-addr> and <rel-port>: convey transport addresses related to the candidate, useful for ;diagnostics and other purposes. <rel-addr> and <rel-port> MUST be present for server reflexive, ;peer reflexive, and relayed candidates. If a candidate is server or peer reflexive, <rel-addr> and ;<rel-port> are equal to the base for that server or peer reflexive candidate. If the candidate is ;relayed, <rel-addr> and <rel-port> are equal to the mapped address in the Allocate response ;that provided the client with that relayed candidate (see Section 3.2.24.3 for a discussion of its ;purpose). If the candidate is a host candidate, <rel-;addr> and <rel-port> MUST be omitted.

;The candidate attribute can itself be extended. The grammar allows for new name/value pairs ;to be added at the end of the attribute. An implementation MUST ignore any name/value pairs ;it doesn't understand.

; "remote-candidates" Attribute

;The syntax of the "remote-candidates" attribute is defined using ABNF as defined in ;RFC 5234. The remote-candidates attribute is a media level only.

```
remote-candidate-att   =      "remote-candidates" ":" remote-candidate
                                 0*(SP remote-candidate)
remote-candidate       =      component-ID SP connection-address SP
                                 port
```

;The attribute contains a connection-address and port for each component. The ordering of ;components is irrelevant. However, a value MUST be present for each component of a media ;stream. This attribute MUST be included in an offer by a controlling agent for a media stream ;that is Completed and MUST NOT be included in any other case.

;"ice-lite" and "ice-mismatch" Attributes (RFC 5245, see Section 3.2)

;The syntax of the "ice-lite" and "ice-mismatch" attributes, both of which are flags, is

```
ice-lite       =   "ice-lite"
ice-mismatch   =   "ice-mismatch"
```

;"ice-lite" is a session-level attribute only and indicates that an agent is a lite implementation. ;"ice-mismatch" is a media-level attribute only and when present in an answer indicates that the ;offer arrived with a default destination for a media component that didn't have a corresponding ;candidate attribute.

;"ice-ufrag" and "ice-pwd" Attributes (RFC 5245, see Section 3.2)

;The "ice-ufrag" and "ice-pwd" attributes convey the username fragment and password used by ;ICE for message integrity. Their syntax is

```
ice-pwd-att    =  "ice-pwd" ":" password
ice-ufrag-att  =  "ice-ufrag" ":" ufrag
password          =   22*256ice-char
ufrag             =   4*256ice-char
```

;The "ice-pwd" and "ice-ufrag" attributes can appear at either the session level or media level. ;When present in both, the value in the media level takes precedence. Thus, the value at the session ;level is effectively a default that applies to all media streams unless overridden by a media-level

;value. Whether present at the session or media level, there MUST be an ice-pwd and ice-ufrag ;attribute for each media stream. If two media streams have identical ice-ufrags, they MUST have ;identical ice-pwds.

;The ice-ufrag and ice-pwd attributes MUST be chosen randomly at the beginning of a session. ;The ice-ufrag attribute MUST contain at least 24 bits of randomness, and the ice-pwd attribute ;MUST contain at least 128 bits of randomness. This means that the ice-ufrag attribute will be at least ;4 characters long, and the ice-pwd at least 22 characters long, since the grammar for these attributes ;allows for 6 bits of randomness per character. The attributes MAY be longer than 4 and 22 characters, ;respectively, of course, up to 256 characters. The upper limit allows for buffer sizing in implementations. ;Its large upper limit allows for increased amounts of randomness to be added over time.

; "ice-options" Attribute (RFC 5245, see Section 3.2)

;The "ice-options" attribute is a session-level attribute. It contains a series of tokens that identify ;the options supported by the agent. Its grammar is:

```
ice-options      =  "ice-options" ":" ice-option-tag
                        0*(SP ice-option-tag)
ice-option-tag   =  1*ice-char
```

RFC 4574 (Section 5.1):

;RFC 4574 (see Section 5.1) specification defines a new media-level value attribute: "label." Its ;formatting in SDP is described by the following ABNF (RFC 5234):

```
label-attribute   =   "a=label:" pointer
pointer           =   token
token             =   1*(token-char)
token-char        =   %x21 / %x23-27 / %x2A-2B / %x2D-2E / %x30-39
                      / %x41-5A / %x5E-7E
```

;The token-char and token elements are defined in RFC 4566 (this section) but included here to ;provide support for the implementer of this SDP feature. The "label" attribute contains a token ;that is defined by an application and is used in its context. The new attribute can be attached ;to "m=" lines in multiple SDP documents allowing the application to logically group the media ;streams across SDP sessions when necessary.

RFC 5576 (Section 5.3):

;RFC 5576 (see Section 5.3) provides a formal ABNF (RFC 5234) grammar for each of the new ;media and source attributes defined in this document. Grammars for existing session or media ;attributes that have been extended to be source attributes are not included.

```
ssrc-attr         =  "ssrc:" ssrc-id SP attribute
                     ; The base definition of "attribute" is in RFC
                     ; 4566.
                     ; (It is the content of "a=" lines.)
ssrc-id           =  integer ; 0 .. 2**32 - 1
attribute         =  / ssrc-attr
ssrc-group-attr   =  "ssrc-group:" semantics *(SP ssrc-id)
semantics         =  "FEC" / "FID" / token
                              ; Matches RFC 5888 that obsoletes RFC
                              ; 3388
```

```
                                  ; definition and
                                  ; IANA registration rules in this doc.
token                  =          <as defined in RFC 4566>
attribute              =          / ssrc-group-attr
cname-attr             =          "cname:" cname
cname                  =          byte-string
                                  ; Following the syntax conventions for
                                  ; CNAME as defined in RFC 3550.
                                  ; The definition of "byte-string" is in
                                  ; RFC 4566
attribute              =          / cname-attr
previous-ssrc-attr     =          "previous-ssrc:" ssrc-id *(SP ssrc-id)
attribute              =          / previous-ssrc-attr
```

RFC 6236 (Section 8.2):

;RFC 6236 (see Section 8.2) provides the syntax for the image attribute in ABNF (RFC 5234):

```
image-attr             =          "imageattr:" PT 1*2( 1*WSP ( "send" /
                                  "recv" )
                                  1*WSP attr-list )
PT                     =          1*DIGIT / "*"
attr-list              =          ( set *(1*WSP set) ) / "*"
                                     ; WSP and DIGIT defined in [RFC 5234]
set                    =          "[" "x=" xyrange "," "y=" xyrange *( ","
                                  key-value )
                                  "]"
                                     ; x is the horizontal image size
                                     ; range (pixel
                                     ; count)
                                     ; y is the vertical image size range
                                     ; (pixel
                                     ; count)
key-value              =          ( "sar=" srange )
                                  / ( "par=" prange )
                                  / ( "q=" qvalue )
                                     ; Key-value MAY be extended with
                                     ; other
                                     ; keyword
                                     ; parameters.
                                     ; At most, one instance each of sar,
                                     ; par, or q
                                     ; is allowed in a set.
                                     ; sar (sample aspect ratio) is the
                                     ; sample
                                     ; aspect ratio
                                     ; associated with the set (optional,
                                     ; MAY be
                                     ; ignored)
                                     ; par (picture aspect ratio) is the
                                     ; allowed
                                     ; ratio between the display's x and y
                                     ; physical
                                     ; size (optional)
```

```
                                  ; q (optional, range [0.0..1.0],
                                  ; default value
                                  ; 0.5)
                                  ; is the preference for the given set,
                                  ; a higher value means a higher
                                  ; preference
onetonine           =           "1" / "2" / "3" / "4" / "5" / "6" / "7"
                                / "8" / "9"
                                  ; Digit between 1 and 9
xyvalue             =           onetonine *5DIGIT
                                  ; Digit between 1 and 9 that is
                                  ; followed by 0 to 5 other digits
step                =           xyvalue
xyrange             =           ( "[" xyvalue ":" [ step ":" ]
                                xyvalue "]" )
                                  ; Range between a lower and an upper
                                  ; value
                                  ; with an optional step, default
                                  ; step = 1
                                  ; The rightmost occurrence of xyvalue
                                  ; MUST ; have
                                  ; a
                                  ; higher value than the leftmost
                                  ; occurrence.
                                  ; / ( "[" xyvalue 1*( "," xyvalue )
                                  ; "]" )
                                  ; Discrete values separated by ','
                                  ; / ( xyvalue )
                                  ; A single value
spvalue             =           ( "0" "." onetonine *3DIGIT )
                                  ; Values between 0.1000 and 0.9999
                                  ; / ( onetonine "." 1*4DIGIT )
                                  ; Values between 1.0000 and 9.9999
srange              =           ( "[" spvalue 1*( "," spvalue ) "]" )
                                  ; Discrete values separated by ','.
                                  ; Each occurrence of spvalue MUST be
                                  ; greater than the previous
                                  ; occurrence.
                                / ( "[" spvalue "-" spvalue "]" )
                                  ; Range between a lower and an upper
                                  ; level
                                  ; (inclusive)
                                  ; The second occurrence of spvalue
                                  ; MUST
                                  ; have a higher
                                  ; value than the first
                                / ( spvalue )
                                  ; A single value
prange              =           ( "[" spvalue "-" spvalue "]" )
                                  ; Range between a lower and an upper
                                  ; level
                                  ; (inclusive)
                                  ; The second occurrence of spvalue
                                  ; MUST
                                  ; have a
```

```
                                        ; higher
                                        ; value than the first
qvalue                =           ( "0" "." 1*2DIGIT ) / ( "1" "."
                                  1*2("0") )
                                        ; Values between 0.00 and 1.00
```

;The attribute typically contains a "send" and a "recv" keyword. These specify the preferences for ;the media once the session is set up, in the send and receive direction, respectively, from the point ;of view of the sender of the session description. One of the keywords ("send" or "recv") MAY be ;omitted; see Sections 8.2.3.2.4 and 8.2.3.2 for a description of cases when this may be appropriate.

;The "send" keyword and corresponding attribute list (attr-list) MUST NOT occur more than ;once per image attribute.

;The "recv" keyword and corresponding attribute list (attr-list) MUST NOT occur more than ;once per image attribute.

;PT is the payload type number; it MAY be set to "*" (wild card) to indicate that the attribute ;applies to all payload types in the media description.

;For sendrecv streams, the send and recv directions SHOULD be present in the SDP.

;For inactive streams, it is RECOMMENDED that both the send and recv directions are ;present in the SDP.

RFC 5939 (Section 8.1):

;In RFC 5939 (see Section 8.1), the "csup" attribute adheres to the RFC 4566 "attribute" ;production, with an att-value defined as follows:

```
att-value          =  option-tag-list
option-tag-list    =  option-tag *("," option-tag)
option-tag         =  token ; defined in RFC 4566
```

;A special base option tag with a value of "cap-v0" is defined for the basic SDP capability negotiation ;framework defined in this document. Entities can use this option-tag with the "a=csup" attribute ;to indicate support for the SDP capability negotiation framework specified in this document. ;Please note that white space is not allowed in this rule.

;In RFC 5939 (see Section 8.1), the attribute can be provided at the session level and the media ;level. The "acap" attribute adheres to the RFC 4566 "attribute" production, with an att-value ;defined as follows:

```
att-value      =   att-cap-num 1*WSP att-par
att-cap-num    =   1*10(DIGIT) ;defined in RFC 5234
att-par        =   attribute ;defined in RFC 4566
```

;Note that white space is not permitted before the att-cap-num. When the attribute capability ;contains a session-level attribute, that "acap" attribute can only be provided at the session level.

;In RFC 5939 (see Section 8.1), attribute capabilities are used in a potential configuration by ;use of the attribute-config-list parameter, which is defined by the following ABNF:

```
attribute-config-list   =  "a=" delete-attributes
attribute-config-list   =  / "a=" [delete-attributes ":"]
                             mo-att-cap-list *(BAR mo-att-cap-list)
delete-attributes       =  DELETE ( "m" ; media attributes
                             / "s" ; session attributes
```

```
                              / "ms" )  ; media and session attributes
mo-att-cap-list                     =     mandatory-optional-att-
                                          cap-list /
                                          mandatory-att-cap-list /
                                          optional-att-cap-list
mandatory-optional-att-cap-list     =     mandatory-att-cap-list
                                          "," optional-att-cap-list
mandatory-att-cap-list              =     att-cap-list
optional-att-cap-list               =     "[" att-cap-list "]"
att-cap-list                        =     att-cap-num *(","
                                          att-cap-num)
att-cap-num                         =     1*10(DIGIT) ;defined in
                                          [RFC 5234]
BAR                                 =     "|"
DELETE                              =     "-"
```

;Note that white space is not permitted within the attribute-configlist rule. Each attribute ;configuration list can optionally begin with instructions for how to handle attributes that are ;part of the actual configuration SDP session description (i.e., the "a=" lines present in the original ;SDP session description).

;In RFC 5939 (see Section 8.1), the transport protocol configuration lists are included in a ;potential configuration by use of the transport-protocol-config-list parameter, which is defined ;by the following ABNF:

```
transport-protocol-config-list      =     "t=" trpr-cap-num
                                          *(BAR trpr-cap-num)
trpr-cap-num                        =     1*10(DIGIT) ; defined in
                                          RFC 5234
```

;Note that white space is not permitted within this rule. The trpr-cap-num refers to transport protocol ;capability numbers defined above and hence MUST be between 1 and 2^31-1 (both included).

;In RFC 5939 (see Section 8.1), extension capabilities are included in a potential configuration ;as well by use of extension configuration lists. Extension configuration lists MUST adhere to the ;following ABNF:

```
extension-config-list               =     ["+"] ext-cap-name "="
                                          ext-cap-list
ext-cap-name                        =     1*(ALPHA / DIGIT)
ext-cap-list                        =     1*VCHAR ; defined in RFC
                                          5234
```

;Note that white space is not permitted within this rule. The ext-cap-name refers to the name of the ;extension capability, and the ext-cap-list is here merely defined as a sequence of visible characters.

;In RFC 5939 (see Section 8.1), the "acfg" attribute adheres to the RFC 4566 "attribute" ;production, with an att-value defined as follows:

```
att-value                           =     config-number [1*WSP
                                          sel-cfg-list]
                                          ;config-number defined in
                                          ;Section 8.1.3.5.1.
sel-cfg-list                        =     sel-cfg *(1*WSP sel-cfg)
sel-cfg                             =     sel-attribute-config /
```

```
                                          sel-transport-protocol-config /
                                          sel-extension-config
sel-attribute-config              =       "a=" [delete-attributes ":"]
                                          mo-att-cap-list
                                          ; defined in Section 3.5.1.
sel-transport-protocol-config     =           "t=" trpr-cap-num
                                              ; defined in Section
                                              ; 8.1.3.5.1.
sel-extension-config              =           ext-cap-name
                                              "=" 1*VCHAR
                                              ; defined in Section
                                              ; 8.1.3.5.1.
```

;Note that white space is not permitted before the config-number. The actual configuration ("a=acfg") attribute can be provided only at the media level.

RFC 6871 (Section 8.3):

;RFC 6871 (see Section 8.3) defines the attributes that allow multiple capability numbers to be ;defined for the media format in question by specifying a range of media capability numbers. ;This permits the media format to be associated with different media parameters in different ;configurations. When a range of capability numbers is specified, the first (leftmost) capability ;number MUST be strictly smaller than the second (rightmost) (i.e., the range increases and covers ;at least two numbers). In ABNF (RFC 5234), we have:

```
media-capability-line             =       rtp-mcap / non-rtp-mcap
rtp-mcap                          =       "a=rmcap:" media-cap-num-list
                                          1*WSP encoding-name "/"
                                          clock-rate
                                          ["/" encoding-parms]
non-rtp-mcap                      =       "a=omcap:" media-cap-num-list
                                          1*WSP
                                          format-name
media-cap-num-list                =       media-cap-num-element
                                          *("," media-cap-num-element)
media-cap-num-element             =       media-cap-num
                                          / media-cap-num-range
media-cap-num-range               =       media-cap-num "-" media-cap-num
media-cap-num                     =       NonZeroDigit *9(DIGIT)
encoding-name                     =       token ;defined in RFC 4566
clock-rate                        =       NonZeroDigit *9(DIGIT)
encoding-parms                    =       token
format-name                       =       token ;defined in RFC 4566
NonZeroDigit                      =       %x31-39 ; 1-9
```

;The encoding-name, clock-rate, and encoding-params are as defined to appear in an "rtpmap" ;attribute for each media type/subtype. Thus, it is easy to convert an "rmcap" attribute line into ;one or more "rtpmap" attribute lines, once a payload type number is assigned to a media-cap-num ;(see Section 8.3.3.3.5). The format-name is a media format description for non-RTP-based media ;as defined for the <fmt> part of the media description ("m=" line) in SDP (RFC 4566). In simple ;terms, it is the name of the media format (e.g., "t38"). This form can also be used in cases such

;as the Binary Floor Control Protocol (BFCP) (RFC 4585) where the fmt list in the "m=" line is ;effectively ignored (BFCP uses "*").

;In RFC 6871 (see Section 8.3), the media capability numbers may include a wildcard ("*"), ;which will be used instead of any payload type mappings in the resulting SDP (see, e.g., RFC 4585 ;and the example below). In ABNF, we have:

```
media-specific-capability = "a=mscap:" media-caps-star
                                    1*WSP att-field ; from RFC 4566
                                    1*WSP att-value ; from RFC 4566
media-caps-star = media-cap-star-element
                                    *("," media-cap-star-element)
 media-cap-star-element = (media-cap-num [wildcard])
                                    / (media-cap-num-range [wildcard])
                                    wildcard = "*"
```

;RFC 6871 (see Section 8.3) defines the payload type number mapping parameter, payload- ;number-config-;list, in accordance with the extension-config-list format defined in RFC 5939.

;In ABNF:

```
payload-number-config-list = ["+"] "pt=" media-map-list
media-map-list = media-map *("," media-map)
media-map = media-cap-num ":" payload-type-number
         ; media-cap-num is defined in Section 3.3.1
payload-type-number = NonZeroDigit *2(DIGIT) ; RTP payload
         ; type number
```

;In RFC 6871 (see Section 8.3), the media configuration parameter is used to specify the media ;format(s) and related parameters for a potential, actual, or latent configuration. Adhering to the ;ABNF for extension-config-list in RFC 5939 with

```
ext-cap-name = "m"
ext-cap-list = media-cap-num-list
                 [*(BAR media-cap-num-list)]
```

;we have

```
media-config-list  =  ["+"] "m=" media-cap-num-list
                      *(BAR media-cap-num-list)
                      ;BAR is defined in RFC 5939
                      ;media-cap-num-list is defined above
```

;Alternative media configurations are separated by a vertical bar ("|").

;The example in Section 8.3.3.3.7 shows how the parameters from the rmcap line are mapped ;to payload type numbers from the "pcfg" "pt" parameter. The use of the plus sign ("+") is described ;in RFC 5939.

;In RFC 6871 (see Section 8.3), session capability attributes may be used to determine whether ;a latent configuration may be used to form an offer for an additional simultaneous stream or to ;reconfigure an existing stream in a subsequent offer/answer exchange. The latent configuration ;attribute is of the form:

```
;a=lcfg:<config-number> <latent-cfg-list>
```

;which adheres to the SDP (RFC 4566) "attribute" production with att-field and att-value ;defined as:

```
att-field = "lcfg"
att-value = config-number 1*WSP lcfg-cfg-list
config-number = NonZeroDigit *9(DIGIT) ;DIGIT defined in RFC 5234
lcfg-cfg-list = media-type 1*WSP pot-cfg-list
                ; as defined in RFC 5939
                ; and extended herein
```

;The media-type (mt=) parameter identifies the media type (audio, video, etc.) to be associated ;with the latent media stream, and it MUST be present. The pot-cfg-list MUST contain a ;transport-protocolconfig-list (t=) parameter and a media-config-list (m=) parameter.

;In RFC 6871 (see Section 8.3), session capability attributes may be used to determine whether ;a latent configuration may be used to form an offer for an additional simultaneous stream or to ;reconfigure an existing stream in a subsequent offer/answer exchange. The latent configuration ;attribute is of the form

```
;a=lcfg:<config-number> <latent-cfg-list>
```

;which adheres to the SDP (RFC 4566) "attribute" production with att-field and att-value ;defined as:

```
att-field = "lcfg"
att-value = config-number 1*WSP lcfg-cfg-list
config-number = NonZeroDigit *9(DIGIT) ;DIGIT defined in RFC 5234
lcfg-cfg-list = media-type 1*WSP pot-cfg-list
                ; as defined in RFC 5939
                ; and extended herein
```

;The media-type (mt=) parameter identifies the media type (audio, video, etc.) to be associated ;with the latent media stream, and it MUST be present. The pot-cfg-list MUST contain a ;transport-protocolconfig-list (t=) parameter and a media-config-list (m=) parameter.

;In RFC 6871 (see Section 8.3), the session capability attribute is a session-level attribute ;described by

```
"a=sescap:" <session num> <list of configs>
```

;which corresponds to the standard value attribute definition with

```
att-field = "sescap"
att-value = session-num 1*WSP list-of-configs
                   [1*WSP optional-configs]
                   session-num = NonZeroDigit *9(DIGIT) ; DIGIT
                   defined
                   ; in RFC 5234
list-of-configs = alt-config *("," alt-config)
optional-configs = "[" list-of-configs "]"
alt-config = config-number *("|" config-number)
```

;The session-num identifies the session: a lower-number session is preferred over a higher-number ;session, and leading zeroes are not permitted.

RFC 7006 (Section 8.4):

;In RFC 7006 (see Section 8.4), from the subrules of the attribute ("a=") line in SDP (RFC 4566), ;SDP attributes are of the form:

```
attribute = (att-field ":" att-value) / att-field
att-field = token
att-value = byte-string
```

;Capability attributes use only the "att-field:att-value" form. The new capabilities may be referenced ;in potential configurations ("a=pcfg") or in latent configurations ("a=lcfg") as productions ;conforming to the <extension-config-list>, as respectively defined in RFC 5939 (see Section 8.1) ;and RFC 6871 (see Section 8.3).

```
extension-config-list = ["+"] ext-cap-name "=" ext-cap-list
ext-cap-name = 1*(ALPHA / DIGIT)
;; ALPHA and DIGIT defined in RFC 5234
ext-cap-list = 1*VCHAR ;; VCHAR defined in RFC 5234
```

;The optional "+" is used to indicate that the extension is mandatory and MUST be supported ;in order to use that particular configuration. The new capabilities may also be referenced in ;actual configurations ("a=acfg") as productions conforming to the <sel-extension-config> defined ;in RFC 5939 (see Section 8.1).

```
sel-extension-config = ext-cap-name "=" 1*VCHAR
```

;The specific parameters are defined in the individual description of each capability below. The "bcap," ;"ccap," and "icap" capability attributes can be provided at the SDP session and/or media level.

;In RFC 7006 (see Section 8.4), the other elements are as defined in RFC 4566. This format ;corresponds to the RFC 4566 attribute production rules, according to the following ABNF ;specified in RFC 5234 syntax:

```
att-field =/ "ccap"
att-value =/ conn-cap-num 1*WSP nettype SP addrtype
SP connection-address
conn-cap-num = 1*10(DIGIT) ;; 1 to 2^31-1, inclusive
; DIGIT defined in RFC 5234
```

;The "ccap" capability attribute allows for expressing alternative connections address ("c=") lines ;in SDP as part of the SDP capability negotiation process.

;In RFC 7006 (see Section 8.4), the concept is extended by the SDP media capabilities negotiation ;(RFC 6871, see Section 8.3)) with an "lcfg" attribute that conveys latent configurations. In this ;document, we define a <connection-config> parameter to be used to specify a connection data ;capability in a potential or latent configuration attribute. The parameter follows the form of an:

```
<extension-config-list>
```

;with

```
ext-cap-name = "c"
ext-cap-list = conn-cap-list
```

;where, according to the following ABNF (RFC 5234) syntax:

```
extension-config-list =/ conn-config-list
conn-config-list = ["+"] "c=" conn-cap-list
conn-cap-list = conn-cap-num *(BAR conn-cap-num)
conn-cap-num = 1*10(DIGIT) ;; 1 to 2^32-1 inclusive
```

;Each capability configuration alternative contains a single connection data capability attribute ;number and refers to the conn-cap-num capability number defined explicitly earlier in this document; ;hence, the values MUST be between 1 and 2^31-1 (both included). The connection data capability ;allows the expression of only a single capability in each alternative, rather than a list of capabilities, ;since no more than a single connection data field is permitted per media block. Nevertheless, it is ;allowed to express alternative potential connection configurations separated by a vertical bar ("|"). ;An endpoint includes a plus sign ("+") in the configuration attribute to mandate support for this ;extension. An endpoint that receives this parameter prefixed with a plus sign and does not support ;this extension MUST treat that potential configuration as not valid. The connection data parameter ;to the actual configuration attribute ("a=acfg") is formulated as a <sel-extension-config> with

```
ext-cap-name = "c"
```

;hence

```
sel-extension-config =/ sel-connection-config
sel-connection-config = "c=" conn-cap-num ;;
```

;as defined above.

;In RFC 7006 (see Section 8.4), this format satisfies the general attribute production rules in ;SDP (RFC 4566), according to the following ABNF (RFC 5234) syntax:

```
att-field =/ "bcap"
att-value =/ bw-cap-num 1*WSP bwtype ":" bandwidth
bw-cap-num = 1*10(DIGIT) ;; DIGIT defined in RFC 5234
```

;Negotiation of bandwidth per media stream can be useful when negotiating media encoding ;capabilities with different bandwidths.

;RFC 7006 (see Section 8.4) extends the <extension-config-list> field to be able to convey lists ;of bandwidth capabilities in latent or potential configurations, according to the following ABNF ;specified in RFC 5234 syntax:

```
extension-config-list =/ bandwidth-config-list
bandwidth-config-list = ["+"] "b=" bw-cap-list *(BAR bw-cap-list)
; BAR defined in RFC 5939 (see Section 8.1).
bw-cap-list = bw-cap-num *("," bw-cap-num)
bw-cap-num = 1*10(DIGIT) ;; DIGIT defined in RFC 5234
```

;Each bandwidth capability configuration is a comma-separated list of bandwidth capability attribute ;numbers where <bw-cap-num> refers to the <bw-cap-num> bandwidth capability numbers defined ;explicitly earlier in this document, and hence MUST be between 1 and 2^31 – 1 (both included).

;The bandwidth parameter to the actual configuration attribute ("a=acfg") is formulated as a ;<sel-extension-config> with

```
ext-cap-name            =       "b"
```

;hence

```
sel-extension-config  =       / sel-bandwidth-config
sel-bandwidth-config  =       "b=" bw-cap-list ;; bw-cap-list as
                              above.
```

;This format corresponds to the RFC 4566 (this Section 2) attribute production rules, according ;to the following ABNF (RFC 5234) syntax:

```
att-field             =       / "ccap"
att-value             =       / conn-cap-num 1*WSP nettype SP addrtype
                                   SP connection-address
conn-cap-num          =       1*10(DIGIT) ;; 1 to 2^31-1, inclusive
                              ; DIGIT defined in RFC 5234
```

;In RFC 7006 (see Section 8.4), the concept is extended by the SDP media capabilities negotiation ;(RFC 6871, see Section 8.3) with an "lcfg" attribute that conveys latent configurations. In this ;document, we define a <connection-config> parameter to be used to specify a connection data ;capability in a potential or latent configuration attribute. The parameter follows the form of an

```
<extension-config-list>
```

;with

```
ext-cap-name = "c"
ext-cap-list = conn-cap-list
```

;where, according to the following ABNF described in RFC 5234 syntax:

```
extension-config-list =/ conn-config-list
conn-config-list = ["+"] "c=" conn-cap-list
conn-cap-list = conn-cap-num *(BAR conn-cap-num)
conn-cap-num = 1*10(DIGIT) ;; 1 to 2^32-1 inclusive
```

;Each capability RFC 7006 (see Section 8.4) configuration alternative contains a single connection ;data capability attribute number and refers to the conn-cap-num capability number defined ;explicitly earlier in this document; hence, the values MUST be between 1 and 2^31 - 1 (both ;included). The connection data capability allows the expression of only a single capability in each ;alternative, rather than a list of capabilities, since no more than a single connection data field is ;permitted per media block. Nevertheless, it is allowed to express alternative potential connection ;configurations separated by a vertical bar ("|"). An endpoint includes a plus sign ("+") in the ;configuration attribute to mandate support for this extension. An endpoint that receives this ;parameter prefixed with a plus sign and does not support this extension MUST treat that potential ;configuration as not valid. The connection data parameter to the actual configuration attribute ;("a=acfg") is formulated as a:

```
<sel-extension-config>
```

;with

```
ext-cap-name = "c"
```

;hence

```
sel-extension-config =/ sel-connection-config
sel-connection-config = "c=" conn-cap-num ;;
```

;as defined above.

;The title capability RFC 7006 (see Section 8.4) attribute satisfies the general attribute ;production rules in SDP (RFC 4566), according to the following ABNF (RFC 5234) syntax:

```
att-field =/ "icap"
att-value =/ title-cap-num 1*WSP text
;; text defined in RFC 4566
title-cap-num = 1*10(DIGIT) ;; DIGIT defined in RFC 5234
```

;In RFC 7006 (see Section 8.4), the SDP capability negotiation framework (RFC 5939, see ;Section 8.1) provides for the existence of the "pcfg" and "acfg" attributes. The concept is extended ;by the SDP media capabilities negotiation (RFC 6871) with an "lcfg" attribute that conveys ;latent configurations. In this document, we define a <title-config-list> parameter to be used to ;convey title capabilities in a potential or latent configuration. This parameter is defined as an ;<extension-config-list> with the following associations:

```
ext-cap-name = "i"
ext-cap-list = title-cap-list
```

;This leads to the following definition for the title capability parameter:

```
extension-config-list =/ title-config-list
title-config-list = ["+"] "i=" title-cap-list
title-cap-list = title-cap-num *(BAR title-cap-num)
; BAR defined in RFC 5939 (see Section 8.1)
title-cap-num = 1*10(DIGIT) ;; DIGIT defined in RFC 5234
```

;Each potential capability configuration contains a single title capability attribute number where ;"title-cap-num" is the title capability number defined explicitly earlier in this document and must ;be between 1 and 2^31-1 (both included). The title capability allows the expression of only a ;single capability in each alternative, since no more than a single title field is permitted per block. ;Nevertheless, it is allowed to express alternative potential title configurations separated by a vertical ;bar ("|"). An endpoint includes a plus sign ("+") in the configuration attribute to mandate support ;for this extension. An endpoint that receives this parameter prefixed with a plus sign and does not ;support this extension MUST treat that potential configuration as not valid. The title parameter to ;the actual configuration attribute ("a=acfg") is formulated as a <sel-extension-config> with:

```
ext-cap-name = "I"
```

;hence

```
sel-extension-config =/ sel-title-config
sel-title-config = "i=" title-cap-num ;;
```

;as defined above.

RFC 7195 (Section 9.5):

;In RFC 7195 (see Section 9.5), the following is the formal ABNF (RFC 5234) syntax that supports ;the extensions defined in this specification. The syntax is built above the SDP (RFC 4566) and ;the tel URI (RFC 3966) grammars. Implementations that are compliant with this specification ;MUST be compliant with this syntax. The following shows the formal syntax of the extensions ;defined in this memo.

; extension to the connection field originally specified in RFC 4566

```
connection-field = [%x63 "=" nettype SP addrtype SP
connection-address CRLF]
```

;CRLF defined in RFC 5234

;nettype and addrtype are defined in RFC 4566

```
connection-address =/ global-number-digits / "-"
```

;global-number-digits specified in RFC 3966

;subrules for correlation attribute

```
attribute =/ cs-correlation-attr
```

;attribute defined in RFC 4566

```
cs-correlation-attr = "cs-correlation:" corr-mechanisms
corr-mechanisms = corr-mech *(SP corr-mech)
corr-mech = caller-id-mech / uuie-mech /
dtmf-mech / external-mech /
ext-mech
caller-id-mech = "callerid" [":" caller-id-value]
caller-id-value = "+" 1*15DIGIT
```

;DIGIT defined in RFC 5234

```
uuie-mech = "uuie" [":" uuie-value]
uuie-value = 1*65(HEXDIG HEXDIG)
```

;This represents up to 130 HEXDIG

; (65 octets)

;HEXDIG defined in RFC 5234

;HEXDIG defined as 0-9, A-F

```
dtmf-mech = "dtmf" [":" dtmf-value]
dtmf-value = 1*32(DIGIT / %x41-44 / %x23 / %x2A )
```

;0-9, A-D, "#" and "*"

```
external-mech = "external"
ext-mech = ext-mech-name [":" ext-mech-value]
ext-mech-name = token
ext-mech-value = token
```

;token is specified in RFC 4566

RFC 6849 (Section 11.1):

;This specification (RFC 6849, see Section 11.1) defines a new "loopback" attribute, which ;indicates that the agent wishes to perform loopback and the type of loopback the agent is able to ;do. The loopback-type is a value media attribute (RFC 4566) with the following syntax:

```
a                      =      loopback:<loopback-type>
```

;Following is the ABNF for loopback-type:

```
attribute              =      / loopback-attr
                              ; attribute defined in RFC 4566
loopback-attr          =      "loopback:" SP loopback-type
loopback-type          =      loopback-choice [1*SP loopback-choice]
loopback-choice        =      loopback-type-pkt / loopback-type-media
loopback-type-pkt      =      "rtp-pkt-loopback"
loopback-type-media    =      "rtp-media-loopback"
```

;The loopback-type is used to indicate the type of loopback. The loopback-type values are ;rtp-pkt-loopback and rtp-media-loopback. rtp-pkt-loopback: In this mode, the RTP packets are ;looped back to the sender at a point before the encoder/decoder function in the receive direction ;to a point after the encoder/decoder function in the send direction.

;The loopback role (RFC 6849, see Section 11.1) defines two property media attributes ;(RFC 4566) that are used to indicate the role of the agent generating the SDP offer or answer. ;The syntax of the two loopback-role media attributes is as follows:

```
a                      =      loopback-source
```

;and

```
a                      =      loopback-mirror
```

;Following is the ABNF (RFC 5234) for loopback-source and loopback-mirror:

```
attribute              =      / loopback-source / loopback-mirror
                              ; attribute defined in RFC 4566
loopback-source        =      "loopback-source"
loopback-mirror        =      "loopback-mirror"
```

;loopback-source: This attribute specifies that the entity that generated the SDP is the media ;source and expects the receiver of the SDP message to act as a loopback mirror.

;loopback-mirror: This attribute specifies that the entity that generated the SDP will mirror ;(echo) all received media back to the sender of the RTP stream. No media is generated ;locally by the looping-back entity for transmission in the mirrored stream.

;The "m=" line in the SDP includes all the payload types that will be used during the loopback ;session. The complete payload space for the session is specified in the "m=" line, and the rtpmap ;attribute is used to map from the payload-type number to an encoding name denoting the payload ;format to be used.

RFC 6947 (Section 9.2):

;The "altc" attribute of RFC 6947 (Section 9.2) adheres to the RFC 4566 "attribute" production. ;The ABNF syntax (RFC 5234) of altc is provided below.

```
altc-attr          =     "altc" ":" att-value
att-value          =     altc-num SP addrtype
                         SP connection-address SP port
                           ["/" rtcp-port]
altc-num           =     1*DIGIT
rtcp-port          =     port
```

;The meanings of the fields are as follows:

;altc-num: digit to uniquely refer to an address alternative. It indicates the preference order, ;with "1" indicated the most preferred address.

;addrtype: the addrtype field as defined in RFC 4566 for connection data.

;connection-address: a network address as defined in RFC 4566 corresponding to the address ;type specified by addrtype.

;port: the port number to be used, as defined in RFC 4566. Distinct port numbers may be used ;for each IP address type. If the specified address type does not require a port number, a ;value defined for that address type should be used.

;rtcp-port: including an RTP Control Protocol (RTCP) port is optional. An RTCP port may ;be indicated in the alternative "c=" line when the RTCP port cannot be derived from the ;RTP port. The "altc" attribute is applicable only in an SDP offer. The "altc" attribute is a ;media-level-only attribute and MUST NOT appear at the SDP session level. (Because it ;defines a port number, it is inherently tied to the media level.) There MUST NOT be more ;than one "altc" attribute per addrtype within each media description. This restriction is ;necessary so that the addrtype of the reply may be used by the offerer to determine which ;alternative was accepted. The "addrtype"s of the altc MUST correspond to the "nettype" of ;the current connection ("c=") line.

RFC 4570 (Section 9.3):

;The "dest-address" value in each source-filter field MUST match an existing connection-field ;value, unless the wildcard connection-address value "*" is specified.

```
source-filter   =     "source-filter" ":" SP filter-mode SP
                      filter-spec
                      ; SP is the ASCII 'space' character
                      ; (0x20, defined in RFC 5234 that obsoletes
                      ; RFC 4234).
filter-mode     =     "excl" / "incl"
                      ; either exclusion or inclusion mode.
filter-spec     =     nettype SP address-types SP dest-address SP
                      src-list
                      ; nettype is as defined in RFC 4566 (see
                      ; Section 2).
address-types   =     "*" / addrtype
                      ; "*" for all address types (both IP4 and
                      ; IP6),
                      ; but only when <dest-address> and <src-list>
                      ; reference FQDNs.
                      ; addrtype is as defined in RFC 4566 (see
                      ; Section 2).
dest-address    =     "*" / basic-multicast-address /
                      unicast-address
```

```
                            ; "*" applies to all connection-address
                            ; values.
                            ; unicast-address is as defined in RFC 4566
                            ; (see Section 2).
src-list          =         *(unicast-address SP) unicast-address
                            ; one or more unicast source addresses (in
                            ; standard IPv4 or IPv6 ASCII-notation form)
                            ; or FQDNs.
                            ; unicast-address is as defined in RFC 4566
                            ; (see Section 2).
basic-multicast-address      =     basic-IP4-multicast /
                                   basic-IP6-multicast
                                   / FQDN / extn-addr
                                   ; i.e., the same as
                                   ; multicast-address
                                   ; defined in RFC 4566 (see
                                   ; Section 2), except
                                   ; that the
                                   ; /<ttl> and /<number of
                                   ; addresses>
                                   ; fields are not included.
                                   ; FQDN and extn-addr are as
                                   ; defined
                                   ; in RFC 4566 (see Section 2).
basic-IP4-multicast          =     m1 3( "." decimal-uchar )
                                   ; m1 and decimal-uchar are as
                                   ; defined
                                   ; in RFC 4566 (see Section 2)
basic-IP6-multicast          =     hexpart
                                   ; hexpart is as defined in RFC
                                   ; 4566
                                   ; (see Section 2).
```

RFC 5432 (Section 12.1):

;RFC 5432 (see Section 12.1) defines the "qos-mech-send" and "qos-mech-recv" session and ;media-level SDP (RFC 4566) attributes. The following is the ABNF (RFC 5234) syntax, which ;is based on the SDP (RFC 4566) grammar:

```
attribute                    =     / qos-mech-send-attr
attribute                    =     / qos-mech-recv-attr
qos-mech-send-attr           =     "qos-mech-send" ":"
                                   [[SP] qos-mech *(SP qos-mech)]
qos-mech-recv-attr           =     "qos-mech-recv" ":"
                                    [[SP] qos-mech *(SP qos-mech)]
qos-mech                     =     "rsvp" / "nsis" / extension-mech
extension-mech               =     token
```

;The "qos-mech" token identifies a QOS mechanism that is supported by the entity generating the ;session description. A token that appears in a "qos-mech-send" attribute identifies a QOS mechanism ;that can be used to reserve resources for traffic sent by the entity generating the session description. A ;token that appears in a "qos-mech-recv" attribute identifies a QOS mechanism that can be used to reserve ;resources for traffic received by the entity generating the session description. The "qos-mech-send"' and ;"qos-;mech-recv" attributes are not interdependent; one can be used without the other.

RFC 3605 (Section 12.3):

;The formal description (RFC 3605, see Section 12.3) of the attribute is defined by the following ;ABNF (RFC 5234 that obsoletes RFC 4234 while RFC 4234 obsoletes RFC 2234) syntax:

```
rtcp-attribute          =      "a=rtcp:" port [nettype space addrtype
                               space
                               connection-address] CRLF
```

;In this description, the "port," "nettype," "addrtype," and "connection-address" tokens are defined.

RFC 3890 (Section 12.4):

;This chapter (RFC 3890, see Section 12.4) defines in ABNF from RFC 2234 (that was made ;obsolete by RFC 4234, while RFC 4234 was made obsolete by RFC 5234) the bandwidth ;modifier and the packet rate attribute. The bandwidth modifier:

```
TIAS-bandwidth-def      =      "b" "=" "TIAS" ":" bandwidth-value CRLF
bandwidth-value         =      1*DIGIT
```

;The maximum packet rate attribute:

```
max-p-rate-def          =      "a" "=" "maxprate" ":" packet-rate CRLF
packet-rate             =      1*DIGIT ["." 1*DIGIT]
```

RFC 4145 (Section 10.1):

;The following (RFC 4145, see Section 10.1) is the ABNF for an "m=" line, as specified by RFC ;4566 that obsoletes RFC 2327.

```
media-field             =      "m=" media space port ["/" integer]
                               space proto 1*(space fmt) CRLF
```

;This document defines a new value for the proto field: "TCP." The "TCP" protocol identifier ;is similar to the "UDP" protocol identifier in that it only describes the transport protocol and ;not the upper-layer protocol. An "m=" line that specifies "TCP" MUST further qualify the ;application-layer protocol using an fmt identifier. Media described using an "m=" line containing ;the "TCP" protocol identifier are carried using TCP.

;The "setup" attribute (RFC 4145, see Section 10.1) indicates which of the endpoints should ;initiate the TCP connection establishment (i.e., send the initial TCP SYN). The "setup" attribute ;is charset-independent and can be a session-level or a media-level attribute. The following is the ;ABNF of the "setup" attribute:

```
setup-attr     =     "a=setup:" role
role           =     "active" / "passive" / "actpass"
                     / "holdconn"
```

;"active": The endpoint will initiate an outgoing connection.
;"passive": The endpoint will accept an incoming connection.
;"actpass": The endpoint is willing to accept an incoming connection or to initiate an outgoing connection.
;"holdconn": The endpoint does not want the connection to be established for the time being.

;The media-level "'connection" attribute (RFC 4145, see Section 10.1), which is charset-independent, ;is used to disambiguate these two scenarios. The following is the ABNF of the connection ;attribute:

```
connection-attr        =     "a=connection:" conn-value
conn-value             =     "new" / "existing"
```

RFC 5547 (Section 13.1):

;In (RFC 5547, see Section 13.1), we define a number of new SDP (RFC 4566) attributes that ;provide the required information to describe the transfer of a file with Message Session Relay ;Protocol (MSRP). These are all media-level-only attributes in SDP. The following is the formal ;ABNF syntax (RFC 5234) of these new attributes. It is built above the SDP (RFC 4566) gram ;mar, RFC 2045, RFC 2183, RFC 2392, and RFC 5322.

```
attribute              =     / file-selector-attr / file-disp-attr /
                             file-tr-id-attr / file-date-attr /
                             file-icon-attr / file-range-attr
                             ; attribute is defined in RFC 4566 (see
                             ; above)
file-selector-attr     =     "file-selector" [":" selector *(SP
                             selector)]
selector               =     filename-selector / filesize-selector /
                             filetype-selector / hash-selector
filename-selector      =     "name:" DQUOTE filename-string DQUOTE
                             ; DQUOTE defined in RFC 5234
filename-string        =     1*(filename-char/percent-encoded)
filename-char          =     %x01-09/%x0B-0C/%x0E-21/%x23-24/%x26-FF
                             ; any byte except NUL, CR, LF,
                             ; double quotes, or percent
percent-encoded        =     "%" HEXDIG HEXDIG
                             ; HEXDIG defined in RFC 5234
filesize-selector      =     "size:" filesize-value
filesize-value         =     integer ;integer defined in RFC 4566
filetype-selector      =     "type:" type "/" subtype *(";"
                             ft-parameter)
ft-parameter           =     attribute "=" DQUOTE value-string DQUOTE
                             ; attribute is defined in RFC 2045
                             ; free insertion of linear-white-space
                             ; is not
                             ; permitted in this context.
                             ; note: value-string has to be
                             ; re-encoded
                             ; when translating between this and a
                             ; Content-Type header.
value-string           =     filename-string
hash-selector          =     "hash:" hash-algorithm ":" hash-value
hash-algorithm         =     token ; see IANA Hash Function
                             ; Textual Names registry
                             ; only "sha-1" currently supported
hash-value             =     2HEXDIG *(":" 2HEXDIG)
                             ; Each byte in upper-case hex, separated
                             ; by colons.
                             ; HEXDIG defined in RFC 5234
```

```
file-tr-id-attr      =    "file-transfer-id:" file-tr-id-value
file-tr-id-value     =    token
file-disp-attr       =    "file-disposition:" file-disp-value
file-disp-value      =    token
file-date-attr       =    "file-date:" date-param *(SP date-param)
date-param           =    c-date-param / m-date-param /
                          r-date-param
c-date-param         =    "creation:" DQUOTE date-time DQUOTE
m-date-param         =    "modification:" DQUOTE date-time DQUOTE
r-date-param         =    "read:" DQUOTE date-time DQUOTE
                          ; date-time is defined in RFC 5322
                          ; numeric timezones (+HHMM or -HHMM)
                          ; must be used
                          ; DQUOTE defined in RFC 5234 files.
file-icon-attr       =    "file-icon:" file-icon-value
file-icon-value      =    cid-url ; cid-url defined in RFC 2392
file-range-attr      =    "file-range:" start-offset "-"
                          stop-offset
start-offset         =    integer ; integer defined in RFC 4566
stop-offset          =    integer / "*"
```

;When used for capability query (see Section 13.1.8.5), the "file-selector" attribute MUST NOT ;contain any selector, because its presence merely indicates compliance to this specification. When ;used in an SDP offer or answer, the "file-selector"' attribute MUST contain at least one selector. ;Selectors characterize the file to be transferred. There are four selectors in this attribute: "name," ;"size," "type," and "hash." The "name" selector in the "file-selector" attribute contains the filename ;of the content enclosed in double quotes. The filename is encoded in UTF-8 (RFC 3629). Its value ;SHOULD be the same as the "filename" parameter of the Content-Disposition header field (RFC ;2183) that would be signaled by the actual file transfer. If a file name contains double quotes or any ;other character that the syntax does not allow in the "name" selector, they MUST be percent-encoded.

RFC 4583 (Section 13.2):

;In RFC 4583 (see Section 13.2), we define the "floorctrl" SDP media-level attribute to perform ;floor control determination. Its ABNF syntax is:

```
floor-control-attribute     =    "a=floorctrl:" role *(SP role)
role                        =    "c-only" / "s-only" / "c-s"
```

;The offerer includes this attribute to state all the roles it would be willing to perform:

;c-only: The offerer would be willing to act as a floor control client only.
;s-only: The offerer would be willing to act as a floor control server only.
;c-s: The offerer would be willing to act both as a floor control client and
;as a floor control server.

;In RFC 4583 (see Section 13.2), we define the "confid" and the "userid" SDP media-level ;attributes. These attributes are used by a floor control server to provide a client with a conference ;ID and a user ID, respectively. Their ABNF syntax is

```
confid-attribute     =    "a=confid:" conference-id
conference-id        =    token
```

```
userid-attribute        =       "a=userid:" user-id
user-id                 =       token
```

;The "confid" and the "userid" attributes carry the integer representation of a conference ID and ;a user ID, respectively. Endpoints that use the offer/answer model to establish BFCP connections ;MUST support the "confid" and "userid" attributes. A floor control server acting as an offerer or ;as an answerer SHOULD include these attributes in its session descriptions.

RFC 4583 (Section 10.1):

;In RFC 4583 (RFC 4145, see Section 10.1), we define the "floorid" SDP media-level attribute. Its ;ABNF syntax is

```
floor-id-attribute    =       "a=floorid:" token [" mstrm:" token *(SP
                              token)]
```

;The "floorid" attribute is used in BFCP "m=" lines.

RFC 2848 (Section 14.3):

;Note: RFC 2848 uses RFC 2543 that was made obsolete by RFC 3261 (also see Section 1.2) and ;RFC 2327 that was made obsolete by RFC 4566. We have not updated ABNF syntaxes of RFC ;2848 here due to changes that have been updated by RFCs 3261 and 4566. The same is true for ;RFCs 2396, 0822, and 2234 that were made obsolete by RFCs 3986, 5322, and 5226, respectively.

;The Telephone Network (TN) network type (RFC 2848, see Section 14.3) is used to indicate ;that the terminal is connected to a telephone network. The address types allowed for network ;type TN are "RFC 3261 that obsoletes RFC 2543" and private address types, which MUST ;begin with an "X-". 0 Address type RFC 2543 is followed by a string conforming to a subset of ;the "telephone-subscriber" Backus-Naur Form (BNF) specified in SIP (RFC 2543). Note that this ;BNF is NOT identical to the BNF that defines the "phone-number" within the "p=" field of SDP.

;Examples:

```
;c= TN RFC 2543 +1-201-406-4090
;c= TN RFC 2543 12014064090
```

;A telephone-subscriber string is of one of two types: global phone number or local phone number. ;These are distinguished by preceding a global phone number with a "plus" sign ("+"). A global ;phone number is by default to be interpreted as an internationally significant E.164 Number Plan ;Address, as defined by E.164 (ITU-T Study Group 2, June 1997), whilst a local phone number is ;a number specified in the default dialing plan within the context of the recipient Public Switched ;Telephone Network (PSTN)/Internet Inter-Networking (PINT) gateway. An implementation ;MAY use private addressing types, which can be useful within a local domain. These address types ;MUST begin with an "X-" and SHOULD contain a domain name after the X-, (e.g., "Xmytype. ;mydomain.com"). An example of such a connection line is as follows:

```
c= TN X-mytype.mydomain.com A*8-HELEN
```

;where "X-mytype.mydomain.com" identifies this private address type, and "A*8-HELEN" is the ;number in this format. Such a format is defined as an "OtherAddr" in the ABNF of Appendix A of

;this book. Note that most dialable telephone numbers are expressible as local phone numbers within ;address RFC 2543; new address types SHOULD only be used for formats that cannot be so written.

;To support this, two extensions (RFC 2848, see Section 14.3) to the session description format ;are specified. These are some new allowed values for the media field, and a description of the ;"fmtp" parameter when used with the media field values (within the context of the contact field ;network type "TN"). An addition is also made to the SIP message format to allow the inclusion ;of data objects as subparts within the request message itself. The original SDP syntax (RFC 2327 ;that was made obsolete by RFC 4566) for media-field is given as:

```
media-field            =      "m=" media space port ["/" integer]
                              space proto 1*(space fmt) CRLF
```

;When used within PINT requests, the definition of the subfields is expanded slightly. The Media ;subfield definition is relaxed to accept all of the discrete "top-level" media types defined in RFC ;2046. In the milestone services, the discrete type "video" is not used, and the extra types "data" ;and "control" are likewise not needed. The use of these types is not precluded, but the behavior ;expected of a PINT gateway receiving a request including such a type is not defined here. The ;port subfield has no meaning in PINT requests as the destination terminals are specified using ;"TN" addressing, so the value of the port subfield in PINT requests is normally set to "1." A value ;of "0" may be used as in SDP to indicate that the terminal is not receiving media. This is useful to ;indicate that a telephone terminal has gone "on hold" temporarily. Likewise, the optional integer ;subfield is not used in PINT.

;In RFC 2848 (see Section 14.3), for each element of the fmt subfield, there MUST be a ;following fmtp attribute. When used within PINT requests, the fmtp attribute has a general ;structure as defined here:

```
"a        =      fmtp:" <subtype> <space> resolution
                 *(<space> resolution)
                 (<space> ";" 1(<attribute>)
                 *(<space> <attribute>))
```

;where

```
<resolution> : =    (<uri-ref> | <opaque-ref> | <sub-part-ref>)
```

;A fmtp attribute describes the sources used with a given fmt entry in the media field. The entries ;in a fmt subfield are alternatives (with the preferred one first in the list). Each entry will have a ;matching fmtp attribute.

;The general (RFC 2848, see Section 14.3) syntax of a reference to an Internet-based external ;data object in a fmtp line within a PINT session description is

```
<uri-ref>          :=       ("uri:" URI-reference)
```

;where URI reference is as defined in Appendix A of RFC 2396.

;These data form an opaque reference, in that they are sent "untouched" through the PINT ;infrastructure. A reference to some data object held on the GSTN has the general definition

```
<opaque-ref> := ("opr:" *uric)
```

;where uric is as defined in Appendix A of RFC 2396.

;In RFC 2848 (see Section 14.3), as an alternative to pointing to the data via a URI or an ;opaque reference to a data item held on the GSTN, it is possible to include the content data within ;the SIP request itself. This is done by using multipart MIME for the SIP payload. The first MIME ;part contains the SDP description of the telephone network session to be executed. The other ;MIME parts contain the content data to be transported. Format specific attribute lines within ;the session description are used to indicate which other MIME part within the request contains ;the content data. Instead of a URI or opaque reference, the format-specific attribute indicates the ;Content-ID of the MIME part of the request that contains the actual data and is defined as:

```
<sub-part-ref> := ("spr:" Content-ID)
```

;where Content-ID is as defined in Appendix A of RFC 2396 and in RFC 0822.

;Within the telephone network (RFC 2848 see Section 14.3), the "local context" is provided ;by the physical connection between the subscriber's terminal and the central office. An analogous ;association between the PINT client and the PINT server that first receives the request may not ;exist, which is why it may be necessary to supply this missing "telephone ;network context." This ;attribute is defined as follows:

```
a                       =     phone-context: <phone-context-ident>
phone-context-ident     =     network-prefix / private-prefix
network-prefix          =     intl-network-prefix /
                              local-network-prefix
intl-network-prefix     =     "+" 1*DIGIT
local-network-prefix    =     1*DIGIT
excldigandplus          =     (0x21-0x2d,0x2f,0x40-0x7d))
private-prefix          =     1*excldigandplus 0*uric
```

;An intl-network-prefix and local-network-prefix MUST be bona fide network prefixes, and a ;network-prefix that is an intl-network-prefix MUST begin with an E.164 service code ("country ;code"). It is possible to register new private-prefixes with IANA to avoid collisions. Prefixes that ;are not so registered MUST begin with an "X-" to indicate their private, nonstandard nature (see ;Appendix C of RFC 2248).

;It is, however, defined here (RFC 2248, see Section 14.3) for use where there are regulatory ;restrictions on GSTN operation, and in that case the executive system can use it to honor the ;originator's request. The attribute is specified as follows:

```
a=clir:<"true" | "false">
```

;This Boolean value is needed within the attribute as it may be that the GSTN address is, by ;default, set to NOT present its identity to correspondents, and the originator wants to do so for ;this particular call.

;In RFC 2848 (see Section 14.3), according to the SDP specification, a PINT server is allowed ;simply to ignore attribute parameters that it does not understand. In order to force a server to ;decline a request if it does not understand one of the PINT attributes, a client SHOULD use the ;"require" attribute, specified as follows:

```
a=require:<attribute-list>
```

;where the attribute-list is a comma-separated list of attributes that appear elsewhere in the session ;description. In order to process the request successfully, the PINT server must BOTH understand

;the attribute AND ALSO fulfill the request implied by the presence of the attribute, for each ;attribute appearing within the attribute-list of the require attribute.

;In RFC 2848 (see Section 14.3)

;; --(ABNF is specified in RFC 2234)
;; --Variations on SDP definitions

```
connection-field      =     ["c=" nettype space addrtype space
                            connection-address CRLF]
```

; -- this is the original definition from SDP, included for completeness
; -- the following are PINT interpretations and modifications

```
nettype               =     ("IN"/"TN")
```

; -- redefined as a superset of the SDP definition

```
addrtype              =     (INAddrType / TNAddrType)
```

; -- redefined as a superset of the SDP definition

```
INAddrType            =     ("IP4"/"IP6")
```

; -- this nonterminal added to hold original SDP address types

```
TNAddrType            =     ("RFC 2543"/OtherAddrType)
OtherAddrType         =     (<X-Token>)
```

; -- X-token is as defined in RFC 2045

```
addr                  =     (<FQDN> / <unicast-address> / TNAddr)
```

; -- redefined as a superset of the original SDP definition
; -- FQDN and unicast address as specified in SDP

```
TNAddr                =     (RFC 2543Addr/OtherAddr)
```

; -- TNAddr defined only in context of nettype == "TN"

```
RFC 2543Addr          =     (INPAddr/LDPAddr)
INPAddr               =     "+" <POS-DIGIT> 0*(("-"
                            <DIGIT>)/<DIGIT>)
```

; -- POS-DIGIT and DIGIT as defined in SDP

```
LDPAddr               =     <DIGIT> 0*(("-" <DIGIT>)/<DIGIT>)
OtherAddr             =     1*<uric>
```

; -- OtherAdd defined in the context of OtherAddrType
; -- uric is as defined in RFC 2396

```
media-field           =     "m=" media <space> port <space> proto
```

```
                              1*(<space> fmt) <CRLF>
```

; -- NOTE redefined as subset/relaxation of original SDP definition
; -- space and CRLF as defined in SDP

```
media                    =       ("application"/"audio"/"image"/"text")
```

; -- NOTE redefined as a subset of the original SDP definition
; -- This could be any MIME discrete type; only those listed are
; -- used in PINT 1.0

```
port                     =       ("0" / "1")
```

; -- NOTE redefined from the original SDP definition;
; -- 0 retains usual sdp meaning of "temporarily no media"
; -- (i.e., "line is on hold")
; -- (1 means there is media)

```
proto                    =       (INProto/TNProto)
```

; -- redefined as a superset of the original SDP definition

```
INProto                  =       1* (<alpha-numeric>)
```

; -- this is the "classic" SDP protocol, defined if nettype == "IN"
; -- alpha-numeric is as defined in SDP

```
TNProto                  =       ("voice"/"fax"/"pager")
```

; -- this is the PINT protocol, defined if nettype == "TN"

```
fmt                      =       (<subtype> / "-")
```

; -- NOTE redefined as a subset of the original SDP definition
; -- subtype as defined in RFC 2046, or "-." MUST be a subtype of type
; -- in associated media subfield or the special value "-."

```
attribute-fields         =       *("a=" attribute-list <CRLF>)
```

; -- redefined as a superset of the definition given in SDP
; -- CRLF is as defined in SDP

```
attribute-list           =       1(PINT-attribute / <attribute>)
```

; -- attribute is as defined in SDP

```
PINT-attribute           =       (clir-attribute /
                                 q763-nature-attribute /
                                 q763plan-attribute /
                                 q763-INN-attribute /
                                 phone-context-attribute /
                                 tsp-attribute /
```

```
                                           pint-fmtp-attribute /
                                           strict-attribute)
clir-attribute                    =        clir-tag ":" ("true" / "false")
clir-tag                          =        "clir"
q763-nature-attribute             =        Q763-nature-tag ":" q763-natures
q763-nature-tag                   =        "Q763-nature"
q763-natures                      =        ("1" / "2" / "3" / "4")
q763-plan-attribute               =        Q763-plan-tag ":" q763-plans
q763-plan-tag                     =        "Q763-plan"
q763-plans                        =        ("1" / "2" / "3" / "4" / "5" /
                                           "6" / "7")
```

; -- of these, the meanings of 1, 3, and 4 are defined in the text

```
q763-INN-attribute                =        Q763-INN-tag ":" q763-INNs
q763-INN-tag                      =        "Q763-INN"
q763-INNs                         =        ("0" / "1")
phone-context-attribute           =        phone-context-tag ":"
                                           phone-context-ident
phone-context-tag                 =        "phone-context"
phone-context-ident               =        network-prefix / private-prefix
network-prefix                    =        intl-network-prefix /
                                           local-network-prefix
intl-network-prefix               =        "+" 1*<DIGIT>
local-network-prefix              =        1*<DIGIT>
private-prefix                    =        1*excldigandplus 0*<uric>
excldigandplus                    =        (0x21-0x2d,0x2f,0x40-0x7d))
tsp-attribute = tsp-tag "         =        " provider-domainname
tsp-tag                           =        "tsp"
provider-domainname               =        <domain>
```

; -- domain is defined in RFC 1035
; -- NOTE the following is redefined relative to the normal use in SDP

```
pint-fmtp-attribute               =        "fmtp:" <subtype> <space>
                                           resolution
                                           *(<space> resolution)
                                           (<space> ";" 1(<attribute>)
                                           *(<space>
                                           <attribute>))
```

; -- subtype as defined in RFC 2046.
; -- NOTE that this value MUST match a fmt on the ultimately preceding
; -- media-field
; -- attribute is as defined in SDP

```
resolution                        =        (uri-ref / opaque-ref /
                                           sub-part-ref)
uri-ref                           =        uri-tag ":" <URI-Reference>
```

; -- URI-reference defined in RFC 3986 that obsoletes RFC 2396

```
uritag                            =        "uri"
```

```
opaque-ref               =      opr-tag ":" 0*<uric>
opr-tag                  =      "opr"
sub-part-ref             =      spr-tag ":" <Content-ID>
```

; -- Content-ID is as defined in RFC 2046 and RFC 822

```
spr-tag                  =      "spr"
strict-attribute         =      "require:" att-tag-list
att-tag-list             =      1(PINT-att-tag-list / <att-field> /
                                pint-fmtp-tag-list)
                                *(","
                                (PINT-att-tag-list / <att-field> /
                                pint-fmtp-tag-list)
                                )
```

; -- att-field as defined in SDP

```
PINT-att-tag-list        =      (phone-context-tag / clir-tag /
                                q763-nature-tag / q763-plan-tag /
                                q763-INN-tag)
pint-fmtp-tag-list       =      (uri-tag / opr-tag / spr-tag)
```

;; --Variations on SIP definitions

```
clir-parameter           =      clir-tag "=" ("true" / "false")
q763-nature-parameter    =      Q763-nature-tag "=" Q763-natures
q763plan-parameter       =      Q763-plan-tag "=" q763plans
q763-INN-parameter       =      Q763-INN-tag "=" q763-INNs
tsp-parameter            =      tsp-tag "=" provider-domainname
phone-context-parameter  =   phone-context-tag "=" phone-context-ident
SIP-param                =      (<transport-param>/<user-param>/<method-
                                param> / <ttl-param> / <maddr-param> /
                                <other-param> )
```

; -- the values in this list are all as defined in SIP

```
PINT-param               =      ( clir-parameter /
                                q763-nature-parameter /
                                q763plan-parameter / q763-INN-parameter/
                                tsp-parameter /
                                phone-context-parameter )
URL-parameter            =      (SIP-param / PINT-param)
```

; -- redefined SIP's URL-parameter to include ones defined in PINT

```
Require-header           =      "require:" 1(required-extensions)
                                *("," required-extensions)
```

; -- NOTE this is redefined as a subset of the SIP definition
; -- (from RFC 2543/section 6.30)

```
required-extensions      =      ("org.ietf.sip.subscribe" /
                                "org.ietf.sdp.require")
```

RFC 4567 (Section 16.1):

;The key management protocol (RFC 4567, see Section 16.1) data contains the necessary ;information to establish the security protocol (e.g., keys and cryptographic parameters). All ;parameters and keys are protected by the key management protocol. The key management data ;SHALL be base64 (RFC 4648 that obsoletes RFC 3548) encoded and comply with the base64 ;grammar as defined in RFC 4566. The key management protocol identifier, KMPID, is defined ;here in ABNF grammar (RFC 5234 made obsolete by RFC 4234).

```
KMPID             =       1*(ALPHA / DIGIT)
```

;Values for the identifier, KMPID, are registered and defined in accordance to Section 16.1.9. ;Note that the KMPID is case sensitive, and it is RECOMMENDED that values registered are ;lowercase letters.

;RFC 4567 (see Section 16.1) provides an ABNF grammar (as used in RFC 4566) for the key ;management extensions to SDP. Note that the new definitions are compliant with the definition ;of an attribute field, that is,

```
attribute = (att-field ":" att-value) / att-field
```

;The ABNF for the key management extensions (conforming to the att-field and att-value) are as ;follows:

```
key-mgmt-attribute    =    key-mgmt-att-field ":"
                           key-mgmt-att-value
key-mgmt-att-field    =    "key-mgmt"
key-mgmt-att-value    =    0*1SP prtcl-id SP keymgmt-data
prtcl-id              =    KMPID
                           ; e.g., "mikey"
keymgmt-data          =    base64
SP                    =    %x20
```

;where KMPID is as defined in Section 16.1.3 of this memo, and base64 is as defined in SDP RFC ;4566. Prtcl-id refers to the set of values defined for KMPID in Section 16.1.9.

;The attribute MAY be used at session level, media level, or at both levels. An attribute defined at ;media level overrides an attribute defined at session level. In other words, if the media-level attribute ;is present, the session-level attribute MUST be ignored for this media. Section 16.1.4.1 describes ;in detail how the attributes are used and how the SDP is handled in different usage scenarios. ;The choice of the level depends, for example, on the particular key management protocol. Some ;protocols may not be able to derive enough key material for all the sessions furthermore, possibly a ;different protection to each session could be required. The particular protocol might achieve this only ;by specifying it at the media level. Other protocols, such as MIKEY, have instead those capabilities ;(as it can express multiple security policies and derive multiple keys), so it may use the session level.

RFC 4796 (Section 5.2):

;This specification defines (RFC 4796, see Section 5.2) a new media-level value attribute, "content." ;Its formatting in SDP is described by the following ABNF (RFC 5234 that obsoletes RFC 4234):

```
content-attribute    =    "a=content:" mediacnt-tag
mediacnt-tag         =     mediacnt *("," mediacnt)
```

```
mediacnt              =      "slides" / "speaker" / "sl" / "main"
                             / "alt" / mediacnt-ext
mediacnt-ext          =      token
```

;The "content" attribute contains one or more tokens, which MAY be attached to a media stream ;by a sending application. An application MAY attach a "content" attribute to any media stream ;it describes. This document provides a set of predefined values for the "content" attribute. Other ;values can be defined in the future.

RFC 4568 (Section 16.2):

;In RFC 4568 (see Section 16.2), we first provide the ABNF grammar for the generic crypto ;attribute, and then we provide the ABNF grammar for the SRTP-specific use of the crypto attribute.

;Generic "Crypto" Attribute Grammar

;The ABNF grammar for the crypto attribute is defined as follows:

```
"a=crypto:" tag 1*WSP crypto-suite 1*WSP key-params
*(1*WSP session-param)
tag                   =      1*9DIGIT
crypto-suite          =      1*(ALPHA / DIGIT / "_")
key-params            =      key-param *(";" key-param)
key-param             =      key-method ":" key-info
key-method            =      "inline" / key-method-ext
key-method-ext        =      1*(ALPHA / DIGIT / "_")
key-info              =      1*(%x21-3A / %x3C-7E) ; visible
                             (printing) chars
                             ; except semi-colon
session-param         =      1*(VCHAR) ; visible (printing)
                             characters
                             ; where WSP, ALPHA, DIGIT, and
                             ; VCHAR
                             ; are defined in RFC 5234 that obsoletes
                             ; RFC 4234.
```

;SRTP "Crypto" Attribute Grammar

;This section provides an ABNF (RFC 5234 that obsoletes RFC 4234) grammar for the SRTP- ;specific use of the SDP crypto attribute:

```
crypto-suite          =      srtp-crypto-suite key-method =
                             srtp-key-method
key-info              =      srtp-key-info
session-param         =      srtp-session-param
srtp-crypto-suite     =      "AES_CM_128_HMAC_SHA1_32" /
                             "F8_128_HMAC_SHA1_32" /
                             "AES_CM_128_HMAC_SHA1_80" /
                             srtp-crypto-suite-ext
srtp-key-method       =      "inline"
srtp-key-info         =      key-salt ["|" lifetime] ["|" mki]
key-salt              =      1*(base64) ; binary key and salt values
                                   ; concatenated together, and then
                                   ; base64 encoded [section 3 of
                                   ; RFC 4648 that obsoletes RFC 3548
```

```
lifetime              =     ["2^"] 1*(DIGIT) ; see section 6.1 for
                            "2^"
mki                   =     mki-value ":" mki-length
mki-value             =     1*DIGIT
mki-length            =     1*3DIGIT ; range 1..128.
srtp-session-param    =     kdr /
                            "UNENCRYPTED_SRTP" /
                            "UNENCRYPTED_SRTCP" /
                            "UNAUTHENTICATED_SRTP" /
                            fec-order /
                            fec-key /
                            wsh /
                            srtp-session-extension
kdr = "KDR            =     " 1*2(DIGIT) ; range 0..24,
                                   ; power of two
fec-order             =     "FEC_ORDER=" fec-type
fec-type              =     "FEC_SRTP" / "SRTP_FEC"
fec-key               =     "FEC_KEY=" key-params
wsh                   =     "WSH=" 2*DIGIT ; minimum value is 64
base64                =     ALPHA / DIGIT / "+" / "/" / "="
srtp-crypto-suite-ext =     1*(ALPHA / DIGIT / "_")
srtp-session-extension =    ["-"] 1*(VCHAR) ;visible chars
                                   ; (RFC 5234 obsoletes RFC 4234)
                                   ; first character must not be
                                   ; dash ("-")
```

RFC 4572 (Section 16.3):

;A fingerprint is represented (RFC 4572, see Section 16.3) in SDP as an attribute (an "a" line). It ;consists of the name of the hash function used, followed by the hash value itself. The hash value ;is represented as a sequence of uppercase hexadecimal bytes, separated by colons. The number ;of bytes is defined by the hash function. (This is the syntax used by openssl and by the browsers' ;certificate managers. It is different from the syntax used to represent hash values in, for example, ;HTTP digest authentication (RFC 2617 made obsolete by RFCs 7235, 7615–7617), which uses ;unseparated lowercase hexadecimal bytes. It was felt that consistency with other applications ;of fingerprints was more important.) The formal syntax of the fingerprint attribute is given in ;ABNF (RFC 5234 that obsoletes RFC 4234). This syntax extends the BNF syntax of SDP (RFC ;4566).

```
attribute              =    / fingerprint-attribute
fingerprint-attribute =     "fingerprint" ":" hash-func SP
                            fingerprint
hash-func              =    "sha-1" / "sha-224" / "sha-256" /
                            "sha-384" / "sha-512" /
                            "md5" / "md2" / token
                                   ; Additional hash functions can
                                   ; only come
                                   ; from updates to RFC 3279
fingerprint            =    2UHEX *(":" 2UHEX)
                                   ; Each byte in upper-case hex,
                                   ; separated
                                   ; by colons.
UHEX                   =    DIGIT / %x41-46 ; A-F uppercase
```

;A certificate fingerprint MUST be computed using the same one-way hash function used in ;the certificate's signature algorithm. (This ensures that the security properties required for the ;certificate also apply for the fingerprint.)

RFC 6193 (Section 16.5):

;RFC 6191 (see Section 16.5) defines a new attribute "ike-setup," which can be used when the ;protocol identifier is "udp" and the "fmt" field is "ike-esp" or "ike-esp-udpencap," in order to ;describe how endpoints should perform the Internet Key Exchange (IKE) session setup procedure. ;The "ikesetup" attribute indicates which of the endpoints should initiate the establishment of an ;IKE session. The "ike-setup" attribute is charset-independent and can be a session- or media-level ;attribute. The following is the ABNF of the "ike-setup" attribute.

```
ike-setup-attr          =      "a=ike-setup:" role
role                    =      "active" / "passive" / "actpass"
```

;"active": The endpoint will initiate an outgoing session.
;"passive": The endpoint will accept an incoming session.
;"actpass": The endpoint is willing to accept an incoming session or to initiate an outgoing session.

;Both endpoints use the SDP offer/answer model to negotiate the value of "ike-setup," following the ;procedures determined for the "setup" attribute defined in Section 10.1.4.1 of RFC 4145 (see Section ;10.1). However, "holdconn," as defined in RFC 4145, is not defined for the "ike-setup" attribute.

;If a preshared key (RFC 6193, see Section 16.5) for IKE authentication is installed in both ;endpoints in advance, they need not exchange the fingerprints of their certificates. However, they ;may still need to specify which preshared key they will use in the following IKE authentication in ;SDP because they may have several preshared keys. Therefore, a new attribute, "psk-fingerprint," ;is defined to exchange the fingerprint of a preshared key over SDP. This attribute also has the role ;of making authorization in SIP consistent with authentication in IKE. Attribute "psk-fingerprint" ;is applied to preshared keys as the "fingerprint" defined in RFC 4572 (see Section 16.3) is ;applied to certificates. The following is the ABNF of the "psk-fingerprint" attribute. The use of ;"psk-fingerprint" is OPTIONAL.

```
attribute                     =    / psk-fingerprint-attribute
psk-fingerprint-attribute     =    "psk-fingerprint" ":"
                                   hash-func SP
                                   psk-fingerprint
hash-func                     =    "sha-1" / "sha-224" /
                                   "sha-256" /
                                   "sha-384" / "sha-512" / token
                                         ; Additional hash
                                         ; functions can only
                                         ; come
                                         ; from updates to RFC 3279
psk-fingerprint               =    2UHEX *(":" 2UHEX)
                                         ; Each byte in upper-case
                                         ; hex,
                                         ; separated
                                         ; by colons.
UHEX                          =    DIGIT / %x41-46 ; A-F uppercase
```

;An example of SDP negotiation for IKE with preshared key authentication without IPsec NAT-
;Traversal is as follows.

RFC 5576 (Section 5.3):

;This memo (RFC 5576, see Section 5.3) defines a new media-level attribute, "depend," with the
;following ABNF (RFC 5234). The identification-tag is defined in RFC 5888 that obsoletes RFC
;3388. In the following ABNF, fmt, token, SP, and CRLF are used as defined in RFC 4566.

```
depend-attribute        =     "a=depend:" dependent-fmt SP dependency-
                              tag *(";" SP dependent-fmt SP dependency-
                              tag) CRLF
dependency-tag          =     dependency-type *1( SP
                              identification-tag
                              ":"
                              fmt-dependency *("," fmt-dependency ))
dependency-type         =     "lay" / "mdc" / token
dependent-fmt           =     fmt
fmt-dependency          =     fmt
```

;dependency-tag indicates one or more dependencies of one dependent fmt in the media description.
;These dependencies are signaled as fmt-dependency values, which indicate fmt values of other
;media descriptions. These other media descriptions are identified by their identification-tag
;values in the depend-attribute. There MUST be exactly one dependency-tag indicated per
;dependent-fmt. dependent-fmt indicates the media format description, as defined in RFC 4566
;(that is, this Section 2), that depends on one or more media format descriptions in the media
;description indicated by the value of the identification-tag within the dependency-tag.

RFC 3108 (Section 9.4):

See Section 9.4 for ABNF syntaxes. We have not repeated it here for sake of brevity.

2.10 Changes in RFC 4566 by RFC 4566bis

RFC 4566bis (draft-ietf-mmusic-rfc4566bis-17) that is yet to be standardized in the IETF will make some changes in RFC 4566 described in this section. The summary of changes is as follows:

- The ABNF rule for IP6-address has been corrected. As a result, the ABNF rule for IP6-multicast has changed, and the (now unused) rules for hexpart, hexseq, and hex4 have been removed.
- IP4 unicast and multicast addresses in the example SDP descriptions have been revised per RFCs 5735 and 5771.
- Text in Section 2.5.2 (Section 5.2 of RFC 4566) has been revised to clarify the use of local addresses in case of ICE-like SDP extensions.
- Normative and informative references have been updated.
- The text regarding the session- vs. media-level attribute usage has been clarified.
- The case-insensitivity rules from RFC 4855 have been included in this document.

We are providing the changes that have been proposed in the following sections.

2.10.1 Revised ABNF for SDP Grammar including IPv6/IPv4

This section provides an ABNF grammar for SDP. ABNF is defined in RFCs 5234 and 7405.
;SDP Syntax

```
session-description     =     proto-version
                              origin-field
                              session-name-field
                              [information-field]
                              [uri-field]
                              *email-field
                              *phone-field
                              [connection-field]
                              *bandwidth-field
                              1*time-field
                              [key-field]
                              *attribute-field
                              *media-description
proto-version           =     %s"v" "=" 1*DIGIT CRLF
                                ;this memo describes version 0

origin-field            =     %s"o" "=" username SP sess-id SP
                              sess-version SP
                               nettype SP addrtype SP unicast-address
                              CRLF

session-name-field      =     %s"s" "=" text CRLF

information-field       =     %s"i" "=" text CRLF

uri-field               =     %s"u" "=" uri CRLF

email-field             =     %s"e" "=" email-address CRLF

phone-field             =     %s"p" "=" phone-number CRLF

connection-field        =     %s"c" "=" nettype SP addrtype SP
                              connection-address CRLF
                              ;a connection field must be present
                              ;in every media description or at the
                              ;session-leve

bandwidth-field         =     %s"b" "=" bwtype ":" bandwidth CRLF

time-field              =     %s"t" "=" start-time SP stop-time
                              *(CRLF repeat-fields) CRLF
                               [zone-adjustments CRLF]

repeat-fields           =     %s"r" "=" repeat-interval SP typed-time
                              1*(SP typed-time)

zone-adjustments        =     %s"z" "=" time SP ["-"] typed-time
                              *(SP time SP ["-"] typed-time)

key-field               =     %s"k" "=" key-type CRLF
```

```
attribute-field        =    %s"a" "=" attribute CRLF

media-description      =    media-field
                            [information-field]
                            *connection-field
                            *bandwidth-field
                            [key-field]
                            *attribute-field

media-field            =    %s"m" "=" media SP port ["/" integer]
                            SP proto 1*(SP fmt) CRLF
                            ; sub-rules of 'o='

username               =    non-ws-string
                            ;pretty wide definition, but doesn't
                            ;include space

sess-id                =    1*DIGIT
                            ;should be unique for this username/host

sess-version           =    1*DIGIT

nettype                =    token
                            ;typically "IN"

addrtype               =    token
                            ;typically "IP4" or "IP6"
                            ; sub-rules of 'u='
uri                         =     URI-reference
                            ; see RFC 3986
                            ; sub-rules of 'e=', see RFC 5322 for
                            definitions

email-address          =    address-and-comment /
                            dispname-and-address
                            / addr-spec

address-and-comment    =    addr-spec 1*SP "(" 1*email-safe ")"

dispname-and-address   =    1*email-safe 1*SP "<" addr-spec ">"
                            ; sub-rules of 'p='
phone-number                =     phone *SP "(" 1*email-safe ")" /
                            1*email-safe "<" phone ">" /
                            phone

phone                  =    ["+"] DIGIT 1*(SP / "-" / DIGIT)
                                  ; sub-rules of 'c='

connection-address     =    multicast-address / unicast-address
                            ; sub-rules of 'b='

bwtype                 =    token

bandwidth              =    1*DIGIT
                            ; sub-rules of 't='
```

```
start-time          =      time / "0"

stop-time           =      time / "0"

time                =      POS-DIGIT 9*DIGIT
                           ; Decimal representation of NTP time in
                           ; seconds since 1900.  The representation
                           ; of NTP time is an unbounded length field
                           ; containing at least 10 digits.  Unlike the
                           ; 64-bit representation used elsewhere, time
                           ; in SDP does not wrap in the year 2036.
                           ; sub-rules of 'r=' and 'z='

repeat-interval     =      POS-DIGIT *DIGIT [fixed-len-time-unit]

typed-time          =      1*DIGIT [fixed-len-time-unit]

fixed-len-time-unit =      %s"d" / %s"h" / %s"m" / %s"s"
                                  ; NOTE: These units are case-sensitive.
                                  ; sub-rules of 'k='

key-type            =      %s"prompt" /
                           %s"clear:" text /
                           %s"base64:" base64 /
                           %s"uri:" uri
                           ; NOTE: These names are case-sensitive.

base64              =      *base64-unit [base64-pad]

base64-unit         =      4base64-char

base64-pad          =      2base64-char "==" / 3base64-char "="

base64-char         =      ALPHA / DIGIT / "+" / "/"
                           ; sub-rules of 'a='

attribute           =      (att-field ":" att-value) / att-field

att-field           =      token

att-value           =      byte-string
                                  ; sub-rules of 'm='

media               =      token
                            ;typically "audio", "video", "text", "image"
                            ;or "application"

fmt                 =      token
                                  typically an RTP payload type for audio
                                  ;and video media

proto               =      token *("/" token)
                                  ;typically "RTP/AVP" or "udp"
```

```
port                  =    1*DIGIT
                           ; generic sub-rules: addressing

unicast-address       =    IP4-address / IP6-address / FQDN / extn-addr

multicast-address     =    IP4-multicast / IP6-multicast / FQDN
                           / extn-addr

IP4-multicast         =    m1 3( "." decimal-uchar )
                           "/" ttl [ "/" numaddr ]
                           ; IP4 multicast addresses may be in the
                           ; range 224.0.0.0 to 239.255.255.255

m1                    =     ("22" ("4"/"5"/"6"/"7"/"8"/"9")) /
                           ("23" DIGIT )

IP6-multicast         =    IP6-address [ "/" numaddr ]
                           ; IP6 address starting with FF

numaddr               =    integer

ttl                   =    (POS-DIGIT *2DIGIT) / "0"

FQDN                  =    4*(alpha-numeric / "-" / ".")
                                  ; fully qualified domain name as
                                  specified
                                  ; in RFC 1035 (and updates)

IP4-address           =    b1 3("." decimal-uchar)

b1                    =    decimal-uchar
                           ; less than "224"

IP6-address           =    6( h16 ":" ) ls32
                                  /"::" 5( h16 ":" ) ls32
                           / [             h16 ] "::" 4( h16 ":" ) ls32
                           / [ *1( h16 ":" ) h16 ] "::" 3( h16 ":" ) ls32
                           / [ *2( h16 ":" ) h16 ] "::" 2( h16 ":" ) ls32
                           / [ *3( h16 ":" ) h16 ] "::"    h16 ":"   ls32
                           / [ *4( h16 ":" ) h16 ] "::"              ls32
                           / [ *5( h16 ":" ) h16 ] "::"              h16
                           / [ *6( h16 ":" ) h16 ] "::"

h16                   =    1*4HEXDIG

ls32                  =    ( h16 ":" h16 ) / IP4-address
                                  ; Generic for other address families

extn-addr             =    non-ws-string
                           ; generic sub-rules: datatypes

text                  =    byte-string
                           ;default is to interpret this as UTF8 text.
```

```
                          ;ISO 8859-1 requires "a=charset:ISO-8859-1"
                          ;session-level attribute to be used

byte-string         =     1*(%x01-09/%x0B-0C/%x0E-FF)
                          ;any byte except NUL, CR, or LF

non-ws-string       =     1*(VCHAR/%x80-FF)
                          ;string of visible characters

token-char          =     ALPHA / DIGIT
                          / "!" / "#" / "$" / "%" / "&"
                          / "'" ; (single quote)
                          / "*" / "+" / "-" / "." / "^" / "_"
                          / "`" ; (Grave accent)
                          / "{" / "|" / "}" / "~"

token               =     1*(token-char)

email-safe          =     %x01-09/%x0B-0C/%x0E-27/%x2A-3B/%x3D/%x3F-FF
                          ;any byte except NUL, CR, LF, or the quoting
                          ;characters ()<>

integer             =     POS-DIGIT *DIGIT

zero-based-integer        =     "0" / integer

non-zero-int-or-real      =     integer / non-zero-real

non-zero-real             =     zero-based-integer "." *DIGIT POS-DIGIT
                                ; generic sub-rules: primitives

alpha-numeric             =     ALPHA / DIGIT

POS-DIGIT                 =     %x31-39 ; 1 - 9

decimal-uchar             =     DIGIT
                                / POS-DIGIT DIGIT
                                / ("1" 2*(DIGIT))
                                / ("2" ("0"/"1"/"2"/"3"/"4") DIGIT)
                                / ("2" "5" ("0"/"1"/"2"/"3"/"4"/"5"))
                                      ; external references:
                                      ; ALPHA, DIGIT, CRLF, SP, VCHAR:
                                      ; from RFC 5234
                                      ; URI-reference: from RFC 3986
                                      ; addr-spec: from RFC 5322
```

2.10.2 Revised Texts in Section 2.5.2 (Section 5.2 of RFC 4566)

Origin ("o=")

```
o=<username> <sess-id> <sess-version> <nettype> <addrtype>
<unicast-address>
```

The "o=" line gives the originator of the session (his/her username and the address of the user's host) plus a session identifier and version number:

<username> is the user's login on the originating host, or it is "-" if the originating host does not support the concept of user IDs. The <username> MUST NOT contain spaces.

<sess-id> is a numeric string such that the tuple of <username>, <sess-id>, <nettype>, <addrtype>, and <unicast-address> forms a globally unique identifier for the session. The method of <sessid> allocation is up to the creating tool, but it has been suggested that a NTP format timestamp be used to ensure uniqueness (RFC 5905).

<sess-version> is a version number for this session description. Its usage is up to the creating tool, as long as <sess-version> is increased when a modification is made to the session data. Again, it is RECOMMENDED that an NTP format timestamp is used.

<nettype> is a text string giving the type of network. Initially "IN" is defined to have the meaning "Internet," but other values MAY be registered in the future (see Section 2.8).

<addrtype> is a text string giving the type of the address that follows. Initially "IP4" and "IP6" are defined, but other values MAY be registered in the future (see Section 2.8).

<unicast-address> is an address of the machine from which the session was created. For an address type of IP4, this is either a FQDN of the machine or the dotted-decimal representation of an IP version 4 address of the machine. For an address type of IP6, this is either a FQDN of the machine or the compressed textual representation of an IP version 6 address of the machine. For both IP4 and IP6, the FQDN is the form that SHOULD be given unless this is unavailable, in which case a globally unique address MAY be substituted. Unless an SDP extension for NAT-traversal is used (e.g., ICE - RFC 5245, ICE TCP - RFC 6544), a local IP address MUST NOT be used in any context where the SDP description might leave the scope in which the address is meaningful (for example, a local address MUST NOT be included in an application-level referral that might leave the scope).

In general, the "o=" line serves as a globally unique identifier for this version of this session description, and the subfields excepting the version taken together identify the session irrespective of any modifications.

For privacy reasons, it is sometimes desirable to obfuscate the username and IP address of the session originator. If this is a concern, an arbitrary <username> and private <unicast-address> MAY be chosen to populate the "o=" line, provided that these are selected in a manner that does not affect the global uniqueness of the field.

2.10.3 RFC 4855 Case-Insensitivity Rules

To take care of the case-insensitivity rules, RFC 4566bis (draft-ietf-mmusic-rfc4566bis-22) has proposed to add the following texts in <fmt> attributes of media descriptions (see Section 2.5.14):

"Section 3 of RFC 4855 states that the payload format (encoding) names defined in the RTP Profile are commonly shown in upper case, while media subtype names are commonly shown in lower case. It also states that both of these names are case-insensitive in both places, similar to parameter names which are case-insensitive both in media type strings and in the default mapping to the SDP a=fmtp attribute."

2.11 Summary

We have described the basic SDP specification along with semantics and syntaxes of various features/capabilities/functionalities of audio, video, and/or data sharing application. The SDP attributes that deal with features/capabilities/functionalities of multimedia sessions are expressed with some simple abbreviated texts using grammar of ABNF syntaxes. Although simple, these standardized SDP ABNF texts need to be understood in relation to bigger context of complex features/capabilities/functionalities of multimedia applications. The inner meaning of adding and modifying session parameters during negotiations between conferencing parties is not so obvious from the simple SDP textual attributes unless the original descriptive texts of RFCs are known. That is, it requires a certain level of expertise in knowing the hidden meaning of each standardized SDP attribute from the elaborate original texts of RFCs. For example, a given multimedia application, say videoconferencing, may support many different videos.

A given kind of video may have many streaming parameters that may need to be negotiated among different parties. In addition to these streaming attributes of video, the network, QOS/grouping, and/or security attributes, on which streaming attributes may depend, may also need to be negotiated. The dynamic negations of all videos along with all individual parameters of each video between different parties may create a serious challenge. Because the simplicity of SDP attributes may not be capable of complying with all negotiation requirements simultaneously unless the requirements of negotiations are simplified. In subsequent sections, we will see how many more SDP extensions have been made to meet.

2.12 Problems

1. What is the definition of SDP? How does the SDP characteristic differ from that of the other protocols? Why do the call control protocols need to carry SDP for session negotiation?
2. Describe the basic rules for ABNF. Provide some examples of how the ABNF for the SDP keeps the bigger contexts of session negotiations hidden in the simple user-friendly texts unless one is familiar with the original texts of RFCs.
3. Provide some conceptual descriptions of how SDP can be used in session negotiations by SIP, WebRTC, streaming media, email, WWW, and multicast session announcement applications.
4. What are the major SDP semantic components for describing a multimedia session? Describe the functions of each semantic that describes a given session.
5. What requirements does SDP need to satisfy in order to carry the session negotiation information related to media and transport, timing, private sessions, more detail session information, categorization in session description, and internationalization using many different languages?
6. Create an end-to-end call flow message for session negotiation that includes the basic SDP parameters such as protocol version, origin, session name, session information, URI, email address and phone number, connection data, bandwidth, timing, repeat times, time zones, encryption keys, attributes, and media descriptions for both SIP and WebRTC call control protocol.
7. What are the SDP attributes defined in RFC 4566? Create examples using each of those attributes for both SIP and WebRTC.
8. What are the recommendations provided for security considerations in SDP for SIP, SAP, and WWW? Create an example using call flows that follow RFC 4566 recommendations for SIP, SAP, and WWW.
9. Provide the call flow example that includes recommended SDP attributes and security features provided in RFC 4566; should it be standardized for WebRTC?

10. Describe conceptually how security-related semantics for SDP can support encryption, authentication, authorization, integrity, and nonrepudiation.

References

1. C. Holmberg, H. Alvestrand, and C. Jennings, "Negotiating Media Multiplexing Using the Session Description Protocol (SDP)," draft-ietf-mmusic-sdp-bundle-negotiation-29.txt, IETF Draft (Work in Progress), April 15, 2016.
2. M. Handley, et al, "SDP: Session Description Protocol, draft-ietf-mmusic-rfc4566bis-22 (Work in Progress), January 1, 2018.

Chapter 3

Negotiations Model in SDP

Abstract
This section describes the Session Description Protocol (SDP) offer/answer model based on Request for Comments (RFC) 3264. The SDP provides a negotiations mechanism used by communicating entities to arrive at a common view of a multimedia session. In the offer/answer negotiation model that SDP uses, one participant offers the other a description of the desired session from his/her perspective, and the other answers with the desired session from his/her perspective. This offer/answer model is most useful in unicast sessions where information from both participants is needed for the complete view of the session. However, some simple examples of SDP offer/answer model are provided in Section 18.1 following RFC 4317 and using Session Initiation Protocol (SIP). By the same token, additional examples of SDP offer/answer model using web real-time communication are provided in Section 18.2 following Internet Engineering Task Force drafts. These examples are related to codec negotiation and selection, hold and resume, and addition and deletion of media streams, multiple media types, bidirectional, unidirectional, inactive streams, and dynamic payload types, and Third Party Call Control using the offer/answer model. In addition, this section describes a protocol using RFC 5245 for network address translator traversal for User Datagram Protocol-based multimedia sessions established with the offer/answer model. This protocol is called Interactive Connectivity Establishment (ICE). ICE makes use of the simple traversal of UDP around NAT protocol and its extension, traversal using relays around NAT. ICE can be used by any protocol utilizing the offer/answer model, such as the SIP.

3.1 Offer/Answer Model with SDP

This section describes Request for Comments (RFC) 3264 that defines a mechanism by which two entities can make use of the Session Description Protocol (SDP) to arrive at a common view of a multimedia session. In the model, one participant offers the other a description of the desired session from his/her perspective, and the other participant answers with the desired session from his/her perspective. This offer/answer model is most useful in unicast sessions where information from both participants is needed for the complete view of the session. The offer/answer model is used by call control protocols like the Session Initiation Protocol (SIP) and web real-time communication (WebRTC).

3.1.1 Introduction

The SDP (RFC 4566 that made RFC 2327 obsolete) was originally conceived as a way to describe multicast sessions carried on the Mbone. The Session Announcement Protocol (RFC 2974) was devised as a multicast mechanism to carry SDP messages. Although the SDP specification allows for unicast operation, it is not complete. Unlike multicast, where there is a global view of the session used by all participants, unicast sessions involve two participants, and a complete view of the session requires information from both and agreement on parameters between them.

As an example, a multicast session requires conveying a single multicast address for a particular media stream. However, for a unicast session, two addresses are needed—one for each participant. As another example, a multicast session requires an indication of which codecs will be used in the session. However, for unicast, the set of codecs needs to be determined by finding an overlap in the set supported by each participant. As a result, even though SDP has the expressiveness to describe unicast sessions, it is missing the semantics and operational details of how it is actually done. In this offer/answer model with SDP, we remedy that by defining a simple offer/answer model based on SDP. In this model, one participant in the session generates an SDP message that constitutes the offer—the set of media streams and codecs the offerer wishes to use, along with the Internet Protocol (IP) addresses and ports the offerer would like to use to receive the media.

The offer is conveyed to the other participant, called the answerer. The answerer generates an answer, which is an SDP message that responds to the offer provided by the offerer. The answer has a matching media stream for each stream in the offer, indicating whether the stream is accepted, along with the codecs that will be used and the IP addresses and ports the answerer wants to use to receive media. It is also possible for a multicast session to work similarly to a unicast one; its parameters are negotiated between a pair of users as in the unicast case, but both sides send packets to the same multicast address, rather than unicast ones. This chapter also discusses the application of the offer/answer model to multicast streams. We define guidelines for how the offer/answer model is used to update a session after an initial offer/answer exchange. The means by which the offers and answers are conveyed are outside the scope of this document. The offer/answer model defined here is the mandatory baseline mechanism used by the SIP (RFC 3261 also see Section 1.2).

3.1.2 Terminology

Some terminologies used in this section are provided in Table 2.1.

3.1.3 Definitions

Definitions of terminologies used in this section are provided in Table 2.1.

3.1.4 Protocol Operation

The offer/answer exchange assumes the existence of a higher-layer protocol (such as SIP) which is capable of exchanging SDP for the purposes of session establishment between agents. Protocol operation begins when one agent sends an initial offer to another agent. An offer is initial if it is outside of any context that may have already been established through the higher-layer protocol. It is assumed that the higher layer protocol provides maintenance of some kind of context that allows the various SDP exchanges to be associated together.

The agent receiving the offer MAY generate an answer, or it MAY reject the offer. The means for rejecting an offer are dependent on the higher-layer protocol. The offer/answer exchange is atomic; if the answer is rejected, the session reverts to the state prior to the offer (which may be absence of a session). At any time, either agent MAY generate a new offer that updates the session. However, it MUST NOT generate a new offer if it has received an offer it has not yet answered or rejected.

Furthermore, it MUST NOT generate a new offer if it has generated a prior offer for which it has not yet received an answer or a rejection. If an agent receives an offer after having sent one, but before receiving an answer to it, this is considered a "glare" condition. The term glare was originally used in circuit-switched telecommunications networks to describe the condition in which two switches both attempt to seize the same available circuit on the same trunk at the same time. Here, it means both agents have attempted to send an updated offer at the same time. The higher-layer protocol needs to provide a means for resolving such conditions. The higher layer protocol will need to provide a means for ordering messages in each direction. SIP meets these requirements specified in RFC 3261 (see Section 1.1).

3.1.5 Generating the Initial Offer

The offer (and answer) MUST be a valid SDP message, as defined by RFC 4566 (see Section 2), with one exception. RFC 4566 mandates that either an "e=" or a "p=" line be present in the SDP message. This specification relaxes that constraint; an SDP formulated for an offer/answer application MAY omit both the e and p lines. The numeric value of the session id and version in the o line MUST be representable with a 64-bit signed integer. The initial value of the version MUST be less than (2**62) – 1, to avoid rollovers. Although the SDP specification allows for multiple session descriptions to be concatenated together into a large SDP message, an SDP message used in the offer/answer model MUST contain exactly one session description.

The SDP "s=" line conveys the subject of the session, which is reasonably defined for multicast, but ill defined for unicast. For unicast sessions, it is RECOMMENDED that it consist of a single space character (0x20) or a dash (-). Unfortunately, SDP does not allow the "s=" line to be empty. The SDP "t=" line conveys the time of the session. Generally, streams for unicast sessions are created and destroyed through external signaling means, such as SIP. In that case, the "t=" line SHOULD have a value of "0 0."

The offer will contain zero or more media streams (each media stream is described by an "m=" line and its associated attributes). Zero media streams imply that the offerer wishes to communicate but that the streams for the session will be added at a later time through a modified offer. The streams MAY be for a mix of unicast and multicast; the latter obviously implies a multicast address in the relevant "c=" line(s). Construction of each offered stream depends on whether the stream is multicast or unicast.

3.1.5.1 Unicast Streams

If the offerer wishes to only send media on a stream to its peer, it MUST mark the stream as sendonly with the "a=sendonly" attribute. We refer to a stream as being marked with a certain direction if a direction attribute was present as either a media stream attribute or a session attribute. If the offerer wishes to only receive media from its peer, it MUST mark the stream as recvonly. If the offerer wishes to communicate, but wishes to neither send nor receive media at this time, it MUST mark the stream with an "a=inactive" attribute. The inactive direction attribute is specified in RFC 3108 (see Section 9.4). Note that in the case of the Real-Time Transport Protocol described in RFC 3550 (that obsoletes RFC 1889) Real-Time Transport Control Protocol (RTCP) is still sent and received for sendonly, recvonly, and inactive streams. That is, the directionality of the media stream has no impact on the RTCP usage. If the offerer wishes to both send and receive media with its peer, it MAY include an "a=sendrecv" attribute, or it MAY omit it, since sendrecv is the default.

For recvonly and sendrecv streams, the port number and address in the offer indicate where the offerer would like to receive the media stream. For sendonly RTP streams, the address and port number indirectly indicate where the offerer wants to receive RTCP reports. Unless there is an explicit indication otherwise, reports are sent to the port number one higher than the number indicated. The IP address and port present in the offer indicate nothing about the source IP address and source port of RTP and RTCP packets that will be sent by the offerer. A port number of zero in the offer indicates that the stream is offered but MUST NOT be used. This has no useful semantics in an initial offer but is allowed for reasons of completeness, since the answer can contain a zero port indicating a rejected stream (Section 3.1.6). Furthermore, existing streams can be terminated by setting the port to zero (Section 3.1.8). In general, a port number of zero indicates that the media stream is not wanted.

The list of media formats for each media stream conveys two pieces of information, namely the set of formats (codecs and any parameters associated with the codec, in the case of RTP) that the offerer is capable of sending and/or receiving (depending on the direction attributes), and, in the case of RTP, the RTP payload type numbers used to identify those formats. If multiple formats are listed, it means that the offerer is capable of making use of any of those formats during the session. In other words, the answerer MAY change formats in the middle of the session, making use of any of the formats listed, without sending a new offer. For a sendonly stream, the offer SHOULD indicate those formats the offerer is willing to send for this stream.

For a recvonly stream, the offer SHOULD indicate those formats the offerer is willing to receive for this stream. For a sendrecv stream, the offer SHOULD indicate those codecs the offerer is willing to send and receive with. For recvonly RTP streams, the payload type numbers indicate the value of the payload type field in RTP packets the offerer is expecting to receive for that codec. For sendonly RTP streams, the payload type numbers indicate the value of the payload type field in RTP packets the offerer is planning to send for that codec. For sendrecv RTP streams, the payload type numbers indicate the value of the payload type field the offerer expects to receive and would prefer to send.

However, for sendonly and sendrecv streams, the answer might indicate different payload type numbers for the same codecs, in which case, the offerer MUST send with the payload type numbers from the answer. Different payload type numbers may be needed in each direction because of interoperability concerns with H.323. As per RFC 4566 (see Section 2) that obsoletes RFC 2327, fmtp parameters MAY be present to provide additional parameters of the media format. In the case of RTP streams, all media descriptions SHOULD contain "a=rtpmap" mappings from RTP payload types to encodings. If there is no "a=rtpmap," the default payload type mapping, as defined by the current profile in use (e.g., RFC 3551 that obsoletes RFC 1890) is to be used.

This allows easier migration away from static payload types. In all cases, the formats in the "m=" line MUST be listed in order of preference, with the first format listed being preferred. In this case, preferred means that the recipient of the offer SHOULD use the format with the highest preference acceptable to it. If the ptime attribute is present for a stream, it indicates the desired packetization interval that the offerer would like to receive. The ptime attribute MUST be greater than zero. If the bandwidth attribute is present for a stream, it indicates the desired bandwidth that the offerer would like to receive. A value of zero is allowed but discouraged. It indicates that no media should be sent. In the case of RTP, it would also disable all RTCP.

If multiple media streams of different types are present, it means that the offerer wishes to use those streams at the same time. A typical case is an audio and a video stream as parts of a videoconference. If multiple media streams of the same type are present in an offer, it means that the offerer wishes to send (and/or receive) multiple streams of that type at the same time. When sending multiple streams of the same type, it is a matter of local policy as to how each media source of that type (e.g., a video camera and videocassette recorder in the case of video) is mapped to each stream. When a user has a single source for a particular media type, only one policy makes sense: The source is sent to each stream of the same type. Each stream MAY use different encodings. When receiving multiple streams of the same type, it is a matter of local policy as to how each stream is mapped to the various media sinks for that particular type (e.g., speakers or a recording device in the case of audio).

There are a few constraints on the policies, however. First, when receiving multiple streams of the same type, each stream MUST be mapped to at least one sink for the purpose of presentation to the user. In other words, the intent of receiving multiple streams of the same type is that they should all be presented in parallel, rather than choosing just one. Another constraint is that when multiple streams are received and sent to the same sink, they MUST be combined in some media-specific way. For example, in the case of two audio streams, the received media from each might be mapped to the speakers. In that case, the combining operation would be to mix them. In the case of multiple instant messaging streams, where the sink is the screen, the combining operation would be to present all of them to the user interface. The third constraint is that if multiple sources are mapped to the same stream, those sources MUST be combined in some media-specific way before they are sent on the stream. Although policies beyond these constraints are flexible, an agent won't generally want a policy that will copy media from its sinks to its sources unless it is a conference server (i.e., don't copy media received on one stream to another stream).

A typical usage example for multiple media streams of the same type is a prepaid calling card application, where the user can press and hold the pound ("#") key at any time during a call to hang up and make a new call on the same card. This requires media from the user to two destinations—the remote gateway, and the dual-tone multifrequency (DTMF) processing application, which looks for the pound. This could be accomplished with two media streams, one sendrecv to the gateway and the other sendonly (from the perspective of the user) to the DTMF application.

Once the offerer has sent the offer, it MUST be prepared to receive media for any recvonly streams described by that offer. It MUST be prepared to send and receive media for any sendrecv streams in the offer and send media for any sendonly streams in the offer (of course, it cannot actually send until the peer provides an answer with the needed address and port information). In the case of RTP, even though it may receive media before the answer arrives, it will not be able to send RTCP receiver reports until the answer arrives.

3.1.5.2 Multicast Streams

If a session description contains a multicast media stream listed as receive (or send) only, it means that the participants, including the offerer and answerer, can only receive (or send) on that stream. This differs from the unicast view, where the directionality refers to the flow of media between offerer and answerer. Beyond that clarification, the semantics of an offered multicast stream are exactly as described in RFC 4566 (see Section 2) that obsoletes RFC 2327.

3.1.6 Generating the Answer

The answer to an offered session description is based on the offered session description. If the answer is different from the offer in any way (different IP addresses, ports, etc.), the origin line MUST be different in the answer, since the answer is generated by a different entity. In that case, the version number in the "o=" line of the answer is unrelated to the version number in the "o=" line of the offer. For each "m=" line in the offer, there MUST be a corresponding "m=" line in the answer. The answer MUST contain exactly the same number of "m=" lines as the offer. This allows streams to be matched up based on their order. This implies that if the offer contained zero "m=" lines, the answer MUST contain zero "m=" lines.

The "t=" line in the answer MUST equal that of the offer. The time of the session cannot be negotiated. An offered stream MAY be rejected in the answer, for any reason. If a stream is rejected, the offerer and answerer MUST NOT generate media (or RTCP packets) for that stream. To reject an offered stream, the port number in the corresponding stream in the answer MUST be set to zero. Any media formats listed are ignored. At least one MUST be present, as specified by SDP. Constructing an answer for each offered stream differs for unicast and multicast.

3.1.6.1 Unicast Streams

If a stream is offered with a unicast address, the answer for that stream MUST contain a unicast address. The media type of the stream in the answer MUST match that of the offer. If a stream is offered as sendonly, the corresponding stream MUST be marked as recvonly or inactive in the answer. If a media stream is listed as recvonly in the offer, the answer MUST be marked as sendonly or inactive in the answer. If an offered media stream is listed as sendrecv (or if there is no direction attribute at the media or session level, in which case the stream is sendrecv by default), the corresponding stream in the answer MAY be marked as sendonly, recvonly, sendrecv, or inactive. If an offered media stream is listed as inactive, it MUST be marked as inactive in the answer.

For streams marked as recvonly in the answer, the "m=" line MUST contain at least one media format the answerer is willing to receive with from amongst those listed in the offer. The stream MAY indicate additional media formats, not listed in the corresponding stream in the offer, that the answerer is willing to receive. For streams marked as sendonly in the answer, the "m=" line MUST contain at least one media format the answerer is willing to send from amongst those listed in the offer. For streams marked as sendrecv in the answer, the "m=" line MUST contain at least one codec the answerer is willing to both send and receive, from amongst those listed in the offer.

The stream MAY indicate additional media formats, not listed in the corresponding stream in the offer, that the answerer is willing to send or receive (of course, it will not be able to send them at this time, since it was not listed in the offer). For streams marked as inactive in the answer, the list of media formats is constructed based on the offer. If the offer was sendonly, the list is constructed as if the answer were recvonly. Similarly, if the offer was recvonly, the list is constructed as if the answer were sendonly, and if the offer was sendrecv, the list is constructed as if the answer were sendrecv. If the offer was inactive, the list is constructed as if the offer were actually sendrecv and the answer were sendrecv. The connection address and port in the answer indicate the address where the answerer wishes to receive media (in the case of RTP, RTCP will be received on the port one higher unless there is an explicit indication otherwise). This address and port MUST be present even for sendonly streams; in the case of RTP, the port one higher is still used to receive RTCP.

In the case of RTP, if a particular codec was referenced with a specific payload type number in the offer, that same payload type number SHOULD be used for that codec in the answer. Even if the same payload type number is used, the answer MUST contain rtpmap attributes to define the payload type mappings for dynamic payload types and SHOULD contain mappings for static payload types. The media formats in the "m=" line MUST be listed in order of preference, with the first format listed being preferred. In this case, preferred means that the offerer SHOULD use the format with the highest preference from the answer.

Although the answerer MAY list the formats in their desired order of preference, it is RECOMMENDED that unless there is a specific reason, the answerer list formats in the same relative order they were present in the offer. In other words, if a stream in the offer lists audio codecs 8, 22, and 48, in that order, and the answerer only supports codecs 8 and 48, it is RECOMMENDED that, if the answerer has no reason to change it, the ordering of codecs in the answer be 8, 48, and not 48, 8. This helps assure that the same codec is used in both directions.

The interpretation of fmtp parameters in an offer depends on the parameters. In many cases, those parameters describe specific configurations of the media format and should therefore be processed as the media format value itself would be. This means that the same fmtp parameters with the same values MUST be present in the answer if the media format they describe is present in the answer. Other fmtp parameters are more like parameters, for which it is perfectly acceptable for each agent to use different values. In that case, the answer MAY contain fmtp parameters, and those MAY have the same values as those in the offer, or they MAY be different. SDP extensions that define new parameters SHOULD specify the proper interpretation in offer/answer.

The answerer MAY include a nonzero ptime attribute for any media stream; this indicates the packetization interval the answerer would like to receive. There is no requirement that the packetization interval be the same in each direction for a particular stream. The answerer MAY include a bandwidth attribute for any media stream; this indicates the bandwidth that the answerer would like the offerer to use when sending media. The value of zero is allowed, interpreted as described in Section 3.1.5. If the answerer has no media formats in common for a particular offered stream, the answerer MUST reject that media stream by setting the port to zero.

If there are no media formats in common for all streams, the entire offered session is rejected. Once the answerer has sent the answer, it MUST be prepared to receive media for any recvonly streams described by that answer. It MUST be prepared to send and receive media for any sendrecv streams in the answer, and it MAY send media immediately. The answerer MUST be prepared to receive media for recvonly or sendrecv streams using any media formats listed for those streams in the answer, and it MAY send media immediately. When sending media, it SHOULD use a packetization interval equal to the value of the ptime attribute in the offer, if any was present. It SHOULD send media using a bandwidth no higher than the value of the bandwidth attribute in the offer, if any was present. The answerer MUST send using a media format in the offer that is also listed in the answer, and SHOULD send using the most preferred media format in the offer that is also listed in the answer. In the case of RTP, it MUST use the payload type numbers from the offer, even if they differ from those in the answer.

3.1.6.2 Multicast Streams

Unlike unicast, where there is a two-sided view of the stream, there is only a single view of the stream for multicast. As such, generating an answer to a multicast offer generally involves modifying a limited set of aspects of the stream. If a multicast stream is accepted, the address and port information in the answer MUST match that of the offer. Similarly, the directionality information in the answer (sendonly, recvonly, or sendrecv) MUST equal that of the offer. This is because all participants in a multicast session need to have equivalent views of the parameters of the session, an underlying assumption of the multicast bias of RFC 2327 (made obsolete by RFC 4566, see Section 2). The set of media formats in the answer MUST be equal to or a subset of those in the offer. Removing a format is a way for the answerer to indicate that the format is not supported. The ptime and bandwidth attributes in the answer MUST equal those in the offer, if present. If not present, a nonzero ptime MAY be added to the answer.

3.1.7 Offerer Processing of the Answer

When the offerer receives the answer, it MAY send media on the accepted stream(s) (assuming it is listed as sendrecv or recvonly in the answer). It MUST send using a media format listed in the answer, and it SHOULD use the first media format listed in the answer when it does send. The reason this is a SHOULD, and not a MUST (it's also a SHOULD, and not a MUST, for the answerer), is because there will oftentimes be a need to change codecs on the fly. For example, during silence periods, an agent might like to switch to a comfort noise codec. Or, if the user presses a number on the keypad, the agent might like to send that using RFC 2833 (that is made obsolete by RFCs 4733). Congestion control might necessitate changing to a lower rate codec based on feedback. The offerer SHOULD send media according to the value of any ptime and bandwidth attribute in the answer. The offerer MAY immediately cease listening for media formats that were listed in the initial offer but are not present in the answer.

3.1.8 Modifying the Session

At any point during the session, either participant MAY issue a new offer to modify characteristics of the session. It is fundamental to the operation of the offer/answer model that the exact same offer/answer procedure defined previously be used for modifying parameters of an existing session. The offer MAY be identical to the last SDP provided the other party (which may have been provided in an offer or an answer), or it MAY be different. We refer to the last SDP provided as the "previous SDP." If the offer is the same, the answer MAY be the same as the previous SDP from the answerer, or it MAY be different. If the offered SDP is different from the previous SDP, some constraints are placed on its construction, discussed in the next section. Nearly all aspects of the session can be modified. New streams can be added, existing streams can be deleted, and parameters of existing streams can change. When issuing an offer that modifies the session, the "o=" line of the new SDP MUST be identical to that in the previous SDP, except that the version in the origin field MUST increment by one from the previous SDP. If the version in the origin line does not increment, the SDP MUST be identical to the SDP with that version number. The answerer MUST be prepared to receive an offer that contains SDP with a version that has not changed; this is effectively a no-op. However, the answerer MUST generate a valid answer (which MAY be the same as the previous SDP from the answerer, or MAY be different), according to the procedures defined in Section 3.1.6.

If an SDP is offered and is different from the previous SDP, the new SDP MUST have a matching media stream for each media stream in the previous SDP. In other words, if the previous SDP had N "m=" lines, the new SDP MUST have at least N "m=" lines. The i-th media stream in the previous SDP, counting from the top, matches the i-th media stream in the new SDP, counting from the top. This matching is necessary in order for the answerer to determine which stream in the new SDP corresponds to a stream in the previous SDP. Because of these requirements, the number of "m=" lines in a stream never decreases; it either stays the same or increases. Deleted media streams from a previous SDP MUST NOT be removed in a new SDP; however, attributes for these streams need not be present.

3.1.8.1 Adding a Media Stream

New media streams are created by new additional media descriptions below the existing ones or by reusing the "slot" used by an old media stream that had been disabled by setting its port to zero. Reusing its slot means that the new media description replaces the old one but retains its positioning relative to other media descriptions in the SDP. New media descriptions MUST appear below any existing media sections. The rules for formatting these media descriptions are identical to those described in Section 3.1.3. When the answerer receives an SDP with more media descriptions than the previous SDP from the offerer, or it receives an SDP with a media stream in a slot where the port was previously zero, the answerer knows that new media streams are being added. These can be rejected or accepted by placing an appropriately structured media description in the answer. The procedures for constructing the new media description in the answer are described in Section 3.1.6.

3.1.8.2 Removing a Media Stream

An existing media stream is removed by creating a new SDP with the port number for that stream set to zero. The stream description MAY omit all attributes present previously and MAY list just a single media format. A stream that is offered with a port of zero MUST be marked with port zero in the answer. Like the offer, the answer MAY omit all attributes present previously and MAY list just a single media format from amongst those in the offer. Removal of a media stream implies that media is no longer sent for that stream, and any media that is received is discarded. In the case of RTP, RTCP transmission also ceases, as does processing of any received RTCP packets. Any resources associated with it can be released. The user interface might indicate that the stream has terminated, by closing the associated window on a PC, for example.

3.1.8.3 Modifying a Media Stream

Nearly all characteristics of a media stream can be modified.

3.1.8.3.1 Modifying Address, Port, or Transport

The port number for a stream MAY be changed. To do this, the offerer creates a new media description, with the port number in the "m=" line different from the corresponding stream in the previous SDP. If only the port number is to be changed, the rest of the media stream description SHOULD remain unchanged. The offerer MUST be prepared to receive media on both the old and new ports as soon as the offer is sent. The offerer SHOULD NOT cease listening for media on the old port until the answer is received and media arrives on the new port. Doing so could result in loss of media during the transition. Received, in this case, means that the media is passed to a media sink. This means that if there is a playout buffer, the agent will continue to listen on the old port until the media on the new port reaches the top of the playout buffer. At that time, it MAY cease listening for media on the old port.

The corresponding media stream in the answer MAY be the same as the stream in the previous SDP from the answerer, or it MAY be different. If the updated stream is accepted by the answerer, the answerer SHOULD begin sending traffic for that stream to the new port immediately. If the answerer changes the port from the previous SDP, it MUST be prepared to receive media on both the old and new ports as soon as the answer is sent. The answerer MUST NOT cease listening for media on the old port until media arrives on the new port. At that time, it MAY cease listening for media on the old port. The same is true for an offerer that sends an updated offer with a new port; it MUST NOT cease listening for media on the old port until media arrives on the new port. Of course, if the offered stream is rejected, the offerer can cease being prepared to receive using the new port as soon as the rejection is received. To change the IP address where media is sent to, the same procedure is followed for changing the port number. The only difference is that the connection line is updated; the port number is not. The transport for a stream MAY be changed. The process for doing this is identical to changing the port, except the transport is updated, not the port.

3.1.8.3.2 Changing the Set of Media Formats

The list of media formats used in the session MAY be changed. To do this, the offerer creates a new media description, with the list of media formats in the "m=" line different from the corresponding media stream in the previous SDP. This list MAY include new formats and MAY remove formats present in the previous SDP. However, in the case of RTP, the mapping from a particular dynamic payload type number to a particular codec within that media stream MUST NOT change for the duration of a session. For example, if A generates an offer with G.711 assigned to dynamic payload type number 46, payload type number 46 MUST refer to G.711 from that point forward in any offers or answers for that media stream within the session. However, it is acceptable for multiple payload type numbers to be mapped to the same codec, so that an updated offer could also use payload type number 72 for G.711.

The mappings need to remain fixed for the duration of the session because of the loose synchronization between signaling exchanges of SDP and the media stream. The corresponding media stream in the answer is formulated as described in Section 3.1.6 and may result in a change in media formats as well. Similarly, as described in Section 3.1.6, as soon as it sends its answer, the answerer MUST begin sending media using any formats in the offer that were also present in the answer, SHOULD use the most preferred format in the offer that was also listed in the answer (assuming the stream allows for sending), and MUST NOT send using any formats that are not in the offer, even if they were present in a previous SDP from the peer. Similarly, when the offerer receives the answer, it MUST begin sending media using any formats in the answer, SHOULD use the most preferred one (assuming the stream allows for sending), and MUST NOT send using any formats that are not in the answer, even if they were present in a previous SDP from the peer.

When an agent ceases using a media format (by not listing that format in an offer or answer, even though it was in a previous SDP) the agent will still need to be prepared to receive media with that format for a brief time. How does it know when it can be prepared to stop receiving with that format? If it needs to know, there are three techniques that can be applied. First, the agent can change ports in addition to changing formats. When media arrives on the new port, it knows that the peer has ceased sending with the old format, and it can cease being prepared to receive with it. This approach has the benefit of being media-format independent. However, changes in ports may require changes in resource reservation or rekeying of security protocols. The second approach is to use a totally new set of dynamic payload types for all codecs when one is discarded. When media is received with one of the new payload types, the agent knows that the peer has ceased sending with the old format. This approach doesn't affect reservations or security contexts, but it is RTP specific and wasteful of a very small payload type space. A third approach is to use a timer. When the SDP from the peer is received, the timer is set. When it fires, the agent can cease being prepared to receive with the old format. A value of 1 minute would typically be more than sufficient. In some cases, an agent may not care and thus continually be prepared to receive with the old formats. Nothing need be done in this case. Of course, if the offered stream is rejected, the offer can cease being prepared to receive using any new formats as soon as the rejection is received.

3.1.8.3.3 Changing Media Types

The media type (audio, video, etc.) for a stream MAY be changed. It is RECOMMENDED that the media type be changed (as opposed to adding a new stream), when the same logical data is being conveyed, but just in a different media format. This is particularly useful for changing between voiceband fax and fax in a single stream, which are separate media types. To do this, the offerer creates a new media description, with a new media type, in place of the description in the previous SDP, which is to be changed. The corresponding media stream in the answer is formulated as described in Section 3.1.6. Assuming the stream is acceptable, the answerer SHOULD begin sending with the new media type and formats as soon as it receives the offer. The offerer MUST be prepared to receive media with both the old and new types until the answer is received, and media with the new type is received, and it reaches the top of the playout buffer.

3.1.8.3.4 Changing Attributes

Any other attributes in a media description MAY be updated in an offer or answer. Generally, an agent MUST send media (if the directionality of the stream allows) using the new parameters once the SDP with the change is received.

3.1.8.4 Putting a Unicast Media Stream on Hold

If a party in a call wants to put the other party "on hold" (i.e., request that it temporarily stops sending one or more unicast media streams) a party offers the other an updated SDP. If the stream to be placed on hold was previously a sendrecv media stream, it is placed on hold by marking it as sendonly. If the stream to be placed on hold was previously a recvonly media stream, it is placed on hold by marking it inactive. This means that a stream is placed "on hold" separately in each direction. Each stream is placed "on hold" independently. The recipient of an offer for a stream on hold SHOULD NOT automatically return an answer with the corresponding stream on hold. An SDP with all streams "on hold" is referred to as held SDP. Certain third party call control scenarios do not work when an answerer responds to held SDP.

Typically, when a user "presses" hold, the agent will generate an offer with all streams in the SDP indicating a direction of sendonly, and it will also locally mute, so that no media is sent to the far end, and no media is played out. RFC 2543 (that was made obsolete by RFC 3261, also see Section 1.2) specified that placing a user on hold was accomplished by setting the connection address to 0.0.0.0. Its usage for putting a call on hold is no longer recommended, since it doesn't allow for RTCP to be used with held streams, doesn't work with IPv6, and breaks with connection-oriented media. However, it can be useful in an initial offer when the offerer knows it wants to use a particular set of media streams and formats but doesn't know the addresses and ports at the time of the offer. Of course, when used, the port number MUST NOT be zero, which would specify that the stream has been disabled. An agent MUST be capable of receiving SDP with a connection address of 0.0.0.0, in which case it means that neither RTP nor RTCP should be sent to the peer.

3.1.9 Indicating Capabilities

Before an agent sends an offer, it is helpful to know if the media formats in that offer would be acceptable to the answerer. Certain protocols, like SIP, provide a means to query for such capabilities. SDP can be used in responses to such queries to indicate capabilities. This section describes how such an SDP message is formatted. Since SDP has no way to indicate that the message is for the purpose of capability indication, this is determined from the context of the higher-layer protocol. The ability of baseline SDP to indicate capabilities is very limited. It cannot express allowed parameter ranges or values and cannot be in parallel with an offer/answer itself. Extensions might address such limitations in the future.

An SDP constructed to indicate media capabilities is structured as follows. It MUST be a valid SDP, except that it MAY omit both "e=" and "p=" lines. The "t=" line MUST be equal to "0 0."

For each media type supported by the agent, there MUST be a corresponding media description of that type. The session ID in the origin field MUST be unique for each SDP constructed to indicate media capabilities. The port MUST be set to zero, but the connection address is arbitrary. The usage of port zero makes sure that an SDP formatted for capabilities does not cause media streams to be established if it is interpreted as an offer or answer. The transport component of the "m=" line indicates the transport for that media type. For each media format of that type supported by the agent, there SHOULD be a media format listed in the "m=" line. In the case of RTP, if dynamic payload types are used, an rtpmap attribute MUST be present to bind the type to a specific format. There is no way to indicate constraints, such as how many simultaneous streams can be supported for a particular codec, and so on.

The SDP of Figure 3.1 indicates that the agent can support three audio codecs (Pulse Code Modulation mu-law [PCMU], 1016, and GSM) and two video codecs (H.261 and H.263).

3.1.10 Example Offer/Answer Exchanges

This section provides example offer/answer exchanges.

3.1.10.1 Basic Exchange

Assume that the caller, Alice, has included the following description in her offer. It includes a bidirectional audio stream and two bidirectional video streams, using H.261 (payload type 31) and MPEG (payload type 32). The offered SDP is:

```
v=0
o=alice 2890844526 2890844526 IN IP4 host.anywhere.com
s=
c=IN IP4 host.anywhere.com
t=0 0
m=audio 49170 RTP/AVP 0
a=rtpmap:0 PCMU/8000
m=video 51372 RTP/AVP 31
a=rtpmap:31 H261/90000
m=video 53000 RTP/AVP 32
a=rtpmap:32 MPV/90000
```

The callee, Bob, does not want to receive or send the first video stream, so he returns this SDP as the answer:

```
v=0
o=carol 28908764872 28908764872 IN IP4 100.3.6.6
s=-
t=0 0
c=IN IP4 192.0.2.4
m=audio 0 RTP/AVP 0 1 3
a=rtpmap:0 PCMU/8000
a=rtpmap:1 1016/8000
a=rtpmap:3 GSM/8000
m=video 0 RTP/AVP 31 34
a=rtpmap:31 H261/90000
a=rtpmap:34 H263/90000
```

Figure 3.1 SDP indicating capabilities.

```
v=0
o=bob 2890844730 2890844730 IN IP4 host.example.com
s=
c=IN IP4 host.example.com
t=0 0
m=audio 49920 RTP/AVP 0
a=rtpmap:0 PCMU/8000
m=video 0 RTP/AVP 31
m=video 53000 RTP/AVP 32
a=rtpmap:32 MPV/90000
```

At some point later, Bob decides to change the port where he will receive the audio stream (from 49920 to 65422) and at the same time add an additional audio stream as receive only, using the RTP payload format for events (RFC 2833 that was made obsolete by RFCs 4733 & 4734). Bob offers the following SDP in the offer:

```
v=0
o=bob 2890844730 2890844731 IN IP4 host.example.com
s=
c=IN IP4 host.example.com
t=0 0
m=audio 65422 RTP/AVP 0
a=rtpmap:0 PCMU/8000
m=video 0 RTP/AVP 31
m=video 53000 RTP/AVP 32
a=rtpmap:32 MPV/90000
m=audio 51434 RTP/AVP 110
a=rtpmap:110 telephone-events/8000
a=recvonly
```

Alice accepts the additional media stream and so generates the following answer:

```
v=0
o=alice 2890844526 2890844527 IN IP4 host.anywhere.com
s=
c=IN IP4 host.anywhere.com
t=0 0
m=audio 49170 RTP/AVP 0
a=rtpmap:0 PCMU/8000
m=video 0 RTP/AVP 31
a=rtpmap:31 H261/90000
m=video 53000 RTP/AVP 32
a=rtpmap:32 MPV/90000
m=audio 53122 RTP/AVP 110
a=rtpmap:110 telephone-events/8000
a=sendonly
```

3.1.10.2 One of N Codec Selection

A common occurrence in embedded phones is that the Digital Signal Processor (DSP) used for compression can support multiple codecs at a time, but once that codec is selected, it cannot be readily changed on the fly. This example shows how a session can be set up using an initial offer/answer exchange, followed immediately by a second one to lock down the set of codecs. The initial offer from Alice to Bob indicates a single audio stream with the three audio codecs that are available in the DSP. The stream is marked as inactive, since media cannot be received until a codec is locked down:

```
v=0
o=alice 2890844526 2890844526 IN IP4 host.anywhere.com
s=
c=IN IP4 host.anywhere.com
t=0 0
m=audio 62986 RTP/AVP 0 4 18
a=rtpmap:0 PCMU/8000
a=rtpmap:4 G723/8000
a=rtpmap:18 G729/8000
a=inactive
```

Bob can support dynamic switching between PCMU and G.723. So, he sends the following answer:

```
v=0
o=bob 2890844730 2890844731 IN IP4 host.example.com
s=
c=IN IP4 host.example.com
t=0 0
m=audio 54344 RTP/AVP 0 4
a=rtpmap:0 PCMU/8000
a=rtpmap:4 G723/8000
a=inactive
Alice can then select any one of these two codecs. So, she sends an
updated offer with a sendrecv stream:
v=0
o=alice 2890844526 2890844527 IN IP4 host.anywhere.com
s=
c=IN IP4 host.anywhere.com
t=0 0
m=audio 62986 RTP/AVP 4
a=rtpmap:4 G723/8000
a=sendrecv
Bob accepts the single codec:
v=0
o=bob 2890844730 2890844732 IN IP4 host.example.com
s=
c=IN IP4 host.example.com
t=0 0
m=audio 54344 RTP/AVP 4
a=rtpmap:4 G723/8000
a=sendrecv
```

If the answerer (Bob) was only capable of supporting one-of-N codecs, Bob would select one of the codecs from the offer, and place that in his answer. In this case, Alice would do a re-INVITE to activate that stream with that codec. As an alternative to using "a=inactive" in the first exchange, Alice can list all codecs, and as soon as she receives media from Bob, generate an updated offer locking down the codec to the one just received. Of course, if Bob only supports one-of-N codecs, there will be one codec in his answer, and in this case, there is no need for a re-INVITE to lock down a single codec.

3.1.11 Security Considerations

The security specific to SDP offer/answer model described in RFC 3264 is provide in Section 17.2.

3.1.12 Internet Assigned Numbers Authority (IANA) Considerations

There are no Internet Assigned Numbers Authority (IANA) considerations with this specification.

3.2 ICE for Network Address Translator Traversal by SDP Offer/Answer Protocols

This section describes the SDP offer/answer model for network address translator (NAT) traversal for User Datagram Protocol (UDP)-based multimedia sessions (RFC 5245). Note that NATs are troublesome devices in network paths that not only hamper multimedia session negotiations; they are also vulnerable for security attacks. The NAT crossing for multimedia sessions is so complicated, in fact, RFC 5245 has used Interactive Connectivity Establishment (ICE) protocol. It uses traversal using relays around NAT (TURN) and simple traversal of UDP around NAT (STUN) and its extensions for crossing NATs by UDP-based multimedia sessions that are established with the SDP offer/answer model.

3.2.1 Introduction

RFC 3264 (see Section 3.1) defines a two-phase exchange of SDP messages (RFC 4566, see Section 2) for the purposes of establishment of multimedia sessions. This offer/answer mechanism is used by protocols such as the Session Initiation NATs. Because their purpose is to establish a flow of media packets, they tend to carry the IP addresses and ports of media sources and sinks within their messages, known to be problematic through NAT (RFC 3235). The protocols also seek to create a media flow directly between participants, so that there is no application layer intermediary between them. This is done to reduce media latency, decrease packet loss, and reduce the operational costs of deploying the application. However, this is difficult to accomplish through NAT. A full treatment of the reasons for this is beyond the scope of this specification.

Numerous solutions have been defined for allowing these protocols to operate through NAT. These include Application Layer Gateways (ALGs), the Middlebox Control Protocol (RFC 3303), the original Simple Traversal of UDP Through NAT (RFC 3489) specification, and Realm Specific IP (RFCs 3102 and 3103) along with session description extensions needed to make them work, such as the SDP (RFC 4566, see Section 2) attribute for the RTCP (RFC 3605, see Section 12.3) Unfortunately, these techniques all have pros and cons that make each one optimal in some network topologies but a poor choice in others. The result is that administrators and implementers make assumptions about the topologies of the networks in which their solutions will be deployed. This introduces complexity and brittleness into the system. What is needed is a single solution flexible enough to work well in all situations.

This specification defines ICE as a technique for NAT traversal for UDP-based media streams (though ICE can be extended to handle other transport protocols, such as Transmission Control Protocol (TCP), RFC 6544) established by the offer/answer model. ICE is an extension to the offer/answer model and works by including a multiplicity of IP addresses and ports in SDP offers and answers, which are then tested for connectivity by peer-to-peer connectivity checks. The IP addresses and ports included in the SDP and the connectivity checks are performed using the revised STUN specification (RFC 5389), now renamed Session Traversal Utilities for NAT. The new name and specification reflect its new role as a tool that is used with other NAT traversal techniques (namely ICE) rather than a standalone NAT traversal solution, as the original STUN specification was. ICE also makes use of TURN (RFC 5766), an extension to STUN. Because ICE exchanges a multiplicity of IP addresses and ports for each media stream, it also allows for address selection for multihomed and dual-stack hosts; for this reason it deprecates RFCs 4091 and 4092.

3.2.2 Overview of ICE

In a typical ICE deployment, we have two endpoints (known as AGENTS in RFC 3264, see Section 3.1, terminology) that want to communicate. They are able to communicate indirectly via some signaling protocol (such as SIP), by which they can perform an offer/answer exchange of SDP (RFC 3264, see Section 3.1) messages. Note that ICE is not intended for NAT traversal for SIP, which is assumed to be provided via another mechanism (RFC 5626). At the beginning of the ICE process, the agents are ignorant of their own topologies. In particular, they might or might not be behind a NAT (or multiple tiers of NATs). ICE allows the agents to discover enough information about their topologies to potentially find one or more paths by which they can communicate.

Figure 3.2 shows a typical environment for ICE deployment. The two endpoints are labelled L and R (for left and right, which helps visualize call flows). Both L and R are behind their own respective NATs, though they may not be aware of this. The type of NAT and its properties are also unknown. Agents L and R are capable of engaging in an offer/answer exchange by which they can exchange SDP messages, whose purpose is to set up a media session between L and R. Typically, this exchange will occur through a SIP server.

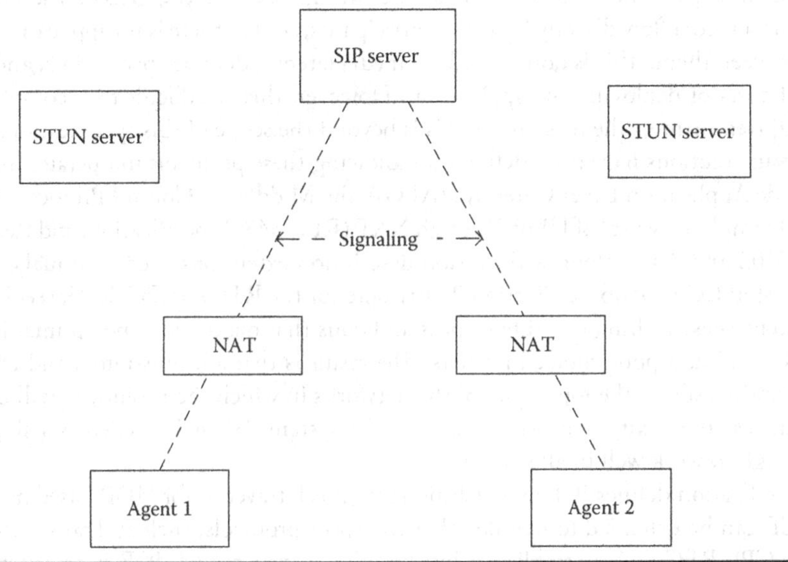

Figure 3.2 ICE deployment scenario (Copyright: Internet Engineering Task Force (IETF).

In addition to the agents, a SIP server and NATs, ICE is typically used in concert with STUN or TURN servers in the network. Each agent can have its own STUN or TURN server, or they can be the same.

The basic idea behind ICE is as follows: Each agent has a variety of candidate TRANSPORT ADDRESSES (combination of IP address and port for a particular transport protocol, which is always UDP in this specification) it could use to communicate with the other agent. These might include:

- A transport address on a directly attached network interface
- A translated transport address on the public side of a NAT (a "server reflexive" address)
- A transport address allocated from a TURN server (a "relayed address")

Potentially, any of L's candidate transport addresses can be used to communicate with any of R's candidate transport addresses. In practice, however, many combinations will not work. For instance, if L and R are both behind NATs, their directly attached interface addresses are unlikely to be able to communicate directly (this is why ICE is needed, after all!). The purpose of ICE is to discover which pairs of addresses will work. The way that ICE does this is to systematically try all possible pairs (in a carefully sorted order) until it finds one or more that work.

3.2.2.1 Gathering Candidate Addresses

In order to execute ICE, an agent has to identify all of its address candidates. A CANDIDATE is a transport address—a combination of IP address and port for a particular transport protocol (with only UDP specified here). This document defines three types of candidates, some derived from physical or logical network interfaces, others discoverable via STUN and TURN. Naturally, one viable candidate is a transport address obtained directly from a local interface. Such a candidate is called a HOST CANDIDATE. The local interface could be Ethernet or Wi-Fi, or it could be obtained through a tunnel mechanism, such as a Virtual Private Network (VPN) or Mobile IP. In all cases, such a network interface appears to the agent as a local interface from which ports (and thus candidates) can be allocated.

If an agent is multihomed, it obtains a candidate from each IP address. Depending on the location of the PEER (the other agent in the session) on the IP network relative to the agent, the agent may be reachable by the peer through one or more of those IP addresses. Consider, for example, an agent that has a local IP address on a private net 10 network (I1), and a second connected to the public Internet (I2). A candidate from I1 will be directly reachable when communicating with a peer on the same private net 10 network, while a candidate from I2 will be directly reachable when communicating with a peer on the public Internet. Rather than trying to guess which IP address will work prior to sending an offer, the offering agent includes both candidates in its offer.

Next, the agent uses STUN or TURN to obtain additional candidates. These come in two flavors: translated addresses on the public side of a NAT (SERVER REFLEXIVE CANDIDATES) and addresses on TURN servers (RELAYED CANDIDATES). When TURN servers are utilized, both types of candidates are obtained from the TURN server. If only STUN servers are utilized, only server reflexive candidates are obtained from them. The relationship of these candidates to the host candidate is shown in Figure 3.3. In this figure, both types of candidates are discovered using TURN. In the figure, the notation X:x means IP address X and UDP port x.

When the agent sends the TURN Allocate request from IP address and port X:x, the NAT (assuming there is one) will create a binding X1':x1', mapping this server reflexive candidate to the host candidate X:x. Outgoing packets sent from the host candidate will be translated by the NAT

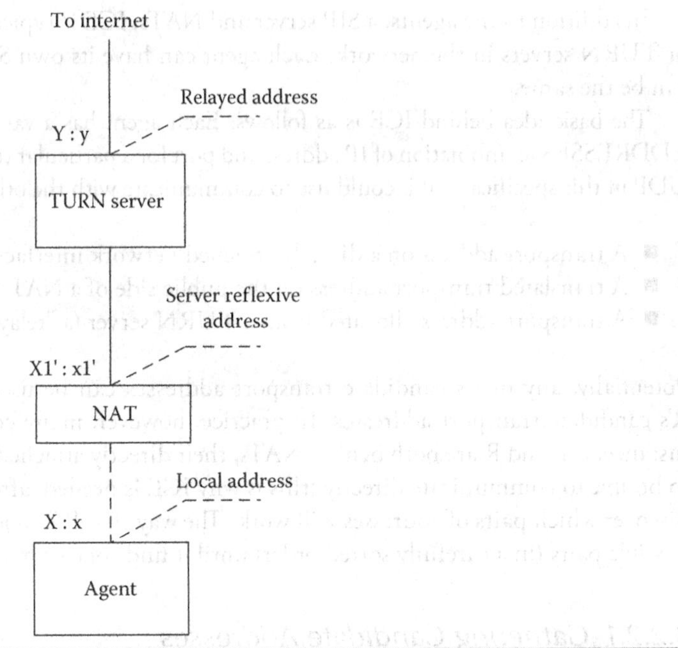

Figure 3.3 Candidate relationships (Copyright: IETF).

to the server reflexive candidate. Incoming packets sent to the server reflexive candidate will be translated by the NAT to the host candidate and forwarded to the agent. We call the host candidate associated with a given server reflexive candidate the BASE.

Note: "Base" refers to the address an agent sends from for a particular candidate. Thus, as a degenerate case, a host candidate also has a base, but it's the same as the host candidate. When there are multiple NATs between the agent and the TURN server, the TURN request will create a binding on each NAT, but only the outermost server reflexive candidate (the one nearest the TURN server) will be discovered by the agent. If the agent is not behind a NAT, then the base candidate will be the same as the server reflexive candidate; the server reflexive candidate is redundant and will be eliminated.

The Allocate request then arrives at the TURN server. The TURN server allocates a port y from its local IP address Y and generates an Allocate response, informing the agent of this relayed candidate. The TURN server also informs the agent of the server reflexive candidate, X1':x1' by copying the source transport address of the Allocate request into the Allocate response. The TURN server acts as a packet relay, forwarding traffic between L and R. In order to send traffic to L, R sends traffic to the TURN server at Y:y, and the TURN server forwards that to X1':x1', which passes through the NAT where it is mapped to X:x and delivered to L. When only STUN servers are utilized, the agent sends a STUN Binding request (RFC 5389) to its STUN server. The STUN server will inform the agent of the server reflexive candidate X1':x1' by copying the source transport address of the Binding request into the Binding response.

3.2.2.2 Connectivity Checks

Once L has gathered all of its candidates, it orders them in highest to lowest priority and sends them to R over the signaling channel. The candidates are carried in attributes in the SDP offer.

When R receives the offer, it performs the same gathering process and responds with its own list of candidates. At the end of this process, each agent has a complete list of both its candidates and its peer's candidates. It pairs them up, resulting in CANDIDATE PAIRS. To see which pairs work, each agent schedules a series of CHECKS. Each check is a STUN request/response transaction that the client will perform on a particular candidate pair by sending a STUN request from the local candidate to the remote candidate. The basic principle of the connectivity checks is simple:

1. Sort the candidate pairs in priority order.
2. Send checks on each candidate pair in priority order.
3. Acknowledge checks received from the other agent.

With both agents performing a check on a candidate pair, the result is a four-way handshake (Figure 3.4):

It is important to note that the STUN requests are sent to and from the exact same IP addresses and ports that will be used for media (e.g., RTP and RTCP). Consequently, agents demultiplex STUN and RTP/RTCP using contents of the packets, rather than the port on which they are received. Fortunately, this demultiplexing is easy to do, especially for RTP and RTCP. Because a STUN Binding request is used for the connectivity check, the STUN Binding response will contain the agent's translated transport address on the public side of any NATs between the agent and its peer. If this transport address is different from other candidates the agent already learned, it represents a new candidate, called a PEER REFLEXIVE CANDIDATE, which then gets tested by ICE just the same as any other candidate. As an optimization, as soon as R gets L's check message, R schedules a connectivity check message to be sent to L on the same candidate pair. This accelerates the process of finding a valid candidate and is called a TRIGGERED CHECK. At the end of this handshake, both L and R know that they can send (and receive) messages end to end in both directions.

3.2.2.3 Sorting Candidates

Because the algorithm above searches all candidate pairs, if a working pair exists it will eventually find it no matter what order the candidates are tried in. In order to produce faster (and better) results, the candidates are sorted in a specified order. The resulting list of sorted candidate pairs is

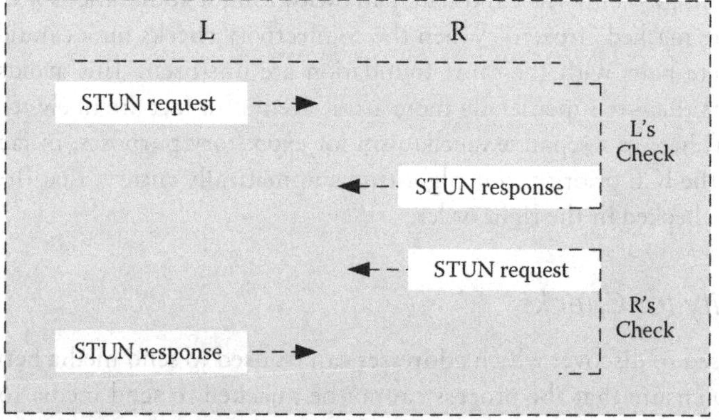

Figure 3.4 Basic connectivity check (Copyright: IETF).

called the CHECKLIST. The algorithm is described in Section 3.2.4.1.2 but follows two general principles:

- Each agent gives its candidates a numeric priority, which is sent along with the candidate to the peer.
- The local and remote priorities are combined so that each agent has the same ordering for the candidate pairs.

The second property is important for getting ICE to work when there are NATs in front of L and R. Frequently, NATs will not allow packets in from a host until the agent behind the NAT has sent a packet toward that host. Consequently, ICE checks in each direction will not succeed until both sides have sent a check through their respective NATs. The agent works through this checklist by sending a STUN request for the next candidate pair on the list periodically. These are called ORDINARY CHECKS.

In general, the priority algorithm is designed so that candidates of similar type get similar priorities so that more direct routes (that is, fewer media relays and fewer NATs) are preferred over indirect ones (more media relays and more NATs). Within those guidelines, however, agents have a fair amount of discretion about how to tune their algorithms.

3.2.2.4 Frozen Candidates

The previous description only addresses the case where the agents wish to establish a media session with one COMPONENT (a piece of a media stream requiring a single transport address; a media stream may require multiple components, each of which has to work for the media stream as a whole to work). Typically (e.g., with RTP and RTCP), the agents actually need to establish connectivity for more than one flow. The network properties are likely to be very similar for each component (especially because RTP and RTCP are sent and received from the same IP address). It is usually possible to leverage information from one media component in order to determine the best candidates for another. ICE does this with a mechanism called "frozen candidates."

Each candidate is associated with a property called its FOUNDATION. Two candidates have the same foundation when they are "similar," of the same type and obtained from the same host candidate and STUN server using the same protocol. Otherwise, their foundation is different. A candidate pair has a foundation, too, which is just the concatenation of the foundations of its two candidates. Initially, only the candidate pairs with unique foundations are tested. The other candidate pairs are marked "frozen." When the connectivity checks for a candidate pair succeed, the other candidate pairs with the same foundation are unfrozen. This avoids repeated checking of components that are superficially more attractive but in fact are likely to fail. While we've described "frozen" here as a separate mechanism for expository purposes, in fact it is an integral part of ICE, and the ICE prioritization algorithm automatically ensures that the right candidates are unfrozen and checked in the right order.

3.2.2.5 Security for Checks

Because ICE is used to discover which addresses can be used to send media between two agents, it is important to ensure that the process cannot be hijacked to send media to the wrong location. Each STUN connectivity check is covered by a message authentication code (MAC) computed using a key exchanged in the signaling channel. This MAC provides message integrity and data origin authentication, thus stopping an attacker from forging or modifying connectivity check messages. Furthermore, if the SIP (RFC 3261, also see Section 1.2) caller is using ICE, and the call forks, the ICE exchanges happen independently with each forked recipient. In such a case, the keys exchanged in the signaling help associate each ICE exchange with each forked recipient.

3.2.2.6 Concluding ICE

ICE checks are performed in a specific sequence, so that high priority candidate pairs are checked first, followed by lower priority ones. One way to conclude ICE is to declare victory as soon as a check for each component of each media stream completes successfully. Indeed, this is a reasonable algorithm, and details for it are provided below in paragraphs explaining with figures. However, it is possible that a packet loss will cause a higher-priority check to take longer to complete. In that case, allowing ICE to run a little longer might produce better results.

More fundamentally, however, the prioritization defined by this specification may not yield "optimal" results. As an example, if the aim is to select low-latency media paths, usage of a relay is a hint that latencies may be higher, but it is nothing more than a hint. An actual round-trip time measurement could be made, and it might demonstrate that a pair with lower priority is actually better than one with higher priority. Consequently, ICE assigns one of the agents in the role of the CONTROLLING AGENT and the other of the CONTROLLED AGENT. The controlling agent gets to nominate which candidate pairs will get used for media amongst the ones that are valid. It can do this in one of two ways: using REGULAR NOMINATION or AGGRESSIVE NOMINATION.

With regular nomination, the controlling agent lets the checks continue until at least one valid candidate pair for each media stream is found. Then, it picks amongst those that are valid and sends a second STUN request on its NOMINATED candidate pair, but this time with a flag set to tell the peer that this pair has been nominated for use. This is shown in Figure 3.5.

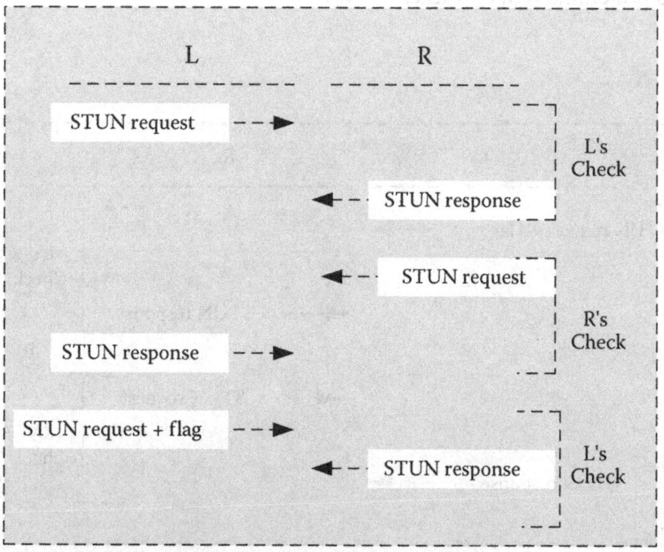

Figure 3.5 Regular nomination (Copyright: IETF).

Once the STUN transaction with the flag completes, both sides cancel any future checks for that media stream. ICE will now send media using this pair. The pair an ICE agent is using for media is called the SELECTED PAIR. In aggressive nomination, the controlling agent puts the flag in every STUN request it sends. This way, once the first check succeeds, ICE processing is complete for that media stream, and the controlling agent doesn't have to send a second STUN request. The selected pair will be the highest-priority valid pair whose check succeeded. Aggressive nomination is faster than regular nomination but gives less flexibility. Aggressive nomination is shown in Figure 3.6.

Once all of the media streams are completed, the controlling endpoint sends an updated offer if the candidates in the "m=" and "c=" lines for the media stream (called the DEFAULT CANDIDATES) don't match ICE's SELECTED CANDIDATES. Once ICE is concluded, it can be restarted at any time for one or all of the media streams by either agent. This is done by sending an updated offer indicating a restart.

3.2.2.7 Lite Implementations

In order for ICE to be used in a call, both agents need to support it. However, certain agents will always be connected to the public Internet and have a public IP address at which it can receive packets from any correspondent. To make it easier for these devices to support ICE, ICE defines a special type of implementation called LITE (in contrast to the normal FULL implementation). A lite implementation doesn't gather candidates; it includes only host candidates for any media stream. Lite agents do not generate connectivity checks or run the state machines, though they need to be able to respond to connectivity checks. When a lite implementation connects with a full implementation, the full agent takes the role of the controlling agent, and the lite agent takes on the controlled role. When two lite implementations connect, no checks are sent. For guidance on when a lite implementation is appropriate, see the discussion in Section 3.2.23. It is important to note that the lite implementation was added to this specification to provide a stepping stone to full implementation. Even for devices that are always connected to the public Internet, a full implementation is preferable if achievable.

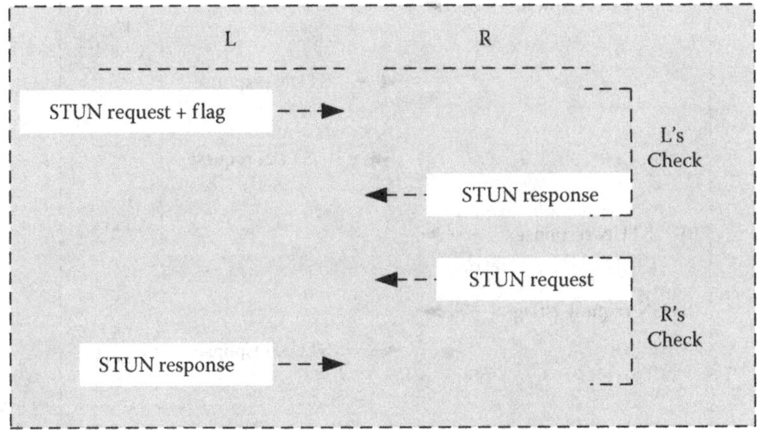

Figure 3.6 Aggressive nomination (Copyright: IETF).

3.2.3 Terminology

Readers should be familiar with the terminology defined in the offer/answer model (RFC 3264), STUN (RFC 5389), and NAT Behavioral requirements for UDP (RFC 4787). This specification makes use of some additional terminology that can be found in Table 2.1.

3.2.4 Sending the Initial Offer

In order to send the initial offer in an offer/answer exchange, an agent must (1) gather candidates, (2) prioritize them, (3) eliminate redundant candidates, (4) choose default candidates, and then (5) formulate and send the SDP offer. All but the last of these five steps differ for full and lite implementations.

3.2.4.1 Full Implementation Requirements

3.2.4.1.1 Gathering Candidates

An agent gathers candidates when it believes that communication is imminent. An offerer can do this based on a user interface cue or on an explicit request to initiate a session. Every candidate is a transport address. It also has a type and a base. Four types are defined and gathered by this specification: host candidates, server reflexive candidates, peer reflexive candidates, and relayed candidates. The server reflexive candidates are gathered using STUN or TURN, and relayed candidates are obtained through TURN. Peer reflexive candidates are obtained in later phases of ICE, due to connectivity checks. The base of a candidate is the candidate from which an agent must send when using that candidate.

3.2.4.1.1.1 Host Candidates The first step is to gather host candidates. Host candidates are obtained by binding to ports (typically ephemeral) on an IP address attached to an interface (physical or virtual, including VPN interfaces) on the host. For each UDP media stream the agent wishes to use, the agent SHOULD obtain a candidate for each component of the media stream on each IP address the host has. It obtains each candidate by binding to a UDP port on the specific IP address. A host candidate (and indeed every candidate) is always associated with a specific component for which it is a candidate. Each component has an ID assigned to it, called the component ID. For RTP-based media streams, the RTP itself has a component ID of 1, and RTCP a component ID of 2. If an agent is using RTCP, it MUST obtain a candidate for it. If an agent is using both RTP and RTCP, it would end up with 2*K host candidates if an agent has K IP addresses. The base for each host candidate is set to the candidate itself.

3.2.4.1.1.2 Server Reflexive and Relayed Candidates Agents SHOULD obtain relayed candidates and SHOULD obtain server reflexive candidates. These requirements are at SHOULD strength to allow for provider/implementation variation. Use of STUN and TURN servers may be unnecessary in closed networks where agents are never connected to the public Internet or to endpoints outside of the closed network. In such cases, a full implementation would be used for agents that are dual stack or multihomed to select a host candidate. Use of TURN servers is expensive, and when ICE is being used, they will only be utilized when both endpoints are behind NATs that perform address- and port-dependent mapping. Consequently, some deployments might consider this use case to be marginal and elect not to use TURN servers. If an agent does not gather server reflexive or relayed candidates, it is RECOMMENDED that the functionality be implemented and just disabled through configuration, so that it can be re-enabled through configuration if conditions change in the future.

If an agent is gathering both relayed and server-reflexive candidates, it uses a TURN server. If it is gathering just server-reflexive candidates, it uses a STUN server. The agent next pairs each host candidate with the STUN or TURN server with which it is configured or that it has discovered by some means. If a STUN or TURN server is configured, it is RECOMMENDED that a domain name be configured, and the DNS procedures in (RFC 5389) (using Service Records (SRV) with the "stun" service) be used to discover the STUN server, and the DNS procedures in (RFC 5766) (using SRV records with the "turn" service) be used to discover the TURN server.

This specification only considers usage of a single STUN or TURN server. When there are multiple choices for that single STUN or TURN server (when, for example, they are learned through DNS records and multiple results are returned), an agent SHOULD use a single STUN or TURN server (based on its IP address) for all candidates for a particular session. This improves the performance of ICE. The result is a set of pairs of host candidates with STUN or TURN servers. The agent then chooses one pair and sends a Binding or Allocate request to the server from that host candidate. Binding requests to a STUN server are not authenticated, and any ALTERNATESERVER attribute in a response is ignored. Agents MUST support the backwards compatibility mode for the Binding request defined in RFC 5389. Allocate requests SHOULD be authenticated using a long-term credential obtained by the client through some other means.

Every *Ta* milliseconds thereafter, the agent can generate another new STUN or TURN transaction. This transaction can be a retry of a previous transaction that failed with a recoverable error (such as authentication failure) or a transaction for a new host candidate and STUN or TURN server pair. The agent SHOULD NOT generate transactions more frequently than one every Ta milliseconds. See Section 3.2.16 for guidance on how to set *Ta* and the STUN retransmit timer, RTO (Retransmission Timeout).

The agent will receive a Binding or Allocate response. A successful Allocate response will provide the agent with a server reflexive candidate (obtained from the mapped address) and a relayed candidate in the XOR-RELAYED-ADDRESS attribute. If the Allocate request is rejected because the server lacks resources to fulfill it, the agent SHOULD instead send a Binding request to obtain a server reflexive candidate. A Binding response will provide the agent with only a server reflexive candidate (also obtained from the mapped address). The base of the server reflexive candidate is the host candidate from which the Allocate or Binding request was sent. The base of a relayed candidate is that candidate itself. If a relayed candidate is identical to a host candidate (which can happen in rare cases), the relayed candidate MUST be discarded.

3.2.4.1.1.3 Computing Foundations

Finally, the agent assigns each candidate a foundation. The foundation is an identifier, scoped within a session. Two candidates MUST have the same foundation ID when all of the following are true:

- They are of the same type (host, relayed, server reflexive, or peer reflexive).
- Their bases have the same IP address (the ports can be different).
- For reflexive and relayed candidates, the STUN or TURN servers used to obtain them have the same IP address.
- They were obtained using the same transport protocol (TCP, UDP, etc.).

Similarly, two candidates MUST have different foundations if their types are different, their bases have different IP addresses, the STUN or TURN servers used to obtain them have different IP addresses, or their transport protocols are different.

3.2.4.1.1.4 Keeping Candidates Alive Once server reflexive and relayed candidates are allocated, they MUST be kept alive until ICE processing is completed, as described in Section 3.2.8.3. For server reflexive candidates learned through a Binding request, the bindings MUST be kept alive by additional Binding requests to the server. Refreshes for allocations are done using the Refresh transaction, as described in RFC 5766. The Refresh requests will also refresh the server reflexive candidate.

3.2.4.1.2 Prioritizing Candidates

The prioritization process results in the assignment of a priority to each candidate. Each candidate for a media stream MUST have a unique priority that MUST be a positive integer between 1 and (2**31 - 1). This priority will be used by ICE to determine the order of the connectivity checks and the relative preference for candidates. An agent SHOULD compute this priority using the formula in Section 3.2.4.1.2.1 and choose its parameters using the guidelines in Section 3.2.4.1.2.2. If an agent elects to use a different formula, ICE will take longer to converge since both agents will not be coordinated in their checks.

3.2.4.1.2.1 Recommended Formula When using the formula, an agent computes the priority by determining a preference for each type of candidate (server reflexive, peer reflexive, relayed, and host), and, when the agent is multihomed, choosing a preference for its IP addresses. These two preferences are then combined to compute the priority for a candidate. That priority is computed using the following formula:

$$\text{priority} = (2 \wedge 24) * (\text{type preference}) + (2 \wedge 8) * (\text{local preference}) \\ + (2 \wedge 0) * (256 - \text{componed ID})$$

The type preference MUST be an integer from 0 to 126 inclusive and represents the preference for the type of the candidate (where the types are local, server reflexive, peer reflexive, and relayed). 126 is the highest preference, and 0 is the lowest. Setting the value to 0 means that candidates of this type will only be used as a last resort. The type preference MUST be identical for all candidates of the same type and MUST be different for candidates of different types. The type preference for peer reflexive candidates MUST be higher than that of server reflexive candidates. Note that candidates gathered based on the procedures of Section 3.2.4.1.1 will never be peer reflexive candidates; candidates of this type are learned from the connectivity checks performed by ICE.

The local preference MUST be an integer from 0 to 65535 inclusive. It represents a preference for the particular IP address from which the candidate was obtained, in cases where an agent is multihomed. 65535 represents the highest preference; 0, the lowest. When there is only a single IP address, this value SHOULD be set to 65535. More generally, if there are multiple candidates for a particular component for a particular media stream that have the same type, the local preference MUST be unique for each one. In this specification, this only happens for multihomed hosts. If a host is multihomed because it is dual stack, the local preference SHOULD be set equal to the precedence value for IP addresses described in RFC 3484 (that was made obsolete by RFC 6724). The component ID is the component ID for the candidate and MUST be between 1 and 256 inclusive.

3.2.4.1.2.2 Guidelines for Choosing Type and Local Preferences

One criterion for selection of the type and local preference values is the use of a media intermediary, such as a TURN server, VPN server, or NAT. With a media intermediary, if media is sent to that candidate, it will first transit the media intermediary before being received. Relayed candidates are one type of candidate that involves a media intermediary. Host candidates obtained from a VPN interface comprise another type. When media is transited through a media intermediary, it can increase the latency between transmission and reception. It can increase the packet losses, because of the additional router hops that may be taken. It may increase the cost of providing service, since media will be routed in and right back out of a media intermediary run by a provider. If these concerns are important, the type preference for relayed candidates SHOULD be lower than that of host candidates.

The RECOMMENDED values are 126 for host candidates, 100 for server reflexive candidates, 110 for peer reflexive candidates, and 0 for relayed candidates. Furthermore, if an agent is multihomed and has multiple IP addresses, the local preference for host candidates from a VPN interface SHOULD have a priority of 0. Another criterion for selection of preferences is IP address family. ICE works with both IPv4 and IPv6. It therefore provides a transition mechanism that allows dual-stack hosts to prefer connectivity over IPv6 but to fall back to IPv4 in case the v6 networks are disconnected (due, for example, to a failure in a 6to4 relay) (RFC 3056). It can also help with hosts that have both a native IPv6 address and a 6to4 address. In such a case, higher local preferences could be assigned to the v6 addresses, followed by the 6to4 addresses, followed by the v4 addresses. This allows a site to obtain and begin using native v6 addresses immediately, yet still fall back to 6to4 addresses when communicating with agents in other sites that do not yet have native v6 connectivity.

Another criterion for selecting preferences is security. If a user is a telecommuter, and therefore connected to a corporate network and a local home network, the user may prefer their voice traffic to be routed over the VPN in order to keep it on the corporate network when communicating within the enterprise, but use the local network when communicating with users outside of the enterprise. In such a case, a VPN address would have a higher local preference than any other address. Another criterion for selecting preferences is topological awareness. This is most useful for candidates that make use of intermediaries. In those cases, if an agent has preconfigured or dynamically discovered knowledge of the topological proximity of the intermediaries to itself, it can use that to assign higher local preferences to candidates obtained from closer intermediaries.

3.2.4.1.3 Eliminating Redundant Candidates

Next, the agent eliminates redundant candidates. A candidate is redundant if its transport address equals that of another candidate and its base equals the base of that other candidate. Note that two candidates can have the same transport address yet have different bases and not be considered redundant. Frequently, a server reflexive candidate and a host candidate will be redundant when the agent is not behind a NAT. The agent SHOULD eliminate the redundant candidate with the lower priority.

3.2.4.1.4 Choosing Default Candidates

A candidate is said to be default if it would be the target of media from a non-ICE peer; that target is called the DEFAULT DESTINATION. If the default candidates are not selected by the ICE algorithm when communicating with an ICE-aware peer, an updated offer/answer will be required after ICE processing completes in order to "fix up" the SDP so that the default destination for media matches the candidates selected by ICE. If ICE happens to select the default candidates, no updated offer/answer is required.

An agent MUST choose a set of candidates, one for each component of each in-use media stream, to be default. A media stream is in use if it does not have a port of 0 (which is used in RFC 3264 (see Section 3.1) to reject a media stream). Consequently, a media stream is in use even if it is marked as a=inactive (RFC 4566, see Section 2) or has a bandwidth value of 0. It is RECOMMENDED that default candidates be chosen based on the likelihood of those candidates to work with the peer that is being contacted. It is RECOMMENDED that the default candidates are the relayed candidates (if relayed candidates are available), server reflexive candidates (if server reflexive candidates are available), and finally host candidates.

3.2.4.2 Lite Implementation Requirements

Lite implementations only utilize host candidates. A lite implementation MUST, for each component of each media stream, allocate zero or one IPv4 candidate. It MAY allocate zero or more IPv6 candidates, but no more than one per each IPv6 address utilized by the host. Since there can be no more than one IPv4 candidate per component of each media stream, if an agent has multiple IPv4 addresses, it MUST choose one for allocating the candidate. If a host is dual stack, it is RECOMMENDED that it allocate one IPv4 candidate and one global IPv6 address. With the lite implementation, ICE cannot be used to dynamically choose amongst candidates. Therefore, including more than one candidate from a particular scope is NOT RECOMMENDED, since only a connectivity check can truly determine whether to use one address or the other.

Each component has an ID assigned to it, called the component ID. For RTP-based media streams, the RTP itself has a component ID of 1, and RTCP a component ID of 2. If an agent is using RTCP, it MUST obtain candidates for it. Each candidate is assigned a foundation. The foundation MUST be different for two candidates allocated from different IP addresses and MUST be the same otherwise. A simple integer that increments for each IP address will suffice. In addition, each candidate MUST be assigned a unique priority amongst all candidates for the same media stream. It SHOULD be equal to:

$$\text{priority} = (2 \wedge 24) * (126) + (2 \wedge 8) * (\text{IP precedence}) + (2 \wedge 0) * (256 - \text{componed ID})$$

If a host is v4 only, it SHOULD set the IP precedence to 65535. If a host is v6 or dual stack, the IP precedence SHOULD be the precedence value for IP addresses described in RFC 3484 (that was made obsolete by RFC 6724).

Next, an agent chooses a default candidate for each component of each media stream. If a host is IPv4 only, there will only be one candidate for each component of each media stream; therefore, that candidate is the default. If a host is IPv6 or dual stack, the selection of default is a matter of local policy. This default SHOULD be chosen such that it is the candidate most likely to be used with a peer. For IPv6-only hosts, this will typically be a globally scoped IPv6 address. For dual-stack hosts, the IPv4 address is RECOMMENDED.

3.2.4.3 Encoding the SDP

The process of encoding the SDP is identical between full and lite implementations. The agent will include an "m=" line for each media stream it wishes to use. The ordering of media streams in the SDP is relevant for ICE. ICE will perform its connectivity checks for the first m line first, and consequently media will be able to flow for that stream first. Agents SHOULD place their most important media stream, if there is one, first in the SDP. There will be a candidate attribute for each candidate for a particular media stream. Section 3.2.15 provides detailed rules for constructing this attribute. The attribute carries the IP address, port, and transport protocol for the candidate, in addition to its properties that need to be signaled to the peer for ICE to work: the priority, foundation, and component ID. The candidate attribute also carries information about the candidate that is useful for diagnostics and other functions: its type and related transport addresses.

STUN connectivity checks between agents are authenticated using the short-term credential mechanism defined for STUN (RFC 5389). This mechanism relies on a username and password that are exchanged through protocol machinery between the client and server. With ICE, the offer/answer exchange is used to exchange them. The username part of this credential is formed by concatenating a username fragment from each agent, separated by a colon. Each agent also provides a password, used to compute the message integrity for requests it receives. The username fragment and password are exchanged in the ice-ufrag and ice-pwd attributes, respectively. In addition to providing security, the username provides disambiguation and correlation of checks to media streams. See Section 3.2.24.4 or motivation.

If an agent is a lite implementation, it MUST include an "a=ice-lite" session-level attribute in its SDP. If an agent is a full implementation, it MUST NOT include this attribute. The default candidates are added to the SDP as the default destination for media. For streams based on RTP, this is done by placing the IP address and port of the RTP candidate into the "c=" and "m=" lines, respectively. If the agent is utilizing RTCP, it MUST encode the RTCP candidate using the a=rtcp attribute as defined in RFC 3605 (see Section 12.3). If RTCP is not in use, the agent MUST signal that using b=RS:0 and b=RR:0 as defined in RFC 3556.

The transport addresses that will be the default destination for media when communicating with non-ICE peers MUST also be present as candidates in one or more a=candidate lines.

ICE provides for extensibility by allowing an offer or answer to contain a series of tokens that identify the ICE extensions used by that agent. If an agent supports an ICE extension, it MUST include the token defined for that extension in the ice-options attribute. The following is an example SDP message that includes ICE attributes (lines folded for readability):

```
v=0
o=jdoe 2890844526 2890842807 IN IP4 10.0.1.1
s=
c=IN IP4 192.0.2.3
t=0 0
a=ice-pwd:asd88fgpdd777uzjYhagZg
a=ice-ufrag:8hhY
m=audio 45664 RTP/AVP 0
b=RS:0
b=RR:0
a=rtpmap:0 PCMU/8000
a=candidate:1 1 UDP 2130706431 10.0.1.1 8998 typ host
a=candidate:2 1 UDP 1694498815 192.0.2.3 45664 typ srflx raddr
  10.0.1.1 rport 8998
```

Once an agent has sent its offer or its answer, that agent MUST be prepared to receive both STUN and media packets on each candidate. As discussed in Section 3.2.11.1, media packets can be sent to a candidate prior to its appearance as the default destination for media in an offer or answer.

3.2.5 Receiving the Initial Offer

When agent receives an initial offer, it will check whether the offerer supports ICE, determine its own role, gather candidates, prioritize them, choose default candidates, encode and send an answer, and for full implementations, form the checklists and begin connectivity checks.

3.2.5.1 Verifying ICE Support

The agent will proceed with the ICE procedures defined in this specification if, for each media stream in the SDP it received, the default destination for each component of that media stream appears in a candidate attribute. For example, in the case of RTP, the IP address and port in the "c=" and "m=" lines, respectively, appear in a candidate attribute, and the value in the rtcp attribute appears in a candidate attribute. If this condition is not met, the agent MUST process the SDP based on normal RFC 3264 (see Section 3.1) procedures, without using any of the ICE mechanisms described in the remainder of this specification with the following exceptions:

1. The agent MUST follow the rules of Section 10, which describe keepalive procedures for all agents.
2. If the agent is not proceeding with ICE because there were a=candidate attributes, but none that matched the default destination of the media stream, the agent MUST include an a=icemismatch attribute in its answer.
3. If the default candidates were relayed candidates learned through a TURN server, the agent MUST create permissions in the TURN server for the IP addresses learned from its peer in the SDP it just received. If this is not done, initial packets in the media stream from the peer may be lost.

3.2.5.2 Determining Role

For each session, each agent takes on a role. There are two roles: controlling and controlled. The controlling agent is responsible for the choice of the final candidate pairs used for communications. For a full agent, this means nominating the candidate pairs that can be used by ICE for each media stream and for generating the updated offer based on ICE's selection, when needed. For a lite implementation, being the controlling agent means selecting a candidate pair based on those in the offer and answer (for IPv4, there is only ever one pair), and then generating an updated offer reflecting that selection, when needed (it is never needed for an IPv4-only host). The controlled agent is told which candidate pairs to use for each media stream and does not generate an updated offer to signal this information. The next paragraphs below describe in detail the actual procedures followed by controlling and controlled nodes. The rules for determining the role and the impact on behavior are as follows:

Both agents are full: The agent that generated the offer that started the ICE processing MUST take the controlling role, and the other MUST take the controlled role. Both agents will form checklists, run the ICE state machines, and generate connectivity checks. The controlling agent will execute the logic in Section 3.2.8.1 to nominate pairs that will be selected by ICE, and then both agents end ICE as described in Section 3.2.8.1.2. In unusual cases, described in Section 3.2.24.11 of this book (that is, Appendix B.11 of RFC 5245), it is possible for both agents to mistakenly believe they are controlled or controlling. To resolve this, each agent MUST select a random number, called the tie-breaker, uniformly distributed between 0 and (2**64) – 1 (that is, a 64-bit positive integer). This number is used in connectivity checks to detect and repair this case, as described in Section 3.2.7.1.2.2.

One agent full, one lite: The full agent MUST take the controlling role, and the lite agent MUST take the controlled role. The full agent will form checklists, run the ICE state machines, and generate connectivity checks. That agent will execute the logic in Section 3.2.8.1 to nominate pairs that will be selected by ICE and use the logic in Section 3.2.8.1.2 to end ICE. The lite implementation will just listen for connectivity checks, receive them and respond to them, and then conclude ICE as described in Section 3.2.8.2. For the lite implementation, the state of ICE processing for each media stream is considered to be Running, and the state of ICE overall is Running.

Both lite: The agent that generated the offer that started the ICE processing MUST take the controlling role, and the other MUST take the controlled role. In this case, no connectivity checks are ever sent. Rather, once the offer/answer exchange completes, each agent performs the processing described in Section 3.2.8 without connectivity checks. It is possible that both agents will believe they are controlled or controlling. In the latter case, the conflict is resolved through glare detection capabilities in the signaling protocol carrying the offer/answer exchange. The state of ICE processing for each media stream is considered to be Running, and the state of ICE overall is Running. Once roles are determined for a session, they persist unless ICE is restarted. An ICE restart (Section 3.2.9.1) causes a new selection of roles and tie-breakers.

3.2.5.3 Gathering Candidates

The process for gathering candidates at the answerer is identical to the process at the offerer as described in Section 3.2.4.1.1 for full implementations and Section 3.2.4.2 for lite implementations. It is RECOMMENDED that this process begin immediately on receipt of the offer, prior to alerting the user. Such gathering MAY begin when an agent starts.

3.2.5.4 Prioritizing Candidates

The process for prioritizing candidates at the answerer is identical to the process followed by the offerer, as described in Section 3.2.4.1.2 for full implementations and Section 3.2.4.2 for lite implementations.

3.2.5.5 Choosing Default Candidates

The process for selecting default candidates at the answerer is identical to the process followed by the offerer, as described in Section 3.2.4.1.4 for full implementations and Section 3.2.4.2 for lite implementations.

3.2.5.6 Encoding the SDP

The process for encoding the SDP at the answerer is identical to the process followed by the offerer for both full and lite implementations, as described in Section 3.2.4.3.

3.2.5.7 *Forming the Checklists*

Forming checklists is done only by full implementations. Lite implementations MUST skip the steps defined in this section. There is one checklist per in-use media stream resulting from the offer/answer exchange. To form the checklist for a media stream, the agent forms candidate pairs, computes a candidate pair priority, orders the pairs by priority, prunes them, and sets their states. These steps are described in this section.

3.2.5.7.1 Forming Candidate Pairs

First, the agent takes each of its candidates for a media stream (called LOCAL CANDIDATES) and pairs them with the candidates it received from its peer (called REMOTE CANDIDATES) for that media stream. In order to prevent the attacks described in Section 3.2.18.5.2, agents MAY limit the number of candidates they'll accept in an offer or answer. A local candidate is paired with a remote candidate if and only if the two candidates have the same component ID and have the same IP address version. It is possible that some of the local candidates won't get paired with remote candidates and some of the remote candidates won't get paired with local candidates. This can happen if one agent doesn't include candidates for the all of the components for a media stream. If this happens, the number of components for that media stream is effectively reduced and considered to be equal to the minimum across both agents of the maximum component ID provided by each agent across all components for the media stream.

In the case of RTP, this would happen when one agent provides candidates for RTCP and the other does not. As another example, the offerer can multiplex RTP and RTCP on the same port and signal that it can do that in the SDP through an SDP attribute (RFC 5761). However, since the offerer doesn't know if the answerer can perform such multiplexing, the offerer includes candidates for RTP and RTCP on separate ports so that the offer has two components per media stream. If the answerer can perform such multiplexing, it will include just a single component for each candidate, for the combined RTP/RTCP mux. ICE will end up acting as if there were just a single component for this candidate. The candidate pair whose local and remote candidates are both the default candidates for a particular component is called, unsurprisingly, the default candidate pair for that component. This is the pair that would be used to transmit media if both agents had not been ICE aware.

In order to aid understanding, Figure 3.7 shows the relationships among several key concepts—transport addresses, candidates, candidate pairs, and checklists—in addition to indicating the main properties of candidates and candidate pairs.

3.2.5.7.2 Computing Pair Priority and Ordering Pairs

Once the pairs are formed, a candidate pair priority is computed. Let G be the priority for the candidate provided by the controlling agent. Let D be the priority for the candidate provided by the controlled agent. The priority for a pair is computed as

$$\text{pair priority} = 2 \wedge 32 * \text{MIN}(G,D) + 2 * \text{MAX}(G,D) + (G > D?1:0)$$

Where G>D?1:0 is an expression whose value is 1 if G is greater than D, and 0 otherwise. Once the priority is assigned, the agent sorts the candidate pairs in decreasing order of priority. If two pairs have identical priority, the ordering amongst them is arbitrary.

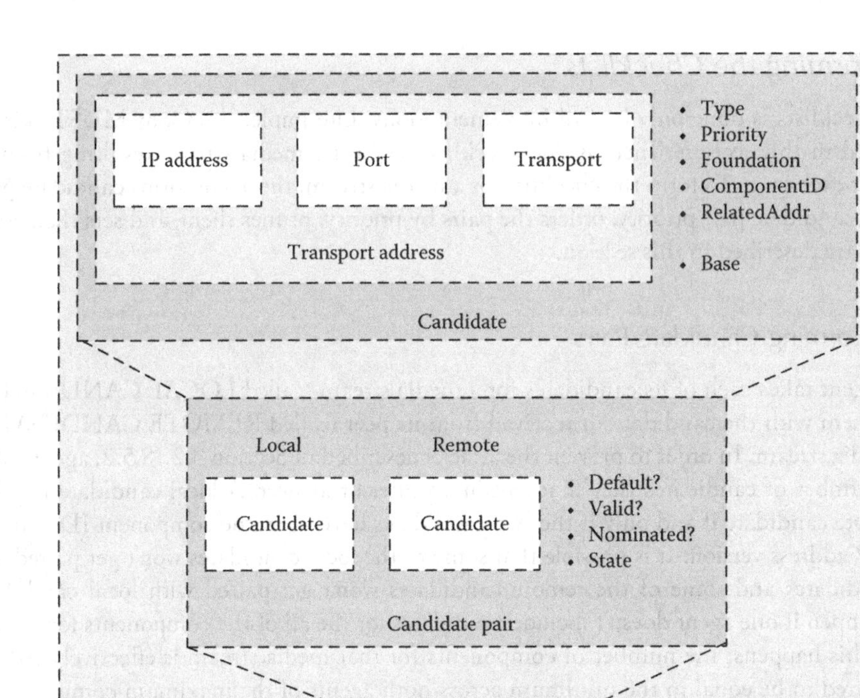

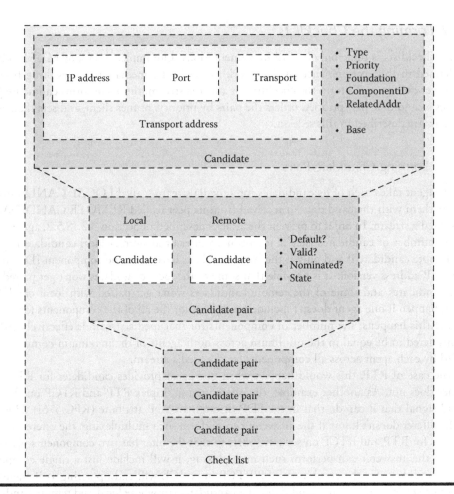

Figure 3.7 Conceptual diagram of a checklist (Copyright: IETF).

3.2.5.7.3 Pruning the Pairs

This sorted list of candidate pairs is used to determine a sequence of connectivity checks that will be performed. Each check involves sending a request from a local candidate to a remote candidate. Since an agent cannot send requests directly from a reflexive candidate, but only from its base, the agent next goes through the sorted list of candidate pairs. For each pair where the local candidate is server reflexive, the server reflexive candidate MUST be replaced by its base. Once this has been done, the agent MUST prune the list. This is done by removing a pair if its local and remote candidates are identical to the local and remote candidates of a pair higher up on the priority list. The result is a sequence of ordered candidate pairs, called the checklist for that media stream. In addition, in order to limit the attacks described in Section 3.2.18.5.2, an agent MUST limit the total number of connectivity checks the agent performs across all checklists to a specific value, and this value MUST be configurable. A default of 100 is RECOMMENDED. This limit is enforced by discarding the lower-priority candidate pairs until there are less than 100. It is RECOMMENDED that a lower value be utilized when possible, set to the maximum number of plausible checks that might be seen in an actual deployment configuration. The requirement for configuration is meant to provide a tool for fixing this value in the field if, once deployed, it is found to be problematic.

3.2.5.7.4 Computing States

Each candidate pair in the checklist has a foundation and a state. The foundation is the combination of the foundations of the local and remote candidates in the pair. The state is assigned once the checklist for each media stream has been computed. There are five potential values that the state can have:

- **Waiting**: A check has not been performed for this pair and can be performed as soon as it is the highest-priority Waiting pair on the checklist.
- **In-Progress**: A check has been sent for this pair, but the transaction is in progress.
- **Succeeded**: A check for this pair was already done and produced a successful result.
- **Failed**: A check for this pair was already done and failed, either never producing any response or producing an unrecoverable failure response.
- **Frozen**: A check for this pair hasn't been performed, and it can't yet be performed until some other check succeeds, allowing this pair to unfreeze and move into the Waiting state.

As ICE runs, the pairs will move between states as shown in Figure 3.8.

The initial states for each pair in a checklist are computed by performing the following sequence of steps:

1. The agent sets all of the pairs in each checklist to the Frozen state.
2. The agent examines the checklist for the first media stream (a media stream is the first media stream when it is described by the first "m=" line in the SDP offer and answer). For that media stream:
   - For all pairs with the same foundation, it sets the state of the pair with the lowest component ID to Waiting. If there is more than one such pair, the one with the highest priority is used.

One of the checklists will have some number of pairs in the Waiting state, and the other checklists will have all of their pairs in the Frozen state. A checklist with at least one pair that is Waiting is called an active checklist, and a checklist with all pairs Frozen is called a frozen checklist.

The checklist itself is associated with a state that captures the state of ICE checks for that media stream. There are three states:

- Running: In this state, ICE checks are still in progress for this media stream.
- Completed: In this state, ICE checks have produced nominated pairs for each component of the media stream. Consequently, ICE has succeeded and media can be sent.
- Failed: In this state, the ICE checks have not completed successfully for this media stream.

When a checklist is first constructed as the consequence of an offer/answer exchange, it is placed in the running state. ICE processing across all media streams also has a state associated with it. This state is equal to Running while ICE processing is under way. The state is Completed when ICE processing is complete and Failed if it ended without success. Rules for transitioning between states are described in the next section.

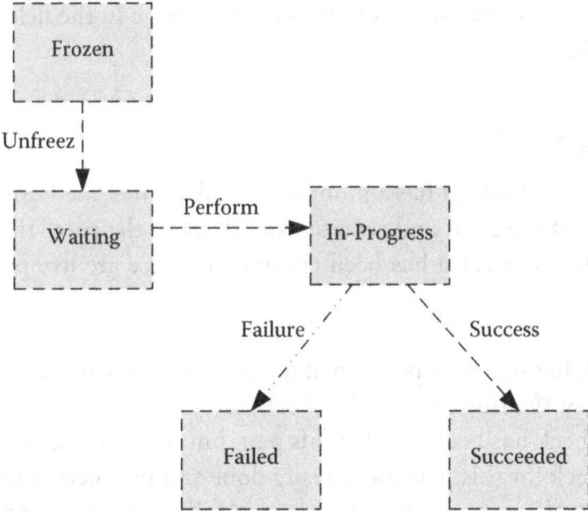

Figure 3.8 Pair state FSM (Copyright: IETF).

3.2.5.8 Scheduling Checks

Checks are generated only by full implementations. Lite implementations MUST skip the steps described in this section. An agent performs ordinary checks and triggered checks. The generation of both checks is governed by a timer that fires periodically for each media stream. The agent maintains a first-in-first-out queue, called the triggered check queue, which contains candidate pairs for which checks are to be sent at the next available opportunity. When the timer fires, the agent removes the top pair from the triggered check queue, performs a connectivity check on that pair, and sets the state of the candidate pair to In-Progress. If there are no pairs in the triggered check queue, an ordinary check is sent.

Once the agent has computed the checklists as described in Section 3.2.5.7, it sets a timer for each active checklist. The timer fires every $Ta * N$ seconds, where N is the number of active checklists (initially, there is only one active checklist). Implementations MAY set the timer to fire less frequently. Implementations SHOULD take care to spread out these timers so that they do not fire at the same time for each media stream. Ta and the retransmit timer RTO are computed as described in Section 3.2.16. Multiplying by N allows this aggregate check throughput to be split among all active checklists. The first timer fires immediately, so that the agent performs a connectivity check the moment the offer/answer exchange has been done, followed by the next check Ta seconds later (since there is only one active checklist). When the timer fires and there is no triggered check to be sent, the agent MUST choose an ordinary check as follows:

- Find the highest-priority pair in the Waiting state in that checklist.
- If there is such a pair:
  - Send a STUN check from the local candidate of that pair to the remote candidate of that pair. The procedures for forming the STUN request for this purpose are described in Section 3.2.7.1.2.
  - Set the state of the candidate pair to In-Progress.
- If there is no such pair:
  - Find the highest-priority pair in the Frozen state in that checklist.

- If there is such a pair:
  - Unfreeze the pair.
  - Perform a check for that pair, causing its state to transition to In-Progress.
- If there is no such pair:
  - Terminate the timer for that checklist.

To compute the message integrity for the check, the agent uses the remote username fragment and password learned from the SDP from its peer. The local username fragment is known directly by the agent for its own candidate.

3.2.6 Receipt of the Initial Answer

This section describes the procedures that an agent follows when it receives the answer from the peer. It verifies that its peer supports ICE, determines its role, and for full implementations, forms the checklist and begins performing ordinary checks. When ICE is used with SIP, forking may result in a single offer generating a multiplicity of answers. In that case, ICE proceeds completely in parallel and independently for each answer, treating the combination of its offer and each answer as an independent offer/answer exchange, with its own set of pairs, checklists, states, and so on. The only case in which processing of one pair impacts another is freeing of candidates, discussed in Section 3.2.8.3.

3.2.6.1 Verifying ICE Support

The logic at the offerer is identical to that of the answerer as described in Section 3.2.5.1, with the exception that an offerer would not ever generate a=ice-mismatch attributes in an SDP. In some cases, the answer may omit a=candidate attributes for the media streams and instead include an a=ice-mismatch attribute for one or more of the media streams in the SDP. This signals to the offerer that the answerer supports ICE but that ICE processing was not used for the session because a signaling intermediary modified the default destination for media components without modifying the corresponding candidate attributes. See Section 3.2.18 for a discussion of cases in which this can happen. This specification provides no guidance on how an agent should proceed in such a failure case.

3.2.6.2 Determining Role

The offerer follows the same procedures described for the answerer in Section 3.2.5.2.

3.2.6.3 Forming the Checklist

Formation of checklists is performed only by full implementations. The offerer follows the same procedures described for the answerer in Section 3.2.5.7.

3.2.6.4 Performing Ordinary Checks

Ordinary checks are performed only by full implementations. The offerer follows the same procedures described for the answerer in Section 3.2.5.8.

3.2.7 Performing Connectivity Checks

This section describes how connectivity checks are performed. All ICE implementations are required to be compliant to RFC 5389, as opposed to the older RFC 3489. However, whereas a full implementation will both generate checks (acting as a STUN client) and receive them (acting as a STUN server), a lite implementation will only receive checks and thus will only act as a STUN server.

3.2.7.1 STUN Client Procedures

These procedures define how an agent sends a connectivity check, whether an ordinary or a triggered check. These procedures are only applicable to full implementations.

3.2.7.1.1 Creating Permissions for Relayed Candidates

If the connectivity check is being sent using a relayed local candidate, the client MUST create a permission first if it has not already created one. It would have created one if it had told the TURN server to create a permission for the given relayed candidate toward the IP address of the remote candidate. To create the permission, the agent follows the procedures defined in RFC 5766. The permission MUST be created toward the IP address of the remote candidate. It is RECOMMENDED that the agent defer creation of a TURN channel until ICE completes, in which case permissions for connectivity checks are normally created using a CreatePermission request. Once established, the agent MUST keep the permission active until ICE concludes.

3.2.7.1.2 Sending the Request

The check is generated by sending a Binding request from a local candidate to a remote candidate. RFC 5389 describes how Binding requests are constructed and generated. A connectivity check MUST utilize the STUN short-term credential mechanism. Support for backwards compatibility with RFC 3489 MUST NOT be used or assumed with connectivity checks. The FINGERPRINT mechanism MUST be used for connectivity checks. ICE extends STUN by defining several new attributes, including PRIORITY, USE-CANDIDATE, ICE-CONTROLLED, and ICE-CONTROLLING. These new attributes are formally defined in Section 3.2.19.1, and their usage is described in the subsections below. These STUN extensions are applicable only to connectivity checks used for ICE.

3.2.7.1.2.1 PRIORITY and USE-CANDIDATE An agent MUST include the PRIORITY attribute in its Binding request. The attribute MUST be set equal to the priority that would be assigned, based on the algorithm described in Section 3.2.4.1.2, to a peer reflexive candidate, should one be learned as a consequence of this check (see Section 3.2.7.1.3.2.1) for how peer reflexive candidates are learned). This priority value will be computed identically to how the priority for the local candidate of the pair was computed, except that the type preference is set to the value for peer reflexive candidate types.

The controlling agent MAY include the USE-CANDIDATE attribute in the Binding request. The controlled agent MUST NOT include it in its Binding request. This attribute signals that the controlling agent wishes to cease checks for this component and use the candidate pair resulting from the check for this component. Section 3.2.8.1.1 provides guidance on determining when to include it.

3.2.7.1.2.2 Ice-Controlled and Ice-Controlling The agent MUST include the ICE-CONTROLLED attribute in the request if it is in the controlled role and MUST include the ICE-CONTROLLING attribute in the request if it is in the controlling role. The content of either attribute MUST be the tie-breaker that was determined in Section 3.2.5.2. These attributes are defined fully in Section 3.2.19.1.

3.2.7.1.2.3 Forming Credentials A Binding request serving as a connectivity check MUST utilize the STUN short-term credential mechanism. The username for the credential is formed by concatenating the username fragment provided by the peer with the username fragment of the agent sending the request, separated by a colon (":"). The password is equal to the password provided by the peer. For example, consider the case where agent L is the offerer, and agent R is the answerer. Agent L included a username fragment of LFRAG for its candidates and a password of LPASS. Agent R provided a username fragment of RFRAG and a password of RPASS. A connectivity check from L to R utilizes the username RFRAG:LFRAG and a password of RPASS. A connectivity check from R to L utilizes the username LFRAG:RFRAG and a password of LPASS. The responses utilize the same usernames and passwords as the requests (note that the USERNAME attribute is not present in the response).

3.2.7.1.2.4 DiffServ Treatment If the agent is using DiffServ (Differentiated Service) Codepoint markings (RFC 2475) in its media packets, it SHOULD apply those same markings to its connectivity checks.

3.2.7.1.3 Processing the Response

When a Binding response is received, it is correlated to its Binding request using the transaction ID, as defined in RFC 5389, which then ties it to the candidate pair for which the Binding request was sent. This section defines additional procedures for processing Binding responses specific to this usage of STUN.

3.2.7.1.3.1 Failure Cases If the STUN transaction generates a 487 (Role Conflict) error response, the agent checks whether it included the ICE-CONTROLLED or ICE-CONTROLLING attribute in the Binding request. If the request contained the ICE-CONTROLLED attribute, the agent MUST switch to the controlling role if it has not already done so. If the request contained the ICE-CONTROLLING attribute, the agent MUST switch to the controlled role if it has not already done so. Once it has switched, the agent MUST enqueue the candidate pair whose check generated the 487 into the triggered check queue. The state of that pair is set to Waiting. When the triggered check is sent, it will contain an ICE-CONTROLLING or ICE-CONTROLLED attribute reflecting its new role.

Note, however, that the tie-breaker value MUST NOT be reselected. A change in roles will require an agent to recompute pair priorities (Section 5.7.2), since those priorities are a function of controlling and controlled roles. The change in role will also impact whether the agent is responsible for selecting nominated pairs and generating updated offers upon conclusion of ICE. Agents MAY support receipt of ICMP (Internet Control Message Protocol) errors for connectivity checks. If the STUN transaction generates an ICMP error, the agent sets the state of the pair to Failed. If the STUN transaction generates a STUN error response that is unrecoverable (as defined in RFC 5389) or times out, the agent sets the state of the pair to Failed. The agent MUST check that the source IP address and port of the response equal the destination IP address and port to which the Binding request was sent and that the destination IP address and port of the response match the source IP address and port from which the Binding request was sent. In other words, the source and destination transport addresses in the request and responses are symmetric. If they are not symmetric, the agent sets the state of the pair to Failed.

3.2.7.1.3.2 Success Cases A check is considered a success if all of the following are true:

- The STUN transaction generated a success response.
- The source IP address and port of the response equals the destination IP address and port to which the Binding request was sent.
- The destination IP address and port of the response match the source IP address and port from which the Binding request was sent.

3.2.7.1.3.2.1 Discovering Peer Reflexive Candidates The agent checks the mapped address from the STUN response. If the transport address does not match any of the local candidates the agent knows about, the mapped address represents a new candidate—a peer reflexive candidate. Like other candidates, it has a type, base, priority, and foundation. They are computed as follows:

- Its type is equal to peer reflexive.
- Its base is set equal to the local candidate of the candidate pair from which the STUN check was sent.
- Its priority is set equal to the value of the PRIORITY attribute in the Binding request.
- Its foundation is selected as described in Section 3.2.4.1.1.3.

This peer reflexive candidate is then added to the list of local candidates for the media stream. Its username fragment and password are the same as all other local candidates for that media stream.

However, the peer reflexive candidate is not paired with other remote candidates. This is not necessary; a valid pair will be generated from it momentarily based on the procedures in Section 3.2.7.1.3.2.2. If an agent wishes to pair the peer reflexive candidate with remote candidates besides the one in the valid pair that will be generated, the agent MAY generate an updated offer that includes the peer reflexive candidate. This will cause it to be paired with all other remote candidates.

3.2.7.1.3.2.2 Constructing a Valid Pair The agent constructs a candidate pair whose local candidate equals the mapped address of the response and whose remote candidate equals the destination address to which the request was sent. This is called a valid pair, since it has been validated by a STUN connectivity check. The valid pair may equal the pair that generated the check, may equal a different pair in the checklist, or may be a pair not currently on any checklist. If the pair equals the pair that generated the check or is on a checklist currently, it is also added to the VALID LIST, which is maintained by the agent for each media stream. This list is empty at the start of ICE processing and fills as checks are performed, resulting in valid candidate pairs.

It will be very common that the pair will not be on any checklist. Recall that the checklist has pairs whose local candidates are never server reflexive; those pairs had their local candidates converted to the base of the server reflexive candidates, and then pruned if they were redundant. When the response to the STUN check arrives, the mapped address will be reflexive if there is a NAT between the two.

In that case, the valid pair will have a local candidate that doesn't match any of the pairs in the checklist. If the pair is not on any checklist, the agent computes the priority for the pair based on the priority of each candidate, using the algorithm in Section 3.2.5.7. The priority of the local candidate depends on its type. If it is not peer reflexive, it is equal to the priority signaled for that candidate in the SDP. If it is peer reflexive, it is equal to the PRIORITY attribute the agent placed in the Binding request that just completed. The priority of the remote candidate is taken from the SDP of the peer. If the candidate does not appear there, then the check must have been a triggered check to a new remote candidate. In that case, the priority is taken as the value of the PRIORITY attribute in the Binding request that triggered the check that just completed. The pair is then added to the VALID LIST.

3.2.7.1.3.2.3 Updating Pair States The agent sets the state of the pair that *generated* the check to Succeeded. Note that, the pair that *generated* the check may be different than the valid pair constructed in Section 3.2.7.1.3.2.2 as a consequence of the response. The success of this check might cause the state of other checks to change as well. The agent MUST perform the following two steps:

1. Change the states for all other Frozen pairs for the same media stream and same foundation to Waiting. Typically, but not always, these other pairs will have different component IDs.
2. If there is a pair in the valid list for every component of this media stream (where this is the actual number of components being used, in cases where the number of components signaled in the SDP differs from offerer to answerer), the success of this check may unfreeze checks for other media streams. Note that this step is followed not just the first time the valid list under consideration has a pair for every component, but every subsequent time a check succeeds and adds yet another pair to that valid list. Examine the checklist for each other media stream in turn:
   - If the checklist is active, the agent changes the state of all Frozen pairs in that checklist whose foundation matches a pair in the valid list under consideration to Waiting.
   - If the checklist is frozen, and there is at least one pair in the checklist whose foundation matches a pair in the valid list under consideration, the state of all pairs in the checklist whose foundation matches a pair in the valid list under consideration is set to Waiting. This will cause the checklist to become active, and ordinary checks will begin for it, as described in Section 3.2.5.8.
   - If the checklist is frozen, and there are no pairs in the checklist whose foundation matches a pair in the valid list under consideration, the agent
     - Groups together all of the pairs with the same foundation and
     - For each group, sets the state of the pair with the lowest component ID to Waiting. If there is more than one such pair, the one with the highest priority is used.

3.2.7.1.3.2.4 Updating the Nominated Flag If the agent was a controlling agent, and it had included a USE-CANDIDATE attribute in the Binding request, the valid pair generated from that check has its nominated flag set to true. This flag indicates that this valid pair should be used for media, if it is the highest-priority one amongst those whose nominated flag is set. This may conclude ICE processing for this media stream or all media streams; see Section 3.2.8. If the agent is the controlled agent, the response may be the result of a triggered check that was sent in response to a request that itself had the USE-CANDIDATE attribute. This case is described in Section 3.2.7.2.1.5 and may now result in setting the nominated flag for the pair learned from the original request.

3.2.7.1.3.3 Checklist and Timer State Updates Regardless of whether the check was successful, the completion of the transaction may require updating of checklist and timer states. If all of the pairs in the checklist are now in either the Failed or Succeeded state:

- If there is not a pair in the valid list for each component of the media stream, the state of the checklist is set to Failed.
- For each frozen checklist, the agent
  - Groups together all of the pairs with the same foundation, and
  - For each group, sets the state of the pair with the lowest component ID to Waiting. If there is more than one such pair, the one with the highest priority is used.

If none of the pairs in the checklist is in the Waiting or Frozen state, the checklist is no longer considered active and will not count toward the value of N in the computation of timers for ordinary checks as described in Section 3.2.5.8.

3.2.7.2 STUN Server Procedures

An agent MUST be prepared to receive a Binding request on the base of each candidate it included in its most recent offer or answer. This requirement holds even if the peer is a lite implementation. The agent MUST use a short-term credential to authenticate the request and perform a message integrity check. The agent MUST consider the username valid if it consists of two values separated by a colon, where the first value is equal to the username fragment generated by the agent in an offer or answer for a session in-progress. It is possible (and in fact very likely) that an offerer will receive a Binding request prior to receiving the answer from its peer. If this happens, the agent MUST immediately generate a response (including computation of the mapped address as described in Section 3.2.7.2.1.2). The agent has sufficient information at this point to generate the response; the password from the peer is not required. Once the answer is received, it MUST proceed with the remaining steps required, namely, 3.2.7.2.1.3, 3.2.7.2.1.4, and 3.2.7.2.1.5 for full implementations. In cases where multiple STUN requests are received before the answer, this may cause several pairs to be in the triggered check queue.

An agent MUST NOT utilize the ALTERNATE-SERVER mechanism and MUST NOT support the backwards compatibility mechanisms to RFC 3489. It MUST utilize the FINGERPRINT mechanism. If the agent is using DiffServ Codepoint markings (RFC 2475) in its media packets, it SHOULD apply those same markings to its responses to Binding requests. The same would apply to any layer 2 markings the endpoint might be applying to media packets.

3.2.7.2.1 Additional Procedures for Full Implementations

This subsection defines the additional server procedures applicable to full implementations.

3.2.7.2.1.1 Detecting and Repairing Role Conflicts Normally, the rules for selection of a role in Section 3.2.5.2 will result in each agent selecting a different role—one controlling and one controlled. However, in unusual call flows, typically utilizing third party call control, it is possible for both agents to select the same role. This section describes procedures for checking for this case and repairing it. An agent MUST examine the Binding request for either the ICE-CONTROLLING or ICE-CONTROLLED attribute. It MUST follow these procedures:

- If neither ICE-CONTROLLING nor ICE-CONTROLLED is present in the request, the peer agent may have implemented a previous version of this specification. There may be a conflict, but it cannot be detected.
- If the agent is in the controlling role, and the ICE-CONTROLLING attribute is present in the request:
  - If the agent's tie-breaker is larger than or equal to the contents of the ICE-CONTROLLING attribute, the agent generates a Binding error response and includes an ERROR-CODE attribute with a value of 487 (Role Conflict) but retains its role.
  - If the agent's tie-breaker is less than the contents of the ICE-CONTROLLING attribute, the agent switches to the controlled role.
- If the agent is in the controlled role, and the ICE-CONTROLLED attribute is present in the request:
  - If the agent's tie-breaker is larger than or equal to the contents of the ICE-CONTROLLED attribute, the agent switches to the controlling role.
  - If the agent's tie-breaker is less than the contents of the ICE-CONTROLLED attribute, the agent generates a Binding error response and includes an ERROR-CODE attribute with a value of 487 (Role Conflict) but retains its role.
- If the agent is in the controlled role and the ICE-CONTROLLING attribute was present in the request, or the agent was in the controlling role and the ICE-CONTROLLED attribute was present in the request, there is no conflict.

A change in roles will require an agent to recompute pair priorities (Section 3.2.5.7.2), since those priorities are a function of controlling and controlled roles. The change in role will also impact whether the agent is responsible for selecting nominated pairs and generated updated offers upon conclusion of ICE. The remaining sections in Section 3.2.7.2.1 are followed if the server generated a successful response to the Binding request, even if the agent changed roles.

3.2.7.2.1.2 Computing Mapped Address For requests being received on a relayed candidate, the source transport address used for STUN processing (namely, generation of the XOR-MAPPED-ADDRESS attribute) is the transport address as seen by the TURN server. That source transport address will be present in the XOR-PEER-ADDRESS attribute of a Data Indication message, if the Binding request was delivered through a Data Indication. If the Binding request was delivered through a ChannelData message, the source transport address is the one that was bound to the channel.

3.2.7.2.1.3 Learning Peer Reflexive Candidates If the source transport address of the request does not match any existing remote candidates, it represents a new peer reflexive remote candidate. This candidate is constructed as follows:

- The priority of the candidate is set to the PRIORITY attribute from the request.
- The type of the candidate is set to peer reflexive.
- The foundation of the candidate is set to an arbitrary value, different from the foundation for all other remote candidates. If any subsequent offer/answer exchanges contain this peer reflexive candidate in the SDP, it will signal the actual foundation for the candidate.
- The component ID of this candidate is set to the component ID for the local candidate to which the request was sent.

This candidate is added to the list of remote candidates. However, the agent does not pair this candidate with any local candidates.

3.2.7.2.1.4 Triggered Checks Next, the agent constructs a pair whose local candidate is equal to the transport address on which the STUN request was received and a remote candidate equal to the source transport address from which the request came (which may be the peer reflexive remote candidate that was just learned). The local candidate will be either a host candidate (for cases where the request was not received through a relay) or a relayed candidate (for cases where it was received through a relay). The local candidate can never be a server reflexive candidate. Since both candidates are known to the agent, it can obtain their priorities and compute the candidate pair priority. This pair is then looked up in the checklist. There can be one of several outcomes:

- If the pair is already on the checklist:
  - If the state of that pair is Waiting or Frozen, a check for that pair is put into the triggered check queue if not already present.
  - If the state of that pair is In-Progress, the agent cancels the in-progress transaction. Cancellation means that the agent will not retransmit the request, will not treat the lack of response to be a failure, but will wait the duration of the transaction timeout for a response. In addition, the agent MUST create a new connectivity check for that pair (representing a new STUN Binding request transaction) by placing the pair in the triggered check queue. The state of the pair is then changed to Waiting.
  - If the state of the pair is Failed, it is changed to Waiting, and the agent MUST create a new connectivity check for that pair (representing a new STUN Binding request transaction), by placing the pair in the triggered check queue.
  - If the state of that pair is Succeeded, nothing further is done. These steps are done to facilitate rapid completion of ICE when both agents are behind NAT.
- If the pair is not already on the checklist:
  - The pair is inserted into the checklist based on its priority.
  - Its state is set to Waiting.
  - The pair is placed into the triggered check queue.

When a triggered check is to be sent, it is constructed and processed as described in Section 3.2.7.1.2. These procedures require the agent to know the transport address, username fragment, and password for the peer. The username fragment for the remote candidate is equal to the part after the colon of the USERNAME in the Binding request that was just received. Using that username fragment, the agent can check the SDP messages received from its peer (there may be more than one in cases of forking) and find this username fragment. The corresponding password is then selected.

3.2.7.2.1.5 Updating the Nominated Flag If the Binding request received by the agent had the USE-CANDIDATE attribute set, and the agent is in the controlled role, the agent looks at the state of the pair computed in Section 3.2.7.2.1.4:

- If the state of this pair is Succeeded, the check generated by this pair produced a successful response. This would have caused the agent to construct a valid pair when that success response was received (see Section 3.2.7.1.3.2.2). The agent now sets the nominated flag in the valid pair to true. This may end ICE processing for this media stream; see Section 3.2.8.

- If the state of this pair is In-Progress, if its check produces a successful result, the resulting valid pair has its nominated flag set when the response arrives. This may end ICE processing for this media stream when it arrives; see Section 3.2.8.

3.2.7.2.2 Additional Procedures for Lite Implementations

If the check that was just received contained a USE-CANDIDATE attribute, the agent constructs a candidate pair whose local candidate is equal to the transport address on which the request was received and whose remote candidate is equal to the source transport address of the request that was received. This candidate pair is assigned an arbitrary priority and placed into a list of valid candidates called the valid list. The agent sets the nominated flag for that pair to true. ICE processing is considered complete for a media stream if the valid list contains a candidate pair for each component.

3.2.8 Concluding ICE Processing

This section describes how an agent completes ICE.

3.2.8.1 Procedures for Full Implementations

Concluding ICE involves nominating pairs by the controlling agent and updating of state machinery.

3.2.8.1.1 Nominating Pairs

The controlling agent nominates pairs to be selected by ICE by using one of two techniques: regular nomination or aggressive nomination. If its peer has a lite implementation, an agent MUST use a regular nomination algorithm. If its peer is using ICE options (present in an ice-options attribute from the peer) that the agent does not understand, the agent MUST use a regular nomination algorithm. If its peer is a full implementation and isn't using any ICE options or is using ICE options understood by the agent, the agent MAY use either the aggressive or the regular nomination algorithm. However, the regular algorithm is RECOMMENDED since it provides greater stability.

3.2.8.1.1.1 Regular Nomination With regular nomination, the agent lets some number of checks complete, each of which omits the USE-CANDIDATE attribute. Once one or more checks complete successfully for a component of a media stream, valid pairs are generated and added to the valid list. The agent lets the checks continue until some stopping criterion is met and then picks from the valid pairs based on an evaluation criterion. The criterion for stopping the checks and for evaluating the valid pairs is entirely a matter of local optimization.

When the controlling agent selects the valid pair, it repeats the check that produced this valid pair (by putting the pair that generated the check into the triggered check queue), this time with the USE-CANDIDATE attribute. This check should succeed (since the previous did), causing the nominated flag of that and only that pair to be set. Consequently, there will be only a single nominated pair in the valid list for each component, and when the state of the checklist moves to completed, that exact pair is selected by ICE for sending and receiving media for that component.

Regular nomination provides the most flexibility, since the agent has control over the stopping and selection criteria for checks. The only requirement is that the agent MUST eventually pick one and only one candidate pair and generate a check for that pair with the USECANDIDATE attribute present. Regular nomination improves ICE's resilience to variations in implementation (see Section 3.2.14). Regular nomination is also more stable, allowing both agents to converge on a single pair for media without any transient selections, which can happen with the aggressive algorithm. The drawback of regular nomination is that it is guaranteed to increase latencies because it requires an additional check be done.

3.2.8.1.1.2 Aggressive Nomination With aggressive nomination, the controlling agent includes the USE-CANDIDATE attribute in every check it sends. Once the first check for a component succeeds, the component will be added to the valid list and have its nominated flag set. When all components have a nominated pair in the valid list, media can begin to flow using the highest-priority nominated pair. However, because the agent included the USE-CANDIDATE attribute in all of its checks, another check may yet complete, causing another valid pair to have its nominated flag set. ICE always selects the highest-priority nominated candidate pair from the valid list as the one used for media. Consequently, the selected pair may actually change briefly as ICE checks complete, resulting in a set of transient selections until it stabilizes.

3.2.8.1.2 Updating States

For both controlling and controlled agents, the state of ICE processing depends on the presence of nominated candidate pairs in the valid list and on the state of the checklist. Note that, at any time, more than one of the following cases can apply:

- If there are no nominated pairs in the valid list for a media stream and the state of the checklist is Running, ICE processing continues.
- If there is at least one nominated pair in the valid list for a media stream and the state of the checklist is Running
  - The agent MUST remove all Waiting and Frozen pairs in the checklist and triggered check queue for the same component as the nominated pairs for that media stream.
  - If an In-Progress pair in the checklist is for the same component as a nominated pair, the agent SHOULD cease retransmissions for its check if its pair priority is lower than the lowest-priority nominated pair for that component.
- Once there is at least one nominated pair in the valid list for every component of at least one media stream and the state of the checklist is Running
  - The agent MUST change the state of processing for its checklist for that media stream to Completed.
  - The agent MUST continue to respond to any checks it may still receive for that media stream and MUST perform triggered checks if required by the processing of Section 3.2.7.2.
  - The agent MUST continue retransmitting any In-Progress checks for that checklist.
  - The agent MAY begin transmitting media for this media stream as described in Section 3.2.11.1.
- Once the state of each checklist is Completed
  - The agent sets the state of ICE processing overall to Completed.

- If an agent is controlling, it examines the highest-priority nominated candidate pair for each component of each media stream. If any of those candidate pairs differ from the default candidate pairs in the most recent offer/answer exchange, the controlling agent MUST generate an updated offer as described in Section 3.2.9. If the controlling agent is using an aggressive nomination algorithm, this may result in several updated offers as the pairs selected for media change. An agent MAY delay sending the offer for a brief interval (one second is RECOMMENDED) in order to allow the selected pairs to stabilize.

- If the state of the checklist is Failed, ICE has not been able to complete for this media stream. The correct behavior depends on the state of the checklists for other media streams:
  - If all checklists are Failed, ICE processing overall is considered to be in the Failed state, and the agent SHOULD consider the session a failure, SHOULD NOT restart ICE, and the controlling agent SHOULD terminate the entire session.
  - If at least one of the checklists for other media streams is Completed, the controlling agent SHOULD remove the failed media stream from the session in its updated offer.
  - If none of the checklists for other media streams is Completed, but at least one is Running, the agent SHOULD let ICE continue.

3.2.8.2 Procedures for Lite Implementations

Concluding ICE for a lite implementation is relatively straightforward. There are two cases to consider:

The implementation is lite, and its peer is full.

The implementation is lite, and its peer is lite.

The effect of ICE concluding is that the agent can free any allocated host candidates that were not utilized by ICE, as described in Section 3.2.8.3.

3.2.8.2.1 Peer Is Full

In this case, the agent will receive connectivity checks from its peer. When an agent has received a connectivity check that includes the USE-CANDIDATE attribute for each component of a media stream, the state of ICE processing for that media stream moves from Running to Completed. When the state of ICE processing for all media streams is Completed, the state of ICE processing overall is Completed. The lite implementation will never itself determine that ICE processing has failed for a media stream; rather, the full peer will make that determination and then remove or restart the failed media stream in a subsequent offer.

3.2.8.2.2 Peer Is Lite

Once the offer/answer exchange has completed, both agents examine their candidates and those of their peer. For each media stream, each agent pairs up its own candidates with the candidates of its peer for that media stream. Two candidates are paired up when they are for the same component, utilize the same transport protocol (UDP in this specification), and are from the same IP address family (IPv4 or IPv6).

- If there is a single pair per component, that pair is added to the Valid list. If all of the components for a media stream have one pair, the state of ICE processing for that media stream is set to Completed. If all media streams are Completed, the state of ICE processing is set to Completed overall. This will always be the case for implementations that are IPv4 only.

- If there is more than one pair per component
  - The agent MUST select a pair based on local policy. Since this case only arises for IPv6, it is RECOMMENDED that an agent follow the procedures of RFC 3484, made obsolete by RFC 6724, to select a single pair.
  - The agent adds the selected pair for each component to the valid list. As described in Section 3.2.11.1, this will permit media to begin flowing. However, it is possible (and in fact likely) that the two agents have chosen different pairs.
  - To reconcile this, the controlling agent MUST send an updated offer as described in Section 3.1.9.1.3, which will include the remote-candidates attribute.
  - The agent MUST NOT update the state of ICE processing when the offer is sent. If this subsequent offer completes, the controlling agent MUST change the state of ICE processing to Completed for all media streams and the state of ICE processing overall to Completed. The states for the controlled agent are set based on the logic in Section 3.2.9.2.3.

3.2.8.3 Freeing Candidates

3.2.8.3.1 Full Implementation Procedures

The procedures in Section 3.2.8 require that an agent continue to listen for STUN requests and continue to generate triggered checks for a media stream, even once processing for that stream completes. The rules in this section describe when it is safe for an agent to cease sending or receiving checks on a candidate that was not selected by ICE and then free the candidate.

When ICE is used with SIP, and an offer is forked to multiple recipients, ICE proceeds in parallel with and independently of each answerer, all using the same local candidates. Once ICE processing has reached the Completed state for all peers for media streams using those candidates, the agent SHOULD wait an additional three seconds, and then it MAY cease responding to checks or generating triggered checks on those candidates. It MAY free the candidates at that time. Freeing of server reflexive candidates is never explicit; it happens by lack of a keepalive. The three-second delay handles cases in which aggressive nomination is used, and the selected pairs can quickly change after ICE has completed.

3.2.8.3.2 Lite Implementation Procedures

A lite implementation MAY free candidates not selected by ICE as soon as ICE processing has reached the Completed state for all peers for all media streams using those candidates.

3.2.9 Subsequent Offer/Answer Exchanges

Either agent MAY generate a subsequent offer at any time allowed by RFC 3264 (see Section 3.1). The rules in Section 3.2.8 will cause the controlling agent to send an updated offer at the conclusion of ICE processing when ICE has selected different candidate pairs from the default pairs. This section defines rules for construction of subsequent offers and answers. Should a subsequent offer be rejected, ICE processing continues as if the subsequent offer had never been made.

3.2.9.1 Generating the Offer

3.2.9.1.1 Procedures for All Implementations

3.2.9.1.1.1 ICE Restarts An agent MAY restart ICE processing for an existing media stream. An ICE restart, as the name implies, will cause all previous states of ICE processing to be flushed and checks to start anew. The only difference between an ICE restart and a brand new media session is that, during the restart, media can continue to be sent to the previously validated pair.

An agent MUST restart ICE for a media stream if:

- The offer is being generated for the purposes of changing the target of the media stream. In other words, if an agent wants to generate an updated offer that, had ICE not been in use, would result in a new value for the destination of a media component.
- An agent is changing its implementation level. This typically only happens in third-party call control use cases, where the entity performing the signaling is not the entity receiving the media and it has changed the target of media mid-session to another entity that has a different ICE implementation.

These rules imply that setting the IP address in the "c=" line to 0.0.0.0 will cause an ICE restart. Consequently, ICE implementations MUST NOT utilize this mechanism for call hold, and instead MUST use a=inactive and a=sendonly as described in RFC 3264 (see Section 3.1). To restart ICE, an agent MUST change both the ice-pwd and the ice-ufrag for the media stream in an offer. Note that it is permissible to use a session-level attribute in one offer but provide the same ice-pwd or ice-ufrag as a media-level attribute in a subsequent offer. This is not a change in password, just a change in its representation and does not cause an ICE restart.

An agent sets the rest of the fields in the SDP for this media stream as it would in an initial offer of this media stream (see Section 3.2.4.3). Consequently, the set of candidates MAY include some, none, or all of the previous candidates for that stream and MAY include a totally new set of candidates gathered as described in Section 3.2.4.1.1.

3.2.9.1.1.2 Removing a Media Stream If an agent removes a media stream by setting its port to 0, it MUST NOT include any candidate attributes for that media stream and SHOULD NOT include any other ICE-related attributes defined in Section 3.2.15 for that media stream.

3.2.9.1.1.3 Adding a Media Stream If an agent wishes to add a new media stream, it sets the fields in the SDP for this media stream as if this were an initial offer for that media stream (see Section 3.2.4.3). This will cause ICE processing to begin for this media stream.

3.2.9.1.2 Procedures for Full Implementations

This section describes additional procedures for full implementations, covering existing media streams. The username fragments, password, and implementation level MUST remain the same as used previously. If an agent needs to change one of these, it MUST restart ICE for that media stream. Additional behavior depends on the state ICE processing for that media stream.

3.2.9.1.2.1 Existing Media Streams with ICE Running

If an agent generates an updated offer including a media stream that was previously established, and for which ICE checks are in the Running state, the agent follows the procedures defined here.

An agent MUST include candidate attributes for all local candidates it had signaled previously for that media stream. The properties of that candidate as signaled in SDP—the priority, foundation, type, and related transport address—SHOULD remain the same. The IP address, port, and transport protocol, which fundamentally identify that candidate, MUST remain the same (if they change, it would be a new candidate). The component ID MUST remain the same. The agent MAY include additional candidates it did not offer previously, which it has gathered since the last offer/answer exchange, including peer reflexive candidates.

The agent MAY change the default destination for media. As with initial offers, there MUST be a set of candidate attributes in the offer matching this default destination.

3.2.9.1.2.2 Existing Media Streams with ICE Completed

If an agent generates an updated offer including a media stream that was previously established, and for which ICE checks are in the Completed state, the agent follows the procedures defined here.

The default destination for media (i.e., the values of the IP addresses and ports in the "m-" and "c=" lines used for that media stream) MUST be the local candidate from the highest-priority nominated pair in the valid list for each component. This "fixes" the default destination for media to equal the destination ICE has selected for media.

The agent MUST include candidate attributes for candidates matching the default destination for each component of the media stream and MUST NOT include any other candidates.

In addition, if the agent is controlling, it MUST include the a=remote-candidates attribute for each media stream whose checklist is in the Completed state. The attribute contains the remote candidates from the highest-priority nominated pair in the valid list for each component of that media stream. It is needed to avoid a race condition whereby the controlling agent chooses its pairs, but the updated offer beats the connectivity checks to the controlled agent, which doesn't even know these pairs are valid, let alone selected. See Section 3.2.24.6 (that is, Appendix B.6 of RFC 5245) for elaboration on this race condition.

3.2.9.1.3 Procedures for Lite Implementations

3.2.9.1.3.1 Existing Media Streams with ICE Running

This section describes procedures for lite implementations for existing streams for which ICE is running.

A lite implementation MUST include all of its candidates for each component of each media stream in an a=candidate attribute in any subsequent offer. These candidates are formed identically to the procedures for initial offers, as described in Section 3.2.4.2.

A lite implementation MUST NOT add additional host candidates in a subsequent offer. If an agent needs to offer additional candidates, it MUST restart ICE. The username fragments, password, and implementation level MUST remain the same as used previously. If an agent needs to change one of these, it MUST restart ICE for that media stream.

3.2.9.1.3.2 Existing Media Streams with ICE Completed

If ICE has completed for a media stream, the default destination for that media stream MUST be set to the remote candidate of the candidate pair for that component in the valid list. For a lite implementation, there is always just a single candidate pair in the valid list for each component of a media stream. Additionally, the agent MUST include a candidate attribute for each default destination.

Additionally, if the agent is controlling (which only happens when both agents are lite), the agent MUST include the a=remote-candidates attribute for each media stream. The attribute contains the remote candidates from the candidate pairs in the valid list (one pair for each component of each media stream).

3.2.9.2 Receiving the Offer and Generating an Answer

3.2.9.2.1 Procedures for All Implementations

When receiving a subsequent offer within an existing session, an agent MUST reapply the verification procedures in Section 3.2.5.1 without regard for the results of verification from any previous offer/answer exchanges. Indeed, it is possible that a previous offer/answer exchange resulted in ICE not being used, but it is now used because of a subsequent exchange.

3.2.9.2.1.1 Detecting ICE Restart If the offer contained a change in the a=ice-ufrag or a=ice-pwd attributes compared to the previous SDP from the peer, it indicates that ICE is restarting for this media stream. If all media streams are restarting, then ICE is restarting overall.

If ICE is restarting for a media stream

- The agent MUST change the a=ice-ufrag and a=ice-pwd attributes in the answer.
- The agent MAY change its implementation level in the answer.

An agent sets the rest of the fields in the SDP for this media stream as it would in an initial answer to this media stream (see Section 3.2.4.3). Consequently, the set of candidates MAY include some, none, or all of the previous candidates for that stream and MAY include a totally new set of candidates gathered as described in Section 3.2.4.1.1.

3.2.9.2.1.2 New Media Stream If the offer contains a new media stream, the agent sets the fields in the answer as if it had received an initial offer containing that media stream (see Section 3.2.4.3). This will cause ICE processing to begin for this media stream.

3.2.9.2.1.3 Removed Media Stream If an offer contains a media stream whose port is 0, the agent MUST NOT include any candidate attributes for that media stream in its answer and SHOULD NOT include any other ICE-related attributes defined in Section 3.2.15 for that media stream.

3.2.9.2.2 Procedures for Full Implementations

Unless the agent has detected an ICE restart from the offer, the username fragments, password, and implementation level MUST remain the same as used previously. If an agent needs to change one of these, it MUST restart ICE for that media stream by generating an offer; ICE cannot be restarted in an answer. Additional behaviors depend on the state of ICE processing for that media stream.

3.2.9.2.2.1 Existing Media Streams with ICE Running and No Remote-Candidates If ICE is running for a media stream, and the offer for that media stream lacks the remote-candidates attribute, the rules for construction of the answer are identical to those for the offerer as described in Section 3.2.9.1.2.1.

3.2.9.2.2.2 Existing Media Streams with ICE Completed and No Remote-Candidates

If ICE is Completed for a media stream, and the offer for that media stream lacked the remote-candidates attribute, the rules for construction of the answer are identical to those for the offerer as described in Section 3.2.9.1.2.2, except that the answerer MUST NOT include the a=remote-candidates attribute in the answer.

3.2.9.2.2.3 Existing Media Streams and Remote-Candidates

A controlled agent will receive an offer with the a=remote-candidates attribute for a media stream when its peer has concluded ICE processing for that media stream. This attribute is present in the offer to deal with a race condition between the receipt of the offer and the receipt of the Binding response that tells the answerer the candidate that will be selected by ICE. See Section 3.2.24.9 (that is, Appendix B.6 of RFC 5245) for an explanation of this race condition. Consequently, processing of an offer with this attribute depends on the winner of the race.

The agent forms a candidate pair for each component of the media stream by

- Setting the remote candidate equal to the offerer's default destination for that component (e.g., the contents of the "m=" and "c=" lines for RTP and the a=rtcp attribute for RTCP)
- Setting the local candidate equal to the transport address for that same component in the a=remote-candidates attribute in the offer.

The agent then sees if each of these candidate pairs is present in the valid list. If a particular pair is not in the valid list, the check has "lost" the race. Call such a pair a "losing pair."

The agent finds all the pairs in the checklist whose remote candidates equal the remote candidate in the losing pair.

- If none of the pairs is In-Progress and at least one is Failed, it is most likely that a network failure, such as a network partition or serious packet loss, has occurred. The agent SHOULD generate an answer for this media stream as if the remotecandidates attribute had not been present and then restart ICE for this stream.
- If at least one of the pairs is In-Progress, the agent SHOULD wait for those checks to complete and as each completes, redo the processing in this section until there are no losing pairs.

Once there are no losing pairs, the agent can generate the answer. It MUST set the default destination for media to the candidates in the remote-candidates attribute from the offer (each of which will now be the local candidate of a candidate pair in the valid list). It MUST include a candidate attribute in the answer for each candidate in the remote-candidates attribute in the offer.

3.2.9.2.3 Procedures for Lite Implementations

If the received offer contains the remote-candidates attribute for a media stream, the agent forms a candidate pair for each component of the media stream by

- Setting the remote candidate equal to the offerer's default destination for that component (e.g., the contents of the "m=" and "c=" lines for RTP and the a=rtcp attribute for RTCP).
- Setting the local candidate equal to the transport address for that same component in the a=remote-candidates attribute in the offer.

It then places those candidates into the Valid list for the media stream. The state of ICE processing for that media stream is set to Completed.

Furthermore, if the agent believed it was controlling, but the offer contained the remote-candidates attribute, both agents believe they are controlling. In this case, both would have sent updated offers around the same time. However, the signaling protocol carrying the offer/answer exchanges will have resolved this glare condition, so that one agent is always the "winner" by having its offer received before its peer has sent an offer. The winner takes the role of controller, so that the loser (the answerer under consideration in this section) MUST change its role to controlled. Consequently, if the agent was going to send an updated offer, because based on the rules in Section 3.2.8.2.2 it was controlling, it no longer needs to.

Besides the potential role change, change in the Valid list, and state changes, the construction of the answer is performed identically to the construction of an offer as described in Section 3.2.9.1.3.

3.2.9.3 Updating the Check and Valid Lists

3.2.9.3.1 Procedures for Full Implementations

3.2.9.3.1.1 ICE Restarts The agent MUST remember the highest-priority nominated pairs in the Valid list for each component of the media stream, called the previous selected pairs, prior to the restart. The agent will continue to send media using these pairs, as described in Section 3.2.11.1. Once these destinations are noted, the agent MUST flush the valid list and checklist, and then recompute the checklist and its states as described in Section 3.2.5.7.

3.2.9.3.1.2 New Media Stream If the offer/answer exchange added a new media stream, the agent MUST create a new checklist for it (and an empty Valid list to start, of course), as described in Section 3.2.5.7.

3.2.9.3.1.3 Removed Media Stream If the offer/answer exchange removed a media stream, or an answer rejected an offered media stream, an agent MUST flush the Valid list for that media stream. It MUST terminate any STUN transactions in progress for that media stream. An agent MUST remove the checklist for that media stream and cancel any pending ordinary checks for it.

3.2.9.3.1.4 ICE Continuing for Existing Media Stream The valid list is not affected by an updated offer/answer exchange unless ICE is restarting.

If an agent is in the Running state for that media stream, the checklist is updated (the checklist is irrelevant if the state is completed). To do that, the agent recomputes the checklist using the procedures described in Section 3.2.7. If a pair on the new checklist was also on the previous checklist, and its state was Waiting, In-Progress, Succeeded, or Failed, its state is copied over. Otherwise, its state is set to Frozen.

If none of the checklists is active (meaning that the pairs in each checklist are Frozen), the full-mode agent sets the first pair in the checklist for the first media stream to Waiting and then sets the state of all other pairs in that checklist for the same component ID and with the same foundation to Waiting as well.

Next, the agent goes through each checklist, starting with the highest-priority pair. If a pair has a state of Succeeded, and it has a component ID of 1, then all Frozen pairs in the same checklist with the same foundation whose component IDs are not 1 have their states set to Waiting.

If, for a particular checklist, there are pairs for each component of that media stream in the Succeeded state, the agent moves the state of all Frozen pairs for the first component of all other media streams (and thus in different checklists) with the same foundation to Waiting.

3.2.9.3.2 Procedures for Lite Implementations

If ICE is restarting for a media stream, the agent MUST start a new Valid list for that media stream. It MUST remember the pairs in the previous Valid list for each component of the media stream, called the previous selected pairs, and continue to send media there as described in Section 3.2.11.1. The state of ICE processing for each media stream MUST change to Running, and the state of ICE processing MUST change to Running.

3.2.10 Keepalives

All endpoints MUST send keepalives for each media session. These keepalives serve the purpose of keeping NAT bindings alive for the media session. These keepalives MUST be sent regardless of whether the media stream is currently inactive, sendonly, recvonly, or sendrecv, and regardless of the presence or value of the bandwidth attribute. These keepalives MUST be sent even if ICE is not being utilized for the session at all. The keepalive SHOULD be sent using a format that is supported by its peer. ICE endpoints allow for STUN-based keepalives for UDP streams, and as such, STUN keepalives MUST be used when an agent is a full ICE implementation and is communicating with a peer that supports ICE (lite or full). An agent can determine that its peer supports ICE by the presence of a=candidate attributes for each media session. If the peer does not support ICE, the choice of a packet format for keepalives is a matter of local implementation. A format that allows packets to easily be sent in the absence of actual media content is RECOMMENDED. Examples of formats that readily meet this goal are RTP No-Op [1] and, in cases where both sides support it, RTP comfort noise (RFC 3389). If the peer doesn't support any formats that are particularly well suited for keepalives, an agent SHOULD send RTP packets with an incorrect version number or some other form of error that would cause them to be discarded by the peer.

If no packet has been sent on the candidate pair ICE is using for a media component for *Tr* seconds (where packets include those defined for the component (RTP or RTCP) and previous keepalives), an agent MUST generate a keepalive on that pair. *Tr* SHOULD be configurable and SHOULD have a default of 15 seconds. *Tr* MUST NOT be configured to less than 15 seconds. Alternatively, if an agent has a dynamic way to discover the binding lifetimes of the intervening NATs, it can use that value to determine *Tr*. Administrators deploying ICE in more controlled networking environments SHOULD set *Tr* to the longest duration possible in their environment.

If STUN is being used for keepalives, a STUN Binding Indication is used (RFC 5389). The Indication MUST NOT utilize any authentication mechanism. It SHOULD contain the FINGERPRINT attribute to aid in demultiplexing, but SHOULD NOT contain any other attributes. It is used solely to keep the NAT bindings alive. The Binding Indication is sent using the same local and remote candidates that are being used for media. Though Binding Indications are used for keepalives, an agent MUST be prepared to receive a connectivity check as well. If a connectivity check is received, a response is generated as discussed in RFC 5389, but there is no impact on ICE processing otherwise.

An agent MUST begin the keepalive processing once ICE has selected candidates for usage with media, or media begins to flow, whichever happens first. Keepalives end once the session terminates or the media stream is removed.

3.2.11 Media Handling

3.2.11.1 Sending Media

Procedures for sending media differ for full and lite implementations.

3.2.11.1.1 Procedures for Full Implementations

Agents always send media using a candidate pair, called the selected candidate pair. An agent will send media to the remote candidate in the selected pair (setting the destination address and port of the packet equal to that remote candidate) and will send it from the local candidate of the selected pair. When the local candidate is server or peer reflexive, media is originated from the base. Media sent from a relayed candidate is sent from the base through that TURN server, using procedures defined in RFC 5766.

If the local candidate is a relayed candidate, it is RECOMMENDED that an agent create a channel on the TURN server toward the remote candidate. This is done using the procedures for channel creation as defined in Section 11 of RFC 5766.

The selected pair for a component of a media stream is

- Empty if the state of the checklist for that media stream is Running, and there is no previously selected pair for that component due to an ICE restart
- Equal to the previously selected pair for a component of a media stream if the state of the checklist for that media stream is Running, and there was a previously selected pair for that component due to an ICE restart
- Equal to the highest-priority nominated pair for that component in the valid list if the state of the checklist is Completed

If the selected pair for at least one component of a media stream is empty, an agent MUST NOT send media for any component of that media stream. If the selected pair for each component of a media stream has a value, an agent MAY send media for all components of that media stream.

Note that the selected pair for a component of a media stream may not equal the default pair for that same component from the most recent offer/answer exchange. When this happens, the selected pair is used for media, not the default pair. When ICE first completes, if the selected pairs aren't a match for the default pairs, the controlling agent sends an updated offer/answer exchange to remedy this disparity. However, until that updated offer arrives, there will not be a match. Furthermore, in very unusual cases, the default candidates in the updated offer/answer will not be a match.

3.2.11.1.2 Procedures for Lite Implementations

A lite implementation MUST NOT send media until it has a Valid list that contains a candidate pair for each component of that media stream. Once that happens, the agent MAY begin sending media packets. To do that, it sends media to the remote candidate in the pair (setting the destination address and port of the packet equal to that remote candidate) and will send it from the local candidate.

3.2.11.1.3 Procedures for All Implementations

ICE has interactions with jitter buffer adaptation mechanisms. An RTP stream can begin using one candidate and switch to another one, though this happens rarely with ICE. The newer candidate may result in RTP packets taking a different path through the network—one with different delay characteristics. As discussed below, agents are encouraged to readjust jitter buffers when there are changes in the source or destination address of media packets. Furthermore, many audio codecs use the marker bit to signal the beginning of a talkspurt, for the purposes of jitter buffer adaptation. For such codecs, it is RECOMMENDED that the sender set the marker bit (RFC 3550) when an agent switches transmission of media from one candidate pair to another.

3.2.11.2 Receiving Media

ICE implementations MUST be prepared to receive media on each component on any candidates provided for that component in the most recent offer/answer exchange (in the case of RTP, this would include both RTP and RTCP if candidates were provided for both).

It is RECOMMENDED that, when an agent receives an RTP packet with a new source or destination IP address for a particular media stream, the agent readjust its jitter buffers.

Section 8.2 of RFC 3550 describes an algorithm for detecting synchronization source (SSRC) collisions and loops. These algorithms are based, in part, on seeing different source transport addresses with the same SSRC. However, when ICE is used, such changes will sometimes occur as the media streams switch between candidates. An agent will be able to determine that a media stream is from the same peer as a consequence of the STUN exchange that proceeds media transmission. Thus, if there is a change in source transport address, but the media packets come from the same peer agent, this SHOULD NOT be treated as an SSRC collision.

3.2.12 Usage with SIP

3.2.12.1 Latency Guidelines

ICE requires a series of STUN-based connectivity checks to take place between endpoints. These checks start from the answerer on generation of its answer and start from the offerer when it receives the answer. These checks can take time to complete, and as such, the selection of messages to use with offers and answers can affect perceived user latency. Two latency figures are of particular interest. These are the post-pickup delay and the post-dial delay. The post-pickup delay refers to the time between when a user "answers the phone" and when any speech he/she utters can be delivered to the caller. The post-dial delay refers to the time between when a user enters the destination address for the user and ringback begins as a consequence of having successfully started ringing the phone of the called party. Two cases can be considered—one where the offer is present in the initial INVITE and one where it is in a response.

3.2.12.1.1 Offer in INVITE

To reduce post-dial delays, it is RECOMMENDED that the caller begin gathering candidates prior to actually sending its initial INVITE. This can be started upon user interface cues that a call is pending, such as activity on a keypad or the phone going offhook.

If an offer is received in an INVITE request, the answerer SHOULD begin to gather its candidates on receipt of the offer and then generate an answer in a provisional response once it has completed that process. ICE requires that a provisional response with an SDP be transmitted reliably. This can be done through the existing Provisional Response Acknowledgment (PRACK) mechanism (RFC 3262) or through an optimization that is specific to ICE. With this optimization, provisional responses containing an SDP answer that begins ICE processing for one or more media streams can be sent reliably without RFC 3262. To do this, the agent retransmits the provisional response with the exponential backoff timers described in RFC 3262. Retransmits MUST cease on receipt of a STUN Binding request for one of the media streams signaled in that SDP (because receipt of a Binding request indicates the offerer has received the answer) or on transmission of the answer in a 2xx response.

If the peer agent is lite, there will never be a STUN Binding request. In such a case, the agent MUST cease retransmitting the 18x after sending it four times (ICE will actually work even if the peer never receives the 18x; however, experience has shown that sending it is important for middleboxes and firewall traversal). If no Binding request is received prior to the last retransmit, the agent does not consider the session terminated. Despite the reliable delivery of the provisional response, the rules for when an agent can send an updated offer or answer do not change from those specified in RFC 3262. Specifically, if the INVITE contained an offer, the same answer appears in all of the 1xx and 2xx responses to the INVITE. Only after the 2xx has been sent can an updated offer/answer exchange occur. This optimization SHOULD NOT be used if both agents support PRACK. Note that the optimization is very specific to provisional response carrying answers that start ICE processing; it is not a general technique for 1xx reliability.

Alternatively, an agent MAY delay sending an answer until the 200 OK; however, this results in a poor user experience and is NOT RECOMMENDED.

Once the answer has been sent, the agent SHOULD begin its connectivity checks. Once candidate pairs for each component of a media stream enter the valid list, the answerer can begin sending media on that media stream.

However, prior to this point, any media that needs to be sent toward the caller (such as SIP early media (RFC 3960) MUST NOT be transmitted. For this reason, implementations SHOULD delay alerting the called party until candidates for each component of each media stream have entered the valid list. In the case of a Public Switched Telephone Network (PSTN) gateway, this would mean that the setup message into the PSTN is delayed until this point. Doing this increases the post-dial delay, but has the effect of eliminating "ghost rings," cases where the called party hears the phone ring, picks up, but hears nothing and cannot be heard. This technique works without requiring support for, or usage of, preconditions (RFC 3312), since it's a localized decision. It also has the benefit of guaranteeing that not a single packet of media will get clipped, so that post-pickup delay is zero. If an agent chooses to delay local alerting in this way, it SHOULD generate a 180 response once alerting begins.

3.2.12.1.2 Offer in Response

In addition to uses where the offer is in an INVITE, and the answer is in the provisional and/or 200 OK response, ICE works with cases where the offer appears in the response. In such cases, which are common in third-party call control (RFC 3725), ICE agents SHOULD generate their offers in a reliable provisional response (which MUST utilize RFC 3262) and not alert the user on receipt of the INVITE.

The answer will arrive in a PRACK. This allows for ICE processing to take place prior to alerting, so that there is no post-pickup delay, at the expense of increased call setup delays. Once ICE completes, the callee can alert the user and then generate a 200 OK when he/she answers. The 200 OK would contain no SDP, since the offer/answer exchange has completed.

Alternatively, agents MAY place the offer in a 2xx instead (in which case the answer comes in the ACK). When this happens, the callee will alert the user on receipt of the INVITE, and the ICE exchanges will take place only after the user answers. This has the effect of reducing call setup delay but can cause substantial post-pickup delays and media clipping.

3.2.12.2 SIP Option-Tags and Media Feature-Tags

RFC 5768 specifies a SIP option-tag and media feature-tag for usage with ICE. ICE implementations using SIP SHOULD support this specification, which uses a feature-tag in registrations to facilitate interoperability through signaling intermediaries.

3.2.12.3 Interactions with Forking

ICE interacts very well with forking. Indeed, ICE fixes some of the problems associated with forking. Without ICE, when a call forks and the caller receives multiple incoming media streams, it cannot determine which media stream corresponds to which callee. With ICE, this problem is resolved. The connectivity checks that occur prior to transmission of media carry username fragments, which in turn are correlated to specific callees. Subsequent media packets that arrive on the same candidate pair as the connectivity check will be associated with that same callee. Thus, the caller can perform this correlation as long as it has received an answer.

3.2.12.4 Interactions with Preconditions

Quality of Service (QOS) preconditions, which are defined in RFCs 3312 and 4032, apply only to the transport addresses listed as the default targets for media in an offer/answer.

If ICE changes the transport address where media is received, this change is reflected in an updated offer that changes the default destination for media to match ICE's selection. As such, it appears like any other re-INVITE would and is fully treated in RFCs 3312 and 4032, which apply without regard to the fact that the destination for media is changing due to ICE negotiations occurring "in the background." Indeed, an agent SHOULD NOT indicate that QOS preconditions have been met until the checks have completed and selected the candidate pairs to be used for media. ICE also has (purposeful) interactions with connectivity preconditions (RFC 5898). Those interactions are described there.

Note that the procedures in Section 3.2.12.1 describe their own type of "preconditions," albeit with less functionality than those provided by the explicit preconditions in RFC 5898.

3.2.12.5 Interactions with Third-Party Call Control

ICE works with Flows I, III, and IV as described in RFC 3725. Flow I works without the controller supporting or being aware of ICE. Flow IV will work as long as the controller passes along the ICE attributes without alteration. Flow II is fundamentally incompatible with ICE; each agent will believe itself to be the answerer and thus never generate a re-INVITE.

The flows for continued operation, as described in Section 7 of RFC 3725, require additional behavior of ICE implementations to support. In particular, if an agent receives a mid-dialog re-INVITE that contains no offer, it MUST restart ICE for each media stream and go through the process of gathering new candidates. Furthermore, that list of candidates SHOULD include those currently being used for media.

3.2.13 Relationship with Alternative Network Address Types

RFC 4091 (made obsolete by RFC 5245, described here), the Alternative Network Address Types (ANAT) Semantics for the SDP grouping framework, and this RFC 5245, its usage with SIP, define a mechanism for indicating that an agent can support both IPv4 and IPv6 for a media stream; it does so by including two "m=" lines, one for v4 and one for v6. This is similar to ICE, which allows an agent to indicate multiple transport addresses using the candidate attribute. However, ANAT relies on static selection to pick between choices, rather than a dynamic connectivity check used by ICE. This specification (that is, this RFC 5245 described here) deprecates RFCs 4091 and 4092. Instead, agents wishing to support dual stack will utilize ICE.

3.2.14 Extensibility Considerations

This specification makes very specific choices about how both agents in a session coordinate to arrive at the set of candidate pairs that are selected for media. It is anticipated that future specifications will want to alter these algorithms, whether they are simple changes like timer tweaks or larger changes like a revamp of the priority algorithm. When such a change is made, providing interoperability between the two agents in a session is critical.

First, ICE provides the a=ice-options SDP attribute. Each extension or change to ICE is associated with a token. When an agent supporting such an extension or change generates an offer or an answer, it MUST include the token for that extension in this attribute. This allows each side to know what the other side is doing. This attribute MUST NOT be present if the agent doesn't support any ICE extensions or changes. At this time, no IANA registry or registration procedures are defined for these option-tags. At time of writing, it is unclear whether ICE changes and extensions will be sufficiently common to warrant a registry.

One of the complications in achieving interoperability is that ICE relies on a distributed algorithm running on both agents to converge on an agreed-upon set of candidate pairs. If the two agents run different algorithms, it can be difficult to guarantee convergence on the same candidate pairs. The regular nomination procedure described in Section 3.2.8 eliminates some of the tight coordination by delegating the selection algorithm completely to the controlling agent.

Consequently, when a controlling agent is communicating with a peer that supports options it doesn't know about, the agent MUST run a regular nomination algorithm. When regular nomination is used, ICE will converge perfectly even when both agents use different pair-prioritization algorithms. One of the keys to such convergence is triggered checks, which ensure that the nominated pair is validated by both agents. Consequently, any future ICE enhancements MUST preserve triggered checks.

ICE is also extensible to other media streams beyond RTP and for transport protocols beyond UDP. Extensions to ICE for non-RTP media streams need to specify how many components they utilize and assign component IDs to them, starting at 1 for the most important component ID. Specifications for new transport protocols must define how, if at all, various steps in the ICE processing differ from UDP.

3.2.15 Grammar

This specification defines seven new SDP attributes: the "candidate," "remote-candidates," "ice-lite," "ice-mismatch," "iceufrag," "ice-pwd," and "ice-options" attributes. We have included the grammar/ Augmented Backus-Naur Form (ABNF) syntax for these SDP attributes in Section 2.9.2, but we have repeated here because of convenience.

3.2.15.1 "candidate" Attribute

The candidate attribute is a media-level attribute only. It contains a transport address for a candidate that can be used for connectivity checks. The syntax of this attribute is defined using ABNF as defined in RFC 5234:

```
candidate-attribute   = "candidate" ":" foundation SP component-id SP
                        transport SP
                        priority SP
                        connection-address SP ;from RFC 4566
                        port ;port from RFC 4566
                        SP cand-type
                        [SP rel-addr]
                        [SP rel-port]
                        *(SP extension-att-name SP
                             extension-att-value)
foundation            = 1*32ice-char
component-id          = 1*5DIGIT
transport             = "UDP" / transport-extension
transport-extension   = token ; from RFC 3261
priority              = 1*10DIGIT
cand-type             = "typ" SP candidate-types
candidate-types       = "host" / "srflx" / "prflx" / "relay" / token
rel-addr              = "raddr" SP connection-address
rel-port              = "rport" SP port
extension-att-name    = byte-string ;from RFC 4566
extension-att-value   = byte-string
ice-char              = ALPHA / DIGIT / "+" / "/"
```

This grammar encodes the primary information about a candidate: its IP address, port and transport protocol, and its properties—the foundation, component ID, priority, type, and related transport address.

<connection-address>: is taken from RFC 4566 (see Section 2). It is the IP address of the candidate, allowing for IPv4 addresses, IPv6 addresses, and fully qualified domain names (FQDNs). When parsing this field, an agent can differentiate an IPv4 address and an IPv6 address by the presence of a colon in its value; the presence of a colon indicates IPv6. An agent MUST ignore candidate lines that include candidates with IP address versions that are not supported or recognized. An IP address SHOULD be used, but an FQDN MAY be used in place of an IP address. In that case, when receiving an offer or answer containing an FQDN in an a=candidate attribute, the FQDN is looked up in the DNS first using an AAAA record (assuming the agent supports IPv6), and if no result is found or the agent only supports IPv4, using an A. If the DNS query returns more than one IP address, one is chosen and then used for the remainder of ICE processing.

<port>: is also taken from RFC 4566 (see Section 2). It is the port of the candidate.

<transport>: indicates the transport protocol for the candidate. This specification only defines UDP. However, extensibility is provided to allow for future transport protocols to be used with ICE, such as TCP or the Datagram Congestion Control Protocol (RFC 4340).

<foundation>: is composed of 1 to 32 <ice-char>s. It is an identifier equivalent for two candidates that are of the same type, share the same base, and come from the same STUN server. The foundation is used to optimize ICE performance in the Frozen algorithm.

<component-id>: is a positive integer between 1 and 256 that identifies the specific component of the media stream for which this is a candidate. It MUST start at 1 and MUST increment by 1 for each component of a particular candidate. For media streams based on RTP, candidates for the actual RTP media MUST have a component ID of 1, and candidates for RTCP MUST have a component ID of 2. Other types of media streams that require multiple components MUST develop specifications that define the mapping of components to component IDs. See Section 3.2.14 for additional discussion on extending ICE to new media streams.

<priority>: is a positive integer between 1 and (2**31 - 1).

<cand-type>: encodes the type of candidate. This specification defines the values "host," "srflx," "prflx," and "relay" for host, server reflexive, peer reflexive, and relayed candidates, respectively. The set of candidate types is extensible for the future.

<rel-addr> and <rel-port>: convey transport addresses related to the candidate, useful for diagnostics and other purposes. <rel-addr> and <rel-port> MUST be present for server reflexive, peer reflexive, and relayed candidates. If a candidate is server or peer reflexive, <rel-addr> and <rel-port> are equal to the base for that server or peer reflexive candidate. If the candidate is relayed, <rel-addr> and <rel-port> are equal to the mapped address in the Allocate response that provided the client with that relayed candidate (see Section 3.2.24.3, that is, Appendix B.3 of RFC 5245, for a discussion of its purpose). If the candidate is a host candidate, <rel-addr> and <rel-port> MUST be omitted.

The candidate attribute can itself be extended. The grammar allows for new name/value pairs to be added at the end of the attribute. An implementation MUST ignore any name/value pairs it doesn't understand.

3.2.15.2 "remote-candidates" Attribute

The syntax of the "remote-candidates" attribute is defined using ABNF as defined in RFC 5234. The remote-candidates attribute is media level only.

```
remote-candidate-att  = "remote-candidates" ":" remote-candidate
                0*(SP remote-candidate)
remote-candidate  = component-ID SP connection-address SP port
```

The attribute contains a connection-address and port for each component. The ordering of components is irrelevant. However, a value MUST be present for each component of a media stream. This attribute MUST be included in an offer by a controlling agent for a media stream that is Completed and MUST NOT be included in any other case.

3.2.15.3 "ice-lite" and "ice-mismatch" Attributes

The syntax of the "ice-lite" and "ice-mismatch" attributes, both of which are flags, is

```
ice-lite          = "ice-lite"
ice-mismatch      = "ice-mismatch"
```

"ice-lite" is a session-level attribute only and indicates that an agent is a lite implementation. "ice-mismatch" is a media-level attribute only and when present in an answer, indicates that the offer arrived with a default destination for a media component that didn't have a corresponding candidate attribute.

3.2.15.4 "ice-ufrag" and "ice-pwd" Attributes

The "ice-ufrag" and "ice-pwd" attributes convey the username fragment and password used by ICE for message integrity. Their syntax is

```
ice-pwd-att         = "ice-pwd" ":" password
ice-ufrag-att       = "ice-ufrag" ":" ufrag
password            = 22*256ice-char
ufrag               = 4*256ice-char
```

The "ice-pwd" and "ice-ufrag" attributes can appear at either the session level or media level. When present in both, the value in the media level takes precedence. Thus, the value at the session level is effectively a default that applies to all media streams, unless overridden by a media-level value. Whether present at the session or media level, there MUST be an ice-pwd and ice-ufrag attribute for each media stream. If two media streams have identical ice-ufrags, they MUST have identical ice-pwds.

The ice-ufrag and ice-pwd attributes MUST be chosen randomly at the beginning of a session. The ice-ufrag attribute MUST contain at least 24 bits of randomness, and the ice-pwd attribute MUST contain at least 128 bits of randomness. This means that the ice-ufrag attribute will be at least 4 characters long, and the ice-pwd at least 22 characters long, since the grammar for these attributes allows for 6 bits of randomness per character. The attributes MAY be longer than 4 and 22 characters, respectively, of course, up to 256 characters. The upper limit allows for buffer sizing in implementations. Its large upper limit allows for increased amounts of randomness to be added over time.

3.2.15.5 "ice-options" Attribute

"ice-options" is a session-level attribute. It contains a series of tokens that identify the options supported by the agent. Its grammar is

```
ice-options         = "ice-options" ":" ice-option-tag
                      0*(SP ice-option-tag)
ice-option-tag      = 1*ice-char
```

3.2.16 Setting Ta and RTO

During the gathering phase of ICE (Section 3.2.4.1.1) and while ICE is performing connectivity checks (Section 3.2.7), an agent sends STUN and TURN transactions. These transactions are paced at a rate of one every Ta milliseconds and utilize a specific RTO. This section describes how the values of Ta and RTO are computed. This computation depends on whether ICE is used with a real-time media stream (such as RTP). When ICE is used for a stream with a known maximum bandwidth, the computation in Section 3.2.16.1 MAY be followed to rate-control the ICE exchanges. For all other streams, the computation in Section 3.2.16.2 MUST be followed.

3.2.16.1 RTP Media Streams

The values of RTO and Ta change during the lifetime of ICE processing. One set of values applies during the gathering phase and the other, for connectivity checks.

The value of Ta SHOULD be configurable, and SHOULD have a default of, for each media stream *i*

$$Ta\_i = (stun\_packet\_size / rtp\_packet\_size) * rtp\_ptime$$

$$Ta = \max\left[20ms, \frac{1}{\sum_{i=1}^{k} \frac{1}{Ta\_i}}\right]$$

where *k* is the number of media streams. During the gathering phase, *Ta* is computed based on the number of media streams the agent has indicated in its offer or answer, and the RTP packet size and RTP ptime are those of the most preferred codec for each media stream. Once an offer and answer have been exchanged, the agent recomputes *Ta* to pace the connectivity checks. In that case, the value of *Ta* is based on the number of media streams that will actually be used in the session, and the RTP packet size and RTP ptime are those of the most preferred codec with which the agent will send.

In addition, the retransmission timer for the STUN transactions, *RTO*, defined in RFC 5389, SHOULD be configurable and during the gathering phase, SHOULD have a default of

$$RTO = \max(100ms, Ta * (number\ of\ pairs))$$

where the number of pairs refers to the number of pairs of candidates with STUN or TURN servers.

For connectivity checks, RTO SHOULD be configurable and SHOULD have a default of

$$RTO = \max(100ms, Ta * N * (Num-Waiting + Num-In-Progress))$$

where $Num-Waiting$ is the number of checks in the checklist in the Waiting state, and $Num-In-Progress$ is the number of checks in the In-Progress state. Note that the *RTO* will be different for each transaction as the number of checks in the Waiting and In-Progress states change.

These formulas are aimed at causing STUN transactions to be paced at the same rate as media. This ensures that ICE will work properly under the same network conditions needed to support the media as well. See Section 3.2.24.1 (that is, Appendix B.1 of RFC 5245) for additional discussion and motivations. Because of this pacing, it will take a certain amount of time to obtain all of the server reflexive and relayed candidates. Implementations should be aware of the time required to do this and if the application requires a time budget, limit the number of candidates that are gathered.

The formulas result in a behavior whereby an agent will send its first packet for every single connectivity check before performing a retransmit. This can be seen in the formulas for the *RTO* (which represents the retransmit interval). Those formulas scale with *N*, the number of checks to be performed. As a result, ICE maintains a nicely constant rate, but becomes more sensitive to packet loss. The loss of the first single packet for any connectivity check is likely to cause that pair to take a long time to be validated, and instead, a lower-priority check (but one for which there was no packet loss) is much more likely to complete first. This results in ICE performing suboptimally, choosing lower-priority pairs over higher-priority pairs. Implementers should be aware of this consequence, but still should utilize the timer values described here.

3.2.16.2 Non-RTP Sessions

In cases in which ICE is used to establish some kind of session that is not real time, and has no fixed rate associated with it that is known to work on the network in which ICE is deployed, Ta and RTO revert to more conservative values. Ta SHOULD be configurable, SHOULD have a default of 500 ms, and MUST NOT be configurable to be less than 500 ms.

In addition, the retransmission timer for the STUN transactions, RTO, SHOULD be configurable and during the gathering phase SHOULD have a default of

$$RTO = \max(500\ ms, Ta * (number\ of\ pairs))$$

where the number of pairs refers to the number of pairs of candidates with STUN or TURN servers.

For connectivity checks, RTO SHOULD be configurable and SHOULD have a default of

$$RTO = \max(500 \text{ ms}, Ta * N * (Num - Waiting + Num - In\_Progress))$$

3.2.17 Example

This example is based on the simplified topology of Figure 3.9.

Two agents, L and R, are using ICE. Both are full-mode ICE implementations and use aggressive nomination when they are controlling. Both agents have a single IPv4 address. For agent L, it

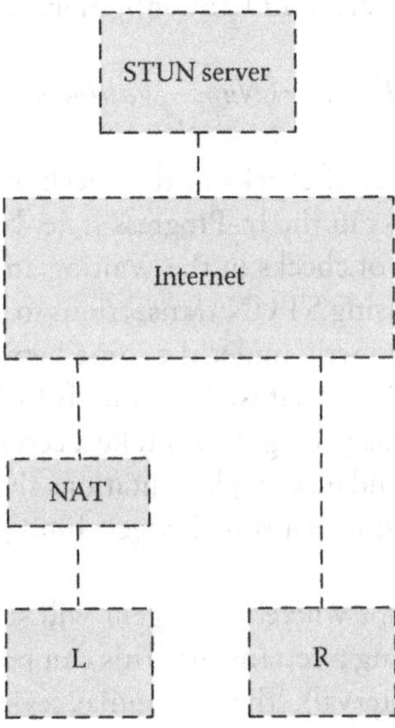

Figure 3.9 Example topology (Copyright: IETF).

is 10.0.1.1 in private address space (RFC 1918), and for agent R, 192.0.2.1 on the public Internet. Both are configured with the same STUN server (shown in this example for simplicity, although in practice the agents do not need to use the same STUN server), which is listening for STUN Binding requests at an IP address of 192.0.2.2 and port 3478. TURN servers are not used in this example. Agent L is behind a NAT, and agent R is on the public Internet. The NAT has an endpoint independent mapping property and an address dependent filtering property. The public side of the NAT has an IP address of 192.0.2.3.

To facilitate understanding, transport addresses are listed using variables that have mnemonic names. The format of the name is entity-type-seqno, where entity refers to the entity whose IP address the transport address is on, and is one of "L," "R," "STUN," or "NAT." The type is either "PUB" for transport addresses that are public or "PRIV" for transport addresses that are private. Finally, seq-no is a sequence number that is different for each transport address of the same type on a particular entity. Each variable has an IP address and port, denoted by varname.IP and varname.PORT, respectively, where varname is the name of the variable.

The STUN server has advertised transport address STUN-PUB-1 (which is 192.0.2.2:3478).

In the call flow itself (Figure 3.10), STUN messages are annotated with several attributes. The "S=" attribute indicates the source transport address of the message. The "D=" attribute indicates the destination transport address of the message. The "MA=" attribute is used in STUN Binding response messages and refers to the mapped address. "USE-CAND" implies the presence of the USE-CANDIDATE attribute.

The call flow (Figure 3.10) examples omit STUN authentication operations and RTCP and focus on RTP for a single media stream between two full implementations.

First, agent L obtains a host candidate from its local IP address (not shown) and from that, sends a STUN Binding request to the STUN server to get a server reflexive candidate (messages F1–F4). Recall that the NAT has the address and port independent mapping property. Here, it creates a binding of NAT-PUB-1 for this UDP request, and this becomes the server reflexive candidate for RTP.

Agent L sets a type preference of 126 for the host candidate and 100 for the server reflexive. The local preference is 65535. Based on this, the priority of the host candidate is 2130706431 and for the server reflexive candidate, 1694498815. The host candidate is assigned a foundation of 1; the server reflexive, a foundation of 2. It chooses its server reflexive candidate as the default candidate and encodes it into the “m=” and “c=” lines. The resulting offer (message F5) looks like (lines folded for clarity)

```
v=0
o=jdoe 2890844526 2890842807 IN IP4 $L-PRIV-1.IP
s=
c=IN IP4 $NAT-PUB-1.IP
t=0 0
a=ice-pwd:asd88fgpdd777uzjYhagZg
a=ice-ufrag:8hhY
m=audio $NAT-PUB-1.PORT RTP/AVP 0
b=RS:0
b=RR:0
a=rtpmap:0 PCMU/8000
a=candidate:1 1 UDP 2130706431 $L-PRIV-1.IP $L-PRIV-1.PORT typ
host
a=candidate:2 1 UDP 1694498815 $NAT-PUB-1.IP $NAT-PUB-1.PORT typ
srflx raddr $L-PRIV-1.IP rport $L-PRIV-1.PORT
```

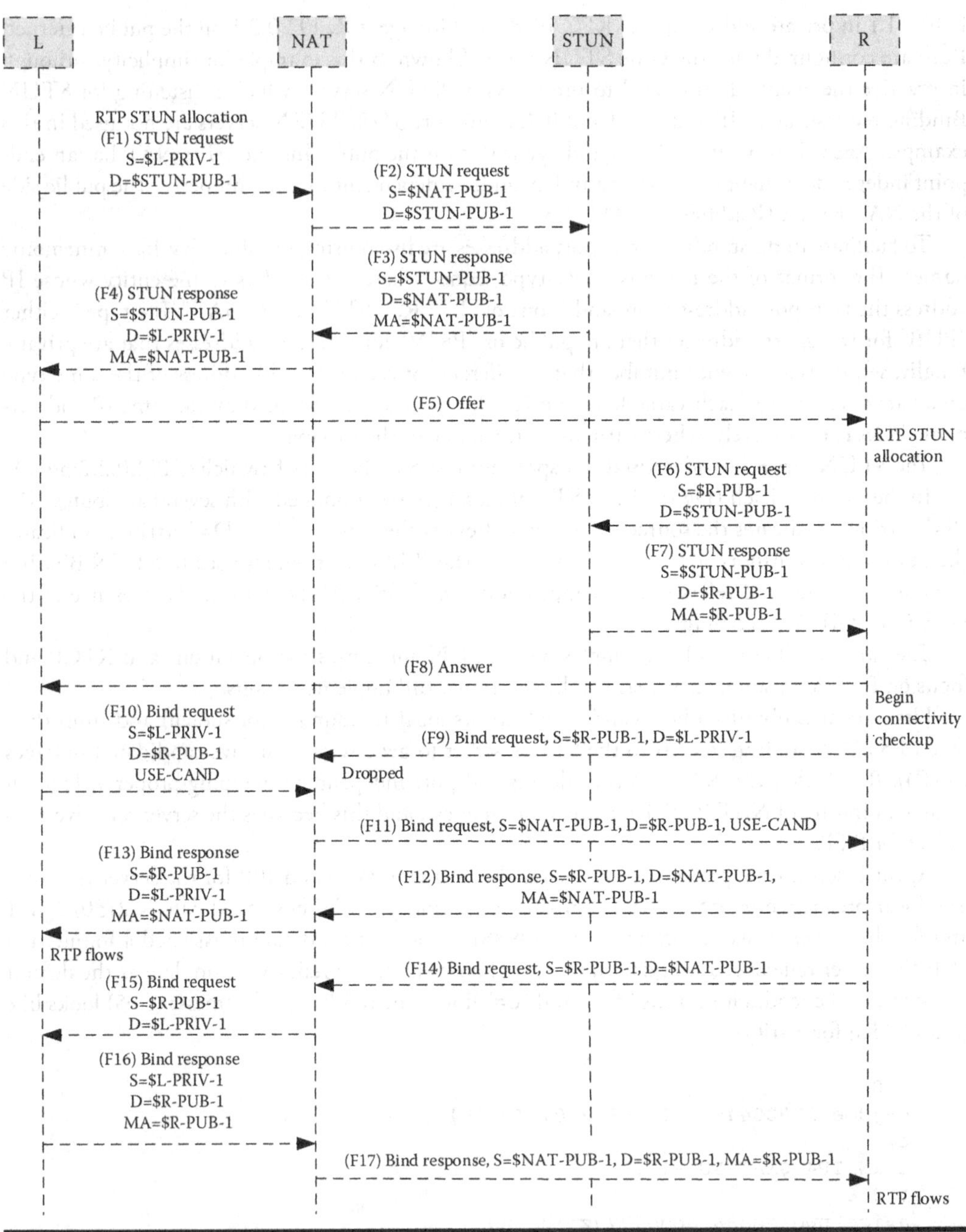

Figure 3.10 Example flow (Copyright: IETF).

The offer, with the variables replaced with their values, will look like (lines folded for clarity)

```
v=0
o=jdoe 2890844526 2890842807 IN IP4 10.0.1.1
s=
```

```
c=IN IP4 192.0.2.3
t=0 0
a=ice-pwd:asd88fgpdd777uzjYhagZg
a=ice-ufrag:8hhY
m=audio 45664 RTP/AVP 0
b=RS:0
b=RR:0
a=rtpmap:0 PCMU/8000
a=candidate:1 1 UDP 2130706431 10.0.1.1 8998 typ host
a=candidate:2 1 UDP 1694498815 192.0.2.3 45664 typ srflx raddr
10.0.1.1 rport 8998
```

This offer is received at agent R. Agent R will obtain a host candidate and from it obtain a server reflexive candidate (messages F6–F7). Since R is not behind a NAT, this candidate is identical to its host candidate, and they share the same base. It therefore discards this redundant candidate and ends up with a single host candidate. With identical type and local preferences as L, the priority for this candidate is 2130706431. It chooses a foundation of 1 for its single candidate. Its resulting answer looks like

```
v=0
o=bob 2808844564 2808844564 IN IP4 $R-PUB-1.IP
s=
c=IN IP4 $R-PUB-1.IP
t=0 0
a=ice-pwd:YH75Fviy6338Vbrhrlp8Yh
a=ice-ufrag:9uB6
m=audio $R-PUB-1.PORT RTP/AVP 0
b=RS:0
b=RR:0
a=rtpmap:0 PCMU/8000
a=candidate:1 1 UDP 2130706431 $R-PUB-1.IP $R-PUB-1.PORT typ host
```

With the variables filled in:

```
v=0
o=bob 2808844564 2808844564 IN IP4 192.0.2.1
s=
c=IN IP4 192.0.2.1
t=0 0
a=ice-pwd:YH75Fviy6338Vbrhrlp8Yh
a=ice-ufrag:9uB6
m=audio 3478 RTP/AVP 0
b=RS:0
b=RR:0
a=rtpmap:0 PCMU/8000
a=candidate:1 1 UDP 2130706431 192.0.2.1 3478 typ host
```

Since neither side indicated that it is lite, the agent that sent the offer that began ICE processing (agent L) becomes the controlling agent.

Agents L and R pair up the candidates. They both initially have two pairs. However, agent L will prune the pair containing its server reflexive candidate, resulting in just one. At agent L, this pair has a local candidate of $L_PRIV_1 and remote candidate of $R_PUB_1, and has a candidate pair priority of 4.57566E+18 (note that an implementation would represent this as a 64-bit integer so as not to lose precision). At agent R, there are two pairs. The highest priority has a local candidate of $R_PUB_1 and remote candidate of $L_PRIV_1 and has a priority of 4.57566E+18, and the second has a local candidate of $R_PUB_1 and remote candidate of $NAT_PUB_1 and priority 3.63891E+18.

Agent R begins its connectivity check (message F9) for the first pair (between the two host candidates). Since R is the controlled agent for this session, the check omits the USE-CANDIDATE attribute. The host candidate from agent L is private and behind a NAT, and thus this check won't be successful, because the packet cannot be routed from R to L.

When agent L gets the answer, it performs its one and only connectivity check (messages F10–F13). It implements the aggressive nomination algorithm and thus includes a USE-CANDIDATE attribute in this check. Since the check succeeds, agent L creates a new pair, whose local candidate is from the mapped address in the Binding response (NAT-PUB-1 from message F13) and whose remote candidate is the destination of the request (R-PUB-1 from message F10). This is added to the valid list. In addition, it is marked as selected since the Binding request contained the USE-CANDIDATE attribute. Since there is a selected candidate in the Valid list for the one component of this media stream, ICE processing for this stream moves into the Completed state. Agent L can now send media if it so chooses.

Soon after receipt of the STUN Binding request from agent L (message F11), agent R will generate its triggered check. This check happens to match the next one on its checklist—from its host candidate to agent L's server reflexive candidate. This check (messages F14–F17) will succeed. Consequently, agent R constructs a new candidate pair using the mapped address from the response as the local candidate (R-PUB-1) and the destination of the request (NAT-PUB-1) as the remote candidate. This pair is added to the Valid list for that media stream. Since the check was generated in the reverse direction of a check that contained the USE-CANDIDATE attribute, the candidate pair is marked as selected. Consequently, processing for this stream moves into the Completed state, and agent R can also send media.

3.2.18 Security Considerations

There are several types of attacks possible in an ICE system. This section considers these attacks and their countermeasures. These countermeasures include

- Using ICE in conjunction with secure signaling techniques, such as SIP Security (SIPS) and
- Limiting the total number of connectivity checks to 100, and optionally limiting the number of candidates they'll accept in an offer or answer.

The details of security for SDP offer/answer model crossing NAT intermediaries using ICE are described in Section 17.3.

3.2.19 STUN Extensions

3.2.19.1 New Attributes

This specification defines four new attributes, PRIORITY, USECANDIDATE, ICE-CONTROLLED, and ICE-CONTROLLING.

The PRIORITY attribute indicates the priority that is to be associated with a peer reflexive candidate, should one be discovered by this check. It is a 32-bit unsigned integer and has an attribute value of 0x0024.

The USE-CANDIDATE attribute indicates that the candidate pair resulting from this check should be used for transmission of media. The attribute has no content (the length field of the attribute is zero); it serves as a flag. It has an attribute value of 0x0025.

The ICE-CONTROLLED attribute is present in a Binding request and indicates that the client believes it is currently in the controlled role. The content of the attribute is a 64-bit unsigned integer in network byte order, which contains a random number used for tiebreaking of role conflicts.

The ICE-CONTROLLING attribute is present in a Binding request and indicates that the client believes it is currently in the controlling role. The content of the attribute is a 64-bit unsigned integer in network byte order, which contains a random number used for tie-breaking of role conflicts.

3.2.19.2 New Error Response Codes

This specification defines a single error response code: 487 (Role Conflict): The Binding request contained either the ICE-CONTROLLING or ICE-CONTROLLED attribute, indicating a role that conflicted with the server. The server ran a tie-breaker based on the tie-breaker value in the request and determined that the client needs to switch roles.

3.2.20 Operational Considerations

This section discusses issues relevant to network operators looking to deploy ICE.

3.2.20.1 NAT and Firewall Types

ICE was designed to work with existing NAT and firewall equipment. Consequently, it is not necessary to replace or reconfigure existing firewall and NAT equipment in order to facilitate deployment of ICE. Indeed, ICE was developed to be deployed in environments in which the Voice over IP (VoIP) operator has no control over the IP network infrastructure, including firewalls and NAT.

That said, ICE works best in environments in which the NAT devices are "behave" compliant, meeting the recommendations defined in RFCs 4787 and 5766. In networks with behave-compliant NAT, ICE will work without the need for a TURN server, thus improving voice quality, decreasing call setup times, and reducing the bandwidth demands on the network operator.

3.2.20.2 Bandwidth Requirements

Deployment of ICE can have several interactions with available network capacity that operators should take into consideration.

3.2.20.2.1 STUN and TURN Server Capacity Planning

First and foremost, ICE makes use of TURN and STUN servers, which would typically be located in the network operator's data centers. The STUN servers require relatively little bandwidth. For each component of each media stream, there will be one or more STUN transactions from each client to the STUN server. In a basic voice-only IPv4 VoIP deployment, there will be four transactions per call (one for RTP and one for RTCP, for both caller and callee). Each transaction is a single request and a single response, the former being 20 bytes long, and the latter, 28. Consequently, if a system has N users, and each makes four calls in a busy hour, this would require $N*$ 1.7 bps. For one million users, this is 1.7 Mbps, a very small number (relatively speaking).

TURN traffic is more substantial. The TURN server will see traffic volume equal to the STUN volume (indeed, if TURN servers are deployed, there is no need for a separate STUN server), in addition to the traffic for the actual media traffic. The amount of calls requiring TURN for media relay is highly dependent on network topologies, and can and will vary over time. In a network with 100% behave-compliant NAT, it is exactly zero. At time of writing, large-scale consumer deployments were seeing between 5% and 10% of calls requiring TURN servers. Considering a voice-only deployment using G.711 (so 80 kbps in each direction), with 0.2 erlangs during the busy hour, this is $N*$ 3.2 kbps. For a population of one million users, this is 3.2 Gbps, assuming a 10% usage of TURN servers.

3.2.20.2.2 Gathering and Connectivity Checks

The process of gathering of candidates and performing of connectivity checks can be bandwidth intensive. ICE has been designed to pace both of these processes. The gathering phase and the connectivity check phase are meant to generate traffic at roughly the same bandwidth as the media traffic itself. This was done to ensure that, if a network is designed to support multimedia traffic of a certain type (voice, video, or just text), it will have sufficient capacity to support the ICE checks for that media. Of course, the ICE checks will cause a marginal increase in the total utilization; however, this will typically be an extremely small increase.

Congestion due to the gathering and check phases has proven to be a problem in deployments that did not utilize pacing. Typically, access links became congested as the endpoints flooded the network with checks as fast as they can send them. Consequently, network operators should make sure that their ICE implementations support the pacing feature. Though this pacing does increase call setup times, it makes the ICE network friendly and easier to deploy.

3.2.20.2.3 Keepalives

STUN keepalives (in the form of STUN Binding Indications) are sent in the middle of a media session. However, they are sent only in the absence of actual media traffic. In deployments that are not utilizing Voice Activity Detection (VAD), the keepalives are never used and there is no increase in bandwidth usage. When VAD is being used, keepalives will be sent during silence periods. This involves a single packet every 15–20 seconds, far less than the packet every 20–30 ms that is sent when there is voice. Therefore, keepalives don't have any real impact on capacity planning.

3.2.20.3 ICE and ICE-lite

Deployments utilizing a mix of ICE and ICE-lite interoperate perfectly. They have been explicitly designed to do so, without loss of function. However, ICE-lite can only be deployed in limited use cases. Those cases, and the caveats involved in doing so, are documented in Section 3.2.23 (that is, Appendix A of RFC 5245).

3.2.20.4 Troubleshooting and Performance Management

ICE utilizes end-to-end connectivity checks and places much of the processing in the endpoints. This introduces a challenge to the network operator—how can they troubleshoot ICE deployments? How can they know how ICE is performing?

ICE has built-in features to help deal with these problems. SIP servers on the signaling path, typically deployed in the data centers of the network operator, will see the contents of the offer/answer exchanges that convey the ICE parameters. These parameters include the type of each candidate (host, server reflexive, or relayed), along with their related addresses. Once ICE processing has completed, an updated offer/answer exchange takes place, signaling the selected address (and its type). This updated re-INVITE is performed exactly for the purposes of educating network equipment (such as a diagnostic tool attached to a SIP server) about the results of ICE processing.

As a consequence, through the logs generated by the SIP server, a network operator can observe what types of candidates are being used for each call and what address was selected by ICE. This is the primary information that helps evaluate how ICE is performing.

3.2.20.5 Endpoint Configuration

ICE relies on several pieces of data being configured into the endpoints. This configuration data includes timers, credentials for TURN servers, and hostnames for STUN and TURN servers. ICE itself does not provide a mechanism for this configuration. Instead, it is assumed that this information is attached to whatever mechanism is used to configure all of the other parameters in the endpoint. For SIP phones, standard solutions such as the configuration framework (RFC 6080) have been defined.

3.2.21 IANA Considerations

This specification registers new SDP attributes, four new STUN attributes, and one new STUN error response. We have not included all these IANA registration here for the sake of brevity. The detail can be found in RFC 5245.

3.2.22 Internet Architecture Board Considerations

The Internet Architecture Board (IAB) has studied the problem of "UNilateral Self-Address Fixing" (UNSAF), which is the general process by which an agent attempts to determine its address in another realm on the other side of a NAT through a collaborative protocol reflection mechanism (RFC 3424). ICE is an example of a protocol that performs this type of function. Interestingly, the process for ICE is not unilateral, but bilateral, and the difference has a significant impact on the issues raised by IAB. Indeed, ICE can be considered a Bilateral Self-Address Fixing protocol, rather than an UNSAF protocol. Regardless, the IAB has mandated that any protocols developed for this purpose document a specific set of considerations. This section meets those requirements.

3.2.22.1 Problem Definition

From RFC 3424, any UNSAF proposal must provide

"Precise definition of a specific, limited-scope problem that is to be solved with the UNSAF proposal. A short-term fix should not be generalized to solve other problems; this is why 'short-term fixes usually aren't'."

The specific problems being solved by ICE are

"Provide a means for two peers to determine the set of transport addresses that can be used for communication. Provide a means for an agent to determine an address that is reachable by another peer with which it wishes to communicate."

3.2.22.2 Exit Strategy

From RFC 3424, any UNSAF proposal must provide

"Description of an exit strategy/transition plan. The better short-term fixes are the ones that will naturally see less and less use as the appropriate technology is deployed."

ICE itself doesn't easily get phased out. However, it is useful even in a globally connected Internet, to serve as a means for detecting whether a router failure has temporarily disrupted connectivity, for example. ICE also helps prevent certain security attacks that have nothing to do with NAT. However, ICE does help phase out other UNSAF mechanisms. ICE effectively selects amongst those mechanisms, prioritizing those that are better and deprioritizing those that are worse. Local IPv6 addresses can be preferred. As NATs begin to dissipate as IPv6 is introduced, server reflexive and relayed candidates (both forms of UNSAF addresses) simply never get used, because higher-priority connectivity exists for the native host candidates. Therefore, the servers get used less and less and can eventually be remove when their usage goes to zero.

Indeed, ICE can assist in the transition from IPv4 to IPv6. It can be used to determine whether to use IPv6 or IPv4 when two dual-stack hosts communicate with SIP (IPv6 gets used). It can also allow a network with both 6to4 and native v6 connectivity to determine which address to use when communicating with a peer.

3.2.22.3 Brittleness Introduced by ICE

From RFC 3424, any UNSAF proposal must provide

"Discussion of specific issues that may render systems more 'brittle'. For example, approaches that involve using data at multiple network layers create more dependencies, increase debugging challenges, and make it harder to transition."

ICE actually removes brittleness from existing UNSAF mechanisms. In particular, classic STUN (as described in RFC 3489 that was made obsolete by RFC 5389) has several points of brittleness. One of them is the discovery process that requires an agent to try to classify behind which type of NAT it is. This process is error-prone. With ICE, that discovery process is simply not used. Rather than unilaterally assessing the validity of the address, its validity is dynamically determined by measuring connectivity to a peer. The process of determining connectivity is very robust.

Another point of brittleness in classic STUN and any other unilateral mechanism is its absolute reliance on an additional server. ICE makes use of a server for allocating unilateral addresses but allows agents to directly connect if possible. Therefore, in some cases, the failure of a STUN server would still allow a call to progress when ICE is used.

Another point of brittleness in classic STUN is that it assumes that the STUN server is on the public Internet. Interestingly, with ICE, that is not necessary. There can be a multitude of STUN servers in a variety of address realms. ICE will discover the one that has provided a usable address.

The most troubling point of brittleness in classic STUN is that it doesn't work in all network topologies. In cases in which there is a shared NAT between each agent and the STUN server, traditional STUN may not work. With ICE, that restriction is removed.

Classic STUN also introduces some security considerations. Fortunately, those security considerations are also mitigated by ICE. Consequently, ICE serves to repair the brittleness introduced in classic STUN and does not introduce any additional brittleness into the system.

The penalty of these improvements is that ICE increases session establishment times.

3.2.22.4 Requirements for a Long-Term Solution

From RFC 3424, any UNSAF proposal must provide: "... requirements for longer term, sound technical solutions -- contribute to the process of finding the right longer term solution. Our conclusions from RFC 3489 (obsoleted by RFC 5890) remain unchanged. However, we feel ICE actually helps because we believe it can be part of the long-term solution."

3.2.22.5 Issues with Existing Network Address Port Translation (NAPT) Boxes

From RFC 3424, any UNSAF proposal must provide:

"Discussion of the impact of the noted practical issues with existing, deployed NA[P]Ts and experience reports."

A number of NAT boxes are now being deployed into the market that try to provide "generic" Application Layer Gateway (ALG) functionality. These generic ALGs hunt for IP addresses, either in text or binary form within a packet, and rewrite them if they match a binding. This interferes with classic STUN. However, the update to STUN (RFC 5389) uses an encoding that hides these binary addresses from generic ALGs.

Existing Network Address Port Translation (NAPT) boxes have nondeterministic and typically short expiration times for UDP-based bindings. This requires implementations to send periodic keepalives to maintain those bindings. ICE uses a default of 15 seconds, which is a very conservative estimate. Eventually, over time, as NAT boxes become compliant to behave (RFC 4787), this minimum keepalive will become deterministic and well known, and the ICE timers can be adjusted. Having a way to discover and control the minimum keepalive interval would be far better still.

3.2.23 Lite and Full Implementations (Appendix A of RFC 5245)

ICE allows for two types of implementations. A full implementation supports the controlling and controlled roles in a session and can perform address gathering. In contrast, a lite implementation is a minimalist implementation that does little but respond to STUN checks.

Because ICE requires both endpoints to support it in order to bring benefits to either endpoint, incremental deployment of ICE in a network is more complicated. Many sessions involve an endpoint that is, by itself, not behind a NAT and not one that would worry about NAT traversal. A very common case is to have one endpoint that requires NAT traversal (such as a VoIP hard phone or soft phone) make a call to one of these devices. Even if the phone supports a full ICE implementation, ICE won't be used at all if the other device doesn't support it. The lite implementation allows for a low-cost entry point for these devices. Once they support the lite implementation, full implementations can connect to them and get the full benefits of ICE.

Consequently, a lite implementation is only appropriate for devices that will *always* be connected to the public Internet and have a public IP address at which it can receive packets from any correspondent. ICE will not function when a lite implementation is placed behind a NAT.

ICE allows a lite implementation to have a single IPv4 host candidate and several IPv6 addresses. In that case, candidate pairs are selected by the controlling agent using a static algorithm, such as the one in RFC 3484, which is recommended by this specification. However, static mechanisms for address selection are always prone to error, since they cannot ever reflect the actual topology and can never provide actual guarantees on connectivity. They are always heuristics. Consequently, if an agent is implementing ICE just to select between its IPv4 and IPv6 addresses, and none of its IP addresses are behind NAT, usage of full ICE is still RECOMMENDED in order to provide the most robust form of address selection possible.

It is important to note that the lite implementation was added to this specification to provide a stepping-stone to full implementation. Even for devices that are always connected to the public Internet with just a single IPv4 address, a full implementation is preferable if achievable. A full implementation will reduce call setup times, since ICE's aggressive mode can be used. Full implementations also obtain the security benefits of ICE unrelated to NAT traversal; in particular, the voice hammer attack described in Section 3.2.18 is prevented only for full implementations, not lite. Finally, it is often the case that a device that finds itself with a public address today will be placed in a network tomorrow where it will be behind a NAT. It is difficult to definitively know, over the lifetime of a device or product, that it will always be used on the public Internet. Full implementation provides assurance that communications will always work.

3.2.24 Design Motivations (Appendix B of RFC 5245)

ICE contains a number of normative behaviors that may themselves be simple but derive from complicated or nonobvious thinking or use cases that merit further discussion. Since these design motivations are not necessary to understand for purposes of implementation, they are discussed here in an appendix to the specification. This section is non-normative.

3.2.24.1 Pacing of STUN Transactions

STUN transactions used to gather candidates and to verify connectivity are paced out at an approximate rate of one new transaction every Ta milliseconds. Each transaction, in turn, has a retransmission timer RTO that is a function of Ta as well. Why are these transactions paced, and why are these formulas used?

Sending of these STUN requests will often have the effect of creating bindings on NAT devices between the client and the STUN servers. Experience has shown that many NAT devices have upper limits on the rate at which they will create new bindings. Experiments have shown that once every 20 ms is well supported, but not much lower than that. This is why Ta has a lower bound of 20 ms. Furthermore, transmission of these packets on the network makes use of bandwidth and needs to be rate limited by the agent. Deployments based on earlier draft versions of this document tended to overload rate-constrained access links and perform poorly overall, in addition to negatively impacting the network. Consequently, the pacing ensures that the NAT device does not get overloaded and that traffic is kept at a reasonable rate.

The definition of a "reasonable" rate is that STUN should not use more bandwidth than the RTP itself will use, once media starts flowing. The formula for Ta is designed so that, if a STUN packet were sent every Ta seconds, it would consume the same amount of bandwidth as RTP packets, summed across all media streams. Of course, STUN has retransmits, and the desire is to pace those as well. For this reason, RTO is set such that the first retransmit on the first transaction happens just as the first STUN request on the last transaction occurs. Pictorially

Three transactions will be sent (e.g., in the case of candidate gathering, there are three host candidate/STUN server pairs). These are transactions A, B, and C. The retransmit timer is set so that the first retransmission on the first transaction (packet A2) is sent at time 3Ta.

Subsequent retransmits after the first will occur even less frequently than Ta milliseconds apart, since STUN uses an exponential back-off on its retransmissions.

3.2.24.2 Candidates with Multiple Bases

Section 3.2.4.1.3 talks about eliminating candidates that have the same transport address and base. However, candidates with the same transport addresses but different bases are not redundant. When can an agent have two candidates that have the same IP address and port, but different bases? Consider the topology of Figure 3.11:

In this case, the offerer is multihomed. It has one IP address, 10.0.1.100, on network C, which is a net 10 private network. The answerer is on this same network. The offerer is also connected to network A, which is 192.168/16. The offerer has an IP address of 192.168.1.100 on this network. There is a NAT on this network, natting into network B, which is another net 10 private network, but not connected to network C. There is a STUN server on network B. The offerer obtains a

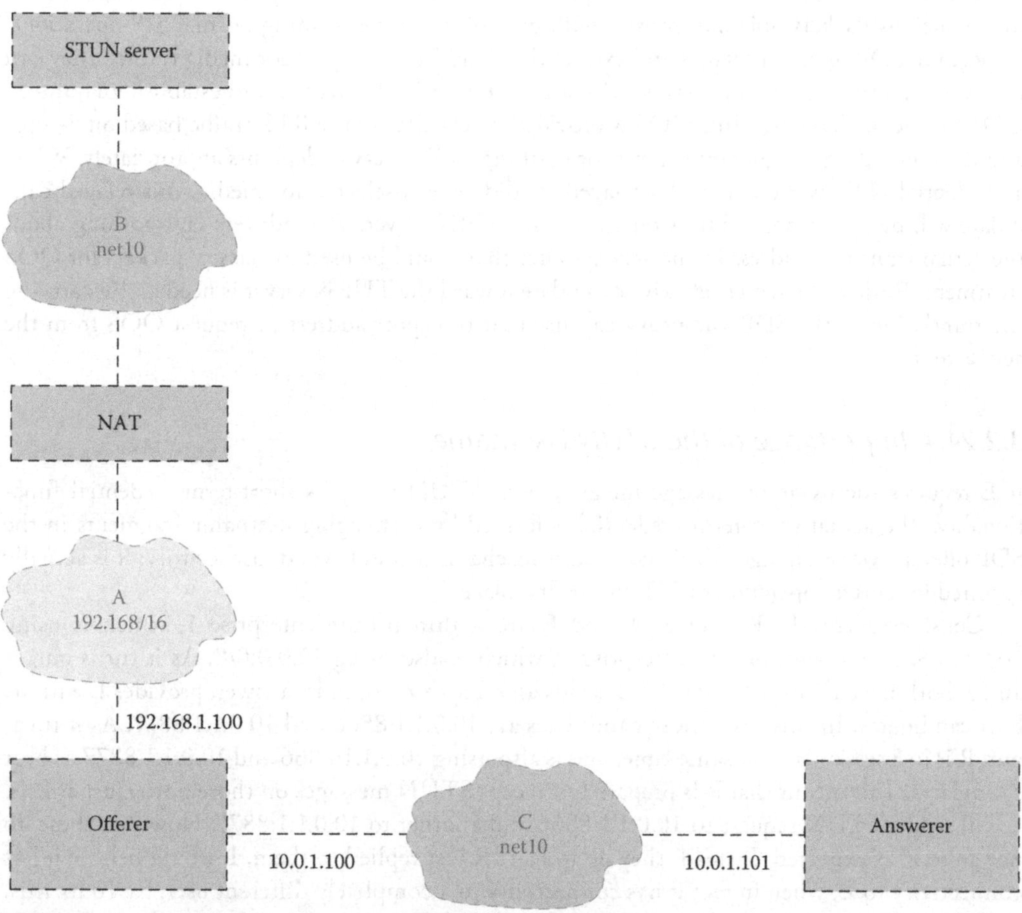

Figure 3.11 Identical candidates with different bases (Copyright: IETF).

host candidate on its IP address on network C (10.0.1.100:2498) and a host candidate on its IP address on network A (192.168.1.100:3344). It performs a STUN query to its configured STUN server from 192.168.1.100:3344. This query passes through the NAT, which happens to assign the binding 10.0.1.100:2498. The STUN server reflects this in the STUN Binding response. Now, the offerer has obtained a server reflexive candidate with a transport address that is identical to a host candidate (10.0.1.100:2498). However, the server reflexive candidate has a base of 192.168.1.100:3344, and the host candidate has a base of 10.0.1.100:2498.

3.2.24.3 Purpose of the <rel-addr> and <rel-port> Attributes

The candidate attribute contains two values that are not used at all by ICE itself: <rel-addr> and <rel-port>. Why are they present?

There are two motivations for their inclusion. The first is diagnostic. It is very useful to know the relationship between the different types of candidates. By including them, an agent can know which relayed candidate is associated with which reflexive candidate, which in turn is associated with a specific host candidate. When checks for one candidate succeed and not for others, this provides useful diagnostics on what is going on in the network.

The second reason has to do with off-path QOS mechanisms. When ICE is used in environments such as PacketCable 2.0, proxies will, in addition to performing normal SIP operations, inspect the SDP in SIP messages and extract the IP address and port for media traffic. They can then interact, through policy servers, with access routers in the network, to establish guaranteed QOS for the media flows. This QOS is provided by classifying the RTP traffic based on 5-tuple and then providing it a guaranteed rate, or marking its DiffServ codepoints appropriately. When a residential NAT is present, and a relayed candidate gets selected for media, this relayed candidate will be a transport address on an actual TURN server. That address says nothing about the actual transport address in the access router that would be used to classify packets for QOS treatment. Rather, the server reflexive candidate toward the TURN server is needed. By carrying the translation in the SDP, the proxy can use that transport address to request QOS from the access router.

3.2.24.4 Importance of the STUN Username

ICE requires the usage of message integrity with STUN using its short-term credential functionality. The actual short-term credential is formed by exchanging username fragments in the SDP offer/answer exchange. The need for this mechanism goes beyond just security; it is actually required for correct operation of ICE in the first place.

Consider agents L, R, and Z. L and R are within private enterprise 1, which is using 10.0.0.0/8. Z is within private enterprise 2, which is also using 10.0.0.0/8. As it turns out, R and Z both have IP address 10.0.1.1. L sends an offer to Z. Z, in its answer, provides L with its host candidates. In this case, those candidates are 10.0.1.1:8866 and 10.0.1.1:8877. As it turns out, R is in a session at that same time, and is also using 10.0.1.1:8866 and 10.0.1.1:8877 as host candidates. This means that R is prepared to accept STUN messages on those ports, just as Z is. L will send a STUN request to 10.0.1.1:8866 and another to 10.0.1.1:8877. However, these do not go to Z as expected. Instead, they go to R! If R just replied to them, L would believe it has connectivity to Z, when in fact it has connectivity to a completely different user, R. To fix this, the STUN short-term credential mechanisms are used. The username fragments are sufficiently random that it is highly unlikely that R would be using the same values as Z. Consequently, R would reject the STUN request since the credentials were invalid. In essence, the STUN username fragments provide a form of transient host identifiers, bound to a particular offer/answer session.

An unfortunate consequence of the nonuniqueness of IP addresses is that, in this example, R might not even be an ICE agent. It could be any host, and the port to which the STUN packet is directed could be any ephemeral port on that host. If there is an application listening on this socket for packets, and it is not prepared to handle malformed packets for whatever protocol is in use, the operation of that application could be affected. Fortunately, since the ports exchanged in SDP are ephemeral and usually drawn from the dynamic or registered range, the odds are good that the port is not used to run a server on host R, but rather is the agent side of some protocol. This decreases the probability of hitting an allocated port, due to the transient nature of port usage in this range.

However, the possibility of a problem does exist, and network deployers should be prepared for it. Note that this is not a problem specific to ICE; stray packets can arrive at a port at any time for any type of protocol, especially ones on the public Internet. As such, this requirement is just restating a general design guideline for Internet applications: Be prepared for unknown packets on any port.

3.2.24.5 The Candidate Pair Priority Formula

The priority for a candidate pair has an odd form. It is

$$\text{pair priority} = 2 \wedge 32 * \min(G, D) + 2 * \max(G, D) + (G > D ? 1 : 0)$$

Why is this? When the candidate pairs are sorted based on this value, the resulting sorting has the MAX/MIN property. This means that the pairs are first sorted by decreasing value of the minimum of the two priorities. For pairs that have the same value of the minimum priority, the maximum priority is used to sort them. If the max and the min priorities are the same, the controlling agent's priority is used as the tie-breaker in the last part of the expression. The factor of 2*32 is used since the priority of a single candidate is always less than 2*32, resulting in the pair priority being a "concatenation" of the two component priorities. This creates the MAX/MIN sorting. MAX/MIN ensures that, for a particular agent, a lower-priority candidate is never used until all higher-priority candidates have been tried.

3.2.24.6 The Remote-Candidates Attribute

The a=remote-candidates attribute exists to eliminate a race condition between the updated offer and the response to the STUN Binding request that moved a candidate into the Valid list. This race condition is shown in Figure 3.12. On receipt of message F4, agent L adds a candidate pair to the valid list. If there was only a single media stream with a single component, agent L could now send an updated offer. However, the check from agent R has not yet generated a response, and agent R receives the updated offer (message F7) before getting the response (message F9). Thus, it does not yet know that this particular pair is valid. To eliminate this condition, the actual candidates at R that were selected by the offerer (the remote candidates) are included in the offer itself, and the answerer delays its answer until those pairs validate.

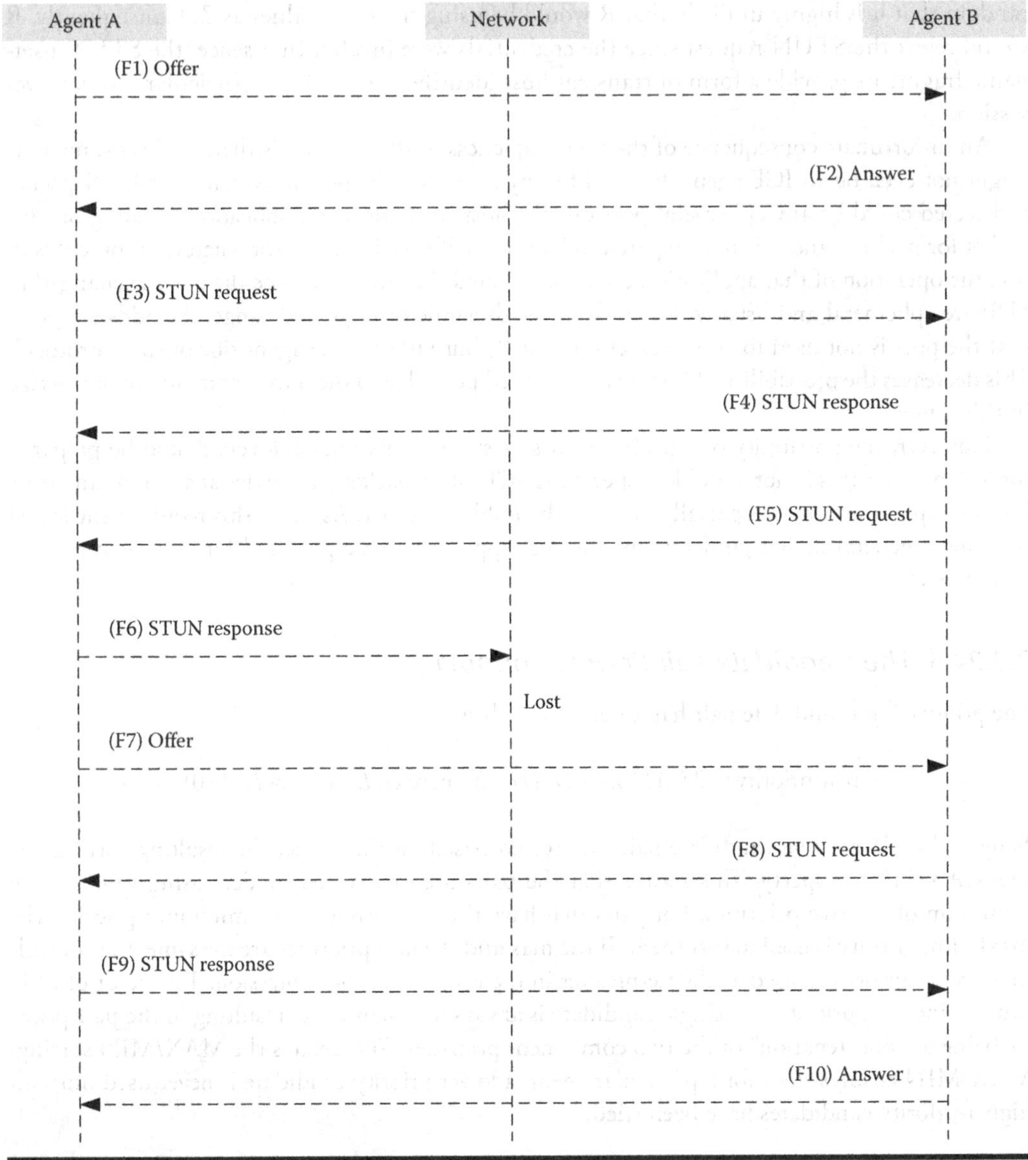

Figure 3.12 Race condition flow (Copyright: IETF).

3.2.24.7 Why Are Keepalives Needed?

Once media begins flowing on a candidate pair, it is still necessary to keep the bindings alive at intermediate NATs for the duration of the session. Normally, the media stream packets themselves (e.g., RTP) meet this objective. However, several cases merit further discussion. First, in some RTP usages, such as SIP, the media streams can be "put on hold." This is accomplished by using the SDP "sendonly" or "inactive" attributes, as defined in RFC 3264. RFC 3264 directs implementations to cease transmission of media in these cases. However, doing so may cause NAT bindings to time out, and media won't be able to come off hold.

Second, some RTP payload formats, such as the payload format for text conversation (RFC 4103), may send packets so infrequently that the interval exceeds the NAT binding timeouts.

Third, if silence suppression is in use, long periods of silence may cause media transmission to cease sufficiently long for NAT bindings to time out.

For these reasons, the media packets themselves cannot be relied upon. ICE defines a simple periodic keepalive utilizing STUN Binding indications. This makes its bandwidth requirements highly predictable and thus amenable to QOS reservations.

3.2.24.8 Why Prefer Peer Reflexive Candidates?

Section 3.2.4.1.2 describes procedures for computing the priority of candidate based on its type and local preferences. That section requires that the type preference for peer reflexive candidates always be higher than server reflexive. Why is that? The reason has to do with the security considerations in Section 3.2.18. It is much easier for an attacker to cause an agent to use a false server reflexive candidate than it is for an attacker to cause an agent to use a false peer reflexive candidate. Consequently, attacks against address gathering with Binding requests are thwarted by ICE by preferring the peer reflexive candidates.

3.2.24.9 Why Send an Updated Offer?

Section 3.2.11.1 describes rules for sending media. Both agents can send media once ICE checks complete, without waiting for an updated offer. Indeed, the only purpose of the updated offer is to "correct" the SDP so that the default destination for media matches where media is being sent based on ICE procedures (which will be the highest priority nominated candidate pair).

This begs the question: Why is the updated offer/answer exchange needed at all? Indeed, in a pure offer/answer environment, it would not be. The offerer and answerer will agree on the candidates to use through ICE and then can begin using them. As far as the agents themselves are concerned, the updated offer/answer provides no new information. However, in practice, numerous components along the signaling path look at the SDP information. These include entities performing off-path QOS reservations, NAT traversal components such as ALGs and Session Border Controllers (SBCs), and diagnostic tools that passively monitor the network. For these tools to continue to function without change, the core property of SDP—that the existing, pre-ICE definitions of the addresses used for media (the "m=" and "c=" lines and the rtcp attribute)—must be retained. For this reason, an updated offer must be sent.

3.2.24.10 Why Are Binding Indications Used for Keepalives?

Media keepalives are described in Section 3.2.10. These keepalives make use of STUN when both endpoints are ICE capable. However, rather than using a Binding request transaction (which generates a response), the keepalives use an Indication. Why is that?

The primary reason has to do with network QOS mechanisms. Once media begins flowing, network elements will assume that the media stream has a fairly regular structure, making use of periodic packets at fixed intervals, with the possibility of jitter. If an agent is sending media packets, and then receives a Binding request, it would need to generate a response packet along with its media packets. This will increase the actual bandwidth requirements for the 5-tuple carrying the media packets and introduce jitter in the delivery of those packets. Analysis has shown that this is a concern in certain layer 2 access networks that use fairly tight packet schedulers for media.

Additionally, using a Binding Indication allows integrity to be disabled, allowing for better performance. This is useful for large-scale endpoints, such as PSTN gateways and SBCs.

3.2.24.11 Why Is the Conflict Resolution Mechanism Needed?

When ICE runs between two peers, one agent acts as controlled and the other as controlling. Rules are defined as a function of implementation type and offerer/answerer to determine who is controlling and who is controlled. However, the specification mentions that, in some cases, both sides might believe they are controlling, or both sides might believe they are controlled. How can this happen?

The condition when both agents believe they are controlled shows up in third-party call control cases. Consider the following flow (Figure 3.13):

This flow is a variation on flow III of RFC 3725. In fact, it works better than flow III since it produces fewer messages. In this flow, the controller sends an offerless INVITE to agent A, which responds with its offer, SDP1. The agent then sends an offerless INVITE to agent B, to which it responds with its offer, SDP2. The controller then uses the offer from each agent to generate the answers. When this flow is used, ICE will run between agents A and B, but both will believe they are in the controlling role. With the role conflict resolution procedures, this flow will function properly when ICE is used. At this time, there are no documented flows that can result in the case where both agents believe they are controlled. However, the conflict resolution procedures allow for this case, should a flow arise that would fit into this category.

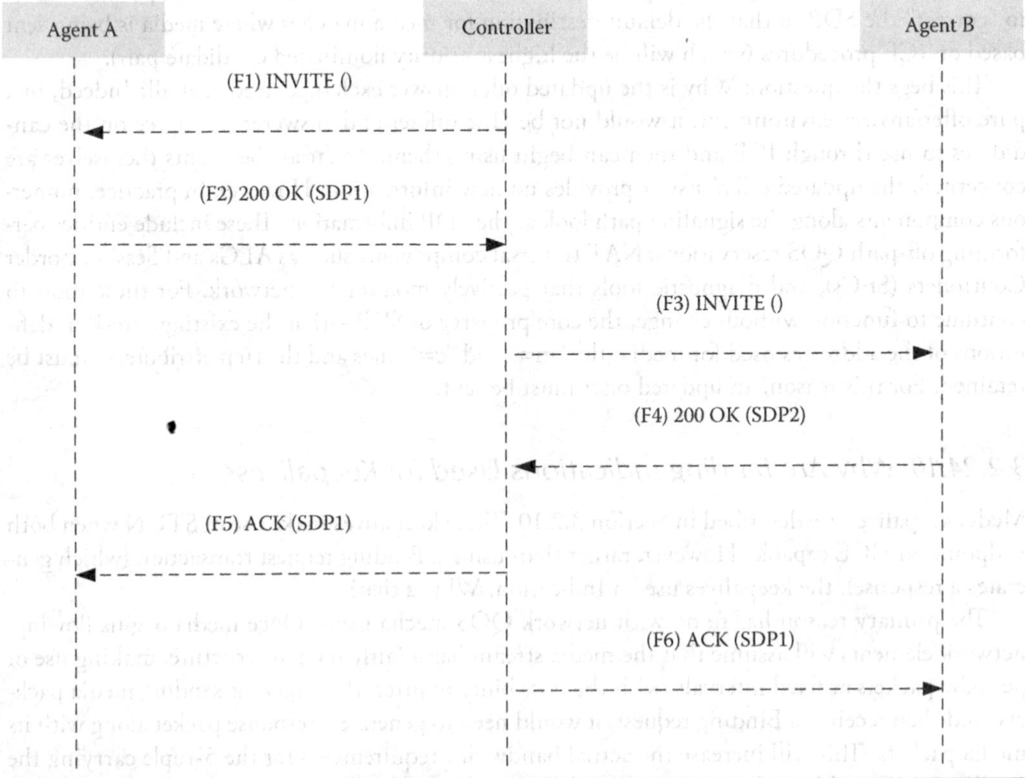

Figure 3.13 Role conflict flow (Copyright: IETF).

3.3 Summary

We have described the most fundamental mechanism of SDP known as the offer/answer model for capability negotiations among the conferencing parties. It can be seen that the basic SDP specification defined in RFC 4566 (see Section 2) does not have any mechanisms for capability negotiations other than the expressions of basic descriptive attributes related to session metadata, stream, QOS/grouping, network, and security. This is because the original SDP specification is devised for supporting the Session Announcement Protocol (SAP) (RFC2974) for one-way multicast/broadcast mode of communications and did not envision that negotiations of capabilities are needed to have common agreed-upon session parameters.

A key observation is that the context of the session negotiation is being maintained by the call control protocol such as SIP and WebRTC that carries SDP offers and answers. It means that acceptance or rejection of any offer is decided by the call control application. In addition, if there is any glare in SDP offer/answer, the upper-layer call control application needs to mitigate it. We have explained the rules of offer/answer model: An agent MAY generate any new offer that updates the session, but MUST NOT generate a new offer if it has received an offer for which it has not yet answered or rejected. Similarly, an agent MUST NOT generate a new offer if it has generated a prior offer for which it has not yet received an answer or a rejection. If an offer is rejected, the system reverses to its original state.

The rules for generation of the initial offer and for creation of answer for both unicast and multicast of a multimedia call are discussed including offerer processing of the answer. At any time any agent MAY modify the session with adding, removing, and modifying a media stream. However, almost all characteristics of a media stream can be modified: address, port, and transport, media format, media type, and any other attributes of a media description. The offer/answer can also be used to keep a media on hold for a unicast media stream. Finally, we have described how to indicate capabilities in the SDP offer/answer model and have provided some examples of offer/answer exchanges.

Finally, we have described UDP-based multimedia session negotiations over NATs that are the most troublesome intermediary entities between end-to-end paths using the SDP offer/answer model. However, it requires the SDP offer/answer model to use ICE protocol. In turn, ICE utilizes the STUN and TURN protocols. Again, as an example, SIP has been used as the call control protocol in conjunction with SDP offer/answer, ICE, STUN, and TURN in performing the session establishment in crossing NATs intermediaries that remain in both call signaling and media end-to-end paths.

3.4 Problems

1. What are the characteristics of the unicast and multicast session? Explain how SDP is used by SAP for session descriptions of the unicast and multicast sessions. How are the SDP session parameters decided between the communicating parties in SAP?
2. What has been missing in SDP capabilities prior to RFC 3264 (see Section 3.1) for session negotiations between conferencing parties?
3. What is meant by offer/answer in SDP and why it is needed? Explain briefly why SDP offer/answer model is needed for both unicast and multicast sessions.
4. How does a glare happen in SDP session negotiations? Describe the glare situation using hypothetical call flows. What is the role of the upper-layer call control application (e.g., SIP or WebRTC) in mitigating glare?

5. Explain the SDP offer protocol operation for both unicast and multicast sessions using hypothetical call flows. What role does an upper-layer call control protocol (e.g., SIP or WebRTC) in offer mode?
6. Explain the SDP answer protocol operation for both unicast and multicast sessions using hypothetical call flows. What role does an upper-layer call control protocol (e.g., SIP or WebRTC) play in answer mode?
7. How can a session be modified in adding and deleting media using the offer/answer model? Use hypothetical call flows in explaining adding and deleting a media.
8. Explain how a given media stream in an existing session can be modified using the offer/answer model. Illustrate all characteristics of the media stream that can be modified.
9. Describe how a unicast media stream of a session can be put on hold using a hypothetical example.
10. Provide a brief description of how session capabilities can be indicated in SDP offer/answer. Use some example offer/answer exchanges that show how the session capabilities are indicated and eventually negotiated between parties of a unicast session.
11. Explain the characteristics of different NATs. What are the basic problems of NAT crossing by UDP-based multimedia sessions? Articulate the problems of the NAT crossing by both call signaling and media.
12. What are the characteristics of ICE protocol if we deploy ICE assuming that there will be NAT intermediaries between source-destination paths? Why do we need the ICE protocol to be used by call control protocol (e.g., SIP) and the SDP offer/answer model for the NAT crossing by UDP-based multimedia applications for session negotiations?
13. How do we formulate the initial SDP offer assuming that NATs will remain as intermediaries for both signaling and media paths in terms of (a) gathering candidates, (b) prioritizing them, (c) eliminating redundant candidates, (d) choosing default candidates, and then (e) formulating and sending the SDP offer?
14. What does the agent do when it receives an initial offer assuming NATs are in the end-to-end path in terms of checking whether the offerer supports ICE, determining its own role, gathering candidates, prioritizing them, choosing default candidates, encoding and sending an answer, and for full implementations, forming the checklists and beginning connectivity checks?
15. Explain in detail what an agent does when it receives the answer from the peer in terms of (a) verifying that its peer supports ICE, (b) determining its role, and (c) for full/lite implementations, forming the checklist and (d) beginning performing ordinary checks.
16. What are the roles of the STUN client and server in performing connectivity checks for both ICE full/lite implementation?
17. What are the roles of the ICE controller for processing signaling messages and freeing candidates in terms of full/lite implementation?
18. How are the SDP offer/answer messages exchanged once initial procedures (Questions 13 through 17) are completed in terms of generating the offer, receiving the offer and generating an answer, updating the check, and validating the list for both full/lite ICE implementation?
19. Why is the keepalive message needed for each media session when end-to-end multimedia sessions are established over NAT intermediaries?
20. Describe how the keepalive message is used for operation/management of multimedia sessions that are established using SDP offer/answer, ICE, and STUN over the NAT intermediaries.

21. Describe how media is handled for multimedia sessions that are established using SDP offer/answer, ICE, and STUN over the NAT intermediaries.
22. Explain in detail the multimedia session establishment and operation/maintenance using SIP call control protocol using SDP offer/answer, ICE, and STUN over the NAT intermediaries.
23. Explain the relationship between ANAT and ICE.
24. Can you explain why and how ICE and STUN can be extended in the future as some of the algorithms used in ICE may not be sufficient in meeting new requirements?
25. Provide detailed call flow examples using a simple topology assuming that two agents are establishing connection over NATs using a STUN server including setting performance parameters for RTP media streams. How does it work over non-RTP media?
26. What problem did the IAB study for UNSAF, and how does it elate ICE? How does ICE differ from this study? Explain the present issues with NAPT boxes.
27. Explain differences between ICE full and lite implementation.
28. What is the design motivation of ICE/STUN in terms of (a) Pacing of STUN Transactions, (b) Candidates with Multiple Bases, (c) Purpose of the <rel-addr> and <rel-port> Attributes, (d) Importance of the STUN Username, (e) Candidate Pair Priority Formula, (f) Remote-candidates Attribute, (g) Necessity for Keepalives, (h) Preference for Peer Reflexive Candidates, (i) Justification for Sending an Updated Offer, (j) Binding Indications Used for Keepalives, and (k) needs for Conflict Resolution Mechanism?

Reference

1. F. Andreasen, D. Oran, and D. Wing, "A No-Op Payload Format for RTP," draft-ietf-avt-rtp-no-op-04, IETF Draft (Work in Progress), November 22, 2007.

Chapter 4

Capability Declaration in SDP

Abstract
The capability declaration in Session Description Protocol (SDP) deals with certain attributes that enable it to provide a minimal and backwards compatible capability declaration mechanism because a session between conferencing parties needs to be established with a minimum set of common capabilities. Such a capability declaration can be used as an input to a subsequent session negotiation, which is done by means outside the scope of Request for Comments (RFC) 3407. This provides a simple and limited solution to the general capability negotiation problem being addressed by the SDP defined in RFC 3407.

4.1 Simple Capability Declaration

This subsection describes Request for Comments (RFC) 3407 that defines the Session Description Protocol (SDP) simple clarity declaration (simcap). RFC 3407 defines a minimal and backwards compatible capability declaration feature in SDP by defining a set of new SDP attributes.

4.1.1 Convention Used in Specification

The convention used here is as per RFC 2026.

4.1.2 Introduction

The SDP (RFC 4566, see Section 2) describes multimedia sessions for the purposes of session announcement, session invitation, and other forms of multimedia session initiation. SDP was not intended to provide capability negotiation. However, several other protocols [e.g., Session Initiation Protocol (SIP)] (RFC 3261; see Section 1.1) and Media Gateway Control Protocol (RFC 2705) use SDP and are likely to continue doing so for the foreseeable future. Nevertheless, in many cases these signaling protocols have an urgent need for some limited form of capability negotiation.

An endpoint may support G.711 audio [over Real-Time Transport Protocol (RTP)] as well as T.38 fax relay [over User Datagram Protocol (UDP) or Transmission Control Protocol (TCP)]. Unless the endpoint is willing to support two media streams at the same time, this cannot currently be expressed in SDP. Another example involves support for multiple codecs. An endpoint indicates this by including all the codecs in the "m=" line in the session description. However, the endpoint thereby also commits to simultaneous support for each of these codecs. In practice, Digital Signal Processor memory and processing power limitations may not make this feasible. The problem with SDP is that media descriptions are used for session parameters as well as capabilities without a clear distinction between the two.

In this specification, we define a minimal and backwards compatible capability declaration feature in SDP by defining a set of new SDP attributes. Together, these attributes define a capability set that consists of a capability set sequence number followed by one or more capability descriptions. Each capability description in the set contains information about supported media formats, but the endpoint does not commit to using any of these. In order to actually use a declared capability, session negotiation has to be done by means outside the scope of this specification, for example, using the offer/answer model (RFC 3264, also see Section 3.1). It should be noted that the mechanism is not intended to solve the general capability negotiation problem targeted by RFC 4566 (see Section 2). It is merely intended as a simple and limited solution to the most urgent problems facing current users of SDP.

4.1.3 Simcap Attributes

The SDP simcap is defined by a set of SDP attributes. Together, these attributes form a capability set which describes the complete media capabilities of the endpoint. Any previous capability sets issued by the endpoint for the session in question no longer apply. The capability set consists of a sequence number and one or more capability descriptions. Each such capability-description describes the media type and media formats supported and may include one or more capability parameters to further define the capability. A session description MUST NOT contain more than one capability set, however the capability set can describe capabilities at both the session and media level. Capability descriptions provided at the session level apply to all media streams of the media type indicated, whereas capability descriptions provided at the media level apply to that particular media stream only. We refer to these respectively as session capabilities and media stream capabilities.

A media stream capability may or may not be of the same media type as the media stream to which it applies. The capability set MUST begin with a single sequence number followed by one or more capability descriptions listing all media formats the endpoint is currently able and willing to support. More specifically, if a media format is included in a media ("m=") line, then by definition the media format MUST be included in either a session capability or a media stream capability for that media line. The endpoint MAY include additional media formats in a capability if it is capable of supporting those media formats in a session with its peer. An endpoint MUST NOT include capabilities it knows it cannot use in a particular session. An endpoint receiving a capability set from another endpoint MAY use any of the media formats included in that capability set in a later attempt to negotiate media streams with the other endpoint (e.g., using the offer/answer model) (RFC 3264, see Section 3.1).

If a new capability set is received from the other endpoint, the old capability set MUST NOT continue to be used. Session capabilities can be used for any media streams of the indicated media type, whereas media stream capabilities can only be used for their associated media line. However, an endpoint receiving a capability set with a given media format MUST NOT assume that a subsequent attempt to negotiate a media stream using just this media format will succeed. The individual capability descriptions in a capability set can be provided contiguously, or they can be scattered throughout the session description. The first capability description, however, MUST follow immediately after the sequence number. The sequence number is on the form:

```
a=sqn: <sqn-num>
```

where <sqn-num> is an integer between 0 and 255 (both included). The initial sequence number MUST be 0 (zero), and it MUST be incremented by 1 modulo 256 with each new capability set issued by the endpoint. Receivers may not necessarily see all capability sets issued and hence MUST NOT reject a capability set due to gaps in sequence numbers. The sequence number MUST either be provided as a session-level or media-level attribute; however, there MUST NOT be more than one occurrence of the sequence number attribute in the session description (since there cannot be more than one capability set). Each capability description in the capability set is on the form:

```
a=cdsc: <cap-num> <media> <transport> <fmt list>
```

where <cap-num> is an integer between 1 and 255 (both included) used to number the capabilities, and <media>, <transport>, and <fmt list> are defined as in the SDP "m=" line. The capability description refers to a send- and-receive capability by default. When generating a capability set, the capability number MUST start with 1 in the first capability description and be incremented by the number of media formats in the <fmt list> for each subsequent capability description. The media formats in the <fmt list> are numbered from left to right. Receivers of a capability set MUST NOT, however, reject capability descriptions due to gaps in the capability numbers. The capability number provides a convenient handle within the context of the capability set (as referenced by the sequence number), which may be used to reference a particular capability by means outside of this specification. A capability description can include one or more capability parameter lines on the form:

```
a=cpar: <cap-par>
a=cparmin: <cap-par>
a=cparmax: <cap-par>
```

where <cap-par> is either bandwidth information ("b=") or an attribute ("a=") in its full "<type>=<value>" form (RFC 4566, see Section 2). A capability parameter line provides additional parameters for the preceding "cdsc" attribute line. Capability parameter lines for a capability description SHOULD immediately follow the "cdsc" line to which they refer. Nevertheless, the capability description includes all capability parameter lines until the next description ("cdsc") or media ("m=") line in the session description.

The "cpar" attribute should normally be used when capability parameter values are to be specified. When provided, it means that the endpoint is declaring that it supports the media formats in the preceding "cdsc" line in accordance with the <cap-par> value specified. This can, for example, be used to specify "fmtp" parameters. If a session negotiation is attempted without considering the <cap-par> value, it may fail due to lack of endpoint support. A capability description may contain zero, one, or more "cpar" attribute lines describing either the same or different parameters. The same parameter may be described more than once to specify alternatives.

Where a minimum numerical value is to be specified, the "cparmin" attribute should be used. There may be zero, one, or more "cparmin" attribute lines in a capability description; however, a given parameter MUST NOT be described by a "cparmin" attribute more than once. Where a maximum numerical value is to be specified, the "cparmax" attribute should be used. There may be zero, one, or more "cparmax" attribute lines in a capability description; however, a given parameter MUST NOT be described by a "cparmax" attribute more than once. Ranges of numerical values can be expressed by using a "cparmin" and a "cparmax" attribute for a given parameter. It follows from the previous rules that only a single range can be specified for a given parameter.

Capability descriptions may be provided at both the session level and media level. A capability description provided at the session level applies to all media streams of the indicated media type in the session description. A capability description provided at the media level only applies to that particular media stream (regardless of media type). If a capability description with media type X is provided at the session level, and there are no media streams of type X in the session description, then which of the media streams the capability description applies to is undefined (except if there is only one media stream). It is therefore RECOMMENDED that such capabilities be provided at the media level instead. Below we show an example session description using the above simcap mechanism:

```
v=0
o=- 25678 753849 IN IP4 128.96.41.1
s=
c=IN IP4 128.96.41.1
t=0 0
m=audio 3456 RTP/AVP 18 96
a=rtpmap:96 telephone-event
a=fmtp:96 0-15,32-35
a=sqn: 0
a=cdsc: 1 audio RTP/AVP 0 18 96
a=cpar: a=fmtp:96 0-16,32-35
a=cdsc: 4 image udptl t38
a=cdsc: 5 image tcp t38
```

The sender of this session description is currently prepared to send and receive G.729 audio as well as telephone events 0–15 and 32–35. The sender is furthermore capable of supporting

- Pulse Coded Modulation with μ-Law (PCMU) Algorithm encoding for the audio media stream,
- Telephone events 0–16 and 32–35,
- T.38 fax relay using UDP or TCP [see International Telegraph Union – Telecommunication Recommendation T.38 Annex D, "SIP/SDP Call Establishment Procedures."] as transport protocol.

Note that the first capability number for "cdsc" attribute specified is 1, whereas the next is 4, as three media formats were included in the first capability description. Also note that the rtpmap for payload type 96 was not included in the capability description, as it was already specified for the media ("m=") line. Conversely, it would of course not have been valid to provide the rtpmap in the capability description and then omit the "a=rtpmap" line..Below, we show another example of the simcap mechanism, this time with multiple media streams:

```
v=0
o=- 25678 753849 IN IP4 128.96.41.1
s=
c=IN IP4 128.96.41.1
t=0 0
m=audio 3456 RTP/AVP 18
a=sqn: 0
```

```
a=cdsc: 1 audio RTP/AVP 0 18
m=video 3458 RTP/AVP 31
a=cdsc: 3 video RTP/AVP 31 34
```

The sender of this session description is currently prepared to send and receive G.729 audio and H.261 video. The sender is furthermore capable of supporting

- PCMU Algorithm encoding for the audio media stream
- H.263 for the video media stream

Note that the first capability number specified is 1, whereas the next is 3, as two media formats were included in the first capability description. Also, note that the sequence number applies to the entire capability set (i.e., both audio and video); hence, it is supplied once. Finally, note that the media formats 18 and 31 are listed in both the media lines and the capability set as required. The above session description could have equally been supplied as follows:

```
v=0
o=- 25678 753849 IN IP4 128.96.41.1
s=
c=IN IP4 128.96.41.1
t=0 0
a=sqn: 0
a=cdsc: 1 audio RTP/AVP 0 18
a=cdsc: 3 video RTP/AVP 31 34
m=audio 3456 RTP/AVP 18
m=video 3458 RTP/AVP 31
```

(i.e., with the capability set provided at the session level).

4.1.4 Security Considerations

The security procedures that need to be considered in dealing with the SDP capability negotiation messages described in this section (RFC 3407) are provided in Section 17.4.

4.1.5 Internet Assigned Numbers Authority Considerations

SDP parameters that are described here have been registered with Internet Assigned Numbers Authority (IANA). For the sake of brevity, we have not included the IANA registration procedures here. The procedures for the registration of new SDP parameters are described in RFC 3407.

4.2 Summary

We have described that the initial SDP protocol specification was not designed to support capability negotiations, for a minimum common set of capabilities between conferencing parties. RFC 3407, described here, has mitigated this problem by defining minimal and backwards compatible capability declaration features in SDP by defining a set of new SDP attributes. Furthermore, these attributes define a capability set, which consists of a capability set sequence number followed by one or more capability descriptions. It should be noted that each capability description provided in the set defined here by RFC 3407 merely contains information about supported media formats and that the endpoint is not committing to use any of them. In order to actually use a declared capability, session negotiation will have to be carried out by means of another mechanism such as the SDP offer/answer model defined in RFC 3264 (see Section 3.1). On the other hand, RFC 3407 is not intended to solve the general capability negotiation. Rather, it is simply intended as a simple and limited solution to the most urgent problems for expressing capability features using SDP, which can then be used for negotiations for an agreed-upon minimum common set of capabilities by all communicating parties.

4.3 Problems

1. Why is the capability declaration in SDP not considered in its original specification? Why is the capability declaration in SDP important? Justify in detail and provide examples.
2. How does RFC 3407 mitigate the capability negotiations for having a minimum common set of capabilities between the conferencing parties?
3. Describe differences between session-level and media-level capability sets; provide detailed examples.
4. How are the backward compatibility problems handled in SDP using capability description mechanisms? Describe in detail.
5. Does RFC 3407 offer session negotiation capability between the communication parties? If not, which RFC offers session negotiation capability using SDP? Describe how the capability declaration described by this RFC facilitates session negotiation capabilities.

Chapter 5

Media-Level Attribute Support in SDP

Abstract

This section describes the Session Description Protocol (SDP) media-level attributes defined in Request for Comments (RFCs) 4574, 4796, and 5576. RFC 4574 defines a new SDP media-level attribute: "label." The "label" attribute carries a pointer to a media stream in the context of an arbitrary network application that uses SDP. The sender of the SDP document can attach the "label" attribute to a particular media stream or streams. The application can then use the provided pointer to refer to each particular media stream in its context. RFC 4796 defines another new SDP media-level attribute: "content." The "content" attribute defines the content of the media stream to a more detailed level than the media description line. The sender of an SDP session description can attach the "content" attribute to one or more media streams. The receiving application can then treat each media stream differently (for example, show it on a big or small screen) based on its content. The SDP also provides mechanisms to describe attributes of multimedia sessions and of individual media streams, for example, Real-time Transport Protocol sessions, within a multimedia session, but does not provide any mechanism to describe individual media sources within a media stream. RFC 5576 defines a mechanism, known as a source-specific media attribute, to describe Real-Time Transport Protocol media sources. Source-specific media attributes are identified by their synchronization source identifiers, in SDP, to associate attributes with these sources, and to express relationships among sources. It also defines several source-level attributes that can be used to describe properties of media sources.

5.1 SDP Label Attribute

This section describes Request for Comments (RFC) 4574 that defines a new Session Description Protocol (SDP) media-level attribute: "label." The "label" attribute carries a pointer to a media stream in the context of an arbitrary network application that uses SDP. The sender of the SDP document can attach the "label" attribute to a particular media stream or streams. The application can then use the provided pointer to refer to each particular media stream in its context.

5.1.1 Introduction

SDP is used by a variety of distributed-over-the-network applications. These applications deal with multiple sessions described by SDP (RFC 4566, see Section 2) and serve multiple users or services in the context of a single application instance. Applications of this kind need a means to identify a particular media stream across multiple SDP descriptions exchanged with different users. The Internet Engineering Task Force (IETF) Centralized Conferencing Working Group (XCON) framework is an example of a centralized conference architecture that uses SDP according to the offer/answer mechanism defined in RFC 3264 (see Section 3.1) to establish media streams with each of the conference participants. Additionally, XCON identifies the need to uniquely identify a media stream in terms of its role in a conference regardless of its media type, transport protocol, and media format. This can be accomplished with an external document that points to the appropriate media stream and provides information (e.g., the media stream's role in the conference) about it. The Session Initiation Protocol (SIP) Event Package for Conference State described in RFC 4575 defines and uses a concrete format for such external documents. This specification defines the SDP (RFC 4566, see Section 2) "label" media-level attribute, which provides a pointer to a media stream that is described by an "m=" line in an SDP session description.

5.1.2 Terminology

All terminologies are defined in Table 2.1.

5.1.3 Motivation for the New Label Attribute

Even though SDP and its extensions already provide a few ways to refer to a media stream, none of them is appropriate in the context of external documents that may be created before the session description itself and need to be handled by automata. The "i" SDP attribute, defined in RFC 4566 (see Section 2) that obsoletes RFC 2327, can be used to label media streams. Nevertheless, values of the "I" attribute are intended for human users and not for automata. The "mid" SDP attribute, defined in RFC 3388 (made obsolete by RFC 5888), can be used to identify media streams as well. Nevertheless, the scope of "mid" is too limited to be used by applications dealing with multiple SDP sessions. This is because values of the "mid" attribute are meaningful in the context of a single SDP session, not in the context of a broader application (for example, a multi-party application).

Another way of referring to a media stream is by using the order of the "m=" line in the SDP session document (for example, the 5th media stream in the session description). This is the mechanism used in the offer/answer model (RFC 3264, see Section 3.1). The problem with this mechanism is that it can only be used to refer to media streams in session descriptions that exist already. There are scenarios where a static document needs to refer, using a pointer, to a media stream that will be negotiated by SDP means and created in the future. When the media stream is eventually created, the application needs to label the media stream so that the pointer in the static document points to the proper media stream in the session description.

5.1.4 The Label Attribute

This specification defines a new media-level value attribute: "label." Its Augmented Backus-Naur Form (ABNF) syntax is provided in Section 2.9.2. Its formatting in SDP is described here again by the following ABNF (RFC 4234 that is made obsolete by RFC 5234) for convenience:

```
label-attribute   =    "a=label:" pointer
pointer           =    token
token             =    1*(token-char)
token-char        =    %x21 / %x23-27 / %x2A-2B / %x2D-2E / %x30-39
                       / %x41-5A / %x5E-7E
```

The token-char and token elements are defined in RFC 4566 (see Section 2) but included here to provide support for the implementer of this SDP feature. The "label" attribute contains a token defined by an application and used in its context. The new attribute can be attached to "m=" lines in multiple SDP documents, allowing the application to logically group the media streams across SDP sessions when necessary.

5.1.5 The Label Attribute in the Offer/Answer Model

This specification does not define a means to discover whether the peer endpoint understands the "label" attribute because "label" values are informative only at the offer/answer model level. At the offer/answer level, it means that the lack of label attributes in an offer does not imply that the answer should be without them. It also means that the presence in an offer of label attributes does not imply that the answer should also have them. In addition to the basic offer/answer rule above, applications that use "label" as a pointer to media streams MUST specify its usage constraints. For example, such applications MAY mandate support for "label." In this case, the application will define means for negotiation of the "label" attribute support as part of its specification.

5.1.6 Example

The following is an example of an SDP session description that uses the "label" attribute:

```
v=0
o=bob 280744730 28977631 IN IP4 host.example.com
s=
i=A Seminar on the session description protocol
c=IN IP4 192.0.2.2
t=0 0
m=audio 6886 RTP/AVP 0
a=label:1
m=audio 22334 RTP/AVP 0
a=label:2
```

5.1.7 Security Considerations

The security procedures that need to be considered in dealing with the basic SDP messages described in this section (RFC 4574) are provided in Section 17.4.1.

5.1.8 Internet Assigned Numbers Authority Considerations

SDP parameters that are described here have been registered with Internet Assigned Numbers Authority (IANA). We have not included the IANA registration procedures here for the sake of brevity. The procedures for registration of new SDP parameters are described in RFC 4574.

5.2 SDP Content Attribute

This section describes RFC 4796, which defines a new SDP media-level attribute, "content." The "content" attribute defines the content of the media stream to a more detailed level than the media description line. The sender of an SDP session description can attach the "content" attribute to one or more media streams. The receiving application can then treat each media stream differently (e.g., show it on a big or small screen) based on its content.

5.2.1 Introduction

The SDP (RFC 4566, see Section 2) is a protocol that is intended to describe multimedia sessions for the purposes of session announcement, session invitation, and other forms of multimedia session initiation. One of the most typical use cases of SDP is where it is used with the SIP RFC 3261 (also see Section 1.2). There are situations where one application receives several similar media streams, which are described in an SDP session description. The media streams can be similar in the sense that their content cannot be distinguished just by examining their media description lines (e.g., two video streams). The "content" attribute is needed so that the receiving application can treat each media stream appropriately based on its content. This specification defines the SDP "content" media-level attribute, which provides more information about the media stream than the "m=" line in an SDP session description. The main purpose of this specification is to allow applications to take automated actions based on the "content" attributes. However, this specification does not define those actions. Consequently, two implementations can behave completely differently when receiving the same "content" attribute.

5.2.2 Terminology

See Table 2.1.

5.2.3 Related Techniques

The "label" attribute specified in RFC 4574 (see Section 5.1) enables a sender to attach a pointer to a particular media stream. The namespace of the "label" attribute itself is unrestricted; so, in principle, it could also be used to convey information about the content of a media stream. However, in practice, this is not possible because of the need for backward compatibility. Existing implementations of the "label" attribute already use values from that unrestricted namespace in an application-specific way. So, it is not possible to reserve portions of the "label" attribute's namespace without possible conflict with already used application-specific labels.

It is possible to assign semantics to a media stream with an external document that uses the "label" attribute as a pointer. The downside of this approach is that it requires an external document. Therefore, this kind of mechanism is only applicable to special use cases where such external documents are used (e.g., centralized conferencing). Yet another way to attach semantics to a media stream is to use the "I" SDP attribute, defined in RFC 4566 (see Section 2). However, values of the "I" attribute are intended for human users and not for automata.

5.2.4 Motivation for the New Content Attribute

Currently, SDP does not provide any means for describing the content of a media stream (e.g., speaker's image, slides, sign language) in a form the application can understand. Of course, the end user can see the content of the media stream and read its title, but the application cannot understand what the media stream contains. The application that is receiving multiple similar (e.g., same type and format) media streams needs, in some cases, to know what the contents of those streams are. This kind of situation occurs, for example, in cases where presentation slides, the speaker's image, and sign language are transported as separate media streams. It would be desirable for the receiving application to be able to distinguish them in a way to be able to handle them automatically in an appropriate manner.

Figure 5.1 shows a screen of a typical communication application. The "content" attribute makes it possible for the application to decide where to show each media stream. From an end user's perspective, it is desirable that the user not need to arrange each media stream every time a new media session starts. The "content" attribute could also be used in more complex situations. An example of such a situation is application-controlling equipment in an auditorium. An auditorium can have many different output channels for video (e.g., main screen and two smaller screens) and audio (e.g., main speakers and headsets for the participants). In this kind of environment, a lot of interaction from the end user who operates the application would be required in the absence of cues from a controlling application. The "content" attribute would make it possible, for example, for an end user to specify, only once, which output each media stream of a given session should use. The application could automatically apply the same media layout for subsequent sessions. So, the "content" attribute can help reduce the amount of required end-user interaction considerably.

5.2.5 The Content Attribute

All ABNF syntaxes are shown in Section 2.9.2. This specification defines a new media-level value attribute, "content." Its formatting in SDP is described here again for convenience by the following ABNF (RFC 4234 that is obsoleted by RFC 5234):

```
content-attribute  =  "a=content:" mediacnt-tag
mediacnt-tag       =  mediacnt *("," mediacnt)
mediacnt           =  "slides" / "speaker" / "sl" / "main"
                      / "alt" / mediacnt-ext
mediacnt-ext       =  token
```

The "content" attribute contains one or more tokens, which MAY be attached to a media stream by a sending application. An application MAY attach a "content" attribute to any media stream

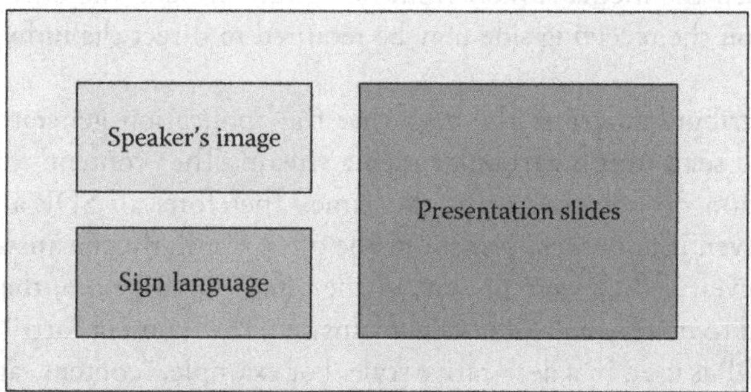

Figure 5.1 Application's screen (Copyright: IETF).

it describes. This document provides a set of predefined values for the "content" attribute. Other values can be defined in the future. The predefined values are:

- **slides:** the media stream includes presentation slides. The media type can be, for example, a video stream or a number of instant messages with pictures. Typical use cases for this are online seminars and courses. This is similar to the "presentation" role in H.239. (International Telegraph Union – Telecommunication, "Infrastructure of audiovisual services – Systems aspects; Role management and additional media channels for H.300-series terminals," Series H.239, July 2003.]
- **speaker:** the media stream contains the image of the speaker. The media can be, for example, a video stream or a still image. Typical use cases for this are online seminars and courses.
- **sl:** the media stream contains sign language. A typical use case for this is an audio stream that is translated into sign language, which is sent over a video stream.
- **main:** the media stream is taken from the main source. A typical use case for this is a concert where the camera is shooting the performer.
- **alt:** the media stream is taken from the alternative source. A typical use case for this is an event where the ambient sound is separated from the main sound. The alternative audio stream could be, for example, the sound of a jungle. Another example is the video of a conference room, while the main stream carries the video of the speaker. This is similar to the "live" role in H.239.

All these values can be used with any media type. We chose not to restrict each value to a particular set of media types in order not to prevent applications from using innovative combinations of a given value with different media types. The application can make decisions on how to handle a single media stream based on both the media type and the value of the "content" attribute. If the application does not implement any special logic for the handling of a given media type and "content" value combination, it applies the application's default handling for the media type. Note: The same "content" attribute value can occur more than once in a single session description.

5.2.6 The Content Attribute in the Offer/Answer Model

This specification does not define a means to discover whether the peer endpoint understands the "content" attribute because "content" values are only informative at the offer/answer model (RFC 3264, see Section 3.1) level. The fact that the peer endpoint does not understand the "content" attribute does not keep the media session from being established. The only consequence is that end user interaction on the receiving side may be required to direct the individual media streams appropriately.

The "content" attribute describes the data that the application generating the SDP session description intends to send over a particular media stream. The "content" values for both directions of a media stream do not need to be the same. Therefore, an SDP answer MAY contain "content" attributes even if none were present in the offer. Similarly, the answer MAY contain no "content" attributes even if they were present in the offer. Furthermore, the values of "content" attributes do not need to match in an offer and an answer. The "content" attribute can also be used in scenarios where SDP is used in a declarative style. For example, "content" attributes can be used in SDP session descriptors that are distributed with Session Announcement Protocol (RFC 2974).

5.2.7 Examples

There are two examples in this section. This first example uses a single "content" attribute value per media stream:

```
v=0
o=Alice 292742730 29277831 IN IP4 131.163.72.4
s=Second lecture from information technology
c=IN IP4 131.164.74.2
t=0 0
m=video 52886 RTP/AVP 31
a=rtpmap:31 H261/9000
a=content:slides
m=video 53334 RTP/AVP 31
a=rtpmap:31 H261/9000
a=content:speaker
m=video 54132 RTP/AVP 31
a=rtpmap:31 H261/9000
a=content:sl
```

The second example is a case where there is more than one "content" attribute value per media stream. The difference with the previous example is that now the conferencing system might automatically mix the video streams from the presenter and slides:

```
v=0
o=Alice 292742730 29277831 IN IP4 131.163.72.4
s=Second lecture from information technology
c=IN IP4 131.164.74.2
t=0 0
m=video 52886 RTP/AVP 31
a=rtpmap:31 H261/9000
a=content:slides,speaker
m=video 54132 RTP/AVP 31
a=rtpmap:31 H261/9000
a=content:sl
```

5.2.8 Operation with Synchronized Multimedia Integration Language

The values of "content" attribute, defined in Section 5.2.5, can also be used with Synchronized Multimedia Integration Language (SMIL) (Michel, T. and J. Ayars, "SMIL 2.0," W3C, January 2005). SMIL contains a "param" element, which is used for describing the content of a media flow. However, this "param" element, like the "content" attribute, provides an application-specific description of the media content. Details on how to use the values of the "content" attribute with SMIL's "param" element are outside the scope of this specification.

5.2.9 Security Considerations

The security procedures that need to be considered in dealing with the SDP capability negotiation messages described in this section (RFC 4796) are provided in Section 17.4.2.

5.2.10 IANA Considerations

SDP parameters that are described here have been registered with IANA. We have not included the IANA registration procedures here for the sake of brevity. The procedures for registration of new SDP parameters are described in RFC 4796.

5.3 Source-Specific Media Attributes in SDP

This section describes RFC 5576 that defines a mechanism to describe Real-Time Transport Protocol (RTP) media sources, which are identified by their SSRC (Synchronization Source) identifiers, in SDP, to associate attributes with these sources and to express relationships among sources. This mechanism is needed because RTP does not provide any mechanism to describe individual media sources within a media stream. It also defines several source-level attributes that can be used to describe properties of media sources.

The SDP provides mechanisms to describe attributes of multimedia sessions and of individual media streams (e.g., RTP sessions) within a multimedia session but does not provide any mechanism to describe individual media sources within a media stream. RFC 5576 defines a mechanism to describe RTP media sources, which are identified by their SSRC identifiers, in SDP, to associate attributes with these sources and to express relationships among sources. It also defines several source-level attributes that can be used to describe properties of media sources.

5.3.1 Introduction

The SDP (RFC 4566, see Section 2) provides mechanisms to describe attributes of multimedia sessions and of media streams (e.g., RTP specified in RFC 3550) sessions within a multimedia session but does not provide any mechanism to describe individual media sources within a media stream. Several recently proposed protocols, notably RTP control protocol single-source multicast (RFC 5760), have found it useful to describe specific media sources in SDP messages. Single-source multicast, in particular, needs to ensure that receivers' RTP SSRC identifiers do not collide with those of media senders, as the RTP specification (RFC 3550) requires that colliding sources change their SSRC values after a collision has been detected. Earlier work has used mechanisms specific to each protocol to describe the individual sources of an RTP session.

Moreover, whereas the RTP (RFC 3550) is defined as allowing multiple sources in an RTP session (e.g., if a user has more than one camera), SDP has no existing mechanism for an endpoint to indicate that it will be using multiple sources or to describe their characteristics individually. To address all these problems, this document defines a mechanism to describe RTP sources, identified by their SSRC identifier, in SDP, to associate attributes with these sources and to express relationships among individual sources. It also defines a number of new SDP attributes that apply to individual sources ("source-level" attributes), describes how a number of existing media stream ("media-level") attributes can also be applied at the source level, and establishes IANA registries for source-level attributes and source grouping semantics.

5.3.2 Terminology

See Table 2.1.

5.3.3 Overview

In the RTP (RFC 3550), an association among a group of communicating participants is known as an RTP session. An RTP session is typically associated with a single transport address (in the case of multicast) or communication flow (in the case of unicast), though RTP translators and Real-Time Transport Control Protocol (RTCP) single-source multicast (RFC 5760) can make the situation more complex. RTP topologies are discussed in more detail in RFC 5117 (that was made obsolete by RFC 7667). Within an RTP session, the source of a single stream RTP packet is known as a SSRC. Every synchronization source is identified by a 32-bit numeric identifier. In addition, receivers (who may never send RTP packets) also have source identifiers, which are used to identify their RTP control protocol receiver reports and other feedback messages.

Messages of the SDP (RFC 4566, see Section 2), known as session descriptions, describe multimedia sessions. A multimedia session is a set of multimedia senders and receivers as well as the data streams flowing from senders to receivers. A multimedia session contains a number of media streams, which are the individual RTP sessions or other media paths over which one type of multimedia data is carried. Information that applies to an entire multimedia session is called session-level information, while information pertaining to one media stream is called media-level information. The collection of all the information describing a media stream is known as a media description. (Media descriptions are also sometimes known informally as SDP "m=" lines, after the SDP syntax that begins a media description.) Several standard information elements are defined at both the session level and the media level. Extended information can be included at both levels through attributes. (The term "media stream" does not appear in the SDP specification itself but is used by a number of SDP extensions, for instance, Interactive Connectivity Establishment (RFC 5245), to denote the object described by an SDP media description. This term is unfortunately rather confusing, as the RTP specification (RFC 3550) uses the term "media stream" to refer to an individual media source or RTP packet stream, identified by an SSRC, whereas an SDP media stream describes an entire RTP session, which can contain any number of RTP sources. In RFC 4796, the term "media stream" means an SDP media stream (i.e., the thing described by an SDP media description), whereas "media source" is used for a single source of media packets (i.e., an RTP media stream). The core SDP specification does not have any way of describing individual media sources, particularly RTP synchronization sources, within a media stream. To address this problem, in this RFC 4796 (Section 5.2), we introduce a third level of information, called source-level information. Syntactically, source-level information is described by a new SDP media-level attribute, "ssrc," which identifies specific synchronization sources within an RTP session and acts as a meta attribute mapping source-level attribute information to these sources. This document also defines an SDP media-level attribute, "ssrc-group," which can represent relationships among media sources within an RTP session in much the same way as the "group" attribute (RFC 5888, that obsoletes RFC 3388, see Section 7.1) represents relationships among media streams within a multimedia session.

5.3.4 Media Attributes

This section defines two media-level attributes, "ssrc" and "ssrc-group." The ABNF syntaxes are provided here for convenience.

5.3.4.1 The "ssrc" Media Attribute

```
a=ssrc:<ssrc-id> <attribute>
a=ssrc:<ssrc-id> <attribute>:<value>
```

The SDP media attribute "ssrc" indicates a property (known as a "source-level attribute") of a media source (RTP stream) within an RTP session. <ssrc-id> is the SSRC ID of the source being described, interpreted as a 32-bit unsigned integer in network byte order and represented in decimal. <attribute> or <attribute>:<value> represents the source-level attribute specific to the given media source. The source-level attribute follows the syntax of the SDP "a=" line. It thus consists of either a single attribute name (a flag) or an attribute name and value (e.g., "cname:user@example.com" Note: CNAME=Canonical Name). No attributes of the former type are defined by this document.

Within a media stream, "ssrc" attributes with the same value of <ssrc-id> describe different attributes of the same media sources. Across media streams, <ssrc-id> values are not correlated (unless correlation is indicated by media-stream grouping or some other mechanism) and MAY be repeated. Each "ssrc" media attribute specifies a single source-level attribute for the given <ssrc-id>. For each source mentioned in SDP, the source-level attribute "cname", defined in Section 5.3.6.1, MUST be provided. Any number of other source-level attributes for the source MAY also be provided.

The "ssrc" media attribute MAY be used for any RTP-based media transport. It is not defined for other transports. If any other SDP attributes also mention RTP SSRC values (e.g., Multimedia Internet KEYing) (RFCs 3830 and 4567), the values used MUST be consistent. (These attributes MAY provide additional information about a source described by an "ssrc" attribute or MAY describe additional sources.) Though the source-level attributes specified by the ssrc property follow the same syntax as session-level and media-level attributes, they are defined independently. All source-level attributes MUST be registered with IANA. The "ssrc" media attribute is not dependent on charset.

5.3.4.2 The "ssrc-group" Media Attribute

```
a=ssrc-group:<semantics> <ssrc-id> ...
```

The SDP media attribute "ssrc-group" expresses a relationship among several sources of an RTP session. It is analogous to the "group" session-level attribute (RFC 3388 that is made obsolete by RFC 5888, see Section 7.1), which expresses a relationship among media streams in an SDP multimedia session (i.e., a relationship among several logically related RTP sessions). As sources are already identified by their SSRC IDs, no analogous property to the "mid" attribute is necessary; groups of sources are identified by their SSRC IDs directly. The <semantics> parameter is taken from the specification of the "group" attribute (RFC 5888, Section 7.1). The initial semantic values defined for the "ssrc-group" attribute are Flow Identification. RFC 5888 and Forward Error Correction (RFC 4756). In each case, the relationship among the grouped sources is the same as the relationship among corresponding sources in media streams grouped using the SDP "group" attribute.

Though the "ssrc-group" semantic values follow the same syntax as "group" semantic values, they are defined independently. All "ssrc-group" semantic values MUST be registered with IANA. Lip Synchronization (RFC 5888) is redundant for sources within a media stream as RTP sources with the same CNAME are implicitly synchronized in RTP. Single Reservation Flow (RFC 3524) and Alternative Network Address Types (RFC 4091 made obsolete by RFC 5245, see Section 3.3) refer specifically to the media stream's transport characteristics. Composite Session termed "FLUTE" is used to group FLUTE sessions, and so is not applicable to RTP. The "ssrc-group" attribute indicates the sources in a group by listing the <ssrc-id>s of the sources in the group. It

MUST list at least one <ssrc-id> for a group and MAY list any number of additional ones. Every <ssrc-id> listed in an "ssrc-group" attribute MUST be defined by a corresponding "ssrc:" line in the same media description. The "ssrc-group" media attribute is not dependent on charset.

5.3.5 Usage of Identified Source Identifiers in RTP

The synchronization source identifiers used in an RTP session are chosen randomly and independently by endpoints. As such, it is possible for two RTP endpoints to choose the same SSRC identifier. Though the probability of this is low, the RTP specification (RFC 3550) requires that all RTP endpoints MUST be prepared to detect and resolve collisions. As a result, all endpoints MUST be prepared for the fact that information about specific sources identified in a media stream might be out of date. The actual binding between SSRCs and source CNAMEs can only be identified by the Source Description (SDES) RTCP packets transmitted on the RTP session.

When endpoints choose their own local SSRC values for media streams for which source-level attributes have been specified, they MUST NOT use for themselves any SSRC identifiers mentioned in media descriptions they have received for the media stream.

However, sources identified by SDP source-level attributes do not otherwise affect RTP transport logic. Specifically, sources that are only known through SDP, for which neither RTP nor RTCP packets have been received, MUST NOT be counted for RTP group size estimation, and report blocks MUST NOT be sent for them in Sender Report or Receiver Report RTCP messages. Endpoints MUST NOT assume that only the sources mentioned in SDP will be present in an RTP session; additional sources, with previously unmentioned SSRC IDs, can be added at any time, and endpoints MUST be prepared to receive packets from these sources. (How endpoints handle such packets is not specified here; they SHOULD be handled in the same manner as packets from additional sources would be handled had the endpoint not received any a=ssrc: attributes at all.)

An endpoint that observes an SSRC collision between its explicitly signaled source and another entity that has not explicitly signaled an SSRC MAY delay its RTP collision-resolution actions (RFC 3550) by, where is the deterministic, calculated, reporting interval for receivers defined in Section 6.3.1 of the RTP specification (RFC 3550), to see whether the conflict still exists. (This gives precedence to explicitly signaled sources and places the burden of collision resolution on nonsignaled sources.) SSRC collisions between multiple explicitly signaled sources, however, MUST be acted upon immediately.

If, following RTP's collision-resolution procedures (RFC 3550), a source identified by source-level attributes has been forced to change its SSRC identifier, the author of the SDP containing the source-level attributes for these sources SHOULD send out an updated SDP session description with the new SSRC if the mechanism by which SDP is distributed for the multimedia session has a mechanism to distribute updated SDP. This updated SDP MUST include a "previous-ssrc" source-level attribute, described in Section 5.3.6.2, listing the source's previous SSRC ID. (If only a single source with a given CNAME has collided, the other RTP session members can infer a correspondence between the sources' old and new SSRC IDs without requiring an updated session description. However, if more than one source collides at once, or if sources are leaving and rejoining, this inference is not possible. To avoid confusion, therefore, sending updated SDP messages is always RECOMMENDED.)

Endpoints MUST NOT reuse the same SSRC ID for identified sources with the same CNAME for at least the duration of the RTP session's participant timeout interval (see Section 6.3.5 of RFC 3550). They SHOULD NOT reuse any SSRC ID ever mentioned in SDP (either by themselves or by other endpoints) for the entire lifetime of the RTP session. Endpoints MUST be prepared for the possibility that other parties in the session do not understand SDP source-level attributes, unless some higher-level mechanism normatively requires them. See Section 5.3.9 for more discussion of this.

5.3.6 Source Attributes

This section describes specific source attributes that can be applied to RTP sources. The ABNF syntaxes are provided here for convenience.

5.3.6.1 The "cname" Source Attribute

```
a=ssrc:<ssrc-id> cname:<cname>
```

The "cname" source attribute associates a media source with its CNAME, SDES item. This MUST be the CNAME value that the media sender will place in its RTCP SDES packets; it therefore MUST follow the syntax conventions of CNAME defined in the RTP specification (RFC 3550). If a session participant receives an RTCP SDES packet associating this SSRC with a different CNAME, it SHOULD assume there has been an SSRC collision and that the description of the source that was carried in the SDP description is not applicable to the actual source being received. This source attribute is REQUIRED to be present if any source attributes are present for a source. The "cname" attribute MUST NOT occur more than once for the same ssrc-id within a given media stream. The "cname" source attribute is not dependent on charset. Figure 5.6 in Section 5.3.10 gives a formal ABNF (RFC 5234, also see Section 2.9.2 and Appendix A of this book) grammar for the "cname" attribute.

5.3.6.2 The "previous-ssrc" Source Attribute

```
a=ssrc:<ssrc-id> previous-ssrc:<ssrc-id> ...
```

The "previous-ssrc" source attribute associates a media source with previous source identifiers used for the same media source. Following an SSRC change due to an SSRC collision involving a media source described in SDP, the updated session description of the source's new SSRC (described in Section 5.3.5) MUST include the "previous-ssrc" attribute associating the new SSRC with the old one. If further updated SDP descriptions are published for the media source, the "previous-ssrc" attribute SHOULD be included if the session description was generated before the participant timeout of the old SSRC, and MAY be included after that point. This attribute, if present, MUST list at least one previous SSRC and MAY list any number of additional SSRCs for the source if the source has collided more than once. This attribute MUST be present only once for each source. The "previous-ssrc" source attribute is not dependent on charset. Figure 5.8 in Section 5.3.10 gives a formal ABNF (RFC 5234) grammar for the previous ssrc attribute.

5.3.6.3 The "fmtp" Source Attribute

```
a=ssrc:<ssrc> fmtp:<format> <format specific parameters>
```

The "fmtp" source attribute allows format-specific parameters to be conveyed about a given source. The <format> parameter MUST be one of the media formats (i.e., RTP payload types) specified for the media stream. The meaning of the <format specific parameters> is unique for each media type. This parameter MUST only be used for media types for which source-level format parameters have explicitly been specified; media-level format parameters MUST NOT be carried over blindly. The "fmtp" source attribute is not dependent on charset.

5.3.6.4 Other Source Attributes

This section only defines source attributes that are necessary or useful for an endpoint to decode and render the sources in a media stream. It does not include any attributes that would contribute to an endpoint's decision to accept or reject a stream (e.g., in an offer/answer exchange). Such attributes are for future consideration.

5.3.7 Examples

This section gives several examples of SDP descriptions of media sessions containing source attributes. For brevity, only the media sections of the descriptions are given.

The example in Figure 5.2 shows an audio stream advertising a single source.

The example in Figure 5.3 shows a video stream where one participant (identified by a single CNAME) has several cameras. The sources could be further distinguished by RTCP SDES information.

The example in Figure 5.4 shows how the relationships among sources used for RTP retransmission (RFC 4588) can be explicitly signaled. This prevents the complexity of associating original sources with retransmission sources when SSRC multiplexing is used for RTP retransmission, as is described in Section 5.3 of RFC 4588.

```
m=audio 49168 RTP/AVP 0
a=ssrc:314159 cname:user@example.com
```

Figure 5.2 Example of a declaration of a single synchronization source.

```
m=video 49170 RTP/AVP 96
a=rtpmap:96 H264/90000
a=ssrc:12345 cname:another-user@example.com
a=ssrc:67890 cname:another-user@example.com
```

Figure 5.3 Example of a media stream containing several independent sources from a single session member.

```
m=video 49174 RTP/AVPF 96 98
a=rtpmap:96 H.264/90000
a=rtpmap:98 rtx/90000
a=fmtp:98 apt=96;rtx-time=3000
a=ssrc-group:FID 11111 22222
a=ssrc:11111 cname:user3@example.com
a=ssrc:22222 cname:user3@example.com
a=ssrc-group:FID 33333 44444
a=ssrc:33333 cname:user3@example.com
a=ssrc:44444 cname:user3@example.com

Note: Audio-Visual-Profile with Feedback (AVPF)
```

Figure 5.4 Example of the relationships among several retransmission sources (Copyright: IETF).

5.3.8 Usage with the Offer/Answer Model

When used with the SDP Offer/Answer Model (RFC 3264, see Section 3.1), SDP source-specific attributes describe only the sources each party is willing to send (whether it is sending RTP data or RTCP report blocks). No mechanism is provided by which an answer can accept or reject individual sources within a media stream; if the set of sources in a media stream is unacceptable, the answerer's only option is to reject the media stream or the entire multimedia session. The SSRC IDs for sources described by an SDP answer MUST be distinct from the SSRC IDs for sources of that media stream in the offer. Similarly, new SSRC IDs in an updated offer MUST be distinct from the SSRC IDs for that media stream established in the most recent offer/answer exchange for the session and SHOULD be distinct from any SSRC ID ever used by either party within the multimedia session (whether or not it is still being used).

5.3.9 Backward Compatibility

According to the definition of SDP, interpreters of SDP SDESs ignore unknown attributes. Thus, endpoints MUST be prepared for recipients of their RTP media session not to understand their explicit SDESs, unless some external mechanism indicates that they were understood. In some cases (such as RTP Retransmission - RFC 4588), this may constrain some choices about the bit-streams that are transmitted. SDESs are specified in this document such that RTP endpoints that are compliant with the RTP specification (RFC 3550) will be able to decode the media streams they describe, regardless of whether they support explicit SDESs. However, some deployed RTP implementations may not actually support multiple media sources in a media stream. Media senders MAY wish to restrict themselves to a single source at a time unless they have some means of concluding that the receivers of the media stream support source multiplexing.

5.3.10 Formal Grammar

Section 2.9.2 provides all syntaxes. We have provided ABNF syntaxes here again for convenience. This section gives a formal ABNF (RFC 5234) grammar for each of the new media and source attributes defined in this specification. Grammars for existing sessions or media attributes that have been extended to be source attributes are not included (Figures 5.5–5.8).

```
ssrc-attr =     "ssrc:" ssrc-id SP attribute
                ; The base definition of "attribute" is in RFC 4566 (see
                ; Section 2).
                ; (It is the content of "a=" lines.)
ssrc-id   =     integer ; 0 .. 2**32 -1
attribute =     / ssrc-attr
```

Figure 5.5 Syntax of the "ssrc" media attribute (Copyright: IETF).

```
ssrc-group-attr   =   "ssrc-group:" semantics *(SP ssrc-id)
semantics         =   "FEC" / "FID" / token
                      ; Matches RFC 3388 definition and
                      ; IANA registration rules in this doc.
token             =   <as defined in RFC 4566>
attribute         =   / ssrc-group-attr
```

Figure 5.6 Syntax of the "ssrc-group" media attribute (Copyright: IETF).

```
cname-attr      =     "cname:" cname
cname           =     byte-string
                      ; Following the syntax conventions for
                            CNAME as defined in RFC 3550.
                      ; The definition of "byte-string" is in RFC
                          4566.
attribute       =     / cname-attr
```

Figure 5.7 Syntax of the "cname" source attribute (Copyright: IETF).

```
previous-ssrc-attr =     "previous-ssrc:" ssrc-id *(SP ssrc-id)
attribute          =     / previous-ssrc-attr
```

Figure 5.8 Syntax of the "previous-ssrc" source attribute (Copyright: IETF).

5.3.11 Security Considerations

Security procedures that need to be considered in dealing with the SDP capability negotiation messages described in this section (RFC 5576) are provided in Section 17.4.3.

5.3.12 IANA Considerations

SDP parameters that are described here have been registered with IANA. We have not included the IANA registration procedures here for the sake of brevity. The procedures for registration of new SDP parameters are described in RFC 5576.

5.4 Summary

We have described two new media-level attributes termed as "label" (RFC 4574) and "content" (RFC 4796). In addition, new source-specific media attributes identified by SSRC identifiers namely "ssrc" and "ssrc-group" are defined (RFC 5576). The label attribute is a pointer to a particular media stream (or streams). Once the pointer is placed, an application can use the specific pointer to refer to each particular media stream (or streams) on its context. On the other hand, the "content" attribute defines the content of a media stream (or streams) to a more detailed level than the media description line. Similar to the "label" attribute, the "content" attribute can also enable the handling of each media stream differently by the receiving application. We have also explained that the source-specific media attributes identified by their synchronization source identifiers such as "ssrc" and "ssrc-group" describe individual media sources within a media stream. The "ssrc" attribute is used to identify characteristics of media sources within a media stream. The "ssrc-group" attribute is used to identify relationships among media sources within a media stream. Moreover, several source-level attributes such as "cname," "previous-ssrc," and "fmtp" are specified to describe properties of media sources. The formats of all media-level attributes are defined in this section.

5.5 Problems

1. Why are the SDP media-level attributes defined by "a=" (RFC 4566), "mid" (RFC 5888) and "m=" line (RFC 4566) not sufficient to meet the requirements for multimedia conferencing applications?

2. Why does a new SDP medial-level attribute termed as “label” defined by RFC 4574? Explain in detail using an example of how the “label” attribute is used in expressing a new capability in multimedia conferencing.
3. Why do we need a new SDP media-level attribute “content” as defined by RFC 4796? Explain in detail using an offer/answer capability negotiation model. What are the problems in using the “label” attribute for representing “content” by SDP?
4. Why are the two "ssrc" and "ssrc-group" source-specific media attributes defined in SDP? What are the new capabilities addressed by these two source-specific media attributes in the context of RTP media streams in the context of multimedia conferencing?
5. What capabilities do the SDP source-level media attributes “cname,” “previous-ssrc,” and “fmtp” represent? Explain the utility of these new attributes using offer/answer model for capability negotiations in the context of multimedia conferencing.

Chapter 6

Semantics for Signaling Media Decoding Dependency in SDP

Abstract

This section defines semantics that allow for signaling the decoding dependency (DDP) of different media descriptions with the same media type in the Session Description Protocol (SDP) specified in Request for Comments (RFC) 5583. If multiple "m=" lines exist indicating the same media type, a receiver cannot identify a specific relationship between those media. The relationship between different layers of coding for the same media type needs to be defined. For example, if media data is separated and transported in different network streams due to the use of a layered or multiple descriptive media coding process, the receiver should have a mechanism to know the dependency between different SDP media streams for each media type. A new grouping type "DDP" is defined, to be used in conjunction with RFC 3388 (that is made obsolete by RFC 5888, see Section 7.1), which defines SDP attributes such as "group" and "mid" in addition to others. Further, an attribute is specified describing the relationship of the media streams in a "DDP" group indicated by media identification attribute(s) and media format description(s). The layered dependencies between different coding layers are aimed at increasing the coding efficiency, and SDP is enhanced to carry this information.

6.1 Decoding Dependency with Layered or Multiple Descriptive Media

This section describes Request for Comments (RFC) 5583 that defines semantics that allow for signaling the decoding dependency ("DDP") of different media descriptions with the same media type in the Session Description Protocol (SDP). This is required, for example, if media data is separated and transported in different network streams due to the use of a layered or multiple descriptive media coding process. A new grouping type—("DDP")—is defined to be used in conjunction with RFC 5888 (see Section 7.1). In addition, an attribute is specified describing the relationship of the media streams in a "DDP" group indicated by media identification attribute(s) and media format description(s).

6.1.1 Introduction

An SDP session description may contain one or more media descriptions, each identifying a single media stream. A media description is identified by one "m=" line. Today, if multiple "m=" lines exist indicating the same media type, a receiver cannot identify a specific relationship between those media. A Multiple Description Coding (MDC) or layered Media Bitstream contains, by definition, one or more Media Partitions that are conveyed in their own media streams. The cases we are interested in are layered and MDC Bitstreams with two or more Media Partitions. Carrying more than one Media Partition in its own session is one of the key use cases for employing layered or MDC-coded media. Senders, network elements, or receivers can suppress sending/forwarding/subscribing/decoding individual Media Partitions and still preserve perhaps suboptimal, but still useful, media quality.

One property of all Media Bitstreams relevant to this dependency with layers codec is that their Media Partitions have a well-defined usage relationship. For example, in layered coding, "higher" Media Partitions are useless without "lower" ones. In MDC coding, Media Partitions are complementary—the more Media Partitions one receives, the better a reproduced quality may be. The specification described here defines an SDP extension to indicate such a "DDP." The trigger for the present specification has been the standardization process of the Real-Time Transport Protocol (RTP) payload format for the Scalable Video Coding (SVC) extension to International Telegraph Union – Telecommunication (ITU-T) Rec. H.264/MPEG-4 AVC (RFC 6190). In RFC 6190, it was observed that the aforementioned lack in signaling support is not specific to SVC, but applies to all layered or MDC codecs. Therefore, this specification presents a generic solution. Likely, the second technology utilizing the mechanisms of this specification will be Multiview Video Coding (MVC). In MVC specified in RFC 6190, layered dependencies between views are used to increase coding efficiency; therefore, the properties of MVC with respect to SDP signaling are comparable to those of SVC. The mechanisms defined herein are media transport protocol dependent and applicable only in conjunction with the use of RTP (RFC 3550). The SDP grouping of Media Lines of different media types is out of the scope of this RFC 5583.

6.1.2 Terminology

See Section 2.2/Table 2.1.

6.1.3 Definitions

See Section 2.2/Table 2.1.

6.1.4 Motivation, Use Cases, and Architecture

6.1.4.1 Motivation

This section is concerned with two types of "DDP": layered and MDC. The transport of layered and MDC share as key motivators the desire for media adaptation to network conditions (i.e., related to bandwidth, error rates, connectivity of endpoints in multicast or broadcast scenarios, and the like).

6.1.4.1.1 Layered "DDP"

In layered coding, the partitions of a Media Bitstream are known as media layers or simply layers. One or more layers may be transported in different media streams in the sense of (RFC 4566, see Section 2). A classic use case is known as receiver-driven layered multicast, in which a receiver selects a combination of media streams in response to quality or bit-rate requirements. Back in the mid-1990s, the then-available layered media formats and codecs envisioned primarily (or even exclusively) a one-dimensional hierarchy of layers. That is, each so-called enhancement layer referred to exactly one layer "below." The single exception has been the base layer, which is self-contained.

Therefore, the identification of one enhancement layer fully specifies the Operation Point of a layered coding scheme, including knowledge about all the other layers that need to be decoded. SDP (RFC 4566, see Section 2) contains rudimentary support for exactly this use case and media formats, in that it allows for signaling a range of transport addresses in a certain media description. By definition, a higher transport address identifies a higher layer in the one-dimensional hierarchy. A receiver needs only to decode data conveyed over this transport address and lower transport addresses to decode this Operation Point.

Newer media formats depart from this simple one-dimensional hierarchy, in that highly complex (at least tree-shaped) dependency hierarchies can be implemented. Compelling use cases for these complex hierarchies have been identified by industry. Support for it is therefore desirable. However, SDP, in its current form, does not allow for the signaling of these complex relationships. Therefore, receivers cannot make an informed decision on which layers to subscribe (in cases of layered multicast).

Layered "DDPs" may also exist in an MVC environment. Views may be coded using inter-view dependencies to increase coding efficiency. This results in Media Bitstreams that logically may be separated into Media Partitions representing different views of the reconstructed video signal. These Media Partitions cannot be decoded independently; therefore, other Media Partitions are required for reconstruction. To express this relationship, the signaling needs to express the dependencies of the views, which in turn are Media Partitions in the sense of this document.

6.1.4.1.2 Multiple Descriptive "DDP"

In the most basic form of MDC, each Media Partition forms an independent representation of the media. That is, decoding of any of the Media Partitions yields useful reproduced media data. When more than one Media Partition is available, then a decoder can process them jointly, and the resulting media quality increases. The highest reproduced quality is available if all original Media Partitions are available for decoding. More complex forms of MDC can also be envisioned (i.e., where, as a minimum, N-out-of-M total Media Partitions need to be available to allow meaningful decoding). MDC has not yet been embraced heavily by the media standardization community, though it is the subject of a lot of academic research. As an example, we refer to Reference [1]. In this specification, we cover MDC because we (1) envision that MDC media formats will come into practical use within the lifetime of this specification, and (2) the solution for its signaling is very similar to that of layered coding.

6.1.4.1.3 Other "DDP" Relationships

At the time of writing, no "DDP" relationships beyond the two already mentioned have been identified and warrant standardization. However, the mechanisms of this specification could be extended by introducing new code points for new DDP types. If such an extension becomes necessary, as formally required in Section 6.1.5.2.2, the new "DDP" type MUST be documented in an Internet Engineering Task Force (IETF) Standards-Track document.

6.1.4.2 Use Cases

- Receiver-driven layered multicast: This technology is discussed in RFC 3550 and references therein. We refrain from elaborating further; the subject is well known and understood.
- Multiple end-to-end transmission with different properties: Assume a unicast and point-to-point topology, wherein one endpoint sends media to another. Assume further that different forms of media transmission are available. The difference may lie in the cost of the transmission (free, charged), in the available protection (unprotected/secure), in the quality of service (QOS) (guaranteed quality/best effort), or other factors. Layered and MDC coding allows matching of the media characteristics to the available transmission path(s). For example, in layered coding, it makes sense to convey the base layer over high QOS. Enhancement layers, on the other hand, can be conveyed over best effort, as they are "optional" in their characteristic—nice to have, but nonessential for media consumption. In a different scenario, the base layer may be offered in a nonencrypted session as a free preview. An encrypted enhancement layer references this base layer and allows optimal quality playback; however, it is only accessible to users who have the key, which may have been distributed by a conditional access mechanism.

6.1.5 Signaling Media Dependencies

Note that all Augmented Backus-Naur Form (ABNF) syntaxes are shown in Section 2.9.2. We have reproduced ABNF syntaxes here for convenience.

6.1.5.1 Design Principles

Dependency signaling is only feasible between media descriptions described with an "m=" line and with an assigned media identification attribute ("mid"), as defined in RFC 3388 (that is made obsolete by RFC 5888, see Section 7.1). All media descriptions grouped according to this specification MUST have the same media type. Other dependencies relations expressed by the SDP grouping have to be addressed in other specifications. A media description MUST NOT be part of more than one group of the grouping type defined in this specification.

6.1.5.2 Semantics

6.1.5.2.1 SDP Grouping Semantics for "DDP"

This specification defines a new grouping semantic DDP: DDP associates a media stream, identified by its mid attribute, with a DDP group. Each media stream MUST be composed of an integer number of Media Partitions. A media stream is identified by a session unique media format description (RTP payload type number) within a media description. In a DDP group, all media streams MUST have the same type of DDP (as signaled by the attribute defined in Section 6.1.5.2.2). All media streams MUST contain at least one Operation Point. The DDP group type informs a receiver about the requirement for handling the media streams of the group according to the new media-level attribute "depend," as defined in Section 6.1.5.2.2. When using multiple codecs (e.g., for the Offer/Answer model), the media streams MUST have the same dependency structure, regardless of which media format description (RTP payload type number) is used.

6.1.5.2.2 Attribute ("Depend") for Dependency Signaling per Media Stream

This specification defines a new media-level attribute, "depend," with the following ABNF (RFC 5234). The identification-tag is defined in RFC 3388 (that is made obsolete by RFC 5888, see Section 7.1). In the following, ABNF, fmt, token, single-space (SP), and CRLF are used as defined in RFC 4566 (see Section 2).

```
depend-attribute = "a=depend:" dependent-fmt SP dependency-tag
                   *(";" SP dependent-fmt SP dependency-tag) CRLF
dependency-tag   = dependency-type *1(SP identification-tag ":"
                   fmt-dependency *("," fmt-dependency))
dependency-type  = "lay" / "mdc" / token
dependent-fmt    = fmt
fmt-dependency   = fmt
```

dependency-tag indicates one or more dependencies of one dependent fmt in the media description. These dependencies are signaled as fmt-dependency values, which indicate fmt values of other media descriptions. These other media descriptions are identified by their identification-tag values in the depend-attribute. There MUST be exactly one dependency-tag indicated per dependent-fmt. dependent-fmt indicates the media format description, as defined in RFC 4566 (see Section 2), that depends on one or more media format descriptions in the media description indicated by the value of the identification-tag within the dependency-tag.

fmt-dependency indicates the media format description in the media description identified by the identification-tag within the dependency-tag, on which the dependent-fmt of the dependent media description depends. In case a list of fmt-dependency values is given, any element of the list is sufficient to satisfy the dependency, at the choice of the decoding entity. The depend-attribute describes the DDP. The depend-attribute MUST be followed by a sequence of dependent-fmt and the corresponding dependency-tag fields, which identify all related media format descriptions in all related media descriptions of the dependent-fmt. The attribute MAY be used with multicast as well as with unicast transport addresses. The following dependency-type values are defined in this specification:

- **lay:** Layered DDP identifies the described media stream as one or more Media Partitions of a layered Media Bitstream. When "lay" is used, all media streams required for decoding the Operation Point MUST be identified by identification-tag and fmt-dependency following the "lay" string.
- **mdc:** Multidescriptive DDP signals that the described media stream is part of a set of a MDC Media Bitstream. By definition, at least N-out-of-M media streams of the group need to be available to from an Operation Point. The values of N and M depend on the properties of the Media Bitstream and are not signaled within this context. When "mdc" is used, all required media streams for the Operation Point MUST be identified by identification-tag and fmt-dependency following the "mdc" string. Further, dependency types MUST be defined in a Standards-Track document.

6.1.6 Usage of New Semantics in SDP

6.1.6.1 Usage with the SDP Offer/Answer Model

The backward compatibility in Offer/Answer is generally handled as specified in Section 8.4 of RFC 3388 (that is made obsolete by RFC 5888, see Section 9.4 of this book), as summarized here. Depending on the implementation, a node that does not understand DDP grouping (either does not understand line grouping at all or just does not understand the DDP semantics) SHOULD respond to an offer containing DDP grouping either (1) with an answer that ignores the grouping attribute or (2) with a refusal to the request (e.g., 488 Not acceptable here or 606 Not acceptable in Session Initiation Protocol).

In case (1), if the original sender of the offer still wishes to establish communications, it SHOULD generate a new offer with a single media stream that represents an Operation Point. Note: In most cases, this will be the base layer of a layered Media Bitstream, equally possible are Operation Points containing a set of enhancement layers as long as all are part of a single media stream. In case (2), if the sender of the original offer has identified that the refusal to the request is caused by the use of DDP grouping, and if the sender of the offer still wishes to establish the session, it SHOULD retry the request with an offer including only a single media stream.

If the answerer understands the DDP semantics, it is necessary to take the "depend" attribute into consideration in the Offer/Answer procedure. The main rule for the "depend" attribute is that the offerer decides the number of media streams and the dependency between them. The answerer cannot change the dependency relations. For unicast sessions where the answerer receives media (i.e., for offers including media streams that have a directionality indicated by "sendonly," "sendrecv," or have no directionality indicated, the answerer MAY remove media Operation Points. The answerer MUST use the dependency relations provided in the offer when sending media.

The answerer MAY send according to all of the Operation Points present in the offer, even if the answerer has removed some of those Operation Points. Thus, an answerer can limit the number of Operation Points being delivered to the answerer while the answerer can still send media to the offerer using all of the Operation Points indicated in the offer. For multicast sessions, the answerer MUST accept all Operation Points and their related DDPs or MUST remove nonaccepted Operation Points completely. Due to the nature of multicast, the receiver can select which Operation Points it actually receives and processes. For multicast sessions that allow the answerer to also send data, the answerer MAY send all of the offered Operation Points.

In any case, if the answerer cannot accept one or more offered Operation Points and/or the media stream's dependencies, the answerer MAY re-invite with an offer including acceptable Operation Points and/or dependencies. Note: Applications may limit the possibility of performing a re-invite. The previous offer is also a good hint to the capabilities of the other agent.

6.1.6.2 Declarative Usage

If a Real-Time Streaming Protocol (RTSP) receiver understands signaling according to this specification, it SHALL set up all media streams that are required to decode the Operation Point of its choice. If an RTSP receiver does not understand the signaling defined within this specification, it falls back to normal SDP processing. Two likely cases have to be distinguished:

1. If at least one of the media types included in the SDP is within the receiver's capabilities, it selects among those candidates according to implementation specific criteria for setup, as usual.
2. If none of the media types included in the SDP can be processed, then obviously no setup can occur.

6.1.6.3 Usage with AVP and SAVP RTP Profiles

The signaling mechanisms defined in this specification MUST NOT be used to negotiate between using the attribute-value pair (AVP) (RFC 3551) and SAVP (RFC 3711) profiles for RTP. However, both profiles MAY be used separately or jointly with the signaling mechanism defined in this specification.

6.1.6.4 Usage with Capability Negotiation

This RFC 5583 does not cover the interaction with capability negotiation. This issue is for further study and will need to be addressed in a different document.

6.1.6.5 Examples

6.1.6.5.1 Example for Signaling Layered DDP

The example provided after a couple of paragraphs shows a session description with three media descriptions, all of type video and with layered DDP ("lay"). Each of the media descriptions includes two possible media format descriptions with different encoding parameters as (e.g., "packetization-mode" (not shown in the example) for the media subtypes "H264" and "H264-SVC" given by the "a=rtpmap:" line). The first media description includes two H264 payload types as media format descriptions, "96" and "97" as defined in RFC 3984 and represents the base layer Operation Point (identified by "mid:L1"). The two other media descriptions (identified by "mid:L2" and "mid:L3") include H264-SVC payload types as defined in RFC 6190, which contain enhancements to the base layer Operation Point or the first enhancement layer Operation Point (media description identified by "mid:L2").

The example shows the dependencies of the media format descriptions of the different media descriptions indicated by the "DDP" grouping, "mid", and "depend" attributes. The "depend" attribute is used with the DDP type "lay" indicating layered DDP. For example, the third media description ("m=video 40004...") identified by "mid:L3" has different dependencies on the media format descriptions of the two other media descriptions: Media format description "100" depends on media format description "96" or "97" of the media description identified by "mid:L1". This is an exclusive-OR (i.e., payload type "100" may be used with payload type "96" or with "97"; one of the two combinations is required for decoding payload type "100").

For media format description "101", it is different. This one depends on two of the other media descriptions at the same time (i.e., it depends on media format description "97" of the media description identified by "mid:L1"; it also depends on media format description "99" of the media description identified by "mid:L2"). For decoding media format description "101" both media format description "97" and media format description "99" are required by definition.

```
v=0
o=svcsrv 289083124 289083124 IN IP4 host.example.com
s=LAYERED VIDEO SIGNALING Seminar
t=0 0
c=IN IP4 192.0.2.1/127
a=group:DDP L1 L2 L3
m=video 40000 RTP/AVP 96 97
b=AS:90
a=framerate:15
a=rtpmap:96 H264/90000
a=rtpmap:97 H264/90000
a=mid:L1
m=video 40002 RTP/AVP 98 99
b=AS:64
a=framerate:15
a=rtpmap:98 H264-SVC/90000
a=rtpmap:99 H264-SVC/90000
a=mid:L2
a=depend:98 lay L1:96,97; 99 lay L1:97
m=video 40004 RTP/AVP 100 101
b=AS:128
a=framerate:30
a=rtpmap:100 H264-SVC/90000
a=rtpmap:101 H264-SVC/90000
a=mid:L3
a=depend:100 lay L1:96,97; 101 lay L1:97 L2:99
```

6.1.6.5.2 Example for Signaling of Multidescriptive DDP

The example shows a session description with three media descriptions, all of type video with multidescriptive DDP. Each of the media descriptions includes one media format description. The example shows the dependencies of the media format descriptions of the different media descriptions indicated by "DDP" grouping, "mid," and "depend" attributes. The "depend" attribute is used with the DDP type "mdc" indicating layered DDP. For example, media format description "104" in the media description ("m=video 40000...") with "mid:M1" depends on the two other media descriptions. It depends on media format description "105" of media description with "mid:M2", and it also depends on media format description "106" of media description with "mid:M3". In case of the multidescriptive DDP, media format description "105" and "106" can be used by definition to enhance the decoding process of media format description "104", but they are not required for decoding.

```
v=0
o=mdcsrv 289083124 289083124 IN IP4 host.example.com
s=MULTI DESCRIPTION VIDEO SIGNALING Seminar
t=0 0
c=IN IP4 192.0.2.1/127
a=group:DDP M1 M2 M3
m=video 40000 RTP/AVP 104
a=mid:M1
```

```
a=depend:104 mdc M2:105 M3:106
m=video 40002 RTP/AVP 105
a=mid:M2
a=depend:105 mdc M1:104 M3:106
m=video 40004 RTP/AVP 106
a=mid:M3
a=depend:106 mdc M1:104 M2:105
```

6.1.7 Security Considerations

The security procedures that need to be considered in dealing with the SDP messages described in this section (RFC 5583) are provided in Section 17.5.

6.1.8 IANA Considerations

SDP parameters that are described here have been registered with Internet Assigned Numbers Authority (IANA). We have not included the IANA registration procedures here for the sake of brevity. The procedures for registration of new SDP parameters are described in RFC 5583.

6.1.9 Informative Note on "The SDP Grouping Framework"

The grouping mechanism is extended in RFC 3388 (that is made obsolete by RFC 5888, see Section 7.1) in a way that a media description can be part of more than one group of the same grouping type in the same session description. However, media descriptions grouped by this specification must be at most part of one group of the type "DDP" in the same session description.

6.2 Summary

The layered or MDC provides opportunity for variable video quality, and SDP needs to have the mechanisms to express the variable qualities of different streams that relate to the same media. The layered or multiple DDP between views of video streams are used to increase the coding efficiency. The semantics for layered or multiple media streams of the same video are defined here to facilitate the negotiations between the communicating parties using SDP. More importantly, the dependency of different layers of coding is also expressed in SDP media descriptions. However, the mechanisms defined herein are media transport protocol dependent and applicable only in conjunction with the use of RTP defined in RFC 3550.

6.3 Problems

1. What are the salient differences between features of the layered and the multiple description video coding, which needs to be included in the session description? Can the feature sets of multiview and layered video coding be included in the same generic session description? If so, justify your answer.
2. Why is the use of a SDP media description identified by one "m=" line in SDP not sufficient in expressing the layered or multiple descriptive media coding? Explain using an example of the known codec such as ITU-T Rec. H.264/MPEG-4.

3. Explain the SDP grouping semantics for DDP using the offer/answer model one for layered coding and another for multi-descriptive decoding with the use of RTP.
4. Can you propose any solution for the grouping of multiple media streams of different media types that can be used in SDP for capability negotiations using the offer/answer model?

Reference

1. Vitali, A., Borneo, A., Fumagalli, M., and R. Rinaldo, "Video over IP using Standard-Compatible Multiple Description Coding: an IETF proposal," Packet Video Workshop, April 2006, Hangzhou, China.

Chapter 7

Media Grouping Support in SDP

Abstract

This chapter defines a framework based on Request for Comments (RFC) 5888 that obsoletes RFC 3388 to group media lines commonly known as "m" lines in the Session Description Protocol (SDP) for different purposes, such as expressing a particular relationship between two or more of them. This framework uses the "group" and "mid" SDP attributes, both of which are defined in RFC 5888 specification. Additionally, it specifies how to use the framework for two more important purposes: lip synchronization and for receiving a media flow consisting of several media streams on different transport addresses.

This section also describes the Internet Engineering Task Force (IETF) draft (draft-ietf-mmusic-sdp-bundle-negotiation-29.txt) [1]. This specification defines a new Session Description Protocol (SDP) grouping framework extension, "BUNDLE." The extension can be used with the SDP Offer/Answer mechanism to negotiate the usage of a single address:port combination (BUNDLE address) for receiving media, referred to as bundled media, specified by multiple SDP media descriptions ("m=" lines). To assist endpoints in negotiating the use of bundle, this specification defines a new SDP attribute, "bundle-only," which can be used to request that specific media is only used if bundled. There are multiple ways to correlate the bundled Real-Time Transport Protocol (RTP) packets with the appropriate media descriptions. This specification defines a new RTP Security Descriptions (SDES) item and a new RTP header extension that provides an additional way to do this correlation by using attributes to carry a value that associates the RTP/Real-Time Transport Control Protocol (RTCP) packets with a specific media description.

Another SDP grouping mechanism for RTP media streams that can be used to specify relations between media streams is described here based on the IETF draft (draft-ietf-mmusic-msid-13) [5]. This mechanism is used to signal the association between the SDP concept of "media description" and the WebRTC concept of "MediaStream"/"MediaStreamTrack" using SDP signaling.

7.1 SDP Grouping Framework

In this section, we describe Request for Comments (RFC) 5888 that provides a framework to group "m" lines in the SDP for different purposes. This framework uses the "group" and "mid" SDP attributes, both of which are defined in this specification. Additionally, we specify how to use the framework for two different purposes: lip synchronization and for receiving a media flow consisting of several media streams on different transport addresses. This RFC 5888 obsoletes RFC 3388.

7.1.1 Introduction

RFC 5888 has specified a media-line grouping framework for SDP (RFC 4566, see Section 2) and obsoletes RFC 3388. An SDP as specified in RFC 4566, a session description typically contains one or more media lines, which are commonly known as "m" lines. When a session description contains more than one "m" line, SDP does not provide any means to express a particular relationship between two or more of them. When an application receives an SDP session description with more than one "m" line, it is up to the application to determine what to do with them. SDP does not carry any information about grouping media streams. While in some environments this information can be carried out of band, it is necessary to have a mechanism in SDP to express how different media streams within a session description relate to each other. The framework defined in this specification is such a mechanism.

7.1.2 Terminology

See Table 2.1.

7.1.3 Overview of Operation

This section provides a nonnormative description of how the SDP grouping framework defined in this specification works. In a given session description, each "m" line is identified by a token, which is carried in a "mid" attribute below the "m" line. The session description carries session-level "group" attributes that group different "m" lines (identified by their tokens) using different group semantics. The semantics of a group describe the purpose for which the "m" lines are grouped. For example, the "group" line in the session description below indicates that the "m" lines identified by tokens 1 and 2 (the audio and the video "m" lines, respectively) are grouped for the purpose of lip synchronization.

```
v=0
o=Laura 289083124 289083124 IN IP4 one.example.com
c=IN IP4 192.0.2.1
t=0 0
a=group:LS 1 2
m=audio 30000 RTP/AVP 0
a=mid:1
m=video 30002 RTP/AVP 31
a=mid:2
```

7.1.4 Media Stream Identification Attribute

This document defines the "media stream identification" media attribute, which is used for identifying media streams within a session description. Its formatting in SDP (RFC 4566, see Section 2) is described by the following Augmented Backus-Naur Form (ABNF) (RFC 5234):

```
mid-attribute       = "a=mid:" identification-tag
identification-tag  = token
                      ; token is defined in RFC 4566
```

The identification-tag MUST be unique within an SDP session description.

7.1.5 Group Attribute

This document defines the "group" session-level attribute, which is used for grouping together different media streams. Its formatting in SDP is described by the following ABNF (RFC 5234):

```
group-attribute      = "a=group:" semantics
                       *(SP identification-tag)
semantics            = "LS" / "FID" / semantics-extension
semantics-extension  = token
                       ; token is defined in RFC 4566
                       ; (see Section 2)
```

This document defines two standard semantics: Lip Synchronization and Flow Identification (FID). Semantics extensions follow the Standards Action policy (RFC 5226).

7.1.6 Use of "group" and "mid"

All of the "m" lines of a session description that uses "group" MUST be identified with a "mid" attribute regardless of whether they appear in the group line(s). If a session description contains at least one "m" line that has no "mid" identification, the application MUST NOT perform any grouping of media lines. "a=group" lines are used to group together several "m" lines that are identified by their "mid" attribute. "a=group" lines that contain identification-tags that do not correspond to any "m" line within the session description MUST be ignored. The application acts as if the "a=group" line did not exist. The behavior of an application receiving an SDP description with grouped "m" lines is defined by the semantics field in the "a=group" line. There MAY be several "a=group" lines in a session description. The "a=group" lines of a session description can use the same or different semantics. An "m" line identified by its "mid" attribute MAY appear in more than one "a=group" line.

7.1.7 Lip Synchronization

An application that receives a session description that contains "m=" lines that are grouped together using lip-synchronization (LS) semantics MUST synchronize the playout of the corresponding media streams. Note that LS semantics apply not only to a video stream that has to be synchronized with an audio stream; the playout of two streams of the same type can be synchronized as well. For RTP streams, synchronization is typically performed using the RTCP, which provides enough information to map time stamps from the different streams into a local absolute time value. However, the concept of media stream synchronization MAY also apply to media streams that do not make use of RTP. If this is the case, the application MUST recover the original timing relationship between the streams using whatever mechanism is available.

7.1.7.1 Example of LS

The following example shows a session description of a conference that is being multicast. The first media stream (mid:1) contains the voice of the speaker who speaks in English. The second media stream (mid:2) contains the video component, and the third (mid:3) media stream carries the translation to Spanish of what the speaker is saying. The first and second media streams have to be synchronized.

```
v=0
o=Laura 289083124 289083124 IN IP4 two.example.com
c=IN IP4 233.252.0.1/127
t=0 0
a=group:LS 1 2
m=audio 30000 RTP/AVP 0
a=mid:1
m=video 30002 RTP/AVP 31
a=mid:2
m=audio 30004 RTP/AVP 0
i=This media stream contains the Spanish translation
a=mid:3
```

Note: Although the third media stream is not present in the group line, it still has to contain a "mid" attribute (mid:3), as stated before.

7.1.8 Flow Identification

An "m" line in an SDP session description defines a media stream. However, SDP does not define what a media stream is. This definition can be found in the Real-Time Streaming Protocol (RTSP) specification. The RTSP (RFC 2326) defines a media stream as "a single media instance, e.g., an audio stream or a video stream as well as a single whiteboard or shared application group. When using RTP, a stream consists of all RTP and RTCP packets created by a source within an RTP session." This definition assumes that a single audio (or video) stream maps into an RTP session. The RTP RFC 1889 (that is made obsolete by RFC 3550) used to define an RTP session as follows: "For each participant, the session is defined by a particular pair of destination transport addresses (one network address plus a port pair for RTP and RTCP)."

While the previous definitions cover the most common cases, there are situations where a single media instance (e.g., an audio stream or a video stream) is sent using more than one RTP session. Two examples (among many others) of this kind of situation are cellular systems using the Session Initiation Protocol (SIP) (RFC 3261, also see Section 2.1) and systems receiving dual-tone multifrequency (DTMF) tones on a host different from the voice.

7.1.8.1 SIP and Cellular Access

Systems using a cellular access and SIP as the signaling protocol need to receive media over the air. During a session, the media can be encoded using different codecs. The encoded media has to traverse the radio interface. The radio interface is generally characterized as being prone to bit errors and associated with relatively high packet transfer delays. In addition, radio interface resources in a cellular environment are scarce and thus expensive, which calls for special measures in providing a highly efficient transport. In order to get an appropriate speech quality in combination with an efficient transport, precise knowledge of codec properties is required so that a proper radio bearer for the RTP session can be configured before transferring the media. These radio bearers are dedicated bearers per media type (i.e., codec). Cellular systems typically configure different radio bearers on different port numbers. Therefore, incoming media has to have different destination port numbers for the different possible codecs in order to be routed properly to the correct radio bearer. Thus, this is an example in which several RTP sessions are used to carry a single media instance (the encoded speech from the sender).

7.1.8.2 DTMF Tones

Some voice sessions include DTMF tones. Sometimes, the voice handling is performed by a host different from the DTMF handling. It is common to have an application server in the network gathering DTMF tones for the user while the user receives the encoded speech on his user agent. In this situation, it is necessary to establish two RTP sessions: one for the voice and the other for the DTMF tones. Both RTP sessions are logically part of the same media instance.

7.1.8.3 Media Flow Definition

The previous examples show that the definition of a media stream in RFC 2326 does not cover some scenarios. It cannot be assumed that a single media instance maps into a single RTP session. Therefore, we introduce the definition of a media flow: A media flow consists of a single media instance, e.g., an audio stream or a video stream as well as a single whiteboard or shared application group. When using RTP, a media flow comprises one or more RTP sessions.

7.1.8.4 FID Semantics

Several "m" lines grouped together using FID semantics form a media flow. A media agent handling a media flow that comprises several "m" lines MUST send a copy of the media to every "m" line that is part of the flow as long as the codecs and the direction attribute present in a particular "m" line allow it. It is assumed that the application uses only one codec at a time to encode the media produced. This codec MAY change dynamically during the session, but at any particular moment, only one codec is in use. The application encodes the media using the current codec and checks, one by one, all of the "m" lines that are part of the flow. If a particular "m" line contains the codec being used and the direction attribute is "sendonly" or "sendrecv," a copy of the encoded media is sent to the address/port specified in that particular media stream. If either the "m" line does not contain the codec being used or the direction attribute is neither "sendonly" nor "sendrecv," nothing is sent over this media stream. The application typically ends up sending media to different destinations (IP address/port number) depending on the codec used at any moment.

7.1.8.4.1 Examples of FID

The session description below might be sent by an SIP user agent using cellular access. The user agent supports GSM (Global System for Mobile communications) on port 30000 and Adaptive Multi-Rate (AMR) on port 30002. When the remote party sends GSM, it will send RTP packets to port number 30000. When AMR is the codec chosen, packets will be sent to port 30002. Note that the remote party can switch between codecs dynamically in the middle of the session. However, in this example, only one media stream at a time carries voice. The other remains "muted" while its corresponding codec is not in use.

```
v=0
o=Laura 289083124 289083124 IN IP4 three.example.com
c=IN IP4 192.0.2.1
```

```
t=0 0
a=group:FID 1 2
m=audio 30000 RTP/AVP 3
a=rtpmap:3 GSM/8000
a=mid:1
m=audio 30002 RTP/AVP 97
a=rtpmap:97 AMR/8000
a=fmtp:97 mode-set=0,2,5,7; mode-change-period=2;
mode-change-neighbor; maxframes=1
a=mid:2
```

(The linebreak in the fmtp line accommodates RFC formatting restrictions; SDP does not have continuation lines.)

In the previous example, a system receives media on the same IP address on different port numbers. The following example shows how a system can receive different codecs on different IP addresses.

```
v=0
o=Laura 289083124 289083124 IN IP4 four.example.com
c=IN IP4 192.0.2.1
t=0 0
a=group:FID 1 2
m=audio 20000 RTP/AVP 0
c=IN IP4 192.0.2.2
a=rtpmap:0 PCMU/8000
a=mid:1
m=audio 30002 RTP/AVP 97
a=rtpmap:97 AMR/8000
a=fmtp:97 mode-set=0,2,5,7; mode-change-period=2;
mode-change-neighbor; maxframes=1
a=mid:2
```

(The linebreak in the fmtp line accommodates RFC formatting restrictions; SDP does not have continuation lines.)

The cellular terminal in this example only supports the AMR codec. However, many current IP phones only support Pulse-Code Modulation (PCM) with payload 0. In order to be able to interoperate with them, the cellular terminal uses a transcoder whose IP address is 192.0.2.2. The cellular terminal includes the transcoder IP address in its SDP description to provide support for PCM. Remote systems will send AMR directly to the terminal, but PCM will be sent to the transcoder. The transcoder will be configured (using whatever method is preferred) to convert the incoming PCM audio to AMR and send it to the terminal. The next example shows how the "group" attribute used with FID semantics can indicate the use of two different codecs in the two directions of a bidirectional media stream.

```
v=0
o=Laura 289083124 289083124 IN IP4 five.example.com
c=IN IP4 192.0.2.1
t=0 0
a=group:FID 1 2
m=audio 30000 RTP/AVP 0
a=mid:1
m=audio 30002 RTP/AVP 8
```

```
a=recvonly
a=mid:2
```

A user agent that receives the previous SDP description knows that, at a certain moment, it can send either PCM μ-law to port number 30000 or PCM A-law to port number 30002. However, the media agent also knows that the other end will only send PCM μ-law (payload 0). The following example shows a session description with different "m" lines grouped together using FID semantics that contain the same codec.

```
v=0
o=Laura 289083124 289083124 IN IP4 six.example.com
c=IN IP4 192.0.2.1
t=0 0
a=group:FID 1 2 3
m=audio 30000 RTP/AVP 0
a=mid:1
m=audio 30002 RTP/AVP 8
a=mid:2
m=audio 20000 RTP/AVP 0 8
c=IN IP4 192.0.2.2
a=recvonly
a=mid:3
```

At a particular point in time, if the media agent receiving the SDP message above is sending PCM μ-law (payload 0), it sends RTP packets to 192.0.2.1 on port 30000 and to 192.0.2.2 on port 20000 (first and third "m" lines). If it is sending PCM A-law (payload 8), it sends RTP packets to 192.0.2.1 on port 30002 and to 192.0.2.2 on port 20000 (second and third "m" lines). The system that generated the previous SDP description supports PCM with μ-law algorithm on port 30000 and PCM with A-law algorithm (PCMA) on port 30002. Besides, it uses an application server, whose IP address is 192.0.2.2, that records the conversation. The application server does not need to understand the media content, so it always receives a copy of the media stream, regardless of the codec and payload type (PT) being used. That is why the application server always receives a copy of the audio stream regardless of the codec being used at any given moment (it actually performs an RTP dump, so it can effectively receive any codec). Remember that if several "m" lines grouped together using the FID semantics contain the same codec, the media agent MUST send copies of the same media stream as several RTP sessions at the same time. The last example in this section deals with DTMF tones. DTMF tones can be transmitted using a regular voice codec or can be transmitted as telephony events. The RTP payload for DTMF tones treated as telephone events is described in RFC 4733. Following is an example of an SDP session description using FID semantics and this PT.

```
v=0
o=Laura 289083124 289083124 IN IP4 seven.example.com
c=IN IP4 192.0.2.1
t=0 0
a=group:FID 1 2
m=audio 30000 RTP/AVP 0
a=mid:1
m=audio 20000 RTP/AVP 97
c=IN IP4 192.0.2.2
a=rtpmap:97 telephone-events
a=mid:2
```

The remote party would send PCM encoded voice (payload 0) to 192.0.2.1 and DTMF tones encoded as telephony events to 192.0.2.2. Note that only voice or DTMF is sent at a particular point in time. When DTMF tones are sent, the first media stream does not carry any data and, when voice is sent, there is no data in the second media stream. FID semantics provide different destinations for alternative codecs.

7.1.8.5 Scenarios That FID Does Not Cover

It is worthwhile mentioning some scenarios where the "group" attribute using existing semantics (particularly FID) might seem to be applicable but is not.

7.1.8.5.1 Parallel Encoding Using Different Codecs

FID semantics are useful when the application only uses one codec at a time. An application that encodes the same media using different codecs simultaneously MUST NOT use FID to group those media lines. Some systems that handle DTMF tones are a typical example of parallel encoding using different codecs. Some systems implement the RTP payload defined in RFC 4733, but when they send DTMF tones, they do not mute the voice channel. Therefore, in effect they are sending two copies of the same DTMF tone: encoded as voice and encoded as a telephony event. When the receiver gets both copies, it typically uses the telephony event rather than the tone encoded as voice. FID semantics MUST NOT be used in this context to group both media streams, since such a system does not use alternative codecs but rather uses different parallel encodings for the same information.

7.1.8.5.2 Layered Encoding

Layered encoding schemes encode media in different layers. The quality of the media stream at the receiver varies depending on the number of layers received. SDP provides a means to group together contiguous multicast addresses that transport different layers. The "c" line below:

```
c=IN IP4 233.252.0.1/127/3 is equivalent to the following three
  "c" lines:

c=IN IP4 233.252.0.1/127
c=IN IP4 233.252.0.2/127
c=IN IP4 233.252.0.3/127
```

FID MUST NOT be used to group "m" lines that do not represent the same information. Therefore, FID MUST NOT be used to group "m" lines that contain the different layers of layered encoding schemes. Besides, we do not define new group semantics to provide a more flexible way of grouping different layers, because the already existing SDP mechanism covers the most useful scenarios. Since the existing SDP mechanism already covers the most useful scenarios, we do not define a new group semantics to define a more flexible way of grouping different layers.

7.1.8.5.3 Same IP Address and Port Number

If media streams using several different codecs have to be sent to the same IP address and port, the traditional SDP syntax of listing several codecs in the same "m" line MUST be used. FID MUST

NOT be used to group "m" lines with the same IP address/port. Therefore, an SDP description like the following MUST NOT be generated.

```
v=0
o=Laura 289083124 289083124 IN IP4 eight.example.com
c=IN IP4 192.0.2.1
t=0 0
a=group:FID 1 2
m=audio 30000 RTP/AVP 0
a=mid:1
m=audio 30000 RTP/AVP 8
a=mid:2
```

The correct SDP description for the session above would be the following:

```
v=0
o=Laura 289083124 289083124 IN IP4 nine.example.com
c=IN IP4 192.0.2.1
t=0 0
m=audio 30000 RTP/AVP 0 8
```

If two "m" lines are grouped using FID, they MUST differ in their transport addresses (i.e., IP address plus port).

7.1.9 Usage of the "group" Attribute in SIP

SDP descriptions are used by several different protocols, SIP among them. We include a section about SIP, because the "group" attribute will most likely be used mainly by SIP systems. SIP (RFC 3261, also see Section 1.2) is an application layer protocol for establishing, terminating, and modifying multimedia sessions. SIP carries session descriptions in the bodies of the SIP messages but is independent from the protocol used for describing sessions. SDP (RFC 4566, see Section 2) can be used for this purpose. At session establishment, SIP provides a three-way handshake (INVITE-200 OK-ACK) between end systems. However, just two of these three messages carry SDP, as described in RFC 3264 (see Section 3.1).

7.1.9.1 Mid Value in Answers

The "mid" attribute is an identifier for a particular media stream. Therefore, the "mid" value in the offer MUST be the same as the "mid" value in the answer. Besides, subsequent offers (e.g., a re-INVITE) SHOULD use the same "mid" value for the already existing media streams. RFC 3264 (see Section 3.1) describes the usage of SDP in text of SIP. The offerer and the answerer align their media description so that the nth media stream ("m=" line) in the offerer's session description corresponds to the nth media stream in the answerer's description. The presence of the "group" attribute in an SDP session description does not modify this behavior. Since the "mid" attribute provides a means to label "m" lines, it would be possible to perform media alignment using "mid" labels rather than matching nth "m" lines. However, this would not bring any gain and would add complexity to implementations. Therefore, SIP systems MUST perform media alignment matching nth lines regardless of the presence of the "group" or "mid" attributes. If a media stream that contained a particular "mid" identifier in the offer contains a different identifier in the answer, the application ignores all of the "mid" and "group" lines that might appear in the session description. The following example illustrates this scenario.

7.1.9.1.1 Example

Two SIP entities exchange SDPs during session establishment. The INVITE contains the SDP description

```
v=0
o=Laura 289083124 289083124 IN IP4 ten.example.com
c=IN IP4 192.0.2.1
t=0 0
a=group:FID 1 2
m=audio 30000 RTP/AVP 0 8
a=mid:1
m=audio 30002 RTP/AVP 0 8
a=mid:2
```

The 200 OK response contains the following SDP description:

```
v=0
o=Bob 289083122 289083122 IN IP4 eleven.example.com
c=IN IP4 192.0.2.3
t=0 0
a=group:FID 1 2
m=audio 25000 RTP/AVP 0 8
a=mid:2
m=audio 25002 RTP/AVP 0 8
a=mid:1
```

Since alignment of "m" lines is performed based on the matching of nth lines, the first stream had "mid:1" in the INVITE and "mid:2" in the 200 OK. Therefore, the application ignores every "mid" and "group" line contained in the SDP description. A well-behaved SIP user agent would have returned the SDP description below in the 200 OK response.

```
v=0
o=Bob 289083122 289083122 IN IP4 twelve.example.com
c=IN IP4 192.0.2.3
t=0 0
a=group:FID 1 2
m=audio 25002 RTP/AVP 0 8
a=mid:1
m=audio 25000 RTP/AVP 0 8
a=mid:2
```

7.1.10 Group Value in Answers

A SIP entity that receives an offer that contains an "a=group" line with semantics that it does not understand MUST return an answer without the "group" line. Note that, as described in the previous Section 7.1.8.9, the "mid" lines MUST still be present in the answer. A SIP entity that receives an offer that contains an "a=group" line with semantics that are understood MUST return an answer that contains an "a=group" line with the same semantics. The identification-tags contained in this "a=group" line MUST be the same as those received in the offer or a subset of them (zero identification-tags is a valid subset). When the identification-tags in the answer are a subset, the "group" value to be used in the session MUST be the one present in the answer. SIP entities refuse media streams by setting the port to zero in the corresponding "m" line. "a=group" lines MUST NOT contain identification-tags that correspond to "m" lines with the port set to zero. Note that grouping of "m" lines MUST always be requested by the offerer but never by the answerer. Since SIP provides a two-way SDP exchange, an answerer that requested grouping would not know whether the "group" attribute was accepted by the offerer. An answerer that wants to group media lines issues another offer after having responded to the first one (in a re-INVITE, for instance).

7.1.10.1 Example

The following example shows how the callee refuses a media stream offered by the caller by setting its port number to zero. The "mid" value corresponding to that media stream is removed from the "group" value in the answer. SDP description in the INVITE from caller to callee:

```
v=0
o=Laura 289083124 289083124 IN IP4 thirteen.example.com
c=IN IP4 192.0.2.1
t=0 0
a=group:FID 1 2 3
m=audio 30000 RTP/AVP 0
a=mid:1
m=audio 30002 RTP/AVP 8
a=mid:2
m=audio 30004 RTP/AVP 3
a=mid:3
```

SDP description in the INVITE from callee to caller:

```
v=0
o=Bob 289083125 289083125 IN IP4 fourteen.example.com
c=IN IP4 192.0.2.3
t=0 0
a=group:FID 1 3
m=audio 20000 RTP/AVP 0
a=mid:1
m=audio 0 RTP/AVP 8
a=mid:2
m=audio 20002 RTP/AVP 3
a=mid:3
```

7.1.10.2 Capability Negotiation

A client that understands "group" and "mid" but does not want to use these SDP features in a particular session may still want to indicate that it supports these features. To indicate this support, a client can add an "a=3Dgroup" line with no identification-tags for every semantics value it understands. If a server receives an offer that contains empty "a=group" lines, it SHOULD add its capabilities also in the form of empty "a=group" lines to its answer.

7.1.10.2.1 Example

A system that supports both LS and FID semantics but does not want to group any media stream for this particular session generates the following SDP description:

```
v=0
o=Bob 289083125 289083125 IN IP4 fifteen.example.com
c=IN IP4 192.0.2.3
t=0 0
a=group:LS
a=group:FID
m=audio 20000 RTP/AVP 0 8
```

The server that receives that offer supports FID but not LS. It responds with this SDP description:

```
v=0
o=Laura 289083124 289083124 IN IP4 sixteen.example.com
c=IN IP4 192.0.2.1
t=0 0
a=group:FID
m=audio 30000 RTP/AVP 0
```

7.1.10.3 Backward Compatibility

This document does not define any SIP "Require" header field. Therefore, if one of the SIP user agents does not understand the "group" attribute, the standard SDP fallback mechanism MUST be used, namely, attributes that are not understood are simply ignored.

7.1.10.3.1 Offerer Does Not Support "group"

This situation does not represent a problem, because grouping requests are always performed by offerers and not by answerers. If the offerer does not support "group," this attribute will simply not be used.

7.1.10.3.2 Answerer Does Not Support "group"

The answerer will ignore the "group" attribute since it does not understand it and will also ignore the "mid" attribute. For LS semantics, the answerer might decide to perform, or not to perform, synchronization between media streams. For FID semantics, the answerer will consider the session to consist of several media streams. Different implementations will behave in different ways. In the case of audio and different "m" lines for different codecs, an implementation might decide to act as a mixer with the different incoming RTP sessions, which is the correct behavior.

An implementation might also decide to refuse the request (e.g., 488 Not Acceptable Here or 606 Not Acceptable) because it contains several "m" lines. In this case, the server does not support the type of session that the caller wanted to establish. In case the client is willing to establish a simpler session anyway, the client can retry the request without the "group" attribute and with only one "m" line per flow.

7.1.11 Changes from RFC 3388

Section 7.1.3 (Overview of Operation) has been added for clarity. The AMR and GSM acronyms are now expanded on their first use. The examples now use IP addresses in the range suitable for examples. The grouping mechanism is now defined as an extensible framework. Earlier, RFC 3388 used to discourage extensions to this mechanism in favor of new SDPs. Given a semantics value, RFC 3388 used to restrict "m" line identifiers to only appear in a single group using that semantic.

That restriction has been lifted in this specification. From conversations with implementers, existing (i.e., legacy) implementations enforce this restriction on a per-semantics basis. That is, they only enforce this restriction for supported semantics. Because of the nature of existing semantics, implementations will only use a single "m" line identifier across groups using a given semantic even after the restriction has been lifted by this specification. Consequently, the lifting of this restriction will not cause backward compatibility problems because implementations supporting new semantics will be updated to not enforce this restriction at the same time as they are updated to support the new semantics.

7.1.12 Security Considerations

The security procedures that need to be considered in dealing with the SDP messages described in this section (RFC 5888) are provided in Section 17.6.1.

7.1.13 Internet Assigned Numbers Authority Considerations

SDP parameters that are described here have been registered with Internet Assigned Numbers Authority (IANA). We have not included the IANA registration procedures here for the sake of brevity. The procedures for registration of new SDP parameters are described in RFC 5888.

7.2 Negotiating Media Multiplexing/Bundling Using SDP

This section describes the IETF draft [1] (draft-ietf-mmusic-sdp-bundle-negotiation-29.txt) that defines a new SDP grouping framework extension, "BUNDLE."

The extension can be used with the SDP Offer/Answer mechanism to negotiate the usage of a single address:port combination (BUNDLE address) for receiving media, referred to as bundled media, specified by multiple SDP media descriptions ("m=" lines).

To assist endpoints in negotiating the use of bundle, this specification defines a new SDP attribute, "bundle-only," which can be used to request that specific media is only used if bundled. There are multiple ways to correlate the bundled RTP packets with the appropriate media descriptions. This specification defines a new RTP SDES item and a new RTP header extension that provides an additional way to do this correlation by using these attributes to carry a value that associates the RTP/RTCP packets with a specific media description.

7.2.1 Introduction

This specification [1] (draft-ietf-mmusic-sdp-bundle-negotiation-29.txt) defines a way to use a single address:port combination (BUNDLE address) for receiving media specified by multiple SDP media descriptions ("m=" lines). This specification defines a new SDP grouping framework (RFC 5888, see Section 7.1) extension called "BUNDLE." The extension can be used with the SDP Offer/Answer mechanism (RFC 3264, see Section 3.1) to negotiate the usage of a BUNDLE group. Within the BUNDLE group, a BUNDLE address is used for receiving media specified by multiple "m=" lines. This is referred to as bundled media. The offerer and answerer in RFC 3264 (see Section 3.1) use the BUNDLE extension to negotiate the BUNDLE addresses, one for the offerer (offerer BUNDLE address) and one for the answerer (answerer BUNDLE address), to be used for receiving the bundled media specified by a BUNDLE group. Once the offerer and the answerer have negotiated a BUNDLE group, they associate their respective BUNDLE address with each "m=" line in the BUNDLE group. The BUNDLE addresses are used to receive all media specified by the BUNDLE group.

The use of a BUNDLE group and a BUNDLE address also allows the usage of a single set of Interactive Connectivity Establishment (ICE) [2] [I-D.ietf-ice-rfc5245bis] candidates for multiple "m=" lines. This specification also defines a new SDP attribute, "bundle-only," which can be used to request that specific media is only used if kept within a BUNDLE group. As defined in RFC 4566 (see Section 2), the semantics of assigning the same port value to multiple "m=" lines are undefined, and there is no grouping defined by such means. Instead, an explicit grouping mechanism needs to be used to express the intended semantics. This specification provides such an extension. This specification also updates Sections 3.1.5.1, 3.1.8.1 and 3.1.8.2 of RFC 3264 (see Section 3.1). The update allows an answerer to assign a nonzero port value to an "m=" line in an SDP answer, even if the "m=" line in the associated SDP offer contained a zero port value.

This specification also defines a new RTP (RFC 3550) SDES item and a new RTP header extension that can be used to carry a value that associates RTP/RTCP packets with a specific media description. This can be used to correlate a RTP packet with the correct media. SDP bodies can contain multiple BUNDLE groups. A given BUNDLE address MUST only be associated with a single BUNDLE group. The procedures in this specification apply independently to a given BUNDLE group. All RTP based media flows described by a single BUNDLE group belong to a single RTP session of RFC 3550. The BUNDLE extension is backward compatible. Endpoints that do not support the extension are expected to generate offers and answers without an SDP "group:BUNDLE" attribute and are expected to associate a unique address with each "m=" line within an offer and answer, according to the procedures in RFC 4566 (see Section 2) and RFC 3264 (see Section 3.1).

7.2.2 Terminology

See Table 2.1.

7.2.3 Conventions

Conventions that are used are from IETF RFC 2119.

7.2.4 Applicability Statement

The mechanism in this specification only applies to the SDP (RFC 4566, see Section 2), when used together with the SDP Offer/Answer mechanism (RFC 3264, see Section 3.1). Declarative usage of SDP is out of scope of this document and is thus undefined.

7.2.5 SDP Grouping Framework BUNDLE Extension

This section defines a new SDP grouping framework extension (RFC 5888, see Section 7.1), "BUNDLE." The BUNDLE extension can be used with the SDP Offer/Answer mechanism to negotiate the usage of a single address:port combination (BUNDLE address) for receiving bundled media. A single address:port combination is also used for sending bundled media. The address:port combination used for sending bundled media MAY be the same as the BUNDLE address, used to receive bundled media, depending on whether symmetric RTP (RFC 4961) is used. All media specified by a BUNDLE group share a single 5-tuple (i.e. in addition to using a single address:port combination all bundled media MUST be transported using the same transport-layer protocol [e.g. User Datagram Protocol {UDP} or Transmission Control Protocol {TCP}]).

The BUNDLE extension is indicated using an SDP "group" attribute with a "BUNDLE" semantics value (RFC 5888, see Section 7.1). An identification-tag is associated with each bundled "m=" line, and each identification-tag is listed in the SDP "group:BUNDLE" attribute identification-tag list. Each "m=" line whose identification-tag is listed in the identification-tag list is associated with a given BUNDLE group. SDP bodies can contain multiple BUNDLE groups. Any given bundled "m=" line MUST NOT be associated with more than one BUNDLE group. Section 7.2.8 defines the detailed SDP Offer/Answer procedures for the BUNDLE extension.

7.2.6 SDP "bundle-only" Attribute

This section defines a new SDP media-level attribute (RFC 4566, see Section 2), "bundle-only." "bundle-only" is a property attribute (RFC 4566) and hence has no value.

```
Name: bundle-only
Value: N/A
Usage Level: media
Charset Dependent: no
```

Example:

```
a=bundle-only
```

In order to ensure that an answerer that does not support the BUNDLE extension always rejects a bundled "m=" line, the offerer can assign a zero port value to the "m=" line. According to (RFC 4566, see Section 2), an answerer will reject such "m=" line. By associating an SDP "bundle-only" attribute with such "m=" line, the offerer can request that the answerer accept the "m=" line if the answerer supports the BUNDLE extension and if the answerer keeps the "m=" line within the associated BUNDLE group.

NOTE: Once the offerer BUNDLE address has been selected, the offerer does not need to include the "bundle-only" attribute in subsequent offers. By associating the offerer BUNDLE address with an "m=" line of a subsequent offer, the offerer will ensure that the answerer will either keep the "m=" line within the BUNDLE group or the answerer will have to reject the "m=" line. The usage of the "bundle-only" attribute is only defined for a bundled "m=" line with a zero port value, within an offer. Other usage is unspecified. Section 7.2.8 defines the detailed SDP Offer/Answer procedures for the "bundle-only" attribute.

7.2.7 SDP Information Considerations

7.2.7.1 General

This section describes restrictions associated with the usage of SDP parameters within a BUNDLE group. It also describes, when parameter and attribute values have been associated with each bundled "m=" line, how to calculate a value for the whole BUNDLE group.

7.2.7.2 Connection Data (c=)

The "c=" line nettype value (RFC 4566, see Section 2) associated with a bundled "m=" line MUST be "IN." The "c=" line addrtype value (RFC 4566) associated with a bundled "m=" line MUST be "IP4" or "IP6." The same value MUST be associated with each "m=" line.

NOTE: Extensions to this specification can specify usage of the BUNDLE mechanism for nettype and addrtype values other than those listed above.

7.2.7.3 Bandwidth (b=)

An offerer and answerer MUST use the rules and restrictions defined in [3] [I-D. ietf-mmusic-sdp-mux-attributes] when associating the SDP bandwidth (b=) line with bundled "m=" lines.

7.2.7.4 Attributes (a=)

An offerer and answerer MUST use the rules and restrictions defined in [3] [I-D.ietf-mmusic-sdp-mux-attributes] for when associating SDP attributes with bundled "m=" lines.

7.2.8 SDP Offer/Answer Procedures

7.2.8.1 General

This section describes the SDP Offer/Answer (RFC 3264, see Section 3.1) procedures for

- Negotiating and creating of a BUNDLE group
- Selecting the BUNDLE addresses (offerer BUNDLE address and answerer BUNDLE address)
- Adding an "m=" line to a BUNDLE group
- Moving an "m=" line out of a BUNDLE group
- Disabling an "m=" line within a BUNDLE group

The generic rules and procedures defined in RFC 3264 (see Section 3.1) and RFC 4566 (see Section 2) also apply to the BUNDLE extension. For example, if an offer is rejected by the answerer, the previously negotiated SDP parameters and characteristics (including those associated with a BUNDLE group) apply. Hence, if an offerer generates an offer in which the offerer wants to create a BUNDLE group, and the answerer rejects the offer, the BUNDLE group is not created.

The procedures in this section are independent of the media type or "m=" line proto value represented by a bundled "m=" line. Section 7.2.10 defines additional considerations for RTP based media. Section 7.2.6 defines additional considerations for the usage of the SDP "bundle-only" attribute. Section 7.2.11 defines additional considerations for the usage of ICE [2] [I-D.ietf-ice-rfc5245bis] mechanism. SDP offers and answers can contain multiple BUNDLE groups. The procedures in this section apply independently to a given BUNDLE group.

7.2.8.2 Mux Category Considerations

When an offerer associates a shared address with a bundled "m=" line, the offerer shall associate IDENTICAL and TRANSPORT mux category SDP attributes [3] [I-D.ietf-mmusic-sdp-mux-attributes] with the "m=" line only if the "m=" line is associated with the offerer BUNDLE-tag. Otherwise the offerer MUST NOT associate such SDP attributes with the "m=" line.

When an answerer associates a shared address with a bundled "m=" line, the answerer shall associate IDENTICAL and TRANSPORT category SDP attributes with the "m=" line only if the "m=" line is associated with the answerer BUNDLE-tag. Otherwise the answerer MUST NOT associate such SDP attributes with the "m=" line.

NOTE: As bundled "m=" lines associated with a shared address will share the same IDENTICAL and TRANSPORT mux category SDP attributes and attribute values, there is no need to associate such SDP attributes with each "m=" line.

7.2.8.3 Generating the Initial SDP Offer

7.2.8.3.1 General

When an offerer generates an initial offer, in order to create a BUNDLE group, it MUST

- Assign a unique address to each "m=" line within the offer, following the procedures in RFC 3264 (see Section 3.1), unless the media line is a "bundle-only" "m=" line (see Sections 7.2.8.3.2–7.2.8.5);
- Add an SDP "group:BUNDLE" attribute to the offer
- Place the identification-tag of each bundled "m=" line in the SDP "group:BUNDLE" attribute identification-tag list
- Indicate which unique address the offerer suggests as the offerer BUNDLE address (Section 7.2.8.3.2)

If the offerer wants to request that the answerer accept a given bundled "m=" line only if the answerer keeps the "m=" line within the BUNDLE group, the offerer MUST

- Associate an SDP "bundle-only" attribute (Section 7.2.8.3.2) with the "m=" line
- Assign a zero port value to the "m=" line

NOTE: If the offerer assigns a zero port value to an "m=" line, but does not also associate an SDP "bundle-only" attribute with the "m=" line, it is an indication that the offerer wants to disable the "m=" line (Section 7.2.8.6.5). Section 7.2.17.1 shows an example of an initial offer.

7.2.8.3.2 Suggesting the Offerer BUNDLE Address

In the offer, the address associated with the "m=" line associated with the offerer BUNDLE-tag indicates the address that the offerer suggests as the offerer BUNDLE address. The "m=" line associated with the offerer BUNDLE-tag MUST NOT contain a zero port value or an SDP "bundle-only" attribute.

7.2.8.4 Generating the SDP Answer

7.2.8.4.1 General

When an answerer generates an answer that contains a BUNDLE group, the following general SDP grouping framework restrictions, defined in RFC 5888 (see Section 7.1), also apply to the BUNDLE group:

- The answerer MUST NOT include a BUNDLE group in the answer, unless the offerer requested the BUNDLE group to be created in the corresponding offer.

- The answerer MUST NOT include an "m=" line within a BUNDLE group, unless the offerer requested the "m=" line to be within that BUNDLE group in the corresponding offer.

If the answer contains a BUNDLE group, the answerer MUST

- Select an Offerer BUNDLE Address (Section 7.2.8.4.2)
- Select an Answerer BUNDLE Address (Section 7.2.8.4.3)

The answerer is allowed to select a new answerer BUNDLE address each time it generates an answer to an offer. If the answerer does not want to keep an "m=" line within a BUNDLE group, it MUST

- Move the "m=" line out of the BUNDLE group (Section 7.2.8.4.4) or
- Reject the "m=" line (Section 7.2.8.4.5);

If the answerer keeps a bundle-only "m=" line within the BUNDLE group, it follows the procedures (associates the answerer BUNDLE address with the "m=" line, etc.) for any other "m=" line kept within the BUNDLE group. If the answerer does not want to keep a bundle-only "m=" line within the BUNDLE group, it MUST reject the "m=" line (Section 7.2.8.4.5).

The answerer MUST NOT associate an SDP "bundle-only" attribute with any "m=" line in an answer.

NOTE: If a bundled "m=" line in an offer contains a zero port value, but the "m=" line does not contain an SDP "bundle-only" attribute, it is an indication that the offerer wants to disable the "m=" line (Section 7.2.8.6.5).

7.2.8.4.2 Answerer Selection of Offerer Bundle Address

In an offer, the address (unique or shared) associated with the bundled "m=" line associated with the offerer BUNDLE-tag indicates the address that the offerer suggests as the offerer BUNDLE address (Section 7.2.8.3.2). The answerer MUST check whether that "m=" line fulfils the following criteria:

- The answerer will not move the "m=" line out of the BUNDLE group (Section 7.2.8.4.4).
- The answerer will not reject the "m=" line (Section 7.2.8.4.5).
- The "m=" line does not contain a zero port value.

If all of these criteria are fulfilled, the answerer MUST select the address associated with the "m=" line as the offerer BUNDLE address. In the answer, the answerer BUNDLE-tag represents the "m=" line, and the address associated with the "m=" line in the offer becomes the offerer BUNDLE address. If one or more of the criteria are not fulfilled, the answerer MUST select the next identification-tag in the identification-tag list and perform the same criteria check for the "m=" line associated with that identification-tag. If there are no more identification-tags in the identification-tag list, the answerer MUST NOT create the BUNDLE group. In addition, unless the answerer rejects the whole offer, the answerer MUST apply the answerer procedures for moving an "m=" line out of a BUNDLE group (Section 7.2.8.4.4) to each bundled "m=" line in the offer when creating the answer. Section 7.2.17.1 shows an example of an offerer BUNDLE address selection.

7.2.8.4.3 Answerer Selection of Answerer BUNDLE Address

When the answerer selects a BUNDLE address for itself, referred to as the answerer BUNDLE address, it MUST associate that address with each bundled "m=" line within the created BUNDLE group in the answer. The answerer MUST NOT associate the answerer BUNDLE address with an "m=" line that is not within the BUNDLE group or to an "m=" line that is within another BUNDLE group. Section 7.2.17.1 shows an example of an answerer BUNDLE address selection.

7.2.8.4.4 Moving a Media Description Out of a BUNDLE Group

When an answerer wants to move an "m=" line out of a BUNDLE group, it MUST first check the following criteria:

- In the corresponding offer, the "m=" line is associated with a shared address (e.g., a previously selected offerer BUNDLE address) or
- In the corresponding offer, if an SDP "bundle-only" attribute is associated with the "m=" line and if the "m=" line contains a zero port value.

If either criteria is fulfilled, the answerer MUST reject the "m=" line (Section 7.2.8.4.5).

Otherwise, if in the corresponding offer the "m=" line is associated with a unique address, the answerer MUST associate a unique address with the "m=" line in the answer (the answerer does not reject the "m=" line). In addition, in either case, the answerer MUST NOT place the identification-tag, associated with the moved "m=" line, in the SDP "group" attribute identification-tag list associated with the BUNDLE group.

7.2.8.4.5 Rejecting a Media Description in a BUNDLE Group

When an answerer rejects an "m=" line, it MUST associate an address with a zero port value with the "m=" line in the answer, according to the procedures in RFC 4566 (see Section 2). In addition, the answerer MUST NOT place the identification-tag, associated with the rejected "m=" line, in the SDP "group" attribute identification-tag list associated with the BUNDLE group.

7.2.8.5 Offerer Processing of the SDP Answer

When an offerer receives an answer, if the answer contains a BUNDLE group, the offerer MUST check that any bundled "m=" line in the answer was indicated as bundled in the corresponding offer. If there is no mismatch, the offerer MUST use the offerer BUNDLE address, selected by the answerer (Section 7.2.8.4.2), as the address for each bundled "m=" line.

NOTE: As the answerer might reject one or more bundled "m=" lines, or move a bundled "m=" line out of a BUNDLE group, each bundled "m=" line in the offer might not be indicated as bundled in the answer. If the answer does not contain a BUNDLE group, the offerer MUST process the answer as a normal answer.

7.2.8.6 Modifying the Session

7.2.8.6.1 General

When an offerer generates a subsequent offer, it MUST associate the previously selected offerer BUNDLE address (Section 7.2.8.4.2) with each bundled "m=" line (including any bundle-only "m=" line), except if

- The offerer suggests a new offerer BUNDLE address (Section 7.2.8.6.2) or
- The offerer wants to add a bundled "m=" line to the BUNDLE group (Section 7.2.8.6.3) or
- The offerer wants to move a bundled "m=" line out of the BUNDLE group (Section 7.2.8.6.4) or
- The offerer wants to disable the bundled "m=" line (Section 7.2.8.6.5)

In addition, the offerer MUST select an offerer BUNDLE-tag (Section 7.2.8.3.2) associated with the previously selected offerer BUNDLE address, unless the offerer suggests a new offerer BUNDLE address.

7.2.8.6.2 Suggesting a New Offerer BUNDLE Address

When an offerer generates an offer, in which it suggests a new offerer BUNDLE address (Section 7.2.8.3.2), the offerer MUST:

- Assign the address (shared address) to each "m=" line within the BUNDLE group or
- Assign the address (unique address) to one bundled "m=" line

In addition, the offerer MUST indicate that the address is the new suggested offerer BUNDLE address (Section 7.2.8.3.2).

NOTE: Unless the offerer associates the new suggested offerer BUNDLE address with each bundled "m=" line, it can associate unique addresses with any number of bundled "m=" lines (and the previously selected offerer BUNDLE address to any remaining bundled "m=" line) if it wants to suggest multiple alternatives for the new offerer BUNDLE address.

7.2.8.6.3 Adding a Media Description to a BUNDLE Group

When an offerer generates an offer, in which it wants to add a bundled "m=" line to a BUNDLE group, the offerer MUST:

- Assign a unique address to the added "m=" line or
- Assign the previously selected offerer BUNDLE address to the added "m=" line or
- If the offerer associates a new (shared address) suggested offerer BUNDLE address with each bundled "m=" line (Section 7.2.8.6.2), also associate that address with the added "m=" line.

In addition, the offerer MUST extend the SDP "group:BUNDLE" attribute identification-tag list with the BUNDLE group (Section 7.2.8.3.2) by adding the identification-tag associated with the added "m=" line to the list.

NOTE: Assigning a unique address to the "m=" line allows the answerer to move the "m=" line out of the BUNDLE group (Section 7.2.8.4.4) without having to reject the "m=" line. If the offerer associates a unique address with the added "m=" line, and if the offerer suggests that address as the new offerer BUNDLE address (Section 7.2.8.6.2), the offerer BUNDLE-tag MUST represent the added "m=" line (Section 7.2.8.3.2).

If the offerer associates a new suggested offerer BUNDLE address with each bundled "m=" line (Section 7.2.8.6.2), including the added "m=" line, the offerer BUNDLE-tag MAY represent the added "m=" line (Section 7.2.8.3.2). Section 7.2.17.3 shows an example in which an offerer sends an offer in order to add a bundled "m=" line to a BUNDLE group.

7.2.8.6.4 Moving a Media Description Out of a BUNDLE Group

When an offerer generates an offer in which it wants to move a bundled "m=" line out of a BUNDLE group it was added to in a previous offer/answer transaction, the offerer

- MUST associate a unique address with the "m=" line and
- MUST NOT place the identification-tag associated with the "m=" line in the SDP "group:BUNDLE" attribute identification-tag list associated with the BUNDLE group.

NOTE: If the removed "m=" line is associated with the previously selected BUNDLE-tag, the offerer needs to suggest a new BUNDLE-tag (Section 7.2.8.3.2).

NOTE: If an "m=" line, when being moved out of a BUNDLE group, is added to another BUNDLE group, the offerer applies the procedures in Section 7.2.8.6.3 to the "m=" line. Section 7.2.17.4 shows an example of an offer for moving an "m=" line out of a BUNDLE group.

7.2.8.6.5 A Media Description in a BUNDLE Group

When an offerer generates an offer, in which it wants to disable a bundled "m=" line (added to the BUNDLE group in a previous offer/answer transaction), the offerer

- MUST associate an address with a zero port value with the "m=" line, following the procedures in RFC 4566 and
- MUST NOT place the identification-tag associated with the "m=" line in the SDP "group:BUNDLE" attribute identification-tag list associated with the BUNDLE group. Section 7.2.17.5 shows an example of an offer for disabling an "m=" line within a BUNDLE group.

7.2.9 Protocol Identification

7.2.9.1 General

Each "m=" line within a BUNDLE group MUST use the same transport layer protocol. If bundled "m=" lines use different protocols on top of the transport-layer protocol, there MUST exist a publicly available specification describing a mechanism for this particular protocol combination, how to associate received data with the correct protocol. In addition, if received data can be associated with more than one bundled "m=" line, there MUST exist a publicly available specification that describes a mechanism for associating the received data with the correct "m=" line.

This document describes a mechanism to identify the protocol of received data among the simple traversal of UDP around NAT (STUN), Datagram Transport Layer Security (DTLS), and Secure RTP (SRTP) protocols (in any combination), when UDP is used as transport-layer protocol, but does not describe how to identify different protocols transported on DTLS. While the mechanism is generally applicable to other protocols and transport-layers protocols, any such use requires further specification around how to multiplex multiple protocols on a given transport-layer protocol and how to associate received data with the correct protocols.

7.2.9.2 STUN, DTLS, SRTP

Section 5.1.2 of RFC 5764 describes a mechanism to identify the protocol of a received packet among the STUN, DTLS, and SRTP protocols (in any combination). If an offer or answer includes bundled "m=" lines that represent these protocols, the offerer or answerer MUST support the mechanism described in RFC 5764, and no explicit negotiation is required in order to indicate support and usage of the mechanism. RFC 5764 does not describe how to identify different protocols transported on DTLS, only how to identify the DTLS protocol itself. If multiple protocols are transported on DTLS, there MUST exist a specification describing a mechanism for identifying each individual protocol. In addition, if a received DTLS packet can be associated with more than one "m=" line, there MUST exist a specification that describes a mechanism for associating the received DTLS packet with the correct "m=" line. Section 7.2.10.2 describes how to associate a received (S)RTP packet with the correct "m=" line.

7.2.10 RTP Considerations

7.2.10.1 Single RTP Session

7.2.10.1.1 General

All RTP-based media within a single BUNDLE group belong to a single RTP session (RFC 3550). Disjoint BUNDLE groups will form multiple RTP sessions, one per BUNDLE group. Since a single RTP session is used for each bundle group, all "m=" lines representing RTP-based media in a bundle group will share a single SSRC numbering space RFC 3550. The following rules and restrictions apply for a single RTP session:

- A specific PT value can be used in multiple bundled "m=" lines if each codec associated with the PT number shares an identical codec configuration (Section 7.2.10.1.2).
- The proto value in each bundled RTP-based "m=" line MUST be identical (e.g., RTP/Audio-Visual-Profile with Feedback.
- The RTP MID header extension MUST be enabled, by associating an SDP "extmap" attribute (RFC 5285), with a "urn:ietf:params:rtphdrext: sdes:mid" Uniform Resource Identifier (URI) value, with each bundled RTP-based "m=" line in every offer and answer.
- A given SSRC MUST NOT transmit RTP packets using PTs that originate from different bundled "m=" lines.

NOTE: This last bullet intends to avoid sending multiple media types from the same SSRC. If transmission of multiple media types is done with time overlap, RTP and RTCP fail to function. Even if done in proper sequence, this causes RTP Timestamp rate switching issues (RFC 7160). However, once an SSRC has left the RTP session (by sending an RTCP BYE packet), that SSRC value can later be reused by another source (possibly associated with a different bundled "m=" line.

7.2.10.1.2 PT Value Reuse

Multiple bundled "m=" lines might represent RTP based media. As all RTP based media specified by a BUNDLE group belong to the same RTP session, in order for a given PT value to be used inside more than one bundled "m=" line, all codecs associated with the PT number MUST share an identical codec configuration. This means that the codecs MUST share the same media type, encoding name, clock rate, and any parameter that can affect the codec configuration and packetization. IETF draft [3] [I-D.ietf-mmusic-sdp-mux-attributes] lists SDP attributes, whose attribute values must be identical for all codecs that use the same PT value.

7.2.10.2 Associating RTP/RTCP Packets with Correct SDP Media Description

Multiple mechanisms can be used by an endpoint to associate received RTP/RTCP packets with a bundled "m=" line. Such mechanisms include using the PT value carried inside the RTP packets, the SSRC values carried inside the RTP packets, and other "m=" line specific information carried inside the RTP packets. As all RTP/RTCP packets associated with a BUNDLE group are received (and sent) using single address:port combinations, the local address:port combination cannot be used to associate received RTP packets with the correct "m=" line.

As described in Section 7.2.10.1.2, the same PT value might be used inside RTP packets described by multiple "m=" lines. In such cases, the PT value cannot be used to associate received RTP packets with the correct "m=" line. An offerer and answerer can inform each other of which SSRC values they will use for RTP and RTCP by using the SDP "ssrc" attribute (RFC 5576, see Section 5.3). To allow for proper association with this mechanism, the "ssrc" attribute needs to be associated with each "m=" line that shares a PT with any other "m=" line in the same bundle.

As the SSRC values will be carried inside the RTP/RTCP packets, the offerer and answerer can then use that information to associate received RTP packets with the correct "m=" line. However, an offerer will not know which SSRC values the answerer will use until it has received the answer providing that information. Thus, before the offerer has received the answer, the offerer will not be able to associate received RTP/RTCP packets with the correct "m=" line using the SSRC values. In order for an offerer and answerer to always be able to associate received RTP and RTCP packets with the correct "m=" line, an offerer and answerer using the BUNDLE extension MUST support the mechanism defined in Section 7.2.14, where the remote endpoint inserts the identification-tag associated with an "m=" line in RTP and RTCP packets associated with that "m=" line.

7.2.10.3 RTP/RTCP Multiplexing

7.2.10.3.1 General

Within a BUNDLE group, the offerer and answerer MUST enable RTP/RTCP multiplexing (RFC 5761) for the RTP-based media specified by the BUNDLE group. When RTP/RTCP multiplexing is enabled, the same address:port combination will be used for sending all RTP packets and the RTCP packets associated with the BUNDLE group. Each endpoint will send the packets toward the BUNDLE address of the other endpoint. The same address:port combination MAY be used for receiving RTP packets and RTCP packets.

7.2.10.3.2 SDP Offer/Answer Procedures

7.2.10.3.2.1 General This section describes how an offerer and answerer use the SDP "rtcp-mux" attribute (RFC 5761) and the SDP "rtcp-mux-only" attribute [3] [I-D.ietf-mmusic-mux-exclusive] to negotiate usage of RTP/RTCP multiplexing for RTP-based media specified by a BUNDLE group. The procedures in this section only apply to RTP-based "m=" lines.

7.2.10.3.2.2 Generating the Initial SDP Offer When an offerer generates an initial offer, the offerer MUST associate either an SDP "rtcp-mux" attribute (RFC 5761) or an SDP "rtcp-mux-only" attribute [3] [I-D.ietf-mmusic-mux-exclusive] with each bundled RTP-based "m=" line in the offer. The offerer MUST associate an SDP "rtcp-mux-only" attribute with each bundle-only "m=" line. If the offerer associates an "rtcp-mux-only" attribute with an "m=" line, the offerer may also associate an "rtcp-mux" attribute with the same "m=" line, as described in [3] [I-D. ietf-mmusic-mux-exclusive].

NOTE: Within a BUNDLE group, the offerer can associate the SDP "rtcp-mux" attribute with some of the RTP-based "m=" lines, while it associates the SDP "rtcp-mux-only" attribute with other RTP-based "m=" lines, depending on whether the offerer supports fallback to usage of a separate port for RTCP in case the answerer does not include the "m=" line in the BUNDLE group.

NOTE: If the offerer associates an SDP "rtcp-mux" attribute with an "m=" line, the offerer can also associate an SDP "rtcp" attribute (RFC 3605, see Section 12.3) with a bundled "m=" line, excluding a bundle-only "m=" line, in order to provide a fallback port for RTCP, as described in RFC 5761. However, the fallback port will only be used in case the answerer does not include the "m=" line in the BUNDLE group in the associated answer. In the initial offer, the address:port combination for RTCP MUST be unique in each bundled RTP-based "m=" line (excluding a "bundle-only" "m=" line), similar to RTP.

7.2.10.3.2.3 Generating the SDP Answer

When an answerer generates an answer, if the answerer accepts one or more RTP-based "m=" lines within a BUNDLE group, the answerer MUST enable usage of RTP/RTCP multiplexing. The answerer MUST associate an SDP "rtcp-mux" attribute with each RTP-based "m=" line in the answer. In addition, if an "m=" line in the corresponding offer contained an SDP "rtcp-mux-only" attribute, the answerer MUST also associate an SDP "rtcp-mux-only" attribute with the "m=" line in the answer. If an RTP-based "m=" line in the corresponding offer did not contain an SDP "rtcp-mux" attribute or an SDP "rtcp-mux-only" attribute, the answerer MUST NOT include the "m=" line within a BUNDLE group in the answer.

If an RTP-based "m=" line in the corresponding offer contained an SDP "rtcp-mux-only" attribute, and if the answerer moves the "m=" line out of the BUNDLE group in the answer Section 7.2.8.4.4, the answerer MUST still either enable RTP/RTCP multiplexing for the media associated with the "m=" line, or reject the "m=" line Section 7.2.8.4.5. The answerer MUST NOT associate an SDP "rtcp" attribute with any bundled "m=" line in the answer. The answerer will use the port value of the selected offerer BUNDLE address for sending RTP and RTCP packets associated with each RTP-based bundled "m=" line toward the offerer. If the usage of RTP/RTCP multiplexing within a BUNDLE group has been negotiated in a previous offer/answer transaction, the answerer MUST associate an SDP "rtcp-mux" attribute with each bundled RTP-based "m=" line in the answer.

7.2.10.3.2.4 Offerer Processing of the SDP Answer

When an offerer receives an answer, if the answerer has accepted the usage of RTP/RTCP multiplexing (see Section 7.2.10.3.2.3), the answerer follows the procedures for RTP/RTCP multiplexing defined in RFC 5761. The offerer will use the port value associated with the answerer BUNDLE address for sending RTP and RTCP packets associated with each RTP-based bundled "m=" line toward the answerer.

NOTE: It is considered a protocol error if the answerer has not accepted the usage of RTP/RTCP multiplexing for RTP-based "m=" lines that the answerer included in the BUNDLE group.

7.2.10.3.2.5 Modifying the Session

When an offerer generates a subsequent offer, it MUST associate an SDP "rtcp-mux" attribute or an SDP "rtcp-mux-only" attribute with each RTP-based bundled "m=" line within the BUNDLE group (including any bundled RTP-based "m="

line that the offerer wants to add to the BUNDLE group), unless the offerer wants to disable or remove the "m=" line from the BUNDLE group.

7.2.11 ICE Considerations

7.2.11.1 General

This section describes how to use the BUNDLE grouping extension together with the ICE mechanism [2] [I-D.ietf-ice-rfc5245bis]. The generic procedures for negotiating usage of ICE using SDP, defined in [4] [I-D.ietf-mmusic-ice-sip-sdp], also apply to usage of ICE with BUNDLE, with the following exceptions:

- When BUNDLE addresses for a BUNDLE group have been selected for both endpoints, ICE connectivity checks and keepalives only need to be performed for the whole BUNDLE group, instead of per bundled "m=" line.
- Among bundled "m=" lines with which the offerer has associated a shared address, the offerer only associates ICE-related media level SDP attributes with the "m=" line associated with the offerer BUNDLE-tag.
- Among bundled "m=" lines with which the answerer has associated a shared address, the answerer only associates ICE-related media level SDP attributes with the "m=" line associated with the answerer BUNDLE-tag. Support and usage of ICE mechanism together with the BUNDLE extension is OPTIONAL.

7.2.11.2 SDP Offer/Answer Procedures

When an offerer associates a unique address with a bundled "m=" line (excluding any bundle-only "m=" line), the offerer MUST associate SDP "candidate" attributes (and other applicable ICE-related media-level SDP attributes), containing unique ICE properties (candidates, etc.), with the "m=" line, according to the procedures in [4] [I-D.ietf-mmusic-ice-sip-sdp]. When an offerer associates a shared address with a bundled "m=" line, if the "m=" line is associated with the offerer BUNDLE-tag, the offerer MUST associate SDP "candidate" attributes (and other applicable ICE-related media-level SDP attributes), containing shared ICE properties, with the "m=" line. If the "m=" line is not associated with the offerer BUNDLE-tag, the offerer MUST NOT associate ICE-related SDP attributes with the "m=" line.

When an answerer associates a shared address with a bundled "m=" line, if the "m=" line is associated with the answerer BUNDLE-tag, the answerer MUST associate SDP "candidate" attributes (and other applicable ICE-related media-level SDP attributes), containing shared ICE properties, with the "m=" line. If the "m=" line is not associated with the answerer BUNDLE-tag, the answerer MUST NOT associate ICE-related SDP attributes with the "m=" line.

NOTE: As most ICE-related media-level SDP attributes belong to the TRANSPORT mux category [3] [I-D.ietf-mmusic-sdp-mux-attributes], the offerer and answerer follow the rules in Section 7.2.8.2. However, in the case of ICE-related media-level attributes, the rules apply to all attributes (see note), even if they belong to a different mux category.

NOTE: The following ICE-related media-level SDP attributes are defined in [4] [I-D.ietf-mmusic-ice-sip-sdp]: "candidate," "remotecandidates," "ice-mismatch," "ice-ufrag," "ice-pwd," and "icepacing."

7.2.11.2.1 Generating the Initial SDP Offer

When an offerer generates an initial offer, the offerer MUST associate ICE-related media-level SDP attributes with each bundled "m=" line, according to Section 7.2.11.2.1.

7.2.11.2.2 Generating the SDP Answer

When an answerer generates an answer that contains a BUNDLE group, the answerer MUST associated ICE-related SDP attributes with the "m=" line associated with the answerer BUNDLE-tag, according to Section 7.2.11.2.1.

7.2.11.2.3 Offerer Processing of the SDP Answer

When an offerer receives an answer, if the answerer supports and uses the ICE mechanism and the BUNDLE extension, the offerer MUST associate the ICE properties associated with the offerer BUNDLE address, selected by the answerer (Section 7.2.8.4.2), with each bundled "m=" line.

7.2.11.2.4 Modifying the Session

When an offerer generates a subsequent offer, it MUST associate unique or shared ICE properties to one or more bundled "m=" lines, according to Section 7.2.11.2.1.

7.2.12 DTLS Considerations

One or more media streams within a BUNDLE group might use the DTLS protocol (RFC 6347) in order to encrypt the data or to negotiate encryption keys if another encryption mechanism is used to encrypt media. When DTLS is used within a BUNDLE group, the following rules apply:

- There can only be one DTLS association (RFC 6347) associated with the BUNDLE group.
- Each usage of the DTLS association within the BUNDLE group MUST use the same mechanism for determining which endpoints (the offerer or answerer) becomes DTLS client and DTLS server.
- Each usage of the DTLS association within the BUNDLE group MUST use the same mechanism for determining whether an offer or answer will trigger the establishment of a new DTLS association or whether an existing DTLS association will be used.
- If the DTLS client supports DTLS-SRTP (RFC 5764), it MUST include the "use_srtp" extension (RFC 5764) in the DTLS ClientHello message (RFC 5764); the client MUST include the extension even if the usage of DTLS-SRTP is not negotiated as part of the multimedia session (e.g., SIP session; RFC 3261, see also Section 1.2).

NOTE: The inclusion of the "use_srtp" extension during the initial DTLS handshake ensures that a DTLS renegotiation will not be required in order to include the extension, in case DTLS-SRTP encrypted media is added to the BUNDLE group later during the multimedia session.

7.2.13 Update to RFC 3264

7.2.13.1 General

This section replaces the text of the following sections of RFC 3264:

- Section 5.1 (Unicast Streams). See Section 7.2.5.1 of this book.
- Section 8.2 (Removing a Media Stream). See Section 7.2.8.2 of this book.
- Section 8.4 (Putting a Unicast Media Stream on Hold). See Section 7.2.8.4 of this book.

7.2.13.2 Original Text of Section 5.1 (2nd Paragraph) of RFC 3264

For recvonly and sendrecv streams, the port number and address in the offer indicate where the offerer would like to receive the media stream. For sendonly RTP streams, the address and port number indirectly indicate where the offerer wants to receive RTCP reports. Unless there is an explicit indication otherwise, reports are sent to the port number one higher than the number indicated. The IP address and port present in the offer indicate nothing about the source IP address and source port of RTP and RTCP packets that will be sent by the offerer. A port number of zero in the offer indicates that the stream is offered but MUST NOT be used. This has no useful semantics in an initial offer but is allowed for reasons of completeness, since the answer can contain a zero port indicating a rejected stream (Section 6 of RFC 3264; see Section 3.1.6 of this book). Furthermore, existing streams can be terminated by setting the port to zero (Section 8 of RFC 3264; see Section 3.1.6 of this book). In general, a port number of zero indicates that the media stream is not wanted.

7.2.13.3 New Text Replacing Section 5.1 (2nd Paragraph) of RFC 3264

Note: For Section 5.1 (2nd paragraph) of RFC 3264, see Section 3.1.5.1 (2nd paragraph) of this book.

For recvonly and sendrecv streams, the port number and address in the offer indicate where the offerer would like to receive the media stream. For sendonly RTP streams, the address and port number indirectly indicate where the offerer wants to receive RTCP reports. Unless there is an explicit indication otherwise, reports are sent to the port number one higher than the number indicated. The IP address and port present in the offer indicate nothing about the source IP address and source port of RTP and RTCP packets that will be sent by the offerer. A port number of zero in the offer by default indicates that the stream is offered but MUST NOT be used, but an extension mechanism might specify different semantics for the usage of a zero port value. Furthermore, existing streams can be terminated by setting the port to zero (Section 3.1.8). In general, a port number of zero by default indicates that the media stream is not wanted.

7.2.13.4 Original Text of Section 8.2 (2nd Paragraph) of RFC 3264

Note: For Section 8.2 (2nd paragraph) of RFC 3264, see Section 3.1.8.2 (2nd paragraph) of this book.

A stream that is offered with a port of zero MUST be marked with port zero in the answer. Like the offer, the answer MAY omit all attributes present previously and MAY list just a single media format from amongst those in the offer.

7.2.13.5 New Text Replacing Section 8.2 (2nd Paragraph) of RFC 3264

Note: For Section 8.2 (2nd paragraph) of RFC 3264, see Section 3.1.8.2 (2nd paragraph) of this book.

A stream that is offered with a port of zero MUST by default be marked with port zero in the answer, unless an extension mechanism, which specifies semantics for the usage of a nonzero port value, is used. If the stream is marked with port zero in the answer, the answer MAY omit all attributes present previously and MAY list just a single media format from amongst those in the offer.

7.2.13.6 Original Text of Section 8.4 (6th Paragraph) of RFC 3264

RFC 2543, that is made obsolete by RFC 3261 (also see Section 1.2), specified that placing a user on hold was accomplished by setting the connection address to 0.0.0.0. Its usage for putting a call on hold is no longer recommended, since it doesn't allow for RTCP to be used with held streams, doesn't work with IPv6, and breaks with connection-oriented media. However, it can be useful in an initial offer when the offerer knows it wants to use a particular set of media streams and formats but doesn't know the addresses and ports at the time of the offer. Of course, when used, the port number MUST NOT be zero, which would specify that the stream has been disabled. An agent MUST be capable of receiving SDP with a connection address of 0.0.0.0, in which case it means that neither RTP nor RTCP should be sent to the peer.

7.2.13.7 New Text Replacing Section 8.4 (6th Paragraph) of RFC 3264

Note: For Section 8.2 (6th paragraph) of RFC 3264, see Section 3.1.8.2 (6th paragraph) of this book.

RFC 2543, that is made obsolete by RFC 3261 (also see Section 1.2), specified that placing a user on hold was accomplished by setting the connection address to 0.0.0.0. Its usage for putting a call on hold is no longer recommended, since it doesn't allow for RTCP to be used with held streams, doesn't work with IPv6, and breaks with connection-oriented media. However, it can be useful in an initial offer when the offerer knows it wants to use a particular set of media streams and formats but doesn't know the addresses and ports at the time of the offer. Of course, when used, the port number MUST NOT be zero, if it would specify that the stream has been disabled. However, an extension mechanism might specify different semantics of the zero port number usage. An agent MUST be capable of receiving SDP with a connection address of 0.0.0.0, in which case it means that neither RTP nor RTCP should be sent to the peer.

7.2.14 RTP/RTCP Extensions for Identification-Tag Transport

7.2.14.1 General

SDP offerers and answerers (RFC 3264, see Section 3.1) can associate identification-tags with "m=" lines within SDP offers and answers, using the procedures in RFC 5888 (see Section 7.1). Each identification-tag uniquely represents an "m=" line. This section defines a new RTCP SDES item (RFC 3550), "MID," which is used to carry identification-tags within RTCP SDES packets. This section also defines a new RTP header extension (RFC 5285), which is used to carry identification-tags in RTP packets. The SDES item and RTP header extension make it possible for a receiver to associate received RTCP and RTP packets with a specific "m=" line, with which the receiver has associated an identification-tag, even if those "m=" lines are part of the same RTP session. A media recipient informs the media sender about the identification-tag associated with an "m=" line through the use of a "mid" attribute (RFC 5888, see Section 7.1). The media sender then inserts the identification-tag in RTCP and RTP packets sent to the media recipient.

NOTE: This text defines how identification-tags are carried in SDP offers and answers. The usage of other signaling protocols for carrying identification-tags is not prevented, but the usage of such protocols is outside the scope of this document. RFC 3550 defines general procedures regarding the RTCP transmission interval. The RTCP MID SDES item SHOULD be sent in the first few RTCP packets sent on joining the session, and SHOULD be sent regularly thereafter. The exact number of RTCP packets in which this SDES item is sent is intentionally not specified here, as it will depend on the expected packet loss rate, the RTCP reporting interval, and the allowable overhead.

The RTP MID header extension SHOULD be included in some RTP packets at the start of the session and whenever the SSRC changes. It might also be useful to include the header extension in RTP packets that comprise random access points in the media (e.g., with video I-frames). The exact number of RTP packets in which this header extension is sent is intentionally not specified here, as it will depend on expected packet loss rate and loss patterns, the overhead the application can tolerate, and the importance of immediate receipt of the identification-tag. For robustness purpose, endpoints need to be prepared for situations where the reception of the identification-tag is delayed and SHOULD NOT terminate sessions in such cases, as the identification-tag is likely to arrive soon.

7.2.14.2 RTCP MID SDES Item

```
 0                   1                   2                   3
 0 1 2 3 4 5 6 7 8 9 0 1 2 3 4 5 6 7 8 9 0 1 2 3 4 5 6 7 8 9 0 1
-----------------------------------------------------------------
| MID = assigned SDES |     Length     |                            |
|  identifier value   |                |  Identification-Tag  ....  |
-----------------------------------------------------------------
```

The identification-tag payload is UTF-8 encoded, as in SDP. The identification-tag is not zero terminated.

7.2.14.3 RTP MID Header Extension

The payload, containing the identification-tag, of the RTP MID header extension element can be encoded using either the one-byte or two-byte header (RFC 5285). The identification-tag payload is UTF-8 encoded, as in SDP. The identification-tag is not zero terminated. Note that the set of header extensions included in the packet needs to be padded to the next 32-bit boundary using zero bytes (RFC 5285). As the identification-tag is included in either an RTCP SDES item or an RTP header extension, or both, there should be some consideration about the packet expansion caused by the identification-tag. To avoid Maximum Transmission Unit issues for the RTP packets, the header extension's size needs to be taken into account when encoding the media.

It is recommended that the identification-tag be kept short. Due to the properties of the RTP header extension mechanism, when using the one-byte header, a tag that is 1 to 3 bytes will result in a minimal number of 32-bit words used for the RTP header extension, in case no other header extensions are included at the same time.

Note: It does take into account that some single characters, when UTF-8 encoded, will result in multiple octets.

7.2.15 IANA Considerations

SDP parameters that are described here have been registered with IANA. We have not included the IANA registration procedures here for the sake of brevity. The procedures for registration of new SDP parameters are described in the IETF draft [1] [draft-ietf-mmusic-sdp-bundle-negotiation-29.txt].

7.2.16 Security Considerations

The security procedures that need to be considered in dealing with the SDP messages described in this section [1] (draft-ietf-mmusic-sdp-bundle-negotiation-29.txt) are provided in Section 17.6.2.

7.2.17 Examples

7.2.17.1 Example: Bundle Address Selection

The following example shows

- An offer, in which the offerer associates a unique address with each bundled "m=" line within the BUNDLE group
- An answer, in which the answerer selects the offerer BUNDLE address and that selects its own BUNDLE address (the answerer BUNDLE address) and associates it with each bundled "m=" line within the BUNDLE group.

```
SDP Offer (1)
v=0
o=alice 2890844526 2890844526 IN IP4 atlanta.example.com
s=
c=IN IP4 atlanta.example.com
t=0 0
a=group:BUNDLE foo bar
m=audio 10000 RTP/AVP 0 8 97
b=AS:200
a=mid:foo
a=rtpmap:0 PCMU/8000
a=rtpmap:8 PCMA/8000
a=rtpmap:97 iLBC/8000
a=extmap 1 urn:ietf:params:rtp-hdrext:sdes:mid
m=video 10002 RTP/AVP 31 32
b=AS:1000
a=mid:bar
a=rtpmap:31 H261/90000
a=rtpmap:32 MPV/90000
a=extmap 1 urn:ietf:params:rtp-hdrext:sdes:mid
SDP Answer (2)
v=0
```

```
o=bob 2808844564 2808844564 IN IP4 biloxi.example.com
s=
c=IN IP4 biloxi.example.com
t=0 0
a=group:BUNDLE foo bar
m=audio 20000 RTP/AVP 0
b=AS:200
a=mid:foo
a=rtpmap:0 PCMU/8000
a=extmap 1 urn:ietf:params:rtp-hdrext:sdes:mid
m=video 20000 RTP/AVP 32
b=AS:1000
a=mid:bar
a=rtpmap:32 MPV/90000
a=extmap 1 urn:ietf:params:rtp-hdrext:sdes:mid
```

7.2.17.2 Example: BUNDLE Extension Rejected

The following example shows

- An offer, in which the offerer associates a unique address with each bundled "m=" line within the BUNDLE group
- An answer, in which the answerer rejects the offered BUNDLE group and associates a unique address with each "m=" line (following normal RFC 3264, see Section 3.1 procedures).

```
SDP Offer (1)
v=0
o=alice 2890844526 2890844526 IN IP4 atlanta.example.com
s=
c=IN IP4 atlanta.example.com
t=0 0
a=group:BUNDLE foo bar
m=audio 10000 RTP/AVP 0 8 97
b=AS:200
a=mid:foo
a=rtpmap:0 PCMU/8000
a=rtpmap:8 PCMA/8000
a=rtpmap:97 iLBC/8000
a=extmap 1 urn:ietf:params:rtp-hdrext:sdes:mid
m=video 10002 RTP/AVP 31 32
b=AS:1000
a=mid:bar
a=rtpmap:31 H261/90000
a=rtpmap:32 MPV/90000
a=extmap 1 urn:ietf:params:rtp-hdrext:sdes:mid
SDP Answer (2)
v=0
o=bob 2808844564 2808844564 IN IP4 biloxi.example.com
s=
c=IN IP4 biloxi.example.com
t=0 0
```

```
m=audio 20000 RTP/AVP 0
b=AS:200
a=rtpmap:0 PCMU/8000
m=video 30000 RTP/AVP 32
b=AS:1000
a=rtpmap:32 MPV/90000
```

7.2.17.3 Example: Offerer Adds a Media Description to a BUNDLE Group

The following example shows

- A subsequent offer (the BUNDLE group has been created as part of a previous offer/answer transaction), in which the offerer adds a new "m=" line, represented by the "zen" identification-tag, to a previously negotiated BUNDLE group, associates a unique address with the added "m=" line, and associates the previously selected offerer BUNDLE address with each of the other bundled "m=" lines within the BUNDLE group
- An answer, in which the answerer associates the answerer BUNDLE address with each bundled "m=" line (including the newly added "m=" line) within the BUNDLE group.

```
SDP Offer (1)
v=0
o=alice 2890844526 2890844526 IN IP4 atlanta.example.com
s=
c=IN IP4 atlanta.example.com
t=0 0
a=group:BUNDLE foo bar zen
m=audio 10000 RTP/AVP 0 8 97
b=AS:200
a=mid:foo
a=rtpmap:0 PCMU/8000
a=rtpmap:8 PCMA/8000
a=rtpmap:97 iLBC/8000
a=extmap 1 urn:ietf:params:rtp-hdrext:sdes:mid
m=video 10000 RTP/AVP 31 32
b=AS:1000
a=mid:bar
a=rtpmap:31 H261/90000
a=rtpmap:32 MPV/90000
a=extmap 1 urn:ietf:params:rtp-hdrext:sdes:mid
m=video 20000 RTP/AVP 66
b=AS:1000
a=mid:zen
a=rtpmap:66 H261/90000
a=extmap 1 urn:ietf:params:rtp-hdrext:sdes:mid
SDP Answer (2)
v=0
o=bob 2808844564 2808844564 IN IP4 biloxi.example.com
s=
c=IN IP4 biloxi.example.com
t=0 0
a=group:BUNDLE foo bar zen
m=audio 20000 RTP/AVP 0
```

```
b=AS:200
a=mid:foo
a=rtpmap:0 PCMU/8000
Holmberg, et al. Expires October 17, 2016 [Page 35]
Internet-Draft Bundled media April 2016
a=extmap 1 urn:ietf:params:rtp-hdrext:sdes:mid
m=video 20000 RTP/AVP 32
b=AS:1000
a=mid:bar
a=rtpmap:32 MPV/90000
a=extmap 1 urn:ietf:params:rtp-hdrext:sdes:mid
m=video 20000 RTP/AVP 66
b=AS:1000
a=mid:zen
a=rtpmap:66 H261/90000
a=extmap 1 urn:ietf:params:rtp-hdrext:sdes:mid
```

7.2.17.4 Example: Offerer Moves a Media Description out of a BUNDLE Group

The following example shows

- A subsequent offer (the BUNDLE group has been created as part of a previous offer/answer transaction) in which the offerer moves a bundled "m=" line out of a BUNDLE group, associates a unique address with the moved "m=" line, and associates the offerer BUNDLE address with each other bundled "m=" line within the BUNDLE group
- An answer, in which the answerer moves the "m=" line out of the BUNDLE group, associates a unique address with the moved "m=" line, and associates the answerer BUNDLE address with each of the remaining bundled "m=" lines within the BUNDLE group

```
SDP Offer (1)
v=0
o=alice 2890844526 2890844526 IN IP4 atlanta.example.com
s=
c=IN IP4 atlanta.example.com
t=0 0
a=group:BUNDLE foo bar
m=audio 10000 RTP/AVP 0 8 97
b=AS:200
a=mid:foo
a=rtpmap:0 PCMU/8000
a=rtpmap:8 PCMA/8000
a=rtpmap:97 iLBC/8000
a=extmap 1 urn:ietf:params:rtp-hdrext:sdes:mid
m=video 10000 RTP/AVP 31 32
b=AS:1000

a=mid:bar
a=rtpmap:31 H261/90000
a=rtpmap:32 MPV/90000
a=extmap 1 urn:ietf:params:rtp-hdrext:sdes:mid
m=video 50000 RTP/AVP 66
b=AS:1000
```

```
a=mid:zen
a=rtpmap:66 H261/90000
SDP Answer (2)
v=0
o=bob 2808844564 2808844564 IN IP4 biloxi.example.com
s=
c=IN IP4 biloxi.example.com
t=0 0
a=group:BUNDLE foo bar
m=audio 20000 RTP/AVP 0
b=AS:200
a=mid:foo
a=rtpmap:0 PCMU/8000
a=extmap 1 urn:ietf:params:rtp-hdrext:sdes:mid
m=video 20000 RTP/AVP 32
b=AS:1000
a=mid:bar
a=rtpmap:32 MPV/90000
a=extmap 1 urn:ietf:params:rtp-hdrext:sdes:mid
m=video 60000 RTP/AVP 66
b=AS:1000
a=mid:zen
a=rtpmap:66 H261/90000
```

7.2.17.5 Example: Offerer Disables a Media Description within a BUNDLE Group

The following example shows

- A subsequent offer (the BUNDLE group has been created as part of a previous offer/answer transaction) in which the offerer disables a bundled "m=" line within a BUNDLE group, assigns a zero port number to the disabled "m=" line, and associates the offerer BUNDLE address with each of the other bundled "m=" lines within the BUNDLE group
- An answer in which the answerer moves the disabled "m=" line out of the BUNDLE group, assigns a zero port value to the disabled "m=" line, and associates the answerer BUNDLE address with each of the remaining bundled "m=" lines within the BUNDLE group.

```
SDP Offer (1)
v=0
o=alice 2890844526 2890844526 IN IP4 atlanta.example.com
s=
c=IN IP4 atlanta.example.com
t=0 0
a=group:BUNDLE foo bar
m=audio 10000 RTP/AVP 0 8 97
b=AS:200
a=mid:foo
a=rtpmap:0 PCMU/8000
a=rtpmap:8 PCMA/8000
a=rtpmap:97 iLBC/8000
a=extmap 1 urn:ietf:params:rtp-hdrext:sdes:mid
m=video 10000 RTP/AVP 31 32
```

```
b=AS:1000
a=mid:bar
a=rtpmap:31 H261/90000
a=rtpmap:32 MPV/90000
a=extmap 1 urn:ietf:params:rtp-hdrext:sdes:mid
m=video 0 RTP/AVP 66
a=mid:zen
a=rtpmap:66 H261/90000
SDP Answer (2)
v=0
o=bob 2808844564 2808844564 IN IP4 biloxi.example.com
s=
c=IN IP4 biloxi.example.com
t=0 0
a=group:BUNDLE foo bar
m=audio 20000 RTP/AVP 0
b=AS:200
a=mid:foo
a=rtpmap:0 PCMU/8000
a=extmap 1 urn:ietf:params:rtp-hdrext:sdes:mid
m=video 20000 RTP/AVP 32
b=AS:1000
a=mid:bar

a=rtpmap:32 MPV/90000
a=extmap 1 urn:ietf:params:rtp-hdrext:sdes:mid
m=video 0 RTP/AVP 66
a=mid:zen
a=rtpmap:66 H261/90000
```

7.3 RTP Media Streams Grouping in SDP

This section describes the IETF draft [5] (draft-ietf-mmusic-msid-13) that defines a SDP grouping mechanism for RTP media streams that can be used to specify relations between media streams. This mechanism is used to signal the association between the SDP concept of "media description" and the WebRTC concept of "MediaStream"/"MediaStreamTrack" using SDP signaling.

7.3.1 Introduction

7.3.1.1 Structure of This Document

In this section we describe the RTP media stream grouping in SDP specified in the IETF draft [5] (draft-ietf-mmusic-msid-13). This document adds a new SDP (RFC 4566, see Section 2) mechanism that can associate application layer identifiers with the binding between media streams, attaching identifiers to the media streams, and attaching identifiers to the groupings they form. It is designed for use with WebRTC [6] [I-D.ietf-rtcweb-overview]. Section 7.3.1.2 gives the background on why a new mechanism is needed. Section 7.3.2 gives the definition of the new mechanism. Section 7.3.3 gives the necessary semantic information and procedures for using the media Stream ID (msid or MSID) attribute to signal the association of MediaStreamTracks to MediaStreams in support of the WebRTC application programming interface (API) [7] [W3C. WD-webrtc-20150210].

7.3.1.2 Why a New Mechanism Is Needed

When media is carried by RTP (RFC 3550), each RTP media stream is distinguished inside an RTP session by its SSRC; each RTP session is distinguished from all other RTP sessions by being on a different transport association (strictly speaking, two transport associations, one used for RTP and one used for RTCP, unless RTP/RTCP multiplexing (RFC 5761) is used). SDP gives a description based on media descriptions. According to the model used in [8] [I-D.ietf-rtcweb-jsep], each media description describes exactly one media source; if multiple media sources are carried in an RTP session, this is signaled using BUNDLE [1] [I-D.ietf-mmusic-sdp-bundle-negotiation]; if BUNDLE is not used, each media source is carried in its own RTP session. The SDP grouping framework (RFC 5888, see Section 7.1) can be used to group media descriptions. However, for the use case of WebRTC, there is the need for an application to specify some application-level information about the association between the media description and the group. This is not possible using the SDP grouping framework.

7.3.1.3 The WebRTC MediaStream

The W3C WebRTC API specification [7] [W3C.WD-webrtc-20150210] specifies that communication between WebRTC entities is done via MediaStreams, which contain MediaStreamTracks. A MediaStreamTrack is generally carried using a single SSRC in an RTP session (forming an RTP media stream. The collision of terminology is unfortunate.) There might possibly be additional SSRCs, possibly within additional RTP sessions, in order to support functionality like forward error correction or simulcast. This complication is ignored below throughout this specification. MediaStreamTracks are unidirectional; they carry media in one direction only. In the RTP specification, media streams are identified using the SSRC field. Streams are grouped into RTP Sessions, and carry a CNAME. Neither CNAME nor RTP session corresponds to a MediaStream.

Therefore, the association of an RTP media stream to MediaStreams needs to be explicitly signaled. WebRTC defines a mapping (documented in [8] [I-D.ietf-rtcweb-jsep]) where one SDP media description is used to describe each MediaStreamTrack, and the BUNDLE mechanism [1] [I-D.ietf-mmusic-sdp-bundle-negotiation] is used to group MediaStreamTracks into RTP sessions. Therefore, the need is to specify the ID of a MediaStreamTrack and its associated MediaStream for each media description, which can be accomplished with a media-level SDP attribute. This usage is described in Section 7.3.3.

7.3.2 The Msid Mechanism

This document defines a new SDP (RFC 4566, see Section 2) media-level "msid" attribute. This new attribute allows endpoints to associate RTP media streams that are described in different media descriptions with the same MediaStreams as defined in [7] [W3C.WD-webrtc-20150210] and to carry an identifier for each MediaStreamTrack in its "appdata" field. The value of the "msid" attribute consists of an identifier and an optional "appdata" field. The name of the attribute is "msid." The value of the attribute is specified by the following ABNF (RFC 5234) grammar:

```
msid-value=msid-id [ SP msid-appdata ]
msid-id=1*64token-char ; see RFC 4566 (see Section 2)
msid-appdata=1*64token-char ; see RFC 4566 (see Section 2)
```

An example msid value for a group with the identifier "examplefoo" and application data "examplebar" might look like this: msid:examplefoo examplebar The identifier is a string of ASCII characters that are legal in a "token" consisting of between 1 and 64 characters.

Application data (msid-appdata) is carried on the same line as the identifier, separated from the identifier by a space. The identifier (msid-id) uniquely identifies a group within the scope of an SDP description. There may be multiple msid attributes in a single media description. This represents the case where a single MediaStreamTrack is present in multiple MediaStreams; the value of "msid-appdata" MUST be identical for all occurrences.

Multiple media descriptions with the same value for msid-id and msidappdata is not permitted. Endpoints can update the associations between RTP media streams as expressed by msid attributes at any time. The msid attributes depend on the association of RTP media streams with media descriptions but do not depend on the association of RTP media streams with RTP transports; therefore, its mux category is NORMAL—the process of deciding on MSID attributes doesn't have to take into consideration whether the media streams are bundled.

7.3.3 Procedures

This section describes the procedures for associating media descriptions representing MediaStreamTracks within MediaStreams as defined in [7] [W3C.WD-webrtc-20150210]. In the JavaScript API, each MediaStream and MediaStreamTrack has an "id" attribute, which is a DOMString. The value of the "msid-id" field in the msid consists of the "id" attribute of a MediaStream, as defined in its WebIDL specification. The value of the "msid-appdata" field in the msid consists of the "id" attribute of a MediaStreamTrack, as defined in its WebIDL specification.

When an SDP is updated, a specific "msid-id" continues to refer to the same MediaStream, and a specific "msidappdata" to the same MediaStreamTrack. There is no memory apart from the currently valid SDP descriptions; if an msid "identifier" value disappears from the SDP and appears in a later negotiation, it will be taken to refer to a new MediaStream. The following is a high-level description of the rules for handling SDP updates. Detailed procedures are in Section 7.3.3.2.

- When a new msid "identifier" value occurs in a session description, the recipient can signal to its application that a new MediaStream has been added.
- When a session description is updated to have media descriptions with an msid "identifier" value, with one or more different "appdata" values, the recipient can signal to its application that new MediaStreamTracks have been added to the MediaStream. This is done for each different msid "identifier" value.
- When a session description is updated to no longer list any msid attribute on a specific media description, the recipient can signal to its application that the corresponding MediaStreamTrack has ended.

In addition to signaling that the track is closed when its msid attribute disappears from the SDP, the track will also be signaled as being closed when all associated SSRCs have disappeared by the rules of RFC 3550, Section 6.3.4 (BYE packet received) and Section 6.3.5 (timeout) and when the corresponding media description is disabled by setting the port number to zero. Changing the direction of the media description (by setting "sendonly," "recvonly," or "inactive" attributes) will not close the MediaStreamTrack. The association between SSRCs and media descriptions is specified in [8] [I-D.ietf-rtcweb-jsep].

7.3.3.1 Handling of Nonsignaled Tracks

Entities that do not use msid will not send msid. This means that the recipient has no predefined MediaStream id value for some incoming RTP packets. Note that this handling is triggered by incoming RTP packets, not by SDP negotiation. When MSID is used, the only time this can happen is when, at a time subsequent to the initial negotiation, a negotiation is performed where the answerer adds a MediaStreamTrack to an already established connection and starts sending data before the answer is received by the offerer. For initial negotiation, packets won't flow until the ICE candidates and fingerprints have been exchanged, so this is not an issue. The recipient of those packets will perform the following steps:

- When RTP packets are initially received, it will create an appropriate MediaStreamTrack based on the type of the media (carried in PT), and use the MID RTP header extension [1] [I-D.ietf-mmusic-sdp-bundle-negotiation] (if present) to associate the RTP packets with a specific media section. If the connection is not in the RTCSignalingState "stable," it will wait at this point.
- When the connection is in the RTCSignalingState "stable," it will look at the relevant media section to find the msid attribute.
- If there is an msid attribute, it will use that attribute to populate the "id" field of the MediaStreamTrack and associated MediaStreams, as described above.
- If there is no msid attribute, the identifier of the MediaStreamTrack will be set to a randomly generated string, and it will be signaled as being part of a MediaStream with the WebIDL "label" attribute set to "Non-WebRTC stream."
- After deciding on the "id" field to be applied to the MediaStreamTrack, the track will be signaled to the user. This process may involve a considerable amount of buffering before the stable state is entered. If the implementation wishes to limit this buffering, it MUST signal to the user that media has been discarded.

It follows that media stream tracks in the "default" media stream cannot be closed by removing the msid attribute; the application must instead signal these as closed when the SSRC disappears according to the rules of RFC 3550, Sections 6.3.4 and 6.3.5, or by disabling the media description by setting its port to zero.

7.3.3.2 Detailed Offer/Answer Procedures

These procedures are given in terms of RFC 3264 (see Section 3.1)-recommended sections. They describe the actions to be taken in terms of MediaStreams and MediaStreamTracks; they do not include event signaling inside the application, which is described in Javascript Session Establishment Protocol (JSEP).

7.3.3.2.1 Generating the Initial Offer

For each media description in the offer, if there is an associated outgoing MediaStreamTrack, the offerer adds one "a=msid" attribute to the section for each MediaStream with which the MediaStreamTrack is associated. The "identifier" field of the attribute is set to the WebIDL "id" attribute of the MediaStream, and the "appdata" field is set to the WebIDL "id" attribute of the MediaStreamTrack.

7.3.3.2.2 Answerer Processing of the Offer

For each media description in the offer, and for each "a=msid" attribute in the media description, the receiver of the offer will perform the following steps:

- Extract the "appdata" field of the "a=msid" attribute
- Check if a MediaStreamTrack with the same WebIDL "id" attribute as the "appdata" field already exists, and is not in the "ended" state. If it is not found, create it.
- Extract the "identifier" field of the "a=msid" attribute
- Check if a MediaStream with the same WebIDL "id" attribute already exists. If not, create it.
- Add the MediaStreamTrack to the MediaStream
- Signal to the user that a new MediaStreamTrack is available

7.3.3.2.3 Generating the Answer

The answer is generated in exactly the same manner as the offer. "a=msid" values in the offer do not influence the answer.

7.3.3.2.4 Offerer Processing of the Answer

The answer is processed in exactly the same manner as the offer.

7.3.3.2.5 Modifying the Session

On subsequent exchanges, precisely the same procedure as for the initial offer/answer is followed, but with one additional step in the parsing of the offer and answer:

- For each MediaStreamTrack that has been created as a result of previous offer/answer exchanges, and is not in the "ended" state, check to see if there is still an "a=msid" attribute in the present SDP whose "appdata" field is the same as the WebIDL "id" attribute of the track.
- If no such attribute is found, stop the MediaStreamTrack. This will set its state to "ended."

7.3.3.3 Example SDP Description

The following SDP description shows the representation of a WebRTC PeerConnection with two MediaStreams, each of which has one audio and one video track. Only the parts relevant to the MSID are shown. Line wrapping, empty lines, and comments are added for clarity. They are not part of the SDP.

First MediaStream – id is 4701…

```
m=audio 56500 UDP/TLS/RTP/SAVPF 96 0 8 97 98
a=msid:47017fee-b6c1-4162-929c-a25110252400
f83006c5-a0ff-4e0a-9ed9-d3e6747be7d9
m=video 56502 UDP/TLS/RTP/SAVPF 100 101
a=msid:47017fee-b6c1-4162-929c-a25110252400
b47bdb4a-5db8-49b5-bcdc-e0c9a23172e0
```

Note: Secure Audio-Visual-Profile with Feedback (SAVPF)
Second MediaStream – id is 6131...

```
m=audio 56503 UDP/TLS/RTP/SAVPF 96 0 8 97 98
a=msid:61317484-2ed4-49d7-9eb7-1414322a7aae
b94006c5-cade-4e0a-9ed9-d3e6747be7d9
m=video 56504 UDP/TLS/RTP/SAVPF 100 101
a=msid:61317484-2ed4-49d7-9eb7-1414322a7aae
f30bdb4a-1497-49b5-3198-e0c9a23172e0
```

7.3.4 IANA Considerations

SDP parameters that are described here have been registered with IANA. We have not included the IANA registration procedures here for the sake of brevity. The procedures for registration of new SDP parameters are described in the IETF Draft [5] (draft-ietf-mmusic-msid-13).

7.3.5 Security Considerations

The security procedures that need to be considered in dealing with the SDP messages described in this section (IETF Draft [5]) are provided in Section 17.6.3.

7.4 Summary

We have described new SDP semantics for media groupings that are used to meet some specific needs of upper-layer multimedia applications. First, "group" and "mid" (media stream identification) SDP attributes were defined expressing a particular relationship between two or more of media lines commonly known as "m=" lines. As an example, we have described how this relationship (i.e., "group/mid") can be used for lip synchronization and for receiving a media flow consisting of several media streams on different transport addresses. The media flow identification ("FID") media flow attribute was defined for identifying the DTMF tones that are used by a variety of upper-layer applications. "bundle-only" is another attribute that extends the grouping relationship to deal with the SDP Offer/Answer mechanism to negotiate the usage of a single address:port combination (BUNDLE address) for receiving media, referred to as bundled media, specified by multiple SDP media descriptions ("m=" lines). Finally, the media attribute (i.e., "msid-value") has been introduced to express the association between different media stream descriptions and the group (i.e., "group/mid") itself. The msid-value attribute that specifies the application-specific information is needed by WebRTC applications.

7.5 Problems

1. Why are the SDP grouping attributes such as "group" and "mid" needed? How is lip synchronization performed using the "group/mid" attributes by upper-layer applications?
2. Why is the flow identification ("FID") SDP attribute needed? Provide an example of using the "FID" attribute for transferring DTMF tones of PCM μ-law and A-law codec over cellular networks.

3. Why are the "FID" attributes not used for parallel encoding for different codecs, layered encoding, and same IP address & port number? Explain in details using examples.
4. Provide some call flows in SIP using "group," "mid," and "FID" attributes in setting up the calls, answering the calls, and capability negotiations including backward compatibility where offerer and answerer may not support SDP group framework.
5. What is the SDP grouping framework with BUNDLE extension? Explain with examples. Why is the "bundle-only" attribute needed for extending the SDP group framework ("group/mid/FID")? How does this attribute influence the SDP media description parameters such as connection data (c=), bandwidth (b=), and attributes (a=) along with IP address:port?
6. What is SDP media streaming multiplexing? Provide examples of different SDP multiplexing categories.
7. What are the dependencies between the SDP multiplexing category and the bundle-only parameter? Explain in detail using examples.
8. Explain the use of the "bundle-only" attribute SDP offer/answer negotiations model in detail using an example. Explain how a session established using a bundle-only attribute is modified.
9. What are the 5-tuple parameters in the context of ICE used for IP network address translator (NAT) crossing by all media flows transparently? Provide examples with detail call flows using the SDP offer/answer model where ICE is used including the usage of the SDP "bundle-only" attribute.
10. Why is a new mechanism needed to provide association between the SDP media description and the group ("group/mid") attribute for some applications such as WebRTC? Explain using examples of WebRTC API parameters "MediaStream"/"MediaStreamTrack."
11. Describe call flows related to general procedures, handling of signaling and nonsignaling tracks, offer/answer procedures, of handling a specific "a=misid:examplefoo examplebar" including session modifications for WebRTC Java API where "ms-id=examplefoo," "application data=examplebar." (Note that a specific "ms-id" value continues to refer to the same MediaStream and a specific "msid-appdata" to the same MediaStreamTrack.)
12. Repeat Example 11 using different categories of SDP multiplexers.

References

1. C. Holmberg, H. Alvestrand, and C. Jennings, "Negotiating Media Multiplexing Using the Session Description Protocol," draft-ietf-mmusic-sdp-bundle-negotiation-29.txt, Work-in-Progress, April 15, 2016.
2. A. Keranen, C. Holmberg, and J. Rosenberg, "Interactive Connectivity Establishment (ICE): A Protocol for Network Address Translator (NAT) Traversal," draft-ietf-ice-rfc5245bis-10, Work-in-Progress, May 26, 2017.
3. S. Nandakumar, "A Framework for SDP Attributes when Multiplexing," draft-ietf-mmusic-sdp-mux-attributes-16, Work-in-Progress, December 19, 2016.
4. M. Petit-Huguenin, A. Keranen, and S. Nandakumar, "Session Description Protocol (SDP) Offer/Answer procedures for Interactive Connectivity Establishment (ICE)," draft-ietf-mmusic-ice-sip-sdp-13, Work-in-Progress, December 30, 2017.
5. H. Alvestrand, "WebRTC MediaStream Identification in the Session Description Protocol," draft-ietf-mmusic-msid-13, Work-in-Progress, 2016.

6. H. Alvestrand, "Overview: Real Time Protocols for Browser-based Applications," draft-ietf-rtcweb-overview-15, Work-in-Progress, July 24, 2016.
7. WebRTC 1.0: Real-time Communication between Browsers, W3C Working Draft: W3C. WD-webrtc-20150210, 05 June 2017.
8. J. Uberti, C. Jennings, and E. Rescorla, Ed., "Javascript Session Establishment Protocol," draft-ietf-rtcweb-jsep-14, Work-in-Progress, September 22, 2016.

Chapter 8

Generic Capability Attribute Negotiations in SDP

Abstract
This section describes Request for Comments (RFCs) 5939, 6236, 6871, and 7006 that deal with how the session description protocol is used for negotiation of capabilities between conference participants for session establishment based on agreed-upon session, media, and transport attributes. First, we describe RFC 5939 that addresses the Session Description Protocol (SDP) offer/answer procedures providing a general SDP capability negotiation framework. It also specifies how to provide attributes and transport protocols as capabilities and then negotiate them using the framework. Extensions for other types of capabilities (e.g., media types and media formats) are provided throughout this book. Second, a new generic session setup attribute is described here (RFC 6236) to make it possible to negotiate different image attributes such as image size. A possible use case is to make it possible for a low-end handheld terminal to display video without the need to rescale the image, something that may consume large amounts of memory and processing power. This chapter also explains how to maintain an optimal bit rate for video as only the image size that is desired by the receiver is transmitted. Third, the capability negotiating framework has been extended further by defining media capabilities (RFC 6871) that can be used to negotiate media types and their associated parameters. Finally, we describe RFC 7006 that extends the SDP capability negotiation framework to allow endpoints to negotiate additional SDP capabilities, namely, bandwidth ("b=" line), connection data ("c=" line), and session or media titles ("i=" line for each session or media).

8.1 SDP Capability Negotiation

This section describes Request for Comments (RFC) 5939 that extends SDP with capability negotiation parameters and associated offer/answer procedures to use those parameters in a backwards compatible manner. This RFC defines a general SDP capability negotiation framework. It also specifies how to provide attributes and transport protocols as capabilities and negotiate them using the framework. This capability of SDP will allow the negotiation of one or more alternative transport protocols (e.g., Real-Time Transport Protocol [RTP] profiles) or attributes. As a result, this facilitates the deployment of new RTP profiles such as Secure Real-Time Protocol (SRTP) or RTP with Real-Time Transport Control Protocol (RTCP)-based feedback, the negotiation of the use of different security keying mechanisms, and other actions.

8.1.1 Introduction

We are describing SDP capability negotiation examples for the Session Initiation Protocol (SIP) specified in RFC 3264 (see Section 3.1). The SDP was intended to describe multimedia sessions for the purposes of session announcement, session invitation, and other forms of multimedia session initiation. An SDP session description contains one or more media stream descriptions with information such as Internet Protocol (IP) address and port, type of media stream (e.g., audio or video), transport protocol, possibly including profile information (e.g., RTP/Audio Video Profile [AVP] or RTP/Secure Audio Video Profile [SAVP]), media formats (e.g., codecs), and various other session and media stream parameters that define the session. Simply providing media stream descriptions is sufficient for session announcements for a broadcast application, where the media stream parameters are fixed for all participants. When a participant wants to join the session, he/she obtains the session announcement and uses the media descriptions provided (e.g., joins a multicast group and receives media packets in the encoding format specified). If the media stream description is not supported by the participant, he/she is unable to receive the media.

Such restrictions are not generally acceptable to multimedia session invitations, where two or more entities attempt to establish a media session that uses a set of media stream parameters acceptable to all participants. First of all, each entity must inform the other of its receive address; second, the entities need to agree on media stream parameters to use for the session (e.g., transport protocols and codecs). To solve this, RFC 3264 (see Section 3.1) defined the offer/answer model, whereby an offerer constructs an offer SDP session description that lists the media streams, codecs, and other SDP parameters the offerer is willing to use. This offer session description is sent to the answerer, who chooses from among the media streams, codecs, and other session description parameters provided and generates an answer session description with his parameters, based on that choice. The answer session description is sent back to the offerer, thereby completing the session negotiation and enabling the establishment of the negotiated media streams.

Taking a step back, we can make a distinction between the capabilities supported by each participant, the way in which those capabilities can be supported, and the parameters that can actually be used for the session. More generally, we can say that we have the following:

- A set of capabilities for the session and its associated media stream components, supported by each side. The capability indications by themselves do not imply a commitment to use the capabilities in the session. Capabilities can, for example, be that the "RTP/SAVP" profile is supported, that the Pulse Code Modulation mu-law (PCMU) codec is supported, or that the "crypto" attribute is supported with a particular value.
- A set of potential configurations indicating which combinations of those capabilities can be used for the session and its associated media stream components. Potential configurations are not ready for use. Instead, they provide an alternative that may be used, subject to further negotiation. A potential configuration can, for example, indicate that the "PCMU" codec and the "RTP/SAVP" transport protocol are not only supported (i.e., listed as capabilities), but they are offered for potential use in the session.

- An actual configuration for the session and its associated media stream components, that specifies which combinations of session parameters and media stream components can be used currently and with what parameters. Use of an actual configuration does not require any further negotiation. An actual configuration can, for example, be that the "PCMU" codec and the "RTP/SAVP" transport protocol are offered for use currently.
- A negotiation process that takes the set of actual and potential configurations (combinations of capabilities) as input and provides the negotiated actual configurations as output. SDP by itself was designed to provide only one of these, namely, listing of the actual configurations; however, over the years, use of SDP has been extended beyond its original scope. Of particular importance are the session negotiation semantics that were defined by the offer/answer model in RFC 3264 (see Section 3.1). In this model, both the offer and answer contain actual configurations; separate capabilities and potential configurations are not supported.

Other relevant extensions have been defined as well. RFC 3407 (see Section 4.1) defined simple capability declarations, which extend SDP with a simple and limited set of capability descriptions. Grouping of media lines, which defines how media lines in SDP can have other semantics than the traditional "simultaneous media streams" semantics, was defined in RFC 5888 (see Section 7.1), etc. Each of these extensions was designed to solve a specific limitation of SDP. Since SDP had already been stretched beyond its original intent, a more comprehensive capability declaration and negotiation process was intentionally not defined. Instead, work on a "next generation" of a protocol to provide session description and capability negotiation was initiated (RFC 4566, see Section 2). RFC 4566 (see Section 2) defined a comprehensive capability negotiation framework and protocol that was not bound by existing SDP constraints.

Existing real-time multimedia communication protocols such as SIP, Real-Time Streaming Protocol, Megaco, and Media Gateway Control Protocol continue to use SDP. However, SDP does not address an increasingly important problem: the ability to negotiate one or more alternative transport protocols (e.g., RTP profiles) and associated parameters (e.g., SDP attributes).

This makes it difficult to deploy new RTP profiles such as SRTP (RFC 3711), RTP with RTCP-based feedback (RFC 4585), etc. The problem is exacerbated by the fact that RTP profiles are defined independently. When a new profile is defined and N other profiles already exist, there is a potential need for defining N additional profiles, since profiles cannot be combined automatically. For example, in order to support the plain and SRTP versions of RTP with and without RTCP-based feedback, four separate profiles (and hence profile definitions) are needed: RTP/AVP (RFC 3551), RTP/SAVP (RFC 3711), RTP/Audio Video Profile Feedback (AVPF) (RFC 4585), and RTP/Secure Audio Video Profile Feedback (SAVPF) (RFC 5124). In addition to the pressing profile negotiation problem, other important real-life limitations have been found. Keying material and other parameters, for example, need to be negotiated with some of the transport protocols but not others.

Similarly, some media formats and types of media streams need to negotiate a variety of different parameters. The purpose of this document is to define a mechanism that enables SDP to provide limited support for indicating capabilities and their associated potential configurations and negotiate the use of those potential configurations as actual configurations. It is not the intent to provide a full-fledged capability indication and negotiation mechanism along the lines of RFC 4566 (see Section 2) or ITU-T H.245. Instead, the focus is on addressing a set of well-known real-life limitations. More specifically, the solution provided in this specification provides a general SDP capability negotiation framework that is backwards compatible with existing SDP. It also defines specifically how to provide attributes and transport protocols as capabilities and negotiate them using the framework. Extensions for other types of capabilities (e.g., media types and formats) may be provided in other documents.

As mentioned, SDP is used by several protocols; hence, the mechanism should be usable by all of these. One particularly important protocol for this problem is SIP (RFC 3261, also see Section 1.2). SIP uses the offer/answer model (RFC 3264, see Section 3.1), which is not specific to SIP, to negotiate sessions; hence, the mechanism defined here provides the offer/answer procedures to be used for the capability negotiation framework. The rest of the specification is structured as follows. In Section 8.1.3, we present the SDP capability negotiation solution, which consists of new SDP attributes and associated offer/answer procedures. In Section 8.1.4, we provide examples illustrating its use. In Section 8.1.5, we provide security considerations.

8.1.2 Conventions Used in This Specification

The convention that is being used in this document is as per Internet Engineering Task Force (IETF) RFC 2119.

8.1.3 SDP Capability Negotiation Solution

In this section, we present the conceptual model behind the SDP capability negotiation framework followed by an overview of the SDP capability negotiation solution. We then define new SDP attributes for the solution and provide its associated updated offer/answer procedures.

8.1.3.1 SDP Capability Negotiation Model

Our model uses the concepts of capabilities, potential configurations, actual configurations, and negotiation process as defined in Section 8.1.1. Conceptually, we want to offer not just the actual configuration SDP session description (which is done with the offer/answer model defined in RFC 3264, see Section 3.1), but the actual configuration SDP session description as well as one or more alternative SDP session descriptions (i.e., potential configurations). The answerer must choose either the actual configuration or one of the potential configurations and generate an answer SDP session description based on that. The offerer may need to perform processing on the answer, which depends on the offer that was chosen (actual or potential configuration). The answerer therefore informs the offerer which configuration the answerer chose. The process can be viewed "conceptually" as follows:

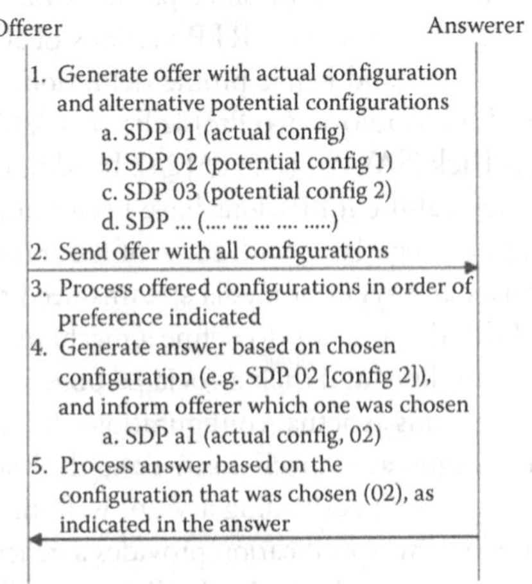

Inline Figure 1 illustrates the conceptual model; the actual solution uses a single SDP session description, which contains the actual configuration (as with existing SDP session descriptions and the offer/answer model defined in RFC 3264, see Section 3.1) and several new attributes and associated procedures that encode the capabilities and potential configurations. A more accurate depiction of the actual offer SDP session description is therefore as follows:

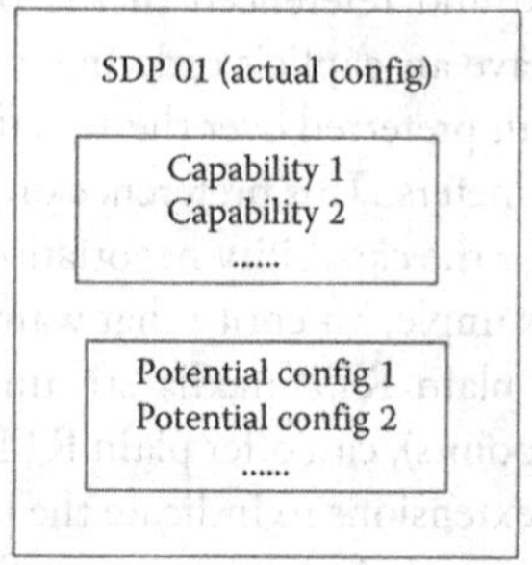

The structure in Inline Figure 2 is used for two reasons:

- Backwards compatibility: As already noted, support for multipart Multipurpose Internet Mail Extensions (MIME) is not ubiquitous. By encoding both capabilities and potential configurations in SDP attributes, we can represent everything in a single SDP session description, thereby avoiding any multipart MIME support issues. Furthermore, since unknown SDP attributes are ignored by the SDP recipient, we ensure that entities that do not support the framework simply perform the regular RFC 3264 (see Section 3.1) offer/answer procedures. This provides us with seamless backwards compatibility.
- Message size efficiency: When we have multiple media streams, each of which may potentially use two or more different transport protocols with a variety of different associated parameters, the number of potential configurations can be large. If each possible alternative is represented as a complete SDP session description in an offer, we can easily end up with large messages. By providing a more compact encoding, we get more efficient message sizes.

In the next Section 8.1.3.2, we describe the exact structure and specific SDP parameters used to represent this.

8.1.3.2 Solution Overview

The solution consists of the following:

- Two new SDP attributes to support extensions to the framework itself as follows:
- A new attribute ("a=csup") that lists the supported base (optionally) and any supported extension options to the framework.
- A new attribute ("a=creq") that lists the extensions to the framework that are required to be supported by the entity receiving the SDP session description in order to perform capability negotiation.
- Two new SDP attributes used to express capabilities as follows (additional attributes can be defined as extensions):
- A new attribute ("a=acap") that defines how to list an attribute name and its associated value (if any) as a capability.
- A new attribute ("a=tcap") that defines how to list transport protocols (e.g., "RTP/AVP") as capabilities.

- Two new SDP attributes to negotiate configurations as follows:
  - A new attribute ("a=pcfg") that lists potential supported configurations. This is done by reference to the capabilities from the SDP session description in question. Extension capabilities can be defined and referenced in the potential configurations. Alternative potential configurations have an explicit ordering associated with them. Also, potential configurations are by default preferred over the actual configuration included in the "m=" line and its associated parameters. This preference order was chosen to provide maximum backwards compatibility for the capability negotiation framework and the possible values offered for a session. For example, an entity that wants to establish a SRTP media stream but is willing to accept a plain RTP media stream (assumed to be the least common denominator for most endpoints), can offer plain RTP in the actual configuration and use the capability negotiation extensions to indicate the preference for SRTP. Entities that do not support the capability negotiation extensions or SRTP will then default to plain RTP.
  - A new attribute ("a=acfg") to be used in an answer SDP session description. The attribute identifies a potential configuration from an offer SDP session description that was used as an actual configuration to form the answer SDP session description. Extension capabilities can be included as well.
- Extensions to the offer/answer model that allow for capabilities and potential configurations to be included in an offer. Capabilities can be provided at the session level and the media level. Potential configurations can be included only at the media level, where they constitute alternative offers that may be accepted by the answerer instead of the actual configuration(s) included in the "m=" line(s) and associated parameters.

  The mechanisms defined in this document enable potential configurations to change the transport protocol, add new attributes, and remove all existing attributes from the actual configuration. The answerer indicates which (if any) of the potential configurations it used to form the answer by including the actual configuration attribute ("a=acfg") in the answer. Capabilities may be included in answers as well, where they can aid in guiding a subsequent new offer.

The mechanism is illustrated by the offer/answer exchange in Inline Figure 3, where Alice sends an offer to Bob:

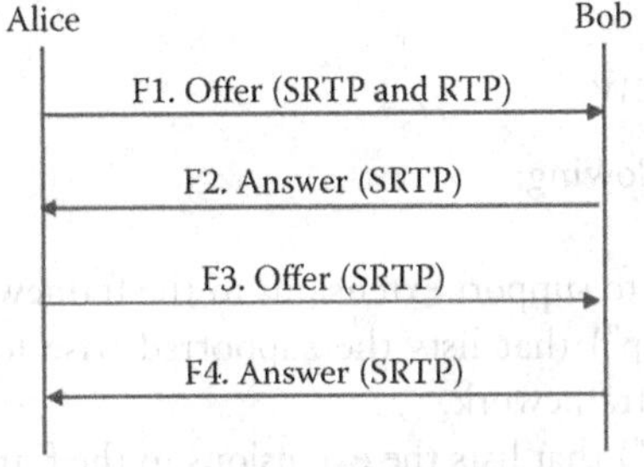

Alice's offer includes RTP and Secure Real-Time Protocol (SRTP) as alternatives, where RTP is the default (actual configuration), but SRTP is the preferred one (potential configuration):

```
v=0
o=- 25678 753849 IN IP4 192.0.2.1
s=
c=IN IP4 192.0.2.1
t=0 0
m=audio 53456 RTP/AVP 0 18
```

```
a=tcap:1 RTP/SAVP
a=acap:1 crypto:1 AES_CM_128_HMAC_SHA1_80
     inline:WVNfX19zZW1jdGwgKCkgewkyMjA7fQp9CnVubGVz|2^20|1:4
a=pcfg:1 t=1 a=1
```

The "m=" line indicates that Alice is offering to use plain RTP with PCMU or G.729. The capabilities are provided by the "a=tcap" and "a=acap" attributes. The transport capability attribute ("a=tcap") indicates that SRTP under the AVP profile ("RTP/SAVP") is supported with an associated transport capability handle of 1. The "acap" attribute provides an attribute capability with a handle of 1.

The attribute capability is a "crypto" attribute, which provides the keying material for SRTP using SDP security descriptions (RFC 4568, see Section 16.2). The "a=pcfg" attribute provides the potential configuration included in the offer by reference to the capability parameters. One alternative is provided; it has a configuration number of 1, and it consists of transport protocol capability 1 (i.e., the RTP/SAVP profile, SRTP) and the attribute capability 1 (i.e., the "crypto" attribute provided). Potential configurations are preferred over the actual configuration included in the offer SDP session description, and hence Alice is expressing a preference for using SRTP.

Bob receives the SDP session description offer from Alice. Bob supports SRTP and the SDP capability negotiation framework; hence, he accepts the (preferred) potential configuration for SRTP provided by Alice and generates the following answer SDP session description:

```
v=0
o=- 24351 621814 IN IP4 192.0.2.2
s=
c=IN IP4 192.0.2.2
t=0 0
m=audio 54568 RTP/SAVP 0 18
a=crypto:1 AES_CM_128_HMAC_SHA1_80
inline:PS1uQCVeeCFCanVmcjkpPywjNWhcYD0mXXtxaVBR|2^20|1:4
a=acfg:1 t=1 a=1
```

Bob includes the "a=acfg" attribute in the answer to inform Alice that he based his answer on an offer using potential configuration 1 with transport protocol capability 1 and attribute capability 1 from the offer SDP session description (i.e., the RTP/SAVP profile using the keying material provided). Bob also includes his keying material in a "crypto" attribute. If Bob had supported one or more extensions to the capability negotiation framework, he would have included option-tags for those in the answer as well (in an "a=csup" attribute).

When Alice receives Bob's answer, session negotiation is complete; however, Alice nevertheless generates a new offer using the negotiated configuration as the actual configuration. This is done purely to assist any intermediaries that may reside between Alice and Bob but do not support the SDP capability negotiation framework and may not understand the negotiation that just took place. Alice's updated offer includes only SRTP, and it does not use the SDP capability negotiation framework (Alice could have included the capabilities as well if she had wanted):

```
v=0
o=- 25678 753850 IN IP4 192.0.2.1
s=
c=IN IP4 192.0.2.1
t=0 0
m=audio 53456 RTP/SAVP 0 18
a=crypto:1 AES_CM_128_HMAC_SHA1_80
     inline:WVNfX19zZW1jdGwgKCkgewkyMjA7fQp9CnVubGVz|2^20|1:4
```

The "m=" line now indicates that Alice is offering to use SRTP with PCMU or G.729. The "crypto" attribute, which provides the SRTP keying material, is included with the same value again. Bob receives the SDP session description offer from Alice, which he accepts, and then generates an answer to Alice:

```
v=0
o=- 24351 621815 IN IP4 192.0.2.2
s=
c=IN IP4 192.0.2.2
t=0 0
m=audio 54568 RTP/SAVP 0 18
a=crypto:1 AES_CM_128_HMAC_SHA1_80
        inline:PS1uQCVeeCFCanVmcjkpPywjNWhcYD0mXXtxaVBR|2^20|1:4
```

Bob includes the same "crypto" attribute as before, and the session proceeds without change. Although Bob did not include any capabilities in his answer, he could have done so if he had wanted. Note that in this particular example, the answerer supported the capability negotiation extensions defined here. Had he not, he would simply have ignored the new attributes and accepted the (actual configuration) offer to use normal RTP. In that case, the following answer would have been generated instead:

```
v=0
o=- 24351 621814 IN IP4 192.0.2.2
s=
c=IN IP4 192.0.2.2
t=0 0
m=audio 54568 RTP/AVP 0 18
```

8.1.3.3 Version and Extension Indication Attributes

In this section, we present the new attributes associated with indicating the SDP capability negotiation extensions supported and required.

8.1.3.3.1 Supported Capability Negotiation Extensions Attribute

The SDP capability negotiation solution allows capability negotiation extensions to be defined. Associated with each such extension is an option-tag that identifies the extension in question. Option-tags MUST be registered with Internet Assigned Numbers Authority (IANA) per the procedures defined in Section 8.1.6.2. The supported capability negotiation extensions attribute ("a=csup") contains a comma-separated list of option-tags identifying the SDP capability negotiation extensions supported by the entity that generated the SDP session description. All SDP Augmented Backus-Naur Form (ABNF) syntaxes are provided in Section 2.9. However, we have repeated them here for convenience. The attribute can be provided at the session level and the media level, and it is defined as follows:

```
a=csup: <option-tag-list>
```

RFC 4566, Section 2.9, provides the ABNF (RFC 5234) for SDP attributes. The "csup" attribute adheres to the RFC 4566 "attribute" production, with an att-value defined as follows:

```
att-value       = option-tag-list
option-tag-list = option-tag *("," option-tag)
option-tag      = token ; defined in RFC 4566 (see Section 2)
```

A special base option-tag with a value of "cap-v0" is defined for the basic SDP capability negotiation framework defined in this document. Entities can use this option-tag with the "a=csup" attribute to indicate support for the SDP capability negotiation framework specified in this document. Please note that white space is not allowed in this rule. The following examples illustrate use of the "a=csup" attribute with the "cap-v0" option-tag and two hypothetical option-tags, "foo" and "bar" (note the lack of white space):

```
a=csup:cap-v0
a=csup:foo
a=csup:bar
a=csup:cap-v0,foo,bar
```

The "a=csup" attribute can be provided at the session and the media level. When provided at the session level, it applies to the entire SDP session description. When provided at the media level, it applies only to the media description in question (option-tags provided at the session level apply as well). There MUST NOT be more than one "a=csup" attribute at the session level and one at the media level (one per media description in the latter case).

Whenever an entity that supports one or more extensions to the SDP capability negotiation framework generates an SDP session description, it SHOULD include the "a=csup" attribute with option-tags for the extensions it supports at the session and/or media level, unless those option-tags are already provided in one or more "a=creq" attribute (see Section 8.1.3.3.2) at the relevant levels. Inclusion of the base option-tag is OPTIONAL; support for the base framework can be inferred from presence of the "a=pcfg" attribute defined in Section 8.1.3.5.1. Use of the base option-tag may still be useful in some scenarios (e.g., when using SIP OPTIONS (RFC 3261, also see Section 1.2) or generating an answer to an offer that did not use the SDP capability negotiation framework).

8.1.3.3.2 Required Capability Negotiation Extensions Attribute

The required capability negotiation extensions attribute ("a=creq") contains a comma-separated list of option-tags (see Section 8.1.3.3.1) specifying the SDP capability negotiation extensions that MUST be supported by the entity receiving the SDP session description, in order for that entity to properly process the SDP capability negotiation attributes and associated procedures. There is no need to include the base option-tag ("cap-v0") with the "creq" attribute, since any entity that supports the "creq" attribute in the first place also supports the base option-tag. Still, it is permissible to do so. Such functionality may be important if a future version of the capability negotiation framework were not backwards compatible. The attribute can be provided at the session level and the media level, and it is defined as follows:

```
a=creq: <option-tag-list>
```

The "creq" attribute adheres to the RFC 4566 (see Section 2) "attribute" production, with an att-value defined as follows:

```
att-value = option-tag-list
```

The following examples illustrate use of the "a=creq" attribute with the "cap-v0" base option-tag and two hypothetical option-tags, "foo" and "bar" (note the lack of white space):

```
a=creq:cap-v0
a=creq:foo
```

```
a=creq:bar
a=creq:cap-v0,foo,bar
```

The "a=creq" attribute can be provided at the session level and the media level. When provided at the session level, it applies to the entire SDP session description. When provided at the media level, it applies only to the media description in question (required option-tags provided at the session level apply as well). There MUST NOT be more than one "a=creq" attribute at the session level and one "a=creq" attribute at the media level (one per media description in the latter case).

When an entity generates an SDP session description and it requires the recipient of that SDP session description to support one or more SDP capability negotiation extensions (except for the base) at the session level or media level in order to properly process the SDP capability negotiation, the "a=creq" attribute MUST be included with option-tags that identify the required extensions at the session and/or media level. If support for an extension is needed only in one or more specific potential configurations, the potential configuration provides a way to indicate that instead (see Section 8.1.3.5.1). Support for the basic negotiation framework is implied by the presence of an "a=pcfg" attribute (see Section 8.1.3.5.1); hence, it does not have to include the "a=creq" attribute with the base option-tag ("cap-v0").

A recipient that receives an SDP session description and does not support one or more of the required extensions listed in a "creq" attribute MUST NOT perform the SDP capability negotiation defined in this specification; instead, the recipient MUST proceed as if the SDP capability negotiation attributes were not included in the first place (i.e., the capability negotiation attributes are ignored). In that case, if the SDP session description recipient is an SDP answerer (RFC 3264, see Section 3.1), the recipient SHOULD include a "csup" attribute in the resulting SDP session description answer listing the SDP capability negotiation extensions it actually supports.

This ensures that introduction of the SDP capability negotiation mechanism by itself does not lead to session failures For nonsupported extensions provided at the session level, this implies that SDP capability negotiation MUST NOT be performed at all. For nonsupported extensions at the media level, this implies that SDP capability negotiation MUST NOT be performed for the media stream in question. An entity that does not support the SDP capability negotiation framework at all will ignore these attributes (as well as the other SDP capability negotiation attributes) and not perform any SDP capability negotiation in the first place.

8.1.3.4 Capability Attributes

In this section, we present the new attributes associated with indicating the capabilities for use by the SDP capability negotiation.

8.1.3.4.1 Attribute Capability Attribute

Attributes and their associated values can be expressed as capabilities by use of a new attribute capability attribute ("a=acap"), which is defined as follows:

```
a=acap: <att-cap-num> <att-par>
```

where <att-cap-num> is an integer between 1 and 2^31-1 (both included) used to number the attribute capability, and <att-par> is an attribute ("a=") in its "<attribute>" or "<attribute>:<value>" form (i.e., excluding the "a=" part) (see RFC 4566, Section 2.9). The attribute can be provided at the session level and the media level. The "acap" attribute adheres to the RFC 4566 "attribute" production, with an att-value defined as follows:

```
att-value     = att-cap-num 1*WSP att-par
att-cap-num   = 1*10(DIGIT) ;defined in RFC 5234
att-par       = attribute ;defined in RFC 4566 (see Section 2.9)
```

Note that white space is not permitted before the att-cap-num. When the attribute capability contains a session-level attribute, that "acap" attribute can only be provided at the session level. Conversely, media-level attributes can be provided in attribute capabilities at either the media level or session level. The base SDP capability negotiation framework only defines procedures for use of media-level attribute capabilities at the media level. Implementations that conform only to the base framework MUST NOT generate media-level attribute capabilities at the session level; however, extensions may change this (see, e.g., RFC 6871 and Section 8.3, for one such extension) and hence all implementations MUST still be prepared to receive such capabilities (see Section 8.1.3.6.2 for processing rules).

Each occurrence of the "acap" attribute in the entire session description MUST use a different value of <att-cap-num>. Consecutive numbering of the <att-cap-num> values is not required. There is a need to be able to reference both session-level and media-level attributes in potential configurations at the media level, and this provides a simple solution to avoiding overlap between the references (handles) to each attribute capability.

The <att-cap-num> values provided are independent of similar <cap-num> values provided for other types of capabilities (i.e., they form a separate name-space for attribute capabilities). The following examples illustrate use of the "acap" attribute:

```
a=acap:1 ptime:20
a=acap:2 ptime:30
a=acap:3 key-mgmt:mikey
       AQAFgM0XflABAAAAAAAAAAAAAsAyONQ6gAAAAAGEEoo2pee4hp2
       UaDX8ZE22YwKAAAPZG9uYWxkQGR1Y2suY29tAQAAAAAAAQAk0JKpg
       aVkDaawi9whVBtBt0KZ14ymNuu62+Nv3ozPLygwK/GbAV9iemnGUIZ19fW
       QUOSrzKTAv9zV
a=acap:4 crypto:1 AES_CM_128_HMAC_SHA1_32
       inline:NzB4d1BINUAvLEw6UzF3WSJ+PSdFcGdUJShpX1Zj|2^20|1:32
```

The first two attribute capabilities provide attribute values for the ptime attribute. The third provides SRTP parameters by using Multimedia Internet KEYing (MIKEY) (RFC 3830) with the "key-mgmt" attribute (RFC 4567, see Section 16.1). The fourth provides SRTP parameters by use of security descriptions with the "crypto" attribute (RFC 4568, see Section 16.2). Note that the line wrapping and new lines in example three and four are provided for formatting reasons only; they are not permitted in actual SDP session descriptions. Readers familiar with RFC 3407 (see Section 4.1) may notice the similarity between the RFC 3407 "cpar" attribute and what has just been described. There are, however, a couple of important differences, notably that the "acap" attribute contains a handle that enables its reference it and it furthermore supports only attributes (the "cpar" attribute defined in RFC 3407 supports bandwidth information as well). The "acap" attribute also is not automatically associated with any particular capabilities. See Section 8.1.3.14 for the relationship to RFC 3407.

Attribute capabilities MUST NOT embed any capability negotiation parameters. This restriction applies to all the capability negotiation parameters defined in this document ("csup," "creq," "acap," "tcap," "pcfg," and "acfg") as well as any capability negotiation extensions defined. The following examples are thus invalid attribute capabilities and MUST NOT be used:

```
a=acap:1 acap:2 foo:a ;Not allowed to embed "acap"
a=acap:2 a=pcfg:1 t=1 a=1 ;Not allowed to embed "pcfg"
```

The reason for this restriction is to avoid overly complex processing rules resulting from the expansion of such capabilities into potential configurations (see Section 8.1.3.6.2 for further details).

8.1.3.4.2 Transport Protocol Capability Attribute

Transport protocols can be expressed as capabilities by use of a new transport protocol capability attribute ("a=tcap") defined as follows:

```
a=tcap: <trpr-cap-num> <proto-list>
```

where <trpr-cap-num> is an integer between 1 and 2^31-1 (both included) used to number the transport address capability for later reference, and <proto-list> is one or more <proto>, separated by white space, as defined in the SDP "m=" line. The attribute can be provided at the session level and the media level. The "tcap" attribute adheres to the RFC 4566 (see Section 2.9) "attribute" production, with an att-value defined as follows:

```
att-value      =    trpr-cap-num 1*WSP proto-list
trpr-cap-num   =    1*10(DIGIT) ;defined in RFC 5234
proto-list     =    proto *(1*WSP proto) ;defined in RFC 4566 (see
                              ;Section 2.9)
```

Note that white space is not permitted before the trpr-cap-num. The "tcap" attribute can be provided at the session level and the media level. There MUST NOT be more than one "a=tcap" attribute at the session level and one at the media level (one per media description in the latter case). Each occurrence of the "tcap" attribute in the entire session description MUST use a different value of <trpr-cap-num>. When multiple <proto> values are provided, the first is associated with the value <trpr-cap-num>, the second with the value one higher, etc. There MUST NOT be any capability number overlap between different "tcap" attributes in the entire SDP session description. The <trpr-cap-num> values provided are independent of similar <cap-num> values provided for other capability attributes (i.e., they form a separate name space for transport protocol capabilities). Consecutive numbering of the <trpr-cap-num> values in different "tcap" attributes is not required. The following are examples of the "a=tcap" attribute:

```
a=tcap:1 RTP/AVP
a=tcap:2 RTP/AVPF
a=tcap:3 RTP/SAVP RTP/SAVPF
a=tcap:5 UDP/TLS/RTP/SAVP
```

The first provides a capability for the "RTP/ attribute-value pair (AVP)" profile defined in RFC 3551, and the second provides a capability for the RTP with RTCP-based feedback profile defined in (RFC 4585). The third provides capabilities for the "RTP/SAVP" (transport capability number 3) and "RTP/SAVPF" profiles (transport protocol capability number 4). The last provides capabilities for "User Datagram Protocol (UDP)/Transport Layer Security (TLS)/RTP/SAVP" (i.e., Datagram Transport Layer Security [DTLS]-SRTP (RFC 5764) (transport capability number 5)). The "tcap" attribute by itself can only specify transport protocols as defined by <proto> in RFC 4566 (see Section 2.9); however, full specification of a media stream requires further qualification of the transport protocol by one or more media format descriptions, which themselves often depend on the transport protocol. As an example, RFC 3551 defines the "RTP/AVP" transport for use with audio and video codecs (media formats), whereas RFC 4145 (see Section 10.1) defines the "Transmission Control Protocol (TCP)" transport, which, for example, may be used to negotiate T.38 fax ("image/t38"), etc.

In a non-SDP context, some media formats could be viewed as transports themselves (e.g., T.38); however, in the context of SDP and SDP capability negotiation, they are not. If capability negotiation is required for such media formats, they MUST all be valid under the transport protocol indicated in the "m=" line included for the media stream description, or a suitable extension must be used (e.g., SDP media capabilities) (RFC 6871, see Section 8.3).

The ability to use a particular transport protocol is inherently implied by including it in the "m=" line, regardless of whether it is provided in a "tcap" attribute. However, if a potential configuration needs to reference that transport protocol as a capability, the transport protocol MUST be included explicitly in a "tcap" attribute. This may seem redundant (and indeed it is from the offerer's point of view); however, it is done to protect against intermediaries (e.g., middleboxes) that may modify "m=" lines while passing unknown attributes through.

If an implicit transport capability were used instead (e.g., a reserved transport capability number could be used to refer to the transport protocol in the "m=" line), and an intermediary were to modify the transport protocol in the "m=" line (e.g., to translate between plain RTP and SRTP), then the potential configuration referencing that implicit transport capability may no longer be correct. With explicit capabilities, we avoid this pitfall; however, the potential configuration preference (see Section 8.1.3.5.1) may not reflect that of the intermediary (which some may view as a feature). Note that a transport protocol capability may be provided, irrespective of whether it is referenced in a potential configuration (just like any other capability).

8.1.3.4.3 Extension Capability Attributes

The SDP capability negotiation framework allows new types of capabilities to be defined as extensions and used with the general capability negotiation framework. The syntax and semantics of such new capability attributes are not defined here; however, in order to be used with potential configurations, they SHOULD allow a numeric handle to be associated with each capability. This handle can be used as a reference within the potential and actual configuration attributes (see Sections 8.1.3.5.1 and 8.1.3.5.2). The definition of such extension capability attributes MUST also state whether they can be applied at the session level, media level, or both. Note that extensions can have option-tags defined for them, and option-tags MUST be registered with the IANA in accordance with the procedures specified in Section 8.1.6.2.

Extension capabilities SHOULD NOT embed any capability negotiation parameters. This applies to all the capability negotiation parameters defined in this document as well as any extensions that are defined. The reason for this restriction is to avoid overly complex processing rules resulting from the expansion of such capabilities into potential configurations (see Section 8.1.3.6.2 for further details). If an extension does not follow the above "SHOULD NOT" recommendation, the extension MUST provide a careful analysis of why such behavior is both necessary and safe.

8.1.3.5 Configuration Attributes

8.1.3.5.1 Potential Configuration Attribute

Potential configurations can be expressed by use of a new potential configuration attribute ("a=pcfg") defined as follows:

```
a=pcfg: <config-number> [<pot-cfg-list>]
```

where <config-number> is an integer between 1 and 2^31-1 (both included). The attribute can be provided only at the media level. The "pcfg" attribute adheres to the RFC 4566 (see Section 2.9) "attribute" production, with an att-value defined as follows:

```
att-value       =    config-number [1*WSP pot-cfg-list]
config-number   =    1*10(DIGIT) ;defined in RFC 5234
pot-cfg-list    =    pot-config *(1*WSP pot-config)
pot-config      =    attribute-config-list /
                     transport-protocol-config-list /
                     extension-config-list
```

The missing productions are defined after a couple of paragraphs below. Note that white space is not permitted before the config-number. The potential configuration attribute can be provided only at the media level, and there can be multiple instances of it within a given media description. The attribute includes a configuration number, which is an integer between 1 and 2^31-1 (both included). The configuration number MUST be unique within the media description (i.e., it has only media-level scope). The configuration number also indicates the relative preference of potential configurations; lower numbers are preferred over higher numbers. Consecutive numbering of the configuration numbers in different "pcfg" attributes in a media description is not required.

A potential configuration list is normally provided after the configuration number. When the potential configuration list is omitted, the potential configuration equals the actual configuration. The potential configuration list contains one or more attribute, transport, and extension configuration lists. A potential configuration may, for example, include attribute capabilities and transport capabilities, transport capabilities only, or some other combination of capabilities. If transport capabilities are not included in a potential configuration, the default transport for that media stream is used.

The potential configuration lists generally reference one or more capabilities (extension configuration lists MAY use a different format). Those capabilities are (conceptually) used to construct a new internal version of the SDP session description by use of purely syntactic add and (possibly) delete operations on the original SDP session description (actual configuration). This provides an alternative potential configuration SDP session description that can be used by conventional SDP and offer/answer procedures if selected. This document defines attribute configuration lists and transport protocol configuration lists. Each of these MUST NOT be present more than once in a particular potential configuration attribute.

Attribute capabilities referenced by the attribute configuration list (if included) are added to the actual configuration, whereas a transport capability referenced by the transport protocol configuration list (if included) replaces the default transport protocol from the actual configuration. Extension configuration lists can be included as well. There can be more than one extension configuration list; however, each particular extension MUST NOT be present more than once in a given "a=pcfg" attribute. Together, the various configuration lists define a potential configuration. There can be multiple potential configurations in a media description. Each of these indicates not only a willingness but in fact a desire to use the potential configuration. The following example SDP session description contains two potential configurations:

```
v=0
o=- 25678 753849 IN IP4 192.0.2.1
s=
c=IN IP4 192.0.2.1
t=0 0
m=audio 53456 RTP/AVP 0 18
```

```
a=tcap:1 RTP/SAVP RTP/SAVPF
a=acap:1 crypto:1 AES_CM_128_HMAC_SHA1_32
      inline:NzB4d1BINUAvLEw6UzF3WSJ+PSdFcGdUJShpX1Zj|2^20|1:32
a=pcfg:1 t=1 a=1
a=pcfg:2 t=2 a=1
```

Potential configuration 1 contains a transport protocol configuration list that references transport capability 1 ("RTP/SAVP") and an attribute configuration list that references attribute capability 1 ("a=crypto:..."). Potential configuration 2 contains a transport protocol configuration list that references transport capability 2 ("RTP/SAVPF") and an attribute configuration list that references attribute capability 1 ("a=crypto:..."). Attribute capabilities are used in a potential configuration by use of the attribute-config-list parameter, which is defined by the following ABNF:

```
attribute-config-list     =  "a=" delete-attributes
attribute-config-list     =  / "a=" [delete-attributes ":"]
                             mo-att-cap-list *(BAR mo-att-cap-list)
delete-attributes         =  DELETE ( "m" ; media attributes
                             / "s" ; session attributes
                             / "ms" ) ; media and session attributes
mo-att-cap-list           =  mandatory-optional-att-cap-list /
                                   mandatory-att-cap-list /
                                   optional-att-cap-list
mandatory-optional-att-cap-list  =  mandatory-att-cap-list
                                   "," optional-att-cap-list
mandatory-att-cap-list    =  att-cap-list
optional-att-cap-list     =  "[" att-cap-list "]"
att-cap-list              =  att-cap-num *("," att-cap-num)
att-cap-num               =  1*10(DIGIT) ;defined in [RFC 5234]
BAR                       =  "|"
DELETE                    =  "-"
```

Note that white space is not permitted within the attribute-config-list rule. Each attribute configuration list can optionally begin with instructions for how to handle attributes that are part of the actual configuration SDP session description (i.e., the "a=" lines present in the original SDP session description). By default, such attributes will remain as part of the potential configuration in question. However, if delete-attributes indicate "-m," then all attribute lines within the media description in question will be deleted in the resulting potential configuration SDP session description (i.e., all "a=" lines under the "m=" line in question).

If delete-attributes indicates "-s", then all attribute lines at the session level will be deleted (i.e., all "a=" lines before the first "m=" line). If delete-attributes indicates "-ms," then all attribute lines within this media description ("m=" line) and all attribute lines at the session level will be deleted. The attribute capability list comes next (if included). It contains one or more alternative lists of attribute capabilities. The alternative attribute capability lists are separated by a vertical bar ("|"), and each list contains one or more attribute capabilities separated by commas (","). The attribute capabilities are either mandatory or optional. Mandatory attribute capabilities MUST be supported in order to use the potential configuration, whereas optional attribute capabilities MAY be supported in order to use the potential configuration.

Within each attribute capability list, all the mandatory attribute capabilities (if any) are listed first, and all the optional attribute capabilities (if any) are listed last. The optional attribute capabilities are contained within a pair of square brackets ("[" and "]"). Each attribute capability is merely an attribute capability number (att-cap-num) that identifies a particular attribute capability by referring to the attribute capability numbers defined above and hence MUST be between 1 and 2^31 -1 (both included). The following example illustrates this concept:

```
a=pcfg:1 a=-m:1,2,[3,4]|1,7,[5]
```

where

- "a=-m:1,2,[3,4]|1,7,[5]" is the attribute configuration list
- "-m" indicates to delete all attributes from the media description of the actual configuration
- "1,2,[3,4]" and "1,7,[5]" are both attribute capability lists.

The two lists are alternatives, since they are separated by a vertical bar above

- "1," "2," and "7" are mandatory attribute capabilities
- "3," "4," and "5" are optional attribute capabilities

Note that in this example, we have a single handle ("1") for the potential configuration(s), but there are actually two different potential configurations (separated by a vertical bar). This is done for message size efficiency reasons, which is especially important when we add other types of capabilities to the potential configuration. If there is a need to provide a unique handle for each, then separate "a=pcfg" attributes with different handles MUST be used instead.

Each referenced attribute capability in the potential configuration will result in the corresponding attribute name and its associated value (contained inside the attribute capability) being added to the resulting potential configuration SDP session description.

Alternative attribute capability lists are separated by a vertical bar ("|"), the scope of which extends to the next alternative (i.e., "," has higher precedence than "|"). The alternatives are ordered by preference with the most preferred listed first. In order for a recipient of the SDP session description (e.g., an answerer receiving this in an offer) to use this potential configuration, exactly one of the alternative lists MUST be selected in its entirety. This requires that all mandatory attribute capabilities referenced by the potential configuration are supported with the attribute values provided.

Transport protocol configuration lists are included in a potential configuration by use of the transport-protocol-config-list parameter, which is defined by the following ABNF:

```
transport-protocol-config-list  =   "t=" trpr-cap-num
                                    *(BAR trpr-cap-num)
trpr-cap-num                    =   1*10(DIGIT) ; defined in RFC 5234
```

Note that white space is not permitted within this rule. The trpr-cap-num refers to the already defined transport protocol capability numbers and MUST be between 1 and 2^31-1 (both included). Alternative transport protocol capabilities are separated by a vertical bar ("|"). The alternatives are ordered by preference with the most preferred listed first. If there are no transport protocol capabilities included in a potential configuration at the media level, the transport protocol information from the associated "m=" line MUST be used. In order for a recipient of the SDP session description (e.g., an answerer receiving this in an offer) to use this potential configuration, exactly one of the alternatives MUST be selected. This requires that the transport protocol in question is supported.

In the presence of intermediaries (the existence of which may not be known), care should be taken in assuming that the transport protocol in the "m=" line will not be modified by an intermediary.

Use of an explicit transport protocol capability will guard against capability negotiation implications of that. Extension capabilities can be included in a potential configuration as well by use of extension configuration lists. Extension configuration lists MUST adhere to the following ABNF:

```
extension-config-list   =   ["+"] ext-cap-name "=" ext-cap-list
ext-cap-name            =   1*(ALPHA / DIGIT)
ext-cap-list            =   1*VCHAR ; defined in RFC 5234
```

Note that white space is not permitted within this rule. The ext-cap-name refers to the name of the extension capability and the ext-cap-list is here merely defined as a sequence of visible characters. The actual extension supported MUST refine both of these further. For extension capabilities that merely need to be referenced by a capability number, it is RECOMMENDED to follow a structure similar to what has been specified above. Unsupported or unknown potential extension configuration lists in a potential configuration attribute MUST be ignored, unless they are prefixed with the plus ("+") sign, which indicates that the extension is mandatory and MUST be supported in order to use that potential configuration.

The "creq" attribute and its associated rules can be used to ensure that required extensions are supported in the first place. Extension configuration lists define new potential configuration parameters and hence they MUST be registered with IANA per the procedures defined in Section 8.1.6.3. Potential configuration attributes can be provided only at the media level; however, it is possible to reference capabilities provided at either the session or media level. There are certain semantic rules and restrictions associated with this:

A (media-level) potential configuration attribute in a given media description MUST NOT reference a media-level capability provided in a different media description; doing so invalidates that potential configuration (note that a potential configuration attribute can contain more than one potential configuration by use of alternatives). A potential configuration attribute can however reference a session-level capability. The semantics of doing so depends on the type of capability. In the case of transport protocol capabilities, it has no particular implication. In the case of attribute capabilities, however, it does. More specifically, the attribute name and value (provided within that attribute capability) will be considered part of the resulting SDP for that particular configuration at the *session* level. In other words, it will be as if that attribute were provided with that value at the session level in the first place. As a result, the base SDP capability negotiation framework REQUIRES that potential configurations not reference any session-level attribute capabilities that contain media-level attributes (since that would place a media-level attribute at the session level). Extensions may modify this behavior, as long as it is fully backwards compatible with the base specification. Individual media streams perform capability negotiation individually; hence, it is possible that one media stream (where the attribute was part of a potential configuration) chose a configuration without a session-level attribute that was chosen by another media stream. The session-level attribute however remains "active" and applies to the entire resulting potential configuration SDP session description.

In theory, this is problematic if one or more session-level attributes either conflict with or potentially interact with another session-level or media-level attribute in an undefined manner. In practice, such examples seem to be rare (at least with the SDP attributes that had been defined at time of publication of this document). A related set of problems can occur if we need coordination between session-level attributes from multiple media streams in order for a particular functionality to work. The grouping framework (RFC 5888, see Section 7.1) is an example of this. If we use the SDP capability negotiation framework to select a session-level group attribute (provided as an attribute capability), and we require two media descriptions to do this consistently, we could have a problem. The Forward Error Correction grouping semantics (RFC 5956 that obsoletes RFC 4756) is one example in which this in theory could cause problems, however in practice, it is unclear whether there is a significant problem with the grouping semantics that had been defined at time of publication of this document.

Resolving the above issues in general requires intermedia stream constraints and synchronized potential configuration processing; this would add considerable complexity to the overall solution. In practice, with the SDP attributes defined at time of publication of this document, it does not seem to be a significant problem; hence, the base SDP capability negotiation solution does not provide a solution to this issue. Instead, it is RECOMMENDED that use of session-level attributes in a potential configuration be avoided when possible and when not, that such use is examined closely for any potential interaction issues. If interaction is possible, the entity generating the SDP session description SHOULD NOT assume that well defined operation will occur at the receiving entity. This implies that mechanisms that might have such interactions cannot be used in security critical contexts. The session-level operation of extension capabilities is undefined. Consequently, each new session-level extension capability defined MUST specify the implication of making it part of a configuration at the media level. Following, we provide an example of the "a=pcfg" attribute in a complete media description in order to properly indicate the supporting attributes:

```
v=0
o=- 25678 753849 IN IP4 192.0.2.1
s=
c=IN IP4 192.0.2.1
t=0 0
m=audio 53456 RTP/AVPF 0 18
a=acap:1 crypto:1 AES_CM_128_HMAC_SHA1_32
      inline:NzB4d1BINUAvLEw6UzF3WSJ+PSdFcGdUJShpX1Zj|2^20|1:32
a=tcap:1 RTP/AVPF RTP/AVP RTP/SAVP RTP/SAVPF
a=pcfg:1 t=4|3 a=1
a=pcfg:8 t=1|2
```

We have two potential configuration attributes listed here. The first one (and most preferred, since its configuration number is "1") indicates that either of the profiles RTP/SAVPF or RTP/SAVP (specified by the transport protocol capability numbers 4 and 3) can be supported with attribute capability 1 (the "crypto" attribute); RTP/SAVPF is preferred over RTP/SAVP since its capability number (4) is listed first in the preferred potential configuration. Note that although we have a single potential configuration attribute and associated handle, we have two potential configurations. The second potential configuration attribute indicates that the RTP/AVPF or RTP/AVP profiles can be used, with RTP/AVPF being the preferred one. This non-SRTP alternative is the less preferred one since its configuration number is "8." Again, note that we have two potential configurations here; hence, a total of four potential configurations in the SDP session description above.

8.1.3.5.2 Actual Configuration Attribute

The actual configuration attribute identifies which of the potential configurations from an offer SDP session description was selected and used as the actual configuration to generate an answer SDP session description. This is done by including the configuration number and the configuration lists (if any) from the offer that were selected and used by the answerer in his offer/answer procedure as follows:

- A selected attribute configuration MUST include the delete attributes and the known and supported parameters from the selected alternative mo-att-cap-list (i.e., containing all mandatory and all known and supported optional capability numbers from the potential configuration). If delete-attributes were not included in the potential configuration, they will of course not be present here either.

- A selected transport protocol configuration MUST include the selected transport protocol capability number.
- A selected potential extension configuration MUST include the selected extension configuration parameters as specified for that particular extension.
- When a configuration list contains alternatives (separated by "|"), the selected configuration only MUST be provided.

Note that the selected configuration number and all selected capability numbers used in the actual configuration attribute refer to those from the offer, not the answer. The answer may, for example, include capabilities as well to inform the offerer of the answerer's capabilities above and beyond the negotiated configuration. The actual configuration attribute does not refer to any of those answer capabilities. The actual configuration attribute ("a=acfg") is defined as follows:

```
a=acfg: <config-number> [<sel-cfg-list>]
```

where <config-number> is an integer between 1 and 2^31-1 (both included) that refers to the selected potential configuration. The attribute can be provided only at the media level.

The "acfg" attribute adheres to the RFC 4566 "attribute" production, with an att-value defined as follows:

```
att-value              =  config-number [1*WSP sel-cfg-list]
                          ;config-number defined in Section 8.1.3.5.1.
sel-cfg-list           =  sel-cfg *(1*WSP sel-cfg)
sel-cfg                =     sel-attribute-config /
                          sel-transport-protocol-config /
                          sel-extension-config
sel-attribute-config   =  "a=" [delete-attributes ":"]mo-att-cap-list
                          ; defined in Section 3.5.1.
sel-transport-protocol-config  =  "t=" trpr-cap-num
                                  ; defined in Section 8.1.3.5.1.
sel-extension-config           =  ext-cap-name
                                  "=" 1*VCHAR
                                  ; defined in Section 8.1.3.5.1.
```

Note that white space is not permitted before the config-number. The actual configuration ("a=acfg") attribute can be provided only at the media level. There MUST NOT be more than one occurrence of an actual configuration attribute within a given media description. Below, we provide an example of the "a=acfg" attribute (building on the previous example with the potential configuration attribute):

```
v=0
o=- 24351 621814 IN IP4 192.0.2.2
s=
c=IN IP4 192.0.2.2
t=0 0
m=audio 54568 RTP/SAVPF 0
a=crypto:1 AES_CM_128_HMAC_SHA1_32
      inline:WSJ+PSdFcGdUJShpX1ZjNzB4d1BINUAvLEw6UzF3|2^20|1:32
a=acfg:1 t=4 a=1
```

It indicates that the answerer used an offer consisting of potential configuration number 1 with transport protocol capability 4 from the offer (RTP/SAVPF) and attribute capability 1 (the "crypto" attribute). The answerer includes his own "crypto" attribute as well.

8.1.3.6 Offer/Answer Model Extensions

In this section, we define extensions to the offer/answer model defined in RFC 3264 (see Section 3.1) to allow for potential configurations to be included in an offer, where they constitute alternative offers that may be accepted by the answerer instead of the actual configuration(s) included in the "m=" line(s). The procedures defined in the following subsections apply to both unicast and multicast streams.

8.1.3.6.1 Generating the Initial Offer

An offerer that wants to use the SDP capability negotiation defined in this document MUST include the following in the offer:

- Zero or more attribute capability attributes. There MUST be an attribute capability attribute ("a=acap") as defined in Section 8.1.3.4.1 for each attribute name and associated value (if any) that needs to be indicated as a capability in the offer. Attribute capabilities may be included irrespective of whether they are referenced by a potential configuration. Session-level attributes and associated values MUST be provided in attribute capabilities only at the session level, whereas media-level attributes and associated values can be provided in attribute capabilities at either the media level or session level. Attributes that are allowed at either the session or the media level can be provided in attribute capabilities at either level.
- Zero or more transport protocol capability attributes. There MUST be transport protocol capabilities as defined in Section 8.1.3.4.2 with values for each transport protocol that needs to be indicated as a capability in the offer. Transport protocol capabilities may be included irrespective of whether they are referenced by a potential configuration. Transport protocols that apply to multiple media descriptions SHOULD be provided as transport protocol capabilities at the session level whereas transport protocols that apply only to a specific media description ("m=" line), SHOULD be provided as transport protocol capabilities within that particular media description. In either case, there MUST NOT be more than a single "a=tcap" attribute at the session level and a single "a=tcap" attribute in each media description.
- Zero or more extension capability attributes. There MUST be one or more extension capability attributes (as outlined in Section 8.1.3.4.3) for each extension capability that is referenced by a potential configuration. Extension capability attributes that are not referenced by a potential configuration can be provided as well.
- Zero or more potential configuration attributes. There MUST be one or more potential configuration attributes ("a=pcfg"), as defined in Section 8.1.3.5.1, in each media description where alternative potential configurations are to be negotiated. Each potential configuration attribute MUST adhere to the rules provided in Section 8.1.3.5.1 and the additional rules provided below. If the offerer requires support for one or more extensions (besides the base protocol defined here), then the offerer MUST include one or more "a=creq" attributes as follows:
- If support for one or more capability negotiation extensions is required for the entire session description, then option-tags for those extensions MUST be included in a single session-level "creq" attribute.

- For each media description that requires support for one or more capability negotiation extensions not listed at the session level, a single "creq" attribute containing all the required extensions for that media description MUST be included within the media description (in accordance with Section 8.1.3.3.2). Note that extensions that only need to be supported by a particular potential configuration can use the "mandatory" extension prefix ("+") within the potential configuration (see Section 8.1.3.5.1).

The offerer SHOULD furthermore include the following:

- A supported capability negotiation extension attribute ("a=csup") at the session level and/or media level as defined in Section 8.1.3.3.2 for each capability negotiation extension supported by the offerer and not included in a corresponding "a=creq" attribute (i.e., at the session level or in the same media description).

Option-tags provided in a "a=csup" attribute at the session level indicate extensions supported for the entire session description, whereas option-tags provided in a "a=csup" attribute in a media description indicate extensions supported for only that particular media description. Capabilities provided in an offer merely indicate what the offerer is capable of doing. They do not constitute a commitment or even an indication to use them. In contrast, each potential configuration constitutes an alternative offer that the offerer would like to use. The potential configurations MUST be used by the answerer to negotiate and establish the session.

The offerer MUST include one or more potential configuration attributes ("a=pcfg") in each media description where the offerer wants to provide alternative offers (in the form of potential configurations). Each potential configuration attribute in a given media description MUST contain a unique configuration number and zero, one, or more potential configuration lists, as described in Section 8.1.3.5.1. Each potential configuration list MUST refer to capabilities that are provided at the session level or within that particular media description; otherwise, the potential configuration is considered invalid. The base SDP capability negotiation framework REQUIRES that potential configurations not reference any session-level attribute capabilities that contain media-level-only attributes; however, extensions may modify this behavior, as long as it is fully backwards compatible with the base specification. Furthermore, it is RECOMMENDED that potential configurations avoid use of session-level capabilities whenever possible; refer to Section 3.5.1.

The current actual configuration is included in the "m=" line (as defined by RFC 3264, see Section 3.1) and any associated parameters for the media description (e.g., attribute ("a=") and bandwidth ("b=") lines). Note that the actual configuration is by default the least-preferred configuration; hence, the answerer will seek to negotiate use of one of the potential configurations instead. If the offerer wishes a different preference for the actual configuration, the offerer MUST include a corresponding potential configuration with the relevant configuration number (which indicates the relative preference between potential configurations); this corresponding potential configuration should simply duplicate the actual configuration.

This can be done implicitly (by not referencing any capabilities) or explicitly (by providing and using capabilities for the transport protocol and all the attributes that are part of the actual configuration). The latter may help detect intermediaries that modify the actual configuration but are not SDP capability negotiation aware. Per RFC 3264 (see Section 3.1), once the offerer generates the offer, he/she must be prepared to receive incoming media in accordance with that offer. That rule applies here as well, but only for the actual configurations provided in the offer. Media received by the offerer according to one of the potential configurations MAY be discarded, until the offerer receives an answer indicating what the actual selected configuration is. Once that answer is received, incoming media MUST be processed in accordance with the actual selected configuration indicated and the answer received (provided the offer/answer exchange completed successfully).

This rule assumes that the offerer can determine whether incoming media adheres to the actual configuration offered or one of the potential configurations instead; this may not always be the case. If the offerer wants to ensure he/she does not play out any garbage, the offerer SHOULD discard all media received before the answer SDP session description is received. Conversely, if the offerer wants to avoid clipping, he/she SHOULD attempt to play any incoming media as soon as it is received (at the risk of playing out garbage). In either case, please note that this document does not place any requirements on the offerer to process and play media before answer. For further details, please refer to Section 8.1.3.9.

8.1.3.6.2 Generating the Answer

When receiving an offer, the answerer MUST check for the presence of a required capability negotiation extension attribute ("a=creq") provided at the session level. If one is found, then capability negotiation MUST be performed. If none is found, then the answerer MUST check each offered media description for the presence of a required capability negotiation extension attribute ("a=creq") and one or more potential configuration attributes ("a=pcfg"). Capability negotiation MUST be performed for each media description where either of those is present in accordance with the procedures described below. The answerer MUST first ensure that it supports any required capability negotiation extensions:

- If a session-level "creq" attribute is provided, and it contains an option-tag that the answerer does not support, then the answerer MUST NOT use any of the potential configuration attributes provided for any of the media descriptions. Instead, the normal offer/answer procedures MUST continue as per RFC 3264 (see Section 3.1). Furthermore, the answerer MUST include a session-level supported capability negotiation extensions attribute ("a=csup") with option-tags for the capability negotiation extensions supported by the answerer.
- If a media-level "creq" attribute is provided, and it contains an option-tag that the answerer does not support, then the answerer MUST NOT use any of the potential configuration attributes provided for that particular media description. Instead, the offer/answer procedures for that media description MUST continue as per RFC 3264 (see Section 3.1) (SDP capability negotiation is still performed for other media descriptions in the SDP session description).

Furthermore, the answerer MUST include a supported capability negotiation extensions attribute ("a=csup") in that media description with option-tags for the capability negotiation extensions supported by the answerer for that media description. Assuming all required capability negotiation extensions are supported, the answerer now proceeds as follows. For each media description where capability negotiation is to be performed (i.e., all required capability negotiation extensions are supported and at least one valid potential configuration attribute is present), the answerer MUST perform capability negotiation by using the most preferred potential configuration that is valid to the answerer, subject to any local policies. A potential configuration is valid to the answerer if:

1. It is in accordance with the syntax and semantics provided in Section 8.1.3.5.1.
2. It contains a configuration number that is unique within that media description.

3. All attribute capabilities referenced by the potential configuration are valid themselves (as defined in Section 8.1.3.4.1), and each of them is provided either at the session level or within this particular media description. For session-level attribute capabilities referenced, the attributes contained inside them MUST NOT be media-level-only attributes. Note that the answerer can only determine this for attributes supported by the answerer. If an attribute is not supported, it will simply be ignored by the answerer, and it will not trigger an "invalid" potential configuration.
4. All transport protocol capabilities referenced by the potential configuration are valid themselves (as defined in Section 8.1.3.4.2), and each of them is furthermore provided either at the session level or within this particular media description.
5. All extension capabilities referenced by the potential configuration and supported by the answerer are valid themselves (as defined by that particular extension) and each of them is furthermore provided either at the session level or within this particular media description. Unknown or unsupported extension capabilities MUST be ignored, unless they are prefixed with the plus ("+") sign, which indicates that the extension MUST be supported in order to use that potential configuration. If the extension is not supported, that potential configuration is not valid to the answerer.

The most preferred valid potential configuration in a media description is the valid potential configuration with the lowest configuration number. The answerer MUST now process the offer for that media stream based on the most preferred valid potential configuration. Conceptually, this entails the answerer constructing an (internal) offer as follows. First, all capability negotiation parameters from the offer SDP session description are removed, thereby yielding an offer SDP session description with the actual configuration as if SDP capability negotiation were not done in the first place. Second, this actual configuration SDP session description is modified as follows for each media stream offered, based on the capability negotiation parameters included originally:

- If a transport protocol capability is included in the potential configuration, then it replaces the transport protocol provided in the "m=" line for that media description.
- If attribute capabilities are present with a delete-attributes session indication ("-s") or media and session indication ("-ms"), then all session-level attributes from the actual configuration SDP session description MUST be deleted in the resulting potential configuration SDP session description in accordance with the procedures in Section 8.1.3.5.1. If attribute capabilities are present with a delete-attributes media indication ("-m") or media and session indication ("-ms"), then all attributes from the actual configuration SDP session description inside this media description MUST be deleted.
- If a session-level attribute capability is included, the attribute (and its associated value, if any) contained in it MUST be added to the resulting SDP session description. All such added session-level attributes MUST be listed before the session-level attributes that were initially present in the SDP session description. Furthermore, the added session-level attributes MUST be added in the order they were provided in the potential configuration (see also Section 8.1.3.5.1). This allows for attributes with implicit preference ordering to be added in the desired order; the "crypto" attribute (RFC 4568, see Section 16.2) is one such example.
- If a media-level attribute capability is included, then the attribute (and its associated value, if any) MUST be added to the resulting SDP session description within the media description in question. All such added media-level attributes MUST be listed before the media-level attributes that were initially present in the media description in question. Furthermore, the

added media-level attributes MUST be added in the order they were provided in the potential configuration (see also Section 8.1.3.5.1).

- If a supported extension capability is included, then it MUST be processed in accordance with the rules provided for that particular extension capability.

These steps MUST be performed exactly once per potential configuration (i.e., there MUST NOT be any recursive processing of any additional capability negotiation parameters that may (illegally) have been nested inside capabilities themselves). As an example of this, consider the (illegal) attribute capability a=acap:1 acap:2 foo:a. The resulting potential configuration SDP session description will, after this processing has been done, contain the attribute capability a=acap:2 foo:a.

However, since we do not perform any recursive processing of capability negotiation parameters, this second attribute capability parameter will not be processed by the offer/answer procedure. Instead, it will simply appear as a (useless) attribute in the SDP session description that will be ignored by further processing. Note that a transport protocol from the potential configuration replaces the transport protocol in the actual configuration, but an attribute capability from the potential configuration is simply added to the actual configuration. In some cases, this can result in one or more meaningless attributes in the resulting potential configuration SDP session description or worse, ambiguous or potentially even illegal attributes. Delete-attributes MUST be used for session- and/or media-level attributes to avoid such scenarios. Nevertheless, it is RECOMMENDED that implementations ignore meaningless attributes that may result from potential configurations.

For example, if the actual configuration was using SRTP and included an "a=crypto" attribute for the SRTP keying material, then use of a potential configuration that uses plain RTP would make the "crypto" attribute meaningless. The answerer may or may not ignore such a meaningless attribute. The offerer can here ensure correct operation by using delete-attributes to remove the "crypto" attribute (but will then need to provide attribute capabilities to reconstruct the SDP session description with the necessary attributes deleted, e.g., rtpmaps).

Also note, that while it is permissible to include media-level attribute capabilities at the session level, the base SDP capability negotiation framework defined here does not define any procedures for use of them (i.e., the answerer effectively ignores them). Please refer to Section 8.1.3.6.2.1 for examples of how the answerer may conceptually "see" the resulting offered alternative potential configurations. The answerer MUST check that he/she supports all mandatory attribute capabilities from the potential configuration (if any), the transport protocol capability (if any) from the potential configuration, and all mandatory extension capabilities from the potential configuration (if any). If he/she does not, the answerer MUST proceed to the second most preferred valid potential configuration for the media description, etc.

- In the case of attribute capabilities, support implies that the attribute name contained in the capability is supported and it can (and will) be negotiated successfully in the offer/answer exchange with the value provided. This does not necessarily imply that the value provided is supported in its entirety. For example, the "a=fmtp" parameter is often provided with one or more values in a list, where the offerer and answerer negotiate use of some subset of the values provided. Other attributes may include mandatory and optional parts to their values; support for the mandatory part is all that is required here. A side effect of the above rule is that whenever an "fmtp" or "rtpmap" parameter is provided as a mandatory attribute capability, the corresponding media format (codec) must be supported and use of it negotiated successfully. If this is not the offerer's intent, the corresponding attribute capabilities must be listed as optional instead.

- In the case of transport protocol capabilities, support implies that the transport protocol contained in the capability is supported and the transport protocol can (and will) be negotiated successfully in the offer/answer exchange.
- In the case of extension capabilities, the extension MUST define the rules for when the extension capability is considered supported and those rules MUST be satisfied. If the answerer has exhausted all potential configurations for the media description, without finding a valid one that is also supported, then the answerer MUST process the offered media stream based on the actual configuration plus any session-level attributes added by a valid and supported potential configuration from another media description in the offered SDP session description.

This process describes potential configuration selection as a per-media-stream process. Intermedia stream coordination of selected potential configurations however is required in some cases. First of all, session-level attributes added by a potential configuration for one media description MUST NOT cause any problems for potential configurations selected by other media descriptions in the offer SDP session description. If the session-level attributes are mandatory, then those session-level attributes MUST furthermore be supported by the session as a whole (i.e., all the media descriptions if relevant). As mentioned earlier, this adds additional complexity to the overall processing; hence, it is RECOMMENDED not to use session-level attribute capabilities in potential configurations, unless absolutely necessary.

Once the answerer has selected a valid and supported offered potential configuration for all of the media streams (or has fallen back to the actual configuration plus any added session attributes), the answerer MUST generate a valid virtual answer SDP session description based on the selected potential configuration SDP session description, as "seen" by the answerer using normal offer/answer rules (see Section 8.1.3.6.2.1 for examples). The actual answer SDP session description is formed from the virtual answer SDP session description as follows: If the answerer selected one of the potential configurations in a media description, the answerer MUST include an actual configuration attribute ("a=acfg") within that media description. The "a=acfg" attribute MUST identify the configuration number for the selected potential configuration as well as the actual parameters that were used from that potential configuration; if the potential configuration included alternatives, only the selected alternatives MUST be included. Only the known and supported parameters will be included. Unknown or unsupported parameters MUST NOT be included in the actual configuration attribute. In the case of attribute capabilities, only the known and supported capabilities are included; unknown or unsupported attribute capabilities MUST NOT be included.

If the answerer supports one or more capability negotiation extensions that were not included in a required capability negotiation extensions attribute in the offer, then the answerer SHOULD furthermore include a supported capability negotiation attribute ("a=csup") at the session level with option-tags for the extensions supported across media streams. Also, if the answerer supports one or more capability negotiation extensions for only particular media descriptions, then a supported capability negotiation attribute with those option-tags SHOULD be included within each relevant media description. The required capability negotiation attribute ("a=creq") MUST NOT be used in an answer.

The offerer's originally provided actual configuration is contained in the offer media description's "m=" line (and associated parameters). The answerer MAY send media to the offerer in accordance with that actual configuration as soon as it receives the offer; however, it MUST NOT send media based on that actual configuration if it selects an alternative potential configuration. If the answerer selects one of the potential configurations, then the answerer MAY immediately start to send media to the offerer in accordance with the selected potential configuration; however, the offerer MAY discard such media or play out garbage until the offerer receives the answer. Please refer to Section 8.1.3.9, for additional considerations and possible alternative solutions outside the base SDP capability negotiation framework.

If the answerer selected a potential configuration instead of the actual configuration, then it is RECOMMENDED that the answerer send back an answer SDP session description as soon as possible. This minimizes the risk of having media discarded or played out as garbage by the offerer. In the case of SIP (RFC 3261, also see Section 1.2) without any extensions, this implies that if the offer was received in an INVITE message, then the answer SDP session description should be provided in the first non-100 provisional response sent back (per RFC 3261, the answer would need to be repeated in the 200 response as well, unless a relevant extension such as RFC 3262 is being used).

8.1.3.6.2.1 Example Views of Potential Configurations The following examples illustrate how the answerer may conceptually "see" a potential configuration. Consider the following offered SDP session description:

```
v=0
o=alice 2891092738 2891092738 IN IP4 lost.example.com
s=
t=0 0
c=IN IP4 lost.example.com
a=tool:foo
a=acap:1 key-mgmt:mikey AQAFgM0XflABAAAAAAAAAAAAAAsAyO...
a=tcap:1 RTP/SAVP RTP/AVP
m=audio 59000 RTP/AVP 98
a=rtpmap:98 AMR/8000
a=acap:2 crypto:1 AES_CM_128_HMAC_SHA1_32
inline:NzB4d1BINUAvLEw6UzF3WSJ+PSdFcGdUJShpX1Zj|2^20|1:32
a=pcfg:1 t=1 a=1|2
m=video 52000 RTP/AVP 31
a=rtpmap:31 H261/90000
a=acap:3 crypto:1 AES_CM_128_HMAC_SHA1_80
      inline:d0RmdmcmVCspeEc3QGZiNWpVLFJhQX1cfHAwJSoj|2^20|1:32
a=pcfg:1 t=1 a=1|3
```

This particular SDP session description offers an audio stream and a video stream, each of which can either use plain RTP (actual configuration) or SRTP (potential configuration). Furthermore, two different keying mechanisms are offered, namely session-level key management extensions using MIKEY (attribute capability 1) and media-level SDP security descriptions (attribute capabilities 2 and 3). There are several potential configurations here, however, below we show the one the answerer "sees" when using potential configuration 1 for both audio and video, and furthermore using attribute capability 1 (MIKEY) for both (we have removed all the capability negotiation attributes for clarity):

```
v=0
o=alice 2891092738 2891092738 IN IP4 lost.example.com
s=
```

```
t=0 0
c=IN IP4 lost.example.com
a=tool:foo
a=key-mgmt:mikey AQAFgM0XflABAAAAAAAAAAAAAAsAyO...
m=audio 59000 RTP/SAVP 98
a=rtpmap:98 AMR/8000
m=video 52000 RTP/SAVP 31
a=rtpmap:31 H261/90000
```

Note that the transport protocol in the media descriptions indicate use of SRTP. Following, we show the offer the answerer "sees" when using potential configuration 1 for both audio and video and furthermore using attribute capability 2 and 3, respectively (SDP security descriptions), for the audio and video stream; note the order in which the resulting attributes are provided:

```
v=0
o=alice 2891092738 2891092738 IN IP4 lost.example.com
s=
t=0 0
c=IN IP4 lost.example.com
a=tool:foo
m=audio 59000 RTP/SAVP 98
a=crypto:1 AES_CM_128_HMAC_SHA1_32
inline:NzB4d1BINUAvLEw6UzF3WSJ+PSdFcGdUJShpX1Zj|2^20|1:32
a=rtpmap:98 AMR/8000
m=video 52000 RTP/SAVP 31
a=crypto:1 AES_CM_128_HMAC_SHA1_80
      inline:d0RmdmcmVCspeEc3QGZiNWpVLFJhQX1cfHAwJSoj|2^20|1:32
a=rtpmap:31 H261/90000
```

Again, note that the transport protocol in the media descriptions indicates use of SRTP. Finally, we show the offer the answerer "sees" when using potential configuration 1 with attribute capability 1 (MIKEY) for the audio stream and potential configuration 1 with attribute capability 3 (SDP security descriptions) for the video stream:

```
v=0
o=alice 2891092738 2891092738 IN IP4 lost.example.com
s=
t=0 0
c=IN IP4 lost.example.com
a=key-mgmt:mikey AQAFgM0XflABAAAAAAAAAAAAAAsAyO...
a=tool:foo
m=audio 59000 RTP/SAVP 98
a=rtpmap:98 AMR/8000
m=video 52000 RTP/SAVP 31
a=crypto:1 AES_CM_128_HMAC_SHA1_80
      inline:d0RmdmcmVCspeEc3QGZiNWpVLFJhQX1cfHAwJSoj|2^20|1:32
a=rtpmap:31 H261/90000
```

8.1.3.6.3 Offerer Processing of the Answer

When the offerer attempts to use SDP capability negotiation in the offer, the offerer MUST examine the answer for the actual use of SDP capability negotiation. For each media description in which the offerer included a potential configuration attribute ("a=pcfg"), the offerer MUST first examine that media description for the presence of a valid actual configuration attribute ("a=acfg"). An actual configuration attribute is valid if:

- It refers to a potential configuration that was present in the corresponding offer, and
- It contains the actual parameters that were used from that potential configuration; if the potential configuration included alternatives, the selected alternatives only MUST be included. Note that the answer will include only parameters and attribute capabilities that are known and supported by the answerer, as described in Section 8.1.3.6.2. If a valid actual configuration attribute is not present in a media description, then the offerer MUST process the answer SDP session description for that media stream per the normal offer/answer rules defined in RFC 3264 (see Section 3.1). However, if a valid one is found, the offerer MUST instead process the answer as follows:
  - The actual configuration attribute specifies which of the potential configurations was used by the answerer to generate the answer for this media stream. This includes all the supported attribute capabilities and the transport capabilities referenced by the potential configuration selected, where the attribute capabilities have any associated delete-attributes included. Extension capabilities supported by the answerer are included as well.
  - The offerer MUST now process the answer in accordance with the rules in RFC 3264 (see Section 3.1), except that it must be done as if the offer consisted of the selected potential configuration instead of the original actual configuration, including any transport protocol changes in the media ("m=") line(s), attributes added and deleted by the potential configuration at the media and session level, and any extensions used. If this derived answer is not a valid answer to the potential configuration offer selected by the answerer, the offerer MUST instead continue further processing as it would have for a regular offer/answer exchange, where the answer received does not adhere to the rules of RFC 3264 (see Section 3.1).

If the offer/answer exchange was successful, and if the answerer selected one of the potential configurations from the offer as the actual configuration, and the selected potential configuration differs from the actual configuration in the offer (the "m=," "a=," etc., lines), then the offerer SHOULD initiate another offer/answer exchange. This second offer/answer exchange will not modify the session in any way; however, it will help intermediaries (e.g., middleboxes), which look at the SDP session description but do not support the capability negotiation extensions, understand the details of the media stream(s) that were actually negotiated. This new offer MUST contain the selected potential configuration as the actual configuration (i.e., with the actual configuration used in the "m=" line and any other relevant attributes, bandwidth parameters, etc.).

Note that, per normal offer/answer rules, the second offer/answer exchange still needs to update the version number in the "o=" line (<sess-version> in RFC 4566, see Section 2.9). Attribute lines carrying keying material SHOULD repeat the keys from the previous offer, unless rekeying is necessary (e.g., due to a previously forked SIP INVITE request). Please refer to Section 8.1.3.12 for additional considerations related to intermediaries.

8.1.3.6.4 Modifying the Session

Capabilities and potential configurations may be included in subsequent offers as defined in RFC 3264 (see Section 3.1.8). The procedure for doing so is similar to that described above with the answer including an indication of the actual selected configuration used by the answerer. If the answer indicates use of a potential configuration from the offer, then the guidelines provided in Section 8.1.3.6.3 for doing a second offer/answer exchange using that potential configuration as the actual configuration apply.

8.1.3.7 Interactions with Interactive Connectivity Establishment

Interactive Connectivity Establishment (ICE) (RFC 5245, see Section 3.2) provides a mechanism for verifying connectivity between two endpoints by sending STUN messages directly between the media endpoints. The basic ICE specification (RFC 5245) is defined only to support UDP-based connectivity; however, it allows extensions to support other transport protocols, such as TCP, which is specified in RFC 6544. ICE defines a new "a=candidate" attribute, which, among other things, indicates the possible transport protocol(s) to use and then associates a priority with each of them. The most preferred transport protocol that *successfully* verifies connectivity will end up being used.

When using ICE, it is thus possible that the transport protocol that will be used differs from what is specified in the "m=" line. Since both ICE and SDP capability negotiation may specify alternative transport protocols, there is a potentially unintended interaction when using these together. We provide the following guidelines for addressing that. There are two basic scenarios to consider:

1. A particular media stream can run over different transport protocols (e.g., UDP, TCP, or TCP/TLS), and the intent is simply to use the one that works (in the preference order specified).
2. A particular media stream can run over different transport protocols (e.g., UDP, TCP, or TCP/TLS), and the intent is to have the negotiation process decide which one to use (e.g., T.38 over TCP or UDP).

In scenario 1, there should be ICE "a=candidate" attributes for UDP, TCP, etc., but otherwise nothing special in the potential configuration attributes to indicate the desire to use different transport protocols (e.g., UDP, or TCP). The ICE procedures essentially cover the capability negotiation required (by having the answerer select something it supports and then use of trial and error connectivity checks). Scenario 2 does not require a need to support or use ICE. Instead, we simply use transport protocol capabilities and potential configuration attributes to indicate the desired outcome.

The scenarios may be combined, for example, by offering potential configuration alternatives where some of them can support only one transport protocol (e.g., UDP), whereas others can support multiple transport protocols (e.g., UDP or TCP). In that case, there is a need for tight control over the ICE candidates that will be used for a particular configuration, yet the actual configuration may want to use all of the ICE candidates. In that case, the ICE candidate attributes can be defined as attribute capabilities, and the relevant ones should then be included in the proper potential configurations (e.g., candidate attributes for UDP only for potential configurations that are restricted to UDP), whereas there could be candidate attributes for UDP, TCP, and TCP/TLS for potential configurations that can use all three. Furthermore, use of the delete-attributes in a potential configuration can be used to ensure that ICE will not end up using a transport protocol that is not desired for a particular configuration.

SDP capability negotiation recommends use of a second offer/answer exchange when the negotiated actual configuration was one of the potential configurations from the offer (see Section

8.1.3.6.3). Similarly, ICE requires use of a second offer/answer exchange if the chosen candidate is not the same as the one in the "m="/"c="-line from the offer. When ICE and capability negotiation are used at the same time, the two secondary offer/answer exchanges SHOULD be combined into a single one.

8.1.3.8 Interactions with SIP Option-Tags

SIP (RFC 3261, also see Section 1.3) allows for SIP extensions to define a SIP option-tag that identifies the SIP extension. Support for one or more such extensions can be indicated by use of the SIP supported header, and required support for one or more such extensions can be indicated by use of the SIP require header. The "a=csup" and "a=creq" attributes defined by the SDP capability negotiation framework are similar, except that support for these two attributes by themselves cannot be guaranteed (since they are specified as extensions to the SDP specification, RFC 4566, see Section 2, itself).

SIP extensions with associated option-tags can introduce enhancements to not only SIP, but also SDP. This is, for example, the case for SIP preconditions defined in RFC 3312. When using SDP capability negotiation, some potential configurations may include certain SDP extensions, whereas others may not. Since the purpose of the SDP capability negotiation is to negotiate a session based on the features supported by both sides, use of the SIP require header for such extensions may not produce the desired result. For example, if one potential configuration requires SIP preconditions support, another does not, and the answerer does not support preconditions, then use of the SIP require header for preconditions would result in a session failure, in spite of the fact that a valid and supported potential configuration was included in the offer.

In general, this can be alleviated by use of mandatory and optional attribute capabilities in a potential configuration. There are cases, however, where permissible SDP values are tied to the use of the SIP require header. SIP preconditions (RFC 3312) is one such example, where preconditions with a "mandatory" strength-tag can only be used when a SIP Require header with the SIP option-tag "precondition" is included. Future SIP extensions that may want to use the SDP capability negotiation framework should avoid such coupling.

8.1.3.9 Processing Media before Answer

The offer/answer model (RFC 3264, see Section 3.1) requires an offerer to be able to receive media in accordance with the offer prior to receiving the answer. This property is retained with the SDP capability negotiation extensions defined here, but only when the actual configuration is selected by the answerer. If a potential configuration is chosen, the offerer may decide not to process any media received before the answer is received. This may lead to clipping. Consequently, the SDP capability negotiation framework recommends sending back an answer SDP session description as soon as possible.

The issue can be resolved by introducing a three-way handshake. In the case of SIP, for example, this can be done by defining a precondition (RFC 3312) for capability negotiation (or by using an existing precondition that is known to generate a second offer/answer exchange before proceeding with the session). However, preconditions are often viewed as complicated to implement, and they may add to overall session establishment delay by requiring an extra offer/answer exchange.

An alternative three-way handshake can be performed by use of ICE (RFC 5245, see Section 3.2). When ICE is used, and the answerer receives a STUN Binding request for any one of the accepted media streams from the offerer, the answerer knows the offer has received his/her answer. At that point, the answerer knows that the offerer will be able to process incoming media according to the negotiated configuration and hence he/she can start sending media without the risk of the offerer either discarding it or playing garbage.

Please note that, these considerations notwithstanding, this document does not place any requirements on the offerer to process and play media before answer; it merely provides recommendations for how to ensure that media sent by the answerer and received by the offerer prior to receiving the answer can in fact be rendered by the offerer. In some use cases, a three-way handshake is not needed, for example, when the offerer does not need information from the answer, such as keying material in the SDP session description, in order to process incoming media. The SDP capability negotiation framework does not define any such solutions; however, extensions may do so. For example, one technique proposed for best-effort SRTP in Reference [1] is to provide different RTP payload type mappings for different transport protocols used, outside of the actual configuration, while still allowing them to be used by the answerer (exchange of keying material is still needed, e.g., inband). The basic SDP capability negotiation framework defined here does not include the ability to do so; however, extensions that enable that may be defined.

8.1.3.10 Indicating Bandwidth Usage

The amount of bandwidth used for a particular media stream depends on the negotiated codecs, transport protocol, and other parameters. For example, the use of SRTP (RFC 3711) with integrity protection requires more bandwidth than plain RTP (RFC 3551). SDP defines the bandwidth ("b=") parameter to indicate the proposed bandwidth for the session or media stream. In SDP, as defined by RFC 4566 (see Section 2), each media description contains one transport protocol and one or more codecs. When specifying the proposed bandwidth, the worst-case scenario must be taken into account (i.e., use of the highest bandwidth codec provided, the transport protocol indicated, and the worst-case (bandwidth-wise) parameters that can be negotiated (e.g., a 32-bit Hashed Message Authentication Code (HMAC) or an 80-bit HMAC).

The base SDP capability negotiation framework does not provide a way to negotiate bandwidth parameters. The issue thus remains; however, it is potentially worse than with SDP per RFC 4566 (see Section 2), since it is easier to negotiate additional codecs and furthermore possible to negotiate different transport protocols. The recommended approach for addressing this is the same as for plain SDP; the worst case (now including potential configurations) needs to be taken into account when specifying the bandwidth parameters in the actual configuration. This can make the bandwidth value less accurate than in SDP per RFC 4566 (see Section 2) (due to potential greater variability in the potential configuration bandwidth use). Extensions can be defined to address this shortcoming.

Note that when using RTP retransmission (RFC 4588) with the RTCP-based feedback profile (RFC 4585) (RTP/AVPF), the retransmitted packets are part of the media stream bandwidth when using SSRC multiplexing. If a feedback-based protocol is offered as the actual configuration transport protocol, a nonfeedback-based protocol is offered as a potential configuration transport protocol and ends up being used, the actual bandwidth usage may be lower than the indicated bandwidth value in the offer (and vice versa).

8.1.3.11 Dealing with a Large Number of Potential Configurations

When using the SDP capability negotiation, it is easy to generate offers that contain a large number of potential configurations. For example, in the offer:

```
v=0
o=- 25678 753849 IN IP4 192.0.2.1
s=
c=IN IP4 192.0.2.1
t=0 0
m=audio 53456 RTP/AVP 0 18
a=tcap:1 RTP/SAVPF RTP/SAVP RTP/AVPF
a=acap:1 crypto:1 AES_CM_128_HMAC_SHA1_80
      inline:WVNfX19zZW1jdGwgKCkgewkyMjA7fQp9CnVubGVz|2^20|1:4
FEC_ORDER=FEC_SRTP
a=acap:2 key-mgmt:mikey AQAFgM0XflABAAAAAAAAAAAAAAsAyO...
a=acap:3 rtcp-fb:0 nack
a=pcfg:1 t=1 a=1,3|2,3
a=pcfg:2 t=2 a=1|2
a=pcfg:3 t=3 a=3
```

We have 5 potential configurations on top of the actual configuration for a single media stream. Adding an extension capability with just two alternatives for each would double that number (to 10), and doing the equivalent with two media streams would again double that number (to 20). While it is easy (and inexpensive) for the offerer to generate such offers, processing them at the answering side may not be. Consequently, it is RECOMMENDED that offerers do not create offers with unnecessarily large number of potential configurations in them.

On the answering side, implementers MUST take care to avoid excessive memory and CPU consumption. For example, a naive implementation that first generates all the valid potential configuration SDP session descriptions internally could find itself being memory exhausted, especially if it supports a large number of endpoints. Similarly, a naive implementation that simply performs iterative trial-and-error processing on each possible potential configuration SDP session description (in the preference order specified) could find itself being CPU constrained. An alternative strategy is to prune the search space first by discarding the set of offered potential configurations where the transport protocol indicated (if any) is not supported, and/or one or more mandatory attribute capabilities (if any) are either not supported or not valid. Potential configurations with unsupported mandatory extension configurations in them can be discarded as well.

8.1.3.12 SDP Capability Negotiation and Intermediaries

An intermediary is here defined as an entity, between a SIP user agent A and a SIP user agent B, that needs to perform some kind of processing on the SDP session descriptions exchanged between A and B, in order for the session establishment to operate as intended. Examples of such intermediaries include session border controllers that may perform media relaying, proxy call session control functions that may authorize use of a certain amount of network resources (bandwidth), etc. The presence and design of such intermediaries may not follow the "Internet" model or the SIP requirements for proxies (which are not supposed to look in message bodies such as SDP session descriptions); however, they are a fact of life in some deployment scenarios and hence deserve consideration.

If the intermediary needs to understand the characteristics of the media sessions being negotiated (e.g., the amount of bandwidth used or the transport protocol negotiated), then use of the SDP capability negotiation framework may impact them. For example, some intermediaries are known to disallow answers where the transport protocol differs from the one in the offer. Use of the SDP capability negotiation framework in the presence of such intermediaries could lead to session failures. Intermediaries that need to authorize use of network resources based on the negotiated media stream parameters are affected as well. If they inspect only the offer, then they may authorize parameters assuming a different transport protocol, codecs, etc., than what is actually being negotiated. For these, and other, reasons it is RECOMMENDED that implementers of intermediaries add support for the SDP capability negotiation framework.

The SDP capability negotiation framework itself attempts to help these intermediaries as well, by recommending a second offer/answer exchange when use of a potential configuration has been negotiated (see Section 8.1.3.6.3). However, there are several limitations with this approach. First, although the second offer/answer exchange is RECOMMENDED, it is not required and hence may not be performed. Second, the intermediary may refuse the initial answer (e.g., due to perceived transport protocol mismatch). Third, the strategy is not foolproof since the offer/answer procedures (RFC 3264, see Section 3.1) leave the original offer/answer exchange in effect when a subsequent one fails. Consider the following example:

1. Offerer generates an SDP session description offer with the actual configuration specifying a low-bandwidth configuration (e.g., plain RTP) and a potential configuration specifying a high(er) bandwidth configuration (e.g., SRTP with integrity).
2. An intermediary (e.g., an Session Border Controller or Proxy-Call Session Control Function), that does not support SDP capability negotiation, authorizes the session based on the actual configuration it sees in the SDP session description.
3. The answerer chooses the high(er) bandwidth potential configuration and generates an answer SDP session description based on that.
4. The intermediary passes through the answer SDP session description.
5. The offerer sees the accepted answer and generates an updated offer that contains the selected potential configuration as the actual configuration. In other words, the high(er) bandwidth configuration (which has already been negotiated successfully) is now the actual configuration in the offer SDP session description.
6. The intermediary sees the new offer; however, it does not authorize the use of the high(er) bandwidth configuration and consequently generates a rejection message to the offerer.
7. The offerer receives the rejected offer.

After step 7, per RFC 3264 (see Section 3.1), the offer/answer exchange that was completed in step 5 remains in effect; however, the intermediary may not have authorized the necessary network resources; hence, the media stream may experience quality issues. The solution to this problem is to upgrade the intermediary to support the SDP capability negotiation framework.

8.1.3.13 Considerations for Specific Attribute Capabilities

8.1.3.13.1 The "rtpmap" and "fmtp" Attributes

The base SDP capability negotiation framework defines transport capabilities and attribute capabilities. Media capabilities, which can be used to describe media formats and their associated parameters, are not defined in this document; however, the "rtpmap" and "fmtp" attributes can be used as attribute capabilities. Using such attribute capabilities in a potential configuration requires a bit of care. The rtpmap parameter binds an RTP payload type to a media format (e.g., codec). While it is possible to provide rtpmaps for payload types not found in the corresponding "m=" line, such rtpmaps provide no value in normal offer/answer exchanges, since only the payload types found in the "m=" line are part of the offer (or answer). This applies to the base SDP capability negotiation framework as well.

Only the media formats (e.g., RTP payload types) provided in the "m=" line are actually offered; inclusion of "rtpmap" attributes with other RTP payload types in a potential configuration does not change this fact; hence, they do not provide any useful information there. They may still be useful as pure capabilities (outside a potential configuration) in order to inform a peer of additional codecs supported. It is possible to provide an "rtpmap" attribute capability with a payload type mapping to a different codec than a corresponding actual configuration "rtpmap" attribute for the media description has. Such practice is permissible as a way of indicating a capability. If that capability is included in a potential configuration, then delete attributes (see Section 8.1.3.5.1) MUST be used to ensure that there are not multiple "rtpmap" attributes for the same payload type in a given media description (which would not be allowed by SDP; RFC 4566, see Section 2).

Similar considerations and rules apply to the "fmtp" attribute. An "fmtp" attribute capability for a media format not included in the "m=" line is useless in a potential configuration (but may be useful as a capability by itself). An "fmtp" attribute capability in a potential configuration for a media format that already has an "fmtp" attribute in the actual configuration may lead to multiple fmtp format parameters for that media format, and that is not allowed by SDP (RFC 4566, see Section 2). The delete-attributes MUST be used to ensure that there are not multiple "fmtp" attributes for a given media format in a media description. Extensions to the base SDP capability negotiation framework may change this behavior.

8.1.3.14 Direction Attributes

SDP defines the "inactive," "sendonly," "recvonly," and "sendrecv" direction attributes. The direction attributes can be applied at either the session level or the media level. In either case, it is possible to define attribute capabilities for these direction capabilities; if used by a potential configuration, the normal offer/answer procedures still apply. For example, if an offered potential configuration includes the "sendonly" direction attribute, and it is selected as the actual configuration, then the answer MUST include a corresponding "recvonly" (or "inactive") attribute.

8.1.3.15 Relationship to RFC 3407

RFC 3407 defines capability descriptions with limited abilities to describe attributes, bandwidth parameters, transport protocols, and media formats. RFC 3407 does not define any negotiation procedures for actually using those capability descriptions. This document defines new attributes for describing attribute capabilities and transport capabilities. It also defines procedures for using those capabilities as part of an offer/answer exchange. In contrast to RFC 3407, this document does not define bandwidth parameters, and it does not define how to express ranges of values. Extensions to this document may be defined in order to fully cover all the capabilities provided by RFC 3407 (see Section 4.1) (for example, more general media capabilities).

It is RECOMMENDED that implementations use the attributes and procedures defined in this document instead of those defined in RFC 3407 (see Section 4.1). If capability description interoperability with legacy RFC 3407 implementations is desired, implementations MAY include both RFC 3407 capability descriptions and capabilities defined by this document. The offer/answer negotiation procedures defined in this document will not use the RFC 3407 capability descriptions.

8.1.4 Examples

In this section, we provide examples showing how to use SDP capability negotiation.

8.1.4.1 Multiple Transport Protocols

The following example illustrates how to use the SDP capability negotiation extensions to negotiate use of one out of several possible transport protocols. The offerer uses the expected least common denominator (plain RTP) as the actual configuration, and the alternative transport protocols as the potential configurations. The example is illustrated by the following offer/answer exchange, in which Alice sends an offer to Bob:

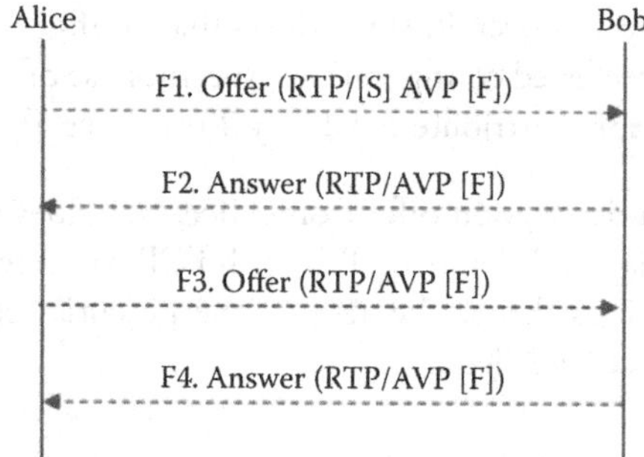

Alice's offer includes plain RTP (RTP/AVP), RTP with RTCP-based feedback (RTP/AVPF), SRTP (RTP/SAVP), and SRTP with RTCP-based feedback (RTP/SAVPF (Secure Audio Visual Profile for RTCP Feedback)) as alternatives. RTP is the default, with RTP/SAVPF, RTP/SAVP, and RTP/AVPF as the alternatives and preferred in the order listed:

```
v=0
o=- 25678 753849 IN IP4 192.0.2.1
s=
c=IN IP4 192.0.2.1
t=0 0
m=audio 53456 RTP/AVP 0 18
a=tcap:1 RTP/SAVPF RTP/SAVP RTP/AVPF
a=acap:1 crypto:1 AES_CM_128_HMAC_SHA1_80
     inline:WVNfX19zZW1jdGwgKCkgewkyMjA7fQp9CnVubGVz|2^20|1:4
FEC_ORDER=FEC_SRTP
a=acap:2 rtcp-fb:0 nack
a=pcfg:1 t=1 a=1,[2]
a=pcfg:2 t=2 a=1
a=pcfg:3 t=3 a=[2]
```

The "m=" line indicates that Alice is offering to use plain RTP with PCMU or G.729. The capabilities are provided by the "a=tcap" and "a=acap" attributes. The "tcap" capability indicates that SRTP with RTCP-based feedback (RTP/SAVPF), SRTP (RTP/SAVP), and RTP with

RTCP-based feedback are supported. The first "acap" attribute provides an attribute capability with a handle of 1. The capability is a "crypto" attribute, which provides the keying material for SRTP using SDP security descriptions (RFC 4568, see Section 16.2). The second "acap" attribute provides an attribute capability with a handle of 2. The capability is an "rtcp-fb" attribute, which is used by the RTCP-based feedback profiles to indicate that payload type 0 (PCMU) supports feedback type "nack." The "a=pcfg" attributes provide the potential configurations included in the offer by reference to the capabilities. There are three potential configurations:

- Potential configuration 1, which is the most preferred potential configuration specifies use of transport protocol capability 1 (RTP/SAVPF) and attribute capabilities 1 (the "crypto" attribute) and 2 (the "rtcp-fb" attribute). Support for the first is mandatory whereas support for the second is optional.
- Potential configuration 2, which is the second most preferred potential configuration, specifies use of transport protocol capability 2 (RTP/SAVP) and mandatory attribute capability 1 (the "crypto" attribute).
- Potential configuration 3, which is the least preferred potential configuration (but the second least preferred configuration overall, since the actual configuration provided by the "m=" line is always the least preferred configuration), specifies use of transport protocol capability 3 (RTP/AVPF) and optional attribute capability 2 (the "rtcp-fb" attribute).

Bob receives the SDP session description offer from Alice. Bob does not support any SRTP profiles; however, he supports plain RTP and RTP with RTCP-based feedback, as well as the SDP capability negotiation extensions; hence, he accepts the potential configuration for RTP with RTCP-based feedback provided by Alice:

```
v=0
o=- 24351 621814 IN IP4 192.0.2.2
s=
c=IN IP4 192.0.2.2
t=0 0
m=audio 54568 RTP/AVPF 0 18
a=rtcp-fb:0 nack
a=acfg:1 t=3 a=[2]
```

Bob includes the "a=acfg" attribute in the answer to inform Alice that he based his answer on an offer containing the potential configuration with transport protocol capability 3 and optional attribute capability 2 from the offer SDP session description (i.e., the RTP/AVPF profile using the "rtcp-fb" value provided). Bob also includes an "rtcp-fb" attribute with the value "nack" value for RTP payload type 0.

When Alice receives Bob's answer, session negotiation is complete; however, Alice chooses to generate a new offer using the actual configuration. This is done purely to assist any intermediaries that may reside between Alice and Bob but do not support the SDP capability negotiation framework (and may not understand the negotiation that just took place):

Alice's updated offer includes only RTP/AVPF and does not use the SDP capability negotiation framework (Alice could have included the capabilities as well if she had wanted):

```
v=0
o=- 25678 753850 IN IP4 192.0.2.1
s=
c=IN IP4 192.0.2.1
```

```
t=0 0
m=audio 53456 RTP/AVPF 0 18
a=rtcp-fb:0 nack
```

The "m=" line now indicates that Alice is offering to use RTP with RTCP-based feedback and PCMU or G.729. The "rtcp-fb" attribute provides the feedback type "nack" for payload type 0 again (but as part of the actual configuration). Bob receives the SDP session description offer from Alice, which he accepts, and then generates an answer to Alice:

```
v=0
o=- 24351 621815 IN IP4 192.0.2.2
s=
c=IN IP4 192.0.2.2
t=0 0
m=audio 54568 RTP/AVPF 0 18
a=rtcp-fb:0 nack
```

Bob includes the same "rtcp-fb" attribute as before, and the session proceeds without change. Although Bob did not include any capabilities in his answer, he could have done so if he had wanted. Note that in this particular example, the answerer supported the SDP capability negotiation framework and hence the attributes and procedures defined here; however, had he not, the answerer would simply have ignored the new attributes received in step 1 and accepted the offer to use normal RTP. In that case, the following answer would have been generated in step 2 instead:

```
v=0
o=- 24351 621814 IN IP4 192.0.2.2
s=
c=IN IP4 192.0.2.2
t=0 0
m=audio 54568 RTP/AVP 0 18
```

8.1.4.2 DTLS-SRTP or SRTP with Media-Level Security Descriptions

The following example illustrates how to use the SDP capability negotiation framework to negotiate use of SRTP using either SDP security descriptions or DTLS-SRTP. The offerer (Alice) wants to establish a SRTP audio stream but is willing to use plain RTP. Alice prefers to use DTLS-SRTP as the key management protocol but supports SDP security descriptions as well (note that RFC 5763 contains additional DTLS-SRTP examples). The example is illustrated by the offer/answer exchange in Inline Figure 5, where Alice sends an offer to Bob:

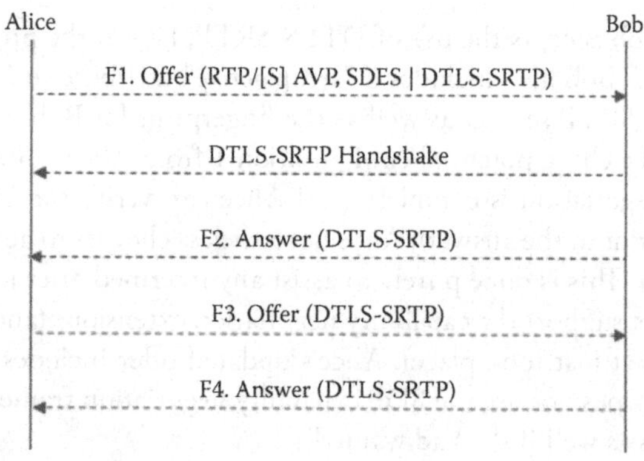

Alice's offer includes an audio stream that offers use of plain RTP and SRTP as alternatives. For the SRTP stream, it can be established using either DTLS-SRTP or SDP security descriptions.

```
v=0
o=- 25678 753849 IN IP4 192.0.2.1
s=
t=0 0
c=IN IP4 192.0.2.1
a=acap:1 setup:actpass
a=acap:2 fingerprint: SHA-1 \
4A:AD:B9:B1:3F:82:18:3B:54:02:12:DF:3E:5D:49:6B:19:E5:7C:AB
a=tcap:1 UDP/TLS/RTP/SAVP RTP/SAVP
m=audio 59000 RTP/AVP 98
a=rtpmap:98 AMR/8000
a=acap:3 crypto:1 AES_CM_128_HMAC_SHA1_32
      inline:NzB4d1BINUAvLEw6UzF3WSJ+PSdFcGdUJShpX1Zj|2^20|1:32
a=pcfg:1 t=1 a=1,2
a=pcfg:2 t=2 a=3
```

The first (and preferred) potential configuration for the audio stream specifies use of transport capability 1 (UDP/TLS/RTP/SAVP) (i.e., DTLS-SRTP) and attribute capabilities 1 and 2 (active/passive mode and certificate fingerprint), both of which must be supported to choose this potential configuration). The second (and less preferred) potential configuration specifies use of transport capability 2 (RTP/SAVP) and mandatory attribute capability 3 (i.e., the SDP security description). Bob receives the SDP session description offer from Alice. Bob supports DTLS-SRTP as preferred by Alice, and Bob now initiates the DTLS-SRTP handshake to establish the DTLS-SRTP session (see RFC 5764 for details). Bob also sends back an answer to Alice as follows:

```
v=0
o=- 24351 621814 IN IP4 192.0.2.2
s=
a=setup:active
a=fingerprint: SHA-1 \
      FF:FF:FF:B1:3F:82:18:3B:54:02:12:DF:3E:5D:49:6B:19:E5:7C:AB
t=0 0
c=IN IP4 192.0.2.2
m=audio 54568 UDP/TLS/RTP/SAVP 98
a=rtpmap:98 AMR/8000
a=acfg:1 t=1 a=1,2
```

For the audio stream, Bob accepts the use of DTLS-SRTP; hence, the profile in the "m=" line is "UDP/TLS/RTP/SAVP." Bob also includes a "setup:active" attribute to indicate he is the active endpoint for the DTLS-SRTP session as well as the fingerprint for Bob's certificate. Bob's "acfg" attribute indicates that he chose potential configuration 1 from Alice's offer. When Alice receives Bob's answer, session negotiation is complete (and Alice can verify the DTLS handshake using Bob's certificate fingerprint in the answer); Alice nevertheless chooses to generate a new offer using the actual configuration. This is done purely to assist any intermediaries that may reside between Alice and Bob but do not support the capability negotiation extensions (and hence may not understand the negotiation that just took place). Alice's updated offer includes only DTLS-SRTP for the audio stream, and it does not use the SDP capability negotiation framework (Alice could have included the capabilities as well if she had wanted):

```
v=0
o=- 25678 753850 IN IP4 192.0.2.1
s=
t=0 0
c=IN IP4 192.0.2.1
a=setup:actpass
a=fingerprint: SHA-1 \
      4A:AD:B9:B1:3F:82:18:3B:54:02:12:DF:3E:5D:49:6B:19:E5:7C:AB
m=audio 59000 UDP/TLS/RTP/AVP 98
a=rtpmap:98 AMR/8000
```

The "m=" line for the audio stream now indicates that Alice is offering to use DTLS-SRTP in active/passive mode using her certificate fingerprint provided. Bob receives the SDP session description offer from Alice, which he accepts, and then generates an answer to Alice:

```
v=0
o=- 24351 621814 IN IP4 192.0.2.2
s=
a=setup:active
a=fingerprint: SHA-1 \
      FF:FF:FF:B1:3F:82:18:3B:54:02:12:DF:3E:5D:49:6B:19:E5:7C:AB
t=0 0
c=IN IP4 192.0.2.2
m=audio 54568 UDP/TLS/RTP/SAVP 98
a=rtpmap:98 AMR/8000
a=acfg:1 t=1 a=1,2
```

Bob includes the same "setup:active" and fingerprint attributes as before, and the session proceeds without change. Although Bob did not include any capabilities in his answer, he could have done so if he had wanted. Note that in this particular example, the answerer supported the capability extensions defined here; however, had he not, the answerer would simply have ignored the new attributes received in step 1 and accepted the offer to use normal (without security features) RTP. In that case, the following answer would have been generated in step 2 instead:

```
v=0
o=- 24351 621814 IN IP4 192.0.2.2
s=
t=0 0
c=IN IP4 192.0.2.2
m=audio 54568 RTP/AVP 98
a=rtpmap:98 AMR/8000
```

Finally, if Bob had chosen to use SDP security descriptions instead of DTLS-SRTP, the following answer would have been generated:

```
v=0
o=- 24351 621814 IN IP4 192.0.2.2
s=
t=0 0
c=IN IP4 192.0.2.2
m=audio 54568 RTP/SAVP 98
a=rtpmap:98 AMR/8000
a=crypto:1 AES_CM_128_HMAC_SHA1_32
```

```
    inline:WSJ+PSdFcGdUJShpX1ZjNzB4d1BINUAvLEw6UzF3|2^20|1:32
a=acfg:2 t=2 a=3
```

8.1.4.3 Best-Effort SRTP with Session-Level MIKEY and Media-Level Security Descriptions

The following example illustrates how to use the SDP capability negotiation extensions to support so-called Best-Effort SRTP as well as alternative keying mechanisms, more specifically MIKEY (RFC 3830) and SDP security descriptions. The offerer (Alice) wants to establish an audio and video session. Alice prefers to use session-level MIKEY as the key management protocol but supports SDP security descriptions as well. The example is illustrated by the offer/answer exchange in Inline Figure 6, where Alice sends an offer to Bob:

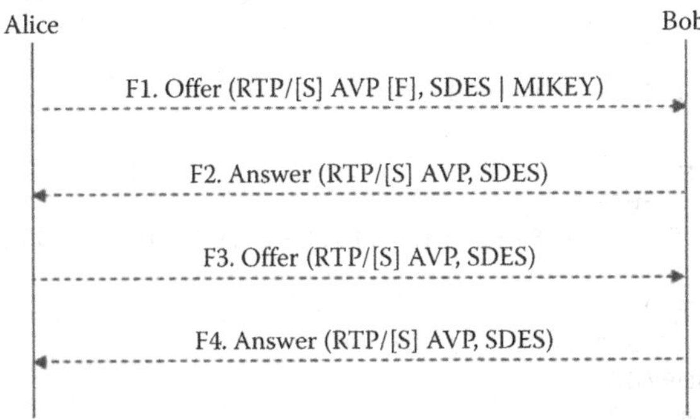

Alice's offer includes an audio and a video stream. The audio stream offers use of plain RTP and SRTP as alternatives, whereas the video stream offers use of plain RTP, RTP with RTCP-based feedback, SRTP, and SRTP with RTCP-based feedback as alternatives:

```
v=0
o=- 25678 753849 IN IP4 192.0.2.1
s=
t=0 0
c=IN IP4 192.0.2.1
a=acap:1 key-mgmt:mikey AQAFgM0XflABAAAAAAAAAAAAAAsAyO...
a=tcap:1 RTP/SAVPF RTP/SAVP RTP/AVPF
m=audio 59000 RTP/AVP 98
a=rtpmap:98 AMR/8000
a=acap:2 crypto:1 AES_CM_128_HMAC_SHA1_32
      inline:NzB4d1BINUAvLEw6UzF3WSJ+PSdFcGdUJShpX1Zj|2^20|1:32
a=pcfg:1 t=2 a=1|2
m=video 52000 RTP/AVP 31
a=rtpmap:31 H261/90000
a=acap:3 crypto:1 AES_CM_128_HMAC_SHA1_80
      inline:d0RmdmcmVCspeEc3QGZiNWpVLFJhQX1cfHAwJSoj|2^20|1:32
a=acap:4 rtcp-fb:* nack
a=pcfg:1 t=1 a=1,4|3,4
a=pcfg:2 t=2 a=1|3
a=pcfg:3 t=3 a=4
```

The potential configuration for the audio stream specifies use of transport capability 2 (RTP/SAVP) and either attribute capability 1 (session-level MIKEY as the keying mechanism) or 2 (SDP security descriptions as the keying mechanism). Support for either of these attribute capabilities is mandatory. There are three potential configurations for the video stream.

- The first configuration with configuration number 1 uses transport capability 1 (RTP/SAVPF) with either attribute capabilities 1 and 4 (session-level MIKEY and the "rtcp-fb" attribute) or attribute capabilities 3 and 4 (SDP security descriptions and the "rtcp-fb" attribute). In this example, the offerer insists not only on the keying mechanism being supported, but also that the "rtcp-fb" attribute is supported with the value indicated. Consequently, all the attribute capabilities are marked as mandatory in this potential configuration.
- The second configuration with configuration number 2 uses transport capability 2 (RTP/SAVP) and either attribute capability 1 (session-level MIKEY) or attribute capability 3 (SDP security descriptions). Both attribute capabilities are mandatory in this configuration.
- The third configuration with configuration number 3 uses transport capability 3 (RTP/AVPF) and mandatory attribute capability 4 (the "rtcp-fb" attribute).

Bob receives the SDP session description offer from Alice. Bob supports SRTP, SRTP with RTCP-based feedback, and the SDP capability negotiation extensions. Bob also supports SDP security descriptions, but not MIKEY; hence, he generates the following answer:

```
v=0
o=- 24351 621814 IN IP4 192.0.2.2
s=
t=0 0
c=IN IP4 192.0.2.2
m=audio 54568 RTP/SAVP 98
a=rtpmap:98 AMR/8000
a=crypto:1 AES_CM_128_HMAC_SHA1_32
      inline:WSJ+PSdFcGdUJShpX1ZjNzB4d1BINUAvLEw6UzF3|2^20|1:32
a=acfg:1 t=2 a=2
m=video 55468 RTP/SAVPF 31
a=rtpmap:31 H261/90000
a=crypto:1 AES_CM_128_HMAC_SHA1_80
      inline:AwWpVLFJhQX1cfHJSojd0RmdmcmVCspeEc3QGZiN|2^20|1:32
a=rtcp-fb:* nack
a=acfg:1 t=1 a=3,4
```

For the audio stream, Bob accepts the use of SRTP; hence, the profile in the "m=" line is "RTP/SAVP." Bob also includes a "crypto" attribute with his own keying material and an "acfg" attribute identifying actual configuration 1 for the audio media stream from the offer, using transport capability 2 (RTP/SAVP) and attribute capability 2 (the "crypto" attribute from the offer). For the video stream, Bob accepts the use of SRTP with RTCP-based feedback; hence, the profile in the "m=" line is "RTP/SAVPF." Bob also includes a "crypto" attribute with his own keying material and an "acfg" attribute identifying actual configuration 1 for the video stream from the offer, using transport capability 1 (RTP/SAVPF) and attribute capabilities 3 (the "crypto" attribute from the offer) and 4 (the "rtcp-fb" attribute from the offer).

When Alice receives Bob's answer, session negotiation is complete; however, Alice nevertheless chooses to generate a new offer using the actual configuration. This is done purely to assist any intermediaries that may reside between Alice and Bob but do not support the capability negotiation extensions (and hence may not understand the negotiation that just took place). Alice's updated offer includes only SRTP for the audio stream SRTP with RTCP-based feedback for the video stream, and it does not use the SDP capability negotiation framework (Alice could have included the capabilities as well if she had wanted):

```
v=0
o=- 25678 753850 IN IP4 192.0.2.1
s=
t=0 0
c=IN IP4 192.0.2.1
m=audio 59000 RTP/SAVP 98
a=rtpmap:98 AMR/8000
a=crypto:1 AES_CM_128_HMAC_SHA1_32
      inline:NzB4d1BINUAvLEw6UzF3WSJ+PSdFcGdUJShpX1Zj|2^20|1:32
m=video 52000 RTP/SAVPF 31
a=rtpmap:31 H261/90000
a=crypto:1 AES_CM_128_HMAC_SHA1_80
      inline:d0RmdmcmVCspeEc3QGZiNWpVLFJhQX1cfHAwJSoj|2^20|1:32
a=rtcp-fb:* nack
```

The "m=" line for the audio stream now indicates that Alice is offering to use SRTP with PCMU or G.729, whereas the "m=" line for the video stream indicates that Alice is offering to use SRTP with RTCP-based feedback and H.261. Each media stream includes a "crypto" attribute, which provides the SRTP keying material, with the same value again. Bob receives the SDP session description offer from Alice, which he accepts, and then generates an answer to Alice:

```
v=0
o=- 24351 621815 IN IP4 192.0.2.2
s=
t=0 0
c=IN IP4 192.0.2.2
m=audio 54568 RTP/SAVP 98
a=rtpmap:98 AMR/8000
a=crypto:1 AES_CM_128_HMAC_SHA1_32
      inline:WSJ+PSdFcGdUJShpX1ZjNzB4d1BINUAvLEw6UzF3|2^20|1:32
m=video 55468 RTP/SAVPF 31
a=rtpmap:31 H261/90000
a=crypto:1 AES_CM_128_HMAC_SHA1_80
      inline:AwWpVLFJhQX1cfHJSojd0RmdmcmVCspeEc3QGZiN|2^20|1:32
a=rtcp-fb:* nack
```

Bob includes the same "crypto" attribute as before, and the session proceeds without change. Although Bob did not include any capabilities in his answer, he could have done so if he had wanted. Note that in this particular example, the answerer supported the capability extensions defined here; however, had he not, the answerer would simply have ignored the new attributes received in step 1 and accepted the offer to use normal RTP. In that case, the following answer would have been generated in step 2 instead:

```
v=0
o=- 24351 621814 IN IP4 192.0.2.2
s=
t=0 0
c=IN IP4 192.0.2.2
m=audio 54568 RTP/AVP 98
a=rtpmap:98 AMR/8000
```

```
m=video 55468 RTP/AVP 31
a=rtpmap:31 H261/90000
a=rtcp-fb:* nack
```

Finally, if Bob had chosen to use session-level MIKEY instead of SDP security descriptions, the following answer would have been generated:

```
v=0
o=- 24351 621814 IN IP4 192.0.2.2
s=
t=0 0
c=IN IP4 192.0.2.2
a=key-mgmt:mikey AQEFgM0XflABAAAAAAAAAAAAAAAYAyO...
m=audio 54568 RTP/SAVP 98
a=rtpmap:98 AMR/8000
a=acfg:1 t=2 a=1
m=video 55468 RTP/SAVPF 31
a=rtpmap:31 H261/90000
a=rtcp-fb:* nack
a=acfg:1 t=1 a=1,4
```

It should be noted that, although Bob could have chosen session-level MIKEY for one media stream and SDP security descriptions for another media stream, there are no well-defined offerer processing rules of the resulting answer for this; hence, the offerer may incorrectly assume use of MIKEY for both streams. To avoid this, if the answerer chooses session-level MIKEY, then all SRTP-based media streams SHOULD use MIKEY (this applies irrespective of whether SDP capability negotiation is being used). Use of media-level MIKEY does not have a similar constraint.

8.1.4.4 SRTP with Session-Level MIKEY and Media-Level Security Descriptions as Alternatives

The following example illustrates how to use the SDP capability negotiation framework to negotiate use of either MIKEY or SDP security descriptions, when one of them is included as part of the actual configuration and the other one is being selected. The offerer (Alice) wants to establish an audio and video session. Alice prefers to use session-level MIKEY as the key management protocol but supports SDP security descriptions as well. The example is illustrated by the offer/answer exchange in Inline Figure 7, where Alice sends an offer to Bob:

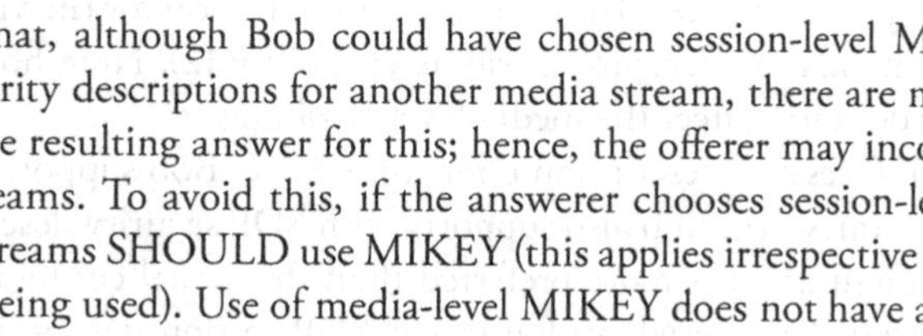

Alice Bob

F1. Offer (RTP/[S] AVP [F], SDES | MIKEY)

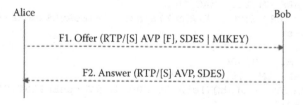

F2. Answer (RTP/[S] AVP, SDES)

Alice's offer includes an audio and a video stream. Both the audio and the video stream offer use of SRTP:

```
v=0
o=- 25678 753849 IN IP4 192.0.2.1
s=
```

```
t=0 0
c=IN IP4 192.0.2.1
a=key-mgmt:mikey AQAFgM0XflABAAAAAAAAAAAAAAsAyO...
m=audio 59000 RTP/SAVP 98
a=rtpmap:98 AMR/8000
a=acap:1 crypto:1 AES_CM_128_HMAC_SHA1_32
      inline:NzB4d1BINUAvLEw6UzF3WSJ+PSdFcGdUJShpX1Zj|2^20|1:32
a=pcfg:1 a=-s:1
m=video 52000 RTP/SAVP 31
a=rtpmap:31 H261/90000
a=acap:2 crypto:1 AES_CM_128_HMAC_SHA1_80
      inline:d0RmdmcmVCspeEc3QGZiNWpVLFJhQX1cfHAwJSoj|2^20|1:32
a=pcfg:1 a=-s:2
```

Alice does not know whether Bob supports MIKEY or SDP security descriptions. She could include attributes for both; however, the resulting procedures and potential interactions are not well defined. Instead, she places a session-level "key-mgmt" attribute for MIKEY in the actual configuration with SDP security descriptions as an alternative in the potential configuration. The potential configuration for the audio stream specifies that all session-level attributes are to be deleted (i.e., the session-level "a=key-mgmt" attribute) and that mandatory attribute capability 2 is to be used (i.e., the "crypto" attribute). The potential configuration for the video stream is similar, except it uses its own mandatory "crypto" attribute capability (2). Note how the deletion of the session-level attributes does not affect the media-level attributes.

Bob receives the SDP session description offer from Alice. Bob supports SRTP and the SDP capability negotiation framework. Bob also supports both SDP security descriptions and MIKEY. Since the potential configuration is more preferred than the actual configuration, Bob (conceptually) generates an internal potential configuration SDP session description that contains the "crypto" attributes for the audio and video stream, but not the "key-mgmt" attribute for MIKEY, thereby avoiding any ambiguity between the two keying mechanisms. As a result, he generates the following answer:

```
v=0
o=- 24351 621814 IN IP4 192.0.2.2
s=
t=0 0
c=IN IP4 192.0.2.2
m=audio 54568 RTP/SAVP 98
a=rtpmap:98 AMR/8000
a=crypto:1 AES_CM_128_HMAC_SHA1_32
      inline:WSJ+PSdFcGdUJShpX1ZjNzB4d1BINUAvLEw6UzF3|2^20|1:32
a=acfg:1 a=-s:1
m=video 55468 RTP/SAVP 31
a=rtpmap:31 H261/90000
a=crypto:1 AES_CM_128_HMAC_SHA1_80
      inline:AwWpVLFJhQX1cfHJSojd0RmdmcmVCspeEc3QGZiN|2^20|1:32
a=acfg:1 a=-s:2
```

For the audio stream, Bob accepts the use of SRTP using SDP security descriptions. Bob therefore includes a "crypto" attribute with his own keying material and an "acfg" attribute identifying the actual configuration 1 for the audio media stream from the offer, with the delete-attributes ("-s") and attribute capability 1 (the "crypto" attribute from the offer). For the video stream, Bob also accepts the use of SRTP using SDP security descriptions. Bob therefore includes a "crypto" attribute with his own keying material, and an "acfg" attribute identifying actual configuration 1 for the video stream from the offer, with the delete-attributes ("-s") and attribute capability 2. Following, we illustrate the offer SDP session description, when Bob instead offers the "crypto" attribute as the actual configuration keying mechanism and "key-mgmt" as the potential configuration:

```
v=0
o=- 25678 753849 IN IP4 192.0.2.1
s=
t=0 0
c=IN IP4 192.0.2.1
a=acap:1 key-mgmt:mikey AQAFgM0XflABAAAAAAAAAAAAAAsAyO...
m=audio 59000 RTP/SAVP 98
a=rtpmap:98 AMR/8000
a=crypto:1 AES_CM_128_HMAC_SHA1_32
      inline:NzB4d1BINUAvLEw6UzF3WSJ+PSdFcGdUJShpX1Zj|2^20|1:32
a=acap:2 rtpmap:98 AMR/8000
a=pcfg:1 a=-m:1,2
m=video 52000 RTP/SAVP 31
a=rtpmap:31 H261/90000
a=acap:3 crypto:1 AES_CM_128_HMAC_SHA1_80
      inline:d0RmdmcmVCspeEc3QGZiNWpVLFJhQX1cfHAwJSoj|2^20|1:32
a=acap:4 rtpmap:31 H261/90000
a=pcfg:1 a=-m:1,4
```

Note how we this time need to perform delete-attributes at the media level instead of at the session level. When doing that, all attributes from the actual configuration SDP session description, including the rtpmaps provided, are removed. Consequently, we had to include these rtpmaps as capabilities as well and then include them in the potential configuration, thereby effectively recreating the original "rtpmap" attributes in the resulting potential configuration SDP session description.

8.1.5 Security Considerations

The security procedures that need to be considered in dealing with the SDP messages described in this section (RFC 5939) are provided in Section 17.7.1.

8.1.6 IANA Considerations

SDP parameters that are described here have been registered with IANA. We have not included the IANA registration procedures here for the sake of brevity. The procedures for registration of new SDP parameters are described in RFC 5939.

8.2 Negotiation of Generic Image Attributes in SDP

This section describes RFC 6236, which proposes a new generic session setup attribute to make it possible to negotiate different image attributes such as image size. A possible use case is to make it possible for a low-end handheld terminal to display video without having to rescale the image, something that may consume large amounts of memory and processing power. The document also helps to maintain an optimal bit rate for video as only the image size that is desired by the receiver is transmitted.

8.2.1 Introduction

In this subsection, we describe the capability negotiations of generic image attributes using SDP specified in RFC 6236. This document proposes a new SDP attribute to make it possible to negotiate different image attributes, such as image size. The term image size is defined here, as it may differ from the physical screen size of, for instance, a hand-held terminal. As an example, it may be beneficial to display a video image on a part of the physical screen and leave space on the screen for other features such as menus and other info. Allowing negotiation of the image size provides a number of benefits:

- Less image distortion: Rescaling of images introduces additional distortion, something that can be avoided (at least on the receiver side) if the image size can be negotiated.
- Reduced receiver complexity: Image rescaling can be quite computation intensive. For low-end devices, this can be a problem.
- Optimal quality for the given bit rate: The sender does not need to encode an entire CIF (352 × 288) image if only an image size of 288 × 256 is displayed on the receiver screen.
- Memory requirement: The receiver device will know the size of the image and can then allocate memory accordingly.
- Optimal aspect ratio: The indication of the supported image sizes and aspect ratio allows the receiver to select the most appropriate combination based on its rescaling capabilities and the desired rendering. For example, if a sender proposes three resolutions in its SDP offer (100 × 200, 200 × 100, and 100 × 100) with sar=1.0 (1:1) etc., then the receiver can select the option that fits the receiver screen best. In cases where rescaling is not implemented (e.g., rescaling is not mandatory to implement in H.264 (ITU-T H.264 standard), the indication of the image attributes may still provide an optimal use of bandwidth because the attribute will give the encoder a better indication of what image size is preferred anyway and will thus help to avoid wasting bandwidth by encoding with an unnecessarily large resolution. For implementers that are considering rescaling issues, it is worth noting that there are several benefits to doing it on the sender side:
  - Rescaling on the sender/encoder side is likely to be easier to do as the camera-related software/hardware already contains the necessary functionality for zooming/cropping/trimming/sharpening the video signal. Moreover, rescaling is generally done in RGB or YUV domains and should not depend on the codecs used.
  - The encoder may be able to encode in a number of formats but may not know which format to choose as, without the image attribute, it does not know the receiver's performance or preference.
  - The quality drop due to digital domain rescaling using interpolation is likely to be lower if it is done before the video encoding rather than after the decoding especially when low bit rate video coding is used.
  - If low-complexity rescaling operations such as simple cropping must be performed, the benefit with having this functionality on the sender side is that it is then possible to present a miniature "what you send" image on the display to help the user to frame the image correctly.

Several of the existing standards such as H.263 (ITU-T standard), H.264 (ITU-T standard), and MPEG-4 (ISO Standard) have support for different resolutions at different frame rates. The purpose of this document is to provide a generic mechanism, which is targeted mainly at the negotiation of the image size. However, to make it more general, the attribute is named "image-attr." This document is limited to point-to-point unicast communication scenarios. The attribute may be used in centralized conferencing scenarios as well, but due to the abundance of configuration options, it may then be difficult to come up with a configuration that fits all parties.

8.2.1.1 Requirements

The design of the image attribute is based on the following requirements, which are listed only for informational purposes:

1. Support the indication of one or more set(s) of image attributes that the SDP endpoint wishes to receive or send. Each image attribute set must include a specific image size.
2. Support setup/negotiation of image attributes, meaning that each side in the offer/answer should be able to negotiate the image attributes it prefers to send and receive.
3. Interoperate with codec-specific parameters such as sproppparameter-sets in H.264 or config in MPEG-4.
4. Make the attribute generic with as few codec specific details/tricks as possible in order to be codec agnostic.

   Besides these requirements, the following may be applicable.
5. The image attribute should support the description of image-related attributes for various types of media, including video, pictures, images, etc.

8.2.2 Conventions Used in This Document

The convention that is being used here is per IETF RFC 2119.

8.2.3 Specification of the "imageattr" SDP Attribute

This section defines the SDP image attribute "imageattr," which can be used in an SDP offer/answer exchange to indicate various image attribute parameters. In this document, we define the following image attribute parameters: image resolution, sample aspect ratio (sar), allowed range in picture aspect ratio (par) and the preference of a given parameter set over another (q). The attribute is extensible, and guidelines for defining additional parameters are provided in Section 8.2.3.2.10.

8.2.3.1 Attribute Syntax

All SDP syntaxes are provided in Section 2.9. However, we have repeated this here for convenience. In this section, the syntax of the "imageattr" attribute is described. The "imageattr" attribute is a media-level attribute. The section is split up in two parts: the first gives an overall view of the syntax, and the second describes how the syntax is used.

8.2.3.1.1 Overall View of Syntax

The syntax for the image attribute is in ABNF (RFC 5234):

```
image-attr    =    "imageattr:" PT 1*2( 1*WSP ( "send" / "recv" )
                   1*WSP attr-list )
```

```
PT            =   1*DIGIT / "*"
attr-list     =   ( set *(1*WSP set) ) / "*"
                      ; WSP and DIGIT defined in [RFC 5234]
set           =   "[" "x=" xyrange "," "y=" xyrange *( "," key-value )
                  "]"
                           ; x is the horizontal image size range
                           (pixel
                           ; count)
                           ; y is the vertical image size range
                           (pixel
                           ; count)
key-value     =   ( "sar=" srange )
                  / ( "par=" prange )
                  / ( "q=" qvalue )
                           ; Key-value MAY be extended with other
                           ; keyword
                           ; parameters.
                           ; At most, one instance each of sar, par,
                           or q
                           ; is allowed in a set.
                           ;
                           ; sar (sample aspect ratio) is the sample
                           ; aspect ratio
                           ; associated with the set (optional, MAY be
                           ; ignored)
                           ; par (picture aspect ratio) is the allowed
                           ; ratio between the display's x and y
                           ; physical
                           ; size (optional)
                           ; q (optional, range [0.0..1.0], default
                           value
                           ; 0.5)
                           ; is the preference for the given set,
                           ; a higher value means a higher preference
onetonine     =   "1" / "2" / "3" / "4" / "5" / "6" / "7" / "8" / "9"
                           ; Digit between 1 and 9
xyvalue       =   onetonine *5DIGIT
                           ; Digit between 1 and 9 that is
                           ; followed by 0 to 5 other digits
step          =   xyvalue
xyrange       =   ( "[" xyvalue ":" [ step ":" ] xyvalue "]" )
                           ; Range between a lower and an upper value
                           ; with an optional step, default step = 1
                           ; The rightmost occurrence of xyvalue MUST
                           have
                           ; a
                           ; higher value than the leftmost occurrence.
                           ; / ( "[" xyvalue 1*( "," xyvalue ) "]" )
                           ; Discrete values separated by ','
                           ; / ( xyvalue )
                           ; A single value
spvalue       =   ( "0" "." onetonine *3DIGIT )
                           ; Values between 0.1000 and 0.9999
                           ; / ( onetonine "." 1*4DIGIT )
                           ; Values between 1.0000 and 9.9999
```

```
srange      =    ( "[" spvalue 1*( "," spvalue ) "]" )
                      ; Discrete values separated by ','.
                      ; Each occurrence of spvalue MUST be
                      ; greater than the previous occurrence.
                      ; / ( "[" spvalue "-" spvalue "]" )
                      ; Range between a lower and an upper level
                      ; (inclusive)
                      ; The second occurrence of spvalue MUST
                      ; have a higher
                      ; value than the first / ( spvalue )
                      ; A single value
prange      =    ( "[" spvalue "-" spvalue "]" )
                      ; Range between a lower and an upper level
                      ; (inclusive)
                      ; The second occurrence of spvalue MUST
                      ; have a
                      ; higher
                      ; value than the first
qvalue      =    ( "0" "." 1*2DIGIT ) / ( "1" "." 1*2("0") )
                      ; Values between 0.00 and 1.00
```

1. The attribute typically contains a "send" and a "recv" keyword. These specify the preferences for the media once the session is set up, in the send and receive direction respectively from the point of view of the sender of the session description. One of the keywords ("send" or "recv") MAY be omitted; see Sections 8.2.3.2.4 and 8.2.3.2.2 for a description of cases when this may be appropriate.
2. The "send" keyword and corresponding attribute list (attr-list) MUST NOT occur more than once per image attribute.
3. The "recv" keyword and corresponding attribute list (attr-list) MUST NOT occur more than once per image attribute.
4. PT is the payload type number; it MAY be set to "*" (wild card) to indicate that the attribute applies to all payload types in the media description.
5. For sendrecv streams, both of the send and recv directions SHOULD be present in the SDP.
6. For inactive streams it is RECOMMENDED that both of the send and recv directions are present in the SDP.

8.2.3.1.1.1 Parameter Rules

The following rules apply for the parameters.

- Payload type number (PT): The image attribute is bound to a specific codec by means of the payload type number. A wild card (*) can be specified for the payload type number to indicate that it applies to all payload types in the media description. Several image attributes can be defined, for instance for different video codec alternatives. This however requires that the payload type numbers differ. Note that the attribute is associated to the codec(s), for instance an SDP offer may specify payload type number 101 while the answer may specify 102, this may make it troublesome to specify a payload type number with the "imageattr" attribute. See Section 8.2.3.2.2 for a discussion and recommendation for solving this.
- Preference (q): The preference for each set is 0.5 by default; setting the optional q parameter to another value makes it possible to set different preferences for the sets. A higher value gives a higher preference for the given set.

- sar: The sar (storage aspect ratio) parameter specifies the sample aspect ratio associated with the given range of *x* and *y* values. The sar parameter is defined as *dx*/*dy* where *dx* and *dy* are the physical size of the pixels. Square pixels give a sar=1.0. The parameter sar MAY be expressed as a range or as a single value. If this parameter is not present, a default sar value of 1.0 is assumed. The interpretation of sar differs between the send and the receive directions.
  - In the send direction, sar defines a specific sample aspect ratio associated with a given *x* and *y* image size (range).
  - In the recv direction, sar expresses that the receiver of the given medium prefers to receive a given *x* and *y* resolution with a given sample aspect ratio. See Section 8.2.3.2.5 for a more detailed discussion.

  The sar parameter will likely not solve all the issues that are related to different sample aspect ratios, but it can help solve them and reduce aspect ratio distortion. The response MUST NOT include a sar parameter if there is no acceptable value given. The reason for this is that if the response includes a sar parameter it is interpreted as "sar parameter accepted," while removal of the sar parameter is treated as "sar parameter not accepted.". For this reason, it is safer to remove an unacceptable sar parameter altogether.
- par: The par (width/height = *x*/*y* ratio) parameter indicates a range of allowed ratios between *x* and *y* physical size (picture aspect ratio). This is used to limit the number of x and y image size combinations; par is given as:

  ```
  par=[ratio_min-ratio_max]
  ```

  where ratio_min and ratio_max are the min and max allowed picture aspect ratios.

  If sar and the sample aspect ratio that the receiver actually uses in the display are the same (or close), the relation between the *x* and *y* pixel resolution and the physical size of the image is straightforward. If however sar differs from the sample aspect ratio of the receiver display, this must be taken into consideration when the *x* and *y* pixel resolution alternatives are sorted out. See Section 8.2.4.2.4 for an example of this.

8.2.3.1.1.2 Offer/Answer Rules In accordance with RFC 3264 (see Section 3.1), offer/answer exchange of the image attribute is as follows.

- Offerer sending the offer:
  - The offerer must be able to support the image attributes that it offers, unless the offerer has expressed a wild card (*) in the attribute list.
  - It is recommended that a device that sees no reason to use the image attribute includes the attribute with wild cards (*) in the attribute lists anyway for the send and recv directions.

  An example of this looks like:

  ```
  a=imageattr:97 send * recv *
  ```

  This gives the answerer the possibility of expressing its preferences. The use of wild cards introduces a risk that the message size can increase in an uncontrolled way. To reduce this risk, these wild cards SHOULD only be replaced by as small set as possible.

- Answerer receiving the offer and sending the answer:
  - The answerer may choose to keep the image attribute but is not required to do so.
  - The answerer may, for its receive and send direction, include one or more entries that it can support from the set of entries proposed in the offer.
  - The answerer may also, for its receive and send direction, replace the entries with a complete new set of entries different from the original proposed by the offerer. The implementer of this feature should however be aware that this may cause extra offer/answer exchanges.
  - The answerer may also remove its send direction completely if it is deemed that it cannot support any of the proposed entries.
  - The answerer should not include an image attribute in the answer if it was not present in the offer.
- Offerer receiving the answer:
  - If the image attribute is not included in the SDP answer the offerer SHOULD continue to process the answer as if this mechanism had not been offered.
  - If the image attribute is included in the SDP answer but none of the entries is usable or acceptable, the offerer MUST resort to other methods to determine the appropriate image size. In this case, the offerer must also issue a new offer/answer without the image attribute to avoid misunderstandings between the offerer and answerer. This will avoid the risk of infinite negotiations.

8.2.3.2 Considerations

8.2.3.2.1 No imageat tr in First Offer

When the initial offer does not contain the "imageattr" attribute, the rules in Section 8.2.3.1.1.2 require the attribute to be absent in the answer. The reasons for this are:

- The offerer of the initial SDP is not likely to understand the image attribute if it did not include it in the offer, bearing in mind that Section 8.2.3.1.1 recommends that the offerer provide the attribute with wild carded parameters if it has no preference.
- Inclusion of the image attribute in the answer may come in conflict with the rules in Section 8.2.3.1.1.2, especially the rules that apply to "offerer receiving the answer." For these reasons, it is RECOMMENDED that a device that sees no reason to use the image attribute include the attribute with wild cards (*) in the attribute lists anyway for the send and recv directions.

8.2.3.2.2 Different Payload Type Numbers in Offer and Answer

In some cases, the answer may specify a different media payload type number than the offer. As an example, the offer SDP may have the "m=" line:

```
m=video 49154 RTP/AVP 99
```

while the answer SDP may have the "m=" line:

```
m=video 49154 RTP/AVP 100
```

If the image attribute in the offer specifies payload type number 99, this attribute will then have no meaning in the answerers receive direction, as the "m=" line specifies media payload type number 100. There are a few ways to solve this.

1. Use a wild card "*" as the payload type number in the image attribute in the offer SDP. The answer SDP also uses the wild card. The drawback with this approach is that this attribute then applies to all payload type numbers in the media description.
2. Specify a wild card "*" as the payload type number in the image attribute in the answer SDP. The offer SDP may contain a defined payload type number in the image attribute but the answer SDP replaces this with a wild card. The drawback here is similar to what is listed above.
3. The image attribute is split in two parts in the SDP answer. For example, the offer SDP (only the parts of interest in this discussion) looks like:

```
m=video 49154 RTP/AVP 99
a=imageattr:99 send ... recv ...
The answer SDP looks like:
m=video 49154 RTP/AVP 100
a=imageattr:99 send ...
a=imageattr:100 recv ...
```

This alternative does not pose any drawbacks. Moreover, it allows specification of different image attributes if more than one payload type is specified in the offer SDP. Of the alternatives listed above, the last one MUST be used as it is the safest. The other alternatives MUST NOT be used.

8.2.3.2.3 Asymmetry

While the image attribute supports asymmetry, there are some limitations. One important limitation is that the codec being used can only support up to a given maximum resolution for a given profile level. As an example, H.264 with profile level 1.2 does not support higher resolution than 352 × 288 Common Interchange Format (CIF). The offer/answer rules imply that the same profile level must be used in both directions. This means that in an asymmetric scenario where Alice wants an image size of 580 × 360 and Bob wants 150 × 120, profile level 2.2 is needed in both directions even though profile level 1 would have been sufficient in one direction. Currently, the only solution to this problem is to specify two unidirectional media descriptions. Note, however, that the asymmetry issue for the H.264 codec is solved by means of the level-asymmetry allowed parameter in RFC 6184.

8.2.3.2.4 sendonly and recvonly

If the directional attributes a=sendonly or a=recvonly are given for a medium, there is of course no need to specify the image attribute for both directions. Therefore, one of the directions in the attribute may be omitted. However, it may be good to do the image attribute negotiation in both directions in case the session is updated for media in both directions at a later stage.

8.2.3.2.5 Sample Aspect Ratio

The relationship between the sar parameter and the x and y pixel resolution deserves some extra discussion. Consider the offer from Alice to Bob (we set the recv direction aside for the moment):

```
a=imageattr:97 send [x=720,y=576,sar=1.1]
```

If the receiver display has square pixels, the 720 × 576 image would need to be rescaled to, for example, 792 × 576 or 720 × 524 to ensure a correct image aspect ratio. This in practice means that rescaling would need to be performed on the receiver side, something that is contrary to the spirit of this document. To avoid this problem, Alice may specify a range of values for the sar parameter like:

```
a=imageattr:97 send [x=720,y=576,sar=[0.91,1.0,1.09,1.45]]
```

Meaning that Alice can encode with any of the mentioned sample aspect ratios, leaving Bob to decide which one he prefers.

8.2.3.2.6 SDP Capability Negotiation Support

The image attribute can be used within the SDP capability negotiation (RFC 5939, see Section 8.1) framework, and its use is then specified using the "a=acap" parameter. An example is:

```
a=acap:1 imageattr:97 send [x=720,y=576,sar=[0.91,1.0,1.09,1.45]]
```

For use with SDP media capability negotiation extension (RFC 6871, see Section 8.3), where it is no longer possible to specify payload type numbers, it is possible to use the parameter substitution rule, an example of this is

```
...
a=mcap:1 video H264/90000
a=acap:1 imageattr:%1% send [x=720,y=576,sar=[0.91,1.0,1.09,1.45]]
...
```

where %1% maps to media capability number 1.

It is also possible to use the a=mscap attribute as in the following example.

```
...
a=mcap:1 video H264/90000
a=mscap:1 imageattr send [x=720,y=576,sar=[0.91,1.0,1.09,1.45]]
...
```

8.2.3.2.7 Interaction with Codec Parameters

As the SDP for most codecs already specifies some kind of indication of, for example, the image size, at session setup, measures must be taken to avoid conflicts between the image attribute and this already existing information. The following subsections describe the best-known codecs and how they define image-size related information. Section 8.2.3.2.7.4 outlines a few possible solutions, but this document does not make a recommendation for any of them.

8.2.3.2.7.1 H.263 The payload format for H.263 is described in RFC 4629. H.263 defines (on the fmtp line) a list of image sizes and their maximum frame rates (profiles) that the offerer can receive. The answerer is not allowed to modify this list and must reject a payload type that contains an unsupported profile. The CUSTOM profile may be used for image size negotiation, but support for asymmetry requires the specification of two unidirectional media descriptions using the sendonly/recvonly attributes.

8.2.3.2.7.2 H.264 The payload format for H.264 is described in RFC 6184. H.264 defines information related to image size in the fmtp line by means of sprop-parameter-sets. According to the specification, several sprop-parameter-sets may be defined for one payload type. The sprop-parameter-sets describe the image size (+ more) that the offerer sends in the stream and need not be complete. This means that sprop-parameter-sets do not represent any negotiation, and the answerer is not allowed to change the sprop-parameter-sets. This configuration may be changed later inband if, for instance, image sizes need to be changed or added.

8.2.3.2.7.3 MPEG-4 The payload format for MPEG-4 is described in RFC 3016. MPEG-4 defines a config parameter on the fmtp line, which is a hexadecimal representation of the MPEG-4 visual configuration information. This configuration does not represent any negotiation, and the answerer is not allowed to change the parameter. It is not possible to change the configuration using inband signaling.

8.2.3.2.7.4 Possible Solutions All the subsections described above clearly indicate that this kind of information must be aligned well with the image attribute to avoid conflicts. There are a number of possible solutions, listed without any preference:

- Ignore payload format parameters: This may not work well in the presence of bad channel conditions, especially in the beginning of a session. Moreover, this is not a good option for MPEG-4.
- Second session-wide offer/answer round: In the second offer/answer, the parameters specific to codec payload format are defined based on the outcome of the "imageattr" negotiation. The drawback with this is that setup of the entire session (including audio) may be delayed considerably, especially as the "imageattr" negotiation can already itself cost up to two offer/answer rounds. Also, the conflict between the "imageattr" negotiation and the parameters specific to payload format is still present after the first offer/answer round, and a fuzzy/buggy implementation may start media before the second offer/answer is completed with unwanted results.
- Second session-wide offer/answer round only for video: This is similar to the alternative above with the exception that setup time for audio is not increased; moreover, the port number for video is set to 0 during the first offer answer round to avoid the flow of media. This has the effect that video will blend in some time after the audio is started (up to 2 seconds delay). This alternative is likely the most clean-cut and failsafe. The drawback is that, as the port number in the first offer is always zero, the media startup will always be delayed even though it would in fact have been possible to start media after the first offer/answer round.

Note that according to RFC 3264 (see Section 3.1), a port number of zero means that the whole media line is rejected and that a new offer for the same port number should be treated as a completely new stream and not an update. The safest way to solve this problem is to use preconditions; this is however outside the scope of this document.

8.2.3.2.8 Change of Display in the Middle of a Session

A very likely scenario is that a user switches to another phone during a video telephony call or plugs a cellphone into an external monitor. In both cases, it is very likely that a renegotiation is initiated using the SIP-REFER (RFC 3515) or SIP-UPDATE (RFC 3311) method. It is RECOMMENDED that one negotiate the image size during this renegotiation.

8.2.3.2.9 Use with Layered Codecs

As the image attribute is a media-level attribute, its use with layered codecs causes some concern. If the layers are transported in different RTP streams, the layers are specified on different media descriptions, and the relation is specified using the grouping framework (RFC 5888, see Section 7.1) and the depend attribute (RFC 5583, see Section 6.1). As it is not possible to specify only one image attribute for several media descriptions, the solution is either to specify the same image attribute for each media description or to only specify the image attribute for the base layer.

8.2.3.2.10 Addition of Parameters

The image attribute allows for the addition of parameters in the future. To make backwards adaptation possible, an entity that processes the attribute MUST ignore any unknown parameters in the offer and MUST NOT include them in the answer it generates. The addition of future parameters that are not understood by the receiving endpoint may lead to ambiguities if mutual dependencies between parameters exist; therefore, addition of parameters must be done with great care.

8.2.4 Examples

This section gives some more information on how to use the attribute by means of a high-level example and a few detailed examples.

8.2.4.1 A High-Level Example

Assume that Alice wishes to set up a session with Bob and that Alice takes the first initiative. The syntactical white-space delimiters (1*WSP) and double-quotes are removed to make reading easier. In the offer, Alice provides information for both the send and receive (recv) directions. For the send direction, Alice provides a list from which the answerer can select. For the receive direction, Alice may either specify a desired image size range right away or a * to instruct Bob to reply with a list of image sizes that Bob can support for sending. Using the overall high-level syntax, the image attribute may then look like:

```
a=imageattr:PT send attr-list recv attr-list
```

or

```
a=imageattr:PT send attr-list recv *
```

In the first alternative, the recv direction may be a full list of desired image size formats. It may however (and most likely) just be a list with one alternative for the preferred x and y resolution. If Bob supports an x and y resolution in at least one of the X and Y ranges given in the send attr-list and in the recv attr-list of the offer, the answer from Bob will look like:

```
a=imageattr:PT send attr-list recv attr-list
```

and the offer/answer negotiation is done. Note that the attr-list will likely be pruned in the answer. While it may contain many different alternatives in the offer, it may in the end contain just one or two alternatives. If Bob does not support any x and y resolution in one of the provided send or recv ranges given in the send attr-list or in the recv attr-list, the corresponding part is removed completely. For instance, if Bob doesn't support any of the offered alternatives in the recv attr-list in the offer, the answer from Bob will look like:

```
a=imageattr:PT recv attr-list
```

8.2.4.2 Detailed Examples

This section gives a few detailed examples. It is assumed where needed that Alice initiates a session with Bob.

8.2.4.2.1 Example 1

Two image resolution alternatives are offered with 800 × 640 with sar=1.1 having the highest preference. It is also indicated that Alice wishes to display video with a resolution of 330 × 250 on her display.

```
a=imageattr:97 send [x=800,y=640,sar=1.1,q=0.6] [x=480,y=320] \
                                   recv [x=330,y=250]
```

If Bob accepts the "recv [x=330,y=250]", the answer may look like

```
a=imageattr:97 recv [x=800,y=640,sar=1.1] \
                                   send [x=330,y=250]
```

indicating that the receiver (Bob) wishes the encoder (on Alice's side) to compensate for a sample aspect ratio of 1.1 (11:10) and desires an image size on its screen of 800 × 640. There is however a possibility that "recv [x=330,y=250]" is not supported. If this is the case, Bob may completely remove this part or replace it with a list of supported image sizes.

```
a=imageattr:97 recv [x=800,y=640,sar=1.1] \
                                   send [x=[320:16:640],y=[240:16:480],
                                   par=[1.2-1.3]]
```

Alice can then select a valid image size that is closest to that originally desired (336 × 256) and performs a second offer/answer.

```
a=imageattr:97 send [x=800,y=640,sar=1.1] \
                              recv [x=336,y=256]
```

Bob replies with:

```
a=imageattr:97 recv [x=800,y=640,sar=1.1] \
                              send [x=336,y=256]
```

8.2.4.2.2 Example 2

Two image resolution sets are offered, with the first having a higher preference (q=0.6).

```
a=imageattr:97 \
          send [x=[480:16:800],y=[320:16:640],par=[1.2-1.3],q=0.6] \
          [x=[176:8:208],y=[144:8:176],par=[1.2-1.3]] \
          recv *
```

The x-axis resolution can take the values 480–800 in 16-pixel steps and 176–208 in 8-pixel steps. The par parameter limits the set of possible x and y screen resolution combinations such that 800 × 640 (ratio=1.25) is a valid combination while 720 × 608 (ratio=1.18) and 800 × 608 (ratio=1.31) are invalid combinations. For the recv direction (Bob->Alice), Bob is requested to provide a list of supported image sizes.

8.2.4.2.3 Example 3

In this example, more of the SDP offer is shown. A complicating factor is that the answerer changes the media payload type number in the offer/answer exchange.

```
m=video 49154 RTP/AVP 99
a=rtpmap:99 H264/90000
a=fmtp:99 packetization-mode=0;profile-level-id=42e011; \
      sprop-parameter-sets=Z0LgC5ZUCg/I,aM4BrFSAa
a=imageattr:99 \
      send [x=176,y=144] [x=224,y=176] [x=272,y=224] [x=320,y=240] \
      recv [x=176,y=144] [x=224,y=176] [x=272,y=224,q=0.6]
      [x=320,y=240]
```

In the send direction, sprop-parameter-sets is defined for a resolution of 320 × 240, which is the largest image size offered in the send direction. This means that if 320 × 240 is selected, no additional offer/answer is necessary. In the receive direction, four alternative image sizes are offered, with 272 × 224 being the preferred choice. The answer may look like:

```
m=video 49154 RTP/AVP 100
a=rtpmap:100 H264/90000
a=fmtp:100 packetization-mode=0;profile-level-id=42e011; \
sprop-parameter-sets=Z0LgC5ZUCg/I,aM4BrFSAa
a=imageattr:99 send [x=320,y=240]
a=imageattr:100 recv [x=320,y=240]
```

indicating (in this example) that the image size is 320 × 240 in both directions. Although the offerer preferred 272 × 224 for the receive direction, the answerer might not be able to offer 272 × 224 or not allow encoding and decoding of video of different image sizes simultaneously. The answerer sets new sprop-parameter-sets, constructed for both send and receive directions at the restricted conditions and image size of 320 × 240. Note also that, because the payload type number is changed by the answerer, the image attribute is also split into two parts according to the recommendation in Section 8.2.3.2.2.

8.2.4.2.4 Example 4

This example illustrates in more detail how compensation for different sample aspect ratios can be negotiated with the image attribute. We set up a session between Alice and Bob; Alice is the offerer of the session. The offer (from Alice) contains this image attribute:

```
a=imageattr:97 \
      send [x=400:16:800],y=[320:16:640],sar=[1.0-1.3],par=[1.2-1.3]] \
      recv [x=800,y=600,sar=1.1]
```

First, we consider the recv direction: The offerer (Alice) explicitly states that she wishes to receive the screen resolution 800 × 600. However, she also indicates that the screen on her display does not use square pixels; the sar value=1.1 means that Bob must (preferably) compensate for this. So, if Bob's video camera produces square pixels, and if Bob wishes to satisfy Alice's sar requirement, the image processing algorithm must rescale a 880 × 600 pixel image (880=800*1.1) to 800 × 600 pixels (could be done other ways). ... and now the send direction: Alice indicates that she can (in the image processing algorithms) rescale the image for sample aspect ratios in the range 1.0–1.3. She can also provide a number of different image sizes (in pixels) ranging from 400 × 320 to 800 × 640. Bob inspects the offered sar and image sizes and responds with the modified image attribute.

```
a=imageattr:97 \
                  recv [x=464,y=384,sar=1.15] \
                  send [x=800,y=600,sar=1.1]
```

Alice will (in order to satisfy Bob's request) need to rescale the image from her video camera from 534 × 384 (534=464*1.15) to 464 × 384.

8.2.5 IANA Considerations

SDP parameters that are described here have been registered with IANA. We have not included the IANA registration procedures here for the sake of brevity. The procedures for registration of new SDP parameters are described in RFC 6236.

8.2.6 Security Considerations

The security procedures that need to be considered in dealing with the SDP negotiation messages for generic image attributes described in this section (RFC 6236) are provided in Section 17.7.2.

8.3 SDP Media Capabilities Negotiation

This section describes RFC 6871 that extends the SDP generic capability framework for negotiating transport protocols and attributes by defining media capabilities that can be used to negotiate media types and their associated parameters. In addition, this RFC updates the IANA Considerations of RFC 5939 (see Section 8.1).

8.3.1 Introduction

In this section, we describe the SDP media capability negotiations specified in RFC 6871. "SDP Capability Negotiation" – RFC 5939 (see Section 8.1) provides a general framework for indicating and negotiating capabilities in SDP (RFC 4566, see Section 2). The base framework defines only capabilities for negotiating transport protocols and attributes. RFC 5939 (see Section 8.1) lists some of the issues with the current SDP capability negotiation process. An additional real-life problem is to be able to offer one media stream (e.g., audio) but list the capability to support another media stream (e.g., video) without actually offering it concurrently. In this document, we extend the framework by defining media capabilities that can be used to indicate and negotiate media types and their associated format parameters.

This document also adds the ability to declare support for media streams, the use of which can be offered and negotiated later, and the ability to specify session configurations as combinations of media stream configurations. The definitions of new attributes for media capability negotiation are chosen to make the translation from these attributes to "conventional" SDP (RFC 4566, see Section 2) media attributes as straightforward as possible in order to simplify implementation. This goal is intended to reduce processing in two ways: Each proposed configuration in an offer may be easily translated into a conventional SDP media stream record for processing by the receiver, and the construction of an answer based on a selected proposed configuration would be straightforward. This document updates RFC 5939 (see Section 8.1) by updating the IANA considerations. All other extensions defined in this document are considered extensions above and beyond RFC 5939.

8.3.2 Terminology

See Section 2.2/Table 2.1.

8.3.3 SDP Media Capabilities

The SDP capability negotiation (RFC 5939, see Section 8.1) discusses the use of any SDP (RFC 4566, see Section 2) attribute (a=) under the attribute capability "acap." The limitations of using "acap" for "fmtp" and "rtpmap" in a potential configuration are described in RFC 5939; for example, they can be used only at the media level since they are media-level attributes. RFC 5939 does not provide a way to exchange media-level capabilities prior to the actual offer of the associated media stream. This section provides an overview of extensions providing an SDP media capability negotiation solution offering more robust capabilities negotiation. This is followed by definitions of new SDP attributes for the solution and its associated updated offer/answer procedures (RFC 3264, see Section 3.1).

8.3.3.1 Requirements

The capability negotiation extensions requirements considered herein are as follows:

1. Support the specification of alternative (combinations of) media formats (codecs) in a single media block.
2. Support the specification of alternative media format parameters for each media format.
3. Retain backwards compatibility with conventional SDP. Ensure that each and every offered configuration can be easily translated into a corresponding SDP media block expressed with conventional SDP lines.
4. Ensure that the scheme operates within the offer/answer model in such a way that media formats and parameters can be agreed upon with a single exchange.
5. Provide the ability to express offers in such a way that the offerer can receive media as soon as the offer is sent. (Note that the offerer may not be able to render received media prior to exchange of keying material.)
6. Provide the ability to offer latent media configurations for future negotiation.

7. Provide reasonable efficiency in the expression of alternative media formats and/or format parameters, especially in those cases in which many combinations of options are offered.
8. Retain the extensibility of the base capability negotiation mechanism.
9. Provide the ability to specify acceptable combinations of media streams and media formats. For example, offer a PCMU audio stream with an H.264 video stream or a G.729 audio stream with an H.263 video stream. This ability would give the offerer a means to limit processing requirements for simultaneous streams. This would also permit an offer to include the choice of an audio/T38 stream or an image/T38 stream, but not both.

Other possible extensions have been discussed but have not been treated in this document. They may be considered in the future. Three such extensions are:

1. Provide the ability to mix, or change, media types within a single media block. Conventional SDP does not support this capability explicitly; the usual technique is to define a media subtype that represents the actual format within the nominal media type. For example, T.38 facsimile (FAX) as an alternative to audio/PCMU within an audio stream is identified as audio/T38; a separate FAX stream would use image/T38.
2. Provide the ability to support multiple transport protocols within an active media stream without reconfiguration. This is not explicitly supported by conventional SDP.
3. Provide capability negotiation attributes for all media-level SDP line types in the same manner as already done for the attribute type, with the exception of the media line type itself. The media line type is handled in a special way to permit compact expression of media coding/format options. The line types are bandwidth ("b="), information ("i="), connection data ("c="), and, possibly, the deprecated encryption key ("k=").

8.3.3.2 Solution Overview

The solution consists of new capability attributes corresponding to conventional SDP line types, new parameters for the "pcfg," "acfg," and the new "lcfg" attributes extending the base attributes from RFC 5939 (see Section 8.1) and a use of the "pcfg" attribute to return capability information in the SDP answer. Several new attributes are defined in a manner that can be related to the capabilities specified in a media line and its corresponding "rtpmap" and "fmtp" attributes.

- A new attribute ("a=rmcap") defines RTP-based media format capabilities in the form of a media subtype (e.g., "PCMU") and its encoding parameters (e.g., "/8000/2"). Each resulting media format type/subtype capability has an associated handle called a media capability number. The encoding parameters are as specified for the "rtpmap" attribute defined in SDP (RFC 4566, see Section 2), without the payload type number part.
- A new attribute ("a=omcap") defines other (non-RTP-based) media format capabilities in the form of a media subtype only (e.g., "T38"). Each resulting media format type/subtype capability has an associated handle called a media capability number.
- A new attribute ("a=mfcap") specifies media format parameters associated with one or more media format capabilities. The "mfcap" attribute is used primarily to associate the media format parameters normally carried in the "fmtp" attribute. Note that media format parameters can be used with RTP and non-RTP-based media formats.

- A new attribute ("a=mscap") specifies media parameters associated with one or more media format capabilities. The "mscap" attribute is used to associate capabilities with attributes other than "fmtp" or "rtpmap," for example, the "rtcp-fb" attribute defined in RFC 4585.
- A new attribute ("a=lcfg") specifies latent media stream configurations when no corresponding media line ("m=") is offered. An example is the offer of latent configurations for video even though no video is currently offered. If the peer indicates support for one or more offered latent configurations, the corresponding media stream(s) may be added via a new offer/answer exchange.
- A new attribute ("a=sescap") is used to specify an acceptable combination of simultaneous media streams and their configurations as a list of potential and/or latent configurations. New parameters are defined for the potential configuration ("pcfg"), latent configuration ("lcfg"), and accepted configuration ("acfg") attributes to associate the new attributes with particular configurations.
- A new parameter type ("m=") is added to the potential configuration ("a=pcfg:") attribute and the actual configuration ("a=acfg:") attribute defined in RFC 5939 (see Section 8.1) and to the new latent configuration ("a=lcfg:") attribute. This permits specification of media capabilities (including their associated parameters) and combinations thereof for the configuration. For example, the "a=pcfg:" line might specify PCMU and telephone events (RFC 4733) or G.729B and telephone events as acceptable configurations. The "a=acfg:" line in the answer would specify the configuration chosen.
- A new parameter type ("pt=") is added to the potential configuration, actual configuration, and latent configuration attributes. This parameter associates RTP payload type numbers with the referenced RTP-based media format capabilities and is appropriate only when the transport protocol uses RTP.
- A new parameter type ("mt=") is used to specify the media type for latent configurations. Special processing rules are defined for capability attribute arguments in order to reduce the need to replicate essentially identical attribute lines for the base configuration and potential configurations.
- A substitution rule is defined for any capability attribute to permit the replacement of the (escaped) media capability number with the media format identifier (e.g., the payload type number in audio/video profiles).
- Replacement rules are defined for the conventional SDP equivalents of the "mfcap" and "mscap" capability attributes. This reduces the necessity of using the deletion qualifier in the "a=pcfg" parameter in order to ignore "rtpmap," "fmtp," and certain other attributes in the base configuration.
- An argument concatenation rule is defined for "mfcap" attributes that refer to the same media capability number. This makes it convenient to combine format options concisely by associating multiple mfcap lines with multiple media format capabilities. This document extends the base protocol extensions to the offer/answer model that allow for capabilities and potential configurations to be included in an offer. Media capabilities constitute capabilities that can be used in potential and latent configurations. Whereas potential configurations constitute alternative offers that may be accepted by the answerer instead of the actual configuration(s) included in the "m=" line(s) and associated parameters, latent configurations merely inform the other side of possible configurations supported by the entity. Those latent configurations may be used to guide subsequent offer/answer exchanges, but they are not part of the current offer/answer exchange.

The mechanism is illustrated by the offer/answer exchange in Inline Figure 8, where Alice sends an offer to Bob:

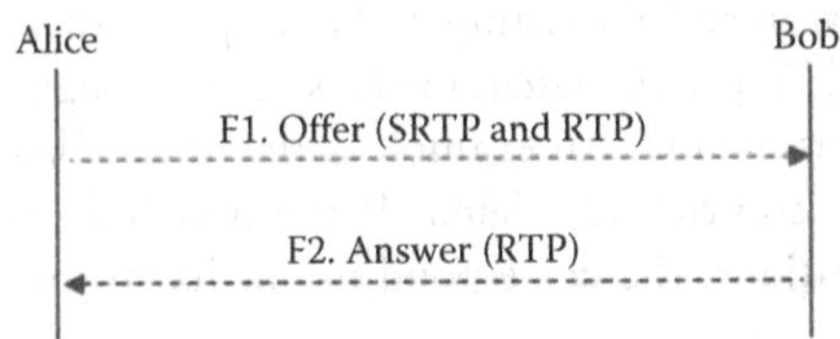

Alice's offer includes RTP and SRTP as alternatives. RTP is the default, but SRTP is preferred (long lines are folded to fit the margins):

```
v=0
o=- 25678 753849 IN IP4 192.0.2.1
s=
c=IN IP4 192.0.2.1
t=0 0
a=creq:med-v0
m=audio 3456 RTP/AVP 0 18
a=tcap:1 RTP/SAVP RTP/AVP
a=rtpmap:0 PCMU/8000/1
a=rtpmap:18 G729/8000/1
a=fmtp:18 annexb=yes
a=rmcap:1,4 G729/8000/1
a=rmcap:2 PCMU/8000/1
a=rmcap:5 telephone-event/8000
a=mfcap:1 annexb=no
a=mfcap:4 annexb=yes
a=mfcap:5 0-11
a=acap:1 crypto:1 AES_CM_128_HMAC_SHA1_32 \
       inline:NzB4d1BINUAvLEw6UzF3WSJ+PSdFcGdUJShpX1Zj|2^20|1:32
a=pcfg:1 m=4,5|1,5 t=1 a=1 pt=1:100,4:101,5:102
a=pcfg:2 m=2 t=1 a=1 pt=2:103
a=pcfg:3 m=4 t=2 pt=4:18
```

The required base and extensions are provided by the "a=creq" attribute defined in RFC 5939 (see Section 8.1), with the option-tag "med-v0," which indicates that the extension framework defined here must be supported. The base-level capability negotiation support ("cap-v0" – RFC 5939) is implied, since it is required for the extensions. The "m=" line indicates that Alice is offering to use plain RTP with PCMU or G.729B. The media line implicitly defines the default transport protocol (RTP/AVP in this case) and the default actual configuration.

The "a=tcap:1" line, specified in the SDP capability negotiation base protocol (RFC 5939, see Section 8.1), defines transport protocol capabilities, in this case SRTP (SAVP profile) as the first option and RTP (AVP profile) as the second option. The "a=rmcap:1,4" line defines two G.729 RTP-based media format capabilities, numbered 1 and 4, and their encoding rate. The capabilities are of media type "audio" and subtype G729. Note that the media subtype is explicitly specified here, rather than RTP payload type numbers. This permits the assignment of payload type numbers in the media stream configuration specification. In this example, two G.729 subtype capabilities are defined. This permits the declaration of two sets of formatting parameters for G.729. The "a=rmcap:2" line defines a G.711 mu-law capability, numbered 2. The "a=rmcap:5" line defines an audio telephone-event capability, numbered 5.

The "a=mfcap:1" line specifies the "fmtp" formatting parameters for capability 1 (offerer will not accept G.729 Annex B packets). The "a=mfcap:4" line specifies the "fmtp" formatting parameters for capability 4 (offerer will accept G.729 Annex B packets). The "a=mfcap:5" line specifies the "fmtp" formatting parameters for capability 5 (the dual-tone multifrequency [DTMF] touchtones 0-9,*,#). The "a=acap:1" line specified in the base protocol provides the "crypto" attribute that provides the keying material for SRTP using SDP security descriptions. The "a=pcfg:" attributes provide the potential configurations included in the offer by reference to the media capabilities, transport capabilities, attribute capabilities, and specified payload type number mappings. Three explicit alternatives are provided; the lowest-numbered one is the preferred one. The "a=pcfg:1 ..." line specifies media capabilities 4 and 5 (i.e., G.729B and DTMF), including their associated media format parameters) or media capability 1 and 5 (i.e., G.729 and DTMF), including their associated media format parameters). Furthermore, it specifies transport protocol capability 1 (i.e., the RTP/SAVP profile, SRTP) and the attribute capability 1 (i.e., the "crypto" attribute provided).

Last, it specifies a payload type number mapping for (RTP-based) media capabilities 1, 4, and 5, thereby permitting the offerer to distinguish between encrypted media and unencrypted media received prior to receipt of the answer. Use of unique payload type numbers in alternative configurations is not required; codecs such as adaptive multi-rate wideband (AMR-WB) (RFC 4867) have the potential for so many combinations of options that it may be impractical to define unique payload type numbers for all supported combinations. If unique payload type numbers cannot be specified, then the offerer will be obliged to wait for the SDP answer before rendering received media. For SRTP using security descriptions (SDES) inline keying (RFC 4568, see Section 16.2), the offerer will still need to receive the answer before being able to decrypt the stream. The second alternative ("a=pcfg:2 ...") specifies media capability 2 (i.e., PCMU, under the RTP/SAVP profile, with the same SRTP key material).

The third alternative ("a=pcfg:3 ...") offers G.729B unsecured; its only purpose in this example is to show a preference for G.729B over PCMU. Per RFC 5939 (see Section 8.1), the media line, with any qualifying attributes such as "fmtp" or "rtpmap," is itself considered a valid configuration (the current actual configuration); it has the lowest preference (per RFC 5939). Bob receives the SDP offer from Alice. Bob supports G.729B, PCMU, and telephone events over RTP, but not SRTP, hence he accepts the potential configuration 3 for RTP provided by Alice. Bob generates the following answer:

```
v=0
o=- 24351 621814 IN IP4 192.0.2.2
s=
c=IN IP4 192.0.2.2
t=0 0
a=csup:med-v0
m=audio 4567 RTP/AVP 18
a=rtpmap:18 G729/8000
a=fmtp:18 annexb=yes
a=acfg:3 m=4 t=2 pt=4:18
```

Bob includes the "a=csup" and "a=acfg" attributes in the answer to inform Alice that he can support the med-v0 level of capability negotiations. Note that in this particular example the answerer supported the capability extensions defined here; however, had he/she not, he/she would simply have processed the offer based on the offered PCMU and G.729 codecs under the RTP/AVP profile only. Consequently, the answer would have omitted the "a=csup" attribute line and chosen one or both of the PCMU and G.729 codecs instead. The answer carries the accepted configuration in the "m=" line along with corresponding "rtpmap" and/or "fmtp" parameters, as appropriate. Note that per the base protocol, after the above, Alice MAY generate a new offer with an actual configuration ("m=" line, etc.) corresponding to the actual configuration referenced in Bob's answer (not shown here).

8.3.3.3 New Capability Attributes

In this section, we present the new attributes associated with indicating the media capabilities for use by the SDP capability negotiation. The approach taken is to keep things similar to the existing media capabilities defined by the existing media descriptions ("m=" lines) and the associated "rtpmap" and "fmtp" attributes. We use media subtypes and "media capability numbers" to link the relevant media capability parameters. This permits the capabilities to be defined at the session level and to be used for multiple streams, if desired. For RTP-based media formats, payload types are then specified at the media level (see Section 8.3.3.3.4.2).

A media capability merely indicates possible support for the media type and media format(s) and parameters in question. In order to actually use a media capability in an offer/answer exchange, it MUST be referenced in a potential configuration. Media capabilities (i.e., the attributes associated with expressing media capability formats, parameters, etc.) can be provided at the session level and/or the media level. Media capabilities provided at the session level may be referenced in any "pcfg" or "lcfg" attribute at the media level (consistent with the media type), whereas media capabilities provided at the media level may be referenced only by the "pcfg" or "lcfg" attribute within that media stream. In either case, the scope of the <med-cap-num> is the entire session description. This enables each media capability to be uniquely referenced across the entire session description (e.g., in a potential configuration).

8.3.3.3.1 The Media Format Capability Attributes

Media subtypes can be expressed as media format capabilities by use of the "a=rmcap" and "a=omcap" attributes. The "a=rmcap" attribute MUST be used for RTP-based media, whereas the "a=omcap" attribute MUST be used for non-RTP-based (other) media formats. The two attributes are defined as follows:

```
a=rmcap:<media-cap-num-list> <encoding-name>/<clock-rate>
      [/<encoding-parms>]
a=omcap:<media-cap-num-list> <format-name>
```

where <media-cap-num-list> is a (list of) media capability number(s) used to number a media format capability, the <encoding name> or <format-name> is the media subtype (e.g., H.263-1998, PCMU, or T38), <clock rate> is the encoding rate, and <encoding parms> are the media encoding parameters for the media subtype. All media format capabilities in the list are assigned to the same media type/subtype. Each occurrence of the "rmcap" and "omcap" attribute MUST use unique values in its <media-cap-num-list>; the media capability numbers are shared between the two attributes, and the numbers MUST be unique across the entire SDP session. In short, the "rmcap" and "omcap" attributes define media format capabilities and associate them with a media capability number in the same manner as the "rtpmap" attribute defines them and associates them with a payload type number.

All SDP ABNF syntaxes are provided in Section 2.9. However, we are repeating this here for convenience. Additionally, the attributes allow multiple capability numbers to be defined for the media format in question by specifying a range of media capability numbers. This permits the media format to be associated with different media parameters in different configurations. When a range of capability numbers is specified, the first (leftmost) capability number MUST be strictly smaller than the second (rightmost) (i.e., the range increases and covers at least two numbers). In ABNF (RFC 5234), we have

```
media-capability-line   =  rtp-mcap / non-rtp-mcap
rtp-mcap                =  "a=rmcap:" media-cap-num-list
                           1*WSP encoding-name "/" clock-rate
                           ["/" encoding-parms]
non-rtp-mcap            =  "a=omcap:" media-cap-num-list 1*WSP
                           format-name
media-cap-num-list      =  media-cap-num-element
                           *("," media-cap-num-element)
media-cap-num-element   =  media-cap-num
                           / media-cap-num-range
media-cap-num-range     =  media-cap-num "-" media-cap-num
media-cap-num           =  NonZeroDigit *9(DIGIT)
encoding-name           =  token ;defined in RFC 4566 (see Section 2.9)
clock-rate              =  NonZeroDigit *9(DIGIT)
encoding-parms          =  token
format-name             =  token ;defined in RFC 4566 (see Section 2.9)
NonZeroDigit            =  %x31-39 ; 1-9
```

The encoding-name, clock-rate, and encoding-params are as defined to appear in an "rtpmap" attribute for each media type/subtype. Thus, it is easy to convert an "rmcap" attribute line into one or more "rtpmap" attribute lines, once a payload type number is assigned to a media-cap-num (see Section 8.3.3.3.5). The format-name is a media format description for non-RTP-based media as defined for the <fmt> part of the media description ("m=" line) in SDP (RFC 4566, see Section 2). In simple terms, it is the name of the media format, e.g., "t38." This form can also be used in cases such as Binary Floor Control Protocol (BFCP) (RFC 4583, see Section 13.2) where the fmt list in the "m=" line is effectively ignored (BFCP uses "*").

The "rmcap" and "omcap" attributes can be provided at the session level and/or the media level. There can be more than one "rmcap" and more than one "omcap" attribute at both the session and media levels (i.e., more than one of each at the session level and more than one of each in each media description). Media capability numbers cannot include leading zeroes, and each media-cap-num MUST be unique within the entire SDP record; it is used to identify that media capability in potential, latent, and actual configurations and in other attribute lines as explained below. Note that media-cap-num values are shared between the "rmcap" and "omcap" attributes; hence, the uniqueness requirement applies to the union of them. When the media capabilities are used in a potential, latent, or actual configuration, the media formats referred by those configurations apply at the media level, irrespective of whether the media capabilities themselves were specified at the session or media level. In other words, the media capability applies to the specific media description associated with the configuration that invokes it. For example

```
v=0
o=- 24351 621814 IN IP4 192.0.2.2
s=
c=IN IP4 192.0.2.2
t=0 0
a=rmcap:1 L16/8000/1
```

```
a=rmcap:2 L16/16000/2
a=rmcap:3 H263-1998/90000
a=omcap:4 example
m=audio 54320 RTP/AVP 0
a=pcfg:1 m=1|2, pt=1:99,2:98
m=video 66544 RTP/AVP 100
a=rtpmap:100 H264/90000
a=pcfg:10 m=3 pt=3:101
a=tcap:1 TCP
a=pcfg:11 m=4 t=1
```

8.3.3.3.2 The Media Format Parameter Capability Attribute

This attribute is used to associate media format specific parameters with one or more media format capabilities. The form of the attribute is

```
a=mfcap:<media-caps> <list of parameters>
```

where <media-caps> permits the list of parameters to be associated with one or more media format capabilities and the format parameters are specific to the type of media format. The mfcap lines map to a single traditional SDP "fmtp" attribute line (one for each entry in <media-caps>) of the form:

```
a=fmtp:<fmt> <list of parameters>
```

where <fmt> is the media format parameter defined in RFC 4566 (see Section 2), as appropriate for the particular media stream. The "mfcap" attribute MUST be used to encode attributes for media capabilities, which would conventionally appear in an "fmtp" attribute. The existing "acap" attribute MUST NOT be used to encode "fmtp" attributes. The "mfcap" attribute adheres to SDP (RFC 4566, see Section 2) attribute production rules with:

```
media-format-parameter-capability="a=mfcap:"media-cap-num-list 1*WSP fmt-
                                  specific-param-list
fmt-specific-param-list= text ; defined in RFC 4566 (see Section 2)
```

Note that media format parameters can be used with RTP-based and non-RTP-based media formats.

8.3.3.3.2.1 Format Parameter Concatenation Rule

The appearance of media subtypes with a large number of formatting options (e.g., AMR-WB, RFC 4867), coupled with the restriction that only a single "fmtp" attribute can appear per media format, suggests that it is useful to create a combining rule for "mfcap" parameters that are associated with the same media capability number. Therefore, different mfcap lines MAY include the same media-cap-num in their media-cap-num-list. When a particular media capability is selected for processing, the parameters from each mfcap line that references the particular capability number in its media-cap-num-list are concatenated together via ";", in the order the "mfcap" attributes appear in the SDP record, to form the equivalent of a single "fmtp" attribute line. This permits one to define a separate mfcap line for a single parameter and value that is to be applied to each media capability designated in the media-cap-num-list. This provides a compact method to specify multiple combinations of format parameters when using codecs with multiple format options.

Note that order-dependent parameters SHOULD be placed in a single mfcap line to avoid possible problems with line rearrangement by a middlebox. Format parameters are not parsed by SDP; their content is specific to the media type/subtype. When format parameters for a specific media capability are combined from multiple "a=mfcap" lines that reference that media capability, the format-specific parameters are concatenated together and separated by ";" for construction of the corresponding format attribute ("a=fmtp"). The resulting format attribute will look something like the following (without line breaks):

```
a=fmtp:<fmt> <fmt-specific-param-list1>;
       <fmt-specific-param-list2>;
       ...
```

where <fmt> depends on the transport protocol in the manner defined in RFC 4566 (see Section 2). SDP cannot assess the legality of the resulting parameter list in the "a=fmtp" line; the user must take care to ensure that legal parameter lists are generated. The "mfcap" attribute can be provided at the session level and the media level. There can be more than one "mfcap" attribute at the session or media level. The unique media-cap-num is used to associate the parameters with a media capability. As a simple example, a G.729 capability is, by default, considered to support comfort noise as defined by Annex B of ITU-T G.729 specification. Capabilities for G.729 with and without comfort noise support may thus be defined by

```
a=rmcap:1,2 G729/8000
a=mfcap:2 annexb:no
      Media capability 1 supports G.729 with Annex B, whereas media
      capability 2 supports G.729 without Annex B.
```

Example for H.263 video:

```
a=rmcap:1 H263-1998/90000
a=rmcap:2 H263-2000/90000
a=mfcap:1 CIF=4;QCIF=2;F=1;K=1
a=mfcap:2 profile=2;level=2.2
```

Finally, for eight format combinations of the AMR codec

```
a=rmcap:1-3 AMR/8000/1
a=rmcap:4-6 AMR-WB/16000/1
a=mfcap:1,2,3,4 mode-change-capability=1
a=mfcap:5,6 mode-change-capability=2
a=mfcap:1,2,3,5 max-red=220
a=mfcap:3,4,5,6 octet-align=1
a=mfcap:1,3,5 mode-set=0,2,4,7
a=mfcap:2,4,6 mode-set=0,3,5,6
```

So that AMR codec #1, when specified in a "pcfg" attribute within an audio stream block (and assigned payload type number 98) as in:

```
a=pcfg:1 m=1 pt=1:98
```

is essentially equivalent to the following:

```
m=audio 49170 RTP/AVP 98
a=rtpmap:98 AMR/8000/1
```

```
a=fmtp:98 mode-change-capability=1; \
max-red=220; mode-set=0,2,4,7
```

and AMR codec #4 with payload type number 99, depicted by the potential configuration:

```
a=pcfg:4 m=4, pt=4:99
```

is equivalent to the following:

```
m=audio 49170 RTP/AVP 99
a=rtpmap:99 AMR-WB/16000/1
a=fmtp:99 mode-change-capability=1; octet-align=1; \
mode-set=0,3,5,6
```

and so on for the other four combinations. SDP could thus convert the media capabilities specifications into one or more alternative media stream specifications, one of which can be chosen for the answer.

8.3.3.3.3 The Media-Specific Capability Attribute

Attributes and parameters associated with a media format are typically specified using the "rtpmap" and "fmtp" attributes in SDP and the similar "rmcap" and "mfcap" attributes in SDP media capabilities. Some SDP extensions define other attributes that need to be associated with media formats, for example, the "rtcp-fb" attribute defined in RFC 4585. Such media-specific attributes, beyond the "rtpmap" and "fmtp" attributes, may be associated with media capability numbers via a new media-specific attribute, "mscap," of the following form:

```
a=mscap:<media caps star> <att field> <att value>
```

where <media caps star> is a (list of) media capability number(s), <att field> is the attribute name, and <att value> is the value field for the named attribute. Note that the media capability numbers refer to media format capabilities specified elsewhere in the SDP ("rmcap" and/or "omcap"). If a range of capability numbers is specified, the first (leftmost) capability number MUST be strictly smaller than the second (rightmost). The media capability numbers may include a wildcard ("*"), which will be used instead of any payload type mappings in the resulting SDP (see, e.g., RFC 4585 and the following example). In ABNF, we have:

```
media-specific-capability = "a=mscap:" media-caps-star
                            1*WSP att-field ; from RFC 4566 (see
                            Section 2)
                            1*WSP att-value ; from RFC 4566 (see
                            Section 2)
media-caps-star = media-cap-star-element
                            *("," media-cap-star-element)
media-cap-star-element = (media-cap-num [wildcard])
                            / (media-cap-num-range [wildcard])
                            wildcard = "*"
```

Given an association between a media capability and a payload type number as specified by the "pt=" parameters in a "pcfg" attribute line, a mscap line may be translated easily into a conventional SDP attribute line of the form

```
a=<att field>":"<fmt> <att value> ; <fmt> defined in SDP (RFC 4566 - see
                                   ; Section 2)
```

A resulting attribute that is not a legal SDP attribute, as specified by RFC 4566, MUST be ignored by the receiver. If a media capability number (or range) contains a wildcard character at the end, any payload type mapping specified for that media-specific capability (or range of capabilities) will use the wildcard character in the resulting SDP instead of the payload type specified in the payload type mapping ("pt" parameter) in the configuration attribute.

A single mscap line may refer to multiple media capabilities by use of a capability number range; this is equivalent to multiple mscap lines, each with the same attribute values (but different media capability numbers), one line per media capability. Multiple mscap lines may refer to the same media capability, but, unlike the "mfcap" attribute, no concatenation operation is defined. Hence, multiple mscap lines applied to the same media capability are equivalent to multiple lines of the specified attribute in a conventional media record. Here is an example with the "rtcp-fb" attribute, modified from an example in RFC 5104 (with the session level and audio media omitted). If the offer contains a media block like the following (note the wildcard character),

```
m=video 51372 RTP/AVP 98
a=rtpmap:98 H263-1998/90000
a=tcap:1 RTP/AVPF
a=rmcap:1 H263-1998/90000
a=mscap:1 rtcp-fb ccm tstr
a=mscap:1 rtcp-fb ccm fir
a=mscap:1* rtcp-fb ccm tmmbr smaxpr=120
a=pcfg:1 t=1 m=1 pt=1:98
```

and if the proposed configuration is chosen, then the equivalent media block will look like the following:

```
m=video 51372 RTP/AVPF 98
a=rtpmap:98 H263-1998/90000
a=rtcp-fb:98 ccm tstr
a=rtcp-fb:98 ccm fir
a=rtcp-fb:* ccm tmmbr smaxpr=120
```

8.3.3.3.4 New Configuration Parameters

Along with the new attributes for media capabilities, new extension parameters are defined for use in the potential configuration, the actual configuration, and/or the new latent configuration defined in Section 8.3.3.3.5.

8.3.3.3.4.1 The Media Configuration Parameter (m=) The media configuration parameter is used to specify the media format(s) and related parameters for a potential, actual, or latent configuration. Adhering to the ABNF for extension-config-list in RFC 5939 (see Section 8.1) with

```
ext-cap-name = "m"
ext-cap-list = media-cap-num-list
               [*(BAR media-cap-num-list)]
```

we have

```
media-config-list = ["+"] "m=" media-cap-num-list
                    *(BAR media-cap-num-list)
                    ;BAR is defined in RFC 5939 (see Section 8.1)
                    ;media-cap-num-list is defined above
```

Alternative media configurations are separated by a vertical bar ("|"). The alternatives are ordered by preference, most-preferred first. When media capabilities are not included in a potential configuration at the media level, the media type and media format from the associated "m=" line will be used. The use of the plus sign ("+") is described in RFC 5939 (see Section 8.1).

8.3.3.3.4.2 The Payload Type Number Mapping Parameter (pt=) The payload type number mapping parameter is used to specify the payload type number to be associated with each RTP-based media format in a potential, actual, or latent configuration. We define the payload type number mapping parameter, payload-number-config-list, in accordance with the extension-config-list format defined in RFC (see Section 8.1). In ABNF

```
payload-number-config-list = ["+"] "pt=" media-map-list
media-map-list = media-map *("," media-map)
media-map = media-cap-num ":" payload-type-number
          ; media-cap-num is defined in Section 3.3.1
payload-type-number = NonZeroDigit *2(DIGIT) ; RTP payload
          ; type number
```

The example in Section 8.3.3.3.7 shows how the parameters from the rmcap line are mapped to payload type numbers from the "pcfg" "pt" parameter. The use of the plus sign ("+") is described in RFC 5939 (see Section 8.1). A latent configuration represents a future capability; hence, the "pt=" parameter is not directly meaningful in the "lcfg" attribute because no actual media session is being offered or accepted. It is permitted in order to tie any payload type number parameters within attributes to the proper media format. A primary example is the case of format parameters for the redundant audio data (RFC 2198) payload, which are payload type numbers. Specific payload type numbers used in a latent configuration MAY be interpreted as suggestions to be used in any future offer based on the latent configuration, but they are not binding; the offerer and/or answerer may use any payload type numbers each deems appropriate. The use of explicit payload type numbers for latent configurations can be avoided by use of the parameter substitution rule of Section 8.3.3.3.7. Future extensions are also permitted. Note that leading zeroes are not permitted.

8.3.3.3.4.3 The Media Type Parameter When a latent configuration is specified (always at the media level), indicating the ability to support an additional media stream, it is necessary to specify the media type (audio, video, etc.) as well as the format and transport type. The media type parameter is defined in ABNF as media-type = ["+"] "mt=" media; media defined in RFC 4566 (see Section 2). At present, the media type parameter is used only in the latent configuration attribute, and the use of the "+" prefix to specify that the entire attribute line is to be ignored if the mt= parameter is not understood is unnecessary. However, if the media type parameter is later added to an existing capability attribute such as "pcfg," then the "+" would be useful. The media format(s) and transport type(s) are specified using the media configuration parameter ("+m=") defined above and the transport parameter ("t=") defined in RFC 5939 (see Section 8.1), respectively.

8.3.3.3.5 The Latent Configuration Attribute

One of the goals of this work is to permit the exchange of supportable media configurations in addition to those offered or accepted for immediate use. Such configurations are referred to as "latent configurations." For example, a party may offer to establish a session with an audio stream and, at the same time, announce its ability to support a video stream as part of the same session. The offerer can supply its video capabilities by offering one or more latent video configurations with the media stream for audio; the responding party may indicate its ability and willingness to support such a video session by returning a corresponding latent configuration.

Latent configurations returned in SDP answers MUST match offered latent configurations (or parameter subsets thereof). Therefore, it is appropriate for the offering party to announce most, if not all, of its capabilities in the initial offer. This choice has been made in order to keep the size of the answer more compact by not requiring acap, rmcap, tcap, etc. lines in the answer. Latent configurations may be announced by use of the latent configuration attribute, which is defined in a manner very similar to the potential configuration attribute. The latent configuration attribute combines the properties of a media line and a potential configuration. A latent configuration MUST include a media type (mt=) and a transport protocol configuration parameter since the latent configuration is independent of any media line present. In most cases, the media configuration (m=) parameter needs to be present as well (see Section 8.3.4 for examples).

The "lcfg" attribute is a media-level attribute. The "lcfg" attribute is defined as a media-level attribute since it specifies a possible future media stream. However, the "lcfg" attribute is not necessarily related to the media description within which it is provided. Session capability attributes ("a=sescap") may be used to indicate supported media stream configurations. Each media line in an SDP description represents an offered simultaneous media stream, whereas each latent configuration represents an additional stream that may be negotiated in a future offer/answer exchange. Session capability attributes may be used to determine whether a latent configuration may be used to form an offer for an additional simultaneous stream or to reconfigure an existing stream in a subsequent offer/answer exchange. The latent configuration attribute is of the form

```
a=lcfg:<config-number> <latent-cfg-list>
```

which adheres to the SDP (RFC 4566, see Section 2) "attribute" production with att-field and att-value defined as

```
att-field = "lcfg"
att-value = config-number 1*WSP lcfg-cfg-list
config-number = NonZeroDigit *9(DIGIT) ;DIGIT defined in RFC 5234
lcfg-cfg-list = media-type 1*WSP pot-cfg-list
                ; as defined in RFC 5939 (see Section 8.1)
                ; and extended herein
```

The media type (mt=) parameter identifies the media type (audio, video, etc.) to be associated with the latent media stream, and it MUST be present. The pot-cfg-list MUST contain a transport-protocolconfig-list (t=) parameter and a media-config-list (m=) parameter. The pot-cfg-list MUST NOT contain more than one instance of each type of parameter list. As specified in RFC 5939 (see Section 8.1), the use of the "+" prefix with a parameter indicates that the entire configuration MUST be ignored if the parameter is not understood; otherwise, the parameter itself may be ignored.

Media stream payload numbers are not assigned by a latent configuration. Assignment will take place if and when the corresponding stream is actually offered via an "m=" line in a later exchange. The payload-number-config-list is included as a parameter to the "lcfg" attribute in case it is necessary to tie payload numbers in attribute capabilities to specific media capabilities. If an "lcfg" attribute invokes an "acap" attribute that appears at the session level, then that attribute will be expected to appear at the session level of a subsequent offer when and if a corresponding media stream is offered. Otherwise, "acap" attributes that appear at the media level represent media-level attributes. Note, however, that "rmcap," "omcap," "mfcap," "mscap," and "tcap" attributes may appear at the session level because they always result in media-level attributes or "m=" line parameters.

The configuration numbers for latent configurations do not imply a preference; the offerer will imply a preference when actually offering potential configurations derived from latent configurations negotiated earlier. Note, however, that the offerer of latent configurations MAY specify preferences for combinations of potential and latent configurations by use of the "sescap" attribute defined in Section 8.3.3.3.8. For example, if an SDP offer contains, say, an audio stream with "pcfg:1," and two latent video configurations, "lcfg:2" and "lcfg:3," then a session with one audio stream and one video stream could be specified by including "a=sescap:1 1,2|3." One audio stream and two video streams could be specified by including "a=sescap:2 1,2,3" in the offer. In order to permit combinations of latent and potential configurations in session capabilities, latent configuration numbers MUST be different from those used for potential configurations.

This restriction is especially important if the offerer does not require cmed-v0 capability and the recipient of the offer doesn't support it. If the "lcfg" attribute is not recognized, the capability attributes intended to be associated with it may be confused with those associated with a potential configuration of some other media stream. Note also that leading zeroes are not permitted in configuration numbers. If a cryptographic attribute, such as the SDES "a=crypto:" attribute (RFC 4568, see Section 16.2), is referenced by a latent configuration through an "acap" attribute, any keying material required in the conventional attribute, such as the SDES key/salt string, MUST be included in order to satisfy formatting rules for the attribute. Since the keying material will be visible but not actually used at this stage (since it's a latent configuration), the value(s) of the keying material MUST NOT be a real value used for real exchange of media, and the receiver of the "lcfg" attribute MUST ignore the value(s).

8.3.3.3.6 Enhanced Potential Configuration Attribute

The present work requires new extensions (parameters) for the "pcfg" attribute defined in the SDP capability negotiation base protocol (RFC 5939, see Section 8.1). The parameters and their definitions are "borrowed" from the definitions provided for the latent configuration attribute in Section 8.3.3.3.5. The expanded ABNF definition of the "pcfg" attribute is

```
a=pcfg: <config-number> [<pot-cfg-list>]
```

where

```
config-number = 1*DIGIT ;defined in (RFC 5234)
pot-cfg-list = pot-config *(1*WSP pot-config)
pot-config = attribute-config-list / ;def in RFC 5939 - see Section 8.1)
               transport-protocol-config-list / ;defined in RFC 5939
               - see Section 8.1
```

```
extension-config-list / ;RFC 5939 (see Section 8.1)
media-config-list / ; Section 8.3.3.3.4.1
payload-number-config-list ; Section 8.3.3.3.4.2
```

Except for the extension-config-list, the pot-cfg-list MUST NOT contain more than one instance of each parameter list.

8.3.3.3.6.1 Returning Capabilities in the Answer

Potential and/or latent configuration attributes may be returned within an answer SDP to indicate the ability of the answerer to support alternative configurations of the corresponding stream(s). For example, an offer may include multiple potential configurations for a media stream and/or latent configurations for additional streams. The corresponding answer will indicate (via an "acfg" attribute) the configuration accepted and used to construct the base configuration for each active media stream in the reply, but the reply MAY also contain potential and/or latent configuration attributes, with parameters, to indicate which other offered configurations would be acceptable. This information is useful if it becomes desirable to reconfigure a media stream (e.g., to reduce resource consumption).

When potential and/or latent configurations are returned in an answer, all numbering MUST refer to the configuration and capability attribute numbering of the offer. The offered capability attributes need not be returned in the answer. The answer MAY include additional capability attributes and/or configurations (with distinct numbering). The parameter values of any returned "pcfg" or "lcfg" attributes MUST be a subset of those included in the offered configurations and/or those added by the answerer; values MAY be omitted only if they were indicated as alternative sets, or optional, in the original offer. The parameter set indicated in the returned "acfg" attribute need not be repeated in a returned "pcfg" attribute. The answerer MAY return more than one "pcfg" attribute with the same configuration number if it is necessary to describe selected combinations of optional or alternative parameters.

Similarly, one or more session capability attributes ("a=sescap") MAY be returned to indicate which of the offered session capabilities is/are supportable by the answerer (see Section 8.3.3.3.8). Note that, although the answerer MAY return capabilities beyond those included by the offerer, these capabilities MUST NOT be used to form any base level media description in the answer. For this reason, it is advisable for the offerer to include most, if not all, potential and latent configurations it can support in the initial offer, unless the size of the resulting SDP is a concern. Either party MAY later announce additional capabilities by renegotiating the session in a second offer/answer exchange.

8.3.3.3.6.2 Payload Type Number Mapping

When media format capabilities defined in "rmcap" attributes are used in potential configuration lines, the transport protocol uses RTP and it is necessary to assign payload type numbers. In some cases, it is desirable to assign different payload type numbers to the same media format capability when used in different potential configurations. One example is when configurations for AVP and SAVP are offered: The offerer would like the answerer to use different payload type numbers for encrypted and unencrypted media, so the offerer can decide whether to render early media that arrives before the answer is received.

For example, if use of AVP was selected by the answerer, then media received by the offerer is not encrypted; hence, it can be played out prior to receiving the answer. Conversely, if SAVP was selected, cryptographic parameters and keying material present in the answer may be needed to decrypt received media. If the offer configuration indicated that AVP media uses one set of payload types and SAVP a different set, then the offerer will know whether media received prior to the answer is encrypted or not by simply looking at the RTP payload type number in the received packet. This association of distinct payload type number(s) with different transport protocols requires a separate pcfg line for each protocol. Clearly, this technique cannot be used if the number of potential configurations exceeds the number of possible payload type numbers.

8.3.3.3.6.3 Processing of Media-Format-Related Conventional Attributes for Potential Configurations When media capabilities negotiation is employed, SDP records are likely to contain conventional attributes such as "rtpmap," "fmtp," and other media-format-related lines, as well as capability attributes such as "rmcap," "omcap," "mfcap," and "mscap" that map into those conventional attributes when invoked by a potential configuration. In such cases, it MAY be appropriate to employ the delete-attributes option (RFC 5939, see Section 8.1) in the attribute configuration list parameter in order to avoid the generation of conflicting "fmtp" attributes for a particular configuration. Any media-specific attributes in the media block that refer to media formats not used by the potential configuration MUST be ignored. For example:

```
v=0
o=- 25678 753849 IN IP4 192.0.2.1
s=
c=IN IP4 192.0.2.1
t=0 0
a=creq:med-v0
m=audio 3456 RTP/AVP 0 18 100
a=rtpmap:100 telephone-event
a=fmtp:100 0-11
a=rmcap:1 PCMU/8000
a=rmcap:2 G729/8000
a=rmcap:3 telephone-event/8000
a=mfcap:3 0-15
a=pcfg:1 m=2,3|1,3 a=-m pt=1:0,2:18,3:100
a=pcfg:2
```

In this example, PCMU is media capability 1, G.729 is media capability 2, and telephone-event is media capability 3. The a=pcfg:1 line specifies that the preferred configuration is G.729 with extended DTMF events, second is G.711 mu-law with extended DTMF events, and the base media-level attributes are to be deleted. Intermixing of G.729, G.711, and "commercial" DTMF events is least preferred (the base configuration provided by the "m=" line, which is, by default, the least preferred configuration). The "rtpmap" and "fmtp" attributes of the base configuration are replaced by the "rmcap" and "mfcap" attributes when invoked by the proposed configuration. If the preferred configuration is selected, the SDP answer will look like the following:

```
v=0
o=- 25678 753849 IN IP4 192.0.2.1
s=
c=IN IP4 192.0.2.1
t=0 0
a=csup:med-v0
m=audio 3456 RTP/AVP 18 100
a=rtpmap:100 telephone-event/8000
a=fmtp:100 0-15
a=acfg:1 m=2,3 pt=1:0,2:18,3:100
```

8.3.3.3.7 Substitution of Media Payload Type Numbers in Capability Attribute Parameters

In some cases, for example, when an RFC 2198 redundancy audio subtype capability is defined in an "mfcap" attribute, the parameters to an attribute may contain payload type numbers. Two options are available for specifying such payload type numbers. They may be expressed explicitly, in which case they are bound to actual payload types by means of the payload type number parameter (pt=) in the appropriate potential or latent configuration. For example, the following SDP fragment defines a potential configuration with redundant G.711 mu-law:

```
m=audio 45678 RTP/AVP 0
a=rtpmap:0 PCMU/8000
a=rmcap:1 PCMU/8000
a=rmcap:2 RED/8000
a=mfcap:2 0/0
a=pcfg:1 m=2,1 pt=2:98,1:0
```

The potential configuration is then equivalent to

```
m=audio 45678 RTP/AVP 98 0
a=rtpmap:0 PCMU/8000
a=rtpmap:98 RED/8000
a=fmtp:98 0/0
```

A more general mechanism is provided via the parameter substitution rule. When an "mfcap," "mscap," or "acap" attribute is processed, its arguments will be scanned for a payload type number escape sequence of the following form (in ABNF):

```
ptn-esc ="%m=" media-cap-num "%" ; defined in Section 8.3.3.3.1
```

If the sequence is found, the sequence is replaced by the payload type number assigned to the media capability number, as specified by the "pt=" parameter in the selected potential configuration; only actual payload type numbers are supported—wildcards are excluded. The sequence "%%" (null digit string) is replaced by a single percent sign, and processing continues with the next character, if any. For example, this offer sequence could have been written as

```
m=audio 45678 RTP/AVP 0
a=rtpmap:0 PCMU/8000
a=rmcap:1 PCMU/8000
a=rmcap:2 RED/8000
a=mfcap:2 %m=1%/%m=1%
a=pcfg:1 m=2,1 pt=2:98,1:0
```

and the equivalent SDP is the same as above.

8.3.3.3.8 The Session Capability Attribute

Potential and latent configurations enable offerers and answerers to express a wide range of alternative configurations for current and future negotiation. However, in practice, it may not be possible to support all combinations of these configurations. The session capability attribute provides a means for the offerer and/or the answerer to specify combinations of specific media stream configurations that it is willing and able to support. Each session capability in an offer or answer

MAY be expressed as a list of required potential configurations and MAY include a list of optional potential and/or latent configurations.

The choices of session capabilities may be based on processing load, total bandwidth, or any other criteria of importance to the communicating parties. If the answerer supports media capabilities negotiation, and session configurations are offered, it MUST accept one of the offered configurations or it MUST refuse the session. Therefore, if the offer includes any session capabilities, it SHOULD include all session capabilities the offerer is willing to support. The session capability attribute is a session-level attribute described by

```
"a=sescap:" <session num> <list of configs>
```

which corresponds to the standard value attribute definition with

```
att-field = "sescap"
att-value = session-num 1*WSP list-of-configs
            [1*WSP optional-configs]
            session-num = NonZeroDigit *9(DIGIT) ; DIGIT defined
            ; in RFC 5234
list-of-configs = alt-config *("," alt-config)
optional-configs = "[" list-of-configs "]"
alt-config = config-number *("|" config-number)
```

The session-num identifies the session: A lower-number session is preferred over a higher-number session, and leading zeroes are not permitted. Each alt-config list specifies alternative media configurations within the session; preference is based on config-num as specified in RFC 5939 (see Section 8.1). Note that the session preference order, when present, takes precedence over the individual media stream configuration preference order. Use of session capability attributes requires that configuration numbers assigned to potential and latent configurations MUST be unique across the entire session; RFC 5939 (see Section 8.1) requires only that "pcfg" configuration numbers be unique within a media description. Also, leading zeroes are not permitted.

As an example, consider an endpoint that is capable of supporting an audio stream with either one H.264 video stream or two H.263 video streams with a floor control stream. In the latter case, the second video stream is optional. The SDP offer might look like the following (offering audio, an H.263 video streams, BFCP and another optional H.263 video stream); the empty lines are added for readability only (not part of valid SDP):

```
v=0
o=- 25678 753849 IN IP4 192.0.2.1
s=
c=IN IP4 192.0.2.1
t=0 0
a=creq:med-v0
a=sescap:2 1,2,5,[3]
a=sescap:1 1,4
m=audio 54322 RTP/AVP 0
a=rtpmap:0 PCMU/8000
a=pcfg:1

m=video 22344 RTP/AVP 102
a=rtpmap:102 H263-1998/90000
a=fmtp:102 CIF=4;QCIF=2;F=1;K=1
```

```
i=main video stream
a=label:11
a=pcfg:2
a=rmcap:1 H264/90000
a=mfcap:1 profile-level-id=42A01E; packetization-mode=2
a=acap:1 label:13
a=pcfg:4 m=1 a=1 pt=1:104
m=video 33444 RTP/AVP 103
a=rtpmap:103 H263-1998/90000
a=fmtp:103 CIF=4;QCIF=2;F=1;K=1
i=secondary video (slides)
a=label:12
a=pcfg:3
m=application 33002 TCP/BFCP *
a=setup:passive
a=connection:new
a=floorid:1 m-stream:11 12
a=floor-control:s-only
a=confid:4321
a=userid:1234
a=pcfg:5
```

If the answerer understands media capability negotiations, but cannot support the BFCP, then it would respond with (invalid empty lines in SDP included again for readability)

```
v=0
o=- 25678 753849 IN IP4 192.0.2.1
s=
c=IN IP4 192.0.2.22
t=0 0
a=csup:med-v0
a=sescap:1 1,4
m=audio 23456 RTP/AVP 0
a=rtpmap:0 PCMU/8000
a=acfg:1
m=video 41234 RTP/AVP 104
a=rtpmap:104 H264/90000
a=fmtp:104 profile-level-id=42A01E; packetization-mode=2
a=acfg:4 m=1 a=1 pt=1:104
m=video 0 RTP/AVP 103
a=acfg:3
m=application 0 TCP/BFCP *
a=acfg:5
```

An endpoint that doesn't support media capabilities negotiation, but does support H.263 video, would respond with one or two H.263 video streams. In the latter case, the answerer may issue a second offer to reconfigure the session to one audio and one video channel using H.264 or H.263. Session capabilities can include latent capabilities as well. Here's a similar example in which the offerer wishes to initially establish an audio stream and prefers to later establish two video streams with chair control. If the answerer doesn't understand media capability negotiation, or cannot support the dual video streams or flow control, then it may support a single H.264 video stream. Note that establishment of the most favored configuration will require two offer/answer exchanges.

```
v=0
o=- 25678 753849 IN IP4 192.0.2.1
s=
c=IN IP4 192.0.2.1
t=0 0
a=creq:med-v0
a=sescap:1 1,3,4,5
a=sescap:2 1,2
a=sescap:3 1
a=rmcap:1 H263-1998/90000
a=mfcap:1 CIF=4;QCIF=2;F=1;K=1
a=tcap:1 RTP/AVP TCP/BFCP
m=audio 54322 RTP/AVP 0
a=rtpmap:0 PCMU/8000
a=pcfg:1
m=video 22344 RTP/AVP 102
a=rtpmap:102 H264/90000
a=fmtp:102 profile-level-id=42A01E; packetization-mode=2
a=label:11
a=content:main
a=pcfg:2
a=lcfg:3 mt=video t=1 m=1 a=31,32
a=acap:31 label:12
a=acap:32 content:main
a=lcfg:4 mt=video t=1 m=1 a=41,42
a=acap:41 label:13
a=acap:42 content:slides
a=lcfg:5 mt=application m=51 t=51
a=tcap:51 TCP/BFCP
a=omcap:51 *
a=acap:51 setup:passive
a=acap:52 connection:new
a=acap:53 floorid:1 m-stream:12 13
a=acap:54 floor-control:s-only
a=acap:55 confid:4321
a=acap:56 userid:1234
```

In this example, the default offer, as seen by endpoints that do not understand capabilities negotiation, proposes a PCMU audio stream and an H.264 video stream. Note that the offered lcfg lines for the video streams don't carry "pt=" parameters because they're not needed (payload type numbers will be assigned in the offer/answer exchange that establishes the streams). Note also that the three "rmcap," "mfcap," and "tcap" attributes used by "lcfg:3" and "lcfg:4" are included at the session level so they may be referenced by both latent configurations. As per Section 3.3, the media attributes generated from the "rmcap," "mfcap," and "tcap" attributes are always media-level attributes. If the answerer supports media capability negotiations and supports the most desired configuration, it would return the following SDP:

```
v=0
o=- 25678 753849 IN IP4 192.0.2.1
s=
c=IN IP4 192.0.2.22
t=0 0
a=csup:med-v0
```

```
a=sescap:1 1,3,4,5
a=sescap:2 1,2
a=sescap:3 1
m=audio 23456 RTP/AVP 0
a=rtpmap:0 PCMU/8000
a=acfg:1
m=video 0 RTP/AVP 102
a=pcfg:2
a=lcfg:3 mt=video t=1 m=1 a=31,32
a=lcfg:4 mt=video t=1 m=1 a=41,42
a=lcfg:5 mt=application t=2
```

This exchange supports immediate establishment of an audio stream for preliminary conversation. This exchange would presumably be followed at the appropriate time with a "reconfiguration" offer/answer exchange to add the video and chair control streams.

8.3.3.4 Offer/Answer Model Extensions

In this section, we define extensions to the offer/answer model defined in RFC 3264 [RFC 3264] and RFC 5939 (see Section 8.1) to allow for media format and associated parameter capabilities, latent configurations, and acceptable combinations of media stream configurations to be used with the SDP capability negotiation framework. Note that the procedures defined in this section extend the offer/answer procedures defined in RFC 5939 (see Section 8.1/8.1.6); those procedures form a baseline set of capability negotiation offer/answer procedures that MUST be followed, subject to the extensions defined here.

SDP capability negotiation (RFC 5939, see Section 8.1) provides a relatively compact means to offer the equivalent of an ordered list of alternative configurations for offered media streams (as would be described by separate "m=" lines and associated attributes). The attributes "acap," "mscap," "mfcap," "omcap," and "rmcap" are designed to map somewhat straightforwardly into equivalent "m=" lines and conventional attributes when invoked by a "pcfg," "lcfg," or "acfg" attribute with appropriate parameters. The "a=pcfg:" lines, along with the "m=" line itself, represent offered media configurations. The "a=lcfg:" lines represent alternative capabilities for future use.

8.3.3.4.1 Generating the Initial Offer

The media capabilities negotiation extensions defined in this document cover the following categories of features:

- Media format capabilities and associated parameters ("rmcap," "omcap," "mfcap," and "mscap" attributes)
- Potential configurations using those media format capabilities and associated parameters
- Latent media streams ("lcfg" attribute)
- Acceptable combinations of media stream configurations ("sescap" attribute)

The high-level description of the operation is as follows: When an endpoint generates an initial offer and wants to use the functionality described in the current document, it SHOULD identify and define the media formats and associated parameters it can support via the "rmcap," "omcap," "mfcap," and "mscap" attributes. The SDP media line(s) ("m=") should be made up with the actual configuration to be used if the other party does not understand capability negotiations (by default, this is the least preferred configuration). Typically, the media line configuration will contain the minimum acceptable configuration from the offerer's point of view.

Preferred configurations for each media stream are identified following the media line. The present offer may also include latent configuration ("lcfg") attributes, at the media level, describing media streams and/or configurations the offerer is not now offering but is willing to support in a future offer/answer exchange. A simple example might be the inclusion of a latent video configuration in an offer for an audio stream. Lastly, if the offerer wishes to impose restrictions on the combinations of potential configurations to be used, it will include session capability ("sescap") attributes indicating those. If the offerer requires the answerer to understand the media capability extensions, the offerer MUST include a "creq" attribute containing the value "med-v0." If media capability negotiation is required only for specific media descriptions, the "med-v0" value MUST be provided only in "creq" attributes within those media descriptions, as described in RFC 5939 (see Section 8.1). Below, we provide a more detailed description of how to construct the offer SDP.

8.3.3.4.1.1 Offer with Media Capabilities

For each RTP-based media format the offerer wants to include as a media format capability, the offer MUST include an "rmcap" attribute for the media format as defined in Section 8.3.3.3.1. For each non-RTP-based media format the offer wants to include as a media format capability, the offer MUST include an "omcap" attribute for the media format as defined in Section 3.3.1. Since the media capability number space is shared between the "rmcap" and "omcap" attributes, each media capability number provided (including ranges) MUST be unique in the entire SDP. If an "fmtp" parameter value is needed for a media format (whether or not it is RTP based) in a media capability, then the offer MUST include one or more "mfcap" parameters with the relevant "fmtp" parameter values for that media format as defined in Section 8.3.3.3.2. When multiple "mfcap" parameters are provided for a given media capability, they MUST be provided in accordance with the concatenation rules in Section 8.3.3.3.2.1.

For each of these media format capabilities, the offer MAY include one or more "mscap" parameters with attributes needed for those specific media formats as defined in Section 8.3.3.3.3. Such attributes will be instantiated at the media level; hence, session-level-only attributes MUST NOT be used in the "mscap" parameter. The "mscap" parameter MUST NOT include an "rtpmap" or "fmtp" attribute ("rmcap" and "mfcap" are used instead). If the offerer wants to limit the relevance (and use) of a media format capability or parameter to a particular media stream, the media format capability or parameter MUST be provided within the corresponding media description. Otherwise, the media format capabilities and parameters MUST be provided at the session level.

Note, however, that the attribute or parameter embedded in these will always be instantiated at the media level. This is due to those parameters being effectively media-level parameters. If session-level attributes are needed, the "acap" attribute defined in RFC 5939 (see Section 8.1) can be used; however, it does not provide for media-format-specific instantiation. Inclusion of the above does not constitute an offer to use the capabilities; a potential configuration is needed for that. If the offerer wants to offer one or more of these media capabilities, they MUST be included as part of a potential configuration ("pcfg") attribute as defined in Section 8.3.3.3.4. Each potential configuration MUST include a config-number, and each config-number MUST be unique in the entire SDP (note that this differs from RFC 5939 (see Section 8.1), which only requires uniqueness within a media description). Also, the config-number MUST NOT overlap with any config-number used by a latent configuration in the SDP. As described in RFC 5939, lower config-numbers indicate a higher preference; the ordering still applies within a given media description only though.

For a media capability to be included in a potential configuration, there MUST be an "m=" parameter in the "pcfg" attribute referencing the media capability number in question. When one or more media capabilities are included in an offered potential configuration ("pcfg"), they completely replace the list of media formats offered in the actual configuration ("m=" line). Any attributes included for those formats remain in the SDP though (e.g., "rtpmap," "fmtp," etc.). For non-RTP-based media formats, the format-name (from the "omcap" media capability) is simply added to the "m=" line as a media format (e.g., t38). For RTP-based media, payload type mappings MUST be provided by use of the "pt" parameter in the potential configuration (see Section 8.3.3.3.4.2); payload type escaping may be used in "mfcap," "mscap," and "acap" attributes as defined in Section 8.3.3.3.7. Note that the "mt" parameter MUST NOT be used with the "pcfg" attribute (since it is defined for the "lcfg" attribute only); the media type in a potential configuration cannot be changed from that of the encompassing media description.

8.3.3.4.1.2 Offer with Latent Configuration If the offerer wishes to offer one or more latent configurations for future use, the offer MUST include a latent configuration attribute ("lcfg") for each as defined in Section 8.3.3.3.6. Each "lcfg" attribute

- MUST be specified at the media level
- MUST include a config-number that is unique in the entire SDP (including for any potential configuration attributes). Note that config-numbers in latent configurations do not indicate any preference order
- MUST include a media type ("mt")
- MUST reference a valid transport capability ("t")

Each "lcfg" attribute MAY include additional capability references, which may refer to capabilities anywhere in the session description, subject to any restrictions normally associated with such capabilities. For example, a media-level attribute capability must be present at the media level in some media description in the SDP. Note that this differs from the potential configuration attribute, which cannot validly refer to media-level capabilities in another media description (per RFC 5939, see Section 8.1/8.1.3.5.1). Potential configurations constitute an actual offer and may instantiate a referenced capability. Latent configurations are not actual offers; hence, they cannot instantiate a referenced capability. Therefore, it is safe for those to refer to capabilities in another media description.

8.3.3.4.1.3 Offer with Configuration Combination Restrictions If the offerer wants to indicate restrictions or preferences among combinations of potential and/or latent configurations, a session capability ("sescap") attribute MUST be provided at the session level for each such combination as described in Section 8.3.3.3.8. Each "sescap" attribute MUST include a session-num that is unique in the entire SDP; the lower the session-num the more preferred that combination is. Furthermore, "sescap" preference order takes precedence over any order specified in individual "pcfg" attributes. For example, if we have pcfg-1 and pcfg-2, and sescap-1 references pcfg-2, whereas sescap-2 references pcfg-1, then pcfg-2 will be the most preferred potential configuration. Without the sescap, pcfg-1 would be the most preferred.

8.3.3.4.2 Generating the Answer

When receiving an offer, the answerer MUST check the offer for "creq" attributes containing the value "med-v0"; answerers compliant with this specification will support this value in accordance with the procedures specified in RFC 5939 (see Section 8.1).

The SDP MAY contain

- Media format capabilities and associated parameters ("rmcap," "omcap," "mfcap," and "mscap" attributes)
- Potential configurations using those media format capabilities and associated parameters
- Latent media streams ("lcfg" attribute)
- Acceptable combinations of media stream configurations ("sescap" attribute)

The high-level informative description of the operation is as follows:

When the answering party receives the offer, if it supports the required capability negotiation extensions, it should select the most-preferred configuration it can support for each media stream and build its answer accordingly. The configuration selected for each accepted media stream is placed into the answer as a media line with associated parameters and attributes. If a proposed configuration is chosen for a given media stream, the answer must contain an actual configuration ("acfg") attribute for that media stream to indicate which offered "pcfg" attribute was used to build the answer. The answer should also include any potential or latent configurations the answerer can support, especially any configurations compatible with other potential or latent configurations received in the offer. The answerer should make note of those configurations it might wish to offer in the future. Below we provide a more detailed normative description of how the answerer processes the offer SDP and generates an answer SDP.

8.3.3.4.2.1 Processing Media Capabilities and Potential Configurations

The answerer MUST first determine if it needs to perform media capability negotiation by examining the SDP for valid and preferred potential configuration attributes that include media configuration parameters (i.e., an "m" parameter in the "pcfg" attribute). Such a potential configuration is valid if

1. It is valid according to the rules defined in RFC 5939 (see Section 8.1).
2. It contains a config-number that is unique in the entire SDP and does not overlap with any latent configuration config-numbers.
3. All media format capabilities ("rmcap" or "omcap"), media format parameter capabilities ("mfcap"), and media-specific capabilities ("mscap") referenced by the potential configuration ("m" parameter) are valid themselves (as defined in Sections 8.3.3.3.1, 8.3.3.3.2, and 8.3.3.3.3) and each is provided either at the session level or within this particular media description.
4. All RTP-based media format capabilities ("rmcap") have a corresponding payload type ("pt") parameter in the potential configuration that results in mapping to a valid payload type that is unique within the resulting SDP.
5. Any concatenation (see Section 8.3.3.3.2.1) and substitution (see Section 8.3.3.3.7) applied to any capability ("mfcap," "mscap," or "acap") referenced by this potential configuration results in a valid SDP.

Note that since SDP does not interpret the value of "fmtp" parameters, any resulting "fmtp" parameter value will be considered valid.

Second, the answerer MUST determine the order in which potential configurations are to be negotiated. In the absence of any session capability ("sescap") attributes, this simply follows the rules of RFC 5939 (see Section 8.1), with a lower config-number within a media description being preferred over a higher one. If a valid "sescap" attribute is present, the preference order provided in the "sescap" attribute MUST take precedence. A "sescap" attribute is considered valid if:

1. It adheres to the rules provided in Section 8.3.3.3.8.
2. All the configurations referenced by the "sescap" attribute are valid themselves (note that this can include the actual, potential, and latent configurations).

The answerer MUST now process the offer for each media stream based on the most preferred valid potential configuration in accordance with the procedures specified in RFC 5939 (see Section 8.1/8.1.3.6.2), and further extended as follows:

- If one or more media format capabilities are included in the potential configuration, then they replace all media formats provided in the "m=" line for that media description. For non-RTP-based media formats ("omcap"), the format-name is added. For RTP-based media formats ("rmcap"), the payload type specified in the payload type mapping ("pt") is added and a corresponding "rtpmap" attribute is added to the media description.
- If one or more media format parameter capabilities are included in the potential configuration, then the corresponding "fmtp" attributes are added to the media description. Note that this inclusion is done indirectly via the media format capability.
- If one or more media-specific capabilities are included in the potential configuration, then the corresponding attributes are added to the media description. Note that this inclusion is done indirectly via the media format capability.
- When checking to see if the answerer supports a given potential configuration that includes one or more media format capabilities, the answerer MUST support at least one of the media formats offered. If he does not, the answerer MUST proceed to the next potential configuration based on the preference order that applies.
- If session capability ("sescap") preference ordering is included, then the potential configuration selection process MUST adhere to the ordering provided. Note that this may involve coordinated selection of potential configurations between media descriptions. The answerer MUST accept one of the offered sescap combinations (i.e., all the required potential configurations specified) or it MUST reject the entire session.

Once the answerer has selected a valid and supported offered potential configuration for all of the media streams (or has fallen back to the actual configuration plus any added session attributes), the answerer MUST generate a valid answer SDP as described in RFC 5939 (see Section 8.1/8.1.3.6.2), and further extended as follows:

- Additional answer capabilities and potential configurations MAY be returned in accordance with Section 8.3.3.3.6.1. Capability numbers and configuration numbers for those MUST be distinct from the ones used in the offer SDP.
- Latent configuration processing and answer generation MUST be performed, as specified below.
- Session capability specification for the potential and latent configurations in the answer MAY be included (see Section 8.3.3.3.8).

8.3.3.4.2.2 Latent Configuration Processing

The answerer MUST determine if it needs to perform any latent configuration processing by examining the SDP for valid latent configuration attributes ("lcfg"). An "lcfg" attribute is considered valid if

- It adheres to the description in Section 8.3.3.3.5.
- It includes a config-number that is unique in the entire SDP and does not overlap with any potential configuration config-number.
- It includes a valid media type ("mt=").
- It references a valid transport capability ("t=").
- All other capabilities referenced by it are valid.

For each such valid latent configuration in the offer, the answerer checks to see if it could support the latent configuration in a subsequent offer/answer exchange. If so, it includes the latent configuration with the same configuration number in the answer, similar to the way potential configurations are processed and the selected one returned in an actual configuration attribute (see RFC 5939, see Section 8.1). If the answerer supports only a (nonmandatory) subset of the parameters offered in a latent configuration, the answer latent configuration will include only those parameters supported (similar to "acfg" processing). Note that latent configurations do not constitute an actual offer at this point in time; they merely indicate additional configurations that could be supported.

If a session capability ("sescap") attribute is included, and it references a latent configuration, then the answerer processing of that latent configuration must be done within the constraints specified by that session capability. That is, it must be possible to support it at the same time as any required (i.e., nonoptional) potential configurations in the session capability. The answerer may in turn add his own sescap indications in the answer as well.

8.3.3.4.3 Offerer Processing of the Answer

The offerer MUST process the answer in accordance with Section 8.1.3.6.3 of this book (that is, Section 3.6.3 of RFC 5939 itself) and the further explanation provided here. When the offerer processes the answer SDP based on a valid actual configuration attribute in the answer, and that valid configuration includes one or more media capabilities, the processing MUST furthermore be done as if the offer were sent using those media capabilities instead of the actual configuration. In particular, the media formats in the "m=" line, and any associated payload type mappings ("rtpmap"), "fmtp" parameters ("mfcap"), and media-specific attributes ("mscap") MUST be used. Note that this may involve use of concatenation and substitution rules (see Sections 8.3.3.3.2.1 and 8.3.3.3.7). The actual configuration attribute may also be used to infer the lack of acceptability of higher-preference configurations that were not chosen, subject to any constraints provided by a session capability ("sescap") attribute in the offer. Note that the SDP capability negotiation base specification (RFC 5939, see Section 8.1)] requires the answerer to choose the highest-preference configuration it can support, subject to local policies. When the offerer receives the answer, it SHOULD furthermore make note of any capabilities and/or latent configurations included for future use and any constraints on how those may be combined.

8.3.3.4.4 Modifying the Session

If, at a later time, one of the parties wishes to modify the operating parameters of a session (e.g., by adding a new media stream or by changing the properties used on an existing stream) it can do so via the mechanisms defined for offer/answer (RFC 3264, see Section 3.1). If the initiating party has remembered the codecs, potential configurations, latent configurations, and session capabilities provided by the other party in the earlier negotiation, it MAY use this knowledge to maximize the likelihood of a successful modification of the session. Alternatively, the initiator MAY perform a new capabilities exchange as part of the reconfiguration. In such a case, the new capabilities will replace the previously negotiated capabilities. This may be useful if conditions change on the endpoint.

8.3.4 Examples

In this section, we provide examples showing how to use the media capabilities with the SDP capability negotiation.

8.3.4.1 Alternative Codecs

This example provides a choice of one of six variations of the AMR codec. In this example, the default configuration as specified by the media line is the same as the most preferred configuration. Each configuration uses a different payload type number so the offerer can interpret early media.

```
v=0
o=- 25678 753849 IN IP4 192.0.2.1
s=
c=IN IP4 192.0.2.1
t=0 0
a=creq:med-v0
m=audio 54322 RTP/AVP 96
a=rtpmap:96 AMR-WB/16000/1
a=fmtp:96 mode-change-capability=1; max-red=220; \
mode-set=0,2,4,7
a=rmcap:1,3,5 audio AMR-WB/16000/1
a=rmcap:2,4,6 audio AMR/8000/1
a=mfcap:1,2,3,4 mode-change-capability=1
a=mfcap:5,6 mode-change-capability=2
a=mfcap:1,2,3,5 max-red=220
a=mfcap:3,4,5,6 octet-align=1
a=mfcap:1,3,5 mode-set=0,2,4,7
a=mfcap:2,4,6 mode-set=0,3,5,6
a=pcfg:1 m=1 pt=1:96
a=pcfg:2 m=2 pt=2:97
a=pcfg:3 m=3 pt=3:98
a=pcfg:4 m=4 pt=4:99
a=pcfg:5 m=5 pt=5:100
a=pcfg:6 m=6 pt=6:101
```

In this example, media capability 1 could have been excluded from the first "rmcap" declaration and from the corresponding "mfcap" attributes, and the "pcfg:1" attribute line could have been simply "pcfg:1." The next example offers a video stream with three options of H.264 and four transports. It also includes an audio stream with different audio qualities: four variations of AMR, or AC3. The offer looks something like the following:

```
v=0
o=- 25678 753849 IN IP4 192.0.2.1
s=An SDP Media NEG example
```

```
c=IN IP4 192.0.2.1
t=0 0
a=creq:med-v0
a=ice-pwd:speEc3QGZiNWpVLFJhQX
m=video 49170 RTP/AVP 100
c=IN IP4 192.0.2.56
a=maxprate:1000
a=rtcp:51540
a=sendonly
a=candidate 12345 1 UDP 9 192.0.2.56 49170 host
a=candidate 23456 2 UDP 9 192.0.2.56 51540 host
a=candidate 34567 1 UDP 7 198.51.100.1 41345 srflx raddr \
      192.0.2.56 rport 49170
a=candidate 45678 2 UDP 7 198.51.100.1 52567 srflx raddr \
      192.0.2.56 rport 51540
a=candidate 56789 1 UDP 3 192.0.2.100 49000 relay raddr \
      192.0.2.56 rport 49170
a=candidate 67890 2 UDP 3 192.0.2.100 49001 relay raddr \
      192.0.2.56 rport 51540
b=AS:10000
b=TIAS:10000000
b=RR:4000
b=RS:3000
a=rtpmap:100 H264/90000
a=fmtp:100 profile-level-id=42A01E; packetization-mode=2; \
      sprop-parameter-sets=Z0IACpZTBYmI,aMljiA==; \
      sprop-interleaving-depth=45; sprop-deint-buf-req=64000; \
      sprop-init-buf-time=102478; deint-buf-cap=128000
a=tcap:1 RTP/SAVPF RTP/SAVP RTP/AVPF
a=rmcap:1-3,7-9 H264/90000
a=rmcap:4-6 rtx/90000
a=mfcap:1-9 profile-level-id=42A01E
a=mfcap:1-9 aMljiA==
a=mfcap:1,4,7 packetization-mode=0
a=mfcap:2,5,8 packetization-mode=1
a=mfcap:3,6,9 packetization-mode=2
a=mfcap:1-9 sprop-parameter-sets=Z0IACpZTBYmI
a=mfcap:1,7 sprop-interleaving-depth=45; \
      sprop-deint-buf-req=64000; sprop-init-buf-time=102478; \
      deint-buf-cap=128000
a=mfcap:4 apt=100
a=mfcap:5 apt=99
a=mfcap:6 apt=98
a=mfcap:4-6 rtx-time=3000
a=mscap:1-6 rtcp-fb nack
a=acap:1 crypto:1 AES_CM_128_HMAC_SHA1_80 \
      inline:d0RmdmcmVCspeEc3QGZiNWpVLFJhQX1cfHAwJSoj|220|1:32
a=pcfg:1 t=1 m=1,4 a=1 pt=1:100,4:97
a=pcfg:2 t=1 m=2,5 a=1 pt=2:99,4:96
a=pcfg:3 t=1 m=3,6 a=1 pt=3:98,6:95
a=pcfg:4 t=2 m=7 a=1 pt=7:100
a=pcfg:5 t=2 m=8 a=1 pt=8:99
a=pcfg:6 t=2 m=9 a=1 pt=9:98
a=pcfg:7 t=3 m=1,3 pt=1:100,4:97
a=pcfg:8 t=3 m=2,4 pt=2:99,4:96
```

```
a=pcfg:9 t=3 m=3,6 pt=3:98,6:95
m=audio 49176 RTP/AVP 101 100 99 98
c=IN IP4 192.0.2.56
a=ptime:60
a=maxptime:200
a=rtcp:51534
a=sendonly
a=candidate 12345 1 UDP 9 192.0.2.56 49176 host
a=candidate 23456 2 UDP 9 192.0.2.56 51534 host
a=candidate 34567 1 UDP 7 198.51.100.1 41348 srflx \
raddr 192.0.2.56 rport 49176
a=candidate 45678 2 UDP 7 198.51.100.1 52569 srflx \
raddr 192.0.2.56 rport 51534
a=candidate 56789 1 UDP 3 192.0.2.100 49002 relay \
      raddr 192.0.2.56 rport 49176
a=candidate 67890 2 UDP 3 192.0.2.100 49003 relay \
      raddr 192.0.2.56 rport 51534
b=AS:512
b=TIAS:512000
b=RR:4000
b=RS:3000
a=maxprate:120
a=rtpmap:98 AMR-WB/16000
a=fmtp:98 octet-align=1; mode-change-capability=2
a=rtpmap:99 AMR-WB/16000
a=fmtp:99 octet-align=1; crc=1; mode-change-capability=2
a=rtpmap:100 AMR-WB/16000/2
a=fmtp:100 octet-align=1; interleaving=30
a=rtpmap:101 AMR-WB+/72000/2
a=fmtp:101 interleaving=50; int-delay=160000;
a=rmcap:14 ac3/48000/6
a=acap:23 crypto:1 AES_CM_128_HMAC_SHA1_80 \
      inline:d0RmdmcmVCspeEc3QGZiNWpVLFJhQX1cfHAwJSoj|220|1:32
a=tcap:4 RTP/SAVP
a=pcfg:10 t=4 a=23
a=pcfg:11 t=4 m=14 a=23 pt=14:102
```

This offer illustrates the advantage in compactness that arises if one can avoid deleting the base configuration attributes and recreating them in "acap" attributes for the potential configurations.

8.3.4.2 Alternative Combinations of Codecs (Session Configurations)

If an endpoint has limited signal processing capacity, it might be capable of supporting, say, a G.711 mu-law audio stream in combination with an H.264 video stream or a G.729B audio stream in combination with an H.263-1998 video stream. It might then issue an offer like the following:

```
v=0
o=- 25678 753849 IN IP4 192.0.2.1
s=
c=IN IP4 192.0.2.1
t=0 0
a=creq:med-v0
a=sescap:1 2,4
```

```
a=sescap:2 1,3
m=audio 54322 RTP/AVP 18
a=rtpmap:18 G729/8000
a=fmtp:18 annexb=yes
a=rmcap:1 PCMU/8000
a=pcfg:1 m=1 pt=1:0
a=pcfg:2
m=video 54344 RTP/AVP 100
a=rtpmap:100 H263-1998/90000
a=rmcap:2 H264/90000
a=mfcap:2 profile-level-id=42A01E; packetization-mode=2
a=pcfg:3 m=2 pt=2:101
a=pcfg:4
```

Note that the preferred session configuration (and the default as well) is G.729B with H.263. This overrides the individual media stream preferences that are PCMU and H.264 by the potential configuration numbering rule.

8.3.4.3 Latent Media Streams

Consider a case in which the offerer can support either G.711 mu-law or G.729B, along with DTMF telephony events for the 12 common touchtone signals but is willing to support simple G.711 mu-law audio as a last resort. In addition, the offerer wishes to announce its ability to support video and Message Session Relay Protocol (MSRP) in the future but does not wish to offer a video stream or an MSRP stream at present. The offer might look like the following:

```
v=0
o=- 25678 753849 IN IP4 192.0.2.1
s=
c=IN IP4 192.0.2.1
t=0 0
a=creq:med-v0
m=audio 23456 RTP/AVP 0
a=rtpmap:0 PCMU/8000
a=rmcap:1 PCMU/8000
a=rmcap:2 G729/8000
a=rmcap:3 telephone-event/8000
a=mfcap:3 0-11
a=pcfg:1 m=1,3|2,3 pt=1:0,2:18,3:100
a=lcfg:2 mt=video t=1 m=10|11
a=rmcap:10 H263-1998/90000
a=rmcap:11 H264/90000
a=tcap:1 RTP/AVP
a=lcfg:3 mt=message t=2 m=20
a=tcap:2 TCP/MSRP
a=omcap:20 *
```

The first "lcfg" attribute line ("lcfg:2") announces support for H.263 and H.264 video (H.263 preferred) for future negotiation. The second "lcfg" attribute line ("lcfg:3") announces support for MSRP for future negotiation. The "m=" line and the "rtpmap" attribute offer an audio stream and provide the lowest precedence configuration (PCMU without any DTMF encoding). The rmcap lines define the RTP-based media format capabilities (PCMU, G.729, telephone-event,

H.263-1998, and H.264) and the omcap line defines the non-RTP-based media format capability (wildcard). The "mfcap" attribute provides the format parameters for telephone-event, specifying the 12 commercial DTMF "digits." The "pcfg" attribute line defines the most-preferred media configuration as PCMU plus DTMF events and the next-most-preferred configuration as G.729B plus DTMF events. If the answerer is able to support all the potential configurations and support H.263 video (but not H.264), it would reply with an answer like the following:

```
v=0
o=- 24351 621814 IN IP4 192.0.2.2
s=
c=IN IP4 192.0.2.2
t=0 0
a=csup:med-v0
m=audio 54322 RTP/AVP 0 100
a=rtpmap:0 PCMU/8000
a=rtpmap:100 telephone-event/8000
a=fmtp:100 0-11
a=acfg:1 m=1,3 pt=1:0,3:100
a=pcfg:1 m=2,3 pt=2:18,3:100
a=lcfg:2 mt=video t=1 m=10
```

The "lcfg" attribute line announces the capability to support H.263 video at a later time. The media line and subsequent "rtpmap" and "fmtp" attribute lines present the selected configuration for the media stream. The "acfg" attribute line identifies the potential configuration from which it was taken and the "pcfg" attribute line announces the potential capability to support G.729 with DTMF events as well. If, at some later time, congestion becomes a problem in the network, either party may, with expectation of success, offer a reconfiguration of the media stream to use G.729 in order to reduce packet sizes.

8.3.5 IANA Considerations

SDP parameters that are described here have been registered with IANA. We have not included the IANA registration procedures here for the sake of brevity. The procedures for registration of new SDP parameters are described in RFC 6871.

8.3.6 Security Considerations

The security procedures that need to be considered in dealing with the SDP messages described in this section (RFC 6871) are provided in Section 17.7.3.

8.4 Miscellaneous Capabilities Negotiation in SDP

This section describes RFC 7006 that extends the SDP capability negotiation mechanism transport protocols and attributes framework. RFC 7006 provides a media capabilities negotiation mechanism that allows endpoints to negotiate additional media-related capabilities. This negotiation is embedded into the widely used SDP offer/answer procedures. This memo extends the SDP capability negotiation framework to allow endpoints to negotiate three additional SDP capabilities. In particular, this RFC provides a mechanism to negotiate bandwidth ("b=" line), connection data ("c=" line), and session or media titles ("i=" line for each session or media).

8.4.1 Introduction

In this subsection, we describe the miscellaneous capabilities negotiation in SDP specified in RFC 7006. The SDP (RFC 4566, see Section 2) is intended for describing multimedia sessions for the purposes of session announcement, session invitation, and other forms of multimedia session initiation. SDP has been extended with an SDP capability negotiation mechanism framework (RFC 5939, see Section 8.1) that allows the endpoints to negotiate capabilities, such as support for the RTP (RFC 3550) and the SRTP (RFC 3711). The SDP media capabilities negotiation (RFC 6871, see Section 8.3) provides negotiation capabilities to media lines as well. The capability negotiation is embedded into the widely used SDP offer/answer procedure (RFC 3264, see Section 3.1).

This memo provides the means to negotiate further capabilities than those specified in the SDP capability negotiation mechanism framework (RFC 5939 see Section 8.1) and the SDP media capabilities negotiation (RFC 6871, see Section 8.3). In particular, this memo provides a mechanism to negotiate bandwidth ("b="), connection data ("c="), and session or media titles ("i="). Since the three added capabilities are independent, it is not expected that implementations will necessarily support all of them at the same time. Instead, it is expected that applications will choose their needed capability for their specific purpose. For this reason, the normative part pertaining to each capability is in a self-contained section: Section 8.4.3.1.1 describes the bandwidth capability extension, Section 8.4.3.1.2 describes the connection data capability extension, and Section 8.4.3.1.3 describes the title capability extension. Separate SDP capability negotiation option-tags are defined for each capability, allowing endpoints to indicate and/or require support for these extensions according to procedures defined in SDP capability negotiation (RFC 5939, see Section 8.1).

8.4.2 Conventions Used in This Document

The conventions used in this document are according to IETF RFC 2119.

8.4.3 Protocol Description

All SDP ABNF syntaxes are provided in Section 2.9. However, we have repeated this here for convenience.

8.4.3.1 Extensions to SDP

The SDP capability negotiation framework (RFC 5939, see Section 8.1) and the SDP media capabilities negotiation (RFC 6871, see Section 8.3) specify attributes for negotiating SDP capabilities. These documents specify new attributes (e.g., "acap," "tcap," "rmcap," and "omcap") for achieving their purpose. In this document, we define three new additional capability attributes for SDP lines of the general form:

```
type=value
```

for types "b," "c," and "i." The corresponding capability attributes are respectively defined as

- "bcap": bandwidth capability
- "ccap": connection data capability
- "icap": title capability

From the subrules of the attribute ("a=") line in SDP (RFC 4566, see Section 2), SDP attributes are of the form

```
attribute = (att-field ":" att-value) / att-field
att-field = token
att-value = byte-string
```

Capability attributes use only the "att-field:att-value" form. The new capabilities may be referenced in potential configurations ("a=pcfg") or in latent configurations ("a=lcfg") as productions conforming to the <extension-config-list>, as respectively defined in RFC 5939 (see Section 8.1) and RFC 6871 (see Section 8.3).

```
extension-config-list = ["+"] ext-cap-name "=" ext-cap-list
ext-cap-name = 1*(ALPHA / DIGIT)
;; ALPHA and DIGIT defined in RFC 5234
ext-cap-list = 1*VCHAR ;; VCHAR defined in RFC 5234
```

The optional "+" is used to indicate that the extension is mandatory and MUST be supported in order to use that particular configuration. The new capabilities may also be referenced in actual configurations ("a=acfg") as productions conforming to the <sel-extension-config> defined in RFC 5939 (see Section 8.1).

```
sel-extension-config = ext-cap-name "=" 1*VCHAR
```

The specific parameters are defined in the individual description of each capability below. The "bcap," "ccap," and "icap" capability attributes can be provided at the SDP session and/or media level. According to the SDP capability negotiation (RFC 5939, see Section 8.1), each extension capability must specify the implication of making it part of a configuration at the media level. According to SDP (RFC 4566, see Section 2), "b=," "c=," and "i=" lines may appear at either session or media level. In line with this, the "bcap," "ccap," and "icap" capability attributes, when declared at session level, are to be interpreted as if that attribute were provided with that value at the session level. The "bcap," "ccap," and "icap" capability attributes declared at media level are to be interpreted as if that capability attribute were declared at the media level. For example, extending the example in RFC 6871 (see Section 8.3) with "icap" and "bcap" capability attributes, we get the following SDP (Figure 8.1):

```
v=0
o=-25678 753849 IN IP4 192.0.2.1
s=
c=IN IP4 192.0.2.1
t=0 0
a=bcap:1 CT:200
a=icap:1 Video conference
m=audio 54320 RTP/AVP 0
a=rmcap:1 L16/8000/1
a=rmcap:2 L16/16000/2
a=pcfg:1 m=1|2 pt=1:99,2:98
m=video 66544 RTP/AVP 100
a=rmcap:3 H263-1998/90000
a=rtpmap:100 H264/90000
a=pcfg:10 m=3 pt=3:101 b=1 i=1
```

Figure 8.1 Example SDP offer with bcap and icap defined at session level.

The SDP in Figure 8.1 defines one PCMU audio stream and one H.264 video stream. It also defines two RTP-based media capabilities ("rmcap" numbered 1 and 2), using 16-bit linear (L16) audio at 8 kbps and 16 kbps, respectively, as well as an RTP-based media capability for H.263 video ("rmcap" numbered 3). The RTP-based media capabilities all appear at the media level. The example also contains a single bandwidth capability ("bcap") and a single title capability ("icap"), both defined at session level. According to the definition above, when the capabilities defined in the "bcap" and "icap" attributes are referenced from the potential configuration, in the resulting SDP they are to be interpreted as session-level attributes (but the RTP-based media capabilities are to be interpreted as media-level attributes).

8.4.3.1.1 Bandwidth Capability

According to RFC 4566 (see Section 2), the bandwidth field denotes the proposed bandwidth to be used by the session or media. In this memo, we specify the bandwidth capability attribute, which can also appear at the SDP session and/or media level. The bandwidth field is specified in RFC 4566 (see Section 2) with the following syntax:

```
b=<bwtype>:<bandwidth>
```

where <bwtype> is an alphanumeric modifier giving the meaning of the <bandwidth> figure. In this document, we define a new capability attribute: the bandwidth capability attribute "bcap." This attribute lists bandwidth as capabilities, according to the following definition:

```
"a=bcap:" bw-cap-num 1*WSP bwtype ":" bandwidth CRLF
```

where <bw-cap-num> is a unique integer within all the bandwidth capabilities in the entire SDP, which is used to number the bandwidth capability and can take a value between 1 and 2^31-1 (both included). The other elements are as defined for the "b=" field in SDP (RFC 4566, see Section 2). This format satisfies the general attribute production rules in SDP (RFC 4566), according to the following ABNF (RFC 5234) syntax (Figure 8.2):

Negotiation of bandwidth per media stream can be useful when negotiating media encoding capabilities with different bandwidths.

8.4.3.1.1.1 Configuration Parameters

The SDP capability negotiation framework (RFC 5939, see Section 8.1) provides for the existence of the "pcfg" and "acfg" attributes. The concept is extended by the SDP media capabilities negotiation (RFC 6871, see Section 8.3) with an "lcfg" attribute that conveys latent configurations. Extensions to the "pcfg" and "lcfg" attributes are defined through <extension-config-list>, and extensions to the "acfg" attribute are defined through the <sel-extension-config>, as defined in the SDP capability negotiation (RFC 5939). In this specification, we extend the <extension-config-list> field to be able to convey lists of bandwidth capabilities in latent or potential configurations, according to the following ABNF (RFC 5234) syntax (Figure 8.3):

```
att-field =/ "bcap"
att-value =/ bw-cap-num 1*WSP bwtype ":" bandwidth
bw-cap-num = 1*10(DIGIT) ;; DIGIT defined in RFC 5234
```

Figure 8.2 Syntax of the "bcap" attribute.

```
extension-config-list =/ bandwidth-config-list
bandwidth-config-list = ["+"] "b=" bw-cap-list *(BAR bw-cap-list)
;; BAR defined in RFC 5939 - (see Section 8.1)
bw-cap-list = bw-cap-num *("," bw-cap-num)
bw-cap-num = 1*10(DIGIT) ;; DIGIT defined in RFC 5234
```

Figure 8.3 Syntax of the bandwidth parameter in "lcfg" and "pcfg" attributes.

Each bandwidth capability configuration is a comma-separated list of bandwidth capability attribute numbers where <bw-cap-num> refers to the <bw-cap-num> bandwidth capability numbers defined explicitly earlier in this document and hence MUST be between 1 and 2^31-1 (both included). Alternative bandwidth configurations are separated by a vertical bar ("|"). The above syntax is very flexible, allowing referencing to multiple "b=" lines per configuration, even for the same <bwtype>. While the need for such definitions is not seen, we have not restricted this, as it is not restricted in SDP (RFC 4566, see section 2) either. The bandwidth parameter to the actual configuration attribute ("a=acfg") is formulated as in Figure 8.4:

8.4.3.1.1.2 Option-Tag The SDP capability negotiation framework (RFC 5939, see Section 8.1) allows for capability negotiation extensions to be defined. Associated with each such extension is an option-tag that identifies the extension in question. Hereby, we define a new option-tag "bcap-v0" that identifies support for the bandwidth capability. The endpoints using the "bcap" capability attribute SHOULD add the option-tag to other existing option-tags present in the "csup" and "creq" attributes in SDP, according to the procedures defined in the SDP capability negotiation framework (RFC 5939).

8.4.3.1.2 Connection Data Capability

According to SDP (RFC 4566, see Section 2), the connection data field in SDP contains the connection data, and it has the following syntax:

```
c=<nettype> <addrtype> <connection-address>
```

where <nettype> indicates the network type, <addrtype> indicates the address type, and the <connection-address> is the connection address, which is dependent on the address type. At the moment, network types already defined include "IN," which indicates Internet network type, and "ATM" (RFC 3108, see Section 9.4), used for describing Asynchronous Transfer Mode (ATM) bearer connections. The circuit-switched description in the SDP document (RFC 7195, see Section 9.5) adds a "PSTN" network type for expressing a public switched telephone network (PSTN) circuit switch.

```
            <sel-extension-config> with ext-cap-name = "b"

hence

            sel-extension-config =/ sel-bandwidth-config
            sel-bandwidth-config = "b=" bw-cap-list ;; bw-cap-list

as above.
```

Figure 8.4 Syntax of the bandwidth parameter in "acfg" attributes.

SDP (RFC 4566, see Section 2) permits specification of connection data at the SDP session and/or media level. In order to permit negotiation of connection data at the media level, we define the connection data capability attribute ("a=ccap") in the form:

```
"a=ccap:" conn-cap-num 1*WSP nettype SP addrtype SP connection-
address CRLF
```

where <conn-cap-num> is a unique integer within all the connection capabilities in the entire SDP, which is used to identify the connection data capability and can take a value between 1 and 2^31-1 (both included). The other elements are as defined in RFC 4566 (see Section 2). This format corresponds to the RFC 4566 (see Section 2) attribute production rules, according to the following ABNF (RFC 5234) syntax (Figure 8.5):

The "ccap" capability attribute allows for expressing alternative connections address ("c=") lines in SDP as part of the SDP capability negotiation process. One of the primary use cases for this is offering alternative connection addresses where the <nettype> is "IN" or "PSTN" (i.e., selecting between an IP-based bearer and a circuit-switched bearer). By supporting the "IN" <nettype>, the "ccap" attribute enables the signaling of multiple IPv4 and IPv6 addresses; however, the standards track mechanism for negotiation of alternative IP addresses in SDP is ICE (RFC 5245, Section 3.2). The "ccap" attribute does not change that; hence, the combined set of actual and potential configurations (as defined in RFC 5939, see Section 8.1) for any given media description MUST NOT use the "ccap" attribute to negotiate more than one address with an IN network type (i.e., it is not permissible to select between "IPv4" and "IPv6" address families or different IP addresses within the same IP address family. Figure 8.6 is an example of an SDP offer that includes a "ccap" capability attribute. An audio stream can be set up with an RTP flow or by establishing a circuit-switched audio stream (Figure 8.6):

The example in Figure 8.6 represents an SDP offer indicating an audio flow using RTP, such as the one represented in Figure 8.7, or an audio flow using a circuit-switched connection, such as the one represented in Figure 8.8.

```
att-field =/ "ccap"
att-value =/ conn-cap-num 1*WSP nettype SP addrtype
SP connection-address
conn-cap-num = 1*10(DIGIT) ;; 1 to 2^31-1, inclusive
;; DIGIT defined in RFC 5234
```

Figure 8.5 Syntax of the "ccap" attribute.

```
v=0
o=2987933123 2987933123 IN IP4 198.51.100.7
s=-
t=0 0
a=creq:med-v0,ccap-v0
m=audio 38902 RTP/AVP 0 8
c=IN IP4 198.51.100.7
a=ccap:1 PSTN E164 +15555556666
a=tcap:2 PSTN
a=omcap:1 -
a=acap:1 setup:actpass
a=acap:2 connection:new
a=acap:3 cs-correlation:callerid:+15555556666
a=pcfg:1 c=1 t=2 m=1 a=1,2,3
```

Figure 8.6 Example SDP offer with a "ccap" attribute.

```
v=0
o=2987933123 2987933123 IN IP4 198.51.100.7
s=-
t=0 0
m=audio 38902 RTP/AVP 0 8
c=IN IP4 198.51.100.7
```

Figure 8.7 Equivalent SDP offer with the RTP flow.

```
v=0
o=2987933123 2987933123 IN IP4 198.51.100.7
s=-
t=0 0
m=audio 9 PSTN -
c=PSTN E164 +15555556666
a=setup:actpass
a=connection:new
a=cs-correlation:callerid:+15555556666
```

Figure 8.8 Equivalent SDP offer with the circuit-switched flow.

This document does not define any mechanism for negotiating or describing different port numbers; hence, the port number from the "m=" line MUST be used by default. Exceptions to this default can be provided by extension mechanisms or network type specific rules. This document defines an exception when the network type is "PSTN," in which case the port number is replaced with 9 (the "discard" port), as described in "SDP Extension for Setting Audio and Video Media Streams over Circuit-Switched Bearers in the PSTN" (RFC 7195, see Section 9.5). An endpoint offering alternative IP and PSTN bearers MUST include the IP media description in the actual configuration (IP address in the "c=" line and port number in the "m=" line) and the PSTN media description in the potential configuration. Exceptions for other network types, such as for the "ATM" network type defined in RFC 3108 (see Section 9.4), require additional specifications.

8.4.3.1.2.1 Configuration Parameters The SDP capability negotiation framework (RFC 5939) provides for the existence of the "pcfg" and "acfg" attributes, which can convey one or more configurations to be negotiated. The concept is extended by the SDP media capabilities negotiation (RFC 6871, see Section 8.3) with an "lcfg" attribute that conveys latent configurations. In this document, we define a <connection-config> parameter to be used to specify a connection data capability in a potential or latent configuration attribute. The parameter follows the form of an (Figure 8.9):

```
    <extension-config-list>

with

    ext-cap-name = "c"
    ext-cap-list = conn-cap-list

where, according to the following Augmented Backus-Naur Form (RFC 5234) syntax:

        extension-config-list =/ conn-config-list
        conn-config-list = ["+"] "c=" conn-cap-list
        conn-cap-list = conn-cap-num *(BAR conn-cap-num)
        conn-cap-num = 1*10(DIGIT) ;; 1 to 2^32-1 inclusive
```

Figure 8.9 Syntax of the connection data parameter in "lcfg" and "pcfg" attributes.

Each capability configuration alternative contains a single connection data capability attribute number and refers to the conn-cap-num capability number defined explicitly earlier in this document; hence, the values MUST be between 1 and 2^31-1 (both included). The connection data capability allows the expression of only a single capability in each alternative, rather than a list of capabilities, since no more than a single connection data field is permitted per media block. Nevertheless, it is still allowed to express alternative potential connection configurations separated by a vertical bar ("|"). An endpoint includes a plus sign ("+") in the configuration attribute to mandate support for this extension. An endpoint that receives this parameter prefixed with a plus sign and does not support this extension MUST treat that potential configuration as not valid. The connection data parameter to the actual configuration attribute ("a=acfg") is formulated as in Figure 8.10:

8.4.3.1.2.2 Option-Tag

The SDP capability negotiation framework (RFC 5939, see Section 8.1) solution allows for capability negotiation extensions to be defined. Associated with each such extension is an option-tag that identifies the extension in question. Hereby, we define a new option-tag of "ccap-v0" that identifies support for the connection data capability. This option-tag SHOULD be added to other existing option-tags present in the "csup" and "creq" attributes in SDP, according to the procedures defined in the SDP capability negotiation framework (RFC 5939, see Section 8.1).

8.4.3.1.3 Title Capability

SDP (RFC 4566, see Section 2) provides for the existence of an information field expressed in the format of the "i=" line, which can appear at the SDP session and/or media level. An "i=" line that is present at the session level is known as the "session name," and its purpose is to convey human-readable textual information about the session. The "i=" line in SDP can also appear at the media level, in which case it is used to provide human-readable information about the media stream to which it is related; for example, it may indicate the purpose of the media stream. The "i=" line is not to be confused with the label attribute ("a=label:") (RFC 4574, see Section 5.1), which provides a machine-readable tag. It is foreseen that applications declaring capabilities related to different configurations of a media stream may need to provide different identifying information for each of those configurations.

That is, a party might offer alternative media configurations for a stream, each of which represents a different presentation of the same or similar information. For example, an audio stream might offer English or Spanish configurations, or a video stream might offer a choice of video

```
<sel-extension-config>

with

ext-cap-name = "c"

hence

sel-extension-config =/ sel-connection-config
sel-connection-config = "c=" conn-cap-num ;;

as defined above.
```

Figure 8.10 Syntax of the connection data parameter in "acfg" attributes.

source such as speaker camera, group camera, or document viewer. The title capability is needed to inform the answering user in order to select the proper choice, and the label is used to inform the offering machine which choice the answerer has selected. Hence, there is value in defining a mechanism to provide titles of media streams as capabilities.

As defined in SDP (RFC 4566, see Section 2), the session information field ("i=," referred to as "title" in this document) is subject to the "a=charset" attribute in order to support different character sets and hence internationalization. The title capability attribute itself ("a=icap") is, however, not subject to the "a=charset" attribute as this would preclude the inclusion of alternative session/title information each using different character sets. Instead, the session/title value embedded in an "a=icap" attribute (title capability) will be subject to the "a=charset" value used within a configuration that includes that title capability. This provides for consistent SDP operation while allowing for capabilities and configurations with different session/title information values with different character set encodings (with each such configuration including an "a=charset" value with the relevant character set for the session/title information in question).

According to SDP (RFC 4566, see Section 2), the session information ("i=") line has the following syntax:

```
"i=" text
```

where "text" represents a human-readable text indicating the purpose of the session or media stream. In this document, we define a new capability attribute: the title capability "icap." This attribute lists session or media titles as capabilities, according to the following definition:

```
"a=icap:" title-cap-num 1*WSP text
```

where <title-cap-num> is a unique integer within all the connection capabilities in the entire SDP, which is used to identify the particular title capability and can take a value between 1 and 2^31-1 (both included). <text> is a human-readable text that indicates the purpose of the session or media stream it is supposed to characterize. As an example, one might use:

```
a=icap:1 Document Camera
```

to define a title capability number 1 to identify a particular source of a media stream. Or, in another example, one might offer two title capabilities with different character encodings (using the charset attribute defined in "SDP: Session Description Protocol" (RFC 4566, see Section 2) and the generic attribute capability attribute ("a=acap:") defined in "Session Description Protocol (SDP) Capability Negotiation" RFC 5939, see Section 8.1).

```
a=icap:1 El bar y la cafetería de José
a=acap:1 charset:ISO-8859-1
a=icap:2 Joe's CafÃ© & Bar
a=acap:2 charset:UTF-8
```

The title capability attribute satisfies the general attribute production rules in SDP (RFC 4566, see Section 2), according to the following ABNF (RFC 5234) syntax (Figure 8.11):

8.4.3.1.3.1 Configuration Parameters

The SDP capability negotiation framework (RFC 5939, see Section 8.1) provides for the existence of the "pcfg" and "acfg" attributes. The concept is extended by the SDP media capabilities negotiation (RFC 6871, see Section 8.3) with an "lcfg" attribute that conveys latent configurations. In this document, we define a <title-config-list> parameter to be used to convey title capabilities in a potential or latent configuration. This parameter is defined as an <extension-config-list> with the following associations (Figure 8.12):

Each potential capability configuration contains a single title capability attribute number where "title-cap-num" is the title capability number defined explicitly earlier in this document, and hence must be between 1 and 2^31-1 (both included). The title capability allows the expression of only a single capability in each alternative, since no more than a single-title field is permitted per block. Nevertheless, it is still allowed to express alternative potential title configurations separated by a vertical bar ("|"). An endpoint includes a plus sign ("+") in the configuration attribute to mandate support for this extension. An endpoint that receives this parameter prefixed with a plus sign and does not support this extension MUST treat that potential configuration as not valid. The title parameter to the actual configuration attribute ("a=acfg") is formulated as a <sel-extension-config> with (Figure 8.13):

8.4.3.1.3.2 Option-Tag

At present, it is difficult to envision a scenario in which the "icap" attribute must be supported or the offer must be rejected. In most cases, if the icap attribute or its contents were to be ignored, an offered configuration could still be chosen based on other criteria such as configuration numbering. However, one might imagine an SDP offer that contained English and Spanish potential configurations for an audio stream. The session might be unintelligible if the choice is based on configuration numbering, rather than informed user selection.

```
att-field =/ "icap"
att-value =/ title-cap-num 1*WSP text
;; text defined in RFC 4566
title-cap-num = 1*10(DIGIT) ;; DIGIT defined in RFC 5234
```

Figure 8.11 Syntax of the "icap" attribute.

```
ext-cap-name = "i"
ext-cap-list = title-cap-list

This leads to the following definition for the title capability parameter:

extension-config-list =/ title-config-list
title-config-list = ["+"] "i=" title-cap-list
title-cap-list = title-cap-num *(BAR title-cap-num)
;; BAR defined in RFC 5939
title-cap-num = 1*10(DIGIT) ;; DIGIT defined in RFC 5234
```

Figure 8.12 Syntax of the title capability parameter in "lcfg" and "pcfg" attributes.

```
ext-cap-name = "i"

hence

sel-extension-config =/ sel-title-config
sel-title-config = "i=" title-cap-num ;;

as defined above.
```

Figure 8.13 Syntax of the title parameter in "acfg" attributes.

Based on such considerations, it may well prove useful to announce the ability to use the icap attribute and its contents to select media configurations or to inform the user about the selected configuration(s). Therefore, we define a new option-tag of "icap-v0" that identifies support for the title capability. This option-tag SHOULD be added to other existing option-tags present in the "csup" and/or "creq" attribute in SDP, according to the procedures defined in the SDP capability negotiation framework (RFC 5939, see Section 8.1). The discussion throughout this specification suggests that "icap-v0" will typically appear in a "csup" attribute but rarely in a "creq" attribute.

8.4.3.2 Session Level versus Media Level

The "bcap," "ccap," and "icap" attributes can appear at the SDP session and/or media level. Endpoints MUST interpret capabilities declared at session level as part of the session level in the resulting SDP for that particular configuration. Endpoints MUST interpret capabilities declared at media description as part of the media level in the resulting SDP for that particular configuration. The presence of the "bcap" capability for the same <bwtype> at both the session and media level is subject to the same constraints and restrictions specified in RFC 4566 (see Section 2) for the bandwidth attribute "b=." To avoid confusion, the <type-attr-num> for each "a=bcap," "a=ccap," and "a=icap" line MUST be unique across all capability attributes of the same type within the entire session description.

8.4.3.3 Offer/Answer Model Extensions

In this section, we define extensions to the offer/answer model defined in SDP offer/answer model (RFC 3264, see Section 3.1) and extended in the SDP capability negotiation framework (RFC 5939, see Section 8.1) to allow for bandwidth, connection, and title capabilities to be used with the SDP capability negotiation framework.

8.4.3.3.1 Generating the Initial Offer

When an endpoint generates an initial offer and wants to use the functionality described in the current document, it first defines appropriate values for the bandwidth, connection data, and/or title capability attributes according to the rules defined in RFC 4566 (see Section 2) for "b=," "c=," and "i=" lines. The endpoint then MUST include the respective capability attributes and associated values in the SDP offer. The preferred configurations for each media stream are identified following the media line in a "pcfg" attribute. Bandwidth and title capabilities may also be referenced in latent configurations in an "lcfg" attribute, as defined in the SDP media capabilities negotiation (RFC 6871, see Section 8.3). Implementations are advised to pay attention to the port number that is used in the "m=" line. By default, a potential configuration that includes a connection data capability will use the port number from the "m=" line, unless the network type is "PSTN," in which case the default port number used will be 9. The offer SHOULD include the level of capability negotiation extensions needed to support this functionality in a "creq" attribute.

8.4.3.3.2 Generating the Answer

When the answering party receives the offer, and if it supports the required capability negotiation extensions, it SHOULD select the most preferred configuration it can support for each media stream and build the answer accordingly, as defined in Section 8.1.3.6.2 of this book (that is,

Section 3.6.2 of the SDP capability negotiation RFC 5939). If the connection data capability is used in a selected potential configuration chosen by the answerer, that offer configuration MUST by default use the port number from the actual offer configuration (i.e., the "m=" line), unless the network type is "PSTN," in which case the default port MUST be assumed to be 9. Extensions may be defined to negotiate the port number explicitly instead.

8.4.3.3.3 Offerer Processing of the Answer

When the offerer receives the answer, it MUST process the media lines according to normal SDP processing rules to identify the media stream(s) accepted by the answer, if any, as defined in Section 8.1.3.6.2 of this book (that is, Section 3.6.3 of the SDP capability negotiation RFC 5939). The "acfg" attribute, if present, MUST be used to verify the proposed configuration used to form the answer and to infer the lack of acceptability of higher-preference configurations that were not chosen.

8.4.3.3.4 Modifying the Session

If, at a later time, one of the parties wishes to modify the operating parameters of a session (e.g., by adding a new media stream or by changing the properties used on an existing stream) it may do so via the mechanisms defined for SDP offer/answer (RFC 3264, see Section 3.1) and in accordance with the procedures defined in Section 8.1.3.6.4 of this book (that is, Section 3.6.4 of RFC 5939).

8.4.4 Field Replacement Rules

To simplify the construction of SDP records, given the need to include fields within the media description in question for endpoints that do not support capabilities negotiation, we define some simple field-replacement rules for those fields invoked by potential or latent configurations. In particular, any "i=" or "c=" lines invoked by a configuration MUST replace the corresponding line, if present within the media description in question. Any "b=" line invoked by a configuration MUST replace any "b=" of the same bandwidth type at the media level, but not at the session level.

8.4.5 Security Considerations

The security procedures that need to be considered in dealing with the SDP miscellaneous capability negotiation messages described in this section (RFC 7006) are provided in Section 17.7.4.

8.4.6 IANA Considerations

SDP parameters that are described here have been registered with IANA. We have not included the IANA registration procedures here for the sake of brevity. The procedures for registration of new SDP parameters are described in RFC 7006.

8.5 Summary

We describe the most important enhancement of SDP functionalities that enable capability attribute negotiations between the conferencing parties using offer/answer procedures (RFC 5939, see Section 8.1) for establishment of the session. These capability attributes may include protocol version, media, direction, transport protocol, configuration, session-level and media-level security, and other parameters. In addition, the interaction with ICE protocol that helps for transparent NAT crossing by media in both directions is also considered as a part of session negotiations.

The video application can have many more complex parameters for each video codec-type (e.g. H.263, H.264, and MPEG-4) in addition to layered and nonlayered transmission considerations. We have described a new generic session setup attribute (RFC 6236, see Section 8.2) to make it possible to negotiate different image attributes such as image size using offer/answer procedures along with specific examples.

In addition, the capability negotiating framework has been extended further by defining generic media capabilities (RFC 6871, see Section 8.3) such as media format, media format parameter, media-specific capability, media configuration parameter, latent configuration, enhanced potential configuration, and other attributes that can be used to negotiate media types and their associated parameters including some examples. Finally, we have discussed how detailed performance parameters of each capability attribute of bandwidth, connection data, and session/media titles (RFC 7006, see Section 8.4) can be negotiated using the offer/answer model.

8.6 Problems

1. What is the historical reason in the development of SDP that it did not have the capability-attributes negotiation capabilities? How does the offer/answer model enhance the SDP for capability attributes negotiation?
2. Provide detail call follows in SIP that deal with generic features, transport protocols, and configurations capability attributes negotiation between the two conferencing parties.
3. Expand Exercise 2 to include interactions with ICE, SIP option-tags, and processing media before answer and bandwidth usage.
4. Expand Exercises 1 & 2 involving large number of potential configurations (i.e. media-level description with a large number of instances).
5. Expand Exercises 2–4 to include intermediaries (e.g., session border controllers) between two conferencing parties.
6. Expand Exercises 2–5 with additional capability attributes: rtpmap, fmtp, and direction.
7. Expand Exercises 2–6 for the DTLS-SRTP or SRTP with media-level security description.
8. Expand Exercises 2–6 for the best-effort SRTP with session-level MIKEY and media-level security description.
9. Expand Exercises 2–6 for the SRTP with session-level MIKEY and media-level security description as alternatives.
10. Provide the detail call flows for image codec parameter attributes negotiations using the offer/answer model for H.263, H.264, and MPEG-4 video along with asymmetry, send-only and receive-only, and aspect ratio. Extend this for layered codecs.
11. Describe in detail all features and capabilities that can be negotiated for different media types and their associated parameters standardized in SDP (RFC 6871). Expand Exercises 2–9 for alternative codecs, alternative combination of codecs, and latent media stream streams described in Section 8.3.5.

Reference

1. Kaplan, H. and F. Audet, "Session Description Protocol (SDP) Offer/Answer Negotiation For Best-Effort Secure Real-Time Transport Protocol", Work in Progress, October 2006.

Chapter 9

Network Protocol Support in SDP

Abstract

In this section, we describe Requests for Comments (RFCs) 4566, 6947, 4570, 3108, and 7195 that provide the Session Description Protocol (SDP) capability negotiation capabilities for different network protocols. Specifically, this section describes the SDP's support for network protocols such as Internet Protocol (IP) Version 4 and 6 (IPv4 and IPv6) defined in RFCs 4566 and 6947, Asynchronous Transfer Mode (ATM) specified in RFC 3108, and Public Switched Telephone Network (PSTN) standardized in RFC 7195. Note that RFC 4566 (see Section 2) obsoletes both RFCs 3266 and 2327 for support of IPv4 and IPv6 network address in SDP. More importantly, how a given SDP offer can carry multiple IP addresses of different address families (e.g., IPv4 and IPv6) specified in RFC 6947 is described here. The proposed attribute, the "altc" attribute, solves the backwards compatibility problem that plagued ANAT due to its syntax. This solution is applicable to scenarios where connectivity checks are not required. If connectivity checks are required, Interactive Connectivity Establishment, as specified in RFC 5245, provides such a solution. The SDP's syntax is also extended to express one or more source addresses as a source filter for one or more destination "connection" addresses. RFC 4570 defines the syntax and semantics for an SDP "source-filter" attribute that may reference either IPv4 or IPv6 address(es) as either an inclusive or exclusive source list for either multicast or unicast destinations. In particular, an inclusive source filter can be used to specify a source-specific multicast session. In case of ATM (RFC 3108), this section describes conventions for using the SDP described in RFCs 3266 and 2327 (that are made obsolete by RFC 4566, see Section 2) for controlling ATM Bearer Connections, and any associated ATM Adaptation Layer. In addition, this section deals with use cases, requirements, and protocol extensions for using the SDP offer/answer model for establishing audio and video media streams over circuit-switched bearers in PSTN (RFC 7195).

9.1 Support for IPv4 and IPv6 in SDP

This section describes the use of IPv4/IPv6 addresses in Session Description Protocol (SDP). Note that Request for Comments (RFC) 4566 (see Section 2) made RFC 3266 obsolete.

9.1.1 Introduction

SDP is intended for describing multimedia sessions for the purposes of session announcement, session invitation, and other forms of multimedia session initiation. It is a text-format description that provides many details of a multimedia session including the originator of the session, a Uniform Resource Locator (URL) related to the session, the connection-address for the session media(s), and optional attributes for the session media(s). Each of these pieces of information may involve one or more IPv6 addresses. The Augmented Backus-Naur Form (ABNF) for IPv4 and IPv6 addresses in SDP is defined in RFC 4566 (see Section 2.9.2) that obsoletes RFCs 3266 and 2327. Recently, RFC 4566bis [1] that has not yet been standardized in the Internet Engineering Task Force (IETF) has suggested some changes in ABNF syntaxes of IPv6 and IPv4, and we have included its SDP ABNF syntaxes in Section 2.10.1.

9.1.2 Notation

The notations and conventions that are used here are as per IETF RFC 2119.

9.1.3 Syntax

The detail ABNF for IPv4 and IPv6 addresses for SDP is defined by RFC 4566 (see Section 2.9.2) that obsoletes RFCs 3266 and 2327. Recently, RFC 4566bis [1] that has not yet been standardized in the IETF has suggested some changes in ABNF syntaxes of IPv6 and IPv4, and we have included its SDP ABNF syntaxes in Section 2.10.1.

9.1.4 Example SDP Description with IPv6 Addresses

The following is an example SDP description using the ABNF for IPv6 addresses. In particular, the origin and connection fields contain IPv6 addresses.

```
v=0
o=nasa1 971731711378798081 0 IN IP6 2201:056D::112E:144A:1E24
s=(Almost) live video feed from Mars-II satellite
p=+1 713 555 1234
c=IN IP6 FF1E:03AD::7F2E:172A:1E24
t=3338481189 3370017201
m=audio 6000 RTP/AVP 2
a=rtpmap:2 G726-32/8000
m=video 6024 RTP/AVP 107
a=rtpmap:107 H263-1998/90000
```

9.1.5 Security Considerations

The security procedures that need to be considered for SDP messages dealing with IPv4/IPv6 addresses described in this section (RFC 4566) are provided in Section 17.8.1.

9.1.6 Internet Assigned Numbers Authority (IANA) Considerations

The updated definition of the IP6 addrtype parameter is provided in RFC 4566 (see Section 2) that obsoletes RFC 2327.

9.2 SDP Alternate Connectivity (ALTC) Attribute

This section describes RFC 6947 that proposes a mechanism that allows the same SDP offer to carry multiple IP addresses of different address families (e.g., IPv4 and IPv6). The proposed attribute, the "altc" attribute, solves the backwards compatibility problem that plagued ANAT due to their syntax. The proposed solution is applicable to scenarios where connectivity checks are not required. If connectivity checks are required, Interactive Connectivity Establishment (ICE), as specified in RFC 5245 (see Section 3.2), provides such a solution

9.2.1 Introduction

9.2.1.1 Overall Context

In this section, we describe the SDP ALTC attribute defined in RFC 6947. Due to the IPv4 address exhaustion problem, IPv6 deployment is becoming an urgent need, along with the need to properly handle the coexistence of IPv6 and IPv4. The reality of IPv4-IPv6 coexistence introduces heterogeneous scenarios with combinations of IPv4 and IPv6 nodes, some of which are capable of supporting both IPv4 and IPv6 dual-stack (DS) and some of which are capable of supporting only IPv4 or only IPv6. In this context, Session Initiation Protocol (SIP) (RFC 3261, also see Section 1.2) User Agents (UAs) need to be able to indicate their available IP capabilities in order to increase the ability to establish successful SIP sessions, to avoid invocation of adaptation functions such as Application Layer Gateways (ALGs) and IPv4-IPv6 interconnection functions (e.g., NAT64 - RFC 6146) and to avoid using private IPv4 addresses through consumer network address translators (NATs) or Carrier-Grade NATs (CGNs) (RFC 6888).

In the meantime, service providers are investigating scenarios to upgrade their service offering to be IPv6 capable. The current strategies involve either offering IPv6 only, for example, to mobile devices, or providing both IPv4 and IPv6, but with private IPv4 addresses that are NATed by CGNs. In the latter case, the end device may be using "normal" IPv4 and IPv6 stacks and interfaces, or it may tunnel the IPv4 packets though a Dual-Stack Lite (DS-Lite) stack that is integrated into the host (RFC 6333). In either case, the device has both address families available from an SIP and media perspective. Regardless of the IPv6 transition strategy being used, it is obvious that there will be a need for DS SIP devices to communicate with IPv4-only legacy UAs, IPv6-only UAs, and other DS UAs.

It may not be possible, for example, for a DS UA to communicate with an IPv6-only UA unless the DS UA has a means of providing the IPv6-only UA with an IPv6 address, while clearly it needs to provide a legacy IPv4-only device with an IPv4 address. The communication must be possible in a backwards compatible fashion, such that IPv4-only SIP devices need not support the new mechanism to communicate with DS UAs. The current means by which multiple address families can be communicated are through ANAT (RFC 5245 obsoletes RFC 4091) or ICE (RFC 5245, see Section 3.2). ANAT has serious backwards compatibility problems, as described in RFC 4092, which effectively make it unusable, and it has been deprecated by the IETF (RFC 5245). ICE at least allows interoperability with legacy devices. ICE is a complicated and processing-intensive mechanism; it has seen limited deployment and implementation in SIP applications. ALTC has been implemented as reported in RFC 7225. No issues have been reported in that document.

9.2.1.2 Purpose

This document proposes a new alternative: a backwards compatible syntax for indicating multiple media connection-addresses and ports in an SDP offer, which can immediately be selected from and used in an SDP answer. The proposed mechanism is independent of the model described in RFC 5939 (see Section 8.1) and does not require implementation of SDP capability negotiations to function. It should be noted that "backwards compatible" in this document generally refers to working with legacy IPv4-only devices. The choice has to be made, one way or the other, because to interoperate with legacy devices requires constructing SDP bodies that they can understand and support, such that they detect their local address family in the SDP connection line. It is not possible to support interworking with both legacy IPv4-only and legacy IPv6-only devices with the same SDP offer. Clearly, there are far more legacy IPv4-only devices in existence, and thus those are the ones assumed in this document. However, the syntax allows for a UA to choose which address family to be backwards compatible with, in case it has some means of determining it.

Furthermore, even for cases where both sides support the same address family, there should be a means by which the "best" address family transport is used, based on what the UAs decide. The address family that is "best" for a particular session cannot always be known a priori. For example, in some cases the IPv4 transport may be better, even if both UAs support IPv6. The proposed solution provides the following benefits:

- It allows a UA to signal more than one IP address (type) in the same SDP offer.
- It is backwards compatible. No parsing or semantic errors will be experienced by a legacy UA or by intermediary SIP nodes that do not understand this new mechanism.
- It is as lightweight as possible to achieve the goal, while still allowing and interoperating with nodes that support other similar or related mechanisms.
- It is easily deployable in managed networks.
- It requires minimal increase of the length of the SDP offer (i.e., minimizes fragmentation risks).

ALTC may also be useful for the multicast context (e.g., RFC 6144 or of [2]). More detailed information about ALTC use cases is provided in Section 9.2.7 of this book (that is, Appendix A of RFC 6947).

9.2.1.3 Scope

This specification proposes an alternative scheme, as a replacement to the ANAT procedure (RFC 5245, see Section 3.2, obsoletes RFC 4091), to carry several IP address types in the same SDP offer while preserving backwards compatibility. While two UAs communicating directly at an SIP layer clearly need to be able to support the same address family for SIP itself, current SIP deployments almost always have proxy servers or back-to-back UAs (B2BUAs) in the SIP signaling path, which can provide the necessary interworking of the IP address family at the SIP layer (e.g., RFC 6157). SIP-layer address family interworking is out of the scope of this document. Instead, this document focuses on the problem of communicating media address family capabilities in a backwards compatible fashion. Because media can go directly between two UAs, without a priori knowledge by the UA Client (UAC) of which address family the far-end UA Server (UAS) supports, it has to offer both, in a backwards compatible fashion.

9.2.1.4 Requirements Language

9.2.2 Use Cases

The ALTC mechanism defined in this document is primarily meant for managed networks. In particular, the following use cases were explicitly considered:

- A DS UAC that initiates an SIP session without knowing the address family of the ultimate target UAS
- A UA that receives an SIP session request with a SDP offer and that wishes to avoid using IPv4 or IPv6
- An IPv6-only UA that wishes to avoid using a NAT64 (RFC 6146)
- An SIP UA behind a DS-Lite CGN (RFC 6333)
- An SIP service provider or enterprise domain of an IPv4-only and/or IPv6-only UA that provides interworking by invoking IPv4-IPv6 media relays and that wishes to avoid invoking such functions and to let media go end to end as much as possible
- An SIP service provider or enterprise domain of a UA that communicates with other domains and that wishes either to avoid invoking IPv4-IPv6 interworking or to let media go end to end as much as possible
- An SIP service provider that provides transit peering services for SIP sessions that may need to modify SDP in order to provide IPv4-IPv6 interworking but would prefer to avoid such interworking or to avoid relaying media in general, as much as possible
- SIP sessions that use the new mechanism when crossing legacy SDP-aware middleboxes but that may not understand this new mechanism.

9.2.3 Overview of the ALTC Mechanism

9.2.3.1 Overview

The ALTC mechanism relies solely on the SDP offer/answer mechanism, with specific syntax to indicate alternative connection-addresses. The basic concept is to use a new SDP attribute, "altc," to indicate the IP addresses for potential alternative connection-addresses. The address that is most likely to get chosen for the session is in the normal "c=" line. Typically, in current operational networks, this would be an IPv4 address. The "a=altc" lines contain the alternative addresses offered for this session. This way, a DS UA might encode its IPv4 address in the "c=" line, while possibly preferring to use an IPv6 address by explicitly indicating the preference order in the corresponding "a=altc" line. One of the "a=altc" lines duplicates the address contained in the "c=" line, for reasons explained in Section 9.2.3.2. The SDP answerer would indicate its chosen address by simply using that address family in the "c=" line of its response. An example of an SDP offer using this mechanism is as follows when IPv4 is considered most likely to be used for the session, but IPv6 is preferred:

```
v=0
o=- 25678 753849 IN IP4 192.0.2.1
s=
c=IN IP4 192.0.2.1
t=0 0
m=audio 12340 RTP/AVP 0 8
a=altc:1 IP6 2001:db8::1 45678
a=altc:2 IP4 192.0.2.1 12340
```

If IPv6 were considered more likely to be used for the session, the SDP offer would be as follows:

```
v=0
o=- 25678 753849 IN IP6 2001:db8::1
s=
c=IN IP6 2001:db8::1
t=0 0
m=audio 45678 RTP/AVP 0 8
a=altc:1 IP6 2001:db8::1 45678
a=altc:2 IP4 192.0.2.1 12340
```

Since an alternative address is likely to require an alternative Transmission Control Protocol (TCP)/User Datagram Protocol (UDP) port number as well, the new "altc" attribute includes both an IP address and a transport port number (or multiple port numbers). The ALTC mechanism does not itself support offering a different transport type (i.e., UDP vs. TCP), codec, or any other attribute. It is intended only for offering an alternative IP address and port number.

9.2.3.2 Rationale for the Chosen Syntax

The use of an "a=" attribute line is, according to RFC 4566 (see Section 2), the primary means for extending SDP and tailoring it to particular applications or media. A compliant SDP parser will ignore the unsupported attribute lines. The rationale for encoding the same address and port in the "a=altc" line as in the "m=" and "c=" lines is to provide detection of legacy SDP-changing middleboxes. Such systems may change the connection-address and media transport port numbers but not support this new mechanism, and thus two UAs supporting this mechanism would try to connect to the wrong addresses. Therefore, this document requires the SDP processor to proceed to the matching rules defined in Section 9.2.4.2.1.

9.2.4 ALTC Attribute

All SDP ABNF syntaxes are described in Section 2.9. However, we are repeating this here for convenience.

9.2.4.1 ALTC Syntax

The "altc" attribute adheres to the (RFC 4566, see Section 2) "attribute" production. The ABNF syntax (RFC 5234) of altc is provided as follows (Figure 9.1).

```
altc-attr     =    "altc" ":" att-value
att-value     =    altc-num SP addrtype
                   SP connection-address SP port
                   ["/" rtcp-port]
altc-num      =    1*DIGIT
rtcp-port     =    port
```

Figure 9.1 Connectivity capability attribute ABNF (Copyright: IETF).

```
altc-attr     =     "altc" ":" att-value
att-value     =     altc-num SP addrtype
                    SP connection-address SP port
                    ["/" rtcp-port]
altc-num      =     1*DIGIT
rtcp-port     =     port
```

The meanings of the fields are as follows:

- **altc-num**: digit to uniquely refer to an address alternative. It indicates the preference order, with "1" indicated the most preferred address.
- **addrtype**: the addrtype field as defined in RFC 4566 (see Section 2) for connection data.
- **connection-address**: a network address as defined in RFC 4566 (see Section 2) corresponding to the address type specified by addrtype.
- **port**: the port number to be used, as defined in RFC 4566 (see Section 2). Distinct port numbers may be used for each IP address type. If the specified address type does not require a port number, a value defined for that address type should be used.
- **rtcp-port**: including an RTP Control Protocol (RTCP) port is optional. An RTCP port may be indicated in the alternative "c=" line when the RTCP port cannot be derived from the Real-Time Transport Protocol (RTP) port. The "altc" attribute is applicable only in an SDP offer. The "altc" attribute is a media-level-only attribute and MUST NOT appear at the SDP session level. (Because it defines a port number, it is inherently tied to the media level.) There MUST NOT be more than one "altc" attribute per addrtype within each media description. This restriction is necessary so that the addrtype of the reply may be used by the offerer to determine which alternative was accepted. The "addrtype"s of the altc MUST correspond to the "nettype" of the current connection ("c=") line.

A media description MUST contain two "altc" attributes: the alternative address and an alternative port. It must also contain an address and a port that "duplicate" the address/port information from the current "c=" and "m=" lines. Each media level MUST contain at least one such duplicate "altc" attribute, of the same IP address family, address, and transport port number as those in the SDP connection and media lines of its level. In particular, if a "c=" line appears within a media description, the addrtype and connection-address from that "c=" line MUST be used in the duplicate "altc" attribute for that media description. If a "c=" line appears only at the session level and a given media description does not have its own connection line, then the duplicate "altc" attribute for that media description MUST be the same as the session-level address information.

The "altc" attributes appearing within a media description MUST be prioritized. The explicit preference order is indicated in each line ("1" indicates the address with the highest priority). Given this rule, and the requirement that the address information provided in the "m=" line and "o=" line must be provided in an "altc" attribute as well, it is possible that the addresses in the "m=" line and "o=" line are not the preferred choice. If the addrtype of an "altc" attribute is not compatible with the transport protocol or media format specified in the media description, that "altc" attribute MUST be ignored. Note that "a=altc" lines describe alternative connection-addresses, NOT addresses for parallel connections. When several "altc" lines are present, establishing multiple sessions MUST be avoided. Only one session is to be maintained with the remote party for the associated media description.

9.2.4.2 Usage and Interaction

9.2.4.2.1 Usage

In an SDP offer/answer model, the SDP offer includes "altc" attributes to indicate alternative connection information (i.e., address type, address and port numbers), including the "duplicate" connection information already identified in the "c=" and "m=" lines. Additional, subsequent offers MAY include "altc" attributes again, and they may change the IP address, port numbers, and order of preference, but they MUST include a duplicate "altc" attribute for the connection and media lines in that specific subsequent offer. In other words, every offered SDP media description with an alternative address offer with an "altc" attribute has two "altc" attributes:

- One duplicating the "c=" and "m=" line information for that media description
- One for the alternative

These need not be the same as the original SDP offer. The purpose of encoding a duplicate "altc" attribute is to allow receivers of the SDP offer to detect whether a legacy SDP-changing middlebox has modified the "c=" and/or "m=" line address/port information. If the SDP answerer does not find a duplicate "altc" attribute value for which the address and port exactly match those in the "c=" line and "m=" line, the SDP answerer MUST ignore the "altc" attributes and use the "c=" and "m=" offered address/ports for the entire SDP instead, as if no "altc" attributes were present. The rationale for this is that many SDP-changing middleboxes will end the media sessions if they do not detect media flowing through them. If a middlebox modified the SDP addresses, media MUST be sent using the modified information.

Note that for RTCP, if applicable for the given media types, each side would act as if the chosen "altc" attribute's port number were in the "m=" media line. Typically, this would mean that RTCP is sent to the port number equal to "RTP port + 1," unless some other attribute determines otherwise. For example, the RTP/RTCP multiplexing mechanism defined in RFC 5761 can still be used with ALTC, such that if both sides support multiplexing, they will indicate this using the "a=rtcp-mux" attribute, as defined in RFC 5761, but the IP connection-address and port they use may be chosen using the ALTC mechanism.

If the SDP offerer wishes to use the RTCP attribute defined in RFC 3605 (see Section 12.3), a complication can arise, since it may not be clear which address choice the "a=rtcp" attribute applies to, relative to the choices offered by ALTC. Technically, RFC 3605 (see Section 12.3) allows the address for RTCP to be indicated explicitly in the "a=rtcp" attribute itself, but this is optional and rarely used. For this reason, this document recommends using the "a=rtcp" attribute for the address choice encoded in the "m=" line and including an alternate RTCP port in the "a=altc" attribute corresponding to the alternative address. In other words, if the "a=rtcp" attribute explicitly encodes an address in its attribute, that address applies for ALTC, as per RFC 3605 (see Section 12.3). If it does not, then ALTC assumes that the "a=rtcp" attribute is for the address in the "m=" line, and the alternative "altc" attribute includes an RTCP alternate port number.

9.2.4.2.2 Usage of ALTC in an SDP Answer

The SDP answer SHOULD NOT contain "altc" attributes, because the answer's "c=" line implicitly and definitively chooses the address family from the offer and includes it in "c=" and "m=" lines of the answer. Furthermore, this avoids establishing several sessions simultaneously between the participating peers. Any solution requiring the use of ALTC in the SDP answer SHOULD document its usage, in particular how sessions are established between the participating peers.

9.2.4.2.3 Interaction with ICE

Since ICE (RFC 5245, see Section 3.2) also includes address and port number information in its candidate attributes, a potential problem arises: Which one wins? Since ICE also includes specific ICE attributes in the SDP answer, the problem is easily avoided: If the SDP offerer supports both ALTC and ICE, it may include both sets of attributes in the same SDP offer. A legacy ICE-only answerer will simply ignore the "altc" attributes and use ICE. An ALTC-only answerer will ignore the ICE attributes and reply without them. An answerer that supports both MUST choose to use one and only one of the mechanisms: either ICE or ALTC. However, if the "m=" or "c=" line was changed by a middlebox, the rules for both ALTC and ICE would make the answerer revert to basic SDP semantics.

9.2.4.2.4 Interaction with SDP Capability Negotiation

The ALTC mechanism is orthogonal to SDP capability negotiation (RFC 5939, see Section 8.1). If the offerer supports both ALTC and SDP capability negotiation, it may offer both.

9.2.5 IANA Considerations

SDP parameters that are described here have been registered with IANA. We have not included the IANA registration procedures here for the sake of brevity. The procedures for registration of new SDP parameters are described in RFC 6947.

9.2.6 Security Considerations

The security procedures that need to be considered in dealing with the SDP messages described in this section (RFC 6947) are provided in Section 17.8.2.

9.2.7 ALTC Use Cases (Appendix A of RFC 6947)

9.2.7.1 Terminology

The following terms are used when discussing the ALTC use cases:

- Signaling Path Border Element (SBE) denotes a functional element, located at the boundaries of an IP Telephony Administrative Domain (ITAD) (RFC 2871), that is responsible for intercepting signaling flows received from UAs and relaying them to the core service platform. An SBE may be located at the access segment (i.e., be the service contact point for UAs) or at the interconnection with adjacent domains (RFC 6406). An SBE controls one or more Data Path Border Elements (DBEs). The SBE and DBE may be located in the same device (e.g., the SBC - RFC 5853) or be separated.
- DBE denotes a functional element, located at the boundaries of an ITAD, that is responsible for intercepting media/data flows received from UAs and relaying them to another DBE (or media servers, e.g., an announcement server or Interactive Voice Response). An example of

a DBE is a media gateway that intercepts RTP flows. An SBE may be located at the access segment (i.e., be the service contact point for UAs) or at the interconnection with adjacent domains (RFC 6406).
- Core service platform ("core SPF") is a macro functional block including session routing, interfaces to advanced services, and access control.

Figure 9.2 provides an overview of the overall architecture, including the SBE, DBE, and core Service Platform (SPF).

9.2.7.2 Multicast Use Case

Recently, a significant effort has been undertaken within the IETF to specify new mechanisms to interconnect IPv6-only hosts to IPv4-only servers (e.g., RFC 6146). This effort exclusively covered unicast transfer mode. An ongoing initiative, called "multrans," has been launched to cover multicast issues that are encountered during IPv6 transition. The overall problem statement is documented in [3]. A particular issue encountered in the context of IPv4/IPv6 coexistence and IPv6 transition of multicast services is the discovery of the multicast group and source (refer to Section 3.4 of [3]):

- For an IPv6-only receiver requesting multicast content generated by an IPv4-only source:
  - An ALG is required to help the IPv6 receiver select the appropriate IP address when only the IPv4 address is advertised (e.g., using SDP). Otherwise, access to the IPv4 multicast content cannot be offered to the IPv6 receiver. The ALG may be located downstream of the receiver. As such, the ALG does not know in advance whether the receiver is DS or IPv6-only. The ALG may be tuned to insert both the original IPv4 address and the corresponding IPv6 multicast address using, for instance, the ALTC SDP attribute.
  - To avoid involving an ALG in the path, an IPv4-only source can advertise both its IPv4 address and its IPv4-embedded IPv6 multicast address (RFC 7371) using, for instance, the ALTC SDP attribute.
- For a DS source sending its multicast content over IPv4 and IPv6, both IPv4 and IPv6 addresses need to be inserted in the SDP part. A means (e.g., ALTC) is needed for this purpose.

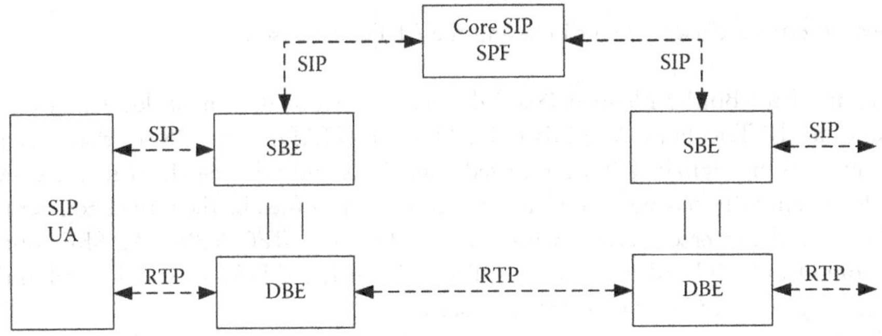

Figure 9.2 Service architecture overview (Copyright: IETF).

9.2.7.3 Introducing IPv6 into SIP-Based Architectures

9.2.7.3.1 Avoiding Crossing CGN Devices

Some service providers are in the process of enabling DS-Lite (RFC 6333) as a means to continue delivering IPv4 services to their customers. To avoiding crossing four levels of NAT when establishing a media session (two NATs in the DS-Lite Address Family Transition Router (AFTR) and two NATs in the DBE), it is recommended to enable IPv6 functions in some SBEs/DBEs. Then, DS-Lite AFTRs will not be crossed for DS-Lite serviced customers if their UA is IPv6-enabled:

- For an SIP UA embedded in the CPE, this is easy to implement since the SIP UA (RFC 3261, also see Section 2) can be tuned to behave as an IPv6-only UA when DS-Lite is enabled. No ALTC is required for this use case.
- For SIP UAs located behind the CPE, a solution to indicate both IPv4 and IPv6 (e.g., ALTC) is required in order to avoid crossing the DS-Lite CGN.

9.2.7.3.2 Basic Scenario for IPv6 SIP Service Delivery

A basic solution to deliver SIP-based services using an IPv4-only core service platform to an IPv6-enabled UA is to enable the IPv4/IPv6 interworking function in the SBE/DBE. Signaling and media between two SBEs and DBEs is maintained over IPv4. IPv6 is used between an IPv6-enabled UA and an SBE/DBE. Figure 9.3 shows the results of session establishment between UAs. In this scenario, the IPv4/IPv6 interworking function is invoked even when both involved UAs are IPv6-enabled.

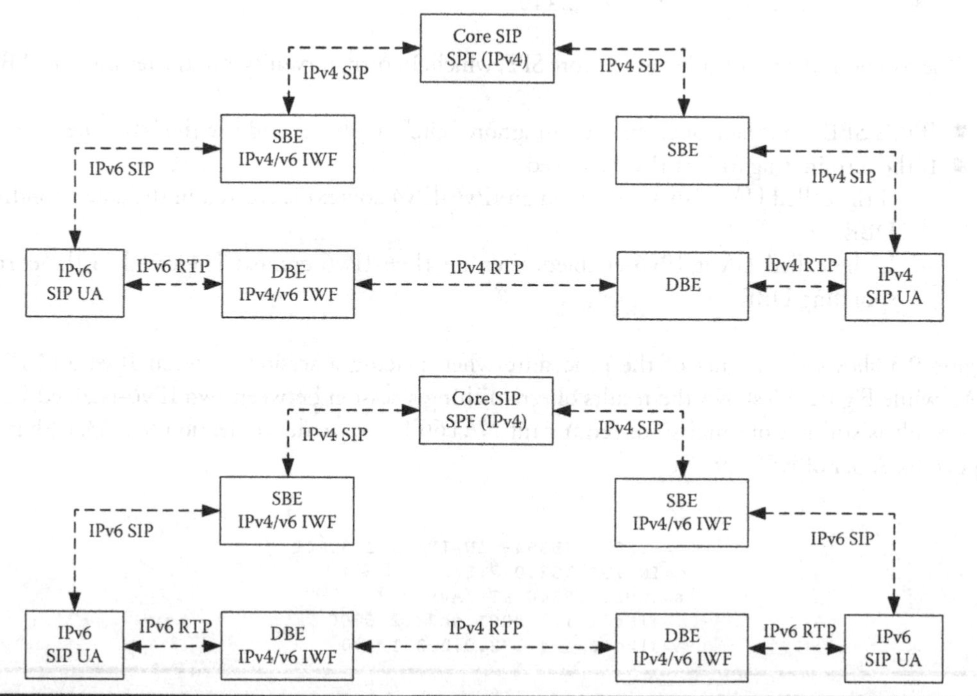

Figure 9.3 Basic scenario (Copyright: IETF).

It may be valuable for service providers to consider solutions that avoid redundant IPv4/IPv6 NATs and avoid involving several DBEs.

9.2.7.3.3 Avoiding IPv4/IPv6 Interworking

A solution to indicate both IPv4 and IPv4 addresses is required for service providers that want the following:

1. A means to promote the invocation of IPv6 transfer capabilities that can be enabled, while no parsing errors are experienced by core service legacy nodes
2. To optimize the cost related to IPv4-IPv6 translation licenses
3. To reduce the DS lifetime
4. To maintain an IPv4-only core
5. To have a set of SBEs/DBEs that are IPv6-enabled.

This section provides an overview of the procedure to avoid IPv4/IPv6 interworking. When an SBE receives an INVITE, it instantiates in its DBE an IPv6-IPv6 context and an IPv6-IPv4 context. Both an IPv6 address and an IPv4 address are returned, together with other information such as port numbers. The SBE builds an SDP offer, including both the IPv4 and IPv6-related information using the "altc" attribute. IPv6 is indicated as the preferred connectivity type; see Figure 9.4.

```
o=- 25678 753849 IN IP4 192.0.2.2
c=IN IP4 192.0.2.2
m=audio 12340 RTP/AVP 0 8
a=altc:1 IP6 2001:db8::2 6000
a=altc:2 IP4 192.0.2.2 12340
```

The request is then forwarded to the core SPF, which, in turn, forwards it to the terminating SBE.

- If this SBE is a legacy one, then it will ignore "altc" attributes and use the "c=" line.
- If the terminating SBE is IPv6-enabled
  - If the called UA is IPv4 only, then an IPv6-IPv4 context is created in the corresponding DBE.
  - If the called UA is IPv6-enabled, then an IPv6-IPv6 context is created in the corresponding DBE.

Figure 9.5 shows the results of the procedure when placing a session between IPv4 and IPv6 UAs, while Figure 9.6 shows the results of establishing a session between two IPv6-enabled UAs. The result is still not optimal since redundant NAT66 is required, see Section 9.2.7.3.4 (that is, Appendix A.3.4 of RFC 6947).

```
o=-25678 753849 IN IP4 192.0.2.2
c=IN IP4 192.0.2.2
m=audio 12340 RTP/AVP 0 8
a=altc:1 IP6 2001:db8::2 6000
a=altc:2 IP4 192.0.2.2 12340
```

Figure 9.4 SDP offer updated by the SBE.

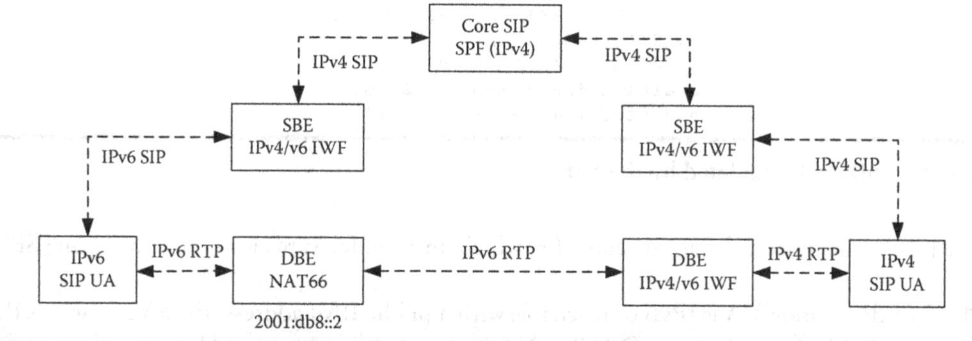

Figure 9.5 Session establishment between IPv4 and IPv6 UAs (Copyright: IETF).

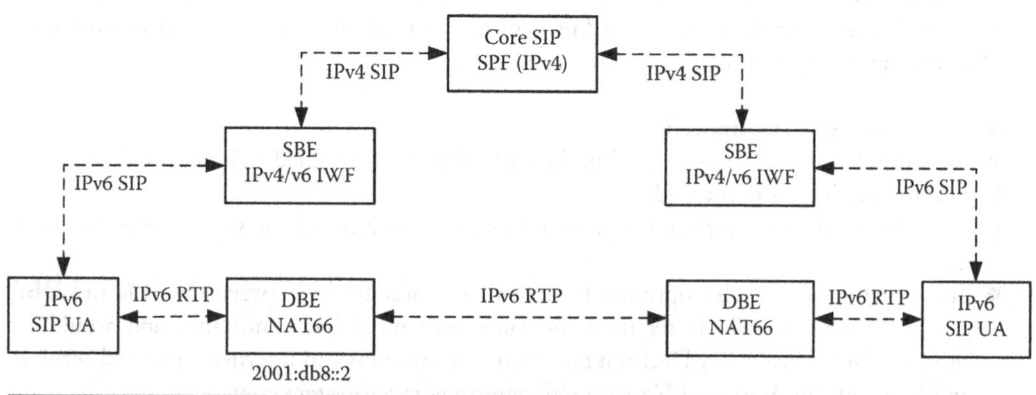

Figure 9.6 Session establishment between IPv6 UAs (Copyright: IETF).

9.2.7.3.4 DBE Bypass Procedure

For service providers wanting to involve only one DBE in the media path when not all SBEs/DBEs and UAs are IPv6-enabled, a means to indicate both IPv4 and IPv6 addresses without inducing session failures is required. This section proposes an example procedure using the "altc" attribute. When the originating SBE receives an INVITE from an IPv6-enabled UA, it instantiates in its DBE both an IPv6-IPv6 context and an IPv6-IPv4 context. Both an IPv6 address and an IPv4 address are returned, together with other information, such as port numbers. The SBE builds an SDP offer, including both IPv4 and IPv6-related information using the "altc" attribute (Figure 9.7). IPv6 is indicated as preferred connectivity type.

```
o=- 25678 753849 IN IP4 192.0.2.2
c=IN IP4 192.0.2.2
m=audio 12340 RTP/AVP 0 8
a=altc:1 IP6 2001:db8::2 6000
a=altc:2 IP4 192.0.2.2 12340
```

```
o=-25678 753849 IN IP4 192.0.2.2
c=IN IP4 192.0.2.2
m=audio 12340 RTP/AVP 0 8
a=altc:1 IP6 2001:db8::2 6000
a=altc:2 IP4 192.0.2.2 12340
```

Figure 9.7 SDP offer updated by the SBE.

The request is then forwarded to the core SPF, which, in turn, forwards it to the terminating SBE:

- If the destination UA is IPv6 or reachable with a public IPv4 address, the SBE forwards the request without altering the SDP offer. No parsing error is experienced by core service nodes since ALTC is backwards compatible.
- If the terminating SBE does not support ALTC, it will ignore this attribute and use the legacy procedure.

As a consequence, only one DBE is maintained in the path when one of the involved parties is IPv6-enabled. Figure 9.8 shows the overall procedure when the involved UAs are IPv6-enabled.

The main advantages of such a solution are as follows:

- DBE resources are optimized.
- No redundant NAT is maintained in the path when IPv6-enabled UAs are involved.
- End-to-end delay is optimized.
- The robustness of the service is optimized since the delivery of the service relies on fewer nodes.
- The signaling path is also optimized since no communication between the SBE and DBE at the terminating side is required for some sessions. (That communication would be through the Service Policy Decision Function in a Telecoms and Internet converged Services and Protocols for Advanced Networks/IP Multimedia Subsystem context.)

9.2.7.3.5 Direct Communications between IPv6-Enabled UAs

For service providers wanting to allow direct IPv6 communications between IPv6-enabled UAs, when not all SBEs/DBEs and UAs are IPv6-enabled, a means to indicate both the IPv4 and IPv6 addresses without inducing session failures is required. Below (Figures 9.9 and 9.10) is an example of

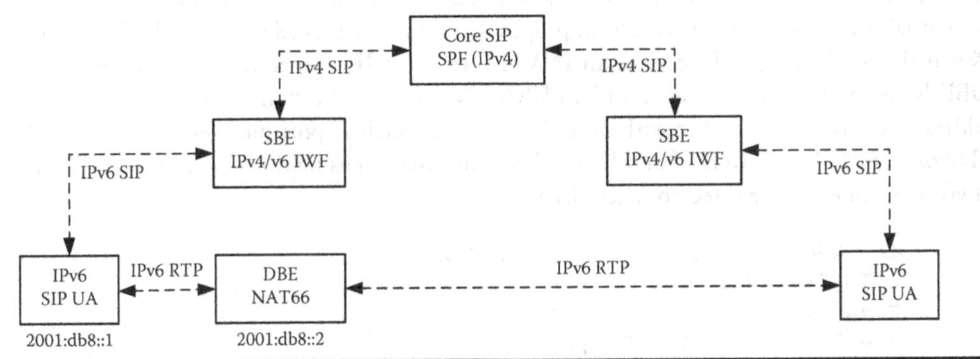

Figure 9.8 DBE bypass overview (Copyright: IETF).

```
o=-25678 753849 IN IP6 2001:db8::1
c=IN IP6 2001:db8::1
m=audio 6000 RTP/AVP 0 8
```

Figure 9.9 SDP offer of the calling UA (Copyright: IETF).

```
o=-25678 753849 IN IP4 192.0.2.2
c=IN IP4 192.0.2.2
m=audio 12340 RTP/AVP 0 8
a=altc:1 IP6 2001:db8::1 6000
a=altc:2 IP4 192.0.2.2 12340
```

Figure 9.10 SDP offer updated by the SBE (Copyright: IETF).

a proposed procedure using the "altc" attribute. At the SBE originating side, when the SBE receives an INVITE from the calling IPv6 UA (Figure 9.9), it uses ALTC to indicate two IP addresses:

1. An IPv4 address belonging to its controlled DBE.
2. The same IPv6 address and port as received in the initial offer made by the calling IPv6.

Figure 9.9 shows an excerpted example of the SDP offer of the calling UA, and Figure 9.10 shows an excerpted example of the updated SDP offer generated by the originating SBE.

```
o=- 25678 753849 IN IP6 2001:db8::1
c=IN IP6 2001:db8::1
m=audio 6000 RTP/AVP 0 8
```

The INVITE message will be routed appropriately to the destination SBE:

```
o=- 25678 753849 IN IP4 192.0.2.2
c=IN IP4 192.0.2.2
m=audio 12340 RTP/AVP 0 8
a=altc:1 IP6 2001:db8::1 6000
a=altc:2 IP4 192.0.2.2 12340
```

1. If the SBE is a legacy device (i.e., IPv4-only), it will ignore IPv6 addresses and will contact its DBE to instantiate an IPv4-IPv4 context.

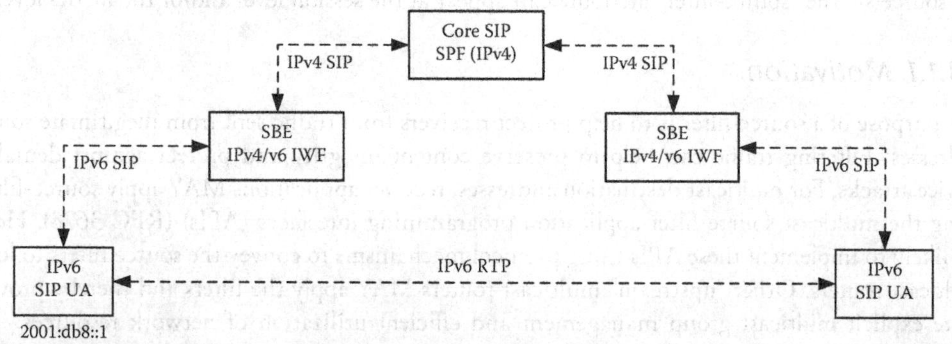

Figure 9.11 Direct IPv6 communication (Copyright: IETF).

2. If the SBE is IPv6-enabled, it will only forward the INVITE to the address of contact of the called party.
   a. If the called party is IPv6-enabled, the communication will be placed using IPv6. As such, no DBE is involved in the data path, as illustrated in Figure 9.11.
   b. Otherwise, IPv4 will be used between the originating DBE and the called UA.

9.3 Source Address Filters in SDP

This section describes RFC 4570 that specifies how to adapt the SDP to express one or more source addresses (e.g., IPv4/IPv6 network addresses) as a source filter for one or more destination "connection" addresses. It defines the syntax and semantics for an SDP "source-filter" attribute that may reference either IPv4 or IPv6 address(es) as either an inclusive or exclusive source list for either multicast or unicast destinations. In particular, an inclusive source-filter can be used to specify a source-specific multicast (SSM) session.

9.3.1 Introduction

In this section, we describe the adoption of SDP to express one or more source addresses as a source filter for one or more destination "connection" addresses defined in RFC 4570. The SDP (RFC 4566, see Section 2) provides a general purpose format for describing multimedia sessions in announcements or invitations. SDP uses an entirely textual data format (the US-ASCII subset of RFC 3629) to maximize portability among transports. SDP does not define a protocol, but only the syntax to describe a multimedia session with sufficient information to discover and participate in that session. Session descriptions may be sent using any number of existing application protocols for transport (e.g., Session Announcement Protocol [SAP], SIP, Real-Time Streaming Protocol [RTSP], email, and HTTP). Typically, session descriptions reference an IP multicast address for the "connection-address" (destination), though unicast addresses or fully qualified domain names MAY also be used.

The "source-filter" attribute defined in this document qualifies the session traffic by identifying the address (FQDN) of legitimate sources (senders). The intent is for receivers to use the source and destination address pair(s) to filter traffic, so that applications receive only legitimate session traffic. Receiver applications are expected to use the SDP source-filter information to identify traffic from legitimate senders, and discard traffic from illegitimate senders. Applications and hosts may also share the source-filter information with network elements (e.g., with routers using RFC 3376) so they can potentially perform the traffic filtering operation further "upstream," closer to the source(s). The "source-filter" attribute can appear at the session level and/or the media level.

9.3.1.1 Motivation

The purpose of a source-filter is to help protect receivers from traffic sent from illegitimate source addresses. Filtering traffic can help to preserve content integrity and protect against denial of service attacks. For multicast destination addresses, receiver applications MAY apply source-filters using the multicast source filter application programming interfaces (APIs) (RFC 3678). Hosts are likely to implement these APIs using protocol mechanisms to convey the source filters to local multicast routers. Other "upstream" multicast routers MAY apply the filters and thereby provide more explicit multicast group management and efficient utilization of network resources. The protocol mechanisms to enable these operations are beyond the scope of this document, but their potential provided motivation for SDP source-filters.

9.3.2 Terminology

See Section 2.2/Table 2-1.

9.3.3 The "source-filter" Attribute

The SDP source-filter attribute does not change any existing SDP syntax or semantics, but defines a format for additional session description information. Specifically, source-filter syntax can prescribe one or more unicast addresses as either legitimate or illegitimate sources for any (or all) SDP session description "connection-address" field values. Note that the unicast source addresses specified by this attribute are those that are seen by a receiver. Therefore, if source addresses undergo translation en-route from the original sender to the receiver (e.g., due to NAT or some tunneling mechanism) then the SDP "source-filter" attribute as presented to the receiver will not be accurate unless the source addresses therein are also translated accordingly. The source-filter attribute has the following syntax:

```
a=source-filter: <filter-mode> <filter-spec>
```

The <filter-mode> is either "incl" or "excl" (for inclusion or exclusion, respectively). The <filter-spec> has four subcomponents:

```
<nettype> <address-types> <dest-address> <src-list>
```

A <filter-mode> of "incl" means that an incoming packet is accepted only if its source address is in the set specified by `<src-list>`. A <filter-mode> of "excl" means that an incoming packet is rejected if its source address is in the set specified by <src-list>. The first subfield, <nettype>, indicates the network type, since SDP is protocol independent. This document is most relevant to the value "IN," which designates the IP. The second subfield, <address-types>, identifies the address family and for the purpose of this document may be <addrtype> value "IP4" or "IP6." Alternately, when <dest-address> is an FQDN, the value MAY be "*" to apply to both address types, since either address type can be returned from a domain name system (DNS) lookup.

The third subfield, <dest-address>, is the destination address, which MUST correspond to one or more of the session's "connection-address" field values. It may be either a unicast or multicast address, an FQDN, or the "*" wildcard to match any/all of the session's "connection-address" values. The fourth subfield, <src-list>, is the list of source hosts/interfaces in the source-filter and consists of one or more unicast addresses or FQDNs, separated by space characters. The format and content of these semantic elements are derived from and compatible with those defined in RFC 4566 (see Section 2). For more detail, see Section 9.3.7 (that is, Appendix A of RFC 4570).

9.3.3.1 Processing Rules

There are a number of details to consider when parsing the SDP source-filter syntax. The <dest-address> value in a "source-filter" attribute MUST correspond to an existing <connection-field> value in the session description. The only exception to this is when a "*" wildcard is used to indicate that the source-filter applies to all <connection-field> values. When the <dest-address> value is a multicast address, the field value MUST NOT include the subfields <ttl> and <number of addresses> from the <connection-address> value. If the <connection-address> specifies more than one multicast address (in the <number of addresses> field), then a source filter, if any, for each such address must be stated explicitly, using a separate "a=source-filter" line for each address (unless a "*" wildcard is used for <dest-address>). See Section 9.3.3.2.4 for an example.

When the <addrtype> value is the "*" wildcard, the <dest-address> MUST be either an FQDN or "*" (i.e., it MUST NOT be an IPv4 or IPv6 address). See Section 9.3.3.2.6 for an example. As has always been the case, the default behavior when a source-filter attribute is not provided in a session description is that all traffic sent to the specified <connection-address> value should be accepted (i.e., from any source address). The source-filter grammar does not include syntax to express either "exclude none" or "include all." Like the standard <connection-field> described in RFC 4566 (see Section 2), the location of the "source-filter" attribute determines whether it applies to the entire session or only to a specific medium (i.e., "session level" or "media level"). A media-level source-filter will always completely override a session-level source-filter.

A "source-filter" need not be located at the same hierarchy level as its corresponding <connection-field>. So, a media-level <source-filter> can reference a session-level <connection-field> value, and a session-level "source-filter" can be applied to all matching media-level <connection-field> values. See Section 9.3.3.2.3 for an example. An SDP description MUST NOT contain more than one session-level "source-filter" attribute that covers the same destination address or more than one media-level "source-filter" attribute that covers the same destination address.

There is no specified limit to the number of entries allowed in the <src-list>; however, there are practical limits that should be considered. For example, depending on the transport to be used for the session description, there may be a limit to the total size of the session description (e.g., as determined by the maximum payload in a single datagram). Also, when the source-filter is applied to control protocols, there may be a limit to the number of source addresses that can be sent. These limits are outside the scope of this document but should be considered when defining source-filter values for SDP.

9.3.3.2 Examples

Here are a number of examples that illustrate how to use the source-filter attribute in some common scenarios. We use the following session description components as the starting point for the examples to follow. For each example, we show the source-filter with additional relevant information and provide a brief explanation.

```
<session-description> =
        v=0
        o=The King <Elvis@example.com>
        s=Elvis Impersonation
        i=All Elvis, all the time
        u=http://www.example.com/ElvisLive/
        t=0 0
        a=recvonly

<media-description 1> =
        m=audio 54320 RTP/AVP 0

<media-description 2> =
        m=video 54322 RTP/AVP 34
```

9.3.3.2.1 SSM Example

Multicast addresses in the SSM range require a single unicast sender address for each multicast destination, so the source-filter specification provides a natural fit. In this example, a session member should receive only traffic sent from 192.0.2.10 to the multicast session address 232.3.4.5.

```
<session-description>

c=IN IP4 232.3.4.5/127
a=source-filter: incl IN IP4 232.3.4.5 192.0.2.10

<media-description 1>
```

This source-filter example uses an inclusion list with a single multicast "connection-address" as the destination and single unicast address as the source. Note that the value of the connection-address matches the value specified in the connection-field. Also, note that since the connection-field is located in the session-description section, the source-filter applies to all media.

Furthermore, if the SDP description specifies an RTP session (e.g., its "m=" line(s) specify "RTP/AVP" as the transport protocol), then the "incl" specification will apply not only to RTP packets, but also to any RTCP packets that are sent to the specified multicast address. This means that, as a side effect of the "incl" specification, the only possible multicast RTCP packets will be "Sender Report" packets sent from the specified source address. Because of this, an SDP description for a SSM RTP session SHOULD also include an

```
a=rtcp-unicast ...
```

attribute, as described in RFC 5760. This specifies that RTCP "Reception Report" packets are to be sent back via unicast.

9.3.3.2.2 Unicast Exclusion Example

Typically, an SDP session <connection-address> value is a multicast address, although it is also possible to use either a unicast address or FQDN. This example illustrates a scenario whereby a session description indicates the unicast source address 192.0.2.10 in an exclusion filter. In effect, this sample source-filter says, "destination 192.0.2.11 should accept traffic from any sender *except* 192.0.2.10."

```
<session-description>

        c = IN IP4 192.0.2.11
        a = source-filter: excl IN IP4 192.0.2.11 192.0.2.10

<media-description 1>
```

9.3.3.2.3 Multiple Session Address Example

This source-filter example uses the wildcard "*" value for <dest-addr> to correspond to any/all <connection-address> values. Hence, the only legitimate source for traffic sent to either 232.2.2.2 or 232.4.4.4 multicast addresses is 192.0.2.10. Traffic sent from any other unicast source address should be discarded by the receiver.

```
<session-description>
        a = source-filter: incl IN IP4 * 192.0.2.10
<media-description 1>
        c = IN IP4 232.2.2.2/127
<media-description 2>
        c = IN IP4 232.4.4.4/63
```

9.3.3.2.4 Multiple Multicast Address Example

In this example, the <connection-address> specifies three multicast addresses: 224.2.1.1, 224.2.1.2, and 224.2.1.3. The first and third of these addresses are given source filters. However, in this example the second address, 224.2.1.2, is *not* given a source filter.

```
<session-description>
      c = IN IP4 224.2.1.1/127/3
      a = source-filter: incl IN IP4 224.2.1.1 192.0.2.10
      a = source-filter: incl IN IP4 224.2.1.3 192.0.2.42

<media-description 1>
```

9.3.3.2.5 IPv6 Multicast Source-Filter Example

This simple example defines a single session-level source-filter that references a single IPv6 multicast destination and source pair. The IP multicast traffic sent to FFOE::11A is valid only from the unicast source address 2001:DB8:1:2:240:96FF:FE25:8EC9.

```
<session-description>
      c  =  IN IP6 FF0E::11A/127
      a  =  source-filter incl IN IP6 FF0E::11A
            2001:DB8:1:2:240:96FF:FE25:8EC9

<media-description 1>
```

9.3.3.2.6 IPv4 and IPv6 FQDN Example

This example illustrates use of the <addrtype> "*" wildcard, along with multicast and source FQDNs that may resolve to either an IPv6 or an IPv4 address or both. Although typically both the multicast and source addresses will be the same (either both IPv4 or both IPv6), using the wildcard for addrtype in the source filter allows asymmetry between the two addresses (so an IPv4 source address may be used with an IPv6 multicast address).

```
<session-description>
      c  =  IN IP4 channel-1.example.com/127
      c  =  IN IP6 channel-1.example.com/127
      a  =  source-filter: incl IN * channel-1.example.com src-1.
            example.com

<media-description 1>
```

9.3.3.3 Offer-Answer Model Considerations

The "source-filter" attribute is not intended to be used as an "offer" in an SDP offer-answer exchange (RFC 3264, see Section 3.1), because sets of source addresses do not represent "capabilities" or "limitations" of the offerer and because the offerer does not, in general, have a priori knowledge of which IP source address(es) will be included in an answer. While an answerer may include the "source-filter" attribute in his/her answer (e.g., to designate a SSM session), the answerer SHOULD ignore any "source-filter" attribute that was present in the original offer.

9.3.4 Interoperability Issues

Defining a list of legitimate sources for a multicast destination address represents a departure from the Any-Source Multicast (ASM) model, as originally described in RFC 1112. The ASM model supports anonymous senders and all types of multicast applications (e.g., many-to-many). Use of a source-filter excludes some (unknown or undesirable) senders, which lends itself more to one-to-many or few-to-few type multicast applications. Although these two models have contrasting operational characteristics and requirements, they can coexist on the same network using the same protocols. Use of source-filters does not corrupt the ASM semantics but provide more control for receivers, at their discretion.

9.3.5 Security Considerations

The security procedures that need to be considered in dealing with the SDP source filter messages described in this section (RFC 4570) are provided in Section 17.8.3.

9.3.6 IANA Considerations

The SDP parameters that are described here have been registered with IANA. We have not included the IANA registration procedures here for the sake of brevity. The procedures for registration of new SDP parameters are described in RFC 4570.

9.3.7 Source-Filter Attribute Syntax (Appendix A of RFC 4570)

This appendix provides an ABNF (RFC 4234 was made obsolete by RFC 5234) grammar for expressing an exclusion or inclusion list of one or more (IPv4 or IPv6) unicast source addresses. It is intended as an extension to the grammar for the SDP, as defined in RFC 4566 (see Section 2). Specifically, it describes the syntax for the new "source-filter" attribute field, which MAY be either a session-level or a media-level attribute. All SDP ABNF syntaxes are provided in Section 2.9. However, we have repeated this here for convenience.

The "dest-address" value in each source-filter field MUST match an existing connection-field value, unless the wildcard connection-address value "*" is specified.

```
source-filter    =   "source-filter" ":" SP filter-mode SP
                     filter-spec
                     ; SP is the ASCII 'space' character
                     ; (0x20, defined in RFC 5234 that
                     ; obsoletes RFC 4234).

filter-mode      =   "excl" / "incl"
                     ; either exclusion or inclusion mode.

filter-spec      =   nettype SP address-types SP dest-address
                     SP src-list
                     ; nettype is as defined in RFC 4566 (see
                     ; Section 2).

address-types    =   "*" / addrtype
                     ; "*" for all address types (both IP4
                     ; and IP6),
                     ; but only when <dest-address> and
                     ; <src-list>
                     ; reference FQDNs.
```

```
                    ; addrtype is as defined in RFC 4566 (see
                    ; Section 2).

dest-address    =   "*" / basic-multicast-address /
                    unicast-address
                    ; "*" applies to all connection-address
                    ; values.
                    ; unicast-address is as defined in RFC
                    ; 4566 (see Section 2).

src-list        =   *(unicast-address SP) unicast-address
                    ; one or more unicast source addresses (in
                    ; standard IPv4 or IPv6 ASCII-notation
                    ; form)
                    ; or FQDNs.
                    ; unicast-address is as defined in RFC
                    ; 4566 (see Section 2).

basic-multicast-address    =    basic-IP4-multicast /
                                basic-IP6-multicast
                                / FQDN / extn-addr
                                ; i.e., the same as
                                ; multicast-address
                                ; defined in RFC 4566 (see
                                ; Section 2), except
                                ; that the
                                ; /<ttl> and /<number of
                                ; addresses>
                                ; fields are not included.
                                ; FQDN and extn-addr are
                                ; as defined
                                ; in RFC 4566 (see
                                ; Section 2).

basic-IP4-multicast        =    m1 3( "." decimal-uchar )
                                ; m1 and decimal-uchar are
                                ; as defined
                                ; in RFC 4566 (see
                                ; Section 2)

basic-IP6-multicast        =    hexpart
                                ; hexpart is as defined in
                                ; RFC 4566
                                ; (see Section 2).
```

9.4 SDP for Asynchronous Transfer Mode (ATM) Bearer Connections

This subsection (RFC 3108) describes conventions for using the SDP described in RFC 4566 that obsoletes RFC 2327 for controlling ATM bearer connections, and any associated ATM adaptation layer (AAL). The AALs addressed are Type 1, Type 2 and Type 5. This list of conventions is meant to be exhaustive. Individual applications can use subsets of these conventions. Further, these conventions are meant to comply strictly with the SDP syntax as defined in RFC 2327 (that was made obsoleted by RFC 4566, see Section 2).

Note: RFC 3108 uses two primary RFCs 2543 and 2327 that have been made obsolete by RFCs 3261 (also see Section 1.2) and 4566 (see Section 2), respectively. The updates due to changes in new RFCs are complicated, and we have not attempted any changes in texts other than mentioning that the new RFCs have obsoleted the old ones.

9.4.1 Introduction

In this section, we describe the SDP for ATM bearer connections specified in RFC 3108. SDP will be used in conjunction with a connection handling /device control protocol such as Megaco (H.248) (RFC 3015 was made obsolete by RFC 3525), SIP (RFC 2543 was made obsolete by RFC 3261, also see Section 1.2) or MGCP (RFC 2705 was made obsolete by RFC 3435) to communicate the information needed to set up ATM and AAL2 bearer connections. These connections include voice connections, voiceband data connections, clear channel circuit emulation connections, video connections, and baseband data connections (e.g., fax relay, modem relay, Service-Specific Connection-Oriented Protocol [SSCOP], frame relay, etc.). These conventions use standard SDP syntax as defined in RFC 2327 (that was made obsolete by RFC 4566, see Section 2) to describe the ATM-level and AAL-level connections, addresses, and other parameters. In general, parameters associated with layers higher than the ATM adaptation layer are included only if they are tightly coupled to the ATM or AAL layers. Since the syntax conforms to RFC 2327 (that was made obsolete by RFC 4566, see Section 2), standard SDP parsers should react in a well-defined and safe manner on receiving session descriptions based on the SDP conventions in this document. This is done by extending the values of fields defined in RFC 2327 (that was made obsolete by RFC 4566, see Section 2) rather than by defining new fields.

This is true for all SDP lines except that of the media attribute lines, in which case new attributes are defined. The SDP protocol allows the definition of new attributes in the media attribute lines, which are free-form. For the remaining lines, the fact that the <networkType> field in an SDP descriptor is set to "ATM" should preclude the misinterpretation of extended parameter values by RFC 2327-compliant SDP parsers. These conventions are meant to address the following ATM applications:

1. Applications in which a new Switched Virtual Circuit (SVC) is set up for each service connection. These SVCs could be AAL1 or AAL5 SVCs or single-channel identifier (CID) AAL2 SVCs.
2. Applications in which existing path resources are assigned to service connections. These resources could be:
   - AAL1/AAL5 permanent virtual circuits (PVCs), Soft PVCs (SPVCs), or cached SVCs
   - AAL2 single-CID PVCs, SPVCs or cached SVCs
   - CIDs within AAL2 SVCs/PVCs/SPVCs that multiplex multiple CIDs
   - Subchannels (identified by CIDs) within AAL1 [4] or AAL2 [5] SVCs/PVCs/SPVCs

Note that the difference between PVCs and SPVCs is in the way the bearer virtual circuit connection is set up. SPVCs are a class of PVCs that use bearer signaling, as opposed to node-by-node provisioning, for connection establishment. This document is limited to the case in which the network type is ATM. This includes raw RTP encapsulation [6] or voice sample encapsulation [7] over AAL5 with no intervening IP layer. It does not address SDP usage for IP, with or without ATM as a lower layer. In some cases, IP connection setup is independent of lower layers, which are configured prior to it. For example, AAL5 PVCs that connect IP routers can be used for VoIP calls. In other cases, VoIP call setup is closely tied to ATM-level connection setup. This might require a chaining of IP and ATM descriptors, as described in Section 9.4.5.6.4.1.

This document makes no assumptions on who constructs the session descriptions (e.g., media gateway, intermediate ATM/AAL2 switch, media gateway controller, etc.). This will be different in different applications. Further, it allows the use of one session description for both directions of a connection (as in SIP and MGCP applications) or the use of separate session descriptions for different directions. It also addresses the ATM multicast and anycast capabilities. This document makes no assumptions about how the SDP description will be coded. Although the descriptions shown here are encoded as text, alternate coding are possible:

- Binary encoding such as ASN.1. This is an option (in addition to text encoding) in the Megaco context.
- Use of extended Integrated Services Digital Network (ISDN) User Part (ISUP) parameters [8] to encode the information in SDP descriptors, with conversion to/from binary/text-based SDP encoding when needed

9.4.2 Representation of Certain Fields within SDP Description Lines

This document conforms to the syntactic conventions of standard SDP as defined in RFC 2327 (that was made obsolete by RFC 4566, see Section 2).

9.4.2.1 Representation of Extension Attributes

The SDP protocol described in RFC 2327 (that was made obsolete by RFC 4566, see Section 2) requires that nonstandard attributes and codec names use an "X-" prefix. In this Internet document, the "X-" prefix is used consistently for codec names (Table 9-2) that have not been registered with the IANA. The IANA-registered codec names do not use this prefix, regardless of whether they are statically or dynamically assigned payload types. Neither is this prefix used for the extension SDP attributes defined in this document. This has been done to enhance legibility. This document suggests that parsers be flexible in the use of the "X-" prefix convention. They should accept codec names and attribute names with or without the "X-" prefix.

9.4.2.2 Representation of Parameter Values

Depending on the format of their representation in SDP, the parameters defined in this document fall into the following classes:

1. Parameters always represented in a decimal format
2. Parameters always represented in a hexadecimal format
3. Parameters always represented as character strings
4. Parameters that can be represented in either decimal or hexadecimal format

No prefixes are needed for classes 1 to 3, since the format is fixed. For class 4, a "0x" prefix shall always be used to differentiate the hexadecimal from the decimal format. For both decimal and hex representations, if the underlying bit field is smaller or larger than the binary equivalent of the SDP representation, then leading 0 bits should be added or removed as needed. Thus, 3 and 0x3 translate into the following five-bit pattern: 0 0011. The SDP representations 0x12 and 18 translate into the following five-bit pattern: 1 0010. Leading 0 digits shall not be used in decimal representations.

Generally, these are also not used in hexadecimal representations. Exceptions are when an exact number of hex digits is expected, as in the case of Network Service Access Point (NSAP) addresses. Parsers shall not reject leading zeros in hex values. Both single-character and multi-character string values are enclosed in double quotes (i.e., "). By contrast, single quotes (i.e., ') are used for emphasizing keywords rather than to refer to characters or strings. In the text representation of decimal and hex numbers, digits to the left are more significant than digits to the right.

9.4.2.3 Directionality Convention

This section defines the meaning of the terms "forward" and "backward" as used in this document. This is especially applicable to parameters that have a specific direction associated with them. In this document, "forward" refers to the direction away from the ATM node under consideration, while "backward" refers to the direction toward the ATM node. This convention must be used in all SDP-based session descriptions regardless of whether the underlying bearer is an SVC, a dynamically allocated PVC/SPVC, or a dynamically allocated CID. This is regardless of which side originates the service connection. If ATM SVC or AAL2 Q.2630.1 signaling is used, the directionality convention is independent of which side originates the SVC or AAL2 connection.

This provides a simple way of identifying the direction in which a parameter is applicable, in a manner that is independent of the underlying ATM or AAL2 bearer. This simplicity comes at a price, described in the following paragraphs. The convention used by all ATM/AAL2 signaling specifications (e.g., ITU-T Standard Q.2931 Section 1.3.3 and ITU-T Standard Q.2630.1) mandates that forward direction is from the end initiating setup/establishment via bearer signaling towards the end receiving the setup/establishment request. The backward direction is in the opposite direction. In some cases, the "forward" and "backward" directions of the ATM signaling convention might be the exact opposite of the SDP convention described in Section 9.4.1, requiring the media gateway to perform the necessary translation. An example case in which this is needed is described below.

Consider an SDP description sent by a media gateway controller to the gateway originating a service-level call. In the backward SVC call set-up model, this gateway terminates (rather than originates) an SVC call. The media gateway refers to the traffic descriptor (and hence the pick cell rate [PCR] in the direction away from this gateway as the forward traffic descriptor and forward PCR). Clearly, this is at odds with ATM SVC signaling, which refers to this very PCR as the backward PCR. The gateway needs to be able to perform the required swap of directions. In this example, the media gateway terminating the service level call (and hence originating the SVC call) does not need to perform this swap.

Certain parameters within attributes are defined exclusively for the forward or backward directions. Examples for the forward direction are the <fsssar> subparameter within the "aal2sscs3661unassured" media attribute line, the <fsssar>, <fsscopsdu> and <fsscopuu> subparameters within the "aal2sscs3661assured" media attribute line, the <fsscopsdu> and <fsscopuu> subparameters within the "aal5sscop" media attribute line, and the <fmaxFrame> parameter within the "aal2sscs3662" media attribute line. Examples for the backward direction are the <bsssar> subparameter within the "aal2sscs3661unassured" media attribute line, the <bsssar>, <bsscopsdu>, and <bsscopuu> subparameters within the "aal2sscs3661assured" media attribute line, the <bsscopsdu> and <bsscopuu> subparameters within the "aal5sscop" media attribute line, and the <bmaxFrame> parameter within the "aal2sscs3662" media attribute line.

9.4.2.4 Case Convention

As defined in RFC 4566, see Section 2, that obsoletes RFC 2327, SDP syntax is case sensitive. Since these ATM conventions conform strictly to SDP syntax, they are also case sensitive. SDP line types (e.g., "c," "m," "o," "a") and fields in the SDP lines should be built according to the case conventions in RFC 4566 (see Section 2) that obsoletes RFC 2327 and in this document. It is suggested, but not required, that SDP parsers for ATM applications be case tolerant where ignoring case does not result in ambiguity. Encoding names, which are defined outside the SDP protocol, are case insensitive.

9.4.2.5 Use of Special Characters in SDP Parameter Values

In general, RFC 2327-conformant (RFC 2327 was made obsolete by RFC 4566, see Section 2) string values of SDP parameters do not include special characters that are neither alphabets nor digits. An exception is the "/" character used in the value "RTP/AVP" of transport subfield of the "m=" line. String values used in SDP descriptions of ATM connections retain this convention, while allowing the use of the special character "/" in a manner commensurate with RFC 2327. In addition, the special characters "$" and "-" are used in the following manner. A "$" value is a wildcard that allows the recipient of the SDP description to select any permitted value of the parameter. A "-" value indicates that it is not necessary to specify the value of the parameter in the SDP description because this parameter is irrelevant for this application or because its value can be known from another source such as provisioning, defaults, another protocol, another SDP descriptor, or another part of the same SDP descriptor. If the use of these special characters is construed as a violation of RFC 2327 (that was made obsolete by RFC 4566, see Section 2) syntax, then reserved string values can be used. The string "CHOOSE" can be used in lieu of "$." The string "OMIT" can be used in lieu of "-" for an omitted parameter.

9.4.3 Capabilities Provided by SDP Conventions

To support the applications listed in Section 9.4.1, the SDP conventions in this document provide the following session control capabilities:

- Identification of the underlying bearer network type as ATM
- Identification by an ATM network element of its own address, in one of several possible formats. A connection peer can initiate SVC setup to this address. A call agent or connection peer can select a pre-established bearer path to this address.
- Identification of the ATM bearer connection that is to be bound to the service-level connection. Depending on the application, this is either a virtual channel connection (VCC) or a subchannel (identified by a CID) within a VCC.
- Identification of media type: audio, video, data
- In AAL1/AAL5 applications, declaration of a set of payload types that can be bound to the ATM bearer connection. The encoding names and payload types defined for use in the RTP context [9] are reused for AAL1 and AAL5, if applicable.
- In AAL2 applications, declaration of a set of profiles that can be bound to the ATM bearer connection. A mechanism for dynamically defining custom profiles within the SDP session description is included. This allows the use of custom profiles for connections that span multinetwork interfaces.
- A means of correlating service-level connections with underlying ATM bearer connections. The backbone network connection identifier or bnc-id specified in ITU Q.1901 [8] standardization work is used for this purpose. In order to provide a common SDP base for applications based on Q.1901 and SIP/SIP+, the neutral term "eecid" is used in lieu of "bnc-id" in the SDP session descriptor.
- A means of mapping codec types and packetization periods into service types (voice, voiceband data, and facsimile). This is useful in determining the encoding to use when the connection is upspeeded in response to modem or facsimile tones.

- A means of describing the adaptation type, Quality-of-Service (QOS) class, ATM transfer capability/service category, broadband bearer class, traffic parameters, Common Part Sublayer (CPS) parameters, and Service Specific Convergence Sublayer (SSCS) parameters related the underlying bearer connection
- Means for enabling or describing special functions such as leaf- initiated-join, anycast, and SVC caching
- For H.323 Annex C applications, a means of specifying the IP address and port number on which the node will receive RTCP messages
- A means of chaining consecutive SDP descriptors so that they refer to different layers of the same connection

9.4.4 Format of the ATM Session Description

The sequence of lines in the session descriptions in this document conforms to RFC 2327 (that was made obsolete by RFC 4566, see Section 2). In general, a session description consists of a session-level part followed by zero or more media-level parts. ATM session descriptions consist of a session-level part followed by one or two media-level parts. The only two media applicable are the ATM bearer medium and RTCP control (where applicable). The session level part consists of the following lines:

```
v= (protocol version, zero or one line)
o= (origin, zero or one line)
s= (session name, zero or one line)
c= (connection information, one line)
b= (bandwidth, zero or more lines)
t= (timestamp, zero or one line)
k= (encryption key, zero or one line)
```

In ATM session descriptions, there are no media attribute lines in the session-level part. These are present in the media-level parts. The media-level part for the ATM bearer consists of the following lines:

```
m= (media information and transport address, one line)
b= (bandwidth, zero or more lines)
k= (encryption key, zero or more lines)
a= (media attribute, zero or more lines)
```

The media-level part for RTCP control consists of the following lines:

```
m= (media information and transport address, one line)
c= (connection information for control only, one line)
```

In general, the "v=," "o=," "s=," and "t=" lines are mandatory. However, in the Megaco (RFC 3105 was made obsolete by RFC 3525; RFC 3525 is designated as historic) context, these lines have been made optional. The "o=," "s=," and "t=" lines are omitted in most MGCP (RFC 2705 was made obsolete by RFC 3435) applications. Note that SDP session descriptors for ATM can contain bandwidth (b=) and encryption key (k=) lines. Like all other lines, these lines should strictly conform to the SDP standard RFC 2327 (that was made obsolete by RFC 4566, see Section 2).

The bandwidth (b=) line is not necessarily redundant in the ATM context since, in some applications, it can be used to convey application-level information that does not map directly into the atmTrfcDesc media attribute line. For instance, the "b=" line can be used in SDP descriptors in RTSP commands to describe content bandwidth.

The encryption key line (k=) can be used to indicate an encryption key for the bearer and a method to obtain the key. At present, the encryption of ATM and AAL2 bearers has not been conventionalized, unlike the encryption of RTP payloads. Nor has the authentication or encryption of ATM or AAL2 bearer signaling. In the ATM and AAL2 contexts, the term "bearer" can include "bearer signaling" as well as "bearer payloads." The order of lines in an ATM session description is exactly in the RFC 2327-conformant (RFC 2327 was made obsolete by RFC 4566, see Section 2) order depicted above. However, there is no order of the media attribute ("a=") lines with respect to other "a" lines.

The SDP protocol version for session descriptions using these conventions is 0. In conformance with standard SDP, it is strongly recommended that the "v=" line be included at the beginning of each SDP session description. In some contexts such as Megaco, the "v" line is optional and may be omitted unless several session descriptions are provided in sequence, in which case the "v" line serves as a delimiter. Depending on the application, sequences of session descriptions might refer to:

- Different connections or sessions
- Alternate ways of realizing the same connection or session
- Different layers of the same session (Section 9.4.5.6.4.1)

The "o=," "s=." and "t=" lines are included for strict conformance with RFC 2327 (that was made obsolete by RFC 4566, see Section 2). It is possible that these lines will not carry useful information in some ATM-based applications. Therefore, some applications might omit these lines, although it is recommended that they not do so. For maximum interoperability, it is preferable that SDP parsers not reject session descriptions that do not contain these lines.

9.4.5 Structure of the Session Description Lines

9.4.5.1 The Origin Line

The origin line for an ATM-based session is structured as follows:

```
o=<username> <sessionID> <version> <networkType> <addressType>
<address>
```

The <username> is set to "-."

The <sessionID> can be set to one of the following:

- a Network Time Protocol (NTP) timestamp referring to the moment when the SDP session descriptor was created
- a Call ID, connection ID, or context ID that uniquely identifies the session within the scope of the ATM node. Since calls can comprise multiple connections (sessions), call IDs are generally not suitable for this purpose.

NTP time stamps can be represented as decimal or hex integers. The part of the NTP timestamp that refers to an integer number of seconds is sufficient. This is a 32-bit field.

On the other hand, call IDs, connection IDs and context IDs can be can be 32 hex digits long.

The <sessionID> field is represented as a decimal or hex number of up to 32 digits. A "0x" prefix is used before the hex representation. The <version> refers to the version of the SDP session descriptor (not that of the SDP protocol). This is can be set to one of the following:

- 0.
- an NTP timestamp referring to the moment when the SDP session descriptor was modified. If the SDP session descriptor has not been modified by an intermediate entity (such as an Media Gateway Controller [MGC]), then the <version> timestamp will be the same as the <sessionId> timestamp, if any. As with the <sessionId>, only the integer part of the NTP timestamp is used.

When equated to the integer part of an NTP timestamp, the <version> field is 10 digits wide. This is more restricted than RFC 2327 (that was made obsolete by RFC 4566, see Section 2), which allows unlimited size. As in RFC 2327 (that was made obsolete by RFC 4566, see Section 2), the most significant digit is non-zero when an NTP timestamp is used.

The <networkType> in SDP session descriptions for ATM applications should be assigned the string value "ATM" or wildcarded to a "$" or "-."

The <addressType> and <address> parameters are identical to those for the connection information ("c=") line. Each of these parameters can be wildcarded per the conventions described for the "c=" line. These parameters should not be omitted since this would violate SDP syntax of RFC 2327 (that was made obsolete by RFC 4566, see Section 2).

As with the "c=" line, SDP parsers are not expected to check the consistency of <networkType> with <addressType>, <address> pairs. The <addressType> and <address> need to be consistent with each other.

9.4.5.2 The Session Name Line

In general, the session name line is structured as follows:

```
s=<sessionName>
```

For ATM-based sessions, the <sessionName> parameter is set to a "-." The resulting line is

```
s=-
```

9.4.5.3 The Connection Information Line

In general, the connection information line [1] is structured as follows:

```
c=<networkType> <addressType> <address>
```

For ATM networks, additional values of <networkType>, <addressType>, and <address> are defined, over and above those listed in RFC 2327 (that was made obsolete by RFC 4566, see Section 2). The ABNF syntax (Section 9.4.9) for ATM SDP does not limit the ways in which <networkType> can be combined with <addressType>, <address> pairs.

However, some combinations will not be valid in certain applications, while others will never be valid. Invalid combinations should be rejected by application-specific functions and not by generic parsers. The ABNF syntax does limit the ways in which <addressType> and <address> can be paired. For ATM networks, the value of <networkType> should be set to "ATM." Further, this may be wildcarded to "$" or "-." If this is done, a node using ATM as the basic transport mechanism will select a value of "ATM." A node that interfaces with multiple network types ("IN," "ATM," etc.) that include ATM can also choose a value of "ATM."

When the SDP description is built by a node such as a media gateway, the <address> refers to the address of the node building the SDP description. When this description is forwarded to another node, it still contains the original node's address. When the media gateway controller builds part or all of the SDP description, the local descriptor contains the address of the local node, while the remote descriptor contains the address of the remote node. If the <address> and/or <addressType> are irrelevant or are known by other means, they can be set to a "$" or a "-," as described below.

Additionally, in all contexts, the "m=" line can have an ATM address in the <virtualConnectionId> subparameter, which, if present, is the remote address if the "c=" line address is local and vice versa.

For ATM networks, the <addressType> can be NSAP, E164, or Gateway Indemnification (GWID) (ALIAS). For ATM networks, the <address> syntax depends on the syntax of the <addressType>. SDP parsers should check the consistency of <addressType> with <address>.

NSAP: If the addressType is NSAP, the address is expressed in the standard dotted hex form. This is a string of 40 hex digits, with dots after the 2nd, 6th, 10th, 14th, 18th, 22nd, 26th, 30th, 34th, and 38th digits. The last octet of the NSAP address is the "selector" field that is available for nonstandard use. An example of a line with an NSAP address is

```
c=ATM NSAP 47.0091.8100.0000.0060.3e64.fd01.0060.3e64.fd01.00
```

A "0x" prefix shall not be used in this case since this is always in hexadecimal format.

E164: If the addressType is E164, the address is expressed as a decimal number with up to 15 digits, for example

```
c=ATM E164 9738294382
```

The use of E.164 numbers in the B-ISDN context is defined in ITU E.191. There is a disparity between the ATM forum and the International Telecommunication Union (ITU) in the use of E.164 numbers for ATM addressing. The ATM forum (e.g., UNI Signaling 4.0) allows only International Format E.164 numbers, while the ITU (e.g., Q.2931) allows private numbering plans. Since the goal of this SDP specification is to interoperate with all bearer signaling protocols, it allows the use of numbers that do not conform to the E.164 International Format. However, to maximize overall consistency, network administrators can restrict the provisioning of numbers to the E.164 International Format. GWID (ALIAS): If the addressType is GWID, it means that the address is a gateway identifier or node alias. This may or may not be globally unique. In this format, the address is expressed as an alphanumeric string ("A"-"Z," "a"-"z," "0" - "9,"".,""-." "_"), for example,

```
c=ATM GWID officeABCmgx101vism12
```

Since these SDP conventions can be used for more than gateways, the string "ALIAS" can be used instead of "GWID" in the "c=" line. Thus, the previous example is equivalent to:

```
c=ATM ALIAS officeABCmgx101vism12
```

An example of a GWID (ALIAS) is the CLLI code used for telecom equipment. For all practical purposes, it should be adequate for the GWID (ALIAS) to be a variable length string with a maximum size of 32 characters.

The connection information line is always present in an SDP session descriptor. However, each of the parameters on this line can be wildcarded to a "$" or a "-" independently of whether other parameters on this line are wildcarded. Not all syntactically legal wildcard combinations are meaningful in a particular application.

Examples of meaningful wildcard combinations in the ATM context are

```
c=- - -
c=$ $ $
c=ATM - -
c=ATM $ $
c=ATM <addressType> -
c=ATM <addressType> $
```

Specifying the ATM address type without specifying the ATM address is useful when the recipient is asked to select an ATM address of a certain type (NSAP, E.164, etc.). Examples of syntactically legal wildcard combinations of dubious utility are

```
c=- $ -
c=- $ $
c=- <addressType> -
c=$ <addressType> $
c=- <addressType> <address>
c=$ <addressType> <address>
```

Note that <addressType> and/or <address> should not omitted without being set to a "-" or "$" since this would violate basic SDP syntax of RFC 2327 (that was made obsolete by RFC 4566, see Section 2).

9.4.5.4 The Timestamp Line

The timestamp line for an SDP session descriptor is structured as follows:

```
t= <startTime> <stopTime>
```

Per Reference [49], NTP time stamps use a 32-bit unsigned representation of seconds and a 32-bit unsigned representation of fractional seconds. For ATM-based sessions, the <startTime> parameter can be made equal to the NTP timestamp referring to the moment when the SDP session descriptor was created. It can also be set to 0 indicating its irrelevance. If it is made equal to the NTP timestamp in seconds, the fractional part of the NTP timestamp is omitted. When equated to the integer part of an NTP timestamp, the <startTime> field is 10 digits wide. This is more restricted than RFC 2327 (that was made obsolete by RFC 4566, see Section 2), which allows unlimited size. As in RFC 2327 (that was made obsolete by RFC 4566, see Section 2), the most significant digit is non-zero when an NTP timestamp is used. The <stopTime> parameter is set to 0 for ATM-based SDP descriptors.

9.4.5.5 Media Information Line for ATM Connections

The general format of the media information line adapted for AAL1 and AAL5 applications is

```
m=<media> <virtualConnectionId> <transport> <format list>
```

The general format of the media information line adapted for AAL2 applications is

```
m=<media> <virtualConnectionId> <transport#1> <format list#1>
<transport#2>
  <format list#2> ... <transport#M> <format list#M>
```

Note that <virtualConnectionId> is equivalent to <port> in RFC 2327 (that was made obsolete by RFC 4566, see Section 2).

The subparameter <media> can take on all the values defined in RFC 2327 (that was made obsolete by RFC 4566, see Section 2). These are "audio," "video," "application," "data," and "control."

When the <transport> parameter has more than one value in the "m=" line, the <transport> <format list> pairs can be arranged in preferential order.

9.4.5.5.1 The Virtual Connection ID

In applications in which the media-level part of a session descriptor is bound to an ATM virtual circuit, the <virtualConnectionId> can be in one of the following formats:

- <ex_vcci>
- <addressType>-<address>/<ex_vcci>
- <address>/<ex_vcci>
- <ex_bcg>/<ex_vcci>
- <ex_portId>/<ex_vpi>/<ex_vci>
- <ex_bcg>/<ex_vpi>/<ex_vci>
- <ex_vpci>/<ex_vci>
- <addressType>-<address>/<ex_vpci>/<ex_vci>
- <address>/<ex_vpci>/<ex_vci>

In applications in which the media-level part of a session descriptor is bound to a subchannel within an ATM virtual circuit, the <virtualConnectionId> can be in one of the following formats:

- <ex_vcci>/<ex_cid>
- <addressType>-<address>/<ex_vcci>/<ex_cid>
- <address>/<ex_vcci>/<ex_cid>
- <ex_bcg>/<ex_vcci>/<ex_cid>
- <ex_portId>/<ex_vpi>/<ex_vci>/<ex_cid>
- <ex_bcg>/<ex_vpi>/<ex_vci>/<ex_cid>
- <ex_vpci>/<ex_vci>/<ex_cid>
- <addressType>-<address>/<ex_vpci>/<ex_vci>/<ex_cid>
- <address>/<ex_vpci>/<ex_vci>/<ex_cid>

Here,

- <ex_vcci> = VCCI-<vcci>
- <ex_vpci> = VPCI-<vpci>
- <ex_bcg> = BCG-<bcg>
- <ex_portId> = PORT-<portId>
- <ex_vpi> = VPI-<vpi>
- <ex_vci> = VCI-<vci>
- <ex_cid> = CID-<cid>

The <vcci>, <vpi>, <vci>, <vpci>, and <cid> are decimal numbers or hexadecimal numbers. An "0x" prefix is used before their values when they are in the hex format.

The <portId> is always a hexadecimal number. An "0x" prefix is not used with it.

The <addressType> and <address> are identical to their definitions above for the connection information line with the difference that this address refers to the remote peer in the media information line. Since the <virtualConnectionId>, as defined here, is meant for use in ATM networks, the values of <addressType> and <address> in the <virtualConnectionId> are limited to ATM-specific values. The <vpi>, <vci>, and <cid> are the virtual path identifier, virtual circuit identifier, and channel identifier, respectively. The <vpi> is an 8- or 12-bit field. The <vci> is a 16-bit field. The <cid> is an 8-bit field ([4] and [5]). For AAL1 applications, it corresponds to the channel number defined in Annex C of [4].

The <vpci> is a 16-bit field defined in Section 4.5.16 of ITU Q.2931 [10]. The <vpci> is similar to the <vpi>, except for its width and the fact that it retains its value across virtual path (VP) crossconnects. In some applications, the size of the <vpci> is the same as the size of the <vpi> (8 or 12 bits). In this case, the most significant 8 or 4 bits are ignored.

The <vcci> is a 16-bit field defined in ITU Recommendation Q.2941.2 [11]. The <vcci> is similar to the <vci>, except for the fact that it retains its value across virtual circuit (VC) crossconnects.

In general, <vpci> and <vcci> values are unique between a pair of nodes. When they are unique between a pair of nodes but not unique within a network, they need to be qualified, at any node, by the ATM address of the remote node. These parameters can be preprovisioned or signaled. When signaled, the <vpci> is encapsulated in the connection identifier information element (IE) of SVC signaling messages.

The <vcci> is encapsulated in the generic information transport (GIT) IE of SVC signaling messages. In an ATM node pair, either node can assign <vcci> values and signal them to the other end via SVC signaling. A glare avoidance scheme is defined in [11] and [12]. This mechanism works in SVC applications. A different glare avoidance technique is needed when a pool of existing PVCs/SPVCs is dynamically assigned to calls. One such scheme for glare reduction is the assignment of <vcci> values from different ends of the <vcci> range, using the lowest or highest available value as applicable.

When <vpci> and <vcci> values are preprovisioned, administrations have the option of provisioning them uniquely in a network. In this case, the ATM address of the far end is not needed to qualify these parameters.

In the AAL2 context, the definition of a VCC implies that there is no CID-level switching between its ends. If either end can assign <cid> values, then a glare reduction mechanism is needed. One such scheme for glare reduction is the assignment of <cid> values from different ends of the <cid> range, using the lowest or highest available value as applicable.

The <portId> parameter is used to identify the physical trunk port on an ATM module. It can be represented as a hexadecimal number of up to 32 hex digits.

In some applications, it is meaningful to bundle a set of connections between a pair of ATM nodes into a bearer connection group. The <bcg> subparameter is an 8-bit field that allows the bundling of up to 255 VPCs or VCCs.

In some applications, it is necessary to wildcard the <virtualConnectionId> parameter, or some elements of this parameter. The "$" wildcard character can be substituted for the entire <virtualConnectionId> parameter, or some of its terms. In the latter case, the constant strings that qualify the terms in the <virtualConnectionId> are retained. The concatenation <addressType>-<address> can be wildcarded in the following ways:

- The entire concatenation, <addressType>-<address>, is replaced with a "$."
- <address> is replaced with a "$," but <addressType> is not.

Examples of wildcarding the <virtualConnectionId> in the AAL1 and AAL5 contexts are $, VCCI-$, BCG-100/VPI-20/VCI-$. Examples of wildcarding the <virtualConnectionId> in the AAL2 context are $, VCCI-40/CID-$, BCG-100/VPI-20/VCI-120/CID-$, NSAP-$/VCCI-$/CID-$, $/VCCI-$/CID-$.

It is also permissible to set the entire <virtualConnectionId> parameter to a "-" indicating its irrelevance.

9.4.5.5.2 The Transport Parameter

The <transport> parameter indicates the method used to encapsulate the service payload. These methods are not defined in this document, which refers to existing Asynchronous Transfer Mode Forum (ATMF) and ITU-T standards, which, in turn, might refer to other standards. For ATM applications, the following <transport> values are defined (Table 9.1):

In H.323 Annex C applications [6], the <transport> parameter has a value of "RTP/AVP." This is because these applications use the RTP Protocol (RFC 1889 was made obsolete by RFC 3550) and audio/video profile (RFC 1890 was made obsolete by RFC 3551). The fact that RTP is carried directly over AAL5 per [6] can be indicated explicitly via the aalApp media attribute.

A value of "AAL1/custom," "AAL2/custom," or "AAL5/custom" for the <transport> parameter can indicate nonstandard or semi-standard encapsulation schemes defined by a corporation or a multivendor agreement. Since there is no standard administration of this convention, care should be taken to preclude inconsistencies within the scope of a deployment.

Table 9.1 List of Transport Parameter Values Used in SDP in the ATM Context (Copyright: IETF).

Transport	*Controlling Document for Encapsulation of Service Payload*
AAL1/ATMF	af-vtoa-0078.000 [15]
AAL1/ITU	ITU-T H.222.1 (RFC 2833) [47]
AAL5/ATMF	af-vtoa-0083.000 [13]
AAL5/ITU	ITU-T H.222.1 (RFC 2833)
AAL2/ATMF	af-vtoa-0113.000 [25] and af-vmoa-0145.000 [23]
AAL2/ITU	ITU-T I.366.2 [10]
AAL1/custom AAL2/custom AAL5/custom	Corporate document or application-specific interoperability statement
AAL1/<corporateName> AAL2/<corporateName> AAL5/<corporateName> AAL1/IEEE:<oui> AAL2/IEEE:<oui> AAL5/IEEE:<oui>	Corporate document
RTP/AVP	Annex C of H.323 [6]

The use of <transport> values "AAL1/<corporateName>," "AAL2/<corporateName>," "AAL5/<corporateName>," "AAL1/IEEE:<oui>," "AAL2/IEEE:<oui>," and "AAL5/IEEE:<oui>" is similar. These indicate nonstandard transport mechanisms or AAL2 profiles, which should be used consistently within the scope of an application or deployment. The parameter <corporateName> is the registered, globally unique name of a corporation (e.g., Cisco, Telcordia, etc.). The parameter <oui> is the hex representation of a three-octet field identical to the Organizationally Unique Identifier (OUI) maintained by the Institute of Electrical and Electronics Engineers (IEEE). Since this is always represented in hex, the "0x" prefix shall not be used. Leading zeros can be omitted. For example, "IEEE:00000C" and "IEEE:C" both refer to Cisco Systems, Inc.

9.4.5.5.3 The Format List for AAL1 and AAL5 Applications

In the AAL1 and AAL5 contexts, the <format list> is a list of payload types:

```
<payloadType#1> <payloadType#2> . . . <payloadType#n>
```

In most AAL1 and AAL5 applications, the ordering of payload types implies a preference (preferred payload types before less favored ones). The payload type can be statically assigned or dynamically mapped. Although the transport is not the same, SDP in the ATM context leverages the encoding names and payload types registered with IANA [9] for RTP. Encoding names not listed in [9] use an "X-" prefix. Encodings that are not statically mapped to payload types in [9] are to be dynamically mapped at the time of connection establishment to payload types in the decimal range 96-127. The SDP "atmmap" attribute (similar to "rtpmap") is used for this purpose. In addition to listing the IANA-registered encoding names and payload types found in [9], Table 9.2 defines a few nonstandard encoding names (with "X-" prefixes).

9.4.5.5.4 The Format List for AAL2 Applications

In the AAL2 context, the <format list> is a list of AAL2 profile types:

```
<profile#1> <profile#2>...<profile#n>
```

In most applications, the ordering of profiles implies a preference (preferred profiles before less favored ones). The <profile> parameter is expressed as a decimal number in the range 1-255.

9.4.5.5.5 Media Information Line Construction

Using the parameter definitions above, the "m=" for AAL1-based audio media can be constructed as follows:

```
m=audio <virtualConnectionId> AAL1/ATMF <payloadType#1>
  <payloadType#2>...<payloadType #n>
```

Note that only those payload types, whether statically mapped or dynamically assigned, that are consistent with af-vtoa-78 [13] can be used in this construction.

Backwards compatibility note: The transport value "AAL1/AVP" used in previous versions of this document should be considered equivalent to the value "AAL1/ATMF" defined above. "AAL1/AVP" is unsuitable because the AVP profile is closely tied to RTP.

An example "m=" line use for audio media over AAL1 is:

```
m=audio VCCI-27 AAL1/ATMF 0
```

This indicates the use of an AAL1 VCC with VCCI=24 to carry Pulse Code Modulation mu-law (PCMU) audio that is encapsulated according to ATMF's af-vtoa-78 [13]. Another example of the use of the "m=" line use for audio media over AAL1 is:

```
m=audio $ AAL1/ATMF 0 8
```

This indicates that any AAL1 VCC may be used. If it exists already, then its selection is subject to glare rules. The audio media on this VCC is encapsulated according to ATMF's af-vtoa-78 [13]. The encodings to be used are either Pulse Coded Modulation mu-law (PCMU) or Pulse Coded Modulation a-law (PCMA), in preferential order.

The "m" for AAL5-based audio media can be constructed as follows:

```
m=audio <virtualConnectionId> AAL5/ATMF <payloadType#1>
  <payloadType#2>...<payloadType #n>
```

An example "m" line use for audio media over AAL5 is:

```
m=audio PORT-2/VPI-6/$ AAL5/ITU 9 15
```

implies that any VCI on VPI= 6 of trunk port #2 may be used. The identities of the terms in the virtual connection ID are implicit in the application context. The audio media on this VCC is encapsulated according to ITU-T H.222.1 [14]. The encodings to be used are either ITU-T G.722 or ITU-T G.728 (LD-CELP), in preferential order.

The "m=" for AAL5-based H.323 Annex C audio [6] can be constructed as follows:

```
m=audio <virtualConnectionId> RTP/AVP <payloadType#1>
  <payloadType#2>...<payloadType #n>
```

For example:

```
m=audio PORT-9/VPI-3/VCI-$ RTP/AVP 2 96
a=rtpmap:96 X-G727-32
a=aalType:AAL5
a=aalApp:itu_h323c - -
```

implies that any VCI on VPI= 3 of trunk port #9 may be used. This VC encapsulates RTP packets directly on AAL5 per [6]. The "rtpmap" (rather than the "atmmap") attribute is used to dynamically map the payload type of 96 into the codec name X-G727-32 (Table 9.2). This name represents 32 kbps EADPCM.

The "m=" line for AAL5-based video media can be constructed as follows:

```
m=video <virtualConnectionId> AAL5/ITU <payloadType#1>
  <payloadType#2>...<payloadType #n>
```

In this case, the use of AAL5/ITU as the transport points to H.222.1 as the controlling standard [41]. An example "m=" line use for video media is:

```
m=video PORT-9/VPI-3/VCI-$ AAL5/ITU 33
```

This indicates that any VCI on VPI= 3 of trunk port #9 may be used. The video media on this VCC is encapsulated according to ITU-T H.222.1 [14]. The encoding scheme is an MPEG 2 transport stream ("MP2T" in Table 9.1). This is statically mapped per [9] to a payload type of 33.

Using the parameter definitions in the previous subsections, the media information line for AAL2-based audio media can be constructed as follows:

```
m=<media> <virtualConnectionId> <transport#1> <format list#1>
  <transport#2> <format list#2> ... <transport#M> <format list#M>
```

where <format list#i> has the form <profile#i_1>...<profile#i_N>

Unlike the "m=" line for AAL1 or AAL5 applications, the "m=" line for AAL2 applications can have multiple <transport> parameters, each followed by a <format list>. This is because it is possible to consider definitions from multiple sources (ATMF, ITU, and nonstandard documents) when selecting an AAL2 profile to be bound to a connection.

In most applications, the ordering of profiles implies a preference (preferred profiles before less favored ones). Therefore, there can be multiple instances of the same <transport> value in the same "m= line.

An example "m=" line use for audio media over AAL2 is

```
m=audio VCCI-27/CID-19 AAL2/ITU 7 AAL2/custom 100 AAL2/ITU 1
```

This indicates the use of CID #19 on VCCI #27 to carry audio. It provides a preferential list of profiles for this connection: profile AAL2/ITU 7 defined in [15], AAL2/custom 100 defined in an application-specific or interoperability document, and profile AAL2/ITU 1 defined in [15].

Another example of the use of the "m" line use for audio media over AAL2 is

```
m=audio VCCI-$/CID-$ AAL2/ATMF 6 8
```

This indicates that any AAL2 CID may be used, subject to any applicable glare avoidance/reduction rules. The profiles that can be bound to this connection are AAL2/ATMF 6 defined in af-vtoa-0113.000 [12] and AAL2/ATMF 8 defined in af-vmoa-0145.000 [35]. These sources use nonoverlapping profile number ranges. The profiles they define fall under the <transport> category "AAL2/ATMF." This application does not order profiles preferentially. This rule is known a priori. It is not embedded in the "m=" line.

Another example of the use of the "m" line for audio media over AAL2 is

```
m=audio VCCI-20/CID-$ AAL2/xyzCorporation 11
```

AAL2 VCCs in this application are single-CID VCCs. Therefore, it is possible to wildcard the CID. The single-CID VCC with VCCI=20 is selected. The AAL2 profile to be used is AAL2/xyzCorporation 11 defined by xyzCorporation.

In some applications, a "-" can be used in lieu of

- <format list>
- <transport> and <format list>

This implies that these parameters are irrelevant or are known by other means (such as defaults). For example

```
m=audio VCCI-234 - -
a=aalType:AAL1
```

indicates the use of VCCI=234 with AAL1 adaptation and unspecified encoding.

In another example application, the "aal2sscs3662" attribute can indicate <faxDemod> = "on" and any other competing options as "off," and the <aalType> attribute can indicate AAL2. Thus

```
m=audio VCCI-123/CID-5 - -
a=aalType:AAL2
a=aal2sscs3662:audio off off on off on off off off - - -
```

Besides indicating an audio medium, a VCCI of 123 and a CID of 5, the "m=" line indicates an unspecified profile. The media attribute lines indicate an adaptation layer of AAL2 and the use of the audio SAP [15] to carry demodulated facsimile.

The media information line for "data" media has one of the following the following formats:

```
m=data <virtualConnectionId> - -
m=data - - -
```

The data could be circuit emulation data carried over AAL1 or AAL2 or packet data carried over AAL5. Media attribute lines, rather than the "m=" line, are used to indicate the adaptation type for the data media. Examples of the representation of data media are listed below.

```
m=data PORT-7/VPI-6/VCI-$ - -
a=aalApp:AAL5_SSCOP- -
```

implies that any VCI on VPI= 6 of trunk port #7 may be used. This VC uses SSCOP on AAL5 to transport data.

```
m=data PORT-7/VPI-6/VCI-50 - -
a=aalType:AAL1_SDT
a=sbc:6
```

implies that VCI 50 on VPI 6 on port 7 uses structured AAL1 to transfer 6 x 64 kbps circuit emulation data. This may be alternately represented as:

```
m=data PORT-7/VPI-6/VCI-50 - -
b=AS:384
a=aalType:AAL1_SDT
```

The following lines:

```
m=data VCCI-123/CID-5 - -
a=aalType:AAL2
a=sbc:2
```

imply that CID 5 of VCCI 123 is used to transfer 2 x 64 kbps circuit emulation data.

In the AAL1 context, it is also permissible to represent circuit mode data as an "audio" codec. If this is done, the codec types used are X-CCD or X-CCD-CAS. These encoding names are dynamically mapped into payload types through the "atmmap" attribute. For example,

```
m=audio VCCI-27 AAL1/AVP 98
a=atmmap:98 X-CCD
a=sbc:6
```

implies that AAL1 VCCI=27 is used for 6 x 64 transmission.

In the AAL2 context, the X-CCD codec can be assigned a profile type and number. Even though it is not possible to construct a profile table as described in ITU I.366.2 for this "codec," it is preferable to adopt the common AAL2 profile convention in its case.

An example AAL2 profile mapping for the X-CCD codec could be as follows: PROFILE TYPE PROFILE NUMBER "CODEC" (ONLY ONE) "custom" 200 X-CCD. The profile does not identify the number of subchannels ("n" in nx64). This is known by other means such as the "sbc" media attribute line.

For example, the media information line

```
m=audio $ AAL2/custom 200
a=sbc:6
```

implies 384 kbps circuit emulation using AAL2 adaptation.

It is not necessary to define a profile with the X-CCD-CAS codec, since this method of channel associated signaling (CAS) transport [13] is not used in AAL2 applications.

9.4.5.6 The Media Attribute Lines

In an SDP line sequence, the media information line "m" is followed by one or more media attribute or "a" lines. Media attribute lines are per the following format:

```
a=<attribute>:<value>
```

or

```
a=<value>
```

In general, media attribute lines are optional except when needed to qualify the media information line. This qualification is necessary when the "m" line for an AAL1 or AAL5 session specifies a payload type that needs to be dynamically mapped. The "atmmap" media attribute line defined in the next paragraphs is used for this purpose.

In attribute lines, subparameters that are meant to be left unspecified are set to a "-." These are generally inapplicable or, if applicable, are known by other means such as provisioning. In some cases, a media attribute line with all parameters set to "-" carries no information and should be preferably omitted. In other cases, such as the "lij" media attribute line, the very presence of the media attribute line conveys meaning.

There are no restrictions placed by RFC 2327 (that was made obsolete by RFC 4566, see Section 2) regarding the order of "a" lines with respect to other "a" lines. However, these lines must not contradict each other or the other SDP lines. Inconsistencies are not to be ignored and should be flagged as errors. Repeated media attribute lines can carry additional information. These should not be inconsistent with each other.

Applications will selectively use the optional media attribute lines listed below. This is meant to be an exhaustive list for describing the general attributes of ATM bearer networks. The base specification for SDP, RFC 2327 (that was made obsolete by RFC 4566, see Section 2), allows the definition of new attributes. In keeping with this spirit, some of the attributes defined in this document can also be used in SDP descriptions of IP and other non-ATM sessions, for example, the "vsel," "dsel," and "fsel" attributes defined below refer generically to codec-s. These can be bad for service-specific codec negotiation and assignment in non-ATM as well as ATM applications.

SDP media attributes defined in this document for use in the ATM context are classified as

- ATM bearer connection attributes (Section 9.4.5.6.1)
- AAL attributes (Section 9.4.5.6.2)
- Service attributes (Section 9.4.5.6.3)
- Miscellaneous media attributes, that cannot be classified as ATM, AAL, or service attributes (Section 9.4.5.6.4). In addition to these, the SDP attributes defined in RFC 4566 (see Section 2) that obsoletes RFC 2327 can also be used in the ATM context. Examples are
- The attributes defined in RFC 4566 (see Section 2) that obsoletes RFC 2327; they allow indication of the direction in which a session is active. These are a=sendonly, a=recvonly, a=sendrecv, a=inactive.
- The "ptime" attribute defined in RFC 4566 (see Section 2) that obsoletes RFC 2327. It indicates the packet period. It is not recommended that this attribute be used in ATM applications since packet period information is provided with other parameters (e.g., the profile type and number in the "m" line, and the "vsel," "dsel," and "fsel" attributes). Also, for AAL1 applications, "ptime" is not applicable and should be flagged as an error. If used in AAL2 and AAL5 applications, "ptime" should be consistent with the rest of the SDP description.
- The "fmtp" attribute used to designate format-specific parameters

9.4.5.6.1 ATM Bearer Connection Attributes

The following is a summary list of the SDP media attributes that can be used to describe ATM bearer connections. These are detailed in subsequent subsections.

- The "eecid" attribute. This stands for "end-to-end connection identifier" and provides a means of correlating service-level connections with underlying ATM bearer connections. In the Q.1901 [8] context, the eecid is synonymous with the bnc-id (backbone network connection identifier).
- The "aalType" attribute. This is used to indicate the nature of the ATM adaptation layer (AAL).
- The "capability" attribute, which indicates the ATM transfer capability (ITU nomenclature), synonymous with the ATM Service Category (ATMF nomenclature)
- The "qosClass" attribute, which indicates the QOS class of the ATM bearer connection
- The "bcob" attribute, which indicates the broadband connection oriented bearer class and whether end-to-end timing is required
- The "stc" attribute, which indicates susceptibility to clipping
- The "upcc" attribute, which indicates the user plane connection configuration.
- The "atmQOSparms" attribute, which is used to describe certain key ATM QOS parameters
- The "atmTrfcDesc" attribute, which is used to describe ATM traffic descriptor parameters

- The "abrParms" attribute, which is used to describe ABRspecific parameters. These parameters are per the UNI 4.0 signaling specification [16]
- The "abrSetup" attribute, which is used to indicate the ABR parameters needed during call/connection establishment
- The "bearerType" attribute, which is used to indicate whether the underlying bearer is an ATM PVC/SPVC, an ATM SVC, or a subchannel within an existing ATM SVC/PVC/SPVC
- The "lij" attribute, which is used to indicate the presence of a connection that uses the leaf-initiated-join capability described in UNI 4.0 [16] and to optionally describe parameters associated with this capability
- The "anycast" attribute, which is used to indicate the applicability of the anycast function described in UNI 4.0 [16] and to optionally qualify it with certain parameters.
- The "cache" attribute, which is used to enable SVC caching and to specify an inactivity timer for SVC release
- The "bearerSigIE" attribute, which can be used to represent ITU Q-series IEs in bit-map form. This is useful in describing parameters that are not closely coupled to the ATM and AAL layers. Examples are the B-HLI and B-LLI IEs specified in ITU Q.2931 [10] and the user-to-user IE described in ITU Q.2957 [17].

9.4.5.6.1.1 The "eecid" Attribute The "eecid" attribute is synonymous with the 4-byte "bnc-id" parameter used by T1SI, the ATM forum and the ITU (Q.1901) standardization effort. The term "eecid" stands for "end-to-end connection identifier," while "bnc-id" stands for "backbone network connection identifier." The name "backbone" is slightly misleading since it refers to the entire ATM network including the ATM edge and ATM core networks. In Q.1901 terminology, an ATM "backbone" connects TDM or analog edges.

While the term "bnc-id" might be used in the bearer-signaling plane and in an ISUP (Q.1901) call control plane, SDP session descriptors use the neutral term "eecid." This provides a common SDP baseline for applications that use ISUP (Q.1901) and applications that use SIP/SIP+.

Section 9.4.5.6.6 depicts the use of the eecid in call establishment procedures. In these procedures, the eecid is used to correlate service-level calls with SVC set-up requests.

In the forward SVC establishment model, the call-terminating gateway selects an eecid and transmits it via SDP to the call-originating gateway. The call-originating gateway transmits this eecid to the call-terminating gateway via the bearer set-up message (SVC set-up or Q.2630.1 establish request).

In the backward SVC establishment model, the call-originating gateway selects an eecid and transmits it via SDP to the call-terminating gateway. The call-terminating gateway transmits this eecid to the call-originating gateway via the bearer set-up message (SVC set-up or Q.2630.1 establish request).

The value of the eecid attribute needs to be unique within the node terminating the SVC setup but not across multiple nodes. Hence, the SVC-terminating gateway has complete control over using and releasing values of this parameter. The eecid attribute is used to correlate, one-to-one, received bearer set-up requests with service-level call control signaling.

Within an SDP session description, the eecid attribute is used as follows:

```
a=eecid:<eecid>
```

where <eecid> consists of up to 8 hex digits (equivalent to 4 octets). Since this is always represented in hex, the "0x" prefix shall not be used.

Within the text representation of the <eecid> parameter, hex digits to the left are more significant than hex digits to the right (Section 9.4.2.2).

This SDP document does not specify how the eecid (synonymous with bnc-id) is to be communicated through bearer signaling (Q.931, UNI, PNNI, AINI, IISP, proprietary signaling equivalent, Q.2630.1). This is a task of these bearer-signaling protocols. However, the following informative statements are made to convey a sense of the interoperability that is a goal of current standardization efforts:

- ITU Q.2941.3 and the ATMF each recommend the use of the GIT IE for carrying the eecid (synonymous with bnc-id) in the set-up message of ATM signaling protocols (Q.2931, UNI 4.0, PNNI, AINI, and IISP). The coding for carrying the eecid (bnc-id) in the GIT IE is defined in ITU Q.2941.3 and accepted by the ATM forum.
- Another alternate method is to use the called party subaddress IE. In some networks, this might be considered a protocol violation and is not the recommended means of carrying the eecid (bnc-id). The GIT IE is the preferred method of transporting the eecid (bnc-id) in ATM signaling messages.
- The establish request message of the Q.2630.1 [18] signaling protocol can use the served user generated reference IE to transport the eecid (bnc-id).

The node assigning the eecid can release and reuse it when it receives a Q.2931 [10] set-up message or a Q.2630.1 [18] establish request message containing the eecid.

However, in both cases (backward and forward models), it is recommended that this eecid be retained until the connection terminates. Since the eecid space is large enough, it is not necessary to release it as soon as possible.

9.4.5.6.1.2 The "aalType" Attribute

When present, the "aalType" attribute is used to indicate the ATM adaptation layer. If this information is redundant with the "m" line, it can be omitted. The format of the "aalType" media attribute line is as follows:

```
a=aalType: <aalType>
```

Here, <aalType> can take on the following string values: "AAL1," "AAL1_SDT," "AAL1_UDT (Unstructured Data Transfer)," "AAL2," "AAL3/4," "AAL5," and "USER_DEFINED_AAL." Note that "AAL3/4" and "USER DEFINED AAL" are not addressed in this document.

9.4.5.6.1.3The "capability" Attribute

When present, the "capability" attribute indicates the ATM transfer capability described in ITU I.371 [19], equivalent to the ATM service category described in the UNI 4.1 traffic management specification [20].

The "capability" media attribute line is structured in one of the following ways:

```
a=capability:<asc> <subtype>
a=capability:<atc> <subtype>
```

Possible values of the <asc> are enumerated as follows:

```
"CBR," "nrt-VBR," "rt-VBR," "UBR," "ABR," "GFR"
```

Possible values of the <atc> are enumerated as follows:

```
"DBR,""SBR,""ABT/IT,""ABT/DT,""ABR"
```

Some applications might use nonstandard <atc> and <asc> values that are not listed above. Equipment designers will need to agree on the meaning and implications of nonstandard transfer capabilities/service capabilities.

The <subtype> field essentially serves as a subscript to the <asc> and <atc> fields. In general, it can take on any integer value or the "-" value indicating that it does not apply or that the underlying data is to be known by other means, such as provisioning.

For an <asc> value of CBR and an <atc> value of DBR, the <subtype> field can be assigned values from Table 4.6 of ITU Q.2931 [10]. These are:

<asc>/<atc>	<subtype>	Meaning
"CBR"/"DBR"	1	Voiceband signal transport (ITU G.711, G.722, I.363)
"CBR"/"DBR"	2	Circuit transport (ITU I.363)
"CBR"/"DBR"	4	High-quality audio signal transport (ITU I.363)
"CBR"/"DBR"	5	Video signal transport (ITU I.363)

Note that [10] does not define a <subtype> value of 3. For other values of the <asc> and <atc> parameters, the following values can be assigned to the <subtype> field, based on [20] and [19].

<asc>/<atc>	<subtype>	Meaning
nrt-VBR	1	nrt-VBR.1
nrt-VBR	2	nrt-VBR.2
nrt-VBR	3	nrt-VBR.3
rt-VBR	1	rt-VBR.1
rt-VBR	2	rt-VBR.2
rt-VBR	3	rt-VBR.3
UBR	1	UBR.1
UBR	2	UBR.2
GFR	1	GFR.1
GFR	2	GRR.2
SBR	1	SBR1
SBR	2	SBR2
SBR	3	SBR3

It is beyond the scope of this specification to examine the equivalence of some of the ATMF and ITU definitions. These need to be recognized from the ATMF and ITU source specifications and exploited, as much as possible, to simplify ATM node design.

When the bearer connection is a single AAL2 CID connection within a multiplexed AAL2 VC, the "capability" attribute does not apply.

9.4.5.6.1.4 The "qosClass" Attribute

When present, the "qosClass" attribute indicates the QOS class specified in ITU I.2965.1 [21]. The "qosClass" media attribute line is structured as follows:

```
a=qosClass:<qosClass>
```

Here, <qosClass> is an integer in the range 0-5.

<qosClass>	Meaning
0	Default QOS
1	Stringent
2	Tolerant
3	Bi-level
4	Unbounded
5	Stringent bi-level

9.4.5.6.1.5 The "bcob" Attribute When present, the "bcob" attribute represents the broadband connection-oriented bearer class defined in [16], [10], and [22]. It can also be used to indicate whether end-to-end timing is required. The "bcob" media attribute line is structured as follows:

```
a=bcob:<bcob> <eetim>
```

Here, <bcob> is the decimal or hex representation of a 5-bit field. The following values are currently defined:

<bcob>	Meaning
0x01	BCOB-A
0x03	BCOB-C
0x05	Frame relaying bearer service
0x10	BCOB-X
0x18	BCOB-VP (transparent VP service)

The <eetim> parameter can be assigned a value of "on" or "off," depending on whether end-to-end timing is required (Table 4.8 of [10]).

Either of these parameters can be left unspecified by setting it to a "-." A "bcob" media attribute line with all parameters set to "-" carries no information and should be omitted.

9.4.5.6.1.6 The "stc" Attribute When present, the "stc" attribute represents susceptibility to clipping. The "stc" media attribute line is structured as follows:

```
a=stc:<stc>
```

Here, <stc> is the decimal equivalent of a 2-bit field. Currently, all values are unused and reserved with the following exceptions:

<stc> value	Binary Equivalent	Meaning
0	00	Not susceptible to clipping
1	01	Susceptible to clipping

9.4.5.6.1.7 The "upcc" Attribute When present, the "upcc" attribute represents the user plane connection configuration. The "upcc" media attribute line is structured as follows:

```
a=upcc:<upcc>
```

Here, <upcc> is the decimal equivalent of a 2-bit field. Currently, all values are unused and reserved with the following exceptions:

```
<upcc> value        Binary Equivalent          Meaning
     0                    00                   Point to point
     1                    01                   Point to multipoint
```

9.4.5.6.1.8 The "atmQOSparms" Attribute When present, the "atmQOSparms" attribute is used to describe certain key ATM QOS parameters.

The "atmQOSparms" media attribute line is structured as follows:

```
a=atmQOSparms:<directionFlag><cdvType><acdv><ccdv><eetd><cmtd><aclr>
```

The <directionFlag> can be assigned the following string values: "f," "b," and "fb." "f" and "b" indicate the forward and backward directions respectively. "fb" refers to both directions (forward and backward). Conventions for the forward and backward directions are per section 9.4.2.3.

The <cdvType> parameter can take on the string values of "PP" and "2P." These refer to the peak-to-peak and two-point CDV as defined in UNI 4.0 [16] and ITU Q.2965.2 [23] respectively.

The CDV parameters, <acdv> and <ccdv>, refer to the acceptable and cumulative CDVs respectively. These are expressed in units of microseconds and represented as the decimal equivalent of a 24-bit field. These use the cell loss ratio, <aclr>, as the "alpha" quantiles defined in the ATMF TM 4.1 specification [20] and in ITU I.356 [24].

The transit delay parameters, <eetd> and <cmtd>, refer to the end-to-end and cumulative transit delays respectively in milliseconds. These are represented as the decimal equivalents of 16-bit fields. These parameters are defined in Q.2965.2 [23], UNI 4.0 [16] and Q.2931 [10].

The <aclr> parameter refers to forward and backward acceptable cell loss ratios. These are the ratios between the number of cells lost and the number of cells transmitted. It is expressed as the decimal equivalent of an 8-bit field. This field expresses an order of magnitude n, where n is an integer in the range 1-15. The cell loss ratio takes on the value 10 raised to the power of minus n.

The <directionFlag> is always specified. Except for the <directionFlag>, the remaining parameters can be set to "-" to indicate that they are not specified, inapplicable, or implied. However, there must be some specified parameters for the line to be useful in an SDP description.

There can be several "atmQOSparms" lines in an SDP description.

An example use of these attributes for an rt-VBR, single-CID AAL2 voice VC is:

```
a=atmQOSparms:f PP 8125 3455 32000 - 11
a=atmQOSparms:b PP 4675 2155 18000 - 12
```

This implies a forward acceptable peak-to-peak CDV of 8.125 ms, a backward acceptable peak-to-peak CDV of 4.675 ms, forward cumulative peak-to-peak CDV of 3.455 ms, a backward cumulative peak-to-peak CDV of 2.155 ms, a forward end-to-end transit delay of 32 ms, a backward end-to-end transit delay of 18 ms, an unspecified forward cumulative transit delay, an unspecified backward cumulative transit delay, a forward cell loss ratio of 10 raised to minus 11, and a backward cell loss ratio of 10 to the minus 12.

An example of specifying the same parameters for the forward and backward directions is

```
a=atmQOSparms:fb PP 8125 3455 32000 - 11
```

This implies a forward and backward acceptable peak-to-peak CDV of 8.125 ms, a forward and backward cumulative peak-to-peak CDV of 3.455 ms, a forward and backward end-to-end transit delay of 32 ms, an unspecified cumulative transit delay in the forward and backward directions, and a cell loss ratio of 10 raised to minus 11 in the forward and backward directions.

9.4.5.6.1.9 The "atmTrfcDesc" Attribute When present, the "atmTrfcDesc" attribute is used to indicate ATM traffic descriptor parameters. There can be several "atmTrfcDesc" lines in an SDP description.

The "atmTrfcDesc" media attribute line is structured as follows:

```
a=atmTrfcDesc:<directionFlag><clpLvl><pcr><scr><mbs>
  <cdvt><mcr><mfs><fd><te>
```

The <directionFlag> can be assigned the following string values: "f," "b," and "fb." "f" and "b" indicate the forward and backward directions respectively. "fb" refers to both directions (forward and backward). Conventions for the forward and backward directions are per Section 9.4.2.3.

The <directionFlag> is always specified. Except for the <directionFlag>, the remaining parameters can be set to "-" to indicate that they are not specified, inapplicable, or implied. However, there must be some specified parameters for the line to be useful in an SDP description.

The <clpLvl> (CLP level) parameter indicates whether the rates and bursts described in these media attribute lines apply to CLP values of 0 or (0+1). It can take on the following string values: "0," "0+1," and "-." If rates and bursts for both <clpLvl> values are to be described, then it is necessary to use two separate media attribute lines for each direction in the same session descriptor.

If the <clpLvl> parameter is set to "-," then it implies that the CLP parameter is known by other means such as default, Management Information Base (MIB) provisioning, etc.

The meaning, units, and applicability of the remaining parameters are per [20] and [19]:

PARAMETER	MEANING	UNITS	APPLICABILITY
<pcr>	PCR	Cells/ second	CBR, rt-VBR, nrt-VBR, ABR, UBR, GFR; CLP=0,0+1
<scr>	SCR Cells/ second		rt-VBR, nrt-VBR; CLP=0,0+1
<mbs>	MBS	Cells	rt-VBR, nrt-VBR, GFR; CLP=0,0+1
<cdvt>	CDVT	Microsec.	CBR, rt-VBR, nrt-VBR, ABR, UBR, GFR; CLP=0,0+1
<mcr>	MCR	Cells/ second	ABR,GFR; CLP=0+1
<mfs>	MFS	Cells	GFR; CLP=0,0+1
<fd>	Frame Discard Allowed	"on"/"off"	CBR, rt-VBR, nrt-VBR, ABR, UBR, GFR; CLP=0+1
<te>	CLP tagging Enabled	"on"/"off"	CBR, rt-VBR, nrt-VBR, ABR, UBR, GFR; CLP=0

<fd> indicates that frame discard is permitted. It can take on the string values of "on" or "off." Note that, in the GFR case, frame discard is always enabled. Hence, this subparameter can be set to "-" in the case of GFR. Since the <fd> parameter is independent of CLP, it is meaningful in the case when <clpLvl> = "0+1." It should be set to "-" for the case when <clpLvl> = "0."

<te> (tag enable) indicates that CLP tagging is allowed. These can take on the string values of "on" or "off." Since the <te> parameter applies only to cells with a CLP of 0, it is meaningful in the case when <clpLvl> = "0." It should be set to "-" for the case when <clpLvl> = "0+1."

An example use of these media attribute lines for an rt-VBR, single-CID AAL2 voice VC is:

```
a=atmTrfcDesc:f 0+1 200 100 20 - - - on -
a=atmTrfcDesc:f 0 200 80 15 - - - - off
a=atmTrfcDesc:b 0+1 200 100 20 - - - on -
a=atmTrfcDesc:b 0 200 80 15 - - - - off
```

This implies a forward and backward PCR of 200 cells per second all cells regardless of CLP, forward and backward PCR of 200 cells per second for cells with CLP=0, a forward and backward SCR of 100 cells per second for all cells regardless of CLP, a forward and backward SCR of 80 cells per second for cells with CLP=0, a forward and backward MBS of 20 cells for all cells regardless of CLP, a forward and backward MBS of 15 cells for cells with CLP=0, an unspecified CDVT, which can be known by other means and an MCR and MFS, which are unspecified because they are inapplicable. Frame discard is enabled in both the forward and backward directions. Tagging is not enabled in either direction.

The <pcr>, <scr>, <mbs>, <cdvt>, <mcr>, and <mfs> are represented as decimal integers, with range as defined in Section 9.4.6. See Section 9.4.2.2 regarding the omission of leading zeros in decimal representations.

9.4.5.6.1.10 The "abrParms" Attribute

When present, the "abrParms" attribute is used to indicate the "additional" ABR parameters specified in the UNI 4.0 signaling specification [16]. There can be several "abrParms" lines in an SDP description.

The "abrParms" media attribute line is structured as follows:

```
a=abrParms:<directionFlag><nrm><trm><cdf><adtf>
```

The <directionFlag> can be assigned the following string values: "f," "b," and "fb." "f" and "b" indicate the forward and backward directions, respectively. "fb" refers to both directions (forward and backward). Conventions for the forward and backward directions are per Section 9.4.2.3.

The <directionFlag> is always specified. Except for the <directionFlag>, the remaining parameters can be set to "-" to indicate that they are not specified, inapplicable, or implied. However, there must be some specified parameters for the line to be useful in an SDP description.

These parameters are mapped into the ABR service parameters in [20] in the manner described in the table. These parameters can be represented in SDP as decimal integers, with fractions permitted for some. Details of the meaning, units, and applicability of these parameters are in [16] and [20].

In SDP, these parameters are represented as the decimal or hex equivalent of the binary fields mentioned in the table.

PARAMETER	*MEANING*	*FIELD SIZE*
<nrm>	Maximum number of cells per forward resource management cell	3 bits
<trm>	Maximum time between forward resource management cells	3 bits
<cdf>	Cutoff decrease factor	3 bits
<adtf>	Allowed cell rate decrease time factor	10 bits

9.4.5.6.1.11 The "abrSetup" Attribute

When present, the "abrSetup" attribute is used to indicate the ABR parameters needed during call/connection establishment (Section 10.1.2.2 of the UNI 4.0 signaling specification [16]). This line is structured as follows:

```
a=abrSetup:<ficr><bicr><ftbe><btbe><crmrtt><frif><brif><frdf><brdf>
```

These parameters are defined as follows:

PARAMETER	*MEANING*	*REPRESENTATION*
<ficr>	Forward initial cell rate (Cells per second)	Decimal equivalent of 24-bit field
<bicr>	Backward initial cell rate (Cells per second)	Decimal equivalent of 24-bit field
<ftbe>	Forward transient buffer exposure (Cells per second)	Decimal equivalent of 24-bit field
<btbe>	Backward transient buffer exposure (Cells)	Decimal equivalent of 24-bit field
<crmrtt>	Cumulative RM round-trip time (Microseconds)	Decimal equivalent of 24-bit field
<frif>	Forward rate increase factor (used to derive cell count)	Decimal integer 0-15
<brif>	Backward rate increase factor (used to derive cell count)	Decimal integer 0-15
<frdf>	Forward rate decrease factor (used to derive cell count)	Decimal integer 0-15
<brdf>	Backward rate decrease factor (used to derive cell count)	Decimal integer 0-15

See Section 9.4.2.3 for a definition of the terms "forward" and "backward."

If any of these parameters in the "abrSetup" media attribute line is not specified, is inapplicable, or is implied, then it is set to h "-."

9.4.5.6.1.12 The "bearerType" Attribute When present, the "bearerType" attribute is used to indicate whether the underlying bearer is an ATM PVC/SPVC, an ATM SVC, or a subchannel within an existing ATM SVC/PVC/SPVC. Additionally, for ATM SVCs and AAL2 CID connections, the "bearerType" attribute can be used to indicate whether the media gateway initiates connection setup via bearer signaling (Q.2931 based or Q.2630.1 based). The format of the "bearerType" media attribute line is as follows:

```
a=bearerType: <bearerType> <localInitiation>
```

The <bearerType> field can take on the following string values: "PVC," "SVC," "CID," with semantics as defined in Section 9.4.5.5. Here, "PVC" includes both the PVC and SPVC cases.

In the case when the underlying bearer is a PVC/SPVC, or a CID assigned by the MGC rather than through bearer signaling, the <localInitiation> flag can be omitted or set to "-." In the case when bearer signaling is used, this flag can be omitted when it is known, by default or by other means, whether the media gateway initiates the connection setup via bearer signaling. Only when this is to be indicated explicitly that the <localInitiation> flag takes on the values of "on" or "off." An "on" value indicates that the media gateway is responsible for initiating connection setup via bearer signaling (SVC signaling or ITU-T Standard Q.2630.1 signaling), an "off" value indicates otherwise.

9.4.5.6.1.13 The "lij" Attribute When present, the "lij" attribute is used to indicate the presence of a connection that uses the leaf-initiated-join capability described in UNI 4.0 [16] and to optionally describe parameters associated with this capability. The format of the "lij" media attribute line is as follows:

```
a=lij: <sci><lsn>
```

The <sci> (screening indication) is a 4-bit field expressed as a decimal or hex integer. It is defined in the UNI 4.0 signaling specification [16]. It is possible that the values of this field will be defined later by the ATMF and/or ITU. Currently, all values are reserved with the exception of 0, which indicates a "network join without root notification."

The <lsn> (leaf sequence number) is a 32-bit field expressed as a decimal or hex integer. Per the UNI 4.0 signaling specification [16], it is used by a joining leaf to associate messages and responses during LIJ (leaf initiated join) procedures.

Each of these fields can be set to a "-" when the intention is to not specify it in an SDP descriptor.

9.4.5.6.1.14 The "anycast" Attribute When present, the "anycast" attribute line is used to indicate the applicability of the anycast function described in UNI 4.0 [16]. Optional parameters to qualify this function are provided. The format of the "anycast" attribute is:

```
a=anycast: <atmGroupAddress> <cdStd> <conScpTyp> <conScpSel>
```

The <atmGroupAddress> is per Annex 5 of UNI 4.0 [16]. Within an SDP descriptor, it can be represented in one of the formats (NSAP, E.164, GWID/ALIAS) described elsewhere in this document.

The remaining subparameters mirror the connection scope selection IE in UNI 4.0 [16]. Their meaning and representation is as shown in the following:

PARAMETER	*MEANING*	*REPRESENTATION*
<cdStd>	Coding standard for the connection scope selection IE Definition: UNI 4.0 [16]	Decimal or hex equivalent of 2 bits
<conScpTyp>	Type of connection scope Definition: UNI 4.0 [16]	Decimal or hex equivalent of 4 bits
<conScpSel>	Connection scope selection Definition: UNI 4.0 [16]	Decimal or hex equivalent of 8 bits

Currently, all values of <cdStd> and <conScpTyp> are reserved with the exception of <cdStd> = 3 (ATMF coding standard) and <conScpTyp> = 1 (connection scope type of "organizational").

Each of these fields can be set to a "-" when the intention is to not specify it in an SDP descriptor.

9.4.5.6.1.15 The "cache" Attribute This attribute is used to enable SVC caching. This attribute has the following format:

```
a=cache:<cacheEnable><cacheTimer>
```

The <cacheEnable> flag indicates whether caching is enabled, corresponding to the string values of "on" and "off," respectively.

The <cacheTimer> indicates the period of inactivity following which the SVC is to be released by sending an SVC release message into the network. This is specified as the decimal or hex equivalent of a 32-bit field, indicating the timeout in seconds. As usual, leading zeros can be omitted. For instance,

```
a=cache:on 7200
```

implies that the cached SVC is to be deleted if it is idle for 2 hours.

The <cacheTimer> can be set to "-" if it is inapplicable or implied.

9.4.5.6.1.16 The "bearerSigIE" Attribute ATM signaling standards provide "escape mechanisms" to represent, signal, and negotiate higher-layer parameters. Examples are the B-HLI and B-LLI IEs specified in ITU Q.2931 [10], and the user-to-user IE described in ITU Q.2957 [17].

The "bearerSigIE" (bearer signaling IE) attribute is defined to allow a similar escape mechanism that can be used with these ATM SDP conventions. The format of this media attribute line is as follows:

```
a=bearerSigIE: <bearerSigIEType> <bearerSigIELng> <bearerSigIEVal>
```

When a "bearerSigIE" media attribute line is present, all its subparameters are mandatory. The "0x" prefix is not used since these are always represented in hex.

The <bearerSigIEType> is represented as exactly 2 hex digits. It is the unique IE identifier as defined in the ITU Q-series standards. Leading zeros are not omitted. Some pertinent values are 7E (Useruser IE per ITU Q.2957 [17]), 5F (B-LLI IE), and 5D (B-HLI IE). B-LLI and B-HLI, which stand for broadband low-layer information and broadband high-layer information, respectively, are defined in ITU Q.2931 [10]. Both of these refer to layers above the ATM adaptation layer.

The <bearerSigIELng> consists of 1-4 hex digits. It is the length of the IE in octets. Leading zeros may be omitted.

The <bearerSigIEVal> is the value of the IE, represented as a hexadecimal bit map. Although the size of this bit map is network/service dependent, setting an upper bound of 256 octets (512 hex digits) is adequate. Since this a bit map, leading zeros should not be omitted. The number of hex digits in this bit map is even.

9.4.5.6.2 ATM AAL Attributes

The following is a summary of the SDP media attributes that can be used to describe the ATM AAL. These are detailed in subsequent subsections.

- The "aalApp" attribute, which is used to point to the controlling standard for an AAL above the ATM AAL
- The "cbrRate" attribute, which represents the constant bit rate (CBR) rate octet defined in Table 4.6 of ITU Q.2931 [10]
- The "sbc" attribute, which denotes the subchannel count in the case of n x 64 clear channel communication
- The "clkrec" attribute, which indicates the clock recovery method for AAL1 UDT
- The "fec" attribute, which indicates the use of forward error correction
- The "prtfl" attribute, which indicates the fill level of partially filled cells
- The "structure" attribute, which is used to indicate the presence or absence of AAL1 SDT and the size of the SDT blocks
- The "cpsSDUsize" attribute, which is used to indicate the maximum size of the CPCS SDU payload
- The "aal2CPS" attribute, which is used to indicate that an AAL2 CPS sublayer as defined in ITU I.363.2 [15] is associated with the VCC referred to in the "m=" line. Optionally, it can be used to indicate selected CPS options and parameter values for this VCC.
- The "aal2CPSSDUrate" attribute, which is used to place an upper bound on the SDU bit rate for an AAL2 CID.
- The "aal2sscs3661unassured" attribute, which is used to indicate the presence of an AAL2 SSCS sublayer with unassured transmission as defined in ITU I.366.1 [25]. Optionally, it can be used to indicate selected options and parameter values for this SSCS.
- The "aal2sscs3661assured" attribute, which is used to indicate the presence of an AAL2 SSCS sublayer with assured transmission as defined in ITU I.366.1 [25]. Optionally, it can be used to indicate selected options and parameter values for this SSCS.

- The “aal2sscs3662” attribute, which is used to indicate the presence of an AAL2 SSCS sublayer as defined in ITU I.366.2. Optionally, it can be used to indicate selected options and parameter values for this SSCS.
- The “aal5sscop” attribute, which is used to indicate the existence of an SSCOP protocol layer over an AAL5 CPS layer, and the parameters that pertain to this SSCOP layer.

9.4.5.6.2.1 The “aalApp” Attribute When present, the “aalApp” attribute is used to point to the controlling standard for an application layer above the ATM adaptation layer. The format of the “aalApp” media attribute line is as follows:

```
a=aalApp: <appClass> <oui> <appId>
```

If any of the subparameters, <appClass>, <oui>, or <appId>, is meant to be left unspecified, it is set to "-." However, an “aalApp” attribute line with all subparameters set to "-" carries no information and should be omitted.

The <appClass>, or application class, field can take on the string values as listed below here.

This list is not exhaustive. An "X-" prefix should be used with <appClass> values not listed here.

```
<appClass>                 Meaning

"itu_h323c"                Annex C of H.323 which specifies direct
                           RTP on AAL5 [6].
"af83"                     af-vtoa-0083.001, which specifies
                           variable size AAL5 PDUs with PCM voice
                           and a null SSCS [7].
"AAL5_SSCOP"               SSCOP as defined in ITU Q.2110 [26]
                           running over an AAL5 CPS [27].
                           No information is provided regarding
                           any layers above SSCOP such as Service
                           Specific Coordination Function (SSCF)
                           layers.
"itu_i3661_unassured"      SSCS with unassured transmission,
                           per ITU I.366.1 [25].
"itu_i3661_assured"        SSCS with assured transmission,
                           per ITU I.366.1 [25]. This uses SSCOP [26].
"itu_i3662"                SSCS per ITU I.366.2 [15].
"itu_i3651"                Frame relay SSCS per ITU I.365.1 [28].
"itu_i3652"                Service-specific coordination function,
                           as defined in ITU I.365.2, for Connection
                           Oriented Network Service (SSCF-CONS) [29].
                           This uses SSCOP [26].
"itu_i3653"                Service-specific coordination function,
                           as defined in ITU I.365.3, for Connection
                           Oriented Transport Service
                           (SSCF-COTS) [30].
                           This uses SSCOP [26].
"itu_i3654"                HDLC Service-specific coordination
                           function,
                           as defined in ITU I.365.4 [31].
"FRF5"                     Use of the FRF.5 frame relay standard [32],
                           which references ITU I.365.1 [28].
```

```
"FRF8"                  Use of the FRF.8.1 frame relay
                        standard [33].
                        This implies a null SSCS and the mapping of
                        the frame relay header into the ATM header.
"FRF11"                 Use of the FRF.11 frame relay
                        standard [34].
"itu_h2221"             Use of the ITU standard H.222.1 for
                        audiovisual communication over AAL5 [14].
```

The <oui>, or organizationally unique identifier, refers to the organization responsible for defining the <appId>, or application identifier. The <oui> is maintained by the IEEE. One of its uses is in 802 medium access control (MAC) addresses. It is a 3-octet field represented as 1 to 6 hex digits. Since this is always represented in hex, the "0x" prefix is not used. Leading zeros may be omitted.

The <appId> subparameter refers to the application ID, a hex number consisting of up to 8 digits. Leading zeros may be omitted. The "0x" prefix is not used, since the representation is always hexadecimal. Currently, the only organization that has defined application identifiers is the ATM forum. These have been defined in the context of AAL2 ([12], [35], Section 5 of [36]). Within SDP, these can be used with <appClass> = itu_i3662. The <oui> value for the ATM forum is 0x00A03E.

In the following example, the aalApp media attribute line is used to indicate "Loop Emulation." Service using CAS (POTS only) without the Emulated Loop Control Protocol (ELCP) [52]. The Application ID is defined by the ATM forum [36]. The SSCS used is per ITU I.366.2 [15].

```
a=aalApp:itu_i3662 A03E A
```

If leading zeros are not dropped, this can be represented as

```
a=aalApp:itu_i3662 00A03E 0000000A
```

Since application identifiers have been specified only in the context of the AAL2 SSCS defined in ITU I.366.2 [15], the <appClass> can be set to "-" without ambiguity. The "aalApp" media attribute line can be reduced to

```
a=aalApp:- A03E A
```

or

```
a=aalApp:- 00A03E 0000000A
```

9.4.5.6.2.2 The "cbrRate" Attribute When present, the "cbrRate" attribute is used to represent the CBR rate octet defined in Table 4.6 of ITU Q.2931 [10]. The format of this media attribute line is

```
a=cbrRate: <cbrRate>
```

Here, <cbrRate> is represented as exactly two hex digits. The "0x" prefix is omitted since this parameter is always represented in hex. Values currently defined by the ITU are:

It is preferable that the "cbrRate" attribute be omitted rather than set to an unspecified value of "-" since it conveys no information in the latter case.

VALUE (hex)	*MEANING*
01	64 kbps
04	1544 kbps
05	6312 kbps
06	32064 kbps
07	44736 kbps
08	97728 kbps
10	2048 kbps
11	8448 kbps
12	34368 kbps
13	139264 kbps
40	n x 64 kbps
41	n x 8 kbps

9.4.5.6.2.3 The "sbc" Attribute

The "sbc" media attribute line denotes the subchannel count and is meaningful only in the case of n x 64 clear channel communication. A clear n x 64 channel can use AAL1 (ATM forum af-vtoa-78) or AAL2 adaptation (ITU I.366.2). Although no such standard definition exists, it is also possible to use AAL5 for this purpose. An n x 64 clear channel is represented by the encoding names of "X-CCD" and "X-CCD-CAS" in Table 9.2.

The format of the "sbc" media attribute line is as follows:

```
a=sbc:<sbc>
```

Here, <sbc> can be expressed as a decimal or hex integer. This attribute indicates how many DS0s in a T1 or E1 frame are aggregated for transmitting clear channel data. For T1-based applications, it can take on integral values in the inclusive range [1...24]. For E1-based applications, it can take on integral values in the inclusive range [1...31]. When omitted, other means are to be used to determine the subchannel count.

Use of the "sbc" attribute provides a direct way to indicate the number of 64 kbps subchannels bundled into an n x 64 clear channel. An alternate mechanism to indicate this exists within the SDP bandwidth information, or "b," line (RFC 4566, see Section 2, that obsoletes RFC 2327). In this case, instead of specifying the number of subchannels, the aggregate bandwidth in kbps is specified. The syntax of the "b" line, copied verbatim from RFC 2327 (that was made obsolete by RFC 4566, see Section 2), is as follows:

```
b=<modifier>:<bandwidth-value>
```

In the case of n x 64 clear channels, the <modifier> is assigned a text string value of "AS," indicating that the "b" line is application-specific. The <bandwidth-value> parameter, which is a decimal number indicating the bandwidth in kbps, is limited to one of the following values in the n x 64 clear channel application context:

64, 128, 192, 256, 320, 384, 448, 512, 576, 640, 704, 768, 832, 896, 960, 1024, 1088, 1152, 1216, 1280, 1344, 1408, 1472, 1600, 1664, 1728, 1792, 1856, 1920, 1984

Thus, for n x 64 circuit mode data service,

```
a=sbc:6
```

is equivalent to

```
b=AS:384
```

The media attribute line

```
a=sbc:2
```

is equivalent to

```
b=AS:128
```

9.4.5.6.2.4 The "clkrec" Attribute When present, the "clkrec" attribute is used to indicate the clock recovery method. This attribute is meaningful in the case of AAL1 UDT. The format of the "clkrec" media attribute line is as follows:

```
a=clkrec:<clkrec>
```

The <clkrec> field can take on the following string values: "NULL," "SRTS," or "ADAPTIVE." SRTS and adaptive clock recovery are defined in ITU I.363.1 [37]. "NULL" indicates that the stream (e.g., T1/E1) encapsulated in ATM is synchronous to the ATM network or is retimed, before AAL1 encapsulation, via slip buffers.

9.4.5.6.2.5 The "fec" Attribute When present, the "fec" attribute is used to indicate the use of forward error correction. Currently, there exists a forward error correction method defined for AAL1 in ITU I.363.1 [37]. The format of the "fec" media attribute line is as follows:

```
a=fec:<fecEnable>
```

The <fecEnable> flag indicates the presence of absence of forward error correction. It can take on the string values of "NULL," "LOSS_SENSITIVE," and "DELAY_SENSITIVE." An "NULL" value implies disabling this capability. FEC can be enabled differently for delay-sensitive and loss-sensitive connections.

9.4.5.6.2.6 The "prtfl" Attribute When present, the "prtfl" attribute is used to indicate the fill level of cells. When this attribute is absent, then other means (such as provisionable defaults) are used to determine the presence and level of partial fill.

This attribute indicates the number of nonpad payload octets, not including any AAL SAR or convergence sublayer octets. For example, in some AAL1 applications that use partially filled cells with padding at the end, this attribute indicates the number of leading payload octets not including any AAL overhead.

The format of the "prtfl" media attribute line is as follows:

```
a=prtfl:<partialFill>
```

Here, <partialFill> can be expressed as a decimal or a hex integer.

In general, permitted values are integers in the range 1–48 inclusive. However, this upper bound is different for different adaptations since the AAL overhead, if any, is different. If the specified partial fill is greater than or equal to the maximum fill, then complete fill is used. Using a "partial" fill of 48 always disables partial fill.

In the AAL1 context, this media attribute line applies uniformly to both P and non-P cells. In AAL1 applications that do not distinguish between P and non-P cells, a value of 47 indicates complete fill (i.e., the absence of partial fill). In AAL1 applications that distinguish between P and non-P cells, a value of 46 indicates no padding in P-cells and a padding of one in non-P cells.

If partial fill is enabled (i.e., there is padding in at least some cells), then AAL1 structures must not be split across cell boundaries. These shall fit into any cell. Hence, their size shall be less than or equal to the partial fill size. Further, the partial fill size is preferably an integer multiple of the structure size. If not, then the partial fill size stated in the SDP description shall be truncated to an integer multiple (e.g., a partial fill size of 40 is truncated to 36 to support six 6 x 64 channels).

9.4.5.6.2.7 The "structure" Attribute

This attribute applies to AAL1 connections only. When present, the "structure" attribute is used to indicate the presence or absence of SDT and the size in octets of the SDT blocks. The format of the "structure" media attribute line is as follows:

```
a=structure: <structureEnable> <blksz>
```

where the <structureEnable> flag indicates the presence of absence of SDT. It can take on the values of "on" or "off." An "on" value implies AAL1 SDT, while an "off" value implies AAL1 UDT.

The block size field, <blksz>, is an optional 16-bit field [10] that can be represented in decimal or hex. It is set to a "-" when not applicable, as in the case of UDT. For SDT, it can be set to a "-" when <blksz> is known by other means. For instance, af-vtoa-78 [13] fixes the structure size for n x 64 service, with or without CAS. The theoretical maximum value of <blksz> is 65,535, although most services use much less.

9.4.5.6.2.8 The "cpsSDUsize" Attribute

When present, the "cpsSDUsize" attribute is used to indicate the maximum size of the CPCS SDU payload. There can be several "cpsSDUsize" lines in an SDP description.

The format of this media attribute line is as follows:

```
a=cpsSDUsize:<directionFlag><cpcs>
```

The <directionFlag> can be assigned the following string values: "f," "b," and "fb." "f" and "b" indicate the forward and backward directions, respectively. "fb" refers to both directions (forward and backward). Conventions for the forward and backward directions are per Section 9.4.2.3.

The <cpcs> field is a 16-bit integer that can be represented in decimal or in hex. The meaning and values of these fields are as follows:

Application	*Field*	*Meaning*	*Values*
AAL5	<cpcs>	Maximum CPCS-SDU size	1-65,535
AAL2	<cpcs>	Maximum CPCS-SDU size	45 or 64

9.4.5.6.2.9 The "aal2CPS" Attribute When present, the "aal2CPS" attribute is used to describe parameters associated with the AAL2 CPS layer.

The format of the "aal2CPS" media attribute line is as follows:

```
a=aal2CPS:<cidLowerLimit><cidUpperLimit><timerCU> <simplifiedCPS>
```

Each of these fields can be set to a "-" when the intention is to not specify them in an SDP descriptor.

The <cidLowerLimit> and <cidUpperLimit> can be assigned integer values between 8 and 255 [5], with the limitation that <cidUpperLimit> be greater than or equal to <cidLowerLimit>. For instance, for POTS applications based on [35], <cidLowerLimit> and <cidUpperLimit> can have values of 16 and 223, respectively.

The <timerCU> integer represents the "combined use" timerCU defined in ITU I.363.2. This timer is represented as an integer number of microseconds. It is represented as the decimal integer equivalent of 32 bits.

The <simplifiedCPS> parameter can be assigned the values "on" or "off." When it is "on," the AAL2 CPS simplification described in [35] is adopted. Under this simplification, each ATM cell contains exactly one AAL2 packet. If necessary, octets at the end of the cell are padded with zeros. Since the <timerCU> value in this context is always 0, it can be set to "-."

9.4.5.6.2.10 The "aal2CPSSDUrate" Attribute When present, the "aal2CPSSDUrate" attribute is used to place an upper bound on the SDU bit rate for an AAL2 CID. This is useful for limiting the bandwidth used by a CID, specially if the CID is used for frame mode data defined in [15] or with the SSSAR defined in [25]. The format of this media attribute line is as follows:

```
a=aal2CPSSDUrate: <fSDUrate><bSDUrate>
```

The fSDUrate and bSDUrate are the maximum forward and backward SDU rates in bits/second. These are represented as decimal integers, with range as defined in Section 9.4.6. If any of these parameters in these media attribute lines is not specified, is inapplicable, or is implied, then it is set to "-."

9.4.5.6.2.11 The "aal2sscs3661unassured" Attribute When present, the "aal2sscs3661unassured" attribute is used to indicate the options that pertain to the unassured transmission SSCS defined in ITU I.366.1 [25]. This SSCS can be selected via the

aalApp attribute defined here or by virtue of the presence of the "aal2sscs3661unassured" attribute. The format of this media attribute line is as follows:

```
a=aal2sscs3661unassured: <ted> <rastimer> <fsssar> <bsssar>
```

Each of these fields can be set to a "-" when the intention is to not specify it in an SDP descriptor.

The <ted> flag indicates the presence or absence of transmission error detection as defined in I.366.1. It can be assigned the values of "on" or "off." An "on" value indicates presence of the capability.

The <rastimer> subparameter indicates the SSSAR reassembly timer in microseconds. It is represented as the decimal equivalent of 32 bits.

The <fsssar> and <bsssar> fields are 24-bit integers that can be represented in decimal or in hex. The meaning and values of the <fsssar> and <bsssar> fields are as follows:

Field	*Meaning*	*Values*
<fsssar>	Maximum SSSAR-SDU size forward direction	1- 65,568
<bsssar>	Maximum SSSAR-SDU size backward direction	1- 65,568

If present, the service-specific transmission error detection (SSTED) sublayer is above the service-specific segmentation and reassembly (SSSAR) sublayer [25]. Since the maximum size of the SSTED-SDUs can be derived from the maximum SSSAR-SDU size, it need not be specified separately.

9.4.5.6.2.12 The "aal2sscs3661assured" Attribute When present, the "aal2sscs3661assured" attribute is used to indicate the options that pertain to the assured transmission SSCS defined in ITU I.366.1 [25] on the basis of ITU Q.2110 [26]. This SSCS can be selected via the "aalApp" attribute or by virtue of the presence of the "aal2sscs3661assured" attribute. The format of this media attribute line is as follows:

```
a=aal2sscs3661assured: <rastimer> <fsssar> <bsssar> <fsscopsdu>
<bsscopsdu><fsscopuu> <bsscopuu>
```

Each of these fields can be set to a "-" when the intention is to not specify it in an SDP descriptor.

The <rastimer> subparameter indicates the SSSAR reassembly timer in microseconds. It is represented as the decimal equivalent of 32 bits.

The <fsssar> and <bsssar> fields are 24-bit integers that can be represented in decimal or in hex. The <fsscopsdu>, <bsscopsdu>, <fsscopuu>, and <bsscopuu> fields are 16-bit integers that can be represented in decimal or in hex. The meaning and values of these fields is as follows:

Field	*Meaning*	*Values*
<fsssar>	Maximum SSSAR-SDU size forward direction	1- 65,568
<bsssar>	Maximum SSSAR-SDU size backward direction	1- 65,568
<fsscopsdu>	Maximum SSCOP-SDU size forward direction	1- 65,528

(*Continued*)

Field	*Meaning*	*Values*
<bsscopsdu>	Maximum SSCOP-SDU size backward direction	1- 65,528
<fsscopuu>	Maximum SSCOP-UU field size forward direction	1- 65,524
<bsscopuu>	Maximum SSCOP-UU field size backward direction	1- 65,524

The SSTED sublayer is above the SSSAR sublayer [25]. The service-specific assured data transfer (SSADT) sublayer is above the SSTED sublayer. Since the maximum size of the SSTED-SDUs and SSADT-SDUs can be derived from the maximum SSSAR-SDU size, they need not be specified separately.

The SSCOP protocol defined in [26] is used by the assured data transfer service defined in [25]. In the context of the ITU I.366.1 SSCS, it is possible to use the "aal2sscs3661assured" attribute to limit the maximum sizes of the SSCOP SDUs and user-to-user (UU) fields in either direction. Note that it is necessary for the parameters on the "aal2sscs3661assured" media attribute line to be consistent with each other.

9.4.5.6.2.13 The "aal2sscs3662" Attribute When present, the "aal2sscs3662" attribute is used to indicate the options that pertain to the SSCS defined in ITU I.366.2 [7]. This SSCS can be selected via the aalApp attribute defined here or by the presence of the "aal2sscs3662" attribute. The format of this media attribute line is as follows:

```
a=aal2sscs3662: <sap> <circuitMode> <frameMode> <faxDemod>
        <cas> <dtmf> <mfall> <mfr1> <mfr2>
        <PCMencoding> <fmaxFrame> <bmaxFrame>
```

Each of these fields can be set to a "-" when the intention is to not specify them in an SDP descriptor. Additionally, the values of these fields need to be consistent with each other. Inconsistencies should be flagged as errors.

The <sap> field can take on the following string values: "AUDIO" and "MULTIRATE." These correspond to the audio and multirate SAPs defined in ITU I.366.2.

For the multirate SAP, the following parameters on the aal2sscs3662 attribute line do not apply: <faxDemod>,<cas>, <dtmf>, <mfall>, <mfr1>, <mfr2>, and <PCMencoding>. These are set to "-" for the multirate SAP.

The <circuitMode> flag indicates whether the transport of circuit mode data is enabled or disabled, corresponding to the string values of "on" and "off," respectively. For the multirate SAP, it cannot have a value of "off." For the audio SAP, it can be assigned a value of "on," "off," or "-." Note that the <sbc> attribute, defined elsewhere in this document, can be used to specify the number of 64-kbps subchannels bundled into a circuit mode data channel.

The <frameMode> flag indicates whether the transport of frame mode data is enabled or disabled, corresponding to the string values of "on" and "off," respectively.

The <faxDemod> flag indicates whether facsimile demodulation and remodulation are enabled or disabled, corresponding to the string values of "on" and "off," respectively.

The <cas> flag indicates whether the transport of CAS bits in AAL2 type 3 packets is enabled or disabled, corresponding to the string values of "on" and "off," respectively.

The <dtmf> flag indicates whether the transport of dual-tone multifrequency (DTMF) dialed digits in AAL2 type 3 packets is enabled or disabled, corresponding to the string values of "on" and "off," respectively.

The <mfall> flag indicates whether the transport of multi-frequency (MF) dialled digits in AAL2 type 3 packets is enabled or disabled, corresponding to the string values of "on" and "off," respectively. This flag enables MF dialled digits in a generic manner, without specifying type (e.g., R1, R2, etc.).

The <mfr1> flag indicates whether the transport, in AAL2 type 3 packets, of MF dialled digits for signaling system R1 is enabled or disabled, corresponding to the string values of "on" and "off," respectively.

The <mfr2> flag indicates whether the transport, in AAL2 type 3 packets, of MF dialled digits for signaling system R2 is enabled or disabled, corresponding to the string values of "on" and "off," respectively.

The <PCMencoding> field indicates whether pulse codec modulation (PCM) encoding, if used, is based on the A-law or the Mu-law. This can be used to qualify the "generic PCM" codec stated in some of the AAL2 profiles. The <PCMencoding> field can take on the string values of "PCMA" and "PCMU."

The <fmaxFrame> and <bmaxFrame> fields are 16-bit integers that can be represented in decimal or in hex. The meaning and values of the <fmaxFrame> and <bmaxFrame> fields are as follows:

Field	*Meaning*	*Values*
<fmaxFrame>	Maximum length of a frame mode data unit forward direction	1- 65,535
<bmaxFrame>	Maximum length of a frame mode data unit backward direction	1- 65,535

9.4.5.6.2.14 The "aal5sscop" Attribute When present, the "aal5sscop" attribute is used to indicate the existence of an SSCOP [26] protocol layer over an AAL5 CPS layer [27] and the parameters that pertain to this SSCOP layer. SSCOP over AAL5 can also be selected via the aalApp attribute defined here. The format of the "aal5sscop" media attribute line is as follows:

```
a=aal5sscop: <fsscopsdu> <bsscopsdu> <fsscopuu> <bsscopuu>
```

Each of these fields can be set to a "-" when the intention is to not specify them in an SDP descriptor.

The representation, meaning, and values of the <fsscopsdu>, <bsscopsdu>, <fsscopuu>, and <bsscopuu> fields are identical to those for the "aal2sscs3661assured" media attribute line (Section 9.4.5.6.2.12). Note that it is necessary for the parameters on the "aal5sscop" media attribute line to be consistent with each other.

9.4.5.6.3 Service Attributes

The following is a summary of the SDP media attributes that can be used to describe the services that use the ATM AAL. These attributes are detailed in subsequent subsections.

- The "atmmap" attribute. In the AAL1 and AAL5 contexts, this is used to dynamically map payload types into codec strings.
- The "silenceSupp" attribute, used to indicate the use of voice activity detection for silence suppression and to optionally parameterize the silence suppression function
- The "ecan" attribute, used to indicate the use of echo cancelation and to parameterize the this function
- The "gc" attribute, used to indicate the use of gain control, and to parameterize this function
- The "profileDesc" attribute, which can be used to describe AAL2 profiles. Although any AAL2 profile can be so described, this attribute is useful for describing, at connection establishment time, custom profiles that might not be known to the far end. This attribute applies in the AAL2 context only.
- The "vsel" attribute, which indicates a prioritized list of 3-tuples for voice service. Each 3-tuple indicates a codec, an optional packet length, and an optional packetization period. This complements the "m=" line information and should be consistent with it.
- The "dsel" attribute, which indicates a prioritized list of 3-tuples for voiceband data service. Each 3-tuple indicates a codec, an optional packet length, and an optional packetization period. This complements the "m=" line information and should be consistent with it.
- The "fsel" attribute, which indicates a prioritized list of 3-tuples for facsimile service. Each 3-tuple indicates a codec, an optional packet length, and an optional packetization period. This complements the "m=" line information and should be consistent with it.
- The "onewaySel" attribute, which indicates a prioritized list of 3-tuples for one direction of an asymmetric connection. Each 3-tuple indicates a codec, an optional packet length, and an optional packetization period. This complements the "m=" line information and should be consistent with it.
- The "codecconfig" attribute, which is used to represent the contents of the single codec IE defined in ITU Q.765.5 [38]
- The "isup_usi" attribute, which is used to represent the bearer capability IE defined in Section 4.5.5 of ITU Q.931 [59] and reiterated as the user service IE in Section 3.57 of ITU Q.763 [39].
- The "uiLayer1_Prot" attribute, which is used to represent the "User Information Layer 1 protocol" field within the bearer capability IE defined in Section 4.5.5 of ITU Q.931 [40].

9.4.5.6.3.1 The "atmmap" Attribute The "atmmap" attribute is defined on the basis of the "rtpmap" attribute used in RFC 4566 (see Section 2) that obsoletes RFC 2327.

```
a=atmmap:<payloadType> <encodingName>
```

The "atmmap" attribute is used to dynamically map encoding names into payload types. This is necessary for those encoding names that have not been assigned a static payload type through IANA [9]. Payload types and encoding techniques that have been registered with IANA for RTP are retained for AAL1 and AAL5.

The range of statically defined payload types is in the range 0-95. All static assignments of payload types to codecs are listed in [9]. The range of payload types defined dynamically via the "atmmap" attribute is 96-127.

In addition to reiterating the payload types and encoding names in [9], Table 9.2 defines non-standard encoding names (with "X-" prefixes). Note that [9], rather than Table 9.2, is the authoritative list of standard codec names and payload types in the ATM context.

9.4.5.6.3.2 The "silenceSupp" Attribute When present, the "silenceSupp" attribute is used to indicate the use or nonuse of silence suppression. The format of the "silenceSupp" media attribute line is as follows:

```
a=silenceSupp: <silenceSuppEnable> <silenceTimer> <suppPref>
<sidUse> <fxnslevel>
```

Table 9.2 Encoding Names and Payload Types (Copyright: IETF)

Encoding Technique	*Encoding Name*	*Payload Type*
PCM - Mu law	"PCMU"	0 (Statically mapped)
32-kbps ADPCM	"G726-32"	2 (Statically mapped)
Dual rate 5.3/6.3kbps	"G723"	4 (Statically mapped)
PCM - A law	"PCMA"	8 (Statically mapped)
7 KHz audio coding within 64 kbps	"G722"	9 (Statically mapped)
LD-CELP	"G728"	15 (Statically mapped)
CS-ACELP (normal/low-complexity)	"G729"	18 (Statically mapped)
Low-complexity CS-ACELP	"X-G729a"	None, map dynamically
Normal CS-ACELP w/ ITU defined silence suppression	"X-G729b"	None, map dynamically
Low-complexity CS-ACELP w/ ITU defined silence suppression	"X-G729ab"	None, map dynamically
16-kbps ADPCM	"X-G726-16"	None, map dynamically
24-kbps ADPCM	"X-G726-24"	None, map dynamically
40-kbps ADPCM	"X-G726-40"	None, map dynamically
Dual rate 5.3/6.3 kbps, high rate	"X-G7231-H"	None, map dynamically
Dual rate 5.3/6.3 kbps, low rate	"X-G7231-L"	None, map dynamically
Dual rate 5.3/6.3 kbps, high rate w/ ITU-defined silence suppression	"X-G7231a-H"	None, map dynamically
Dual rate 5.3/6.3 kbps, high rate w/ ITU-defined silence suppression	"X-G7231a-L"	None, map dynamically
16-kbps EADPCM	"X-G727-16"	None, map dynamically
24-kbps EADPCM	"X-G727-24"	None, map dynamically

(Continued)

Table 9.2 (*Continued*) Encoding Names and Payload Types (Copyright: IETF)

Encoding Technique	*Encoding Name*	*Payload Type*
32-kbps EADPCM	"X-G727-32	None, map dynamically
n x 64 kbps Clear Channel without CAS per af-vtoa-78 [13]	"X-CCD"	None, map dynamically
n x 64 kbps Clear Channel with CAS per af-vtoa-78 [13]	"X-CCD-CAS"	None, map dynamically
GSM Full Rate	"GSM"	3 (Statically mapped)
GSM Half Rate	"GSM-HR"	None, map dynamically
|GSM-Enhanced Full Rate	"GSM-EFR"	None, map dynamically
GSM-Enhanced Half Rate	"GSM-EHR"	None, map dynamically
Group 3 fax demodulation	"X-FXDMOD-3"	None, map dynamically
Federal Standard FED-STD 1016 CELP	"1016"	1 (Statically mapped)
DVI4, 8 KHz (RFC 3551 that obsoletes RFC 1890)	"DVI4"	5 (Statically mapped)
DVI4, 16 KHz (RFC 3551 that obsoletes RFC 1890)	"DVI4"	6 (Statically mapped)
LPC (RFC 3551 that obsoletes RFC 1890), linear Predictive Coding	"LPC"	7 (Statically mapped)
L16 (RFC 3551 that obsoletes RFC 1890) 16-bit linear PCM, double channel	"L16"	10 (Statically mapped)
L16 (RFC 3551 that obsoletes RFC 1890), 16-bit linear PCM, single channel	"L16"	11 (Statically mapped)
QCELP (RFC 3551 that obsoletes RFC 1890)	"QCELP"	12 (Statically mapped)
MPEG1/MPEG2 audio	"MPA"	14 (Statically mapped)
DVI4, 11.025 KHz (RFC 3551 that obsoletes RFC 1890)	"DVI4"	16 (Statically mapped)
DVI4, 22.05 KHz (RFC 3551 that obsoletes RFC 1890)	"DVI4"	17 (Statically mapped)
MPEG1/MPEG2 video	"MVP"	32 (Statically mapped)
MPEG 2 audio/video transport stream	"MP2T"	33 (Statically mapped)
ITU H.261 video	"H261"	31 (Statically mapped)
ITU H.263 video	"H263"	33 (Statically mapped)

(*Continued*)

Table 9.2 (*Continued*) Encoding Names and Payload Types (Copyright: IETF)

Encoding Technique	*Encoding Name*	*Payload Type*
ITU H.263 video 1998 version	"H263-1998"	None, map dynamically
MPEG 1 system stream	"MP1S"	None, map dynamically
MPEG 2 program stream	"MP2P"	None, map dynamically
Redundancy	"RED"	None, map dynamically
Variable rate DVI4	"VDVI"	None, map dynamically
Cell-B	"CelB"	25
JPEG	"JPEG"	26
nv	"nv"	28
L8, 8-bitit Linear PCM	"L8"	None, map dynamically
ITU-R Recommendation BT.656-3 for |digital video	"BT656"	None, map dynamically
Adaptive Multirate-Full Rate (3GPP) [41]	"FR-AMR"	None, map dynamically
Adaptive Multirate-Half Rate (3GPP) [41]	"HR-AMR"	None, map dynamically
ARM-UMTS(3GPP) [41]	"UMTS-AMR"	None, map dynamically
Adaptive Multirate-Generic [41]	"AMR"	None, map dynamically

If any of the parameters in the silenceSupp media attribute line is not specified, is inapplicable, or is implied, then it is set to "-."

The <silenceSuppEnable> can take on values of "on" or "off." If it is "on," then silence suppression is enabled.

The <silenceTimer> is a 16-bit field that can be represented in decimal or hex. Each increment (tick) of this timer represents a millisecond. The maximum value of this timer is between 1 and 3 minutes. This timer represents the time lag before silence suppression kicks in. Even though this can, theoretically, be as low as 1 ms, most Digital Signal Processor algorithms take more than that to detect silence. Setting <silenceTimer> to a large value (say 1 minute> is equivalent to disabling silence suppression within a call. However, idle channel suppression between calls on the basis of silence suppression is still operative in nonswitched, trunking applications if <silenceSuppEnable> = "on" and <silenceTimer> is a large value.

The <suppPref> specifies the preferred silence suppression method that is preferred or already selected. It can take on the string values of "standard" and "custom." If its value is "standard," then a standard method (e.g., ITU-defined) is preferred to custom methods if such a standard is defined. Otherwise, a custom method may be used. If <suppPref> is set to "custom," then a custom method, if available, is preferred to the standard method.

The <sidUse> indicates whether silence insertion descriptors (SIDs) are to be used and whether they use fixed comfort noise or sampled background noise. It can take on the string values of "No SID," "Fixed Noise," and "Sampled Noise."

If the value of <sidUse> is "Fixed Noise," then <fxnslevel> provides its level. It can take on integer values in the range 0-127, as follows:

<fxnslevel> value	*Meaning*
0-29	Reserved
30	-30 dBm0
31	-31dBm0
....	....
77	-77 dBm0
78	-78 dBm0
79-126	reserved
127	Idle Code (no noise)

In addition to the decimal representation of <fxnslevel>, a hex representation, preceded by a "0x" prefix, is also allowed.

9.4.5.6.3.3 The "ecan" Attribute When present, the "ecan" attribute is used to indicate the use or nonuse of echo cancelation. There can be several "ecan" lines in an SDP description.

The format of the "ecan" media attribute line is as follows:

```
a=ecan:<directionFlag><ecanEnable><ecanType>
```

The <directionFlag> can be assigned the following string values: "f," "b," and "fb." "f" and "b" indicate the forward and backward directions, respectively. "fb" refers to both directions (forward and backward). Conventions for the forward and backward directions are per Section 9.4.2.3.

The <directionFlag> is always specified. Except for the <directionFlag>, the remaining parameters can be set to "-" to indicate that they are not specified, inapplicable, or implied. However, there must be some specified parameters for the line to be useful in an SDP description.

If the "ecan" media attribute lines is not present, then means other than the SDP descriptor must be used to determine the applicability and nature of echo cancelation for a connection direction. Examples of such means are MIB provisioning, the local connection options structure in MGCP, etc.

The <ecanEnable> parameter can take on values of "on" or "off." If it is "on," then echo cancelation is enabled. If it is "off," then echo cancelation is disabled.

The <ecanType> parameter can take on the string values "G165" and "G168," respectively.

When SDP is used with some media gateway control protocols such as MGCP and Megaco, there exist means outside SDP descriptions to specify the echo cancelation properties of a connection. Nevertheless, this media attribute line is included for completeness. As a result, the SDP can be used for describing echo cancelation in applications where alternate means for this are unavailable.

9.4.5.6.3.4 The "gc" Attributes When present, the "gc" attribute is used to indicate the use or nonuse of gain control. There can be several "gc" lines in an SDP description.

The format of the "gc" media attribute line is as follows:

```
a=gc:<directionFlag><gcEnable><gcLvl>
```

The <directionFlag> can be assigned the following string values: "f," "b," and "fb." "f" and "b" indicate the forward and backward directions respectively. "fb" refers to both directions (forward and backward). Conventions for the forward and backward directions are per Section 9.4.2.3.

The <directionFlag> is always specified. Except for the <directionFlag>, the remaining parameters can be set to "-" to indicate that they are not specified, inapplicable, or implied. However, there must be some specified parameters for the line to be useful in an SDP description.

If the "gc" media attribute lines is not present, then means other than the SDP descriptor must be used to determine the applicability and nature of gain control for a connection direction. Examples of such means are MIB provisioning, the local connection options structure in MGCP, etc.

The <gcEnable> parameter can take on values of "on" or "off." If it is "on," then gain control is enabled. If it is "off," then gain control is disabled.

The <gcLvl> parameter is represented as the decimal or hex equivalent of a 16-bit binary field. A value of 0xFFFF implies automatic gain control. Otherwise, this number indicates the number of decibels of inserted loss. The upper bound, 65,535 dB (0xFFFE) of inserted loss, is a large number and is a carryover from Megaco. In practical applications, the inserted loss is much lower.

When SDP is used with some media gateway control protocols such as MGCP and Megaco, there exist means outside SDP descriptions to specify the gain control properties of a connection. Nevertheless, this media attribute line is included for completeness. As a result, the SDP can be used for describing gain control in applications where alternate means for this are unavailable.

9.4.5.6.3.5 The "profileDesc" Attribute

There is one "profileDesc" media attribute line for each AAL2 profile that is intended to be described. The "profileDesc" media attribute line is structured as follows:

```
a=profileDesc: <aal2transport> <profile> <uuiCodeRange#1>
  <encodingName#1> <packetLength#1> <packetTime#1>
  <uuiCodeRange#2> <encodingName#2> <packetLength#2>
  <packetTime#2>... <uuiCodeRange#N> <encodingName#N>
  <packetLength#N> <packetTime#N>
```

Here, <aal2transport> can have those values of <transport> (Table 9.1) that pertain to AAL2. These are:

```
AAL2/ATMF
AAL2/ITU
AAL2/custom
AAL2/<corporateName>
AAL2/IEEE:<oui>
```

The parameter <profile> is identical to its definition for the "m=" line (Section 9.4.5.5.4).

The profile elements (rows in the profile tables of ITU I.366.2 or AF-VTOA-0113) are represented as 4-tuples following the <profile> parameter in the "profileDesc" media attribute line. If a member of one of these 4-tuples is not specified or is implied, then it is set to "-."

The <uuiCodeRange> parameter is represented by D1-D2, where D1 and D2 are decimal integers in the range 0 through 15.

The <encodingName> parameter can take one of the values in column 2 of Table 9.2. Additionally, it can take on the following descriptor strings: "PCMG," "SIDG," and "SID729." These stand for generic PCM, generic SID, and G.729 SID respectively.

The <packetLength> is a decimal integer representation of the AAL2 packet length in octets.

The <packetTime> is a decimal integer representation of the AAL2 packetization interval in microseconds.

For instance, the "profileDesc" media attribute line in the next section defines the AAL2/custom 100 profile. This profile is reproduced in Table 9.3 for a description of the parameters in this profile such as M and the sequence number interval, see ITU I.366.2 [13].

```
a=profileDesc:AAL2/custom 100 0-7 PCMG 40 5000 0-7 SIDG 1
5000 8-15
G726-32 40 10000 8-15 SIDG 1 5000
```

If the <packetTime> parameter is to be omitted or implied, then the same profile can be represented as follows:

```
a=profileDesc:AAL2/custom 100 0-7 PCMG 40 - 0-7 SIDG 1 - 8-15
G726-32 40 - 8-15 SIDG 1 -
```

If a gateway has a provisioned or hard-coded definition of a profile, then any definition provided via the "profileDesc" line overrides it.

The exception to this rule is with regard to standard profiles such as ITU-defined profiles and ATMF-defined profiles. In general, these should not be defined via a "profileDesc" media attribute line. If they are, then the definition needs to be consistent with the standard definition else the SDP session descriptor should be rejected with an appropriate error code.

9.4.5.6.3.6 The "vsel" Attribute The "vsel" attribute indicates a prioritized list of one or more 3-tuples for voice service. Each 3-tuple indicates a codec, an optional packet length, and an optional packetization period. This complements the "m=" line information and should be consistent with it.

Table 9.3 Example of a Custom AAL2 Profile (Copyright: IETF)

User-user information (UUI) code point range	*Packet length (octets)*	*Encoding per ITU I.366.22/99 version*	*Description of algorithm*	*M*	*Packet time (ms)*	*Sequence number interval (ms)*
0-7	40	Figure B-1	PCM, G.811-64, generic	1	5	5
0-7	1	Figure I-1	Generic SID	1	5	5
8-15	40	Figure E-2	ADPCM, G.726-32	2	10	5
8-15	1	Figure I-1	Generic SID	1	5	5

The "vsel" attribute refers to all directions of a connection. For a bidirectional connection, these are the forward and backward directions. For a unidirectional connection, this can be either the backward or the forward direction.

The "vsel" attribute is not meant to be used with bidirectional connections that have asymmetric codec configurations described in a single SDP descriptor. For these, the "onewaySel" attribute (Section 9.4.5.6.3.9) should be used. See Section 5.6.3.9 for the requirement for not using the "vsel" and "onewaySel" attributes in the same SDP descriptor.

The "vsel" line is structured as follows:

```
a=vsel:<encodingName #1> <packetLength #1><packetTime #1>
       <encodingName #2> <packetLength #2><packetTime #2>
       ...
       <encodingName #N> <packetLength #N><packetTime #N>
```

where the <encodingName> parameter can take one of the values in column 2 of Table 9.2. The <packetLength> is a decimal integer representation of the packet length in octets.

The <packetTime> is a decimal integer representation of the packetization interval in microseconds. The parameters <packetLength> and <packetTime> can be set to "-" when not needed. Also, the entire "vsel" media attribute line can be omitted when not needed. For example,

```
a=vsel:G729 10 10000 G726-32 40 10000
```

indicates first preference of G.729 or G.729a (both are interoperable) as the voice encoding scheme. A packet length of 10 octets and a packetization interval of 10 ms are associated with this codec. G726-32 is the second preference stated in this line, with an associated packet length of 40 octets and a packetization interval of 10 ms. If the packet length and packetization interval are intended to be omitted, then this media attribute line becomes

```
a=vsel:G729 - - G726-32 - -
```

The media attribute line

```
a=vsel:G726-32 40 10000
```

indicates preference for or selection of 32-kbps ADPCM with a packet length of 40 octets and a packetization interval of 10 ms.

This media attribute line can be used in ATM as well as non-ATM contexts. Within the ATM context, it can be applied to the AAL1, AAL2, and AAL5 adaptations.

The <packetLength> and <packetTime> are not meaningful in the AAL1 case and should be set to "-." In the AAL2 case, this line determines the use of some or all of the rows in a given profile table. If multiple 3-tuples are present, they can indicate a hierarchical assignment of some rows in that profile to voice service (e.g., row A preferred to row B, etc.). If multiple profiles are present on the "m=" line, the profile qualified by this attribute is the first profile. If a single profile that has been selected for a connection is indicated in the "m=" line, the "vsel" attribute qualifies the use, for voice service, of codecs within that profile.

With most of the encoding names in Figure 2 of ITU I.366.2 2/99 version, the packet length and packetization period can be derived from each other. One of them can be set to "-" without a loss of information. There are some exceptions such as the IANA-registered encoding names

G723, DVI4, and L16 for which this is not true. Therefore, there is a need to retain both the packet length and packetization period in the definition of the "vsel" line.

9.4.5.6.3.7 The "dsel" Attribute The "dsel" attribute indicates a prioritized list of one or more 3-tuples for voiceband data service. The <fxIncl> flag indicates whether this definition of voiceband data includes fax ("on" value) or not ("off" value). If <fxIncl> is "on," then the "dsel" line must be consistent with any "fse"' line in the session description. In this case, an error event is generated in the case of inconsistency.

Each 3-tuple indicates a codec, an optional packet length, and an optional packetization period. This complements the "m" line information and should be consistent with it. The "dsel" attribute refers to all directions of a connection. For a bidirectional connection, these are the forward and backward directions. For a unidirectional connection, this can be either the backward or the forward direction.

The "dsel" attribute is not meant to be used with bidirectional connections that have asymmetric codec configurations described in a single SDP descriptor. For these, the "onewaySel" attribute (Section 9.4.5.6.3.9) should be used. See Section 9.4.5.6.3.9 for the requirement to not use the "dsel" and "onewaySel" attributes in the same SDP descriptor.

The "dsel" line is structured as follows:

```
a=dsel:<fxIncl> <encodingName #1> <packetLength #1><packetTime #1>
        <encodingName #2> <packetLength #2><packetTime #2>
        ...
        <encodingName #N> <packetLength #N><packetTime #N>
```

where the <encodingName> parameter can take one of the values in column 2 of Table 9.2. The <packetLength> and <packetTime> parameters are per their definition, above, for the "vsel" media attribute line.

The parameters <packetLength> and <packetTime>) can be set to "-" when not needed. The <fxIncl> flag is presumed to be "off" if it is set to "-." Also, the entire "dsel" media attribute line can be omitted when not needed.

For example,

```
a=dsel:- G726-32 20 5000 PCMU 40 5000
```

indicates that this line does not address facsimile and that the first preference for the voiceband data codes is 32-kbps ADPCM, while the second preference is PCMU. The packet length and the packetization interval associated with G726-32 are 20 octets and 5 ms, respectively. For PCMU, they are 40 octets and 5 ms, respectively.

This media attribute line can be used in ATM as well as non-ATM contexts. Within the ATM context, it can be applied to the AAL1, AAL2, and AAL5 adaptations. The <packetLength> and <packetTime> are not meaningful in the AAL1 case and should be set to "-." In the AAL2 case, this line determines the use of some or all of the rows in a given profile table. If multiple 3-tuples are present, they can indicate a hierarchical assignment of some rows in that profile to voiceband data service (e.g., row A preferred to row B, etc.) If multiple profiles are present on the "m=" line, the profile qualified by this attribute is the first profile. If a single profile that has been selected for a connection is indicated in the "m=" line, the "dsel" attribute qualifies the use, for voiceband data service, of codecs within that profile.

With most of the encoding names in Figure 2 of ITU I.366.2 2/99 version, the packet length and packetization period can be derived from each other. One of them can be set to "-" without a loss of information. There are some exceptions such as the IANA-registered encoding names G723, DVI4, and L16 for which this is not true. Therefore, there is a need to retain both the packet length and packetization period in the definition of the "dsel" line.

9.4.5.6.3.8 The "fsel" Attribute

The "fsel" attribute indicates a prioritized list of one or more 3-tuples for facsimile service. If an "fsel" line is present, any "dsel" line with <fxIncl> set to "on" in the session description must be consistent with it. In this case, an error event is generated in the case of inconsistency. Each 3-tuple indicates a codec, an optional packet length, and an optional packetization period. This complements the "m=" line information and should be consistent with it.

The "fsel" attribute refers to all directions of a connection. For a bidirectional connection, these are the forward and backward directions. For a unidirectional connection, this can be either the backward or the forward direction.

The "fsel" attribute is not meant to be used with bidirectional connections that have asymmetric codec configurations described in a single SDP descriptor. For these, the "onewaySel" attribute (Section 9.4.5.6.3.9) should be used. See Section 9.4.5.6.3.9 for the requirement for not using the "fsel" and "onewaySel" attributes in the same SDP descriptor.

The "fsel" line is structured as follows:

```
a=fsel:<encodingName #1> <packetLength #1><packetTime #1>
       <encodingName #2> <packetLength #2><packetTime #2>
       ...
       <encodingName #N> <packetLength #N><packetTime #N>
```

where the <encodingName> parameter can take one of the values in column 2 of Table 9.2. The <packetLength> and <packetTime> parameters are per their definition for the "vsel" media attribute line.

The parameters <packetLength> and <packetTime> can be set to "-" when not needed. Also, the entire "fsel" media attribute line can be omitted when not needed.

For example,

```
a=fsel:FXDMOD-3 - -
```

indicates demodulation and remodulation of ITU-T group 3 fax at the gateway.

```
a=fsel:PCMU 40 5000 G726-32 20 5000
```

indicates a first and second preference of Mu-law PCM and 32-kbps ADPCM as the facsimile encoding scheme. The packet length and the packetization interval associated with G726-32 are 20 octets and 5 ms, respectively. For PCMU, they are 40 octets and 5 ms, respectively.

This media attribute line can be used in ATM as well as non-ATM contexts. Within the ATM context, it can be applied to the AAL1, AAL2, and AAL5 adaptations. The <packetLength> and <packetTime> are not meaningful in the AAL1 case and should be set to "-." In the AAL2 case, this line determines the use of some or all of the rows in a given profile table. If multiple 3-tuples are present, they can indicate a hierarchical assignment of some rows in that profile to facsimile service (e.g., row A preferred to row B, etc.). If multiple profiles are present on the "m" line, the profile qualified by this attribute is the first profile. If a single profile that has been selected for a connection is indicated in the "m" line, the "fsel" attribute qualifies the use, for facsimile service, of codecs within that profile.

With most of the encoding names in Figure 2 of ITU I.366.2 2/99 version, the packet length and packetization period can be derived from each other. One of them can be set to "-" without a loss of information. There are some exceptions such as the IANA-registered encoding names G723, DVI4, and L16 for which this is not true. Therefore, there is a need to retain both the packet length and packetization period in the definition of the "fsel" line.

9.4.5.6.3.9 The "onewaySel" Attribute

The "onewaySel" (one way select) attribute can be used with connections that have asymmetric codec configurations. There can be several "onewaySel" lines in an SDP description. The "onewaySel" line is structured as follows:

```
a=onewaySel:<serviceType> <directionFlag>
      <encodingName #1> <packetLength #1><packetTime #1>
      <encodingName #2> <packetLength #2><packetTime #2>
      ...
      <encodingName #N> <packetLength #N><packetTime #N>
```

The <serviceType> parameter can be assigned the following string values: "v," "d," "f," "df," and "all." These indicate voice, voiceband data (fax not included), fax, voiceband data (fax included), and all services, respectively.

The <directionFlag> can be assigned the following string values: "f," "b," and "fb." "f" and "b" indicate the forward and backward directions, respectively. "fb" refers to both directions (forward and backward) and shall not be used with the "onewaySel" line. Conventions for the forward and backward directions are per Section 9.4.2.3.

Following <directionFlag>, there is a prioritized list of one or more 3-tuples. Each 3-tuple indicates a codec, an optional packet length, and an optional packetization period. This complements the "m=" line information and should be consistent with it.

Within each 3-tuple, the <encodingName> parameter can take one of the values in column 2 of Table 9.2. The <packetLength> is a decimal integer representation of the packet length in octets.

The <packetTime> is a decimal integer representation of the packetization interval in microseconds.

The "onewaySel" attribute must not be used in SDP descriptors that have one or more of the following attributes: "vsel," "dsel," "fsel." If it is present, then command containing the SDP description may be rejected. An alternate response to such an ill-formed SDP descriptor might be the selective ignoring of some attributes, which must be coordinated via an application-wide policy.

The <serviceType>, <directionFlag>, and <encodingName> parameters may not be set to "-." However, the parameters <packetLength> and <packetTime> can be set to "-" when not needed.

For example,

```
a=onewaySel:v f G729 10 10000
a=onewaySel:v b G726-32 40 10000
```

indicates that for voice service, the codec to be used in the forward direction is G.729 or G.729a (both are interoperable), and the codec to be used in the backward direction is G726-32. A packet length of 10 octets and a packetization interval of 10 ms are associated with the G.729/G.729a codec. A packet length of 40 octets and a packetization interval of 10 ms are associated with the G726-32 codec.

For example,

```
a=onewaySel:d f G726-32 20 5000
a=onewaySel:d b PCMU 40 5000
```

indicates that for voiceband service (fax not included), the codec to be used in the forward direction is G726-32), and the codec to be used in the backward direction is PCMU. A packet length of 20 octets and a packetization interval of 5 ms are associated with the G726-32 codec. A packet length of 40 octets and a packetization interval of 5 ms are associated with the PCMU codec.

This media attribute line can be used in ATM as well as non-ATM contexts. Within the ATM context, it can be applied to the AAL1, AAL2, and AAL5 adaptations. The <packetLength> and <packetTime> are not meaningful in the AAL1 case and should be set to "-." In the AAL2 case, these lines determine the use of some or all of the rows in a given profile table. If multiple 3-tuples are present, they can indicate a hierarchical assignment of some rows in that profile to voice service (e.g., row A preferred to row B, etc.). If multiple profiles are present on the "m" line, the profile qualified by this attribute is the first profile.

With most of the encoding names in Figure 2 of ITU I.366.2 2/99 version, the packet length and packetization period can be derived from each other. One of them can be set to "-" without a loss of information. There are some exceptions such as the IANA-registered encoding names G723, DVI4, and L16 for which this is not true. Therefore, there is a need to retain both the packet length and packetization period in the definition of the "onewaySel" line.

9.4.5.6.3.10 The "codecconfig" Attribute

When present, the "codecconfig" attribute is used to represent the contents of the single codec IE defined in [38]. The contents of this IE are a single-octet organizational identifier (OID) field, followed by a single-octet codec type field, followed by zero or more octets of a codec configuration bit map. The semantics of the codec configuration bit map are specific to the organization [38, 41]. The "codecconfig" attribute is represented as follows:

```
a=codecconfig:<q7655scc>
```

The <q7655scc> (Q.765.5 single codec IE contents) parameter is represented as a string of hex digits. The number of hex digits is even (range 4-32). The "0x" prefix shall be omitted since this value is always hexadecimal. As with other hex values in Section 9.4.2.2, digits to the left are more significant than digits to the right. Leading zeros shall not be omitted.

An example of the use of this media attribute is

```
a=codecconfig:01080C
```

The first octet indicates an OID of 0x01 (the ITU-T). Using ITU Q.765.5 [38], the second octet (0x08) indicates a codec type of G.726 (ADPCM). The last octet, 0x0C indicates that 16-kbps and 24-kbps rates are NOT supported, while the 32-kbps and 40-kbps rates ARE supported.

9.4.5.6.3.11 The "isup_usi" Attribute

When present, the "isup_usi" attribute is used to represent the bearer capability IE defined in Section 4.5.5 of ITU Q.931 [40] (excluding the IE identifier and length). This IE is reiterated as the user service IE in Section 3.57 of ITU Q.763 [39]. The "isup_usi" attribute is represented as follows:

```
a=isup_usi:<isupUsi>
```

The <isupUsi> parameter is represented as a string of hex digits. The number of hex digits is even (allowed range 4-24). The "0x" prefix shall be omitted since this value is always hexadecimal. As with other hex values (Section 2.2), digits to the left are more significant than digits to the right. Leading zeros shall not be omitted.

9.4.5.6.3.12 The "uiLayer1_Prot" Attribute

When present, the "uiLayerl_Prot" attribute is used to represent the "User Information Layer 1 protocol" field within the bearer capability IE defined in Section 4.5.5 of [40] and reiterated as the user service IE in Section 3.57 of [39]. The "User Information Layer 1 protocol" field consists of the five least significant bits of Octet 5 of this IE.

Within SDP, the "uiLayerl_Prot" attribute is represented as follows:

```
a='uiLayer1_Prot':<uiLayer1Prot>
```

The <uiLayerlProt> parameter is represented as a string of two hex digits. The "0x" prefix shall be omitted since this value is always hexadecimal. As with other hex values (Section 2.2), digits to the left are more significant than digits to the right. These hex digits are constructed from an octet with three leading "0" bits and last five bits equal to the "User Information Layer 1 protocol" field described above. As specified in [40] and [42], bit 5 of this field is the most significant. The resulting values of the <uiLayerlProt> parameter are as follows:

VALUE	*MEANING*
0x01	CCITT standardized rate adaption V.110 and X.30
0x02	Recommendation G.711 Mu-law
0x03	Recommendation G.711 A-law
0x04	Recommendation G.721 32-kbps ADPCM and Recommendation I.460
0x05	Recommendations H.221 and H.242
0x06	Recommendation H.223 and H.245
0x07	Non-ITU-T standardized rate adaption
0x08	ITU-T standardized rate adaption V.120
0x09	CCITT standardized rate adaption X.31 HDLC flag stuffing

9.4.5.6.4 Miscellaneous Media Attributes

The “chain” media attribute line, which is used to chain consecutive SDP descriptions, cannot be classified as an ATM, AAL, or service attribute. It is detailed in the following subsection.

9.4.5.6.4.1 The “chain” Attribute

The start of an SDP descriptor is marked by a “v=” line. In some applications, consecutive SDP descriptions are alternative descriptions of the same session. In others, these describe different layers of the same connection (e.g., IP, ATM, frame relay). This is useful when the connectivity at these layers is established at the same time (e.g., an IP-based session over an ATM SVC). To distinguish between the alternation and concatenation of SDP descriptions, a “chain” attribute can be used in the case of concatenation.

When present, the “chain” attribute binds an SDP description to the next or previous SDP description. The next or previous description is separated from the current one by a “v” line. It is not necessary for this description to also have a “chain” media attribute line. Chaining averts the need to set up a single SDP description for a session that is simultaneously created at multiple layers. It allows the SDP descriptors for different layers to remain simple and clean. Chaining is not needed in the Megaco context, where it is possible to create separate terminations for the different layers of a connection.

The “chain” media attribute line has the following format:

```
a=chain:<chainPointer>
```

The <chainPointer> field can take on the following string values: "NEXT," "PREVIOUS," and "NULL." The value "NULL" is not equivalent to omitting the chain attribute from a description since it expressly precludes the possibility of chaining. If the “chain” attribute is absent in an SDP description, chaining can still be realized by the presence of a chain media attribute line in the previous or next description.

9.4.5.6.5 Use of the Second Media-Level Part in H.323 Annex C applications

Section 4 mentions that H.323 annex C applications have a second media level part for the ATM session description. This is used to convey information about the RTCP stream. Although the RTP stream is encapsulated in AAL5 with no intervening IP layer, the RTCP stream is sent to an IP address and RTCP port. This media-level part has the following format:

```
m= control <rtcpPortNum> H323c -
c= IN IP4 <rtcpIPaddr>
```

Consistency with RFC 2327 (that was made obsolete by RFC 4566, see Section 2) is maintained in the location and format of these lines. The <fmt list> in the “m” line is set to "-." The “c” line in the second media-level part pertains to RTCP only.

The <rtcpPortNum> and <rtcpIPaddr> subparameters indicate the port number and IP address on which the media gateway is prepared to receive RTCP packets.

Any of the subparameters on these lines can be set to "-" if it is known by other means.

The range and format of the <rtcpPortNum> and <rtcpIPaddr> subparameters is per [1]. The <rtcpPortNum> is a decimal number between 1024 and 65535. It is an odd number. If an even number in this range is specified, the next odd number is used. The <rtcpIPaddr> is expressed in the usual dotted decimal IP address representation, from 0.0.0.0 to 255.255.255.255.

9.4.5.6.6 Use of the eecid Media Attribute in Call Establishment Procedures

This informative section supplements the definition of the eecid attribute (Section 9.4.5.6.1.1) by describing example procedures for its use. These procedures assume a bearer-signaling mechanism for connection setup that is independent of service-level call control. These procedures are independent of the media gateway control protocol (MGCP, Megaco, SIP, etc.), the protocol used between media gateway controllers (ITU Q.1901, SIP, etc.), and the protocol used for bearer connection setup (Q.2931, UNI, PNNI, AINI, IISP, Q.2630.1, etc.).

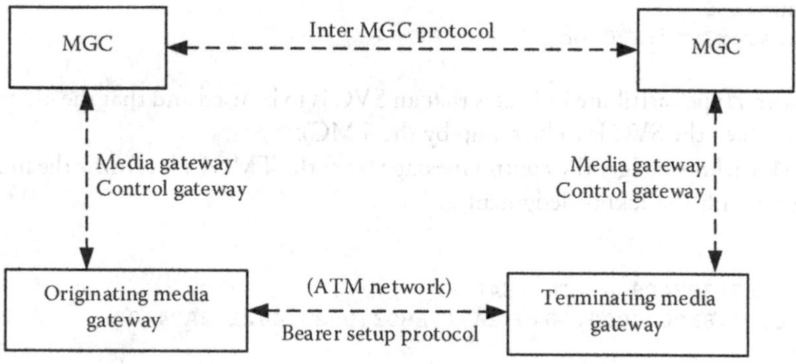

In the diagram in the Inline Figure, the originating media gateway originates the service-level call. The terminating media gateway terminates it. In the forward bearer connection setup model, the originating media gateway initiates bearer connection setup. In the backward bearer connection setup model, the terminating gateway initiates bearer connection setup.

Example use of the Backward Bearer Connection Set-up Model:

1. The originating media gateway controller (OMGC) initiates service-level call establishment by sending the appropriate control message to the originating media gateway (OMG).
2. The OMG provides its NSAP address and an eecid value to the OMGC, using the following SDP description:

```
v=0
o=- 2873397496 0 ATM NSAP
        47.0091.8100.0000.0060.3E64.FD01.0060.3E64.FD01.00
s=-
c=ATM NSAP
        47.0091.8100.0000.0060.3E64.FD01.0060.3E64.FD01.00
t=0 0
m=audio $ AAL2/ITU 8
a=eecid:B3D58E32
```

3. The OMGC signals the terminating media gateway controller (TMGC) through the appropriate mechanism (ISUP with Q.1901 extensions, SIP, etc.). It provides the TMGC with the NSAP address and the eecid provided by the OMG.
4. The TMGC sends the appropriate control message to the TMG. This includes the session descriptor received from the OMG. This descriptor contains the NSAP address of the OMG and the EECID assigned by the OMG. Additionally, the TMGC instructs the TMG to set up an SVC to the OMG. It also requests the TMG to notify the TMGC when SVC setup

is complete. Depending on the control protocol used, this can be done through a variety of means. In the Megaco context, the request to set up an SVC (not the notification request for the SVC setup event) can be made through the following local descriptor:

```
v=0
o=- 2873397497 0 ATM - -
s=-
c=ATM - -
t=0 0
m=audio $ - -
a=bearerType:SVC on
```

The "bearerType" attribute indicates that an SVC is to be used and that the <localInitiation> flag is on (i.e., the SVC is to be set up by the TMG).

5. The TMG acknowledges the control message from the TMGC. It returns the following SDP descriptor with the acknowledgment

```
v=0
o=- 2873397498 0 ATM NSAP
47.0091.8100.0000.0040.2A74.EB03.0020.4421.2A04.00
s=-
c=ATM NSAP
47.0091.8100.0000.0040.2A74.EB03.0020.4421.2A04.00
t=0 0
m=audio $ AAL2/ITU 8
```

   The NSAP address information provided in this descriptor is not needed. It can be omitted (by setting it to "- -").
6. The TMG sends an SVC setup message to the OMG. Within the GIT IE, it includes eecid (B3D58E32) received from the OMG.
7. The OMG uses the eecid to correlate the SVC setup request with service-level control message received before from the OMGC.
8. The OMG returns an SVC connect message to the TMG. On receiving this message, the TMG sends an event notification to the TMGC indicating successful SVC set-up.

Note that, for this example, the "v=," "o=," "s=," and "t=" lines can be omitted in the Megaco context.

Example use of the forward bearer connection setup model:

1. The OMGC initiates service-level call establishment by sending the appropriate controls message to the OMG.
2. The OMG provides its NSAP address to the OMGC, using the following SDP description:

```
v=0
o=- 2873397496 0 ATM NSAP
47.0091.8100.0000.0060.3E64.FD01.0060.3E64.FD01.00
s=-
c=ATM NSAP
47.0091.8100.0000.0060.3E64.FD01.0060.3E64.FD01.00
t=0 0
m=audio $ AAL2/ITU 8
```

The NSAP address information provided in this descriptor is not needed. It can be omitted (by setting it to "- -").

3. The OMGC signals the TMGC through the appropriate mechanism (ISUP with Q.1901 extensions, SIP, etc.). Although this is not necessary, it can provide the TMGC with the NSAP address provided by the OMG.
4. The TMGC sends the appropriate control message to the TMG. This includes the session descriptor received from the OMG. This descriptor contains the NSAP address of the OMG.
5. The TMG acknowledges the control message from the TMGC. Along with the acknowledgement, it provides an SDP descriptor with a locally assigned eecid.

```
v=0
o=- 2873397714 0 ATM NSAP
47.0091.8100.0000.0040.2A74.EB03.0020.4421.2A04.00
s=-
c=ATM NSAP
47.0091.8100.0000.0040.2A74.EB03.0020.4421.2A04.00
t=0 0
m=audio $ AAL2/ITU 8
a=eecid:B3D58E32
```

6. The TMGC signals the OMGC through the appropriate mechanism (ISUP with Q.1901 extensions, SIP, etc.). It provides the OMGC with the NSAP address and the eecid provided by the TMG.
7. The OMGC sends the appropriate control message to the OMG. This includes the session descriptor received from the TMG. This descriptor contains the NSAP address of the TMG and the EECID assigned by the TMG. Additionally, the OMGC instructs the OMG to set up an SVC to the TMG. It also requests the OMG to notify the OMGC when SVC setup is complete. Depending on the control protocol used, this can be done through a variety of means. In the Megaco context, the request to set up an SVC (not the notification request for the SVC setup event) can be made through the following local descriptor:

```
v=0
o=- 2873397874 0 ATM - -
s=-
c=ATM - -
t=0 0
m=audio $ - -
a=bearerType:SVC on
```

The bearerType" attribute indicates that an SVC is to be used and that the <localInitiation> flag is on (i.e., the SVC is to be set up by the TMG).

8. The OMG acknowledges the control message from the OMGC.
9. The OMG sends an SVC setup message to the TMG. Within the GIT IE, it includes eecid (B3D58E32) received from the TMG.
10. The TMG uses the eecid to correlate the SVC set-up request with the service-level control message received before from the TMGC.
11. The TMG returns an SVC connect message to the OMG. On receiving this message, the OMG sends an event notification to the OMGC indicating successful SVC setup.

Note that, for this example, the "v=," "o=," "s=," and "t=" lines can be omitted in the Megaco context.

9.4.6 List of Parameters with Representations

This section provides a list of the parameters used in this document and the formats used to represent them in SDP descriptions. In general, a "-" value can be used for any field that is not specified, is inapplicable, or is implied.

PARAMETER	*MEANING*	*REPRESENTATION*
<username>	User name	Constant "-"
<sessionID>	Session ID	Up to 32 decimal or hex digits
<version>	Version of SDP descriptor	"0" or 10 decimal digits
<networkType>	Network type	Constant "ATM" for ATM transport
<addressType>	Address type	String values:
		"NSAP," "E164," "GWID,"
		"ALIAS"
<address>	Address	"NSAP": 40 hex digits, dotted
		"E164": up to 15 decimal digits
		"GWID": up to 32 characters
		"ALIAS": up to 32 characters
<sessionName>	Session name	Constant "-"
<startTime>	Session start time	"0" or 10 decimal digits
<stopTime>	Session stop time	Constant "0"
<vcci>	Virtual circuit connection identifier	Decimal or hex equivalent of 16 bits
<ex_vcci>	Explicit representation of <vcci>	"VCCI-" prefixed to <vcci>
<bcg>	Bearer connection group	Decimal or hex equivalent of 8 bits
<ex_bcg>	Explicit representation of <bcg>	"BCG-" prefixed to <bcg>
<portId>	Port ID	Hex number of up to 32 digits
<ex_portId>	Explicit representation of <portId>	"PORT-" prefixed to <portId>
<vpi>	Virtual Path Identifier	Decimal or hex equivalent of 8 or 12 bits
<ex_vpi>	Explicit representation of <vpi>	"VPI-" prefixed to <vpi>
<vci>	Virtual Circuit Identifier	Decimal or hex equivalent of 16 bits
<ex_vci>	Explicit representation of <vci>	"VCI-" prefixed to <vci>

(Continued)

PARAMETER	*MEANING*	*REPRESENTATION*
<vpci>	Virtual Path Connection Identifier	Decimal or hex equivalent of 16 bits
<ex_vpci>	Explicit representation of <vpci>	"VPCI-" prefixed to <vpci>
<cid>	Channel identifier	Decimal or hex equivalent of 8 bits
<ex_cid>	Explicit representation of <cid>	"CID-" prefixed to <cid>
<payloadType>	Payload type	Decimal integer 0-127
<transport>	Transport	Values listed in Table 9.1.
<profile>	Profile	Decimal integer 1-255
<eecid>	End-to-end connection Identifier	Up to 8 hex digits
<aalType>	AAL type	String values:
		"AAL1,""AAL1_SDT,""AAL1_UDT,"
		"AAL2," "AAL3/4,"
		"AAL5," "USER_DEFINED_AAL"
<asc>	ATM service category defined by the ATMF	String values:
		"CBR," "nrt-VBR," "rt-VBR,"
		"UBR," "ABR," "GFR"
<atc>	TM transfer capability defined by the ITU	String values:
		"DBR,""SBR,""ABT/IT,""ABT/DT,"
		"ABR"
<subtype>	<asc>/<atc> subtype	Decimal integer 1-10
<qosClass>	QOS class	Decimal integer 0-5
<bcob>	Broadband bearer class	Decimal or hex representation of 5-bit field
<eetim>	End-to-end timing	String values: "on" required "off"
<stc>	Susceptibility to clipping	Decimal equivalent of a 2-bit field
<upcc>	User plane connection a configuration	Decimal equivalent of 2-bit field
<directionFlag>	Direction flag	String values: "f," "b," "fb"
<cdvType>	CDV type	String values:
		"PP," "2P"

(*Continued*)

PARAMETER	*MEANING*	*REPRESENTATION*
<acdv>	Acceptable CDV	Decimal equivalent of 24-bit field
<ccdv>	Cumulative CDV	Decimal equivalent of 24-bit field
<eetd>	End-to-end transit delay	Decimal equivalent of 16-bit field
<cmtd>	Cumulative transit delay	Decimal equivalent of 16-bit field
<aclr>	Acceptable cell loss ratio	Decimal equivalent of 8-bit field
<clpLvl>	CLP level	String values:
		"0" "0+1"
<pcr>	Peak cell rate	Decimal equivalent of a 24-bit field
<scr>	Sustained cell rate	Decimal equivalent of a 24-bit field
<mbs>	Maximum burst size	Decimal equivalent of 16-bit field
<cdvt>	CDVT	Decimal equivalent of 24-bit field.
<mcr>	Minimum cell rate	Decimal equivalent of a 24-bit field
<mfs>	Maximum frame size	Decimal equivalent of a 16-bit field
<fd>	Frame discard allowed	String values:
		"on" "off"
<te>	CLP tagging	String Values: "on" "off"
<nrm>	NRM	Decimal/hex equivalent of 3 bit field
<trm>	TRM	-ditto-
<cdf>	CDF	-ditto-
<adtf>	ADTF	Decimal/Hex equivalent of 10 bit field
<ficr>	Forward initial cell rate	Decimal equivalent of 24-bit field
<bicr>	Backward initial cell rate	Decimal equivalent of 24-bit field
<ftbe>	Forward transient buffer exposure	Decimal equivalent of 24-bit field
<btbe>	Backward transient buffer exposure	Decimal equivalent of 24-bit field
<crmrtt>	Cumulative RM round-trip time (microseconds)	Decimal equivalent of 24-bit field
<frif>	Forward rate increase factor	Decimal integer 0-15
<brif>	Backward rate increase factor	Decimal integer 0-15
<frdf>	Forward rate decrease factor	Decimal integer 0-15
<brdf>	Backward rate decrease factor	Decimal integer 0-15

(Continued)

PARAMETER	MEANING	REPRESENTATION
<bearerType>	Bearer type	String values: "PVC," "SVC," "CID"
<localInitiation>	Local initiation	String values: "on" "off"
<sci>	Screening indication	Decimal or hex equivalent of 4 bits
<lsn>	Leaf sequence number	Decimal or hex equivalent of 32 bits
<cdStd>	Coding standard for connection scope selection IE definition: UNI 4.0 [16]	Decimal or hex equivalent of 2 bits
<conScpTyp>	Type of connection scope definition: UNI 4.0 [16]	Decimal or hex equivalent of 4 bits
<conScpSel>	Connection scope selection definition: UNI 4.0 [16]	Decimal or hex equivalent of 8 bits
<cacheEnable>	Enable SVC caching	String values: "on" "off"
<cacheTimer>	Timer for cached SVC deletion	Decimal or hex equivalent of 32-bit field
<bearerSigIEType>	Bearer signaling IE type	2 hex digits
<bearerSigIELng>	Bearer signaling IE length	1-4 hex digits
<bearerSigIEVal>	Bearer signaling IE value	Even number of hex digits, 2-512
<appClass>	Application specification	String values: "itu_h323c,""af83," "AAL5_SSCOP," "itu_i3661_unassured," "itu_i3661_assured," "itu_i3662," "itu_i3651," "itu_i3652," "itu_i3653," "itu_i3654," "FRF5," "FRF8,""FRF11," "itu_h2221"
<oui>	Organizationally unique identifier	1 to 6 hex digits
<appId>	Application identifier	1 to 8 digits
<cbrRate>	CBR rate	Two hex digits
<sbc>	Subchannel count	T1: Decimal integer 1-24 or hex equivalent E1: Decimal integer 1-31 or hex equivalent
<clkrec>	Clock recovery method	String values: "NULL," "SRTS," "ADAPTIVE"

(Continued)

PARAMETER	MEANING	REPRESENTATION
<fecEnable>	Forward error correction enable	String values: "NULL," "LOSS_SENSITIVE," "DELAY_SENSITIVE"
<partialFill>	Partial fill	Decimal integer 1-48 or hex equivalent
<structureEnable>	Structure present	String values: "on" "off"
<blksz>	Block size	Decimal or hexadecimal equivalent of 16 bits
<cpcs>	Maximum CPCS SDU	AAL5: Decimal or hex size equivalent of 16 bits AAL2: 45 or 64, decimal or hex representation
<cidLowerLimit>	AAL2 CID lower limit	Decimal integer 8-255 or hex equivalent
<cidUpperLimit>	AAL2 CID	Decimal integer 8-255 upper limit or hex equivalent
<timerCU>	Timer, combined use (microseconds)	Integer decimal; range determined by application Use decimal equivalent of 32 bits.
<simplifiedCPS>	Simplified CPS [48]	String values: "on" "off"
<fSDUrate>	Forward SDU rate (bits per second)	Decimal equivalent of 24-bit field
<bSDUrate>	Backward SDU rate (bits per second)	Decimal equivalent of 24-bit field
<ted>	Transmission error detection enable	String values: "on" "off"
<rastimer>	SSSAR reassembly (microseconds)	Integer decimal, range determined by application. Use decimal equivalent of 32 bits.
<fsssar>	Maximum SSSAR-SDU size, forward direction	Decimal 1-65568 or hex equivalent
<bsssar>	Maximum SSSAR-SDU size, backward direction	Decimal 1-65568 or hex equivalent
<fsscopsdu>	Maximum SSCOP-SDU size, forward direction	Decimal 1-65528 or hex equivalent
<bsscopsdu>	Maximum SSCOP-SDU size, backward direction	Decimal 1-65528 or hex equivalent
<fsscopuu>	Maximum SSCOP-UU field size, forward direction	Decimal 1-65524 or hex equivalent
<bsscopuu>	Maximum SSCOP-UU field size, backward direction	Decimal 1-65524 or hex equivalent
<sap>	SAP	String values: "AUDIO" "MULTIRATE"

(Continued)

PARAMETER	*MEANING*	*REPRESENTATION*
<circuitMode>	Circuit mode enable	String values: "on" "off"
<frameMode>	Frame mode enable	String values: "on" "off"
<faxDemod>	Fax demodulation enable	String values: "on" "off"
<cas>	Enable CAS transport via Type 3 packets	String values: "on" "off"
<dtmf>	Enable DTMF transport via Type 3 packets	String values: "on" "off"
<mfall>	Enable MF transport via Type 3 packets	String values: "on" "off"
<mfr1>	Enable MF (R1) transport via Type 3 packets	String values: "on" "off"
<mfr2>	Enable MF (R2) transport via Type 3 packets	String values: "on" "off"
<PCMencoding>	PCM encoding	String values: "PCMA" "PCMU"
<fmaxFrame>	Maximum length of a frame mode, data unit forward direction	Decimal or hex equivalent of 16-bit field
<bmaxFrame>	Maximum length of a frame mode data unit, backward direction	–ditto
<silenceSuppEnable> Silence suppression Enable	String values: "on" "off"	
<silenceTimer>	Kick-in timer for silence suppression	Decimal or hex representation of 16-bit field
<suppPref>	Preferred silence suppression method	String values: "standard" "custom"
<sidUse>	SID use method	String values: "No SID," "Fixed Noise," "Sampled Noise"
<fxnslevel>	Fixed noise level	Decimal or hex representation of a 7-bit field
<ecanEnable>	Enable echo cancelation	String values: "on" "off"
<ecanType>	Type of echo Cancelation	String values: "G165" "G168"
<gcEnable>	Enable gain control	String values: "on" "off"
<gcLvl>	Level of inserted loss	Decimal or hex equivalent of 16-bit field
<aal2transport>	AAL2 transport	Values listed in Table 9.1 that begin with the string "AAL2"
<uuiCodeRange>	UUI code range	Decimal integer 0-15

(*Continued*)

PARAMETER	*MEANING*	*REPRESENTATION*
<encodingName>	Encoding name	String values: "PCMG," "SIDG," "SID729," any value from column 2 of Table 9.2
<packetLength>	Packet length	Decimal integer 0-45
<packetTime>	Packetization interval in microsec	Decimal integer 1-65,536
<fxIncl>	Facsimile included	String values: "on" "off"
<serviceType>	Service type	String values: "v," "d," "f," "df," "all"
<q7655scc>	Contents of the Q.765.5 codec IE	Even number of hex Single digits (4-32)
<isupUsi>	ISUP user service information	Even number of hex digits (4-24)
<uiLayer1Prot>	User Information Layer 1 Protocol	Two hex digits
<chainPointer>	Chain pointer	String values: "NEXT," "PREVIOUS," "NULL"
<rtcpPortNum>	RTCP port number for H.323 Annex C applications	Odd decimal in range 1,024 to 65,535 Preferred: odd number in the range 49,152 to 65,535
<rtcpIPaddr>	IP address for receipt of RTCP packets	Dotted decimal, 7-15 chars

9.4.7 Examples of ATM Session Descriptions Using SDP

An example of a complete AAL1 session description in SDP is

```
v=0
o=- A3C47F21456789F0 0 ATM NSAP
47.0091.8100.0000.0060.3e64.fd01.0060.3e64.fd01.00
s=-
c=ATM NSAP
47.0091.8100.0000.0060.3e64.fd01.0060.3e64.fd01.00
t=0 0
m=audio $ AAL1/AVP 18 0 96
a=atmmap:96 X-G727-32
a=eecid:B3D58E32
```

An example of a complete AAL2 session description in SDP is

```
v=0
o=- A3C47F21456789F0 0 ATM NSAP
47.0091.8100.0000.0060.3e64.fd01.0060.3e64.fd01.00
s=-
c=ATM NSAP
47.0091.8100.0000.0060.3e64.fd01.0060.3e64.fd01.00
t=0 0
```

```
m=audio $ AAL2/ITU 8 AAL2/custom 100 AAL2/ITU 1
a=eecid:B3E32
```

The AAL2 session descriptor here is the same as the one above except that it states an explicit preference for a voice codec, a voiceband data codec, and a voiceband fax codec. Further, it defines the profile AAL2/custom 100 rather than assume that the far end is cognizant of the elements of this profile.

```
v=0
o=- A3C47F21456789F0 0 ATM NSAP
47.0091.8100.0000.0060.3e64.fd01.0060.3e64.fd01.00
s=-
c=ATM NSAP
47.0091.8100.0000.0060.3e64.fd01.0060.3e64.fd01.00
t=0 0
m=audio $ AAL2/ITU 8 AAL2/custom 100 AAL2/ITU 1
a=eecid:B3E32
a=profileDesc:AAL2/custom 100 0-7 PCMG 40 5000 0-7 SIDG 1
5000 8-15 G726-32 40 10000 8-15 SIDG 1 5000
a=vsel:G726-32 40 10000
a=dsel:off PCMU - -
a=fsel:G726-32 40 10000
```

An example of an SDP session descriptor for an AAL5 SVC for delivering MPEG-2 video:

```
v=0
o=- A3C47F21456789F0 0 ATM NSAP
47.0091.8100.0000.0060.3e64.fd01.0060.3e64.fd01.00
s=-
c=ATM NSAP 47.0091.8100.0000.0060.3e64.fd01.0060.3e64.fd01.00
t=0 0
m=video $ AAL5/ITU 33
a=eecid:B3E32
a=aalType:AAL5
a=bearerType:SVC on
a=atmTrfcDesc:f 0+1 7816 - - - - - off -
a=atmTrfcDesc:b 0+1 0 - - - - - on -
a=cpsSDUsize:f 20680
a=aalApp:itu_h2221 - -
```

An example of an SDP session descriptor for an AAL5 PVC for delivering MPEG-2 video:

```
v=0
o=- A3C47F21456789F0 0 ATM - -
s=-
c=ATM - -
t=0 0
m=video PORT-$/VPI-0/VCI-$ AAL5/ITU 33
a=bearerType:PVC -
a=atmTrfcDesc:f 0+1 7816 - - - - - off -
a=atmTrfcDesc:b 0+1 0 - - - - - on -
a=cpsSDUsize:f 20680
a=aalApp:itu_h2221 - -
```

9.4.8 Security Considerations

The security procedures that need to be considered in dealing with the basic SDP messages described in this section (RFC 3108) are provided in Section 17.8.4.

9.4.9 ATM SDP Grammar

This appendix provides an ABNF grammar for the ATM conventions for SDP. ABNF is defined in RFC 2234 (that was made obsolete by RFC 4234; again RFC 4234 was made obsolete by RFC 5234). This is not a complete ABNF description of SDP. Readers are referred to RFC 4566 (see Section 2) that obsoletes RFC 2327 for an ABNF description of the SDP base line protocol and to RFC 2848 (see Section 14.3), RFC 3261 (also see Section 1.2) that obsoletes RFC 2543, RFC 2045, and RFC 2326 for application-specific conventions of SDP use. For case conventions, see Section 9.4.2.4.

```
; Constant definitions

safe   =   alpha-numeric / "'" / "-" / "." / "/" / ":" / "?" /
           DQUOTE /
           "#" / "$" / "&" / "*" / ";" / "=" / "@" / "[" / "]" /
           "^" / "_" /
           "`" / "{" / "|" / "}" / "+" / "~"
DQUOTE =   %x22 ; double quote
alpha-numeric =      ALPHA / DIGIT
ALPHA  =   "a" / "b" / "c" / "d" / "e" / "f" / "g" / "h" / "i" /
            "j" /
           "k" / "l" / "m" / "n" / "o" / "p" / "q" / "r" / "s" /
           "t" /
           "u" / "v" / "w" / "x" / "y" / "z" /
           "A" / "B" / "C" / "D" / "E" / "F" / "G" / "H" / "I" /
           "J" /
           "K" / "L" / "M" / "N" / "O" / "P" / "Q" / "R" / "S" /
           "T" /
           "U" / "V" / "W" / "X" / "Y" / "Z"
DIGIT  =   "0" / POS-DIGIT
POS-DIGIT =   "1" / "2" / "3" / "4" / "5" / "6" / "7" / "8" / "9"
hex-prefix    =       "0" ("x" / "X")
HEXDIG=    DIGIT / "a" / "b" / "c" / "d" / "e" / "f" /
        "A" / "B" / "C" / "D" / "E" / "F"
space  =   %d32
EOL    =   (CR / LF / CRLF) ; as per Megaco RFC
CR     =   %d13
LF     =   %d10

decimal-uchar  =   DIGIT
                       / POS-DIGIT DIGIT
                       / ("1" 2*(DIGIT))
                       / ("2" ("0"/"1"/"2"/"3"/"4") DIGIT)
                       / ("2" "5" ("0"/"1"/"2"/"3"/"4"/"5"))
generic-U8     =   (hex-prefix hex-U8) / decimal-uchar
generic-U12    =   (hex-prefix hex-U12) / 1*4 (DIGIT)
generic-U16    =   (hex-prefix hex-U16) / 1*5(DIGIT)
generic-U24    =   (hex-prefix hex-U24) / 1*8(DIGIT)
```

```
generic-U32    =    (hex-prefix hex-U32) / 1*10(DIGIT)
hex-U8     =    1*2(HEXDIG)
hex-U12    =    1*3(HEXDIG)
hex-U16    =    1*4(HEXDIG)
hex-U24    =    1*6(HEXDIG)
hex-U32    =    1*8(HEXDIG)
generic-U8-or-null    =    generic-U8 / "-"
generic-U12-or-null  =    generic-U12 / "-"
generic-U16-or-null  =    generic-U16 / "-"
generic-U24-or-null  =    generic-U24 / "-"
generic-U32-or-null  =    generic-U32 / "-"
decimal-U8-or-null    =   decimal-uchar / "-"
decimal-U12-or-null  =    1*4(DIGIT) / "-"
decimal-U16-or-null  =    1*5(DIGIT)  / "-"
decimal-U24-or-null  =    1*8 (DIGIT) / "-"
decimal-U32-or-null  =    1*10(DIGIT) / "-"
on-off-or-null        =    "on" / "off" / "-"

; ABNF definition of SDP with ATM conventions

SDP-infoset = 1*(announcement)announcement = proto-version
origin-field     session-name-field    information-field
                 uri-field
email-fields     phone-fields    connection-field bandwidth-fields
time-fields key-field attribute-fields media-descriptions

proto-version = ["v=" 1*4(DIGIT) EOL] ; use "v=0" for ATM SDP

origin-field = ["o=" username space sess-id space sess-version
                space net-type-addr EOL]

username = 1* safe ; for ATM use "-"

sess-id = (1*32 DIGIT) / (hex-prefix 1*32 HEXDIG)
sess-version = (1*10 DIGIT) / (hex-prefix 1*8 HEXDIG)

net-type-addr= nettype space addrtype-addr

netttype = "ATM" / "IN" / "TN" / "-" / "$"

; Other nettype values may be defined in the future in other
  documents
; Validity of nettype and addrtype-addr combination to be
  checked at
; application level, not protocol syntax level

addrtype-addr = atm-addrtype-addr / ip-addrtype-addr
/ tn-addrtype-addr
; ip-addrtype-addr per rfc2327
; tn-addrtype-addr per rfc2848

; ATM address definition

atm-addrtype-addr = atm-nsap-addr / atm-e164-addr / atm-alias-addr
atm-nsap-addr = ("NSAP" / "-" / "$") space (nsap-addr / "-" / "$")
```

```
atm-e164-addr = ("E164" / "-" / "$") space (e164-addr / "-" / "$")
atm-alias-addr = ("GWID" / "ALIAS" / "-" / "$") space
(alias-addr /
"-" / "$")
nsap-addr = 2(HEXDIG) "." 9(4(HEXDIG) ".") 2(HEXDIG)

e164-addr = 1*15 (DIGIT)
alias-addr = 1*32(alpha-numeric / "-" / "." / "_")

session-name-field = ["s=" text EOL] ; for ATM use "s=-"
text = byte-string
byte-string = 1*(byte-string-char) ; definition per rfc2327
byte-string-char = %x01-09/ %x0B/ %x0C/ %x0E-FF ; all ASCII except
                    NUL, CR & LF
; Definitions of information-field, uri-field, email-fields,
; phone-fields per rfc2327. These fields are omitted in
; ATM SDP descriptions. If received, they are ignored in the ATM
; context

connection-field = ["c=" c-net-type-addr]
; connection-field required, not optional, in ATM

c-net-type-addr = nettype space c-addrtype-addr

c-addrtype-addr = atm-addrtype-addr / c-ip-addrtype-addr /
                                      tn-addrtype-addr
; atm-addrtype-addr defined above
; c-ip-addrtype-addr per rfc2327
; difference in address usage between 'o' and 'c' lines per RFC
  2327 (that is obsoleted by
;RFC 4566 (see Section 2)
; tn-addrtype-addr per rfc2848

bandwidth-fields = *("b=" bwtype ":" bandwidth EOL)

bwtype = 1*(alpha-numeric)

bandwidth = 1*(DIGIT)

time-fields = *( "t=" start-time space stop-time
               *(EOL repeat-fields) EOL)
               [zone-adjustments EOL]
start-time = time / "0"

stop-time = time / "0" ; always "0" in ATM

time = POS-DIGIT 9*(DIGIT) ; same as RFC 2327 (that is obsoleted
by RFC 4566
; see Section 2)
; repeat-fields and zone-adjustments per rfc2327, not used in ATM
; Definition of optional key-field per RFC 2327 (that is obsoleted
  by RFC 4566
; see Section 2)
;
```

```
attribute-fields = *("a=" attribute EOL)

; SDP descriptors for ATM do not have session-level media
  attribute
; lines. If these are provided, they should be ignored.

media-descriptions = *(media-description)
media-description = media-field information-field
                    *(connection-field)
                    bandwidth-fields key-field attribute-fields
; Definitions of information-field per RFC 2327 (that is obsoleted
  by RFC 4566
; see Section 2). These fields are
; omitted in ATM SDP descriptions. If received, they are
  ignored in
; the ATM context
;
; In ATM, the connection-field is used in media-description to
  indicate
; the IP address associated with the RTCP control protocol
  in H.323.C
; applications. In this case, the connection field is per the
  RFC 2327
; definition for IP v4-based connections. Otherwise, it is not
  used in
; media-description. If received as part of media-description,
; it is ignored.
;
; Definition of optional bandwidth-fields as above.
: Definition of optional key-field as in RFC 2327
media-field = rfc2327-media-field / rfc2848-media-field /
              atm-media-field
; rfc2327-media-field and rfc2848-media-field defined in
  those rfc's
; Note: RFC 2327 is obsoleted by RFC 4566 (see Section 2)

atm-media-field = "m=" media space vcId space transport-fmts EOL
; superset of rfc2327 definition
; Note: RFC 2327 is obsoleted by RFC 4566 (see Section 2)

media = "audio" / "video" / "data" / "application" / "control"
              /1*(alpha-numeric)
vcId = "$" / "-" / ex-vcci / (ex-vcci "/" ex-cid) /
       (atm-type-addr-m "/" ex-vcci) /
       (atm-type-addr-m "/" ex-vcci "/" ex-cid) /
       (ex-bcg "/" ex-vcci) / (ex-bcg "/" ex-vcci "/" ex-cid)
       (ex-portid "/" ex-vpi "/" ex-vci) /
       (ex-portid "/" ex-vpi "/" ex-vci "/" ex-cid) /
       (ex-bcg "/" ex-vpi "/" ex-vci) /
       (ex-bcg "/" ex-vpi "/" ex-vci "/" ex-cid) /
       (ex-vpci "/" ex-vci) /
       (ex-vpci "/" ex-vci "/" ex-cid) /
       (atm-type-addr-m "/" ex-vpci "/" ex-vci) /
       (atm-type-addr-m "/" ex-vpci "/" ex-vci "/" ex-cid)
```

```
atm-type-addr-m    =  atm-nsap-addr-m / atm-e164-addr-m /
                      atm-alias-addr-m
atm-nsap-addr-m    =  ["NSAP-"] (nsap-addr / "$")
atm-e164-addr-m    =  ["E164-"] (e164-addr / "$")
atm-alias-addr-m =  ["GWID-" / "ALIAS-"] (alias-addr / "$")

; The -m at the end indicates use in the media field
; Wildcarding rules different from ATM address on 'o' and 'c'
  lines

ex-vcci            =  "VCCI-" vcci
ex-cid             =  "CID-" cid
ex-bcg             =  "BCG-" bcg
ex-portid          =  "PORT-" portid
ex-vpi             =  "VPI-" vpi
ex-vci             =  "VCI-" vci
ex-vpci            =  "VPCI-" vpci
vcci               =  generic-U16
cid                =  generic-U8
bcg                =  generic-U8
portid             =  1*32 (HEXDIG)
vpi                =  generic-U12
vci                =  generic-U16
vpci               =  generic-U16

transport-fmts = generic-transport-fmts / known-transport-fmts
     / "- -"
generic-transport-fmts = generic-transport 1*(space fmt)
generic-transport = 1*(alpha-numeric / "/")
fmt = 1*(alpha-numeric)

known-transport-fmts = aal1-transport space aal1-fmt-list /
                           aal2-transport space aal2-fmt-list
                           *(space aal2-transport space aal2-fmt-
                           list) /
                           aal5-transport space aal5-fmt-list /
                           rtp-transport space rtp-fmt-list /
                           tn-proto space tn-fmt-list /
                           h323c-proto "-"
h323c-proto = "H323c"

; h323c-proto used for RTCP control ports in H.323 annex C
; applications. tn-proto and tn-fmt-list per rfc2848

 aal1-transport        =    "AAL1" "/" aal1-transport-list
 aal1-transport-list   =    "ATMF" / "ITU" / "custom" / "IEEE:" oui /
                            corporate-name
 corporate-name        =    1*(safe)
 aal2-transport        =    "AAL2" "/" aal2-transport-list
 aal2-transport-list   =    aal1-transport-list
 aal5-transport        =    "AAL5" "/" aal5-transport-list
 aal5-transport-list   =    aal1-transport-list
 rtp-transport         =    "RTP" "/" rtp-transport-list
 rtp-transport-list    =    "AVP"
```

```
aal1-fmt-list        =    (payload-type *(space payload-type))
                          / "-"
payload-type         =    decimal-uchar
aal5-fmt-list        =    aal1-fmt-list
rtp-fmt-list         =    aal1-fmt-list
aal2-fmt-list        =    (profile *(space profile)) / "-"
profile              =    decimal-uchar
attribute-fields     =    *("a=" attribute EOL)
attribute            =    known-attribute / (generic-att-field ":"
                          att-value) /
                          generic-att-field
generic-att-field    =    1*(alpha-numeric)
att-value            =    byte-string
known-attribute      =    atm-attribute / PINT-attribute /
                          rfc2327-attribute
; PINT-attribute as defined in RFC 2848 (see Section 14.3)
; rfc2327 attribute as defined in that rfc
; Note: RFC 2327 is obsoleted by RFC 4566 (see Section 2)

atm-attribute  =
                "eecid" ":" eecid /
                "aalType" ":" aalType /
                "capability" ":" (asc / atc) space subtype /
                "qosclass" ":" qosclass /
                "bcob" ":" bcob space eetim /
                "stc" ":" stc /
                "upcc" ":" upcc /
                "atmQOSparms" ":" directionFlag space cdvType
                                   space acdv space ccdv space
                                   eetd space cmtd
                                  space aclr /
                "atmTrfcDesc" ":" directionFlag space clpLvl
                                   space pcr space scr space mbs
                                   space cdvt space
                                   mcr space mfs space fd space te /
                "abrParms" ":" directionFlag space nrm space trm
                                   space cdf
                                   space adtf /
                "abrSetup" ":" ficr space bicr space ftbe space
                                  btbe space
                                  crmrtt space frif space brif
                                  space frdf space brdf /
                "bearertype" ":" bearerType space
                                  localInitiation /
                "lij" ":" sci space lsn /
                "anycast" ":" atmGroupAddress space cdStd space
                                   conScpTyp space conScpSel /
                "cache" ":" cacheEnable space cacheTimer /
                "bearerSigIE" ":" bearerSigIEType space
                                   bearerSigIELng space
                                   bearerSigIEVal /
                "aalApp" ":" appClass space oui space appId /
                "cbrRate" ":" cbrRate /
                "sbc" ":" sbc /
```

```
"clkrec" ":" clkrec /
"fec" ":" fecEnable /
"prtfl" ":" partialFill /
"structure" ":" structureEnable space blksz /
"cpsSDUsize" ":" directionFlag space cpcs /
"aal2CPS" ":" cidLowerLimit space cidUpperLimit
              space
              timerCU space simplifiedCPS /
"aal2CPSSDUrate" ":" fSDUrate space bSDUrate /
"aal2sscs3661unassured" ":" ted space rastimer
                            space fsssar
                            space bsssar /
"aal2sscs3661assured" ":" rastimer space fsssar
                          space bsssar
                        space fsscopsdu space
                        bsscopsdu space fsscopuu
                        space bsscopuu /
"aal2sscs3662" ":" sap space circuitMode space
frameMode
                 space faxDemod space cas space
                 dtmf space mfall space mfr1
                 space mfr2 space PCMencoding
                 space fmaxFrame
                 space bmaxFrame /
"aal5sscop" ":" fsscopsdu space bsscopsdu space
fsscopuu
                space bsscopuu /
"atmmap" ":" payload-type space encoding-name /
"silenceSupp" ":" silenceSuppEnable space
                  silenceTimer
                  space suppPref space sidUse
                  space fxnslevel /
"ecan" ":" directionFlag space ecanEnable space
           ecanType /
"gc" ":" directionFlag space gcEnable space
           gcLvl /
"profileDesc" ":" aal2-transport space profile
                  space
                  1*(profile-row) /
"vsel" ":" 1*(encoding-name space packet-length
           space
           packet-time space) /
"dsel" ":" fxIncl space
           1*(encoding-name space packet-length
           space
           packet-time space) /
"fsel" ":" 1*(encoding-name space packet-length
           space
           packet-time space) /
"onewaySel" ":" serviceType space directionFlag
           space
           1*(encoding-name space packet-length
           space
           packet-time space) /
```

```
                    "codecconfig" ":" q7655scc /
                    "isup_usi" ":" isupUsi /

                    "uiLayer1_Prot" ":" uiLayer1Prot /
                    "chain" ":" chainPointer
eecid = 8 (HEXDIG)
aalType = "AAL1" / "AAL2" / "AAL3/4" / "AAL5" / "USER_DEFINED_AAL"

asc    =      "CBR" / "nrt-VBR" / "rt-VBR" / "UBR" / "ABR" / "GFR"
atc    =      "DBR" / "SBR" / "ABT/IT" / "ABT/DT" / "ABR"
subtype        =      decimal-U8-or-null
qosclass       =      decimal-U8-or-null
bcob           =      generic-U8
eetim          =      on-off-or-null
stc            =      decimal-uchar
upcc           =      decimal-uchar
directionFlag=        "f" / "b" / "fb"
cdvType        =      "PP" / "2P" / "-"
acdv           =      decimal-U32-or-null
ccdv           =      decimal-U32-or-null
eetd           =      decimal-U16-or-null
cmtd           =      decimal-U16-or-null
aclr           =      decimal-U8-or-null
clpLvl         =      "0" / "0+1" / "-"
pcr            =      decimal-U24-or-null
scr            =      decimal-U24-or-null
mbs            =      decimal-U16-or-null
cdvt           =      decimal-U24-or-null
mcr            =      decimal-U24-or-null
mfs            =      decimal-U16-or-null
fd             =      on-off-or-null
te             =      on-off-or-null
nrm            =      generic-U8-or-null
trm            =      generic-U8-or-null
cdf            =      generic-U8-or-null
adtf           =      generic-U16-or-null
ficr           =      decimal-U24-or-null
bicr           =      decimal-U24-or-null
ftbe           =      decimal-U24-or-null
btbe           =      decimal-U24-or-null
crmrtt         =      decimal-U24-or-null
frif           =      1*2 (DIGIT)
brif           =      1*2 (DIGIT)
frdf           =      1*2 (DIGIT)
brdf           =      1*2 (DIGIT)
bearerType     =      "PVC" / "SVC" / "CID"
localInitiation       =      on-off-or-null
sci            =      generic-U8-or-null
lsn            =      generic-U32-or-null

atmGroupAddress       =     atm-type-addr

cdStd          =      generic-U8-or-null
```

```
conScpTyp     =      generic-U8-or-null
conScpSel     =      generic-U8-or-null
cacheEnable   =      on-off-or-null
cacheTimer    =      generic-U32-or-null

bearerSigIEType      =      2 * (HEXDIG)
bearerSigIELng       =      1*4 (HEXDIG)
bearerSigIEVal       =      2*512 (HEXDIG)

appClass = "-" /
               "itu_h323c" / "af83" / "AAL5_SSCOP" / "itu_i3661_
                 unassured" /
               "itu_ i3661_assured"/ "itu_i3662"/ "itu_i3651" /
               "itu_i3652" / "itu_i3653" / "itu_i3654" / "FRF11" /
                 "FRF5" /
               "FRF8" / "itu_h2221"

oui           =      "-" / 1*6 (HEXDIG)
appId         =      "-" / 1*8 (HEXDIG)
cbrRate       =      2 (HEXDIG)
sbc           =      generic-U8
clkrec        =      "NULL" / "SRTS" / "ADAPTIVE"
fecEnable     =      "NULL" / "LOSS_SENSITIVE" / "DELAY_SENSITIVE"
partialFill   =      generic-U8

structureEnable      =      on-off-or-null

blksz         =      generic-U16-or-null
cpcs          =      generic-U16

cidLowerLimit =      generic-U8-or-null
cidUpperLimit =      generic-U8-or-null
timerCU       =      decimal-U32-or-null
simplifiedCPS =      on-off-or-null

fSDUrate      =      decimal-U24-or-null
bSDUrate      =      decimal-U24-or-null
ted           =      on-off-or-null
rastimer      =      decimal-U32-or-null
fsssar        =      generic-U24-or-null
bsssar        =      generic-U24-or-null
fsscopsdu     =      generic-U16-or-null
bsscopsdu     =      generic-U16-or-null
fsscopuu      =      generic-U16-or-null
bsscopuu      =      generic-U16-or-null
sap           =      "AUDIO" / "MULTIRATE" / "-"
circuitMode   =      on-off-or-null
frameMode     =      on-off-or-null
faxDemod      =      on-off-or-null
cas           =      on-off-or-null
dtmf          =      on-off-or-null
mfall         =      on-off-or-null
```

```
mfr1               =    on-off-or-null
mfr2               =    on-off-or-null

PCMencoding             =       "PCMA" / "PCMU" / "-"
fmaxframe               =       generic-U16-or-null
bmaxframe               =       generic-U16-or-null

silenceSuppEnable       =       on-off-or-null
silenceTimer            =       generic-U16-or-null
suppPref                =       "standard" / "custom" / "-"

sidUse             =    "No SID" / "Fixed Noise" / "Sampled
                        Noise" / "-"
fxnslevel          =    generic-U8-or-null
ecanEnable         =    on-off-or-null
ecanType           =    "G165" / "G168" / "-"
gcEnable           =    on-off-or-null
gcLvl              =    generic-U16-or-null
profile-row        =    uuiCodeRange space encoding-name space
                        packet-length
                        space packet-time space

uuiCodeRange =          decimal-uchar "-" decimal-uchar / "-"

encoding-name           =       "-" /
                        "PCMG" / "SIDG" / "SID729" /
                        "PCMU" / "G726-32" / "G723" / "PCMA" / "G722"
                         / "G728" /
                        "G729" / "X-G729a" / "X-G729b" / "X-G729ab" /
                        "X-G726-16" / "X-G726-24" / "X-G726-40" /
                          "X-G7231-H" /
                        "X-G7231-L" / "X-G7231a-H" / "X-G7231a-L" /
                        "X-G727-16" / "X-G727-24" / "X-G727-32" /
                        "X-CCD" / "X-CCD-CAS" / "GSM" / "GSM-HR" /
                        "GSM-EFR" / "GSM-EHR" / "X-FXDMOD-3" /
                        "1016" / "DVI4" / "L16" /
                        "LPC" / "MPA" / "QCELP" / "H263" /
                          "H263-1998" /
                        "JPEG" / "H261" / "MPV" / "MP2T" / "nv" /
                          "RED" /
                        "CelB" / "L8" / "VDVI" / "MP1S" / "MP2P" /
                          "BT656" /
                        "FR-AMR" / "HR-AMR" / "UMTS-AMR" / "AMR"

packet-length           =       decimal-U8-or-null
packet-time             =       decimal-U16-or-null
fxIncl                  =       on-off-or-null
serviceType             =       "v" / "d" / "f" / "df" / "all"
q7655scc                =       4*32 (HEXDIG)
isupUsi                 =       4*24 (HEXDIG)
uiLayer1Prot            =       2 (HEXDIG)
chainPointer            =       "NEXT" / "PREVIOUS" / "NULL"
```

9.5 Audio and Video Media Streams over Public Switched Telephone Network (PSTN) in SDP

This section describes RFC 7195 that specifies the use cases, requirements, and protocol extensions for using the SDP offer/answer model for establishing audio and video media streams over circuit-switched (CS) bearers in the PSTN.

9.5.1 Introduction

In this section, we describe the SDP for setting audio and video media streams over CS bearers in PSTN specified in RFC 7195. The SDP (RFC 4566, see Section 2) is intended for describing multimedia sessions for the purposes of session announcement, session invitation, and other forms of multimedia session initiation. SDP is most commonly used for describing media streams that are transported over the RTP (RFC 3550), using the profiles for audio and video media defined in "RTP Profile for Audio and Video Conferences with Minimal Control" (RFC 3551). However, SDP can be used to describe media transport protocols other than RTP. Previous work includes SDP conventions for describing ATM bearer connections (RFC 3108, see Section 9.4) and the Message Session Relay Protocol (RFC 4975).

SDP is commonly carried in SIP (RFC 3261, also see Section 1.2) messages in order to agree on a common media description among the endpoints. "An Offer/Answer Model with the SDP" (RFC 3264, see Section 3.1) defines a framework by which two endpoints can exchange SDP media descriptions and come to an agreement as to which media streams should be used, along with the media-related parameters.

In some scenarios, it might be desirable to establish the media stream over a CS bearer connection even if the signaling for the session is carried over an IP bearer. An example of such a scenario is illustrated with two mobile devices capable of both CS and packet-switched communication over a low bandwidth radio bearer. The radio bearer may not be suitable for carrying real-time audio or video media, and using a CS bearer would offer a perceived QOS. So, according to this scenario, SDP and its higher-layer session control protocol (e.g., SIP - RFC 3261, also see Section 1.2) are used over regular IP connectivity, while the audio or video is received through the classical CS bearer. This document addresses only the use of CS bearers in the PSTN, not a generic CS network.

The mechanisms presented here require a call signaling protocol of the PSTN to be used (such as ITU-T Q.931 [43] or 3GPP TS 24.008 [44]). Setting up a signaling relationship in the IP domain instead of just setting up a CS call also offers the possibility of negotiating, in the same session, other IP-based media that is not sensitive to jitter and delay, for example, text messaging or presence information. At a later point in time, the mobile device might move to an area where a high-bandwidth packet-switched bearer, for example, a Wireless Local Area Network connection, is available. At this point, the mobile device may perform a handover and move the audio or video media streams over to the high-speed bearer. This implies a new exchange of SDP offer/answer that leads to a renegotiation of the media streams.

Other use cases exist. For example, an endpoint might have at its disposal CS and packet-switched connectivity, but the same audio or video codecs are not feasible for both access networks. For example, the CS audio or video stream supports narrow-bandwidth codecs, while the packet-switched access allows any other audio or video codec implemented in the endpoint. In this case, it might be beneficial for the endpoint to describe different codecs for each access type and get an agreement on the bearer together with the remote endpoint.

There are additional use cases related to third-party call control where the session setup time is improved when the CS bearer in the PSTN is described together with one or more codecs. The rest of the document is structured as follows: Section 9.5.2 provides the document conventions, Section 9.5.3 introduces the requirements, Section 9.5.4 presents an overview of the proposed solutions, and Section 9.5.5 contains the protocol description. Section 9.5.6 provides examples of CS audio or video streams in SDP. Sections 9.5.7 and 9.5.8 contain the security and IANA considerations, respectively.

9.5.2 Conventions Used in This Document

The conventions used in this document are per IETF RFC 2119.

9.5.3 Requirements

This section presents the general requirements that are specific for the audio or video media streams over CS bearers.

1. A mechanism for endpoints to negotiate and agree on an audio or video media stream established over a CS bearer MUST be available.
2. The mechanism MUST allow the endpoints to combine CS audio or video media streams with other complementary media streams, for example, text messaging.
3. The mechanism MUST allow the endpoint to negotiate the direction of the CS bearer (i.e., which endpoint is active when initiating the CS bearer).
4. The mechanism MUST be independent of the type of the CS access (e.g., ISDN, GSM, etc.).
5. There MUST be a mechanism that helps an endpoint correlate an incoming CS bearer with the one negotiated in SDP, as opposed to another incoming call that is not related. In case correlation by programmatic means is not possible, correlation may also be performed by the human user.
6. It MUST be possible for endpoints to advertise lists of audio or video codecs in the CS audio or video stream different from those used in a packet-switched audio or video stream.
7. It MUST be possible for endpoints to not advertise the list of available codecs for CS audio or video streams.

9.5.4 Overview of Operation

The mechanism defined in this memo extends SDP (RFC 4566, see Section 2) and allows describing an audio or video media stream established over a CS bearer. A new network type ("PSTN") and a new protocol type ("PSTN") are defined for the "c=" and "m=" lines to be able to describe a media stream over a CS bearer. These SDP extensions are described in Section 9.5.5.2. Since CS bearers are connection-oriented media streams, the mechanism reuses the connection-oriented extensions defined in RFC 4145 to negotiate the active and passive sides of a connection setup. This is further described in Section 9.5.5.3.1.

9.5.4.1 Example Call Flow

Consider the example presented in Figure 9.12. In this example, Endpoint A is located in an environment where it has access to both IP and CS bearers for communicating with other endpoints.

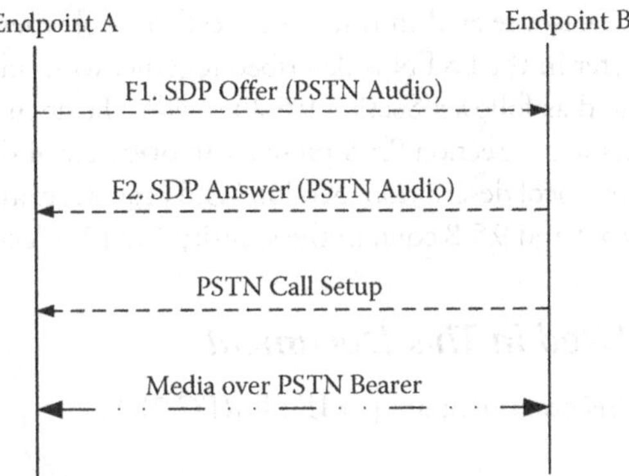

Figure 9.12 Example flow (Copyright: IETF).

Endpoint A decides that the CS bearer offers a perceived better QOS for voice and issues an SDP offer containing the description of an audio media stream over a CS bearer.

Endpoint B receives the SDP offer and determines that it is located in an environment where the IP-based bearer is not suitable for real-time audio media. However, Endpoint B also has a PSTN CS bearer available for audio. Endpoint B generates an SDP answer containing a description of the audio media stream over a CS bearer. During the offer/answer exchange, Endpoints A and B also agree upon the direction in which the CS bearer should be established. In this example, Endpoint B becomes the active party; in other words, it establishes the CS call to the other endpoint. The offer/answer exchange contains identifiers or references that can be used on the CS network for addressing the other endpoint, as well as information that is used to determine that the incoming CS bearer establishment is related to the ongoing session between the two endpoints.

Endpoint B establishes a CS bearer toward Endpoint A using whatever mechanisms are defined for the network type in question. When receiving the incoming CS connection attempt, Endpoint A is able to determine that the attempt is related to the session it is just establishing with B. Endpoint A accepts the CS connection; the CS bearer setup is complete. The two endpoints can now use the CS connection for two-way audio media. If, for some reason, Endpoint B would like to reject the offered stream, it would set the port number of the specific stream to 0, as specified in RFC 3264 (see Section 3.1). Also, if B does not understand some of the SDP attributes specified in this document, it would ignore them, as specified in RFC 4566 (see Section 2).

9.5.5 Protocol Description

9.5.5.1 Level of Compliance

Implementations that are compliant with this specification MUST implement the SDP extensions described in Section 9.5.5.2 and MUST implement the considerations discussed in Sections 9.5.5.3, 9.5.5.4, and 9.5.5.6.

9.5.5.2 Extensions to SDP

This section provides the syntax and semantics of the extensions required for providing a description of audio or video media streams over CS bearers in SDP.

9.5.5.2.1 Connection Data

According to SDP (RFC 4566, see Section 2), the connection data line in SDP has the following syntax:

```
c=<nettype> <addrtype> <connection-address>
```

where <nettype> indicates the network type, <addrtype> indicates the address type, and <connection-address> is the connection-address, which is dependent from the address type.

At the moment, the only network type defined is "IN," which indicates Internet network type. The address types "IP4" and "IP6" indicate the type of IP addresses. This memo defines a new network type for describing a CS bearer network type in the PSTN. The mnemonic "PSTN" is used for this network type. For the address type, we initially considered the possibility of describing E.164 telephone numbers. We define a new "E164" address type to be used within the context of a "PSTN" network type. The "E164" address type indicates that the connection-address contains an E.164 number represented according to the ITU-T E.164 [46] recommendation.

It is a common convention that an international E.164 number contains a leading "+" sign. For consistency's sake, we also require the E.164 telephone is prepended with a "+," even if that is not necessary for routing of the call in the PSTN network. There are cases, though, when the endpoint is merely aware of a CS bearer, without having further information about the E.164 number allocated to it. In these cases, a dash ("-") is used to indicate an unknown connection-address. This makes the connection data line consistent with SDP syntax. Please note that the "E164" address type defined in this memo is exclusively defined to be used in conjunction with the "PSTN" network type in accordance with regular offer/answer procedures (RFC 4566, see Section 2).

Note: RFC 3108 (see Section 9.4) also defines address type "E.164." This definition is distinct from the one defined by this memo and shall not be used with <nettype> "PSTN." This memo exclusively uses the international representation of E.164 numbers (i.e., those including a country code and, as described above, prepended with a "+" sign. Implementations conforming to this specification and using the "E164" address type together with the "PSTN" network type MUST use the "global-number-digits" construction specified in RFC 3966 for representing international E.164 numbers. This representation requires the presence of the "+" sign and additionally allows for the presence of one or more "visualseparator" constructions for easier human readability (see Section 9.5.5.7). Note that <connection-address> MUST NOT be omitted when unknown since this would violate basic syntax of SDP (RFC 4566, see Section 2). In such cases, it MUST be set to a "-." The following are examples of the extension to the connection data line:

```
c=PSTN E164 +441134960123
c=PSTN E164 -
```

When the <addrtype> is E164, the connection-address is defined as follows:

- An international E.164 number (prepended with a "+" sign)
- The value "-" signifying that the address is unknown
- Any other value resulting from the production rule of connection-address in RFC 4566 (see Section 2), but in all cases any value encountered will be ignored.

9.5.5.2.2 Media Descriptions

According to SDP (RFC 4566, see Section 2), the media description line in SDP has the following syntax:

```
m=<media> <port> <proto> <fmt> ...
```

The <media> subfield carries the media type. For establishing an audio bearer, the existing "audio" media type is used. For establishing a video bearer, the existing "video" media type is used. The <port> subfield is the transport port to which the media stream is sent. CS access lacks the concept of a port number; therefore, the <port> subfield does not carry any meaningful value. In order to be compliant with SDP syntax, implementations SHOULD set the <port> subfield to the discard port value "9" and MUST ignore it on reception.

According to RFC 3264 (see Section 3.1), a port number of 0 in the offer of a unicast stream indicates that the stream is offered but must not be used. If a port number of 0 is present in the answer of a unicast stream, it indicates that the stream is rejected. These rules are still valid when the media line in SDP represents a CS bearer.

The <proto> subfield is the transport protocol. The CS bearer uses whatever transport protocol it has available. This subfield SHOULD be set to the mnemonic "PSTN" to be syntactically correct with SDP (RFC 4566, see Section 2) and to indicate the usage of CS protocols in the PSTN.

The <fmt> subfield is the media format description. In the classical usage of SDP to describe RTP-based media streams, when the <proto> subfield is set to "RTP/AVP" or "RTP/SAVP," the <fmt> subfield contains the payload types as defined in the RTP audio profile (RFC 3551).

When "RTP/AVP" is used in the <proto> field, the <fmt> subfield contains the RTP payload type numbers. We use the <fmt> subfield to indicate the list of available codecs over the CS bearer, by reusing the conventions and payload type numbers defined for RTP/AVP. The RTP audio and video media types, when applied to PSTN CS bearers, represent merely an audio or video codec. If the endpoint is able to determine the list of available codecs for CS media streams, it MUST use the corresponding payload type numbers in the <fmt> subfield.

In some cases, the endpoint is not able to determine the list of available codecs for CS media streams. In this case, in order to be syntactically compliant with SDP (RFC 4566, see Section 2), the endpoint MUST include a single dash ("-") in the <fmt> subfield. As per RFC 4566, the media format descriptions are listed in priority order. Examples of media descriptions for CS audio streams are

```
m=audio 9 PSTN 3 0 8
m=audio 9 PSTN -
```

Similarly, an example of a media description for CS video stream is

```
m=video 9 PSTN 34
m=video 9 PSTN -
```

9.5.5.2.3 Correlating the PSTN CS Bearer with SDP

The endpoints should be able to correlate the CS bearer with the session negotiated with SDP in order to avoid ringing for an incoming CS bearer that is related to the session controlled with SDP (and SIP). Several alternatives exist for performing this correlation. This memo provides three mutually nonexclusive correlation mechanisms. Additionally, we define a fourth mechanism where correlation may be performed by external means, typically by the human user, in case using other correlation mechanisms is not possible or does not succeed. Other correlation mechanisms may exist, and their usage will be specified when need arises.

All mechanisms share the same principle: Some unique information is sent in the SDP and in the CS signaling protocol. If these pieces of information match, then the CS bearer is part of the session described in the SDP exchange. Otherwise, there is no guarantee that the CS bearer is related to such session. The first mechanism is based on the exchange of PSTN caller ID between the endpoints. The caller ID is also available as the calling party number in the CS signaling.

The second mechanism is based on the inclusion in SDP of a value that is also sent in the user-user IE (UUIE) that is part of the bearer setup signaling in the PSTN. The third mechanism is based on sending in SDP a string that represents DTMF digits that will be sent right after the CS bearer is established. The fourth correlation mechanism declares support for cases where correlation is done by external means. Typically, this means that the decision is left to the human user. This is how some current conferencing systems operate: After logging on to the conference, the system calls back to the user's phone number to establish audio communications, and it is up to the human user to accept or reject the incoming call. By declaring explicit support for this mechanism, endpoints can use it only when such a possibility exists. Endpoints may opt to implement any combination of the correlation mechanisms specified in Sections 9.5.5.2.3.2, 9.5.5.2.3.3,9.5.5.2.3.4, and 9.5.5.2.3.5, including the option to implement none at all.

9.5.5.2.3.1 The "cs-correlation" Attribute

In order to provide support for the correlation mechanisms, we define a new media-level SDP attribute called "cs-correlation." There MUST be at most one "cs-correlation" attribute per media description. This "cs-correlation" attribute MAY contain zero or more subfields: "callerid," "uuie," "dtmf," or "external" to specify additional information required by the caller ID, UUIE, DTMF, or external correlation mechanisms, respectively. The list of correlation mechanisms may be extended by other specifications; see Section 9.5.5.2.3.6 for more details. The following sections provide more detailed information about these subfields. The values "callerid," "uuie," "dtmf," and "external" refer to the correlation mechanisms defined in Sections 9.5.5.2.3.2, 9.5.5.2.3.3, 9.5.5.2.3.4, and 9.5.5.2.3.5, respectively. The ABNF syntax of the "cs-correlation" attribute is presented in Section 9.5.5.7.

9.5.5.2.3.2 Caller ID Correlation Mechanism

The caller ID correlation mechanism consists of an exchange of the calling party number as an international E.164 number in SDP, followed by the availability of the calling party number IE in the call setup signaling of the CS connection. If both pieces of information match, the CS bearer is correlated to the session described in SDP. An example of inclusion of an international E.164 number in the "cs-correlation" attribute is

```
a=cs-correlation:callerid:+441134960123
```

The presence of the "callerid" subfield indicates that the endpoint supports use of the calling party number as a means of correlating a PSTN call with the session being negotiated. The "callerid" subfield MAY be accompanied by the international E.164 number of the party inserting the parameter. Note that there are no guarantees that this correlation mechanism works or is even available, due a number of problems:

- The endpoint might not be aware of its own E.164 number, in which case it cannot populate the SDP appropriately.

- The calling party number IE in the CS signaling might not be available (e.g., due to policy restrictions of the network operator or caller restriction) due to privacy.
- The calling party number IE in the CS signaling might be available, but the digit representation of the E.164 number might differ from the one expressed in the SDP, due to, for example, lack of country code. To mitigate this problem, implementations should consider only some of the rightmost digits from the E.164 number for correlation. For example, the numbers +44-113-496-0123 and 0113-496-0123 could be considered as the same number. This is also the behavior of some cellular phones, which correlate the incoming calling party with a number stored in the phone book, for the purpose of displaying the caller's name. Please refer to ITU-T E.164 recommendation [46] for consideration of the relevant number of digits to consider.

9.5.5.2.3.3 UUIE Correlation Mechanism A second correlation mechanism is based on including in SDP a string that represents the UUIE that is part of the call setup signaling of the CS bearer. The UUIE is specified in ITU-T Q.931 [43] and 3GPP TS 24.008 [44], among others. The UUIE has a maximum size of 35 or 131 octets, depending on the actual message of the PSTN protocol where it is included and the network settings. The mechanism works as follows. An endpoint creates a UUIE, according to the requirements of the call setup signaling protocol. The same value is included in the SDP offer or SDP answer, in the "uuie" subfield of the "cs-correlation" attribute.

When the SDP offer/answer exchange is complete, each endpoint has become aware of the value that will be used in the UUIE of the call setup message of the PSTN protocol. The endpoint that initiates the call setup attempt includes this value in the UUIE. The recipient of the call setup attempt can extract the UUIE and correlate it with the value previously received in the SDP. If both values match, then the call setup attempt corresponds to that indicated in the SDP.

According to ITU-T Q.931 [43], the UUIE identifier is composed of a first octet identifying this as a UUIE, a second octet containing the length of the user-user contents, a third octet containing a protocol discriminator, and a value of up to 32 or 128 octets (depending on network settings) containing the actual user information (see Figure 4.36 in [43]). The first two octets of the UUIE MUST NOT be used for correlation; only the octets carrying the protocol discriminator and the user information value are input to the creation of the value of the "uuie" subfield in the "cs-correlation" attribute. Therefore, the value of the "uuie" subfield in the "cs-correlation" attribute MUST start with the protocol discriminator octet, followed by the user information octets. The value of the protocol discriminator octet is not specified in this document; it is expected that organizations using this technology will allocate a suitable value for the protocol discriminator.

Once the binary value of the "uuie" subfield in the "cs-correlation" attribute is created, it MUST be base 16 (also known as "hex") encoded before it is inserted in SDP. Please refer to RFC 4648 for a detailed description of base 16 encoding. The resulting encoded value needs to have an even number of hexadecimal digits and MUST be considered invalid if it has an odd number.

Note: The encoding of the "uuie" subfield of the "cs-correlation" attribute is largely inspired by the encoding of the same value in the user-to-user header field in SIP, according to "A Mechanism for Transporting User to User Call Control Information in SIP" (RFC 7433).

As an example, an endpoint willing to send a UUIE containing a protocol discriminator with the hexadecimal value of %x56 and an hexadecimal user information value of %xA390F3D2B7310023 would include an "a=cs-correlation" attribute line as follows:

```
a=cs-correlation:uuie:56A390F3D2B7310023
```

Note that the value of the UUIE is considered an opaque string and only used for correlation purposes. Typically, call signaling protocols impose requirements on the creation of a UUIE for end-user protocol exchange. The details regarding the generation of the UUIE are outside the scope of this specification.

Please note that there are no guarantees that this correlation mechanism works. On one side, policy restrictions might not make the user-user information available end to end in the PSTN. On the other hand, the generation of the UUIE is controlled by the PSTN CS call protocol, which might not offer enough freedom for generating different values from one endpoint to another one or from one call to another in the same endpoint. This might result in the same value of the UUIE for all calls.

9.5.5.2.3.4 DTMF Correlation Mechanism

We introduce a third mechanism for correlating the CS bearer with the session described with SDP. This is based on agreeing on a sequence of digits that are negotiated in the SDP offer/answer exchange and sent as DTMF tones as described in ITU-T Recommendation Q.23 [45] over the CS bearer once this bearer is established. If the DTMF digit sequence received through the CS bearer matches the digit string negotiated in the SDP, the CS bearer is correlated with the session described in the SDP. The mechanism is similar to many voice conferencing systems that require the user to enter a PIN code using DTMF tones in order to be accepted in a voice conference.

The mechanism works as follows. An endpoint selects a DTMF digit sequence. The same sequence is included in the SDP offer or SDP answer, in a "dtmf" subfield of the "cs-correlation" attribute. When the SDP offer/answer exchange is complete, each endpoint has become aware of the DTMF sequence that will be sent right after the CS bearer is set up. The endpoint that initiates the call setup attempt sends the DTMF digits according to the procedures defined for the CS bearer technology used. The recipient (passive side of the bearer setup) of the call setup attempt collects the digits and compares them with the value previously received in the SDP. If the digits match, then the call setup attempt corresponds to that indicated in the SDP.

Note: Implementations are advised to select a number of DTMF digits that provide enough assurance that the call is related but do not prolong the bearer setup time unnecessarily. A number of 5 to 10 digits is a good compromise.

As an example, an endpoint willing to send DTMF tone sequence "14D*3" would include an "a=cs-correlation" attribute line as follows:

```
a=cs-correlation:dtmf:14D*3
```

If the endpoints successfully agree on the usage of the DTMF digit correlation mechanism but the passive side does not receive any DTMF digits after successful CS bearer setup or receives a set of DTMF digits that do not match the value of the "dtmf" attribute (including receiving too many digits), the passive side SHOULD consider that this DTMF mechanism has failed to correlate the incoming call.

9.5.5.2.3.5 External Correlation Mechanism

The fourth correlation mechanism relies on external means for correlating the incoming call to the session. Since endpoints can select which correlation mechanisms they support, it may happen that no other common correlation mechanism is found or that the selected correlation mechanism does not succeed due to the required feature not being supported by the underlying PSTN network. In these cases, the human user can make the decision to accept or reject the incoming call, thus "correlating" the call with the session.

Since not all endpoints are operated by a human user and since there may be no other external means implemented by the endpoint for the correlation function, we explicitly define support for such an external correlation mechanism.

Endpoints wishing to use this external correlation mechanism would use the "external" subfield in the "cs-correlation" attribute. Unlike the other three correlation mechanisms, the "external" subfield does not accept a value. The following is an example of an "a=cs-correlation" attribute line:

```
a=cs-correlation:external
```

Endpoints that are willing to only use the three explicit correlation mechanisms defined in this document ("callerid," "uuie," and/or "dtmf") would not include the "external" mechanism in the offer/answer exchange.

The external correlation mechanism typically relies on the human user to make the decision on whether the call is related to the ongoing session. After the user accepts the call, that bearer is considered related to the session. There is a small chance that the user receives at the same time another CS call that is not related to the ongoing session. The user may reject this call if he/she is able to determine (e.g., based on the calling line identification) that the call is not related to the session and continue waiting for another call attempt. If the user accepts the incoming CS call, but it turns out to be not related to the session, the endpoints need to rely on the human user to take appropriate action (typically, the user would hang up).

9.5.5.2.3.6 Extensions to Correlation Mechanisms New values for the "cs-correlation" attribute may be specified. The registration policy for new values is "Specification Required"; see Section 9.5.8. Any such specification MUST include a description of how the SDP offer/answer mechanism is used to negotiate the use of the new values, taking into account how endpoints determine which side will become active or passive (see Section 9.5.5.3 for more details). If, during the offer/answer negotiation, either endpoint encounters an unknown value in the "cs-correlation" attribute, it MUST consider that mechanism as unsupported and MUST NOT include that value in subsequent offer/answer negotiation.

9.5.5.3 Negotiating the Correlation Mechanisms

The four correlation mechanisms just presented (based on called party number, UUIE, DTMF digit sending, and external) are nonexclusive and can be used independently of each other. In order to know how to populate the "cs-correlation" attribute, the endpoints need to agree which endpoint will become the active party (i.e., the one that will set up the CS bearer).

9.5.5.3.1 Determining the Direction of the CS Bearer Setup

In order to avoid a situation where both endpoints attempt to initiate a connection simultaneously, the direction in which the CS bearer is set up MUST be negotiated during the offer/answer exchange. The framework defined in RFC 4145 (see Section 10.1) allows the endpoints to agree which endpoint acts as the active endpoint when initiating a TCP connection. While RFC 4145 (see Section 10.1) was originally designed for establishing TCP connections, it can be easily extrapolated to the connection establishment of CS bearers. This specification uses the concepts specified in RFC 4145 (see Section 10.1) for agreeing on the direction of establishment of a CS bearer. RFC 4145 (see Section 10.1) defines two new attributes in SDP: "setup" and "connection." The "setup" attribute indicates which of the endpoints should initiate the connection establishment of the PSTN CS bearer. Four values are defined in Section 10.1.4 of this book (that is, Section 4 of RFC 4145): "active," "passive," "actpass," and "holdconn." Please refer to Section 10.1.4 of this book (that is, Section 4 of RFC 4145) for a detailed description of this attribute.

The "connection" attribute indicates whether a new connection is needed or an existing connection is reused. The attribute can take the values "new" or "existing." Please refer to Section 10.1.5 of this book (that is, Section 5 of RFC 4145) for a detailed description of this attribute. Implementations that are compliant with this specification MUST support the "setup" and "connection" attributes specified in RFC 4145 (see Section 10.1) but applied to CS bearers in the PSTN. We define the active party as the one that initiates the CS bearer after the offer/answer exchange. The passive party is the one receiving the CS bearer. Either party may indicate its desire to become the active or passive party during the offer/answer exchange using the procedures described in Section 9.5.5.6.

9.5.5.3.2 Populating the "cs-correlation" Attribute

By defining values for the subfields in the "cs-correlation" attribute, the endpoint indicates that it is willing to become the active party and that it can use those values in the calling party number, in the UUIE, or as DTMF tones during the CS bearer setup. Thus, the following rules apply:

- An endpoint that can only become the active party in the CS bearer setup MUST include all correlation mechanisms it supports in the "cs-correlation" attribute and MUST also specify values for the "callerid," "uuie," and "dtmf" subfields. Notice that the "external" subfield does not accept a value.
- An endpoint that can only become the passive party in the CS bearer setup MUST include all correlation mechanisms it supports in the "cs-correlation" attribute but MUST NOT specify values for the subfields.
- An endpoint that is willing to become either the active or passive party (by including the "a=setup:actpass" attribute in the offer) MUST include all correlation mechanisms it supports in the "cs-correlation" attribute and MUST also specify values for the "callerid," "uuie," and "dtmf" subfields. Notice that the "external" subfield does not accept a value.

9.5.5.3.3 Considerations for Correlations

Passive endpoints should expect an incoming CS call for setting up the audio bearer. Passive endpoints MAY suppress the incoming CS alert during certain time periods. Additional restrictions can be applied, such as the passive endpoint not alerting incoming calls originated from the number that was observed during the offer/answer negotiation. There may be cases when an endpoint is not willing to include one or more correlation mechanisms in the "a=cs-correlation" attribute line even if it supports it.

For example, some correlation mechanisms can be omitted if the endpoint is certain that the PSTN network does not support carrying the correlation identifier. Also, since using the DTMF-based correlation mechanism requires the call to be accepted before DTMF tones can be sent, some endpoints may enforce a policy restricting this due to, for example, cost associated with received calls, making the DTMF-based mechanism unusable. Note that it cannot be guaranteed that the correlation mechanisms relying on caller identification, UUIE, and DTMF sending will succeed even if the usage of those was agreed beforehand. This is due to the fact that correlation mechanisms require support from the CS bearer technology used.

Therefore, even a single positive indication using any of these mechanisms SHOULD be interpreted by the passive endpoint so that the CS bearer establishment is related to the ongoing session, even if the other correlation mechanisms fail. If, after successfully negotiating any of the "callerid," "uuie," or "dtmf" correlation mechanisms in the SDP offer/answer exchange, an endpoint receives an incoming establishment of a CS bearer with no correlation information present, the endpoint first checks whether the offer/answer exchange was also used to successfully negotiate the "external" correlation mechanism. If it was, the endpoint should let the decision be made by external means, typically the human user. If the "external" correlation mechanism was not successfully negotiated, the endpoint should treat the call as unrelated to the ongoing session in the IP domain.

9.5.5.4 Considerations for Usage of Existing SDP

9.5.5.4.1 Originator of the Session

According to SDP (RFC 4566, see Section 2), the origin line in SDP has the following syntax:

```
o=<username> <sess-id> <sess-version> <nettype> <addrtype>
<unicast-address>
```

Of interest here are the <nettype> and <addrtype> fields, which indicate the type of network and type of address, respectively. Typically, this field carries the IP address of the originator of the session. Even if the SDP was used to negotiate an audio or video media stream transported over a CS bearer, the originator is using SDP over an IP bearer. Therefore, <nettype> and <addrtype> fields in the "o=" line should be populated with the IP address identifying the source of the signaling.

9.5.5.4.2 Contact Information

SDP (RFC 4566, see Section 2) defines the "p=" line, which may include the phone number of the person responsible for the conference. Even though this line can carry a phone number, it is not suited for the purpose of defining a connection-address for the media. Therefore, we have selected to define the PSTN-specific connection-addresses in the "c=" line.

9.5.5.5 Considerations for Usage of Third-Party Call Control

"Best Current Practices for Third Party Call Control (3PCC) in the Session Initiation Protocol (SIP)" (RFC 3725) outlines several flows that are possible in third-party call control scenarios and recommends some flows for specific situations. One of the assumptions in RFC 3725 is that an SDP offer may include a "black hole" connection-address, which has the property that packets sent to it will never leave the host that sent them. For IPv4, this "black hole" connection-address is 0.0.0.0 or a domain name within the .invalid DNS top-level domain. When using an E.164 address scheme in the context of third-party call control, when the UA needs to indicate an unknown phone number, it MUST populate the <addrtype> of the SDP "c=" line with a "-" string.

Note: This may result in the recipient of the initial offer rejecting such offer if the recipient of the offer was not aware of its own E.164 number. Consequently, it will not be possible to establish a CS bearer, since neither party is aware of its E.164 number.

9.5.5.6 Offer/Answer Mode Extensions

In this section, we define extensions to the offer/answer model defined in "An Offer/Answer Model with the SDP" (RFC 3264, see Section 3.1) to allow PSTN addresses to be used with the offer/answer model.

9.5.5.6.1 Generating the Initial Offer

The offerer, wishing to use PSTN audio or video stream, MUST populate the "c=" and "m=" lines as follows. The endpoint MUST set the <nettype> in the "c=" line to "PSTN" and the <addrtype> to "E164." Furthermore, the endpoint SHOULD set the <connection-address> field to its own international E.164 number (with a leading "+"). If the endpoint is not aware of its own E.164 number, it MUST set the <connection-address> to "-." In the "m=" line, the endpoint MUST set the <media> subfield to "audio" or "video," depending on the media type, and the <proto> subfield to "PSTN." The <port> subfield SHOULD be set to "9" (the discard port). The values "audio" or "video" in the <media> subfield MUST NOT be set by the endpoint unless it has knowledge that these bearer types are available on the CS network.

The <fmt> subfield carries the payload type number(s) the endpoint is wishing to use. Payload type numbers in this case refer to the codecs that the endpoint wishes to use on the PSTN media stream. For example, if the endpoint wishes to use the GSM codec, it would add payload type number 3 in the list of codecs. The list of payload types MUST only contain those codecs the endpoint is able to use on the PSTN bearer. In case the endpoint is not aware of the codecs available for the CS media streams, it MUST include a dash ("-") in the <fmt> subfield.

The mapping table of static payload types numbers to payload types is initially specified in RFC 3551 and maintained by IANA. For dynamic payload types, the endpoint MUST define the set of valid encoding names and related parameters using the "a=rtpmap" attribute line. See Section 2.6 of this book (that is, Section 6 of RFC 4566) for details. When generating the offer, the offerer MUST include an "a=cs-correlation" attribute line in the SDP offer. The offerer MUST NOT include more than one "cs-correlation" attribute per media description. The "a=cs-correlation" line SHOULD contain an enumeration of all the correlation mechanisms supported by the offerer, in the format of subfields. See Section 9.5.5.3.3 for more information on usage of the correlation mechanisms.

The current list of subfields include "callerid," "uuie," "dtmf," and "external," and they refer to the correlation mechanisms defined in Sections 9.5.5.2.3.2, 9.5.5.2.3.3, 9.5.5.2.3.4, and 9.5.5.2.3.5, respectively.

If the offerer supports any of the correlation mechanisms defined in this memo and is willing to become the active party, the offerer MUST add the "callerid," "uuie," "dtmf," and/or "external" subfields and MUST specify values for them as follows:

- The international E.164 number as the value in the "callerid" subfield
- The contents of the UUIE as the value of the "uuie" subfield
- The DTMF tone string as the value of the "dtmf" subfield
- The endpoint MUST NOT specify any value for the "external" subfield

If the offerer is only able to become the passive party in the CS bearer setup, it MUST add at least one of the possible correlation mechanisms but MUST NOT specify values for those subfields. For example, if the offerer is willing to use the UUIE and DTMF digit-sending mechanisms but can only become the passive party and is also able to let the human user decide whether the correlation should be done, it includes the following lines in the SDP:

```
a=cs-correlation:uuie dtmf external
a=setup:passive
```

If, on the other hand, the offerer is willing to use the UUIE and the DTMF correlation mechanisms and is able to become the active or passive side, and is also able to let the human user decide whether the correlation should be done or not, it includes the following lines in the SDP:

```
a=cs-correlation:uuie:56A390F3D2B7310023 dtmf:14D*3 external
a=setup:actpass
```

The negotiation of the value of the "setup" attribute takes place as defined in Section 10.1.4.1 of this book (that is, Section 4.1 of RFC 4145). The offerer states which role or roles it is willing to perform; the answerer, taking the offerer's willingness into consideration, chooses which roles both endpoints will actually perform during the CS bearer setup. By "active" endpoint, we refer to an endpoint that will establish the CS bearer; by "passive" endpoint, we refer to an endpoint that will receive a CS bearer. If an offerer does not know its international E.164 number, it MUST set the "setup" attribute to the value "active." If the offerer knows its international E.164 number, it SHOULD set the value to either "actpass" or "passive." Also "holdconn" is a permissible value in the "setup" attribute. It indicates that the connection should not be established for the time being. The offerer uses the "connection" attribute to decide whether a new CS bearer is to be established. For the initial offer, the offerer MUST use value "new."

9.5.5.6.2 Generating the Answer

If the offer contained a CS audio or video stream, the answerer first determines whether it is able to accept and use such streams on the CS network. If the answerer does not support or is not willing to use CS media for the session, it MUST construct an answer where the port number for such media stream(s) is set to 0, according to Section 3.1.6 of this book (that is, Section 6 of RFC 3264). If the answerer is willing to use CS media for the session, it MUST ignore the received port number (unless the port number is set to 0). If the offer included a "-" as the payload type number, it indicates that the offerer is not willing or able to define any specific payload type. Most often, a "-" is expected to be used instead of the payload type when the endpoint is not aware of or not willing to define the codecs that will eventually be used on the CS bearer. The CS signaling protocols have their own means of negotiating or indicating the codecs; therefore, an answerer SHOULD accept such offers and SHOULD set the payload type to "-" in the answer. If the answerer explicitly wants to specify a codec for the CS media, it MAY set the respective payload numbers in the <fmt> subfield in the answer. This behavior, however, is NOT RECOMMENDED. When receiving the offer, the answerer MUST determine whether it becomes the active or passive party.

If the SDP in the offer indicates that the offerer is only able to become the active party, the answerer needs to determine whether it is able to become the passive party. If this is not possible (e.g., due to the answerer not knowing its international E.164 number) the answerer MUST reject the CS media by setting the port number to 0 on the answer. If the answerer is aware of its international E.164 number, it MUST include the "setup" attribute in the answer and set it to value "passive" or "holdconn." The answerer MUST also include its E.164 number in the "c=" line. If the SDP in the offer indicates that the offerer is only able to become the passive party, the answerer MUST verify that the offerer's E.164 number is included in the "c=" line of the offer. If the number is included, the answerer MUST include the "setup" attribute in the answer and set it to value "active" or "holdconn." If the number is not included, the recipient of the offer is not willing to establish a connection the E.164 based on a priori knowledge of cost, or other reasons, call establishment is not possible, and the answerer MUST reject the CS media by setting the port number to zero in the answer.

If the SDP in the offer indicates that the offerer is able to become either the active or passive party, the answerer determines which role it will take. If the offer includes an international E.164 number in the "c=" line, the answerer SHOULD become the active party. If the answerer does not become the active party and if the answerer is aware of its E.164 number, it MUST become the passive party. If the answerer does not become the active or the passive party, it MUST reject the CS media by setting the port number to 0 in the answer. For each media description where the offer includes a "cs-correlation" attribute, the answerer MUST select from the offer those correlation mechanisms it supports and include in the answer one "a=cs-correlation" attribute line containing those mechanisms it is willing to use. The answerer MUST only add one "cs-correlation" attribute in those media descriptions where also the offer included a "cs-correlation" attribute. The answerer MUST NOT add any mechanisms that were not included in the offer. If there is more than one "cs-correlation" attribute per media description in the offer, the answerer MUST discard all but the first for any media description. Also, the answerer MUST discard all unknown "cs-correlation" attribute values. If the answerer becomes the active party, it MUST add a value to any of the possible subfields.

If the answerer becomes the passive party, it MUST NOT add any values to the subfields in the "cs-correlation" attribute. After generating and sending the answer,

- If the answerer became the active party, it MUST extract the E.164 number from the "c=" line of the offer and MUST establish a CS bearer to that address.
- If the SDP answer contained a value for the "callerid" subfield, it MUST set the calling party number IE to that number.
- If the SDP answer contained a value for the "uuie" subfield, it MUST send the UUIE according to the rules defined for the CS technology used and set the value of the IE to that received in the SDP offer.
- If the SDP answer contained a value for the "dtmf" subfield, it MUST send those DTMF digits according to the CS technology used.

If, on the other hand, the answerer became the passive party, it MUST be prepared to receive a CS bearer.

- If the offer contained a value for the "callerid" subfield, it MUST compare that value to the calling party number IE of the CS bearer. If the received calling party number IE matches the value of the "callerid" subfield, the call SHOULD be treated as correlated to the ongoing session.
- If the offer contained a value for the "dtmf" subfield, it MUST be prepared to receive and collect DTMF digits once the CS bearer is set up. The answerer MUST compare the received DTMF digits to the value of the "dtmf" subfield. If the received DTMF digits match the

value of the "dtmf" subfield in the "cs-correlation" attribute, the call SHOULD be treated as correlated to the ongoing session.

- If the offer contained a value for the "uuie" subfield, it MUST be prepared to receive a UUIE once the CS bearer is set up. The answerer MUST compare the received UUIE to the value of the "uuie" subfield. If the value of the received UUIE matches the value of the "uuie" subfield, the call SHOULD be treated as correlated to the ongoing session.
- If the offer contained an "external" subfield, it MUST be prepared to receive a CS call and use the external means (typically, the human user) for accepting or rejecting the call.

If the answerer becomes the active party, generates an SDP answer, and then it finds out that the CS call cannot be established, then the answerer MUST create a new SDP offer where the CS stream is removed from the session (actually, by setting the corresponding port in the "m=" line to 0) and send it to its counterpart. This is to synchronize both parties (and potential intermediaries) on the state of the session.

9.5.5.6.3 Offerer Processing the Answer

When receiving the answer, if the SDP does not contain an "a=cs-correlation" attribute line, the offerer should take that as an indication that the other party does not support or is not willing to use the procedures defined in the document for this session and MUST revert to normal processing of SDP. When receiving the answer, the offerer MUST first determine whether it becomes the active or passive party, as described in Section 9.5.5.3.1. If the offerer becomes the active party, it

- MUST extract the E.164 number from the "c=" line and MUST establish a CS bearer to that address.
- If the SDP answer contained a value for the "uuie" subfield, it MUST send the UUIE according to the rules defined for the CS technology used and set the value of the IE to that received in the SDP answer.
- If the SDP answer contained a value for the "dtmf" subfield, it MUST send those DTMF digits according to the CS technology used.

If the offerer becomes the passive party:

- It MUST be prepared to receive a CS bearer.
- Note that if delivery of the answer is delayed for some reason, the CS call attempt may arrive at the offerer before the answer has been processed. In this case, since the correlation mechanisms are negotiated as part of the offer/answer exchange, the answerer cannot know whether the incoming CS call attempt is correlated with the session being negotiated; thus, the offerer SHOULD answer the CS call attempt only after it has received and processed the answer.
- If the answer contained a value for the "dtmf" subfield, the offerer MUST be prepared to receive and collect DTMF digits once the CS bearer is set up. The offerer SHOULD compare the received DTMF digits to the value of the "dtmf" subfield. If the received DTMF digits match the value of the "dtmf" subfield in the "cs-correlation" attribute, the call SHOULD be treated as correlated to the ongoing session.
- If the answer contained a value for the "uuie" subfield, the offerer MUST be prepared to receive a UUIE once the CS bearer is set up. The offerer SHOULD compare the received

UUIE to the value of the "uuie" subfield. If the value of the received UUIE matches the value of the "uuie" subfield, the call SHOULD be treated as correlated to the ongoing session.

- If the answer contained an "external" subfield, the offerer MUST be prepared to receive a CS call and use the external means (typically, the human user) for accepting or rejecting the call.

According the "An Offer/Answer Model with the Session Description Protocol (SDP)" (RFC 3264, see Section 3.1), the offerer needs to be ready to receive media as soon as the offer has been sent. It may happen that the answerer, if it became the active party, will initiate a CS bearer setup that will arrive at the offerer before the answer has arrived. However, the offerer needs to receive the answer and examine the information about the correlation mechanisms in order to successfully perform correlation of the CS call to the session. Therefore, if the offerer receives an incoming CS call, it MUST NOT accept the call before the answer has been received. If no answer is received during an implementation-specific time, the offerer MUST either modify the session according to RFC 3264 (see Section 3.1) or terminate it according to the session signaling procedures in question (for terminating a SIP session, see Section 15 of RFC 3261).

9.5.5.6.4 Modifying the Session

If, at a later time, one of the parties wishes to modify the session (e.g., by adding a new media stream or by changing properties used on an existing stream, it may do so via the mechanisms defined in "An Offer/Answer Model with the Session Description Protocol (SDP)" (RFC 3264, see Section 3.1). If there is an existing CS bearer between the endpoints and the offerer wants to reuse that, the offerer MUST set the value of the "connection" attribute to "existing." If either party removes the CS media from the session (by setting the port number to 0), it MUST terminate the CS bearer using whatever mechanism is appropriate for the technology in question.

If either party wishes to drop and reestablish an existing call, that party MUST first remove the CS media from the session by setting the port number to 0 and then use another offer/answer exchange where it MUST set the "connection" attribute to "new." If the media types are different (for example, a different codec will be used for the CS bearer), the media descriptions for terminating the existing bearer and the new bearer can be in the same offer.

If either party would like to remove existing RTP-based media from the session and replace that with a CS bearer, it would create a new offer to add the CS media as described in Section 9.5.5.6.1, replacing the RTP-based media description with the CS media description, as specified in RFC 3264 (see Section 3.1). Once the offer/answer exchange is done, but the CS bearer is not yet established, there may be a period of time when no media is available. Also, it may happen that correlating the CS call fails for reasons discussed in Section 9.5.5.3.3.

In this case, even if the offer/answer exchange was successful, endpoints are not able to receive or send media. It is up to the implementation to decide the behavior in this case; if nothing else is done, the user most likely hangs up after a while if there is no other media in the session. Note that this may also happen when switching from one RTP media to another RTP media (e.g., when firewall blocks the new media stream). If either party would like to remove existing CS media from the session and replace that with RTP-based media, it would modify the media description as per the procedures defined in RFC 3264 (see Section 3.1). The endpoint MUST then terminate the CS bearer using whatever mechanism is appropriate for the technology in question.

9.5.5.7 Formal Syntax

All ABNF syntaxes for SDP are provided in Section 2.9. However, we are repeating this here for convenience. The following is the formal ABNF (RFC 5234) syntax that supports the extensions defined in this specification. The syntax is built above the SDP (RFC 4566, see Section 2) and the tel URI (RFC 3966) grammars. Implementations that are compliant with this specification MUST be compliant with this syntax. Figure 9.13 shows the formal syntax of the extensions defined in this memo.

```
; extension to the connection field originally specified
; in RFC 4566 (see Section 2/2.9)

connection-field        =    [%x63 "=" nettype SP addrtype SP
                             connection-address CRLF]

; CRLF defined in RFC 5234
;nettype and addrtype are defined in RFC 4566 (see Section 2/2.9)

connection-address      =    / global-number-digits / "-"

; global-number-digits specified in RFC 3966
;subrules for correlation attribute

attribute               =    / cs-correlation-attr

; attribute defined in RFC 4566 (see Section 2/2.9)

cs-correlation-attr     =    "cs-correlation:" corr-mechanisms
corr-mechanisms         =    corr-mech *(SP corr-mech)
corr-mech               =    caller-id-mech / uuie-mech /
                             dtmf-mech / external-mech /
                             ext-mech
caller-id-mech          =    "callerid" [":" caller-id-value]
caller-id-value         =    "+" 1*15DIGIT

; DIGIT defined in RFC 5234

uuie-mech               =    "uuie" [":" uuie-value]
uuie-value              =    1*65(HEXDIG HEXDIG)

;This represents up to 130 HEXDIG
; (65 octets)
;HEXDIG defined in RFC 5234
;HEXDIG defined as 0-9, A-F

dtmf-mech               =    "dtmf" [":" dtmf-value]
dtmf-value              =    1*32(DIGIT / %x41-44 / %x23 / %x2A )

;0-9, A-D, '#' and '*'

external-mech           =    "external"
ext-mech                =    ext-mech-name [":" ext-mech-value]
ext-mech-name           =    token
ext-mech-value          =    token

; token is specified in RFC 4566 (see Section 2/2.9)
```

Figure 9.13 Syntax of the SDP extensions.

```
; extension to the connection field originally specified
; in RFC 4566 (see Section 2/2.9)

connection-field      =  [%x63 "=" nettype SP addrtype SP
connection-address CRLF]
; CRLF defined in RFC 5234
;nettype and addrtype are defined in RFC 4566 (see Section
;2/2.9)

connection-address    =  / global-number-digits / "-"

; global-number-digits specified in RFC 3966
;subrules for correlation attribute

attribute             =  / cs-correlation-attr

; attribute defined in RFC 4566 (see Section 2/2.9)

cs-correlation-attr   =  "cs-correlation:" corr-mechanisms
corr-mechanisms       =  corr-mech *(SP corr-mech)
corr-mech             =  caller-id-mech / uuie-mech
                         /dtmf-mech / external-mech /
ext-mech
caller-id-mech        =  "callerid" [":" caller-id-value]
caller-id-value       =  "+" 1*15DIGIT

; DIGIT defined in RFC 5234

uuie-mech             =  "uuie" [":" uuie-value]
uuie-value            =  1*65(HEXDIG HEXDIG)

;This represents up to 130 HEXDIG
; (65 octets)
;HEXDIG defined in RFC 5234
;HEXDIG defined as 0-9, A-F

dtmf-mech             =  "dtmf" [":" dtmf-value]
dtmf-value            =  1*32(DIGIT / %x41-44 / %x23 / %x2A )

;0-9, A-D, '#' and '*'

external-mech         =  "external"
ext-mech              =  ext-mech-name [":" ext-mech-value]
ext-mech-name         =  token
ext-mech-value        =  token
; token is specified in RFC 4566 (see Section 2/2.9)
```

9.5.6 Examples

In these examples, where an SDP line is too long to be displayed as a single line, a breaking character "\" indicates continuation in the following line. Note that this character is included for display purposes only. Implementations MUST write a single line without breaks.

9.5.6.1 Single PSTN Audio Stream

Figure 9.14 shows a basic example that describes a single audio media stream over a CS bearer. Endpoint A generates an SDP offer, which is shown in Figure 9.15. The offer describes a PSTN CS bearer in the "m=" and "c=" lines where it also indicates its international E.164 number format.

Additionally, Endpoint A expresses that it can initiate the CS bearer or be the recipient of it in the "a=setup" attribute line. The SDP offer also includes correlation identifiers that this endpoint will insert in the calling party number and/or UUIE of the PSTN call setup if eventually this endpoint initiates the PSTN call. Endpoint A also includes "external" as one correlation mechanism, indicating that it can use the human user to perform correlation in case other mechanisms fail.

```
v=0
o=alice 2890844526 2890842807 IN IP4 192.0.2.5
s=
t=0 0
m=audio 9 PSTN -
c=PSTN E164 +441134960123
a=setup:actpass
a=connection:new
a=cs-correlation:callerid:+441134960123 \
uuie:56A390F3D2B7310023 external
```

Endpoint B generates an SDP answer (Figure 9.16), describing a PSTN audio media on port 9 without information on the media subtype on the "m=" line. The "c=" line contains B's international

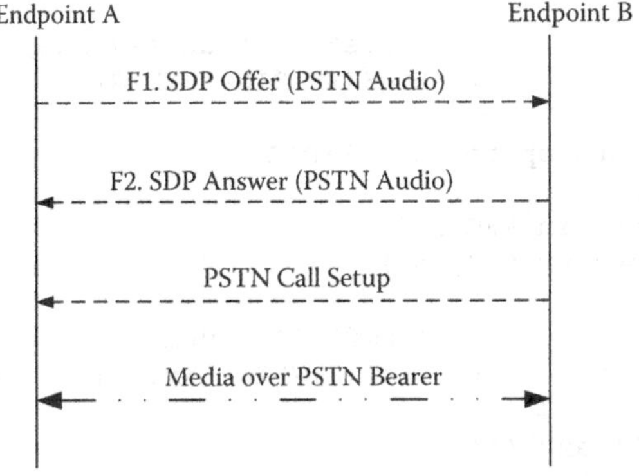

Figure 9.14 Basic flow (Copyright: IETF).

```
v=0
o=alice 2890844526 2890842807 IN IP4 192.0.2.5
s=
t=0 0
m=audio 9 PSTN -
c=PSTN E164 +441134960123
a=setup:actpass
a=connection:new
a=cs-correlation:callerid:+441134960123 \
uuie:56A390F3D2B7310023 external
```

Figure 9.15 SDP offer (1) (Copyright: IETF).

```
v=0
o=-2890973824 2890987289 IN IP4 192.0.2.7
s=
t=0 0
m=audio 9 PSTN -
c=PSTN E164 +441134960124
a=setup:active
a=connection:new
a=cs-correlation:callerid:+441134960124 \
uuie:74B9027A869D7966A2 external
```

Figure 9.16 SDP answer with CS media (Copyright: IETF).

E.164 number. In the "a=setup" line, Endpoint B indicates that it is willing to become the active endpoint when establishing the PSTN call, and it also includes the "a=cs-correlation" attribute line containing the values it is going to include in the calling party number and UUIE of the PSTN call establishment. Endpoint B is also able to perform correlation by external means, in case other correlation mechanisms fail.

```
v=0
o=- 2890973824 2890987289 IN IP4 192.0.2.7
s=
t=0 0
m=audio 9 PSTN -
c=PSTN E164 +441134960124
a=setup:active
a=connection:new
a=cs-correlation:callerid:+441134960124 \
uuie:74B9027A869D7966A2 external
```

When Endpoint A receives the answer, it discovers that B is willing to become the active endpoint when setting up the PSTN call. Endpoint A temporarily stores B's E.164 number and the UUIE value of the "cs-correlation" attribute and waits for a CS bearer establishment. Endpoint B initiates a CS bearer using whatever CS technology is available for it. The called party number is set to A's number, and the calling party number is set to B's own number. Endpoint B also sets the UUIE value to the one contained in the SDP answer.

When Endpoint A receives the CS bearer establishment, it examines the UUIE and the calling party number and, by comparing those received during the offer/answer exchange, determines that the call is related to the SDP session. It may also be that neither the UUIE nor the calling party number is received by the called party or the format of the calling party number is changed by the PSTN. Implementations may still accept such call establishment attempts as being related to the session that was established in the IP network. As it cannot be guaranteed that the values used for correlation are always passed intact through the network, they should be treated as additional hints that the CS bearer is actually related to the session.

9.5.6.2 Advanced SDP Example: CS Audio and Video Streams

Figure 9.17 shows an example of negotiating audio and video media streams over CS bearers.

```
v=0
o=alice 2890844526 2890842807 IN IP4 192.0.2.5
s=
```

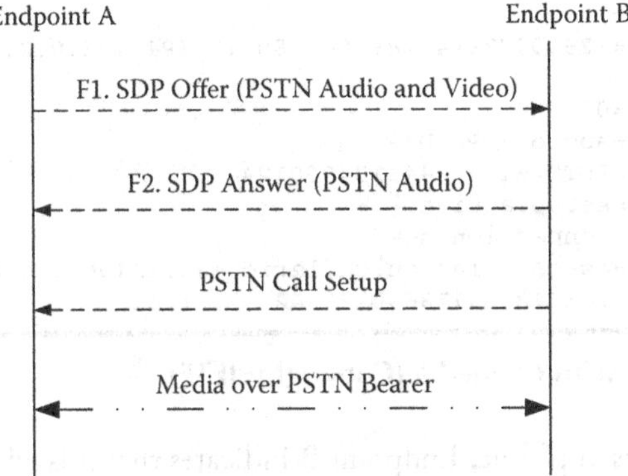

Figure 9.17 CS audio and video streams (Copyright: IETF).

```
t=0 0
a=setup:actpass
a=connection:new
c=PSTN E164 +441134960123
m=audio 9 PSTN -
a=cs-correlation:dtmf:1234536
m=video 9 PSTN 34
a=rtpmap:34 H263/90000
a=cs-correlation:callerid:+441134960123
```

Upon receiving the SDP offer described in Figure 9.18, Endpoint B rejects the video stream as the device does not currently support video, but it accepts the CS audio stream. As Endpoint A indicated that it is able to become either the active or passive party, Endpoint B gets to select which role it would like to take. Since the offer contained the international E.164 number of Endpoint A, Endpoint B decides that it becomes the active party in setting up the CS bearer. B includes a new value in the "dtmf" subfield of the "cs-correlation" attribute, which it is going to send as DTMF tones once the bearer setup is complete. The answer is described in Figure 9.19.

```
v=0
o=- 2890973824 2890987289 IN IP4 192.0.2.7
s=
t=0 0
a=setup:active
a=connection:new
c=PSTN E164 +441134960124
m=audio 9 PSTN -
a=cs-correlation:dtmf:654321
m=video 0 PSTN 34
a=cs-correlation:callerid:+441134960124
```

9.5.7 Security Considerations

The security procedures that need to be considered in dealing with the basic SDP messages described in this section (RFC 7195) are provided in Section 17.8.5.

```
v=0
o=alice 2890844526 2890842807 IN IP4 192.0.2.5
s=
t=0 0
a=setup:actpass
a=connection:new
c=PSTN E164 +441134960123
m=audio 9 PSTN -
a=cs-correlation:dtmf:1234536
m=video 9 PSTN 34
a=rtpmap:34 H263/90000
a=cs-correlation:callerid:+441134960123
```

Figure 9.18 SDP offer with CS audio and video (1) (Copyright: IETF).

```
v=0
o=-2890973824 2890987289 IN IP4 192.0.2.7
s=
t=0 0
a=setup:active
a=connection:new
c=PSTN E164 +441134960124
m=audio 9 PSTN -
a=cs-correlation:dtmf:654321
m=video 0 PSTN 34
a=cs-correlation:callerid:+441134960124
```

Figure 9.19 SDP answer with CS audio and video (2) (Copyright: IETF).

9.5.8 *IANA Considerations*

SDP parameters that are described here have been registered with IANA. We have not included the IANA registration procedures here for the sake of brevity. The procedures for registration of new SDP parameters are described in RFC 7195.

9.6 Summary

We have described syntaxes showing how SDP can support IPv4, IPv6, ATM, and PSTN network protocols. A given multimedia session may include multiple media such as audio, video, and data (application) as well as source and destination address. As a result, multiple network addresses may be needed for supporting a given session. In IPv4 and IPv6, the support for SDP syntaxes for all attributes has been straightforward. Moreover, we have discussed the DS IPv4/IPv6 networking situations where a SDP offer can carry multiple IP addresses of different address families (e.g., IPv4 and IPv6). The SDP's "altc" attribute offers a generic solution including backwards compatibility. This "altc" solution is only applicable where an end-to-end connectivity-check is not required. For example, if network address translators remain between the end-to-end session establishment path, the network connectivity-check is needed. In that situation, ICE defined in RFC 5245 needs to be used. In ATM, SDP syntaxes need to be used for controlling ATM bearer connections and any associated ATM adaptation layer. The AALs addressed can be the exhaustive list of Type 1, Type 2, and Type 5 where individual applications can use subsets of these conventions. Finally,

SDP syntaxes are extended for supporting audio or video stream established over a CS bearer. In this connection, a new network type ("PSTN") and a new protocol type ("PSTN") are defined for the "c=" and "m=" lines that describe a media stream over a CS bearer.

9.7 Problems

1. Provide examples in SDP that use IPv4 addresses for different media for a given session setup following the syntaxes provided in this section. Repeat the same example using IPv6 addresses.
2. Provide an example where SDP will carried out over a DS IPv4-IPv6 network, that is, one party is connected to IPv4 network while the other party is in the IPv6 network; identify all assumptions and end-to-end network configurations.
3. What are the fundamental differences of SDP syntaxes in supporting multimedia (audio, video) for IPv4/IPv6 and ATM network?
4. What is the purpose of SDP syntax "altc" creation? How does it solve the creation of session establishment over the DS IPv4/IPv6 network? Provide a detailed example for capability negotiations over the DS network for SIP session establishments.
5. Define the following: SBE, DBE, and core service platform. Create a generic SIP-based service architecture that contains core service platform, SBE, and DBE.
6. Specify a SIP-based service architecture along with detailed SDP calls flows that support multicast over the IP network interconnecting IPv6-only hosts to IPv4-only server.
7. Create a SIP-based IP service architecture along with SDP call flows using an IPv4-only core service platform to an IPv6-enabled UA that enables the IPv4/IPv6 interworking function in the SBE/DBE signaling and media between two SBEs and an SBE/DBE.
8. Develop a SIP-based service architecture along with SDP call flows that avoid IPv4/IPv6 interworking. Do the same with DBE bypass procedure. Do the same with direct communications between IPv6-enabled users.
9. Provide an example with call flows where SDP's "source-filter" is used for offer/answer for both unicast and multicast session over IPv4 and IPv6 network. Explain how the SDP's "source-filter" attribute prevents denial-of-service attacks in unicast and multicast sessions.
10. Does the SDP's "source-filter" attribute cause any interoperability issue with the ASM model? Explain in detail using examples.
11. Provide examples of when and how SDP syntaxes are used for ATM bearer connections associated with ATM adaptation layer of Type 1, Type 2, and Type 5.
12. How does it differ in SDP syntaxes supporting audio and video media stream over packet-switched bearers in IPv4/IPv6 and CS bearers in PSTN?
13. How do you create correlation of PSTN CS bearers with SDP for UUIE and DTMF? How do you negotiate the correlation mechanisms using SDP over the PSTN?
14. Provide detailed call flows for the third-party call control mechanism for audio and video over the CS PSTN.
15. Why is the SDP offer/answer model extended for supporting capability negotiations over the PSTN? Describe the offer/answer model for audio and video capability negotiations over the PSTN starting with initial offer, answer, processing the answer, modifying the session, and establishment of the session.

References

1. M. Handley, C. Perkins, and V. Jacobson, "SDP: Session Description Protocol", draft-ietf-mmusic-rfc4566bis-22 (Work in Progress), January 1, 2018.
2. Tsou, T., Clauberg, A., Boucadair, M., Venaas, S., and Q. Sun, "Address Acquisition For Multicast Content When Source and Receiver Support Differing IP Versions", IETF Draft, Work in Progress, January 2013.
3. C. Jacquenet, M. Boucadair, Y. Lee, J. Qin, T. Tsou, Q. Sun, "IPv4-IPv6 Multicast: Problem Statement and Use Cases", draft-ietf-mboned-v4v6-mcast-ps-04, Work-in-Progress, March 10, 2014.
4. ATMF Voice and Telephony over ATM—ATM Trunking using AAL1 for Narrowband Services, version 1.0, af-vtoa-0089.000, July 1997.
5. ITU-T I.363.2, B-ISDN ATM Adaptation Layer Specification: Type 2 AAL, September 1997.
6. H.323-2, Packet-Based Multimedia Communications Systems.
7. af-vtoa-0083.000, Voice and Telephony over ATM to the Desktop.
8. ITU Q.1901, Bearer Independent Call Control Protocol.
9. http://www.iana.org/assignments/rtp-parameters for a list of codecs with static payload types.
10. ITU-T Q.2931, B-ISDN Application Protocol for Access Signaling.
11. ITU Q.2941-2, Digital Subscriber Signaling System No. 2 (DSS 2): Generic identifier transport extensions.
12. af-vtoa-0113.000, ATM Trunking Using AAL2 for Narrowband Services.
13. ATMF Circuit Emulation Service (CES) Interoperability Specification, version 2.0, af-vtoa-0078.000, January 1997.
14. ITU-T H.222.1, Multimedia multiplex and synchronization for audiovisual communication in ATM environments.
15. ITU-T I.366.2, AAL Type 2 Reassembly Service Specific Convergence Sublayer for Trunking, February 1999.
16. ATMF UNI 4.0, Signaling Specification, af-sig-0061.000.
17. ITU Q.2957, Digital Subscriber Signaling System No. 2, User to user signaling.
18. ITU Q.2630.1, AAL Type 2 Signaling Protocol—Capability Set 1.
19. ITU I.371, Traffic Control and Congestion Control in the BISDN.
20. ATMF Traffic Management Specification, Version 4.1, af-tm- 0121.000.
21. ITU Q. 2965.1, Digital Subscriber Signaling System No.2 (DSS 2)—Support of Quality of Service classes.
22. ITU Q.2961, Digital Subscriber Signaling System No.2 (DSS 2)—Additional Traffic Parameters. Also, Amendment 2 to Q.2961.
23. ITU Q. 2965.2, Digital Subscriber Signaling System No.2 (DSS 2)—Signaling of Individual Quality of Service Parameters.
24. I.356, BISDN ATM Layer Cell Transfer Performance.
25. ITU-T I.366.1, Segmentation and Reassembly Service Specific Convergence Sublayer for AAL Type 2, June 1998.
26. Q.2110, B-ISDN ATM Adaptation Layer—Service Specific Connection Oriented Protocol (SSCOP).
27. ITU-T I.363.5, B-ISDN ATM Adaptation Layer Specification: Type 5 AAL, August 1996.
28. I.365.1, Frame Relaying Service Specific Convergence Sublayer (FR-SSCS).
29. I.365.2, B-ISDN ATM Adaptation Layer Sublayers: Service Specific Coordination Function to Provide the Connection Oriented Network Service.
30. I.365.3, B-ISDN ATM Adaptation Layer Sublayers: Service Specific Coordination Function to Provide the Connection-Oriented Transport Service.
31. I.365.4, B-ISDN ATM Adaptation Layer Sublayers: Service Specific Convergence Sublayer for HDLC Applications.
32. FRF.5, Frame Relay/ATM PVC Network Interworking Implementation Agreement.
33. FRF.8.1, Frame Relay/ATM PVC Service Interworking Implementation Agreement.
34. FRF.11, Voice over Frame Relay Implementation Agreement.

35. af-vmoa-0145.000, Voice and Multimedia over ATM, Loop Emulation Service using AAL2.
36. http://www.atmforum.com/atmforum/specs/specs.html, ATM Forum, Well-Known Addresses and Assigned Codes.
37. ITU-T I.363.1, B-ISDN ATM Adaptation Layer Specification: Type 1 AAL, August 1996.
38. ITU Q.765.5, Application Transport Mechanism—Bearer Independent Call Control.
39. ITU Q.763, SS7—ISUP Formats and Codes.
40. ITU Q.931, Digital Subscriber Signaling System No. 1: Network Layer.
41. http://www.3gpp.org/ftp/Specs for specifications related to 3GPP, including AMR codecs.
42. Cuervo, F., Greene, N., Rayhan, A., Huitema, C., Rosen, B. and J. Segers, "Megaco Protocol Version 1.0", RFC 3015, November 2000.
43. International Telecommunications Union, "Digital Subscriber Signaling System No. 1—ISDN User-Network Interface Layer 3 Specification for Basic Call Control," ITU-T Recommendation Q931, May 1998.
44. 3GPP, "Mobile Radio Interface Layer 3 Specification; Core Network Protocols; Stage 3", 3GPP TS 24.008 3.20.0, December 2005.
45. International Telecommunications Union, "Technical Features of Push-Button Telephone Sets", ITU-T Technical Recommendation Q.23, 1988.
46. International Telecommunications Union, "The International Public Telecommunication Numbering Plan", ITU-T Recommendation E.164, 2010.
47. ITU-T H.222.1, Multimedia Multiplex and Synchronization for Audiovisual Communication in ATM Environments.
48. af-vmoa-0145.000, Voice and Multimedia over ATM, Loop Emulation Service using AAL2.
49. Mills, D., "Network Time Protocol (Version 3) Specification, Implementation and Analysis", RFC 1305, March 1992.

Chapter 10

Transport Protocol Support in SDP

Abstract

The real-time audio and video applications are usually sent over Real-Time Transport Protocol (RTP) rather than User Datagram Protocol (UDP) because of suitability of connectionless transport mechanisms; these real-time traffics can tolerate some errors. The support for the RTP and UDP transport protocols are well taken care of in RFC 4566 (see Section 2), and we do not need to address these transport protocols separately here. However, texts and other data applications that are considered nonreal-time applications need connection-oriented transport protocols because they cannot tolerate errors. In this chapter, we discuss the support of connection-oriented media transport (e.g., Transmission Control Protocol (TCP)) in Session Description Protocol (SDP). RFC 4145 describes how to express media transport over TCP using the SDP. It defines the SDP "TCP" protocol identifier, the SDP "setup" attribute, which describes the connection setup procedure, and the SDP "connection" attribute, which handles connection reestablishment.

10.1 TCP-Based Media Transport in SDP

This section describes RFC 4145 that defined how to express media transport over Transmission Control Protocol (TCP) using the Session Description Protocol (SDP). It defines the SDP "TCP" protocol identifier, the SDP "setup" attribute, which describes the connection setup procedure, and the SDP "connection" attribute, which handles connection reestablishment.

10.1.1 Introduction

In this subsection, we describe TCP-Based Media Transport in the SDP specified in RFC 4115. The SDP described in RFC 2327 (that is made obsolete by RFC 4566, see Section 2) provides a general-purpose format for describing multimedia sessions in announcements or invitations. SDP uses an entirely textual data format (the US-ASCII subset of UTF-8 RFC 3629) to maximize portability among transports. SDP does not define a protocol; it defines the syntax to describe a multimedia session with sufficient information to participate in that session. Session descriptions may be sent using arbitrary existing application protocols for transport (e.g., SAP (RFC 2974), Session Initiation Protocol (SIP) (RFC 3261, also see Section 1.2), Real-Time Streaming Protocol (RTSP) (RFC 2326), email, HTTP (RFC 2616 is made obsolete by RFC 7230, RFC 7231, RFC 7232, RFC 7233, RFC 7234, and RFC 7235, etc.)). SDP in RFC 2327 (that is made obsolete by RFC 4566, see Section 2) defines two protocol identifiers: Real-Time Transport Protocol (RTP)/Audio Video Profile (AVP) and User Datagram Protocol (UDP), both of which represent unreliable, connectionless protocols. While these transports are appropriate choices for multimedia streams, there are applications for which TCP is more appropriate. This specification defines a new protocol identifier, "TCP," to describe TCP connections in SDP. TCP introduces two new factors when describing a session: how and when should endpoints perform the TCP connection setup procedure. This document defines two new attributes to describe TCP connection setups: "setup" and "connection." All SDP Augmented Backus-Naur Form (ABNF) syntaxes are shown in Section 2.9. However, we have repeated this here for convenience.

10.1.2 Terminology

See Table 2.1.

10.1.3 Protocol Identifier

The following is the ABNF for an "m=" line, as specified by RFC 2327 (that is made obsolete by RFC 4566, see Section 2).

```
media-field       =    "m=" media space port ["/" integer]
                       space proto 1*(space fmt) CRLF
```

This specification defines a new value for the proto field: "TCP." The "TCP" protocol identifier is similar to the "UDP" protocol identifier in that it only describes the transport protocol, and not the upper-layer protocol. An "m=" line that specifies "TCP" MUST further qualify the application-layer protocol using an fmt identifier. Media described using an "m=" line containing the "TCP" protocol identifier are carried using TCP (RFC 0793).

10.1.4 Setup Attribute

The "setup" attribute indicates which of the endpoints should initiate the TCP connection establishment (i.e., send the initial "TCP SYN"). The "setup" attribute is charset-independent and can be a session-level or a media-level attribute. The following is the ABNF of the "setup" attribute:

```
setup-attr  =  "a=setup:" role
role        =  "active" / "passive" / "actpass"
               / "holdconn"
```

"active": The endpoint will initiate an outgoing connection.
"passive": The endpoint will accept an incoming connection.
"actpass": The endpoint is willing to accept an incoming connection or to initiate an outgoing connection.
"holdconn": The endpoint does not want the connection to be established for the time being.

10.1.4.1 The Setup Attribute in the Offer/Answer Model

The offer/answer model, defined in RFC 3264 (see Section 3.1), provides endpoints with a means to obtain shared view of a session. Some session parameters are negotiated (e.g., codecs to use), while others are simply communicated from one endpoint to the other (e.g., IP addresses). The value of the "setup" attribute falls into the first category. That is, both endpoints negotiate its value using the offer/answer model. The negotiation of the value of the "setup" attribute takes places as follows. The offerer states which role or roles it is willing to perform; and the answerer, taking the offerer's willingness into consideration, chooses which roles both endpoints will actually perform during connection establishment. The following are the values the "setup" attribute can take in an offer/answer exchange:

Offer	Answer
Active	passive/holdconn
Passive	active/holdcom
Actpass	Active/passive/holdconn
Holdconn	Holdconn

The active endpoint SHOULD initiate a connection to the port number on the "m=" line of the other endpoint. The port number on its own "m=" line is irrelevant, and the opposite endpoint MUST NOT attempt to initiate a connection to the port number specified there. Nevertheless, since the "m=" line must contain a valid port number, the endpoint using the value "active" SHOULD specify a port number of 9 (the discard port) on its "m=" line. The endpoint MUST NOT specify a port number of 0, except to denote an "m=" line that has been or is being refused.

The passive endpoint SHOULD be ready to accept a connection on the port number specified in the "m=" line. A value of "actpass" indicates that the offerer can either initiate a connection to the port number on the "m"= line in the answer or accept a connection on the port number specified in the "m=" line in the offer. That is, the offerer has no preference as to whether it accepts or initiates the connection and, so, is letting the answerer choose. A value of "holdconn" indicates that the connection should not be established for the time being. The default value of the setup attribute in an offer/answer exchange is "active" in the offer and "passive" in the answer.

10.1.5 The Connection Attribute

The preceding description of the "setup" attribute is placed in the context of using SDP to initiate a session. Still, SDP may be exchanged between endpoints at various stages of a session to accomplish tasks such as terminating a session, redirecting media to a new endpoint, or renegotiating the media parameters for a session. After the initial session has been established, it may be ambiguous whether a subsequent SDP exchange represents a confirmation that the endpoint is to continue using the current TCP connection unchanged or is a request to make a new TCP connection. The media-level "connection" attribute, which is charset-independent, is used to disambiguate these two scenarios. The following is the ABNF of the connection attribute:

```
connection-attr   =   "a=connection:" conn-value
conn-value        =   "new" / "existing"
```

10.1.5.1 Offerer Behavior

Offerers and answerers use the "connection" attribute to decide whether a new transport connection needs to be established or, on the other hand, the existing TCP connection should still be used. When an offerer generates an "m=" line that uses TCP, it SHOULD provide a connection attribute for the "m=" line unless the application using the "m" line has other means to deal with connection reestablishment. After the initial offer/answer exchange, any of the endpoints can generate a new offer to change some characteristics of the session (e.g., the direction attribute). If such an offerer wants to continue using the previously established transport-layer connection for the "m=" line, the offerer MUST use a connection value of "existing" for the "m=" line. If, on the other hand, the offerer wants to establish a new transport-layer connection for the "m" line, it MUST use a connection value of "new."

Note that, according to the rules in this section, an offer that changes the transport address (IP address or port number) of an "m=" line will have a connection value of "new." Similarly, the "connection" attribute in an initial offer (i.e., no transport connection has been established yet) takes the value of "new." The "connection" value resulting from an offer/answer exchange is the "connection" value in the answer. If the "connection" value in the answer is "new," the endpoints SHOULD establish a new connection. If the connection value in the answer is "existing," the endpoints SHOULD continue using the existing connection. Taking into consideration the rules in Section 5.2, the following are the values that the "connection" attribute can take in an offer/answer exchange:

Offer	*Answer*
New	New
existing	existing/new

If the connection value resulting from an offer/answer exchange is "existing," the endpoints continue using the existing connection. Consequently, the port numbers, IP addresses, and "setup" attributes negotiated in the offer/answer exchange are ignored because there is no need to establish a new connection. The previous rule implies that an offerer generating an offer with a connection value of "existing" and a setup value of "passive" needs to be ready (i.e., needs to allocate resources) to receive a connection request from the answerer just in case the answerer chooses a connection value of "new" for the answer. However, if the answerer uses a connection value of "existing" in the answer, the offerer needs to deallocate the previously allocated resources that were never used because no connection request was received.

To avoid allocating resources unnecessarily, offerers using a connection value of "existing" in their offers may choose to use a setup value of "holdconn." Nevertheless, offerers using this strategy should be aware that if the answerer chooses a connection value of "new," a new offer/answer exchange (typically initiated by the previous offerer) with a setup value other than "holdconn" will be needed to establish the new connection. This may, of course, cause delays in the application using the TCP connection. The default value of the connection attribute in both offers and answers is "new."

10.1.5.2 Answerer Behavior

The connection value for an "m=" line is negotiated using the offer/ answer model. The resulting connection value after an offer/answer exchange is the connection value in the answer. If the connection value in the offer is "new," the answerer MUST also use a value of "new" in the answer. If the connection value in the offer is "existing," the answerer uses a value of "existing" in the answer if it wishes to continue using the existing connection and a value of "new" if it wants a new connection to be established. In some scenarios where third party call control (RFC 3261, see Section 1.2) is used, an endpoint may receive an initial offer with a connection value of "existing." Following the previous rules, such an answerer would use a connection value of "new" in the answer. If the connection value for an "m=" line resulting from an offer/answer exchange is "new," the endpoints SHOULD establish a new TCP connection as indicated by the "setup" attribute. If a previous TCP connection is still up, the endpoints SHOULD close it as soon as the offer/answer exchange is completed. It is up to the application to ensure proper data synchronization between the two TCP connections. If the connection value for an "m=" line resulting from an offer/answer exchange is "existing," the endpoints SHOULD continue using the existing TCP connection.

10.1.6 Connection Management

This section addresses connection establishment, connection reestablishment, and connection termination.

10.1.6.1 Connection Establishment

An endpoint that according to an offer/answer exchange is supposed to initiate a new TCP connection SHOULD initiate it as soon as it is able to, even if the endpoint does not intend to immediately begin sending media to the remote endpoint. This allows media to flow from the remote endpoint if needed. Note that some endpoints need to wait for some event to happen before being able to establish the connection. For example, a wireless terminal may need to set up a radio bearer before being able to initiate a TCP connection.

10.1.6.2 Connection Reestablishment

If an endpoint determines that the TCP for an "m=" line has been closed and should be reestablished, it SHOULD perform a new offer/answer exchange using a connection value of "new" for this "m=" line. Note that the SDP direction attribute (e.g., "a=sendonly") deals with the media sent over the TCP connection but has no impact on the TCP connection itself.

10.1.6.3 Connection Termination

Typically, endpoints do not close the TCP connection until the session has expired, has been explicitly terminated, or a new connection value has been provided for the "m=" line. Additionally, specific applications can describe further scenarios where an endpoint may close a given TCP connection (e.g., whenever a connection is in the half-closed state). As soon as an endpoint notices that it needs to terminate a TCP connection, it SHOULD do so. In any case, individual applications may provide further considerations on how to achieve a graceful connection termination. For example, a file application using TCP to receive a FIN (a flag that confirms reception) from the remote endpoint may need to finish the ongoing transmission of a file before sending its own FIN.

10.1.7 Examples

The following examples show the most common usage of the "setup" attribute combined with TCP-based media descriptions. For the purpose of brevity, the main portion of the session description is omitted in the examples, which only show "m=" lines and their attributes (including "c" lines).

10.1.7.1 Passive/Active

An offerer at 192.0.2.2 signals its availability for a T.38 fax session at port 54111:

```
m=image 54111 TCP t38
c=IN IP4 192.0.2.2
a=setup:passive
a=connection:new
```

An answerer at 192.0.2.1 receiving this offer responds with the following answer:

```
m=image 9 TCP t38
c=IN IP4 192.0.2.1
a=setup:active
a=connection:new
```

The endpoint at 192.0.2.1 then initiates the TCP connection to port 54111 at 192.0.2.2.

10.1.7.2 Actpass/Passive

In another example, an offerer at 192.0.2.2 signals its availability for a T.38 fax session at TCP port 54111. Additionally, this offerer is also willing to set up the media stream by initiating the TCP connection:

```
m=image 54111 TCP t38
c=IN IP4 192.0.2.2
a=setup:actpass
a=connection:new
```

The endpoint at 192.0.2.1 responds with the following description:

```
m=image 54321 TCP t38
c=IN IP4 192.0.2.1
a=setup:passive
a=connection:new
```

This will cause the offerer (at 192.0.2.2) to initiate a connection to port 54321 at 192.0.2.1.

10.1.7.3 Existing Connection Reuse

Subsequent to the exchange in Section 10.1.7.2, another offer/answer exchange is initiated in the opposite direction. The endpoint at 192.0.2.1 wishes to continue using the existing connection:

```
m=image 54321 TCP t38
c=IN IP4 192.0.2.1
a=setup:passive
a=connection:existing
```

The endpoint at 192.0.2.2 also wishes to use the existing connection and responds with the following description:

```
m=image 9 TCP t38
c=IN IP4 192.0.2.2
a=setup:active
a=connection:existing
```

The existing connection from 192.0.2.2 to 192.0.2.1 will be reused. Note that the endpoint at 192.0.2.2 uses "setup:active" in response to the offer of "setup:passive" and uses port 9 because it is active.

10.1.7.4 Existing Connection Refusal

Subsequent to the exchange in Section 10.1.7.3, another offer/answer exchange is initiated by the endpoint at 192.0.2.2, again wishing to reuse the existing connection:

```
m=image 54111 TCP t38
c=IN IP4 192.0.2.2
a=setup:passive
a=connection:existing
```

However, this time the answerer is unaware of the old connection and thus wishes to establish a new one. (This could be the result of a transfer via third-party call control.) It is unable to act in the "passive" mode and thus responds as "active":

```
m=image 9 TCP t38
c=IN IP4 192.0.2.3
a=setup:active
a=connection:new
```

The endpoint at 192.0.2.3 then initiates the TCP connection to port 54111 at 192.0.2.2, and the endpoint at 192.0.2.2 closes the old connection. Note that the endpoint at 192.0.2.2, while using a connection value of "existing," has used a setup value of "passive." Had it not done this and instead used a setup value of "holdconn" (probably to avoid allocating resources as described in Section 10.1.5.1), a new offer/answer exchange would have been needed in order to establish the new connection.

10.1.8 Other Connection-Oriented Transport Protocols

This specification specifies how to describe TCP-based media streams using SDP. Still, some of the attributes defined here could possibly be used to describe media streams based on other connection-oriented transport protocols as well. This section provides advice to authors of specifications of SDP extensions that deal with connection-oriented transport protocols other than TCP. It is recommended that documents defining new SDP protocol identifiers that involve extra protocol layers between TCP and the media itself, for example, Transport Layer Security (TLS) (RFC 2246 that is obsoleted by RFC 4346) over TCP start with the string "TCP/" (e.g., "TCP/TLS"). The "setup" and the "connection" attributes are specified in Sections 10.1.4 and 10.1.5 respectively. While both attributes are applicable to "m=" lines that use the "TCP" protocol identifier, they are general enough to be reused in "m=" lines with other connection-oriented transport protocols. Therefore, it is recommended that the "setup" and "connection" attributes be reused, as long as possible, for new proto values associated with connection-oriented transport protocols.

Section 10.1.6 deals with TCP connection management. It should be noted that while in TCP both endpoints need to close a connection, other connection-oriented transport protocols may not have the concept of half-closed connections. In such a case, a connection would be terminated as soon as one of the endpoints closed it, making it unnecessary for the other endpoint to perform any further action to terminate the connection. So, specifications dealing with such transport protocols may need to specify slightly different procedures regarding connection termination.

10.1.9 Security Considerations

The security procedures that need to be considered in dealing with the basic SDP messages described in this section (RFC 4145) are provided in Section 17.9.1.

10.1.10 IANA Considerations

SDP parameters that are described here have been registered with Internet Assigned Numbers Authority (IANA). We have not included the IANA registration procedures here for the sake of brevity. The procedures for registration of new SDP parameters are described in RFC 4145.

10.2 Summary

We have described the support of transport protocols in SDP and specifically for connection-oriented protocols (e.g., TCP and TCP/TLS). The support of RTP and UDP transport protocols by SDP are provided in RFC 2327 (that is made obsolete by RFC 4566, see Section 2). We have described setup and connection attribute and connection management in SDP for support of the connection-oriented transport such as TCP. In addition, examples for passive/active, actpass/passive, existing connection reuse, and existing connection refusal are provided. Some guidance for other kinds of connection-oriented transport are provided in light of TCP-based media transport specifically described in this section.

10.3 Problems

1. How does the connection-oriented TCP-based media transport differ from that of the connectionless UDP-based media transport?
2. Describe in detail call flows for TCP-based media transport with the SDP offer/answer model that takes care of passive/active, actpass/passive, existing connection reuse, and existing connection refusal.
3. Describe in detail the behavior of the offerer and answerer of Problem 1 along with the connection management (establishment, reestablishment, and termination).
4. Describe the call flows and offer/answer like those of Problems 2 and 3 with TLS-over-TCP. Show call flows in detail.
5. What are the specific recommendations made for supporting other connection-oriented media transport in SDP in the light of TCP? Describe in detail.

Chapter 11

RTP Media Loopback Support in SDP

Abstract

This chapter describes the Real-Time Transport Protocol media loopback-support in Session Description Protocol (SDP) specified in RFC (Request for Comments) 6849. The wide deployment of Voice-over IP (VoIP), real-time text, and Video-over IP services has introduced new challenges in managing and maintaining real-time voice/text/video quality, reliability, and overall performance. In particular, media delivery is an area that needs attention. One method of meeting these challenges is monitoring the media delivery performance by looping media back to the transmitter. This is typically referred to as "active monitoring" of services. Media loopback is especially popular in ensuring the quality of transport to the edge of a given VoIP, real-time text, or Video-over IP service. Today, in networks that deliver real-time media, short of running "ping" and "traceroute" to the edge, administrators are left without the necessary tools to actively monitor, manage, and diagnose quality issues with their service. The extension defined herein adds new SDP media types and attributes that enable establishment of media sessions where the media is looped back to the transmitter. Such media sessions will serve as monitoring and troubleshooting tools by providing the means for measurement of more advanced VoIP, real-time text, and Video-over IP performance metrics.

11.1 Support for Media Loopback in SDP and RTP

This section describes the support of media loopback in both Session Description Protocol (SDP) and Real-Time Transport Protocol (RTP) specified in RFC (Request for Comments) 6849. The extension defined herein adds new SDP media types and attributes that enable the establishment of media sessions where the media is looped back to the transmitter. Thereby, this capability facilitates monitoring and troubleshooting of media sessions.

11.1.1 Introduction

The overall quality, reliability, and performance of Voice-over IP (VoIP), real-time text, and Video-over IP services rely on the performance and quality of the media path using RTP. In order to assure the quality of the delivered media, there is a need to monitor the performance of the media transport. One method of monitoring and managing the overall quality of real-time VoIP, real-time text, and Video-over IP services is through monitoring the quality of the media in an active session. This type of "active monitoring" of services is a method of proactively managing the performance and quality of VoIP-based services. The goal of active monitoring is to measure the media quality of a VoIP, real-time text, or Video-over IP session. A way to achieve this goal is to request an endpoint to loop media back to the other endpoint and to provide media statistics, for example, Real-Time Transport Control Protocol (RTCP) (RFC 3550) and RTCP Extended Reports (RTCP-XR) (RFC 3611) information. Another method involves deployment of special endpoints that always loop incoming media back for all sessions.

Although the latter method has been used and is functional, it does not scale to support large networks and introduces new network management challenges. Further, it does not offer the granularity of testing a specific endpoint that may be exhibiting problems. The extension defined in this document introduces new SDP media types and attributes that enable the establishment of media sessions where the media is looped back to the transmitter. The SDP offer/answer model (RFC 3264, see Section 3.1) is used to establish a loopback connection. Furthermore, this extension provides guidelines on handling RTP (RFC 3550), as well as usage of RTCP (RFC 3550) and RTCP-XR (RFC 3611) for reporting media-related measurements.

11.1.1.1 Use Cases Supported

As a matter of terminology in this specification, packets flow from one peer acting as a "loopback-source," to the other peer acting as a "loopback-mirror," which in turn returns packets to the loopback-source. In advance of the session, the peers negotiate to determine which acts in which role, using the SDP offer/answer exchange. The negotiation also includes details such as the type of loopback to be used. This specification supports three use cases: "encapsulated packet loopback," "direct loopback," and "media loopback." These are distinguished by the treatment of incoming RTP packets at the loopback-mirror.

11.1.1.1.1 Encapsulated Packet Loopback

In the encapsulated packet loopback case, the entire incoming RTP packet is encapsulated as payload within an outer RTP packet that is specific to this use case and specified in Section 11.1.7.1. The encapsulated packet is returned to the loopback-source. The loopback-source can generate statistics for one-way path performance up to the RTP level for each direction of travel by examining sequence numbers and timestamps in the encapsulating outer RTP header and the encapsulated RTP packet payload. The loopback-source can also play back the returned media content for evaluation. Because the encapsulating RTP packet header extends the packet size, it could encounter difficulties in an environment where the original RTP packet size is close to the path Maximum Transmission Unit (MTU) size. The encapsulating payload format therefore offers the possibility of RTP-level fragmentation of the returned packets. The use of this facility could affect statistics derived for the return path. In addition, the increased bit rate required in the return direction may affect these statistics more directly in a restricted bandwidth situation.

11.1.1.1.2 Direct Loopback

In the direct loopback-case, the loopback-mirror copies the payload of the incoming RTP packet into a new RTP packet, using a payload format specific to this use case and specified in Section 11.1.7.2. The loopback-mirror returns the new packet to the packet source. There is no provision in this case for RTP-level fragmentation. This use case has the advantage of keeping the packet size the same in both directions. The packet source can compute only two-way path statistics from the direct loopback-packet header but can play back the returned media content. It has been suggested that the loopback-source, knowing that the incoming packet will never be passed to a decoder, can store a timestamp and sequence number inside the payload of the packet it sends to the mirror then extract that information from the returned direct loopback-packet and compute one-way path statistics as in the previous case. Obviously, playout of returned content is no longer possible if this is done.

11.1.1.1.3 Media Loopback

In the media loopback-case, the loopback-mirror submits the incoming packet to a decoder appropriate to the incoming payload type (PT). The packet is taken as close as possible to the analog level then re-encoded according to an outgoing format determined by SDP negotiation. The re-encoded content is returned to the loopback-source as an RTP packet with PT corresponding to the re-encoding format. This usage allows troubleshooting at the codec level. The capability for path statistics is limited to what is available from RTCP reports.

11.1.2 Terminology

The SDP is as defined in RFC 4566 (see Section 2). This specification assumes that the SDP offer/answer model is followed, per RFC 3264 (see Section 3.1), but does not assume any specific signaling protocol for carrying the SDP. The following terms are borrowed from RFC 3264 (see Section 3.1) definitions: offer, offerer, answer, answerer, and agent.

11.1.3 Overview of Operation

This specification defines two loopback "types," two "roles," and two encoding formats for loopback. For any given SDP offerer or answerer pair, one side is the source of RTP packets, while the other is the mirror looping packets/media back. Those define the two loopback roles. As the mirror, two "types" of loopback can be performed: packet level or media level. When media level is used, there is no further choice of encoding format; there is only one format: whatever is indicated for normal media, since the "looping" is performed at the codec level. When packet-level looping is performed, however, the mirror can send back RTP in an encapsulated format or in direct loopback format. The rest of this specification describes these loopback types, roles, and encoding formats and the SDP offer/answer rules for indicating them.

11.1.3.1 SDP Offerer Behavior

An SDP offerer compliant to this specification and attempting to establish a media session with media loopback will include "loopback" media attributes for each individual media description in the offer message it wishes to have looped back. Note that the offerer may choose to request loopback for some media descriptions/streams but not for others. For example, it might wish to request loopback for a video stream but not audio, or vice versa. The offerer will look for the "loopback" media attributes in the media description(s) of the response from the SDP answer for confirmation that the request is accepted.

11.1.3.2 SDP Answerer Behavior

In order to accept a loopback offer (that is, an offer containing "loopback" in the media description), an SDP answerer includes the "loopback" media attribute in each media description for which it desires loopback. An answerer can reject an offered stream (either with loopback-source or loopback-mirror) if the loopback-type is not specified, the specified loopback-type is not supported, or the endpoint cannot honor the offer for any other reason. The loopback request is rejected by setting the stream's media port number to zero in the answer as defined in RFC 3264 (see Section 3.1) or by rejecting the entire offer (i.e., by rejecting the session request entirely). Note that an answerer that is not compliant with this specification and that receives an offer with the "loopback" media attributes would ignore the attributes and treat the incoming offer as a normal request. If the offerer does not wish to establish a "normal" RTP session, it needs to terminate the session upon receiving such an answer.

11.1.4 New SDP Attributes

Three new SDP media-level attributes are defined: One indicates the type of loopback, and the other two define the role of the agent. All SDP syntaxes are provided in Section 2.9. However, we have repeated this here for convenience.

11.1.4.1 Loopback-Type Attribute

This specification defines a new "loopback" attribute, which indicates that the agent wishes to perform loopback and the type of loopback that the agent is able to do. The loopback-type is a value media attribute (RFC 4566, see Section 2) with the following syntax:

```
a=loopback:<loopback-type>
```

Following is the Augmented Backus-Naur Form (BNF) (RFC 5234) for loopback-type:

```
attribute            =    / loopback-attr
                          ; attribute defined in RFC 4566
                          (see Section 2/2.9)
loopback-attr        =    "loopback:" SP
loopback-type
loopback-type        =    loopback-choice [1*SP
loopback-choice]
loopback-choice      =    loopback-type-pkt /
loopback-type-media
loopback-type-pkt    =    "rtp-pkt-loopback"
loopback-type-media  =    "rtp-media-loopback"
```

The loopback-type is used to indicate the type of loopback. The loopback-type values are rtp-pkt-loopback and rtp-media-loopback.

- **rtp-pkt-loopback**: In this mode, the RTP packets are looped back to the sender at a point before the encoder/decoder function in the receive direction to a point after the encoder/decoder function in the send direction. This effectively re-encapsulates the RTP payload with the RTP/UDP (User Datagram Protocol)/IP headers appropriate for sending it in the reverse direction. Any type of encoding-related functions, such as packet loss concealment, MUST NOT be part of this type of loopback path. In this mode, the RTP packets are looped back with a new PT and format. Section 11.1.7 describes the payload formats that are to be used for this type of loopback. This type of loopback applies to the encapsulated and direct loopback use cases described in Section 11.1.1.1.

- **rtp-media-loopback:** This loopback is activated as close as possible to the analog interface and after the decoder so that the RTP packets are subsequently re-encoded prior to transmission back to the sender. This type of loopback applies to the media loopback use case described in Section 11.1.1.1.3.

11.1.4.2 Loopback-Role Attributes: Loopback-Source and Loopback-Mirror

The loopback role defines two property media attributes (RFC 4566 – see Section 2) that are used to indicate the role of the agent generating the SDP offer or answer. The syntax of the two loopback-role media attributes is as follows:

```
a=loopback-source
```

and

```
a=loopback-mirror
```

Following is the Augmented BNF (RFC 5234) for loopback-source and loopback-mirror:

```
attribute          =  / loopback-source / loopback-mirror
                      ;attribute defined in RFC 4566
                      (see Section 2/2.9)
loopback-source    =  "loopback-source"
loopback-mirror    =  "loopback-mirror"
```

- **loopback-source:** This attribute specifies that the entity that generated the SDP is the media source and expects the receiver of the SDP message to act as a loopback-mirror.
- **loopback-mirror:** This attribute specifies that the entity that generated the SDP will mirror (echo) all received media back to the sender of the RTP stream. No media is generated locally by the looping-back entity for transmission in the mirrored stream.

The "m=" line in the SDP includes all the PTs that will be used during the loopback session. The complete payload space for the session is specified in the "m=" line, and the rtpmap attribute is used to map from the PT number to an encoding name denoting the payload format to be used.

11.1.5 Rules for Generating the SDP Offer/Answer

11.1.5.1 Generating the SDP Offer for Loopback Session

If an offerer wishes to make a loopback request, it includes both the loopback-type and loopback-role attributes in a valid SDP offer:

Example:

```
m=audio 41352 RTP/AVP 0 8 100
a=loopback:rtp-media-loopback
```

```
a=loopback-source
a=rtpmap:0 pcmu/8000
a=rtpmap:8 pcma/8000
a=rtpmap:100 G7221/16000/1
```

Since media loopback requires bidirectional RTP, its normal direction mode is "sendrecv"; the "sendrecv" direction attribute MAY be encoded in SDP or not, as per Section 3.1.5.1 of this book (that is, Section 5.1 of RFC 3264), since it is implied by default. If either the loopback-source or mirror wishes to disable loopback use during a session, the direction mode attribute "inactive" MUST be used as per RFC 3264 (see Section 3.1). The direction mode attributes "recvonly" and "sendonly" are incompatible with the loopback mechanism and MUST NOT be indicated when generating an SDP offer or answer. When receiving an SDP offer or answer, if "recvonly" or "sendonly" is indicated for loopback, the SDP-receiving agent SHOULD treat it as a protocol failure of the loopback negotiation and terminate the session through its normal means (e.g., by sending a Session Initiation Protocol (SIP) BYE if SIP is used) or reject the offending media stream.

The offerer may offer more than one loopback-type in the SDP offer. The port number and the address in the offer (m/c = lines) indicate where the offerer would like to receive the media stream(s). The PT numbers indicate the value of the payload the offerer expects to receive. However, the answer might indicate a subset of PT numbers from those given in the offer. In that case, the offerer MUST only send the PTs received in the answer, per normal SDP offer/answer rules.

If the offer indicates rtp-pkt-loopback attribute support, the offer MUST also contain either an encapsulated or direct loopback encoding format encoding name, or both, as defined in Sections 11.1.7.1 and 11.1.2 of this book. If the offer only indicates rtp-media-loopback support, then neither encapsulated nor direct loopback encoding format applies and MUST NOT be in the offer. If loopback-type is rtp-pkt-loopback, the loopback mirror MUST send, and the loopback source MUST receive, the looped-back packets encoded in one of the two payload formats (encapsulated RTP or direct loopback) as defined in Section 11.1.7.

Example:

```
m=audio 41352 RTP/AVP 0 8 112
a=loopback:rtp-pkt-loopback
a=loopback-source
a=rtpmap:112 encaprtp/8000
```

Example:

```
m=audio 41352 RTP/AVP 0 8 112
a=loopback:rtp-pkt-loopback
a=loopback-source
a=rtpmap:112 rtploopback/8000
```

11.1.5.2 Generating the SDP Answer for Loopback Session

As with the offer, an SDP answer for loopback follows SDP offer/answer rules for the direction attribute, but directions of "sendonly" or "recvonly" do not apply for loopback operation. The port number and the address in the answer ("m="/"c=" lines) indicate where the answerer would like to receive the media stream. The PT numbers indicate the value of the PTs the answerer expects to send and receive. An answerer includes both the loopback-role and loopback-type attributes in the answer to indicate that it will accept the loopback request. When a stream is offered with the loopback-source attribute, the corresponding stream in the response will be loopback-mirror and vice versa, provided the answerer is capable of supporting the requested loopback type. For example, if the offer contains the loopback-source attribute:

```
m=audio 41352 RTP/AVP 0 8
a=loopback:rtp-media-loopback
a=loopback-source
```

The answer that is capable of supporting the offer must contain the loopback-mirror attribute:

```
m=audio 12345 RTP/AVP 0 8
a=loopback:rtp-media-loopback
a=loopback-mirror
```

If a stream is offered with multiple loopback-type attributes, the answer MUST include only one of the loopback types that are accepted by the answerer. The answerer SHOULD give preference to the first loopback-type in the SDP offer. For example, if the offer contains:

```
m=audio 41352 RTP/AVP 0 8 112
a=loopback:rtp-media-loopback rtp-pkt-loopback
a=loopback-source
a=rtpmap:112 encaprtp/8000
```

The answer that is capable of supporting the offer and chooses to loopback the media using the rtp-media-loopback type must contain:

```
m=audio 12345 RTP/AVP 0 8
a=loopback:rtp-media-loopback
a=loopback-mirror
```

As specified in Section 11.1.7, if the loopback-type is rtp-pkt-loopback, the encapsulated RTP payload format or direct loopback RTP payload format MUST be used for looped-back packets. For example, if the offer contains:

```
m=audio 41352 RTP/AVP 0 8 112 113
a=loopback:rtp-pkt-loopback
a=loopback-source
a=rtpmap:112 encaprtp/8000
a=rtpmap:113 rtploopback/8000
```

The answer that is capable of supporting the offer must contain one of the following:

```
m=audio 12345 RTP/AVP 0 8 112
a=loopback:rtp-pkt-loopback
a=loopback-mirror
a=rtpmap:112 encaprtp/8000
m=audio 12345 RTP/AVP 0 8 113
a=loopback:rtp-pkt-loopback
a=loopback-mirror
a=rtpmap:113 rtploopback/8000
```

The previous examples used the "encaprtp" and "rtploopback" encoding names, which will be defined in Sections 11.1.7.1.3 and 11.1.7.2.3.

11.1.5.3 Offerer Processing of the SDP Answer

If the received SDP answer does not contain an a=loopback-mirror or a=loopback-source attribute, it is assumed that the loopback extensions are not supported by the remote agent. This is not a protocol failure and instead merely completes the SDP offer/answer exchange with whatever normal rules apply; the offerer MAY decide to end the established RTP session (if any) through normal means of the upper-layer signaling protocol (e.g., by sending a SIP BYE).

11.1.5.4 Modifying the Session

At any point during the loopback session, either participant MAY issue a new offer to modify the characteristics of the previous session, as defined in Section 3.1.8 of this book (that is, Section 8 of RFC 3264). This also includes transitioning from a normal media processing mode to loopback mode, and vice versa.

11.1.5.5 Establishing Sessions between Entities behind Network Address Translators

Interactive Connectivity Establishment (ICE) (RFC 5245, see Section 3.2), traversal using relays around NAT (TURN) (RFC 5766), and simple traversal of UDP around NAT (STUN) (RFC 5389) provide a general solution to establishing media sessions between entities that are behind Network Address Translators (NATs). Loopback sessions that involve one or more endpoints behind NATs can also use these general solutions wherever possible.

If ICE is not supported, then in the case of loopback, the mirroring entity will not send RTP packets and therefore will not automatically create the NAT pinhole in the way that other SIP sessions do. Therefore, if the mirroring entity is behind a NAT, it MUST send some packets to the identified address/port(s) of the peer, in order to open the NAT pinhole. Using ICE, this would be accomplished with the STUN connectivity check process or through a TURN server connection. If ICE is not supported, either (RFC 6263) or Section 3.2.10 of this book (that is, Section 10 of RFC 5245) can be followed to open the pinhole and keep the NAT binding alive/refreshed.

Note that for any form of NAT traversal to function, symmetric RTP/RTCP (RFC 4961) MUST be used, unless the mirror can control the NAT(s) in its path to create explicit pinholes. In other words, both agents MUST send packets from the source address and port they receive packets on, unless some mechanism is used to avoid that need (e.g., by using the Port Control Protocol).

11.1.6 RTP Requirements

A loopback source MUST NOT send multiple source streams on the same 5-tuple, since there is no means for the mirror to indicate which is which in its mirrored RTP packets. A loopback-mirror that is compliant to this specification and accepts media with the loopback type rtp-pkt-loopback loops back the incoming RTP packets using either the encapsulated RTP payload format or the direct loopback RTP payload format as defined in Section 11.1.7. A device that is compliant to this specification and performing the mirroring using the loopback type rtp-media-loopback MUST transmit all received media back to the sender, unless congestion feedback or other lower-layer constraints prevent it from doing so. The incoming media is treated as if it were to be played; for example, the media stream may receive treatment from Packet Loss Concealment algorithms. The mirroring entity regenerates the entire RTP header fields as it would when transmitting media. The mirroring entity MAY choose to encode the loopback media according to any of the media descriptions supported by the offering entity. Furthermore, in cases where the same media type is looped back, the mirroring entity can choose to preserve the number of frames/packets and the bit rate of the encoded media according to the received media.

11.1.7 Payload Formats for Packet Loopback

The payload formats described in this section MUST be used by a loopback-mirror when "rtp-pkt-loopback" is the specified loopback type. Two different formats are specified here: an encapsulated RTP payload format and a direct loopback RTP payload format. The encapsulated RTP payload format should be used when the incoming RTP header information needs to be preserved during the loopback operation. This is useful in cases where the loopback source needs to measure performance metrics in both directions. However, this comes at the expense of increased packet size as described in Section 11.1.7.1. The direct loopback RTP payload format should be used when bandwidth requirements prevent the use of the encapsulated RTP payload format.

11.1.7.1 Encapsulated Payload Format

A received RTP packet is encapsulated in the payload section of the RTP packet generated by a loopback-mirror. Each received packet is encapsulated in a separate encapsulating RTP packet; the encapsulated packet would be fragmented only if required (for example, due to MTU limitations).

11.1.7.1.1 Usage of RTP Header Fields

- Payload Type (PT): The assignment of an RTP PT for this packet format is outside the scope of this document; it is either specified by the RTP profile under which this payload format is used or more likely signaled dynamically out-of-band (e.g., using SDP; Section 11.1.7.1.3 defines the name binding).
- Marker (M) bit: If the received RTP packet is looped back in multiple encapsulating RTP packets, the M bit is set to 1 in every fragment except the last packet; otherwise, it is set to 0.
- Extension (X) bit: This bit is defined by the RTP profile used.
- Sequence Number: The RTP sequence number SHOULD be generated by the loopback-mirror in the usual manner with a constant random offset as described in RFC 3550.
- Timestamp: The RTP timestamp denotes the sampling instant for when the loopback-mirror is transmitting this packet to the loopback-source. The RTP timestamp MUST use the same clock rate as that of the encapsulated packet. The initial value of the timestamp SHOULD be random for security reasons (see Section 5.1 of RFC 3550).
- Synchronization source (SSRC): This field is set as described in RFC 3550.
- The CSRC count (CC) and contributing source (CSRC) fields are used as described in RFC 3550.

11.1.7.1.2 RTP Payload Structure

The outer RTP header of the encapsulating packet is followed by the payload header defined in this section, after any header extension(s). If the received RTP packet has to be looped back in multiple encapsulating packets due to fragmentation, the encapsulating RTP header in each packet is followed by the payload header defined in this section. The header is devised so that the loopback-source can decode looped-back packets in the presence of moderate packet loss (RFC 3550). The RTP payload of the encapsulating RTP packet starts with the payload header defined in this section (Figure 11.1).

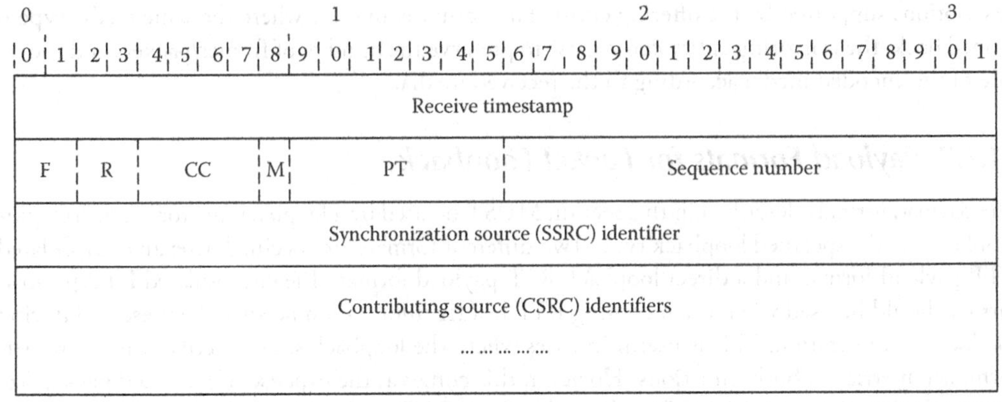

Figure 11.1 Encapsulating RTP packet payload header.

The 12 octets after the receive timestamp are identical to the encapsulated RTP header of the received packet except for the first 2 bits of the first octet. In effect, the received RTP packet is encapsulated by creating a new outer RTP header followed by four new bytes of a receive timestamp, followed by the original received RTP header and payload, except that the first two bits of the received RTP header are overwritten as defined here.

Receive timestamp: 32 bits. The receive timestamp denotes the sampling instant for when the last octet of the received media packet that is being encapsulated by the loopback-mirror is received from the loopback-source. The same clock rate MUST be used by the loopback-source. The initial value of the timestamp SHOULD be random for security reasons (see Section 5.1 of RFC 3550).

- Fragmentation (F): 2 bits

Possible values are First Fragment (00), Last Fragment (01), No Fragmentation (10), or Intermediate Fragment (11). This field identifies how much of the received packet is encapsulated in this packet by the loopback-mirror. If the received packet is not fragmented, this field is set to 10; otherwise, the packet that contains the first fragments sets this field to 00. The packet that contains the last fragment sets this field to 01, and all other packets set this field to 11.

11.1.7.1.3 Usage of SDP

The PT number for the encapsulated stream can be negotiated using SDP. There is no static PT assignment for the encapsulating stream, so dynamic PT numbers MUST be used. The binding to the name is indicated by an rtpmap attribute. The name used in this binding is "encaprtp." The following is an example SDP fragment for encapsulated RTP.

```
m=audio 41352 RTP/AVP 112
a=rtpmap:112 encaprtp/8000
```

11.1.7.2 Direct Loopback RTP Payload Format

The direct loopback RTP payload format can be used in scenarios where the 16-byte overhead of the encapsulated payload format is of concern or simply due to local policy. When using this payload format, the receiver loops back each received RTP packet payload (not header) in a separate RTP packet. Because a direct loopback format does not retain the original RTP headers, there will be no indication of the original payload-type sent to the mirror, in looped-back packets. Therefore, the loopback source SHOULD only send one PT per loopback RTP session if direct mode is used.

11.1.7.2.1 Usage of RTP Header Fields

- Payload Type (PT): The assignment of an RTP PT for the encapsulating packet format is outside the scope of this specification; it is either specified by the RTP profile under which this payload format is used or more likely signaled dynamically out-of-band (e.g., using SDP; Section 11.1.7.2.3 defines the name binding).
- Marker (M) bit: This bit is set to the value in the received packet.
- Extension (X) bit: This bit is defined by the RTP profile used.
- Sequence Number: The RTP sequence number SHOULD be generated by the loopback-mirror in the usual manner with a constant random offset, as per RFC 3550.
- Timestamp: The RTP timestamp denotes the sampling instant for when the loopback-mirror is transmitting this packet to the loopback-source. The same clock rate MUST be used as that of the received RTP packet. The initial value of the timestamp SHOULD be random for security reasons (see Section 5.1 of RFC 3550).
- SSRC: This field is set as described in RFC 3550.
- The CC and CSRC fields are used as described in RFC 3550.

11.1.7.2.2 RTP Payload Structure

This payload format does not define any payload-specific headers. The loopback-mirror simply copies the RTP payload data from the payload portion of the RTP packet received from the loopback-source.

11.1.7.2.3 Usage of SDP

The PT number for the payload loopback stream can be negotiated using a mechanism like SDP. There is no static PT assignment for the stream, so dynamic PT numbers MUST be used. The binding to the name is indicated by an rtpmap attribute. The name used in this binding is "rtploopback." The following is an example SDP fragment for the direct loopback RTP format.

```
m=audio 41352 RTP/AVP 112
a=rtpmap:112 rtploopback/8000
```

11.1.8 SRTP Behavior

Secure RTP (SRTP) (RFC 3711) MAY be used for loopback sessions. SRTP operates at a lower logical layer than RTP, and thus if both sides negotiate to use SRTP, each side uses its own key and performs encryption/decryption, authentication, etc. Therefore, the loopback function on the mirror occurs after the SRTP packet has been decrypted and authenticated, as a normal clear-text RTP packet without a Master Key Identifier or authentication tag; once the clear-text RTP packet or payload is mirrored—either at the media-layer, direct packet-layer, or encapsulated packet-layer—it is encrypted by the mirror using its own key. In order to provide the same level of protection to both forward and reverse media flows (media to and from the mirror), if SRTP is used it MUST be used in both directions with the same properties.

11.1.9 RTCP Requirements

The use of the loopback attribute is intended for the monitoring of media quality of the session. Consequently, the media performance information should be exchanged between the offering and the answering entities. An offering or answering agent that is compliant with this specification SHOULD support RTCP per RFC 3550 and RTCP-XR per RFC 3611. Furthermore, if the offerer or answerer chooses to support RTCP-XR, it SHOULD support the RTCP-XR Loss Run Length Encoding (RLE) Report Block, Duplicate RLE Report Block, Statistics Summary Report Block, and VoIP Metrics Report Block per Sections 4.1, 4.2, 4.6, and 4.7 of RFC 3611. The offerer and the answerer MAY support other RTCP-XR reporting blocks as defined by RFC 3611.

11.1.10 Congestion Control

All participants in a media-level loopback session SHOULD implement congestion control mechanisms as defined by the RTP profile under which the loopback mechanism is implemented. For audio/video profiles, implementations SHOULD conform to the mechanism defined in Section 2 of RFC 3551. For packet-level loopback types, the loopback-source SHOULD implement congestion control. The mirror will simply reflect back the RTP packets it receives (either in encapsulated or direct modes); therefore, the source needs to control the congestion of both forward and reverse paths by reducing its sending rate to the mirror. This keeps the loopback mirror implementation simpler and provides more flexibility for the source performing a loopback test.

11.1.11 Examples

This section provides examples for media descriptions using SDP for different scenarios. The examples are given for SIP-based transactions; for convenience, they are abbreviated and do not show the complete signaling.

11.1.11.1 Offer for Specific Media Loopback Type

An agent sends an SDP offer that looks like

```
v=0
o=alice 2890844526 2890842807 IN IP4 host.atlanta.example.com
s=-
c=IN IP4 host.atlanta.example.com
t=0 0
m=audio 49170 RTP/AVP 0
a=loopback:rtp-media-loopback
a=loopback-source
a=rtpmap:0 pcmu/8000
```

The agent is offering to source the media and expects the answering agent to mirror the RTP stream per the loopback type rtp-media-loopback. An answering agent sends an SDP answer that looks like

```
v=0
o=bob 1234567890 1122334455 IN IP4 host.biloxi.example.com
s=-
c=IN IP4 host.biloxi.example.com
t=0 0
m=audio 49270 RTP/AVP 0
a=loopback:rtp-media-loopback
a=loopback-mirror
a=rtpmap:0 pcmu/8000
```

The answerer agrees to mirror the media from the offerer at the media level.

11.1.11.2 Offer for Choice of Media Loopback Type

An agent sends an SDP offer that looks like

```
v=0
o=alice 2890844526 2890842807 IN IP4 host.atlanta.example.com
s=-
c=IN IP4 host.atlanta.example.com
t=0 0
m=audio 49170 RTP/AVP 0 112 113
a=loopback:rtp-media-loopback rtp-pkt-loopback
a=loopback-source
a=rtpmap:0 pcmu/8000
a=rtpmap:112 encaprtp/8000
a=rtpmap:113 rtploopback/8000
```

The offerer is offering to source the media and expects the answerer to mirror the RTP stream at either the media or RTP level. An answering agent sends an SDP answer that looks like

```
v=0
o=bob 1234567890 1122334455 IN IP4 host.biloxi.example.com
s=-
c=IN IP4 host.biloxi.example.com
t=0 0
m=audio 49270 RTP/AVP 0 112
a=loopback:rtp-pkt-loopback
a=loopback-mirror
a=rtpmap:0 pcmu/8000
a=rtpmap:112 encaprtp/8000
```

The answerer agrees to mirror the media from the offerer at the packet level using the encapsulated RTP payload format.

11.1.11.3 Answerer Rejecting Loopback Media

An agent sends an SDP offer that looks like

```
v=0
o=alice 2890844526 2890842807 IN IP4 host.atlanta.example.com
s=-
```

```
c=IN IP4 host.atlanta.example.com
t=0 0
m=audio 49170 RTP/AVP 0
a=loopback:rtp-media-loopback
a=loopback-source
a=rtpmap:0 pcmu/8000
```

The offerer is offering to source the media and expects the answerer to mirror the RTP stream at the media level. An answering agent sends an SDP answer that looks like

```
v=0
o=bob 1234567890 1122334455 IN IP4 host.biloxi.example.com
s=-
c=IN IP4 host.biloxi.example.com
t=0 0
m=audio 0 RTP/AVP 0
a=rtpmap:0 pcmu/8000
```

Note in this case that the answerer did not indicate loopback support, although it could have and still used a port number of 0 to indicate that it does not wish to accept that media session. Alternatively, the answering agent could have simply rejected the entire SDP offer through some higher-layer signaling protocol means (e.g., by rejecting the SIP INVITE request if the SDP offer was in the INVITE).

11.1.12 Security Considerations

The security procedures that need to be considered in dealing with the basic SDP messages described in this section (RFC 6849) are provided in Section 17.10.1.

11.1.13 Implementation Considerations

The media loopback approach described in this specification is a complete solution that would work under all scenarios. However, it is possible that the solution is not lightweight enough for some implementations. In light of this concern, this section clarifies which features of the loopback proposal MUST be implemented for all implementations and which features MAY be deferred if the complete solution is not desired. All implementations MUST at least support the rtp-pkt-loopback mode for loopback type, with direct media loopback payload encoding. In addition, for the loopback role, all implementations of an SDP offerer MUST at least be able to act as a loopback-source. These requirements are intended to provide a minimal level of interoperability between different implementations.

11.1.14 IANA Considerations

SDP parameters that are described here have been registered with Internet Assigned Numbers Authority (IANA). We have not included the IANA registration procedures here for the sake of brevity. The procedures for registration of new SDP parameters are described in RFC 6849.

11.2 Summary

Media loopback is especially popular in ensuring the quality of transport to the edge of a given VoIP, real-time text, or Video-over IP service. We have discussed the SDP extensions for supporting media loopback to meet new challenges in managing and maintaining real-time voice/text/video quality, reliability, and overall performance. The media loopback approach described here provides a complete solution that is applicable under many scenarios. Three types of media loopback are described: Encapsulated Packet Loopback, Direct Loopback and Media Loopback. We have provided examples of how these SDP extensions can be used creating offer and answer, modifying session, and establishing session with entities behind NATs in dealing with media loopback including specific media-type such as audio, video, text, and application. Furthermore, we have discussed the RTP formats for packet loopback for encapsulated and direct RTP payload format. In addition, the behavior of SRTP, RTCP, and congestion control are described in the context of media loopback. These enhancements in SDP, RTP, SRTP, and RTCP will serve as monitoring and troubleshooting tools by providing the means for measurement of more advanced VoIP, real-time text, and Video-over IP performance metrics.

11.3 Problems

1. Why do we need media loopback capability for real-time applications such as VoIP? What are the different mechanisms for media loopback? Explain each one of those schemes providing examples.
2. What are the new SDP extensions in SDP for supporting the media loop? Explain in detail generating SDP offer/answer, processing, and modifying the session for inclusion of media loopback.
3. Explain in detail how the end-to-end SDP sessions can be set up for media loopback for the entities that are behind NATs?
4. How are the RTP payload formats for packet loopback used in encapsulating payload and direct loopback format?
5. Explain the behavior in detail how SRTP is taking care of the media loopback capability.
6. How will congestion control be taken care of using SDP and RTP/SRTP media loopback capability for real-time applications like VoIP?
7. Explain the following in detail using examples with call-flows: Offer for specific media loopback type, offer for choice of media loopback type, and answerer rejecting loopback media.

What are the specific implementation guidelines for media loopback described in RFC 6849? How does they affect the implementation of media loopback scheme for VoIP services?

Chapter 12

Quality-of-Service Support in SDP

Abstract

This section describes Request for Comments (RFCs) 5432, 3556, 3605, 3890, and 5898 that deal with quality-of-service (QOS) support in Session Description Protocol (SDP). The QOS support in multimedia applications is essential especially for real-time and near-real-time audio and video applications. Although the QOS requirements for nonreal-time applications (e.g., text and data applications) may not be as stringent as that of real-time and near-real-time applications, the support for QOS mechanisms like Resource Reservation Protocol specified in RFC 2205 and Next Steps in Signaling protocol defined in RFC 5974 in SDP are essential for negotiating the specific QOS parameters of multimedia applications among the conferencing parties. RFC 5432 defines a mechanism that allows endpoints to negotiate the QOS mechanism to be used for a particular media stream. It is important to note that QOS parameters negotiated at the multimedia application layer are not sufficient unless the same is supported in the network layer. The network layer QOS in support of the application layer QOS is orthogonal; it is not defined in RFC 5432 and needs to be worked out separately.

In addition, Request for Comments (RFC) 3556 (see Section 12.2) defines an extension to the Session Description Protocol (SDP) to specify two additional modifiers for the bandwidth attribute. These modifiers may be used to specify the bandwidth allowed for Real-Time Transport Control Protocol (RTCP) packets in a Real-Time Transport Protocol (RTP) session. Another problem has been that when a session requires multiple ports, SDP assumes that these ports have consecutive numbers. However, when the session crosses a network address translation (NAT) device that also uses port mapping, the ordering of ports can be destroyed by the translation. To handle this, RFC 3605 (see Section 12.3) defines an extension attribute to SDP. Furthermore, RFC 3890 (see Section 12.4) defines a SDP Transport Independent Application Specific (TIAS) maximum bandwidth modifier that does not include transport overhead; instead, an additional packet rate attribute is defined. The transport independent bit-rate value together with the maximum packet rate can then be used to calculate the real bit rate over the transport actually used. The existing SDP bandwidth modifiers and their values include the bandwidth needed for the transport and IP layers.

Finally, RFC 5898 (see Section 12.5) defines a new connectivity precondition for the SDP precondition framework. A connectivity precondition can be used to delay session establishment or modification until media stream connectivity has been successfully verified. The method of verification may vary depending on the type of transport used for the media. For unreliable datagram transports such as User Datagram Protocol (UDP), verification involves probing the stream with data or control packets. For reliable connection-oriented transports such as Transmission Control Protocol (TCP), verification can be achieved simply by successful connection establishment or by probing the connection with data or control packets, depending on the situation.

12.1 QOS Mechanism Selection SDP

This section describes RFC 5432 that specifies the quality-of-service (QOS) support in SDP assuming the fact that endpoints somehow establish the QOS (e.g., Resource Reservation Protocol (RSVP), DiffServ, and Next Steps in Signaling (NSIS)) required for the media streams they establish.

12.1.1 Introduction

The offer/answer model (RFC 3264, see Section 3.1) for SDP (RFC 4566, see Section 2) does not provide any mechanism for endpoints to negotiate the QOS mechanism to be used for a particular media stream. Even when QOS preconditions (RFC 3312) are used, the choice of the QOS mechanism is left unspecified and is up to the endpoints. Endpoints that support more than one QOS mechanism need a way to negotiate which one to use for a particular media stream. Examples of QOS mechanisms are RSVP (RFC 2205) and NSIS (RFC 5974). This specification defines a mechanism that allows endpoints to negotiate the QOS mechanism to be used for a particular media stream. However, the fact that endpoints agree on a particular QOS mechanism does not imply that that particular mechanism is supported by the network. Discovering which QOS mechanisms are supported at the network layer is out of the scope of this specification. In any case, the information the endpoints exchange to negotiate QOS mechanisms, as defined in this document, can be useful for a network operator to resolve a subset of the QOS interoperability problem—namely, to ensure that a mechanism commonly acceptable to the endpoints is chosen and make it possible to debug potential misconfiguration situations.

12.1.2 Terminology

See Table 2.1.

12.1.3 SDP Attributes Definition

We have shown all SDP Augmented Backus-Naur Form (ABNF) syntaxes in Section 2.9. However, we have repeated this here for convenience. This section defines the "qos-mech-send" and "qos-mech-recv" session and media-level SDP (RFC 4566, see Section 2/2.9) attributes. The following is their ABNF (RFC 5234) syntax, which is based on the SDP (RFC 4566, see Section 2/2.9) grammar:

```
attribute            =    / qos-mech-send-attr
attribute            =    / qos-mech-recv-attr
qos-mech-send-attr   =    "qos-mech-send" ":"
                          [[SP] qos-mech *(SP qos-mech)]
qos-mech-recv-attr   =    "qos-mech-recv" ":"
                          [[SP] qos-mech *(SP qos-mech)]
```

```
qos-mech           =      "rsvp" / "nsis" / extension-mech
extension-mech     =      token
```

The "qos-mech" token identifies a QOS mechanism supported by the entity generating the session description. A token that appears in a "qos-mech-send" attribute identifies a QOS mechanism that can be used to reserve resources for traffic sent by the entity generating the session description. A token that appears in a "qos-mech-recv" attribute identifies a QOS mechanism that can be used to reserve resources for traffic received by the entity generating the session description. The "qos-mech-send" and "qos-mech-recv" attributes are not interdependent; one can be used without the other. The following is an example of an "m=" line with "qos-mech-send" and "qos-mech-recv" attributes:

```
m=audio 50000 RTP/AVP 0
a=qos-mech-send: rsvp nsis
a=qos-mech-recv: rsvp nsis
```

12.1.4 Offer/Answer Behavior

Through the "qos-mech-send" and "qos-mech-recv" attributes, an offer/answer exchange allows endpoints to come up with a list of common QOS mechanisms sorted by preference. However, note that endpoints negotiate in which direction QOS is needed using other mechanisms, such as preconditions (RFC 3312). Endpoints may also use other mechanisms to negotiate, if needed, the parameters to use with a given QOS mechanism (e.g., bandwidth to be reserved).

12.1.4.1 Offerer Behavior

Offerers include a "qos-mech-send" attribute with the tokens corresponding to the QOS mechanisms (in order of preference) that are supported in the send direction. Similarly, offerers include a "qos-mech-recv" attribute with the tokens corresponding to the QOS mechanisms (in order of preference) that are supported in the receive direction.

12.1.4.2 Answerer Behavior

On receiving an offer with a set of tokens in a "qos-mech-send" attribute, the answerer takes those tokens corresponding to QOS mechanisms it supports in the receive direction and includes them, in order of preference, in a "qos-mech-recv" attribute in the answer. On receiving an offer with a set of tokens in a "qos-mech-recv" attribute, the answerer takes those tokens corresponding to QOS mechanisms it supports in the send direction and includes them, in order of preference, in a "qos-mech-send" attribute in the answer.

When ordering the tokens in a "qos-mech-send" or a "qos-mech-recv" attribute by preference, the answerer may take into account its own preferences and those expressed in the offer. However, the exact algorithm to be used to order such token lists is outside the scope of this specification. Note that if the answerer does not have any QOS mechanism in common with the offerer, it will return empty "qos-mech-send" and "qos-mech-recv" attributes.

12.1.4.3 Resource Reservation

Once the offer/answer exchange completes, both offerer and answerer use the token lists in the "qos-mech-send" and "qos-mech-recv" attributes of the answer to perform resource reservations. Offerers and answerers SHOULD attempt to use the QOS mechanism with highest priority in each direction first. If an endpoint (the offerer or the answerer) does not succeed in using the mechanism with highest priority in a given direction, it SHOULD attempt to use the next QOS mechanism in order of priority in that direction, and so on. If an endpoint unsuccessfully tries all the common QOS mechanisms for a given direction, the endpoint MAY attempt to use additional QOS mechanisms not supported by the remote endpoint. This is because there may be network entities out of the endpoint's control (e.g., an RSVP proxy) that make those mechanisms work.

12.1.4.4 Subsequent Offer/Answer Exchanges

If, during an established session for which the QOS mechanism to be used for a given direction was agreed upon using the mechanism defined in this specification, an endpoint receives a subsequent offer that does not contain the QOS selection attribute corresponding to that direction (i.e., the "qos-mech-send" or "qos-mech-recv" attribute is missing), the endpoints SHOULD continue using the same QOS mechanism used up to that moment.

12.1.5 Example

The following is an offer/answer exchange between two endpoints using the "qos-mech-send" and "qos-mech-recv" attributes. Parts of the session descriptions are omitted for clarity. The offerer generates the following session description, listing RSVP and NSIS for both directions. The offerer, preferring to use RSVP, includes it before NSIS.

```
m=audio 50000 RTP/AVP 0
a=qos-mech-send: rsvp nsis
a=qos-mech-recv: rsvp nsis
```

The answerer supports NSIS in both directions, but not RSVP. Consequently, it returns the following session description:

```
m=audio 55000 RTP/AVP 0
a=qos-mech-send: nsis
a=qos-mech-recv: nsis
```

12.1.6 Internet Assigned Numbers Authority Considerations

SDP parameters that are described here have been registered with Internet Assigned Numbers Authority (IANA). We have not included the IANA registration procedures here for the sake of brevity. The procedures for registration of new SDP parameters are described in RFC 5432.

12.1.7 Security Considerations

The security procedures that need to be considered in dealing with the basic SDP messages described in this section (RFC 5432) are provided in Section 17.11.1.

12.2 SDP Bandwidth Modifiers for RTCP Bandwidth

This section describes RFC 3556 that defines an extension to the SDP to specify two additional modifiers for the bandwidth attribute. These modifiers may be used to specify the bandwidth allowed for RTCP packets in an RTP session.

12.2.1 Introduction

The RTP, RFC 3550, includes a control protocol RTCP, which provides synchronization information from data senders and feedback information from data receivers. Normally, the amount of bandwidth allocated to RTCP in an RTP session is 5% of the session bandwidth. For some applications, it may be appropriate to specify the RTCP bandwidth independently of the session bandwidth. Using a separate parameter allows rate-adaptive applications to set an RTCP bandwidth consistent with a "typical" data bandwidth that is lower than the maximum bandwidth specified by the session bandwidth parameter. That allows the RTCP bandwidth to be kept under 5% of the data bandwidth when the rate has been adapted downward. On the other hand, there may be applications that send data at very low rates but need to communicate extra RTCP information, such as Application-Defined RTCP Packets (APP). These applications may need to specify RTCP bandwidth that is higher than 5% of the data bandwidth.

The RTP specification allows a profile to specify that the RTCP bandwidth may be divided into two separate session parameters for those participants that are active data senders and those that are not. Using two parameters allows RTCP reception reports to be turned off entirely for a particular session by setting the RTCP bandwidth for nondata senders to 0 while keeping the RTCP bandwidth for data senders non-0 so that sender reports can still be sent for intermedia synchronization. Turning off RTCP reception reports is not recommended because they are needed for the functions listed in the RTP specification, particularly reception quality feedback and congestion control. However, doing so may be appropriate for systems operating on unidirectional links or for sessions that do not require feedback on the quality of reception or liveness of receivers and that have other means to avoid congestion. This memo defines an extension to the SDP specified in RFC 2327 (that was made obsolete by RFC 4566, see Section 2) to specify RTCP bandwidth for senders and nonsenders (receivers).

12.2.2 SDP Extensions

The SDP includes an optional bandwidth attribute with the following syntax:

```
b=<modifier>:<bandwidth-value>
```

where <modifier> is a single alphanumeric word giving the meaning of the bandwidth figure and where the default units for <bandwidthvalue> are kilobits per second. This attribute specifies the proposed bandwidth to be used by the session or media. A typical use is with the modifier "AS" (for Application Specific Maximum), which may be used to specify the total bandwidth for a single media stream from one site (source). This memo defines two additional bandwidth modifiers:

```
b=RS:<bandwidth-value>
b=RR:<bandwidth-value>
```

where "RS" indicates the RTCP bandwidth allocated to active data senders (as defined by the RTP spec) and "RR" indicates the RTCP bandwidth allocated to other participants in the RTP session (i.e., receivers). The exact behavior induced by specifying these bandwidth modifiers depends upon the algorithm used to calculate the RTCP reporting interval. Different algorithms may be specified by different RTP profiles.

For the RTP A/V Profile (RFC 3551), which specifies that the default RTCP interval algorithm defined in the RTP spec (RFC 3550) is to be used, at least RS/(RS+RR) of the RTCP bandwidth is dedicated to active data senders. If the proportion of senders to total participants is less than or equal to RS/(RS+RR), each sender gets RS divided by the number of senders. When the proportion of senders is greater than RS/(RS+RR), the senders get their proportion of the sum of these parameters, which means that a sender and a nonsender each get the same allocation. Therefore, it is not possible to constrain the data senders to use less RTCP bandwidth than is allowed for nonsenders. A few special cases are worth noting:

- If RR is 0, then the proportion of participants that are senders can never be greater than RS/(RS+RR); therefore, nonsenders never get any RTCP bandwidth independent of the number of senders.
- Setting RS to 0 does not mean that data senders are not allowed to send RTCP packets; it only means that they are treated the same as nonsenders. The proportion of senders (if there are any) is always greater than RS/(RS+RR) if RR is non-0.
- If RS and RR are both 0, it would be unwise to attempt calculation of the fraction RS/(RS+RR). The bandwidth allocation specified by the RS and RR modifiers applies to the total bandwidth consumed by all RTCP packet types, including SR, RR, SDES, BYE, APP, and any new types defined in the future. The <bandwidth-value> for these modifiers is in units of bits per second with an integer value.

NOTE: This specification was in conflict with the initial SDP spec in RFC 2327 (that was made obsolete by RFC 4566, see Section 2), which prescribes that the <bandwidth-value> for all bandwidth modifiers should be an integer number of kilobits per second. This discrepancy was forced by the fact that the desired RTCP bandwidth setting may be less than 1 kb/s. At the 44th Internet Engineering Task Force (IETF) meeting in Minneapolis, two solutions were considered: allow fractional values or specify that the units for these particular modifiers would be in bits per second. The second choice was preferred so that the syntax would not be changed. The SDP spec is being modified by RFC 4566 (see Section 2) and allows this change in semantics.

12.2.3 Default Values

If either or both of the RS and RR bandwidth specifiers are omitted, the default values for these parameters are as specified in the RTP profile in use for the session in question. For the audio/video profile, RFC 3551, the defaults follow the recommendations of the RTP spec:

- The total RTCP bandwidth is 5% of the session bandwidth. If one of these RTCP bandwidth specifiers is omitted, its value is 5% minus the value of the other one, but not less than 0. If both are omitted, the sender and receiver RTCP bandwidths are 1.25% and 3.75% of the session bandwidth, respectively.
- At least RS/(RS+RR) of the RTCP bandwidth is dedicated to active data senders. When the proportion of senders is greater than RS/(RS+RR) of the participants, the senders get their proportion of the sum of these parameters. This memo does not impose limits on the values that may be specified with the RR and RS modifiers, other than that they must be nonnegative. However, the RTP specification and the appropriate RTP profile may specify limits.

12.2.4 Precedence

An SDP description consists of a session-level description (details that apply to the whole session and all media streams) and zero or more media-level descriptions (details that apply only to a single media stream). Bandwidth specifiers may be present either at the session level to specify the total bandwidth shared by all media, in the media sections to specify the bandwidth allocated to each medium, or both. This is true for the two RTCP bandwidth modifiers defined here as well. Since the bandwidth allocated to RTCP is a fraction of the session bandwidth when not specified explicitly using the modifiers defined here, there is an interaction between the session bandwidth and RTCP bandwidth specifiers at the session and media levels of the SDP description. The precedence of these specifiers is as follows, with (1) being the highest precedence:

1. Explicit RR or RS specifier at media level
2. Explicit RR or RS specifier at session level
3. Default based on session bandwidth specifier at media level
4. Default based on session bandwidth specifier at session level

In particular, the relationship of (2) and (3) means that if the RR bandwidth is specified as zero at the session level that turns off RTCP transmission for nondata senders in all media.

12.2.5 Example

An example SDP description is:

```
v=0
o=mhandley 2890844526 2890842807 IN IP4 126.16.64.4
s=SDP Seminar
i=A Seminar on the session description protocol
c=IN IP4 224.2.17.12/127
t=2873397496 2873404696
m=audio 49170 RTP/AVP 0
b=AS:64
b=RS:800
b=RR:2400
m=video 51372 RTP/AVP 31
b=AS:256
b=RS:800
b=RR:2400
```

In this example, the explicit RTCP bandwidths for the audio medium are equal to the defaults and so could be omitted. However, for the video medium, the RTCP bandwidths have been set according to a data bandwidth of 64 kb/s even though the maximum data bandwidth is specified as 256 kb/s. This is based on the assumption that the video data bandwidth will automatically adapt to a lower value based on network conditions.

12.2.6 IANA Considerations

SDP parameters that are described here have been registered with IANA. We have not included the IANA registration procedures here for the sake of brevity. The procedures for registration of new SDP parameters are described in RFC 3556.

12.2.7 Security Considerations

The security procedures that need to be considered in dealing with the basic SDP messages described in this section (RFC 3556) are provided in Section 17.11.2.

12.3 RTCP Attribute in SDP

This section describes RFC 3605. The SDP is used to describe the parameters of media streams used in multimedia sessions. When a session requires multiple ports, SDP assumes that these ports have consecutive numbers. However, when the session crosses a NAT device that also uses port mapping, the ordering of ports can be destroyed by the translation. To handle this, RFC 3605 proposes an extension attribute to SDP.

12.3.1 Introduction

The session invitation protocol such as SIP (RFC 3261, also see Section 1.2) is often used to establish multimedia sessions on the Internet. There are often cases today in which one or both ends of the connection are hidden behind a NAT device (RFC 2766). In this case, the SDP text must document the IP addresses and UDP ports as they appear on the "public Internet" side of the NAT. In this memo, we will suppose that the host located behind a NAT has a way of obtaining these numbers. A possible way to learn these numbers is briefly outlined in Section 3; however, just learning the numbers is not enough. The SIP messages use the encoding defined in SDP (RFC 2327 that is made obsolete by RFC 4566) to describe the IP addresses and TCP or UDP ports used by the various media. Audio and video are typically sent using RTP (RFC 3550), which requires two UDP ports, one for the media and one for the control protocol (RTCP). SDP carries only one port number per media and states, "other ports used by the media application (such as the RTCP port) should be derived algorithmically from the base media port." RTCP port numbers were necessarily derived from the base media port in older versions of RTP, such as RFC 1889 (that is made obsolete by RFC 4566, see Section 2), but now that this restriction has been lifted, there is a need to specify RTCP ports explicitly in SDP. Note, however, that implementations of RTP adhering to the earlier RFC 1889 (that was made obsolete by RFC 4566, see Section 2) specification may not be able to make use of the SDP attributes specified in this document.

When the NAT device performs port mapping, there is no guarantee that the mappings of two separate ports reflect the sequencing and the parity of the original port numbers; in fact, when the NAT manages a pool of IP addresses, it is possible that the RTP and RTCP ports are mapped to different addresses. In order to successfully establish connections despite the misordering of the port numbers and the possible parity switches caused by the NAT, we propose using a specific SDP attribute to document the RTCP port and optionally the RTCP address.

12.3.2 Description of the Solution

The main part of our solution is the declaration of an SDP attribute for documenting the port used by RTCP.

12.3.2.1 The RTCP Attribute

All SDP ABNF syntaxes are provided in Section 2.9. However, we have repeated this here for convenience. The RTCP attribute is used to document the RTCP port used for media stream, when that port is not the next higher (odd) port number following the RTP port described in the media line. The RTCP attribute is a "value" attribute and follows the general syntax specified on page 18 of RFC 2327 (that was made obsolete by RFC 4566, see Section 2/2.9): "a=<attribute>:<value>." For the RTCP attribute:

- The name is the ascii string "rtcp" (lower case),
- The value is the RTCP port number and optional address.

The formal description of the attribute is defined by the following ABNF (RFC 2234 that was made obsolete by RFC 4234; again, RFC 4234 was made obsolete by RFC 5234) syntax:

```
rtcp-attribute    =    "a=rtcp:" port [nettype space addrtype space
                       connection-address] CRLF
```

In this description, the "port," "nettype," "addrtype," and "connection-address" tokens are defined as specified in "Appendix A: SDP Grammar" of RFC 2327 (that was made obsolete by RFC 4566, see Section 2/2.9). Example encodings could be

```
m=audio 49170 RTP/AVP 0
a=rtcp:53020
m=audio 49170 RTP/AVP 0
a=rtcp:53020 IN IP4 126.16.64.4
m=audio 49170 RTP/AVP 0
a=rtcp:53020 IN IP6 2001:2345:6789:ABCD:EF01:2345:6789:ABCD
```

The RTCP attribute MAY be used as a media-level attribute; it MUST NOT be used as a session-level attribute. Though the following examples relate to a method that will return only unicast addresses, both unicast and multicast values are valid.

12.3.3 Discussion of the Solution

The implementation of the solution is fairly straightforward. The questions that have been most often asked regarding this solution are whether this is useful (i.e., whether a host can actually discover port numbers in an unmodified NAT), whether it is sufficient (i.e., whether there is a need to document more than one ancillary port per media type), and why one should not change the media definition instead of adding a new attribute.

12.3.3.1 How Do We Discover Port Numbers?

The proposed solution is only useful if the host can discover the "translated port numbers" (i.e., the value of the ports as they appear on the "external side" of the NAT). One possibility is to ask the cooperation of a well-connected third party that will act as a server according to simple traversal of UDP around NAT (STUN) [RFC 3489]. We thus obtain a four-step process:

1. The host allocates two UDP ports numbers for an RTP/RTCP pair.
2. The host sends a UDP message from each port to the STUN server.
3. The STUN server reads the source address and port of the packet and copies them in the text of a reply.
4. The host parses the reply according to the STUN protocol and learns the external address and port corresponding to each of the two UDP ports.

This algorithm supposes that the NAT will use the same translation for packets sent to the third party and to the "SDP peer" with which the host wants to establish a connection. There is no guarantee that all NAT boxes deployed on the Internet have this characteristic. Implementers are referred to the STUN specification, RFC 3489 (that was made obsolete by RFC 5389), for an extensive discussion of the various types of NAT.

12.3.3.2 Do We Need to Support Multiple Ports?

Most media streams are transmitted using a single pair of RTP and RTCP ports. It is possible, however, to transmit a single media over several RTP flows, for example using hierarchical encoding. In this case, SDP will encode the port number used by RTP on the first flow, and the number of flows, as in:

```
m=video 49170/2 RTP/AVP 31
```

In this example, the media is sent over two consecutive pairs of ports, corresponding respectively to RTP for the first flow (even number, 49170), RTCP for the first flow (odd number, 49171), RTP for the second flow (even number, 49172), and RTCP for the second flow (odd number, 49173). In theory, it is possible to modify SDP and document the many ports corresponding to the separate encoding layers. However, layered encoding is not much used in practice and when used is mostly used in conjunction with multicast transmission. The translation issues documented in this memo apply uniquely to unicast transmission; thus, there is no short-term need for the support of multiple port descriptions. It is more convenient and more robust to focus on the simple case in which a media is sent over exactly one RTP/RTCP stream.

12.3.3.3 Why Not Expand the Media Definition?

The RTP ports are documented in the media description line, and it would seem convenient to document the RTCP port at the same place rather than create an RTCP attribute. We considered this design alternative and rejected it for two reasons: adding an extra port number and an option address in the media description would be awkward; more importantly, it would create problems with existing applications, which would have to reject the entire media description if they did not understand the extension. On the contrary, adding an attribute has a well-defined failure mode: implementations that don't understand the "a=rtcp" attribute will simply ignore it; they will fail to send RTCP packets to the specified address, but they will at least be able to receive the media in the RTP packets.

12.3.4 UNSAF Considerations

The RTCP attribute in SDP is used to enable the establishment of RTP/RTCP flows through NAT. This mechanism can be used in conjunction with an address discovery mechanism such as STUN, RFC 3489 (that was made obsolete by RFC 5389). STUN is a short-term fix to the NAT traversal problem, which requires consideration of the general issues linked to "Unilateral self-address fixing" (UNSAF) (RFC 3424). The RTCP attribute addresses a very specific problem, the documentation of port numbers as they appear after address translation by a port-mapping NAT. The RTCP attribute SHOULD NOT be used for other applications. We expect that, with time, one of two exit strategies can be developed. The IETF may develop an explicit "middlebox control" protocol that will enable applications to obtain a pair of port numbers appropriate for RTP and RTCP. Another possibility is the deployment of IPv6, which will enable use of "end-to-end" addressing and guarantee that the two hosts will be able to use appropriate ports. In both cases, there will be no need for documenting a "nonstandard" RTCP port with the RTCP attribute.

12.3.5 Security Considerations

The security procedures that need to be considered in dealing with the basic SDP messages described in this section (RFC 3605) are provided in Section 17.11.3.

12.3.6 IANA Considerations

SDP parameters that are described here have been registered with IANA. We have not included the IANA registration procedures here for the sake of brevity. The procedures for registration of new SDP parameters are described in RFC 3605.

12.4 Transport Independent Bandwidth Modifier for SDP

This subsection describes RFC 3890 that defines a SDP TIAS maximum bandwidth modifier that does not include transport overhead; instead, an additional packet rate attribute is defined. The transport independent bit-rate value together with the maximum packet rate can then be used to calculate the real bit rate over the transport actually used. The existing SDP bandwidth modifiers and their values include the bandwidth needed for the transport and IP layers. When using SDP with protocols like Session Announcement Protocol (SAP), SIP, and RTSP, and when the involved hosts have different transport overhead, for example, due to different IP versions, the interpretation of what lower layer bandwidths are included is not clear.

12.4.1 Introduction

This specification is structured in the following way: In this section, some information regarding SDP bandwidth modifiers and different mechanisms that affect transport overhead are asserted. In Section 12.4.3, the problems found are described, including problems that are not solved by this specification. In Section 12.4.4, the scope of the problems this specification solves is presented. Section 12.4.5 contains the requirements applicable to the problem scope. Section 12.4.6 defines the solution, which is a new bandwidth modifier, and a new maximum packet rate attribute. Section 12.4.7 looks at the protocol interaction for SIP, RTSP, and SAP. The security considerations are discussed in Section 12.4.8. The remaining sections are the necessary IANA considerations, acknowledgements, reference list, author's address, and copyright and intellectual property right (IPR) notices. Today, the SDP specified in RFC 2327 (that is made obsolete by RFC 4566, see Section 2) is used in several types of applications. The original application is session information and configuration for multicast sessions announced with SAP (RFC 2974). SDP is also a vital component in media negotiation for the SIP (RFC 3261, also see Section 1.2) by using the offer answer model (RFC 3264, see Section 3.1). The Real-Time Streaming Protocol (RTSP) (RFC 2326) also makes use of SDP to declare to the client what media and codec(s) comprise a multimedia presentation.

12.4.1.1 The Bandwidth Attribute

In SDP specified in RFC 2327 (that was made obsolete by RFC 4566, see Section 2), there exists a bandwidth attribute, which has a modifier used to specify what type of bit rate the value refers to. The attribute has the following form:

```
b=<modifier>:<value>
```

Today there are four defined modifiers used for different purposes.

12.4.1.1.1 Conference Total

The conference total is indicated by giving the modifier "CT." Conference total gives a maximum bandwidth a conference session will use. Its purpose is to decide if this session can coexist with any other sessions, defined in RFC 2327 (that was made obsolete by RFC 4566, see Section 2).

12.4.1.1.2 Application Specific Maximum

The Application Specific maximum bandwidth is indicated by the modifier "AS." The interpretation of this attribute is dependent on the application's notion of maximum bandwidth. For an RTP application, this attribute is the RTP session bandwidth as defined in RFC 3550. The session bandwidth includes the bandwidth that the RTP data traffic will consume, including the lower layers, down to the IP layer. Therefore, the bandwidth is in most cases calculated over RTP payload, RTP header, UDP, and IP, defined in RFC 2327 (that was made obsolete by RFC 4566, see Section 2).

12.4.1.1.3 RTCP Report Bandwidth

In RFC 3556 (see Section 12.2), two bandwidth modifiers are defined. These modifiers, "RS" and "RR," define the amount of bandwidth is assigned for RTCP reports by active data senders and RTCP reports by other participants (receivers), respectively.

12.4.1.2 IPv6 and IPv4

Today there are two IP versions, IPv4 specified in RFC 0791 and IPv6 specified in RFC 2460 (that was made obsolete by RFC 8200), used in parallel on the Internet, creating problems. However, there exist a number of possible transition mechanisms.

- The nodes that wish to communicate must share the IP version; typically this is done by deploying dual-stack nodes. For example, an IPv4-only host cannot communicate with an IPv6-only host.
- If communication between nodes that do not share a protocol version is required, use of a translation or proxying mechanism would be required. Work is underway to specify such a mechanism for this purpose (Figure 12.1).
- IPv6 nodes belonging to different domains running IPv6 but lacking IPv6 connectivity between them solve this by tunneling over the IPv4 net, see Figure 12.2. Basically, the IPv6 packets are sent as payload in IPv4 packets between the tunneling end-points at the edge of each IPv6 domain. The bandwidth required over the IPv4 domain will be different from IPv6 domains. However, as the tunneling is normally not performed by the application end-point, this scenario cannot usually be taken into consideration.

IPv4 has a minimum header size of 20 bytes, while the fixed part of the IPv6 header is 40 bytes. The difference in header sizes means that the bit rate required for the two IP versions is different. The significance of the difference depends on the packet rate and payload size of each packet.

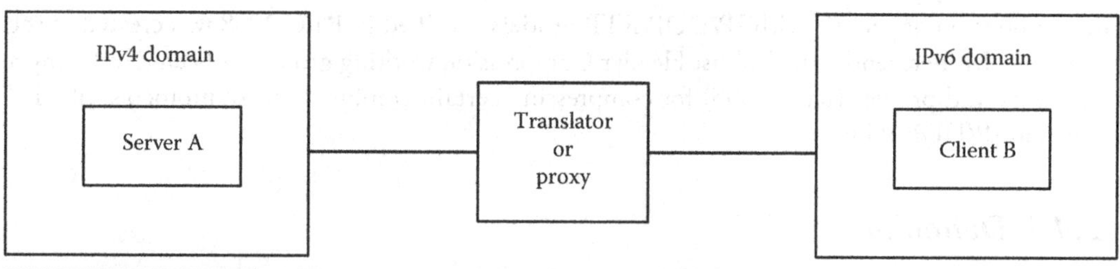

Figure 12.1 Translation or proxying between IPv6 and IPv4 addresses.

12.4.1.3 Further Mechanisms that Change the Bandwidth Utilization

A number of other mechanisms also may change the overhead at layers below media transport. We will briefly cover a few of these here.

12.4.1.3.1 IPsec

IPsec specified in RFC 2401 (that was made obsolete by RFC 4301) can be used between endpoints to provide confidentiality through the application of the IP encapsulating security payload (ESP) described in RFC 2406 (Note: RFC 2406 was made obsolete by RFCs 4303 and 4305; again, RFC 4305 was made obsolete by RFC 4835; furthermore, RFC 4835 was made obsolete by RFC 7321) or integrity protection using the IP Authentication Header (AH) specified in RFC 2402 (Note: RFC 2402 is obsoleted by RFCs 4303 and 4305; again, RFC 4305 was made obsolete by RFC 4835; furthermore, RFC 4835 was made obsolete by RFC 7321) of the media stream. The addition of the ESP and AH headers increases each packet's size. To provide virtual private networks, complete IP packets may be encapsulated between an end node and the private networks security gateway, thus providing a secure tunnel that ensures confidentiality, integrity, and authentication of the packet stream. In this case, the extra IP and ESP header will significantly increase the packet size.

12.4.1.3.2 Header Compression

Another mechanism that alters the actual overhead over links is header compression. Header compression uses the fact that most network protocol headers have either static or predictable values in their fields within a packet stream. Compression is normally only done on a per-hop basis (i.e., on a single link). The normal reason for doing header compression is that the link has fairly limited bandwidth, and significant gain in throughput is achieved. There exist several different header compression standards. For compressing IP headers only, there is RFC 2507. However,

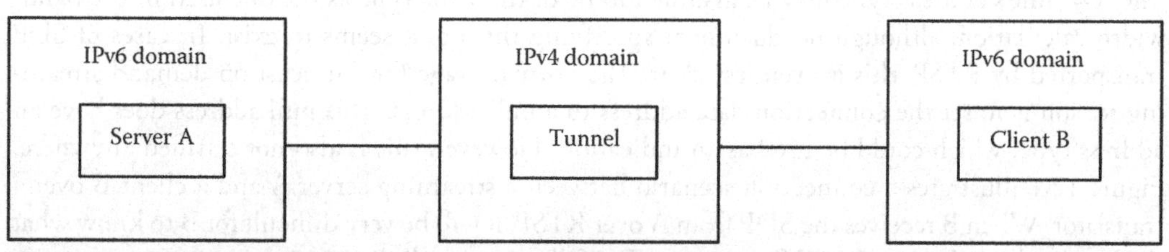

Figure 12.2 Tunneling through an IPv4 domain (Copyright IETF).

the compression of packets with IP/UDP/RTP headers specified in RFC 2508 was created at the same time. More recently, the Robust Header Compression working group has been developing a framework and profiles (RFC 3095) for compressing certain combinations of protocols, like IP/UDP and IP/UDP/RTP.

12.4.2 Definitions

See Table 2.1.

12.4.2.1 Glossary

ALG: Application Level Gateway
bps: bits per second
RTSP: Real-Time Streaming Protocol, see RFC 2326
SDP: Session Description Protocol, see RFC 2327 (was made obsolete by RFC 4566, see Section 2)
SAP: Session Announcement Protocol (see RFC 2974)
SIP: Session Initiation Protocol, see (RFC 3261, also see Section 1.2)
TIAS: Transport Independent Application Specific maximum, a bandwidth modifier

Also see Table 2.1.

12.4.2.2 Terminology

See Table 2.1.

12.4.3 The Bandwidth Signaling Problems

When an application wants to use SDP to signal the bandwidth required for this application, some problems become evident due to the inclusion of the lower layers in the bandwidth values.

12.4.3.1 What IP Version Is Used

If one signals the bandwidth in SDP, for example, using "b=AS:" as an RTP based application, one cannot know if the overhead is calculated for IPv4 or IPv6. An indication of which protocol has been used when calculating the bandwidth values is given by the "c=" connection-address line. This line contains either a multicast group address or a unicast address of the data source or sink. The "c=" line's address type may be assumed to be of the same type as the one used in the bandwidth calculation, although no document specifying this point seems to exist. In cases of SDP transported by RTSP, this is even less clear. The normal usage for a unicast on-demand streaming session is to set the connection data address to a null address. This null address does have an address type, which could be used as an indication. However, this is also not clarified anywhere. Figure 12.1 illustrates a connection scenario between a streaming server A and a client B over a translator. When B receives the SDP from A over RTSP, it will be very difficult for B to know what the bandwidth values in the SDP represent. The following possibilities exist:

1. The SDP is unchanged and the "c=" null address is of type IPv4. The bandwidth value represents the bandwidth needed in an IPv4 network.
2. The SDP has been changed by an Application Level Gateway (ALG). The "c=" address is changed to an IPv6 type. The bandwidth value is unchanged.
3. The SDP is changed, and both "c=" address type and bandwidth value are converted. Unfortunately, this can seldom be done, see Section 12.4.3.3.

In case 1, the client can understand that the server is located in an IPv4 network and that it uses IPv4 overhead when calculating the bandwidth value. The client can almost never convert the bandwidth value, see Section 12.4.3.3. In case 2, the client does not know that the server is in an IPv4 network and that the bandwidth value is not calculated with IPv6 overhead. In cases where a client uses this value to determine if its end of the network has sufficient resources, the client will underestimate the required bit rate, potentially resulting in bad application performance (very rare). If one tries to convert the bandwidth value without further information about the packet rate, significant errors may be introduced into the value.

12.4.3.2 Taking Other Mechanisms into Account

Sections 12.4.1.2 and 12.4.1.3 list a number of factors like header compression and tunnels that would change lower layer header sizes. For these mechanisms, there exist different possibilities for taking them into account. The option of using IPsec directly between endpoints should definitely be known to the application, thus enabling it to take the extra headers into account. However, the same problem exists with the current SDP bandwidth modifiers where a receiver is not able to convert these values, taking the IPsec headers into account. It is less likely that an application would be aware of the existence of a virtual private network. Thus, the generality of the mechanism to tunnel all traffic may prevent the application from even considering whether it would be possible to convert the values.

When using header compression, the actual overhead will be less deterministic, but in most cases an average overhead can be determined for a certain application. If a network node knows that some type of header compression is employed, this can be taken into consideration. For RSVP (RFC 2205), there exists an extension, RFC 3006, that allows the data sender to inform network nodes about the compressibility of the data flow. To be able to do this with any accuracy, the compression factor and packet rate or size is needed, as RFC 3006 provides.

12.4.3.3 Converting Bandwidth Values

If one would like to convert a bandwidth value calculated using IPv4 overhead to IPv6 overhead, the packet rate is required. The new bandwidth value for IPv6 is normally "IPv4 bandwidth" + "packet rate" * 20 bytes, where 20 bytes is the usual difference between IPv6 and IPv4 headers. The overhead difference may be some other value in cases when IPv4 options (RFC 0791) or IPv6 extension headers (RFC 2460 obsoleted by RFC 8200) are used. As converting requires the packet rate of the stream, this is not possible in the general case. Many codecs either have multiple possible packet/frame rates or can perform payload format aggregation, resulting in many possible rates. Therefore, some extra information in the SDP will be required. The "a=ptime:" parameter may be a possible candidate. However, this parameter is normally only used for audio codecs. Its definition as described in RFC 2327 (that was made obsolete by 4566, see Section 2) is that it is only a recommendation, which the sender may disregard. A better parameter is needed.

12.4.3.4 RTCP Problems

When RTCP is used between hosts in IPv4 and IPv6 networks over a translator, similar problems exist. The RTCP traffic going from the IPv4 domain will result in a higher RTCP bit rate than was intended in the IPv6 domain due to the larger headers. This may result in up to a 25% increase in required bandwidth for the RTCP traffic. The largest increase will be for small RTCP packets when the number of IPv4 hosts is much larger than the number of IPv6 hosts. Fortunately, as RTCP has a limited bandwidth compared to RTP, it will result in only a maximum 1.75% increase of the total session bandwidth when RTCP bandwidth is 5% of RTP bandwidth. The RTCP randomization may easily result in short-term effects of the same magnitude, so this increase may be considered tolerable. The increase in bandwidth will be less in most cases. At the same time, this results in unfairness in the reporting between an IPv4 and IPv6 node. In the worst-case scenario, the IPv6 node may report with 25% longer intervals. These problems have been considered insignificant enough to not be worth any complex solutions. Therefore, only a simple algorithm for deriving RTCP bandwidth is defined in this specification.

12.4.3.5 Future Development

Today there is work in the IETF to design a new datagram transport protocol suitable for real-time media. This protocol is called the Datagram Congestion Control Protocol (DCCP). It will most probably have a different header size than UDP, which is the protocol most often used for real-time media today. This results in even more possible transport combinations. This may become a problem if one has the possibility of using different protocols, which will not be determined prior to actual protocol SETUP. Thus, precalculating this value will not be possible, which is one further motivation for the necessity of a transport independent bandwidth modifier needed. DCCP's congestion control algorithms will control how much bandwidth can really be utilized. This may require further work with specifying SDP bandwidth modifiers to declare the dynamic possibilities of an application's media stream. For example, in min and max media bandwidth, the application is capable of producing of all bit rates of media codecs, or for media codecs only capable of producing certain bit rates, enumerating possible rates. However, this is for future study and outside the scope of the present solution.

12.4.3.6 Problem Conclusion

A shortcoming of the current SDP bandwidth modifiers is that they also include the bandwidth needed for lower layers. It is in many cases difficult to determine which lower layers and their versions were included in the calculation, especially in the presence of translation or proxying between different domains. This prevents a receiver from determining if given bandwidth needs to be converted based on the actual lower layers being used. Second, an attribute to give the receiver an explicit determination of the maximum packet rate that will be used does not exist. This value is necessary for accurate conversion of any bandwidth values if the difference in overhead is known.

12.4.4 Problem Scope

The problems described in Section 12.4.3 are common and affect application-level signaling using SDP, other signaling protocols, and RSVPs. However, this document targets the specific problem of signaling the bit rate in SDP. The problems need to be considered in other affected protocols and in new protocols being designed. As this specification only targets carrying the bit rate information within SDP, it will have a limited applicability. As SDP information is normally transported end to end by an application protocol, nodes between the endpoints will not have access to the bit rate information. It will normally only be the endpoints that are able to take this information into account. An interior node will need to receive the information through a means other than SDP, and that is outside the scope of this specification. Nevertheless, the bit rate information provided in this specification is sufficient for cases such as first-hop resource reservation and admission control. It also provides information about the maximum codec rate, which is independent of lower-level protocols. This specification does NOT try to solve the problem of detecting NATs or other middleboxes.

12.4.5 Requirements

The problems outlined in the preceding sections and applicability should meet the following requirements:

- The bandwidth value SHALL be given in a way such that it can be calculated for all possible combinations of transport overhead.

12.4.6 Solution

12.4.6.1 Introduction

This chapter describes a solution for the problems outlined in this document for the AS bandwidth modifier, thus enabling the derivation of the required bit rate for an application, or RTP session's data and RTCP traffic. The solution is based upon the definition of a new TIAS bandwidth modifier and a new SDP attribute for the maximum packet rate (maxprate). The CT is a session-level modifier and cannot easily be dealt with. To address the problems with different overhead, it is RECOMMENDED that the CT value be calculated using reasonable worst-case overhead. An example of how to calculate a reasonable worst case overhead is:

- Take the overhead of the largest transport protocol (using average size if variable) and add that to the largest IP overhead that is expected for use, plus the data traffic rate. Do this for every individual media stream used in the conference and add them together.
- The RR and RS modifiers (RFC 3556, see Section 12.2) will be used as defined and include transport overhead. The small unfairness between hosts is deemed acceptable.

12.4.6.2 The TIAS Bandwidth Modifier

12.4.6.2.1 Usage

A new bandwidth modifier is to be used for the following purposes:

- Resource reservation. A single bit rate can be enough for use as a resource reservation. Some characteristics can be derived from the stream, codec type, etc. In cases where more information is needed, another SDP parameter will be required.
- Maximum media codec rate. With the definition in Section 12.4.6.2.2 of "TIAS," the given bit rate will mostly be from the media codec. Therefore, it gives a good indication of the maximum codec bit rate that must be supported by the decoder.

- Communication bit rate required for the stream. The "TIAS" value together with "maxprate" can be used to determine the maximum communication bit rate the stream will require. Using session-level values or adding all maximum bit rates from the streams in a session together, a receiver can determine whether its communication resources are sufficient to handle the stream. For example, a modem user can determine whether the session fits his modem's capabilities and the established connection.
- Determine the RTP session bandwidth and derive the RTCP bandwidth. The derived transport-dependent attribute will be the RTP session bandwidth in case of RTP based transport. The TIAS value can also be used to determine what RTCP bandwidth to use for implicit allocation. RTP (RFC 3550) specifies that if not explicitly stated, additional bandwidth, equal to 5% of the RTP session bandwidth, shall be used by RTCP. The RTCP bandwidth can be explicitly allocated by using the RR and RS modifiers defined in RFC 3556 (see Section 12.2).

12.4.6.2.2 Definition

A new session and media-level bandwidth modifier is defined:

```
b=TIAS:<bandwidth-value> ; see Section 12.4.6.6 for ABNF definition.
```

The TIAS maximum bandwidth modifier has an integer bit-rate value in bits per second. A fractional bandwidth value SHALL always be rounded up to the next integer. The bandwidth value is the maximum needed by the application (SDP session level) or media stream (SDP media level) without counting IP or other transport layers like TCP or UDP. At the SDP session level, the TIAS value is the maximal amount of bandwidth needed when all declared media streams are used. This MAY be less than the sum of all the individual media streams values. This is due to the possibility that not all streams have their maximum at the same point in time. This can normally only be verified for stored media streams. For RTP transported media streams, TIAS at the SDP media level can be used to derive the RTP "session bandwidth," defined in Section 6.2 of RFC 3550). In the context of RTP transport, the TIAS value is defined as follows:

Only the RTP payload as defined in RFC 3550 SHALL be used in the calculation of the bit rate (i.e., excluding the lower layers (IP/UDP) and RTP headers including RTP header), RTP header extensions, Contributing Source (CSRC) list, and other RTP profile specific fields. Note that the RTP payload includes both the payload format header and the data. This may allow one to use the same value for RTP-based media transport, non-RTP transport, and stored media.

Note 1: The usage of bps is not in accordance with RFC 2327 that is made obsolete by RFC 4566 (see Section 2). This change has no implications on the parser; only the interpreter of the value must be aware. The change is done to allow for better resolution and has been used for the RR and RS bandwidth modifiers, see RFC 3556 (see Section 12.2).

Note 2: RTCP bandwidth is not included in the bandwidth value. In applications using RTCP, the bandwidth used by RTCP is either 5% of the RTP session bandwidth including lower layers or as specified by the RR and RS modifiers (RFC 3556, see Section 12.2). A specification of how to derive the RTCP bit rate when using TIAS is presented in Section 12.4.6.5.

12.4.6.2.3 Usage Rules

"TIAS" is primarily intended to be used at the SDP media level. The "TIAS" bandwidth attribute MAY be present at the session level in SDP, if all media streams use the same transport. In cases where the sum of the media level values for all media streams is larger than the actual maximum bandwidth need for all streams, it SHOULD be included at the session level. However, if present at the session level it SHOULD be present also at the media level. "TIAS" SHALL NOT be present at the session level unless the same transport protocols are used for all media streams. The same transport is used as long as the same combination of protocols is used, like IPv6/UDP/RTP. To allow for backwards compatibility with applications of SDP that do not implement "TIAS," it is RECOMMENDED to also include the "AS" modifier when using "TIAS." The presence of a value including lower-layer overhead, even with its problems, is better than no value. However, an SDP application implementing TIAS SHOULD ignore the "AS" value and use "TIAS" instead when both are present. When using TIAS for an RTP-transported stream, the "maxprate" attribute, if possible to calculate, defined next, SHALL be included at the corresponding SDP level.

12.4.6.3 Packet Rate Parameter

To be able to calculate the bandwidth value including the lower layers actually used, a packet rate attribute is also defined. The SDP session and media level maximum packet rate attribute is defined as

```
a=maxprate:<packet-rate> ; see Section 12.4.6.6 for ABNF definition.
```

The <packet-rate> is a floating-point value for the stream's maximum packet rate in packets per second. If the number of packets is variable, the given value SHALL be the maximum the application can produce in case of a live stream, or for stored on-demand streams, has produced. The packet rate is calculated by adding the number of packets sent within a 1-second window. The maxprate is the largest value produced when the window slides over the entire media stream. In cases for which this can't be calculated (i.e., a live stream), an estimated value of the maximum packet rate the codec can produce for the given configuration and content SHALL be used.

Note: The sliding window calculation will always yield an integer number. However, the attributes field is a floating-point value because the estimated or known maximum packet rate per second may be fractional.

At the SDP session level, the "maxprate" value is the maximum packet rate calculated over all the declared media streams. If this can't be measured (stored media) or estimated (live), the sum of all media level values provides a ceiling value. Note: the value at session level can be less than the sum of the individual media streams due to temporal distribution of media stream's maximums. The "maxprate" attribute MUST NOT be present at the session level if the media streams use different transport. The attribute MAY be present if the media streams use the same transport. If the attribute is present at the session level, it SHOULD also be present at the media level for all media streams. "maxprate" SHALL be included for all transports for which a packet rate can be derived and TIAS is included. For example, if you use TIAS and a transport like IP/UDP/RTP, for which the max packet rate (actual or estimated) can be derived, then "maxprate" SHALL be included. However, if either (a) the packet rate for the transport cannot be derived or (b) TIAS is not included, then, "maxprate" is not required to be included.

12.4.6.4 Converting to Transport-Dependent Values

When converting the transport-independent bandwidth value (bw-value) into a transport-dependent value including the lower layers, the following MUST be done:

1. Determine which lower layers will be used and calculate the sum of the sizes of the headers in bits (h-size). In cases of variable header sizes, the average size SHALL be used. For RTP-transported media, the lower layers SHALL include the RTP header with header extensions, if used, the CSRC list, and any profile-specific extensions.
2. Retrieve the maximum packet rate from the SDP (prate = maxprate).
3. Calculate the transport overhead by multiplying the header sizes by the packet rate (t over = h-size * prate).
4. Round the transport overhead up to nearest integer in bits (t-over = CEIL(t-over)).
5. Add the transport overhead to the transport independent bandwidth value (total bit rate = bw-value + t-over) When this calculation is performed using the "maxprate," the bit-rate value will be the absolute maximum the media stream may use over the transport assumed in the calculations.

12.4.6.5 Deriving RTCP Bandwidth

This chapter does not solve the fairness and possible bit-rate change introduced by IPv4 to IPv6 translation. These differences are considered small enough, and known solutions introduce code changes to the RTP/RTCP implementation. This section provides a consistent way of calculating the bit rate to assign to RTCP, if not explicitly given. First, the transport-dependent RTP session bit rate is calculated, in accordance with Section 12.4.6.4, using the actual transport layers used at the endpoint where the calculation is done. The RTCP bit rate is then derived as usual based on the RTP session bandwidth (i.e., normally equal to 5% of the calculated value).

12.4.6.5.1 Motivation for this Solution

Giving the exact same RTCP bit-rate value to both the IPv4 and IPv6 hosts will result in the IPv4 host having a higher RTCP sending rate. The sending rate represents the number of RTCP packets sent during a given time interval. The sending of RTCP is limited according to rules defined in the RTP specification (RFC 3550). For a 100-byte RTCP packet (including UDP/IPv4), the IPv4 sender has an approximately 20% higher sending rate. This rate falls with larger RTCP packets. For example, 300-byte packets will only give the IPv4 host a 7% higher sending rate.

The rule for deriving RTCP bandwidth gives the same behavior as fixed assignment when the RTP session has traffic parameters giving a large TIAS/maxprate ratio. The two hosts will be fair when the TIAS/maxprate ratio is approximately 40 bytes/packet, given 100-byte RTCP packets. For a TIAS/maxprate ratio of 5 bytes/packet, the IPv6 host will be allowed to send approximately 15%–20% more RTCP packets.

The larger the RTCP packets become, the more they will favor the IPv6 host in its sending rate. The conclusions is that, within the normal useful combination of transport-independent bit rates and packet rates, the difference in fairness between hosts on different IP versions with different overhead is acceptable. For the 20-byte difference in overhead between IPv4 and IPv6 headers, the RTCP bandwidth actually used in a unicast connection case will not be larger than approximately 1% of the total session bandwidth.

12.4.6.6 ABNF Definitions

All SDP ABNF syntaxes are defined in Section 2/2.9. However, we have repeated this here for convenience. This chapter defines in ABNF from RFC 2234 that was made obsolete by RFC 4234 (again, RFC 4234 was made obsolete by RFC 5324) the bandwidth modifier and the packet rate attribute. The bandwidth modifier:

```
TIAS-bandwidth-def      =    "b" "=" "TIAS" ":" bandwidth-value CRLF
bandwidth-value         =    1*DIGIT
```

The maximum packet rate attribute:

```
max-p-rate-def          =    "a" "=" "maxprate" ":" packet-rate CRLF
packet-rate             =    1*DIGIT ["." 1*DIGIT]
```

12.4.6.7 Example

```
v=0
o=Example_SERVER 3413526809 0 IN IP4 server.example.com
s=Example of TIAS and maxprate in use
c=IN IP4 0.0.0.0
b=AS:60
b=TIAS:50780
t=0 0
a=control:rtsp://server.example.com/media.3gp
a=range:npt=0-150.0
a=maxprate:28.0
m=audio 0 RTP/AVP 97
b=AS:12
b=TIAS:8480
a=maxprate:10.0
a=rtpmap:97 AMR/8000
a=fmtp:97 octet-align;
a=control:rtsp://server.example.com/media.3gp/trackID=1
m=video 0 RTP/AVP 99

b=AS:48
b=TIAS:42300
a=maxprate:18.0
a=rtpmap:99 MP4V-ES/90000
a=fmtp:99 profile-level-id=8;
config=000001B008000001B5090000010100000120008840066 82C2090A21F
a=control:rtsp://server.example.com/media.3gp/trackID=3
```

In this SDP example of a streaming session's SDP, there are two media streams, one audio stream encoded with AMR and one video stream encoded with the MPEG-4 Video encoder. AMR is used here to produce a constant rate media stream and uses a packetization resulting in 10 packets per second. This results in a TIAS bandwidth rate of 8480 bits per second and the claimed 10 packets per second. The video stream is more variable. However, it has a measured maximum payload rate of 42,300 bits per second. The video stream also has a variable packet rate, despite the fact that the video is 15 frames per second, where at least one instance in a second-long window contains 18 packets.

12.4.7 Protocol Interaction

12.4.7.1 Real-Time Streaming Protocol

The "TIAS" and "maxprate" parameters can be used with RTSP as currently specified. To be able to calculate the transport-dependent bandwidth, some of the transport header parameters will be required. It should be no problem for a client to calculate the required bandwidth(s) prior to an RTSP SETUP. The reason is that a client supports a limited number of transport setups. The one actually offered to a server in a SETUP request will be dependent on the contents of the SDP description. The "m=" line(s) will signal the desired transport profile(s) to the client.

12.4.7.2 Session Initiation Protocol

The usage of "TIAS" together with "maxprate" should not be different from the handling of the "AS" modifier currently in use. The needed transport parameters will be available in the transport field in the "m=" line. The address class can be determined from the "c=" field and the client's connectivity.

12.4.7.3 Session Announcement Protocol

In the case of SAP, all available information to calculate the transport dependent bit rate should be present in the SDP. The "c=" information gives the address family used for the multicast. The transport layer (e.g., RTP/UDP) for each media is evident in the media line ("m=") and its transport field.

12.4.8 Security Consideration

The security procedures that need to be considered in dealing with the basic SDP messages (RFC 3890) described here are provided in Section 17.11.4.

12.4.9 IANA Considerations

SDP parameters that are described here have been registered with IANA. We have not included the IANA registration procedures here for the sake of brevity. The procedures for registration of new SDP parameters are described in RFC 3890.

12.5 Connectivity Preconditions for SDP Media Streams

This section describes RFC 5898 that defines a new connectivity precondition for the SDP precondition framework. A connectivity precondition can be used to delay session establishment or modification until media stream connectivity has been successfully verified. The method of verification may vary depending on the type of transport used for the media. For unreliable datagram transports such as UDP, verification involves probing the stream with data or control packets. For reliable connection-oriented transports such as TCP, verification can be achieved simply by successful connection establishment or by probing the connection with data or control packets, depending on the situation.

12.5.1 Introduction

The concept of a SDP (RFC 4566, see Section 2) precondition in the SIP (RFC 3261, also see Section 1.2) is defined in RFC 3312 (updated by RFCs 4032 and 5027). A precondition is a condition that has to be satisfied for a given media stream in order for session establishment or modification to proceed. When the precondition is not met, session progress is delayed until the precondition is satisfied or the session establishment fails. For example, RFC 3312 defines the QOS precondition, which is used to ensure availability of network resources prior to establishing a session (i.e., prior to starting to alert the callee).

SIP sessions are typically established in order to set up one or more media streams. Even though a media stream may be negotiated successfully through an SDP offer/answer exchange, the actual media stream itself may fail. For example, when there is one or more NATs or firewalls in the media path, the media stream may not be received by the far end. In cases where the media is carried over a connection-oriented transport such as TCP (RFC 0793), the connection-establishment procedures may fail. The connectivity precondition defined in this document ensures that session progress is delayed until media stream connectivity has been verified. The connectivity precondition type defined in this document follows the guidelines provided in RFC 4032 to extend the SIP preconditions framework.

12.5.2 Terminology

See Table 2.1.

12.5.3 Connectivity Precondition Definition

12.5.3.1 Syntax

The connectivity precondition type is defined by the string "conn"; hence, we modify the grammar found in RFC 3312 and RFC 5027 as follows: precondition-type = "conn" / "sec" / "qos" / token. This precondition tag is registered with the IANA in Section 12.5.8.

12.5.3.2 Operational Semantics

According to RFC 4032, documents defining new precondition types need to describe the behavior of user agents (UAs) from the moment session establishment is suspended due to a set of preconditions until it is resumed when these preconditions are met. An entity that wishes to delay session establishment or modification until media stream connectivity has been established uses this precondition type in an offer. When a mandatory connectivity precondition is received in an offer, session establishment or modification is delayed until the connectivity precondition has been met (i.e., until media stream connectivity has been established in the desired direction or directions).

The delay of session establishment defined here implies that alerting the called party does not occur until the precondition has been satisfied. Packets may be both sent and received on the media streams in question. However, such packets SHOULD be limited to packets that are necessary to verify connectivity between the two endpoints involved on the media stream. That is, the underlying media stream SHOULD NOT be cut through. For example, interactive connectivity establishment (ICE) connectivity checks (RFC 5245, see Section 3.2) and TCP SYN, SYNACK, and ACK packets can be exchanged on media streams that support them as a way of verifying connectivity.

Some media streams are described by a single "m=" line but involve multiple addresses. For example, RFC 5109 specifies how to send forward error correction (FEC) information as a separate stream (the address for the FEC stream is provided in an “a=fmtp” line). When a media stream consists of multiple destination addresses, connectivity to all of them MUST be verified in order for the precondition to be met. In the case of RTP media streams (RFC 3550) that use RTCP, connectivity MUST be verified for both RTP and RTCP; the RTCP transmission interval rules MUST still be adhered to.

12.5.3.3 Status Type

RFC 3312 defines support for two kinds of status types, namely, segmented and end to end. The connectivity precondition type defined here MUST be used with the end-to-end status type; use of the segmented status type is undefined.

12.5.3.4 Direction-Tag

The direction attributes defined in RFC 3312 are interpreted as follows:

- **send:** the party that generated the session description is sending packets on the media stream to the other party, and the other party has received at least one of those packets. That is, there is connectivity in the forward (sending) direction.
- **recv:** the other party is sending packets on the media stream to the party that generated the session description, and this party has received at least one of those packets. That is, there is connectivity in the backward (receiving) direction.
- **sendrecv:** both the send and recv conditions hold. Note that a "send" connectivity precondition from the offerer’s point of view corresponds to a "recv" connectivity precondition from the answerer’s point of view, and vice versa. If media stream connectivity in both directions is required before session establishment or modification continues, the desired status needs to be set to "sendrecv."

12.5.3.5 Precondition Strength

Connectivity preconditions may have a strength-tag of either "mandatory" or "optional." When a mandatory connectivity precondition is offered and the answerer cannot satisfy the connectivity precondition (e.g., because the offer does not include parameters that enable connectivity to be verified without media cut through), the offer MUST be rejected as described in RFC 3312. When an optional connectivity precondition is offered, the answerer MUST generate its answer SDP as soon as possible. Since session progress is not delayed in this case, it is not known whether the associated media streams will have connectivity. If the answerer wants to delay session progress until connectivity has been verified, the answerer MUST increase the strength of the connectivity precondition by using a strength-tag of "mandatory" in the answer.

Note that use of a mandatory precondition requires the presence of a SIP "Require" header with the option-tag "precondition." Any SIP UA that does not support a mandatory precondition will reject such requests. To get around this issue, an optional connectivity precondition and the SIP "Supported" header with the option-tag "precondition" can be used instead. Offers with connectivity preconditions in re-INVITEs or UPDATEs follow the rules given in Section 6 of RFC 3312. That is: Both UAs SHOULD continue using the old session parameters until all the mandatory preconditions are met. At that moment, the UAs can begin using the new session parameters.

12.5.4 Verifying Connectivity

Media stream connectivity is ascertained by use of a connectivity verification mechanism between the media endpoints. A connectivity verification mechanism may be an explicit mechanism, such as ICE (RFC 5245, see Section 3.2) or ICE TCP (RFC 6544), or it may be an implicit mechanism, such as TCP. Explicit mechanisms provide specifications for when connectivity between two endpoints using an offer/answer exchange is ascertained, whereas implicit mechanisms do not. The verification mechanism is negotiated as part of the normal offer/answer exchange; however, it is not identified explicitly. More than one mechanism may be negotiated, but the offerer and answerer need not use the same. The following rules guide which connectivity verification mechanism to use:

- If an explicit connectivity verification mechanism (e.g., ICE) is negotiated, the precondition is met when the mechanism verifies connectivity successfully.
- Otherwise, if a connection-oriented transport (e.g., TCP) is negotiated, the precondition is met when the connection is established.
- Otherwise, if an implicit verification mechanism is provided by the transport itself or the media stream data using the transport, the precondition is met when the mechanism verifies connectivity successfully.
- Otherwise, connectivity cannot be verified reliably, and the connectivity precondition will never be satisfied if requested.

This document does not mandate any particular connectivity verification mechanism; however, in the following, we provide additional considerations for verification mechanisms.

12.5.4.1 Correlation of Dialog to Media Stream

SIP and SDP do not provide any inherent capabilities for associating an incoming media stream packet with a particular dialog. Thus, when an offerer is trying to ascertain connectivity, and an incoming media stream packet is received, the offerer may not know which dialog had its "recv" connectivity verified. Explicit connectivity verification mechanisms therefore typically provide a means to correlate the media stream, whose connectivity is being verified, with a particular SIP dialog. However, some connectivity verification mechanisms may not provide such a correlation. In the absence of a mechanism for the correlation of dialog to media stream (e.g., ICE), a UA Server (UAS) MUST NOT require the offerer to confirm a connectivity precondition.

12.5.4.2 Explicit Connectivity Verification Mechanisms

Explicit connectivity verification mechanisms typically use probe traffic with some sort of feedback to inform the sender of whether reception was successful. We provide two examples of such mechanisms and how they are used with connectivity preconditions: ICE (RFC 5245) provides one or more candidate addresses in signaling between the offerer and the answerer and then uses STUN Binding Requests to determine which pairs of candidate addresses have connectivity. Each STUN Binding Request contains a password that is communicated in the SDP as well; this enables correlation between STUN Binding Requests and candidate addresses for a particular media stream. It also provides correlation with a particular SIP dialog.

ICE implementations may be either full or lite (see RFC 5245, Section 3.2). Full implementations generate and respond to STUN Binding Requests, whereas lite implementations only respond to them. With ICE, one side is a controlling agent, and the other side is a controlled agent. A full implementation can take on either role, whereas a lite implementation can only be a controlled agent. The controlling agent decides which valid candidate to use and informs the controlled agent of it by identifying the pair as the nominated pair. This leads to the following connectivity precondition rules:

- A full implementation ascertains both "send" and "recv" connectivity when it operates as a STUN client and has sent a STUN Binding Request that resulted in a successful check for all the components of the media stream (as defined further in ICE).
- A full or a lite implementation ascertains "recv" connectivity when it operates as a STUN server and has received a STUN Binding Request that resulted in a successful response for all the components of the media stream (as defined further in ICE).
- A lite implementation ascertains "send" and "recv" connectivity when the controlling agent has informed it of the nominated pair for all the components of the media stream.

A simpler and slightly more delay-prone alternative to these rules is for all ICE implementations to ascertain "send" and "recv" connectivity for a media stream when the ICE state for that media stream has moved to Completed. Note that there is never a need for the answerer to request confirmation of the connectivity precondition when using ICE: The answerer can determine the status locally. Also, note that when ICE is used to verify connectivity preconditions, the precondition is not satisfied until connectivity has been verified for all the component transport addresses used by the media stream. For example, with an RTP-based media stream where RTCP is not suppressed, connectivity MUST be ascertained for both RTP and RTCP.

Finally, it should be noted that although connectivity has been ascertained, a new offer/answer exchange may be required before media can flow (per ICE). The above are merely examples of explicit connectivity verification mechanisms. Other techniques can be used as well. It is, however, RECOMMENDED that ICE be supported by entities that support connectivity preconditions. Use of ICE has the benefit of working for all media streams (not just RTP) as well as facilitating NAT and firewall traversal, which may otherwise interfere with connectivity. Furthermore, the ICE recommendation provides a baseline to ensure that all entities that require probe traffic to support the connectivity preconditions have a common way of ascertaining connectivity.

12.5.4.3 Verifying Connectivity for Connection-Oriented Transports

Connection-oriented transport protocols generally provide an implicit connectivity verification mechanism. Connection establishment involves sending traffic in both directions, thereby verifying connectivity at the transport-protocol level. When a three-way (or more) handshake for connection establishment succeeds, bidirectional communication is confirmed and both the "send" and "recv" preconditions are satisfied, regardless of whether they are requested. In the case of TCP, for example, once the TCP three-way handshake has completed (SYN, SYN-ACK, ACK), the TCP connection is established and data can be sent and received by either party (i.e., both a send and a receive connectivity precondition has been satisfied). Stream Control Transmission Protocol (SCTP) (RFC 4960) connections have similar semantics as TCP and SHOULD be treated the same.

When a connection-oriented transport is part of an offer, it may be passive, active, or active/passive (RFC 4145, see Section 10.1). When it is passive, the offerer expects the answerer to initiate the connection establishment, and when it is active, the offerer wants to initiate the connection establishment. When it is active/passive, the answerer decides. As noted earlier, lack of a media-stream-to-dialog correlation mechanism can make it difficult to guarantee with whom connectivity has been ascertained. When the offerer takes on the passive role, the offerer will not necessarily know which SIP dialog originated an incoming connection request. If the offerer instead is active, this problem is avoided.

12.5.5 Connectivity and Other Precondition Types

The role of a connectivity precondition is to ascertain media stream connectivity before establishing or modifying a session. The underlying intent is for the two parties to be able to exchange media packets successfully. However, connectivity by itself may not fully satisfy this. QOS, for example, may be required for the media stream; this can be addressed by use of the "qos" precondition defined in RFC 3312. Successful security parameter negotiation may be another prerequisite; this can be addressed by use of the "sec" precondition defined in RFC 5027 (see Section 16.4).

12.5.6 Examples

The first example uses the connectivity precondition with TCP in the context of a session involving a wireless access medium. Both UAs use a radio access network that does not allow them to send any data (not even a TCP SYN) until a radio bearer has been set up for the connection. Figure 12.3 shows the message flow of this example (the required PRACK transaction has been omitted for clarity, see RFC 3312 for details):

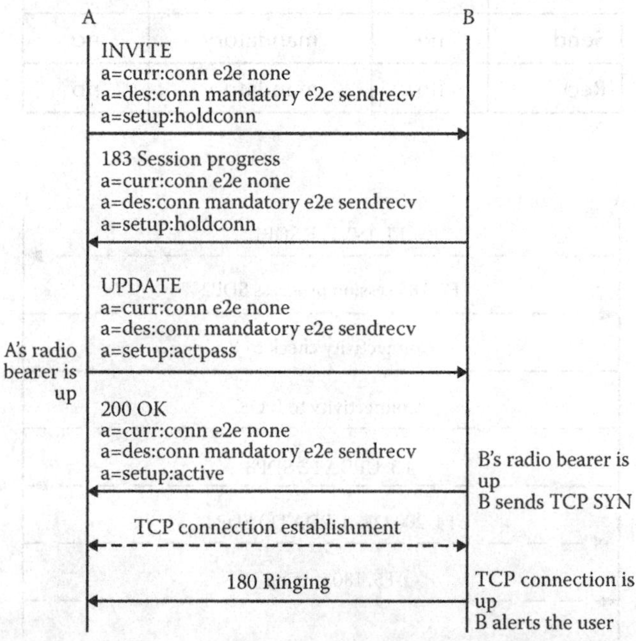

Figure 12.3 Message flow with two types of preconditions (Copyright IETF).

A sends an INVITE requesting connection-establishment preconditions. The setup attribute in the offer is set to hold connection (holdconn) (RFC 4145, see Section 10.1) because A cannot send or receive any data before setting up a radio bearer for the connection. B agrees to use the connectivity precondition by sending a 183 (Session Progress) response. The setup attribute in the answer is also set to hold connection (holdconn) because B, like A, cannot send or receive any data before setting up a radio bearer for the connection. When A's radio bearer is ready, A sends an UPDATE to B with a setup attribute with a value of actpass. This attribute indicates that A can perform an active or a passive TCP open. A is letting B choose which endpoint will initiate the connection.

Since B's radio bearer is not ready yet, B chooses to be the one initiating the connection and indicates this with a setup attribute with a value of active. At a later point, when B's radio bearer is ready, B initiates the TCP connection toward A. Once the TCP connection is established successfully, B knows the "sendrecv" precondition is satisfied, and B proceeds with the session (i.e., alerts the Callee) and sends a 180 (Ringing) response. The second example shows a basic SIP session establishment using SDP connectivity preconditions and ICE (the required PRACK transaction and some SDP details have been omitted for clarity). The offerer (A) is a full ICE implementation whereas the answerer (B) is a lite ICE implementation. The message flow for this scenario is shown in Figure 12.4.

SDP1: A includes a mandatory end-to-end connectivity precondition with a desired status of "sendrecv"; this will ensure media stream connectivity in both directions before continuing with the session setup. Since media stream connectivity in either direction is unknown at this point, the current status is set to "none." A's local status table (see RFC 3312) for the connectivity precondition is as follows:

Direction	*Current*	*Desired Strength*	*Confirm*
Send	no	mandatory	no
Recv	no	mandatory	no

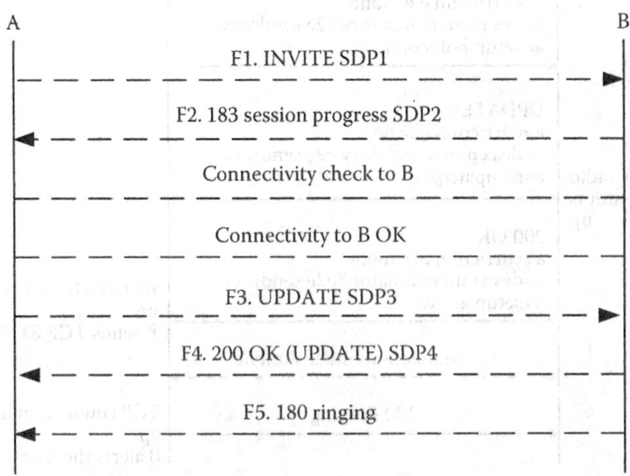

Figure 12.4 Connectivity precondition with ICE connectivity checks (Copyright IETF).

and the resulting offer SDP is

```
a=ice-pwd:asd88fgpdd777uzjYhagZg
a=ice-ufrag:8hhY
m=audio 20000 RTP/AVP 0
c=IN IP4 192.0.2.1
a=rtcp:20001
a=curr:conn e2e none
a=des:conn mandatory e2e sendrecv
a=candidate:1 1 UDP 2130706431 192.0.2.1 20000 typ host
```

SDP2: When B receives the offer, B sees the mandatory sendrecv connectivity precondition. B is a lite ICE implementation and hence B can only ascertain "recv" connectivity (from B's point of view) from A; thus, B wants A to inform it about connectivity in the other direction ("send" from B's point of view). B's local status table therefore looks as follows:

Direction	*Current*	*Desired Strength*	*Confirm*
send	no	mandatory	no
recv	no	mandatory	no

Since B is a lite ICE implementation and B wants to ask A for confirmation about the "send" (from B's point of view) connectivity precondition, the resulting answer SDP becomes

```
a=ice-lite
a=ice-pwd:qrCA8800133321zF9AIj98
a=ice-ufrag:H92p
m=audio 30000 RTP/AVP 0
c=IN IP4 192.0.2.4
a=rtcp:30001
a=curr:conn e2e none
a=des:conn mandatory e2e sendrecv
a=conf:conn e2e send
a=candidate:1 1 UDP 2130706431 192.0.2.4 30000 typ host
```

Since the "send" and the "recv" connectivity precondition (from B's point of view) are still not satisfied, session establishment remains suspended.

SDP3: When A receives the answer SDP, A notes that B is a lite ICE implementation and that confirmation was requested for B's "send" connectivity precondition, which is the "recv" precondition from A's point of view. A performs a successful send and recv connectivity check to B by sending an ICE connectivity check to B and receiving the corresponding response. A's local status table becomes

Direction	*Current*	*Desired Strength*	*Confirm*
send	yes	mandatory	no
recv	yes	mandatory	yes

whereas B's local status table becomes

Direction	*Current*	*Desired Strength*	*Confirm*
send	No	mandatory	no
recv	Yes	mandatory	no

Since B asked for confirmation about the "recv" connectivity (from A's point of view), A now sends an UPDATE (5) to B to confirm the connectivity from A to B:

```
a=ice-pwd:asd88fgpdd777uzjYhagZg
a=ice-ufrag:8hhY
m=audio 20000 RTP/AVP 0
c=IN IP4 192.0.2.1
a=rtcp:20001
a=curr:conn e2e sendrecv
a=des:conn mandatory e2e sendrecv
a=candidate:1 1 UDP 2130706431 192.0.2.1 20000 typ host
```

B knows it has recv connectivity (verified by ICE as well as A's UPDATE) and send connectivity (confirmed by A's UPDATE) at this point. B's local status table becomes

Direction	*Current*	*Desired Strength*	*Confirm*
Send	yes	mandatory	no
Recv	yes	mandatory	no

and the session can continue.

12.5.7 Security Considerations

The security procedures that need to be considered in dealing with the basic SDP messages described in this section (RFC 5898) are provided in Section 17.11.5.

12.5.8 IANA Considerations

SDP parameters that are described here have been registered with IANA. We have not included the IANA registration procedures here for the sake of brevity. The procedures for registration of new SDP parameters are described in RFC 5898.

12.6 Summary

We have described the support of application layer QOS of real-time, near-real-time, and nonreal-time multimedia applications in SDP for negotiations of QOS parameters between the conferencing parties. First, we have explained a mechanism that allows endpoints to negotiate the QOS mechanism to be used for a particular media stream. The SDP offer/answer model is described and how to take care of every performance parameter that needs to be negotiated. Second, two additional modifiers for the bandwidth attribute are defined in the SDP to specify the bandwidth allowed for

RTCP packets in the RTP session. Third, we have addressed another problem in SDP where the ports are not assigned in consecutive numbers due to provisioning of multiple ports, for example, when endpoints are behind the NATs. This problem is solved in extending the RTCP attributes that allow one to discover the medial-level multiple port numbers, and SDP can use these RTCP attributes for port numbers. Fourth, we have explained in detail the SDP extensions for the transport independent bandwidth modifier attributes. We have described the bandwidth for conference total, AS maximum, and RTCP reported bandwidth. Moreover, the change in bandwidth utilization is described, taking the example of IPsec and header compression. The TIAS bandwidth modifier is defined in extending the SDP for taking care of these bandwidth attributes. Moreover, RTSP, SIP, and SAP protocol are described in support of using QOS parameters for negotiations using SDP.

Finally, the connectivity precondition has been defined in SDP to take care of the fact that session establishment can be delayed until the time bandwidth requirements are met. This is a very useful capability for implementation of QOS for multimedia applications (e.g., SIP, SAP, and RTSP). The SDP connectivity precondition attribute includes the syntax, operational semantics, status type, direction-tag, and precondition strength. Moreover, the mechanism allows one to verify the preconditions using attributes like correlation of dialog to media stream and explicit connectivity verification. However, all the support of QOS mechanisms in SDP is done in the application layer between the endpoints. The key issue of network layer QOS needs to be addressed separately. For example, connectivity preconditions attribute defined in SDP is an important capability to make sure that both higher application and lower network layer QOS are in sync for end-to-end connectivity between end users.

12.7 Problems

1. What QOS mechanisms are standardized over the IP network? Describe briefly the characteristics of those mechanisms including their pros and cons.
2. Explain with detail call flows how the SDP "qos-mech" token supports the IP QOS mechanisms for negotiations of QOS between the conferencing parties including offer/answer behavior.
3. If the standardized IP QOS mechanisms are extended, how do the SDP attributes defined in RFC 5432 support this? What are the possible security concerns of SDP QOS attributes? What are the recommended approaches to prevent attacks on QOS attributes?
4. What RTCP bandwidth modifier attributes are defined in RFC 3556? How do the SDP bandwidth modifiers help to define RTCP bandwidth? If the transmitter and receiver bandwidth are not defined, how does the default parameter take care of bandwidth? Explain in detail using call flows.
5. What has been the historical reason for SDP to assume that the media-stream port numbers are consecutive? Why do the NATs fail to follow this rule when media streams pass through them? Explain in detail using examples.
6. How does RFC 3605 help to specify RTCP ports explicitly in SDP through its extensions? What are differences between media- and session-level attributes? Why is it that RTCP attributes MUST NOT be considered as session-level attributes?
7. Provide a detailed example using RTCP attributes in SDP with offer/answer models assuming conferencing parties are behind the NATs.
8. What is the purpose of RFC 3890 that defines the TIAS bandwidth modifier for the SDP? What are the conference total bandwidth, application specific maximum bandwidth in

RTP, and RTCP report bandwidth? How does the bandwidth differ in IPv4, IPv6, and dual-stack IPv4/IPv6? How does the bandwidth change in IPsec and header compression?

9. What are the problems for signaling of bandwidth in SDP described in RFC 3890 related to IP, DCCP, and other mechanisms such as header compression, tunneling, IPsec, converted bandwidth values, and RTCP? What is the conclusion of these problems in terms of considering the bandwidth in SDP? How does this RFC scope the problem?
10. How does RFC 3890 solve the problems of expressing bandwidth in SDP, which takes care of all possible combinations of transport network?
11. Explain in detail the TIAS bandwidth modifier, including usage, definition, and usage rules.
12. How does RFC 3890 provide solutions for taking care of bandwidth in SDP for packet rate parameter, converting to transport-dependent values, and deriving RTCP bandwidth? Show usages of these parameters for RSTP, SIP, and SAP call-control signaling protocol.
13. Why do we need connectivity preconditions for SDP media streams? How do they differ in application between connectionless (e.g., UDP) and connection-oriented (e.g., TCP) transport? Define in detail the connectivity preconditions, including syntax, operational semantics, status type, direction-tag, and precondition strength.
14. Explain in detail using call flows how the connectivity is verified as defined in RFC 5898 including correlation of dialog to media stream, explicit connectivity verification, and verifying connectivity for both connection-oriented and connectionless transport.
15. Describe in detail the differences between the connectivity preconditions defined in RFC 5898, the "qos" precondition defined in RFC 3312, and the "sec" (i.e., security) precondition defined in RFC 5027.

Chapter 13

Application-Specific Extensions in SDP

Abstract
This section describes Requests for Comments (RFCs) 5547 and 4583 that deal with application-specific extensions in Session Description Protocol (SDP). The multimedia applications that are used among conference participants have certain application- and media-level capabilities and features. These specific features need to be agreed upon between the conference participants. We have considered two applications, file (e.g., text, still images, animation, audio, and video) transfer and binary floor control application protocol, that have been standardized in the Internet Engineering Task Force (IETF). In this chapter, we describe a mechanism specified in RFC 5547 to negotiate the transfer of one or more files between two endpoints by using the SDP offer/answer model specified in RFC 3264. SDP is extended to describe the attributes of the files to be transferred. The offerer can describe either the files it wants to send or the files it would like to receive. The answerer can either accept or reject the offer separately for each individual file. The transfer of one or more files is initiated after a successful negotiation. The Message Session Relay Protocol (MSRP) is defined as the default mechanism to actually carry the files between the endpoints. The conventions on how to use MSRP for file transfer are also provided in this chapter. In addition, this chapter specifies SDP syntaxes defined in RFC 4583 for supporting Binary Floor Control Protocol (BFCP) streams in SDP descriptions. User agents using the offer/answer model to establish BFCP streams use this format in their offers and answers.

13.1 SDP Offer/Answer Mechanism to Enable File Transfer

This section describes Request for Comments (RFC) 5547 that provides a mechanism to negotiate the transfer of one or more files between two endpoints by using the Session Description Protocol (SDP) offer/answer model. SDP is extended to describe the attributes of the files to be transferred. The offerer can describe either the files it wants to send or the files it would like to receive. The answerer can either accept or reject the offer separately for each individual file. The transfer of one or more files is initiated after a successful negotiation. The Message Session Relay Protocol (MSRP) is defined as the default mechanism to actually carry the files between endpoints. The conventions on how to use MSRP for file transfer are provided in this section.

13.1.1 Introduction

The SDP offer/answer (RFC 3264, see Section 3.1) provides a mechanism for two endpoints to arrive at a common view of a multimedia session. These sessions often contain real-time media streams such as voice and video but are not limited to those. Any media component type can be supported, as long as there is a specification on how to negotiate it within the SDP offer/answer exchange. The MSRP (RFC 4975) is a protocol for transmitting instant messages (IMs) in the context of a session. The protocol specification describes the usage of SDP for establishing an MSRP session. In addition to plain text messages, MSRP is able to carry arbitrary (binary) Multipurpose Internet Mail Extensions (MIMEs) (RFC 2045) compliant content, such as images or video clips.

There are many cases in which the endpoints involved in a multimedia session would like to exchange files within that session. With MSRP, it is possible to embed files as MIME objects inside the stream of IMs. MSRP has other features that are useful for file transfer. Message chunking enables the sharing of the same transport connection between the transfer of a large file and interactive IM exchange without blocking the IM. MSRP relays (RFC 4976) provide a mechanism for network address translator (NAT) traversal. Finally, Secure MIME (S/MIME) (RFC 3851 was made obsolete by RFC 5751) can be used for ensuring the integrity and confidentiality of the transferred content. However, the baseline MSRP does not readily meet all the requirements for file transfer services within multimedia sessions. There are four main missing features:

- The recipient must be able to distinguish "file transfer" from "file attached to IM," allowing the recipient to treat the cases differently.
- It must be possible for the sender to send the request for a file transfer. It must be possible for the recipient to accept or decline it, using the meta information in the request. The actual transfer must take place only after acceptance by the recipient.
- It must be possible for the sender to pass some meta information on the file before the actual transfer. This must be able to include at least content type, size, hash, and name of the file, as well as a short (human-readable) description.
- It must be possible for the recipient to request a file from the sender, providing meta information about the file. The sender must be able to decide whether to send a file matching the request.

The rest of this document is organized as follows. Section 13.1.3 defines a few terms used in this document. Section 13.1.4 provides the overview of operation. Section 13.1.5 introduces the concept of the file selector. The detailed syntax and semantics of the new SDP attributes and conventions on how the existing ones are used are defined in Section 13.1.6. Section 13.1.7 discusses the file disposition types. Section 13.1.8 describes the protocol operation involving SDP and MSRP. Finally, some examples are given in Section 13.1.9.

13.1.2 Terminology

See Table 2.1.

13.1.3 Definitions

For the purpose of this document, the following definitions specified in RFC 3264 (see Sections 2.2 for definitions and 3.1 for RFC 3264) apply:

- Answer
- Answerer
- Offer
- Offerer

Additionally, see Table 2.1.

13.1.4 Overview of Operation

An SDP offerer creates an SDP body that contains the description of one or more files the offerer wants to send or receive. The offerer sends the SDP offer to the remote endpoint. The SDP answerer can accept or reject the transfer of each of those files separately. The actual file transfer is carried out using the MSRP (RFC 4975). Each SDP "m=" line describes an MSRP media stream used to transfer a single file at a time. That is, the transfer of multiple simultaneous files requires multiple "m=" lines and corresponding MSRP media streams. It should be noted that multiple MSRP media streams can share a single transport layer connection, so this mechanism will not lead to excessive use of transport resources.

Each "m=" line for an MSRP media stream is accompanied by a few attributes describing the file to be transferred. If the file sender generates the SDP offer, the attributes describe a local file to be sent (push), and the file receiver can use this information to either accept or reject the transfer. However, if the SDP offer is generated by the file receiver, the attributes are intended to characterize a particular file the receiver is willing to get (pull) from the sender. It is possible that the sender does not have a matching file or does not want to send the file, in which case the offer is rejected.

The attributes describing each file are provided in SDP by a set of new SDP attributes, most of which have been directly borrowed from MIME. This way, user agents can decide whether to accept a given file transfer based on the file's name, size, description, hash, icon (e.g., if the file is a picture), etc. SDP direction attributes (e.g., "sendonly," "recvonly") are used to indicate the direction of the transfer (i.e., whether the SDP offerer is willing to send or receive the file). Assuming that the answerer accepts the file transfer, the actual transfer of the files takes place with ordinary MSRP. Note that the "sendonly" and "recvonly" attributes refer to the direction of MSRP SEND requests and do not preclude other protocol elements (e.g., 200 responses, REPORT requests, etc.).

In principle the file transfer can work even with an endpoint supporting only regular MSRP without understanding the extensions defined herein, in a particular case where that endpoint is both the SDP answerer and the file receiver. The regular MSRP endpoint answers the offer as it would answer any ordinary MSRP offer without paying attention to the extension attributes. In such a scenario, the user experience would, be reduced, since the recipient would not know (by any protocol means) the reason for the session and would not be able to accept/reject it based on the file attributes.

13.1.5 File Selector

When the file receiver generates the SDP offer, this SDP offer needs to unambiguously identify the requested file at the file sender. For this purpose, we introduce the notion of a file selector, which is a tuple composed of one or more of the following individual selectors: the name, size, type, and hash of the file. The file selector can include any number of selectors, so all four of them do not always need to be present. The purpose of the file selector is to provide enough information about the file to the remote entity, so that both the local and the remote entity can refer to the same file. The file selector is encoded in a "file-selector" media attribute in SDP. The formal syntax of the "file-selector" media attribute is described in Figure 13.1.

```
attribute            =  / file-selector-attr / file-disp-attr /
                        file-tr-id-attr / file-date-attr /
                        file-icon-attr / file-range-attr
                        ; attribute is defined in RFC 4566 (see Section 2/2.9)

file-selector-attr   =  "file-selector" [":" selector *(SP selector)]
selector             =  filename-selector / filesize-selector /
                        filetype-selector / hash-selector

filename-selector    =  "name:" DQUOTE filename-string DQUOTE
                        ; DQUOTE defined in RFC 5234

filename-string      =  1*(filename-char/percent-encoded)

filename-char        =  %x01-09/%x0B-0C/%x0E-21/%x23-24/%x26-FF
                        ; any byte except NUL, CR, LF,
                        ; double quotes, or percent

percent-encoded      =  "%" HEXDIG HEXDIG
                        ; HEXDIG defined in RFC 5234

filesize-selector    =  "size:" filesize-value
filesize-value       =  integer ;integer defined in RFC 4566
                        ; (see Section 2/2.9)
filetype-selector    =  "type:" type "/" subtype *(";" ft-parameter)

ft-parameter         =  attribute "=" DQUOTE value-string DQUOTE
                        ; attribute is defined in RFC 2045
                        ; free insertion of linear-white-space is not
                        ; permitted in this context.
                        ; note: value-string has to be re-encoded
                        ; when translating between this and a
                        ; Content-Type header.

value-string         =  filename-string
hash-selector        =  "hash:" hash-algorithm ":" hash-value
hash-algorithm       =  token ; see IANA Hash Function
                        ; Textual Names registry
                        ; only "sha-1" currently supported

hash-value           =  2HEXDIG *(":" 2HEXDIG)
                        ; Each byte in upper-case hex, separated
                        ; by colons.
                        ; HEXDIG defined in RFC 5234

file-tr-id-attr      =  "file-transfer-id:" file-tr-id-value
file-tr-id-value     =  token
file-disp-attr       =  "file-disposition:" file-disp-value
file-disp-value      =  token
file-date-attr       =  "file-date:" date-param *(SP date-param)
date-param           =  c-date-param / m-date-param / r-date-param
c-date-param         =  "creation:" DQUOTE date-time DQUOTE
```

Figure 13.1 Syntax of the SDP extension.

```
m-date-param      =    "modification:" DQUOTE date-time DQUOTE

r-date-param      =    "read:" DQUOTE date-time DQUOTE
                       ; date-time is defined in RFC 5322
                       ; numeric timezones (+HHMM or -HHMM)
                       ; must be used
                       ; DQUOTE defined in RFC 5234 files.
file-icon-attr    =    "file-icon:" file-icon-value
file-icon-value   =    cid-url ; cid-url defined in RFC 2392
file-range-attr   =    "file-range:" start-offset "-" stop-offset
start-offset      =    integer ; integer defined in RFC 4566
                       ; (see Section 2/2.9)
stop-offset       =    integer / "*"
```

Figure 13.1 (CONTINUED) Syntax of the SDP extension.

The file selection process is applied to all the available files at the host. The process selects those files that match each of the selectors present in the "file-selector" attribute. The result can be zero, one, or more files, depending on the presence of the mentioned selectors in the SDP and depending on the available files in a host. The file transfer mechanism specified in this document requires that a file selector eventually results at most in a single file to be chosen. Typically, if the hash selector is known, it is enough to produce a file selector that points to exactly zero or one file. However, a file selector that selects a unique file is not always known by the offerer. Sometimes only the name, size, or type of file is known, so the file selector may result in selecting more than one file, which is an undesired case. The opposite is also true: If the file selector contains a hash selector and a name selector, there is a risk that the remote host has renamed the file, in which case, although a file whose computed hash equals the hash selector exists, the file name does not match that of the name selector. In this case, the file selection process will result in the selection of zero files.

This specification uses the Secure Hash Algorithm 1 (SHA-1) (RFC 3174). If future needs require adding support for different hashing algorithms, they will be specified as extensions to this document. Implementations according to this specification MUST implement the "file-selector" attribute and MAY implement any of the other attributes specified in this specification. SDP offers and answers for file transfer MUST contain a "file-selector" media attribute that selects the file to be transferred and MAY contain any of the other attributes specified in this specification. The "file-selector" media attribute is also useful when learning the support of the file transfer offer/answer capability that this document specifies. This is further explained in Section 13.1.8.5.

13.1.6 Extensions to SDP

All SDP syntaxes are provided in Section 2/2.9. However, we have repeated this here for convenience. We define a number of new SDP (RFC 4566, see Section 2) attributes that provide the required information to describe the transfer of a file with MSRP. These are all media-level-only attributes in SDP. The following is the formal Augmented Backus-Naur Form (ABNF) syntax (RFC 5234) of these new attributes. It is built above the SDP (RFC 4566, see Section 2.9) grammar (RFC 2045, RFC 2183, RFC 2392, and RFC 5322).

When used for capability query (see Section 13.1.8.5), the "file-selector" attribute MUST NOT contain any selector, because its presence merely indicates compliance to this specification. When used in an SDP offer or answer, the "file-selector" attribute MUST contain at least one selector. Selectors characterize the file to be transferred. There are four selectors in this attribute: "name," "size," "type," and "hash." The "name" selector in the "file-selector" attribute contains the filename of the content enclosed in double quotes. The filename is encoded in UTF-8 (RFC 3629). Its value SHOULD be the same as the "filename" parameter of the Content-Disposition header field (RFC 2183) that would be signaled by the actual file transfer. If a file name contains double quotes or any other character that the syntax does not allow in the "name" selector, they MUST be percent encoded.

The "name" selector MUST NOT contain characters that can be interpreted as directory structure by the local operating system. If such characters are present in the file name, they MUST be percent encoded. Note that the "name" selector might still contain characters that, although not meaningful for the local operating system, might still be meaningful to the remote operating system (e.g., '\', '/', ':'). Therefore, implementations are responsible for sanitizing the input received from the remote endpoint before doing a local operation in the local file system, such as the creation of a local file. Among other things, implementations can percent-encode characters that are meaningful to the local operating system before doing file system local calls.

The "size" selector in the "file-selector" attribute indicates the size of the file in octets. The value of this attribute SHOULD be the same as the "size" parameter of the Content-Disposition header field (RFC 2183) that would be signaled by the actual file transfer. Note that the "size" selector merely includes the file size and does not include any potential overhead added by a wrapper, such as message/cpim (RFC 3862). The "type" selector in the "file-selector" attribute contains the MIME media and submedia types of the content. In general, anything that can be expressed in a Content-Type header field (see RFC 2045) can also be expressed with the "type" selectors. Possible MIME media type values are listed in the Internet Assigned Numbers Authority (IANA) registry for MIME media types. Zero or more parameters can follow. When translating parameters from a Content-Type header and a "type" selector, the parameter has to be re-encoded prior to its accommodation as a parameter of the "type" selector (see the ABNF syntax of "ft-parameter").

The "hash" selector in the "file-selector" attribute provides a hash computation of the file to be transferred. This is commonly used by file transfer protocols. For example, FLUTE (RFC 6726) uses hashes (called message digests) to verify the contents of the transfer. The purpose of the "hash" selector is two-fold: On one side, in pull operations, it allows the file receiver to identify a remote file by its hash rather than by its file name, providing that the file receiver has learned the hash of the remote file by some out-of-band mechanism. On the other side, in either push or pull operations, it allows the file receiver to verify the contents of the received file or even avoid unnecessary transmission of an existing file.

The address space of the SHA-1 algorithm is big enough to avoid any collision in hash computations in between two endpoints. When transferring files, the actual file transfer protocol should provide reliable transmission of data, so verifications of received files should always succeed. However, if endpoints need to protect the integrity of a file, they should use some other mechanism than the "hash" selector specified in this specification. The "hash" selector includes the hash algorithm and its value. Possible hash algorithms are those defined in the IANA registry of Hash Function Textual Names. Implementations according to this specification MUST add a 160-bit string resulting from the computation of US SHA-1 (RFC 3174) if the "hash" selector is present. If need arises, extensions can be drafted to support several hashing algorithms. Therefore, implementations according to this specification MUST be prepared to receive SDP containing more than a single "hash" selector in the "file-selector" attribute.

The value of the "hash" selector is the byte string resulting from applying the hash algorithm to the content of the whole file, even when the file transfer is limited to a number of octets (i.e., the "file-range" attribute is indicated). The "file-transfer-id" attribute provides a randomly chosen globally unique identification to the actual file transfer. It is used to distinguish a new file transfer request from a repetition of the SDP (or the fraction of the SDP that deals with the file description). This attribute is described in much greater detail in Section 13.1.8.1. The "file-disposition" attribute provides a suggestion to the other endpoint about the intended disposition of the file. Section 13.1.7 provides further discussion of the possible values. The value of this attribute SHOULD be the same as the disposition type parameter of the Content-Disposition header field (RFC 2183) that would be signaled by the actual file transfer protocol.

The "file-date" attribute indicates the dates on which the file was created, modified, or last read. This attribute MAY contain a combination of the "creation," "modification," and "read" parameters, but MUST NOT contain more than one of each type. The "creation" parameter indicates the date on which the file was created. The value MUST be a quoted string that contains a representation of the creation date of the file in RFC 5322 "date-time" format. Numeric time zones (+HHMM or -HHMM) MUST be used. The value of this parameter SHOULD be the same as the "creation-date" parameter of the Content-Disposition header field (RFC 2183) that would be signaled by the actual file transfer protocol.

The "modification" parameter indicates the date on which the file was last modified. The value MUST be a quoted string that contains a representation of the last modification date to the file in RFC 5322 "date-time" format. Numeric time zones (+HHMM or -HHMM) MUST be used. The value of this parameter SHOULD be the same as the "modification-date' parameter of the Content-Disposition header field (RFC 2183) that would be signaled by the actual file transfer protocol. The "read" parameter indicates the date on which the file was last read. The value MUST be a quoted string that contains a representation of the last date the file was read in RFC 5322 "date-time" format. Numeric time zones (+HHMM or -HHMM) MUST be used. The value of this parameter SHOULD be the same as the "read-date" parameter of the Content-Disposition header field (RFC 2183) that would be signaled by the actual file transfer protocol.

The "file-icon" attribute can be useful with certain file types such as images. It allows the file sender to include a pointer to a body that includes a small preview icon representing the contents of the file to be transferred, which the file receiver can use to determine whether it wants to receive such file. The "file-icon" attribute contains a Content-ID (CID) URL, which is specified in RFC 2392. Section 13.1.8.8 contains further considerations about the "file-icon: attribute. The "file-range" attribute provides a mechanism to signal a chunk of a file rather than the complete file. This enables use cases where a file transfer can be interrupted and resumed, even perhaps changing one of the endpoints. The "file-range" attribute contains the "start offset" and "stop offset" of the file, separated by a dash "-." The "start offset" value refers to the octet position of the file where the file transfer should start. The first octet of a file is denoted by the ordinal number "1." The "stop offset" value refers to the octet position of the file where the file transfer should stop, inclusive of this octet. The "stop offset" value MAY contain a "*" if the total size of the file is not known in advance.

The absence of this attribute indicates a complete file (i.e., as if the "filerange" attribute would have been present with a value "1-*". The "file-range" attribute must not be confused with the Byte-Range header in MSRP. The former indicates the portion of a file that the application would read and pass onto the MSRP stack for transportation. From the point of view of MSRP, the portion of the file is viewed as a whole message. The latter indicates the number of bytes of that message that are carried in a chunk and the total size of the message. Therefore, MSRP starts counting the delivered message at octet number 1, independently of the position of that octet in the file. The following is an example of an SDP body that contains the extensions defined in this memo (Figure 13.2):

NOTE: The "file-selector" attribute in Figure 13.2 is split in three lines for formatting purposes. Real implementations will encode it in a single line.

```
v=0
o=alice 2890844526 2890844526 IN IP4 host.atlanta.example.com
s=
c=IN IP4 host.atlanta.example.com
t=0 0
m=message 7654 TCP/MSRP *
i=This is my latest picture
a=sendonly
a=accept-types:message/cpim
a=accept-wrapped-types:*
a=path:msrp://atlanta.example.com:7654/jshA7we;tcp
a=file-selector:name:"My cool picture.jpg" type:image/jpeg
     size:32349 hash:sha-1:
     72:24:5F:E8:65:3D:DA:F3:71:36:2F:86:D4:71:91:3E:E4:A2:CE:2E
a=file-transfer-id:vBnG916bdberum2fFEABR1FR3ExZMUrd
a=file-disposition:attachment
a=file-date:creation:"Mon, 15 May 2006 15:01:31 +0300"
a=file-icon:cid:id2@alicepc.example.com
a=file-range:1-32349
```

Figure 13.2 Example of SDP describing a file transfer.

13.1.7 File Disposition Types

The SDP offer/answer for file transfer allows the file sender to indicate a preferred disposition of the file to be transferred in a new "file-disposition" attribute. In principle, any value listed in the IANA registry for Mail Content Disposition Values is acceptable; however, most of them may not be applicable. There are two content dispositions of interest for file transfer operations. On one hand, the file sender may just want the file to be rendered immediately in the file receiver's device. On the other hand, the file sender may just want to indicate to the file receiver that the file should not be rendered at the reception of the file.

The recipient's user agent may want to interact with the user regarding the file disposition, or it may save the file until the user takes an action. In any case, the exact actions are implementation dependent. To indicate that a file should be automatically rendered, this specification uses the existing "render" value of the Content-Disposition type in the new "file-disposition" attribute in SDP. To indicate that a file should not be automatically rendered, this memo uses the existing "attachment" value of the Content-Disposition type in the new "file-disposition" attribute in SDP. The default value is "render" (i.e., the absence of a "file-disposition" attribute in the SDP has the same semantics as "render").

The disposition value "attachment" is specified in RFC 2183 with the following definition: "Body parts can be designated 'attachment' to indicate that they are separate from the main body of the mail message, and that their display should not be automatic, but contingent upon some further action of the user." In the case of this specification, the "attachment" disposition type is used to indicate that the display of the file should not be automatic, but rather contingent upon some further action of the user.

13.1.8 Protocol Operation

This section discusses how to use the parameters defined in Section 13.1.6 in the context of an offer/answer (RFC 3264, see Section 3.1) exchange. Additionally, this section also discusses the behavior of the endpoints using MSRP. A file transfer session is initiated by the offerer sending an SDP offer to the answerer. The answerer either accepts or rejects the file transfer session and sends an SDP answer to the offerer. We can differentiate two use cases, depending on whether the offerer is the file sender or file receiver:

1. The offerer is the file sender (i.e., the offerer wants to transmit a file to the answerer). Consequently, the answerer is the file receiver. In this case, the SDP offer contains a "sendonly" attribute; accordingly, the SDP answer contains a "recvonly" attribute.
2. The offerer is the file receiver (i.e., the offerer wants to fetch a file from the answerer). Consequently, the answerer is the file sender. In this case, the SDP offer contains a session or media "recvonly" attribute; accordingly, the SDP answer contains a session or media "sendonly" attribute.

13.1.8.1 The "file-transfer-id" Attribute

This specification creates an extension to the SDP offer/answer model (RFC 3264, see Section 3.1); because of that, it is assumed that the existing SDP behavior is kept intact. The SDP behavior requires, for example, that SDP is sent again to the remote party in situations where the media description or perhaps other SDP parameters have not changed with respect to a previous offer/answer exchange. Let's consider the Session Initiation Protocol (SIP) session timer, which uses re-INVITE requests to refresh sessions. RFC 4028 recommends sending unmodified SDP in a re-INVITE to refresh the session. Should this re-INVITE contain SDP describing a file transfer operation and occur while the file transfer is still going on, there will be no means to detect whether the SDP creator wanted to abort the current file transfer operation and initiate a new one or the SDP file description was included in the SDP due to other factors (e.g., session timer refresh).

A similar scenario occurs when two endpoints have successfully agreed on a file transfer, which is currently taking place when one of the endpoints wants to add additional media streams to the existing session. In this case, the endpoint sends a re-INVITE request that contains the SDP. The SDP needs to maintain the media descriptions for the current ongoing file transfer and add the new media descriptions. The problem is that the other endpoint is not able to determine whether a new file transfer is requested. In other cases, a file transfer was successfully completed. Then, if an endpoint resends the SDP offer with the media stream for the file transfer, then the other endpoint wouldn't be able to determine whether a new file transfer should start.

To address these scenarios, this specification defines the "filetransfer-id" attribute, which contains a globally unique random identifier allocated to the file transfer operation. The file transfer identifier helps both endpoints determine whether the SDP offer is requesting a new file transfer or is a repetition of the SDP. A new file transfer is one that, in case of acceptance, will provoke the actual transfer of a file. This is typically the case with new offer/answer exchanges or cases in which an endpoint wants to abort the existing file transfer and restart the file transfer once more. On the other hand, the repetition of the SDP does not lead to any actual file to be transferred, potentially because the file transfer is still going on or because it has already finished. This is the case of repeated offer/answer exchanges, which can be due to a number of factors (session timer, addition/removal of other media types in the SDP, updates in SDP due to changes in other session parameters, etc.).

Implementations according to this specification MUST include a "filetransfer-id" attribute in SDP offers and answers. The SDP offerer MUST select a file transfer identifier according to the syntax and add it to the "file-transfer-id" attribute. The SDP answerer MUST copy the value of the "file-transfer-id" attribute in the SDP answer. The file transfer identifier MUST be unique within the current session (never used before in this session), and it is RECOMMENDED to be unique across different sessions. It is RECOMMENDED that one select a relatively big random identifier (e.g., 32 characters) to avoid duplications. The SDP answerer MUST keep track of the proposed file transfer identifiers in each session and copy the value of the received file transfer identifier in the SDP answer. If a file transfer is suspended and resumed later, the resumption is considered a new file transfer (even when the file to be transferred is the same); therefore, the SDP offerer MUST choose a new file transfer identifier.

If an endpoint sets the port number to 0 in the media description of a file transfer, for example, because it wants to reject the file transfer operation, then the SDP answer MUST mirror the value of the "file-transfer-id" attribute included in the SDP offer. This effectively means that setting a media stream to 0 has higher precedence than any value the "file-transfer-id" attribute can take. As a side effect, the "file-transfer-id" attribute can be used for aborting and restarting again an ongoing file transfer. Assume that two endpoints agree on a file transfer and the actual transfer of the file is taking place. At some point in time in the middle of the file transfer, one endpoint sends a new SDP offer, equal to the initial one except for the value of the "file-transfer-id" attribute, which is a new globally unique random value. This indicates that the offerer wants to abort the existing transfer and start a new one, according to the SDP parameters. The SDP answerer SHOULD abort the ongoing file transfer, according to the procedures of the file transfer protocol (e.g., MSRP), and start sending files once more from the initial requested octet. Section 13.1.8.4 further discusses aborting a file transfer.

If an endpoint creates an SDP offer where the "file-transfer-id" attribute value does not change with respect to a previously sent one, but the file selector changes so that a new file is selected, then this is considered an error, and the SDP answerer MUST abort the file transfer operation (e.g., by setting the port number to zero in the SDP answer). Note that endpoints MAY change the "file-selector" attribute as long as the selected file does not change (e.g., by adding a hash selector); however, it is RECOMMENDED that endpoints do not change the value of the "file-selector" attribute if it is requested to transfer the same file described in a previous SDP offer/answer exchange. Figure 13.3 summarizes the relation of the "file-transfer-id" attribute and the file selector in subsequent SDP exchanges.

In another scenario, an endpoint that has successfully transferred a file wants to send an SDP due to other reasons than the transfer of a file. The SDP offerer creates an SDP file description that maintains the media description line corresponding to the file transfer. The SDP offerer MUST then set the port number to 0 and MUST keep the value of the "file-transfer-id" attribute the initial file transfer got.

13.1.8.2 Offerer's Behavior

An offerer who wishes to send or receive one or more files to or from an answerer MUST build an SDP (RFC 4566, see Section 2) description of a session containing one "m=" line per file. When MSRP is used as the transfer mechanism, each "m=" line also describes a single MSRP session, according to the MSRP (RFC 4975) procedures. Any "m=" lines that may have already been present in a previous SDP exchange are normally kept unmodified; the new "m=" lines are added afterwards

'file-transfer-id'	File Selector	
	Different File	Same File
Change	New File Transfer Operation	New File Transfer Operation
Unchanged	Error	Existing File Transfer Operation

Figure 13.3 Relation of the "file-transfer-id" attribute with the selector of the file in a subsequent SDP exchange.

(Section 13.1.8.6 describes cases when "m=" lines are reused). All the media line attributes specified and required by MSRP (RFC 4975) (e.g., "a=path," "a=accept-types," etc.) MUST be included as well.

13.1.8.2.1 The Offerer Is a File Sender

In a push operation, the file sender creates an SDP offer describing the file to be sent. The file sender MUST add a "file-selector" attribute media line containing at least one of the "type," "size," or "hash" selectors in indicating the type, size, or hash of the file, respectively. If the file sender wishes to start a new file transfer, the file sender MUST add a "file-transfer-id" attribute containing a new globally unique random identifier value.

Additionally, the file sender MUST add a session or media "sendonly" attribute to the SDP offer. Then the file sender sends the SDP offer to the file receiver. Not all selectors in the "file-selector" attribute might be known when the file sender creates the SDP offer, for example, because the host is still processing the file.

The "hash" selector in the "file-selector" attribute contains valuable information for the file receiver to identify whether the file is already available and does not need to be transmitted. The file sender MAY also add a "name" selector in the "file-selector" attribute and "file-icon," "file-disposition," and "file-date" attributes further describing the file to be transferred. The "file-disposition" attribute provides a presentation suggestion (e.g., whether the file sender would like the file receiver to render the file). The three date attributes provide the answerer with an indication of the age of the file. The file sender MAY also add a "file-range" attribute indicating the start and stop offsets of the file. When the file sender receives the SDP answer, if the port number of the answer for a file request is non-zero, the file sender starts the transfer of the file according to the negotiated parameters in SDP.

13.1.8.2.2 The Offerer Is a File Receiver

In a pull operation, the file receiver creates the SDP offer and sends it to the file sender. The file receiver MUST include a "file-selector" attribute and MUST include, at least, one of the selectors defined for such attribute (i.e., "name," "type," "size," or "hash"). In many cases, if the hash of the file is known, that is enough to identify the file; therefore, the offerer can include only a "hash" selector. However, particularly in cases where the hash of the file is unknown, the file name, size, and type can provide a description of the file to be fetched. If the file receiver wishes to start a new file transfer, it MUST add a "file-transfer-id" attribute containing a new globally unique random value. The file receiver MAY also add a "file-range" attribute, indicating the start and stop offsets of the file. There is no need for the file receiver to include further file attributes in the SDP offer; thus, it is RECOMMENDED that SDP offerers not include any other file attribute defined by this specification (other than the mandatory ones).

Additionally, the file receiver MUST add a session or media "recvonly" attribute in the SDP offer. Then, the file receiver sends the SDP offer to the file sender. When the file receiver receives the SDP answer, if the port number of the answer for a file request is non-zero, then the file receiver should receive the file using the protocol indicated in the "m=" line. If the SDP answer contains a supported hashing algorithm in the "hash" selectors of the "file-selector" attribute, then the file receiver SHOULD compute the hash of the file after its reception and check it against the hash received in the answer. In case the computed hash does not match the one contained in the SDP answer, the file receiver SHOULD consider that the file transfer failed and SHOULD inform the user. Similarly, the file receiver SHOULD also verify that the other selectors declared in the SDP match the file properties; otherwise, the file receiver SHOULD consider that the file transfer failed and SHOULD inform the user.

13.1.8.2.3 SDP Offer for Several Files

An offerer that wishes to send or receive more than one file generates an "m=" line per file along with the file attributes described in this specification. This way, the answerer can reject individual files by setting the port number of their associated "m=" lines to zero, as per regular SDP (RFC 4566, see Section 2) procedures. Similarly, the answerer can accept each individual file separately by setting the port number of its associated "m=" line to non-zero value. Each file has its own file transfer identifier, which uniquely identifies each file transfer. Using an "m=" line per file implies that different files are transferred using different MSRP sessions. However, all those MSRP sessions can be set up to run over a single Transmission Control Protocol (TCP) connection, as described in Section 8.1 of RFC 4975. The same TCP connection can also be reused for sequential file transfers.

13.1.8.3 Answerer's Behavior

If the answerer wishes to reject a file offered by the offerer, it sets the port number of the "m=" line associated with the file to zero, as per regular SDP (RFC 4566, see Section 2) procedures. The rejected answer MUST contain "file-selector" and "file-transfer-id" attributes whose values mirror the corresponding values of the SDP offer. If the answerer decides to accept the file, it proceeds as per regular MSRP (RFC 4975) and SDP (RFC 3264, see Section 3.1) procedures.

13.1.8.3.1 The Answerer Is a File Receiver

In a push operation, the SDP answerer is the file receiver. When the file receiver gets the SDP offer, it first examines the port number. If the port number is set to zero, the file transfer operation is closed, and no more data is expected over the media stream. Then, if the port number is not zero, the file receiver inspects the "file-transfer-id" attribute. If the value of the "filetransfer-id" attribute has been previously used, then the existing session remains without changes; perhaps the file transfer is still in progress, or perhaps it has concluded, but there are no changes with respect to the status. In any case, independently of the port number, the SDP answerer creates a regular SDP answer and sends it to the offerer.

If the port number is different from zero and the SDP offer contains a new "file-transfer-id" attribute, it is signaling a request for a new file transfer. The SDP answerer extracts the attributes and parameters that describe the file and typically requests permission from the user to accept or reject the reception of the file. If the file transfer operation is accepted, the file receiver MUST create an SDP answer according to the procedures specified in RFC 3264 (see Section 3.1). If the offer contains "name," "type," or "size" selectors in the "file-selector" attribute, the answerer MUST copy them into the answer. The file receiver copies the value of the "file-transfer-id" attribute to the SDP answer. Then the file receiver MUST add a session or media "recvonly" attribute according to the procedures specified in in RFC 3264 (see Section 3.1). The file receiver MUST NOT include "file-icon," "file-disposition," or "file-date" attributes in the SDP answer.

The file receiver can use the hash to find out if a local file with the same hash is already available, which could imply the reception of a duplicate file. It is up to the answerer to determine whether the file transfer is accepted in case of a duplicate file. If the SDP offer contains a "file-range" attribute and the file receiver accepts the receipt of the range of octets declared in there, the file receiver MUST include a "file-range" attribute in the SDP answer with the same range of values. If the file receiver does not accept the receipt of that range of octets, it SHOULD reject the transfer of the file. When the file transfer operation is complete, the file receiver computes the hash of the file and SHOULD verify that it matches the hash declared in the SDP. If they do not match, the file receiver SHOULD consider that the file transfer failed and SHOULD inform the user. Similarly, the file receiver SHOULD verify that the other selectors declared in the SDP match the file properties; otherwise, the file receiver SHOULD consider that the file transfer failed and SHOULD inform the user.

13.1.8.3.2 The Answerer Is a File Sender

In a pull operation, the answerer is the file sender. In this case, the SDP answerer MUST first inspect the value of the "file-transfer-id" attribute. If it has not been previously used throughout the session, then acceptance of the file MUST provoke the transfer of the file over the negotiated protocol. However, if the value has been previously used by another file transfer operation within the session, then the file sender MUST NOT alert the user and MUST NOT start a new transfer of the file. Regardless of whether an actual file transfer is initiated, the file sender MUST create a proper SDP answer that contains the "file-transfer-id" attribute with the same value received in the SDP offer, and then it MUST continue processing the SDP answer.

The file sender MUST always create an SDP answer according to the SDP offer/answer procedures specified in RFC 3264 (se Section 3.1). The file sender inspects the file selector of the received SDP offer, which is encoded in the "file-selector" media attribute line. Then the file sender applies the file selector, which implies selecting those files that match one by one with the "name," "type," "size," and "hash" selectors of the "file-selector" attribute line (if they are present). The file selector identifies zero or more candidate files to be sent. If the file selector is unable to identify any file, then the answerer MUST reject the MSRP stream for file transfer by setting the port number to zero, and then the answerer SHOULD also reject the SDP as per procedures in in RFC 3264 (see Section 3.1), if this is the only stream described in the SDP offer.

If the file selector points to a single file and the file sender decides to accept the file transfer, the file sender MUST create an SDP answer that contains a "sendonly" attribute, according to the procedures described in in RFC 3264 (see Section 3.1). The file sender SHOULD add a "hash" selector in the answer with the locally computed SHA-1 hash over the complete file. If a hash value computed by the file sender differs from that specified by the file receiver, the file sender can either send the file without that hash value or reject the request by setting the port in the media stream to zero. The file sender MAY also include a "type" selector in the "file-selector" attribute line of the SDP answer. The answerer MAY also include "file-icon" and "file-disposition" attributes to further describe the file. Although the answerer MAY also include "name" and "size" selectors in the "file-selector" attribute, and a "file-date" attribute, it is RECOMMENDED not to include them in the SDP answer if the actual file transfer protocol (e.g., MSRP, RFC 4975) can accommodate a Content-Disposition header field (RFC 2183) with the equivalent parameters.

The whole idea of adding file descriptors to SDP is to provide a mechanism where a file transfer can be accepted prior to its start. Adding any SDP attributes that are otherwise signaled later in the file transfer protocol would just duplicate the information but not provide any information to the offerer to accept or reject the file transfer (note that the offerer is requesting a file). Last, if the file selector points to multiple candidate files, the answerer MAY use some local policy (e.g., consulting the user) to choose one of them to be defined in the SDP answer. If that choice cannot be made, the answerer SHOULD reject the MSRP media stream for file transfer (by setting the port number to zero). If the need arises, future specifications could provide a mechanism allowing either the selection of multiple files or, perhaps, the resolution of ambiguities by returning a list of files that match the file selector. If the SDP offer contains a "file-range" attribute and the file sender agrees to send the range of octets declared in there, the file sender MUST include a "file-range" attribute in the SDP answer with the same range of values. If the file sender does not agree to send that range of octets, it SHOULD reject the transfer of the file.

13.1.8.4 Aborting an Ongoing File Transfer Operation

Either the file sender or the file receiver can abort an ongoing file transfer at any time. Unless otherwise noted, the entity that aborts an ongoing file transfer operation MUST follow the procedures at the media level (e.g., MSRP) and at the signaling level (SDP offer/answer), as described in Figures 13.4–13.6). Assume the scenario depicted in Figure 13.4 where a file sender wishes to abort an ongoing file transfer without initiating an alternative file transfer. Assume that an ongoing MSRP SEND request is being transmitted. The file sender aborts the MSRP message by including the "#" character in the continuation field of the end line of a SEND request, according to the MSRP procedures (see Section 7.1 of RFC 4975). Since a file is transmitted as one MSRP message, aborting the MSRP message effectively aborts the file transfer.

The file receiver acknowledges the MSRP SEND request with a 200 response. Then the file sender SHOULD close the MSRP session by creating a new SDP offer that sets the port number to zero in the related "m=" line that describes the file transfer (see Section 3.1.8.2 of this book) (that is, Section 8.2 of RFC 3264). This SDP offer MUST conform to the requirements of Section 13.1.8.2.1 of this book. The "file-transfer-id" attribute MUST be the same attribute that identifies the ongoing transfer. Then the file sender sends this SDP offer to the file receiver. Rather than close the MSRP session by setting the port number to zero in the related "m=" line, the file sender could tear down the whole session (e.g., by sending a SIP BYE request).

Note that it is the responsibility of the file sender to tear down the MSRP session. Implementations should be prepared for misbehaviors and implement measures to avoid hang states. For example, upon expiration of a timer the file receiver can close the aborted MSRP session by using regular MSRP procedures. A file receiver that receives the above SDP offer creates an SDP answer according to the procedures of the SDP offer/answer (RFC 3264, see Section 3.1). This SDP answer MUST conform to the requirements of Section 13.1.8.3.1. Then the file receiver sends this SDP answer to the file sender.

When the file receiver wants to abort the file transfer (Figure 13.4), there are two possible scenarios, depending on the value of the Failure-Report header in the ongoing MSRP SEND request. Assume now the scenario depicted in Figure 13.5 where the MSRP SEND request includes a

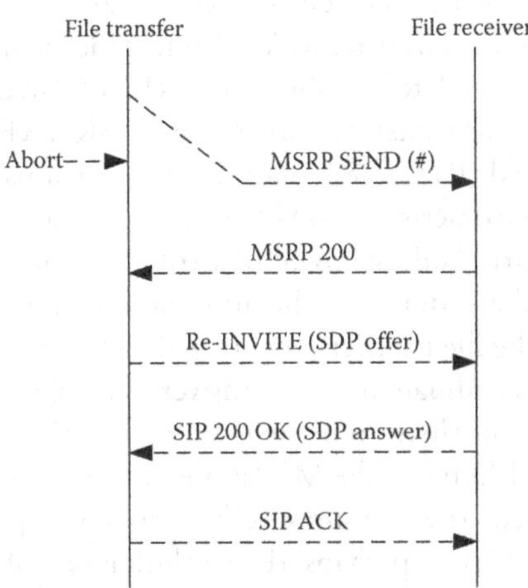

Figure 13.4 File sender aborts an ongoing file transfer (Copyright IETF).

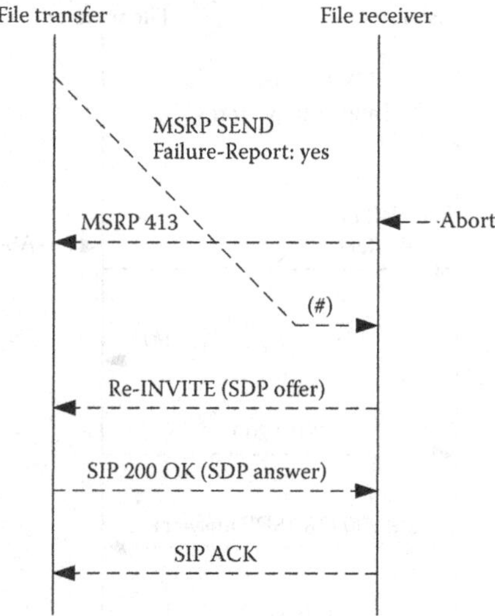

Figure 13.5 File receiver aborts an ongoing file transfer. Failure-Report set to a value different than "no" in MSRP (Copyright IETF).

Failure-Report header set to a value different from "no." When the file receiver wishes to abort the ongoing file transfer, the file receiver generates an MSRP 413 response to the current MSRP SEND request (see Section 10.5 of RFC 4975). Then the file receiver MUST close the MSRP session by generating a new SDP offer that sets the port number to zero in the related "m=" line that describes the file transfer (see Section 13.1.8.2 of this book) (that is, Section 8.2 of RFC 3264). This SDP offer MUST conform to the requirements expressed in Section 13.1.8.2.2. The "file-transfer-id" attribute MUST be the same attribute that identifies the ongoing transfer. Then the file receiver sends this SDP offer to the file sender.

In another scenario, depicted in Figure 13.6, an ongoing file transfer is taking place, where the MSRP SEND request contains a Failure-Report header set to the value "no." When the file receiver wants to abort the ongoing transfer, it MUST close the MSRP session by generating a new SDP offer that sets the port number to zero in the related "m=" line that describes the file transfer (see Section 3.1.8.2) (that is, Section 8.2 of RFC 3264). This SDP offer MUST conform to the requirements expressed in Section 13.1.8.2.2 of this book. The "file-transfer-id" attribute MUST be the same attribute that identifies the ongoing transfer. Then the file receiver sends this SDP offer to the file sender.

A file sender that receives an SDP offer setting the port number to zero in the related "m=" line for file transfer, first, if an ongoing MSRP SEND request is being transmitted, aborts the MSRP message by including the "#" character in the continuation field of the end-line of a SEND request, according to the MSRP procedures (see Section 7.1 of RFC 4975). Since a file is transmitted as one MSRP message, aborting the MSRP message effectively aborts the file transfer. Then the file sender creates an SDP answer according to the procedures of the SDP offer/answer (RFC 3264, see Section 3.1). This SDP answer MUST conform to the requirements of Section 13.1.8.3.2 of this book. Then the file sender sends this SDP answer to the file receiver.

13.1.8.5 Indicating File Transfer Offer/Answer Capability

The SDP offer/answer model (RFC 3264, see Section 3.1) provides provisions for indicating a capability to another endpoint (see Section 3.1.9 of this book) (that is, Section 9 of RFC 3264).

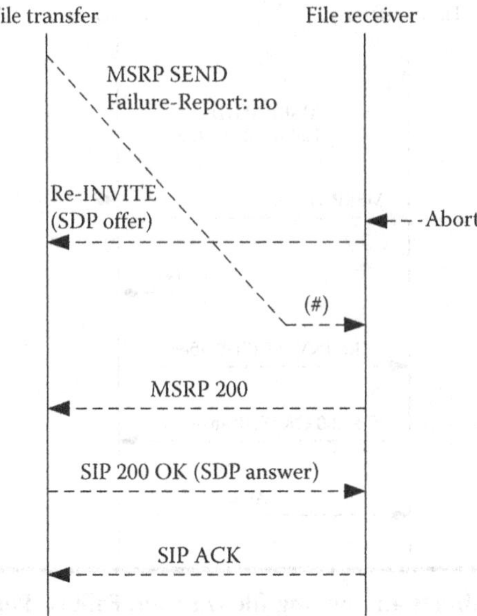

Figure 13.6 File receiver aborts an ongoing file transfer. Failure-Report set to "no" in MSRP (Copyright IETF).

The mechanism assumes a high-level protocol, such as SIP (RFC 3261, also see Section 1.2), that provides a capability query (such as a SIP OPTIONS request). RFC 3264 indicates how to build the SDP that is included in the response to such capability query. As such, RFC 3264 indicates that an endpoint builds an SDP body that contains an "m=" line containing the media type (message, for MSRP). An endpoint that implements the procedures specified in this document SHOULD also add a "file-selector" media attribute for the "m=message" line. The "file-selector" media attribute MUST be empty, i.e., it MUST NOT contain any selector. The endpoint MUST NOT add any of the other file attributes defined in this specification.

13.1.9 Reusage of Existing "m=" Lines in SDP

The SDP offer/answer model (RFC 3264, see Section 3.1) provides rules that allow SDP offerers and answerers to modify an existing media line (i.e., reuse an existing media line with different attributes). The same is also possible when SDP signals a file transfer operation according to the rules of this memo. Therefore, the procedures defined in RFC 3264 (see Section 3.1), in particular those defined in Section 3.1.8.3 of this book, MUST apply for file transfer operations. An endpoint that wants to reuse an existing "m=" line to start the file transfer of another file creates a different "file-selector" attribute and selects a new globally unique random value of the "file-transfer-id" attribute. If the file offerer resends an SDP offer with a port different from zero, then the "file-transfer-id" attribute determines whether a new file transfer will start or whether the file transfer does not need to start. If the SDP answerer accepts the SDP, then file transfer starts from the indicated octet (if a "file-range" attribute is present).

13.1.9.1 MSRP Usage

The file transfer service specified in this document uses "m=" lines in SDP to describe the unidirectional transfer of a file. Consequently, each MSRP session established following the procedures in Sections 13.1.8.2 and 13.1.8.3 is only used to transfer a single file. So, senders MUST only use the dedicated MSRP session to send the file described in the SDP offer or answer. That is, senders MUST NOT send additional files over the same MSRP session. File transfer may be accomplished using a new multimedia session established for the purpose. Alternatively, a file transfer may be conducted within an existing multimedia session, without regard for the media in use within that session. Of particular note, file transfer may be done within a multimedia session containing an MSRP session used for regular instant messaging. If file transfer is initiated within an existing multimedia session, the SDP offerer MUST NOT reuse an existing "m=" line that is still being used by MSRP (either regular MSRP for instant messaging or an ongoing file transfer). Rather, it MUST add an additional "m=" line or reuse an "m=" line that is no longer being used.

Additionally, implementations according to this specification MUST include a single file in a single MSRP message. Notice that the MSRP specification defines "MSRP message" as a complete unit of MIME or text content, which can be split and delivered in more than one MSRP request; each of these portions of the complete message is called a "chunk." So, it is valid to send a file in several chunks, but from the MSRP point of view, all the chunks together form an MSRP message: the common presence and instant messaging (CPIM) message that wraps the file. When chunking is used, it should be noticed that MSRP does not require waiting for a 200-class response for a chunk before sending the following one. Therefore, it is valid to send pipelined MSRP SEND requests containing chunks of the same MSRP message (the file). Section 13.1.9.1 contains an example of a file transfer using pipelined MSRP requests. The MSRP specification (RFC 4975) defines a "max-size" SDP attribute.

This attribute specifies the maximum number of octets of an MSRP message the creator of the SDP is willing to receive (notice once more the definition of "MSRP message"). File receivers MAY add a "max-size" attribute to the MSRP "m=" line that specifies the file, indicating the maximum number of octets of an MSRP message. File senders MUST NOT exceed the "max-size" limit for any message sent in the resulting session. In the absence of a "file-range" attribute in the SDP, the MSRP file transfer MUST start with the first octet of the file and end with the last octet (i.e., the whole file is transferred). If a "file-range" attribute is present in SDP, the file sender application MUST extract the indicated range of octets from the file (start and stop offset octets, both inclusive). Then the file sender application MAY wrap those octets in an appropriate wrapper. MSRP mandates implementations to implement the message/cpim wrapper (RFC 3862).

Usage of a wrapper is negotiated in the SDP (see Section 8.6 in RFC 4975). Last, the file sender application delivers the content (e.g., the message/CPIM body) to MSRP for transportation. MSRP will consider the delivered content a whole message and will start numbering bytes with the number 1. Note that the default content disposition of MSRP bodies is "render." When MSRP is used to transfer files, the MSRP Content-Disposition header can also take the value "attachment" as indicated in Section 13.1.7. Once the file transfer is complete, the file sender SHOULD close the MSRP session and MUST behave according to the MSRP (RFC 4975) procedures with respect to closing MSRP sessions. Note that MSRP session management is not related to TCP connection management. As a matter of fact, MSRP allows multiple MSRP sessions to share the same TCP connection.

13.1.9.2 Considerations about the "file-icon" Attribute

This specification allows a file sender to include a small preview of an image file: an icon. A "file-icon" attribute contains a CID URL (RFC 2392) pointing to an additional body that contains the actual icon. Since the icon is sent as a separate body along the SDP body, the file sender MUST wrap the SDP body and the icon bodies in a MIME multipart/related body. Therefore, implementations according to this specification MUST implement the multipart/related MIME type (RFC 2387). When creating a multipart/related MIME wrapper, the SDP body MUST be the root body, which according to RFC 2387 is identified as the first body in the multipart/related MIME wrapper or explicitly identified by the "start" parameter. According to RFC 2387, the "type" parameter MUST be present and point to the root body (i.e., the SDP body).

Assume that an endpoint behaving according to this specification tries to send a file to a remote endpoint that neither implements this specification nor implements multipart MIME bodies. The file sender sends an SDP offer that contains a multipart/related MIME body that includes an SDP body part and an icon body part. The file receiver, not supporting multipart MIME types, will reject the SDP offer via a higher protocol mechanism (e.g., SIP). In this case, it is RECOMMENDED that the file sender remove the icon body part, create a single SDP body (i.e., without multipart MIME), and resend the SDP offer. This provides some backwards compatibility with file receivers that do not implement this specification and increases the chances of getting the SDP accepted at the file receiver. Since the icon is sent as part of the signaling, it is RECOMMENDED that one keep the size of icons restricted to the minimum number of octets that provide significance.

13.1.10 Examples

13.1.10.1 Offerer Sends a File to the Answerer

This section shows an example flow for a file transfer scenario. The example assumes that SIP (RFC 3261, also see Section 1.2) is used to transport the SDP offer/answer exchange, although the SIP details are briefly shown. Alice, the SDP offerer, wishes to send an image file to Bob (the answerer). Alice's user agent client creates a unidirectional SDP offer that contains the description of the file she wants to send to Bob's user agent server. The description also includes an icon representing the contents of the file to be transferred. The sequence flow is shown in Figure 13.7.

F1: Alice constructs an SDP description of the file to be sent and attaches it to a SIP INVITE request addressed to Bob (Figures 13.7 and 13.8).

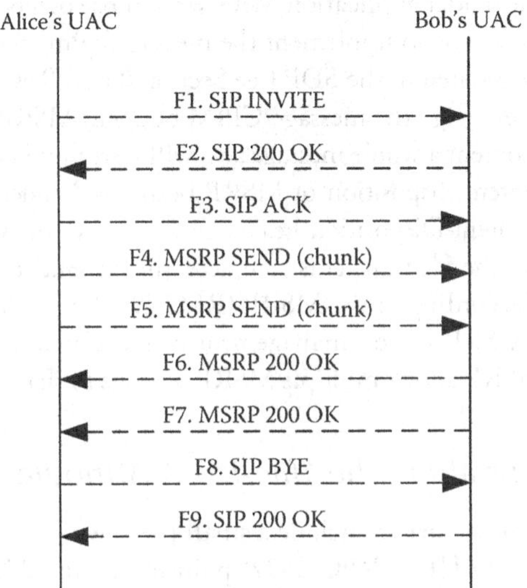

Figure 13.7 Flow diagram of an offerer sending a file to an answerer (Copyright IETF).

```
INVITE sip:bob@example.com SIP/2.0
To: Bob <sip:bob@example.com>
From: Alice <sip:alice@example.com>;tag=1928301774
Call-ID: a84b4c76e66710
CSeq: 1 INVITE
Max-Forwards: 70
Date: Sun, 21 May 2006 13:02:03 GMT
Contact: <sip:alice@alicepc.example.com>
Content-Type: multipart/related; type="application/sdp";
boundary="boundary71"
Content-Length: [length]

--boundary71
Content-Type: application/sdp
Content-Length: [length of SDP]

v=0
o=alice 2890844526 2890844526 IN IP4 alicepc.example.com
s=
c=IN IP4 alicepc.example.com
t=0 0
m=message 7654 TCP/MSRP *
i=This is my latest picture
a=sendonly
a=accept-types:message/cpim
a=accept-wrapped-types:*
a=path:msrp://alicepc.example.com:7654/jshA7we;tcp
a=file-selector:name:"My cool picture.jpg" type:image/jpeg
          size:4092 hash:sha-1:
          72:24:5F:E8:65:3D:DA:F3:71:36:2F:86:D4:71:91:3E:E4:A2:CE:2E
a=file-transfer-id:Q6LMoGymJdh0IKIgD6wD0jkcfgva4xvE
a=file-disposition:render
a=file-date:creation:"Mon, 15 May 2006 15:01:31 +0300"
a=file-icon:cid:id2@alicepc.example.com

--boundary71
Content-Type: image/jpeg
Content-Transfer-Encoding: binary
Content-ID: <id2@alicepc.example.com>
Content-Length: [length of image]
Content-Disposition: icon

[...small preview icon of the file...]

--boundary71--
```

Figure 13.8 INVITE request containing an SDP offer for file transfer (Copyright IETF).

NOTE: The Content-Type header field and the "file-selector" attribute in Figure 13.8 are split in several lines for formatting purposes. Real implementations will encode it in a single line. From now on, we omit the SIP details for the sake of brevity.

F2: Bob receives the INVITE request, inspects the SDP offer and extracts the icon body, checks the creation date and file size, and decides to accept the file transfer. So Bob creates the following SDP answer (Figure 13.9):

NOTE: The "file-selector' attribute in the Figure 13.8 is split in three lines for formatting purposes. Real implementations will encode it in a single line.

F4: Alice opens a TCP connection to Bob and creates an MSRP SEND request. This SEND request contains the first chunk of the file (Figure 13.10).

```
v=0
o=bob 2890844656 2890844656 IN IP4 bobpc.example.com
s=
c=IN IP4 bobpc.example.com
t=0 0
m=message 8888 TCP/MSRP *
a=recvonly
a=accept-types:message/cpim
a=accept-wrapped-types:*
a=path:msrp://bobpc.example.com:8888/9di4ea;tcp
a=file-selector:name:"My cool picture.jpg" type:image/jpeg
          size:4092 hash:sha-1:
          72:24:5F:E8:65:3D:DA:F3:71:36:2F:86:D4:71:91:3E:E4:A2:CE:2E
a=file-transfer-id:Q6LMoGymJdh0IKIgD6wD0jkcfgva4xvE
```

Figure 13.9 SDP answerer accepting the SDP offer for file transfer (Copyright IETF).

```
MSRP d93kswow SEND
To-Path: msrp://bobpc.example.com:8888/9di4ea;tcp
From-Path: msrp://alicepc.example.com:7654/iau39;tcp
Message-ID: 12339sdqwer
Byte-Range: 1-2048/4385
Content-Type: message/cpim

To: Bob <sip:bob@example.com>
From: Alice <sip:alice@example.com>
DateTime: 2006-05-15T15:02:31-03:00
Content-Disposition: render; filename="My cool picture.jpg";
creation-date="Mon, 15 May 2006 15:01:31 +0300";
       size=4092
Content-Type: image/jpeg

... first set of bytes of the JPEG image ...
-------d93kswow+
```

Figure 13.10 MSRP SEND request containing the first chunk of actual file (Copyright IETF).

F5: Alice sends the second and last chunk. Note that MSRP allows one to send pipelined chunks, so there is no need to wait for the 200 (OK) response from the previous chunk (Figure 13.11).

F6: Bob acknowledges the reception of the first chunk (Figure 13.12).

F7: Bob acknowledges the reception of the second chunk (Figure 13.13).

F8: Alice terminates the SIP session by sending a SIP BYE request.

F9: Bob acknowledges the reception of the BYE request and sends a 200 (OK) response.

13.1.10.2 Offerer Requests a File from the Answerer and Second File Transfer

In this example, Alice, the SDP offerer, first wishes to fetch a file from Bob, the SDP answerer. Alice knows that Bob has a specific file she wants to download. She has learned the hash of the file by some out-of-band mechanism. The hash selector is enough to produce a file selector that points to the specific file. So, Alice creates an SDP offer that contains the file descriptor. Bob accepts the file transfer and sends the file to Alice. When Alice has completely received Bob's file, she intends to send a new image file to Bob. Therefore, Alice reuses the existing SDP media line with different attributes and updates the description of the new file she wants to send to Bob's user agent server.

```
MSRP op2nc9a SEND
To-Path: msrp://bobpc.example.com:8888/9di4ea;tcp
From-Path: msrp://alicepc.example.com:7654/iau39;tcp
Message-ID: 12339sdqwer
Byte-Range: 2049-4385/4385
Content-Type: message/cpim

... second set of bytes of the JPEG image ...
-------op2nc9a$
```

Figure 13.11 MSRP SEND request containing the second chunk of actual file (Copyright IETF).

```
MSRP d93kswow 200 OK
To-Path: msrp://alicepc.example.com:7654/iau39;tcp
From-Path: msrp://bobpc.example.com:8888/9di4ea;tcp
Byte-Range: 1-2048/4385

-------d93kswow$
```

Figure 13.12 MSRP 200 OK response (Copyright IETF).

```
MSRP op2nc9a 200 OK
To-Path: msrp://alicepc.example.com:7654/iau39;tcp
From-Path: msrp://bobpc.example.com:8888/9di4ea;tcp
Byte-Range: 2049-4385/4385
-------op2nc9a$
```

Figure 13.13 MSRP 200 OK response (Copyright IETF).

In particular, Alice creates a new file transfer identifier since this is a new file transfer operation. Figure 13.14 shows the sequence flow.

F1: Alice constructs an SDP description of the file she wants to receive and attaches the SDP offer to a SIP INVITE request addressed to Bob (Figure 13.15).

NOTE: The “file-selector” attribute in the Figure 13.5 is split into two lines for formatting purposes. Real implementations will encode it in a single line. From now on, we omit the SIP details for the sake of brevity.

F2: Bob receives the INVITE request, inspects the SDP offer, computes the file descriptor, and finds a local file whose hash equals the one indicated in the SDP. Bob accepts the file transfer and creates an SDP answer as follows (Figure 13.16):

NOTE: The “file-selector” attribute in the above figure is split in two lines for formatting purposes. Real implementations will encode it in a single line.

F4: Alice opens a TCP connection to Bob. Bob then creates an MSRP SEND request that contains the file (Figure 13.17).

F5: Alice acknowledges the reception of the SEND request (Figure 13.18).

F6: Alice reuses the existing SDP media line inserting the description of the file to be sent and attaches it to a SIP re-INVITE request addressed to Bob. Alice reuses the TCP port number for the MSRP stream, but changes the MSRP session and the “file-transfer-id” value according to this specification (Figure 13.19).

NOTE: The Content-Type header field and the “file-selector” attribute in the Figure 13.19 are split in several lines for formatting purposes. Real implementations will encode it in a single line.

F7: Bob receives the re-INVITE request, inspects the SDP offer and extracts the icon body, checks the creation date and the file size, and decides to accept the file transfer. So Bob creates an

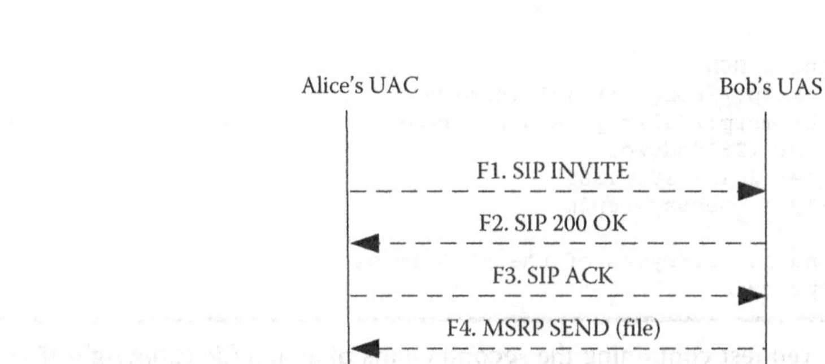

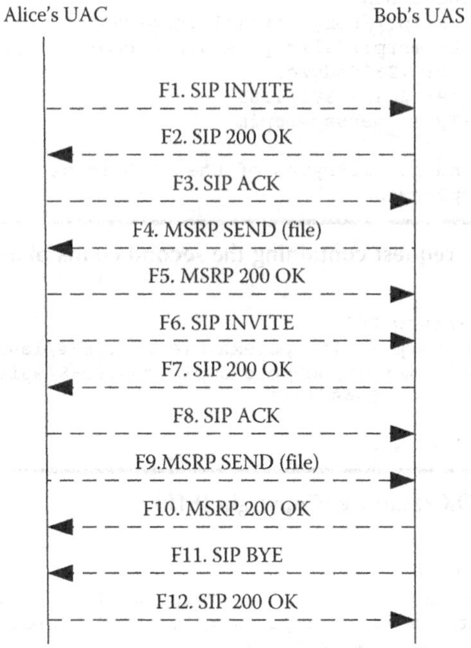

Figure 13.14 Flow diagram of an offerer requesting a file from the answerer and then sending a file to the answerer (Copyright IETF).

```
INVITE sip:bob@example.com SIP/2.0
To: Bob <sip:bob@example.com>
From: Alice <sip:alice@example.com>;tag=1928301774
Call-ID: a84b4c76e66710
CSeq: 1 INVITE
Max-Forwards: 70
Date: Sun, 21 May 2006 13:02:03 GMT
Contact: <sip:alice@alicepc.example.com>
Content-Type: application/sdp
Content-Length: [length of SDP]

v=0
o=alice 2890844526 2890844526 IN IP4 alicepc.example.com
s=
c=IN IP4 alicepc.example.com
t=0 0
m=message 7654 TCP/MSRP *
a=recvonly
a=accept-types:message/cpim
a=accept-wrapped-types:*
a=path:msrp://alicepc.example.com:7654/jshA7we;tcp
a=file-selector:hash:sha-1:
      72:24:5F:E8:65:3D:DA:F3:71:36:2F:86:D4:71:91:3E:E4:A2:CE:2E
a=file-transfer-id:aCQYuBRVoUPGVsFZkCK98vzcX2FXDIk2
```

Figure 13.15 INVITE request containing an SDP offer for file transfer (Copyright IETF).

```
v=0
o=bob 2890844656 2890855439 IN IP4 bobpc.example.com
s=
c=IN IP4 bobpc.example.com
t=0 0
m=message 8888 TCP/MSRP *
a=sendonly
a=accept-types:message/cpim
a=accept-wrapped-types:*
a=path:msrp://bobpc.example.com:8888/9di4ea;tcp
a=file-selector:type:image/jpeg hash:sha-1:
72:24:5F:E8:65:3D:DA:F3:71:36:2F:86:D4:71:91:3E:E4:A2:CE:2E
a=file-transfer-id:aCQYuBRVoUPGVsFZkCK98vzcX2FXDIk2
```

Figure 13.16 SDP answerer accepting the SDP offer for file transfer.

```
MSRP d93kswow SEND
To-Path: msrp://alicepc.example.com:7654/jshA7we;tcp
From-Path: msrp://bobpc.example.com:8888/9di4ea;tcp
Message-ID: 12339sdqwer
Byte-Range: 1-2027/2027
Content-Type: message/cpim

To: Bob <sip:bob@example.com>
From: Alice <sip:alice@example.com>
DateTime: 2006-05-15T15:02:31-03:00
Content-Disposition: render; filename="My cool photo.jpg";
creation-date="Mon, 15 May 2006 15:01:31 +0300";
modification-date="Mon, 15 May 2006 16:04:53 +0300";
read-date="Mon, 16 May 2006 09:12:27 +0300";
size=1931
Content-Type: image/jpeg

...binary JPEG image...

-------d93kswow$
```

Figure 13.17 MSRP SEND request containing the actual file.

```
MSRP d93kswow 200 OK
To-Path: msrp://bobpc.example.com:8888/9di4ea;tcp
From-Path: msrp://alicepc.example.com:7654/jshA7we;tcp
Byte-Range: 1-2027/2027

-------d93kswow$
```

Figure 13.18 MSRP 200 OK response (Copyright IETF).

SDP answer where he reuses the same TCP port number, but changes his MSRP session, according to the procedures of this specification (Figure 13.20).

NOTE: The "file-selector" attribute in the above figure is split in three lines for formatting purposes. Real implementations will encode it in a single line.

F9: If a TCP connection toward Bob is already open, Alice reuses that TCP connection to send an MSRP SEND request that contains the file (Figure 13.21).

F10: Bob acknowledges the reception of the SEND request (Figure 13.22).

F11: Then Bob terminates the SIP session by sending a SIP BYE request.

F12: Alice acknowledges the reception of the BYE request and sends a 200 (OK) response.

```
INVITE sip:bob@example.com SIP/2.0
To: Bob <sip:bob@example.com>;tag=1928323431
From: Alice <sip:alice@example.com>;tag=1928301774
Call-ID: a84b4c76e66710
CSeq: 2 INVITE
Max-Forwards: 70
Date: Sun, 21 May 2006 13:02:33 GMT
Contact: <sip:alice@alicepc.example.com>
Content-Type: multipart/related; type="application/sdp";
boundary="boundary71"
Content-Length: [length of multipart]

--boundary71
Content-Type: application/sdp
Content-Length: [length of SDP]

v=0
o=alice 2890844526 2890844527 IN IP4 alicepc.example.com
s=
c=IN IP4 alicepc.example.com
t=0 0
m=message 7654 TCP/MSRP*
i=This is my latest picture
a=sendonly
a=accept-types:message/cpim
a=accept-wrapped-types:*
a=path:msrp://alicepc.example.com:7654/iau39;tcp
a=file-selector:name:"sunset.jpg" type:image/jpeg
        size:4096 hash:sha-1:
        58:23:1F:E8:65:3B:BC:F3:71:36:2F:86:D4:71:91:3E:E4:B1:DF:2F
a=file-transfer-id:ZVE8MfI9mhAdZ8GyiNMzNN5dpqgzQlCO
a=file-disposition:render
a=file-date:creation:"Sun, 21 May 2006 13:02:15 +0300"
a=file-icon:cid:id3@alicepc.example.com

--boundary71
Content-Type: image/jpeg
Content-Transfer-Encoding: binary
Content-ID: <id3@alicepc.example.com>
Content-Length: [length of image]
Content-Disposition: icon
[..small preview icon...]

--boundary71--
```

Figure 13.19 Reuse of the SDP in a second file transfer (Copyright IETF).

13.1.10.3 Example of a Capability Indication

Alice sends an OPTIONS request to Bob (this request does not contain SDP). Bob answers with a 200 (OK) response that contains the SDP shown in Figure 13.23. The SDP indicates support for CPIM messages that can contain other MIME types. The maximum MSRP message size that the endpoint can receive is 20000 octets. The presence of the "fileselector" attribute indicates support for the file transfer offer/answer mechanism (Figure 13.24).

13.1.11 Security Considerations

The security procedures that need to be considered in dealing with the basic SDP messages described in this section (RFC 5547) are provided in Section 17.12.1.

```
v=0
o=bob 2890844656 2890855440 IN IP4 bobpc.example.com
s=
c=IN IP4 bobpc.example.com
t=0 0
m=message 8888 TCP/MSRP*
a=recvonly
a=accept-types:message/cpim
a=accept-wrapped-types:*
a=path:msrp://bobpc.example.com:8888/eh10dsk;tcp
a=file-selector:name:"sunset.jpg" type:image/jpeg
     size:4096 hash:sha-1:
        58:23:1F:E8:65:3B:BC:F3:71:36:2F:86:D4:71:91:3E:E4:B1:DF:2F
a=file-transfer-id:ZVE8MfI9mhAdZ8GyiNMzNN5dpqgzQlCO
a=file-disposition:render
```

Figure 13.20 SDP answerer accepting the SDP offer for file transfer (Copyright IETF).

```
MSRP d95ksxox SEND
To-Path: msrp://bobpc.example.com:8888/eh10dsk;tcp
From-Path: msrp://alicepc.example.com:7654/iau39;tcp
Message-ID: 13449sdqwer
Byte-Range: 1-2027/2027
Content-Type: message/cpim

To: Bob <sip:bob@example.com>
From: Alice <sip:alice@example.com>
DateTime: 2006-05-21T13:02:15-03:00
Content-Disposition: render; filename="Sunset.jpg";
creation-date="Sun, 21 May 2006 13:02:15 -0300";
size=1931
Content-Type: image/jpeg

...binary JPEG image...

-------d95ksxox+
```

Figure 13.21 MSRP SEND request containing the actual file (Copyright IETF).

```
MSRP d95ksxox 200 OK
To-Path: msrp://alicepc.example.com:7654/iau39;tcp
From-Path: msrp://bobpc.example.com:8888/eh10dsk;tcp
Byte-Range: 1-2027/2027

-------d95ksxox$
```

Figure 13.22 MSRP 200 OK response (Copyright IETF).

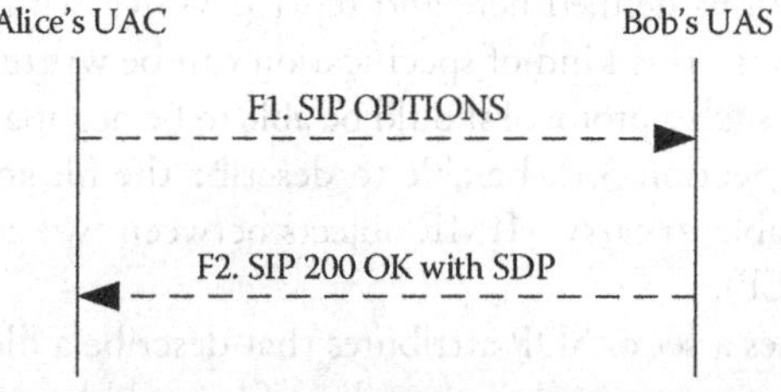

Figure 13.23 Flow diagram of a capability request (Copyright IETF).

```
v=0
o=bob 2890844656 2890855439 IN IP4 bobpc.example.com
s=-
c=IN IP4 bobpc.example.com
t=0 0
m=message 0 TCP/MSRP *
a=accept-types:message/cpim
a=accept-wrapped-types:*
a=max-size:20000
a=file-selector
```

Figure 13.24 SDP of the 200 (OK) response to an OPTIONS request (Copyright IETF).

13.1.12 IANA Considerations

SDP parameters that are described here have been registered with IANA. We have not included the IANA registration procedures here for the sake of brevity. The procedures for registration of new SDP parameters are described in RFC 5547.

13.1.13 Alternatives Considered (Appendix A of RFC 5547)

The requirements are related to the description and negotiation of the session, not to the actual file transfer mechanism. Thus, it is natural that in order to meet them it is enough to define attribute extensions and usage conventions to SDP, while MSRP itself needs no extensions and can be used as it is. This is effectively the approach taken in this specification. Another goal has been to specify the SDP extensions in such a way that a regular MSRP endpoint that does not support them could still in some cases act as an endpoint in a file transfer session, albeit with a somewhat reduced functionality.

In some ways, the aim of this specification is similar to the aim of content indirection mechanism in the SIP (RFC 4483). Both mechanisms allow a user agent to decide whether to download a file based on information about the file. However, there are some differences. With content indirection, it is not possible for the other endpoint to explicitly accept or reject the file transfer. Also, it is not possible for an endpoint to request a file from another endpoint. Furthermore, content indirection is not tied to the context of a media session, which is sometimes a desirable property. Finally, content indirection typically requires some server infrastructure, which may not always be available. It is possible to use content indirection directly between the endpoints, but in that case there is no definition for how it works for endpoints behind NATs. The level of requirements in implementations decides which solution meets the requirements.

Based on the argumentation here, this specification defines the SDP attribute extensions and usage conventions needed for meeting the requirements on file transfer services with the SDP offer/answer model, using MSRP as the transfer protocol within the session. In principle, it is possible to use the SDP extensions defined here and replace MSRP with any other similar protocol that can carry MIME objects. This kind of specification can be written as a separate document if the need arises. Essentially, such a protocol should be able to be negotiated on an SDP offer/answer exchange (RFC 3264, see Section 3.1), be able to describe the file to be transferred in an SDP offer/answer exchange, be able to carry MIME objects between two endpoints, and use a reliable transport protocol (e.g., TCP).

This specification defines a set of SDP attributes that describe a file to be transferred between two endpoints. The information needed to describe a file could be potentially encoded in a few different ways. The MMUSIC working group considered a few alternative approaches before deciding to use the encoding described in Section 13.1.6. In particular, the working group looked at the MIME "external-body" type and the use of a single SDP attribute or parameter. A MIME "external-body" could potentially be used to describe the file to be transferred. In fact, many of the SDP parameters this specification defines are also supported by "external-body" body parts.

The MMUSIC working group decided not to use "external-body" body parts because a number of existing offer/answer implementations do not support multipart bodies. The information carried in the SDP attributes defined in Section 13.1.6 could potentially be encoded in a single SDP attribute. The MMUSIC working group decided not to follow this approach because it is expected that implementations support only a subset of the parameters defined in Section 13.1.6. Those implementations will be able to use regular SDP rules in order to ignore nonsupported SDP parameters. If all information were encoded in a single SDP attribute, those rules, which relate to backwards compatibility, would need to be redefined specifically for that parameter.

13.2 SDP Format for Binary Floor Control Protocol (BFCP) Streams

This section describes RFC 4583 that specifies how to describe Binary Floor Control Protocol (BFCP) streams in SDP descriptions. User agents using the offer/answer model to establish BFCP streams use this format in their offers and answers.

13.2.1 Introduction

As discussed in the BFCP specification (RFC 4582), a given BFCP client needs a set of data in order to establish a BFCP connection to a floor control server. These data include the transport address of the server, the conference identifier, and the user identifier. One way for clients to obtain this information is to use an offer/answer (RFC 3264, see Section 3.1) exchange. This document specifies how to encode this information in the SDP session descriptions that are part of such an offer/answer exchange. User agents typically use the offer/answer model to establish a number of media streams of different types. Following this model, a BFCP connection is described as any other media stream by using an SDP "m=" line, possibly followed by a number of attributes encoded in "a=" lines.

13.2.2 Terminology

See Table 2.1.

13.2.3 Fields in the "m=" Line

This section describes how to generate an "m=" line for a BFCP stream. According to the SDP specification (RFC 4566, see Section 2), the "m=" line format is the following:

```
m=<media> <port> <transport> <fmt> ...
```

The media field MUST have a value of "application." The port field is set following the rules in RFC 4145 (see Section 10.1). Depending on the value of the "setup" attribute (discussed in Section 13.2.7), the port field contains the port to which the remote endpoint will initiate its TCP connection or is irrelevant (i.e., the endpoint will initiate the connection toward the remote endpoint) and should be set to a value of 9, which is the discard port. Since BFCP only runs on top of TCP, the port is always a TCP port. A port field value of zero has the standard SDP meaning (i.e., rejection of the media stream). We define two new values for the transport field: TCP/BFCP and TCP/Transport Layer Security (TLS)/BFCP. The former is used when BFCP runs directly on top of TCP, and the latter is used when BFCP runs on top of TLS, which in turn runs on top of TCP. The fmt (format) list is ignored for BFCP. The fmt list of BFCP "m=" lines SHOULD contain a single "*" character. The following is an example of an "m=" line for a BFCP connection:

```
m=application 50000 TCP/TLS/BFCP *
```

13.2.4 Floor Control Server Determination

When two endpoints establish a BFCP stream, they need to determine which acts as a floor control server. In the most common scenario, a client establishes a BFCP stream with a conference server that acts as the floor control server. Floor control server determination is straightforward because one endpoint can only act as a client and the other can only act as a floor control server. However, there are scenarios where both endpoints could act as a floor control server. For example, in a two-party session that involves an audio stream and a shared whiteboard, the endpoints need to decide which party will be acting as the floor control server. Furthermore, there are situations where both the offerer and the answerer act as both clients and floor control servers in the same session. For example, in a two-party session that involves an audio stream and a shared whiteboard, one party acts as the floor control server for the audio stream and the other acts as the floor control server for the shared whiteboard. All SDP ABNF syntaxes are provided in Section 2.9. However, we have repeated this here for convenience. We define the "floorctrl" SDP media-level attribute to perform floor control determination. Its ABNF syntax (RFC 4234 is obsoleted by RFC 5234) is

```
floor-control-attribute   =   "a=floorctrl:" role *(SP role)
role                      =   "c-only" / "s-only" / "c-s"
```

The offerer includes this attribute to state all the roles it would be willing to perform:

c-only: The offerer would be willing to act as a floor control client only.
s-only: The offerer would be willing to act as a floor control server only.
c-s: The offerer would be willing to act both as a floor control client and as a floor control server.

If an "m=" line in an offer contains a "floorctrl" attribute, the answerer MUST include one in the corresponding "m=" line in the answer. The answerer includes this attribute to state which role the answerer will perform. That is, the answerer chooses one of the roles the offerer is willing to perform and generates an answer with the corresponding role for the answerer. Table 13.1 shows the corresponding roles for an answerer, depending on the offerer's role.

Table 13.1 Roles (Copyright IETF)

Offerer	*Answerer*
c-only	s-only
s-only	c-only

The following are the descriptions of the roles when they are chosen by an answerer:

c-only: The answerer will act as a floor control client. Consequently, the offerer will act as a floor control server.
s-only: The answerer will act as a floor control server. Consequently, the offerer will act as a floor control client.
c-s: The answerer will act both as a floor control client and as a floor control server. Consequently, the offerer will also act both as a floor control client and as a floor control server.

Endpoints that use the offer/answer model to establish BFCP connections MUST support the "floorctrl" attribute. A floor control server acting as an offerer or as an answerer SHOULD include this attribute in its session descriptions. If the "floorctrl" attribute is not used in an offer/answer exchange, by default the offerer and the answerer will act as a floor control client and as a floor control server, respectively. The following is an example of a "floorctrl" attribute in an offer. When this attribute appears in an answer, it only carries one role:

```
a=floorctrl:c-only s-only c-s
```

13.2.5 The "confid" and "userid" SDP Attributes

We define the "confid" and the "userid" SDP media-level attributes. These attributes are used by a floor control server to provide a client with a conference ID and a user ID, respectively. Their ABNF syntax (RFC 4234 is obsoleted by RFC 5234) is

```
confid-attribute  =  "a=confid:" conference-id
conference-id     =  token
userid-attribute  =  "a=userid:" user-id
user-id           =  token
```

The "confid" and the "userid" attributes carry the integer representation of a conference ID and a user ID, respectively. Endpoints that use the offer/answer model to establish BFCP connections MUST support the "confid" and the "userid" attributes. A floor control server acting as an offerer or as an answerer SHOULD include these attributes in its session descriptions.

13.2.6 Association between Streams and Floors

We define the "floorid" SDP media-level attribute. Its ABNF syntax (RFC 4234 was made obsolete by RFC 5234) is

```
floor-id-attribute  =      "a=floorid:" token [" mstrm:" token *(SP
                           token)]
```

The "floorid" attribute is used in BFCP "m=" lines. It defines a floor identifier and, possibly, associates it with one or more media streams. The token representing the floor ID is the integer representation of the floor ID to be used in BFCP. The token representing the media stream is a pointer to the media stream, which is identified by an SDP label attribute (RFC 4574, see Section 5.1). Endpoints that use the offer/answer model to establish BFCP connections MUST support the "floorid" and the "label" attributes. A floor control server acting as an offerer or as an answerer SHOULD include these attributes in its session descriptions.

13.2.7 TCP Connection Management

The management of the TCP connection used to transport BFCP is performed using the "setup" and "connection" attributes, as defined in RFC 4145 (see Section 10.1). The "setup" attribute indicates which of the endpoints (client or floor control server) initiates the TCP connection. The "connection" attribute handles TCP connection reestablishment. The BFCP specification (RFC 4582) describes a number of situations in which the TCP connection between a client and the floor control server needs to be reestablished. However, that specification does not describe the reestablishment process because this process depends on how the connection was established in the first place. BFCP entities using the offer/answer model follow these rules. When the existing TCP connection is reset following the rules in RFC 4582, the client SHOULD generate an offer toward the floor control server in order to reestablish the connection. If a TCP connection cannot deliver a BFCP message and times out, the entity that attempted to send the message (i.e., the one that detected the TCP timeout) SHOULD generate an offer in order to reestablish the TCP connection. Endpoints that use the offer/answer model to establish BFCP connections MUST support the "setup" and "connection" attributes.

13.2.8 Authentication

When a BFCP connection is established using the offer/answer model, it is assumed that the offerer and the answerer authenticate each other using some mechanism. Once this mutual authentication takes place, all the offerer and the answerer need to ensure is that the entity from which they are receiving BFCP messages is the same as the one that generated the previous offer or answer. When SIP is used to perform an offer/answer exchange, the initial mutual authentication takes place at the SIP level. Additionally, SIP uses S/MIME specified in RFC 3850 (that was made obsolete by RFC 5750) to provide an integrity-protected channel with optional confidentiality for the offer/answer exchange. BFCP takes advantage of this integrity-protected offer/answer exchange to perform authentication. Within the offer/answer exchange, the offerer and answerer exchange the fingerprints of their self-signed certificates. These self-signed certificates are then used to establish the TLS connection that will carry BFCP traffic between the offerer and the answerer.

BFCP clients and floor control servers follow the rules in RFC 4572 (see Section 16.3) regarding certificate choice and presentation. This implies that unless a "fingerprint" attribute is included in the session description, the certificate provided at the TLS level MUST either be directly signed by one of the other party's trust anchors or be validated using a certification path that terminates at one of the other party's trust anchors specified in RFC 3280 (that was made obsoleted by RFC 3950). Endpoints that use the offer/answer model to establish BFCP connections MUST support the "fingerprint" attribute and SHOULD include it in their session descriptions. When TLS is used, once the underlying TCP connection is established, the answerer acts as the TLS server regardless of its role (passive or active) in the TCP establishment procedure.

13.2.9 Examples

For the purpose of brevity, the main portion of the session description is omitted in the examples, which only show "m="' lines and their attributes. The following is an example of an offer sent by a conference server to a client.

```
m=application 50000 TCP/TLS/BFCP *
a=setup:passive
```

```
a=connection:new
a=fingerprint:SHA-1 \
      4A:AD:B9:B1:3F:82:18:3B:54:02:12:DF:3E:5D:49:6B:19:E5:7C:AB
a=floorctrl:s-only
a=confid:4321
a=userid:1234
a=floorid:1 m-stream:10
a=floorid:2 m-stream:11
m=audio 50002 RTP/AVP 0
a=label:10
m=video 50004 RTP/AVP 31
a=label:11
```

Note that due to RFC formatting conventions, this document splits SDP across lines whose content would exceed 72 characters. A backslash character marks where this line folding has taken place. This backslash and its trailing CRLF (carriage-return line-feed) and whitespace would not appear in actual SDP content. The following is the answer returned by the client.

```
m=application 9 TCP/TLS/BFCP *
a=setup:active
a=connection:new
a=fingerprint:SHA-1 \
      3D:B4:7B:E3:CC:FC:0D:1B:5D:31:33:9E:48:9B:67:FE:68:40:E8:21
a=floorctrl:c-only
m=audio 55000 RTP/AVP 0
m=video 55002 RTP/AVP 31
```

13.2.10 Security Considerations

The security procedures that need to be considered in dealing with the basic SDP messages described in this section (RFC 4583) are provided in Section 17.12.2.

13.2.11 IANA Considerations

SDP parameters that are described here have been registered with IANA. We have not included the IANA registration procedures here for the sake of brevity. The procedures for registration of new SDP parameters are described in RFC 4583.

13.3 Summary

We have described the application-specific extensions such as file (e.g. text, audio, video, MIME, and others) transfer and conference floor sharing in SDP. The MSRP used for actual file transfer between the conferencing parties is specified for capability negotiations between the communicating parties. However, we have explained in detail that the characteristics/features such as content type, size, hash, and name of files need to be negotiated using the SDP offer/answer model before actual transfer of the file takes place. According, SDP syntaxes are extended to carry those features for negotiations. The details of each feature of a file are explained in this section. In addition, file disposition types, MSRP protocol operation, and file-icon and file-transfer-id attribute are described in great length. Furthermore, offerer's and answerer's behavior in both file sending and

receiving are discussed. We have also described the specific MSRP session establishment specifying in "m=" lines in SDP describing unidirectional transfer of a file. The capability such as aborting an ongoing file transfer operation is also explained here.

We have described another application known as the BFCP that manages joint or exclusive access to shared resources in a (multiparty) conferencing environment. The BFCP is used between floor participants and floor control servers and between floor chairs (i.e., moderators) and floor control servers. Basically, the BFCP provides a means for floor participants to send floor requests to floor control servers, for floor control servers to grant or deny requests to access a given resource from floor participants, and for floor chairs to send floor control servers decisions regarding floor requests. The BFCP client and server exchange the data that includes the transport address of the server, the conference identifier, and the user identifier using SDP offer and answer. A BFCP connection is described as any other media stream by using an SDP 'm=' line, possibly followed by a number of attributes encoded in "a=" lines. We have defined the "floorctrl," "confid," "userid," and "floorid" SDP media-level attributes. The association between media streams and floor, TCP connection management, and authentication for BFCP are specified in detail.

13.4 Problems

1. What different kinds of multimedia applications can used among the conference participants in multimedia conferencing? What are the possible contents in in a MIME file? How do the performance characteristics differ among text, still image, animation, audio, and video file?
2. What are the different types of files that can be transferred using offer/answer mechanisms specified in RFC 5547? Why does RFC 5547 use MSRP for actual transfer of a specific file in multimedia conferencing? What features are not in MSRP for using the SDP offer/answer model for negotiations in choosing a particular file between the conference participants?
3. What SDP-level media attributes defined in RFC 5547 allow one to negotiate a particular file between the conference participants using detail call flows showing the usage of each attribute including offerer's and answerer's behavior for transferring and receiving files, aborting an ongoing file transfer operation, and MSRP usage?
4. What are the possible alternative approaches of RFC 5547 that could offer the same capability for negotiating capabilities using SDP and actual transfer of the file?
5. What is the BFCP application? What are the characteristics of BFCP? What is the role of the floor control server? Describe the features of BFCP in granting access to the floor using detail call flows.
6. What SDP media-level attributes have been standardized in RFC 4583? Show the usage of each attribute including association between media streams and floor, TCP connection management, and authentication of BFCP with conferencing architecture and call flows.

Chapter 14

Service-Specific Extension in SDP

Abstract

In this chapter, we describe Request for Comments (RFCs) 6064, 5139, and 2848 that deal with the service-specific extensions in Session Description Protocol (SDP). Services may usually be created with a combination of software- and network-infrastructure optimizing for any particular function. This chapter describes SDP extensions defined in RFC 6064 for the packet-switched streaming service (PSS) and the multimedia broadcast/multicast service (MBMS) defined by the *Third Generation Partnership Project* (3GPP). 3GPP intends these services for both mobile and fixed users over the networks consisting of both wire-line and mobile cellular networks. RFC 6064 also extends the Real-Time Streaming Protocol for PSS and MBMS services. Similarly, RFC 5159 extends SDP attributes for the Open Mobile Alliance Broadcast Service and Content Protection. The public-switched telephone network (PSTN)/Internet interworking (PINT) is another multimedia service defined in RFC 2848 where the signaling traffic is sent over the Internet, but the media is transferred over the PSTN. That is, the PINT defines a protocol for invoking certain telephone services from an IP network. These services include placing basic calls, sending and receiving faxes, and receiving content over the telephone. The protocol is specified as a set of enhancements and additions to the SDP and Session Initiation Protocol (SIP). Note that we have covered the PINT services here in great detail to clarify how SDP can be used for negotiations of PINT services features between the conferencing parties having telephones connected to both IP and PSTN network.

14.1 SDP and RTSP Extensions for 3GPP Packet-Switched Services

This subsection describes Request for Comments (RFC) 6064 that specifies extensions in both Session Description Protocol (SDP) and Real-Time Streaming Protocol (RTSP). The packet-switched streaming service (PSS) and the multimedia broadcast/multicast service (MBMS) defined by *Third Generation Partnership Project* (3GPP) use the SDP and the RTSP with some extensions. This RFC provides information about these extensions and registers the RTSP and SDP extensions with Internet Assigned Numbers Authority (IANA).

14.1.1 Introduction

3GPP has specified the PSS that uses both RTSP (RFC 2326) and SDP (RFC 4566, see Section 2). The service is specified in technical specifications (TS) 26.233 [1] and [2] in Release 4 and subsequent releases. The basic service defined in Release 4 is enhanced in Release 5 with capability exchange and in Release 6 with a number of features, such as adaptation, digital rights management (DRM), progressive download, and a streaming server file format defined in Reference [3]. Fast start-up and content switching are addressed in Release 7. 3GPP has also specified the MBMS that uses SDP. The IP-layer protocols used by this service are specified in TS 26.346 Release 6 [4]. Release 7 extends the MBMS user service to also work with unicast bearers for interactive and streaming traffic classes.

In the process of defining these services, there has occasionally been a need to extend both SDP and RTSP functionalities. These extensions have mostly been in the form of SDP attributes and RTSP headers and option-tags. 3GPP uses the name "feature-tags" (like RTSP 2.0 for what RTSP 1.0 calls "option-tags"); "option-tag" is the name that will be used in this document. The purpose of this informational document is to register these SDP and RTSP extensions, in order to avoid future conflicts, and to raise the awareness of their existence within Internet Engineering Task Force (IETF).

In Section 14.1.5.4, this document defines three SDP protocol identifiers used in MBMS to enable the usage of block-based forward error correction (FEC). The SDP protocol identifiers require an RFC be defined and registered. As this is an RFC from the IETF stream, any semantic change will require a new IETF-approved RFC. The other SDP and RTSP extensions registered by this document are not normatively defined in this specification. Instead, the normative definitions are referenced by the registrations. 3GPP can update the normative definition in future versions of their specifications. However, to ensure that such a change is visible in the IETF, at minimum, IANA should be notified and the reference to the 3GPP specification updated, and preferably an updated version of this RFC published.

The document begins with two sections presenting the SDP extensions for PSS and MBMS, respectively. They are followed by a section noting that offer/answer considerations are not applicable here. The subsequent section presents the extensions of RTSP for PSS. The IANA registration of SDP attributes and protocol identifiers is given in Section 14.1.8.1, and the RTSP headers and option tags in Section 14.1.8.2. For normative descriptions of all SDP and RTSP extensions, we refer to TS 26.234 [2] and TS 26.346 [4].

14.1.2 Glossary

- 3GP: 3GPP file format, a multimedia file format based on the ISO base media file format, existing in different profiles intended for multimedia messages, direct playback on clients, progressive download, usage on servers to deliver on-demand multimedia sessions in PSS, or servers sending MBMS sessions.
- 3GPP: *Third Generation Partnership Project*; see http://www.3gpp.org for more information about this organization.
- FEC: Forward Error Correction
- MBMS: Multimedia broadcast/multicast service, a service defined by 3GPP that utilizes broadcast or multicast technology in combination with unicast for delivery of a wide range of content to mobile terminals.

- PSS: Packet-switched streaming service, a unicast-based streaming service for delivery of on-demand or live streaming multimedia content to mobile terminals.
- RTSP: Real-Time Streaming Protocol; see RFC 2326.
- SDP: Session Description Protocol; see RFC 4566 (see Section 2).
- SRTP: Secure Real-time Transport Protocol; see RFC 3711.
- QoE: Quality of Experience, the quality level of the user experience of a service. In PSS, this is estimated by a combination of application-level metrics.
- QOS: Quality of Service, the quality (properties) that the network provides toward the upper-layer service.

Additionally, see Section 2.2/Table 2.1.

14.1.3 Applicability Statement

This document describes 3GPP-defined extensions to SDP (RFC 4566, see Section 2) and RTSP (RFC 2326) and registers attributes that are normatively defined in 3GPP TS 26.234, 26.244, and 26.346, up to the referenced versions of the respective documents. The SDP and RTSP extensions have only been defined for usage with the 3GPP service in mind. The applicability for usage outside of these services has not been considered nor addressed. Usage of these attributes in other contexts may require further definitions or clarifications. For example, all SDP attributes lack offer/answer usage rules specified in RFC 3264 (se Section 3.1), which currently makes it impossible to use them with offer/answer. Please note that change control of these SDP and RTSP extensions belongs to 3GPP.

14.1.4 PSS SDP Extensions

The PSS specification [2] defines a number of different SDP attributes for different purposes. They are listed in this specification described here, grouped by their purpose. The text is intentionally not specific enough to allow implementation from this document. The normative definition is in the 3GPP technical specification cited.

14.1.4.1 Video Buffering Attributes

The following attributes are used to provide parameters for the video buffer model provided in Annex G and Section 5.3.3.2 of Reference [2]. The attributes were defined in Release 5 as "X-" attributes and, at the time, were not considered for registration. In hindsight, however, they should not have been "X-" attributes, and they should have been registered, as the registration rules of SDP (RFC 4566, see Section 2) point out. Changing their names today is impossible due to the deployed base of millions of mobile handsets supporting PSS, and therefore they are registered in their current form. All attributes are defined at the media level.

- The "a=X-predecbufsize" attribute provides the size of the predecoder buffer in bytes.
- The "a=X-initpredecbufperiod" attribute provides the time during which a receiver should initially buffer, in 90 kHz ticks, before starting to consume the data in the buffer in order to ensure that underflow does not occur, assuming correct data delivery.
- The "a=X-initpostdecbufperiod" attribute provides the initial buffering period, in 90 kHz ticks, for the post-decoder buffer present in H.263 and MPEG-4 Visual.
- The "a=X-decbyterate" attribute indicates the maximum peak byte decoding rate used in the verification of the Annex G buffer model expressed in bytes per second.

- The "a=3gpp-videopostdecbufsize" attribute is used to indicate the value used in determining the H.264 video post-decoder buffer size.

Note that complete descriptions of these attributes can be found in Section 5.3.3.2 of Reference [2].

14.1.4.2 Video Frame Size Attribute

This media-level attribute provides the receiver with the largest picture size a specific H.263 payload type will carry within the session. The attribute has the following form (see Section 5.3.3.2 of Reference [2]):

```
"a=framesize:<payload type number> <width>-<height>"
```

14.1.4.3 Integrity-Protection Configuration Attributes

These attributes are all used to configure the integrity-protection mechanism defined in Annex K (Sections K.2.2.1, K.2.2.2, and K.2.2.3) of Reference [2].

- The session-level attribute "a=3GPP-Integrity-Key" carries the integrity key used to derive SRTP master keys for integrity protection. The key is protected in different ways depending on a method identifier. When using Open Mobile Alliance (OMA) DRM key management, the key is encrypted using Advanced Encryption Standard [5] before it is base64-encoded (RFC 4648).
- The media-level attribute "a=3GPP-SRTP-Config" is used to configure SRTP for integrity protection and contains an integrity nonce, a key salt used in deriving the SRTP master key from the integrity key, and any SRTP configuration parameters, such as the integrity-tag length.
- The session-level attribute "a=3GPP-SDP-Auth" is used to carry an authentication tag calculated over certain parts of the SDP to prevent manipulation of the security attributes.

14.1.4.4 The Alternative Attributes

Two media-level and one session-level attributes are used in a mechanism for providing alternative SDP lines. One or more SDP lines at media level can be replaced, if desired, by alternatives. The mechanism is backwards compatible in the way that a receiver that does not support the attributes will get the default configuration. The different alternatives can be grouped using different attributes that can be specified hierarchically with a top and a lower level. 3GPP Release 6 supports grouping based on bit rate, according to the SDP bandwidth modifiers AS (RFC 4566, see Section 2) and Transport Independent Application Specific maximum (TIAS) (RFC 3890, see Section 12.4) and language. The SDP attributes (see Sections 5.3.3.3 and 5.3.3.4 of Reference [2]) are

- The media-level attribute "a=alt:<id>:<SDP-Line>" carries any SDP line and an alternative identifier.
- The media-level attribute "a=alt-default-id:<id>" identifies the default configuration to be used in groupings.
- The session-level attribute "a=alt-group" is used to group different recommended media alternatives. This allows providing aggregated properties for the whole group according to the grouping type. Language and bit rate are two defined grouping types.

14.1.4.5 Adaptation Attribute

The media-level SDP attribute "a=3GPP-Adaptation-Support" (see Section 5.3.3.5 in Reference [2]) is defined as part of the negotiation procedure of the PSS adaptation mechanism. The attribute carries a single value indicating how often the Real-Time Transport Control Protocol (RTCP) "Next Application Data Unit" APP packet shall be included in sent RTCP compound packets. The adaptation mechanism allows the client to provide the server with information on the available transmission bit rate and receiver buffer status.

14.1.4.6 Quality of Experience Attribute

The session- and media-level attribute "a=3GPP-QoE-Metrics" (see Section 5.3.3.6 of Reference [2]) is used to negotiate the usage of the QoE metrics. The included parameters indicate which metrics should be used, over which duration there should be measurements, and how often reports should be sent.

14.1.4.7 Asset Information Attribute

The session- and media-level attribute "a=3GPP-Asset-Information" (see Section 5.3.3.7 of Reference [2]) can exist in multiple instances in a description and describes different types of asset information. The different asset classes defined in Release 6 are Title, Description, Copyright, Performer, Author, Genre, Rating, Classification, Keywords, Location, Album, and Recording Year. The different assets are described with a base64-encoded asset box from the 3GPP file format [3].

14.1.5 MBMS SDP Extensions

The MBMS specification [4] defines a number of different SDP attributes for different purposes. They are listed here.

14.1.5.1 MBMS Bearer Mode Declaration Attribute

The session- and media-level attribute "a=mbms-mode" (see Section 7.3.2.7 of Reference [4]) is used to describe MBMS broadcast mode media. The attribute may be used at the session level to set the default for all media and at the media level to specify differences between media. However, the attribute is never used at the session level when the session includes MBMS multicast mode media or at the media level to describe MBMS multicast mode media.

14.1.5.2 FEC Flow ID Attribute

The media-level attribute "a=mbms-flowid" (see Section 8.3.1.9 of Reference [4]) maps one or more FEC source block flow dialogs identification (IDs) to their corresponding destination IP addresses and User Datagram Protocol (UDP) port numbers. It is present in each SDP media block for repair packet streams.

14.1.5.3 MBMS Repair Attribute

The session- and media-level attribute "a=mbms-repair" (see Section 8.3.1.8 of Reference [4]) is used to provide FEC repair packets with non-FEC specific parameters. For Release 6, one such parameter is defined to specify the required minimum receiver buffer time.

14.1.5.4 SDP Protocol Identifiers for FEC

MBMS defines a mechanism to provide block-based FEC for UDP-based traffic. This solution uses the SDP protocol "proto" identifier to identify the media streams that use the FEC shim layer. The media streams may be source streams or repair streams. As required by SDP (RFC 4566, see Section 2), these protocol identifiers are normatively defined in this document in accordance with their usage specified by 3GPP.

14.1.5.4.1 RTP Protocol Identifiers

For FEC-protected Real-Time Transport Protocol (RTP) streams, the following two "proto" identifiers are defined:

- UDP/MBMS-FEC/RTP/AVP
- UDP/MBMS-FEC/RTP/SAVP

They indicate the usage of UDP (RFC 0768) with MBMS FEC source packet formats, as defined in Section 8.2.2.4 of Reference [4], that transport RTP packets in accordance with the Audio Video Profile (AVP) (RFC 3551) or Secure Audio Video Profile (SAVP) (Secure RTP) (RFC 3711) profiles, respectively. These protocol identifiers SHALL use the media formats ("fmt") namespace rules that are used for RTP/AVP and RTP/SAVP, respectively.

14.1.5.4.2 FEC Repair Data Identifier

A media stream carrying MBMS FEC repair information over UDP requires its own "proto" identifier. Protocol identifier "UDP/MBMS-REPAIR" identifies the FEC repair packet containing the protocol combination of UDP (RFC 0768), FEC repair payload ID, and repair symbols as specified in Section 8.2.2.5 of Reference [4]. The "fmt" namespace is not used and SHALL be set to "*."

14.1.5.5 Video Buffering Attribute

The PSS media-level buffer attribute "a=X-initpredecbufperiod" (see Section 14.1.4.1) that specifies an initial buffering time is also used for MBMS in Release 7. It is mainly intended for video streams but may be used for other media types as well (see Section 8.3.1.1 of Reference [4]).

14.1.6 SDP Offer/Answer Consideration

The usage of the SDP attributes in an offer/answer (RFC 3264, see Section 3.1) context is not defined. These SDP attributes are defined for use in a declarative context and for PSS specifically in the RTSP (RFC 2326) context.

14.1.7 PSS RTSP Extensions

The RTSP extensions for PSS consist of a number of new RTSP headers and option tags and a narrowing of Uniform Resource Identifier (URI) usage in regards to 3GP files. The headers and option tags are informatively described here; see Reference [2] for the normative declaration.

14.1.7.1 3GPP-Link-Char Header

The "3GPP-Link-Char" header (see Section 5.3.2.1 of Reference [2]) is used by clients to provide the server with quality of service (QOS) information about the wireless link it is currently using. The header can be used to provide the server with three different QOS parameters:

- Guaranteed bandwidth
- Maximum bandwidth
- Maximum transfer delay

The header may be included in RTSP requests using any of the methods SETUP, PLAY, OPTIONS, or SET_PARAMETER.

14.1.7.2 3GPP-Adaptation Header

The "3GPP-Adaptation" header (see Section 5.3.2.2 of Reference [2]) is used by the client to provide the server with adaptation-related parameters and to indicate support of the adaptation function. The header carries the resource identification as a URI, the client's buffer size, and the desired target time. The header may be included in requests using the methods SETUP, PLAY, OPTIONS, and SET_PARAMETER. The response to a request using this method shall include this header.

14.1.7.3 3GPP-QoE-Metrics Header

The "3GPP-QoE-Metrics" header (see Section 5.3.2.3.1 of Reference [2]) is used to negotiate the usage of the QoE metrics (see Section 11 of Reference [2]). The header may be included in requests and responses using the SETUP, SET_PARAMETER, OPTIONS, or PLAY method.

14.1.7.4 3GPP-QoE-Feedback Header

The "3GPP-QoE-Feedback" header (see Section 5.3.2.3.2 of Reference [2]) is used to carry QoE metrics from the client to the server when it reports, which happens either during or at the end of the media delivery. The header may be included in requests using the SET_PARAMETER, PAUSE, or TEARDOWN method.

14.1.7.5 Video Buffer Headers

PSS uses several headers to provide the client with the different buffer parameters. They provide the buffer status at the point of a stream from which a PLAY request plays. These headers may only be used in PLAY responses. See Section 5.3.2.4 and Annex G of Reference [2] for normative definitions. The three "x-" headers were defined in 3GPP Release 5. When it was realized that they should not have been given "x-" names, it was too late to rename them due to deployment. The RTSP headers are

- x-predecbufsize
- x-initpredecbufperiod
- x-initpostdecbufperiod
- 3gpp-videopostdecbufsize

14.1.7.6 Integrity Protection

The integrity-protection mechanism defined in PSS Annex K uses the "3GPP-Freshness-Token" RTSP header (see Section K.2.2.4 of Reference [2]) to carry a freshness token in DESCRIBE requests.

14.1.7.7 RTSP URI Extension

The PSS specification also defines syntax for referencing tracks within the 3GP file format [PSS-3GP]. The 3GP format is based on the ISO base media file format and is defined in several different profiles, including a streaming-server profile, in Release 6. This syntax is fully contained within the generic URI syntax defined for RTSP URIs. It is a syntax restriction that server manufacturers follow to allow clients or proxies to understand what encodes the track number in the URI. This is provided for information only. To identify a track within a 3GPP file, the last URI segment has to contain a structure that is <alpha string>=<track nr> (see Section 5.3.3.1 of Reference [2]).

14.1.7.8 Fast Start-Up and Content Switching

Release 7 of PSS defines a number of extensions in terms of headers and option tags (see Section 5.5 of Reference [2]) for support of fast start-up and switching of content for on-demand and live applications built on top of PSS. Clients are enabled to reuse the existing RTSP control session and RTP resources while switching to new content. The RTSP headers are

- Switch-Stream
- SDP-Requested
- Pipelined-Requests

The RTSP option tags are:

- 3gpp-pipelined
- 3gpp-switch
- 3gpp-switch-req-sdp
- 3gpp-switch-stream

14.1.8 IANA Considerations

SDP parameters that are described here have been registered with IANA. We have not included the IANA registration procedures here for the sake of brevity. The procedures for registration of new SDP parameters are described in RFC 6064.

14.1.9 Security Considerations

The security procedures that need to be considered in dealing with the basic SDP messages described in this section (RFC 6064) are provided in Section 17.13.1.

14.2 SDP Attributes for OMA Services

This section describes RFC 5139 that provides descriptions of the SDP attributes used by the OMA's Broadcast (BCAST) Service and Content Protection specification.

14.2.1 Introduction

The OMA BCAST group specifies service and content protection mechanisms for broadcast services over wireless networks. As part of that specification, several new SDP attributes are necessary to allow the broadcast server to signal the service and content protection parameters to clients. Section 2.8.2.4 of this book (that is, Section 8.2.4 of RFC 4566) requires that new SDP attributes be registered through IANA with name, contact information, and description (and other similar parameters). A standards track specification is RECOMMENDED if the new attribute(s) will have widespread use and interoperability considerations. OMA BCAST specifications are expected to be used by broadcast wireless systems based on 3GPP MBMS, 3GPP2 Broadcast and Multicast Services, and Digital Video Broadcasting – Handheld. Although this would typically be considered a "widespread" use, in this case IETF chose to use a nonstandards track RFC to register the SDP attributes because OMA maintains change control of the documents that specify the interpretation of the values in the attributes.

This document provides descriptions of the SDP attributes used in the OMA BCAST Service and Content Protection specification [6].

14.2.2 Terminology

See Section 2.2/Table 2.1.

14.2.3 New SDP Attributes

The following new SDP attributes have been specified:

- a=bcastversion:<major>.<minor>
- a=stkmstream:<id of the stkm stream>
- a=SRTPAuthentication:<id for SRTP authentication algorithm value>
- a=SRTPROCTxRate:<ROC transmission rate>

See Section 14.2.5 for details on IANA considerations.

14.2.4 Security Considerations

The security procedures that need to be considered in dealing with the basic SDP messages described in this section (RFC 5159) are provided in Section 17.13.2.

14.2.5 IANA Considerations

SDP parameters that are described here have been registered with IANA. We have not included the IANA registration procedures here for the sake of brevity. The procedures for registration of new SDP parameters are described in RFC 5159.

14.3 SDP and SIP Extensions for PSTN/Internet Interworking Services

This section describes RFC 2848 that specifies a set of extensions and additions in SDP and Session Initiation Protocol (SIP) for the PSTN/Internet Interworking (PINT) service protocols for invoking certain telephone services from an IP network. These services include placing basic calls, sending and receiving faxes, and receiving content over the telephone.

14.3.1 Introduction

The desire to invoke certain telephone call services from the Internet has been identified by many different groups (users, public and private network operators, call center service providers, and equipment vendors, see RFC 2458). The generic scenario is as follows (when the invocation is successful):

1. An IP host sends a request to a server on an IP network.
2. The server relays the request into a telephone network.
3. The telephone network performs the requested call service.

As examples, consider a user who wishes to have a callback placed to his/her telephone. It may be that a customer wants someone in the support department of some business to call back. Similarly, a user may want to hear some announcement of a weather warning sent from a remote automatic weather service in the event of a storm. We use the term PINT service to denote such a complete transaction, starting with the sending of a request from an IP client and including the telephone call itself. PINT services are distinguished by the fact that they always involve two separate networks: an IP network to request the placement of a call, and the global switched telephone network (GSTN) to execute the actual call. It is understood that intelligent network (I.N.) systems, private Private Branch Exchanges (PBXs), cellular phone networks, and the Integrated Switched Digital Network (ISDN) can all be used to deliver PINT services. Also, the request for service might come from within a private IP network disconnected from the whole Internet.

The requirements for the PINT protocol were deliberately restricted to providing the ability to invoke a small number of fixed telephone call services. These "Milestone PINT services" are specified in Section 14.3.2. Great care has been taken, however, to develop a protocol that is aligned with other Internet protocols where possible, so that future extensions to PINT could develop along with Internet conferencing. Within the Internet conference architecture, establishing media calls is done via a combination of protocols. SIP (RFC 2543 was made obsolete by RFC 3261, also see Section 1.2) is used to establish the association between the participants within the call (this association between participants within the call is called a "session"), and SDP (RFC 2327 was made obsolete by RFC 4566, see Section 2) is used to describe the media to be exchanged within the session. The PINT protocol uses these two protocols together, providing some extensions and enhancements to enable SIP clients and servers to become PINT clients and servers.

A PINT user who wishes to invoke a service within the telephone network uses SIP to invite a remote PINT server into a session. The invitation contains an SDP description of the media session that the user would like to take place. This might be a "sending a fax session" or a "telephone call session," for example. In a PINT service execution session the media is transported over the phone system, while in a SIP session the media is normally transported over an Internet. When used to invoke a PINT service, SIP establishes an association between a requesting PINT client and the PINT server responsible for invoking the service within the telephone network. These two entities are not the same as the telephone network entities involved in the telephone network service. The SIP messages carry within their SDP payloads a description of the telephone network media session.

Note that the facts that a PINT server accepts an invitation and that a session is established are no guarantee that the media will be successfully transported. (This is analogous to the fact that if a SIP invitation is accepted successfully is no guarantee against a subsequent failure of audio hardware.) The particular requirements of PINT users lead to some new messages. When a PINT server agrees to send a fax to telephone B, it may be that the fax transmission fails after part of the fax is sent. Therefore, the PINT client may wish to receive information about the status of the actual telephone call session that was invoked as a result of the established PINT session. Three new requests, SUBSCRIBE, UNSUBSCRIBE, and NOTIFY are added here to the initial SIP specification to allow this.

The enhancements and additions specified here are not intended to alter the behavior of baseline SIP or SDP in any way. The purpose of PINT extensions is to extend the usual SIP/SDP services to the telephone world. Apart from integrating well into existing protocols and architectures and the advantages of reuse, this means that the protocol specified here can handle a class of call services wider than just the milestone services. The rest of this chapter is organized as follows: Section 14.3.2 describes the PINT milestone services; Section 14.3.3 specifies the PINT functional and protocol architecture; Section 14.3.4 gives examples of the PINT 1.0 extensions of SIP and SDP; Section 14.3.5 contains some security considerations for PINT. The final section contains descriptions of how the PINT protocol may be used to provide service over the GSTN. For a summary of the extensions to SIP and SDP specified in this document, Section 14.3.3.2 gives a combined list, plus lists describing the extensions to SIP and SDP, respectively.

14.3.1.1 Glossary

- Executive system: A system that interfaces with a PINT Server and a telephone network that executes a PINT service. It need not be directly associated with the Internet and is represented by the PINT server in transactions with Internet entities.
- PINT client: An Internet host that sends requests for invocation of a PINT service, in accordance with this document
- PINT gateway: An Internet host that accepts requests for PINT service and dispatches them onwards toward a telephone network
- PINT service: A service invoked within a phone system in response to a request received from a PINT client
- Requesting User – The initiator of a request for service. This role may be distinct from that of the "party" to any telephone network call that results from the request.
- Requestor: An Internet host from which a request for service originates
- (Service call) party – A person involved in a telephone network call that results from the execution of a PINT service request or a telephone network-based resource that is involved (such as an automatic fax sender or a text-to-speech unit)

14.3.2 PINT Milestone Services

The original motivation for defining this protocol was the desire to invoke the following three telephone network services from within an IP network:

14.3.2.1 Request to Call

A request is sent from an IP host that causes a phone call to be made, connecting party A to some remote party B.

14.3.2.2 Request to Fax Content

A request is sent from an IP host that causes a fax to be sent to fax machine B. The request MAY contain a pointer to the fax data (that could reside in the IP network or in the telephone network)

OR the fax data itself. The content of the fax MAY be text OR some other more general image data. The details of the fax transmission are not accessible to the IP network but remain entirely within the telephone network. Note that this service does not relate to "Fax over IP": the IP network is only used to send the request that a certain fax be sent. Of course, it is possible that the resulting telephone network fax call happens to use a real-time IP fax solution, but this is completely transparent to the PINT transaction.

14.3.2.3 Request to Speak/Send/Play Content

A request is sent from an IP host that causes a phone call to be made to user A and for some sort of content to be spoken out. The request MUST EITHER contain a Uniform Resource Locator (URL) pointing to the content OR include the content itself. The content MAY be text OR some other more general application data. The details of the content transmission are not accessible to the IP network but remain entirely within the telephone network. This service could equally be called "Request to Hear Content"; the user's goal is to hear the content spoken to him/her. The mechanism by which the request is formulated is outside the scope of this document; however, an example might be that a Web page has a button that when pressed causes a PINT request to be passed to the PSTN, resulting in the content of the page (or other details) being spoken to the person.

14.3.2.4 Relation between PINT Milestone Services and Traditional Telephone Services

There are many different versions and variations of each telephone call service invoked by a PINT request. Consider as an example what happens when a user requests to call 1-800-225-5287 via the PINT request-to-call (R2C) service. There may be thousands of agents in the call center, and there may be any number of sophisticated algorithms and pieces of equipment that are used to decide exactly which agent will return the call. Once this choice is made, there may be many different ways to set up the call: The agent's phone might ring first and only then the original user will be called; perhaps the user will be called first and hear some horrible music or prerecorded message while the agent is located.

Similarly, when a PINT request causes a fax to be sent, hundreds of fax protocol details must be negotiated, as well as transmission details within the telephone networks used. PINT requests do not specify too precisely the exact telephone-side service. Operational details of individual events within the telephone network that executes the request are outside the scope of PINT. This does not preclude certain high-level details of the telephone network session from being expressed within a PINT request. For example, it is possible to use the SDP "lang" attribute to express a language preference for the request-to-hear-content (R2HC) service. If a particular PINT system wishes to allow requests to contain details of the telephone-network-side service, it uses the SDP attribute mechanism (see Section 14.3.3.4.2).

14.3.3 PINT Functional and Protocol Architecture

14.3.3.1 PINT Functional Architecture

Familiarity is assumed with SIP 2.0 specified in RFC 2543 (that was made obsolete by RFC 3261, also see Section 1.2) and with SDP specified in RFC 2327 (that was made obsolete by RFC 4566, see Section 2). PINT clients and servers are SIP clients and servers. SIP is used to carry the request over the IP network to the correct PINT server in a secure and reliable manner, and SDP is used to describe the telephone network session to be invoked or whose status is to be returned. A PINT system uses SIP proxy servers and redirect servers for their usual purpose, but at some point there must be a PINT server with the means to relay received requests into a telephone system and to receive acknowledgement of these relayed requests. A PINT server with this capability is called a "PINT gateway." A PINT gateway appears to a SIP system as a user agent server. Notice that a PINT gateway appears to the PINT infrastructure as if it represents a "user," while in fact it represents an entire telephone network infrastructure that can provide a set of telephone network services. So the PINT system might appear to an individual PINT client as follows (Figure 14.1):

The system of PINT servers is represented as a cloud to emphasize that a single PINT request might pass through a series of location servers, proxy servers, and redirect servers before finally reaching the correct PINT gateway that can actually process the request by passing it to the telephone network cloud. The PINT gateway might have a true telephone network interface, or it might be connected via some other protocol or application programming interface to an "Executive System" that is capable of invoking services within the telephone cloud.

As an example, within an I.N. system, the PINT gateway might appear to realize the service control gateway function. In an office environment, it might be a server adjunct to the office PBX, connected to both the office LAN and the office PBX. The executive system that lies beyond the PINT gateway is outside the scope of PINT.

14.3.3.2 PINT Protocol Architecture

This section explains how SIP and SDP work in combination to convey the information necessary to invoke telephone network sessions. The following list summarizes the extension features used in PINT 1.0. Following on, the features are considered separately for SDP and then for SIP:

- Telephony URLs in SDP contact fields
- Refinement of SIP/SDP telephony URLs
  a. Inclusion of private dialing plans
- Specification of telephone service provider (TSP) and/or phone context URL-parameters
- Data objects as session media
  - Protocol Transport formats to indicate the treatment of the media within the GSTN
- Implicit (indirect) media streams and opaque arguments
- In-line data objects using multipart/Multipurpose Internet Mail Extensions (MIME)
- Refinement/clarification of opaque arguments passed onwards to executive systems
  - Framework for presentation restriction indication
  - Framework for Q.763 arguments
- An extension mechanism for SDP to specify strictures and force failure when a recipient does NOT support the specified extensions, using "require" headers

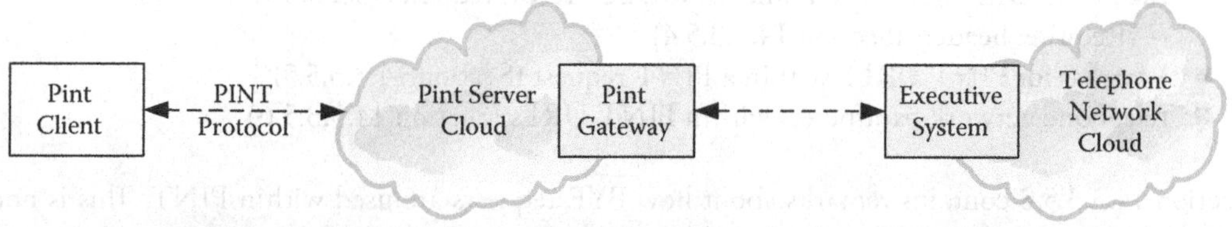

Figure 14.1 PINT functional architecture (Copyright IETF).

- Mandatory support for "Warning" headers to give more detailed information on request disposition
- Mechanism to register interest in the disposition of a requested service and to receive indications on that disposition. Both PINT and SIP rely on features of MIME (RFC 2046). The use of SIP 2.0 specified in RFC 2543 (that was made obsolete by RFC 3261, also see Section 1.2) is implied by PINT 1.0, and this implies compliance with version 1.0 of MIME.

14.3.3.2.1 SDP Operation in PINT

The SDP payload contains a description of the particular telephone network session that the requestor wishes will occur in the GSTN. This information includes such things as the telephone network address (i.e., the "telephone number") of the terminal(s) involved in the call, an indication of the media type to be transported (e.g., audio, text, image, or application data), and an indication of whether the information is to be transported over the telephone network via voice, fax, or pager transport. An indication of the content to be sent to the remote telephone terminal (if there is any) is also included. SDP is flexible enough to convey these parameters independently. For example, a request to send some text via voice transport will be fulfilled by invoking some text-to-speech-over-the-phone service, and a request to send text via fax will be fulfilled by invoking some text-to-fax service. The following is a list of PINT 1.0 enhancements and additions to SDP:

- A new network type "TN" and address types "RFC 2543 (made obsolete by RFC 3261, also see Section 1.2)" and "X-..." (Section 14.3.3.4.1)
- New media types "text," "image," and "application"; new protocol transport keywords "voice," "fax," and "pager"; and the associated format types and attribute tags (Section 14.3.3.4.2)
- New format specific attributes for included content data (Section 14.3.3.4.2.4)
- New attribute tags, used to pass information to the telephone network (Section 14.3.3.4.3)
- A new attribute tag, "require," used by a client to indicate that some attribute requires support in the server (Section 14.3.3.4.4)

14.3.3.2.2 SIP Operation in PINT

SIP is used to carry the request for telephone service from the PINT client to the PINT gateway and may include a telephone number if needed for the particular service. The following is a complete list of PINT enhancements and additions to SIP:

- The multipart MIME payloads (Section 14.3.3.5.1)
- Mandatory support for "Warning:" headers (Section 14.3.3.5.2)
- The SUBSCRIBE, NOTIFY, and UNSUBSCRIBE requests (Section 14.3.3.5.3)
  - Require: headers (Section 14.3.3.5.4)
- A format for PINT URLS within a PINT request (Section 14.3.3.5.5)
- Telephone network parameters within PINT URLs (Section 14.3.3.5.6)

Section 14.3.3.5.8 contains remarks about how BYE requests are used within PINT. This is not an extension to baseline SIP; it is included here only for clarification of the semantics when used with telephone network sessions.

14.3.3.3 REQUIRED and OPTIONAL Elements for PINT Compliance

Of these, only the TN network type (with its associated RFC 2543 address type) (RFC 2543 was made obsolete by RFC 3261, also see Section 1.2) and the "require" attribute MUST be supported by PINT 1.0 clients and servers. In practice, most PINT service requests will use other changes, of which references to data objects in requests are most likely to appear in PINT requests. Each of the other new PINT constructs enables a different function, and a client or server that wishes to enable that particular function MUST do so by the construct specified in this document. For example, building a PINT client and server that provide only the R2C telephone call service, without support for the other milestone services, is allowed. The "Require:" SIP header and the "require" attribute provide a mechanism that can be used by clients and servers to signal their need and/or ability to support specific "new" PINT protocol elements.

It should be noted that many optional features of SIP and SDP make sense as specified in the PINT context. One example is the SDP a=lang: attribute, which can be used to describe the preferred language of the callee. Another example is the use of the "t=" parameter to indicate the time at which the PINT service is to be invoked. This is the normal use of the "t=" field. A third example is the quality attributes. Any SIP or SDP option or facility is available to PINT clients and servers without change. Conversely, support for data objects within Internet conference sessions may be useful, even if the aim is not to provide a GSTN service request. In this case, the extensions covering these items may be incorporated into an otherwise "plain" SIP/SDP invitation. Likewise, support for SDP "require" may be useful, as a framework for addition of features to a "traditional" SIP/SDP infrastructure. Again, these may be convenient to incorporate into SIP/SDP implementations that would not be used for PINT service requests. Such additions are beyond the scope of this document, however.

14.3.3.4 PINT Extensions to SDP

PINT 1.0 adds to SDP the possibility to describe audio, fax, and pager telephone sessions. It is deliberately designed to hide the underlying technical details and complexity of the telephone network. The only network type defined for PINT is the generic "TN." More precise tags such as "ISDN" and "GSM" are not defined. Similarly, the transport protocols are designated simply as "fax," "voice," and "pager"; there are no more specific identifiers for the various telephone network voice, fax, or pager protocols. Similarly, the data to be transported are identified only by a MIME content type, such as "text" data, "image" data, or some more general "application" data. An important example of transporting "application" data is the milestone service "voice access to web content." In this case, the data to be transported are pointed to by a URI, the data content type is application/URI, and the transport protocol would be "voice." Some sort of speech-synthesis facility, speaking out to a phone, will have to be invoked to perform this service. This section gives details of the new SDP keywords.

14.3.3.4.1 Network Type "TN" and Address Type "RFC 2543"

The TN network type is used to indicate that the terminal is connected to a telephone network. The address types allowed for network type TN are "RFC 2543" and private address types, which MUST begin with an "X-." 0 Address type RFC 2543 is followed by a string conforming to a subset of the "telephone-subscriber" Backus-Naur Form (BNF) specified in Figure 4 of SIP, RFC 2543 (made obsolete by RFC 3261, also see Section 1.2). Note that this BNF is NOT identical to the BNF that defines the "phone-number" within the "p=" field of SDP.

Examples:

```
c= TN RFC 2543 +1-201-406-4090
c= TN RFC 2543 12014064090
```

A telephone-subscriber string is of one of two types: global-phone-number or local-phone-number. These are distinguished by preceding a global-phone-number with a "plus" sign ("+"). A global-phone-number is by default to be interpreted as an internationally significant E.164 number plan address, as defined by RFC 2617 (made obsolete by RFCs 7235, 7615, 7616, and 7617), whilst a local-phone number is a number specified in the default dialing plan within the context of the recipient PINT gateway. An implementation MAY use private addressing types, which can be useful within a local domain. These address types MUST begin with an "X-" and SHOULD contain a domain name after the X- (e.g., "Xmytype. mydomain.com"). An example of such a connection line is as follows:

```
c= TN X-mytype.mydomain.com A*8-HELEN
```

where "X-mytype.mydomain.com" identifies this private address type, and "A*8-HELEN" is the number in this format. Such a format is defined as an "OtherAddr" in the Augmented Backus-Naur Form (ABNF) of Appendix A of this book. Note that most dialable telephone numbers are expressible as local-phone-numbers within address RFC 2543 (made obsolete by RFC 3261, also see Section 1.2); new address types SHOULD only be used for formats that cannot be so written.

14.3.3.4.2 Support for Data Objects within PINT

One significant change over traditional SIP/SDP Internet conference sessions with PINT is that a PINT service request may refer to a data object to be used as source information in that request. For example, a PINT service request may specify a document to be processed as part of a GSTN service by which a fax is sent. Similarly, a GSTN service may be take a Web page and result in a vocoder processing that page and speaking the contents over a telephone. The SDP specification does not have explicit support for reference to or carriage of data objects within requests. In order to use SDP for PINT, there is a need to describe such media sessions as "a telephone call to a certain number during which such-and-such an image is sent as a fax."

To support this, two extensions to the session description format are specified. These are some new allowed values for the media-field and a description of the "fmtp" parameter when used with the media-field values (within the context of the contact field network type "TN"). An addition is also made to the SIP message format to allow the inclusion of data objects as subparts within the request message itself. The original SDP syntax from RFC 2327 (that was made obsolete by RFC 4566, see Section 2) for media-field is given as:

```
media-field     =       "m=" media space port ["/" integer]
                        space proto 1*(space fmt) CRLF
```

When used within PINT requests, the definition of the subfields is expanded slightly. The media subfield definition is relaxed to accept all of the discrete "top-level" media types defined in Reference [4]. In the milestone services, the discrete type "video" is not used, and the extra types "data" and "control" are likewise not needed. The use of these types is not precluded, but the behavior expected of a PINT gateway receiving a request including such a type is not defined here. The port subfield has no meaning in PINT requests as the destination terminals are specified using "TN" addressing, so the value of the port subfield in PINT requests is normally set to "1." A value of "0" may be used as in SDP to indicate that the terminal is not receiving media. This is useful to indicate that a telephone terminal has gone "on hold" temporarily. Likewise, the optional integer subfield is not used in PINT.

As mentioned in RFC 2327 (that was made obsolete by RFC 4566, see Section 2), the Transport Protocol subfield is specific to the associated address type. In the case that the address type in the preceding contact field is one of those defined for use with the network type "TN," the following values are defined for the Transport Protocol subfield:

```
"voice," "fax," and "pager."
```

The interpretation of this subfield within PINT requests is the treatment or disposition of the resulting GSTN service. Thus, for Transport Protocol "voice," the intent is that the service will result in a GSTN voice call, whilst for protocol "fax" the result will be a GSTN fax transmission, and protocol "pager" will result in a paper message being sent. Note that this subfield does not necessarily dictate the media type and subtype of any source data; for example, one of the milestone services calls for a textual source to be vocoded and spoken in a resulting telephone service call. The transport protocol value in this case would be "voice," whilst the media type would be "text." The Fmt subfield is described in RFC 2327 (that was made obsolete by RFC 4566, see Section 2) as being Transport Protocol-specific.

When used within PINT requests having one of the above protocol values, this subfield consists of a list of one or more values, each of which is a defined MIME subtype of the associated media subfield value. The special value "-" is allowed, meaning that there is no MIME subtype. This subfield retains from RFC 2543 (that was made obsolete by RFC 4566, see Section 2) its meaning that the list will contain a set of alternative subtypes, the first being the preferred value. For experimental purposes and by mutual consent of the sender and recipient, a subtype value may be specified as an <X-token> (i.e., a character string starting with "X-"). The use of such values is discouraged, and if such a value is expected to find common use then it SHOULD be registered with IANA using the standard content type registration process (see Section 14.3.8). When the fmt parameter is the single character "-" (a dash), this is interpreted as meaning that an unspecified or default subtype can be used for this service. Thus, the media-field value "m=audio 1 voice -<CRLF>" is taken to mean that a voice call is requested, using whatever audio subtype is deemed appropriate by the executive system. PINT service is a special case, in that the request comes from the IP network but the service call is provided within the GSTN. Thus, the service request will not normally be able to define the particular codec used for the resulting GSTN service call. If such an intent IS required, then the quality attribute may be used (see "Suggested Attributes" section of RFC 2327 (that was made obsolete by RFC 4566, see Section 2).

14.3.3.4.2.1 Use of fmtp Attributes in PINT Requests

For each element of the fmt subfield, there MUST be a following fmtp attribute. When used within PINT requests, the fmtp attribute has a general structure as defined here:

```
"a=       fmtp:" <subtype> <space> resolution
          *(<space> resolution)
          (<space> ";" 1(<attribute>)
          *(<space> <attribute>))
```

where:

```
<resolution> :        =         (<uri-ref> | <opaque-ref> |
                                <sub-part-ref>)
```

An fmtp attribute describes the sources used with a given fmt entry in the media-field. The entries in an fmt subfield are alternatives (with the preferred one first in the list). Each entry will have a matching fmtp attribute. The list of resolutions in an fmtp attribute describes the set of sources that resolve the matching Fmt choice; all elements of this set will be used. It should be noted that, for use in PINT services, the elements in such a set will be sent as a sequence; it is unlikely that trying to send them in parallel would be successful. An fmtp attribute can contain a mixture of different kinds of elements. Thus, an attribute might contain a subpart-ref indicating included data held in a subpart of the current message, followed by an opaque-ref referring to some content on the GSTN, followed by an uriref pointing to some data held externally on the IP network. To indicate which form each resolution element takes, each starts with its own literal tag. The detailed syntax of each form is described in the following sections.

14.3.3.4.2.2 Support for Remote Data Object References in PINT

Where data objects stored elsewhere on the IP network are to be used as sources for processing within a PINT service, they may be referred to using the uri-ref form. This is simply a URI, as described in RFC 2396 (that was made obsolete by RFC 3986). Note that the reference SHOULD be an absolute URI, as there may not be enough contextual information for the recipient server to resolve a relative reference; any use of relative references requires some private agreement between the sender and recipient of the message and SHOULD be avoided unless the sender can be sure that the recipient is the one intended and the reference is unambiguous in context. This also holds for partial URIs (such as "uri:http://aNode/index.htm") as these will need to be resolved in the context of the eventual recipient of the message. The general syntax of a reference to an Internet-based external data object in an fmtp line within a PINT session description is

```
<uri-ref>     :=     ("uri:" URI-reference)
```

where URI-reference is as defined in Appendix A of RFC 2396 (that was made obsolete by RFC 3986).

For example:

```
c= TN RFC 2543 +1-201-406-4090
m= text 1 fax plain
a=fmtp:plain uri:ftp://ftp.isi.edu/in-notes/RFC 2468.txt
```

or

```
c= TN RFC 2543 +1-201-406-4090
m= text 1 fax plain
a=fmtp:plain
```

means get this data object from the Internet and use it as a source for the requested GSTN Fax service.

14.3.3.4.2.3 Support for GSTN-Based Data Objects in PINT

PINT services may refer to data that are held not on the IP network but instead within the GSTN. The way in which these items are indicated need have no meaning within the context of the requestor or the PINT gateway; the reference is merely some data that may be used by the executive system to indicate the content intended as part of the request. These data form an opaque reference, in that they are sent "untouched" through the PINT infrastructure. A reference to some data object held on the GSTN has the general definition

```
<opaque-ref> := ("opr:" *uric)
```

where uric is as defined in Appendix A of RFC 2396 (that was made obsolete by RFC 3986).

For example:

```
c= TN RFC 2543 +1-201-406-4090
m= text 1 fax plain
a=fmtp:plain opr:APPL.123.456
```

means send the data that is indexed ON THE GSTN by the reference value "APPL.123.456" to the fax machine on +1-201-406-4090. The executive system may also take the telephone URL held in the To: field of the enclosed IP message into account when deciding the context to be used for the data object dereference. Of course, an opaque reference may also be used for other purposes; it could, for example, be needed to authorize access to a document held on the GSTN rather than being required merely to disambiguate the data object. The purpose to which an opaque reference is put, however, is out of scope for this document. It is merely an indicator carried within a PINT Request.

An opaque reference may have no value in the case where the value to be used is implicit in the rest of the request. For example, suppose some company wishes to use PINT to implement a "fax-back service." In their current implementation, the image(s) to be faxed are entirely defined by the telephone number dialed. Within the PINT request, this telephone number would appear within the "To:" field of the PINT request, so there is no need for an opaque reference value. If there are several resolutions for a PINT service request, and one of these is an opaque reference with no value, then that opaque reference MUST be included in the attribute line, but with an empty value field. For example:

```
c= TN RFC 2543 +1-201-406-4090
m= text 1 fax plain
a=fmtp:plain uri:http://www.sun.com/index.html opr:
```

might be used to precede some data to be faxed with a covering note. In the special case where an opaque reference is the sole resolution of a PINT service request, AND that reference needs no value, there is no need for an fmt list at all; the intent of the service is unambiguous without any further resolution. For example:

```
c= TN RFC 2543 +1-201-406-4090
m= text 1 fax -
```

means that there is an implied content stored on the GSTN, and that this is uniquely identified by the combination of SIP To-URI and the Contact field of the session description.

14.3.3.4.2.4 Session Description Support for Included Data Objects As an alternative to pointing to the data via a URI or an opaque reference to a data item held on the GSTN, it is possible to include the content data within the SIP request itself. This is done by using multipart MIME for the SIP payload. The first MIME part contains the SDP description of the telephone network session to be executed. The other MIME parts contain the content data to be transported.

Format-specific attribute lines within the session description are used to indicate which other MIME part within the request contains the content data. Instead of a URI or opaque reference, the format-specific attribute indicates the Content-ID of the MIME part of the request that contains the actual data and is defined as:

```
<sub-part-ref> := ("spr:" Content-ID)
```

where Content-ID is as defined in Appendix A of RFC 2045 and in RFC 0822. For example:

```
c= TN RFC 2543 +1-201-406-4090
m= text 1 fax plain
a=fmtp:plain spr:<Content-ID>
```

The <Content-ID> parameter is the Content-ID of one of the MIME parts inside the message, and this fragment means that the requesting user would like the data object held in the subpart of this message labeled <Content-ID> to be faxed to the machine at phone number +1-201-406-4090. See also Section 14.3.3.5.1 for a discussion on the support needed in the enclosing SIP request for included data objects.

14.3.3.4.3 Attribute Tags to Pass Information into the Telephone Network

It may be desired that service parameters that can be understood only by some entity in the "Telephone Network Cloud" be included within the PINT request. SDP attribute parameters are used for this purpose. They MAY appear within a particular media description or outside of a media description. These attributes may also appear as parameters within PINT URLS (see Section 14.3.3.5.6) as part of a SIP request. This is necessary so that telephone terminals that require the attributes to be defined can appear within the To: line of a PINT request as well as within PINT session descriptions. The purpose of these attributes is to allow the client to specify extra context within which a particular telephone number is to be interpreted. There are many reasons extra context might be necessary to interpret a given telephone number:

- The telephone number might be reachable in many different ways (such as via competing TSPs), and the PINT client wishes to indicate its selection of service provider.
- The telephone number might be reachable only from a limited number of networks (such as an 800 free phone number).
- The telephone number might be reachable only within a single telephone network (such as the 152 customer service number of BT). Similarly, the number might be an internal corporate extension reachable only within the PBX.

However, as noted above, it is not usually necessary to use SDP attributes to specify the phone context. URLs such as 152@pint.bt.co.il within the To: and From: headers and/or Request-URI, normally offer sufficient context to resolve telephone numbers. If the client wishes the request to fail if the attributes are not supported, these attributes SHOULD be used in conjunction with the "require" attribute (Section 14.3.3.4.4) and the "Require:org.ietf.sdp.require" header (Section 14.3.3.5.4). It is not possible to standardize every possible internal telephone network parameter. PINT 1.0 attributes have been chosen for specification because they are common enough that many different PINT systems will want to use them; therefore, interoperability will be increased by having a single specification.

Proprietary attribute "a=" lines, that by definition are not interoperable, may be nonetheless useful when it is necessary to transport some proprietary internal telephone network variables over the IP network, for example, to identify the order in which service call legs are to be made. These private attributes SHOULD BE, however, subject to the same IANA registration procedures mentioned in the SDP specification provided in RFC 2327 (that was made obsolete by RFC 4566, see Section 2).

14.3.3.4.3.1 The Phone-Context Attribute An attribute is specified to enable "remote local dialing." This is the service that allows a PINT client to reach a number from far outside the area or network that can usually reach the number. It is useful when the sending or receiving address is only dialable within some local context, which may be remote to the origin of the PINT client. For example, if Alice wanted to report a problem with her telephone, she might then dial a "network wide" customer care number; within the British Telecom network in the UK, this is "152." Note that in this case, she doesn't dial any trunk prefix; this is the whole dialable number. If dialed from another operator's network, it will not connect to British Telecom's Engineering Enquiries service; and dialing "+44 152" will not normally succeed. Such numbers are called network-specific service numbers.

Within the telephone network, the "local context" is provided by the physical connection between the subscriber's terminal and the central office. An analogous association between the PINT client and the PINT server that first receives the request may not exist, which is why it may be necessary to supply this missing "telephone network context." This attribute is defined as follows:

```
a                        =     phone-context: <phone-context-ident>
phone-context-ident      =     network-prefix / private-prefix
network-prefix           =     intl-network-prefix /
                               local-network-prefix
intl-network-prefix      =     "+" 1*DIGIT
local-network-prefix     =     1*DIGIT
excldigandplus           =     (0x21-0x2d,0x2f,0x40-0x7d))
private-prefix           =     1*excldigandplus 0*uric
```

An intl-network-prefix and local-network-prefix MUST be bona fide network prefixes, and a network-prefix that is an intl-network-prefix MUST begin with an E.164 service code ("country code"). It is possible to register new private-prefixes with IANA to avoid collisions. Prefixes that are not so registered MUST begin with an "X-" to indicate their private, nonstandard nature.

Example 1:

```
c= TN RFC 2543 1-800-765-4321
a=phone-context:+972
```

This describes a terminal whose address in Israel (E.164 country code 972) is 1-800-765-4321.

Example 2:

```
c= TN RFC 2543 1-800-765-4321
a=phone-context:+1
```

This describes a terminal whose address in North America (E.164 country code 1) is 1-800-765-4321. The two telephone terminals described by examples 1 and 2 are different; in fact they are located in different countries.

Example 3:

```
c=TN RFC 2543 123
a=phone-context:+97252
```

This describes a terminal whose address when dialed from within the network identified by +97252 is the string "123." It so happens that +97252 defines one of the Israeli cell phone providers, and 123 reaches customer service when dialed within that network. It may well be useful or necessary to use the SDP "require" parameter in conjunction with the phone-context attribute.

Example 4:

```
c= TN RFC 2543 321
a=phone-context:X-acme.com-23
```

This might describe the telephone terminal that is at extension 321 of PBX number 23 within the acme.com private PBX network. It is expected that such a description would be understandable by the acme.com PINT server that receives the request. Note that if the PINT server receiving the request is inside the acme.com network, the same terminal might be addressable as follows:

```
c= TN RFC 2543 7-23-321
```

(assuming that "7" is dialed in order to reach the private PBX network from within acme.com)

14.3.3.4.3.2 Presentation Restriction Attribute Although it has no effect on the transport of the service request through the IP Network, there may be a requirement to allow originators of a PINT service request to indicate whether they wish the "B party" in the resulting service call to be presented with the "A party's" calling telephone number. It is a legal requirement in some jurisdictions that a caller be able to select whether the correspondent can find out the calling telephone number (using automatic number indication or caller display or calling line identity presentation equipment). Thus, an attribute may be needed to indicate the originator's preference. Regardless of whether the default behavior of the executive system is to present or not present a party's telephone number to the correspondent GSTN terminal is not specified, and it is not mandatory in all territories for a PINT gateway or executive system to act on this attribute. It is, however, defined here for use where there are regulatory restrictions on GSTN operation, and in that case the executive system can use it to honor the originator's request. The attribute is specified as follows:

```
a=clir:<"true" | "false">
```

This Boolean value is needed within the attribute as it may be that the GSTN address is, by default, set to NOT present its identity to correspondents and that the originator wants to do so for this particular call. It is in keeping with the aim of this attribute to allow the originator to specify what treatment he/she wants for the requested service call. The expected interpretation of this attribute is that, if it is present and the value is "false" then the calling line identity CAN be presented to the correspondent terminal, whilst if it is "true" then if possible the executive system is requested to NOT present the calling line identity.

14.3.3.4.3.3 ITU-T CalledPartyAddress Attributes Parameters These attributes correspond to fields that appear within the International Telecommunication Union-Telecommunication Sector (ITU-T) Q.763 "CalledPartyAddress" field (see Section 3.9 of Reference [7]). PINT clients use these attributes to specify further parameters relating to terminal addresses, in the case when the address indicates a "local-phone-number." In the case that the PINT request contains a reference to a GSTN terminal, the parameters may be required to correctly identify that remote terminal. The general form of this attribute is: "a=Q763-<token>((":" <value>) |"")". Three of the possible elements and their use in SDP attributes are described here. Where other Q.763 elements are to be used, then these should be the subject of further specification to define the syntax of the attribute mapping. It is recommended that any such specification maintain the value sets shown in Q.763. The defined attributes are

```
a=Q763-nature: - indicates the "nature of address indicator."
```

The value MAY be any number between 0 and 127. The following values are specified:

```
"1" a subscriber number
"2" unknown
"3" a nationally significant number
"4" an internationally significant number
```

The values have been chosen to coincide with the values in Q.763. Note that other values are possible, according to national rules or future expansion of Q.763. a=Q763-plan: – indicates the numbering plan to which the address belongs. The value MAY be any number between 0 and 7. The following values are specified:

```
"1" Telephone numbering plan (ITU-T E.164)
"3" Data numbering plan (ITU-T X.121)
"4" Telex numbering plan (ITU-T F.69)
```

The values have been chosen to coincide with the values in Q.763. Other values are allowed, according to national rules or future expansion of Q.763. a=Q763-INN – indicates if routing to the internal network number is allowed. The value MUST be ONE of

```
"0" routing to internal network number allowed
"1" routing to internal network number not allowed
```

The values have been chosen to coincide with the values in Q.763. Note that it is possible to use a local-phone-number and indicate via attributes that the number is in fact an internationally significant E.164 number. Normally this SHOULD NOT be done; an internationally significant E.164 number is indicated by using a "global-phone number" for the address string.

14.3.3.4.4 The "Require" Attribute

According to the SDP specification, a PINT server is allowed simply to ignore attribute parameters that it does not understand. In order to force a server to decline a request if it does not understand one of the PINT attributes, a client SHOULD use the "require" attribute, specified as follows:

```
a=require:<attribute-list>
```

where the attribute-list is a comma-separated list of attributes that appear elsewhere in the session description. In order to process the request successfully, the PINT server must BOTH understand the attribute AND ALSO fulfill the request implied by the presence of the attribute, for each attribute appearing within the attribute-list of the require attribute.

If the server does not recognize the attribute listed, the PINT server MUST return an error status code (such as 420 (Bad Extension) or 400 (Bad Request)), and SHOULD return suitable Warning: lines explaining the problem or an Unsupported: header containing the attribute it does not understand. If the server recognizes the attribute listed but cannot fulfill the request implied by the presence of the attribute, the request MUST be rejected with a status code of 606 (Not Acceptable), along with a suitable Unsupported:

```
header or Warning: line.
```

The "require" attribute may appear anywhere in the session description, and any number of times, but it MUST appear before the use of the attribute marked as required. Since the "require" attribute is itself an attribute, the SIP specification allows a server that does not understand the require attribute to ignore it. In order to ensure that the PINT server will comply with the "require" attribute, a PINT client SHOULD include a Require: header with the tag "org.ietf.sdp.require" (Section 14.3.3.5.4)

Note that the majority of the PINT extensions are "tagged," and these tags can be included in Require strictures. The exception is the use of phone numbers in SDP parts. However, these are defined as a new network and address type, so that a receiving SIP/SDP server should be able to detect whether it supports these forms. The default behavior for any SDP recipient is that it will fail a PINT request if it does not recognize or support the TN and RFC 2543 (that was made obsolete by RFC 3261, also see Section 1.2) or X-token network and address types, as without the contents being recognized no media session could be created. Thus, a separate stricture is not required in this case.

14.3.3.5 PINT Extensions to SIP 2.0

PINT requests are SIP requests; many of the specifications within this document merely explain how to use existing SIP facilities for the purposes of PINT.

14.3.3.5.1 Multi-Part MIME (Sending Data along with SIP Request)

A PINT request can contain a payload, which is multipart MIME. In this case, the first part MUST contain an SDP session description that includes at least one of the format-specific attribute tags for "included content data" specified in Section 14.3.3.4.3. Subsequent parts contain content data that may be transferred to the requested telephone call service. As discussed earlier, within a single PINT request, some of the data MAY be pointed to by a URI within the request, and some of the data MAY be included within the request.

Where included data is carried within a PINT service request, the Content Type entity header of the enclosing SIP message MUST indicate this. To do so, the media type value within this entity header MUST be set to a value of "multipart." A content subtype is intended for situations like this in which subparts are to be handled together. This is the multipart/related type (defined in RFC 2387), and its use is recommended. The enclosed body parts SHOULD include the part-specific Content Type headers as appropriate ("application/sdp" for the first body part holding the session description, with an appropriate content type for each of the subsequent, "included data object," parts). This matches the standard syntax of MIME multipart messages as defined in RFC 2046. For example, in a multipart message where the string "------next-------" is the boundary, the first two parts might be as follows:

```
------next-------

Content-Type: application/sdp
....
c= TN RFC 2543 +1-201-406-4090
m= text 1 pager plain
a=fmtp:plain spr:17@mymessage.acme.com

----------next-------

Content-Type: text/plain
Content-ID: 17@mymessage.acme.com
This is the text that is to be paged to +1-201-406-4090

----------next-----------
```

The ability to indicate different alternatives for the content to be transported is useful, even when the alternatives are included within the request. For example, a request to send a short message to a pager might include the message in Unicode [8] and an alternative version of the same content in text/plain, should the PINT server or telephone network not be able to process the Unicode. PINT clients should be extremely careful when sending included data within a PINT request. Such requests SHOULD be sent via Transmission Control Protocol, to avoid fragmentation and to transmit the data reliably. It is possible that the PINT server is a proxy server that will replicate and fork the request, which could be disastrous if the request contains a large amount of application data. PINT proxy servers should be careful not to create many copies of a request with large amounts of data in it. If the client does not know the actual location of the PINT gateway, is using the SIP location services to find it, and the included data makes the PINT request likely to be transported in several IP datagrams, it is RECOMMENDED that the initial PINT request not include the data object but instead hold a reference to it.

14.3.3.5.2 Warning Header

A PINT server MUST support the SIP "Warning:" header so that it can signal lack of support for individual PINT features. As an example, suppose the PINT request is to send a jpeg picture to a fax machine, but the server cannot retrieve and/or translate jpeg pictures from the Internet into fax transmissions. In such a case, the server fails the request and includes a Warning such as the following:

```
Warning: 305 pint.acme.com Incompatible media format: jpeg
```

SIP servers that do not understand the PINT extensions at all are strongly encouraged to implement "Warning:" headers to indicate that PINT extensions are not understood. Also, "Warning:" headers may be included within NOTIFY requests if it is necessary to notify the client about some condition concerning the invocation of the PINT service described here.

14.3.3.5.3 Mechanism to Register Interest in the Disposition of a PINT Service

We now consider the mechanism to register interest in the disposition of a PINT service and to receive indications on that disposition. It can be very useful to find out whether a requested service has completed, and if so whether it was successful. This is especially true for PINT service, where the person requesting the service is not (necessarily) a party to it, and so may not have an easy way of finding out the disposition of that service. Equally, it may be useful to indicate when the service has changed state, for example, when the service call has started. Arranging a flexible system to provide extensive monitoring and control during a service is nontrivial (see Section 14.3.6.4 for some issues); PINT 1.0 uses a simple scheme that should nevertheless provide useful information. It is possible to expand the scheme in a "backwards compatible" manner, so if required it can be enhanced at a later date.

The PINT 1.0 status registration and indication scheme uses three new methods; SUBSCRIBE, UNSUBSCRIBE, and NOTIFY. These are used to allow a PINT client to register an interest in (or "subscribe" to) the status of a service request, to indicate that a prior interest has lapsed (i.e., "unsubscribe" from the status), and for the server to return service indications. The state machine of SUBSCRIBE/UNSUBSCRIBE is identical to that of INVITE/BYE; just as INVITE signals the beginning and BYE signals the end of participation in a media session, SUBSCRIBE signals the beginning and UNSUBSCRIBE signals the end of participation in a monitoring session. During the monitoring session, NOTIFY messages are sent to inform the subscriber of a change in session state or disposition.

14.3.3.5.3.1 Opening a Monitoring Session with a SUBSCRIBE Request When a SUBSCRIBE request is sent to a PINT Server, it indicates that a user wishes to receive information about the status of a service session. The request identifies the session of interest by including the original session description along with the request, using the SDP global-session-id that forms part of the origin-field to identify the service session uniquely. The SUBSCRIBE request (like any other SIP request about an ongoing session) is sent to the same server that was sent the original INVITE or to a server specified in the Contact field within a subsequent response (this might well be the PINT gateway for the session). Whilst there are situations in which reuse of the Call-ID used in the original INVITE that initiated the session of interest is possible, there are other situations in which it is not possible. In detail, where the subscription is being made by the user who initiated the original service request, the Call-ID may be used as it will be known to the receiver to refer to a previously established session.

However, when the request comes from a user other than the original requesting user, the SUBSCRIBE request constitutes a new SIP call leg, so the Call-ID SHOULD NOT be used; the only common identifier is the origin field of the session description enclosed within the original service request, and so this MUST be used. Rather than have two different methods of identifying the "session of interest," the choice is to use the origin field of the SDP subpart included both in the original INVITE and in this SUBSCRIBE request. Note that the request MUST NOT include any subparts other than the session description, even if these others were present in the original INVITE request. A server MUST ignore whatever subparts are included within a

SUBSCRIBE request with the sole exception of the enclosed session description. The request MAY contain a "Contact:" header, specifying the PINT user agent server to which such information should be sent.

In addition, it SHOULD contain an "Expires:" header, which indicates for how long the PINT requestor wishes to receive notification of the session status. We refer to the period of time before the expiration of the SUBSCRIBE request as the "subscription period." See Section 14.3.5.1.4 for security considerations, particularly privacy implications. A value of 0 within the "Expires:" header indicates a desire to receive one single immediate response (i.e., the request expires immediately). It is possible for a sequence of monitoring sessions to be opened, exist, and complete, all relating to the same service session. A successful response to the SUBSCRIBE request includes the session description, according to the gateway. Normally this will be identical to the last cached response the gateway returned to any request concerning the same SDP global session id specified in RFC 2327 (that was made obsolete by RFC 4566, see Section 2), Section 14.3.6, "o=" field. The " t=" line may be altered to indicate the actual start or stop time, however. The gateway might add an i= line to the session description to indicate such information as how many fax pages were sent. The gateway SHOULD include an "Expires:" header indicating how long it is willing to maintain the monitoring session. If this is unacceptable to the PINT requestor, then it can close the session by sending an immediate UNSUBSCRIBE message (see Section 14.3.3.5.3.3).

In principle, a user might send a SUBSCRIBE request after the telephone network service has completed. This allows, for example, checking up "the morning after" to see if the fax was successfully transmitted. However, a PINT gateway is only required to keep state about a call for as long as it indicated previously in an "Expires:" header sent within the response to the original INVITE message that triggered the service session, within the response to the SUBSCRIBE message, within the response to any UNSUBSCRIBE message, or within its own UNSUBSCRIBE message (but see Section 14.3.3.5.8, point 3). If the server no longer has a record of the session to which a requestor has SUBSCRIBEd, it returns "606 Not Acceptable" along with the appropriate Warning: 307 header indicating that the SDP session ID is no longer valid. This means that a requesting client that knows that it will want information about the status of a session after the session terminates SHOULD send a SUBSCRIBE request before the session terminates.

14.3.3.5.3.2 Sending Status Indications with a NOTIFY Request

During the subscription period, the gateway may, from time to time, send a spontaneous NOTIFY request to the entity indicated in the Contact header of the "opening" SUBSCRIBE request. Normally this will happen as a result of any change in the status of the service session for which the requestor has subscribed. The receiving user agent server MUST acknowledge this by returning a final response (normally a "200 OK"). In this version of the PINT extensions, the gateway is not required to support redirects (3xx codes) and so may treat them as a failure. Thus, if the response code class is above 2xx then this may be treated by the gateway as a failure of the monitoring session, and in that situation it will immediately attempt to close the session described in Section 14.3.3.5.3.3. The NOTIFY request contains the modified session description. For example, the gateway may be able to indicate a more accurate start or stop time. The gateway may include a Warning: header to describe some problem with the invocation of the service and may indicate within an i= line some information about the telephone network session itself.

Example:

```
NOTIFY sip:petrack@pager.com SIP/2.0
To: sip:petrack@pager.com
```

```
From: sip:R2F.pint.com@service.com
Call-ID: 19971205T234505.56.78@pager.com
CSeq: 4711 SUBSCRIBE
Warning: xxx fax aborted, will try for the next hour.
Content-Type:application/sdp
c=...
i=3 pages of 5 sent
t=...
```

14.3.3.5.3.3 Closing a Monitoring Session with an UNSUBSCRIBE Request At some point, either the client's representative user agent server or the gateway may decide to terminate the monitoring session. This is achieved by sending an UNSUBSCRIBE request to the correspondent server. Such a request indicates that the sender intends to close the monitoring session immediately, and, on receipt of the final response from the receiving server, the session is deemed over. Note that unlike the SUBSCRIBE request, which is never sent by a PINT gateway, an UNSUBSCRIBE request can be sent by a PINT gateway to the user agent server to indicate that the monitoring session is closed. (This is analogous to the fact that a gateway never sends an INVITE, although it can send a BYE to indicate that a telephone call has ended.)

If the gateway initiates closure of the monitoring session by sending an UNSUBSCRIBE message, it SHOULD include an "Expires:" header showing for how much longer after this monitoring session is closed it is willing to store information on the service session. This acts as a minimum time within which the client can send a new SUBSCRIBE message to open another monitoring session; after the time indicated in the Expires: header the gateway is free to dispose of any record of the service session, so that subsequent SUBSCRIBE requests can be rejected with a "606" response. If the subscription period specified by the client has expired, then the gateway may send an immediate UNSUBSCRIBE request to the client's representative user agent server. This ensures that the monitoring session always completes with a UNSUBSCRIBE/response exchange and that the representative user agent server can avoid maintaining state in certain circumstances.

14.3.3.5.3.4 Timing of SUBSCRIBE Requests As it relies on the gateway having a copy of the INVITEd session description, the SUBSCRIBE message is limited in when it can be issued. The gateway must have received the service request to which this monitoring session is to be associated, which from the client's perspective happens as soon as the gateway has sent a 1xx response back to it. However, once this has been done, there is no reason the client should not send a monitoring request. It does not have to wait for the final response from the gateway, and it can certainly send the SUBSCRIBE request before sending the ACK for the service request final response. Beyond this point, the client is free to send a SUBSCRIBE request when it decides, unless the gateway's final response to the initial service request indicated a short Expires: time.

However, there are good reasons (see Section 14.3.6.4) it may be appropriate to start a monitoring session immediately before the service is confirmed by the PINT client sending an ACK. At this point, the gateway will have decided whether it can handle the service request but will not have passed the request on to the executive system. It is therefore in a good position to ask the executive system to enable monitoring when it sends the service request onwards. In practical implementations, it is likely that more information on transient service status will be available if this is indicated as being important BEFORE or AS the service execution phase starts; once execution has begun, the level of information that can be returned may be difficult to change. Thus, whilst it is free to send a SUBSCRIBE request at any point after receiving an interim response from the gateway to its service request, it is recommended that the client should send such a monitoring request immediately prior to sending an ACK message confirming the service if it is interested in transient service status messages.

14.3.3.5.4 The "Require:" Header for PINT

PINT clients use the "Require:" header to signal to the PINT server that a certain PINT extension of SIP is required. PINT 1.0 defines two strings that can go into the "Require:" header:

```
org.ietf.sip.subscribe -- the server can fulfill SUBSCRIBE requests
and associated methods (see Section 14.3.3.5.3)
org.ietf.sdp.require -- the PINT server (or the SDP parser
associated to it) understands the "require" attribute defined in
Section 14.3.3.4.4
```

Example:

```
Require:org.ietf.sip.subscribe,org.ietf.sdp.require
```

A client SHOULD only include a "Require:" header where it truly requires the server to reject the request if the option is not supported.

14.3.3.5.5 PINT URLs within PINT Requests

Normally the hostnames and domain names that appear in the PINT URLs are the internal affair of each individual PINT system. A client uses the appropriate SDP payload to indicate the particular service it wishes to invoke; it is not necessary to use a particular URL to identify the service. A PINT URL is used in two different ways within PINT requests: within the Request-URI, and within the To: and From: headers. Use within the Request-URI requires clarification in order to ensure smooth interworking with the telephone network serviced by the PINT infrastructure, and this is covered next.

14.3.3.5.5.1 PINT URLS within Request-URIs There are some occasions when it may be useful to indicate service information within the URL in a standardized way:

a. It may not be possible to use SDP information to route the request if it is encrypted.
b. It allows implementation that makes use of I.N. "service indicators."
c. It enables multiple competing PINT gateways to REGISTER with a single "broker" server (proxy or redirect) (see Section 14.3.6.3).

For these reasons, the following conventions for URLs are offered for use in PINT requests:

1. The user portion of a sip URL indicates the service to be requested. At present, the following services are defined:

```
R2C (for Request-to-Call)
R2F (for Request-to-Fax)
R2HC (for Request-to-Hear-Content)
```

The user portions "R2C," "R2F," and "R2HC" are reserved for the PINT milestone services. Other user portions MUST be used in case the requested service is not one of the milestone services. See Section 14.3.6.2 for some related considerations concerning registrations by competing PINT systems to a single PINT proxy server acting as a service broker.

2. The host portion of a sip URL contains the domain name of the PINT service provider.
3. A new url-parameter is defined to be "tsp" (for "telephone service provider"). This can be used to indicate the actual telephone network provider to be used to fulfill the PINT request.

Thus, for example,

```
INVITE sip:R2C@pint.pintservice.com SIP/2.0
INVITE sip:R2F@pint.pintservice.com;tsp=telco.com SIP/2.0
INVITE sip:R2HC@pint.mycom.com;tsp=pbx23.mycom.com SIP/2.0
INVITE sip:13@pint.telco.com SIP/2.0
```

14.3.3.5.6 Telephony Network Parameters within PINT URLs

Any legal SIP URL can appear as a PINT URL within the Request-URI or To: header of a PINT request. But if the address is a telephone address, we indicated in Section 14.3.3.4.3 that it may be necessary to include more information in order to correctly identify the remote telephone terminal or service. PINT clients MAY include these attribute tags within PINT URLs if they are necessary or a useful complement to the telephone number within the SIP URL. These attribute tags MUST be included as URL parameters as defined in RFC 2543 (that was made obsolete by 3261; also see Section 1.2) (i.e., in the semi-colon separated manner). The following is an example of a PINT URL containing extra attribute tags:

```
sip:+9725228808@pint.br.com;user=phone;require=Q763-plan;a=Q763-plan:4
```

As we noted in Section 14.3.3.4.3, these extra attribute parameters will not normally be needed within a URL, because there is a great deal of context available to help the server interpret the phone number correctly. In particular, there is the SIP URL within the To: header, and there is also the Request-URI. In most cases, this provides sufficient information for the telephone network. The SDP attributes defined in Section 14.3.3 will normally only be used when they are needed to supply necessary context to identify a telephone terminal.

14.3.3.5.7 REGISTER Requests within PINT

A PINT gateway is a SIP user agent server. A user agent server uses the REGISTER request to tell a proxy or redirect server that it is available to "receive calls" (i.e., to service requests). Thus, a PINT gateway registers with a proxy or redirect server the service that is accessible via itself, whilst in SIP, a user is registering his/her presence at a particular SIP server. Competing PINT servers able to offer the same PINT service may be trying to register at a single PINT server. The PINT server might act as a "broker" among the various PINT gateways that can fulfill a request. A format for PINT URLs was specified in Section 14.3.3.5.5 that enables independent PINT systems to REGISTER an offer to provide the same service. The registrar can apply its own mechanisms and policies to decide how to respond to INVITEs from clients seeking service

(See Section 14.3.6.3 for some possible deployment options). There is no change between SIP and PINT REGISTER semantics or syntax. Of course, the information in the PINT URLs within the REGISTER request may not be sufficient to completely define the service a gateway can offer. The use of SIP and SDP within PINT REGISTER requests to enable a gateway to specify in more detail the services it can offer is the subject of future study.

14.3.3.5.8 BYE Requests in PINT

The semantics of BYE requests within PINT requires some extra precision. One issue concerns conferences that "cannot be left," and the other concerns keeping call state after the BYE. The BYE request (RFC 2543 was made obsolete by RFC 3261, also see Section 1.2) is normally used to indicate that the originating entity no longer wishes to be involved in the specified call. The request terminates the call and the media session. Applying this model to PINT, if a PINT client makes a request that results in invocation of a telephone call from A to B, a BYE request from the client, if accepted, should result in a termination of the phone call. One might expect this to be the case if the telephone call has not started when the BYE request is received. For example, if a request to fax is sent with a "t=" line indicating that the fax is to be sent tomorrow at 4 AM, the requestor might wish to cancel the request before the specified time.

However, even if the call has yet to start, it may not be possible to terminate the media session on the telephone system side. For example, the fax call may be in progress when the BYE arrives, and perhaps it is just not possible to cancel the fax in session. Another possibility is that the entire telephone-side service might be completed before the BYE is received. In the above R2F example, the BYE might be sent the following morning, and the entire fax has been sent before the BYE was received. It is too late to send the BYE.

In the case where the telephone network cannot terminate the call, the server MUST return a "606 Not Acceptable" response to the BYE, along with a session description that indicates the telephone network session that is causing the problem. Thus, in PINT, a "Not Acceptable" response MAY be returned both to INVITE and BYE requests. It indicates that some aspect of the session description makes the request unacceptable. By allowing a server to return a "Not Acceptable" response to BYE requests, we are not changing its semantics, just enlarging its use. A combination of "Warning:" headers and "i=" lines within the session description can be used to indicate the precise nature of the problem.

Example:

```
SIP/2.0 606 Not Acceptable
From: ...
To: .......
.....
Warning: 399 pint.mycom.com Fax in progress, service cannot be aborted
Content-Type: application/sdp
Content-Length: ...
v=0
...
...
i=3 of 5 pages sent OK
c=TN RFC 2543 +12014064090
m=image 1 fax tif
a=fmtp:tif uri:http://tifsRus.com/yyyyyy.tif
```

Note that the server might return an updated session description within a successful response to a BYE as well. This can be used, for example, to indicate the actual start times and stop times of the telephone session or how many pages were sent in the fax transmission. The second issue concerns how long must a server keep call state after receiving a BYE. A question arises because other clients might still wish to send queries about the telephone network session that was the subject of the PINT transaction. Ordinary SIP semantics have three important implications for this situation:

1. A BYE indicates that the requesting client will clear out all call state as soon as it receives a successful response. A client SHOULD NOT send a SUBSCRIBE request after it has sent a BYE.
2. A server may return an "Expires:" header within a successful response to a BYE request. This indicates for how long the server will retain session state about the telephone network session. At any point during this time, a client may send a SUBSCRIBE request to the server to learn about the session state (although as explained in the previous paragraph, a client that has sent a BYE will not normally send a SUBSCRIBE).
3. When engaged in a SUBSCRIBE/NOTIFY monitoring session, PINT servers that send UNSUBSCRIBE to a URL listed in the Contact: header of a client request SHOULD not clear session state until after the successful response to the UNSUBSCRIBE message is received. For example, it may be that the requesting client host is turned off (or in a low power mode) when the telephone service is executed (and is therefore not available at the location previously specified in the "Contact:" attribute) to receive the PINT server's UNSUBSCRIBE. Of course, it is possible that the UNSUBSCRIBE request will simply time out.

14.3.4 Examples of PINT Requests and Responses

14.3.4.1 A Request to a Call Center from an Anonymous User

We are considering a PINT request to a call center from an anonymous user to receive a phone call.

```
C->S:
     INVITE sip:R2C@pint.mailorder.com SIP/2.0
     Via: SIP/2.0/UDP 169.130.12.5
     From: sip:anon-1827631872@chinet.net
     To: sip:+1-201-456-7890@iron.org;user=phone
     Call-ID: 19971205T234505.56.78@pager.com
     CSeq: 4711 INVITE
     Subject: Sale on Ironing Boards
     Content-type: application/sdp
     Content-Length: 174

     v=0
     o=- 2353687637 2353687637 IN IP4 128.3.4.5
     s=R2C
     i=Ironing Board Promotion
     e=anon-1827631872@chinet.net
     t=2353687637 0
     m=audio 1 voice -
     c=TN RFC 2543 +1-201-406-4090
```

In this example, the context that is required to interpret the To: address as a telephone number is not given explicitly; it is implicitly known to the R2C@pint.mailorder.com server. But the telephone of the person who wishes to receive the call is explicitly identified as an internationally significant E.164 number that falls within the North American numbering plan (because of the "+1" within the “c=” line).

14.3.4.2 John Jones to Receive a Phone Call from Mary James

Let us imagine a scenario: A request from a nonanonymous customer (John Jones) to receive a phone call from a particular sales agent (Mary James) concerning the defective ironing board that was purchased.

```
C->S:
    INVITE sip:marketing@pint.mailorder.com SIP/2.0
    Via: SIP/2.0/UDP 169.130.12.5
    From: sip:john.jones.3@chinet.net
    To: sip:mary.james@mailorder.com
    Call-ID: 19971205T234505.56.78@pager.com
    CSeq: 4712 INVITE
    Subject: Defective Ironing Board - want refund
    Content-type: application/sdp
    Content-Length: 150

    v=0
    o=- 2353687640 2353687640 IN IP4 128.3.4.5
    s=marketing
    e=john.jones.3@chinet.net
    c= TN RFC 2543 +1-201-406-4090
    t=2353687640 0
    m=audio 1 voice -
```

The To: line might include Mary James’s phone number instead of an email-like address. An implementation that cannot accept email-like URLs in the "To:" header must decline the request with a 606 Not Acceptable. Note that the sending PINT client "knows" that the PINT gateway contacted with the "marketing@pint.mailorder.com" Request-URI is capable of processing the client request as expected. (See Section 14.3.3.5.5.1 for a discussion on this.) Note also that such a telephone call service could be implemented on the phone side with different details. For example, it might be that first the agent’s phone rings and then the customer’s phone rings, or it might be that first the customer’s phone rings and he hears silly music until the agent comes on line. If necessary, such service parameter details might be indicated in "a=" attribute lines within the session description. The specification of such attribute lines for service consistency is beyond the scope of the PINT 1.0 specifications.

14.3.4.3 A Request from the Same User to Get a Fax Back on How to Assemble the Ironing Board

```
C->S:
    INVITE sip:faxback@pint.mailorder.com SIP/2.0
    Via: SIP/2.0/UDP 169.130.12.5
    From: sip:john.jones.3@chinet.net
```

```
To: sip:1-800-3292225@steam.edu;user=phone;phone-context=+1
Call-ID: 19971205T234505.66.79@chinet.net
CSeq: 4713 INVITE
Content-type: application/sdp
Content-Length: 218

v=0
o=- 2353687660 2353687660 IN IP4 128.3.4.5
s=faxback
e=john.jones.3@chinet.net
t=2353687660 0
m=application 1 fax URI
c=TN RFC 2543 1-201-406-4091
a=fmtp:URI uri:http://localstore/Products/IroningBoards/2344.html
```

In this example, the fax to be sent is stored on some local server (localstore), whose name may be only resolvable, or that may only be reachable, from within the IP network on which the PINT server sits. The phone number to be dialed is a "local phone number" as well. There is no "phone-context" attribute, so the context (in this case, for which nation the number is "nationally significant") must be supplied by the faxback@pint.mailorder.com PINT server. If the server that receives it does not understand the number, it SHOULD decline the request and include a "Network Address Not Understood" warning. Note that no "require" attribute was used here, since it is very likely that the request can be serviced even by a server that does not support the "require" attribute.

14.3.4.4 A Request from Same User to Have that Same Information Read Out over the Phone

```
C->S:
    INVITE sip:faxback@pint.mailorder.com SIP/2.0
    Via: SIP/2.0/UDP 169.130.12.5
    From: sip:john.jones.3@chinet.net
    To: sip:1-800-3292225@steam.edu;user=phone;phone-context=+1
    Call-ID: 19971205T234505.66.79@chinet.net
    CSeq: 4713 INVITE
    Content-type: application/sdp
    Content-Length: 220

    v=0
    o=- 2353687660 2353687660 IN IP4 128.3.4.5
    s=faxback
    e=john.jones.3@chinet.net
    t=2353687660 0
    m=application 1 voice URI
    c=TN RFC 2543 1-201-406-4090
    a=fmtp:URI uri:http://localstore/Products/IroningBoards/2344.html
```

14.3.4.5 A Request to Send an Included Text Page to a Friend's Pager

In this example, the text to be paged out is included in the request.

```
C->S:
    INVITE sip:R2F@pint.pager.com SIP/2.0
    Via: SIP/2.0/UDP 169.130.12.5
    From: sip:scott.petrack@chinet.net
    To: sip:R2F@pint.pager.com
    Call-ID: 19974505.66.79@chinet.net
    CSeq: 4714 INVITE

    Content-Type: multipart/related; boundary=--next

    ----next

    Content-Type: application/sdp
    Content-Length: 236
    v=0
    o=- 2353687680 2353687680 IN IP4 128.3.4.5
    s=R2F
    e=scott.petrack@chinet.net
    t=2353687680 0
    m=text 1 pager plain
    c= TN RFC 2543 +972-9-956-1867
    a=fmtp:plain spr:2@53655768

    ----next

    Content-Type: text/plain
    Content-ID: 2@53655768
    Content-Length:50
    Hi Joe! Please call me asap at 555-1234.

    ----next--
```

14.3.4.6 A Request to Send an Image as a Fax to Phone Number +972-9-956-1867

```
C->S:
    INVITE sip:faxserver@pint.vocaltec.com SIP/2.0
    Via: SIP/2.0/UDP 169.130.12.5
    From: sip:scott.petrack@chinet.net
    To: sip:faxserver@pint.vocaltec.com
    Call-ID: 19971205T234505.66.79@chinet.net
    CSeq: 4715 INVITE
    Content-type: application/sdp
    Content-Length: 267

    v=0
    o=- 2353687700 2353687700 IN IP4 128.3.4.5
    s=faxserver
    e=scott.petrack@chinet.net
    t=2353687700 0
    m=image 1 fax tif gif
    c= TN RFC 2543 +972-9-956-1867
    a=fmtp:tif uri:http://petrack/images/tif/picture1.tif
    a=fmtp:gif uri:http://petrack/images/gif/picture1.gif
```

The image is available as tif or as gif. The tif is the preferred format. Note that the http server where the pictures reside is local and the PINT server is also local (because it can resolve machine name "petrack").

14.3.4.7 A Request to Read Out over the Phone Two Pieces of Content in Sequence

First, some included text is read out by text-to-speech. Then some text that is stored at some URI on the Internet is read out.

```
C->S:
     INVITE sip:R2HC@pint.acme.com SIP/2.0
     Via: SIP/2.0/UDP 169.130.12.5
     From: sip:scott.petrack@chinet.net
     To: sip:R2HC@pint.acme.com
     Call-ID: 19974505.66.79@chinet.net
     CSeq: 4716 INVITE
     Content-Type: multipart/related; boundary=next

     --next

     Content-Type: application/sdp
     Content-Length: 316

     v=0
     o=- 2353687720 2353687720 IN IP4 128.3.4.5
     s=R2HC
     e=scott.petrack@chinet.net
     c= TN RFC 2543 +1-201-406-4091
     t=2353687720 0
     m=text 1 voice plain
     a=fmtp:plain spr:2@53655768
     m=text 1 voice plain
     a=fmtp:plain uri:http://www.your.com/texts/stuff.doc

     --next

     Content-Type: text/plain
     Content-ID: 2@53655768
     Content-Length: 172
```

Hello!! I am about to read out to you the document you requested, "uri:http://www.your.com/texts/stuff.doc." We hope you like acme.com's new speech synthesis server.

```
     --next--
```

14.3.4.8 Request for the Prices for ISDN to Be Sent to My Fax Machine

```
     INVITE sip:R2FB@pint.bt.co.uk SIP/2.0
     Via: SIP/2.0/UDP 169.130.12.5
```

```
To: sip:0345-12347-01@pint.bt.co.uk;user=phone;phone-context=+44
From: sip:hank.wangford@newts.demon.co.uk
Call-ID: 19981204T201505.56.78@demon.co.uk
CSeq: 4716 INVITE
Subject: Price List
Content-type: application/sdp
Content-Length: 169

v=0
o=- 2353687740 2353687740 IN IP4 128.3.4.5
s=R2FB
i=ISDN Price List
e=hank.wangford@newts.demon.co.uk
t=2353687740 0
m=text 1 fax -
c=TN RFC 2543 +44-1794-8331010
```

14.3.4.9 Request for a Callback

```
INVITE sip:R2C@pint.bt.co.uk SIP/2.0
Via: SIP/2.0/UDP 169.130.12.5
To: sip:0345-123456@pint.bt.co.uk;user=phone;phone-context=+44
From: sip:hank.wangford@newts.demon.co.uk
Call-ID: 19981204T234505.56.78@demon.co.uk
CSeq: 4717 INVITE
Subject: It costs HOW much?
Content-type: application/sdp
Content-Length: 176

v=0
o=- 2353687760 2353687760 IN IP4 128.3.4.5
s=R2C
i=ISDN pre-sales query
e=hank.wangford@newts.demon.co.uk
c=TN RFC 2543 +44-1794-8331013
t=2353687760 0
m=audio 1 voice -
```

14.3.4.10 Sending a Set of Information in Response to an Enquiry

```
INVITE sip:R2FB@pint.bt.co.uk        SIP/2.0
Via: SIP/2.0/UDP 169.130.12.5
To: sip:0345-12347-01@pint.bt.co.uk;user=phone;phone-context=+44
From: sip:colin.masterton@sales.hh.bt.co.uk
Call-ID: 19981205T234505.56.78@sales.hh.bt.co.uk
CSeq: 1147 INVITE
Subject: Price Info, as requested
Content-Type: multipart/related; boundary=next

--next

Content-type: application/sdp
```

```
Content-Length: 325

v=0
o=- 2353687780 2353687780 IN IP4 128.3.4.5
s=R2FB
i=Your documents
e=colin.masterton@sales.hh.bt.co.uk
t=2353687780 0
m=application 1 fax octet-stream
c=TN RFC 2543 +44-1794-8331010
a=fmtp:octet-stream uri:http://www.bt.co.uk/imgs/pipr.gif opr:
spr:2@53655768

--next

Content-Type: text/plain
Content-ID: 2@53655768
Content-Length: 352
```

Dear Sir,

Thank you for your enquiry. I have checked availability in your area, and we can provide service to your cottage. I enclose a quote for the costs of installation, together with the ongoing rental costs for the line. If you want to proceed with this, please quote job reference isdn/hh/123.45.9901.

Yours Sincerely,
Colin Masterton

```
--next--
```

Note that the "implicit" faxback content is given by an EMPTY opaque reference in the middle of the fmtp line in this example.

14.3.4.11 Sportsline "Headlines" Message Sent to Your Phone/Pager/Fax

(i) phone

```
INVITE sip:R2FB@pint.wwos.skynet.com SIP/2.0
Via: SIP/2.0/UDP 169.130.12.5
To:
sip:1-900-123-456-7@wwos.skynet.com;user=phone;phone-context=+1
From: sip:fred.football.fan@skynet.com
Call-ID: 19971205T234505.56.78@chinet.net
CSeq: 4721 INVITE
Subject: Wonderful World Of Sports NFL Final Scores
Content-type: application/sdp
Content-Length: 220

v=0
o=- 2353687800 2353687800 IN IP4 128.3.4.5
s=R2FB
i=NFL Final Scores
```

```
e=fred.football.fan@skynet.com
c=TN RFC 2543 +44-1794-8331013
t=2353687800 0
m=audio 1 voice x-pay
a=fmtp:x-pay opr:mci.com/md5:<crypto signature>
```

(ii) fax

```
INVITE sip:R2FB@pint.wwos.skynet.com SIP/2.0
Via: SIP/2.0/UDP 169.130.12.5
To: sip:1-900-123-456-7@wwos.skynet.com;user=phone;
phone-context=+1
From: sip:fred.football.fan@skynet.com
Call-ID: 19971205T234505.56.78@chinet.net
CSeq: 4722 INVITE
Subject: Wonderful World Of Sports NFL Final Scores
Content-type: application/sdp
Content-Length: 217

v=0
o=- 2353687820 2353687820 IN IP4 128.3.4.5
s=R2FB
i=NFL Final Scores
e=fred.football.fan@skynet.com
c=TN RFC 2543 +44-1794-8331010
t=2353687820 0
m=text 1 fax x-pay
a=fmtp:x-pay opr:mci.com/md5:<crypto signature>
```

(iii) pager

```
INVITE sip:R2FB@pint.wwos.skynet.com SIP/2.0
Via: SIP/2.0/UDP 169.130.12.5
To: sip:1-900-123-456-7@wwos.skynet.com;user=phone;
phone-context=+1
From: sip:fred.football.fan@skynet.com
Call-ID: 19971205T234505.56.78@chinet.net
CSeq: 4723 INVITE
Subject: Wonderful World Of Sports NFL Final Scores
Content-type: application/sdp
Content-Length: 219
v=0
o=- 2353687840 2353687840 IN IP4 128.3.4.5
s=R2FB
i=NFL Final Scores
e=fred.football.fan@skynet.com
c=TN RFC 2543 +44-1794-8331015
t=2353687840 0
m=text 1 pager x-pay
a=fmtp:x-pay opr:mci.com/md5:<crypto signature>
```

Note that these are all VERY similar.

14.3.4.12 Automatically Giving Someone a Fax Copy of Your Phone Bill

```
INVITE sip:BillsRUs@pint.sprint.com SIP/2.0
Via: SIP/2.0/UDP 169.130.12.5
To: sip:+1-555-888-1234@fbi.gov;user=phone
From: sip:agent.mulder@fbi.gov
Call-ID: 19991231T234505.56.78@fbi.gov
CSeq: 911 INVITE
Subject: Itemised Bill for January 98
Content-type: application/sdp
Content-Length: 247

v=0
o=- 2353687860 2353687860 IN IP4 128.3.4.5
s=BillsRUs
i=Joe Pendleton's Phone Bill
e=agent.mulder@fbi.gov
c=TN RFC 2543 +1-202-833-1010
t=2353687860 0
m=text 1 fax x-files-id
a=fmtp:x-files-id opr:fbi.gov/jdcn-123@45:3des;base64,<signature>
```

Note: in this case the opaque reference is a collection of data used to convince the executive system that the requestor has the right to get this information, rather than selecting the particular content (the A party in the "To:" field of the SIP "wrapper" does that alone).

14.3.5 Security Considerations

The security procedures that need to be considered in dealing with the basic SDP messages described in this section (RFC 2848) are provided in Section 17.13.3.

14.3.6 Deployment Considerations and the Relationship PINT to I.N.

14.3.6.1 Web Front End to PINT Infrastructure

It is possible that some other protocol may be used to communicate a requesting user's requirements. Due to the high numbers of available web browsers and servers, it seems likely that some PINT systems will use HTML/HTTP as a "front end." In this scenario, HTTP will be used over a connection from the requesting user's web browser client to an Intermediate Web Server. This will be closely associated with a PINT Client (using some unspecified mechanism to transfer the data from the web server to the PINT client). The PINT client will represent the requesting user to the PINT gateway and thus to the executive system that carries out the required action (Figure 14.2).

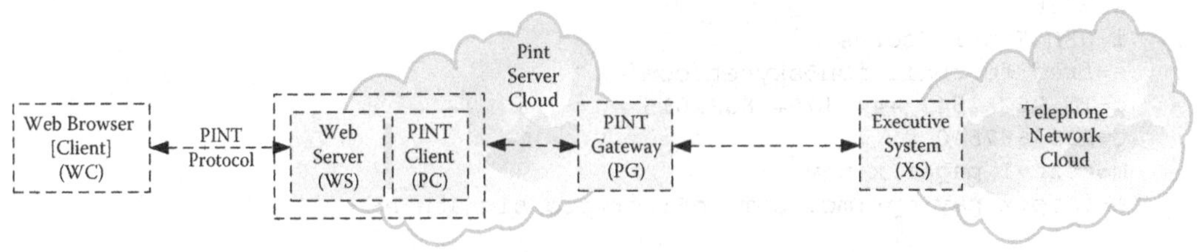

Figure 14.2 Basic "Web-fronted" configuration (Copyright IETF).

14.3.6.2 Redirects to Multiple Gateways

It is quite possible that a given PINT gateway is associated with an executive system (or systems) that can connect to the GSTN at different places. Equally, if there is a chain of PINT servers, then each of these intermediate or proxy servers may be able to route PINT requests to executive systems that connect at specific points to the GSTN. The result of this is that there may be more than one PINT gateway or executive system that can deal with a given request. The mechanisms by which the choice on where to deliver a request are outside the scope of this document (Figure 14.3).

However, there do seem to be two approaches. Either a server that acts as a proxy or redirect will select the appropriate gateway itself and will cause the request to be sent on accordingly, or a list of possible locations will be returned to the requesting user from which they can select their choice for PINT services. In SIP, the implication is that if a proxy cannot resolve to a single unique match for a request destination, then a response containing a list of the choices should be returned to the requesting user for selection. This is not too likely a scenario within the normal use of SIP. However, within PINT, such ambiguity may be quite common; it implies that there are a number of possible providers of a given service.

14.3.6.3 Competing PINT Gateways REGISTERing to Offer the Same Service

With PINT, the registration is not for an individual but instead for a service that can be handled by a service provider. Thus, one can envisage a registration by the PINT Server of the domain telcoA.com of its ability to support the service R2C as "R2C@telcoA.com" sent to an intermediary server that acts as registrar for the "broker.telcos.com" domain from "R2C@pint.telcoA.com" as follows:

```
REGISTER sip:registrar@broker.telcos.com SIP/2.0
To: sip:R2C@pint.telcoA.com
From: sip:R2C@pint.telcoA.com
...
```

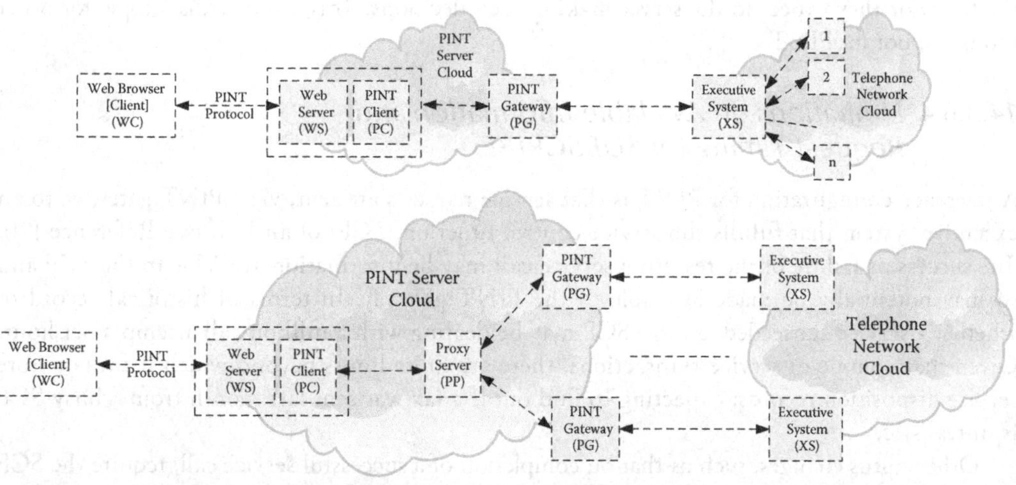

Figure 14.3 Multiple access configurations (Copyright IETF).

This is the standard SIP registration service. However, what happens if there are a number of different service providers, all of whom support the "R2C" service? Suppose there is a PINT system at domain "broker.com." PINT clients requesting a R2C service from broker.com might be very willing to be redirected or proxied to any one of the various service providers that had previously registered with the registrar. PINT servers might also be interested in providing service for requests that did not specify the service provider explicitly, as well as those requests that were directed "at them." To enable such service, PINT servers would REGISTER at the broker PINT server registrations of the form:

```
REGISTER sip:registrar@broker.com SIP/2.0
To: sip:R2C@broker.com
From: sip:R2C@pint.telcoA.com
```

When several such REGISTER messages appear at the registrar, each differing only in the URL in the "From:" line, the registrar has many possibilities, for example:

- It overwrites the prior registration for R2C@broker.telcos.com when the next comes in
- It rejects the subsequent registration for "R2C@broker.telcos.com"
- It maintains all such registrations. In this last case, on receiving an Invitation for the "general" service, either
  - It passes on the invitation to all registered service providers, returning a collated response with all acceptances, using multiple Location: headers, or
  - It silently selects one of the registrations (using, for example, a "round robin" approach) and routes the invitation and response onwards without further comment. As an alternative to all of the above approaches, it:
    - May choose to not allow registrations for the "general" service, rejecting all such REGISTER requests.

The algorithm by which such a choice is made will be implementation-dependent, and is outside the scope of PINT. Where a behavior is to be defined by requesting users, then some sort of call processing language might be used to allow those clients, as a preservice operation, to download the behavior they expect to the server making such decisions. This, however, is a topic for other protocols, not for PINT.

14.3.6.4 Limitations on Available Information and Request Timing for SUBSCRIBE

A reference configuration for PINT is that service requests are sent, via a PINT gateway, to an executive system that fulfills the service control function (SCF) of an I.N. (see Reference [9]). The success or failure of the resulting service call may be information available to the SCF and so may potentially be made available to the PINT gateway. In terms of historical record of whether a service succeeded, a large SCF may be dealing with a million call attempts per hour. Given that volume of service transactions, there are finite limits beyond which it cannot store service disposition records; expecting to find out if a fax was sent last month from a busy SCF is unrealistic.

Other status changes, such as that on completion of a successful service call, require the SCF to arrange monitoring of the service call in a way that the service may not do normally, for performance reasons. In most implementations, it is difficult efficiently to interrupt a service to change it once it has begun execution, so it may be necessary to have two different services: one that sets GSTN resources to monitor service call termination, and one that doesn't. It is unlikely to be possible to decide that monitoring is required once the service has started.

These factors can have implications both on the information that is potentially available at the PINT gateway and when a request to register interest in the status of a PINT service can succeed. The alternative to using a general SCF is to provide a dedicated service node just for PINT services. As this node is involved in placing all service calls, it is in a position to collect the information needed.

However, it may well still not be able to respond successfully to a registration of interest in call state changes once a service logic program instance is running. Thus, although a requesting user may register an interest in the status of a service request, the PINT gateway may not be in a position to comply with that request. Although this does not affect the protocol used between the requestor and the PINT gateway, it may influence the response returned. To avoid the problem of changing service logic once running, any registration of interest in status changes should be made at or before the time at which the service request is made.

Conversely, if a historical request is made on the disposition of a service, this should be done within a short time after the service has completed; the executive system is unlikely to store the results of service requests for long; these will have been processed as automatic message accounting records quickly, after which the executive system has no reason to keep them, and so they may be discarded. Where the PINT gateway and the executive system are intimately linked, the gateway can respond to status subscription requests that occur while a service is running. It may accept these requests and simply not even try to query the executive system until it has information that a service has completed, merely returning the final status. Thus the PINT requestor may be in what it believes is a monitoring state, whilst the PINT gateway has not even informed the executive system that a request has been made. This will increase the internal complexity of the PINT gateway in that it will have a complex set of interlocking state machines but does mean that status registration and indication CAN be provided in conjunction with an I.N. system.

14.3.6.5 Parameters Needed for Invoking Traditional GSTN Services within PINT

This section describes how parameters needed to specify certain traditional GSTN services can be carried within PINT requests.

14.3.6.5.1 Service Identifier

When a requesting user asks for a service to be performed, he or she will, of course, have to specify in some way which service. This can be done in the URLs within the "To:" header and the Request-URI (see Section 14.3.3.5.5.1).

14.3.6.5.2 A and B Parties

With the R2C service, they will also need to specify the A and B parties they want to be engaged in the resulting service call. The A party could identify, for example, the call center from which they want a call back, whilst the B party is their own specific telephone number (i.e., who the call center agent is to call). The R2F and R2HC services require the B party to be specified (respectively the telephone number of the destination fax machine or the telephone to which spoken content is to be delivered), but the A party is a telephone network-based resource (either a fax or speech transcoder/sender) and is implicit; the requesting user does not (and cannot) specify it. With the "Fax-Back" variant of the R2F service, (i.e., where the content to be delivered resides on the GSTN) they will also have to specify two parties. As before, the B party is the telephone number of the fax machine to which they want a fax to be sent.

However, within this variant, the A party identifies the "document context" for the GSTN-based document store from which a particular document is to be retrieved; the analogy here is to a GSTN user dialing a particular telephone number and then entering the document number to be returned using "touch tone" digits. The telephone number dialed is that of the document store or A party, with the "touch tone" digits selecting the document within that store.

14.3.6.5.3 Other Service Parameters

In terms of the extra parameters to the request, the services again differ. The R2C service needs only the A and B parties. Also, it is convenient to assert that the resulting service call will carry voice, as the executive system within the destination GSTN may be able to check that assertion against the A and B party numbers specified and may treat the call differently. With the R2F and R2HC services, the source information to be transcoded is held on the Internet. That means either that this information is carried along with the request itself or that a reference to the source of this information is given.

In addition, it is convenient to assert that the service call will carry fax or voice, and, where possible, to specify the format for the source information. The GSTN-based content or "Fax-Back" variant of the R2F service needs to specify the document store number and the fax machine number to which the information is to be delivered. It is convenient to assert that the call will carry fax data, as the destination executive system may be able to check that assertion against the document store number and that of the destination fax machine. Further, the document number may also need to be sent. This parameter is an opaque reference that is carried through the Internet but has significance only within the GSTN. The document store number and document number together uniquely specify the actual content to be faxed.

14.3.6.5.4 Service Parameter Summary

The following table summarizes the information needed in order to specify fully the intent of a GSTN service request. Note that it excludes any other parameters (such as authentication or authorization tokens, "Expires:," or "CallId:" headers) that may be used in a request.

Service	*ServiceID*	*A-Party*	*B-Party*	*CallFmt*	*Source*	*SourceFm*
R2C	x	x	x	voice	-	-
R2F	x	-	x	fax	URI/IL	ISF/ILSF
R2FB	x	x	x	fax	OR	-
R2HC	x	-	x	voice	URI/IL	ISF/ILSF

In this table, "x" means that the parameter is required, whilst "-" means that the parameter is not required. The services listed are R2C, R2F, the GSTN-based content or "Fax-Back" variant of request-to-fax (R2FB), and R2HC. The call format parameter values "voice" or "fax" indicate the kind of service call that results. The source indicator "URI/IL" implies that the information is either an Internet source reference (a URI) or is carried "in-line" with the message. The source indicator "OR" means that the value passed is an opaque reference that should be carried along with the rest of the message but is to be interpreted only within the destination GSTN context. As an alternative, it could be given as a "local" reference with the "file" style or even using a partial reference with the "http" style.

However, the way in which such reference is interpreted is a matter for the receiving PINT server and executive system; it remains, in effect, an opaque reference. The source format value "ISF/ILSF" means either that the format of the source is specified in terms of the URI or that it is carried "in-line." Note that, for some data, the format can be detected by inspection or, if all else fails, be assumed from the URI (for example, by assuming that the file extension part of a URL indicates the data type). For an opaque reference, the source format is not available on the Internet, and so is not given.

14.3.6.6 Parameter Mapping to PINT Extensions

This section describes the way in which the parameters needed to specify a GSTN service request fully might be carried within a "PINT extended" message. There are other choices, and these are not precluded. However, in order to ensure that the requesting user receives the service he/she expects, it is necessary to have some shared understanding of the parameters passed and the behavior expected of the PINT server and its attendant executive system. The service identifier can be sent as the userinfo element of the Request-URI. Thus, the first line of a PINT Invitation would be of the form

```
INVITE <serviceID>@<pint-server>.<domain> SIP/2.0
```

The A party for the R2C and R2FB service can be held in the "To:" header field. In this case the "To:" header value will be different from the Request-URI. In the services where the A party is not specified, the "To:" field is free to repeat the value held in the Request-URI. This is the case for R2F and R2HC services. The B party is needed in all these milestone services and can be held in the enclosed SDP subpart, as the value of the "c=" field. The call format parameter can be held as part of the "m=" field value. It maps to the "transport protocol" element as described in Section 14.3.3.4.2 of this document.

The source format specifier is held in the "m=," as a type and either "-" or subtype. The latter is normally required for all services except R2C or "Fax Back," where the "-" form may be used. As shown earlier, the source format and source are not always required when generating requests for services. However, the inclusion in all requests of a source format specifier can make parsing the request simpler and allows other services to be specified in the future, so values are always given. The source format parameter is covered in Section 14.3.3.4.2 as the "media type" element.

The source itself is identified by an "a=fmtp:" field value, where needed. With the exception of the R2C service, all invitations will normally include such a field. From the perspective of the SDP extensions, it can be considered as qualifying the media subtype, as if to say, for example, "when I say jpeg, what I mean is the following." In summary, the parameters needed by the different services are carried in fields as shown in the following table:

Service	*Svc Param*	*PINT/SIP or SDP field used*	*Example value*
R2C			
	ServiceID:	<SIP Request-URI userinfo>	R2C
	AParty:	<SIP To: field>	sip:123@p.com
	BParty:	<SDP c= field>	TN RFC 2543 4567
	CallFormat:	<SDP transport protocol sub-field of m= field> voice	voice
	SourceFmt:	<SDP media type sub-field of m= field>	audio
		(--- only "-" sub-type sub-field value used)	----
	Source:	(--- No source specified)	----
R2F			
	ServiceID:	<SIP Request-URI userinfo>	R2F
	AParty:	(--- SIP To: field not used)	sip:R2F@pint.xxx.net
	BParty:	<SDP c= field>	TN RFC xxx +441213553
	CallFormat:	<SDP transport protocol sub-field of m= field>	fax
	SourceFmt:	<SDP media type sub-field of m= field>	image
		<SDP media sub-type sub-field of m= field>	jpeg
	Source:	<SDP a=fmtp: field qualifying preceding m= field>	a=fmtp:jpeg<uri-ref>
R2FB			
	ServiceID:	<SIP Request-URI userinfo>	R2FB
	AParty:	<SIP To: field>	sip:1-730-1234@p.com
	BParty:	<SDP c= field>	TN RFC xxx +441213553
	CallFormat:	<SDP transport protocol sub-field of m= field>	fax
	SourceFmt:	<SDP media type sub-field of m= field>	image
		<SDP media type sub-field of m= field>	image
	Source:	<SDP a=fmtp: field qualifying preceding m= field>	a=fmtp:jpeg opr:1234

(Continued)

R2HC			
	ServiceID:	<SIP Request-URI userinfo>	R2HC
	AParty:	(--- SIP To: field not used)	sip:R2HC@pint.ita.il
	BParty:	<SDP c= field>	TN RFC xxx +441213554
	CallFormat:	: <SDP transport protocol sub-field of m= field> z	voice
	SourceFmt:	<SDP media type sub-field of m= field>	text
		<SDP media sub-type sub-field of m= field>	html
	Source:	<SDP a=fmtp: field qualifying preceding m=field>	a=fmtp:html<uri-ref>

14.3.7 Collected ABNF for PINT Extensions (Appendix A of RFC 2848)

All SDP ABNF syntaxes are provided in Section 2.9. However, we have repeated here for convenience.

```
;; --(ABNF is specified in RFC 2234 that is obsoleted by RFC 4234
(RFC 4234 that is obsoleted by RFC 5234)
;; --Variations on SDP definitions

connection-field      =      ["c=" nettype space addrtype space
                             connection-address CRLF]
; -- this is the original definition from SDP, included for
completeness
; -- the following are PINT interpretations and modifications

nettype               =      ("IN"/"TN")

; -- redefined as a superset of the SDP definition

addrtype              =      (INAddrType / TNAddrType)

; -- redefined as a superset of the SDP definition

INAddrType            =      ("IP4"/"IP6")

; -- this non-terminal added to hold original SDP address types

TNAddrType            =      ("RFC 2543"/OtherAddrType)
OtherAddrType         =      (<X-Token>)

; -- X-token is as defined in RFC 2045

addr                  =      (<FQDN> / <unicast-address> / TNAddr)

; -- redefined as a superset of the original SDP definition
; -- FQDN and unicast address as specified in SDP

TNAddr                =      (RFC 2543Addr/OtherAddr)
```

```
; -- TNAddr defined only in context of nettype == "TN"

RFC 2543Addr          =       (INPAddr/LDPAddr)
INPAddr               =       "+" <POS-DIGIT>
                              0*(("-" <DIGIT>)/<DIGIT>)

; -- POS-DIGIT and DIGIT as defined in SDP
LDPAddr               =       <DIGIT> 0*(("-" <DIGIT>)/<DIGIT>)
OtherAddr             =       1*<uric>

; -- OtherAdd defined in the context of OtherAddrType
; -- uric is as defined in RFC 2396 that is obsoleted by RFC 3986

media-field           =       "m=" media <space> port <space> proto
                              1*(<space> fmt) <CRLF>
; -- NOTE redefined as subset/relaxation of original SDP definition
; -- space and CRLF as defined in SDP

media                 =       ("application"/"audio"/"image"/"text")

; -- NOTE redefined as a subset of the original SDP definition
; -- This could be any MIME discrete type; Only those listed are
; -- used in PINT 1.0

port                  =       ("0" / "1")

; -- NOTE redefined from the original SDP definition;
; -- 0 retains usual sdp meaning of "temporarily no media"
; -- (i.e. "line is on hold")
; -- (1 means there is media)

proto                 =       (INProto/TNProto)

; -- redefined as a superset of the original SDP definition

INProto               =       1* (<alpha-numeric>)

; -- this is the "classic" SDP protocol, defined if nettype == "IN"
; -- alpha-numeric is as defined in SDP

TNProto               =       ("voice"/"fax"/"pager")

; -- this is the PINT protocol, defined if nettype == "TN"

fmt                   =       (<subtype> / "-")

; -- NOTE redefined as a subset of the original SDP definition
; -- subtype as defined in RFC 2046, or "-". MUST be a subtype
of type held

; -- in associated media sub-field or the special value "-".

attribute-fields      =       *("a=" attribute-list <CRLF>)
```

```
; -- redefined as a superset of the definition given in SDP
; -- CRLF is as defined in SDP
attribute-list          =       1(PINT-attribute / <attribute>)

; -- attribute is as defined in SDP

PINT-attribute          =       (clir-attribute /
                                q763-nature-attribute /
                                q763plan-attribute /
                                q763-INN-attribute /
                                phone-context-attribute /
                                tsp-attribute /
                                pint-fmtp-attribute / strict-attribute)
clir-attribute          =       clir-tag ":" ("true" / "false")
clir-tag                =       "clir"
q763-nature-attribute =         Q763-nature-tag ":" q763-natures
q763-nature-tag         =       "Q763-nature"
q763-natures            =       ("1" / "2" / "3" / "4")
q763-plan-attribute     =       Q763-plan-tag ":" q763-plans
q763-plan-tag           =       "Q763-plan"
q763-plans              =       ("1" / "2" / "3" / "4" / "5" / "6"
                                / "7")

; -- of these, the meanings of 1, 3, and 4 are defined in the text

q763-INN-attribute      =       Q763-INN-tag ":" q763-INNs
q763-INN-tag            =       "Q763-INN"
q763-INNs               =       ("0" / "1")
phone-context-attribute =       phone-context-tag ":"
                                phone-context-ident
phone-context-tag       =       "phone-context"
phone-context-ident     =       network-prefix / private-prefix
network-prefix          =       intl-network-prefix /
local-network-prefix
intl-network-prefix     =       "+" 1*<DIGIT>
local-network-prefix    =       1*<DIGIT>
private-prefix          =       1*excldigandplus 0*<uric>
excldigandplus          =       (0x21-0x2d,0x2f,0x40-0x7d))
tsp-attribute           =       tsp-tag " = " provider-domainname
tsp-tag                 =       "tsp"
provider-domainname     =       <domain>

; -- domain is defined in RFC 1035
; -- NOTE the following is redefined relative to the normal use in
SDP

pint-fmtp-attribute     =       "fmtp:" <subtype> <space> resolution
                                *(<space> resolution)
                                (<space> ";" 1(<attribute>) *(<space>
                                <attribute>))

; -- subtype as defined in RFC 2046.
; -- NOTE that this value MUST match a fmt on the ultimately
preceding
```

```
; -- media-field
; -- attribute is as defined in SDP

resolution              =      (uri-ref / opaque-ref / sub-part-ref)
uri-ref                 =      uri-tag ":" <URI-Reference>

; -- URI-Reference defined in RFC 2396 that is obsoleted by RFC 3986

uritag                  =      "uri"
opaque-ref              =      opr-tag ":" 0*<uric>
opr-tag                 =      "opr"
sub-part-ref            =      spr-tag ":" <Content-ID>

; -- Content-ID is as defined in RFC 2046 and RFC 0822

spr-tag                 =      "spr"
strict-attribute        =      "require:" att-tag-list
att-tag-list            =      1(PINT-att-tag-list / <att-field> /
                               pint-fmtp-tag-list)
                               *(","
                                (PINT-att-tag-list / <att-field> /
                               pint-fmtp-tag-list)
                               )

; -- att-field as defined in SDP

PINT-att-tag-list       =      (phone-context-tag / clir-tag /
                               q763-nature-tag / q763-plan-tag /
                               q763-INN-tag)
pint-fmtp-tag-list      =      (uri-tag / opr-tag / spr-tag)

;; --Variations on SIP definitions

clir-parameter          =      clir-tag "=" ("true" / "false")
q763-nature-parameter =        Q763-nature-tag "=" Q763-natures
q763plan-parameter      =      Q763-plan-tag "=" q763plans
q763-INN-parameter      =      Q763-INN-tag "=" q763-INNs
tsp-parameter           =      tsp-tag "=" provider-domainname
phone-context-parameter  =  phone-context-tag "=" phone-context-ident
SIP-param               =      (<transport-param>/<user-param>/<method-
                               param> / <ttl-param> / <maddr-param> /
                               <other-param> )

; -- the values in this list are all as defined in SIP

PINT-param              =      ( clir-parameter /
                               q763-nature-parameter /
                               q763plan-parameter / q763-INN-parameter/
                               tsp-parameter /
                               phone-context-parameter )

URL-parameter                  =      (SIP-param / PINT-param)

; -- redefined SIP's URL-parameter to include ones defined in PINT
```

```
Require-header         =        "require:" 1(required-extensions)
                                *("," required-extensions)

; -- NOTE this is redefined as a subset of the SIP definition
; -- (from RFC 2543/section 6.30); RFC 2543 is obsoleted by RFC 3261
; (also see Section 1.2)

required-extensions    =        ("org.ietf.sip.subscribe" /
                                "org.ietf.sdp.require")
```

14.3.8 IANA Considerations (Appendix B of RFC 2848)

SDP parameters that are described here have been registered with IANA. We have not included the IANA registration procedures here for the sake of brevity. The procedures for registration of new SDP parameters are described in RFC 2848.

14.4 Summary

We have described the service-specific extensions in SDP for the 3GPP-defined PSS and MBMS. We have defined the new SDP attributes such as video buffering, video frame size, integration-protection configuration, alternative, adaptation, quality-of-experience, and asset information that are used for PSS services. Similarly, the MBMS service-specific SDP attributes such as MBMS bearer mode declaration, FEC flow ID, MBMS repair, SDP protocol identifiers to FEC (RTP protocol identifiers and FEC repair data identifier), and video buffering attribute are defined. Although not related to SDP, we have described RSTP extensions for PSS services here just for the completeness of RFC 6064. We have also covered the SDP extensions for OMA's BCAST service that is similar to 3GPP's MBMS; they are not the same because of the services are invoked differently because of SDP attributes for this service such as major, minor, id for of the stream, id for STRP authentication algorithm value, and ROC transmission rate.

The PINT is a detailed multimedia service where the phone has dual interfaces: IP and PSTN network. First, the PINT service function and protocol architecture are defined. The new SDP attributes that are defined for PINT are network type and address type, data objects in PINT (use of fmtp, remote data, GSTN-based data, session description included in data objects), attribute tags to pass information into the telephone (phone context, presentation restriction, and ITU-T called party), and require. In addition, PINT's extensions to SIP are defined considering the fact that SIP/SDP goes hand-to-hand as well as the usages of SDP extensions where SIP is used as the call control signaling protocol. In this conferencing environment, we have also discussed the PINT request and responses in great length. Finally, the deployment of PINT in the context of I.N. services is described. The key is that the I.N. features over the PSTN/ISDN network show how well those intelligent services can be easily deployed with combinations of Internet and PSTN/ISDN network eliminating the need for a separate costly Signaling System No. 7 (SS7) signaling network.

14.5 Problems

1. What is the motivation of service-specific extensions in SDP? How do you justify that the SDP extensions that accommodate PSS, MBMS, BCAST, and PINT are service-specific? Does mobility play any role in defining these services? Explain with detailed examples.

2. Justify that the features and characteristics of the PSS and MBMS service need new SDP extensions that are service-specific. Explain using detailed call flows for both services using each SDP attribute.
3. Use detailed technical explanations to show that SDP attributes that are defined for accommodating MBMS and BCAST services are really two different services.
4. What are the RTSP extensions made by PSS services? Explain the service using call flows.
5. What is the PINT service protocol? What has been the motivation behind the PINT service?
6. Define the PINT functional and protocol architecture with call flows using SIP/SDP.
7. What are the SDP attributes that have been extended? Show with detailed call flows the usage of each new SDP attribute.
8. Describe each new extension of SIP to accommodate PINT. Describe each extension in SIP with detailed call flows including SDP negotiations.
9. Describe PINT requests and responses in detail using call flows for negotiation services using SDP.
10. Explain in detail using call flows how PINT can be used for I.N. services in view of Web front, redirection to multiple gateways, PINT gateways to offer the same services, usage of SIP's SUBSCRIBE method, and GSTN services using PINT.

References

1. 3GPP TS 26.233 version 7.0.0 (2007-06), "Transparent end-to-end packet switched streaming service (PSS) General Description".
2. 3GPP TS 26.234 version 7.7.0 (2009-03), "Transparent end-to-end Packet-switched Streaming Service (PSS); Protocols and codecs".
3. 3GPP TS 26.244 version 7.3.0 (2007-12), "Transparent end-to-end packet switched streaming service (PSS); 3GPP file format (3GP)".
4. 3GPP TS 26.346 version 7.10.0 (2009-03), "Multimedia Broadcast/Multicast Service (MBMS); Protocols and codecs".
5. NIST, "Advanced Encryption Standard (AES)", FIPS PUB 197, http://www.nist.gov/itl/fipscurrent.cfm.
6. Open Mobile Alliance, "Service and Content Protection for Mobile Broadcast Services", OMA OMA-TS-BCAST_SvcCntProtection-V1_0-20071218-D, 2007.
7. ITU-T Study Group XI, "Q.763 – Formats and Codes for the ISDN User Part of SS No7", ITU-T, August 1994.
8. The Unicode Consortium, "The Unicode Standard -- Version 2.0", Addison-Wesley, 1996.
9. ITU-T Study Group XI, "Q.1204 – IN Distributed Functional Plane Architecture", ITU-T, February 1994.

Chapter 15

Compression Support in SDP

Abstract
This chapter describes Request for Comments (RFC) 3485 that deals with the compression support in Session Description Protocol (SDP) and Session Initiation Protocol (SIP). The SDP is a text-based protocol intended for describing multimedia sessions and may carry files like Multipurpose Internet Mail Extensions and others. The call control protocols like SIP (see Section 1.2) or web real-time communication (see Section 1.3) that uses SDP for session negotiations between the conferencing parties is also a text-based protocol. In a combination of SDP and call control protocol, the bandwidth may be a problem to carry the signaling traffic, especially in resource-constraints wireless networks. RFC 3485 specifies the SIP/SDP-specific static dictionary that the signaling compression scheme defined in RFC 3320 may use in order to achieve higher bandwidth efficiency as well as better response time for the call setup. The dictionary is compression algorithm independent.

15.1 SDP and SIP Static Dictionary for Signaling Compression

This section describes Request for Comments (RFC) 3485 that specifies Session Description Protocol (SDP) and Session Initiation Protocol (SIP) static dictionary for SIP and SDP for signaling compression (SigComp). SIP is a text-based protocol for initiating and managing communication sessions. The protocol can be compressed by using SigComp. Similarly, the SDP is a text-based protocol intended for describing multimedia sessions for the purposes of session announcement, session invitation, and other forms of multimedia session initiation. This RFC defines the SIP/SDP-specific static dictionary that SigComp may use in order to achieve higher efficiency. The dictionary is compression algorithm independent.

15.1.1 Introduction

SIP (RFC 3261, also see Section 1.2) and SDP (RFC 4566, see Section 2) are text-based protocols that use the UTF-8 charset specified in RFC 2279 (that was made obsolete by RFC 3629). SIP and SDP were designed for rich bandwidth links. However, when SIP/SDP is run over narrow bandwidth links, such as radio interfaces or low speed serial links, the session setup time increases substantially, compared to an operation over a rich bandwidth link. The session setup time can decrease dramatically if the SIP/SDP signaling is compressed. The SigComp mechanisms specified in SigComp described in RFC 3320 provide a multiple compression/decompression algorithm framework to compress and decompress text-based protocols such as SIP and SDP. When compression is used in SIP/SDP, the compression achieves its maximum rate once a few message exchanges have taken place. This is due to the fact that the first message the compressor sends to the decompressor is only partially compressed, as there is not a previous stored state against which to compress. As the goal is to reduce the session setup time as much as possible, it seems sensible to investigate a mechanism to boost the compression rate from the first message. In this specification, we introduce the static dictionary for SIP and SDP. The dictionary is to be used in conjunction with SIP, SDP, and SigComp. The static SIP/SDP dictionary constitutes a SigComp state that can be referenced in the first SIP message that the compressor sends out.

15.1.2 Design Considerations

The static SIP/SDP dictionary is a collection of well-known strings that appear in most of the SIP and SDP messages. The dictionary is not a comprehensive list of reserved words, but it includes many of the strings that appear in SIP and SDP signaling. The static dictionary is unique and MUST be available in all SigComp implementations for SIP/SDP. The dictionary is not intended to evolve as SIP or SDP evolves. It is defined once, and stays as is forever. This solves the problems of updating, upgrading, and finding out the dictionary that is supported at the remote end when several versions of the same dictionary coexist.

Section 15.1.4 of this book (that is, Appendix A of RFC 3485) contains the collection of strings that SIP contributed to the static dictionary and includes references to the documents that define those strings. Section 15.1.5 of this book (that is, Appendix B of RFC 3485) contains the collection of strings that SDP contributed to the static dictionary and includes references to the documents that define those strings. While these appendices are of an informative nature, Section 15.1.3 gives the normative binary form of the SIP/SDP dictionary. This dictionary is included in the SigComp implementation. This dictionary has been formed from the collection of individual dictionaries given in Sections 15.1.4 and 15.1.5 of this book (that is, Appendices A and B of RFC 3485, respectively).

The two input collections are collections of UTF-8 encoded character strings. In order to facilitate the readability, the appendices describe them in one table for each collection. In these tables, each row represents an entry. Each entry contains the string that actually occurs in the dictionary, its priority (see bullets below), its offset from the first octet and its length (both in hexadecimal), and one or more references that elucidate why this string is expected to occur in SIP/SDP messages. Note: Length in this document always refers to octets. The columns in the tables are described as follows:

- **String**: represents the UTF-8 string that is inserted into the dictionary. Note that the quotes (") are not part of the string itself. Note also that the notation "carriage-return line-feed" ("CRLF") represents a carriage return character (ASCII code 0x0D) followed by a line feed character (ASCII code 0x0A).
- **Pr**: indicates the priority of this string within the dictionary. Some compression algorithms, such as DEFLATE, offer an increased efficiency when the most commonly used strings are located at the bottom of the dictionary. To facilitate generating a dictionary that has the most frequently occurring strings further down at the bottom, we have decided to allocate a

priority to each string in the dictionary. Priorities range from 1 until 5. A low number in the priority column (e.g., 1) indicates that we believe in a high probability of finding the string in SIP or SDP messages. A high number in the priority column (e.g., 5) indicates lower probability of finding the string in a SIP or SDP message. This is typically the case for less frequent error codes or optional infrequent tags.

- **Off:** indicates the hexadecimal offset of the entry with respect to the first octet in the dictionary. Note that several strings in the collections can share space in the dictionary if they exhibit suitable common substrings.
- **Len:** the length of the string (in octets, in hexadecimal). References: contains one or more references to the specification and the section within the specification where the string is defined.

Note that the strings stored in the dictionary are case sensitive. (Again, the strings do not comprise the quotes ("), they are just shown here to increase the readability.) Where the string is a header field, we also included the colon ":" and the amount of white space expected to occur. Note that this means that not all messages that conform to the SIP Augmented Backus-Naur Form (ABNF), which allows other combinations (e.g., a white space or horizontal tabulator before the colon (":") sign), will benefit as much from the dictionary—the best increase in compression performance is to be expected for messages that use the recommended formatting guidelines for SIP.

Some strings appear followed by an equal sign, and some others do not. This depends on whether the string is part of a parameter name or a parameter value. In a SIP message, all the SIP headers terminate with a CRLF pair of characters. As these characters are appended to the end of each SIP header line, right after the header values, and because the header values are typically not part of the static SIP dictionary, we cannot include the terminating CRLF as part of the SIP static dictionary. Instead, the approach we have taken is to include in each header field entry the CRLF from the previous line that prefixes every header field. We have represented CRLF by the notation "[CRLF]."

Therefore, in generating the actual binary dictionary, an entry in the dictionary represented as "[CRLF]From: " has been interpreted as an entry whose value is CR, LF, the word From, a colon and a whitespace.

Note that most SIP header field names are included with the full string from CRLF to the colon-blank pair. However, in certain situations, when the likelihood of occurrence is not considered high (as indicated by a priority value of 3–5), and when there are common substrings shared by a number of headers, we have added one entry with the common substring and several entries with the non-common substrings remaining. An example is the "Proxy-Authenticate" and "Proxy-Authorization" headers. There are three entries in the dictionary: the common substring "[CRLF] Proxy-" and the uncommon substrings "Authenticate: " and "Authorization: ."

This allows the reuse of the uncommon substrings by other entries and may save a number of bytes in the binary form of the dictionary. Note that this splitting mechanism does not apply with strings that are likely to occur very often (those whose priority is set to 1 or 2). SIP (RFC 3261, also see Section 1.2) responses start with a status code (e.g., "302") and a reason phrase (e.g., "Moved Temporarily"). The status code is a normative part, whereas the reason phrase is not normative, it is just a suggested text. For instance, both "302 Moved Temporarily" and "302 Redirect" are valid beginnings of SIP responses.

In the SIP dictionary we have included two entries per response code, one including only the status code and a space (e.g., "302 ") and another including both the status code and the suggested reason phrase (e.g., "302 Moved Temporarily"). The former can be used when the SIP response changed the suggested reason phrase to another one. The latter can be used when the suggested reason phrase is part of the response. In this way, we accommodate both alternatives. (Note that in the actual dictionary, both strings occupy the same space in the string subset, but have two separate entries in the table subset.)

15.1.3 Binary Representation of the SIP/SDP Dictionary

This section contains the result of combining the SIP and the SDP dictionaries described in Sections 15.1.4 and 15.1.5 in order to create a single dictionary that is loaded into SigComp as a state. The binary SigComp dictionary is comprised of two parts, the concatenation of which serves as the state value of the state item: A string subset, which contains all strings in the contributing collections as a substring (roughly ordered such that strings with low priority numbers occur at the end) and a table subset, which contains pairs of length and offset values for all strings in the contributing collections. In each of these pairs, the length is stored as a one-byte value, and the offset is stored as a two-byte value that has had 1024 added to the offset (this allows direct referencing from the stored value if the dictionary state has been loaded at address 1024).

The intention is that all compression algorithms will be able to use all or part of the string subset, and some compression methods, notably those that are related to the LZ78 family, will also use the table in order to form an initial set of tokens for that compression method. The text described here therefore gives examples for referencing both the table subset and the string subset of the dictionary state item. As defined in Section 3.3.3 in the SigComp specification of RFC 3320, a SigComp state is characterized by a certain set of information. For the static SIP/SDP dictionary, the information in Table 15.1 fully characterizes the state item.

Note that the string subset of the dictionary can be accessed using

```
STATE-ACCESS (%ps, 6, 0, 0x0D8C, %sa, 0),
```

and the table subset can be accessed using

```
STATE-ACCESS (%ps, 6, 0x0D8C, 0x0558, %sa, 0),
```

where %ps points to Universal Decompressor Virtual Memory (UDVM) containing

```
0xfbe507dfe5e6
```

and %sa is the desired destination address in UDVM (with UDVM byte copying rules applied).

If only a subset of the dictionary up to a specific priority is desired (e.g., to save UDVM space), the values for the third and fourth operand in these STATE-ACCESS instructions can be changed to

```
Priorities  String  String  Table   Table
desired     offset  length  offset  length
==========  ======  ======  ======  ======
1 only      0x0CB2  0x00DA  0x0D8C  0x003F
1..2        0x0920  0x046C  0x0D8C  0x0147
1..3        0x07B8  0x05D4  0x0D8C  0x01A7
1..4        0x0085  0x0D07  0x0D8C  0x044A
1..5        0x0000  0x0D8C  0x0D8C  0x0558
```

The state item consists of the following elements:

Table 15.1 Binary Representation of the Static SIP/SDP Dictionary for SigComp (Copyright IETF)

```
Name:                        Value:
=====================        =======================
state_identifier             0xfbe507dfe5e6aa5af2abb914ceaa05f99ce61b
state_length                 0x12E4
state_address                0 (not relevant for the dictionary)
state_instruction            0 (not relevant for the dictionary)
minimum_access_length        6
state_value                  Representation of the table below.

0000  0d0a 5265 6a65 6374 2d43 6f6e 7461 6374  ..Reject-Contact
0010  3a20 0d0a 4572 726f 722d 496e 666f 3a20  : ..Error-Info:
0020  0d0a 5469 6d65 7374 616d 703a 200d 0a43  ..Timestamp: ..C
0030  616c 6c2d 496e 666f 3a20 0d0a 5265 706c  all-Info: ..Repl
0040  792d 546f 3a20 0d0a 5761 726e 696e 673a  y-To: ..Warning:
0050  200d 0a53 7562 6a65 6374 3a20 3b68 616e   ..Subject: ;han
0060  646c 696e 673d 696d 6167 653b 7075 7270  dling=image;purp
0070  6f73 653d 3b63 6175 7365 3d3b 7465 7874  ose=;cause=;text
0080  3d63 6172 6433 3030 204d 756c 7469 706c  =card300 Multipl
0090  6520 4368 6f69 6365 736d 696d 6573 7361  e Choicesmimessa
00A0  6765 2f73 6970 6672 6167 3430 3720 5072  ge/sipfrag407 Pr
00B0  6f78 7920 4175 7468 656e 7469 6361 7469  oxy Authenticati
00C0  6f6e 2052 6571 7569 7265 6469 6765 7374  on Requiredigest
00D0  2d69 6e74 6567 7269 7479 3438 3420 4164  -integrity484 Ad
00E0  6472 6573 7320 496e 636f 6d70 6c65 7465  dress Incomplete
00F0  6c65 7068 6f6e 652d 6576 656e 7473 3439  lephone-events49
0100  3420 5365 6375 7269 7479 2041 6772 6565  4 Security Agree
0110  6d65 6e74 2052 6571 7569 7265 6465 6163  ment Requiredeac
0120  7469 7661 7465 6434 3831 2043 616c 6c2f  tivated481 Call/
0130  5472 616e 7361 6374 696f 6e20 446f 6573  Transaction Does
0140  204e 6f74 2045 7869 7374 616c 653d 3530   Not Existale=50
0150  3020 5365 7276 6572 2049 6e74 6572 6e61  0 Server Interna
0160  6c20 4572 726f 726f 6275 7374 2d73 6f72  l Errorobust-sor
0170  7469 6e67 3d34 3136 2055 6e73 7570 706f  ting=416 Unsuppo
0180  7274 6564 2055 5249 2053 6368 656d 6572  rted URI Schemer
0190  6765 6e63 7934 3135 2055 6e73 7570 706f  gency415 Unsuppo
01A0  7274 6564 204d 6564 6961 2054 7970 656e  rted Media Typen
01B0  6469 6e67 3438 3820 4e6f 7420 4163 6365  ding488 Not Acce
01C0  7074 6162 6c65 2048 6572 656a 6563 7465  ptable Herejecte
01D0  6434 3233 2049 6e74 6572 7661 6c20 546f  d423 Interval To
01E0  6f20 4272 6965 6672 6f6d 2d74 6167 512e  o Briefrom-tagQ.
01F0  3835 3035 2056 6572 7369 6f6e 204e 6f74  8505 Version Not
0200  2053 7570 706f 7274 6564 3430 3320 466f   Supported403 Fo
0210  7262 6964 6465 6e6f 6e2d 7572 6765 6e74  rbiddenon-urgent
0220  3432 3920 5072 6f76 6964 6520 5265 6665  429 Provide Refe
0230  7272 6f72 2049 6465 6e74 6974 7934 3230  rror Identity420
0240  2042 6164 2045 7874 656e 7369 6f6e 6f72   Bad Extensionor
0250  6573 6f75 7263 650d 0a61 3d6b 6579 2d6d  esource..a=key-m
0260  676d 743a 6d69 6b65 794f 5054 494f 4e53  gmt:mikeyOPTIONS
0270  204c 616e 6775 6167 653a 2035 3034 2053   Language: 504 S
0280  6572 7665 7220 5469 6d65 2d6f 7574 6f2d  erver Time-outo-
0290  7461 670d 0a41 7574 6865 6e74 6963 6174  tag..Authenticat
02A0  696f 6e2d 496e 666f 3a20 4465 6320 3338  ion-Info: Dec 38
02B0  3020 416c 7465 726e 6174 6976 6520 5365  0 Alternative Se
02C0  7276 6963 6535 3033 2053 6572 7669 6365  rvice503 Service
02D0  2055 6e61 7661 696c 6162 6c65 3432 3120   Unavailable421
02E0  4578 7465 6e73 696f 6e20 5265 7175 6972  Extension Requir
```

(*Continued*)

Table 15.1 (*Continued*) Binary Representation of the Static SIP/SDP Dictionary for SigComp (Copyright IETF)

02F0	6564	3430	3520	4d65	7468	6f64	204e	6f74	ed405 Method Not
0300	2041	6c6c	6f77	6564	3438	3720	5265	7175	Allowed487 Requ
0310	6573	7420	5465	726d	696e	6174	6564	6175	est Terminatedau
0320	7468	2d69	6e74	6572	6c65	6176	696e	673d	th-interleaving=
0330	0d0a	6d3d	6170	706c	6963	6174	696f	6e20	..m=application
0340	4175	6720	3531	3320	4d65	7373	6167	6520	Aug 513 Message
0350	546f	6f20	4c61	7267	6536	3837	2044	6961	Too Large687 Dia
0360	6c6f	6720	5465	726d	696e	6174	6564	3330	log Terminated30
0370	3220	4d6f	7665	6420	5465	6d70	6f72	6172	2 Moved Temporar
0380	696c	7933	3031	204d	6f76	6564	2050	6572	ily301 Moved Per
0390	6d61	6e65	6e74	6c79	6d75	6c74	6970	6172	manentlymultipar
03A0	742f	7369	676e	6564	0d0a	5265	7472	792d	t/signed..Retry-
03B0	4166	7465	723a	2047	4d54	6875	2c20	3430	After: GMThu, 40
03C0	3220	5061	796d	656e	7420	5265	7175	6972	2 Payment Requir
03D0	6564	0d0a	613d	6f72	6965	6e74	3a6c	616e	ed..a=orient:lan
03E0	6473	6361	7065	3430	3020	4261	6420	5265	dscape400 Bad Re
03F0	7175	6573	7472	7565	3439	3120	5265	7175	questrue491 Requ
0400	6573	7420	5065	6e64	696e	6735	3031	204e	est Pending501 N
0410	6f74	2049	6d70	6c65	6d65	6e74	6564	3430	ot Implemented40
0420	3620	4e6f	7420	4163	6365	7074	6162	6c65	6 Not Acceptable
0430	3630	3620	4e6f	7420	4163	6365	7074	6162	606 Not Acceptab
0440	6c65	0d0a	613d	7479	7065	3a62	726f	6164	le..a=type:broad
0450	6361	7374	6f6e	6534	3933	2055	6e64	6563	castone493 Undec
0460	6970	6865	7261	626c	650d	0a4d	494d	452d	ipherable..MIME-
0470	5665	7273	696f	6e3a	204d	6179	2034	3832	Version: May 482
0480	204c	6f6f	7020	4465	7465	6374	6564	0d0a	Loop Detected..
0490	4f72	6761	6e69	7a61	7469	6f6e	3a20	4a75	Organization: Ju
04A0	6e20	6d6f	6465	2d63	6861	6e67	652d	6e65	n mode-change-ne
04B0	6967	6862	6f72	3d63	7269	7469	6361	6c65	ighbor=criticale
04C0	7274	6370	2d66	6234	3839	2042	6164	2045	rtcp-fb489 Bad E
04D0	7665	6e74	6c73	0d0a	556e	7375	7070	6f72	ventls..Unsuppor
04E0	7465	643a	204a	616e	2035	3032	2042	6164	ted: Jan 502 Bad
04F0	2047	6174	6577	6179	6d6f	6465	2d63	6861	Gatewaymode-cha
0500	6e67	652d	7065	7269	6f64	3d0d	0a61	3d6f	nge-period=..a=o
0510	7269	656e	743a	7365	6173	6361	7065	0d0a	rient:seascape..
0520	613d	7479	7065	3a6d	6f64	6572	6174	6564	a=type:moderated
0530	3430	3420	4e6f	7420	466f	756e	6433	3035	404 Not Found305
0540	2055	7365	2050	726f	7879	0d0a	613d	7479	Use Proxy..a=ty
0850	6e2d	5374	6174	653a	204e	6f76	200d	0a53	n-State: Nov ..S
0860	6572	7669	6365	2d52	6f75	7465	3a20	5365	ervice-Route: Se
0870	7020	0d0a	416c	6c6f	772d	4576	656e	7473	p ..Allow-Events
0880	3a20	4665	6220	0d0a	613d	696e	6163	7469	: Feb ..a=inacti
0890	7665	5254	502f	5341	5650	2052	5450	2f41	veRTP/SAVP RTP/A
08A0	5650	4620	416e	6f6e	796d	6f75	7369	7073	VPF Anonymousips
08B0	3a0d	0a61	3d74	7970	653a	7465	7374	656c	:..a=type:testel
08C0	3a4d	4553	5341	4745	200d	0a61	3d72	6563	:MESSAGE ..a=rec
08D0	766f	6e6c	790d	0a61	3d73	656e	646f	6e6c	vonly..a=sendonl
08E0	790d	0a63	3d49	4e20	4950	3420	0d0a	5265	y..c=IN IP4 ..Re
08F0	6173	6f6e	3a20	0d0a	416c	6c6f	773a	200d	ason: ..Allow: .
0900	0a45	7665	6e74	3a20	0d0a	5061	7468	3a20	.Event: ..Path:
0910	3b75	7365	723d	0d0a	623d	4153	2043	5420	;user=..b=AS CT
0920	0d0a	5757	572d	4175	7468	656e	7469	6361	..WWW-Authentica
0930	7465	3a20	4469	6765	7374	200d	0a61	3d73	te: Digest ..a=s
0940	656e	6472	6563	7669	6465	6f63	7465	742d	endrecvideooctet-
0950	616c	6967	6e3d	6170	706c	6963	6174	696f	align=applicatio
0960	6e2f	7364	7061	7468	6561	6465	7273	7061	n/sdpatheaderspa

(*Continued*)

Table 15.1 (*Continued*) Binary Representation of the Static SIP/SDP Dictionary for SigComp (Copyright IETF)

0970	7574	683d	0d0a	613d	6f72	6965	6e74	3a70	uth=..a=orient:p
0980	6f72	7472	6169	7469	6d65	6f75	7474	722d	ortraitimeouttr-
0990	696e	7469	636f	6e63	3d34	3833	2054	6f6f	inticonc=483 Too
09A0	204d	616e	7920	486f	7073	6c69	6e66	6f70	Many Hopslinfop
09B0	7469	6f6e	616c	676f	7269	7468	6d3d	3630	tionalgorithm=60
09C0	3420	446f	6573	204e	6f74	2045	7869	7374	4 Does Not Exist
09D0	2041	6e79	7768	6572	6573	706f	6e73	653d	Anywheresponse=
09E0	0d0a	0d0a	5265	7175	6573	742d	4469	7370	....Request-Disp
09F0	6f73	6974	696f	6e3a	204d	4435	3830	2050	osition: MD580 P
0A00	7265	636f	6e64	6974	696f	6e20	4661	696c	recondition Fail
0A10	7572	6570	6c61	6365	7334	3232	2053	6573	ureplaces422 Ses
0A20	7369	6f6e	2049	6e74	6572	7661	6c20	546f	sion Interval To
0A30	6f20	536d	616c	6c6f	6361	6c31	3831	2043	o Smallocal181 C
0A40	616c	6c20	4973	2042	6569	6e67	2046	6f72	all Is Being For
0A50	7761	7264	6564	6f6d	6169	6e3d	6661	696c	wardedomain=fail
0A60	7572	656e	6465	7265	616c	6d3d	5355	4253	urenderealm=SUBS
0A70	4352	4942	4520	7072	6563	6f6e	6469	7469	CRIBE preconditi
0A80	6f6e	6f72	6d61	6c69	7073	6563	2d6d	616e	onormalipsec-man
0A90	6461	746f	7279	3431	3320	5265	7175	6573	datory413 Reques
0AA0	7420	456e	7469	7479	2054	6f6f	204c	6172	t Entity Too Lar
0AB0	6765	3265	3138	3320	5365	7373	696f	6e20	ge2e183 Session
0AC0	5072	6f67	7265	7373	6374	7034	3836	2042	Progressctp486 B
0AD0	7573	7920	4865	7265	6d6f	7465	726d	696e	usy Heremotermin
0AE0	6174	6564	414b	4176	312d	4d44	352d	7365	atedAKAv1-MD5-se
0AF0	7373	696f	6e6f	6e65	0d0a	4175	7468	6f72	ssionone..Author
0B00	697a	6174	696f	6e3a	2036	3033	2044	6563	ization: 603 Dec
0B10	6c69	6e65	7874	6e6f	6e63	653d	3438	3520	linextnonce=485
0B20	416d	6269	6775	6f75	7365	726e	616d	653d	Ambiguousername=
0B30	6175	6469	6f0d	0a43	6f6e	7465	6e74	2d54	audio..Content-T
0B40	7970	653a	204d	6172	200d	0a52	6563	6f72	ype: Mar ..Recor
0B50	642d	526f	7574	653a	204a	756c	2034	3031	d-Route: Jul 401
0B60	2055	6e61	7574	686f	7269	7a65	640d	0a52	Unauthorized..R
0B70	6571	7569	7265	3a20	0d0a	743d	3020	302e	equire: ..t=0 0.
0B80	302e	302e	300d	0a53	6572	7665	723a	2052	0.0.0..Server: R
0B90	4547	4953	5445	5220	0d0a	633d	494e	2049	EGISTER ..c=IN I
0BA0	5036	2031	3830	2052	696e	6769	6e67	3130	P6 180 Ringing10
0BB0	3020	5472	7969	6e67	763d	300d	0a6f	3d55	0 Tryingv=0..o=U
0BC0	5044	4154	4520	4e4f	5449	4659	200d	0a53	PDATE NOTIFY ..S
0BD0	7570	706f	7274	6564	3a20	756e	6b6e	6f77	upported: unknow
0BE0	6e41	4d52	5450	2f41	5650	200d	0a50	7269	nAMRTP/AVP ..Pri
0BF0	7661	6379	3a20	0d0a	5365	6375	7269	7479	vacy: ..Security
0C00	2d0d	0a45	7870	6972	6573	3a20	0d0a	613d	-..Expires: ..a=
0C10	7274	706d	6170	3a0d	0a6d	3d76	6964	656f	rtpmap:..m=video
0C20	200d	0a6d	3d61	7564	696f	200d	0a73	3d20	..m=audio ..s=
0C30	6661	6c73	650d	0a61	3d63	6f6e	663a	3b65	false..a=conf:;e
0C40	7870	6972	6573	3d0d	0a52	6f75	7465	3a20	xpires=..Route:
0C50	0d0a	613d	666d	7470	3a0d	0a61	3d63	7572	..a=fmtp:..a=cur
0C60	723a	436c	6965	6e74	3a20	5665	7269	6679	r:Client: Verify
0C70	3a20	0d0a	613d	6465	733a	0d0a	5241	636b	: ..a=des:..RAck
0C80	3a20	0d0a	5253	6571	3a20	4259	4520	636e	: ..RSeq: BYE cn
0C90	6f6e	6365	3d31	3030	7265	6c75	7269	3d71	once=100reluri=q
0CA0	6f70	3d54	4350	5544	5071	6f73	786d	6c3b	op=TCPUDPqosxml;
0CB0	6c72	0d0a	5669	613a	2053	4950	2f32	2e30	lr..Via: SIP/2.0
0CC0	2f54	4350	2034	3038	2052	6571	7565	7374	/TCP 408 Request
0CD0	2054	696d	656f	7574	696d	6572	7073	6970	Timeoutimerpsip
0CE0	3a0d	0a43	6f6e	7465	6e74	2d4c	656e	6774	:..Content-Lengt

(*Continued*)

Table 15.1 (*Continued*) Binary Representation of the Static SIP/SDP Dictionary for SigComp (Copyright IETF)

```
0CF0  683a 204f 6374 200d 0a56 6961 3a20 5349  h: Oct ..Via: SI
0D00  502f 322e 302f 5544 5020 3b63 6f6d 703d  P/2.0/UDP ;comp=
0D10  7369 6763 6f6d 7072 6f62 6174 696f 6e61  sigcomprobationa
0D20  636b 3b62 7261 6e63 683d 7a39 6847 3462  ck;branch=z9hG4b
0D30  4b0d 0a4d 6178 2d46 6f72 7761 7264 733a  K..Max-Forwards:
0D40  2041 7072 2053 4354 5052 4143 4b20 494e   Apr SCTPRACK IN
0D50  5649 5445 200d 0a43 616c 6c2d 4944 3a20  VITE ..Call-ID:
0D60  0d0a 436f 6e74 6163 743a 2032 3030 204f  ..Contact: 200 O
0D70  4b0d 0a46 726f 6d3a 200d 0a43 5365 713a  K..From: ..CSeq:
0D80  200d 0a54 6f3a 203b 7461 673d 0410 dd10   ..To: ;tag=....
0D90  1131 0d11 0a07 10b9 0c10 fe12 10e1 0611  .1..............
0DA0  4e07 114e 0311 4a04 114a 0710 b208 1179  N..N..J..J.....y
0DB0  0611 810f 1122 0b11 5506 116b 0b11 6013  ....."..U..k..`.
0DC0  10b2 0811 7105 1187 1310 f709 0e8d 080d  ....q...........
0DD0  ae0c 10b9 0710 8e03 0d96 0310 8a04 108a  ................
0DE0  090d d70a 0f12 080f 8f09 0f8f 080d 6c06  ..............l.
0DF0  0e66 090e 6c0a 0e6c 060f c607 0fc6 0511  .f..l..l........
0E00  4806 1148 060f bf07 0fbf 070e 5506 0f16  H..H........U...
0E10  040e f403 0eb1 0310 a609 1050 0310 a30a  ...........P....
0E20  0db4 050e 3606 0ed6 030d f911 0ef8 040c  ....6...........
0E30  d908 0eea 0409 5303 0a4b 040e e410 0f35  ......S..K.....5
0E40  090e e408 0d3f 030f e10b 1001 0310 ac06  .....?..........
0E50  1095 0c0e 760b 0feb 0a0f ae05 102b 0410  ....v........+..
0E60  2b08 107a 100f 4907 0fb8 0910 3e0b 100c  +..z..I.....>...
0E70  070f 780b 0f6d 0910 4708 1082 0b0f f608  ..x..m..G.......
0E80  1062 080f 8708 106a 040f 780d 0fcd 080d  .b.....j..x.....
0E90  ae10 0f5d 0b0f 9814 0d20 1b0d 2004 0de0  ...]..... .. ...
0EA0  140e b40b 0fa3 0b07 340f 0d56 040e f403  ........4..V....
0EB0  10af 070d 3409 0f27 0410 9b04 109f 0910  ....4..'........
0EC0  5908 1072 0910 350a 1021 0a10 1708 0fe3  Y..r..5..!......
0ED0  0310 a905 0cac 040c bd07 0cc1 080c c109  ................
0EE0  0cf6 100c 720c 0c86 040d 640c 0cd5 090c  ....r.....d.....
0EF0  ff1b 0bfc 110c 5d13 0c30 090c a40c 0c24  ......]..0.....$
0F00  0c0d 3b03 0d1a 030d 1d16 0c43 090c 9209  ..;........C....
0F10  0c9b 0d0e cb04 0d16 060d 1005 04f2 0b0c  ................
0F20  e105 0bde 0a0c ec13 0be3 070b d408 0d08  ................
0F30  0c0c c909 0c3a 040a e50c 0a23 080b 3a0e  .....:.....#..:.
0F40  09ab 0f0e fa09 0f6f 0c0a 170f 0976 0c0a  .......o.....v..
0F50  5f17 0de2 0f07 a80a 0f85 0f08 d60e 09b9  _...............
0F60  0b0a 7a03 0bdb 0308 c104 0ec7 0308 d302  ..z.............
0F70  048d 080b 4a05 0b8c 070b 6106 0548 0407  ....J.....a..H..
0F80  f405 1030 0407 1e08 071e 050b 9110 04ca  ...0............
0F90  090a 7109 0e87 0504 9805 0b6e 0b04 9b0f  ..q........n....
0FA0  049b 0704 9b03 04a3 0704 a310 0798 0907  ................
0FB0  9805 0b73 050b 7805 0b7d 0507 b905 0b82  ...s..x..}......
0FC0  050b 8705 0b1d 0508 e405 0c81 050f 4405  ..............D.
0FD0  1140 0508 7805 089d 050f 5805 073f 050c  .@..x.....X..?..
0FE0  6d05 10f2 050c 5805 06a9 0407 b609 058c  m.....X.........
0FF0  0606 1a06 0e81 0a06 160a 0ac4 070b 5a0a  ..............Z.
1000  0aba 030b 1b04 1145 060c 8c07 05ad 0a0e  .......E........
1010  da08 0b42 0d09 f70b 051c 0911 1608 05c9  ...B............
1020  070d 8606 0bcf 0a06 4d04 0ba2 0606 8d08  ........M.......
1030  05e6 080e 110b 0a9b 030a 0403 0bb5 0510  ................
1040  d704 0994 050a e203 0bb2 060d 6704 0d11  ............g...
1050  0808 b71b 0e3b 0a09 a114 0485 1507 8315  .....;..........
1060  076e 0d09 3d17 06ae 0f07 e614 07be 0d06  .n..=...........
```

(*Continued*)

Table 15.1 (*Continued*) Binary Representation of the Static SIP/SDP Dictionary for SigComp (Copyright IETF)

```
1070  0a0d 0930 1606 f212 081e 2104 aa13 10c5  ...0......!.....
1080  080a 0f1c 0e96 180b b81a 0595 1a05 7511  ..............u.
1090  063d 1606 dc1e 0e19 1605 d11d 0620 2305  .=........... #.
10A0  2711 087d 110d 9916 04da 0d0f 1c16 0708  '..}............
10B0  1705 b40d 08c7 1307 f812 0857 1f04 fe19  ...........W....
10C0  054e 1308 0b0f 08e9 1706 c513 067b 1905  .N...........{..
10D0  f115 0744 180d fb0b 0f09 1b0d be12 0830  ...D...........0
10E0  1507 5904 0ba6 040b ae04 0b9e 040b 9604  ..Y.............
10F0  0b9a 0a0a b00b 0a90 080b 320b 096b 080b  ..........2..k..
1100  2a0b 0a85 090b 120a 0aa6 0d09 ea13 0d74  *..............t
1110  1407 d213 090b 1208 4210 095b 1209 1e0d  ........B..[....
1120  0cb1 0e0c 1711 094a 0c0a 530c 0a47 090a  .......J..S..G..
1130  f70e 09c7 0c0a 3b07 0669 0806 6906 09e3  ......;..i..i...
1140  080b 520a 0ad8 1206 570d 0657 0709 e304  ..R.....W..W....
1150  0ae9 1007 3009 0b00 0c0a 2f05 0ae9 050a  ....0...../.....
1160  6b06 0a6b 0a0a ce09 0aee 030b db07 0f7e  k..k...........~
1170  0a09 970a 0671 0e09 d517 0693 070e 5c07  .....q........\.
1180  0fda 0a0f 350d 0dec 0a09 970a 0671 080b  ....5.........q..
1190  220f 0985 060b 680c 0d4a 090b 0913 08f8  ".....h..J......
11A0  1508 a204 0baa 0f05 660d 0723 090a 060b  ........f..#....
11B0  0d4a 0f04 ee06 04f8 0409 2b04 0853 0708  .J........+..S..
11C0  c003 111f 0411 1e07 0d8c 0307 3404 10db  ............4...
11D0  0307 3603 0da9 0d04 200b 0451 0c04 3a04  ..6..... ..Q..:.
11E0  0bb8 040c 2404 0595 0404 7c04 0575 0404  ....$.....|..u..
11F0  8504 096b 0406 3d06 047b 0406 dc04 0783  ...k..=..{......
1200  040e 1912 0400 1008 8e10 0869 0e04 120d  ............i....
1210  042d 0310 b904 05d1 0407 6e04 0620 0704  .-.........n.. ..
1220  7404 0bfc 0a04 5c04 0527 0409 3d04 087d  t.....\..'..=..}
1230  040f ae04 0d99 0406 ae04 04da 0904 0908  ................
1240  1122 040f 1c04 07e6 040e cb05 08bd 0407  ."..............
1250  0804 0fa3 0406 5704 05b4 040f 5d04 08c7  ......W.....]...
1260  080b f404 07f8 0407 3004 07be 0408 5705  ........0.....W.
1270  0d46 0404 fe04 060a 0405 4e04 0e3b 0408  .F........N..;..
1280  0b04 0930 0408 e905 05ee 0406 c504 06f2  ...0............
1290  0406 7b04 09a1 0405 f104 081e 0407 4404  ..{...........D.
12A0  0bdd 040d fb04 04aa 040b e307 0eee 040f  ................
12B0  0904 0eb4 040d be04 10c5 0408 3005 0f30  ............0..0
12C0  0407 5904 0a0f 060e 6104 0481 040d ab04  ..Y.....a.......
12D0  0d93 0411 6b04 0e96 0504 6609 046b 0b04  ....k.....f..k..
12E0  4604 0ce1                                F...
```

15.1.4 SIP Input Strings to the SIP/SDP Static Dictionary (Appendix A of RFC 3485)

For reference, this section lists the SIP input strings used in generating the dictionary, as well as a priority value, the offset of the string in the generated dictionary, the length of the string, and one or more references into the referenced documents that motivate the presence of this string. Note that the notation "[CRLF]" stands for a sequence of two bytes with the values 0x0d and 0x0a, respectively.

The priority value is used for determining the position of the string in the dictionary. Lower priority values (higher priorities) cause the string to occur at a later position in the dictionary, making it more efficient to reference the string in certain compression algorithms. Hence, lower priority values were assigned to strings more likely to occur.

```
String                                 Pr Off  Len  References
====================================== == ==== ==== ==========
"sip:"                                  1 0CDD 0004 [3] 19.1.1
"sips:"                                 3 08AC 0005 [3] 19.1.1
"tel:"                                  3 08BD 0004 [7] 2.2
"SIP/2.0"                               1 0CB9 0007 [3] 25.1
"SIP/2.0/UDP "                          1 0CFE 000C [3] 25.1
"SIP/2.0/TCP "                          2 0CB9 000C [3] 25.1
"INVITE"                                1 0D4E 0006 [3] 25.1
"INVITE "                               1 0D4E 0007 [3] 25.1
"ACK"                                   1 0D4A 0003 [3] 25.1
"ACK "                                  1 0D4A 0004 [3] 25.1
"OPTIONS"                               4 0269 0007 [3] 25.1
"OPTIONS "                              4 0269 0008 [3] 25.1
"BYE"                                   2 0C8A 0003 [3] 25.1
"BYE "                                  2 0C8A 0004 [3] 25.1
"CANCEL"                                4 05E3 0006 [3] 25.1
 "CANCEL "                              4 05E3 0007 [3] 25.1
 "REGISTER"                             2 0B8F 0008 [3] 25.1
 "REGISTER "                            2 0B8F 0009 [3] 25.1
 "INFO"                                 4 06E9 0004 [8] 2
 "INFO "                                4 06E9 0005 [8] 2
 "SUBSCRIBE"                            2 0A6C 0009 [9] 8.1.1
 "SUBSCRIBE "                           2 0A6C 000A [9] 8.1.1
 "NOTIFY"                               2 0BC6 0006 [9] 8.1.2
 "NOTIFY "                              2 0BC6 0007 [9] 8.1.2
 "PRACK"                                2 0D48 0005 [10] 6
 "PRACK "                               2 0D48 0006 [10] 6
 "UPDATE"                               2 0BBF 0006 [11] 7, 10
 "UPDATE "                              2 0BBF 0007 [11] 7, 10
 "REFER"                                4 066B 0005 [13] 2.1, 7
 "REFER "                               4 066B 0006 [13] 2.1, 7
 "MESSAGE"                              3 08C1 0007 [21] 9
 "MESSAGE "                             3 08C1 0008 [21] 9
 "[CRLF]Accept: "                       4 06CE 000A [3] 20.1
 "[CRLF]Accept-"                        4 06EE 0009 [22] 5,
                                          [3] 20.2, 20.3
 "Contact: "                            5 0009 0009 [22] 5
 "Encoding: "                           4 0597 000A [3] 20.2,
                                          [3] 20.12
 "Language: "                           4 0271 000A [3] 20.3,
                                          [3] 20.13
 "[CRLF]Alert-Info: "                   4 05D5 000E [3] 20.4
 "[CRLF]Allow: "                        3 08F6 0009 [3] 20.5
 "[CRLF]Allow-Events: "                 3 0872 0010 [9] 8.2.1
 "[CRLF]Authentication-Info: "          4 0293 0017 [3] 20.6
 "[CRLF]Authorization: "                2 0AF8 0011 [3] 20.7
 "[CRLF]Call-ID: "                      1 0D55 000B [3] 20.8
 "[CRLF]Call-Info: "                    5 002D 000D [3] 20.9
 "[CRLF]Contact: "                      1 0D60 000B [3] 20.10
 "[CRLF]Content-"                       4 0B35 000A [3] 20.11,
                                          20.12, 20.13, [3] 20.14,
                                          20.15
 "Disposition: "                        4 09EC 000D [3] 20.11
 "Encoding: "                           4 0597 000A [3] 20.2,
                                          [3] 20.12
 "Language: "                           4 0271 000A [3] 20.3,
                                          [3] 20.13
 "[CRLF]Content-Length: "               1 0CE1 0012 [3] 20.14
 "[CRLF]Content-Type: "                 2 0B35 0010 [3] 20.15
 "[CRLF]CSeq: "                         1 0D79 0008 [3] 20.16
 "[CRLF]Date: "                         4 0722 0008 [3] 20.17
```

```
"[CRLF]Error-Info: "                    5 0012 000E [3] 20.18
"[CRLF]Event: "                         3 08FF 0009 [9] 8.2.1
"[CRLF]Expires: "                       2 0C01 000B [3] 20.19
"[CRLF]From: "                          1 0D71 0008 [3] 20.20
"[CRLF]In-Reply-To: "                   4 0585 000F [3] 20.21
"[CRLF]Max-Forwards: "                  1 0D31 0010 [3] 20.22
"[CRLF]Min-"                            4 0768 0006 [3] 20.23,
                                          [18] 5
"Expires: "                             4 083A 0009 [3] 20.23
"SE: "                                  4 06E5 0004 [18] 5
"[CRLF]MIME-Version: "                  5 0469 0010 [3] 20.24
"[CRLF]Organization: "                  5 048E 0010 [3] 20.25
"[CRLF]Path: "                          3 0908 0008 [16] 3
"[CRLF]Priority: "                      4 0623 000C [3] 20.26
"[CRLF]Privacy: "                       2 0BEB 000B [33] 4.2
"[CRLF]Proxy-"                          4 073A 0008 [3] 20.27,
                                         20.28, 20.29
"Authenticate: "                         4 05AB 000E [3] 20.27
"Authorization: "                        4 0AFA 000F [3] 20.28
"Require: "                              4 0B6F 0009 [3] 20.29
"[CRLF]RAck: "                           2 0C7A 0008 [10] 7.2
"[CRLF]Reason: "                         3 08EC 000A [17] 2
"[CRLF]Record-Route: "                   2 0B49 0010 [3] 20.30
"[CRLF]Refer-To: "                       4 0617 000C [13] 2.1, 7
"[CRLF]Referred-By: "                    4 0576 000F [34] 9
"[CRLF]Reject-Contact: "                 5 0000 0012 [22] 5
"[CRLF]Replaces: "                       4 065F 000C [14] 3.1
"[CRLF]Reply-To: "                       5 003A 000C [3] 20.31
"[CRLF]Request-Disposition: "            4 09E2 0017 [22] 5
"[CRLF]Require: "                        2 0B6D 000B [3] 20.32
"[CRLF]Retry-After: "                    4 03A8 000F [3] 20.33
"[CRLF]Route: "                          2 0C47 0009 [3] 20.34
"[CRLF]RSeq: "                           2 0C82 0008 [10] 7.1
"[CRLF]Security-"                        2 0BF6 000B [20] 3.3
"Client: "                               2 0C62 0008 [20] 3.3
"Server: "                               2 0B87 0008 [20] 3.3
"Verify: "                               2 0C6A 0008 [20] 3.3
"[CRLF]Server: "                         4 0B85 000A [3] 20.35
"[CRLF]Service-Route: "                  3 085D 0011 [35]
"[CRLF]Session-Expires: "                3 0830 0013 [18] 4
"[CRLF]Subject: "                        5 0051 000B [3] 20.36
"[CRLF]Subscription-State: "             3 0843 0016 [9] 8.2.3
"[CRLF]Supported: "                      2 0BCD 000D [3] 20.37
"[CRLF]Timestamp: "                      5 0020 000D [3] 20.38
"[CRLF]To: "                             1 0D81 0006 [3] 20.39
"[CRLF]Unsupported: "                    4 04D6 000F [3] 20.40
"[CRLF]User-Agent: "                     4 05B9 000E [3] 20.41
"[CRLF]Via: "                            1 0CB2 0007 [3] 20.42
"[CRLF]Via: SIP/2.0/UDP "                1 0CF7 0013 [3] 20.42
"[CRLF]Via: SIP/2.0/TCP "                1 0CB2 0013 [3] 20.42
"[CRLF]Warning: "                        5 0046 000B [3] 20.43
"[CRLF]WWW-Authenticate: "               2 0920 0014 [3] 20.44
"[CRLF]WWW-Authenticate: Digest "        2 0920 001B [3] 20.44
"[CRLF][CRLF]"                           2 09E0 0004 [3] 7
";transport="                            4 067A 000B [3] 25.1
"udp"                                    4 07DB 0003 [3] 25.1,
                                           [24] A, [3] 25.1, [24] A
"tcp"                                    4 04C1 0003 [3] 25.1
"sctp"                                   4 0AC7 0004 [3] 25.1
"tls"                                    4 04D3 0003 [3] 25.1,
                                           [20] 3.3
```

```
";user="                         3 0910 0006 [3] 25.1
"phone"                          3 00F2 0005 [3] 25.1
"ip"                             4 008D 0002 [3] 25.1
";method="                       4 074A 0008 [3] 25.1
";ttl="                          4 078C 0005 [3] 25.1
";lr"                            2 0CAF 0003 [3] 25.1
"Digest "                        2 0934 0007 [6] 3.2.1,
                                   3.2.2
"username="                      2 0B27 0009 [6] 3.2.2
"uri="                           2 0C9B 0004 [6] 3.2.2
"qop="                           2 0C9F 0004 [6] 3.2.1,
                                   3.2.2
"cnonce="                        2 0C8E 0007 [6] 3.2.2
"nc="                            2 0996 0003 [6] 3.2.2
"response="                      2 09D7 0009 [6] 3.2.2
"nextnonce="                     2 0B12 000A [6] 3.2.3
"rspauth="                       2 096C 0008 [6] 3.2.3
"realm="                         2 0A66 0006 [6] 3.2.1
"domain="                        2 0A55 0007 [6] 3.2.1
"nonce="                         2 0B16 0006 [6] 3.2.1
"opaque="                        4 0761 0007 [6] 3.2.1
"stale="                         4 0148 0006 [6] 3.2.1
"true"                           4 03F4 0004 [6] 3.2.1
"false"                          4 0C30 0005 [6] 3.2.1
"algorithm="                     2 09B4 000A [6] 3.2.1,
                                   [19] 3.1
"MD5"                            2 09F9 0003 [6] 3.2.1,
                                   [19] 3.1
"MD5-sess"                       2 0AEA 0008 [6] 3.2.1,
                                   [19] 3.1
"auth"                           4 031E 0004 [6] 3.2.1
"auth-int"                       4 031E 0008 [6] 3.2.1
"AKAv"                           2 0AE4 0004 [19] 3.1, 6
"AKAv1-MD5"                      2 0AE4 0009 [19] 3.1, 6
"auts="                          4 0791 0005 [19] 3.4
"digest-integrity"               4 00CA 0010 [20] 3.3
"ipsec-ike"                      4 0671 0009 [20] 3.3
"ipsec-man"                      4 0A87 0009 [20] 3.3
"smime"                          4 0098 0005 [20] 3.3
";alg="                          4 076E 0005 [20] 3.3
";purpose="                      5 006B 0009 [3] 20.9
"icon"                           5 0993 0004 [3] 20.9, 20.11
"info"                           5 09AB 0004 [3] 20.9
"card"                           5 0081 0004 [3] 20.9
";expires="                      2 0C3E 0009 [3] 25.1,
                                   [9] 8.4
"render"                         5 0A61 0006 [3] 20.11
"session"                        5 0AEE 0007 [3] 20.11,
                                   [33] 4.2
"alert"                          5 04BD 0005 [3] 20.11
";handling="                     5 005C 000A [3] 20.11
"optional"                       2 09AE 0008 [3] 20.11,
                                   [12] 4, [3] 20.11, [12] 4
"required"                       5 07F4 0008 [3] 20.11
"text"                           5 007C 0004 [3] 25.1
"image"                          5 0066 0005 [3] 25.1
"audio"                          5 0B30 0005 [3] 25.1
"video"                          5 0946 0005 [3] 25.1
"application"                    2 0334 000B [3] 25.1
"application/sdp"                2 0956 000F [3] 25.1
```

```
"message/sip"                  4 009B 000B [3] 27.5
"message/sipfrag"              4 009B 000F [15] 2
"message"                      4 009B 0007 [3] 27.5,
                                 [15] 2
"sip"                          4 00A3 0003 [3] 27.5
"sipfrag"                      4 00A3 0007 [15] 2
"multipart/signed"             4 0398 0010 [3] 23.3
"multipart"                    4 0398 0009 [3] 25.1, 7.4.1
"sdp"                          2 064B 0003
"xml"                          2 0CAC 0003
"Mon, "                        4 0773 0005 [3] 25.1
"Tue, "                        4 0778 0005 [3] 25.1
"Wed, "                        4 077D 0005 [3] 25.1
"Thu, "                        4 03B9 0005 [3] 25.1
"Fri, "                        4 0782 0005 [3] 25.1
"Sat, "                        4 0787 0005 [3] 25.1
"Sun, "                        4 071D 0005 [3] 25.1
" Jan "                        4 04E4 0005 [3] 25.1
" Feb "                        4 0881 0005 [3] 25.1
" Mar "                        4 0B44 0005 [3] 25.1
" Apr "                        4 0D40 0005 [3] 25.1
" May "                        4 0478 0005 [3] 25.1
" Jun "                        4 049D 0005 [3] 25.1
" Jul "                        4 0B58 0005 [3] 25.1
" Aug "                        4 033F 0005 [3] 25.1
" Sep "                        4 086D 0005 [3] 25.1
" Oct "                        4 0CF2 0005 [3] 25.1
" Nov "                        4 0858 0005 [3] 25.1
" Dec "                        4 02A9 0005 [3] 25.1
" GMT"                         4 03B6 0004 [3] 25.1
";tag="                        1 0D87 0005 [3] 25.1
"emergency"                    4 018C 0009 [3] 20.26
"urgent"                       4 021A 0006 [3] 20.26
"normal"                       4 0A81 0006 [3] 20.26
"non-urgent"                   4 0216 000A [3] 20.26
";duration="                   4 06C4 000A [3] 20.33
";maddr="                      4 075A 0007 [3] 20.42
";received="                   4 06BA 000A [3] 20.42
";branch="                     5 0D22 0008 [3] 20.42
";branch=z9hG4bK"              1 0D22 000F [3] 8.1.1.7
"SIP"                          5 0CB9 0003 [3] 25.1,
                                 [17] 2
"UDP"                          2 0CA6 0003 [3] 20.42
"TCP"                          2 0CA3 0003 [3] 20.42
"TLS"                          4 071B 0003 [3] 20.42
"SCTP"                         4 0D45 0004 [3] 20.42
"active"                       4 088C 0006 [9] 8.4
"pending"                      4 01AD 0007 [9] 8.4
"terminated"                   4 0ADA 000A [9] 8.4
";reason="                     4 0742 0008 [9] 8.4
";retry-after="                4 05F7 000D [9] 8.4
"deactivated"                  4 011C 000B [9] 8.4
"probation"                    4 0D16 0009 [9] 8.4
"rejected"                     4 01C9 0008 [9] 8.4
"timeout"                      4 0986 0007 [9] 8.4
"giveup"                       4 07CF 0006 [9] 8.4
"noresource"                   4 024D 000A [9] 8.4
";id="                         4 07A2 0004 [9] 8.4
"100rel"                       2 0C95 0006 [10] 8.1
"precondition"                 2 0A76 000C [12] 8
"refer"                        3 07DE 0005 [13] 3.1, 7
```

```
"to-tag"                             4 028D 0006 [14] 3.2
"from-tag"                           4 01E6 0008 [14] 3.2
"replaces"                           4 0A11 0008 [14] 3.4
"Q.850"                              5 01EE 0005 [17] 2
";cause="                            5 0074 0007 [17] 2
";text="                             5 007B 0006 [17] 2
"path"                               3 0964 0004 [16] 3
";refresher="                        4 069B 000B [18] 4
"uac"                                4 0604 0003 [18] 4
"uas"                                4 07B5 0003 [18] 4
"timer"                              4 0CD7 0005 [18] 7.1
"pref"                               5 07DD 0004 [22] 4.1
"TRUE"                               4 0594 0004 [22] 6.2
"FALSE"                              4 06E2 0005 [22] 6.2
";q="                                4 07B2 0003 [3] 25.1,
                                       [22] 6.2, [20] 3.3
";comp=sigcomp"                      1 0D0A 000D [23] 6
"privacy"                            3 07D4 0007 [33] 4.2
"header"                             4 0967 0006 [33] 4.2
"user"                               4 0911 0004 [33] 4.2
"none"                               2 0AF4 0004 [33] 4.2,
                                       [12] 4
"critical"                           4 04B7 0008 [33] 4.2
"100 "                               5 0BAE 0004 [3] 21.1.1
"100 Trying"                         2 0BAE 000A [3] 21.1.1
"180 "                               5 0BA3 0004 [3] 21.1.2
"180 Ringing"                        2 0BA3 000B [3] 21.1.2
"181 "                               5 0A3B 0004 [3] 21.1.3
"181 Call Is Being Forwarded"        4 0A3B 001B [3] 21.1.3
"182 "                               5 05A1 0004 [3] 21.1.4
"182 Queued"                         4 05A1 000A [3] 21.1.4
"183 "                               5 0AB4 0004 [3] 21.1.5
"183 Session Progress"               2 0AB4 0014 [3] 21.1.5
"200 "                               5 0D6B 0004 [3] 21.2.1
"200 OK"                             1 0D6B 0006 [3] 21.2.1
"202 "                               5 0824 0004 [9] 8.3.1
"202 Accepted"                       3 0824 000C [9] 8.3.1
"300 "                               5 0085 0004 [3] 21.3.1
"300 Multiple Choices"               4 0085 0014 [3] 21.3.1
"301 "                               5 0383 0004 [3] 21.3.2
"301 Moved Permanently"              4 0383 0015 [3] 21.3.2
"302 "                               5 036E 0004 [3] 21.3.3
"302 Moved Temporarily"              4 036E 0015 [3] 21.3.3
"305 "                               5 053D 0004 [3] 21.3.4
"305 Use Proxy"                      4 053D 000D [3] 21.3.4
"380 "                               5 02AE 0004 [3] 21.3.5
"380 Alternative Service"            4 02AE 0017 [3] 21.3.5
"400 "                               5 03E6 0004 [3] 21.4.1
"400 Bad Request"                    4 03E6 000F [3] 21.4.1
"401 "                               5 0B5D 0004 [3] 21.4.2
"401 Unauthorized"                   2 0B5D 0010 [3] 21.4.2
"402 "                               5 03BE 0004 [3] 21.4.3
"402 Payment Required"               4 03BE 0014 [3] 21.4.3
"403 "                               5 020A 0004 [3] 21.4.4
"403 Forbidden"                      4 020A 000D [3] 21.4.4
"404 "                               5 0530 0004 [3] 21.4.5
"404 Not Found"                      4 0530 000D [3] 21.4.5
"405 "                               5 02F2 0004 [3] 21.4.6
"405 Method Not Allowed"             4 02F2 0016 [3] 21.4.6
"406 "                               5 041E 0004 [3] 21.4.7
"406 Not Acceptable"                 4 041E 0012 [3] 21.4.7
```

```
"407 "                                   5 00AA 0004 [3] 21.4.8
"407 Proxy Authentication Required"      4 00AA 0021 [3] 21.4.8
"408 "                                   5 0CC5 0004 [3] 21.4.9
"408 Request Timeout"                    4 0CC5 0013 [3] 21.4.9
"410 "                                   5 060F 0004 [3] 21.4.10
"410 Gone"                               4 060F 0008 [3] 21.4.10
"413 "                                   5 0A96 0004 [3] 21.4.11
"413 Request Entity Too Large"           4 0A96 001C [3] 21.4.11
"414 "                                   5 07B8 0004 [3] 21.4.12
"414 Request-URI Too Long"               4 07B8 0018 [3] 21.4.12
"415 "                                   5 0195 0004 [3] 21.4.13
"415 Unsupported Media Type"             4 0195 001A [3] 21.4.13
"416 "                                   5 0175 0004 [3] 21.4.14
"416 Unsupported URI Scheme"             4 0175 001A [3] 21.4.14
"420 "                                   5 023D 0004 [3] 21.4.15
"420 Bad Extension"                      4 023D 0011 [3] 21.4.15
"421 "                                   5 02DC 0004 [3] 21.4.16
"421 Extension Required"                 4 02DC 0016 [3] 21.4.16
"422 "                                   5 0A19 0004 [18] 6, 12.1
"422 Session Interval Too Small"         4 0A19 001E [18] 6, 12.2
"423 "                                   5 01D1 0004 [3] 21.4.17
"423 Interval Too Brief"                 4 01D1 0016 [3] 21.4.17
"429 "                                   5 0220 0004 [34] 9
"429 Provide Referror Identity"          4 0220 001D [34] 9
"480 "                                   5 07FC 0004 [3] 21.4.18
"480 Temporarily Unavailable"            3 07FC 001B [3] 21.4.18
"481 "                                   5 0127 0004 [3] 21.4.19
"481 Call/Transaction Does Not Exist"    4 0127 0023 [3] 21.4.19
"482 "                                   5 047D 0004 [3] 21.4.20
"482 Loop Detected"                      4 047D 0011 [3] 21.4.20
"483 "                                   5 0999 0004 [3] 21.4.21
"483 Too Many Hops"                      4 0999 0011 [3] 21.4.21
"484 "                                   5 00DA 0004 [3] 21.4.22
"484 Address Incomplete"                 4 00DA 0016 [3] 21.4.22
"485 "                                   5 0B1C 0004 [3] 21.4.23
"485 Ambiguous"                          4 0B1C 000D [3] 21.4.23
"486 "                                   5 0ACB 0004 [3] 21.4.24
"486 Busy Here"                          3 0ACB 000D [3] 21.4.24
"487 "                                   5 0308 0004 [3] 21.4.25
"487 Request Terminated"                 4 0308 0016 [3] 21.4.25
"488 "                                   5 01B4 0004 [3] 21.4.26
"488 Not Acceptable Here"                4 01B4 0017 [3] 21.4.26
"489 "                                   5 04C7 0004 [9] 8.3.2
"489 Bad Event"                          4 04C7 000D [9] 8.3.2
"491 "                                   5 03F8 0004 [3] 21.4.27
"491 Request Pending"                    4 03F8 0013 [3] 21.4.27
"493 "                                   5 0457 0004 [3] 21.4.28
"493 Undecipherable"                     4 0457 0012 [3] 21.4.28
"494 "                                   5 00FE 0004 [20] 3.3.1
"494 Security Agreement Required"        4 00FE 001F [20] 3.3.1
"500 "                                   5 014E 0004 [3] 21.5.1
"500 Server Internal Error"              4 014E 0019 [3] 21.5.1
"501 "                                   5 040B 0004 [3] 21.5.2
"501 Not Implemented"                    4 040B 0013 [3] 21.5.2
"502 "                                   5 04E9 0004 [3] 21.5.3
"502 Bad Gateway"                        4 04E9 000F [3] 21.5.3
"503 "                                   5 02C5 0004 [3] 21.5.4
"503 Service Unavailable"                4 02C5 0017 [3] 21.5.4
"504 "                                   5 027B 0004 [3] 21.5.5
"504 Server Time-out"                    4 027B 0013 [3] 21.5.5
```

```
"505 "                                   5 01F1 0004 [3] 21.5.6
"505 Version Not Supported"              4 01F1 0019 [3] 21.5.6
"513 "                                   5 0344 0004 [3] 21.5.7
"513 Message Too Large"                  4 0344 0015 [3] 21.5.7
"580 "                                   5 09FB 0004 [12] 8
"580 Precondition Failure"               4 09FB 0018 [12] 8
"600 "                                   5 07E3 0004 [3] 21.6.1
"600 Busy Everywhere"                    3 07E3 0013 [3] 21.6.1
"603 "                                   5 0B09 0004 [3] 21.6.2
"603 Decline"                            4 0B09 000B [3] 21.6.2
"604 "                                   5 09BE 0004 [3] 21.6.3
"604 Does Not Exist Anywhere"            4 09BE 001B [3] 21.6.3
"606 "                                   5 0430 0004 [3] 21.6.4
"606 Not Acceptable"                     4 0430 0012 [3] 21.6.4
"687 "                                   5 0359 0004 [14] 3.5
"687 Dialog Terminated"                  4 0359 0015 [14] 3.5
"Anonymous"                              3 08A4 0009 [3] 8.1.1.3
```

Note: [3]=RFC 3261, [6]=RFC 2617, [7]=RFC 2806, [8]=RFC 2976, [9]=RFC 3265, [10]=RFC 3262, [11]=RFC 3311, [12]=RFC 3312, [13]=RFC 3515, [14]=RFC 3891, [15]=RFC 3420, [16]=RFC 3327, [17]=RFC 3226, [18]=RFC 4028, [19]=RFC 3310, [20]=RFC 3329, [21]=RFC 3428, [22]=RFC 3841, [33]=RFC 3323, [34]=RFC 3892, [35]=RFC 3608

15.1.5 SDP Input Strings to the SIP/SDP Static Dictionary (Appendix B of RFC 3485)

For reference, this section lists the SDP input strings that were used in generating the dictionary, as well as a priority value, the offset of the string in the generated dictionary, the length of the string, and one or more references into the referenced documents that motivate the presence of this string. Note that the notation "[CRLF]" stands for a sequence of two bytes with the values 0x0d and 0x0a, respectively.

The priority value is used for determining the position of the string in the dictionary. Lower priority values (higher priorities) cause the string to occur at a later position in the dictionary, making it more efficient to reference the string in certain compression algorithms. Hence, lower priority values were assigned to strings more likely to occur.

```
String                                   Pr Off  Len  References
======================================== == ==== ==== ==========
"v=0[CRLF]o="                             2 0BB8 0007 [24] 6
"[CRLF]s="                                2 0C2B 0004 [24] 6
"[CRLF]s= "                               2 0C2B 0005 [32] 5
"[CRLF]i="                                4 07A6 0004 [24] 6
"[CRLF]u="                                4 07AE 0004 [24] 6
"[CRLF]e="                                4 079E 0004 [24] 6
"[CRLF]c=IN IP4 "                         3 08E1 000B [24] 6
"[CRLF]c=IN IP6 "                         2 0B98 000B [24] 6
"[CRLF]c="                                5 08E1 0004 [24] 6
"[CRLF]b="                                3 0916 0004 [24] 6
"[CRLF]t="                                2 0B78 0004 [24] 6
"[CRLF]t=0 0"                             2 0B78 0007 [32] 5
"[CRLF]r="                                4 0796 0004 [24] 6
"[CRLF]z="                                4 079A 0004 [24] 6
"[CRLF]k=clear:"                          4 06B0 000A [24] 6
"[CRLF]k=base64:"                         4 0690 000B [24] 6
"[CRLF]k=uri:"                            4 0732 0008 [24] 6
```

```
"[CRLF]k=prompt:"                      4 056B 000B [24] 6
"[CRLF]k="                             5 056B 0004 [24] 6
"[CRLF]a=cat:"                         4 072A 0008 [24] 6
"[CRLF]a=keywds:"                      4 0685 000B [24] 6
"[CRLF]a=tool:"                        4 0712 0009 [24] 6
"[CRLF]a=ptime:"                       4 06A6 000A [24] 6
"[CRLF]a=maxptime:"                    4 05EA 000D [24] 6
"[CRLF]a=rtpmap:"                      2 0C0C 000B [24] 6, [32] 5
"[CRLF]a=recvonly"                     3 08C9 000C [24] 6
"[CRLF]a=sendrecv"                     3 093B 000C [24] 6
"[CRLF]a=sendonly"                     3 08D5 000C [24] 6
"[CRLF]a=inactive"                     3 0886 000C [24] 6
"[CRLF]a=orient:portrait"              4 0974 0013 [24] 6
"[CRLF]a=orient:landscape"             4 03D2 0014 [24] 6
"[CRLF]a=orient:seascape"              4 050B 0013 [24] 6
"[CRLF]a=type:broadcast"               4 0442 0012 [24] 6
"[CRLF]a=type:meeting"                 4 055B 0010 [24] 6
"[CRLF]a=type:moderated"               4 051E 0012 [24] 6
"[CRLF]a=type:test"                    4 08B1 000D [24] 6
"[CRLF]a=type:H.332"                   4 0817 000E [24] 6
"[CRLF]a=type:recvonly"                4 054A 0011 [24] 6
"[CRLF]a=charset:"                     4 0653 000C [24] 6
"[CRLF]a=sdplang:"                     4 0647 000C [24] 6
"[CRLF]a=lang:"                        4 06F7 0009 [24] 6
"[CRLF]a=framerate:"                   4 05C7 000E [24] 6
"[CRLF]a=quality:"                     4 063B 000C [24] 6
"[CRLF]a=fmtp:"                        2 0C50 0009 [24] 6
"[CRLF]a=curr:"                        2 0C59 0009 [12] 4
"[CRLF]a=des:"                         2 0C72 0008 [12] 4
"[CRLF]a=conf:"                        2 0C35 0009 [12] 4
"[CRLF]a=mid:"                         4 0752 0008 [26] 3
"[CRLF]a=group:"                       4 06D8 000A [26] 3
"[CRLF]a=key-mgmt:mikey"               4 0257 0012 [28] 2.1,
                                         [29] 6
"[CRLF]a=key-mgmt:"                    4 0257 000D [28] 2.1
"[CRLF]a="                             5 0257 0004 [24] 6
"[CRLF]m=audio "                       2 0C21 000A [24] 6
"[CRLF]m=video "                       2 0C17 000A [24] 6
"[CRLF]m=application "                 4 0330 0010 [24] 6
"[CRLF]m=data "                        4 0700 0009 [24] 6
"[CRLF]m=control "                     4 062F 000C [24] 6
"[CRLF]m="                             5 0330 0004 [24] 6
"AS "                                  3 091A 0003 [24] 6
"CT "                                  3 091D 0003 [24] 6
"RTP/AVP "                             2 0BE3 0008 [24] A
"RTP/SAVP "                            3 0892 0009 [30] 12
"RTP/AVPF "                            3 089B 0009 [31] 4.1
"udp"                                  4 07DB 0003 [3] 25.1,
                                         [24] A, [3] 25.1, [24] A
"0.0.0.0"                              4 0B7E 0007 [24] A
"qos"                                  2 0CA9 0003 [12] 4
"mandatory"                            2 0A8D 0009 [12] 4
"optional"                             2 09AE 0008 [3] 20.11,
                                         [12] 4, [3] 20.11, [12] 4
"none"                                 2 0AF4 0004 [33] 4.2,
                                         [12] 4
"failure"                              4 0A5C 0007 [12] 4
"unknown"                              4 0BDA 0007 [12] 4
"e2e"                                  2 0AB1 0003 [12] 4
"local"                                2 0A36 0005 [12] 4
```

```
"remote"                          2 0AD6 0006 [12] 4
"send"                            2 08D9 0004 [12] 4
"recv"                            2 0553 0004 [12] 4
"sendrecv"                        2 093F 0008 [12] 4
"AMR"                             2 0BE1 0003 [25] 8
"octet-align="                    4 094A 000C [25] 8
"mode-set="                       4 0709 0009 [25] 8
"mode-change-period="             4 04F8 0013 [25] 8
"mode-change-neighbor="           4 04A2 0015 [25] 8
"crc="                            4 07AA 0004 [25] 8
"robust-sorting="                 4 0166 000F [25] 8
"interleaving="                   4 0323 000D [25] 8
"channels="                       4 0606 0009 [25] 8
"octet-align"                     4 094A 000B [25] 8
"telephone-event"                 4 00EE 000F [27] 3.3, 6.1
"events"                          4 00F8 0006 [27] 6.1
"rate"                            4 052B 0004 [27] 6.1, 6.2
"tone"                            4 0453 0004 [27] 6.2
"rtcp-fb"                         4 04C0 0007 [31] 4
"ack"                             4 0D1F 0003 [31] 4
"nack"                            4 0D1E 0004 [31] 4
"ttr-int"                         4 098C 0007 [31] 4
"app"                             4 0334 0003 [31] 4
"rpsi"                            4 0CDB 0004 [31] 4
"pli"                             4 0336 0003 [31] 4
"sli"                             4 09A9 0003 [31] 4
```

Note: [12]=RFC 3312, [24]=RFC 4566, [25]=RFC 3267, [26]=RFC 3388, [27]=RFC 2833, [28]=RFC 4567, [30]=RFC 3711, [31]=RFC 3551, [32]=RFC 3264, [33]=RFC 3323

15.2 Summary

We have described the SIP/SDP-specific static dictionary that is used for compression and decompression of the signaling traffic. This lossless compression/decompression scheme defined in RFC 3485 is used with SigComp of RFC 3320. The key is that this static dictionary is loaded only once and independent of the evolution of SIP/SDP standards. The SIP and SDP dictionaries are combined into one dictionary known as the SigComp dictionary using binary representation. This single dictionary is again loaded into SigComp as a state where SigComp is offered as a service between the application and the underlying transport. The SIP and SDP input strings that are used in generating the SIP/SDP static dictionary are also provided along with priority value, the offset of the string in generating the static dictionary, the length of the string, and the references of RFCs that motivate the presence of the string.

15.3 Problems

1. What motivates for compression of the signaling traffic such as SIP and SDP? What are the salient features of the SigComp specified in RFC 3320? Explain using examples.
2. What is the SIP/SDP static dictionary? Will this dictionary change with respect to the future evolution of SIP and/or SDP? Explain in detail to justify your answer. How is this used in the compression and decompression of SIP and SDP signaling traffic?

3. Explain in detail the basic motivation behind the SIP/SDP compression and decompression design scheme.
4. Why do we need binary representation of the SIP/SDP dictionary? Where does the SIP/SDP SigComp stand in terms of logical Open Standard International (OSI) layer architecture? Describe SIP/SDP compression/decompression operation using an example in the context of OSI layer.

Chapter 16

Security Capability Negotiation in SDP

Abstract
This chapter describes Request for Comments (RFCs) 4567, 4568, 4572, 5027, and 6193 that deal with security capability negotiations using Session Description Protocol (SDP). The multimedia conferencing requires a host of security capabilities related to session, media, and transport level, which need to be negotiated between the conferencing parties. In this chapter, we describe how SDP can be used to negotiate all of those security attributes using call control protocols like Session Initiation Protocol (SIP) and others. RFC 4567 defines general extensions for SDP and Real-Time Streaming Protocol (RTSP) to carry messages, as specified by a key management protocol, in order to secure the media. These extensions are presented as a framework to be used by one or more key management protocols. As such, their use is meaningful only when complemented by an appropriate key management protocol. General guidelines are also given on how the framework should be used with SIP and RTSP. The usage with the Multimedia Internet KEYing key management protocol is also defined.

RFC 4568 defines a SDP cryptographic attribute for unicast media streams. The attribute describes a cryptographic key and other parameters that serve to configure security for a unicast media stream in either a single message or a roundtrip exchange. The attribute can be used with a variety of SDP media transports, and this is defines how to use it for the Secure Real-time Transport Protocol (SRTP) unicast media streams. The SDP crypto attribute requires the services of a data security protocol to secure the SDP message.

RFC 4572 specifies how to establish secure connection-oriented media transport sessions over the Transport Layer Security (TLS) Protocol using the SDP. It defines a new SDP protocol identifier, "TCP/TLS." It also defines the syntax and semantics for an SDP "fingerprint" attribute that identifies the certificate that will be presented for the TLS session. This mechanism allows media transport over TLS connections to be established securely, as long as the integrity of session descriptions is assured. RFC 4572 also extends and updates RFC 4145 that describes how to express media transport over Transmission Control Protocol (TCP) using the SDP. In addition, RFC 4145 defines the SDP "TCP protocol identifier," the SDP "setup" attribute, which describes the connection setup procedure, and the SDP "connection" attribute, which handles connection reestablishment.

Furthermore, RFC 5027 defines a new security precondition for the SDP based on the SIP (RFC 3261, also see Section 1.2) precondition framework described in RFCs 3312 and 4032. A security precondition can be used to delay session establishment or modification until media stream security for a secure media stream has been negotiated successfully. Finally, RFC 6193 specifies how to establish a media session that represents a virtual private network using the SIP for the purpose of on-demand media/application sharing between peers. It extends the protocol identifier of the SDP so that it can negotiate the use of the Internet Key Exchange (IKE) Protocol for media sessions in the SDP offer/answer model. It also specifies a method to boot up IKE and generate IPsec security associations using a self-signed certificate.

16.1 Key Management Extensions for SDP and RTSP

This section describes a framework specified by RFC 4567 that describes how SDP can be used to negotiate the generic security messages attributes specified by the key management protocols using call control protocols like SIP and RTSP in order to secure media. In fact, general guidelines are provided on how the framework should be used with SIP and RTSP. Furthermore, a specific example for negotiations of security attributes of the Multimedia Internet KEYing (MIKEY) key management protocol using SDP is provided in this section.

16.1.1 Introduction

There has recently been work to define a security profile for the protection of real-time applications running over SRTP (RFC 3711). However, a security protocol needs a key management solution to exchange keys and security parameters, manage and refresh keys, etc. A key management protocol is executed prior to the security protocol's execution. The key management protocol's main goal is to, in a secure and reliable way, establish a security association for the security protocol. This includes one or more cryptographic keys and the set of necessary parameters for the security protocol (e.g., cipher and authentication algorithms) to be used. The key management protocol has similarities with, for example, SIP (RFC 3261, also see Section 1.2) and RTSP (RFC 2326) in the sense that it negotiates necessary information in order to be able to set up the session. The focus in the following sections is to describe a new SDP attribute and RTSP header extension to support key management and to show how these can be integrated within SIP and RTSP. The resulting framework is completed by one or more key management protocols, which use the extensions provided. Some of the motivations to create a framework with the possibility to include the key management in the session establishment are

- Just as the codec information is a description of how to encode and decode the audio (or video) stream, the key management data is a description of how to encrypt and decrypt the data.
- The possibility of negotiating security for the entire multimedia session at one time.
- Knowledge of the media at session establishment makes it easy to tie key management to the multimedia sessions.
- This approach may be more efficient than setting up the security later, as that approach might force extra roundtrips, possibly also a separate setup for each stream, hence adding more delay to the actual setup of the media session.

- The possibility of negotiating keying material end to end without applying end-to-end protection of the SDP (instead, hop-by-hop security mechanisms can be used, which may be useful if intermediate proxies need access to the SDP).

Currently in SDP (RFC 4566, see Section 2), there exists one field to transport keys, the "k=" field. However, this is not enough for a key management protocol as many more parameters need to be transported, and the "k=" field is not extensible. The approach used is to extend the SDP description through a number of attributes that transport the key management offer/answer and to associate it with media sessions. SIP uses the offer/answer model (RFC 3264, see Section 3.1) whereby extensions to SDP will be enough. However, RTSP does not use the offer/answer model with SDP, so a new RTSP header is introduced to convey key management data. RFC 4568 (see Section 16.2) uses the approach of extending SDP, to carry the security parameters for the media streams. However, the mechanism defined in RFC 4568 requires end-to-end protection of the SDP by some security protocol such as S/ Multipurpose Internet Mail Extensions (MIME), in order to get end-to-end protection. The solution described here focuses only on the end-to-end protection of key management parameters and, as a consequence, does not require external end-to-end protection means. It is important to note, though, and we stress this again, that only the key management parameters are protected. The document also defines the use of the described framework with the key management protocol MIKEY (RFC 3830).

16.1.2 Applicability

RFC 4568 (see Section 16.2) provides similar cryptographic key distribution capabilities, and it is intended for use when keying material is protected along with the signaling. In contrast, this specification expects endpoints to have preconfigured keys or a common security infrastructure. It provides its own security and is independent of the protection of signaling (if any). As a result, it can be applied in environments where signaling protection is not turned on or used hop by hop (i.e., scenarios where the SDP is not protected end to end). This specification will, independently of the signaling protection applied, ensure end-to-end security establishment for the media.

16.1.3 Extensions to SDP and RTSP

This section describes common attributes that can be included in SDP or RTSP when an integrated key management protocol is used. The attribute values follow the general SDP and RTSP guidelines (see RFC 4566, Section 2 and RFC 2326). For both SDP and RTSP, the general method of adding the key management protocol is to introduce new attributes, one identifier to identify the specific key management protocol, and one data field where the key management protocol data is placed. All SDP Augmented Backus-Naur Form (ABNF) syntaxes are provided in Section 2.9. We have repeated this here for convenience.

The key management protocol data contains the necessary information to establish the security protocol (e.g., keys and cryptographic parameters). All parameters and keys are protected by the key management protocol. The key management data SHALL be base64 (RFC 3548 was made obsolete by RFC 4648) encoded and comply with the base64 grammar as defined in RFC 4566 (see Sections 2 and 2.9). The key management protocol identifier (KMPID), is defined here in ABNF grammar (RFC 4234 was made obsolete by RFC 5234).

```
KMPID       =     1*(ALPHA / DIGIT)
```

Values for the identifier, KMPID, are registered and defined in accordance with Section 9. Note that the KMPID is case sensitive, and it is RECOMMENDED that values registered be lowercase letters.

16.1.3.1 SDP Extensions

This section provides an ABNF grammar (as used in RFC 4566, see Sections 2 and 2.9) for the key management extensions to SDP. Note that the new definitions are compliant with the definition of an attribute field, that is,

```
attribute = (att-field ":" att-value) / att-field
```

The ABNF for the key management extensions (conforming to the att-field and att-value) are as follows:

```
key-mgmt-attribute        =    key-mgmt-att-field ":"
                               key-mgmt-att-value
key-mgmt-att-field        =    "key-mgmt"
key-mgmt-att-value        =    0*1SP prtcl-id SP keymgmt-data
prtcl-id                  =    KMPID
                               ; e.g., "mikey"
keymgmt-data              =    base64
SP                        =    %x20
```

where KMPID is as defined in Section 16.1.3 of this specification, and base64 is as defined in SDP (RFC 4566, see Sections 2 and 2.9). Prtcl-id refers to the set of values defined for KMPID in Section 16.1.9.

The attribute MAY be used at session level, media level, or both levels. An attribute defined at media level overrides an attribute defined at session level. In other words, if the media-level attribute is present, the session-level attribute MUST be ignored for this media. Section 16.1.4.1 describes in detail how the attributes are used and how the SDP is handled in different usage scenarios. The choice of the level depends, for example, on the particular key management protocol. Some protocols may not be able to derive enough key material for all sessions; furthermore, possibly a different protection to each session could be required. The particular protocol might achieve this only by specifying it at the media level. Other protocols, such as MIKEY, have those capabilities (and can express multiple security policies and derive multiple keys) and so may use at the session level.

16.1.3.2 RTSP Extensions

To support the key management attributes, the following RTSP header is defined:

```
KeyMgmt          =    "KeyMgmt" ":" key-mgmt-spec
                      0*("," key-mgmt-spec)
key-mgmt-spec    =    "prot" "=" KMPID ";" ["uri" "=" %x22 URI
                      %x22 ";"]
```

where KMPID is as defined in Section 16.1.3 of this memo, "base64" as defined in RFC 4566 (see Sections 2 and 2.9), and "URI" as defined in Section 3 of RFC 3986.

The "uri" parameter identifies the context for which the key management data applies, and the RTSP Uniform Resource Identifier (URI) SHALL match a (session or media) URI present in the description of the session. If the RTSP aggregated control URI is included, it indicates that the key management message is on the session level (the RTSP media control URI, that it applies to the media level). If no "uri" parameter is present in a key-mgmt-spec the specification applies to the context identified by the RTSP request URI. The KeyMgmt header MAY be used in the messages and directions described in this table.

Method	*Direction*	*Requirement*
DESCRIBE response	S->C	RECOMMENDED
SETUP	C->S	REQUIRED
SETUP response	S->C	REQUIRED (error)

Note: Section 16.1.4.2 describes in detail how the RTSP extensions are used.

We define one new RTSP status code to report error due to any failure during the key management processing (Section 16.1.4.2):

```
Status-Code          =       "463" ; Key management failure
```

A 463 response MAY contain a KeyMgmt header with a key management protocol message that further indicates the nature of the error.

16.1.4 Usage with SDP, SIP, RTSP, and SAP

This section gives rules and recommendations of how/when to include the defined key management attribute when SIP and/or RTSP are used with SDP. When a key management protocol is integrated with SIP/SDP and RTSP, the following general requirements are placed on the key management:

- At the current time, it MUST be possible to execute the key management protocol in at most one request-response message exchange. Future relaxation of this requirement is possible but would introduce significant complexity for implementations supporting multiroundtrip mechanisms.
- It MUST be possible from the SIP/SDP and RTSP application, using the key management application programming interface (API), to receive key management data and information of whether a message is accepted. The content of the key management messages depends on the key management protocol that is used. However, the content of such key management messages might be expected to be roughly as follows: the key management Initiator (e.g., the offerer) includes the key management data in a first message, containing the media description to which it should apply. This data in general consists of the security parameters (including key material) needed to secure the communication, together with the necessary authentication information (to ensure that the message is authentic).

At the responder's side, the key management protocol checks the validity of the key management message, with the availability of the parameters offered, and then provides the key management data to be included in the answer. This answer may typically authenticate the responder to the initiator and state whether the initial offer was accepted. Certain protocols might require the responder to include a selection of the security parameters that he/she is willing to support. Again, the actual content of such responses is dependent on the particular key management protocol. Section 16.1.7 describes a realization of the MIKEY protocol using these mechanisms. Procedures to be used when mapping new key management protocols onto this framework are described in Section 16.1.6.

16.1.4.1 Use of SDP

This section describes the processing rules for the different applications that use SDP for the key management.

16.1.4.1.1 General Processing

The processing when SDP is used is slightly different according to the way SDP is transported and whether it uses an offer/answer or an announcement. The processing can be divided into four steps:

1. How to create the initial offer
2. How to handle a received offer
3. How to create an answer
4. How to handle a received answer

It should be noted that the last two steps may not always be applicable, as there are cases in which an answer cannot or will not be sent back. The general processing for creating an initial offer SHALL take the following actions:

- The identifier of the key management protocol used MUST be placed in the prtcl-id field of SDP. A table of legal protocols identifiers is maintained by IANA (see Section 16.1.9).
- The keymgmt-data field MUST be created as follows: the key management protocol MUST be used to create the key management message. This message SHALL be base64 encoded (RFC 3548 was made obsolete by RFC 4648) by the SDP application and then encapsulated in the keymgmt-data attribute. Note though that the semantics of the encapsulated message is dependent on the key management protocol that is used.

The general processing for handling a received offer SHALL follow these actions:

- The key management protocol is identified according to the prtcl-id field. A table of legal protocols identifiers is maintained by Internet Assigned Numbers Authority (IANA) (Section 16.1.9).
- The key management data from the keymgmt-data field MUST be extracted, base64 decoded to reconstruct the original message, and then passed to the key management protocol for processing. Note that depending on key management protocol, some extra parameters might be requested by the specific API, such as the source/destination network address/port(s) for the specified media (however, this will be implementation specific depending on the actual API). The extra parameters that a key management protocol might need (other than those defined here) MUST be documented, describing their use, as well as the interaction of that key management protocol with SDP and RTSP.
- If errors occur, or the key management offer is rejected, the session SHALL be aborted. Possible error messages are dependent on the specific session establishment protocol. At this

stage, the key management will have either accepted or rejected the offered parameters. This MAY cause a response message to be generated, depending on the key management protocol and the application scenario.

If an answer is to be generated, the following general actions SHALL be performed:

- The identifier of the key management protocol used MUST be placed in the prtcl-id field.
- The keymgmt-data field MUST be created as follows. The key management protocol MUST be used to create the key management message. This message SHALL be base64 encoded (RFC 3548 was made obsolete by RFC 4648) by the SDP application and then encapsulated in the keymgmt-data attribute. The semantics of the encapsulated message is dependent on the key management protocol that is used.

The general processing for handling a received answer SHALL follow these actions:

- The key management protocol is identified according to the prtcl-id field.
- The key management data from the keymgmt-data field MUST be extracted, base64 decoded to reconstruct the original message, and then passed to the key management protocol for processing.
- If the key management offer is rejected and the intent is to renegotiate it, it MUST be done through another offer/answer exchange. It is RECOMMENDED to NOT abort the session in that case, but to renegotiate using another offer/answer exchange. For example, in SIP (RFC 3261, also see Section 1.2), the "security precondition" as defined in RFC 5027 (see Section 16.4) solves the problem for a session initiation. The procedures in RFC 5027 (see Section 16.4) are outside the scope of this document. In an established session, an additional offer/answer exchange using a re-INVITE or UPDATE as appropriate MAY be used.
- If errors occur, or the key management offer is rejected and there is no intent to renegotiate it, the session SHALL be aborted. If possible, an error message indicating the failure SHOULD be sent back. Otherwise, if all the steps are successful, the normal setup proceeds.

16.1.4.1.2 Use of SDP with Offer/Answer and SIP

This section defines additional processing rules, to the general rules defined in Section 16.1.4.1.1, applicable only to applications using SDP with the offer/answer model (RFC 3264, see Section 3.1) (and in particular SIP). When an initial offer is created, the following offer/answer-specific procedure SHALL be applied:

- Before creating the key management data field, the list of protocol identifiers MUST be provided by the SDP application to (each) key management protocol, as defined in Section 16.1.4.1.4 (to defeat bidding-down attacks).

For a received SDP offer that contains the key management attributes, the following offer/answer-specific procedure SHALL be applied:

- Before, or in conjunction with, passing the key management data to the key management protocol, the complete list of protocol identifiers from the offer message is provided by the SDP application to the key management protocol (as defined in Section 16.1.4.1.4).

When an answer is created, the following offer/answer-specific procedure SHALL be applied:

- If the key management rejects the offer and the intent is to renegotiate it, the answer SHOULD include the cause of failure in a message from the key management protocol. The renegotiation MUST be done through another offer/answer exchange (e.g., using RFC 5027, see Section 16.4). In an established session, it can also be done through a re-INVITE or UPDATE as appropriate.
- If the key management rejects the offer and the session needs to be aborted, the answerer SHOULD return a "488 Not Acceptable Here" message, optionally also including one or more Warning headers (a "306 Attribute Not Understood" when one of the parameters is not supported, and a "399 Miscellaneous Warning" with arbitrary information to be presented to a human user or logged; see Section 20.43 in RFC 3261 (also see Section 1.2). Further details about the cause of failure MAY be described in an included message from the key management protocol. The session is then aborted (and it is up to local policy or the end user to decide how to continue).

Note that the key management attribute (related to the same key management protocol) MAY be present at both session and media level. Consequently, the process SHALL be repeated for each such key management attribute detected. In case the key management processing of any such attribute does not succeed (e.g., authentication failure, parameters not supported, etc.), on either session or media level, the entire session setup SHALL be aborted, including those parts of the session that successfully completed their parts of the key management. If more than one key management protocol is supported, multiple instances of the key management attribute MAY be included in the initial offer when using the offer/answer model, each transporting a different key management protocol, thus indicating supported alternatives.

If the offerer includes more than one key management protocol attribute at session level (analogous for the media level), these SHOULD be listed in order of preference (the first being the preferred). The answerer selects the key management protocol it wishes to use and processes only it, on either session or media level or both, according to where it is located. If the answerer does not support any of the offerer's suggested key management protocols, the answerer indicates this to the offerer so a new offer/answer can be triggered; alternatively, it may return a "488 Not Acceptable Here" error message, whereby the sender MUST abort the current setup procedure.

Note that the placement of multiple key management offers in a single message has the disadvantage that the message expands and the computational workload for the offerer will increase drastically. Unless the guidelines of Section 16.1.4.1.4 are followed, multiple lines may open up bidding-down attacks. Note also that the multiple-offer option has been added to optimize signaling overhead in case the initiator knows some key (e.g., a public key) the responder has but is unsure what protocol the responder supports. The mechanism is not intended to negotiate options within one and the same protocol.

The offerer MUST include the key management data within an offer that contains the media description to which it applies. Rekeying MUST be handled as a new offer, with the new proposed parameters. The answerer treats this as a new offer where the key management is the issue of change. The rekeying exchange MUST be finalized before the security protocol can change the keys. The same key management protocol used in the original offer SHALL also be used in the new offer carrying rekeying. If the new offer carrying rekeying fails (e.g., the authentication verification fails), the answerer SHOULD send a "488 Not Acceptable Here" message, including one or more Warning headers (at least a 306). The offerer MUST then abort the session.

Note that, in multicast scenarios, unlike unicast, there is only a single view of the stream (RFC 3264, see Section 3.1), hence there MUST be a uniform agreement of the security parameters. After the offer is issued, the offerer SHOULD be prepared to receive media, as the media may arrive prior to the answer. However, this brings issues, as the offerer does not know yet the answerer's choice in terms of, for example, algorithms, or possibly the key is known. This can cause delay, or clipping can occur; if this is unacceptable, the offerer SHOULD use mechanisms outside the scope of this RFC 4567 (e.g., the security preconditions for SIP/SDP described in RFC 5027) (see Section 16.4).

16.1.4.1.3 Use of SDP with SAP

There are cases in which SDP is used without conforming to the offer/answer model; instead, it is a one-way SDP distribution (i.e., without back channel), such as when used with Session Announcement Protocol (SAP)-RFC 2974 and HTTP. The processing follows the two first steps of the general SDP processing (see Section 16.1.4.1.1). It can be noted that the processing in this case differs from the offer/answer case in that only one key management protocol SHALL be offered (i.e., no negotiation will be possible). This implies that the bidding-down attack is not an issue; therefore, the countermeasure is not needed. The key management protocol used MUST support one-way messages.

16.1.4.1.4 Bidding-Down Attack Prevention

The possibility of supporting multiple key management protocols may, unless properly handled, introduce bidding-down attacks. Specifically, a man-in-the-middle could "peel off" cryptographically strong offers (deleting the key management lines from the message), leaving only weaker ones as the responder's choice. To avoid this, the list of identifiers of the proposed key management protocols MUST be authenticated. The authentication MUST be done separately by each key management protocol. Accordingly, it MUST be specified (in the key management protocol specification itself or in a companion document) how the list of key management protocol identifiers can be processed to be authenticated from the offerer to the answerer by the specific key management protocol. Even if only one key management protocol is used it MUST authenticate its own protocol identifier. The list of protocol identifiers MUST then be given to each of the selected (offered) key management protocols by the application with ";" separated identifiers. All offered protocol identifiers MUST be included in the same order as they appear in the corresponding SDP description.

The protocol list can formally be described as:

```
prtcl-list      =       KMPID *(";" KMPID)
```

where KMPID is as defined in Section 16.1.3. For example, if the offered protocols are MIKEY and two yet-to-be invented protocols KEYP1, KEYP2, the SDP is

```
v=0
o=alice 2891092738 2891092738 IN IP4 lost.example.com
s=Secret discussion
t=0 0
c=IN IP4 lost.example.com
a=key-mgmt:mikey AQAFgM0Xf1ABAAAAAAAAAAAAAAsAyO...
a=key-mgmt:keyp1 727gkdOshsuiSDF9sdhsdKnD/dhsoSJokdo7eWD...
a=key-mgmt:keyp2 DFsnuiSDSh9sdh Kksd/dhsoddo7eOok727gWsJD...
m=audio 39000 RTP/SAVP 98
a=rtpmap:98 AMR/8000
```

```
m=video 42000 RTP/SAVP 31
a=rtpmap:31 H261/90000
```

The protocol list, "mikey;keypl;keyp2," would be generated from the SDP description and used as input to each specified key management protocol (with the data for that protocol). Each of the three protocols includes this protocol identifier list in its authentication coverage (according to its protocol specification). If more than one protocol is supported by the offerer, it is RECOMMENDED that all acceptable protocols be included in the first offer, rather than making single, subsequent alternative offers in response to error messages; see "Security Considerations." End-to-end integrity protection of the key-mgmt attributes altogether, provided externally to the key management itself, protects against this bidding-down attack. This is, for example, the case if SIP uses S/MIME (RFC 3851 was made obsolete by RFC 5751) to end-to-end integrity protect the SDP description. However, as this end-to-end protection is not an assumption of the framework, the mechanisms defined in this section SHALL be applied.

16.1.4.2 RTSP Usage

RTSP does not use the offer/answer model, as SIP does. This causes some problems, as it is not possible (without modifying RTSP) to send back an answer. To solve this, a new header has been introduced (Section 16.1.3.2). This also assumes that the key management has some kind of binding to the media, so that the response to the server will be processed as required. The server SHALL be the Initiator of the key management exchange for sessions in PLAY mode (i.e., transporting media from server to client). The following paragraph describes the behavior for PLAY mode. For any other mode, the behavior is not defined in this specification.

To obtain a session description, the client initially contacts the server via a DESCRIBE message. The initial key management message from the RTSP server is sent to the client in the SDP of the 200 OK in response to the DESCRIBE. Note that only one key management protocol SHALL be used per session/media level. A server MAY allow the SDP with key management attribute(s) to be distributed to the client through other means than RTSP, although this is not specified here.

The "uri" parameter of the KeyMgmt header is used to indicate for the key management protocol on what context the carried message applies. For key management messages on the SDP session level, the answer MUST contain the RTSP aggregated control URL to indicate this. For key management messages initially on SDP media level, the key management response message in the KeyMgmt header MAY use the RTSP media-level URL. For RTSP sessions not using aggregated control (i.e., no session-level control), URI is defined; the key management protocol SHALL only be invoked on individual media streams. In this case also, the key management response SHALL be on individual media streams (i.e., one RTSP key management header per media).

When responding to the initial key management message, the client uses the new RTSP header (KeyMgmt) to send back an answer. How this is done depends on the usage context:

- Key management protocol responses for the initial establishment of security parameters for an aggregated RTSP session SHALL be sent in the first SETUP of the session. This means that if the key management is declared for the whole session but is set up in nonaggregated fashion (i.e., one media per RTSP session), each SETUP MUST carry the same response for the session-level context. When performing a setup of the second or any subsequent media in an RTSP session, the same key management parameters as established for the first media apply to these setups.
- Key management responses for the initial establishment of security parameters for an individual media SHALL only be included in SETUP for the corresponding media stream. If a

server receives a SETUP message in which it expects a key management message, but none is included, a "403 Forbidden" SHOULD be returned to the client, whereby the current setup MUST be aborted. When the server creates an initial SDP message, the procedure SHALL be the same as described in Section 16.1.4.1.1.

The client processing of the initial SDP message from the server SHALL follow the same procedures as described in Section 16.1.4.1.1, except that, if there is an error, the session is aborted (no error message is sent back). The client SHALL create the response, using the key management header in RTSP, as follows:

- The identifier of the key management protocol used (e.g., MIKEY) MUST be placed in the "prot" field of the header. The prot values are maintained by IANA (Section 16.1.9).
- The keymgmt-data field MUST be created as follows: the key management protocol MUST be used to create the key management message. This message SHALL be base64 encoded by the RTSP application and then encapsulated in the "data" field of the header. The semantics of the encapsulated message is dependent on the key management protocol that is used.
- Include, if necessary, the URL to indicate the context in the "uri" parameter. The server SHALL process a received key management header in RTSP as follows:
  - The key management protocol is identified according to the "prot" field.
  - The key management data from the "data" field MUST be extracted, base64 decoded to reconstruct the original message, and then passed to the key management protocol for processing.
  - If the key management protocol is successful, the processing can proceed according to normal rules.
  - Otherwise, if the key management fails (e.g., due to authentication failure or parameter not supported), an error is sent back as the SETUP response using RTSP error code 463 (see Section 16.1.3.2) and the session is aborted. It is up to the key management protocol to specify (within the RTSP status code message or through key management messages) details about the type of error that occurred. Rekeying within RTSP is for further study, given that media updating mechanisms within RTSP are unspecified at the time this document was written.

16.1.5 Example Scenarios

The following examples utilize MIKEY (RFC 3830) as the key management protocol to be integrated into SDP and RTSP.

16.1.5.1 Example 1 (SIP/SDP)

An SIP call is taking place between Alice and Bob. Alice sends an INVITE message consisting of the following offer:

```
v=0
o=alice 2891092738 2891092738 IN IP4 w-land.example.com
s=Cool stuff
e=alice@w-land.example.com
t=0 0
c=IN IP4 w-land.example.com
  a=key-mgmt:mikey
```

```
   AQAFgM0XflABAAAAAAAAAAAAAAsAyONQ6gAAAAAGEEoo2pee4hp2
   UaDX8ZE22YwKAAAPZG9uYWxkQGR1Y2suY29tAQAAAAAAAQAk0JKpga
   VkDaawi9whVBtBt
   0KZ14ymNuu62+Nv3ozPLygwK/GbAV9iemnGUIZ19fWQUOSrzKTAv9zV
m=audio 49000 RTP/SAVP 98
a=rtpmap:98 AMR/8000
m=video 52230 RTP/SAVP 31
a=rtpmap:31 H261/90000
```

That is, Alice proposes to set up one audio stream and one video stream that run over SRTP (signaled by the use of the Secure Audio Video Profile [SAVP] profile). She uses MIKEY to set up the security parameters for SRTP (Section 16.1.7). The MIKEY message contains the security parameters, together with the necessary key material. Note that MIKEY is exchanging the crypto-suite for both streams, as it is placed at the session level. Also, MIKEY provides its own security (i.e., when Bob processes Alice's MIKEY message, he will also find the signaling of the security parameters used to secure the MIKEY exchange). Alice's endpoint's authentication information is also carried within the MIKEY message, to prove that the message is authentic. The above MIKEY message is an example of a message in which the preshared method MIKEY is used. Upon receiving the offer, Bob checks the validity of the received MIKEY message, and, if there is successful verification, he accepts the offer and sends an answer back to Alice (with his authentication information and, if necessary, some key material from his side):

```
v=0
o=bob 2891092897 2891092897 IN IP4 foo.example.com
s=Cool stuff
e=bob@foo.example.com
t=0 0
c=IN IP4 foo.example.com
   a=key-mgmt:mikey
   AQEFgM0XflABAAAAAAAAAAAAAAYAyONQ6gAAAAAJAAAQbWlja2
   V5QG1vdXNlLmNvbQABn8HdGE5BMDXFIuGEga+62AgY5cc=
m=audio 49030 RTP/SAVP 98
a=rtpmap:98 AMR/8000
m=video 52230 RTP/SAVP 31
a=rtpmap:31 H261/90000
```

Upon receiving the answer, Alice verifies its correctness. In case of success, at this point Alice and Bob share the security parameters and the keys needed for a secure RTP (SRTP) communication.

16.1.5.2 Example 2 (SDP)

This example shows what Alice would have done if she had wished to protect only the audio stream. She would have placed the MIKEY line at media level for the audio stream only (also specifying the use of the SRTP profile there, SAVP). The semantics of the MIKEY messages is as in the previous case but applies only to the audio stream.

```
v=0
o=alice 2891092738 2891092738 IN IP4 w-land.example.com
s=Cool stuff
e=alice@w-land.example.com
t=0 0
c=IN IP4 w-land.example.com
m=audio 49000 RTP/SAVP 98
```

```
a=rtpmap:98 AMR/8000
a=key-mgmt:mikey AQAFgM0XflABAAAAAAAAAAAAAAsAy...
m=video 52230 RTP/AVP 31
a=rtpmap:31 H261/90000
```

Bob would then act as described in the previous example, including the MIKEY answer at the media level for the audio stream (as Alice did). Note that even if the key management attribute were specified at the session level, the video part would not be affected (as a security profile is not used, instead the RTP/attribute-value pair profile is signaled).

16.1.5.3 Example 3 (RTSP)

A client wants to set up a streaming session and requests a media description from the streaming server. DESCRIBE rtsp://server.example.com/fizzle/foo RTSP/1.0

```
CSeq: 312
Accept: application/sdp
From: user@example.com
```

The server sends back an OK message including an SDP description with the MIKEY message. The MIKEY message contains the necessary security parameters the server is willing to offer the client, with authentication information (to prove that the message is authentic) and the key material. The SAVP profile also signals the use of SRTP for securing the media sessions.

```
RTSP/1.0 200 OK
CSeq: 312
Date: 23 Jan 1997 15:35:06 GMT
Content-Type: application/sdp
Content-Length: 478

v=0
o=actionmovie 2891092738 2891092738 IN IP4 movie.example.com
s=Action Movie
e=action@movie.example.com
t=0 0
c=IN IP4 movie.example.com
a=control:rtsp://movie.example.com/action
a=key-mgmt:mikey AQAFgM0XflABAAAAAAAAAAAAAAsAy...
m=audio 0 RTP/SAVP 98
a=rtpmap:98 AMR/8000
a=control:rtsp://movie.example.com/action/audio
m=video 0 RTP/SAVP 31
a=rtpmap:31 H261/90000
a=control:rtsp://movie.example.com/action/video
```

The client checks the validity of the received MIKEY message and, in case of successful verification, accepts the message. The client then includes its key management data in the SETUP request going back to the server, the client authentication information (to prove that the message is authentic), and, if necessary, some key material.

```
SETUP rtsp://movie.example.com/action/audio RTSP/1.0
CSeq: 313
Transport: RTP/SAVP/UDP;unicast;client_port=3056-3057
```

```
keymgmt: prot=mikey; uri="rtsp://movie.example.com/action";
data="AQEFgM0XflABAAAAAAAAAAAAAAYAyONQ6g..."
```

The server processes the request including checking the validity of the key management header.

```
RTSP/1.0 200 OK
CSeq: 313
Session: 12345678
Transport: RTP/SAVP/UDP;unicast;client_port=3056-3057;
server_port=5000-5001
```

Note that in this case, the key management line was specified at the session level, and the key management information only goes into the SETUP related to the first stream. The "uri" indicates to the server that the context is for the whole aggregated session the key management applies. The RTSP client then proceeds with setting up the second media (video) in aggregation with the audio. As the two media are run in aggregation and the key context was established in the first exchange, no more key management messages are needed.

16.1.5.4 Example 4 (RTSP)

The use of the MIKEY message at the media level would change the previous example as follows. The 200 OK would contain the two distinct SDP attributes for MIKEY at the media level:

```
RTSP/1.0 200 OK
CSeq: 312
Date: 23 Jan 1997 15:35:06 GMT
Content-Type: application/sdp
Content-Length: 561
v=0
o=actionmovie 2891092738 2891092738 IN IP4 movie.example.com
s=Action Movie
e=action@movie.example.com
t=0 0
c=IN IP4 movie.example.com
a=control:rtsp://movie.example.com/action
m=audio 0 RTP/SAVP 98
a=rtpmap:98 AMR/8000
a=key-mgmt:mikey AQAFgM0XflABAAAAAAAAAAAAAA...
a=control:rtsp://movie.example.com/action/audio
m=video 0 RTP/SAVP 31
a=rtpmap:31 H261/90000
a=key-mgmt:mikey AQAFgM0AdlABAAAAAAAAAAAAAA...
a=control:rtsp://movie.example.com/action/video
```

Each RTSP header is inserted in the SETUP related to the audio and video separately:

```
SETUP rtsp://movie.example.com/action/audio RTSP/1.0
CSeq: 313
Transport: RTP/SAVP/UDP;unicast;client_port=3056-3057
keymgmt: prot=mikey; uri="rtsp://movie.example.com/action/audio";
      data="AQEFgM0XflABAAAAAAAAAAAAAA..."
```

and similarly for the video session:

```
SETUP rtsp://movie.example.com/action/video RTSP/1.0
CSeq: 315
Transport: RTP/SAVP/UDP;unicast;client_port=3058-3059
keymgmt: prot=mikey; uri="rtsp://movie.example.com/action/video";
      data="AQEFgM0AdlABAAAAAAAAAAAAAA..."
```

Note: The "uri" parameter could be excluded from the two SETUP messages in this example.

16.1.6 Adding Further Key Management Protocols

This framework cannot be used with all key management protocols. The key management protocol needs to comply with the requirements described in Section 16.4.4. In addition, the following needs to be defined:

- The key management protocol identifier to be used as the protocol identifier should be registered at IANA according to Section 16.1.9.
- The information that the key management needs from SDP and RTSP, and vice versa, as described in Section 16.1.4. The exact API is implementation specific, but it MUST at least support the exchange of the specified information.
- The key management protocol to be added MUST be such that the processing in Section 16.1.4 (describing its interactions with SDP and RTSP) can be applied. Note in particular, Section 16.1.4.1.4 requires each key management protocol to specify how the list of protocol identifiers is authenticated inside that key management protocol. The key management MUST always be given the protocol identifier(s) of the key management protocol(s) included in the offer in the order they appear.

Finally, it is obviously crucial to analyze possible security implications induced by the introduction of a new key management protocol in the described framework. Today, the MIKEY protocol (RFC 3830) has adopted the key management extensions to work with SIP and RTSP (see Section 16.1.7). Other protocols MAY use the described attribute and header (e.g., Kerberos) (RFC 4120); however, this is subject to future standardization.

16.1.7 Integration of MIKEY

RFC 4120 describes a key management protocol for real-time applications (both for peer-to-peer and group communication). MIKEY carries the security parameters needed for setting up the security protocol (e.g., SRTP) protecting the media stream. MIKEY can be integrated within SDP and RTSP, following the rules and guidelines described in this document. MIKEY satisfies the requirements described in Section 16.1.4. The MIKEY message is formed as defined in RFC 4120; it is then passed from MIKEY to the SDP application that base64 encodes it and encapsulates it in the keymgmt-data attribute. The examples in Section 16.1.5 use MIKEY, where the semantics of the exchange is also briefly explained. The KMPID to be used as the protocol identifier SHALL be "mikey" and is registered at IANA; see Section 16.1.9 for details.

The information the key management needs from SDP and RTSP, and vice versa, follows Section 16.1.4. To avoid bidding-down attacks, the directives in Section 16.1.4.1.4 are followed. The list of protocol identifiers is authenticated within MIKEY by placing the list in a general extension payload (of type "SDP IDs," RFC 4120), which then automatically will be integrity protected/signed. The receiver SHALL then match the list in the general extension payload with the list included in SDP and SHOULD (according to policy), if they differ or if integrity/signature verification fails, reject the offer.

The server will need to be able to know the identity of the client before creating and sending a MIKEY message. To signal the MIKEY identity of the client to the server in the DESCRIBE, it is RECOMMENDED to include the From header field in RTSP. Other methods of establishing identity could be using the IP address or retrieving the identity from the RTSP authentication if used.

16.1.7.1 MIKEY Interface

This subsection describes some aspects implementers SHOULD consider. If the MIKEY implementation is separate from the SDP/SIP/RTSP, an API between MIKEY and those protocols is needed with certain functionality (however, exactly what it looks like is implementation dependent). The following aspects need to be considered:

- The possibility for MIKEY to receive information about the sessions negotiated. This is to some extent implementation dependent. It is RECOMMENDED that, in the case of SRTP streams, the number and direction of SRTP streams be included. It is also RECOMMENDED that one provide the destination addresses and ports to MIKEY. When referring to streams described in SDP, MIKEY SHALL allocate two consecutive numbers for the related crypto session indexes (as each stream can be bidirectional). An example: If the SDP contains two "m=" lines (specifying the direction of the streams) and MIKEY is at the session level, then MIKEY allocates (e.g., the crypto sessions identifiers) (CS IDs; see RFC 4120) "1" and "2" for the first "m=" line, and "3" and "4" for the second "m=" line.
- The possibility for MIKEY to receive incoming MIKEY messages and return a status code from/to the SIP/RTSP application.
- The possibility for the SIP or RTSP applications to receive information from MIKEY. This would typically include the receiving of the crypto session bundle identifier (CSB ID; see RFC 4120, to later be able to identify the active MIKEY session), and the SSRCs and the rollover counter (ROC; see RFC 3711) for SRTP usage. It is also RECOMMENDED that extra information about errors can be received.
- The possibility for the SIP or RTSP application to receive outgoing MIKEY messages.
- The possibility to tear down a MIKEY Crypto Session Bundle (CSB) (e.g., if the SIP session is closed, the CSB SHOULD also be closed).

16.1.8 Security Considerations

The security procedures that need to be considered in dealing with the basic SDP messages described in this section (RFC 4567) are provided in Section 17.14.1.

16.1.9 IANA Considerations

SDP parameters that are described here have been registered with IANA. We have not included the IANA registration procedures here for the sake of brevity. The procedures for registration of new SDP parameters are described in RFC 4567.

16.2 SDP Security Descriptions for Media Streams

This section describes RFC 4568 that specifies a SDP cryptographic key and other parameters for unicast media streams. The security attribute is used to configure security for a unicast media stream in either a single message or a roundtrip exchange. However, the security attribute can be used with a variety of SDP media transports. Specifically, it defines how to use it for the SRTP unicast media streams. It is noted that the SDP crypto attribute requires the services of a data security protocol to secure the SDP message.

16.2.1 Introduction

The SDP (RFC 4566, see Section 2) describes multimedia sessions, which can be audio, video, whiteboard, fax, modem, and other media streams including security services. The SRTP (RFC 3711) provides security services for RTP media and is signaled by the use of SRTP transport (e.g., "RTP/SAVP" or "RTP/SAVPF") in an SDP media ("m=") line. However, there are no means within SDP itself to configure SRTP beyond using default values. This document specifies a new SDP attribute called "crypto," which is used to signal and negotiate cryptographic parameters for media streams in general and for SRTP in particular. The definition of the crypto attribute in this document is limited to two-party unicast media streams where each source has a unique cryptographic key; support for multicast media streams or multipoint unicast streams is for further study.

The crypto attribute is defined in a generic way to enable its use with SRTP and any other secure transports that can establish cryptographic parameters with only a single message or in a single roundtrip exchange using the offer/answer model (RFC 3264, see Section 3.1). Extensions to transports other than SRTP, however, are beyond the scope of this document. Each type of secure media transport needs its own specification for the crypto-attribute parameter. These definitions are frequently unique to the particular type of transport and must be specified in a standards-track RFC and registered with IANA according to the procedures defined in Section 16.2.10. This document defines the security parameters and keying material for SRTP only.

It would be self-defeating not to secure cryptographic keys and other parameters at least as well as the data are secured. Data security protocols such as SRTP rely upon a separate key management system to securely establish encryption and/or authentication keys. Key management protocols provide authenticated key establishment (AKE) procedures to authenticate the identity of each endpoint and protect against man-in-the-middle, reflection/replay, connection hijacking, and some denial-of-service attacks. Along with the key, an AKE protocol such as MIKEY (RFC 3830), Group Domain of Interpretation (GDOI) (RFC 3547 was made obsolete by RFC 6407), Kerberized Internet Negotiation of Keys (KINK) (RFC 4430), IKE (RFC 4306 was made obsolete by RFC 5282), Secure Multiparts (RFC 3851 was made obsolete by RFC 5751, RFC 2015, or TLS (RFC 2246 was made obsolete by RFC 4346; again RFC 4346 was made obsolete by RFC 5246) securely disseminates information describing both the key and the data security session. AKE is needed because it is pointless to provide a key over a medium where an attacker can snoop the key, alter the definition of the key to render it useless, or change the parameters of the security session to gain unauthorized access to session-related information.

SDP, however, was not designed to provide AKE services, and the media security descriptions defined in this document do not add AKE services to SDP. This specification is no replacement for a key management protocol or for the conveyance of key management messages in SDP (RFC 4567). The SDP security descriptions defined here are suitable for restricted cases only where IPsec, TLS, or some other encapsulating data-security protocol (e.g., SIP S/MIME) protects the SDP message. This document adds security descriptions to those encrypted and/or authenticated SDP messages through the new SDP "crypto" attribute, which provides the cryptographic parameters of a media stream.

The "crypto" attribute can be adapted to any media transport, but its precise definition is unique to a particular transport. In Section 16.2.2, we provide notational conventions followed by an applicability statement for the crypto attribute in Section 16.2.3. In Section 16.4.4, we introduce the general SDP crypto attribute, and in Section 16.2.5, we define how it is used with and without the offer/answer model. In Section 16.2.6, we define the crypto attribute details needed for SRTP, and in Section 16.2.7, we define SRTP-specific use of the attribute with and without the offer/answer model. Section 8 recites security considerations, and Section 16.2.9 gives an ABNF grammar for the general crypto attribute as well as the SRTP-specific use of the crypto attribute. IANA considerations are provided in Section 16.2.10.

16.2.2 Notational Conventions

The terminology in this document conforms to RFC 2828 that is obsoleted by RFC 4949, "Internet Security Glossary." n^r is exponentiation, where n is multiplied by itself r times; n and r are integers. 0, ..., k is an integer range of all integers from 0 to k, inclusive. The terms "transport" and "media transport" are used to mean "transport protocol" as defined in RFC 4566 (see Section 2). Explanatory notes are provided in several places throughout the document; these notes are indented three spaces from the surrounding text.

16.2.3 Applicability

RFC 4567 (see Section 16.1) provides similar cryptographic key distribution capabilities; it is intended for use when the signaling is to be confidential and/or integrity-protected separately from the keying material. In contrast, this specification carries the keying material within the SDP message, and it is intended for use when the keying material is protected along with the signaling. Implementations MUST employ security mechanisms that provide confidentiality and integrity for the keying material. When this specification is used in the context of SIP (RFC 3261, also see Section 1.2), the application SHOULD employ either the SIPS URI or S/MIME to provide protection for the SDP message and the keying material it contains. The use of transport layer or IP layer security in lieu of the SIPS URI or S/MIME protection is NOT RECOMMENDED since the protection of the SDP message and the keying material it contains cannot be ensured through all intermediate entities such as SIP proxies.

16.2.4 SDP "Crypto" Attribute and Parameters

A new media-level SDP attribute called "crypto" describes the cryptographic suite, key parameters, and session parameters for the preceding unicast media line. The "crypto" attribute MUST only appear at the SDP media level (not at the session level). The "crypto" attribute follows the format (see Section 16.2.9.1 for the formal ABNF grammar):

```
a=crypto:<tag> <crypto-suite> <key-params> [<session-params>]
```

The fields tag, crypto-suite, key-params, and session-params are described in the following subsections. The values of each of these fields are case-insensitive, unless otherwise noted. However, implementers are encouraged to use the actual case shown in this document and any extensions to it. Note that per normal SDP rules, the "crypto" attribute name itself is case-sensitive. Following is an example of the crypto attribute for the "RTP/SAVP" transport (i.e., the SRTP extension to the audio/video profile (RFC 3711). In the following, new lines are included for formatting reasons only:

```
a=crypto:1 AES_CM_128_HMAC_SHA1_80
       inline:PS1uQCVeeCFCanVmcjkpPywjNWhcYD0mXXtxaVBR|2^20|1:32
```

The crypto-suite is AES_CM_128_HMAC_SHA1_80, key-params is defined by the text starting with "inline:", and session-params is omitted.

16.2.4.1 Tag

The tag is a decimal number used as an identifier for a particular crypto attribute (see Section 16.2.9.1 for details); leading zeroes MUST NOT be used. The tag MUST be unique among all crypto attributes for a given media line. It is used with the offer/answer model to determine which of several offered crypto attributes were chosen by the answerer (see Section 16.2.5.1). In the offer/answer model, the tag is a negotiated parameter.

16.2.4.2 Crypto-Suite

The crypto-suite field is an identifier that describes the encryption and authentication algorithms (e.g., AES_CM_128_HMAC_SHA1_80) for the transport in question (see Section 16.2.9.1 for details). The possible values for the crypto-suite parameter are defined within the context of the transport (i.e., each transport defines a separate namespace for the set of crypto-suites). For example, the crypto-suite "AES_CM_128_HMAC_SHA1_80" defined within the context "RTP/SAVP" transport applies to SRTP only; the string may be reused for another transport (e.g., "RTP/SAVPF"), but a separate definition would be needed. In the offer/answer model, the crypto-suite is a negotiated parameter.

16.2.4.3 Key Parameters

The key-params field provides one or more sets of keying material for the crypto-suite in question. The field consists of a method indicator followed by a colon, and the actual keying information as shown below (the formal grammar is provided in Section 16.2.9.1): key-params = <key-method> ":" <key-info> Keying material might be provided by different means from that for key-params; however, this is out of scope. Only one method is defined in this document, namely, "inline," which indicates that the actual keying material is provided in the key-info field itself. There is a single name space for the key-method, (i.e., the key-method is transport independent). New key-methods (e.g., use of a URL) may be defined in a standards-track RFC in the future. Although the key-method itself may be generic, the accompanying key-info definition is specific not only to the key-method, but also to the transport in question. Key-info encodes keying material for a crypto-suite, which defines that keying material. New key methods MUST be registered with the IANA according to the procedures defined in Section 16.2.10.2.1.

Key-info is defined as a general octet string (see Section 16.2.9.1 for details); further transport and key-method specific syntax and semantics MUST be provided in a standards-track RFC for each combination of transport and key-method that uses it; definitions for SRTP are provided in Section 16.2.6. Note that such definitions are provided within the context of both a particular transport (e.g., "RTP/SAVP") and a specific key-method (e.g., "inline"). IANA will register the list of supported key methods for each transport. When multiple keys are included in the key parameters, it MUST be possible to determine which of the keys is being used in a given media packet by a simple inspection of the media packet received; a trial-and-error approach between the possible keys MUST NOT be performed.

For SRTP, this could be achieved by use of Master Key Identifiers (MKI) (RFC 3711). Use of <"From" and "To"> values are not supported in SRTP security descriptions for reasons explained in Section 16.2.6.1. In the offer/answer model, the key parameter is a declarative parameter.

16.2.4.4 Session Parameters

Session parameters are specific to a given transport, and use of them is OPTIONAL in the security descriptions framework, where they are just defined as general character strings. If session parameters are to be used for a given transport, then transport-specific syntax and semantics MUST be provided in a standards-track RFC; definitions for SRTP are provided in Section 16.2.6. In the offer/answer model, session parameters may be either negotiated or declarative; the definition of specific session parameters MUST indicate whether they are negotiated or declarative. Negotiated parameters apply to data sent in both directions, whereas declarative parameters apply only to media sent by the entity that generated the SDP. Thus, a declarative parameter in an offer applies to media sent by the offerer, whereas a declarative parameter in an answer applies to media sent by the answerer.

16.2.4.5 Example

This example shows use of the crypto attribute for the "RTP/SAVP" media transport type (as defined in Section 16.2.5). The "a=crypto" line is actually one long line; it is shown as two lines due to page formatting.

```
v=0
o=jdoe 2890844526 2890842807 IN IP4 10.47.16.5
s=SDP Seminar
i=A Seminar on the session description protocol
u=http://www.example.com/seminars/sdp.pdf
e=j.doe@example.com (Jane Doe)
c=IN IP4 161.44.17.12/127
t=2873397496 2873404696
m=video 51372 RTP/SAVP 31
a=crypto:1 AES_CM_128_HMAC_SHA1_80
      inline:d0RmdmcmVCspeEc3QGZiNWpVLFJhQX1cfHAwJSoj|2^20|1:32
m=audio 49170 RTP/SAVP 0
a=crypto:1 AES_CM_128_HMAC_SHA1_32
      inline:NzB4d1BINUAvLEw6UzF3WSJ+PSdFcGdUJShpX1Zj|2^20|1:32
m=application 32416 udp wb
a=orient:portrait
```

This SDP message describes three media streams, two of which use the "RTP/SAVP" transport. Each has a crypto attribute for the "RTP/SAVP" transport. These secure-RTP specific descriptions are defined in Section 16.2.6.

16.2.5 General Use of the Crypto Attribute

In this section, we describe the general use of the crypto attribute outside any transport or key-method specific rules.

16.2.5.1 Use with Offer/Answer

The general offer/answer rules for the crypto attribute are in addition to the rules specified in RFC 3264 (see Section 3.1), which MUST be followed unless otherwise noted. RFC 3264 defines operation for both unicast and multicast streams; the following sections describe operation for two-party unicast streams only, since support for multicast streams (and multipoint unicast streams) is for further study.

16.2.5.1.1 Generating the Initial Offer: Unicast Streams

When generating an initial offer for a unicast stream, there MUST be one or more crypto attributes present for each media stream for which security is desired. Each crypto attribute for a given media stream MUST contain a unique tag. The ordering of multiple "a=crypto" lines is significant: the most preferred crypto line is listed first. Each crypto attribute describes the crypto-suite, key(s), and possibly session parameters offered for the media stream. In general, a "more preferred" crypto-suite SHOULD be cryptographically stronger than a "less preferred" crypto-suite.

The crypto-suite always applies to media in the directions supported by the media stream (e.g., send and receive). The key(s), however, apply to data packets (e.g., SRTP and Secure RTCP [SRTCP] packets) that will be sent by the same party that generated the SDP. That is, each endpoint determines its own transmission keys and sends those keys, in SDP, to the other endpoint. This is done for consistency. Also, in the case of SRTP, for example, secure RTCP will still flow in both the send and receive directions for a unidirectional stream.

The inline parameter conveys the keying material used by an endpoint to encrypt the media streams transmitted by that endpoint. The same keying material is used by the recipient to decrypt those streams. The offer may include session parameters. There are no general offer rules for the session parameters; instead, specific rules may be provided as part of the transport-specific definitions of any session parameters.

When issuing an offer, the offerer MUST be prepared to support media security in accordance with any of the crypto attributes included in the offer. There are, however, two problems associated with this. First, the offerer does not know which key the answerer will be using for media sent to the offerer. Second, the offerer may not be able to deduce which of the offered crypto attributes were accepted. Since media may arrive prior to the answer, delay or clipping can occur. If this is unacceptable to the offerer, the offerer SHOULD use a mechanism outside the scope of this document to prevent this problem. For example, in SIP (RFC 3261, also see Section 1.2), a "security" precondition as defined in RFC 5027 (see Section 16.4) could solve this problem.

16.2.5.1.2 Generating the Initial Answer: Unicast Streams

When the answerer receives the initial offer with one or more crypto attributes for a given unicast media stream, the answerer MUST either accept exactly one of the offered crypto attributes, or the offered stream MUST be rejected. If the answerer wishes to indicate support for other crypto attributes, those can be listed by use of the SDP simple capability declaration

(RFC 3407, see Section 4.1) extensions. Only crypto attributes that are valid can be accepted; valid attributes do not violate any of the general rules defined for security descriptions or any specific rules defined for the transport and key-method in question. When selecting one of the valid crypto attributes, the answerer SHOULD select the most preferred crypto attribute it can support (i.e., the first valid supported crypto attribute in the list) according to the answerer's capabilities and security policies.

If there are one or more crypto attributes in the offer, but none of them is valid or none of the valid ones is supported, the offered media stream MUST be rejected. When an offered crypto attribute is accepted, the crypto attribute in the answer MUST contain the following:

- The tag and crypto-suite from the accepted crypto attribute in the offer (the same crypto-suite MUST be used in the send and receive directions).
- The key(s) the answerer will be using for media sent to the offerer. Note that a key MUST be provided, irrespective of any direction attributes in the offer or answer.

Furthermore, any session parameters that are negotiated MUST be included in the answer. Declarative session parameters provided by the offerer are not included in the answer; however, the answerer may provide its own set of declarative session parameters. Once the answerer has accepted one of the offered crypto attributes, the answerer MAY begin sending media to the offerer in accordance with the selected crypto attribute. Note, however, that the offerer may not be able to process such media packets correctly until the answer has been received.

16.2.5.1.3 Processing of the Initial Answer: Unicast Streams

When the offerer receives the answer, the offerer MUST verify that one of the initially offered crypto-suites and its accompanying tag were accepted and echoed in the answer. Also, the answer MUST include one or more keys, which will be used for media sent from the answerer to the offerer. If the offer contained any mandatory negotiated session parameters (see Section 16.2.6.3.7), the offerer MUST verify that said parameters are included in the answer and support them. If the answer contains any mandatory declarative session parameters, the offerer MUST be able to support those. If any of these fails, the negotiation MUST fail.

16.2.5.1.4 Modifying the Session

Once a media stream has been established, it MAY be modified at any time, as described in Section 3.1.8 of this book (that is, Section 8 of RFC 3264). Such a modification MAY be triggered by the security service (e.g., in order to perform a rekeying or change the crypto-suite). If media stream security using the general security descriptions defined here is still desired, the crypto attribute MUST be included in these new offer/answer exchanges. The procedures are similar to those defined in Sections 16.2.5.1.1–16.2.5.1.3 of this book, subject to the considerations provided in Section 3.1.8 of this book (that is, Section 8 of RFC 3264).

16.2.5.2 Use Outside Offer/Answer

The crypto attribute can also be used outside the context of offer/answer where there is no negotiation of the crypto-suite, cryptographic key, or session parameters. In this case, the sender determines security parameters for the stream. Since there is no negotiation mechanism, the sender MUST include exactly one crypto attribute, and the receiver MUST accept it or SHOULD NOT receive the associated stream. The sender SHOULD select the security description that it deems most secure for its purposes.

16.2.5.3 General Backwards Compatibility Considerations

In the offer/answer model, it is possible that the answerer supports a given secure transport (e.g., "RTP/SAVP") and accepts the offered media stream but does not support the crypto attribute defined in this document and hence ignores it. The offerer can recognize this situation by seeing an accepted media stream in the answer that does not include a crypto line. In that case, the security negotiation defined here MUST fail. Similar issues exist when security descriptions are used outside the offer/answer model. But the source of a nonnegotiated security description has no indication that the receiver has ignored the crypto attribute.

16.2.6 SRTP Security Descriptions

In this section, we provide definitions for security descriptions for SRTP media streams. In the next section, we define how to use SRTP security descriptions with and without the offer/answer model. SRTP security descriptions MUST only be used with the SRTP transport (e.g., "RTP/SAVP" or "RTP/SAVPF"). The following specifies security descriptions for the "RTP/SAVP" profile, defined in RFC 3711. However, it is expected that other SRTP profiles (e.g., "RTP/SAVPF") can use the same descriptions, which are in accordance with the SRTP protocol specification (RFC 3711). There is no assurance that an endpoint is capable of configuring its SRTP service with a particular crypto attribute parameter, but SRTP guarantees minimal interoperability among SRTP endpoints through the default SRTP parameters (RFC 3711).

More capable SRTP endpoints support a variety of parameter values beyond the SRTP defaults, and these values can be configured by the SRTP security descriptions defined here. An endpoint that does not support the crypto attribute will ignore it according to the SDP. Such an endpoint will not correctly process the particular media stream. By using the offer/answer model, the offerer and answerer can negotiate the crypto parameters to be used before commencement of the multimedia session (see Section 16.2.7.1). There are over 20 cryptographic parameters listed in the SRTP specification. Many of these parameters have fixed values for particular cryptographic transforms. At the time of session establishment, however, there is usually no need to provide unique settings for many of the SRTP parameters, such as salt length and pseudo-random function (PRF). Thus, it is possible to simplify the list of parameters by defining "cryptographic suites" that fix a set of SRTP parameter values for the security session. This approach is followed by the SRTP security descriptions, which use the general security description parameters as follows:

- crypto-suite: Identifies the encryption and authentication transforms
- key parameter: SRTP keying material and parameters
- session parameters: The following parameters are defined:
  - KDR: The SRTP Key Derivation Rate (KDR) is the rate at which a PRF is applied to a master key.
  - UNENCRYPTED_SRTP: SRTP messages are not encrypted.
  - UNENCRYPTED_SRTCP: SRTCP messages are not encrypted.
  - UNAUTHENTICATED_SRTP: SRTP messages are not authenticated.

- FEC_ORDER: Order of forward error correction (FEC) relative to SRTP services.
- FEC_KEY: Master key for FEC when the FEC stream is sent to a separate address and/or port.
- WSH: Window Size Hint
- Extensions: Extension parameters can be defined.

Please refer to the SRTP specification for a complete list of parameters and their descriptions (see Section 8.2 of RFC 3711). Regarding the UNENCRYPTED_SRTCP parameter, offerers and answerers of SDP security descriptions MUST NOT use the SRTCP E-bit to override UNENCRYPTED_SRTCP or the default, which is to encrypt all SRTCP messages (see Section 16.2.6.3.2). The key parameter, the crypto-suite, and the session parameters shown here are described in detail in the following subsections.

16.2.6.1 SRTP Key Parameter

SRTP security descriptions define the use of the "inline" key method as described in the following. Use of any other keying method (e.g., URL) for SRTP security descriptions is for further study. The "inline" type of key contains the keying material (master key and salt) and all policy related to that master key, including how long it can be used (lifetime) and whether it uses a master key identifier to associate an incoming SRTP packet with a particular master key. Compliant implementations obey the policies associated with a master key and MUST NOT accept incoming packets that violate the policy (e.g., after the master key lifetime has expired). The key parameter contains one or more cryptographic master keys, each of which MUST be a unique cryptographically random (RFC 1750 was made obsolete by RFC 4086) value with respect to other master keys in the entire SDP message (i.e., including master keys for other streams). Each key follows the format (the formal definition is provided in Section 16.2.9.2):

```
"inline:" <key||salt> ["|" lifetime] ["|" MKI ":" length]
```

key||salt concatenated master key and salt, base64 encoded (see RFC 3548 was made obsolete by RFC 4648) lifetime master key (max number of SRTP or SRTCP packets using this master key)

```
MKI:length MKI and length of the MKI field in SRTP packets
```

The following definition provides an example for AES_CM_128_HMAC_SHA1_80:

```
inline:d0RmdmcmVCspeEc3QGZiNWpVLFJhQX1cfHAwJSoj|2^20|1:4
```

The first field ("d0RmdmcmVCspeEc3QGZiNWpVLFJhQX1cfHAwJSoj") of the parameter is the cryptographic master key appended with the master salt; the two are first concatenated and then base64 encoded. The length of the concatenated key and salt is determined by the crypto-suite for which the key applies. If the length (after being decoded from base64) does not match that specified for the crypto-suite, the crypto attribute in question MUST be considered invalid. Each master key and salt MUST be a cryptographically random number and MUST be unique to the entire SDP message. When base64 decoding the key and salt, padding characters (i.e., one or two "=" at the end of the base64-encoded data) are discarded (see RFC 3548, made obsolete by RFC 4648, for details).

Base64 encoding assumes that the base64 encoding input is an integral number of octets. If a given crypto-suite requires the use of a concatenated key and salt with a length that is not an integral number of octets, said crypto-suite MUST define a padding scheme that results in the base64 input being an integral number of octets. For example, if the length defined were 250 bits, then 6 padding bits would be needed, which could be defined to be the last 6 bits in a 256-bit input.

The second field is the OPTIONAL lifetime of the master key as measured in maximum number of SRTP or SRTCP packets using that master key (i.e., the number of SRTP packets and the number of SRTCP packets each has to be less than the lifetime). The lifetime value MAY be written as a non-zero, positive decimal integer or as a power of 2 (see the grammar in Section 16.2.9.2 for details); leading zeroes MUST NOT be used. The "lifetime" value MUST NOT exceed the maximum packet lifetime for the crypto-suite. If the lifetime is too large or otherwise invalid, then the entire crypto attribute MUST be considered invalid. The default MAY be implicitly signaled by omitting the lifetime (note that the lifetime field never includes a colon, whereas the third field always does). This is convenient when the SRTP cryptographic key lifetime is the default value. As a shortcut to avoid long decimal values, the syntax of the lifetime allows using the literal "2^," which indicates "two to the power of."

The example provided in the above paragraphs shows a case where the lifetime is specified as 2^20. The following example, which is for the AES_CM_128_HMAC_SHA1_80 crypto-suite, has a default for the lifetime field, which means that SRTP's and SRTCP's default values will be used (see RFC 3711):

```
inline:YUJDZGVmZ2hpSktMbW9QUXJzVHVWd3l6MTIzNDU2|1066:4
```

The example shows a 30-octet key and concatenated salt that is base64 encoded: The 30-octet key/salt concatenation is expanded to 40 characters (octets) by the three-in-four encoding of base64. The third field, which is also OPTIONAL, is the MKI and its byte length. "MKI" is the master key identifier associated with the SRTP master key. The MKI is here defined as a positive decimal integer that is encoded as a big-endian integer in the actual SRTP packets; leading zeroes MUST NOT be used in the integer representation. If the MKI is given, then the length of the MKI MUST also be given and separated from the MKI by a colon (":"). The MKI length is the size of the MKI field in the SRTP packet, specified in bytes as a decimal integer; leading zeroes MUST NOT be used. If the MKI length is not given or its value exceeds 128 (bytes), then the entire crypto attribute MUST be considered invalid. The substring "1:4" in the first example assigns to the key an MKI of 1 that is 4 bytes long, and the second example assigns a 4-byte MKI of 1066 to the key. One or more master keys with their associated MKIs can be initially defined and then later updated or deleted and new ones defined.

SRTP offers a second feature for specifying the lifetime of a master key in terms of two values, called "From" and "To," which are defined on the SRTP sequence number space RFC 3711. This SRTP security descriptions specification, however, does not support the <"From", "To"> feature since the lifetime of an AES master key is 2^48 SRTP packets, which means that there is no cryptographic reason to replace a master key for practical point-to-point applications. For this reason, there is no need to support two means for signaling key update. The MKI is chosen over <"From", "To"> by this specification for the very few applications that need it since the MKI feature is simpler (though the MKI adds additional bytes to each packet, whereas <"From", "To"> does not). As mentioned earlier, the key parameter can contain one or more master keys. When the key parameter contains more than one master key, all the master keys in that key parameter MUST include an MKI value. When using the MKI, the MKI length MUST be the same for all keys in a given crypto attribute.

16.2.6.2 Crypto-Suites

The SRTP crypto-suites define the encryption and authentication transforms to be used for the SRTP media stream. The SRTP specification has defined three crypto-suites, which are described further in the following subsections in the context of the SRTP security descriptions. This table provides an overview of the crypto-suites and their parameters:

	AES_CM_128_ HMAC_SHA1_80	*AES_CM_128_ HMAC_SHA1_32*	*F8_128_ HMAC_SHA1_80*
Master key length	128 bits	128 bits	128 bits
Master salt length	112 bits	112 bits	112 bits
SRTP lifetime	2^48 packets	2^48 packets	2^48 packets
SRTCP lifetime	2^31 packets	2^31 packets	2^31 packets
Cipher	AES Counter Mode	AES Counter Mode	AES F8 Mode
Encryption key	128 bits	128 bits	128 bits
MAC	HMAC-SHA1	HMAC-SHA1	HMAC-SHA1
SRTP auth. tag	80 bits	32 bits	80 bits
SRTCP auth. tag	80 bits	80 bits	80 bits
SRTP auth. key len.	160 bits	160 bits	160 bits
SRTCP auth. key len.	160 bits	160 bits	160 bits

16.2.6.2.1 AES_CM_128_HMAC_SHA1_80

AES_CM_128_HMAC_SHA1_80 is the SRTP default AES counter mode cipher and HMAC-SHA1 message authentication with an 80-bit authentication tag. The master-key length is 128 bits and has a default lifetime of a maximum of 2^48 SRTP packets or 2^31 SRTCP packets, whichever comes first (see page 39 of RFC 3711). SRTP allows 2^48 SRTP packets or 2^31 SRTCP packets, whichever comes first. However, it is RECOMMENDED that automated key management allow easy and efficient rekeying at intervals far smaller than 2^31 packets given today's media rates or even HDTV media rates.

The SRTP and SRTCP encryption key lengths are 128 bits. The SRTP and SRTCP authentication key lengths are 160 bits (see Security Considerations in Section 16.2.8). The master salt value is 112 bits in length and the session salt value is 112 bits in length. The PRF is the default SRTP PRF that uses Advanced Encryption Standard (AES) Counter Mode with a 128-bit key length. The length of the base64-decoded key and salt value for this crypto-suite MUST be 30 characters (i.e., 240 bits); otherwise, the crypto attribute is considered invalid.

16.2.6.2.2 AES_CM_128_HMAC_SHA1_32

This crypto-suite is identical to AES_CM_128_HMAC_SHA1_80 except that the authentication tag is 32 bits. The length of the base64-decoded key and salt value for this crypto-suite MUST be 30 octets (i.e., 240 bits); otherwise, the crypto attribute is considered invalid.

16.2.6.2.3 F8_128_HMAC_SHA1_80

This crypto-suite is identical to AES_CM_128_HMAC_SHA1_80 except that the cipher is F8 (RFC 3711). The length of the base64-decoded key and salt value for this crypto-suite MUST be 30 octets (i.e., 240 bits); otherwise the crypto attribute is considered invalid.

16.2.6.2.4 Adding New Crypto-Suite Definitions

If new transforms are added to SRTP, new definitions for those transforms SHOULD be given for the SRTP security descriptions and published in a Standards-Track RFC. Sections 16.2.6.2.1–16.2.6.2.3 illustrate how to define crypto-suite values for particular cryptographic transforms. Any new crypto-suites MUST be registered with IANA following the procedures in Section 16.2.10.

16.2.6.3 Session Parameters

SRTP security descriptions define a set of "session" parameters, which OPTIONALLY may be used to override SRTP session defaults for the SRTP and SRTCP streams. These parameters configure an RTP session for SRTP services. The session parameters provide session-specific information to establish the SRTP cryptographic context.

16.2.6.3.1 KDR=n

KDR specifies the key derivation rate, as described in Section 4.3.1 of RFC 3711. The value *n* MUST be a decimal integer in the set {1,2,…,24}, which denotes a power of 2 from 2^1 to 2^24, inclusive; leading zeroes MUST NOT be used. The SRTP key derivation rate controls how frequently a new session key is derived from an SRTP master key (RFC 3711) given in the declaration. When the key derivation rate is not specified (i.e., the KDR parameter is omitted), a single initial key derivation is performed (RFC 3711). In the offer/answer model, KDR is a declarative parameter.

16.2.6.3.2 UNENCRYPTED_SRTCP and UNENCRYPTED_SRTP

SRTP and SRTCP packet payloads are encrypted by default. The UNENCRYPTED_SRTCP and UNENCRYPTED_SRTP session parameters modify the default behavior of the crypto-suites with which they are used:

- UNENCRYPTED_SRTCP signals that the SRTCP packet payloads are not encrypted.
- UNENCRYPTED_SRTP signals that the SRTP packet payloads are not encrypted.

In the offer/answer model, these parameters are negotiated. If UNENCRYPTED_SRTCP is signaled for the session, then the SRTCP E bit MUST be clear (0) in all SRTCP messages. If the default is used, all SRTCP messages are encrypted, and the E bit MUST be set (1) on all SRTCP messages.

16.2.6.3.3 UNAUTHENTICATED_SRTP

SRTP and SRTCP packet payloads are authenticated by default. The UNAUTHENTICATED_SRTP session parameter signals that SRTP messages are not authenticated. The use of

UNAUTHENTICATED_SRTP is NOT RECOMMENDED (see Security Considerations). The SRTP specification requires use of message authentication for SRTCP, but not for SRTP (RFC 3711). In the offer/answer model, this parameter is negotiated.

16.2.6.3.4 FEC_ORDER=order

FEC_ORDER signals the use of FEC for the RTP packets (RFC 2733 was made obsolete by RFC 5109). The FEC values for "order" are FEC_SRTP or SRTP_FEC. FEC_SRTP signals that FEC is applied before SRTP processing by the sender of the SRTP media and after SRTP processing by the receiver of the SRTP media; FEC_SRTP is the default. SRTP_FEC is the reverse processing. In the offer/answer model, FEC_ORDER is a declarative parameter.

16.2.6.3.5 FEC_KEY=key-params

FEC_KEY signals the use of separate master key(s) for a FEC stream. The master key(s) are specified with the exact same format as the SRTP key parameter defined in Section 16.2.6.1, and the semantic rules are the same—in particular, the master key(s) MUST be different from all other master key(s) in the SDP. An FEC_KEY MUST be specified when the FEC stream is sent to a different IP address and/or port than the media stream to which it applies (i.e., the "m=" line) (e.g., as described in Section 11.1 of RFC 2733, made obsolete by RFC 5109). When an FEC stream is sent to the same IP address and port as the media stream to which it applies, an FEC_KEY MUST NOT be specified. If an FEC_KEY is specified in this latter case, the crypto attribute in question MUST be considered invalid. In the offer/answer model, FEC_KEY is a declarative parameter.

16.2.6.3.6 Window Size Hint

SRTP defines the SRTP-WINDOW-SIZE (Section 3.3.2 of RFC 3711) parameter to protect against replay attacks. The minimum value is 64 (RFC 3711); however, this value may be considered too low for some applications (e.g., video). The Window Size Hint (WSH) session parameter provides a hint for how big this window should be to work satisfactorily (e.g., based on sender knowledge of the number of packets per second). However, there might be enough information given in SDP attributes like "a=maxprate" [maxprate] and the bandwidth modifiers to allow a receiver to derive the parameter satisfactorily. Consequently, this value is only considered a hint to the receiver of the SDP that MAY choose to ignore the value provided. The value is a decimal integer; leading zeroes MUST NOT be used. In the offer/answer model, WSH is a declarative parameter.

16.2.6.3.7 Defining New SRTP Session Parameters

New SRTP session parameters for the SRTP security descriptions can be defined in a standards-track RFC and registered with IANA according to the registration procedures defined in Section 16.2.10. New SRTP session parameters are by default mandatory. A newly defined SRTP session parameter that is prefixed with the dash character ("-"), however, is considered optional and MAY be ignored. If an SDP crypto attribute is received with an unknown session parameter that is not prefixed with a "-" character, that crypto attribute MUST be considered invalid.

16.2.6.4 SRTP Crypto Context Initialization

In addition to the various SRTP parameters defined above (Sections 16.2.6.2.1 – 16.2.2.4), three pieces of information are critical to the operation of the default SRTP ciphers:

- SSRC: Synchronization source
- ROC: Roll-over counter for a given SSRC
- SEQ: Sequence number for a given SSRC

In a unicast session, as defined here, there are three constraints on these values. The first constraint is on the SSRC, which makes an SRTP keystream unique from other participants. As explained in SRTP, the keystream MUST NOT be reused on two or more different pieces of plaintext. Keystream reuse makes the ciphertext vulnerable to cryptanalysis. One vulnerability is that known-plaintext fields in one stream can expose portions of the reused keystream, and this could further expose more plaintext in other streams. Since all current SRTP encryption transforms use keystreams, key sharing is a general problem (RFC 3711). SRTP mitigates this problem by including the SSRC of the sender in the keystream. But SRTP does not solve this problem in its entirety because the RTP has SSRC collisions, which although very rare (RFC 3550) are quite possible. During a collision, two or more SSRCs that share a master key will have identical keystreams for overlapping portions of the RTP sequence number space. SRTP security descriptions avoid keystream reuse by making unique master keys REQUIRED for the sender and receiver of the security description. Thus, the first constraint is satisfied. Also note that there is a second problem with SSRC collisions: the SSRC is used to identify the crypto context and thereby the cipher, key, ROC, etc. to process incoming packets. In case of SSRC collisions, crypto context identification becomes ambiguous, and correct packet processing may not occur. Furthermore, if an RTCP BYE packet is to be sent for a colliding SSRC, that packet may also have to be secured. In a (unicast) point-to-multipoint scenario, this can be problematic for the same reasons (i.e., it is not known which of the possible crypto contexts to use). Note that these problems are not unique to the SDP security descriptions; any use of SRTP needs to consider them. The second constraint is that the ROC MUST be zero at the time that each SSRC commences sending packets. Thus, there is no concept of a "late joiner" in SRTP security descriptions, which are constrained to be unicast and pairwise. The ROC and SEQ form a "packet index" in the default SRTP transforms, and the ROC is consistently set to zero at session commencement, according to this document.

The third constraint is that the initial value of SEQ SHOULD be chosen to be within the range of 0.2^15-1; this avoids an ambiguity when packets are lost at the start of the session. If it is at the start of a session, an SSRC source might randomly select a high sequence-number value and put the receiver in an ambiguous situation:

> if initial packets are lost in transit up to the point that the sequence number wraps (i.e., exceeds 2^16-1), then the receiver might not recognize that its ROC needs to be incremented. By restricting the initial SEQ to the range of 0..2^15-1, SRTP packet-index determination will find the correct ROC value, unless all the first 2^15 packets are lost (which seems, if not impossible, rather unlikely). See Section 3.3.1 of the SRTP specification regarding packet-index determination (RFC 3711).

16.2.6.4.1 Late Binding of One or More SSRCs to a Crypto Context

The packet index, therefore, depends on the SSRC, the SEQ of an incoming packet, and the ROC, which is an SRTP crypto context variable. Thus, SRTP has a big security dependency on SSRC uniqueness. Given these constraints, unicast SRTP crypto contexts can be established without the need to negotiate SSRC values in the SRTP security descriptions. Instead, an approach called "late binding" is RECOMMENDED by this specification. When a packet arrives, the SSRC that is contained in it can be bound to the crypto context at the time of session commencement (i.e., SRTP packet arrival) rather than at the time of session signaling (i.e., receipt of an SDP). With the arrival of the packet containing the SSRC, all the data items needed for the SRTP crypto context are held by the receiver. (Note that the ROC value by definition is zero; if nonzero values were to be supported, additional signaling would be required.) In other words, the crypto context for a SRTP session using late binding is initially identified by the SDP as <*, address, port> where "*" is a wildcard SSRC, "address" is the local receive address from the "c=" line, and "port" is the local receive port from the "m=" line. When the first packet arrives with ssrcX in its SSRC field, the crypto context <ssrcX, address, port> is instantiated subject to the following constraints:

- Media packets are authenticated: authentication MUST succeed; otherwise, the crypto context is not instantiated.
- Media packets are not authenticated: crypto context is automatically instantiated.

Note that use of late binding when there is no authentication of the SRTP media packets is subject to numerous security attacks and that consequently it is NOT RECOMMENDED (of course, this can be said for unauthenticated SRTP in general). Note that use of late binding without authentication will result in the creation of local state as a result of receiving a packet from any unknown SSRC. UNAUTHENTICATED_SRTP, therefore, is NOT RECOMMENDED because it invites easy denial-of-service attack. In contrast, late binding with authentication does not suffer from this weakness.

16.2.6.4.2 Sharing Cryptographic Contexts among Sessions or SSRCs

With the constraints and procedures described above (Section 16.2.6.4.1), it is not necessary to explicitly signal the SSRC, ROC, and SEQ for a unicast RTP session. So there are no a=crypto parameters for signaling SSRC, ROC, or SEQ. Thus, multiple SSRCs from the same entity will share a=crypto parameters when late binding is used. Multiple SSRCs from the same entity arise due to either multiple sources (microphones, cameras, etc.) or RTP payloads requiring SSRC multiplexing within that same session. SDP also allows multiple RTP sessions to be defined in the same media description ("m="); these RTP sessions will also share the a=crypto parameters. An application that uses a=crypto in this way serially shares a master key among RTP sessions or SSRCs and MUST replace the master key when the aggregate number of packets among all SSRCs approaches 2^31 packets. SSRCs that share a master key MUST be unique from one another.

16.2.6.5 Removal of Crypto Contexts

The mechanism defined above (Section 16.2.6.4.2) addresses the issue of creating crypto contexts. However, in practice, session participants may want to remove crypto contexts prior to session termination. Since a context contains information that cannot automatically be recovered (e.g., ROC), it is important that the sender and receiver agree on when a crypto context can be removed and perhaps more importantly when it cannot. Even when late binding is used for a unicast stream, the ROC is lost and cannot be recovered automatically (unless it is zero) once the crypto context is removed. We resolve this problem as follows. When SRTP security descriptions are being used, crypto-context removal MUST follow the same rules as SSRC removal from the member table (RFC 3550); note that this can happen as the result of an SRTCP BYE packet or a simple time-out due to inactivity. Inactive session participants that wish to ensure their crypto contexts are not timed out MUST thus send SRTCP packets at regular intervals.

16.2.7 SRTP-Specific Use of the Crypto Attribute

Section 16.2.5 describes general use of the crypto attribute, and this section completes it by describing SRTP-specific use.

16.2.7.1 Use with Offer/Answer

In this section, we describe how the SRTP security descriptions are used with the offer/answer model to negotiate cryptographic capabilities and communicate SRTP master keys. The rules defined below complement the general offer/answer rules defined in Section 16.2.5.1, which MUST be followed unless otherwise specified. Note that the rules below define unicast operation only; support for multicast and multipoint unicast streams is for further study.

16.2.7.1.1 Generating the Initial Offer: Unicast Streams

When the initial offer is generated, the offerer MUST follow the steps in Section 16.2.5.1.1, as well as the following steps. For each unicast media line (m=) using the SRTP transport where the offerer wants to specify cryptographic parameters, the offerer MUST provide at least one valid SRTP security description ("a=crypto" line), as defined in Section 16.2.6. If the media stream includes FEC with a different IP address and/or port from that of media stream itself, an FEC_KEY parameter MUST be included, as described in Section 16.2.6.3.5.

The inline parameter conveys the SRTP master key used by an endpoint to encrypt the SRTP and SRTCP streams transmitted by that endpoint. The same key is used by the recipient to decrypt those streams. However, the receiver MUST NOT use that same key for the SRTP or SRTCP packets it sends to the session because the default SRTP cipher and mode is insecure when the master key is reused across distinct SRTP streams. The offerer MAY include one or more other SRTP session parameters, as defined in Section 16.2.6.3. Note, however, that if any SRTP session parameters are included that are not known to the answerer, but that are nonetheless mandatory (see Section 16.2.6.3.6), the negotiation will fail if the answerer does not support them.

16.2.7.1.2 Generating the Initial Answer: Unicast Streams

When the initial answer is generated, the answerer MUST follow the steps in Section 16.2.5.1.2, as well as the following steps. For each unicast media line that uses the SRTP transport and contains one or more "a=crypto" lines in the offer, the answerer MUST either accept one (and only one) of the crypto lines for that media stream, or it MUST reject the media stream. Only "a=crypto" lines that are considered valid SRTP security descriptions, as defined in Section 16.2.6, can be accepted. Furthermore, all parameters (crypto-suite, key parameter, and mandatory session parameters) MUST be acceptable to the answerer in order for the offered media stream to be accepted. Note that if the media stream includes FEC with a different IP address and/or port from that of the media stream itself, an FEC_KEY parameter MUST be included, as described in Section 16.2.6.3.5. When the answerer accepts an SRTP unicast media stream with a crypto line, the answerer MUST include one or more master keys appropriate for the selected crypto algorithm; the master key(s) included in the answer MUST be different from those in the offer. When the master key(s) are not shared between the offerer and answerer, SSRC collisions between the offerer and answerer will not lead to keystream reuse, and hence SSRC collisions do not necessarily have to be prevented.

If FEC to a separate IP address and/or port is included, the answer MUST include an FEC_KEY parameter, as described in Section 16.2.6.3.5. Declarative session parameters may be added to the answer as usual; however, the answerer SHOULD NOT add any mandatory session parameter (see Section 16.2.6.3.6) that might be unknown to the offerer. If the answerer cannot find any valid crypto line it supports, or if its configured policy prohibits any cryptographic key parameter (e.g., key length) or cryptographic session parameter (e.g., KDR, FEC_ORDER), it MUST reject the media stream, unless it is able to successfully negotiate use of SRTP by other means outside the scope of this document (e.g., by use of MIKEY, RFC 3830).

16.2.7.1.3 Processing of the Initial Answer: Unicast Streams

When the offerer receives the answer, it MUST perform the steps in Section 16.2.5.1.3, as well as the following steps for each SRTP media stream it offered with one or more crypto lines in it. If the media stream was accepted and it contains a crypto line, it MUST be checked that the crypto line is valid according to the constraints specified in Section 16.2.6 (including any FEC constraints). If the offerer either does not support or is not willing to honor one or more of the SRTP parameters in the answer, the offerer MUST consider the crypto line invalid. If the crypto line is not valid, or the offerer's configured policy prohibits any cryptographic key parameter (e.g., key length) or cryptographic session parameter, the SRTP security negotiation MUST be deemed to have failed.

16.2.7.1.4 Modifying the Session

When a media stream using the SRTP security descriptions has been established and a new offer/answer exchange is performed, the offerer and answerer MUST follow the steps in Section 16.2.5.1.4, as well as the following steps. When modifying the session, all negotiated aspects of the SRTP media stream can be modified. For example, a new crypto-suite can be used or a new master key can be established. As described in RFC 3264 (see Section 3.1), when a new offer/answer exchange is made, there will be a window of time where the offerer and the answerer must be prepared to receive media according to both the old and new offer/answer exchanges. This requirement applies here as well; however, the following should be noted:

- When authentication is not being used, it may not be possible for either the offerer or answerer to determine if a given packet is encrypted according to the old or new offer/answer exchange. RFC 3264 (see Section 3.1) defines a couple of techniques to address this problem (e.g., changing the payload types used and/or the transport addresses). Note, however, that a change in transport addresses may have an impact on quality of service as well as on firewall and network address translator (NAT traversal. The SRTP security descriptions use the MKI to deal with this (which adds a few bytes to each SRTP packet), as described in Section 16.2.6.1. For further details on the MKI, please refer to RFC 3711.
- If the answerer changes its master key, the offerer will not be able to process packets secured via this master key until the answer is received. This could be addressed by using a security "precondition" (RFC 5027). If the offerer includes an IP address and/or port that differs

from that used previously for a media stream (or FEC stream), the offerer MUST include a new master key with the offer; in so doing, it will be creating a new crypto context where the ROC is set to zero.

Similarly, if the answerer includes an IP address and/or port that differs from that used previously for a media stream (or FEC stream), the answerer MUST include a new master key with the answer (and hence create a new crypto context with the ROC set to zero). The reason for this is that when the answerer receives an offer or the offerer receives an answer with an updated IP address and/or port, it is not possible to determine if the other side has access to the old crypto context parameters (and in particular the ROC).

For example, if one side is a decomposed media gateway, or if an SIP back-to-back user agent is involved, it is possible that the media endpoint changed and no longer has access to the old crypto context. By always requiring a new master key in this case, the answerer/offerer will know that the ROC is zero for this offer/answer, and any key lifetime constraints will trivially be satisfied too. Another consideration here applies to media relays; if the relay changes the media endpoint on one side transparently to the other side, the relay cannot operate as a simple packet reflector but will have to actively engage in SRTP packet processing and transformation (i.e., decryption and re-encryption, etc.). Finally, note that if the new offer is rejected, the old crypto parameters remain in place.

16.2.7.1.5 Offer/Answer Example

In this example, the offerer supports two crypto-suites (f8 and AES). The a=crypto line is actually one long line, although it is shown as two lines in this document due to page formatting. The f8 example shows two inline parameters; as explained in Section 16.2.6.1, there may be one or more key (i.e., inline) parameters in a crypto attribute. In this way, multiple keys are offered to support key rotation using an MKI.

Offerer sends

```
v=0
o=sam 2890844526 2890842807 IN IP4 10.47.16.5
s=SRTP Discussion
i=A discussion of Secure RTP
u=http://www.example.com/seminars/srtp.pdf
e=marge@example.com (Marge Simpson)
c=IN IP4 168.2.17.12
t=2873397496 2873404696
m=audio 49170 RTP/SAVP 0
a=crypto:1 AES_CM_128_HMAC_SHA1_80
inline:WVNfX19zZW1jdGwgKCkgewkyMjA7fQp9CnVubGVz|2^20|1:4
        FEC_ORDER=FEC_SRTP
a=crypto:2 F8_128_HMAC_SHA1_80
inline:MTIzNDU2Nzg5QUJDREUwMTIzNDU2Nzg5QUJjZGVm|2^20|1:4;
inline:QUJjZGVmMTIzNDU2Nzg5QUJDREUwMTIzNDU2Nzg5|2^20|2:4
        FEC_ORDER=FEC_SRTP
```

Answerer replies

```
v=0
o=jill 25690844 8070842634 IN IP4 10.47.16.5
s=SRTP Discussion
```

```
i=A discussion of Secure RTP
u=http://www.example.com/seminars/srtp.pdf
e=homer@example.com (Homer Simpson)
c=IN IP4 168.2.17.11
t=2873397526 2873405696
m=audio 32640 RTP/SAVP 0
a=crypto:1 AES_CM_128_HMAC_SHA1_80
inline:PS1uQCVeeCFCanVmcjkpPywjNWhcYD0mXXtxaVBR|2^20|1:4
```

In this case, the session would use the AES_CM_128_HMAC_SHA1_80 crypto-suite for the RTP and RTCP traffic. If F8_128_HMAC_SHA1_80 were selected by the answerer, there would be two inline keys associated with the SRTP cryptographic context. One key has an MKI value of 1 and the second has an MKI of 2.

16.2.7.2 SRTP-Specific Use Outside Offer/Answer

Use of SRTP security descriptions outside the offer/answer model is not defined. Use of SRTP security descriptions outside the offer/answer model could have been defined for sendonly media streams; however, there would not be a way to indicate the key to use for SRTCP by the receiver of said media stream.

16.2.7.3 Support for SIP Forking

As mentioned earlier, the security descriptions defined here do not support multicast media streams or multipoint unicast streams. However, in the SIP protocol, it is possible to receive several answers to a single offer due to the use of forking (see SIP, RFC 3261, also see Section 1.2). Receiving multiple answers leads to a couple of problems for the SRTP security descriptions:

- Different answerers may choose different ciphers, keys, etc.; however, there is no way for the offerer to associate a particular incoming media packet with a particular answer.
- Two or more answerers may pick the same SSRC, and hence the SSRC collision problems mentioned earlier may arise.

As stated earlier, the above point-to-multipoint cases are outside the scope of the SDP security descriptions. However, there are still ways of supporting SIP forking (e.g., by changing the multipoint scenario resulting from SIP forking into multiple two-party unicast cases). This can be done as follows:

> For each answer received beyond the initial answer, issue a new offer to that particular answerer using a new receive transport address (IP address and port); note that this requires support for the SIP UPDATE method (RFC 3311). Also, to ensure that two media sessions are not inadvertently established prior to the UPDATE being processed by one of them, use security preconditions (RFC 5027, also see Section 16.4 of this book).

Finally, note that all SIP user agents that received the offer will know the key(s) being proposed by the initial offer. If the offerer wants to ensure security with respect to all other user agents that may have received the offer, a new offer/answer exchange with a new key needs to be performed with the answerer as well. Note that the offerer cannot determine whether a single or multiple SIP user agents received the offer, since intermediate forking proxies may only forward a single answer to the offerer. The above description is intended to suggest one possible way of supporting SIP forking. There are many details missing, and it should not be considered a normative specification. Alternative approaches may also be possible.

16.2.7.4 SRTP-Specific Backwards Compatibility Considerations

It is possible that the answerer supports the SRTP transport and accepts the offered media stream but that it does not support the crypto attribute defined here. The offerer can recognize this situation by seeing an accepted SRTP media stream in the answer that does not include a crypto line. In that case, the security negotiation defined here MUST be deemed to have failed. Also, if a media stream with a given SRTP transport (e.g., "RTP/SAVP") is sent to a device that does not support SRTP, that media stream will be rejected.

16.2.7.5 Operation with KEYMGT= and k=lines

An offer MAY include both "a=crypto" and "a=keymgt" lines [keymgt]. Per SDP rules, the answerer will ignore attribute lines it does not understand. If the answerer supports both "a=crypto" and "a=keymgt," the answer MUST include either "a=crypto" or "a=keymgt" but not both, as including both is undefined. An offer MAY include both "a=crypto" and "k=" lines [RFC 4566]. Per SDP rules, the answerer will ignore attribute lines it does not understand. If the answerer supports both "a=crypto" and "k=," the answer MUST include either "a=crypto" or "k=" but not both, as including both is undefined.

16.2.8 Security Considerations

The security procedures that need to be considered in dealing with the basic SDP messages described in this section (RFC 4568) are provided in Section 17.14.2.

16.2.9 Grammar

All SDP ABNF syntaxes are provided in Section 2.9. However, we are repeating this here for convenience. In this section, we first provide the ABNF grammar for the generic crypto attribute, and then we provide the ABNF grammar for the SRTP-specific use of the crypto attribute.

16.2.9.1 Generic "Crypto" Attribute Grammar

The ABNF grammar for the crypto attribute is defined below:

```
"a=crypto:" tag 1*WSP crypto-suite 1*WSP key-params
              *(1*WSP session-param)
tag                  =    1*9DIGIT
crypto-suite         =    1*(ALPHA / DIGIT / "_")
key-params           =    key-param *(";" key-param)
key-param            =    key-method ":" key-info
key-method           =    "inline" / key-method-ext
key-method-ext       =    1*(ALPHA / DIGIT / "_")
```

```
key-info              =     1*(%x21-3A / %x3C-7E) ; visible
                            (printing) chars
                                   ; except semi-colon
session-param         =     1*(VCHAR) ; visible (printing)
                            characters
                                   ; where WSP, ALPHA, DIGIT, and
                                   ; VCHAR
                                   ; are defined in RFC 4234 obsoleted
                                   ; by RFC 5234.
```

16.2.9.2 SRTP "Crypto" Attribute Grammar

This section provides an ABNF (RFC 4234 was made obsolete by RFC 5234) grammar for the SRTP-specific use of the SDP crypto attribute:

```
crypto-suite          =     srtp-crypto-suite key-method =
                            srtp-key-method
key-info              =     srtp-key-info
session-param         =     srtp-session-param
srtp-crypto-suite     =     "AES_CM_128_HMAC_SHA1_32" /
                            "F8_128_HMAC_SHA1_32" /
                            "AES_CM_128_HMAC_SHA1_80" /
                            srtp-crypto-suite-ext
srtp-key-method       =     "inline"
srtp-key-info         =     key-salt ["|" lifetime] ["|" mki]
key-salt              =     1*(base64) ; binary key and salt values
                                   ; concatenated together, and then
                                   ; base64 encoded [section 3 of
                                   ; RFC 3548
lifetime              =     ["2^"] 1*(DIGIT) ; see section 6.1 for
                            "2^"
mki                   =     mki-value ":" mki-length
mki-value             =     1*DIGIT
mki-length            =     1*3DIGIT ; range 1..128.
srtp-session-param    =     kdr /
                            "UNENCRYPTED_SRTP" /
                            "UNENCRYPTED_SRTCP" /
                            "UNAUTHENTICATED_SRTP" /
                            fec-order /
                            fec-key /
                            wsh /
                            srtp-session-extension
kdr = "KDR            =     " 1*2(DIGIT) ; range 0..24,
                                   ; power of two
fec-order             =     "FEC_ORDER=" fec-type
fec-type              =     "FEC_SRTP" / "SRTP_FEC"
fec-key               =     "FEC_KEY=" key-params
wsh                   =     "WSH=" 2*DIGIT ; minimum value is 64
base64                =     ALPHA / DIGIT / "+" / "/" / "="
srtp-crypto-suite-ext =     1*(ALPHA / DIGIT / "_")
srtp-session-extension=     ["-"] 1*(VCHAR) ;visible chars (RFC 4234
                                   ; obsoleted by RFC 5234)
                                   ; first character must not be
                                   ; dash ("-")
```

16.2.10 IANA Considerations

SDP parameters that are described here have been registered with IANA. We have not included the IANA registration procedures here for the sake of brevity. The procedures for registration of new SDP parameters are described in RFC 4568.

16.2.11 Rationale for Keying Material Directionality (Appendix A of RFC 4568)

SDP security descriptions define the keying material for the sending direction, which is included in the SDP. Thus, the key that is carried in an SDP message is a decryption key for the receiver of that SDP message. This is in contrast to the majority of information included in SDP, which describes information for the receiving (or receiving and sending) direction. This reversed information directionality generates some challenges with using the mechanism in the offer/answer model and in particular with SIP, where early media and forking require special consideration (as described in Section 16.2.7.3). There are however good reasons for doing this, which can be summarized as follows:

First of all, there is the general security philosophy of letting the entity that sends traffic decide what key to use for protecting it. SRTP uses counter mode, which is secure when counters do not overlap among senders who share a master key; the surest way to avoid counter overlap is for each endpoint to generate its own master key.

Secondly, if SDP security descriptions had been designed to keep the normal SDP information directionality, it would have resulted in problems with supporting early media and SIP forking: If an offer generates multiple answers and the keying material was for the receive direction, some of the parameter values (e.g. lifetime) would have to be shared between all the answerers (senders of media), which would lead to considerable complexity, possibly requiring changes or extensions to SRTP. Other problems were discovered as well, which we describe further in the following sections.

In the following scenarios, we analyze what would have occurred if SDP security descriptions had been designed so that the keying material was to receive the keying material (rather than its actual design, where the keying material is sending the keying material).

16.2.11.1 Scenario A: Non-Forking Case

In this scenario, the offer includes the receiving keying material; the answerer receives it and starts sending data packets toward the offerer. If there were a single crypto attribute in the offer, there would be no ambiguity about which crypto-suite was being used and, hence, the incoming packet could be processed. However, in the case where the offer includes multiple alternative crypto-attributes, the offerer would not know which one was chosen, and hence, if the offerer received packets before the answer came back, the offerer would be unable to process those packets (problem 1). (Use of the MKI has been suggested as one possible solution; however, it incurs a per-packet overhead.)

16.2.11.2 Scenario B: Serial Forking Case

In this scenario, Alice generates an offer to Bob, who starts sending (early) media toward Alice (no answer returned yet). In this scenario, we assume we aren't also encountering Scenario A (e.g., the offer includes only a single crypto attribute) and that Bob is using a SSRC value of 1 for his SRTP and SRTCP packets. Alice thus has a crypto context for SSRC 1, including the associated ROC and RTP SEQ. Bob now forwards the call to Carol (Bob has not yet generated an answer). At this point, Bob has Alice's key, which sometimes might be a security weakness. As the exchange proceeds, Carol gets the original offer, including the offered crypto attribute and starts sending media packets toward Alice.

It just so happens that Carol chooses an SSRC value of 1, as did Bob. When Carol starts generating packets, there is a potential for what RFC 3711 calls a "two-time pad" issue (problem2), as well as the potential for the ROC to be out of sync between Alice and Carol (problem 3). Note that since Bob and Carol are (presumably) using different source transport addresses, the SSRC reuse does not constitute an SSRC collision (although it may still be interpreted as such by Alice). Per RFC 3711, since the master key would be shared between Bob and Carol in this case, it is RECOMMENDED that Alice leave the session at that point in order to avoid the two-time pad issue. It should also be noted that RFC 3711 recommends against sharing SRTP master keys, which forking may accidentally introduce when the keying material is for the receiving direction.

If we consider this scenario again, but this time with keying material in the offer (and answer) being the sending keying material (as specified by SDP security descriptions), the scenario instead looks as follows: Bob again chooses SSRC 1, and Bob will need to send back an answer to Alice, since Alice needs to learn Bob's sending key. Bob also starts sending media toward Alice (clipping may occur until Alice receives Bob's answer). Bob again forwards the call to Carol who also starts sending early media using SSRC 1. However, Carol needs to generate a new answer (for the dialog between Alice and Carol) in order for Alice to process Carol's packets. Upon receiving this answer, Alice can initiate a new offer/answer exchange (to move the session to another transport address as described in Section 16.2.7.3). In this case, there is one master key per session and a unique keystream regardless of whether SSRCs collide.

16.2.11.3 Scenario C: Parallel Forking Case

In this scenario, Alice generates an offer (with receive keying material) that gets forked to Bob and Carol in parallel. Bob and Carol both start sending packets (early media) to Alice. If Bob and Carol choose different SSRCs, everything is fine initially. However, one of the crypto context parameters is the master key lifetime, and since Bob and Carol are sharing the same master key (unbeknownst to either), they do not know when they need to rekey (problem 4). If they choose the same SSRC, we have the two-time pad problem again (problem 2). In summary, if keying material were for the receive direction, we would have the following problems:

- Problem 1: Offerer does not know which of multiple crypto offers was chosen by answerer.
- Problem 2: SSRC reuse (or SSRC collisions) between multiple answerers (serial or parallel forking) may lead to the two-time pad issue.
- Problem 3: Part of the crypto context parameters (specifically the ROC) is not communicated but derived, and if we allow multiple entities to use the same SSRC (sequentially), the ROC can be wrong.
- Problem 4: All crypto contexts that share a master key need to maintain a shared set of counters (master key lifetime), and if we allow for multiple entities on different platforms to share a master key, we would need a mechanism to synchronize these counters.

Problem 1 could be addressed by using the MKI as proposed separately; however, it would result in using extra bandwidth for each SRTP media packet. Solving problem 2 implies a need to be able to synchronize SSRC values with the answerer (or abandon the session when SSRC reuse or SSRC collisions occur). Problem 3 implies a need to be able to synchronize ROC values on a per SSRC basis (or abandon the session when SSRC reuse occurs). Problem 4 could be solved by having the offerer (Alice, the entity receiving media) determine how many packets have actually been generated by the total set of senders to Alice and, hence, be the one to initiate the rekeying. In the case of packet losses, etc., this is not foolproof, but in practice it could probably be addressed by use of a reasonable safety margin. In conclusion, it would be expected from an offer/answer and SIP point of view to have the offer (and answer) keying material be the receive keying material; however, doing so would trade security for SIP friendliness (e.g., two-time pad and master key lifetime issues) and violate the RFC 3711 rule for sharing an SRTP master key across SRTP sessions.

16.3 Connection-Oriented Media Transport over TLS Protocol in SDP

This section describes RFC 4572 that specifies how to establish secure connection-oriented media transport sessions over TLS protocol using the SDP. It defines a new SDP protocol identifier, "TCP/TLS." It also defines the syntax and semantics for an SDP "fingerprint" attribute that identifies the certificate that will be presented for the TLS session. This mechanism allows media transport over TLS connections to be established securely, so long as the integrity of session descriptions is assured. In addition, this RFC extends and updates RFC 4145 (see Section 10.1).

16.3.1 Introduction

The SDP (RFC 4566, Section 2) provides a general-purpose format for describing multimedia sessions in announcements or invitations. For many applications, it is desirable to establish, as part of a multimedia session, a media stream that uses a connection-oriented transport. RFC 4145 (see Section 10.1), connection-oriented media transport in the SDP, specifies a general mechanism for describing and establishing such connection-oriented streams; however, the only transport protocol it directly supports is TCP. In many cases, session participants wish to provide confidentiality, data integrity, and authentication for their media sessions. This document therefore extends the connection-oriented media specification to allow session descriptions to describe media sessions that use the TLS protocol (RFC 4346 made obsolete by RFC 5246). The TLS protocol allows applications to communicate over a channel that provides confidentiality and data integrity. The TLS specification, however, does not specify how specific protocols establish and use this secure channel; particularly, TLS leaves the question of how to interpret and validate authentication certificates as an issue for the protocols that run over TLS. This document specifies such usage for the case of connection-oriented media transport.

Complicating this issue, endpoints exchanging media will often be unable to obtain authentication certificates signed by a well-known root certification authority (CA). Most CAs charge for signed certificates, particularly host-based certificates; additionally, there is a substantial administrative overhead for obtaining signed certificates, as CAs must be able to confirm that they are issuing the signed certificates to the correct party. Furthermore, in many cases endpoints' IP addresses and host names are dynamic: They may be obtained from DHCP, for example. It is impractical to obtain a CA-signed certificate valid for the duration of a DHCP lease. For such hosts, self-signed certificates are usually the only option. This specification defines a mechanism that allows self-signed certificates can be used securely, provided that the integrity of the SDP description is assured. It provides for endpoints to include a secure hash of their certificate, known as the "certificate fingerprint," within the session description. As long as the fingerprint of the offered certificate matches the one in the session description, end hosts can trust even self-signed certificates.

The rest of this document is laid out as follows. An overview of the problem and threat model is given in Section 16.3.3. Section 16.3.4 gives the basic mechanism for establishing TLS-based connected-oriented media in SDP. Section 16.3.5 describes the SDP fingerprint attribute, which, assuming that the integrity of SDP content is assured, allows the secure use of self-signed certificates. Section 16.3.6 describes which X.509 certificates are presented and how they are used in TLS. Section 16.3.7 discusses additional security considerations.

16.3.2 Terminology

See Table 2.1.

16.3.3 Overview

This section discusses the threat model that motivates TLS transport for connection-oriented media streams. It also discusses in more detail the need for end systems to use self-signed certificates.

16.3.3.1 SDP Operational Modes

There are two principal operational modes for multimedia sessions: advertised and offer/answer. Advertised sessions are the simpler mode. In this mode, a server publishes, in some manner, an SDP session description of a multimedia session it is making available. The classic example of this mode of operation is the SAP (RFC 2974), in which SDP session descriptions are periodically transmitted to a well-known multicast group. Traditionally, these descriptions involve multicast conferences, but unicast sessions are also possible. (Connection-oriented media, obviously, cannot use multicast.) Recipients of a session description connect to the addresses published in the session description. These recipients may not previously have been known to the advertiser of the session description.

Alternatively, SDP conferences can operate in offer-answer mode (RFC 3264, see Section 3.1). This mode allows two participants in a multimedia session to negotiate the multimedia session. In this model, one participant offers the other a description of the desired session from its perspective, and the other participant answers with the desired session from its own perspective. In this mode, each of the participants in the session has knowledge of the other. This is the mode of operation used by the SIP (RFC 3261, also see Section 1.2).

16.3.3.2 Threat Model

Participants in multimedia conferences often wish to guarantee confidentiality, data integrity, and authentication for their media sessions. This section describes various types of attackers and the ways they attempt to violate these guarantees. It then describes how the TLS protocol can be used to thwart attackers. The simplest type of attacker is one who listens passively to the traffic associated with a multimedia session. This attacker might, for example, be on the same local-area network (LAN) or wireless network as one of the participants in a conference. This sort of attacker does not threaten a connection's data integrity or authentication, and almost any operational mode of TLS can provide media stream confidentiality.

More sophisticated is an attacker who can send his own data traffic over the network, but who cannot modify or redirect valid traffic. In SDP's "advertised" operational mode, this can barely be considered an attack; media sessions are expected to be initiated from anywhere on the network. In SDP's offer-answer mode, however, this type of attack is more serious. An attacker could initiate a connection to one or both of the endpoints of a session, thus impersonating an endpoint, or acting as a man in the middle to listen in on their communications. To thwart these attacks, TLS uses endpoint certificates. As long as the certificates' private keys have not been compromised, the endpoints have an external trusted mechanism (most commonly, a mutually trusted CA) to validate certificates, and the endpoints know what certificate identity to expect, endpoints can be certain that such an attack has not taken place.

Finally, the most serious type of attacker is one who can modify or redirect session descriptions, for example, a compromised or malicious SIP proxy server. Neither TLS itself nor any mechanisms that use it can protect an SDP session against such an attacker. Instead, the SDP description itself must be secured through some mechanism; SIP, for example, defines how S/MIME (RFC 3851 made obsolete by RFC 5750) can be used to secure session descriptions.

16.3.3.3 The Need for Self-Signed Certificates

SDP session descriptions are created by any endpoint that needs to participate in a multimedia session. In many cases, such as SIP phones, such endpoints have dynamically configured IP addresses and host names and must be deployed with nearly zero configuration. For such an endpoint, it is for practical purposes impossible to obtain a certificate signed by a well-known CA. If two endpoints have no prior relationship, self-signed certificates cannot generally be trusted, as there is no guarantee that an attacker is not launching a man-in-the-middle attack. Fortunately, however, if the integrity of SDP session descriptions can be assured, it is possible to consider those SDP descriptions themselves as a prior relationship: certificates can be securely described in the session description itself. This is done by providing a secure hash of a certificate, or "certificate fingerprint," as an SDP attribute; this mechanism is described in Section 16.3.5.

16.3.3.4 Example SDP Description for TLS Connection

Figure 16.1 illustrates an SDP offer that signals the availability of a T.38 fax session over TLS. For the purpose of brevity, the main portion of the session description is omitted in the example, showing only the "m=" line and its attributes. (This example is the same as the first one in RFC 4145 (see Section 10.1), except for the proto parameter and the fingerprint attribute.) See subsequent sections for explanations of the example's TLS-specific attributes. (Note: Due to RFC formatting conventions, this document splits SDP across lines whose content would exceed 72 characters. A backslash character marks where this line folding has taken place. This backslash and its trailing CRLF and whitespace would not appear in actual SDP content.)

```
m=image 54111 TCP/TLS t38
c=IN IP4 192.0.2.2
a=setup:passive
a=connection:new
a=fingerprint:SHA-1 \
        4A:AD:B9:B1:3F:82:18:3B:54:02:12:DF:3E:5D:49:6B:19:E5:7C:AB
```

Figure 16.1 Example SDP description offering a TLS media stream.

16.3.4 Protocol Identifiers

The "m=" line in SDP specifies, among other items, the transport protocol to be used for the media in the session. See the "Media Descriptions" section of SDP (RFC 4566, see Section 2) for a discussion on transport protocol identifiers. This specification defines a new protocol identifier, "TCP/TLS," which indicates that the media described will use the Transport Layer Security protocol (RFC 4346 made obsolete by RFC 5246) over TCP. (Using TLS over other transport protocols is not discussed in this document.) The "TCP/TLS" protocol identifier describes only the transport protocol, not the upper-layer protocol. An "m=" line that specifies "TCP/TLS" MUST further qualify the protocol using a fmt identifier to indicate the application being run over TLS. Media sessions described with this identifier follow the procedures defined in RFC 4145 (see Section 10.1). They also use the SDP attributes defined in that specification, "setup" and "connection."

16.3.5 Fingerprint Attribute

Parties to a TLS session indicate their identities by presenting authentication certificates as part of the TLS handshake procedure. Authentication certificates are X.509 certificates [1], as profiled by RFC 3279, RFC 3280 (made obsolete by RFC 5280), and RFC 4055. In order to associate media streams with connections and to prevent unauthorized barge-in attacks on the media streams, endpoints MUST provide a certificate fingerprint. If the X.509 certificate presented for the TLS connection matches the fingerprint presented in the SDP, the endpoint can be confident that the author of the SDP is indeed the initiator of the connection. A certificate fingerprint is a secure one-way hash of the distinguished encoding rules form of the certificate. (Certificate fingerprints are widely supported by tools that manipulate X.509 certificates; for instance, the command "openssl x509-fingerprint" causes the command-line tool of the openssl package to print a certificate fingerprint, and the certificate managers for Mozilla and Internet Explorer display them when viewing the details of a certificate.)

All SDP ABNF syntaxes are provided in Section 2.9. However, we have repeated this here for convenience. A fingerprint is represented in SDP as an attribute (an "a=" line). It consists of the name of the hash function used, followed by the hash value itself. The hash value is represented as a sequence of uppercase hexadecimal bytes, separated by colons. The number of bytes is defined by the hash function. (This is the syntax used by openssl and by the browsers' certificate managers. It is different from the syntax used to represent hash values in, for example, HTTP digest authentication (RFC 2617 made obsolete by RFC 7230, RFC 7231, RFC 7232, RFC 7233, RFC 7234, and RFC 7235), which uses unseparated lowercase hexadecimal bytes. It was felt that consistency with other applications of fingerprints was more important.) The formal syntax of the fingerprint attribute is given in ABNF (RFC 4234 made obsolete by RFC 5234) in Figure 16.2. This syntax extends the BNF syntax of SDP (RFC 4566, see Section 2/2.9).

A certificate fingerprint MUST be computed using the same one-way hash function used in the certificate's signature algorithm. (This ensures that the security properties required for the certificate also apply for the fingerprint. It also guarantees that the fingerprint will be usable by the other endpoint, as long as the certificate itself is.) Following RFC 3279 as updated by RFC 4055, therefore, the defined hash functions are "SHA-1" [2] (RFC 3174), "SHA-224" [2], "SHA-256" [2], "SHA-384" [2], "SHA-512" [2], "MD5" (RFC 1321), and "MD2" (RFC 1319 was made obsolete by RFC 6149), with "SHA-1" preferred. A new IANA registry of Hash Function Textual Names, specified in Section 16.3.8, allows for the addition of future tokens, but they may only be added if they are included in RFC s that update or obsolete RFC 3279. Self-signed certificates (for which

```
attribute                = / fingerprint-attribute
fingerprint-attribute    = "fingerprint" ":" hash-func SP fingerprint
hash-func                = "sha-1" / "sha-224" / "sha-256" /
                           "sha-384" / "sha-512" /
                           "md5" / "md2" / token
                                ; Additional hash functions can only come
                                ; from updates to RFC 3279
fingerprint              = 2UHEX *(":" 2UHEX)
                                ; Each byte in upper-case hex, separated
                                ; by colons.
UHEX                     = DIGIT / %x41-46 ; A-F uppercase
```

Figure 16.2 ABFN syntax for the fingerprint attribute (Copyright IETF).

legacy certificates are not a consideration) MUST use one of the Federal Information Processing Standards (FIPS) 180 algorithms (SHA-1, SHA-224, SHA-256, SHA-384, or SHA-512) as their signature algorithm, and thus MUST use it to calculate certificate fingerprints. The fingerprint attribute may be either a session-level or a media-level SDP attribute. If it is a session-level attribute, it applies to all TLS sessions for which no media-level fingerprint attribute is defined.

16.3.6 Endpoint Identification

16.3.6.1 Certificate Choice

An X.509 certificate binds an identity and a public key. If SDP describing a TLS session is transmitted over a mechanism that provides integrity protection, a certificate asserting any syntactically valid identity MAY be used. For example, an SDP description sent over HTTP/TLS (RFC 2818) or secured by S/MIME (RFC 3851 was made obsolete by RFC 5751) MAY assert any identity in the certificate securing the media connection. Security protocols that provide only hop-by-hop integrity protection (e.g., the sips protocol RFC 3261, also see Section 1.2, SIP over TLS) are considered sufficiently secure to allow the mode in which any valid identity is accepted. However, see Section 16.3.7 for a discussion of some security implications of this fact. In situations where the SDP is not integrity-protected, however, the certificate provided for a TLS connection MUST certify an appropriate identity for the connection. In these scenarios, the certificate presented by an endpoint MUST certify either the SDP connection address, or the identity of the creator of the SDP message, as follows:

- If the connection address for the media description is specified as an IP address, the endpoint MAY use a certificate with an iPAddress subjectAltName that exactly matches the IP in the connection-address in the session description's "c=" line. Similarly, if the connection address for the media description is specified as a fully qualified domain name, the endpoint MAY use a certificate with a dNSName subjectAltName matching the specified "c=" line connection-address exactly. (Wildcard patterns MUST NOT be used.)
- Alternately, if the SDP session description was transmitted over a protocol (such as SIP, RFC 3261, also see Section 1.2) for which the identities of session participants are defined by uniform resource identifiers (URIs), the endpoint MAY use a certificate with a uniformResourceIdentifier subjectAltName corresponding to the identity of the endpoint that generated the SDP. The details of what URIs are valid are dependent on the transmitting protocol. (For more details on the validity of URIs, see Section 16.3.7). Identity matching is performed using the matching rules specified by RFC 3280 (made obsolete by RFC 5280).

If more than one identity of a given type is present in the certificate (e.g., more than one dNSName name), a match in any one of the set is considered acceptable. To support the use of certificate caches, as described in Section 16.3.7, endpoints SHOULD consistently provide the same certificate for each identity they support.

16.3.6.2 Certificate Presentation

In all cases, an endpoint acting as the TLS server (i.e., one taking the "setup:passive" role, in the terminology of connection-oriented media) MUST present a certificate during TLS initiation, following the rules presented in Section 16.3.6.1. If the certificate does not match the original fingerprint, the client endpoint MUST terminate the media connection with a bad_certificate error. If the SDP offer/answer model (RFC 3264, see Section 3.1) is being used, the client (the endpoint with the 'setup:active' role) MUST also present a certificate following the rules of Section 16.3.6.1. The server MUST request a certificate, and if the client does not provide one or if the certificate does not match the provided fingerprint, the server endpoint MUST terminate the media connection with a bad_certificate error.

Note that when the offer/answer model is being used, it is possible for a media connection to outrace the answer back to the offerer. Thus, if the offerer has offered a "setup:passive" or "setup:actpass" role, it MUST (as specified in RFC 4145, see Section 10.1) begin listening for an incoming connection as soon as it sends its offer.

However, it MUST NOT assume that the data transmitted over the TLS connection is valid until it has received a matching fingerprint in an SDP answer. If the fingerprint, once it arrives, does not match the client's certificate, the server endpoint MUST terminate the media connection with a bad_certificate error, as stated in the previous paragraph. If offer/answer is not being used (e.g., if the SDP was sent over the SAP, RFC 2974), there is no secure channel available for clients to communicate certificate fingerprints to servers. In this case, servers MAY request client certificates, which SHOULD be signed by a well-known CA or MAY allow clients to connect without a certificate.

16.3.7 Security Considerations

The security procedures that need to be considered in dealing with the basic SDP messages described in this section (RFC 4572) are provided in Section 17.14.3.

16.3.8 Internet Assigned Numbers Authority Considerations

SDP parameters that are described here have been registered with IANA. We have not included the IANA registration procedures here for the sake of brevity. The procedures for registration of new SDP parameters are described in RFC 4572.

16.4 Security Preconditions for SDP Media Streams

This section describes RFC 5027 that defines a new security precondition for the SDP precondition framework described in RFCs 3312 and 4032. A security precondition can be used to delay session establishment or modification until media stream security for a secure media stream has been negotiated successfully.

16.4.1 Introduction

The concept of a SDP (RFC 4566, see Section 2) precondition is defined in RFC 3312 that has been updated by RFCs 4032 and 5027. A precondition is a condition that has to be satisfied for a given media stream in order for session establishment or modification to proceed. When a (mandatory) precondition is not met, session progress is delayed until the precondition is satisfied or the session establishment fails. For example, RFC 3312 defines the Quality-of-Service precondition, which is used to ensure availability of network resources prior to establishing (i.e., alerting) a call. Media streams can either be provided in clear text and with no integrity protection, or some kind of media security can be applied (e.g., confidentiality and/or message integrity). For example, the audio/video profile of the RTP (RFC 3551) is normally used without any security services whereas the SRTP (RFC 3711) is always used with security services. When media stream security is being negotiated (e.g., using the mechanism defined in SDP security descriptions) (RFC 4568, see Section 16.2), both the offerer and the answerer (RFC 3264, see Section 3.1) need to know the cryptographic parameters being used for the media stream; the offerer may provide multiple choices for the cryptographic parameters, or the cryptographic parameters selected by the answerer may differ from those of the offerer (e.g., the key used in one direction versus the other). In such cases, to avoid media clipping, the offerer needs to receive the answer prior to receiving any media packets from the answerer. This can be achieved by using a security precondition, which ensures the successful negotiation of media stream security parameters for a secure media stream prior to session establishment or modification.

16.4.2 Notational Conventions

See Table 2.1.

16.4.3 Security Precondition Definition

The semantics for a security precondition are that the relevant cryptographic parameters (cipher, key, etc.) for a secure media stream are known to have been negotiated in the direction(s) required. If the security precondition is used with a nonsecure media stream, the security precondition is by definition satisfied. A secure media stream is here defined as a media stream that uses some kind of security service (e.g., message integrity, confidentiality, or both), regardless of the cryptographic strength of the mechanisms being used. As an extreme example of this, SRTP using the NULL encryption algorithm and no message integrity would be considered a secure media stream whereas use of plain RTP would not. Note, though, that Section 9.5 of RFC 3711 discourages the use of SRTP without message integrity.

Security preconditions do not guarantee that an established media stream will be secure. They merely guarantee that the recipient of the media stream packets will be able to perform any relevant decryption and integrity checking on those media stream packets. Please refer to Section 16.4.5 for further security considerations. The security precondition type is defined by the string "sec"; hence, we modify the grammar found in RFC 3312 as follows:

```
precondition-type = "sec" / "qos" / token
```

RFC 3312 defines support for two kinds of status types, namely, segmented and end to end. The security precondition-type defined here MUST be used with the end-to-end status type; use of the segmented status type is undefined.

A security precondition can use the strength-tag "mandatory," "optional," or "none."

When a security precondition with a strength-tag of "mandatory" is received in an offer, session establishment or modification MUST be delayed until the security precondition has been met (i.e., the relevant cryptographic parameters (cipher, key, etc.) for a secure media stream are known to have been negotiated in the direction(s) required). When a mandatory security precondition is offered, and the answerer cannot satisfy the security precondition (e.g., because the offer was for a secure media stream, but it did not include the necessary parameters to establish the secure media stream keying material), the offered media stream MUST be rejected as described in RFC 3312.

The delay of session establishment defined here implies that alerting the called party MUST NOT occur and media for which security is being negotiated MUST NOT be exchanged until the precondition has been satisfied. In cases where secure media and other nonmedia data is multiplexed on a media stream (e.g., when interactive connectivity establishment [ICE] [RFC 5245] is being used), the nonmedia data is allowed to be exchanged prior to the security precondition being satisfied. When a security precondition with a strength-tag of "optional" is received in an offer, the answerer MUST generate its answer SDP as soon as possible. Since session progress is not delayed in this case, the answerer does not know when the offerer is able to process secure media stream packets and hence clipping may occur. If the answerer wants to avoid clipping and delay session progress until he knows the offerer has received the answer, the answerer MUST increase the strength of the security precondition by using a strength-tag of "mandatory" in the answer.

Note that use of a mandatory precondition in an offer requires the presence of an SIP "Require:" header field containing the option tag "precondition": Any SIP user agent that does not support a mandatory precondition will consequently reject such requests (which also has unintended ramifications for SIP forking that are known as the heterogeneous error response forking problem (see e.g., [3]). To get around this, an optional security precondition and the SIP "Supported:" header field containing the option tag "precondition" can be used instead. When a security precondition with a strength-tag of "none" is received, processing continues as usual. The "none" strength-tag merely indicates that the offerer supports the following security precondition; the answerer MAY upgrade the strength-tag in the answer as described in RFC 3312. The direction tags defined in RFC 3312 are interpreted as follows:

- **send**: Media stream security negotiation is at a stage where it is possible to send media packets to the other party and the other party will be able to process them correctly from a security point of view (i.e., decrypt and/or integrity check them as necessary). The definition of "media packets" includes all packets that make up the media stream. In the case of SRTP, for example, it includes SRTP as well as SRTCP. When media and nonmedia packets are multiplexed on a given media stream (e.g., when ICE is being used), the requirement applies to the media packets only.
- **recv**: Media stream security negotiation is at a stage where it is possible to receive and correctly process media stream packets sent by the other party from a security point of view. The precise criteria for determining when the other party is able to correctly process media stream packets from a security point of view depend on the secure media stream protocol being used as well as the mechanism by which the required cryptographic parameters are negotiated.

We here provide details for SRTP negotiated through SDP security descriptions as defined in RFC 4568 (see Section 16.2):

- When the offerer requests the "send" security precondition, it needs to receive the answer before the security precondition is satisfied. The reason for this is twofold. First, the offerer needs to know where to send the media. Second, in the case where alternative cryptographic parameters are offered, the offerer needs to know which set was selected. The answerer does not know when the answer is actually received by the offerer (which in turn will satisfy the precondition), and hence the answerer needs to use the confirm-status attribute (RFC 3312). This will make the offerer generate a new offer showing the updated status of the precondition.
- When the offerer requests the "recv" security precondition, it also needs to receive the answer before the security precondition is satisfied. The reason for this is straightforward: The answer contains the cryptographic parameters that will be used by the answerer for sending media to the offerer; prior to receipt of these cryptographic parameters, the offerer is unable to authenticate or decrypt such media. When security preconditions are used with the key management extensions for the SDP (RFC 4567, see Section 16.1), the details depend on the actual key management protocol being used.

After an initial offer/answer exchange in which the security precondition is requested, any subsequent offer/answer sequence for the purpose of updating the status of the precondition for a secure media stream SHOULD use the same key material as the initial offer/answer exchange. This means that the key-mgmt attribute lines (RFC 4567, see Section 16.1) or crypto attribute lines (RFC 4568, see Section 16.2) in SDP offers, that are sent in response to SDP answers containing a confirm-status field (RFC 3312) SHOULD repeat the same data as that sent in the previous SDP offer. If applicable to the key management protocol or SDP security description, the SDP answers to these SDP offers SHOULD repeat the same data in the key-mgmt attribute lines (RFC 4567, see Section 16.1) or crypto attribute lines (RFC 4568, see Section 16.2) as that sent in the previous SDP answer.

Of course, this duplication of key exchange during precondition establishment is not to be interpreted as a replay attack. This issue may be solved if, for example, the SDP implementation recognizes that the key management protocol data is identical in the second offer/answer exchange and avoids forwarding the information to the security layer for further processing. Offers with security preconditions in re-INVITEs or UPDATEs follow the rules given in Section 6 of RFC 3312 (i.e.: "Both user agents SHOULD continue using the old session parameters until all the mandatory preconditions are met. At that moment, the user agents can begin using the new session parameters").

At that moment, we furthermore require that user agents MUST start using the new session parameters for media packets being sent. The user agents SHOULD be prepared to process media packets received with either the old or the new session parameters for a short period of time to accommodate media packets in transit. Note that this may involve iterative security processing of the received media packets during that period of time. Section 3.1.8 of this book (that is, Section 8 in RFC 3264) lists several techniques to help alleviate the problem of determining when a received media packet was generated according to the old or new offer/answer exchange.

16.4.4 Examples

16.4.4.1 SDP Security Descriptions Example

The call flow of Figure 16.3 shows a basic session establishment using the SIP (RFC 3261, also see Section 1.2) and SDP security descriptions (RFC 4568, see Section 16.2) with security descriptions for the secure media stream (SRTP in this case).

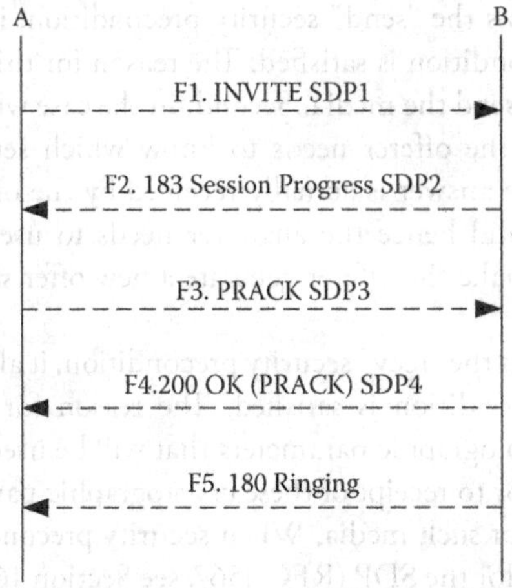

Figure 16.3 Security preconditions with SDP security descriptions example (Copyright IETF).

The SDP descriptions of this example are shown here; we have omitted the details of the SDP security descriptions as well as any SIP details for clarity of the security precondition described here:

SDP1: A includes a mandatory end-to-end security precondition for both the send and receive directions in the initial offer as well as a "crypto" attribute (see (RFC 4568, see Section 16.2), which includes keying material that can be used by A to generate media packets. Since B does not know any of the security parameters yet, the current status (see RFC 3312) is set to "none." A's local status table (see RFC 3312) for the security precondition is as follows:

Direction	*Current*	*Desired Strength*	*Confirm*
send	no	mandatory	no
recv	no	mandatory	no

and the resulting offer SDP is:

```
m=audio 20000 RTP/SAVP 0
c=IN IP4 192.0.2.1
a=curr:sec e2e none
a=des:sec mandatory e2e sendrecv
a=crypto:foo...
```

SDP2: When B receives the offer and generates an answer, B knows the (send and recv) security parameters of both A and B. From a security perspective, B is now able to receive media from A, so B's "recv" security precondition is "yes." However, A does not know any of B's SDP information, so B's "send" security precondition is "no." B's local status table therefore looks as follows:

Direction	*Current*	*Desired Strength*	*Confirm*
send	No	mandatory	no
recv	Yes	mandatory	no

B requests A to confirm when A knows the security parameters used in the send and receive directions (it would suffice for B to ask for confirmation of A's send direction only) and hence the resulting answer SDP becomes:

```
m=audio 30000 RTP/SAVP 0
c=IN IP4 192.0.2.4
a=curr:sec e2e recv
a=des:sec mandatory e2e sendrecv
a=conf:sec e2e sendrecv
a=crypto:bar...
```

SDP3: When A receives the answer, A updates its local status table based on the rules in RFC 3312. A knows the security parameters of both the send and receive directions; hence, A's local status table is updated as follows:

Direction	*Current*	*Desired Strength*	*Confirm*
send	yes	mandatory	yes
recv	yes	mandatory	yes

Since B requested confirmation of the send and recv security preconditions, and both are now satisfied, A immediately sends an updated offer (3) to B showing that the security preconditions are satisfied:

```
m=audio 20000 RTP/SAVP 0
c=IN IP4 192.0.2.1
a=curr:sec e2e sendrecv
a=des:sec mandatory e2e sendrecv
a=crypto:foo...
```

Note that we here use PRACK (RFC 3262, see Section 3.1) instead of UPDATE (RFC 3311) since the precondition is satisfied immediately, and the original offer/answer exchange is complete.

SDP4: Upon receiving the updated offer, B updates its local status table based on the rules in RFC 3312, which yields the following:

Direction	*Current*	*Desired Strength*	*Confirm*
send	yes	mandatory	no
recv	yes	mandatory	no

B responds with an answer (4) that contains the current status of the security precondition (i.e., sendrecv) from B's point of view:

```
m=audio 30000 RTP/SAVP 0
c=IN IP4 192.0.2.4
a=curr:sec e2e sendrecv
```

```
a=des:sec mandatory e2e sendrecv
a=crypto:bar...
```

B's local status table indicates that all mandatory preconditions have been satisfied; hence, session establishment resumes; B returns a 180 (Ringing) response (5) to indicate alerting.

16.4.4.2 Key Management Extension for SDP Example

The call flow of Figure 16.4 shows a basic session establishment using the SIP (RFC 3261, also see Section 1.2) and key management extensions for SDP (RFC 4567, see Section 16.1) with security descriptions for the secure media stream (SRTP in this case):

The SDP descriptions of this example are shown below; we show an example use of MIKEY (RFC 3830) with the key management extensions; however, we have omitted the details of the MIKEY parameters as well as any SIP details for clarity of the security precondition described here:

SDP1: A includes a mandatory end-to-end security precondition for both the send and receive directions in the initial offer as well as a "key-mgmt" attribute (see RFC 4567, Section 16.1), which includes keying material that can be used by A to generate media packets. Since B does not know any of the security parameters yet, the current status (see RFC 3312) is set to "none." A's local status table (see RFC 3312) for the security precondition is as follows:

Direction	*Current*	*Desired Strength*	*Confirm*
send	no	mandatory	no
recv	no	mandatory	no

and the resulting offer SDP is:

```
m=audio 20000 RTP/SAVP 0
c=IN IP4 192.0.2.1
```

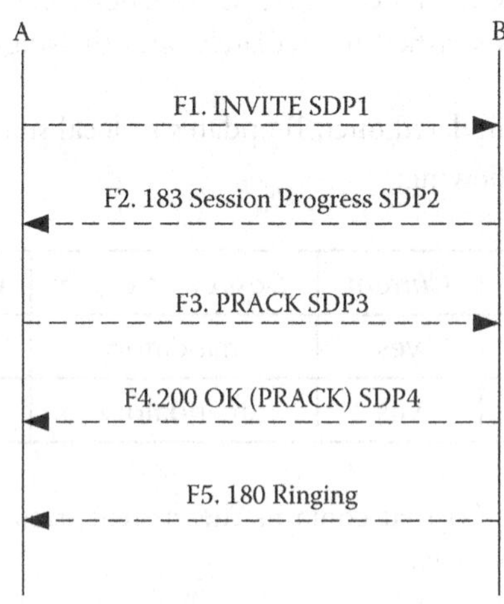

Figure 16.4 Security preconditions with key management extensions for SDP example (Copyright IETF).

```
a=curr:sec e2e none
a=des:sec mandatory e2e sendrecv
a=key-mgmt:mikey AQAFgM0X...
```

SDP2: When B receives the offer and generates an answer, B knows the (send and recv) security parameters of both A and B. B generates keying material for sending media to A; however, A does not know B's keying material, so the current status of B's "send" security precondition is "no." B does know A's SDP information, so B's "recv" security precondition is "yes." B's local status table therefore looks as follows:

Direction	*Current*	*Desired Strength*	*Confirm*
send	no	mandatory	no
recv	yes	mandatory	no

B requests that A confirm when A knows the security parameters used in the send and receive directions; hence, the resulting answer SDP becomes:

```
m=audio 30000 RTP/SAVP 0
c=IN IP4 192.0.2.4
a=curr:sec e2e recv
a=des:sec mandatory e2e sendrecv
a=conf:sec e2e sendrecv
a=key-mgmt:mikey AQAFgM0X...
```

Note that the actual MIKEY data in the answer differs from that in the offer; however, we have only shown the initial and common part of the MIKEY value in the above (Section 16.4.4.2).

SDP3: When A receives the answer, A updates its local status table based on the rules in RFC 3312. A now knows all the security parameters of both the send and receive directions; hence, A's local status table is updated as follows:

Direction	*Current*	*Desired Strength*	*Confirm*
send	yes	mandatory	yes
recv	yes	mandatory	yes

Since B requested confirmation of the send and recv security preconditions, and both are now satisfied, A immediately sends an updated offer (3) to B showing that the security preconditions are satisfied:

```
m=audio 20000 RTP/SAVP 0
c=IN IP4 192.0.2.1
a=curr:sec e2e sendrecv
a=des:sec mandatory e2e sendrecv
a=key-mgmt:mikey AQAFgM0X...
```

SDP4: Upon receiving the updated offer, B updates its local status table based on the rules in RFC 3312, which yields the following:

Direction	*Current*	*Desired Strength*	*Confirm*
Send	yes	mandatory	no
Recv	yes	mandatory	no

B responds with an answer (4) that contains the current status of the security precondition (i.e., sendrecv) from B's point of view:

```
m=audio 30000 RTP/SAVP 0
c=IN IP4 192.0.2.4
a=curr:sec e2e sendrecv
a=des:sec mandatory e2e sendrecv
a=key-mgmt:mikey AQAFgM0X...
```

B's local status table indicates that all mandatory preconditions have been satisfied; hence, session establishment resumes; B returns a 180 (Ringing) response (5) to indicate alerting.

16.4.5 Security Considerations

The security procedures that need to be considered in dealing with the basic SDP messages described in this section (RFC 5027) are provided in Section 17.14.4.

16.4.6 IANA Considerations

IANA has registered an RFC 3312 precondition type called "sec" with the name "Security precondition." The reference for this precondition type is the current document (RFC 5027).

16.5 Media Description for IKE in SDP

This section describes RFC 6193 that extends the protocol identifier of the SDP so that it can negotiate use of the IKE for media sessions in the SDP offer/answer model. In addition, it defines a method to boot up IKE and generate IPsec security associations using a self-signed certificate. In fact, RFC 6193 specifically has shown how to establish a media session that represents a virtual private network using the call control protocol SIP for the purpose of on-demand media/application sharing between peers.

16.5.1 Applicability Statement

This document provides information about a deployed use of the SIP (RFC 3261, also see Section 1.2) for the Internet community. It is not currently an IETF standards track proposal. The mechanisms in this document use SIP as a name resolution and authentication mechanism to initiate an Internet Key Exchange Protocol (RFC 5996 made obsolete by RFC 7296) session. The purpose of this document is to establish an on-demand virtual private network (VPN) to a home router that does not have a fixed IP address using self-signed certificates. It is only applicable under the condition that the integrity of the SDP (RFC 4566, see Section 2) is assured. The method of ensuring this integrity of SDP is outside the scope of this document. This document specifies the process in which a pair of SIP user agents resolve each other's names, exchange the fingerprints of their self-signed certificates securely, and agree to establish an IPsec-based VPN (RFC 4301). However, this document does not make any modifications to the specifications of IPsec/IKE. Despite the limitations of the conditions under which this document can be applied, there are sufficient use cases in which this specification is helpful, such as the following:

- Sharing media using a framework developed by Digital Living Network Alliance (DLNA) or similar protocols over VPN between two user devices
- Accessing remote desktop applications over VPN initiated by SIP call. As an additional function of click-to-call, a customer service agent can access a customer's PC remotely to troubleshoot the problem while talking with the customer over the phone.
- Accessing and controlling medical equipment (medical robotics) remotely to monitor the elderly in a rural area (remote care services)
- Using a LAN-based gaming protocol based on peer-to-peer rather than via a gaming server.

16.5.2 Introduction

This section describes the problem in accessing home networks and provides an overview of the proposed solution.

16.5.2.1 Problem Statement

Home servers and network-capable consumer electronic devices have been widely deployed. People using such devices are willing to share content and applications and are therefore seeking ways to establish multiple communication channels with each other. However, there are several obstacles to remote home access. It is often not possible for a device outside the home network to connect to another device inside the home network because the home device is behind a NAT or firewall that allows outgoing connections but blocks incoming connections. One effective solution for this problem is VPN remote access to the NAT device, which is usually a home router. With this approach, once the external device joins the home network securely, establishing connections with all the devices inside the home will become easy because popular LAN-based communication methods such as DLNA can be used transparently. However, there are more difficult cases in which a home router itself is located behind the NAT. In such cases, it is also necessary to consider NAT traversal of the remote access to the home router. In many cases, because the global IP address of the home router is not always fixed, it is necessary to make use of an effective name resolution mechanism.

In addition, there is the problem of how a remote client and a home router authenticate each other over IKE to establish IPsec for remote access. It is not always possible for the two devices to securely exchange a preshared key in advance. Administrative costs can make it impractical to distribute authentication certificates signed by a well-known root CA to all the devices. In addition, it is inefficient to publish a temporary certificate to a device that does not have a fixed IP address or hostname. To resolve these authentication issues, this document proposes a mechanism that enables the devices to authenticate each other using self-signed certificates.

16.5.2.2 Approach to Solution

This document proposes the use of SIP as a name resolution and authentication mechanism because of three main advantages:

- Delegation of authentication to third party: Devices can be free from managing their signed certificates and whitelists by taking advantage of authentication and authorization mechanisms supported by SIP.
- User Datagram Protocol (UDP) hole punching for IKE/IPsec: SIP has a cross-NAT rendezvous mechanism, and ICE (RFC 5245) has a function to open ports through the NAT. The combination of these effective functions can be used for general applications as well as real-time media. It is difficult to set up a session between devices without SIP if the devices are behind various types of NAT.
- Reuse of existing SIP infrastructure: SIP servers are widely distributed as scalable infrastructure; it is quite practical to reuse them without any modifications. Today, SIP is applied to not only Voice over IP but also various applications and is recognized as a general protocol for session initiation. Therefore, it can also be used to initiate IKE/IPsec sessions.

However, a specification uses a self-signed certificate for authentication in the SIP/SDP framework. "Connection-Oriented Media Transport over the Transport Layer Security (TLS) Protocol in the Session Description Protocol (SDP)" (RFC 4572, see Section 16.3) (hereafter referred to as comedia-tls) specifies a method to exchange the fingerprint of a self-signed certificate to establish a TLS (RFC 5246) connection. This specification defines a mechanism by which self-signed certificates can be used securely, provided that the integrity of the SDP description is assured. Because a certificate itself is used for authentication not only in TLS but also in IKE, this mechanism will be applied to the establishment of an IPsec security association (SA) by extending the protocol identifier of SDP so that it can specify IKE.

One easy method for protecting the integrity of the SDP description, which is the premise of this specification, is to use the SIP identity (RFC 4474) mechanism. This approach is also referred to in RFC 5763. Because the SIP identity mechanism can protect the integrity of a body part as well as the value of the From header in a SIP request by using a valid Identity header, the receiver of the request can establish secure IPsec connections with the sender by confirming that the hash value of the certificate sent during IKE negotiation matches the fingerprint in the SDP. Although SIP identity does not protect the identity of the receiver of the SIP request, SIP-connected identity (RFC 4916) does. Note that the possible deficiencies discussed in [4] could affect this specification if SIP identity is used for the security mechanism.

Considering the above (Sections 16.5.2.1 and 16.5.2.2) background, this specification defines new media formats "ike-esp" and "ike-esp-udpencap," which can be used when the protocol identifier is "udp" to enable the negotiation of using IKE for media sessions over SDP exchange on the condition that the integrity of the SDP description is assured. It also specifies the method to set up an IPsec SA by exchanging fingerprints of self-signed certificates based on comedia-tls, and it notes the example of SDP offer/answer (RFC 3264, see Section 3.1) and the points that should be taken care of by implementation. Because there is a chance that devices are behind NAT, this document also covers the method for combining IKE/IPsec NAT-Traversal (RFCs 3947 and 3948) with ICE. In addition, it defines the attribute "ike-setup" for IKE media sessions, similar to the "setup" attribute for TCP-based media transport defined in RFC 4145 (see Section 10.1). This attribute is used to negotiate the role of each endpoint in the IKE session.

16.5.2.3 Alternative Solution under Prior Relationship between Two Nodes

Under quite limited conditions, certificates signed by trusted third parties or preshared keys between endpoints could be used for authentication in IKE, using SIP servers only for name resolution and authorization of session initiation. Such limited cases are addressed in Section 16.5.8.

16.5.2.4 Authorization Model

In this document, SIP servers are used for the authorization of each SIP call. The actual media sessions of IPsec/IKE are authorized not by SIP servers but by the remote client and the home router based on the information in SIP/SDP. For example, the home router recognizes the remote client with its SIP-URI and IP address in the SDP. If it decides to accept the remote client as a peer of a VPN session, it will accept the following IKE session. Then, during the IKE negotiation, the certificate fingerprint in the SDP is compared with the certificate exchanged in the IKE session. If they match, IKE negotiation continues. Only a successful IKE negotiation establishes an IPsec session with the remote peer.

16.5.2.5 Conventions Used in This Document

The conventions used in RFC 6193 are per IETF RFC 2119.

16.5.3 Protocol Overview

Figure 16.5 shows a case of VPN remote access from a device outside the home to a home router whose IP address is not fixed. In this case, the external device, a remote client, recognizes the address of record of the home router but does not have any information about its contact address and certificate. Generally, establishing an IPsec SA dynamically and securely in this situation is difficult. However, as specified in comedia-tls (RFC 4572, see Section 16.3), if the integrity of SDP session descriptions is assured, it is possible for the home router and the remote client to have a prior relationship with each other by exchanging certificate fingerprints (i.e., secure one-way hashes of the distinguished encoding rules form of the certificates).

1. Both remote client and home router generate secure signaling channels. They may REGISTER to SIP proxy using TLS.
2. Remote client sends an offer SDP with an INVITE request to home router, and home router returns an answer SDP with a reliable response (e.g., 200 OK). Both exchange the fingerprints of their self-signed certificates in SDP during this transaction.
3. Remote client does not accept an answer SDP with an unreliable response as the final response.
4. After the SDP exchange, remote client, which has the active role, initiates IKE with home router, which has the passive role, to establish an IPsec SA. Both validate that the certificate presented in the IKE exchange has a fingerprint that matches the fingerprint from SDP. If they match, IKE negotiation proceeds as normal.
5. Remote client joins the home network. By this method, the self-signed certificates of both parties are used for authentication in IKE, but SDP itself is not concerned with all the negotiations related to key-exchange, such as those of encryption and authentication algorithms.

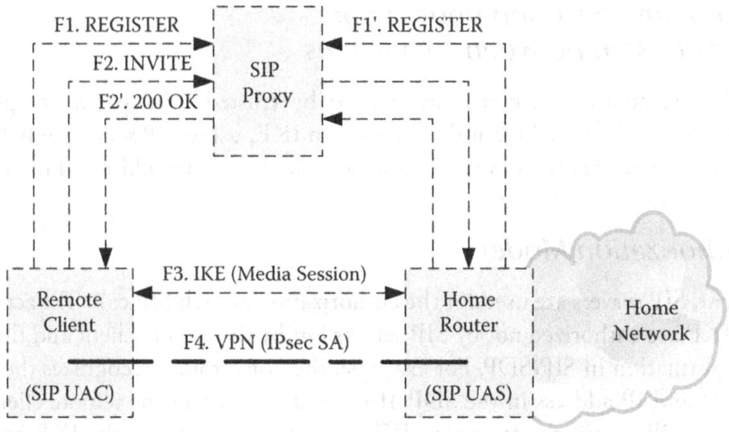

Figure 16.5 Remote access to home network (Copyright IETF).

These negotiations are up to IKE. In many cases where IPsec is used for remote access, a remote client needs to dynamically obtain a private address inside the home network while initiating the remote access. Therefore, the IPsec security policy needs to be set dynamically at the same time. However, such a management function of the security policy is the responsibility of the high-level application. SDP is not concerned with it. The roles of SDP here are to determine the IP addresses of both parties used for IKE connection with c-line in SDP and to exchange the fingerprints of the certificates used for authentication in IKE with the fingerprint attribute in SDP.

16.5.4 Protocol Identifiers

This document defines two SDP media formats for the "udp" protocol under the "application" media type: "ike-esp" and "ike-esp-udpencap." The format "ike-esp" indicates that the media described is IKE for the establishment of an IPsec security association as described in IPsec encapsulating security payload (ESP) (RFC 4303). In contrast, "ike-esp-udpencap" indicates that the media described is IKE, which is capable of NAT traversal for the establishment of UDP encapsulation of IPsec packets through NAT boxes as specified in RFC 3947 and RFC 3948. Even if the offerer and answerer exchange "ike-esp-udpencap," IKE conforming to RFC 3947 and RFC 3948 can end up establishing a normal IPsec tunnel when there is no need to use UDP encapsulation of IPsec. Both the offerer and answerer can negotiate IKE by specifying "udp" in the "proto" field and "ike-esp" or "ike-esp-udpencap" in the "fmt" field in SDP.

All SDP ABNF syntaxes are provided in Section 2.9. However, we have repeated this here for convenience. In addition, this document defines a new attribute "ike-setup," which can be used when the protocol identifier is "udp" and the "fmt" field is "ike-esp" or "ike-esp-udpencap," in order to describe how endpoints should perform the IKE session setup procedure. The "ikesetup" attribute indicates which of the endpoints should initiate the establishment of an IKE session. The "ike-setup" attribute is charset-independent and can be a session- or media-level attribute. The following is the ABNF of the "ike-setup" attribute.

```
ike-setup-attr       =       "a=ike-setup:" role
role                 =       "active" / "passive" / "actpass"
```

- "active": The endpoint will initiate an outgoing session.
- "passive": The endpoint will accept an incoming session.
- "actpass": The endpoint is willing to accept an incoming session or to initiate an outgoing session.

Both endpoints use the SDP offer/answer model to negotiate the value of "ike-setup," following the procedures determined for the "setup" attribute defined in Section 10.1.4.1 (that is, Section 4.1 of RFC 4145). However, "holdconn," as defined in RFC 4145 (see Section 10.1), is not defined for the "ike-setup" attribute.

Offer	*Answer*
active	Passive
passive	Active
actpass	active/passive

The semantics for the "ike-setup" attribute values of "active," "passive," and "actpass" in the offer/answer exchange are the same as those described for the "setup" attribute in Section 10.1.4.1 (that is, Section 4.1 of RFC 4145), except that "ike-setup" applies to an IKE session instead of a TCP connection. The default value of the "ike-setup" attribute is "active" in the offer and "passive" in the answer.

16.5.5 Normative Behavior

In this section, a method of negotiating the use of IKE for media sessions in the SDP offer/answer model is described.

16.5.5.1 SDP Offer and Answer Exchange

An offerer and an answerer negotiate the use of IKE following the usage of the protocol identifiers defined in Section 16.5.4. If IPsec NAT-Traversal is not necessary, the offerer MAY use the media format "ike-esp" to indicate an IKE session. If either of the endpoints that negotiate IKE is behind the NAT, the endpoints need to transmit both IKE and IPsec packets over the NAT. That mechanism is specified in RFC 3947 and RFC 3948: both endpoints encapsulate IPsec-ESP packets with a UDP header and multiplex them into the UDP path that IKE generates. To indicate this type of IKE session, the offerer uses "ike-espudpencap" media lines. In this case, the offerer MAY decide on its transport addresses (combination of IP address and port) before starting IKE, making use of the ICE framework. Because UDPencapsulated ESP packets and IKE packets go through the same UDP hole of a NAT, IPsec NAT-Traversal works if ICE reserves simply one UDP path through the NAT. However, those UDP packets need to be multiplexed with simple traversal of UDP around NAT (STUN) (RFC 5389) packets if ICE is required to use STUN. A method to coordinate IPsec NAT-traversal and ICE is described in Sections 16.5.5.4 and 16.5.5.5.

The offer MAY contain media lines for media other than "ike-esp" or "ike-esp-udpencap.. For example, audio stream may be included in the same SDP to have a voice session when establishing the VPN. This may be useful to verify that the connected device is indeed operated by somebody who is authorized to access it, as described in Section 16.5.9. If that occurs, the negotiation described in this specification occurs only for the "ike-esp" or "ike-esp-udpencap" media lines; other media lines are negotiated and set up normally. If the answerer determines it will refuse the IKE session without beginning the IKE negotiation (e.g., the From address is not on the permitted list), it SHOULD reject the "ike-esp" or "ike-esp-udpencap" media line in the normal manner by setting the port number in the SDP answer to 0 and SHOULD process the other media lines normally (only if it is still reasonable to establish that media without VPN).

If the offerer and the answerer agree to start an IKE session by the offer/answer exchange, they will start the IKE setup. Following the comedia-tls specification (RFC 4572), the fingerprint attribute, which may be either a session- or a media-level SDP attribute, is used to exchange fingerprints of self-signed certificates. If the fingerprint attribute is a session-level attribute, it applies to all IKE sessions and TLS sessions for which no media-level fingerprint attribute is defined. Note that it is possible for an offerer to become the IKE responder and an answerer to become the IKE initiator. For example, when a remote access server (RAS) sends an INVITE to an RAS client, the server may expect the client to become an IKE initiator. In this case, the server sends an offer SDP with ike-setup:passive, and the client returns an answer SDP with ike-setup:active.

16.5.5.2 Maintenance and Termination of VPN Session

If the high-level application recognizes a VPN session as the media session, it MAY discard the IPsec SA and terminate IKE when that media session is terminated by a BYE request. Therefore, the application aware of the VPN session MUST NOT send a BYE request as long as it needs the IPsec SA. On the other hand, if the high-level application detects that a VPN session is terminated, it MAY terminate the media associated with the VPN or the entire SIP session. Session timers in SIP (RFC 4028) MAY be used for the session maintenance of the SIP call, but this does not necessarily ensure that the VPN session is alive. If the VPN session needs session maintenance such as keep-alive and rekeying, it MUST be done utilizing its own maintenance mechanisms. SIP re-INVITE MUST NOT be used for this purpose. Note that each party can cache the certificate of the other party as described in the Security Considerations section of comedia-tls (RFC 4572, see Section 16.3).

16.5.5.3 Forking

Forking to multiple registered instances is outside the scope of this document. At least, it is assumed that a user agent client (UAC) establishes a session with only one user agent server (UAS). Encountering forked answers should be treated as an illegal process, and the UAC should cancel the session.

16.5.5.4 Port Usage

IKE generally uses local UDP port 500, but the IPsec NAT-Traversal specification requires a port transition to local UDP port 4500 during IKE negotiation because IPsec-aware NAT may multiplex IKE sessions using port 500 without changing the port number. If using ICE for IPsec Nat-Traversal, this port transition of IKE means ICE has to generate an additional UDP path for port 4500, and this would be unnecessary overhead. However, IPsec NAT-Traversal allows an IKE session to use local UDP port 4500 from the beginning without using port 500. Therefore, the endpoints SHOULD use their local UDP port 4500 for an IKE session from the beginning, and ICE will only need to generate a UDP path of port 4500. When using ICE, a responder's IKE port observed by an initiator is not necessarily 500 or 4500. Therefore, an IKE initiator MUST allow any destination ports in addition to 500 and 4500 for the IKE packets it sends. An IKE initiator just initiates an IKE session to the port number decided by an SDP offer/answer or ICE.

16.5.5.5 Multiplexing UDP Messages When Using ICE

Conforming to ICE, an offerer and an answerer start a STUN connectivity check after SDP exchange. Then the offerer initiates the IKE session making use of the UDP path generated by STUN packets. In addition, UDP-encapsulated ESP packets are multiplexed into the same UDP path as IKE. Thus, it is necessary to multiplex the three different packets, STUN, IKE, and UDP-encapsulated ESP, into the same UDP path. This section describes how to demultiplex these three packets.

At the first step, the endpoint that received a UDP packet at the multiplexed port MUST check the first 32 bits (bits 0-31) of the UDP payload. If they are all 0, which is defined as a non-ESP marker, that packet MUST be treated as an IKE packet. Otherwise, it is judged as an ESP packet in the IPsec NAT-Traversal specification. It is furthermore necessary to distinguish STUN from ESP. Therefore, the bits 32–63 from the beginning of the UDP payload MUST be checked. If the bits do not match the magic cookie of STUN 0x2112A442 (most packets do not match), the packet is treated as an ESP packet because it is no longer a STUN packet.

However, if the bits do match the magic cookie, an additional test is necessary to determine whether the packet is STUN or ESP. The magic cookie field of STUN overlaps the sequence number field of ESP, so a possibility remains that the sequence number of ESP coincides with 0x2112A442. In this additional test, the validity of the fingerprint attribute of the STUN message MUST be checked. If there is a valid fingerprint in the message, it is judged as a STUN packet; otherwise, it is an ESP packet. The multiplexing of UDP messages logic described here is expressed as follows.

```
if SPI-field-is-all-zeros
     { packet is IKE }
else
     {
         if bits-32-through-63 == stun-magic-cookie-value and
         bits-0-through-1 == 0 and
         bits-2-through-15 == a STUN message type and
         bits-16-through-31 == length of this UDP packet
     {
         fingerprint_found == parse_for_stun_fingerprint();
         if fingerprint_found == 1
         { packet is STUN }
         else
         { packet is ESP }
     }
else
         { packet is ESP }
}
```

16.5.6 Examples

16.5.6.1 Example of SDP Offer and Answer Exchange without IPsec NAT-Traversal

If IPsec NAT-Traversal is not necessary, SDP negotiation to set up IKE is quite simple. Examples of SDP exchange are as follows (Figures 16.6 and 16.7).

(Note: Due to RFC formatting conventions, this document splits SDP across lines whose content would exceed 72 characters. A backslash character marks where this line folding has taken place. This backslash and its trailing CRLF and whitespace would not appear in actual SDP content.)

16.5.6.2 Example of SDP Offer and Answer Exchange with IPsec NAT-Traversal

We consider the following scenario here (Figure 16.8).

```
offer SDP
      ...
      m=application 500 udp ike-esp
      c=IN IP4 192.0.2.10
      a=ike-setup:active
      a=fingerprint:SHA-1 \
      4A:AD:B9:B1:3F:82:18:3B:54:02:12:DF:3E:5D:49:6B:19:E5:7C:AB
      ...
answer SDP
      ...
      m=application 500 udp ike-esp
      c=IN IP4 192.0.2.20
      a=ike-setup:passive
      a=fingerprint:SHA-1 \
      D2:9F:6F:1E:CD:D3:09:E8:70:65:1A:51:7C:9D:30:4F:21:E4:4A:8E
      ...
```

Figure 16.6 SDP example when offerer is an IKE initiator.

```
offer SDP
      ...
      m=application 500 udp ike-esp
      c=IN IP4 192.0.2.10
      a=ike-setup:passive
      a=fingerprint:SHA-1 \
      4A:AD:B9:B1:3F:82:18:3B:54:02:12:DF:3E:5D:49:6B:19:E5:7C:AB
      ...
answer SDP
      ...
      m=application 500 udp ike-esp
      c=IN IP4 192.0.2.20
      a=ike-setup:active
      a=fingerprint:SHA-1 \
      D2:9F:6F:1E:CD:D3:09:E8:70:65:1A:51:7C:9D:30:4F:21:E4:4A:8E
      ...
```

Figure 16.7 SDP example when offerer is an IKE responder.

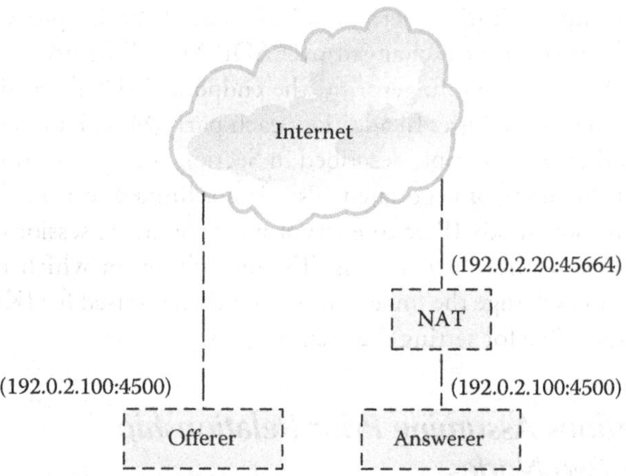

Figure 16.8 NAT-Traversal scenario (Copyright IETF).

As shown above (Figure 16.8), an offerer is on the Internet, but an answerer is behind the NAT. The offerer cannot initiate an IKE session unless the answerer prepares a global routable transport address that accepts IKE packets. In this case, the following offer/answer exchange will take place (Figure 16.9).

16.5.7 Application to IKE

After the fingerprints of both parties are securely shared over the SDP exchange, the IKE initiator MAY start the IKE session with the other party. To follow this specification, a digital signature

```
offer SDP
     ...
     a=ice-pwd:YH75Fviy6338Vbrhrlp8Yh
     a=ice-ufrag:9uB6
     m=application 4500 udp ike-esp-udpencap
     c=IN IP4 192.0.2.10
     a=ike-setup:active
     a=fingerprint:SHA-1 \
     4A:AD:B9:B1:3F:82:18:3B:54:02:12:DF:3E:5D:49:6B:19:E5:7C:AB
     a=candidate:1 1 udp 2130706431 192.0.2.10 4500 typ host
     ...
answer SDP
     ...
     a=ice-pwd:asd88fgpdd777uzjYhagZg
     a=ice-ufrag:8hhY
     m=application 45664 udp ike-esp-udpencap
     c=IN IP4 192.0.2.20
     a=ike-setup:passive
     a=fingerprint:SHA-1 \
     D2:9F:6F:1E:CD:D3:09:E8:70:65:1A:51:7C:9D:30:4F:21:E4:4A:8E
     a=candidate:1 1 udp 2130706431 192.0.2.100 4500 typ host
     a=candidate:2 1 udp 1694498815 192.0.2.20 45664 typ srflx \
     raddr 192.0.2.100 rport 4500
     ...
```

Figure 16.9 SDP example with IPsec NAT-Traversal (Copyright IETF).

MUST be chosen as an authentication method in IKE phase 1. In this process, a certificate whose hash value matches the fingerprint exchanged over SDP MUST be used. If the certificate used in IKE does not match the original fingerprint, the endpoint MUST terminate the IKE session by detecting an authentication failure. In addition, each party MUST present a certificate and be authenticated by the other. The example described in Section 16.5.3 is for tunnel mode IPsec used for remote access, but the mode of negotiated IPsec is not limited to tunnel mode. For example, IKE can negotiate transport mode IPsec to encrypt multiple media sessions between two parties with only a pair of IPsec security associations. The only thing for which the SDP offer/answer model is responsible is to exchange the fingerprints of certificates used for IKE; therefore, the SDP offer/answer is not responsible for setting the security policy.

16.5.8 Specifications Assuming Prior Relationship between Two Nodes

This section describes the specification for the limited cases in which certificates signed by trusted third parties or preshared keys between endpoints can be used for authentication in IKE. Because the endpoints already have a prior relationship in this case, they use SIP servers for only name resolution and authorization. However, even in this case, the integrity of the SDP description MUST be assured.

16.5.8.1 Certificates Signed by Trusted Third Party

The protocol overview in this case is the same as in Section 16.5.3. The SDP offer/answer procedure is also the same as in Sections 16.5.5 and 16.5.6. Both endpoints have a prior relationship through the trusted third parties, and SIP servers are used for name resolution and authorization of session initiation. Even so, they MAY exchange fingerprints in the SDP because one device can have several certificates and it would be necessary to specify in advance which certificate will be used for the following IKE authentication. This process also ensures that the certificate offered in the IKE process is the same as that owned by the peer that has been authorized at the SIP/SDP layer. By this process, authorization in SIP and authentication in IKE become consistent with each other.

16.5.8.2 Configured Preshared Key

If a preshared key for IKE authentication is installed in both endpoints in advance, they need not exchange the fingerprints of their certificates. However, they may still need to specify which preshared key they will use in the following IKE authentication in SDP because they may have several preshared keys. Therefore, a new attribute, "psk-fingerprint," is defined to exchange the fingerprint of a preshared key over SDP. This attribute also has the role of making authorization in SIP consistent with authentication in IKE. Attribute "psk-fingerprint" is applied to preshared keys as the "fingerprint" defined in RFC 4572 (see Section 16.3) is applied to certificates. All SDP ABNF syntaxes are provided in Section 2.9. However, we have repeated this here for convenience. The following is the ABNF of the "psk-fingerprint" attribute. The use of "psk-fingerprint" is OPTIONAL.

```
attribute                    =     / psk-fingerprint-attribute
psk-fingerprint-attribute    =     "psk-fingerprint" ":" hash-func
                                   SP
                                   psk-fingerprint
```

```
hash-func          =     "sha-1" / "sha-224" / "sha-256" /
                         "sha-384" / "sha-512" / token
                               ; Additional hash functions can
                               only
                               ; come
                               ; from updates to RFC 3279
psk-fingerprint =        2UHEX *(":" 2UHEX)
                               ; Each byte in upper-case hex,
                               ; separated
                               ; by colons.
UHEX               =     DIGIT / %x41-46 ; A-F uppercase
```

An example of SDP negotiation for IKE with preshared key authentication without IPsec NAT-Traversal is as follows (Figure 16.10):

16.5.9 Security Considerations

The security procedures that need to be considered in dealing with the basic SDP messages described in this section (RFC 6193) are provided in Section 17.14.5.

16.5.10 IANA Considerations

SDP parameters that are described here have been registered with IANA. We have not included the IANA registration procedures here for the sake of brevity. The procedures for registration of new SDP parameters are described in RFC 6193.

16.6 Summary

We have described extensions of SDP for key management for negotiations of the security attributes for media that is used by call control protocols like SIP and others. In fact, it provides a framework to be used by one or more key management protocols. A key management protocol is executed prior to the security protocol's execution. A security protocol used by the call control

```
offer SDP
     ...
     m=application 500 udp ike-esp
     c=IN IP4 192.0.2.10
     a=ike-setup:active
     a=psk-fingerprint:SHA-1 \
     12:DF:3E:5D:49:6B:19:E5:7C:AB:4A:AD:B9:B1:3F:82:18:3B:54:02
     ...

answer SDP
     ...
     m=application 500 udp ike-esp
     c=IN IP4 192.0.2.20
     a=ike-setup:passive
     a=psk-fingerprint:SHA-1 \
     12:DF:3E:5D:49:6B:19:E5:7C:AB:4A:AD:B9:B1:3F:82:18:3B:54:02
     ...
```

Figure 16.10 SDP example of IKE with preshared key authentication (Copyright IETF).

protocols like SIP needs a key management solution to exchange keys and security parameters and manage and refresh keys and other features before the setup of the session. In fact, we have described how a key management protocol establishes a security association for the security protocol. Similar to SDP, we have described RTSP extensions for supporting key management attributes. Although this SDP extensions are quite generic, not all key management protocols can use this unless they meet certain restrictions that are specified here. The MIKEY key management protocol that is used for both peer-to-peer and group communications is also described here for integration with SDP and RTSP. In addition, the usage of this key management framework with SDP, SIP, RTSP, and SAP is provided.

We have defined the SDP "crypto" attributes and parameters that serve to configure security for the unicast media streams in either a single message or a roundtrip exchange only between the two parties. The SDP offer/answer model is described how to negotiate these attribute and parameters. Although these cryptographic attributes can be used for a variety of SDP media transports, we have described only SRTP unicast media streams in detail. The SRTP key parameter, crypto-suits, and session parameters are described including crypto contexts initialization and removal. Finally, the SRTP-specific usage of crypto attribute for negotiations between two parties using offer/answer model is provided in great detail. However, it should be noted that the SDP crypto attribute requires the services of a data security protocol to secure the SDP message. We have also explained the rationality ofr keying material directionality for the scenarios like non-forking, serial forking, and parallel forking.

The SDP protocol identifier "TCP/TLS" is being defined for the establishment of secure connection-oriented media transport sessions over the TLS. A key security weakness of TLS is that it does not specify how specific protocols establish and use this secure channel. So, it is an issue how to interpret and validate authentication certificates for the protocols that run over TLS. The SDP protocol identifier ("TCP/TLS") resolves this security issue. In addition, the self-signed certificates that can be used securely are defined, as long as the integrity of the SDP description is assured. It allows an endpoint to include a secure hash of its certificate, known as the "certificate fingerprint," within the session description. It requires that the fingerprint of the offered certificate matches the one in the session description; end hosts can trust even self-signed certificates.

We have described the security preconditions for SDP. The multimedia session setup or modifications between the conference participants require a lot of preconditions that need to be satisfied. A precondition is a condition such as quality of service, policy, firewall/NAT crossing, and other features that need to be satisfied for a given media stream in order for session establishment or modification to proceed. A precondition can be optional or mandatory. If the precondition is mandatory, the session progress is delayed until the precondition is satisfied or the session establishment fails. Similarly, the negotiated SDP security preconditions (e.g., cryptographic parameters) that need to be satisfied between the offerer and the answerer ensuring end-to-end security are described. We have also explained the SDP security description and key management using call flows.

In this section, we have specified how to establish a media session that represents a VPN using the call control protocol like SIP for the purpose of on-demand media/application sharing between peers. The home networks with their home servers are being deployed widely in the Internet. Many home networks reside behind NATs. In view of this, home network-based application/media sharing between users located in different home networks is now possible with the help of DLNA framework or similar protocols over VPN between two user devices/servers. We have also addressed another problem: how a remote client and a home router authenticate each other over IKE to establish IPsec for remote access. The ICE framework has been used for UDP hole punching for IKE/IPsec with a call control protocol like SIP over the NAT.

In addition, we have described that the TCP-based media transport over the TLS protocol in SDP uses a self-signed certificate for authentication in the SIP/SDP framework. This certificate itself is used for authentication not only in TLS but also in IKE. This mechanism is applied to the establishment of an IPsec SA by extending the protocol identifier of SDP so that it can specify IKE. We have also defined new media formats "ike-esp" and "ike-esp-udpencap," which can be used when the protocol identifier is "udp," to enable the negotiation of using IKE for media sessions over SDP exchanges. Finally, we have provided examples using call flows using the SDP offer/answer model for security parameter negotiations using SDP including maintenance and termination of the VPN session, forking, port usage, multiple UDP messages when using ICE, offer/answer with and without IPsec-NAT-traversal, certificates signed by trusted third party, and configuring preshared key.

16.7 Problems

1. What is meant by session-, media-, and transport-level attributes in the context of multimedia conference? Explain using some examples including call flows. Justify the fact that some security parameters in all of these levels need to be negotiated.
2. What security parameters are negotiated using the SDP offer/answer model by call control protocols such as SIP in RFC 4567? Describe in detail using call flows.
3. Describe the SDP and RTSP extensions for supporting key management protocol and provide examples using call flows for negotiations security parameters key management protocol by SDP using SIP, RTSP, and SAP including the use of the SDP offer/answer model.
4. How has MIKEY been integrated by RFC 4567 as another key management protocol whose security parameters can be negotiated using SDP? Explain in detail using call flows.
5. What key assumptions are being used in RFC 4561 before and after session establishment? What are the problems in security if the initial offer from the calling party is not acceptable to the responder?
6. Why do we need an SDP cryptographic attribute that has been defined in RFC 4568 for unicast media such as audio, video, fax, whiteboard, modem, and other streams? Justify your answer providing examples using applicability and explaining the fact that S/MIME, TLS, and others are also used for establishing the secured session.
7. Explain why we need security services such as data origin authentication, integrity, and confidentiality of the streams for the session establishment.
8. How do the cryptographic security attributes differ between RFCs 4568 and 4567? Explain in detail.
9. What certain predefined default values provided in SRTP can be used for configuring the SDP for security? What are the generalized SDP cryptographic attributes defined in RFC 4568?
10. What are the SDP crypto attributes and parameters defined in 4568? Provide the use cases of these SDP crypto attributes including SDP offer/answer, generating offer/answer using unicast streams, processing, and modifying the session after negotiation. Explain using the SDP crypto attribute with examples for outside offer/answer and backwards compatibilities.

11. Explain the use of SRTP security attributes including their key parameters, all varieties of crypto-suites, and all session parameters.
12. Provide examples with detailed call flows of SRTP crypto context initialization including late binding of one or more SSRCs to crypto context and sharing cryptographic contexts among sessions or SSRCs. How does one remove the cryptographic contexts?
13. Explain in detail the SRTP-specific use of the crypto attribute including simple offer/answer for unicast streams, generating initial offer, generating initial answer, processing the initial answer, and modifying the session, as well as with call flows of offer/answer.
14. Explain in detail the call flows for SRTP-specific use outside offer/answer, support for SIP forking, SRTP-specific backwards compatibility, and operation with KEYMGT= and k= lines.
15. What is the rationale for keying material for directionality including the scenarios for non-forking, serial forking, and parallel forking? Explain in detail using call flows.
16. What are the main attributes of a connection-oriented transport protocol? Provide examples of the connection-oriented media streams that demand connection-oriented transport protocol. Why does TLS demand the connection-oriented protocol like TCP? What are the security weaknesses of TLS in the context of the establishment of the connection-oriented media session?
17. How do the SDP security attributes "TCP/TLS" and "certificate fingerprints" solve the security holes for setting up a connection-oriented media session?
18. Describe some scenarios in which we need attributes like "precondition" in setting up the multimedia conferencing. Explain in detail using high-level call flows.
19. Which RFC defines preconditions for quality of service in setting up the multimedia session? Explain in detail using call flows rationalizing the precondition for quality of service attributes.
20. What are the basic problems of setting up multimedia sessions between the endpoints that remain behind NATs? Do these NAT-crossing problems affect security? If so, explain in detail.
21. Explain in detail why we need SDP security preconditions as defined in RFC 5027 including the call flows.
22. Devise a scenario in which both SDP security preconditions and quality of service preconditions are required for successful call setup. Do the same for scenarios in which both endpoints remain behind the NATs.
23. What is the DLNA framework referred to in RFC 6193? How does it relate to VPN?
24. Create a scenario in which both remain behind their home networks and NATs and both home networks can be connected via a DLNA or VPN. Furthermore, both uses have their own media applications in their respective home networks, which they share among themselves using a call control protocol like SIP. Define the scenarios with abstract call flows.
25. What are the SDP security attributes known as "protocol identifier" for IKE media sessions defined in RFC 6193? What is the applicability of these SDP security attributes in the context of IPsec/IKE? How does it frame the solution for scenario described in Problem 24?
26. Describe in detail the negotiation of IKE for media sessions using the SDP offer/answer model, maintenance and termination of VPN session, forking, port usage, multiplexing UDP messages over ICE, usage of IKE, prior relationship between the two parties/certificate signed by trusted third party, and configured preshared key including detail call flows for the scenario described in Problem 24.

References

1. International Telecommunications Union, "Information Technology – Open Systems Interconnection—The Directory: Public-key and attribute certificate frameworks", ITU-T Recommendation X.509, ISO Standard 9594-8, March 2000.
2. National Institute of Standards and Technology, "Secure Hash Standard", FIPS PUB 180-2, August 2002, http://csrc.nist.gov/ publications/fips/fips180-2/fips180-2.pdf.
3. Mahy, R., "A Solution to the Heterogeneous Error Response Forking Problem (HERFP) in the Session Initiation Protocol (SIP)", draft-mahy-sipping-herfp-fix-01.txt, Work in Progress, March 2006.
4. Rosenberg, J., "Concerns around the Applicability of RFC 4474", draft-rosenberg-sip-rfc4474-concerns-00, Work in Progress, February 2008.

Chapter 17

Security Considerations in Using SDP Attributes

Abstract
In the earlier sections (Sections 2–16), we have considered the Session Description Protocol (SDP) signaling messages that are used for negotiations of capabilities related to network, transport, session, media, and security attributes. During negotiations, SDP signaling messages are sent back and forth between the endpoints. These endpoints can be humans and/or automat. However, it is critical that the security of these individual SDP signaling messages be provided during negotiations. Each of these Request for Comments (RFCs) related to Sections 2–16 also describes how security needs to be assured for those individual SDP signaling messages. We have described all of the security features described in those RFCs of Sections 2–16 so that the SDP signaling messages described in the subsequent sections create a unified global security picture by virtue of having all of them together in one place. The security parameters for media stream/payload (e.g., audio, video, and/or data [test, graphics, and/or animation]) are negotiated using SDP before the setup of the call. During the actual media transfer between the endpoints, those security algorithms/parameters/features are used as agreed upon before the call setup using SDP.

In this section, we describe the security aspects of how to use individual SDP session-, transport-, and media-signaling messages related to media (e.g., audio, video, and/or data). At first, we describe the general security framework for any SDP signaling messages per RFC 4566. In light of this framework, we consider the security aspects of each SDP message category used for capability negotiations: offer/answer, security grouping, generic capability, image (generic), media loopback, QOS, rtcpbandwidth, transport independent bandwidth modifier, connectivity precondition, tcp-based media transport, source filter, source filter in offer/answer, media stream security, media stream security (packets authentication, key stream reuse, signaling authentication, and encryption), Transport Layer Security, and precondition. The security considerations for each of those SDP attributes are specified in their respective RFCs in subsequent sections.

17.1 Basic SDP Messages

Request for Comments (RFC) 4566 (see Section 2) describes the basic security mechanisms that are applicable to all Session Description Protocol (SDP) messages used for session negotiations. Numerous attacks such as diversion of media streams (enabling eavesdropping), disabling of calls, denial of service (DoS) attacks, and injection of unwanted media streams are possible if adversaries can modify SDP parameters, injecting themselves in the middle of the session negotiations path being initiated by call control protocols. However, in general, SDP is used with multimedia call control protocols such as Session Initiation Protocol (SIP) (RFC 3261, also see Section 1.2), WebRTC (see Section 1.3), Session Announcement Protocol (SAP) (RFC 2974), and others using the offer/answer model (RFC 326 4, see Section 3.1) to agree on parameters for unicast sessions. RFC 4566 clarifies that, when used in this manner, the security considerations of those protocols apply. That is, SDP itself does not need to build any extra security mechanisms to protect itself. SDP is a session description format that describes multimedia sessions containing session metadata, network description, media stream description, security description, and quality of services (QOS) description. Entities receiving and acting upon an SDP message SHOULD be aware that a session description cannot be trusted unless it has been obtained by an authenticated transport protocol from a known and trusted source as explained in RFC 4566 (see Section 2). Many different transport protocols may be used to distribute session descriptions, and the nature of the authentication will differ from transport to transport.

For some transports, security features are often not deployed. When a session description has not been obtained in a trusted manner, the endpoint SHOULD exercise care (RFC 4566) because, among other attacks, the media sessions received may not be the intended ones, the destination of the media may not be the expected one, any of the parameters of the session may be incorrect, or media security may be compromised. It is up to the endpoint to make a sensible decision, taking into account the security risks of the application and the user preferences. The endpoint may decide to ask the user whether to accept the session.

One transport that can be used to distribute session descriptions is SAP (RFC 2974). This provides both encryption and authentication mechanisms, but due to the nature of session announcements it is likely that there are many occasions where the originator of a session announcement cannot be authenticated (RFC 4566) because the originator is previously unknown to the receiver of the announcement and no common public key infrastructure is available. On receiving a session description over an unauthenticated transport mechanism or from an untrusted party, software parsing the session should take a few precautions. Session descriptions contain information required to start software on the receiver's system. Software that parses a session description MUST NOT be able to start other software except that which is specifically configured as appropriate for participating in multimedia sessions.

It is normally considered inappropriate for software parsing a session description to start (RFC 4566), on a user's system, software that is appropriate to participate in multimedia sessions, without the user first understanding and consenting that such software will be started. Thus, a session description arriving by session announcement, email, session invitation, or WWW page MUST NOT deliver the user into an interactive multimedia session unless the user has explicitly preauthorized such action. As it is not always simple to tell whether a session is interactive, applications that are unsure should assume sessions are interactive.

In RFC 4566, no attributes inform the recipient of a session description to start multimedia tools in a mode where they default to transmitting. Under some circumstances, it might be appropriate to define such attributes. If this is done, an application parsing a session description containing such attributes SHOULD either ignore them or inform the user that joining this session will result in the automatic transmission of multimedia data. The default behavior for an unknown attribute is to ignore it.

In certain environments, it has become common for intermediary systems to intercept and analyze session descriptions contained within other signaling protocols. This is done for a range of purposes, including but not limited to opening holes in firewalls to allow media streams to pass and to mark, prioritize, or block traffic selectively. In some cases, such intermediary systems may modify the session description, for example, to have the contents of the session description match network address translator (NAT) bindings dynamically created. Per RFC 4566, these behaviors are NOT RECOMMENDED unless the session description is conveyed in such a manner that it allows the intermediary system to conduct proper checks to establish the authenticity of the session description and the authority of its source to establish such communication sessions.

SDP by itself does not include sufficient information to enable these checks; they depend on the encapsulating protocol (e.g., SIP or Real-Time Streaming Protocol [RTSP]). Use of the "k=" field (see Section 2.5.12, RFC 4566) poses a significant security risk, since it conveys session encryption keys in the clear. SDP MUST NOT be used to convey key material, unless it can be guaranteed that the channel over which the SDP is delivered is both private and authenticated. Moreover, the "k=" line provides no way to indicate or negotiate cryptographic key algorithms. As it provides for only a single symmetric key, rather than separate keys for confidentiality and integrity, its utility is severely limited. The use of the "k=" line is NOT RECOMMENDED, as discussed in Section 2.5.12. The generic security guidelines describe here are applicable for all SDP schemes. In addition to these security considerations, in the subsequent sections we describe a few security mechanisms to use when dealing with some specific SDP parameters.

17.2 Negotiations Model in SDP

17.2.1 Offer/Answer Attributes

RFC 3264 (see Section 3.1) describes the SDP offer/answer model. The SDP offer/answer reveals the fact that session negotiations take a couple of message exchanges between the caller and the callee. As a result, numerous attacks are possible if an attacker can modify offers or answers in transit. Generally, as explained earlier, these include diversion of media streams (enabling eavesdropping), disabling of calls, and injection of unwanted media streams (RFC 3264, see Section 3.1). If a passive listener can construct fake offers and inject those into an exchange, similar attacks are possible. Even if an attacker can simply observe offers and answers, he/she can inject media streams into an existing conversation. Offer/answer relies on transport within an application signaling protocol, such as SIP. It also relies on that protocol for security capabilities. Because of the previously described attacks, that protocol MUST provide a means for end-to-end authentication and integrity protection of offers and answers. It SHOULD offer encryption of bodies to prevent eavesdropping. However, media injection attacks can alternatively be resolved through authenticated media exchange, and therefore the encryption requirement is a SHOULD instead of a MUST. Replay attacks are also problematic. An attacker can replay an old offer, perhaps one that had put media on hold, and thus disable media streams in a conversation. Therefore, the application protocol MUST provide a secure way to sequence offers and answers and to detect and reject old offers or answers.

SDP offer and answer messages (RFC 4317, see Section 18.1 of this book) can contain private information about addresses and sessions to be established between parties. If this information needs to be kept private, some security mechanism in the protocol used to carry the offers and answers must be used. Use of capability renegotiation (RFC 3407, Section 4.1) may make the session susceptible to denial of service (DoS), without design care as to authentication. For SIP, this means using TLS transport and/or S/MIME encryption of the SDP message body. It is important that SDP offer and answer messages be properly authenticated and authorized before they are used to establish a media session. Examples of SIP mechanisms include SIP Digest, certs, and cryptographically verified SIP identity.

17.2.2 ICE: NAT-Traversal with Offer/Answer Attributes

The security specific to NAT-traversal with Offer/Answer Attributes (RFC 5245, see Section 3.2) is described here. Several types of attacks are possible in an Interactive Connectivity Establishment (ICE) system. This section considers these attacks and their countermeasures. These countermeasures include:

- Using ICE in conjunction with secure signaling techniques, such as SIPS
- Limiting the total number of connectivity checks to 100, and optionally limiting the number of candidates they'll accept in an offer or answer

17.2.2.1 Attacks on Connectivity Checks

An attacker might attempt to disrupt the simple traversal of UDP around NAT (STUN) connectivity checks. Ultimately, all of these attacks fool an agent into thinking something incorrect about the results of the connectivity checks. The possible false conclusions an attacker can try to cause are

- False invalid: An attacker can fool a pair of agents into thinking a candidate pair is invalid when it isn't. This can be used to cause an agent to prefer a different candidate (such as one injected by the attacker) or to disrupt a call by forcing all candidates to fail.
- False valid: An attacker can fool a pair of agents into thinking a candidate pair is valid when it isn't. This can cause an agent to proceed with a session but then not be able to receive any media.
- False Peer Reflexive Candidate: An attacker can cause an agent to discover a new peer reflexive candidate when it shouldn't have. This can be used to redirect media streams to a DoS target or to the attacker, for eavesdropping or other purposes.
- False valid on false candidate: An attacker has already deceived an agent into believing the existence of a candidate with an address (for example, by injecting a false peer reflexive candidate or false server reflexive candidate). It must then launch an attack that forces the agents to believe that this candidate is valid.

If an attacker can cause a false peer reflexive candidate or false valid on a false candidate, it can launch any of the attacks described in RFC 5389.

To force the false invalid result, the attacker has to wait for the connectivity check from one of the agents to be sent. When it is, the attacker needs to inject a fake response with an unrecoverable error response, such as a 400. However, since the candidate is, in fact, valid, the original request may reach the peer agent and result in a success response. The attacker needs to force this packet or its response to be dropped, through a DoS attack, layer 2 network disruption, or other technique. If it doesn't do this, the success response will also reach the originator, alerting it to a possible attack. Fortunately, this attack is mitigated completely through the STUN short-term credential mechanism. The attacker needs to inject a fake response, and in order for this response to be processed, the attacker needs the password. If the offer/answer signaling is secured, the attacker will not have the password and its response will be discarded.

Forcing the fake valid result works in a similar way. The agent needs to wait for the Binding request from each agent and inject a fake success response. The attacker won't need to worry about disrupting the actual response since, if the candidate is not valid, it presumably wouldn't be received anyway. However, like the fake invalid attack, this attack is mitigated by the STUN short-term credential mechanism in conjunction with a secure offer/answer exchange.

Forcing the false peer reflexive candidate result can be done either with fake requests or responses or with replays. We consider the fake requests and responses case first. It requires the attacker to send a Binding request to one agent with a source IP address and port for the false candidate. In addition, the attacker must wait for a Binding request from the other agent, and generate a fake response with a XOR-MAPPED-ADDRESS attribute containing the false candidate. Like the other attacks described here, this is mitigated by the STUN message integrity mechanisms and secure offer/answer exchanges.

Forcing the false peer reflexive candidate result with packet replays is different. The attacker waits until one of the agents sends a check. It intercepts this request and replays it toward the other agent with a faked source IP address. It must also prevent the original request from reaching the remote agent, either by launching a DoS attack to cause the packet to be dropped or forcing it to be dropped using layer 2 mechanisms. The replayed packet is received at the other agent and accepted, since the integrity check passes (the integrity check cannot and does not cover the source IP address and port). A response is then triggered; this response will contain a XORMAPPED-ADDRESS with the false candidate and will be sent to that false candidate. The attacker must then receive it and relay it toward the originator.

The other agent will then initiate a connectivity check toward that false candidate. This validation needs to succeed. This requires the attacker to force a false valid on a false candidate. Injecting fake requests or responses to achieve this goal is prevented using the integrity mechanisms of STUN and the offer/answer exchange. Thus, this attack can only be launched through replays; the attacker must intercept the check toward this false candidate and replay it toward the other agent. Then it must intercept the response and replay that back as well.

This attack is very hard to launch unless the attacker is identified by the fake candidate because it requires the attacker to intercept and replay packets sent by two different hosts. If the two agents are on different networks (for example, across the public Internet), this attack can be hard to coordinate: It needs to occur against two different endpoints, on different parts of the network, at the same time.

If the attacker itself is identified by the fake candidate, the attack is easier to coordinate. However, if SRTP is used (RFC 3711), the attacker will not be able to play the media packets but will only be able to discard them, effectively disabling the media stream for the call. However, this attack requires the agent to disrupt packets in order to block the connectivity check from reaching the target. In that case, if the goal is to disrupt the media stream, it's much easier to just disrupt it with the same mechanism, rather than attack ICE.

17.2.2.2 Attacks on Server Reflexive Address Gathering

ICE endpoints make use of STUN Binding requests for gathering server reflexive candidates from a STUN server. These requests are not authenticated in any way. Consequently, an attacker can employ numerous attacks to provide the client with a false server reflexive candidate:

- An attacker can compromise the Domain Name System (DNS), causing DNS queries to return a rogue STUN server address. That server can provide the client with fake server reflexive candidates. This attack is mitigated by DNS security, though Domain Name System Security Extensions (DNS-SEC) is not required to address it.
- An attacker that can observe STUN messages (such as an attacker on a shared network segment, like Wi-Fi) can inject a fake response that is valid and will be accepted by the client.
- An attacker can compromise a STUN server by means of a virus and cause it to send responses with incorrect mapped addresses. A false mapped address learned by these attacks will be used as a server reflexive candidate in the ICE exchange. For this candidate to actually be used for media, the attacker must also attack the connectivity checks and in particular force a false valid on a false candidate. This attack is very hard to launch if the false address identifies a fourth party (neither the offerer, answerer, nor attacker), since it requires attacking the checks generated by each agent in the session and is prevented by Secure Real-Time Protocol (SRTP) if it identifies the attacker itself.

If the attacker elects not to attack the connectivity checks, the worst it can do is prevent the server reflexive candidate from being used. However, if the peer agent has at least one candidate that is reachable by the agent under attack, the STUN connectivity checks themselves will provide a peer reflexive candidate that can be used for the exchange of media. Peer reflexive candidates are generally preferred over server reflexive candidates. As such, an attack solely on the STUN address gathering will normally have no impact on a session at all.

17.2.2.3 Attacks on Relayed Candidate Gathering

An attacker might attempt to disrupt the gathering of relayed candidates, forcing the client to believe it has a false relayed candidate. Exchanges with the traversal using relays around NAT (TURN) server are authenticated using a long-term credential. Consequently, injection of fake responses or requests will not work. In addition, unlike Binding requests, Allocate requests are not susceptible to replay attacks with modified source IP addresses and ports, since the source IP address and port are not utilized to provide the client with its relayed candidate.

However, TURN servers are susceptible to DNS attacks or to viruses aimed at the TURN servers, for purposes of turning them into zombie or rogue servers. These attacks can be mitigated by DNS-SEC and through good box and software security on TURN servers. Even if an attacker has caused the client to believe in a false relayed candidate, the connectivity checks cause such a candidate to be used only if they succeed. Thus, an attacker must launch a false valid on a false candidate, per above paragraph, which is a very difficult attack to coordinate.

17.2.2.4 Attacks on the Offer/Answer Exchanges

An attacker that can modify or disrupt the offer/answer exchanges can readily launch a variety of attacks with ICE. The attacker could direct media to a target of a DoS attack, insert him-/herself into the media stream, and so on. These concerns are similar to the general security considerations for offer/answer exchanges, and the security considerations in RFC 3264 (see Section 3.1) apply. These require techniques for message integrity and encryption for offers and answers, which are satisfied by the SIPS mechanism (RFC 3261, also see Section 1.2) when SIP is used. As such, the usage of SIPS with ICE is RECOMMENDED.

17.2.2.5 Insider Attacks

In addition to attacks where the attacker is a third party trying to insert fake offers, answers, or STUN messages, several attacks are possible with ICE when the attacker is an authenticated and valid participant in the ICE exchange.

17.2.2.5.1 The Voice Hammer Attack

The voice hammer attack is an amplification attack. The attacker initiates sessions to other agents and maliciously includes the IP address and port of a DoS target as the destination for media traffic signaled in the SDP. This causes substantial amplification; a single offer/answer exchange can create a continuing flood of media packets, possibly at high rates (consider video sources). This attack is not specific to ICE, but ICE can help provide remediation.

Specifically, if ICE is used, the agent receiving the malicious SDP will first perform connectivity checks to the target of media before sending media there. If this target is a third-party host, the checks will not succeed, and media is never sent. Unfortunately, ICE doesn't help if it's not used, in which case an attacker can simply send the offer without the ICE parameters. However, in environments where the set of clients is known, and is limited to ones that support ICE, the server can reject any offers or answers that don't indicate ICE support.

17.2.2.5.2 STUN Amplification Attack

The STUN amplification attack is similar to the voice hammer. However, instead of voice packets being directed to the target, STUN connectivity checks are directed to the target. The attacker sends an offer with a large number of candidates, say 50. The answerer receives the offer and starts its checks, which are directed at the target and consequently never generate a response. The answerer will start a new connectivity check every Ta ms (say, Ta = 20 ms).

However, the retransmission timers are set to a large number due to the large number of candidates. As a consequence, packets will be sent at an interval of one every Ta milliseconds and then with increasing intervals after that. Thus, STUN will not send packets at a rate faster than media would be sent, and the STUN packets persist only briefly until ICE fails for the session. Nonetheless, this is an amplification mechanism. It is impossible to eliminate the amplification, but the volume can be reduced through a variety of heuristics. Agents SHOULD limit the total number of connectivity checks they perform to 100. Additionally, agents MAY limit the number of candidates they will accept in an offer or answer.

Frequently, protocols that wish to avoid these kinds of attacks force the initiator to wait for a response prior to sending the next message. However, in the case of ICE, this is not possible. It is not possible to differentiate the following two cases:

- There was no response because the initiator is being used to launch a DoS attack against an unsuspecting target that will not respond.
- There was no response because the IP address and port are not reachable by the initiator.

In the second case, another check should be sent at the next opportunity, while in the former case, no further checks should be sent.

17.2.2.6 Interactions with Application Layer Gateways and SIP

Application layer gateways (ALGs) are functions present in a NAT device that inspect the contents of packets and modify them, in order to facilitate NAT-traversal for application protocols. Session border controllers (SBCs) are close cousins of ALGs but are less transparent since they actually exist as application layer SIP intermediaries. ICE interacts with SBCs and ALGs.

If an Application Layer Gateway (ALG) is SIP aware but not ICE aware, ICE will work through it as long as the ALG correctly modifies the SDP. A correct ALG implementation behaves as follows:

- The ALG does not modify the "m=" and "c=" lines or the rtcp attribute if they contain external addresses.
- If the m and "c=" lines contain internal addresses, the modification depends on the state of the ALG:
  - If the ALG already has a binding established that maps an external port to an internal IP address and port matching the values in the m and c lines or rtcp attribute, the ALG uses that binding instead of creating a new one.
  - If the ALG does not already have a binding, it creates a new one and modifies the SDP, rewriting the m and c lines and rtcp attribute.

Unfortunately, many ALGs are known to work poorly in these corner cases. ICE does not try to work around broken ALGs, as this is outside the scope of its functionality. ICE can help diagnose these conditions, which often show up as a mismatch between the set of candidates and the m and c lines and rtcp attributes. The ICE mismatch attribute is used for this purpose.

ICE works best through ALGs when the signaling is run over TLS. This prevents the ALG from manipulating the SDP messages and interfering with ICE operation. Implementations that are expected to be deployed behind ALGs SHOULD provide for TLS transport of the SDP.

If an SBC is SIP aware but not ICE aware, the result depends on the behavior of the SBC. If it is acting as a proper back-to-back user agent (B2BUA), the SBC will remove any SDP attributes it doesn't understand, including the ICE attributes. Consequently, the call will appear to both endpoints as if the other side doesn't support ICE. This will result in ICE being disabled and media flowing through the SBC, if the SBC has requested it. If, however, the SBC passes the ICE attributes without modification, yet modifies the default destination for media (contained in the m and c lines and rtcp attribute), this will be detected as an ICE mismatch, and ICE processing will be aborted for the call. It is outside of the scope of ICE to act as a tool for "working around" SBCs. If one is present, ICE will not be used and the SBC techniques take precedence.

17.3 Capability Declaration in SDP

17.3.1 Simple Capability Negotiations

RFC 3407 (see Section 4.1) describes capability negotiations using SDP messages. Capability negotiation of security-sensitive parameters is a delicate process; it should not be done without careful evaluation of the design, including the possible susceptibility to downgrade attacks. Use of capability renegotiation may make the session susceptible to DoS, without design care as to authentication.

17.4 Media-Level Attribute Support

17.4.1 Label Attribute (RFC 4574)

In label attribute, defined in RFC 4574 (see Section 5.1), an attacker may attempt to add, modify, or remove "label" attributes from a session description. This could make an application behave in a nondesirable way. It is strongly RECOMMENDED that integrity protection be applied to the SDP session descriptions. For session descriptions carried in SIP (RFC 3261, also see Section 1.2), S/MIME is the natural choice to provide such end-to-end integrity protection, as described in RFC 3261 (also see Section 1.2). Other applications MAY use a different form of integrity protection.

17.4.2 Content Attribute (RFC 4796)

In content attribute specified in RFC 4796 (see Section 5.2), an attacker may attempt to add, modify, or remove "content" attributes from a session description. Depending on how an implementation chooses to react to the presence or absence of a given "content" attribute, this could result in an application behaving in an undesirable way; therefore, it is strongly RECOMMENDED that integrity protection be applied to the SDP session descriptions. Integrity protection can be provided for a session description carried in an SIP (RFC 3261, also see Section 1.2) (e.g., by using S/MIME) (RFC 3851 obsoleted by RFC 5751) or TLS (RFC 4346 obsoleted by RFC 5246). It is assumed that values of "content" attribute do not contain data that would be truly harmful if exposed to a possible attacker. It must be noted that the initial set of values does not contain any data that would require confidentiality protection. However, S/MIME and TLS can be used to protect confidentiality, if needed.

17.4.3 Source-Specific Media Attributes (RFC 5576)

In source-specific media attributes described in RFC 5576 (see Section 5.3), all the security implications of RTP specified in RFC 3550 and of SDP described in RFC 4566 (see Sections 2 and 17.1) apply. Explicitly describing the multiplexed sources of an RTP media stream does not appear to add any further security issues. In accordance with RFC 3550, RTP suffers from the same security liabilities as the underlying protocols. For example, an impostor can fake source or destination network addresses or change the header or payload. Within RTCP, the CNAME and NAME information may be used to impersonate another participant. In addition, RTP may be sent via IP multicast, which provides no direct means for a sender to know all the receivers of the data sent and therefore no measure of privacy. Rightly or not, users may be more sensitive to privacy concerns with audio and video communication than they have been with more traditional forms of network communication (Stubblebine "Security Services for Multimedia Conferencing," in *16th National Computer Security Conference* (Baltimore, MD), pp. 391–395, September 1993). Therefore, the use of security mechanisms with RTP is important. These mechanisms are discussed in Section 9 of RFC 3550. RTP-level translators or mixers may be used to allow RTP traffic to reach hosts behind firewalls. Appropriate firewall security principles and practices, which are beyond the scope of this specification, should be followed in the design and installation of these devices and in the admission of RTP applications for use behind the firewall.

17.5 Signaling Media Decoding Dependency (RFC 5583)

In signaling media decoding dependency specified in RFC 5583 (see Section 6.1), all security implications of SDP apply. There may be a risk of manipulation of the dependency signaling of a session description by an attacker. This may mislead a receiver, or a middlebox (e.g., a receiver) may try to compose a media bit stream out of several RTP packet streams that does not form an operation point, although the signaling made it believe it would form a valid operation point, with potential fatal consequences for the media decoding process. It is recommended that the receiver SHOULD perform an integrity check on SDP and follow the security considerations of SDP to only trust SDP from trusted sources.

17.6 Media Grouping Support

17.6.1 Grouping Framework (RFC 5888)

In the media grouping framework described in RFC 5888 (see Section 7.1), using the "group" parameter with FID semantics, an entity that managed to modify the session descriptions exchanged between the participants to establish a multimedia session could force the participants to send a copy of the media to any destination of its choosing. Integrity mechanisms provided by protocols used to exchange session descriptions and media encryption can be used to prevent this attack. In SIP, S/MIME (RFC 5750) and TLS (RFC 5246) can be used to protect session description exchanges in an end-to-end and a hop-by-hop fashion, respectively.

17.6.2 Negotiating Media Multiplexing/Bundling (IETF Draft)

The security considerations defined in RFC 3264 (see Section 17.2.1) and RFC 5888 (see Section 17.6.1) apply to the BUNDLE extension. Bundle does not change which information flows over the network but only changes the addresses and ports on which information is flowing and thus has very little impact on the security of the RTP sessions. When the BUNDLE extension is used, a single set of security credentials might be used for all media streams specified by a BUNDLE group. When the BUNDLE extension is used, the number of Synchronization Source (SSRC) values within a single RTP session increases, which increases the risk of SSRC collision. RFC 4568 (see Sections 16.2 and 17.14.2) describes how SSRC collision may weaken SRTP and SRTCP encryption in certain situations.

17.6.3 RTP Media Streams Grouping (IETF Draft)

In RTP media stream grouping described in the Internet Engineering Task Force (IETF) draft (see Section 7.3), an adversary with the ability to modify SDP descriptions has the ability to switch around tracks between media streams. This is a special case of the general security consideration that modification of SDP descriptions must be confined to entities trusted by the application. If implementing buffering as mentioned in Section 7.3.3.1, the amount of buffering should be limited to avoid memory exhaustion attacks. No other identified attacks depend on this mechanism.

17.7 Generic Capability Attribute Negotiations

17.7.1 Generic Capability (RFC 5939)

The SDP generic capability negotiation framework (RFC 5939, see Section 8.1) is defined to be used within the context of the offer/answer model; hence, all offer/answer security considerations apply here as well (RFC 3264, see Sections 3.1 and 17.2.1). Similarly, the SIP uses SDP and the offer/answer model; hence, when used in that context, the SIP security considerations apply as well (RFC 3261, also see Section 1.2). However, SDP capability negotiation introduces additional security issues. Its use as a mechanism to enable alternative transport protocol negotiation (secure and nonsecure) as well as its ability to negotiate use of more or less secure keying methods and material warrant further security considerations. Also, the (continued) support for receiving media before answer combined with negotiation of alternative transport protocols (secure and nonsecure) warrants further security considerations. We discuss these issues below.

The SDP capability negotiation framework allows an offered media stream to both indicate and support various levels of security for that media stream. Different levels of security can for example be negotiated by use of alternative attribute capabilities, each indicating more or less secure keying methods as well as more or less strong ciphers. Since the offerer indicates support for each of these alternatives, he/she will presumably accept the answerer's seemingly selecting any of the offered alternatives. If an attacker can modify the SDP session description offer, he can thereby force the negotiation of the weakest security mechanism that the offerer is willing to accept. This may enable the attacker to compromise the security of the negotiated media stream. Similarly, if the offerer wishes to negotiate use of a secure media stream (e.g., Secure RTP (SRTP) but includes a nonsecure media stream (e.g., plain RTP) as a valid (but less preferred) alternative, then an attacker that can modify the offered SDP session description will be able to force the establishment of an insecure media stream. The solution to both of these problems involves the use of integrity protection over the SDP session description.

Ideally, this integrity protection provides end-to-end integrity protection in order to protect from any man-in-the-middle attack; secure multiparts such as S/MIME (RFC 5751) provide one such solution; however, S/MIME requires use and availability of a Public Key Infrastructure (PKI). A slightly less secure alternative when using SIP, but generally much easier to deploy in practice, is to use SIP identity (RFC 4474); this requires the existence of an authentication service (see RFC 4474). Although this mechanism still requires a PKI, it only requires that servers (as opposed to end users) have third-party validatable certificates, which significantly reduces the barrier to entry by ordinary users.

Yet another, and considerably less secure, alternative is to use hop-by-hop security only (e.g., TLS or IPsec) thereby ensuring the integrity of the offered SDP session description on a hop-by-hop basis. This is less secure because SIP allows partially trusted intermediaries on the signaling path, and such intermediaries processing the SIP request at each hop would be able to perform a man-in-the-middle attack by modifying the offered SDP session description. In simple architectures where the two user agents' proxies communicate directly, the security provided by this method is roughly comparable to that provided by the previously discussed signature-based mechanisms.

Per the normal offer/answer procedures, as soon as the offerer has generated an offer, the offerer must be prepared to receive media in accordance with that offer. The SDP capability negotiation preserves that behavior for the actual configuration in the offer; however, the offerer has no way of knowing which configuration (actual or potential) was selected by the answerer until an answer indication is received. This opens up a new security issue where an attacker may be able to interject media toward the offerer until the answer is received. For example, the offerer may use plain RTP as the actual configuration and SRTP as an alternative potential configuration. Even though the answerer selects SRTP, the offerer will not know that until he receives the answer; hence, an attacker will be able to send media to the offerer meanwhile. The easiest protection against such an attack is to not offer the use of the nonsecure media stream in the actual configuration; however, that may in itself have undesirable side effects: If the answerer does not support the secure media stream and also does not support the capability negotiation framework, then negotiation of the media stream will fail. Alternatively, SDP security preconditions (RFC 5027, see Section 16.4) can be used. This will ensure that media is not flowing until session negotiation has completed and the selected configuration is known. Use of preconditions requires both sides to support them. If they don't, and use of them is required, the session will fail. As a (limited) way to work around this, it is RECOMMENDED that SIP entities generate an answer SDP session description and send it to the offerer as soon as possible, for example, in a "183 Session Progress" message. This will limit the time during which an attacker can send media to the offerer. Section 8.1.3.9 presents other alternatives as well.

Additional security considerations apply to the answer SDP session description. The actual configuration attribute tells the offerer on which potential configuration the answer was based; hence, an attacker that can either modify or remove the actual configuration attribute in the answer can cause session failure as well as extend the time window during which the offerer will accept incoming media that does not conform to the actual answer. The solutions to this SDP session description answer integrity problem are the same as for the offer (i.e., use of end-to-end integrity protection, SIP identity, or hop-by-hop protection). The mechanism to use depends on the mechanisms supported by the offerer as well as the acceptable security tradeoffs.

As described in Sections 8.1.3.1 and 8.1.3.11, SDP capability negotiation conceptually allows an offerer to include many different offers in a single SDP session description. This can cause the answerer to process a large number of alternative potential offers, which can consume significant memory and CPU resources. An attacker can use this amplification feature to launch a DoS attack against the answerer. The answerer must protect itself from such attacks. As explained in Section 8.1.3.11, the answerer can help reduce the effects of such an attack by first discarding all potential configurations that contain unsupported transport protocols, unsupported or invalid mandatory attribute capabilities, or unsupported mandatory extension configurations. The answerer should also look out for potential configurations that are designed to pass the above test but still produce a large number of potential configuration SDP session descriptions that cannot be supported.

An attacker can potentially find a valid session-level attribute that causes conflicts or otherwise interferes with individual media description configurations. At the time of publication of this document, we do not know of such an SDP attribute; however, this does not mean it does not exist or will not exist in the future. If such attributes are found to exist, implementers should explicitly protect against them. A significant number of valid and supported potential configurations may remain. However, since all of those contain only valid and supported transport protocols and attributes, it is expected that only a few of them will need to be processed on average. Still, the answerer must ensure that it does not needlessly consume large amounts of memory or CPU resources when processing those as well as being prepared to handle the case where a large number of potential configurations still need to be processed.

17.7.2 Generic Image-Attributes (RFC 6236)

In generic image-attributes described in RFC 6236 (see Section 8.2), the image-attributes and especially the parameters that denote the image size can take on values that may cause memory or CPU exhaustion problems. This may happen either because of a mistake by the sender of the SDP or because of an attack issued by a malicious SDP sender. This issue is similar to the case in which the a=fmtp line(s) may take on extreme values for the same reasons outlined in Section 8.2. A receiver of the SDP containing the image-attribute MUST ensure that the parameters have reasonable values and that the device can handle the implications in terms of memory and CPU usage. Failure to do a sanity check on the parameters may result in memory or CPU exhaustion. In principle, for some SDPs containing the image-attribute and for some deployments, simply checking the parameters may not be sufficient to detect all potential DoS problems. Implementers ought to consider whether any potential DoS attacks would not be detected by simply checking parameters.

17.7.3 Media Capabilities (RFC 6871)

The security considerations of RFC 5939 (see Sections 8.1 and 17.7.1) apply for this SDP media capabilities negotiations (RFC 6871, see Section 8.3). In RFC 5939 (see Sections 8.1 and 17.7.1), it was noted that negotiation of transport protocols (e.g., secure and nonsecure) and negotiation of keying methods and material are potential security issues whose remedy warrant integrity protection. Latent configuration support provides hints to the other side about capabilities supported for further offer/answer exchanges, including transport protocols and attribute capabilities (e.g., for keying methods). If an attacker can remove or alter latent configuration information to suggest that only nonsecure or less-secure alternatives are supported, then he may be able to force negotiation of a less secure session than would otherwise have occurred. While the specific attack, as described here, differs from those described in RFC 5939, the considerations and mitigation strategies are similar to those described in RFC 5939.

Another variation on the above removing or altering latent configuration attack involves the session capability ("sescap") attribute defined in this document. The "sescap" enables a preference order to be specified for all of the potential configurations, and that preference will take precedence over any preference indication provided in individual potential configuration attributes. Consequently, an attacker that can insert or modify a "sescap" attribute may be able to force negotiation of an insecure or less secure alternative than would otherwise have occurred. Again, the considerations and mitigation strategies are similar to those described in RFC 5939.

The addition of negotiable media formats and their associated parameters, defined in this specification (RFC 6871, see Section 8.3) can cause problems for middleboxes that attempt to control bandwidth utilization, media flows, and/or processing resource consumption as part of network policy but that do not understand the media capability negotiation feature. As for the initial SDP capability negotiation work (RFC 5939), the SDP answer is formulated in such a way as to always carry the selected media encoding for every media stream selected. Pending an understanding of capabilities negotiation, the middlebox should examine the answer SDP to obtain the best picture of the media streams being established. As always, middleboxes can best do their job if they fully understand media capabilities negotiation.

17.7.4 Miscellaneous Capabilities (RFC 7006)

This specification (RFC 7006, see Section 8.4) provides an extension on top of the SDP (RFC 4566, see Sections 2 and 17.1), SDP offer/answer model (RFC 3264, see Sections 17.2 and 3.1), SDP capability negotiation framework (RFC 5939, see Sections 8.1 and 17.7.1), and SDP media capabilities negotiation (RFC 6871, see Sections 8.3 and 17.7.3). As such, the security considerations of those documents apply. The bandwidth capability attribute may be used for reserving resources at endpoints and intermediaries that inspect SDP. Modification of the bandwidth value by an attacker can lead to the network being underutilized (too high bandwidth value) or congested (too low bandwidth value).

Similarly, by modifying the alternative connection address(es), an attacker would be able to direct media streams to a desired endpoint, thus launching a version of the well-known voice hammer attack (RFC 5245, see Sections 3.2, 17.2, 17.22, and 17.2.2.5.1). The title capability provides for alternative "i=" line information, which is intended for human consumption. However, manipulating the textual information could be misused to provide false information, leading to a bad user experience or the person using the service making a wrong choice regarding the available media streams. If it is essential to protect the capability attribute values, one of the security mechanisms proposed in RFC 5939 SHOULD be used. The "i=" line, and thus the value carried in the title capability attribute, is intended for human-readable description only. It should not be parsed programmatically.

17.8 Network Protocol Support

17.8.1 IPv4/IPv6 (RFC 4566)

In general, security features described in RFC 4566 (see Sections 2, 9.1, 17.1, and 17.2) are applicable. However, for security details specific to IPv4/IPv6 network addresses, we need to use the recommendations of RFC 4566. The call control protocol's user agent (e.g., SIP RFC 3261, also see Section 1.2, and WebRTC – see Section 1.3) inserts IP4/IPv6 network addresses into SDP for communications over the IPv4/IPv6 network. In accordance to RFC 6157, malicious user agents that may intercept a request can mount a DoS attack targeted to the different network interfaces of the user agent. In such a case, the user agent should use mechanisms that protect the confidentiality and integrity of the messages, such as using the SIPS Uniform Resource Identifier (URI) scheme as described in Section 26.2.2 of RFC 3261, or secure MIME as described in Section 23 of RFC 3261. If HTTP digest is used as an authentication mechanism in SIP, then the user agent should ensure that the quality of protection also includes the SDP payload.

17.8.2 Alternate Connectivity Attribute (RFC 6947)

The security implications for ALTC specified in RFC 6947 (see Section 9.2) are effectively the same as they are for SDP in general described in RFC 4566 (see Sections 2, 17.1, and 17.2).

17.8.3 Source Filters (RFC 4570)

In general, the security features specific to SDP messages described in RFC 4566 (see Sections 2, 17.1, and 17.2) are also applicable for SDP source filters specified in RFC 4570 (see Section 9.3). The central issue relevant to using source address filters is the question of address authenticity. Using the source IP address for authentication is weak, since addresses are often dynamically assigned and it is possible for a sender to "spoof" its source address (i.e., use one other than its own) in datagrams it sends. Proper router configuration, however, can reduce the likelihood of "spoofed" source addresses being sent to or from a network. Specifically, border routers are encouraged to filter traffic so that datagrams with invalid source addresses are not forwarded (e.g., routers drop datagrams if the source address is nonlocal) (RFC 2827). This, however, does not prevent IP source addresses from being spoofed on a local area network (LAN).

Also, as noted in Section 9.3.3, tunneling or NAT mechanisms may require corresponding translation of the addresses specified in the SDP "source-filter" attribute and furthermore may cause a set of original source addresses to be translated to a smaller set of source addresses as seen by the receiver. Use of Fully Qualified Domain Names (FQDNs) for either <dest-address> or <src-list> values provides a layer of indirection that provides great flexibility. However, it also exposes the source-filter to any security inadequacies the DNS system may have. If unsecured, it is conceivable that the DNS server could return illegitimate addresses. In addition, if source-filtering is implemented by sharing the source-filter information with network elements, then the security of the protocol(s) that are used for this (e.g., RFC 3376) becomes important, to ensure that legitimate traffic (and only legitimate traffic) is received. For these reasons, receivers SHOULD NOT treat the SDP "source-filter" attribute as its sole mechanism for protecting the integrity of received content.

17.8.4 ATM Bearer Connections (RFC 3108)

The security implications for Asynchronous Transfer Mode (ATM) bearer connections described in RFC 3108 (see Section 9.4) in carrying SDP messages are described in the following sections.

17.8.4.1 Bearer Security

At present, neither the means of encrypting ATM and AAL2 bearers nor the authentication of ATM or AAL2 bearer signaling is conventionalized in the same manner as are the means of encrypting RTP payloads.

The SDP encryption key line ("k=") defined in RFC 4566 that obsoletes RFC 2327 can be used to represent the encryption key and the method of obtaining the key. In the ATM and AAL2 contexts, the term "bearer" can include "bearer signaling" as well as "bearer payloads."

17.8.4.2 Security of the SDP description

The SDP session descriptions might originate in untrusted areas such as equipment owned by end subscribers or located at end-subscriber premises. SDP relies on the security mechanisms of the encapsulating protocol or layers below the encapsulating protocol. Examples of encapsulating protocols are the SIP, MGCP, and Multimedia Gateway Control Protocol (MEGACO). No additional security mechanisms are needed. SIP, MGCP, and MEGACO can use IPsec authentication as described in RFC 1826. IPsec encryption can be optionally used with authentication to provide an additional, potentially more expensive level of security. IPsec security associations can be made between equipment located in untrusted areas and equipment located in trusted areas through configured shared secrets or the use of a certificate authority.

17.8.5 Audio/Video Media Streams over PSTN (RFC 8195)

RFC 8195, described in Section 9.5, provides an extension to RFC 4566 (see Sections 2, 17.1 and 17.2) and RFC 3264 (see Section 3.1 and 17.2). As such, the security considerations of those documents apply. This memo provides mechanisms to agree on a correlation identifier or identifiers that are used to evaluate whether an incoming circuit switched bearer is related to an ongoing session in the IP domain. If an attacker replicates the correlation identifier and establishes a call within the time window the receiving endpoint is expecting a call, the attacker may be able to hijack the circuit-switched bearer.

These types of attacks are not specific to the mechanisms presented in this memo. For example, caller ID spoofing is a well-known attack in the Public Switched Telephone Network (PSTN). Users are advised to use the same caution before revealing sensitive information as they would on any other phone call. Furthermore, users are advised that mechanisms that may be in use in the IP domain for securing the media, like SRTP (RFC 3711), are not available in the Circuit-Switched domain. For the purposes of establishing a circuit-switched bearer, the active endpoint needs to know the passive endpoint's phone number.

Phone numbers are sensitive information, and some people may choose not to reveal their phone numbers when calling using supplementary services like calling line identification restriction (CLIR) in GSM. Implementations should take the caller's preferences regarding calling line identification into account, if possible, by restricting the inclusion of the phone number in the SDP "c=" line if the caller has chosen to use CLIR. If this is not possible, implementations may present a prompt informing the user that his/her phone number may be transmitted to the other party.

As with IP addresses, if there is a desire to protect the SDP containing phone numbers carried in SIP, implementers are advised to follow the security mechanisms defined in RFC 3261 (also see Section 1.2). It is possible for an attacker to create a circuit-switched session whereby the attacked endpoint should dial a circuit-switched number, perhaps even a premium-rate telephone number. To mitigate the consequences of this attack, endpoints MUST authenticate and trust remote endpoint users who try to remain passive in the circuit-switched connection establishment. It is RECOMMENDED that endpoints have local policies precluding the active establishment of circuit switched connections to certain numbers (e.g., international, premium, and long distance). Additionally, it is strongly RECOMMENDED that the end user be asked for consent prior to the endpoint initiating a circuit-switched connection.

17.9 Transport Protocol Support

17.9.1 TCP-Based Media Support (RFC 4145)

In general, the security and other considerations for TCP-based media support by SDP specified in RFC 4145 (see Section 10.1) are applicable to what has been described for RFC 4566 (see Sections 2, 17.1, and 17.2) that obsoletes RFC 2327. An attacker may attempt to modify the values of the connection and setup attributes in order to have endpoints reestablish connections unnecessarily or to keep them from establishing a connection. So, it is strongly RECOMMENDED that integrity protection be applied to the SDP session descriptions. For session descriptions carried in SIP, S/MIME is the natural choice to provide such end-to-end integrity protection, as described in RFC 3261 (also see Section 1.2). Other applications MAY use a different form of integrity protection.

17.10 Media Loopback Support

17.10.1 RTP Media Loopback (RFC 6849)

The security considerations of RFC 3264 (see Sections 3.1 and 17.2) and RFC 3550 apply. Given that media loopback may be automated without the end user's knowledge, the answerer of the media loopback should be aware of DoS attacks. It is RECOMMENDED that session requests for media loopback be authenticated and the frequency of such sessions be limited by the answerer. If the higher-layer signaling protocol were not authenticated, a malicious attacker could create a session between two parties the attacker wishes to target, each party acting as the loopback mirror to the other, of the rtp-pkt-loopback type. A few RTP packets sent to either party would then infinitely loop between the two, as fast as they could process them, consuming their resources and network bandwidth.

Furthermore, media loopback provides a means of attack indirection, whereby a malicious attacker creates a loopback session as the loopback source and uses the mirror to reflect the attacker's packets against a target—perhaps a target the attacker could not reach directly, such as one behind a firewall, for example. Or the attacker could initiate the session as the loopback mirror, in the hopes of making the peer generate media against another target. If end-user devices such as mobile phones answer loopback requests without authentication and without notifying the end user, then an attacker could cause the battery to drain and possibly deny the end user normal phone service or cause network data usage fees. This could even occur naturally if a legitimate loopback session does not terminate properly and the end device does not have a timeout mechanism for such. For the reasons noted above, end-user devices SHOULD provide a means of indicating to the human user that the device is in a loopback session, even if it is an authenticated session. Devices that answer or generate loopback sessions SHOULD either perform keep-alive/refresh tests of the session state through some means or time out the session automatically.

17.11 QOS Support

17.11.1 QOS Mechanism Selection (RFC 5432)

With respect to security of QOS mechanism selection in SDP described in RFC 5432 (see Section 12.1), an attacker may attempt to add, modify, or remove "qos-mech-send" and "qos-mech-recv" attributes from a session description. This could result in an application behaving in a nondesirable way. For example, the endpoints under attack may not be able to find a common QOS mechanism to use. Consequently, it is strongly RECOMMENDED that integrity and authenticity protection be applied to SDP session descriptions carrying these attributes. For session descriptions carried in SIP (RFC 3261, also see Section 1.2), S/MIME (RFC 3851 made obsolete by RFC 5751) is the natural choice to provide such end-to-end integrity protection, as described in RFC 3261. Other applications MAY use a different form of integrity protection.

17.11.2 Bandwidth Modifier for RTCP Bandwidth (RFC 3556)

RFC 3556 (see Section 12.2) defines bandwidth modifier keywords as an extension to SDP, so the security considerations listed in the SDP specification (RFC 4566, see Sections 2 and 17.1) apply to session descriptions containing these modifiers as with any other. The bandwidth value supplied with one of these modifiers could be unreasonably large and cause the application to send RTCP packets at an excessive rate, resulting in a DoS. This is similar to the risk that an unreasonable bandwidth could be specified for the media data, though encoders generally have a limited bandwidth range. Applications should apply validity checks to all parameters received in an SDP description, particularly one that is not authenticated. This RFC 3556 cannot specify limits because they are dependent on the RTP profile and application.

17.11.3 RTCP Attribute (RFC 3605)

The SDP extension specified in RFC 3605 (see Section 12.3) is not believed to introduce any significant security risk to multimedia applications. One could conceive that a malevolent third party would use the extension to redirect the RTCP fraction of an RTP exchange, but this requires intercepting and rewriting the signaling packet carrying the SDP text; if an interceptor can do that, many more attacks are available, including a wholesale change of the addresses and port numbers to which the media will be sent. In order to avoid attacks of this sort, when SDP is used in a signaling packet where it is of the form application/sdp, end-to-end integrity using S/MIME (RFC 3369) is the technical method to be implemented and applied. This is compatible with SIP (RFC 3261, also see Section 1.2).

17.11.4 Transport Independent Bandwidth Modifier (RFC 3890)

The key feature of transport independent bandwidth modifier in SDP, specified by RFC 3890 (see Section 12.4), is the bandwidth. The bandwidth value supplied by the parameters defined here can be altered, if not integrity protected. By altering the bandwidth value, one can fool a Transporeceiver into reserving either more or less bandwidth than is actually needed. Reserving too much may result in unwanted expenses on behalf of the user, while also blocking resources that other parties could have used. If too little bandwidth is reserved, the receiving user's quality may be affected. Trusting a too-large Transport Independent Application Specific Maximum value may also result in the receiver rejecting the session due to insufficient communication and decoding resources.

Due to these security risks, it is strongly RECOMMENDED that the SDP be integrity protected and source authenticated so tampering cannot be performed and the source can be trusted. It is also RECOMMENDED that any receiver of the SDP perform an analysis of the received bandwidth values to verify that they are reasonable expected values for the application. For example, a single channel AMR-encoded voice stream claiming to use 1000 kbps is not reasonable. Please note that some of the above security requirements are in conflict with that required to make signaling protocols using SDP work through a middlebox, as discussed in the security considerations of RFC 3303.

17.11.5 Connectivity Preconditions for SDP Media Streams (RFC 5898)

The security for connectivity preconditions for SDP media streams described in RFC 5898 (see Section 12.5) is described here. General security considerations for preconditions are discussed in RFC 3312 and RFC 4032. As discussed in RFC 4032, it is strongly RECOMMENDED that S/MIME (RFC 3853) integrity protection be applied to the SDP session descriptions. When the user agent provides identity services (rather than the proxy server), the SIP identity mechanism specified in RFC 4474 provides alternative end-to-end integrity protection. Additionally, the following security issues relate specifically to connectivity preconditions. Connectivity preconditions rely on mechanisms beyond SDP, such as TCP (RFC 0793) connection establishment or ICE connectivity checks (RFC 5245), to establish and verify connectivity between an offerer and an answerer. An attacker that prevents those mechanisms from succeeding (e.g., by keeping ICE connectivity checks from arriving at their destinations) can prevent media sessions from being established.

While this attack relates to connectivity preconditions, it is actually an attack against the connection-establishment mechanisms used by the endpoints. This attack can be performed in the presence or absence of connectivity preconditions. In their presence, the whole session setup will be disrupted. In their absence, only the establishment of the particular stream under attack will be disrupted. This specification does not provide any mechanism against attackers able to block traffic between the endpoints involved in the session because such an attacker will always be able to launch DoS attacks.

Instead of blocking the connectivity checks, the attacker can generate forged connectivity checks that will cause the endpoints to assume that there was connectivity when there was actually no connectivity. This attack would result in the user experience being poor because the session would be established without all the media streams being ready. The same attack can be used, regardless of whether connectivity preconditions are used, to attempt to hijack a connection. The forged connectivity checks would trick the endpoints into sending media to the wrong direction. To prevent these attacks, it is RECOMMENDED that the mechanisms used to check connectivity be adequately secured by message authentication and integrity protection. For example, Section 3.2.2.5 of this book (Section 2.5 of RFC 5245) discusses how message integrity and data origin authentication are implemented in ICE connectivity checks.

17.12 Application-Specific Extensions

17.12.1 Offer/Answer Model Enabling File Transfer (RFC 5547)

The SDP attributes defined in this specification (RFC 5547, see Section 13.1) identify a file to be transferred between two endpoints. An endpoint can offer to send the file to the other endpoint or request to receive the file from the other endpoint. In the former case, an attacker modifying those SDP attributes could cheat the receiver making it think that the file to be transferred was a different one. In the latter case, the attacker could make the sender send a file different from the one requested by the receiver. Consequently, it is RECOMMENDED that integrity protection be applied to the SDP session descriptions carrying the attributes specified in this specification.

Additionally, it is RECOMMENDED that senders verify the properties of the file against the selectors that describe it. The descriptions of the files being transferred between endpoints may reveal information the endpoints may consider confidential. Therefore, it is RECOMMENDED that SDP session descriptions carrying the attributes specified in this specification be encrypted. TLS and S/MIME are the natural choices to provide offer/answer exchanges with integrity protection and confidentiality.

When an SDP offer contains the description of a file to be sent or received, the SDP answerer MUST first authenticate the SDP offerer and then authorize the file transfer operation, usually according to a local policy. Typically, these functions are integrated in the high-level protocol that carries SDP (e.g., SIP) and in the file transfer protocol (e.g., Message Session Relay Protocol [MSRP]). If SIP (RFC 3261, also see Section 1.2) and MSRP (RFC 4975) are used, the standard mechanisms for user authentication and authorization are sufficient.

It is possible that a malicious or misbehaving implementation tries to exhaust the resources of the remote endpoint (e.g., the internal memory or the file system) by sending very large files. To protect from this attack, an SDP answer SHOULD first verify the identity of the SDP offerer, and perhaps only accept file transfers from trusted sources. Mechanisms to verify the identity of the file sender depend on the high-level protocol that carries the SDP, for example, SIP (RFC 3261, also see Section 1.2) and MSRP (RFC 4975). It is also RECOMMENDED that implementations take measures to avoid attacks on resource exhaustion, for example, by limiting the size of received files, verifying that there is enough space in the file system to store the file prior to its reception, or limiting the number of simultaneous file transfers.

File receivers MUST also sanitize all input, such as the local file name, prior to making calls to the local file system to store a file. This is to prevent the existence of meaningful special characters inside the files that could damage the local operating system. Once a file has been transferred, the file receiver must take care of it. Typically, file transfer is a commonly used mechanism for transmitting computer virus, spyware, and other types of malware. File receivers should apply all possible security technologies (e.g., antivirus, antispyware) to mitigate the risk of damage at their host.

17.12.2 Binary Floor Control Protocol (RFC 4583)

The Binary Floor Control Protocol (BFCP) (RFC 4583, see Section 13.2), SDP (RFC 4566, see Sections 2 and 17.1), and offer/answer (RFC 3264, see Section 3.1 and 17.2) specifications discuss security issues related to BFCP, SDP, and offer/answer, respectively. In addition, RFC 4145 (see Sections 10.1 and 17.9.1) and RFC 4572 (see Section 16.3) discuss security issues related to the establishment of TCP and TLS connections using an offer/answer model. BFCP assumes that an initial integrity-protected channel is used to exchange self-signed certificates between a client and the floor control server. For session descriptions carried in SIP (RFC 3261, also see Section 1.2), S/MIME (RFC 3853) is the natural choice for providing such a channel.

17.13 Service-Specific Extensions

17.13.1 3GPP Packet-Switched Services (RFC 6064)

The security related to packet-switched streaming and multimedia broadcast/multicast service specific to 3GPP networks (RFC 6064, see Section 14.1) is described here. SDP attributes are subject to modification by an attacker unless they are integrity protected and authenticated. The security considerations of the SDP specification (RFC 4566, see Sections 2 and 17.1) should be reviewed in this regard. The registered SDP attributes are vulnerable to modification attacks or removal, which may result in problems of a serious nature, including failure to use service and reduced quality. The registered RTSP headers are also vulnerable to insertion, deletion, or modification attacks similar to SDP attributes. Also, in this case, attacks can result in failure of the service or reduced quality of streaming content. The three SDP identifiers by themselves introduce do not any additional security threats that don't already exist for other protocol identifiers in SDP. The media stream and the used protocols identified and configured by the SDP identifier may contain security issues.

17.13.2 Open Mobile Alliance Broadcast Service and Content Protection (RFC 5159)

The security related to Open Mobile Alliance broadcast service and content protection (RFC 5149, see Section 14.2) is described here. In addition to the notes in Sections 2.7 and 17.1 (i.e., Section 7 of RFC 4566), the following considerations may be applicable: The bcastversion parameter indicates the version of the broadcast system used for distribution of broadcast content. If future versions indicated by this parameter allow for enhanced or additional security features, the bcastversion parameter, if unprotected, could be utilized for downgrade attacks. The stkmstream parameter provides references to relevant key management streams so that receivers can map the media streams to key streams and retrieve the necessary keys to decrypt media. As such, this parameter could be utilized, if unprotected, for DoS attacks.

17.13.3 PINT Service Protocol (RFC 2848)

The security related PSTN/Internet Interworking (PINT) service protocol described in RFC 2848 (see Section 14.3) is described here. Note that RFCs 2543 and 2327 have been made obsolete by RFCs 3261 (also see Section 1.2) and 4566 (see Sections 2 and 17.1), respectively. In the context of PINT (RFC 2848, see Section 14.3), we have not updated those obsolete RFCs with new ones for the sake of brevity because it will require major changes in RFC 2848.

17.13.3.1 Basic Principles for PINT Use

A PINT Gateway, and the Executive System(s) with which that Gateway is associated, exist to provide service to PINT Requestors. The aim of the PINT protocol is to pass requests from those users on to a PINT Gateway so an associated Executive System can service those requests.

17.13.3.1.1 Responsibility for Service Requests

The facility of making a General Switched Telephone Network (GSTN)-based call to numbers specified in the PINT request, however, comes with some risks. The request can specify an incorrect telephone or fax number. It is also possible that the Requestor has purposely entered the telephone number of an innocent third party. Finally, the request may have been intercepted on its way through any intervening PINT or SIP infrastructure, and the request may have been altered. In any of these cases, the result may be that a call is placed incorrectly. Where there is intent or negligence, this may be construed as harassment of the person incorrectly receiving the call.

Whilst the regulatory framework for misuse of Internet connections differs throughout the world and is not always mature, the rules under which GSTN calls are made are much more settled. Someone may be liable for mistaken or incorrect calls. Understandably, the GSTN operators would prefer that they are not this someone, so they will need to ensure that any PINT Gateway and Executive System combination not generate incorrect calls through some error in the Gateway or Executive system implementation or GSTN-internal communications fault. Equally, it is important that the Operator be able to show that they act only on requests they have good reason to believe are correct. This means that the Gateway must not pass on requests unless it is sure that they have not been corrupted in transit from the requestor. If a request can be shown to have come from a particular Requestor and to have been acted on in good faith by the PINT service provider, then responsibility for making requests may well fall to the requestor rather than the Operator who executed these requests.

Finally, it may be important for the PINT service provider to be able to show that he/she acts only on requests for which he/she has some degree of assurance of origin. In many jurisdictions, it is a requirement that GSTN Operators place calls only when they can, if required, identify the parties to the call (such as when required to carry out a malicious call trace). It is at least likely that providers of PINT services will have a similar responsibility placed on them. It follows that PINT service providers may require that the identity of the requestor be confirmed. If such confirmation is not available, then they may be forced (or choose) not to provide service. This identification may require personal authentication of the requesting user.

17.13.3.1.2 Authority to Make Requests

Where GSTN resources are used to provide a PINT service, it is at least possible that someone will have to pay for it. This person may not be the Requestor, as, for example, in the case of existing GSTN split-charging services like free phone in which the recipient of a call rather than the originator is responsible for the call cost. This is not, of course, the only possibility; for example, PINT service may be provided on a subscription basis, and there are a number of other models. However, whichever model is chosen, there may be a requirement for confirming the authority of a requestor to make a PINT request. If such confirmation is not available, then, again, the PINT gateway and associated executive system may choose not to provide service.

17.13.3.1.3 Privacy

Even if the identity of the requesting user and the authority under which they make their request is known, there remains the possibility that the request is corrupted, maliciously altered, or even replaced whilst in transit between the Requestor and the PINT Gateway. Similarly, information on the authority under which a request is made may well be carried within that request. This can be sensitive information, as an eavesdropper might steal this and use it within his/her own requests. Such authority SHOULD be treated as if it were financial information such as a credit card number or PIN.

The data authorizing a requesting user to make a PINT request should be known only to him/her and the service provider. However, this information may be in a form that does not match the schemes normally used within the Internet. For example, X.509 certificates (RFC 5280) are commonly used for secured transactions on the Internet both in the IP security architecture (RFC 4346 that obsoletes RFC 2246) and in the TLS protocol (RFC 3801 that obsoletes RFC 2401), but the GSTN provider may only store an account code and PIN (i.e., a fixed string of numbers).

A requesting user has a reasonable expectation that his/her requests for service are confidential. For some PINT services, no content is carried over the Internet; however, the telephone or fax numbers of the parties to a resulting service call may be considered sensitive. As a result, it is likely that the requestor (and the PINT service provider) will require that any request sent across the Internet be protected against eavesdroppers; in short, the requests SHOULD be encrypted.

17.13.3.1.4 Privacy Implications of SUBSCRIBE/NOTIFY

Some special considerations relate to monitoring sessions using the SUBSCRIBE and NOTIFY messages. The SUBSCRIBE message used to register an interest in the disposition of a PINT service transaction uses the original session description carried in the related INVITE message. This current specification does not restrict the source of such a SUBSCRIBE message, so it is possible for an eavesdropper to capture an unprotected session description and use this in a subsequent SUBSCRIBE request. In this way, it is possible to find out details on that transaction that may well be considered sensitive.

The initial solution to this risk is to RECOMMEND that a session description that may be used within a subsequent SUBSCRIBE message be protected. However, there is a further risk: If the origin-field used is "guessable" then it is possible for an attacker to reconstruct the session description and use this reconstruction within a SUBSCRIBE message. SDP "o=" field (see Section 2.5.2 of this book, that is, Section 5.2 of RFC 4566) does not specify the mechanism used to generate the sess-id field and suggests that a method based on timestamps produced by Network Time Protocol (RFC 1305) be used. This is sufficient to guarantee uniqueness but may allow the value to be guessed, particularly if other unprotected requests from the same originator are available.

Thus, to ensure that the session identifier is not guessable, the techniques described in Section 6.2 of RFC 4086 that obsoletes RFC 1750 can be used when generating the origin-field for a session description inside a PINT INVITE message. If all requests from (and responses to) a particular PINT requesting entity are protected, then this is not needed. Where such a situation is not assured, AND where session monitoring is supported, then a method by which an origin-field within a session description is not guessable SHOULD be used.

17.13.3.2 Registration Procedures

Any number of PINT Gateways may register to provide the same service; this is indicated by the Gateways specifying the same "userinfo" part in the "To:" header field of the REGISTER request. Whilst such ambiguity would be unlikely to occur with the scenarios covered by "core" SIP, it is very likely for PINT; there could be any number of service providers willing to support a "Request-To-Fax" service, for example. Unless a request specifies the Gateway name explicitly, an intervening Proxy that acts on a registration database to which several Gateways have all registered is in a position to select from the registrants using whatever algorithm it chooses; in principle, any Gateway that has registered as "R2F" would be appropriate.

However, this opens an avenue for attack, and this is one in which a "rogue" Gateway operator stands to make a significant gain. The standard SIP procedure for releasing a registration is to send a REGISTER request with a Contact field having a wildcard value and an expires parameter with a value of 0. It is important that a PINT Registrar uses authentication of the Registrand, as otherwise one PINT service provider would be able to "spoof" another and remove his/her registration. As this would stop the Proxy passing any requests to that provider, this would both increase requests are sent to the rogue and stop requests going to the victim. Another variant on this attack would be to register a Gateway using a name that has been registered by another provider; thus a rogue Operator might register its Gateway as "R2C@pint.att.com," thereby hijacking requests. The solution is the same; all registrations by PINT Gateways MUST be authenticated; this includes both new or apparent replacement registrations and any cancellation of current registrations. This recommendation is also made in the SIP specification, but for the correct operation of PINT, it is very important indeed.

17.13.3.3 Security Mechanisms and Implications on PINT Service

PINT is a set of extensions to SIP (RFC 2543 made obsolete by RFC 3261, also see Section 1.2) and SDP (RFC 2347 that was made obsolete by RFC 4566, see Section 2), and will use the security procedures described in SIP. There are several implications of this, and these are covered here. For several of the PINT services, the "To:" header field of SIP is used to identify one of the parties to the resulting service call. The PINT Request-To-Call service is an example. As mentioned in the SIP specification, this field is used to route SIP messages through an infrastructure of Redirect and Proxy server between the corresponding user agent servers, and so cannot be encrypted. This means that, although the majority of personal or sensitive data can be protected whilst in transit, the telephone (or fax) number of one of the parties to a PINT service call cannot, and will be "visible" to any interception. For the PINT milestone services this may be acceptable, since the caller named in the To: service is typically a "well known" provider address, such as a Call Center.

Another aspect is that, even if the requesting user does not consider the telephone or fax numbers of the parties to a PINT service private, those parties might. Where PINT servers have reason to believe this, they SHOULD encrypt the requests, even if the requestor has not done so. This could happen, for example, if a Requesting User within a company placed a PINT request and this was carried via the company's Intranet to its proxy/firewall and thence over the Internet to a PINT Gateway at another location. If a request carries data that can be reused by an eavesdropper either to "spoof" the Requestor or to obtain PINT service by inserting the Requestor's authorization token into an eavesdropper's request, then this data MUST be protected. This is particularly important if the authorization token consists of static text (such as an account code and/or PIN).

One approach is to encrypt the whole of the request, using the methods described in the SIP specification. As an alternative, it may be acceptable for the authorization token to be held as an opaque reference (see Section 14.3.3.4.2.3 and examples 14.3.4.11 and 14.3.4.12) (i.e., Section 3.4.2.3 and examples 4.11 and 4.12 of RFC 2848), using some proprietary scheme agreed upon by the requestor and the PINT service provider, as long as this is resistant to interception and reuse. Also, it may be that the authorization token cannot be used outside of a request cryptographically signed by the requestor; if so, this requirement can be relaxed, as in this case the token cannot be reused by another. However, unless both the Requestor and the Gateway are assured this is the case, any authorization token MUST be treated as sensitive, and so MUST be encrypted.

A PINT request may contain data within the SDP message body that can be used more efficiently to route that request. For example, it may be that one Gateway and Executive System combination cannot handle a request that specifies one of the parties as a pager, whilst another can. Both gateways may have registered with a PINT/SIP Registrar, and this information may be available to intervening PINT/SIP Proxies. However, if the message body is encrypted, then the request cannot be decoded at the Proxy server, and so Gateway selection based on contained information cannot be made there. The result is that the Proxy may deliver the request to a Gateway that cannot handle it; the implication is that a PINT/SIP Proxy SHOULD consider its choice of the appropriate Gateway subject to correction and, on receiving a 501 or 415 rejection from the first gateway chosen, try another. In this way, the request will succeed if at all possible, even though it may be delayed (and tie up resources in the inappropriate Gateways).

This opens up an interesting avenue for DoS: sending a valid request that appears to be suitable for a number of different Gateways and simply occupying those Gateways in decrypting a message requesting a service they cannot provide. As mentioned in Section 14.3.3.5.5.1 (i.e., Section 3.5.5.1 of RFC 2848), the choice of service name to be passed in the userinfo portion of the SIP Request-URI is flexible, and it is RECOMMENDED that names be chosen that allow a Proxy to select an appropriate Gateway without having to examine the SDP body part. Thus, in the example given here, the service might be called "Request-To-Page" or "R2P" rather than the more general use of "R2F," if there is a possibility of the SDP body part being protected during transit. A variation on this attack is to provide a request that is syntactically invalid but that, due to the encryption, cannot be detected without expending resources in decoding it. The effects of this form of attack can be minimized in the same way as for any SIP Invitation; the Proxy should detect the 400 rejection returned from the initial Gateway, and not pass the request onwards to another.

Finally, note that the Requesting User may not have a prior relationship with a PINT Gateway, whilst still having a prior relationship with the Operator of the Executive System that fulfills their request. Thus there may be two levels of authentication and authorization; one carried out using the techniques described in the SIP specification (for use between the Requestor and the Gateway), with another being used between the Requesting User or the Requestor and the Executive System. For example, the Requesting User may have an account with the PINT service provider. That provider might require that requests include this identity before they will be convinced to provide service. In addition, to counter attacks on the request whilst it is in transit across the Internet, the Gateway may require a separate X.509-based certification of the request. These are two separate procedures, and data needed for the former would normally be expected to be held in opaque references inside the SDP body part of the request. The detailed operation of this mechanism is, by definition, outside the scope of an Internet Protocol, and so must be considered a private matter. However, one approach to indicating to the Requestor that such "second level" authentication or authorization is required by their Service Provider would be to ask for this inside the textual description carried with a 401 response returned from the PINT Gateway.

17.13.3.4 Summary of Security Implications

From the above discussion, PINT always carries data items that are sensitive, and there may be financial considerations as well as the more normal privacy concerns. As a result, the transactions MUST be protected from interception, modification, and replay in transit. PINT is based on SIP and SDP and can use the security procedures outlined in RFC 2543, Sections 13 and 15, (RFC 2543 was made obsolete by RFC 3261, see also Section 1.2). However, in the case of PINT, the SIP recommendation that requests and responses MAY be protected is not enough. PINT messages MUST be protected, so PINT implementations MUST support SIP Security (as described in RFC 2543, Sections 13 and 15) and be capable of handling such received messages. In some configurations, PINT Clients, Servers, and Gateways can be sure that they operate using the services of network level security (RFC 2401), TLS (RFC 2246), or physical security for all communications between them. In these cases, messages MAY be exchanged without SIP security, since all traffic is protected already. Clients and servers SHOULD support manual configuration to use such lower-layer security facilities. When using network-layer security (RFC 4301 that obsoletes RFC 2401), the Security Policy Database MUST be configured to provide appropriate protection to PINT traffic. When using TLS, a port configured MUST NOT also be configured for non-TLS traffic. When TLS is used, basic authentication MUST be supported, and client-side certificates MAY be supported.

Authentication of the client making the request is required, however, so if this is not provided by the underlying mechanism used, then it MUST be included within the PINT messages using SIP authentication techniques. In contrast with SIP, PINT requests are often sent to parties with which a prior communications relationship exists (such as a Telephone Carrier). In this case, there may be a shared secret between the client and the PINT Gateway. Such PINT systems MAY use authentication based on shared secrets, with HTTP "basic authentication." When this is done, the message integrity and privacy must be guaranteed by some lower-layer mechanism. There are implications on the operation of PINT here, though. If a PINT proxy or redirect server is used, then it must be able to examine the contents of the IP datagrams carried. It follows that an end-to-end approach using network-layer security between the PINT Client and a PINT Gateway precludes the use of an intervening proxy; communication between the Client and Gateway is carried via a tunnel to which any intervening entity cannot gain access, even if the IP datagrams are carried via this node. Conversely, if a "hop-by-hop" approach is used, then any intervening PINT proxies (or redirect servers) are, by implication, trusted entities.

However, if there is any doubt that there is an underlying network or TLS association in place, then the players in a PINT protocol exchange MUST use encryption and authentication techniques within the protocol itself. The techniques described in Section 15 of RFC 2543 (that was made obsolete by RFC 3261, also see Section 1.2) MUST be used, unless there is an alternative protection scheme that is agreed upon by the parties. In either case, the content of any message body (or bodies) carried within a PINT request or response MUST be protected; this has implications on the options for routing requests via proxies (see Section 14.3.5.3) (i.e., Section 5.3 of RFC 2848). Using SIP techniques for protection, the request-URI and "To:" fields headers within PINT requests cannot be protected. In the baseline PINT services, these fields may contain sensitive information. This is a consideration, and if these data ARE considered sensitive, then this will preclude the sole use of SIP techniques; in such a situation, transport (RFC 4346 that obsoletes RFC 2246) or network layer (RFC 4301 that obsoletes RFC 2401) protection mechanisms MUST be used. As a final point, this choice will in turn have an influence on the choice of transport layer protocol that can be used; if a TLS association is available between two nodes, then TCP will have to be used. This is different from the default behavior of SIP (try UDP, then try TCP if that fails).

17.14 Security Capability Negotiation

17.14.1 Key Management Extensions (RFC 4567)

The framework for transfer of key management data as described for RFC 4567 (see Section 16.1) is intended to provide the security parameters for the end-to-end protection of the media session. It is furthermore good practice to secure the session setup (e.g., SDP, SIP, RTSP, and SAP). However, it might be that the security of the session setup is not possible to achieve end to end, but only hop by hop. For example, SIP requires intermediate proxies to have access to part of the SIP message and sometimes also to the SDP description (cf. RFC 4189), although end-to-end confidentiality can hide bodies from intermediaries. General security considerations for the session setup can be found in SDP (RFC 4566, see Sections 2 and 17.1), SIP (RFC 3261, also see Section 1.2), and RTSP (RFC 2326). The framework defined in this memo is useful when the session setup is not protected in an end-to-end fashion, but the media streams need to be end-to-end protected; hence the security parameters (such as keys) are not wanted revealed to or manipulated by intermediaries.

The security will also depend on the level of security the key management protocol offers. It follows that, under the assumption that the key management schemes are secure, the SDP can be passed along unencrypted without affecting the key management as such, and the media streams will still be secure even if some attackers gained knowledge of the SDP contents. Further security considerations can be found for each key management protocol (for MIKEY these can be found in RFC 3830). However, if the SDP messages are not sent integrity protected between the parties, it is possible for an active attacker to change attributes without being detected. As the key management protocol may (indirectly) rely on some of the session information from SDP (e.g., address information), an attack on SDP may have indirect consequences on the key management. Even if the key management protocol does not rely on parameters of SDP and will not be affected by the manipulation of these, different DoS attacks aimed at SDP may lead to undesired interruption in the setup. See also the attacks described at the end of this section.

The only integrity-protected attribute of the media stream is, in the framework proposed here, the set of key management protocols. For instance, it is possible to (1) swap key management offers across SDP messages, or (2) inject a previous key management offer into a new SDP message.

Making the (necessary) assumption that all involved key management protocols are secure, the second attack will be detected by replay protection mechanisms of the key management protocol(s).

Making the further assumption that, according to normal best current practice, the production of each key management offer is done with independent (pseudo)random choices (for session keys and other parameters), the first attack will either be detected in the responder's (now incorrect) verification reply message (if such is used) or be a pure DoS attack, resulting in initiator and responder using different keys. It is RECOMMENDED for the identity at the SPD level to be the one authenticated at the key management protocol level. However, this might need to take into consideration privacy aspects, which are out of scope of this framework.

The use of multiple key management protocols in the same offer may open up the possibility of a bidding-down attack, as specified in Section 16.1.4.1.4. To exclude such possibility, the authentication of the protocol identifier list is used. Note, though, that the security level of the authenticated protocol identifier will be as high (or low), as the "weakest" protocol. Therefore, the offer MUST NOT contain any security protocols (or configurations thereof) weaker than permitted by local security policy.

Note that it is impossible to ensure the authenticity of a declined offer, since even if it comes from the true respondent, the fact that the answerer declines the offer usually means that he does not support the protocol(s) offered, and consequently cannot be expected to authenticate the response either. This means that if the initiator is unsure which protocol(s) the responder supports, we RECOMMEND that the initiator offer all acceptable protocols in a single offer. If not, this opens up the possibility for a "man-in-the-middle" (MITM) to affect the outcome of the eventually agreed-upon protocol, by faking unauthenticated error messages until the initiator eventually offers a protocol "to the liking" of the MITM.

This is not really a security problem, but rather a mild form of DoS that can be avoided by following the above recommendation. Note also that the declined offer could be the result of an attacker who sits on the path and removes all the key management offers. The bidding-down attack prevention, as described above, would not work in this case (as the answerer receives no key management attribute). Also, here it is impossible to ensure the authenticity of a declined offer, though here the reason is the "peeling-off" attack. It is up to the local policy to decide the behavior in the case that the response declines any security (therefore, there is impossibility of authenticating it). For example, if the local policy requires a secure communication and cannot accept an unsecured one, then the session setup SHALL be aborted.

17.14.2 Security Descriptions for Media Streams (RFC 4568)

The security for media streams specified in RFC 4568 (see Section 16.2) is described here. Like all SDP messages, SDP messages containing security descriptions are conveyed in an encapsulating application protocol (e.g., SIP, MGCP). It is the responsibility of the encapsulating protocol to ensure the protection of the SDP security descriptions. Therefore, IT IS REQUIRED that the application invoke its own security mechanisms (e.g., secure multipart such as S/MIME – RFC 3850 made obsolete by RFC 5750) or, alternatively, utilize a lower-layer security service (e.g., TLS or IPsec). IT IS REQUIRED that this security service provide strong message authentication and packet-payload encryption, as well as effective replay protection. "Replay protection" is needed against an attacker that has enough access to the communications channel to intercept messages and to deliver copies to the destination.

A successful replay attack will cause the recipient to perform duplicate processing on a message; the attack is worse when the duped recipient sends a duplicate reply to the initiator. Replay protections are not found in S/MIME or in the other secure multipart standard, PGP/MIME. S/MIME and PGP/MIME, therefore, need to be augmented with some replay-protection mechanism that is appropriate to the encapsulating application protocol (e.g., SIP, MGCP). Three common ways to provide replay protection are to place a sequence number in the message, to use a timestamp, or for the receiver to keep a hash of the message to be compared with incoming messages. There typically needs to be a replay "window" and some policy for keeping state information from previous messages in a "replay table" or list.

The discussion that follows uses "message authentication" and "message confidentiality" in a manner consistent with SRTP (RFC 3711). "Message confidentiality" means that only the holder of the secret decryption key can access the plain-text content of the message. The decryption key is the same key as the encryption key, using the SRTP counter mode and f8 encryption transforms, which are vulnerable to message tampering and need SRTP message authentication to detect such tampering. "Message authentication" and "message integrity validation" generally mean the same thing in IETF security standards:

> An SRTP message is authenticated following a successful HMAC integrity check (RFC 3711), which proves that the message originated from the holder of an SRTP master key and was not altered en-route. Such an "authentic" message, however, can be captured by an attacker and "replayed" when the attacker re-inserts the packet into the channel. A replayed packet can have a variety of bad effects on the session, and SRTP uses the extended sequence number to detect replayed SRTP packets. (RFC 3711)

The SRTP specification identifies which services and features are default values that are normative-to-implement (such as AES_CM_128_80) versus normative-to-use (such as AES_CM_128_32).

17.14.2.1 Authentication of Packets

Security descriptions as defined herein signal security services for RTP packets. RTP messages are vulnerable to a variety of attacks, such as replay and forging. To limit these attacks, SRTP message integrity mechanisms SHOULD be used (SRTP replay protection is always enabled).

17.14.2.2 Keystream Reuse

SRTP security descriptions signal configuration parameters for SRTP sessions. Misconfigured SRTP sessions are vulnerable to attacks on their encryption services when running the crypto suites defined in Sections 6.2.1–6.2.3. An SRTP encryption service is "misconfigured" when two or more media streams are encrypted using the same keystream of AES blocks. When senders and receivers share derived session keys, SRTP requires that the SSRCs of session participants serve to make their corresponding keystreams unique, which is violated in the case of SSRC collision: SRTP SSRC collision drastically weakens SRTP or SRTCP payload encryption during the time that identical keystreams are used (RFC 3711). An attacker, for example, might collect SRTP and SRTCP messages and await a collision. This attack on the AES-CM and AES-f8 encryption is avoided entirely when each media stream has its own unique master key in both the send and receive directions. This specification restricts the use of SDP security description to unicast point-to-point streams so that keys are not shared between SRTP hosts, and the master keys used in the send and receive directions for a given media stream are unique.

17.14.2.3 Signaling Authentication and Signaling Encryption

There is no reason to incur the complexity and computational expense of SRTP, however, when its key establishment is exposed to unauthorized parties. In most cases, the SRTP crypto attribute and its parameters are vulnerable to DoS attacks when they are carried in an unauthenticated SDP message. In some cases, the integrity or confidentiality of the RTP stream can be compromised. For example, if an attacker sets UNENCRYPTED for the SRTP stream in an offer, this could result in the answerer's not decrypting the encrypted SRTP messages. In the worst case, the answerer might itself send unencrypted SRTP and leave its data exposed to snooping. Thus, IT IS REQUIRED that MIME secure multipart, IPsec, TLS, or some other data security service provide message authentication for the encapsulating protocol that carries the SDP messages having a crypto attribute (a=crypto). Furthermore, IT IS REQUIRED that encryption of the encapsulating payload be used whenever a master key parameter (inline) appears in the message. Failure to encrypt the SDP message containing an inline SRTP master key renders the SRTP authentication or encryption service useless in practically all circumstances. Failure to authenticate an SDP message that carries SRTP parameters renders the SRTP authentication or encryption service useless in most practical applications.

When the communication path of the SDP message is routed through intermediate systems that inspect parts of the SDP message, security protocols such as RFC 4301 or TLS SHOULD NOT be used for encrypting and/or authenticating the security description. In the case of intermediate-system processing of a message containing SDP security descriptions, the "a=crypto" attributes SHOULD be protected end to end so that the intermediate system can neither modify the security description nor access the keying material. Network or transport security protocols that terminate at each intermediate system, therefore, SHOULD NOT be used for protecting SDP security descriptions. A security protocol SHOULD allow the security descriptions to be encrypted and authenticated end to end independently of the portions of the SDP message that any intermediate system modifies or inspects: MIME secure multiparts are RECOMMENDED for the protection of SDP messages that are processed by intermediate systems.

17.14.3 Connection-Oriented Media over TLS (RFC 4572)

This entire document (RFC 4572, see Section 16.3) concerns security. The security problem has been presented in Section 16.3.3. See the SDP specification (RFC 4566, see Sections 2 and 17.1) for security considerations applicable to SDP in general. Offering a TCP/TLS connection in SDP (or agreeing to one in the SDP offer/answer mode) does not create an obligation for an endpoint to accept any TLS connection with a given fingerprint. Instead, the endpoint must engage in the standard TLS negotiation procedure to ensure that the TLS stream cipher and MAC algorithm chosen meet the security needs of the higher-level application. (For example, an offered stream cipher of TLS_NULL_WITH_NULL_NULL SHOULD be rejected in almost every application scenario.)

Like all SDP messages, SDP messages describing TLS streams are conveyed in an encapsulating application protocol (e.g., SIP, MGCP, etc.). It is the responsibility of the encapsulating protocol to ensure the integrity of the SDP security descriptions. Therefore, the application protocol SHOULD either invoke its own security mechanisms (e.g., secure multiparts) or, alternatively, utilize a lower-layer security service (e.g., TLS or IPsec). This security service SHOULD provide strong message authentication as well as effective replay protection.

However, such integrity protection is not always possible. For these cases, end systems SHOULD maintain a cache of certificates that other parties have previously presented using this mechanism. If possible, users SHOULD be notified when an unsecured certificate associated with a previously unknown end system is presented and SHOULD be strongly warned if a different unsecured certificate is presented by a party with which it has communicated in the past. In this way, even in the absence of integrity protection for SDP, the security of this document's mechanism is equivalent to that of the Secure Shell protocol (RFC 4251) which is vulnerable to MITM attacks when two parties first communicate but can detect ones that occur subsequently. (Note that a precise definition of the "other party" depends on the application protocol carrying the SDP message.) Users SHOULD NOT, however, under any circumstances be notified about certificates described in SDP descriptions sent over an integrity protected channel.

To aid interoperability and deployment, security protocols that provide only hop-by-hop integrity protection (e.g., the sips protocol, RFC 3261, SIP over TLS) are considered sufficiently secure to allow the mode in which any syntactically valid identity is accepted in a certificate. This decision was made because sips is currently the integrity mechanism most likely to be used in deployed networks in the short to medium term. However, in this mode, SDP integrity is vulnerable to attacks by compromised or malicious middleboxes (e.g., SIP proxy servers). End systems MAY warn users about SDP sessions that are secured in only a hop-by-hop manner, and definitions of media formats running over TCP/TLS MAY specify that only end-to-end integrity mechanisms be used.

Depending on how SDP messages are transmitted, it is not always possible to determine whether a subjectAltName presented in a remote certificate is expected for the remote party. In particular, given call forwarding, third-party call control, or session descriptions generated by endpoints controlled by the Gateway Control Protocol (RFC 5125 that obsoletes RFC 3525), it is not always possible in SIP to determine what entity ought to have generated a remote SDP response. In general, when not using authenticity and integrity protection of SDP descriptions, a certificate transmitted over SIP SHOULD assert the endpoint's SIP Address of Record as a uniformResourceIndicator subjectAltName. When an endpoint receives a certificate over SIP asserting an identity (including an IPAddress or dNSName identity) other than the one to which it placed or received the call, it SHOULD alert the user and ask for confirmation. This applies whether certificates are self-signed or signed by certification authorities; a certificate for sip:bob@example.com may be legitimately signed by a certification authority but may still not be acceptable for a call to sip:alice@example.com. (This issue is not one specific to this specification; the same consideration applies for S/MIME-signed SDP carried over SIP.)

This document does not define any mechanism for securely transporting RTP and RTCP packets over a connection-oriented channel. There was no consensus in the working group as to whether it would be better to send SRTP packets (RFC 3711) over a connection-oriented transport (RFC 4571) or whether it would be better to send standard unsecured RTP packets over TLS using the mechanisms described in this document. The group consensus was to wait until a use case requiring secure connection-oriented RTP was presented. TLS is not always the most appropriate choice for secure connection-oriented media; in some cases, a higher- or lower-level security protocol may be appropriate.

17.14.4 Media Preconditions for SDP Media Streams (RFC 5027)

The security for media preconditions for media streams in SDP specified in RFC 5027 (see Section 16.4) is described here. In addition to the general security considerations for preconditions provided in RFC 3312, the following security issues should be considered. Security preconditions delay session establishment until cryptographic parameters required to send and/or receive media for a media stream have been negotiated. Negotiation of such parameters can fail for a variety of reasons, including policy preventing use of certain cryptographic algorithms, keys, and other security parameters. If an attacker can remove security preconditions or downgrade the strength-tag from an offer/answer exchange, the attacker can thereby cause user-alert for a session that may have no functioning media. This is likely to cause inconvenience to both the offerer and the answerer. Similarly, security preconditions can be used to prevent clipping due to race conditions between offer/answer exchanges and secure media stream packets based on those offer/answer exchanges. If an attacker can remove or downgrade the strength-tag of security preconditions from an offer/answer exchange, the attacker can cause clipping to occur in the associated secure media stream.

Conversely, an attacker might add security preconditions to offers that do not contain them or increase their strength-tag. This in turn may lead to session failure (e.g., if the answerer does not support it), heterogeneous error response forking problems, or a delay in session establishment that was not desired. Use of signaling integrity mechanisms can prevent all of these problems. Where intermediaries on the signaling path (e.g., SIP proxies) are trusted, it is sufficient to use only hop-by-hop integrity protection of signaling (e.g., IPsec or TLS). In all other cases, end-to-end integrity protection of signaling (e.g., S/MIME) MUST be used. Note that the end-to-end integrity protection MUST cover not only the message body, which contains the security preconditions, but also the SIP "Supported" and "Require" headers, which may contain the "precondition" option-tag. If only the message body were integrity protected, removal of the "precondition" option tag could lead to clipping (when a security precondition was otherwise to be used), whereas addition of the option-tag could lead to session failure (if the other side does not support preconditions).

As specified in Section 16.4.3, security preconditions do not guarantee that an established media stream will be secure. They merely guarantee that the recipient of the media stream packets will be able to perform any relevant decryption and integrity checking on those media stream packets. Current SDP (RFC 4566, see Sections 2 and 17.1) and associated offer/answer procedures (RFC 3264) allows only a single type of transport protocol to be negotiated for a given media stream in an offer/answer exchange. Negotiation of alternative transport protocols (e.g., plain and SRTP) is currently not defined. Thus, if the transport protocol offered (e.g., SRTP) is not supported, the offered media stream will simply be rejected. There is however work in progress to address that. For example, the SDP capability negotiation framework (RFCs 5939, see Sections 8.1 and 17.7.1, and 6871, see Sections 8.3 and 17.7.3) defines a method for negotiating the use of a secure or a nonsecure transport protocol by use of SDP and the offer/answer model with various extensions. Such a mechanism introduces a number of security considerations in general; however, use of SDP Security Preconditions with such a mechanism introduces the following security precondition specific security considerations:

A basic premise of negotiating secure and nonsecure media streams as alternatives is that the offerer's security policy allows for nonsecure media. If the offer were to include secure and nonsecure media streams as alternative offers, and media for either alternative may be received prior to the answer, then the offerer may not know whether the answerer accepted the secure alternative. An active attacker thus may be able to inject malicious media stream packets until the answer (indicating the chosen secure alternative) is received. From a security point of view, it is important to note that use of security preconditions (even with a mandatory strength-tag) would not address this vulnerability since security preconditions would effectively apply only to the secure media stream alternatives. If the nonsecure media stream alternative were selected by the answerer, the security precondition would be satisfied by definition, the session could progress and (nonsecure) media could be received prior to the answer being received.

17.14.5 Media Description for IKE (RFC 6193)

The SDP security related to media description for IKE described in RFC 6193 (see Section 16.5) is described here. This entire document (RFC 6193) concerns security, but the security considerations applicable to SDP in general are described in the SDP specification (RFC 4566, see Sections 2 and 17.1). The security issues that should be considered in using comedia-tls are described in Section 16.3.7 in its specification (RFC 4572, see Section 16.3). This section mainly describes the security considerations specific to the negotiation of IKE using comedia-tls. Offering IKE in SDP (or agreeing to one in the SDP offer/answer model) does not create an obligation for an endpoint to accept any IKE session with the given fingerprint. However, the endpoint must engage in the standard IKE negotiation procedure to ensure that the chosen IPsec security associations (including encryption and authentication algorithms) meet the security requirements of the higher-level application. When IKE has finished negotiating, the decision to conclude IKE and establish an IPsec security association with the remote peer is entirely the decision of each endpoint. This procedure is similar to how VPNs are typically established in the absence of SIP. In the general authentication process in IKE, subject DN or subjectAltName is recognized as the identity of the remote party.

However, by using SIP identity and SIP-connected identity mechanisms in this spec, certificates are used simply as carriers for the public keys of the peers, and there is no need for the information about who is the signer of the certificate and who is indicated by subject DN. In this document, the purpose of using IKE is to launch the IPsec Security Association; it is not for the security mechanism of RTP and RTCP (RFC 3550) packets. In fact, this mechanism cannot provide end-to-end security inside the VPN as long as the VPN uses tunnel mode IPsec. Therefore, other security methods such as the SRTP (RFC 3711) must be used to secure the packets. When using the specification defined in this document, it needs to be considered that under the following circumstances, security based on SIP authentication provided by SIP proxy may be breached.

- If a legitimate user's terminal is used by another person, it may be able to establish a VPN with the legitimate identity information. This issue also applies to the general VPN cases based on the shared secret key. Furthermore, in SIP we have a similar problem when file transfer, IM, or comedia-tls where nonvoice/video is used as a means of communication.
- If a malicious user hijacks the proxy, he or she can use whatever credential is on the access control list to gain access to the home network.

For countermeasures to these issues, it is recommended that one use unique information such as a password that only a legitimate user knows for VPN establishment. Validating the originating user by voice or video before establishing VPN would be another method.

17.15 Summary

RFC 4566 that describes the basic SDP provides the security framework considering the fact that SDP itself will not initiate any calls. Rather it will be used in conjunction with call control protocols like SIP (also see Section 1.2), WebRTC (see Section 1.3), and others. So, the call control protocols that use SDP will themselves will have mechanisms for the protection of SDP. However, we have described that RFC 4566 provides detailed descriptions of the security vulnerabilities such as DoS, diversion of media streams that may enable eavesdropping, disabling of calls, injections of unwanted media streams, and other kinds of attacks that are possible if SDP messages are not protected. The conferencing parties must perform authentication and other security features at the time of initiation of the call and nothing to be transferred to an inactive call. The intermediaries (e.g., back-to-back user agent and NAT) may modify the call including SDP, and the call control protocols must take care of the security. It also describes how the "k=" parameter is vulnerable and the channel over which SDP that is delivered itself needs to be authenticated before sending SDP messages.

Over time, the SDP specification described in RFC 4566 has been extended and we have also described security mechanisms for dealing with some SDP parameters for the following SDP attributes: offer/answer, signaling media dependency, security grouping, media multiplexing/bundling, RTP media grouping, generic capability, image (generic), media loopback, QoS, rtcp bandwidth, transport-independent bandwidth modifier, connectivity precondition, TCP-based media transport, source filter, source filter in offer/answer, key management, media stream security (authentication of packets, keystream reuse, signaling authentication and encryption), TLS, and precondition.

17.16 Problems

1. What is the generic security framework defined in RFC 4566 for SDP? Describe in detail using call flows. Does RFC 4566 define any SDP-specific security? If not, why not?
2. Explain in detail how the "k=" parameter is vulnerable to security attacks Describe procedures showing how "k=" can be used securely.
3. Describe in detail including call flows the security vulnerabilities and how these vulnerabilities can be mitigated for each of these SDP attributes: (a) offer/answer, (b) signaling media dependency, (c) security grouping, (d) media multiplexing/bundling, (e) RTP media grouping, (f) generic capability, (g) image (generic), (h) media loopback, (i) QoS, (j) rtcp bandwidth, (k) transport-independent bandwidth modifier, (l) connectivity pre-condition, (m) TCP-based media transport, (n) source filter, (o) source filter in offer/answer, (p) key management, (q) media stream security (authentication of packets, keystream reuse, signaling authentication and encryption), (r) TLS, and (s) pre-condition.
4. Write a complete paper generalizing all security attributes shown in Problem 3 for SDP.

Chapter 18

Capability Negotiations Using SDP by SIP and WebRTC

Abstract

Earlier, we explained that Session Description Protocol (SDP) itself cannot be used for session capability negotiation, because it needs a call control protocol to carry its attributes. The actual application of SDP for capability negotiations between conferencing parties happens when a call control protocol (e.g., Session Initiation Protocol (SIP) or web real-time communication [WebRTC]) initiates the call setup where SDP attributes are carried by the call control protocol in the form of, for example, the offer/answer model. We have chosen SIP and WebRTC call control protocols that will use SDP attributes for capability negotiations for setting up the multimedia conferencing. That is, we provide the use cases for how both call protocols use SDP attributes to meet the common agreed-upon capabilities of the multimedia application/session, transport, and network under a huge variety of conferencing circumstances. For SIP, we have used Request for Comments 4317 that shows the capability negotiations using SDP attributes. Similarly, the Internet Engineering Task Force draft (draft-ietf-rtcweb-sdp-06) provides many scenarios for capability negotiations using SDP under a variety of multimedia conferencing environments.

18.1 Session Negotiations in SIP

This subsection describes Request for Comments (RFC) 4317 that gives examples of the Session Description Protocol (SDP) offer/answer exchanges using the Session Initiation Protocol (SIP) call control protocol. Examples include codec negotiation and selection, hold and resume, and the addition and deletion of media streams. The examples show multiple media types, bidirectional, unidirectional, inactive streams, and dynamic payload types (PTs). The Third Party Call Control (3PCC) examples are also provided here.

18.1.1 Overview

This specification describes offer/answer examples of SDP based on RFC 3264 (see Section 3.1) [1]. The SDP in these examples is defined by RFC 2327 (that was made obsolete by RFC 4566, see Section 2) [2]. The offers and answers are assumed to be transported using a protocol such as SIP (RFC 3261, also see Section 1.2) [3]. Examples include codec negotiation and selection, hold and resume, and the addition and deletion of media streams. The examples show multiple media types, bidirectional, unidirectional, and inactive streams, and dynamic PTs. Common 3PCC (RFC 3725) [5] examples are also given. The following sections contain examples in which two parties, Alice and Bob, exchange SDP offers, answers, and in some cases, additional offers and answers. Note that the subject line ("s=") contains a single space character.

18.1.2 Codec Negotiation and Selection

18.1.2.1 Audio and Video 1

This common scenario shows a video and audio session in which multiple codecs are offered but only one is accepted. As a result of the exchange shown here, Alice and Bob may send only Pulse Code Modulation mu-law (PCMU) audio and MPV video as follows.
Note: Dynamic PT 97 is used for Internet Low Bit Rate codec (iLBC) (RFC 3952) [6].

[Offer]

```
v=0
o=alice 2890844526 2890844526 IN IP4 host.atlanta.example.com
s=
c=IN IP4 host.atlanta.example.com
t=0 0
m=audio 49170 RTP/AVP 0 8 97
a=rtpmap:0 PCMU/8000
a=rtpmap:8 PCMA/8000
a=rtpmap:97 iLBC/8000
m=video 51372 RTP/AVP 31 32
a=rtpmap:31 H261/90000
a=rtpmap:32 MPV/90000
```

[Answer]

```
v=0
o=bob 2808844564 2808844564 IN IP4 host.biloxi.example.com
s=
c=IN IP4 host.biloxi.example.com
t=0 0
m=audio 49174 RTP/AVP 0
a=rtpmap:0 PCMU/8000
m=video 49170 RTP/AVP 32
a=rtpmap:32 MPV/90000
```

18.1.2.2 Audio and Video 2

Alice can support PCMU, PCMA, and iLBC codecs but not more than one at a time. Alice offers all three to maximize chances of a successful exchange, and Bob accepts two of them. An audio-only session is established in the initial exchange between Alice and Bob, using either PCMU or

PCMA codecs (PT in the Real-Time Transport Protocol (RTP) packet tells which is being used). Since Alice only supports one audio codec at a time, a second offer is made with just that one codec to limit the codec choice to just one.

Note: the version number is incremented in both SDP messages in the second exchange. After this exchange, only the PCMU codec may be used for media sessions between Alice and Bob.

Note: The declined video stream is still present in the second exchange of SDP with ports set to zero.

[Offer]

```
v=0
o=alice 2890844526 2890844526 IN IP4 host.atlanta.example.com
s=
c=IN IP4 host.atlanta.example.com
t=0 0
m=audio 49170 RTP/AVP 0 8 97
a=rtpmap:0 PCMU/8000
a=rtpmap:8 PCMA/8000
a=rtpmap:97 iLBC/8000
m=video 51372 RTP/AVP 31 32
a=rtpmap:31 H261/90000
a=rtpmap:32 MPV/90000
```

[Answer]

```
v=0
o=bob 2808844564 2808844564 IN IP4 host.biloxi.example.com
s=
c=IN IP4 host.biloxi.example.com
t=0 0
m=audio 49172 RTP/AVP 0 8
a=rtpmap:0 PCMU/8000
a=rtpmap:8 PCMA/8000
m=video 0 RTP/AVP 31
a=rtpmap:31 H261/90000
```

[Second-Offer]

```
v=0
o=alice 2890844526 2890844527 IN IP4 host.atlanta.example.com
s=
c=IN IP4 host.atlanta.example.com
t=0 0
m=audio 51372 RTP/AVP 0
a=rtpmap:0 PCMU/8000
m=video 0 RTP/AVP 31
a=rtpmap:31 H261/90000
```

[Second-Answer]

```
v=0
o=bob 2808844564 2808844565 IN IP4 host.biloxi.example.com
s=
c=IN IP4 host.biloxi.example.com
t=0 0
```

```
m=audio 49172 RTP/AVP 0
a=rtpmap:0 PCMU/8000
m=video 0 RTP/AVP 31
a=rtpmap:31 H261/90000
```

18.1.2.3 Audio and Video 3

Alice offers three audio and two video codecs, while Bob accepts with a single audio and video codec. As a result of this exchange, Bob and Alice use iLBC for audio and H.261 for video.

Note: change of dynamic PT from 97 to 99 between the offer and the answer is OK since the same codec is referenced.

[Offer]

```
v=0
o=alice 2890844526 2890844526 IN IP4 host.atlanta.example.com
s=
c=IN IP4 host.atlanta.example.com
t=0 0
m=audio 49170 RTP/AVP 0 8 97
a=rtpmap:0 PCMU/8000
a=rtpmap:8 PCMA/8000
a=rtpmap:97 iLBC/8000
m=video 51372 RTP/AVP 31 32
a=rtpmap:31 H261/90000
a=rtpmap:32 MPV/90000
```

[Answer]

```
v=0
o=bob 2808844564 2808844564 IN IP4 host.biloxi.example.com
s=
c=IN IP4 host.biloxi.example.com
t=0 0
m=audio 49172 RTP/AVP 99
a=rtpmap:99 iLBC/8000
m=video 51374 RTP/AVP 31
a=rtpmap:31 H261/90000
```

18.1.2.4 Two Audio Streams

In this example, Alice wishes to establish separate audio streams, one for normal audio and the other for telephone events. Alice offers two separate streams, one audio with two codecs and the other with RFC 2833 (made obsolete by RFCs 4733 and 4734) [4] tones (for dual-tone multifrequency [DTMF]). Bob accepts both audio streams choosing the iLBC codec and telephone events.

[Offer]

```
v=0
o=alice 2890844526 2890844526 IN IP4 host.atlanta.example.com
s=
c=IN IP4 host.atlanta.example.com
t=0 0
m=audio 49170 RTP/AVP 0 97
a=rtpmap:0 PCMU/8000
```

```
a=rtpmap:97 iLBC/8000
m=audio 49172 RTP/AVP 98
a=rtpmap:98 telephone-event/8000
a=sendonly
```

[Answer]

```
v=0
o=bob 2808844564 2808844564 IN IP4 host.biloxi.example.com
s=
c=IN IP4 host.biloxi.example.com
t=0 0
m=audio 49172 RTP/AVP 97
a=rtpmap:97 iLBC/8000
m=audio 49174 RTP/AVP 98
a=rtpmap:98 telephone-event/8000
a=recvonly
```

18.1.2.5 Audio and Video 4

Alice and Bob establish an audio and video session with a single audio and video codec. In a second exchange, Bob changes his address for media and Alice accepts with the same SDP as the initial exchange (and as a result does not increment the version number).

[Offer]

```
v=0
o=alice 2890844526 2890844526 IN IP4 host.atlanta.example.com
s=
c=IN IP4 host.atlanta.example.com
t=0 0
m=audio 49170 RTP/AVP 97
a=rtpmap:97 iLBC/8000
m=video 51372 RTP/AVP 31
a=rtpmap:31 H261/90000
```

[Answer]

```
v=0
o=bob 2808844564 2808844564 IN IP4 host.biloxi.example.com
s=
c=IN IP4 host.biloxi.example.com
t=0 0
m=audio 49174 RTP/AVP 97
a=rtpmap:97 iLBC/8000
m=video 49170 RTP/AVP 31
a=rtpmap:31 H261/90000
```

[Second-Offer]

```
v=0
o=bob 2808844564 2808844565 IN IP4 host.biloxi.example.com
s=
c=IN IP4 newhost.biloxi.example.com
t=0 0
m=audio 49178 RTP/AVP 97
```

```
a=rtpmap:97 iLBC/8000
m=video 49188 RTP/AVP 31
a=rtpmap:31 H261/90000
```

[Second-Answer]

```
v=0
o=alice 2890844526 2890844526 IN IP4 host.atlanta.example.com
s=
c=IN IP4 host.atlanta.example.com
t=0 0
m=audio 49170 RTP/AVP 97
a=rtpmap:97 iLBC/8000
m=video 51372 RTP/AVP 31
a=rtpmap:31 H261/90000
```

18.1.2.6 Audio Only 1

Alice wishes to establish an audio session with Bob using either PCMU codec or iLBC codec with RFC 2833 (made obsolete by RFCs 4733 and 4734) tones, but not both at the same time. The offer contains these two media streams. Bob declines the first one and accepts the second one. If both media streams had been accepted, Alice would have sent a second declining one of the streams, as shown in Section 18.1.4.3.

[Offer]

```
v=0
o=alice 2890844526 2890844526 IN IP4 host.atlanta.example.com
s=
c=IN IP4 host.atlanta.example.com
t=0 0
m=audio 49170 RTP/AVP 0
a=rtpmap:0 PCMU/8000
m=audio 51372 RTP/AVP 97 101
a=rtpmap:97 iLBC/8000
a=rtpmap:101 telephone-event/8000
```

[Answer]

```
v=0
o=bob 2808844564 2808844564 IN IP4 host.biloxi.example.com
s=
c=IN IP4 host.biloxi.example.com
t=0 0
m=audio 0 RTP/AVP 0
a=rtpmap:0 PCMU/8000
m=audio 49170 RTP/AVP 97 101
a=rtpmap:97 iLBC/8000
a=rtpmap:101 telephone-event/8000
```

18.1.2.7 Audio and Video 5

Alice and Bob establish an audio and video session in the first exchange with a single audio and video codec. In the second exchange, Alice adds a second video codec, which Bob accepts. This allows Alice and Bob to switch between the two video codecs without another offer/answer exchange.

[Offer]

```
v=0
o=alice 2890844526 2890844526 IN IP4 host.atlanta.example.com
s=
c=IN IP4 host.atlanta.example.com
t=0 0
m=audio 49170 RTP/AVP 99
a=rtpmap:99 iLBC/8000
m=video 51372 RTP/AVP 31
a=rtpmap:31 H261/90000
```

[Answer]

```
v=0
o=bob 2808844564 2808844564 IN IP4 host.biloxi.example.com
s=
c=IN IP4 host.biloxi.example.com
t=0 0
m=audio 49172 RTP/AVP 99
a=rtpmap:99 iLBC/8000
m=video 51374 RTP/AVP 31
a=rtpmap:31 H261/90000
```

[Second-Offer]

```
v=0
o=alice 2890844526 2890844527 IN IP4 host.atlanta.example.com
s=
c=IN IP4 host.atlanta.example.com
t=0 0
m=audio 49170 RTP/AVP 99
a=rtpmap:99 iLBC/8000
m=video 51372 RTP/AVP 31 32
a=rtpmap:31 H261/90000
a=rtpmap:32 MPV/90000
```

[Second-Answer]

```
v=0
o=bob 2808844564 2808844565 IN IP4 host.biloxi.example.com
s=
c=IN IP4 host.biloxi.example.com
t=0 0
m=audio 49172 RTP/AVP 99
a=rtpmap:99 iLBC/8000
m=video 51374 RTP/AVP 31 32
a=rtpmap:31 H261/90000
a=rtpmap:32 MPV/90000
```

18.1.2.8 Audio and Video 6

This example shows an audio and video offer that is accepted, but the answerer wants the video sent to a different address than that of the audio. This is a common scenario in conferencing where the video and audio mixing utilizes different servers. In this example, Alice offers audio and video, and Bob accepts.

[Offer]

```
v=0
o=alice 2890844526 2890844526 IN IP4 host.atlanta.example.com
s=
c=IN IP4 host.atlanta.example.com
t=0 0
m=audio 49170 RTP/AVP 0 8 97
a=rtpmap:0 PCMU/8000
a=rtpmap:8 PCMA/8000
a=rtpmap:97 iLBC/8000
m=video 51372 RTP/AVP 31 32
a=rtpmap:31 H261/90000
a=rtpmap:32 MPV/90000
```

[Answer]

```
v=0
o=bob 2808844564 2808844564 IN IP4 host.biloxi.example.com
s=
c=IN IP4 host.biloxi.example.com
t=0 0
m=audio 49174 RTP/AVP 0
a=rtpmap:0 PCMU/8000
m=video 49172 RTP/AVP 32
c=IN IP4 otherhost.biloxi.example.com
a=rtpmap:32 MPV/90000
```

18.1.3 Hold and Resume Scenarios

18.1.3.1 Hold and Unhold 1

Alice calls Bob, but when Bob answers, he places Alice on hold. Bob then takes Alice off hold in the second offer. Alice changes port number in the second exchange. The media session between Alice and Bob is now active after Alice's second answer. Note that a=sendrecv could be present in both second offer and answer exchange. This is a common flow in 3PCC (RFC 3725) [5] scenarios.

[Offer]

```
v=0
o=alice 2890844526 2890844526 IN IP4 host.atlanta.example.com
s=
c=IN IP4 host.atlanta.example.com
t=0 0
m=audio 49170 RTP/AVP 0 97
a=rtpmap:0 PCMU/8000
a=rtpmap:97 iLBC/8000
```

[Answer]

```
v=0
o=bob 2808844564 2808844564 IN IP4 host.biloxi.example.com
s=
```

```
c=IN IP4 placeholder.biloxi.example.com
t=0 0
m=audio 49172 RTP/AVP 97
a=rtpmap:97 iLBC/8000
a=sendonly
```

[Second-Offer]

```
v=0
o=bob 2808844564 2808844565 IN IP4 host.biloxi.example.com
s=
c=IN IP4 host.biloxi.example.com
t=0 0
m=audio 49170 RTP/AVP 97
a=rtpmap:97 iLBC/8000
```

[Second-Answer]

```
v=0
o=alice 2890844526 2890844527 IN IP4 host.atlanta.example.com
s=
c=IN IP4 host.atlanta.example.com
t=0 0
m=audio 49178 RTP/AVP 97
a=rtpmap:97 iLBC/8000
```

18.1.3.2 Hold with Two Streams

In this example, two audio streams have been established in the first offer/answer exchange. In this second offer/answer exchange, one of the audio streams is placed on hold. Alice offers two media streams: a bidirectional audio stream and a send-only telephone event stream. Bob accepts both streams. Bob then puts Alice's audio stream on hold but not the tone stream. Alice responds with identical SDP to the initial offer.

[Offer]

```
v=0
o=alice 2890844526 2890844526 IN IP4 host.atlanta.example.com
s=
c=IN IP4 host.atlanta.example.com
t=0 0
m=audio 49170 RTP/AVP 0 97
a=rtpmap:0 PCMU/8000
a=rtpmap:97 iLBC/8000
m=audio 49172 RTP/AVP 98
a=rtpmap:98 telephone-event/8000
a=sendonly
```

[Answer]

```
v=0
o=bob 2808844564 2808844564 IN IP4 host.biloxi.example.com
s=
c=IN IP4 host.biloxi.example.com
t=0 0
```

```
m=audio 49172 RTP/AVP 97
a=rtpmap:97 iLBC/8000
m=audio 49174 RTP/AVP 98
a=rtpmap:98 telephone-event/8000
a=recvonly
```

[Second-Offer]

```
v=0
o=bob 2808844564 2808844565 IN IP4 host.biloxi.example.com
s=
c=IN IP4 host.biloxi.example.com
t=0 0
m=audio 49172 RTP/AVP 97
a=rtpmap:97 iLBC/8000
a=sendonly
m=audio 49174 RTP/AVP 98
a=rtpmap:98 telephone-event/8000
a=recvonly
```

[Second-Answer]

```
v=0
o=alice 2890844526 2890844527 IN IP4 host.atlanta.example.com
s=
c=IN IP4 host.atlanta.example.com
t=0 0
m=audio 49170 RTP/AVP 0 97
a=rtpmap:0 PCMU/8000
a=rtpmap:97 iLBC/8000
m=audio 49172 RTP/AVP 98
a=rtpmap:98 telephone-event/8000
a=sendonly
```

18.1.4 Addition and Deletion of Media Streams

This section shows addition and deletion of media streams.

18.1.4.1 Second Audio Stream Added

In this example, the first offer/answer exchange establishes a single audio stream with a single codec. The second offer/answer exchange adds a second audio stream for telephone events. The second stream is added by Bob's media server (different connection address) to receive RFC 2833 (made obsolete by RFCs 4733 and 4734) telephone events (DTMF digits, typically) from Alice. Alice accepts. Even though the second stream is unidirectional, Alice receives Real-Time Transport Control Protocol (RTCP) packets on port 49173 from the media server.

[Offer]

```
v=0
o=alice 2890844526 2890844526 IN IP4 host.atlanta.example.com
s=
c=IN IP4 host.atlanta.example.com
t=0 0
m=audio 49170 RTP/AVP 0 97
```

```
a=rtpmap:0 PCMU/8000
a=rtpmap:97 iLBC/8000
```

[Answer]

```
v=0
o=bob 2808844564 2808844564 IN IP4 host.biloxi.example.com
s=
c=IN IP4 host.biloxi.example.com
t=0 0
m=audio 49170 RTP/AVP 97
a=rtpmap:97 iLBC/8000
```

[Second-Offer]

```
v=0
o=bob 2808844564 2808844565 IN IP4 host.biloxi.example.com
s=
c=IN IP4 host.biloxi.example.com
t=0 0
m=audio 49170 RTP/AVP 97
a=rtpmap:97 iLBC/8000
m=audio 48282 RTP/AVP 98
c=IN IP4 mediaserver.biloxi.example.com
a=rtpmap:98 telephone-event/8000
a=recvonly
```

[Second-Answer]

```
v=0
o=alice 2890844526 2890844527 IN IP4 host.atlanta.example.com
s=
c=IN IP4 host.atlanta.example.com
t=0 0
m=audio 49170 RTP/AVP 97
a=rtpmap:97 iLBC/8000
m=audio 49172 RTP/AVP 98
c=IN IP4 host.atlanta.example.com
a=rtpmap:98 telephone-event/8000
a=sendonly
```

18.1.4.2 Audio, Then Video Added

An audio-only session is established in the initial exchange between Alice and Bob using PCMU codec. Alice adds a video stream that is accepted by Bob.

[Offer]

```
v=0
o=alice 2890844526 2890844526 IN IP4 host.atlanta.example.com
s=
c=IN IP4 host.atlanta.example.com
t=0 0
m=audio 49170 RTP/AVP 0
a=rtpmap:0 PCMU/8000
```

[Answer]

```
v=0
o=bob 2808844564 2808844564 IN IP4 host.biloxi.example.com
s=
c=IN IP4 host.biloxi.example.com
t=0 0
m=audio 49172 RTP/AVP 0
a=rtpmap:0 PCMU/8000
```

[Second-Offer]

```
v=0
o=alice 2890844526 2890844527 IN IP4 host.atlanta.example.com
s=
c=IN IP4 host.atlanta.example.com
t=0 0
m=audio 49170 RTP/AVP 0
a=rtpmap:0 PCMU/8000
m=video 49172 RTP/AVP 31
a=rtpmap:31 H261/90000
```

[Second-Answer]

```
v=0
o=bob 2808844564 2808844565 IN IP4 host.biloxi.example.com
s=
c=IN IP4 host.biloxi.example.com
t=0 0
m=audio 49172 RTP/AVP 0
a=rtpmap:0 PCMU/8000
m=video 49168 RTP/AVP 31
a=rtpmap:31 H261/90000
```

18.1.4.3 Audio and Video, Then Video Deleted

Alice and Bob establish an audio and video session. In a second exchange, Bob deletes the video session, resulting in an audio-only session.

[Offer]

```
v=0
o=alice 2890844526 2890844526 IN IP4 host.atlanta.example.com
s=
c=IN IP4 host.atlanta.example.com
t=0 0
m=audio 49170 RTP/AVP 97
a=rtpmap:97 iLBC/8000
m=video 51372 RTP/AVP 31
a=rtpmap:31 H261/90000
```

[Answer]

```
v=0
o=bob 2808844564 2808844564 IN IP4 host.biloxi.example.com
s=
```

```
c=IN IP4 host.biloxi.example.com
t=0 0
m=audio 49174 RTP/AVP 97
a=rtpmap:97 iLBC/8000
m=video 49170 RTP/AVP 31
a=rtpmap:31 H261/90000
```

[Second-Offer]

```
v=0
o=bob 2808844564 2808844565 IN IP4 host.biloxi.example.com
s=
c=IN IP4 host.biloxi.example.com
t=0 0
m=audio 49174 RTP/AVP 97
a=rtpmap:97 iLBC/8000
m=video 0 RTP/AVP 31
a=rtpmap:31 H261/90000
```

[Second-Answer]

```
v=0
o=alice 2890844526 2890844527 IN IP4 host.atlanta.example.com
s=
c=IN IP4 host.atlanta.example.com
t=0 0
m=audio 49170 RTP/AVP 97
a=rtpmap:97 iLBC/8000
m=video 0 RTP/AVP 31
a=rtpmap:31 H261/90000
```

18.1.5 Third Party Call Control

This section shows examples common in 3PCC flows (RFC 3725) [5]. Call hold and resume flows are also common in 3PCC.

18.1.5.1 No Media, Then Audio Added

The first offer from Alice contains no media lines, so Bob accepts with no media lines. In the second exchange, Alice adds an audio stream that Bob accepts.

[Offer]

```
v=0
o=alice 2890844526 2890844526 IN IP4 host.atlanta.example.com
s=
c=IN IP4 host.atlanta.example.com
t=0 0
```

[Answer]

```
v=0
o=bob 2808844564 2808844564 IN IP4 host.biloxi.example.com
```

```
s=
c=IN IP4 host.biloxi.example.com
t=0 0
```

[Second-Offer]

```
v=0
o=alice 2890844526 2890844527 IN IP4 host.atlanta.example.com
s=
c=IN IP4 host.atlanta.example.com
t=0 0
m=audio 49170 RTP/AVP 97
a=rtpmap:97 iLBC/8000
```

[Second-Answer]

```
v=0
o=bob 2808844564 2808844565 IN IP4 host.biloxi.example.com
s=
c=IN IP4 host.biloxi.example.com
t=0 0
m=audio 49172 RTP/AVP 97
a=rtpmap:97 iLBC/8000
```

18.1.5.2 Hold and Unhold 2

The first offer from Alice contains the connection address 0.0.0.0 and a random port number, which means that Bob cannot send media to Alice (the media stream is "black holed" or "bh"). Bob accepts with normal SDP. In the second exchange, Alice changes the connection address, Bob accepts, and a media session is established.

[Offer]

```
v=0
o=alice 2890844526 2890844526 IN IP4 host.atlanta.example.com
s=
c=IN IP4 0.0.0.0
t=0 0
m=audio 23442 RTP/AVP 97
a=rtpmap:97 iLBC/8000
```

[Answer]

```
v=0
o=bob 2808844564 2808844564 IN IP4 host.biloxi.example.com
s=
c=IN IP4 host.biloxi.example.com
t=0 0
m=audio 49170 RTP/AVP 97
a=rtpmap:97 iLBC/8000
```

[Second-Offer]

```
v=0
o=alice 2890844526 2890844527 IN IP4 host.atlanta.example.com
s=
```

```
c=IN IP4 host.atlanta.example.com
t=0 0
m=audio 49170 RTP/AVP 97
a=rtpmap:97 iLBC/8000
```

[Second-Answer]

```
v=0
o=bob 2808844564 2808844564 IN IP4 host.biloxi.example.com
s=
c=IN IP4 host.biloxi.example.com
t=0 0
m=audio 49170 RTP/AVP 97
a=rtpmap:97 iLBC/8000
```

18.1.5.3 Hold and Unhold 3

The first offer from Alice contains an audio stream, but the answer from Bob contains the connection address 0.0.0.0 and a random port number, which means that Alice cannot send media to Bob (the media stream is "bh"). In the second exchange, Bob changes the connection address, Alice accepts, and a media session is established.

[Offer]

```
v=0
o=alice 2890844526 2890844526 IN IP4 host.atlanta.example.com
s=
c=IN IP4 host.atlanta.example.com
t=0 0
m=audio 49170 RTP/AVP 97
a=rtpmap:97 iLBC/8000
```

[Answer]

```
v=0
o=bob 2808844564 2808844564 IN IP4 host.biloxi.example.com
s=
c=IN IP4 0.0.0.0
t=0 0
m=audio 9322 RTP/AVP 97
a=rtpmap:97 iLBC/8000
```

[Second-Offer]

```
v=0
o=bob 2808844564 2808844565 IN IP4 host.biloxi.example.com
s=
c=IN IP4 host.biloxi.example.com
t=0 0
m=audio 49172 RTP/AVP 97
a=rtpmap:97 iLBC/8000
```

[Second-Answer]

```
v=0
o=alice 2890844526 2890844526 IN IP4 host.atlanta.example.com
```

```
s=
c=IN IP4 host.atlanta.example.com
t=0 0
m=audio 49170 RTP/AVP 97
a=rtpmap:97 iLBC/8000
```

18.1.6 Security Considerations

SDP offer and answer messages can contain private information about addresses and sessions to be established between parties. If this information needs to be kept private, some security mechanism in the protocol used to carry the offers and answers must be used. For SIP, this means to using Transport Layer Security (TLS) and/or S/Multipurpose Internet Mail Extensions encryption of the SDP message body. It is important that SDP offer and answer messages be properly authenticated and authorized before they are used to establish a media session. Examples of SIP mechanisms include SIP digest, certs, and cryptographically verified SIP identity.

18.2 Session Negotiations in WebRTC

This section describes the Internet Engineering Task Force (IETF) draft [I-D-ietf-rtcweb-sdp] that provides the examples of the session capability negotiations of the WebRTC call control protocol between peers using SDP. Such negotiation happens based on the SDP offer/answer exchange mechanism described in RFC 3264 (see Section 3.1). In addition, this IETF draft provides an informational reference in describing the role of SDP and the offer/answer exchange mechanism for the most common WebRTC use cases. The main purpose of this section is to show how SDP is used for capability negotiations using a call control protocol like WebRTC (see Section 1.3).

18.2.1 Introduction

JavaScript Session Exchange Protocol (JSEP) [I-D. ietf-rtcweb-jsep] specifies a generic protocol needed to generate (RFC 3264, see Section 3.1) offers and answers negotiated between WebRTC peers for setting up, updating, and tearing down a WebRTC session. For this purpose, SDP is used to construct RFC 3264 (see Section 3.1) offers/answers for describing (media and nonmedia) streams as appropriate for the recipients of the session description to participate in the session. The remainder of this document is organized as follows: Sections 18.2.3 and 18.2.4 provide an overview of SDP and the offer/answer exchange mechanism. Section 18.2.5 provides sample SDP generated for the most common WebRTC use cases.

18.2.2 Terminology

See Sections 1.3.2.4 and 2.2.

18.2.3 SDP and the WebRTC

The purpose of this section is to provide a general overview of SDP and its components. For a more in-depth understanding, the readers are advised to refer to RFC 4566 (see Section 2). The SDP (RFC 4566 see Section 2) describes multimedia sessions, which can contain audio, video,

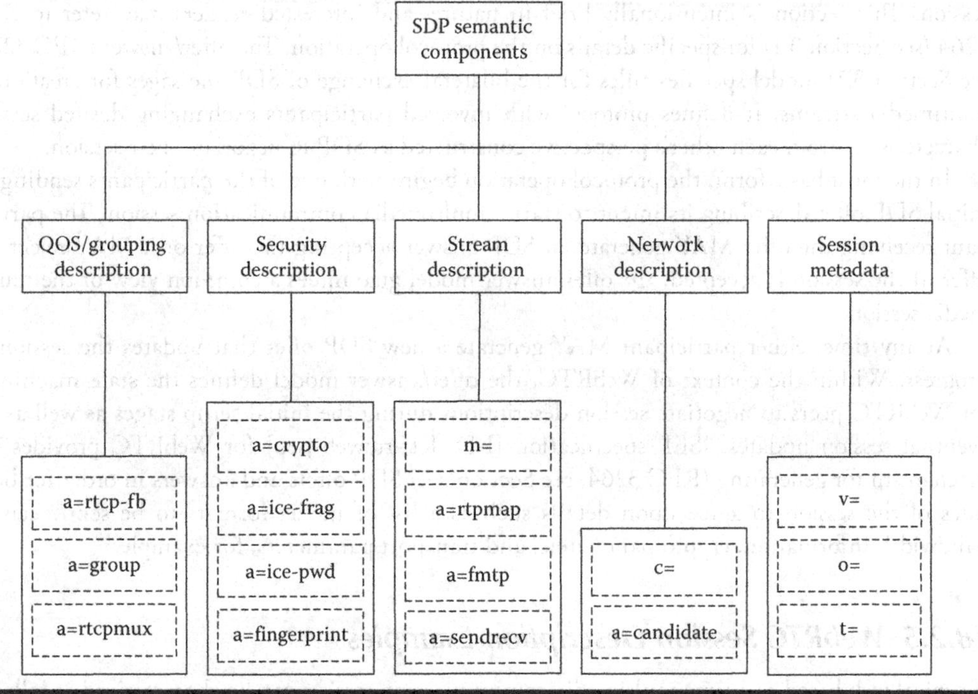

Figure 18.1 Semantic components of SDP.

whiteboard, fax, modem, and other streams. SDP provides a general purpose, standard representation to describe various aspects of multimedia sessions such as media capabilities, transport addresses, and related metadata in a transport agnostic manner, for the purposes of session announcement, session invitation, and parameter negotiation. As of today, SDP is widely used in the context of SIP (RFC 3261, also see Section 1.2), Real-Time Transport Protocol (RTP) (RFC 3550) and Real-Time Streaming Protocol applications (RFC 2326). Figure 18.1 introduces high-level breakup of SDP into components that semantically describe a multimedia session, in our case, a WebRTC session (WebRTC, see Section 1.3). It by no means captures everything about SDP; hence, it should be used for informational purposes only.

WebRTC (see Section 1.3) proposes JavaScript application to fully specify and control the signaling plane of a multimedia session as described in the JSEP specification [I-D. ietf-rtcweb-jsep]. JSEP provides mechanisms for creating session characterization and media definition information to conduct the session based on SDP exchanges. In this context, SDP serves two purposes:

1. Provide grammatical structure syntactically
2. Semantically convey participant's intention and capabilities required to successfully negotiate a session

18.2.4 Offer/Answer and the WebRTC

This section introduces the SDP offer/answer exchange mechanism mandated by WebRTC for negotiating session capabilities while setting up, updating, and tearing down a WebRTC session. This section is intentionally brief in nature, and interested readers may refer to RFC 3264 (see Section 3.1) for specific details on the protocol operation. The offer/answer (RFC 3264, see Section 3.1) model specifies rules for the bilateral exchange of SDP messages for creation of multimedia streams. It defines protocol with involved participants exchanging desired session characteristics from each other's perspective constructed as SDP to negotiate their session.

In the most basic form, the protocol operation begins with one of the participants sending an initial SDP offer describing its intent to start a multimedia communication session. The participant receiving the offer MAY generate an SDP answer accepting the offer or it MAY reject the offer. If the session is accepted, the offer/answer model guarantees a common view of the multimedia session.

At any time, either participant MAY generate a new SDP offer that updates the session in progress. Within the context of WebRTC, the offer/answer model defines the state machinery for WebRTC peers to negotiate session descriptions during the initial setup stages as well as for eventual session updates. JSEP specification [I-D. Ietf-rtcweb-jsep] for WebRTC provides the mechanism for generating (RFC 3264, see Section 3.1) SDP offers and answers in order for both sides of the session to agree upon details such as a list of media formats to be sent/received, bandwidth information, crypto parameters, and transport parameters, for example.

18.2.5 WebRTC Session Description Examples

A typical web based real-time multimedia communication session can be characterized as follows:

- It has zero or more audio only, video only, or audio/video RTP sessions.
- It MAY contain zero or more nonmedia data sessions.
- All the sessions are secured with Datagram Transport Layer Security (DTLS)-Secure Real-Time Protocol (SRTP).
- It supports network address translator (NAT)-traversal using Interactive Connectivity Establishment (ICE) mechanism [7].
- It provides RTCP-based feedback mechanisms.
- Sessions can be over IPv4-only, IPv6-only, or dual-stack based clients.

18.2.5.1 Some Conventions

The examples given in this document follow the conventions listed:

- In all the examples, Alice and Bob are assumed to be WebRTC peers.
- It is assumed that for most of the examples, the support for the call initiation described here [I-D. Ietf-mmusic-sdp-bundle-negotiation] is established apriori either out-of-band or as a consequence of a successful offer/answer negotiation between Alice and Bob, unless explicitly stated otherwise.
- Call-flow diagrams that accompany the use cases capture only prominent aspects of the system behavior; they are intentionally are not detailed to improve readability.
- Even though the call-flow diagrams show SDP being exchanged between the parties, it doesn't represent the only way a WebRTC setup is expected to work. Other approaches may involve WebRTC applications to exchange the media setup information via non-SDP mechanisms as long as they confirm to the [I-D. Ietf-rtcweb-jsep] API specification.

- The SDP examples deviate from actual on-the-wire SDP notation in several ways. This is to facilitate readability and to conform to the restrictions imposed by the RFC formatting rules.
  - Visual markers/empty lines in any SDP example are inserted to make functional divisions in the SDP clearer and are not actually part of SDP syntax.
  - Any SDP line that is indented (compared to the initial line in the SDP block) is a continuation of the preceding line. The line break and indent are to be interpreted as a single-space character.
  - Except for the above two conventions, line endings are to be interpreted as <CR><LF> pairs (that is, an ASCII 13 followed by an ASCII 10).
- Against each SDP line, pointers to the appropriate RFCs are provided for further informational reference. Also, an attempt has been made to provide explanatory notes to enable better understanding of the SDP usage, wherever appropriate.
- Following SDP details are common across all the use cases defined in this document unless mentioned otherwise.
  - DTLS fingerprint for SRTP (a=fingerprint)
  - RTP/RTCP multiplexing (a=rtcp-mux)
  - RTCP feedback support (a=rtcp-fb)
  - Host and server-reflexive candidate lines (a=candidate)
  - SRTP setup framework parameters (a=setup)
  - RTCP attribute (a=rtcp)
  - RTP header extension indicating audio-levels from client to the mixer

For specific details, readers must refer to the [I-D. Ietf-rtcweb-jsep] specification.

- The term "session" is used rather loosely in this document to refer to either a "communication session," an "RTP Session," or a "RTP Stream" depending on the context.
- PT 109 is usually used for OPUS, 0 for PCMU, 8 for PCMA, 99 for H.264, and 120 for VP8 in most of the examples to maintain uniformity.
- The IP Address:Port combinations "192.0.2.4:61665" (host) and "203.0.113.141:54609" (server reflexive) are typically used for Alice.
- The IP Address:Port combinations "198.51.100.7:51556" (host) and "203.0.113.77:49203" (server reflexive) are typically used for Bob.
- The IPv6 addresses 2001:DB8:8101:3a55:4858:a2a9:22ff:99b9 and 2001:DB8:30c:1266:5916:3779:22f6:77f7 are used to represent Alice and Bob host address, respectively.
- In actual use, the values that represent Synchronization Source (SSRC), ICE candidate foundations, WebRTC MediaStream and MediaStreamTrack IDs shall be much larger and/or random than the ones shown in the examples.
- SDP attributes in the examples closely follow the checklist defined in Section 18.2.8.1 (that is, Appendix A.1. of [I-D-ietf-rtcweb-sdp]).

18.2.5.2 Basic Examples

18.2.5.2.1 Audio Only Session

This common scenario shows SDP for secure two-way audio session with Alice offering Opus, PCMU, and PCMA and Bob accepting all the offered audio codecs (Tables 18.1 and 18.2).

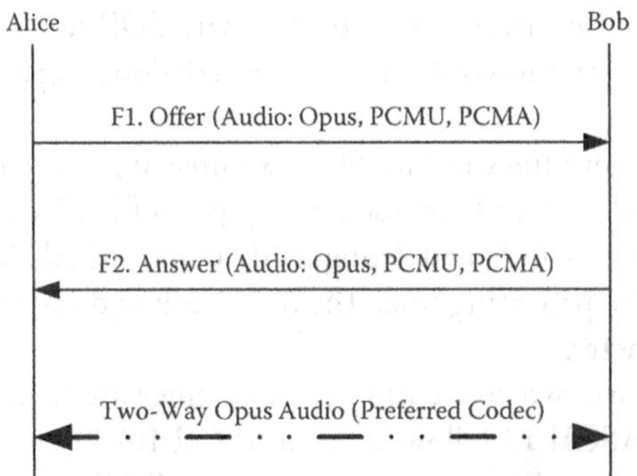

2-Way audio only session

Table 18.1 SDP Offer (Copyright IETF)

```
+---------------------------------------------+----------------------+
| Offer SDP Contents                          | RFC#/Notes           |
+---------------------------------------------+----------------------+
| v=0                                         | [RFC4566]            |
| o=- 20518 0 IN IP4 0.0.0.0                  | [RFC4566] - Session  |
|                                             | Origin Information   |
| s=-                                         | [RFC4566]            |
| t=0 0                                       | [RFC4566]            |
| a=group:BUNDLE audio                        | [I-D.ietf-mmusic-sdp |
|                                             | -bundle-negotiation] |
| a=ice-options:trickle                       | [I-D.ietf-mmusic-tri |
|                                             | ckle-ice]            |
| a=identity:eyJpZHAiOnsiZG9tYWluIjoibmlpZi5  | Section 5.6 of [I-D. |
| odSIsInByb3RvY29sIjoiaWRwLmh0bWwifSwiYXNzZ  | ietf-rtcweb-security |
| XJ0a W9uIjoiZXlKaGJHY2lPaUpTVXpJMU5pSXNJbl  | -arch]               |
| I1Y0NJNklrcFhVeUo5LmV5SmpiMjUwWlc1MGN5STZl  |                      |
| eUptYVc1b lpYSndjbWx1ZENJNlczc2lZV3huYjNKc  |                      |
| GRHaHRJam9pYzJoaExUSTFOaUlzSW1ScFoyVnpkQ0k  |                      |
| 2SWprek9rTXdPa kl6T2pKR09rRXlPakF3T2pBd09q  |                      |
| QkVPalV4T2tGRE9rUXlPalUwT2pZMU9rWTBPak5DT2  |                      |
| pkRU9qa3lPa1JET2pnN E9qTXpPalV4T2pJek9qUXd  |                      |
| PamN5T2preE9qZ3pPalZDT2pBeE9qSkdPalV3T2pjN  |                      |
| E9qTkdJbjFkZlN3aWFXUmxib lJwZEhraU9pSnRhWE  |                      |
| 5wUUc1cGFXWXVhSFVpZlEuSTVQdGhKNFFDT05TOFVX  |                      |
| d25OOUh3MEdaTDl3d0RBVGRrTWtFW llmdlNVTTJ6U  |                      |
| md5R09WSGgzRmpnc2FPZklkRnFsNUx6azBFbndVOTN  |                      |
| QOUlCQ0xZOWtia3V1c0V1S25YRGVNLTNIN WFmdTJv  |                      |
| Zl9CTlZjUnB3MmdBdlNBbVR6SlltcEpqMFEtdmV0Tm  |                      |
| tVT1huZE9HLUIzT3ZGb3QwZVNENlZSNUdhb2wyc Gd  |                      |
| uS3FSTktOd3dacEZ1eUZZbFRodHJIdGNiT19WV3o4Q  |                      |
| nZpTThKS25OdExWd1JxNUhMX2ZLTlRCNzFDYkoyWmh  |                      |
| 5W XU1UEdwWDhXcXJMWC1ybm5YSFY3RnhoTTh5OHdr  |                      |
| LWd5cnRZazVnbFlZeUFrcTVqZklSXzRzWER5d19Qc1  |                      |
| BWTW1aZ XltenVGV3BQTzVFWlJYR0ZpRjFET0o4Q0Q  |                      |
| 3Z3Zta2dUdlBXSWpkemtBIn0=                   |                      |
```

(Continued)

Table 18.1 (*Continued*) SDP Offer (Copyright IETF)

****** Audio m=line *********	*****************************
m=audio 54609 UDP/TLS/RTP/SAVPF 109 0 8	[RFC4566]
c=IN IP4 203.0.113.141	[RFC4566]
a=mid:audio	[RFC5888]
a=msid:ma ta	Identifies RTCMediaStream ID (ma) and RTCMediaStreamTrack ID (ta)
a=sendrecv	[RFC3264] - Alice can send and recv audio
a=rtpmap:109 opus/48000/2	[RFC7587] - Opus Codec 48khz, 2 channels
a=rtpmap:0 PCMU/8000	[RFC3551] PCMU Audio Codec
a=rtpmap:8 PCMA/8000	[RFC3551] PCMA Audio Codec
a=maxptime:120	[RFC4566]
a=ice-ufrag:074c6550	[RFC5245] - ICE user fragment
a=ice-pwd:a28a397a4c3f31747d1ee3474af08a068	[RFC5245] - ICE password
a=fingerprint:sha-256 19:E2:1C:3B:4B:9F:81 :E6:B8:5C:F4:A5:A8:D8:73:04 :BB:05:2F:70:9 F:04:A9:0E:05:E9:26:33:E8:70:88:A2	[RFC5245] - DTLS Fingerprint for SRTP
a=setup:actpass	[RFC4145] - Alice can perform DTLS before Answer arrives
a=dtls-id:1	[I-D.ietf-mmusic-dtls-sdp]
a=rtcp-mux	[RFC5761] - Alice can perform RTP/RTCP Muxing
a=rtcp:60065 IN IP4 203.0.113.141	[RFC3605]
a=rtcp-rsize	[RFC5506] - Alice intends to use reduced size RTCP for this session
a=rtcp-fb:109 nack	[RFC5104] - Indicates NACK RTCP feedback support
a=extmap:1 urn:ietf:params:rtp-hdrext :ssrc-audio-level	[RFC6464] Alice supports RTP header extension to indicate audio levels

(*Continued*)

Table 18.1 (*Continued*) SDP Offer (Copyright IETF)

```
| a=extmap:2 urn:ietf:params:rtp-              | [I-D.ietf-mmusic-sdp |
| hdrext:sdes:mid                              | -bundle-negotiation] |
| a=candidate:0 1 UDP  2122194687 192.0.2.4    | [RFC5245] - RTP Host |
| 61665 typ host                               | Candidate            |
| a=candidate:1 1 UDP  1685987071              | [RFC5245] - RTP      |
| 203.0.113.141 54609 typ srflx raddr          | Server Reflexive ICE |
| 192.0.2.4 rport 61665                        | Candidate            |
| a=candidate:0 2 UDP  2122194687 192.0.2.4    | [RFC5245] - RTCP     |
| 61667 typ host                               | Host Candidate       |
| a=candidate:1 2 UDP  1685987071              | [RFC5245] - RTCP     |
| 203.0.113.141 60065 typ srflx raddr          | Server Reflexive ICE |
| 192.0.2.4 rport 61667                        | Candidate            |
| a=end-of-candidates                          | [I-D.ietf-mmusic-tri |
|                                              | ckle-ice]            |
+----------------------------------------------+----------------------+
```

Table 18.2 SDP Answer (Copyright IETF)

```
+---------------------------------------------+----------------------+
| Answer SDP Contents                         | RFC#/Notes           |
+---------------------------------------------+----------------------+
| v=0                                         | [RFC4566]            |
| o=-  16833 0 IN IP4 0.0.0.0                 | [RFC4566] - Session  |
|                                             | Origin Information   |
| s=-                                         | [RFC4566]            |
| t=0 0                                       | [RFC4566]            |
| a=group:BUNDLE audio                        | [I-D.ietf-mmusic-sdp-|
|                                             | bundle-negotiation]  |
| a=ice-options:trickle                       | [I-D.ietf-mmusic-tric|
|                                             | kle-ice]             |
| a=identity:ew0KICAiaWRwIjp7DQogICAgImRvbW   | Section 5.6 of [I-D.i|
| FpbiI6ICJjaXNjb3NwYXJrLmNvbSIsDQogICAg In   | etf-rtcweb-security-a|
| Byb3RvY29sIjogImRlZmF1bHQiDQogIH0sDQogICJ   | rch]                 |
| hc3NlcnRpb24iOiAibEp3WkVocmFVOXBTblJo V0U   |                      |
| 1d1VVYzFjR0ZYV1hWaFNGVnBabEV1U1RWUWRHaEtO   |                      |
| RkZEVDA1VE9GVlhkMjVPT1VoM01FZGFURGwz ZDBS   |                      |
| QlZHUnJUV3RGVw0KICAgICAgICAgICAgICBsbG1kb   |                      |
| E5WVFRKNlVtZDVSMDlXU0dnelJtcG5jMkZQ Wmtsa   |                      |
| 1JuRnNOVXg2YXpCRmJuZFZPVE5RT1VsQ1EweFpPV3   |                      |
| RpYTNWMWMwVjFTMjVZUkdWTkxUTklODQog ICAgIC   |                      |
| AgICAgICAgIFdGbWRUSnZabDlDVGxaalVuQjNNbWR   |                      |
| CZGxOQmJWUjZTbGx0Y0VwcU1GRXRkbVYw VG10VlQ   |                      |
| xaHVaRTlITFVJelQzWkdiM1F3WlZORU5sWlNOVWRo   |                      |
| YjJ3eWMNCiAgICAgICAgICAgICAgR2R1 UzNGU1Rr   |                      |
| dE9kM2RhY0VaMWVVWlpiRlJvZEhKSWRHTmlUMTlXV   |                      |
| jNvNFFuWnBUVGhLUzI1T2RFeFdkMUp4             |                      |
| TlVoTVgyWkxUbFJDTnpGRFlrb3lXbWg1VyINCn0=    |                      |
| ****** Audio m=line *********               | ******************** |
|                                             | ********             |
```

(*Continued*)

Table 18.2 (*Continued*) SDP Answer (Copyright IETF)

m=audio 49203 UDP/TLS/RTP/SAVPF 109 0 8	[RFC4566]
c=IN IP4 203.0.113.77	[RFC4566]
a=mid:audio	[RFC5888]
a=msid:ma ta	[I-D.ietf-mmusic-msid] Identifies RTCMediaStream ID (ma) and RTCMediaStreamTrack ID (ta)
a=sendrecv	[RFC3264] - Bob can send and recv audio
a=rtpmap:109 opus/48000/2	[RFC7587] Opus Codec
a=rtpmap:0 PCMU/8000	[RFC3551] PCMU Audio Codec
a=rtpmap:8 PCMA/8000	[RFC3551] PCMA Audio Codec
a=maxptime:120	[RFC4566]
a=ice-ufrag:05067423	[RFC5245] - ICE user fragment
a=ice-pwd:1747d1ee3474a28a397a4c3f3af08a068	[RFC5245] - ICE password parameter
a=fingerprint:sha-256 6B:8B:F0:65:5F:78:E2:51:3B:AC:6F:F3:3F:46:1B:35 :DC:B8:5F:64:1A:24:C2:43:F0:A1:58:D0:A1:2C:19:08	[RFC5245] - DTLS Fingerprint for SRTP
a=setup:active	[RFC4145] - Bob carries out DTLS Handshake in parallel
a=dtls-id:1	[I-D.ietf-mmusic-dtls-sdp]
a=rtcp-mux	[RFC5761] - Bob can perform RTP/RTCP Muxing on port 49203
a=rtcp-rsize	[RFC5506] - Bob intends to use reduced size RTCP for this session
a=rtcp-fb:109 nack	[RFC5104] - Indicates NACK RTCP feedback support
a=extmap:1 urn:ietf:params:rtp-hdrext:ssrc-audio-level	[RFC6464] Bob supports audio level RTP header extension as well
a=extmap:2 urn:ietf:params:rtp-hdrext:sdes:mid	[I-D.ietf-mmusic-sdp-bundle-negotiation]
a=candidate:0 1 UDP 2122194687 198.51.100.7 51556 typ host	[RFC5245] - RTP/RTCP Host ICE Candidate
a=candidate:1 1 UDP 1685987071 203.0.113.77 49203 typ srflx raddr 198.51.100.7 rport 51556	[RFC5245] - RTP/RTCP Server Reflexive ICE Candidate
a=end-of-candidates	[I-D.ietf-mmusic-trickle-ice]

18.2.5.2.2 Audio/Video Session

Alice and Bob establish a two-way audio and video session with Opus as the audio codec and H.264 as the video codec.

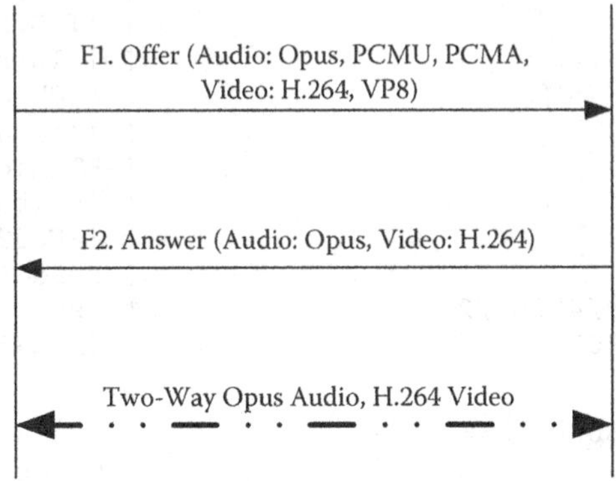

2-way audio, video session

18.2.5.2.2.1 Audio/Video Session

This section shows the IPv4-only offer/answer exchange (Tables 18.3 and 18.4).

Table 18.3 SDP Offer (Copyright IETF)

Offer SDP Contents	RFC#/Notes
v=0	[RFC4566]
o=- 20518 0 IN IP4 0.0.0.0	[RFC4566] - Session Origin Information
s=-	[RFC4566]
t=0 0	[RFC4566]
a=group:BUNDLE audio video	[I-D.ietf-mmusic-sdp-bundle-negotiation]
a=group:LS audio video	[RFC5888] - Alice wants to lip sync her audio and video sreams
a=ice-options:trickle	[I-D.ietf-mmusic-trickle-ice]
****** Audio m=line *********	****************************
m=audio 54609 UDP/TLS/RTP/SAVPF 109 0 8	[RFC4566]
c=IN IP4 203.0.113.141	[RFC4566]
a=mid:audio	[RFC5888]
a=msid:ma ta	[I-D.ietf-mmusic-msid] Identifies RTCMediaStream ID (ma) and RTCMediaStreamTrack ID (ta)
a=sendrecv	[RFC3264] - Alice can send and recv audio

(Continued)

Table 18.3 (*Continued*) SDP Offer (Copyright IETF)

a=rtpmap:109 opus/48000/2	[RFC7587] - Opus Codec 48khz, 2 channels
a=rtpmap:0 PCMU/8000	[RFC3551] PCMU Audio Codec
a=rtpmap:8 PCMA/8000	[RFC3551] PCMA Audio Codec
a=maxptime:120	[RFC4566]
a=ice-ufrag:074c6550	[RFC5245] - ICE user fragment
a=ice-pwd:a28a397a4c3f31747d1ee34 74af08a068	[RFC5245] - ICE password parameter
a=fingerprint:sha-256 19:E2:1C:3B :4B:9F:81:E6:B8:5C:F4:A5:A8:D8:73 :04: BB:05:2F:70:9F:04:A9:0E:05:E 9:26:33:E8:70:88:A2	[RFC5245] - DTLS Fingerprint for SRTP
a=setup:actpass	[RFC4145] - Alice can perform DTLS before Answer arrives
a=dtls-id:1	[I-D.ietf-mmusic-dtls-sdp]
a=rtcp-mux	[RFC5761] - Alice can perform RTP/RTCP Muxing
a=rtcp-mux-only	[I-D.ietf-mmusic-mux-exclusiv e]
a=rtcp-rsize	[RFC5506] - Alice intends to use reduced size RTCP for this session
a=rtcp-fb:109 nack	[RFC5104] - Indicates NACK RTCP feedback support
a=extmap:1 urn:ietf:params:rtp-hdrext:ssrc-audio-level	[RFC6464]
a=extmap:2 urn:ietf:params:rtp-hdrext:sdes:mid	[I-D.ietf-mmusic-sdp-bundle-n egotiation]
a=candidate:0 1 UDP 2122194687 192.0.2.4 61665 typ host	[RFC5245] - RTP/RTCP Host Candidate
a=candidate:1 1 UDP 1685987071 203.0.113.141 54609 typ srflx raddr 192.0.2.4 rport 61665	[RFC5245] - RTP/RTCP Server Reflexive ICE Candidate
a=end-of-candidates	[I-D.ietf-mmusic-trickle-ice]
****** Video m=line *********	******************************
m=video 54609 UDP/TLS/RTP/SAVPF 99 120	[RFC4566]
c=IN IP4 203.0.113.141	[RFC4566]
a=mid:video	[RFC5888]
a=msid:ma tb	Identifies RTCMediaStream ID (ma) and RTCMediaStreamTrack ID (tb)
a=sendrecv	[RFC3264] - Alice can send and recv video
a=rtpmap:99 H264/90000	[RFC6184] - H.264 Video Codec
a=fmtp:99 profile-level-id=4d0028 ;packetization-mode=1	[RFC6184]
a=rtpmap:120 VP8/90000	[RFC7741] - VP8 video codec
a=rtcp-fb:99 nack	[RFC5104] - Indicates NACK

(*Continued*)

Table 18.3 (*Continued*) SDP Offer (Copyright IETF)

	RTCP feedback support
a=rtcp-fb:99 nack pli	[RFC5104] - Indicates support for Picture loss Indication and NACK
a=rtcp-fb:99 ccm fir	[RFC5104] - Full Intra Frame Request-Codec Control Message support
a=rtcp-fb:120 nack	[RFC5104] - Indicates NACK RTCP feedback support
a=rtcp-fb:120 nack pli	[RFC5104] - Indicates support for Picture loss Indication and NACK
a=rtcp-fb:120 ccm fir	[RFC5104] - Full Intra Frame Request-Codec Control Message support
a=extmap:2 urn:ietf:params:rtp-hdrext:sdes:mid	[I-D.ietf-mmusic-sdp-bundle-negotiation]

Table 18.4 SDP Answer (Copyright IETF)

Answer SDP Contents	RFC#/Notes
v=0	[RFC4566]
o=- 16833 0 IN IP4 0.0.0.0	[RFC4566] - Session Origin Information
s=-	[RFC4566]
t=0 0	[RFC4566]
a=group:BUNDLE audio video	[I-D.ietf-mmusic-sdp-bundle-negotiation]
a=group:LS audio video	[RFC5888] - Bob agrees to do the same
a=ice-options:trickle	[I-D.ietf-mmusic-trickle-ice]
****** Audio m=line *********	*****************************
m=audio 49203 UDP/TLS/RTP/SAVPF 109	[RFC4566]
c=IN IP4 203.0.113.77	[RFC4566]
a=mid:audio	[RFC5888]
a=msid:ma ta	Identifies RTCMediaStream ID (ma) and RTCMediaStreamTrack ID (ta)
a=sendrecv	[RFC3264] - Bob can send and recv audio
a=rtpmap:109 opus/48000/2	[RFC7587] - Bob accepts only Opus Codec
a=maxptime:120	[RFC4566]
a=ice-ufrag:c300d85b	[RFC5245] - ICE username frag

(*Continued*)

Table 18.4 (*Continued*) SDP Answer (Copyright IETF)

a=ice-pwd:de4e99bd291c325921d5d47efbabd9a2	[RFC5245] - ICE password
a=fingerprint:sha-256 6B:8B:F0:65:5F:78:E2:51:3B:AC:6F:F3:3F:46:1B:35 :DC:B8:5F:64:1A:24:C2:43:F0:A1:58:D0:A1:2C:19:08	[RFC5245] - DTLS Fingerprint for SRTP
a=setup:active	[RFC4145] - Bob carries out DTLS Handshake in parallel
a=dtls-id:1	[I-D.ietf-mmusic-dtls-sdp]
a=rtcp-mux	[RFC5761] - Bob can perform RTP/RTCP Muxing
a=rtcp-mux-only	[I-D.ietf-mmusic-mux-exclusive]
a=rtcp-rsize	[RFC5506] - Bob intends to use reduced size RTCP for this session
a=extmap:1 urn:ietf:params:rtp-hdrext:ssrc-audio-level	[RFC6464]
a=extmap:2 urn:ietf:params:rtp-hdrext:sdes:mid	[I-D.ietf-mmusic-sdp-bundle-negotiation]
a=candidate:0 1 UDP 3618095783 198.51.100.7 49203 typ host	[RFC5245] - RTP/RTCP Host ICE Candidate
a=candidate:1 1 UDP 565689203 203.0.113.77 49203 typ srflx raddr 198.51.100.7 rport 51556	[RFC5245] - RTP/RTCP Server Reflexive ICE Candidate
a=end-of-candidates	[I-D.ietf-mmusic-trickle-ice]
****** Video m=line *********	*****************************
m=video 49203 UDP/TLS/RTP/SAVPF 99	[RFC4566]
c=IN IP4 203.0.113.77	[RFC4566]
a=mid:video	[RFC5888]
a=msid:ma tb	Identifies RTCMediaStream ID (ma) and RTCMediaStreamTrack ID (tb)
a=sendrecv	[RFC3264] - Bob can send and recv video
a=rtpmap:99 H264/90000	[RFC6184] - Bob accepts H.264 Video Codec.
a=fmtp:99 profile-level-id=4d0028;packetization-mode=1	[RFC6184]
a=rtcp-fb:99 nack	[RFC5104] - Indicates support for NACK based RTCP feedback
a=rtcp-fb:99 nack pli	[RFC5104] - Indicates support for Picture loss Indication and NACK
a=rtcp-fb:99 ccm fir	[RFC5104] - Full Intra Frame Request- Codec Control Message support
a=extmap:2 urn:ietf:params:rtp-hdrext:sdes:mid	[I-D.ietf-mmusic-sdp-bundle-negotiation]

18.2.5.2.2.2 Dual-Stack Audio/Video Session This section captures the offer/answer exchange when Alice and Bob support both IPv4 and IPv6 host addresses (Tables 18.5 and 18.6).

Table 18.5 SDP Offer (Copyright IETF)

Offer SDP Contents	RFC#/Notes
v=0	[RFC4566]
o=- 20518 0 IN IP4 0.0.0.0	[RFC4566] - Session Origin Information
s=-	[RFC4566]
t=0 0	[RFC4566]
a=group:BUNDLE audio video	[I-D.ietf-mmusic-sdp-bundle-negotiation]
a=group:LS audio video	[RFC5888] - Alice wants to lip sync her audio and video sreams
a=ice-options:trickle	[I-D.ietf-mmusic-trickle-ice]
****** Audio m=line *********	****************************
m=audio 54609 UDP/TLS/RTP/SAVPF 109 0 8	[RFC4566]
c=IN IP4 203.0.113.141	[RFC4566]
a=mid:audio	[RFC5888]
a=msid:ma ta	Identifies RTCMediaStream ID (ma) and RTCMediaStreamTrack ID (ta)
a=sendrecv	[RFC3264] - Alice can send and recv audio
a=rtpmap:109 opus/48000/2	[RFC7587] - Opus Codec 48khz, 2 channels
a=rtpmap:0 PCMU/8000	[RFC3551] PCMU Audio Codec
a=rtpmap:8 PCMA/8000	[RFC3551] PCMA Audio Codec
a=maxptime:120	[RFC4566]
a=ice-ufrag:074c6550	[RFC5245] - ICE user fragment
a=ice-pwd:a28a397a4c3f31747d1ee34 74af08a068	[RFC5245] - ICE password parameter
a=fingerprint:sha-256 19:E2:1C:3B :4B:9F:81:E6:B8:5C:F4:A5:A8:D8:73 :04: BB:05:2F:70:9F:04:A9:0E:05:E 9:26:33:E8:70:88:A2	[RFC5245] - DTLS Fingerprint for SRTP
a=setup:actpass	[RFC4145] - Alice can perform DTLS before Answer arrives
a=dtls-id:1	[I-D.ietf-mmusic-dtls-sdp]
a=rtcp-mux	[RFC5761] - Alice can perform RTP/RTCP Muxing
a=rtcp-mux-only	[I-D.ietf-mmusic-mux-exclusive]
a=rtcp-rsize	[RFC5506] - Alice intends to use reduced size RTCP for this session

(Continued)

Table 18.5 (*Continued*) SDP Offer (Copyright IETF)

a=rtcp-fb:109 nack	[RFC5104] - Indicates NACK RTCP feedback support
a=extmap:1 urn:ietf:params:rtp-hdrext:ssrc-audio-level	[RFC6464]
a=extmap:2 urn:ietf:params:rtp-hdrext:sdes:mid	[I-D.ietf-mmusic-sdp-bundle-negotiation]
a=candidate:0 1 UDP 2122194687 192.0.2.4 61665 typ host	[RFC5245] - RTP/RTCP Host Candidate
a=candidate:0 1 UDP 2122194687 2001:DB8:8101:3a55:4858:a2a9:22ff:99b9 61665 typ host	[RFC5245] - RTP/RTCP IPv6 Host Candidate
a=end-of-candidates	[I-D.ietf-mmusic-trickle-ice]
****** Video m=line *********	*****************************
m=video 54609 UDP/TLS/RTP/SAVPF 99 120	[RFC4566]
c=IN IP4 203.0.113.141	[RFC4566]
a=mid:video	[RFC5888]
a=msid:ma tb	Identifies RTCMediaStream ID (ma) and RTCMediaStreamTrack ID (tb)
a=sendrecv	[RFC3264] - Alice can send and recv video
a=rtpmap:99 H264/90000	[RFC6184] - H.264 Video Codec
a=fmtp:99 profile-level-id=4d0028;packetization-mode=1	[RFC6184]
a=rtpmap:120 VP8/90000	[RFC7741] - VP8 video codec
a=rtcp-fb:99 nack	[RFC5104] - Indicates NACK RTCP feedback support
a=rtcp-fb:99 nack pli	[RFC5104] - Indicates support for Picture loss Indication and NACK
a=rtcp-fb:99 ccm fir	[RFC5104] - Full Intra Frame Request-Codec Control Message support
a=rtcp-fb:120 nack	[RFC5104] - Indicates NACK RTCP feedback support
a=rtcp-fb:120 nack pli	[RFC5104] - Indicates support for Picture loss Indication and NACK
a=rtcp-fb:120 ccm fir	[RFC5104] - Full Intra Frame Request-Codec Control Message support
a=extmap:2 urn:ietf:params:rtp-hdrext:sdes:mid	[I-D.ietf-mmusic-sdp-bundle-negotiation]

Table 18.6 SDP Answer (Copyright IETF)

Answer SDP Contents	RFC#/Notes
v=0	[RFC4566]
o=- 16833 0 IN IP4 0.0.0.0	[RFC4566] - Session Origin Information
s=-	[RFC4566]
t=0 0	[RFC4566]
a=group:BUNDLE audio video	[I-D.ietf-mmusic-sdp-bundle-n egotiation]
a=group:LS audio video	[RFC5888] - Bob agrees to do the same
a=ice-options:trickle	[I-D.ietf-mmusic-trickle-ice]
****** Audio m=line *********	***************************
m=audio 49203 UDP/TLS/RTP/SAVPF 109	[RFC4566]
c=IN IP4 203.0.113.77	[RFC4566]
a=mid:audio	[RFC5888]
a=msid:ma ta	Identifies RTCMediaStream ID (ma) and RTCMediaStreamTrack ID (ta)
a=sendrecv	[RFC3264] - Bob can send and recv audio
a=rtpmap:109 opus/48000/2	[RFC7587] - Bob accepts only Opus Codec
a=maxptime:120	[RFC4566]
a=ice-ufrag:c300d85b	[RFC5245] - ICE username frag
a=ice-pwd:de4e99bd291c325921d5d47 efbabd9a2	[RFC5245] - ICE password
a=fingerprint:sha-256 6B:8B:F0:65 :5F:78:E2:51:3B:AC:6F:F3:3F:46:1B :35 :DC:B8:5F:64:1A:24:C2:43:F0:A 1:58:D0:A1:2C:19:08	[RFC5245] - DTLS Fingerprint for SRTP
a=setup:active	[RFC4145] - Bob carries out DTLS Handshake in parallel
a=dtls-id:1	[I-D.ietf-mmusic-dtls-sdp]
a=rtcp-mux	[RFC5761] - Bob can perform RTP/RTCP Muxing
a=rtcp-mux-only	[I-D.ietf-mmusic-mux-exclusiv e]
a=rtcp-rsize	[RFC5506] - Bob intends to use reduced size RTCP for this session
a=extmap:1 urn:ietf:params:rtp- hdrext:ssrc-audio-level	[RFC6464]
a=extmap:2 urn:ietf:params:rtp- hdrext:sdes:mid	[I-D.ietf-mmusic-sdp-bundle-n egotiation]
a=candidate:0 1 UDP 3618095783 198.51.100.7 49203 typ host	[RFC5245] - RTP/RTCP Host ICE Candidate

(*Continued*)

Table 18.6 (*Continued*) SDP Answer (Copyright IETF)

a=candidate:0 1 UDP 3618095783 20 01:DB8:30c:1266:5916:3779:22f6:77 f7 49203 typ host	[RFC5245] - RTP/RTCP IPv6 Host ICE Candidate
a=end-of-candidates	[I-D.ietf-mmusic-trickle-ice]
****** Video m=line *********	*****************************
m=video 49203 UDP/TLS/RTP/SAVPF 99	[RFC4566]
c=IN IP4 203.0.113.77	[RFC4566]
a=mid:video	[RFC5888]
a=msid:ma tb	Identifies RTCMediaStream ID (ma) and RTCMediaStreamTrack ID (tb)
a=sendrecv	[RFC3264] - Bob can send and recv video
a=rtpmap:99 H264/90000	[RFC6184] - Bob accepts H.264 Video Codec.
a=fmtp:99 profile-level-id=4d0028 ;packetization-mode=1	[RFC6184]
a=rtcp-fb:99 nack	[RFC5104] - Indicates support for NACK based RTCP feedback
a=rtcp-fb:99 nack pli	[RFC5104] - Indicates support for Picture loss Indication and NACK
a=rtcp-fb:99 ccm fir	[RFC5104] - Full Intra Frame Request- Codec Control Message support
a=extmap:2 urn:ietf:params:rtp- hdrext:sdes:mid	[I-D.ietf-mmusic-sdp-bundle-n egotiation]

18.2.5.2.3 Data-Only Session

This scenario illustrates the SDP negotiated to set up a data-only session based on the Stream Control Transmission Protocol (SCTP) data channel, thus enabling use cases such as file-transfer and real-time game control, for example (Tables 18.7 and 18.8).

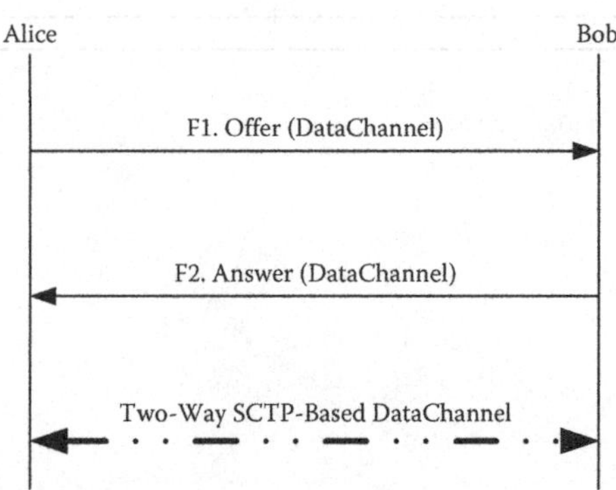

2-way DataChannel session

Table 18.7 SDP Offer (Copyright IETF)

Offer SDP Contents	RFC#/Notes
v=0	[RFC4566]
o=- 20518 0 IN IP4 0.0.0.0	[RFC4566] - Session Origin Information
s=-	[RFC4566]
t=0 0	[RFC4566]
a=group:BUNDLE data	[I-D.ietf-mmusic-sdp-bundle-negotiation]
a=ice-options:trickle	[I-D.ietf-mmusic-trickle-ice]
****** Application m=line *********	****************************
m=application 54609 UDP/DTLS/SCTP webrtc-datachannel	[I-D.ietf-rtcweb-data-channel]
c=IN IP4 203.0.113.141	[RFC4566]
a=mid:data	[RFC5888]
a=sendrecv	[RFC3264] - Alice can send and recv non-media data
a=sctp-port:5000	[I-D.ietf-mmusic-sctp-sdp]
a=max-message-size:100000	[I-D.ietf-mmusic-sctp-sdp]
a=setup:actpass	[RFC4145] - Alice can perform DTLS before Answer arrives
a=dtls-id:1	[I-D.ietf-mmusic-dtls-sdp]
a=ice-ufrag:074c6550	[RFC5245] - Session Level ICE parameter
a=ice-pwd:a28a397a4c3f31747d1ee3474af08a068	[RFC5245] - Session Level ICE parameter
a=fingerprint:sha-256 19:E2:1C:3B:4B:9F:81:E6:B8:5C:F4:A5:A8:D8:73:04 :BB:05:2F:70:9F:04:A9:0E:05:E9:26:33:E8:70:88:A2	[RFC5245] - Session DTLS Fingerprint for SRTP
a=candidate:0 1 UDP 2113667327 192.0.2.4 61665 typ host	[RFC5245]
a=candidate:1 1 UDP 1694302207 203.0.113.141 54609 typ srflx raddr 192.0.2.4 rport 61665	[RFC5245]
a=end-of-candidates	[I-D.ietf-mmusic-trickle-ice]

Table 18.8 SDP Answer (Copyright IETF)

Answer SDP Contents	RFC#/Notes
v=0	[RFC4566]
o=- 16833 0 IN IP4 0.0.0.0	[RFC4566] - Session Origin Information
s=-	[RFC4566]
t=0 0	[RFC4566]
a=group:BUNDLE data	[I-D.ietf-mmusic-sdp-bundle-negotiation]
****** Application m=line *********	*****************************
m=application 49203 UDP/DTLS/SCTP webrtc-datachannel	[I-D.ietf-mmusic-sctp-sdp]
c=IN IP4 203.0.113.77	[RFC4566]
a=mid:data	[RFC5888]
a=sendrecv	[RFC3264] - Bob can send and recv non-media data
a=sctp-port:5000	[I-D.ietf-mmusic-sctp-sdp]
a=max-message-size:100000	[I-D.ietf-mmusic-sctp-sdp]
a=setup:active	[RFC4145] - Bob carries out DTLS Handshake in parallel
a=dtls-id:1	[I-D.ietf-mmusic-dtls-sdp]
a=ice-ufrag:c300d85b	[RFC5245] - Session Level ICE username frag
a=ice-pwd:de4e99bd291c325921d5d47efbabd9a2	[RFC5245] - Session Level ICE password
a=fingerprint:sha-256 6B:8B:F0:65:5F:78:E2:51:3B:AC:6F:F3:3F:46:1B:35 :DC:B8:5F:64:1A:24:C2:43:F0:A1:58:D0:A1:2C:19:08	[RFC5245] - Session DTLS Fingerprint for SRTP
a=candidate:0 1 UDP 2113667327 198.51.100.7 51556 typ host	[RFC5245]
a=candidate:1 1 UDP 1694302207 203.0.113.77 49203 typ srflx raddr 198.51.100.7 rport 51556	[RFC5245]
a=end-of-candidates	[I-D.ietf-mmusic-trickle-ice]

18.2.5.2.4 Audio Call on Hold

Alice calls Bob, but when Bob answers he places Alice on hold by setting the SDP direction attribute to a=inactive in the answer (Tables 18.9 and 18.10).

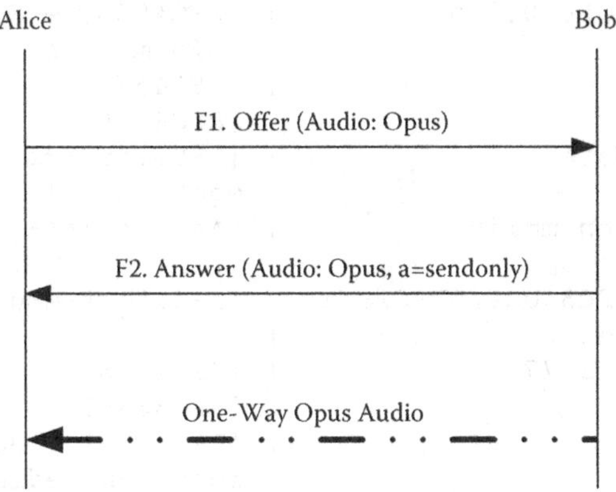

Audio on hold

Table 18.9 SDP Offer (Copyright IETF)

Offer SDP Contents	RFC#/Notes
v=0	[RFC4566]
o=- 20518 0 IN IP4 0.0.0.0	[RFC4566] - Session Origin Information
s=-	[RFC4566]
t=0 0	[RFC4566]
a=group:BUNDLE audio	[I-D.ietf-mmusic-sdp-bundle-negotiation]
a=ice-options:trickle	[I-D.ietf-mmusic-trickle-ice]
****** Audio m=line *********	****************************
m=audio 54609 UDP/TLS/RTP/SAVPF 109	[RFC4566]
c=IN IP4 203.0.113.141	[RFC4566]
a=mid:audio	[RFC5888]
a=msid:ma ta	Identifies RTCMediaStream ID (ma) and RTCMediaStreamTrack ID (ta)
a=sendrecv	[RFC3264] - Alice can send and recv audio
a=rtpmap:109 opus/48000/2	[RFC7587] - Opus Codec 48khz, 2 channels
a=maxptime:120	[RFC4566]
a=ice-ufrag:074c6550	[RFC5245] - ICE user fragment
a=ice-pwd:a28a397a4c3f31747d1ee3474af08a068	[RFC5245] - ICE password

(Continued)

Table 18.9 (*Continued*) SDP Offer (Copyright IETF)

a=fingerprint:sha-256 19:E2:1C:3B :4B:9F:81:E6:B8:5C:F4:A5:A8:D8:73 :04 :BB:05:2F:70:9F:04:A9:0E:05:E 9:26:33:E8:70:88:A2	[RFC5245] - DTLS Fingerprint for SRTP
a=setup:actpass	[RFC4145] - Alice can perform DTLS before Answer arrives
a=dtls-id:1	[I-D.ietf-mmusic-dtls-sdp]
a=rtcp-mux	[RFC5761] - Alice can perform RTP/RTCP Muxing
a=rtcp-mux-only	[I-D.ietf-mmusic-mux-exclusiv e]
a=rtcp-rsize	[RFC5506]
a=rtcp-fb:109 nack	[RFC5104] - Indicates NACK RTCP feedback support
a=extmap:1 urn:ietf:params:rtp- hdrext:ssrc-audio-level	[RFC6464]
a=extmap:2 urn:ietf:params:rtp- hdrext:sdes:mid	[I-D.ietf-mmusic-sdp-bundle-n egotiation]
a=candidate:0 1 UDP 2113667327 192.0.2.4 61665 typ host	[RFC5245]
a=candidate:1 1 UDP 1685987071 203.0.113.141 54609 typ srflx raddr 192.0.2.4 rport 61665	[RFC5245]
a=end-of-candidates	[I-D.ietf-mmusic-trickle-ice]

Table 18.10 SDP Answer

Answer SDP Contents	RFC#/Notes
v=0	[RFC4566]
o=- 16833 0 IN IP4 0.0.0.0	[RFC4566] - Session Origin Information
s=-	[RFC4566]
t=0 0	[RFC4566]
a=group:BUNDLE audio	[I-D.ietf-mmusic-sdp-bundle-n egotiation]
****** Audio m=line *********	*****************************
m=audio 49203 UDP/TLS/RTP/SAVPF 109	[RFC4566]
c=IN IP4 203.0.113.77	[RFC4566]
a=mid:audio	[RFC5888]
a=msid:ma ta	Identifies RTCMediaStream ID (ma) and RTCMediaStreamTrack ID (ta)
a=inactive	[RFC3264] - Bob puts call On Hold

(Continued)

Table 18.10 (*Continued*) SDP Answer

a=rtpmap:109 opus/48000/2	[RFC7587] - Bob accepts Opus Codec
a=maxptime:120	[RFC4566]
a=ice-ufrag:c300d85b	[RFC5245] - ICE username frag
a=ice-pwd:de4e99bd291c325921d5d47efbabd9a2	[RFC5245] - ICE password
a=fingerprint:sha-256 6B:8B:F0:65:5F:78:E2:51:3B:AC:6F:F3:3F:46:1B:35 :DC:B8:5F:64:1A:24:C2:43:F0:A1:58:D0:A1:2C:19:08	[RFC5245] - DTLS Fingerprint for SRTP
a=setup:active	[RFC4145] - Bob carries out DTLS Handshake in parallel
a=dtls-id:1	[I-D.ietf-mmusic-dtls-sdp]
a=rtcp-mux	[RFC5761] - Bob can perform RTP/RTCP Muxing
a=rtcp-mux-only	[I-D.ietf-mmusic-mux-exclusive]
a=rtcp-rsize	[RFC5506]
a=extmap:1 urn:ietf:params:rtp-hdrext:ssrc-audio-level	[RFC6464]
a=extmap:2 urn:ietf:params:rtp-hdrext:sdes:mid	[I-D.ietf-mmusic-sdp-bundle-negotiation]
a=candidate:0 1 UDP 2113667327 198.51.100.7 51556 typ host	[RFC5245] - Host candidate
a=candidate:1 1 UDP 1685987071 203.0.113.141 49203 typ srflx raddr 198.51.100.7 rport 51556	[RFC5245] - Server Reflexive candidate
a=end-of-candidates	[I-D.ietf-mmusic-trickle-ice]

18.2.5.2.5 Audio with DTMF Session

In this example, Alice wishes to establish two separate audio streams, one for normal audio and the other for telephone events. Alice offers first an audio stream with three codecs and then another with (RFC 4733) tones (for DTMF). Bob accepts both audio streams by choosing Opus as the audio codec and telephone event for the other stream (Tables 18.11 and 18.12).

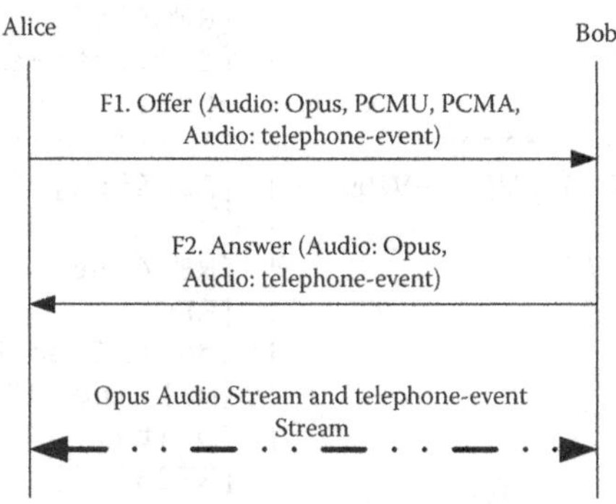

Audio session with DTMF

Table 18.11 SDP Offer (Copyright IETF)

Offer SDP Contents	RFC#/Notes
v=0	[RFC4566]
o=- 20518 0 IN IP4 0.0.0.0	[RFC4566] - Session Origin Information
s=-	[RFC4566]
t=0 0	[RFC4566]
a=group:BUNDLE audio dtmf	[I-D.ietf-mmusic-sdp-bundle-negotiation]
a=ice-options:trickle	[I-D.ietf-mmusic-trickle-ice]
****** Audio m=line *********	*****************************
m=audio 54609 UDP/TLS/RTP/SAVPF 109 0 8	[RFC4566]
c=IN IP4 203.0.113.141	[RFC4566]
a=mid:audio	[RFC5888]
a=msid:ma ta	Identifies RTCMediaStream ID (ma) and RTCMediaStreamTrack ID (ta)
a=sendrecv	[RFC3264] - Alice can send and recv audio
a=rtpmap:109 opus/48000/2	[RFC7587] - Opus Codec 48khz, 2 channels
a=rtpmap:0 PCMU/8000	[RFC3551] PCMU Audio Codec
a=rtpmap:8 PCMA/8000	[RFC3551] PCMA Audio Codec
a=maxptime:120	[RFC4566]
a=ice-ufrag:074c6550	[RFC5245] - ICE user fragment
a=ice-pwd:a28a397a4c3f31747d1ee3474af08a068	[RFC5245] - ICE password parameter
a=fingerprint:sha-256 19:E2:1C:3B:4B:9F:81:E6:B8:5C:F4:A5:A8:D8:73:04 :BB:05:2F:70:9F:04:A9:0E:05:E9:26:33:E8:70:88:A2	[RFC5245] - DTLS Fingerprint for SRTP
a=setup:actpass	[RFC4145] - Alice can perform DTLS before Answer arrives
a=dtls-id:1	[I-D.ietf-mmusic-dtls-sdp]
a=rtcp-mux	[RFC5761] - Alice can perform RTP/RTCP Muxing
a=rtcp-mux-only	[I-D.ietf-mmusic-mux-exclusive]
a=rtcp-rsize	[RFC5506]
a=rtcp-fb:109 nack	[RFC5104] - Indicates NACK RTCP feedback support
a=extmap:1 urn:ietf:params:rtp-hdrext:ssrc-audio-level	[RFC6464]
a=extmap:2 urn:ietf:params:rtp-hdrext:sdes:mid	[I-D.ietf-mmusic-sdp-bundle-negotiation]

(Continued)

Table 18.11 (*Continued*) SDP Offer (Copyright IETF)

a=candidate:0 1 UDP 2122194687 192.0.2.4 61665 typ host	[RFC5245]
a=candidate:1 1 UDP 1685987071 203.0.113.141 54609 typ srflx raddr 192.0.2.4 rport 61665	[RFC5245]
a=end-of-candidates	[I-D.ietf-mmusic-trickle-ice]
****** DTMF m=line *********	******************************
m=audio 54609 UDP/TLS/RTP/SAVPF 126	[RFC4566]
c=IN IP4 203.0.113.141	[RFC4566]
a=mid:dtmf	[RFC5888]
a=msid:ma tb	Identifies RTCMediaStream ID (ma) and RTCMediaStreamTrack ID (tb)
a=sendonly	[RFC3264] - Alice can send DTMF Events
a=rtpmap:126 telephone-event/8000	[RFC4733]
a=rtcp-fb:109 nack	[RFC5104] - Indicates NACK RTCP feedback support
a=extmap:2 urn:ietf:params:rtp-hdrext:sdes:mid	[I-D.ietf-mmusic-sdp-bundle-negotiation]

Table 18.12 SDP Answer (Copyright IETF)

Answer SDP Contents	RFC#/Notes
v=0	[RFC4566]
o=- 16833 0 IN IP4 0.0.0.0	[RFC4566] - Session Origin Information
s=-	[RFC4566]
t=0 0	[RFC4566]
a=group:BUNDLE audio dtmf	[I-D.ietf-mmusic-sdp-bundle-negotiation]
****** Audio m=line *********	******************************
m=audio 49203 UDP/TLS/RTP/SAVPF 109	[RFC4566]
c=IN IP4 203.0.113.77	[RFC4566]
a=mid:audio	[RFC5888]
a=msid:ma ta	Identifies RTCMediaStream ID (ma) and RTCMediaStreamTrack ID (ta)
a=sendrecv	[RFC3264] - Bob can send and receive Opus audio
a=rtpmap:109 opus/48000/2	[RFC7587] - Bob accepts Opus Codec
a=maxptime:120	[RFC4566]

(*Continued*)

Table 18.12 (*Continued*) SDP Answer (Copyright IETF)

a=ice-ufrag:c300d85b	[RFC5245] - ICE username frag
a=ice-pwd:de4e99bd291c325921d5d47 efbabd9a2	[RFC5245] - ICE password
a=fingerprint:sha-256 6B:8B:F0:65 :5F:78:E2:51:3B:AC:6F:F3:3F:46:1B :35 :DC:B8:5F:64:1A:24:C2:43:F0:A 1:58:D0:A1:2C:19:08	[RFC5245] - Fingerprint for SRTP
a=setup:active	[RFC4145] - Bob carries out DTLS Handshake in parallel
a=dtls-id:1	[I-D.ietf-mmusic-dtls-sdp]
a=rtcp-mux	[RFC5761] - Bob can perform RTP/RTCP Muxing on port 49203
a=rtcp-mux-only	[I-D.ietf-mmusic-mux-exclusiv e]
a=rtcp-rsize	[RFC5506] - Alice intends to use reduced size RTCP for this session
a=extmap:1 urn:ietf:params:rtp-hdrext:ssrc-audio-level	[RFC6464]
a=extmap:2 urn:ietf:params:rtp-hdrext:sdes:mid	[I-D.ietf-mmusic-sdp-bundle-n egotiation]
a=candidate:0 1 UDP 2122194687 198.51.100.7 51556 typ host	[RFC5245]
a=candidate:1 1 UDP 1685987071 203.0.113.77 49203 typ srflx raddr 198.51.100.7 rport 51556	[RFC5245]
a=end-of-candidates	[I-D.ietf-mmusic-trickle-ice]
****** DTMF m=line *********	*****************************
m=audio 49203 UDP/TLS/RTP/SAVPF 126	[RFC4566]
c=IN IP4 203.0.113.77	[RFC4566]
a=mid:dtmf	[RFC5888]
a=msid:ma tb	Identifies RTCMediaStream ID (ma) and RTCMediaStreamTrack ID (tb)
a=recvonly	[RFC3264] - Alice can receive DTMF events
a=rtpmap:126 telephone-event/8000	[RFC4733]
a=extmap:2 urn:ietf:params:rtp-hdrext:sdes:mid	[I-D.ietf-mmusic-sdp-bundle-n egotiation]

18.2.5.2.6 One-Way Audio/Video Session: Document Camera

In this scenario, Alice and Bob engage in a one-way audio and video session with Bob receiving Alice's audio and her presentation slides as the video stream (Tables 18.13 and 18.14).

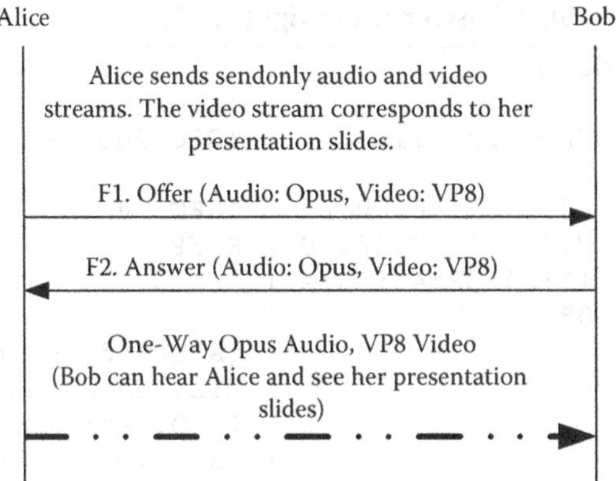

One way audio & video session – document camera

Table 18.13 SDP Offer (Copyright IETF)

Offer SDP Contents	RFC#/Notes
v=0	[RFC4566]
o=- 20519 0 IN IP4 0.0.0.0	[RFC4566]
s=-	[RFC4566]
t=0 0	[RFC4566]
a=group:BUNDLE audio video	[I-D.ietf-mmusic-sdp-bundle-n
	egotiation]
a=group:LS audio video	[RFC5888]
a=ice-options:trickle	[I-D.ietf-mmusic-trickle-ice]
****** Audio m=line *********	*****************************
m=audio 54609 UDP/TLS/RTP/SAVPF	[RFC4566]
109	
c=IN IP4 203.0.113.141	[RFC4566]
a=mid:audio	[RFC5888]
a=msid:ma ta	Identifies RTCMediaStream ID
	(ma) and RTCMediaStreamTrack
	ID (ta)
a=sendonly	[RFC3264] - Send only audio
	stream
a=rtpmap:109 opus/48000/2	[RFC7587]
a=maxptime:120	[RFC4566]
a=ice-ufrag:074c6550	[RFC5245]
a=ice-pwd:a28a397a4c3f31747d1ee34	[RFC5245]
74af08a068	
a=fingerprint:sha-256 19:E2:1C:3B	[RFC5245]
:4B:9F:81:E6:B8:5C:F4:A5:A8:D8:73	
:04 :BB:05:2F:70:9F:04:A9:0E:05:E	
9:26:33:E8:70:88:A2	
a=setup:actpass	[RFC4145] - Alice can perform
	DTLS before Answer arrives

(Continued)

Table 18.13 (*Continued*) SDP Offer (Copyright IETF)

a=dtls-id:1	[I-D.ietf-mmusic-dtls-sdp]
a=rtcp-mux	[RFC5761]
a=rtcp-mux-only	[I-D.ietf-mmusic-mux-exclusive]
a=rtcp-rsize	[RFC5506]
a=rtcp-fb:109 nack	[RFC5104]
a=extmap:1 urn:ietf:params:rtp-hdrext:ssrc-audio-level	[RFC6464]
a=extmap:2 urn:ietf:params:rtp-hdrext:sdes:mid	[I-D.ietf-mmusic-sdp-bundle-negotiation]
a=candidate:0 1 UDP 2122194687 203.0.113.141 54609 typ host	[RFC5245]
a=end-of-candidates	[I-D.ietf-mmusic-trickle-ice]
****** Video m=line *********	****************************
m=video 54609 UDP/TLS/RTP/SAVPF 120	[RFC4566]
c=IN IP4 203.0.113.141	[RFC4566]
a=mid:video	[RFC5888]
a=msid:ma tb	Identifies RTCMediaStream ID (ma) and RTCMediaStreamTrack ID (tb)
a=sendonly	[RFC3264] - Send only video stream
a=rtpmap:120 VP8/90000	[RFC7741]
a=content:slides	[RFC4796] - Alice's presentation video stream
a=rtcp-fb:120 nack	[RFC5104]
a=rtcp-fb:120 nack pli	[RFC5104]
a=rtcp-fb:120 ccm fir	[RFC5104]
a=extmap:2 urn:ietf:params:rtp-hdrext:sdes:mid	[I-D.ietf-mmusic-sdp-bundle-negotiation]

Table 18.14 SDP Answer (Copyright IETF)

Answer SDP Contents	RFC#/Notes
v=0	[RFC4566]
o=- 16833 0 IN IP4 0.0.0.0	[RFC4566]
s=-	[RFC4566]
t=0 0	[RFC4566]
a=group:BUNDLE audio video	[I-D.ietf-mmusic-sdp-bundle-negotiation]
a=group:LS audio video	[RFC5888]
a=ice-options:trickle	[I-D.ietf-mmusic-trickle-ice]
****** Audio m=line *********	****************************
m=audio 49203 UDP/TLS/RTP/SAVPF	[RFC4566]

(*Continued*)

Table 18.14 (*Continued*) SDP Answer (Copyright IETF)

109	
c=IN IP4 203.0.113.77	[RFC4566]
a=mid:audio	[RFC5888]
a=msid:ma ta	Identifies RTCMediaStream ID (ma) and RTCMediaStreamTrack ID (ta)
a=recvonly	[RFC3264] - Receive only audio stream
a=rtpmap:109 opus/48000/2	[RFC7587]
a=maxptime:120	[RFC4566]
a=ice-ufrag:c300d85b	[RFC5245]
a=ice-pwd:de4e99bd291c325921d5d47 efbabd9a2	[RFC5245]
a=fingerprint:sha-256 6B:8B:F0:65 :5F:78:E2:51:3B:AC:6F:F3:3F:46:1B :35 :DC:B8:5F:64:1A:24:C2:43:F0:A 1:58:D0:A1:2C:19:08	[RFC5245]
a=setup:active	[RFC4145] - Bob carries out DTLS Handshake in parallel
a=dtls-id:1	[I-D.ietf-mmusic-dtls-sdp]
a=rtcp-mux	[RFC5761]
a=rtcp-mux-only	[I-D.ietf-mmusic-mux-exclusiv e]
a=rtcp-fb:109 nack	[RFC5104]
a=extmap:1 urn:ietf:params:rtp- hdrext:ssrc-audio-level	[RFC6464]
a=extmap:2 urn:ietf:params:rtp- hdrext:sdes:mid	[I-D.ietf-mmusic-sdp-bundle-n egotiation]
a=candidate:0 1 UDP 2113667327 203.0.113.77 49203 typ host	[RFC5245]
a=end-of-candidates	[I-D.ietf-mmusic-trickle-ice]
****** Video m=line ********	****************************
m=video 49203 UDP/TLS/RTP/SAVPF 120	[RFC4566]
c=IN IP4 203.0.113.77	[RFC4566]
a=mid:video	[RFC5888]
a=msid:ma tb	Identifies RTCMediaStream ID (ma) and RTCMediaStreamTrack ID (tb)
a=recvonly	[RFC3264]
a=rtpmap:120 VP8/90000	[RFC7741]
a=content:slides	[RFC4796] - presentation stream
a=rtcp-fb:120 nack	[RFC5104]
a=rtcp-fb:120 nack pli	[RFC5104]
a=rtcp-fb:120 ccm fir	[RFC5104]
a=extmap:2 urn:ietf:params:rtp- hdrext:sdes:mid	[I-D.ietf-mmusic-sdp-bundle-n egotiation]

18.2.5.2.7 Audio, Video Session with BUNDLE Support Unknown

In this example, since Alice is unsure of Bob's support of the BUNDLE framework, the following steps are performed in order to negotiate and set up a BUNDLE address for the session.

- An SDP offer, in which the Alice assigns unique addresses to each "m=" line in the BUNDLE group and requests the answerer to select the offerer's BUNDLE address.
- An SDP answer, in which Bob indicates his support for BUNDLE, selects the offerer's BUNDLE address, selects its own BUNDLE address and associates it with each BUNDLED "m=" line within the BUNDLE group.

Once the offer/answer exchange completes, both Alice and Bob end up using a single RTP session for both media streams (Tables 18.15 and 18.16).

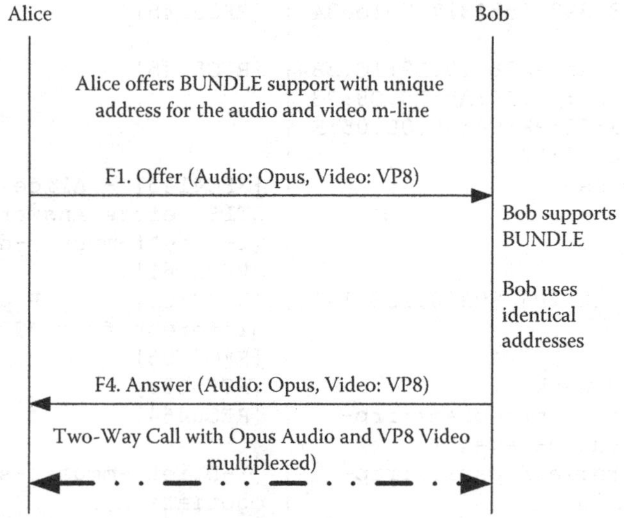

Two-way secure audio, video with BUNDLE support unknown

Table 18.15 SDP Offer w/BUNDLE (Copyright IETF)

Offer SDP Contents	RFC#/Notes
v=0	[RFC4566]
o=- 20518 0 IN IP4 0.0.0.0	[RFC4566]
s=-	[RFC4566]
t=0 0	[RFC4566]
a=group:BUNDLE audio video	[I-D.ietf-mmusic-sdp-bundle-negotiation] Alice supports grouping of m=lines under BUNDLE semantics
a=group:LS audio video	[RFC5888]

(Continued)

Table 18.15 (*Continued*) SDP Offer w/BUNDLE (Copyright IETF)

```
| a=ice-options:trickle                | [I-D.ietf-mmusic-trickle-ice]   |
| ****** Audio m=line ********         | ******************************  |
| m=audio 54609 UDP/TLS/RTP/SAVPF      | [RFC4566]                       |
| 109                                  |                                 |
| c=IN IP4 203.0.113.141               | [RFC4566]                       |
| a=mid:audio                          | [RFC5888] Audio m=line part     |
|                                      | of BUNDLE group with a unique   |
|                                      | port number                     |
| a=msid:ma ta                         | Identifies RTCMediaStream ID    |
|                                      | (ma) and RTCMediaStreamTrack    |
|                                      | ID (ta)                         |
| a=sendrecv                           | [RFC3264]                       |
| a=rtpmap:109 opus/48000/2            | [RFC7587]                       |
| a=maxptime:120                       | [RFC4566]                       |
| a=ice-ufrag:074c6550                 | [RFC5245]                       |
| a=ice-pwd:a28a397a4c3f31747d1ee34    | [RFC5245]                       |
| 74af08a068                           |                                 |
| a=fingerprint:sha-256 19:E2:1C:3B    | [RFC5245]                       |
| :4B:9F:81:E6:B8:5C:F4:A5:A8:D8:73    |                                 |
| :04 :BB:05:2F:70:9F:04:A9:0E:05:E    |                                 |
| 9:26:33:E8:70:88:A2                  |                                 |
| a=setup:actpass                      | [RFC4145] - Alice can perform   |
|                                      | DTLS before Answer arrives      |
| a=dtls-id:1                          | [I-D.ietf-mmusic-dtls-sdp]      |
| a=rtcp-mux                           | [RFC5761]                       |
| a=rtcp:54610 IN IP4 203.0.113.141    | [RFC3605] - RTCP port           |
|                                      | different from RTP Port         |
| a=rtcp-rsize                         | [RFC5506]                       |
| a=rtcp-fb:109 nack                   | [RFC5104]                       |
| a=extmap:1 urn:ietf:params:rtp-      | [RFC6464]                       |
| hdrext:ssrc-audio-level              |                                 |
| a=extmap:2 urn:ietf:params:rtp-      | [I-D.ietf-mmusic-sdp-bundle-n   |
| hdrext:sdes:mid                      | egotiation]                     |
| a=candidate:0 1 UDP  2122194687      | [RFC5245] - RTP host            |
| 192.0.2.4 61665 typ host             | candidate                       |
| a=candidate:1 1 UDP  1685987071      | [RFC5245] - RTP Server          |
| 203.0.113.141 54609 typ srflx        | Reflexive candidate             |
| raddr 192.0.2.4 rport 61665          |                                 |
| a=candidate:0 2 UDP 2122194687       | [RFC5245] - RTCP host           |
| 192.0.2.4 61666 typ host             | candidate                       |
| a=candidate:1 2 UDP  1685987071      | [RFC5245] - RTCP Server         |
| 203.0.113.141 54610 typ srflx        | Reflexive candidate             |
| raddr 192.0.2.4 rport 61666          |                                 |
| ****** Video m=line ********         | ******************************  |
| m=video 62537 UDP/TLS/RTP/SAVPF      | [RFC4566]                       |
| 120                                  |                                 |
| c=IN IP4 203.0.113.141               | [RFC4566]                       |
| a=mid:video                          | [RFC5888] Video m=line part     |
|                                      | of the Bundle group with a      |
|                                      | unique port number              |
```

(*Continued*)

Table 18.15 (*Continued*) SDP Offer w/BUNDLE (Copyright IETF)

a=msid:ma tb	Identifies RTCMediaStream ID (ma) and RTCMediaStreamTrack ID (tb)
a=sendrecv	[RFC3264]
a=rtpmap:120 VP8/90000	[RFC7741]
a=ice-ufrag:6550074c	[RFC5245]
a=ice-pwd:74af08a068a28a397a4c3f3 1747d1ee34	[RFC5245]
a=fingerprint:sha-256 19:E2:1C:3B :4B:9F:81:E6:B8:5C:F4:A5:A8:D8:73 :04 :BB:05:2F:70:9F:04:A9:0E:05:E 9:26:33:E8:70:88:A2	[RFC5245]
a=setup:actpass	[RFC4145] - Alice can perform DTLS before Answer arrives
a=dtls-id:2	[I-D.ietf-mmusic-dtls-sdp]
a=rtcp-mux	[RFC5761]
a=rtcp:62538 IN IP4 203.0.113.141	[RFC3605]
a=rtcp-rsize	[RFC5506]
a=rtcp-fb:120 nack	[RFC5104]
a=rtcp-fb:120 nack pli	[RFC5104]
a=rtcp-fb:120 ccm fir	[RFC5104]
a=extmap:2 urn:ietf:params:rtp-hdrext:sdes:mid	[I-D.ietf-mmusic-sdp-bundle-negotiation]
a=candidate:0 1 UDP 2122194687 192.0.2.4 61886 typ host	[RFC5245] - RTP Host candidate
a=candidate:1 1 UDP 1685987071 203.0.113.141 62537 typ srflx raddr 192.0.2.4 rport 61886	[RFC5245] - RTP Server Reflexive candidate
a=candidate:0 2 2122194687 192.0.2.4 61888 typ host	[RFC5245] - RTCP host candidate
a=candidate:1 2 UDP 1685987071 203.0.113.141 62538 typ srflx raddr 192.0.2.4 rport 61888	[RFC5245] - RTCP Server Reflexive candidate

Table 18.16 SDP Answer w/BUNDLE (Copyright IETF)

Answer SDP Contents	RFC#/Notes
v=0	[RFC4566]
o=- 16833 0 IN IP4 0.0.0.0	[RFC4566]
s=-	[RFC4566]
t=0 0	[RFC4566]
a=group:BUNDLE audio video	[I-D.ietf-mmusic-sdp-bundle-negotiation] Bob supports BUNDLE semantics.
a=group:LS audio video	[RFC5888]
a=ice-options:trickle	[I-D.ietf-mmusic-trickle-ice]
****** Audio m=line *********	****************************

(*Continued*)

Table 18.16 (*Continued*) SDP Answer w/BUNDLE (Copyright IETF)

```
| m=audio 49203 UDP/TLS/RTP/SAVPF    | [RFC4566]                     |
| 109                                |                               |
| c=IN IP4 203.0.113.77              | [RFC4566]                     |
| a=mid:audio                        | [RFC5888] Audio m=line part   |
|                                    | of the BUNDLE group           |
| a=msid:ma ta                       | Identifies RTCMediaStream ID  |
|                                    | (ma) and RTCMediaStreamTrack  |
|                                    | ID (ta)                       |
| a=sendrecv                         | [RFC3264]                     |
| a=rtpmap:109 opus/48000/2          | [RFC7587]                     |
| a=maxptime:120                     | [RFC4566]                     |
| a=ice-ufrag:c300d85b               | [RFC5245]                     |
| a=ice-pwd:de4e99bd291c325921d5d47  | [RFC5245]                     |
| efbabd9a2                          |                               |
| a=fingerprint:sha-256 6B:8B:F0:65  | [RFC5245]                     |
| :5F:78:E2:51:3B:AC:6F:F3:3F:46:1B  |                               |
| :35 :DC:B8:5F:64:1A:24:C2:43:F0:A  |                               |
| 1:58:D0:A1:2C:19:08                |                               |
| a=setup:active                     | [RFC4145] - Bob carries out   |
|                                    | DTLS Handshake in parallel    |
| a=dtls-id:1                        | [I-D.ietf-mmusic-dtls-sdp]    |
| a=rtcp-mux                         | [RFC5761]                     |
| a=rtcp-rsize                       | [RFC5506]                     |
| a=rtcp-fb:109 nack                 | [RFC5104]                     |
| a=extmap:1 urn:ietf:params:rtp-    | [RFC6464]                     |
| hdrext:ssrc-audio-level            |                               |
| a=extmap:2 urn:ietf:params:rtp-    | [I-D.ietf-mmusic-sdp-bundle-n |
| hdrext:sdes:mid                    | egotiation]                   |
| a=candidate:0 1 UDP 2122194687     | [RFC5245]                     |
| 198.51.100.7 49203 typ host        |                               |
| a=candidate:1 1 UDP 1685987071     | [RFC5245]                     |
| 203.0.113.77 51556 typ srflx       |                               |
| raddr 198.51.100.7 rport 49203     |                               |
| ****** Video m=line *********      | ***************************** |
| m=video 49203 UDP/TLS/RTP/SAVPF    | [RFC4566]                     |
| 120                                |                               |
| c=IN IP4 203.0.113.77              | [RFC4566]                     |
| a=mid:video                        | [RFC5888] Video m=line part   |
|                                    | of the BUNDLE group with the  |
|                                    | port from audio line repeated |
| a=msid:ma tb                       | Identifies RTCMediaStream ID  |
|                                    | (ma) and RTCMediaStreamTrack  |
|                                    | ID (tb)                       |
| a=sendrecv                         | [RFC3264]                     |
| a=rtpmap:120 VP8/90000             | [RFC7741]                     |
| a=rtcp-fb:120 nack                 | [RFC5104]                     |
| a=rtcp-fb:120 nack pli             | [RFC5104]                     |
| a=rtcp-fb:120 ccm fir              | [RFC5104]                     |
| a=extmap:2 urn:ietf:params:rtp-    | [I-D.ietf-mmusic-sdp-bundle-n |
| hdrext:sdes:mid                    | egotiation]                   |
+------------------------------------+-------------------------------+
```

18.2.5.2.8 Audio, Video, and Data Session

This example shows SDP for negotiating a session with audio, video and data streams between Alice and Bob with BUNDLE support known (Tables 18.17 and 18.18).

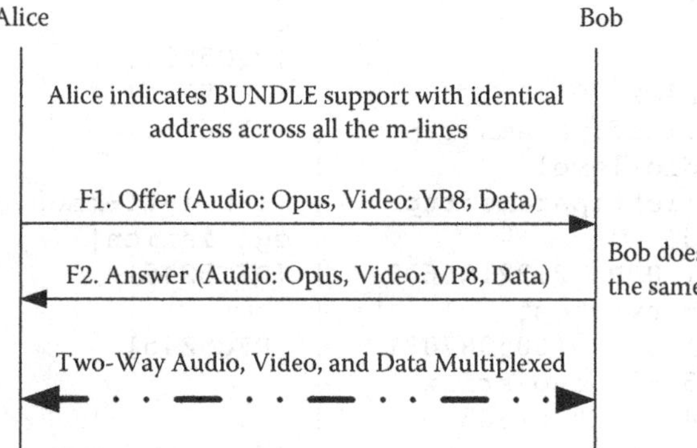

Audio, video, data with BUNDLE support known

Table 18.17 SDP Offer (Copyright IETF)

```
+------------------------------------+---------------------------------+
| Offer SDP Contents                 | RFC#/Notes                      |
+------------------------------------+---------------------------------+
| v=0                                | [RFC4566]                       |
| o=- 20518 0 IN IP4 0.0.0.0         | [RFC4566]                       |
| s=-                                | [RFC4566]                       |
| t=0 0                              | [RFC4566]                       |
| a=group:BUNDLE audio video data    | [I-D.ietf-mmusic-sdp-bundle-n   |
|                                    | egotiation]                     |
| a=group:LS audio video             | [RFC5888]                       |
| a=ice-options:trickle              | [I-D.ietf-mmusic-trickle-ice]   |
| ****** Audio m=line *********      | *****************************   |
| m=audio 54609 UDP/TLS/RTP/SAVPF    | [RFC4566]                       |
| 109                                |                                 |
| c=IN IP4 203.0.113.141             | [RFC4566]                       |
| a=msid:ma ta                       | Identifies RTCMediaStream ID    |
|                                    | (ma) and RTCMediaStreamTrack    |
|                                    | ID (ta)                         |
| a=mid:audio                        | [RFC5888]                       |
| a=sendrecv                         | [RFC3264]                       |
| a=rtpmap:109 opus/48000/2          | [RFC7587]                       |
| a=maxptime:120                     | [RFC4566]                       |
| a=ice-ufrag:074c6550               | [RFC5245]                       |
| a=ice-pwd:a28a397a4c3f31747d1ee34  | [RFC5245]                       |
| 74af08a068                         |                                 |
| a=fingerprint:sha-256 19:E2:1C:3B  | [RFC5245]                       |
| :4B:9F:81:E6:B8:5C:F4:A5:A8:D8:73  |                                 |
| :04 :BB:05:2F:70:9F:04:A9:0E:05:E  |                                 |
| 9:26:33:E8:70:88:A2                |                                 |
```

(Continued)

Table 18.17 (*Continued*) SDP Offer (Copyright IETF)

a=setup:actpass	[RFC4145]
a=dtls-id:1	[I-D.ietf-mmusic-dtls-sdp]
a=rtcp-mux	[RFC5761]
a=rtcp-mux-only	[I-D.ietf-mmusic-mux-exclusive]
a=rtcp-rsize	[RFC5506]
a=rtcp-fb:109 nack	[RFC5104]
a=extmap:1 urn:ietf:params:rtp-hdrext:ssrc-audio-level	[RFC6464]
a=extmap:2 urn:ietf:params:rtp-hdrext:sdes:mid	[I-D.ietf-mmusic-sdp-bundle-negotiation]
a=candidate:0 1 UDP 2122194687 192.0.2.4 61665 typ host	[RFC5245]
a=candidate:1 1 UDP 1685987071 203.0.113.141 54609 typ srflx raddr 192.0.2.4 rport 61665	[RFC5245]
a=end-of-candidates	[I-D.ietf-mmusic-trickle-ice]
****** Video m=line *********	****************************
m=video 54609 UDP/TLS/RTP/SAVPF 120	[RFC4566]
c=IN IP4 203.0.113.141	[RFC4566]
a=mid:video	[RFC5888]
a=msid:ma tb	Identifies RTCMediaStream ID (ma) and RTCMediaStreamTrack ID (tb)
a=sendrecv	[RFC3264]
a=rtpmap:120 VP8/90000	[RFC7741]
a=rtcp-fb:120 nack	[RFC5104]
a=rtcp-fb:120 nack pli	[RFC5104]
a=rtcp-fb:120 ccm fir	[RFC5104]
a=extmap:2 urn:ietf:params:rtp-hdrext:sdes:mid	[I-D.ietf-mmusic-sdp-bundle-negotiation]
****** Application m=line *********	****************************
m=application 54609 UDP/DTLS/SCTP webrtc-datachannel	[I-D.ietf-rtcweb-data-channel]
c=IN IP4 203.0.113.141	[RFC4566]
a=mid:data	[RFC5888]
a=sctp-port:5000	[I-D.ietf-mmusic-sctp-sdp]
a=max-message-size:100000	[I-D.ietf-mmusic-sctp-sdp]
a=sendrecv	[RFC3264]

Table 18.18 SDP Answer (Copyright IETF)

Answer SDP Contents	RFC#/Notes
v=0	[RFC4566]
o=- 16833 0 IN IP4 0.0.0.0	[RFC4566] - Session Origin Information
s=-	[RFC4566]
t=0 0	[RFC4566]
a=group:BUNDLE audio video data	[I-D.ietf-mmusic-sdp-bundle-negotiation]
a=group:LS audio video	[RFC5888]
a=ice-options:trickle	[I-D.ietf-mmusic-trickle-ice]
****** Audio m=line *********	*****************************
m=audio 49203 UDP/TLS/RTP/SAVPF 109	[RFC4566]
c=IN IP4 203.0.113.77	[RFC4566]
a=msid:ma ta	Identifies RTCMediaStream ID (ma) and RTCMediaStreamTrack ID (ta)
a=mid:audio	[RFC5888]
a=sendrecv	[RFC3264]
a=rtpmap:109 opus/48000/2	[RFC7587]
a=maxptime:120	[RFC4566]
a=ice-ufrag:c300d85b	[RFC5245]
a=ice-pwd:de4e99bd291c325921d5d47efbabd9a2	[RFC5245]
a=fingerprint:sha-256 6B:8B:F0:65:5F:78:E2:51:3B:AC:6F:F3:3F:46:1B:35 :DC:B8:5F:64:1A:24:C2:43:F0:A1:58:D0:A1:2C:19:08	[RFC5245]
a=setup:active	[RFC4145]
a=dtls-id:1	[I-D.ietf-mmusic-dtls-sdp]
a=rtcp-mux	[RFC5761]
a=rtcp-mux-only	[I-D.ietf-mmusic-mux-exclusive]
a=rtcp-rsize	[RFC5506]
a=rtcp-fb:109 nack	[RFC5104]
a=extmap:1 urn:ietf:params:rtp-hdrext:ssrc-audio-level	[RFC6464]
a=extmap:2 urn:ietf:params:rtp-hdrext:sdes:mid	[I-D.ietf-mmusic-sdp-bundle-negotiation]
a=candidate:0 1 UDP 2122194687 198.51.100.7 51556 typ host	[RFC5245]
a=candidate:1 1 UDP 1685987071 203.0.113.77 49203 typ srflx raddr 198.51.100.7 rport 51556	[RFC5245]
a=end-of-candidates	[I-D.ietf-mmusic-trickle-ice]

(*Continued*)

Table 18.18 (*Continued*) SDP Answer (Copyright IETF)

```
| ****** Video m=line *********        | ***************************** |
| m=video 49203 UDP/TLS/RTP/SAVPF      | [RFC4566]                     |
| 120                                  |                               |
| c=IN IP4 203.0.113.77                | [RFC4566]                     |
| a=mid:video                          | [RFC5888]                     |
| a=msid:ma tb                         | Identifies RTCMediaStream ID  |
|                                      | (ma) and RTCMediaStreamTrack  |
|                                      | ID (tb)                       |
| a=sendrecv                           | [RFC3264]                     |
| a=rtpmap:120 VP8/90000               | [RFC7741]                     |
| a=rtcp-fb:120 nack                   | [RFC5104]                     |
| a=rtcp-fb:120 nack pli               | [RFC5104]                     |
| a=rtcp-fb:120 ccm fir                | [RFC5104]                     |
| a=extmap:2 urn:ietf:params:rtp-      | [I-D.ietf-mmusic-sdp-bundle-n |
| hdrext:sdes:mid                      | egotiation]                   |
| ****** Application m=line            | ***************************** |
| *********                            |                               |
| m=application 49203 UDP/DTLS/SCTP    | [I-D.ietf-mmusic-sctp-sdp]    |
| webrtc-datachannel                   |                               |
| c=IN IP4 203.0.113.77                | [RFC4566]                     |
| a=mid:data                           | [RFC5888]                     |
| a=sctp-port:5000                     | [I-D.ietf-mmusic-sctp-sdp]    |
| a=max-message-size:100000            | [I-D.ietf-mmusic-sctp-sdp]    |
| a=sendrecv                           | [RFC3264]                     |
+--------------------------------------+-------------------------------+
```

18.2.5.2.9 Audio, Video Session with BUNDLE Unsupported

This use case illustrates the SDP offer/answer exchange where the far end (Bob) either doesn't support media bundling or doesn't want to group "m=" lines over a single 5-tuple.

The same is indicated by dropping the "a=group:BUNDLE" line and BUNDLE RTP header extension in the answer SDP.

On a successful offer/answer exchange, Alice and Bob end up using unique 5-tuples for audio and video media streams (Tables 18.19 and 18.20).

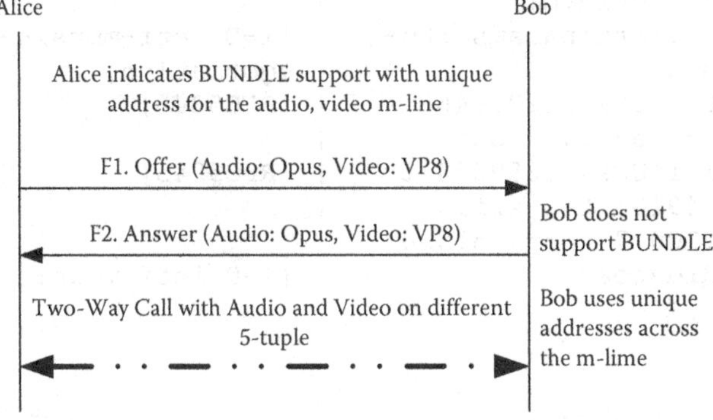

Two-way secure audio, video with BUNDLE unsupported

Table 18.19 SDP Offer w/BUNDLE (Copyright IETF)

Offer SDP Contents	RFC#/Notes
v=0	[RFC4566]
o=- 20518 0 IN IP4 0.0.0.0	[RFC4566]
s=-	[RFC4566]
t=0 0	[RFC4566]
a=group:BUNDLE audio video	[I-D.ietf-mmusic-sdp-bundle-negotiation] Alice supports grouping of m=lines under BUNDLE semantics
a=group:LS audio video	[RFC5888]
a=ice-options:trickle	[I-D.ietf-mmusic-trickle-ice]
****** Audio m=line *********	*****************************
m=audio 54609 UDP/TLS/RTP/SAVPF 109	[RFC4566]
c=IN IP4 203.0.113.141	[RFC4566]
a=mid:audio	[RFC5888] Audio m=line part of BUNDLE group with a unique port number
a=msid:ma ta	Identifies RTCMediaStream ID (ma) and RTCMediaStreamTrack ID (ta)
a=sendrecv	[RFC3264]
a=rtpmap:109 opus/48000/2	[RFC7587]
a=maxptime:120	[RFC4566]
a=ice-ufrag:074c6550	[RFC5245]
a=ice-pwd:a28a397a4c3f31747d1ee3474af08a068	[RFC5245]
a=fingerprint:sha-256 19:E2:1C:3B:4B:9F:81:E6:B8:5C:F4:A5:A8:D8:73:04 :BB:05:2F:70:9F:04:A9:0E:05:E9:26:33:E8:70:88:A2	[RFC5245]
a=setup:actpass	[RFC4145] - Alice can perform DTLS before Answer arrives
a=dtls-id:1	[I-D.ietf-mmusic-dtls-sdp]
a=rtcp-mux	[RFC5761]
a=rtcp:55232 IN IP4 203.0.113.141	[RFC3605] - RTCP port different from RTP port
a=rtcp-rsize	[RFC5506]
a=rtcp-fb:109 nack	[RFC5104]
a=extmap:1 urn:ietf:params:rtp-hdrext:ssrc-audio-level	[RFC6464]
a=extmap:2 urn:ietf:params:rtp-hdrext:sdes:mid	[I-D.ietf-mmusic-sdp-bundle-negotiation]
a=candidate:0 1 UDP 2122194687 192.0.2.4 61665 typ host	[RFC5245]
a=candidate:1 1 UDP 1685987071 203.0.113.141 54609 typ srflx raddr 192.0.2.4 rport 61665	[RFC5245]

(Continued)

Table 18.19 (*Continued*) SDP Offer w/BUNDLE (Copyright IETF)

a=candidate:0 2 UDP 2122194687 192.0.2.4 61666 typ host	[RFC5245]
a=candidate:1 2 UDP 1685987071 203.0.113.141 55232 typ srflx raddr 192.0.2.4 rport 61666	[RFC5245]
a=end-of-candidates	[I-D.ietf-mmusic-trickle-ice]
****** Video m=line *********	*****************************
m=video 54332 UDP/TLS/RTP/SAVPF 120	[RFC4566]
c=IN IP4 203.0.113.141	[RFC4566]
a=mid:video	[RFC5888] Video m=line part of the BUNDLE group with a unique port number
a=msid:ma tb	Identifies RTCMediaStream ID (ma) and RTCMediaStreamTrack ID (tb)
a=sendrecv	[RFC3264]
a=rtpmap:120 VP8/90000	[RFC7741]
a=ice-ufrag:7872093	[RFC5245]
a=ice-pwd:ee3474af08a068a28a397a4 c3f31747d1	[RFC5245]
a=fingerprint:sha-256 19:E2:1C:3B :4B:9F:81:E6:B8:5C:F4:A5:A8:D8:73 :04 :BB:05:2F:70:9F:04:A9:0E:05:E 9:26:33:E8:70:88:A2	[RFC5245]
a=setup:actpass	[RFC4145] - Alice can perform DTLS before Answer arrives
a=dtls-id:2	[I-D.ietf-mmusic-dtls-sdp]
a=rtcp-mux	[RFC5761]
a=rtcp:60052 IN IP4 203.0.113.141	[RFC3605]
a=rtcp-rsize	[RFC5506]
a=rtcp-fb:120 nack	[RFC5104]
a=rtcp-fb:120 nack pli	[RFC5104]
a=rtcp-fb:120 ccm fir	[RFC5104]
a=extmap:2 urn:ietf:params:rtp-hdrext:sdes:mid	[I-D.ietf-mmusic-sdp-bundle-negotiation]
a=candidate:0 1 UDP 2122194687 192.0.2.4 71775 typ host	[RFC5245]
a=candidate:1 1 UDP 1685987071 203.0.113.141 54332 typ srflx raddr 192.0.2.4 rport 71775	[RFC5245]
a=candidate:0 2 2122194687 192.0.2.4 71776 typ host	[RFC5245]
a=candidate:1 2 UDP 1685987071 203.0.113.141 60052 typ srflx raddr 192.0.2.4 rport 71776	[RFC5245]

Table 18.20 SDP Answer without BUNDLE (Copyright IETF)

Answer SDP Contents	RFC#/Notes
v=0	[RFC4566]
o=- 16833 0 IN IP4 0.0.0.0	[RFC4566]
s=-	[RFC4566]
t=0 0	[RFC4566]
a=group:LS audio video	[RFC5888]
a=ice-options:trickle	[I-D.ietf-mmusic-trickle -ice]
****** Audio m=line *********	*********************** *****
m=audio 53214 UDP/TLS/RTP/SAVPF 109	[RFC4566]
c=IN IP4 203.0.113.77	[RFC4566]
a=mid:audio	[RFC5888]
a=msid:ma ta	Identifies RTCMediaStream ID (ma) and RTCMediaStreamTrack ID (ta)
a=sendrecv	[RFC3264]
a=rtpmap:109 opus/48000/2	[RFC7587]
a=maxptime:120	[RFC4566]
a=ice-ufrag:c300d85b	[RFC5245]
a=ice- pwd:de4e99bd291c325921d5d47efbabd9a2	[RFC5245]
a=fingerprint:sha-256 6B:8B:F0:65:5F:7 8:E2:51:3B:AC:6F:F3:3F:46:1B:35 :DC:B8 :5F:64:1A:24:C2:43:F0:A1:58:D0:A1:2C:1 9:08	[RFC5245]
a=setup:active	[RFC4145] - Bob carries out DTLS Handshake in parallel
a=dtls-id:1	[I-D.ietf-mmusic-dtls-sd p]
a=rtcp-mux	[RFC5761]
a=rtcp-rsize	[RFC5506]
a=rtcp-fb:109 nack	[RFC5104]
a=extmap:1 urn:ietf:params:rtp-hdrext :ssrc-audio-level	[RFC6464]
a=candidate:0 1 UDP 2122194687 198.51.100.7 51556 typ host	[RFC5245]
a=candidate:1 1 UDP 1685987071 203.0.113.77 53214 typ srflx raddr 198.51.100.7 rport 51556	[RFC5245]
a=candidate:0 2 UDP 2122194687 198.51.100.7 51558 typ host	[RFC5245]
a=candidate:1 2 UDP 1685987071 203.0.113.77 60065 typ srflx raddr 198.51.100.7 rport 51558	[RFC5245]
****** Video m=line *********	*********************** *****

(*Continued*)

Table 18.20 (*Continued*) SDP Answer without BUNDLE (Copyright IETF)

```
| m=video 58679 UDP/TLS/RTP/SAVPF 120    | [RFC4566]                 |
| c=IN IP4 203.0.113.77                  | [RFC4566]                 |
| a=mid:video                            | [RFC5888]                 |
| a=msid:ma tb                           | Identifies                |
|                                        | RTCMediaStream ID (ma)    |
|                                        | and RTCMediaStreamTrack   |
|                                        | ID (tb)                   |
| a=sendrecv                             | [RFC3264]                 |
| a=rtpmap:120 VP8/90000                 | [RFC7741]                 |
| a=ice-ufrag:85bC300                    | [RFC5245]                 |
| a=ice-                                 | [RFC5245]                 |
| pwd:325921d5d47efbabd9a2de4e99bd291c   |                           |
| a=fingerprint:sha-256 6B:8B:F0:65:5F:7 | [RFC5245]                 |
| 8:E2:51:3B:AC:6F:F3:3F:46:1B:35 :DC:B8 |                           |
| :5F:64:1A:24:C2:43:F0:A1:58:D0:A1:2C:1 |                           |
| 9:08                                   |                           |
| a=setup:active                         | [RFC4145] - Bob carries   |
|                                        | out DTLS Handshake in     |
|                                        | parallel                  |
| a=dtls-id:2                            | [I-D.ietf-mmusic-dtls-sd  |
|                                        | p]                        |
| a=rtcp-mux                             | [RFC5761]                 |
| a=rtcp-rsize                           | [RFC5506]                 |
| a=rtcp-fb:120 nack                     | [RFC5104]                 |
| a=rtcp-fb:120 nack pli                 | [RFC5104]                 |
| a=rtcp-fb:120 ccm fir                  | [RFC5104]                 |
| a=candidate:0 1 UDP 2122194687         | [RFC5245]                 |
| 198.51.100.7 61556 typ host            |                           |
| a=candidate:1 1 UDP 1685987071         | [RFC5245]                 |
| 203.0.113.77 58679 typ srflx raddr     |                           |
| 198.51.100.7 rport 61556               |                           |
| a=candidate:0 1 UDP 2122194687         | [RFC5245]                 |
| 198.51.100.7 61558 typ host            |                           |
| a=candidate:1 1 UDP 1685987071         | [RFC5245]                 |
| 203.0.113.77 56507 typ srflx raddr     |                           |
| 198.51.100.7 rport 61558               |                           |
+----------------------------------------+---------------------------+
```

18.2.5.2.10 Audio, Video BUNDLED, but Data (Not BUNDLED)

This example showcases SDP for negotiating a session with audio, video, and data streams between Alice and Bob with the data stream not being part of the BUNDLE group. This is shown by assigning unique ports for the data media section and not adding the "mid" identification tag to the BUNDLE group (Tables 18.21 and 18.22).

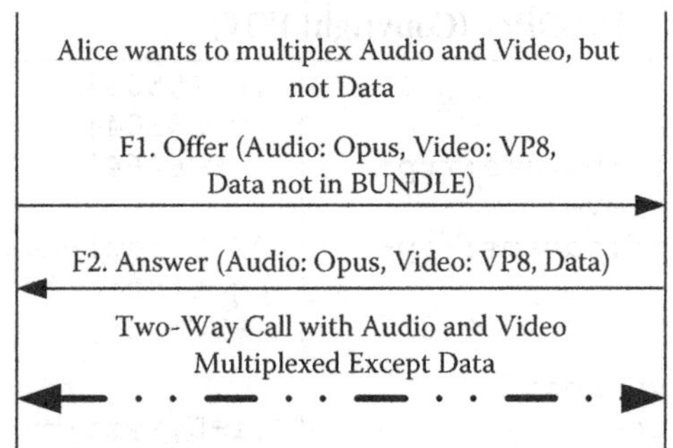

Audio, video, with data (not in BUNDLE)

Table 18.21 SDP Offer (Copyright IETF)

Offer SDP Contents	RFC#/Notes
v=0	[RFC4566]
o=- 20518 0 IN IP4 0.0.0.0	[RFC4566]
s=-	[RFC4566]
t=0 0	[RFC4566]
a=group:BUNDLE audio video	[I-D.ietf-mmusic-sdp-bundle-negotiation] Alice wants to BUNDLE only audio and video media.
a=group:LS audio video	[RFC5888]
a=ice-options:trickle	[I-D.ietf-mmusic-trickle-ice]
****** Audio m=line *********	****************************
m=audio 54609 UDP/TLS/RTP/SAVPF 109	[RFC4566]
c=IN IP4 203.0.113.141	[RFC4566]
a=mid:audio	[RFC5888]
a=msid:ma ta	Identifies RTCMediaStream ID (ma) and RTCMediaStreamTrack ID (ta)
a=sendrecv	[RFC3264]
a=rtpmap:109 opus/48000/2	[RFC7587]
a=maxptime:120	[RFC4566]
a=ice-ufrag:074c6550	[RFC5245]
a=ice-pwd:a28a397a4c3f31747d1ee3474af08a068	[RFC5245]
a=fingerprint:sha-256 19:E2:1C:3B:4B:9F:81:E6:B8:5C:F4:A5:A8:D8:73:04 :BB:05:2F:70:9F:04:A9:0E:05:E9:26:33:E8:70:88:A2	[RFC5245]
a=setup:actpass	[RFC4145]
a=dtls-id:1	[I-D.ietf-mmusic-dtls-sdp]
a=rtcp-mux	[RFC5761]
a=rtcp-mux-only	[I-D.ietf-mmusic-mux-exclusive]

(Continued)

Table 18.21 (*Continued*) SDP Offer (Copyright IETF)

a=rtcp-rsize	[RFC5506]
a=rtcp-fb:109 nack	[RFC5104]
a=extmap:1 urn:ietf:params:rtp-hdrext:ssrc-audio-level	[RFC6464]
a=extmap:2 urn:ietf:params:rtp-hdrext:sdes:mid	[I-D.ietf-mmusic-sdp-bundle-negotiation]
a=candidate:0 1 UDP 2113667327 192.0.2.4 54609 typ host	[RFC5245]
a=end-of-candidates	[I-D.ietf-mmusic-trickle-ice]
****** Video m=line *********	*****************************
m=video 54609 UDP/TLS/RTP/SAVPF 120	[RFC4566]
c=IN IP4 203.0.113.141	[RFC4566]
a=mid:video	[RFC5888]
a=msid:ma tb	Identifies RTCMediaStream ID (ma) and RTCMediaStreamTrack ID (tb)
a=sendrecv	[RFC3264]
a=rtpmap:120 VP8/90000	[RFC7741]
a=rtcp-fb:120 nack	[RFC5104]
a=rtcp-fb:120 nack pli	[RFC5104]
a=rtcp-fb:120 ccm fir	[RFC5104]
a=extmap:2 urn:ietf:params:rtp-hdrext:sdes:mid	[I-D.ietf-mmusic-sdp-bundle-negotiation]
****** Application m=line *********	*****************************
m=application 10000 UDP/DTLS/SCTP webrtc-datachannel	[I-D.ietf-rtcweb-data-channel]
c=IN IP4 203.0.113.141	[RFC4566]
a=mid:data	[RFC5888]
a=sctp-port:5000	[I-D.ietf-mmusic-sctp-sdp]
a=max-message-size:100000	[I-D.ietf-mmusic-sctp-sdp]
a=sendrecv	[RFC3264]
a=setup:actpass	[RFC4145]
a=ice-ufrag:89819013	[RFC5245]
a=ice-pwd:1747d1ee3474af08a068a28a397a4c3f3	[RFC5245]
a=fingerprint:sha-256 29:E2:1C:3B:4B:9F:81:E6:B8:5C:F4:A5:A8:D8:73:04: BB:05:2F:70:9F:04:A9:0E:05:E9:26:33:E8:70:88:A2	[RFC5245]
a=candidate:0 1 UDP 2113667327 192.0.2.4 10000 typ host	[RFC5245]
a=end-of-candidates	[I-D.ietf-mmusic-trickle-ice]

Table 18.22 SDP Answer (Copyright IETF)

```
+-----------------------------------+-----------------------------+
| Answer SDP Contents               | RFC#/Notes                  |
+-----------------------------------+-----------------------------+
| v=0                               | [RFC4566]                   |
| o=-  16833 0 IN IP4 0.0.0.0       | [RFC4566] - Session Origin  |
|                                   | Information                 |
| s=-                               | [RFC4566]                   |
| t=0 0                             | [RFC4566]                   |
| a=group:BUNDLE audio video        | [I-D.ietf-mmusic-sdp-bundle-n |
|                                   | egotiation]                 |
| a=group:LS audio video            | [RFC5888]                   |
| a=ice-options:trickle             | [I-D.ietf-mmusic-trickle-ice] |
| ****** Audio m=line *********     | ***************************** |
| m=audio 49203 UDP/TLS/RTP/SAVPF   | [RFC4566]                   |
| 109                               |                             |
| c=IN IP4 203.0.113.77             | [RFC4566]                   |
| a=mid:audio                       | [RFC5888]                   |
| a=msid:ma ta                      | Identifies RTCMediaStream ID |
|                                   | (ma) and RTCMediaStreamTrack |
|                                   | ID (ta)                     |
| a=sendrecv                        | [RFC3264]                   |
| a=rtpmap:109 opus/48000/2         | [RFC7587]                   |
| a=maxptime:120                    | [RFC4566]                   |
| a=ice-ufrag:c300d85b              | [RFC5245]                   |
| a=ice-pwd:de4e99bd291c325921d5d47 | [RFC5245]                   |
| efbabd9a2                         |                             |
| a=fingerprint:sha-256 6B:8B:F0:65 | [RFC5245]                   |
| :5F:78:E2:51:3B:AC:6F:F3:3F:46:1B |                             |
| :35 :DC:B8:5F:64:1A:24:C2:43:F0:A |                             |
| 1:58:D0:A1:2C:19:08               |                             |
| a=setup:active                    | [RFC4145]                   |
| a=dtls-id:1                       | [I-D.ietf-mmusic-dtls-sdp]  |
| a=rtcp-mux                        | [RFC5761]                   |
| a=rtcp-mux-only                   | [I-D.ietf-mmusic-mux-exclusiv |
|                                   | e]                          |
| a=rtcp-rsize                      | [RFC5506]                   |
| a=rtcp-fb:109 nack                | [RFC5104]                   |
| a=extmap:1 urn:ietf:params:rtp-   | [RFC6464]                   |
| hdrext:ssrc-audio-level           |                             |
| a=extmap:2 urn:ietf:params:rtp-   | [I-D.ietf-mmusic-sdp-bundle-n |
| hdrext:sdes:mid                   | egotiation]                 |
| a=candidate:0 1 UDP 2113667327    | [RFC5245]                   |
| 198.51.100.7 49203 typ host       |                             |
| a=end-of-candidates               | [I-D.ietf-mmusic-trickle-ice] |
| ****** Video m=line *********     | ***************************** |
| m=video 49203 UDP/TLS/RTP/SAVPF   | [RFC4566]                   |
| 120                               |                             |
| c=IN IP4 203.0.113.77             | [RFC4566]                   |
| a=mid:video                       | [RFC5888]                   |
| a=msid:ma tb                      | Identifies RTCMediaStream ID |
|                                   | (ma) and RTCMediaStreamTrack |
|                                   | ID (tb)                     |
| a=sendrecv                        | [RFC3264]                   |
| a=rtpmap:120 VP8/90000            | [RFC7741]                   |
| a=rtcp-fb:120 nack                | [RFC5104]                   |
| a=rtcp-fb:120 nack pli            | [RFC5104]                   |
```

(Continued)

Table 18.22 (*Continued*) SDP Answer (Copyright IETF)

```
| a=rtcp-fb:120 ccm fir                | [RFC5104]                     |
| a=extmap:2 urn:ietf:params:rtp-      | [I-D.ietf-mmusic-sdp-bundle-n |
| hdrext:sdes:mid                      | egotiation]                   |
| ****** Application m=line            | ***************************** |
| *********                            |                               |
| m=application 20000 UDP/DTLS/SCTP    | [I-D.ietf-mmusic-sctp-sdp]    |
| webrtc-datachannel                   |                               |
| c=IN IP4 203.0.113.77                | [RFC4566]                     |
| a=mid:data                           | [RFC5888]                     |
| a=sctp-port:5000                     | [I-D.ietf-mmusic-sctp-sdp]    |
| a=max-message-size:100000            | [I-D.ietf-mmusic-sctp-sdp]    |
| a=setup:active                       | [RFC4145]                     |
| a=sendrecv                           | [RFC3264]                     |
| a=ice-ufrag:991Ca2a5e                | [RFC5245]                     |
| a=ice-pwd:921d5d47efbabd9a2de4e99    | [RFC5245]                     |
| bd291c325                            |                               |
| a=fingerprint:sha-256 7B:8B:F0:65    | [RFC5245]                     |
| :5F:78:E2:51:3B:AC:6F:F3:3F:46:1B    |                               |
| :35: DC:B8:5F:64:1A:24:C2:43:F0:A    |                               |
| 1:58:D0:A1:2C:19:08                  |                               |
| a=candidate:0 1 UDP 2113667327       | [RFC5245]                     |
| 198.51.100.7 20000 typ host          |                               |
| a=end-of-candidates                  | [I-D.ietf-mmusic-trickle-ice] |
+--------------------------------------+-------------------------------+
```

18.2.5.2.11 Audio Only, Add Video to BUNDLE

This example involves two offer/answer exchanges. First, one is used to negotiate and set up BUNDLE support for an audio-only session followed by an updated offer/answer exchange to add video stream to the ongoing session. Also, the newly added video stream is BUNDLED with the audio stream (Tables 18.23–18.26).

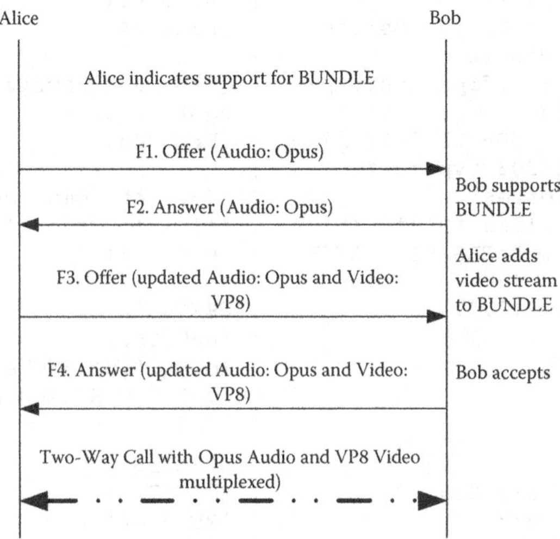

Audio only, add video and BUNDLE

Table 18.23 SDP Offer (Copyright IETF)

Offer SDP Contents	RFC#/Notes
v=0	[RFC4566]
o=- 20518 0 IN IP4 0.0.0.0	[RFC4566]
s=-	[RFC4566]
t=0 0	[RFC4566]
a=group:BUNDLE audio	[I-D.ietf-mmusic-sdp-bundle-negotiation] Alice adds audio m=line to the BUNDLE group
a=ice-options:trickle	[I-D.ietf-mmusic-trickle-ice]
****** Audio m=line *********	****************************
m=audio 54609 UDP/TLS/RTP/SAVPF 109	[RFC4566]
c=IN IP4 203.0.113.141	[RFC4566]
a=mid:audio	[RFC5888]
a=msid:ma ta	Identifies RTCMediaStream ID (ma) and RTCMediaStreamTrack ID (ta)
a=sendrecv	[RFC3264]
a=rtpmap:109 opus/48000/2	[RFC7587]
a=maxptime:120	[RFC4566]
a=ice-ufrag:074c6550	[RFC5245]
a=ice-pwd:a28a397a4c3f31747d1ee3474af08a068	[RFC5245]
a=fingerprint:sha-256 19:E2:1C:3B:4B:9F:81:E6:B8:5C:F4:A5:A8:D8:73:04 :BB:05:2F:70:9F:04:A9:0E:05:E9:26:33:E8:70:88:A2	[RFC5245]
a=setup:actpass	[RFC4145]
a=dtls-id:1	[I-D.ietf-mmusic-dtls-sdp]
a=rtcp-mux	[RFC5761]
a=rtcp-mux-only	[I-D.ietf-mmusic-mux-exclusive]
a=rtcp-rsize	[RFC5506]
a=rtcp-fb:109 nack	[RFC5104]
a=extmap:1 urn:ietf:params:rtp-hdrext:ssrc-audio-level	[RFC6464]
a=extmap:2 urn:ietf:params:rtp-hdrext:sdes:mid	[I-D.ietf-mmusic-sdp-bundle-negotiation]
a=candidate:0 1 UDP 2113667327 192.0.2.4 61665 typ host	[RFC5245]
a=candidate:1 1 UDP 694302207 203.0.113.141 54609 typ srflx raddr 192.0.2.4 rport 61665	[RFC5245]
a=end-of-candidates	[I-D.ietf-mmusic-trickle-ice]

Table 18.24 SDP Answer (Copyright IETF)

Answer SDP Contents	RFC#/Notes
v=0	[RFC4566]
o=- 16833 0 IN IP4 0.0.0.0	[RFC4566] - Session Origin Information
s=-	[RFC4566]
t=0 0	[RFC4566]
a=group:BUNDLE audio	[I-D.ietf-mmusic-sdp-bundle-negotiation]
a=ice-options:trickle	[I-D.ietf-mmusic-trickle-ice]
****** Audio m=line *********	****************************
m=audio 49203 UDP/TLS/RTP/SAVPF 109	[RFC4566]
c=IN IP4 203.0.113.77	[RFC4566]
a=mid:audio	[RFC5888]
a=msid:ma ta	Identifies RTCMediaStream ID (ma) and RTCMediaStreamTrack ID (ta)
a=sendrecv	[RFC3264]
a=rtpmap:109 opus/48000/2	[RFC7587]
a=maxptime:120	[RFC4566]
a=ice-ufrag:c300d85b	[RFC5245]
a=ice-pwd:de4e99bd291c325921d5d47efbabd9a2	[RFC5245]
a=fingerprint:sha-256 6B:8B:F0:65:5F:78:E2:51:3B:AC:6F:F3:3F:46:1B:35 :DC:B8:5F:64:1A:24:C2:43:F0:A1:58:D0:A1:2C:19:08	[RFC5245]
a=setup:active	[RFC4145]
a=dtls-id:1	[I-D.ietf-mmusic-dtls-sdp]
a=rtcp-mux	[RFC5761]
a=rtcp-mux-only	[I-D.ietf-mmusic-mux-exclusive]
a=rtcp-rsize	[RFC5506]
a=rtcp-fb:109 nack	[RFC5104]
a=extmap:1 urn:ietf:params:rtp-hdrext:ssrc-audio-level	[RFC6464]
a=extmap:2 urn:ietf:params:rtp-hdrext:sdes:mid	[I-D.ietf-mmusic-sdp-bundle-negotiation]
a=candidate:0 1 UDP 2113667327 198.51.100.7 51556 typ host	[RFC5245]
a=candidate:1 1 UDP 1694302207 203.0.113.77 49203 typ srflx raddr 198.51.100.7 rport 51556	[RFC5245]
a=end-of-candidates	[I-D.ietf-mmusic-trickle-ice]

Table 18.25 SDP Updated Offer (Copyright IETF)

Updated Offer SDP Contents	RFC#/Notes
v=0	Version number incremented [RFC4566]
o=- 20518 1 IN IP4 0.0.0.0	[RFC4566]
s=-	[RFC4566]
t=0 0	[RFC4566]
a=group:BUNDLE audio video	[I-D.ietf-mmusic-sdp-bundle-negotiation]
a=group:LS audio video	[RFC5888]
a=ice-options:trickle	[I-D.ietf-mmusic-trickle-ice]
****** Audio m=line *********	****************************
m=audio 54609 UDP/TLS/RTP/SAVPF 109	[RFC4566]
c=IN IP4 203.0.113.141	[RFC4566]
a=mid:audio	[RFC5888]
a=msid:ma ta	Identifies RTCMediaStream ID (ma) and RTCMediaStreamTrack ID (ta)
a=sendrecv	[RFC3264]
a=rtpmap:109 opus/48000/2	[RFC7587]
a=maxptime:120	[RFC4566]
a=ice-ufrag:074c6550	[RFC5245]
a=ice-pwd:a28a397a4c3f31747d1ee3474af08a068	[RFC5245]
a=fingerprint:sha-256 19:E2:1C:3B:4B:9F:81:E6:B8:5C:F4:A5:A8:D8:73:04 :BB:05:2F:70:9F:04:A9:0E:05:E9:26:33:E8:70:88:A2	[RFC5245]
a=setup:actpass	[RFC4145]
a=dtls-id:1	[I-D.ietf-mmusic-dtls-sdp]Alice want's to use the same DTLS association
a=rtcp-mux	[RFC5761]
a=rtcp-mux-only	[I-D.ietf-mmusic-mux-exclusive]
a=rtcp-rsize	[RFC5506]
a=rtcp-fb:109 nack	[RFC5104]
a=extmap:1 urn:ietf:params:rtp-hdrext:ssrc-audio-level	[RFC6464]
a=extmap:2 urn:ietf:params:rtp-hdrext:sdes:mid	[I-D.ietf-mmusic-sdp-bundle-negotiation]
a=candidate:0 1 UDP 2113667327 192.0.2.4 61665 typ host	[RFC5245]
a=candidate:1 1 UDP 694302207 203.0.113.141 54609 typ srflx raddr 192.0.2.4 rport 61665	[RFC5245]
a=end-of-candidates	[I-D.ietf-mmusic-trickle-ice]
****** Video m=line *********	****************************
m=video 54609 UDP/TLS/RTP/SAVPF	[RFC4566]

(Continued)

Table 18.25 (*Continued*) SDP Updated Offer (Copyright IETF)

```
| 120                                 |                              |
| c=IN IP4 203.0.113.141              | [RFC4566]                    |
| a=mid:video                         | [RFC5888]                    |
| a=msid:ma tb                        | Identifies RTCMediaStream ID |
|                                     | (ma) and RTCMediaStreamTrack |
|                                     | ID (tb)                      |
| a=sendrecv                          | [RFC3264]                    |
| a=rtpmap:120 VP8/90000              | [RFC7741]                    |
| a=rtcp-fb:120 nack                  | [RFC5104]                    |
| a=rtcp-fb:120 nack pli              | [RFC5104]                    |
| a=rtcp-fb:120 ccm fir               | [RFC5104]                    |
| a=extmap:2 urn:ietf:params:rtp-     | [I-D.ietf-mmusic-sdp-bundle-n |
| hdrext:sdes:mid                     | egotiation]                  |
+-------------------------------------+------------------------------+
```

Table 18.26 SDP Updated Answer (Copyright IETF)

```
+-------------------------------------+------------------------------+
| Updated Answer SDP Contents         | RFC#/Notes                   |
+-------------------------------------+------------------------------+
| v=0                                 | [RFC4566] Version number     |
|                                     | incremented                  |
| o=-  16833 1 IN IP4 0.0.0.0         | [RFC4566] - Session Origin   |
|                                     | Information                  |
| s=-                                 | [RFC4566]                    |
| t=0 0                               | [RFC4566]                    |
| a=group:BUNDLE audio video          | [I-D.ietf-mmusic-sdp-bundle-n |
|                                     | egotiation]                  |
| a=group:LS audio video              | [RFC5888]                    |
| a=ice-options:trickle               | [I-D.ietf-mmusic-trickle-ice] |
| ****** Audio m=line *********       | **************************** |
| m=audio 49203 UDP/TLS/RTP/SAVPF     | [RFC4566]                    |
| 109                                 |                              |
| c=IN IP4 203.0.113.77               | [RFC4566]                    |
| a=mid:audio                         | [RFC5888]                    |
| a=msid:ma ta                        | Identifies RTCMediaStream ID |
|                                     | (ma) and RTCMediaStreamTrack |
|                                     | ID (ta)                      |
| a=sendrecv                          | [RFC3264]                    |
| a=rtpmap:109 opus/48000/2           | [RFC7587]                    |
| a=maxptime:120                      | [RFC4566]                    |
| a=ice-ufrag:c300d85b                | [RFC5245]                    |
| a=ice-pwd:de4e99bd291c325921d5d47   | [RFC5245]                    |
| efbabd9a2                           |                              |
| a=fingerprint:sha-256 6B:8B:F0:65   | [RFC5245]                    |
| :5F:78:E2:51:3B:AC:6F:F3:3F:46:1B   |                              |
| :35 :DC:B8:5F:64:1A:24:C2:43:F0:A   |                              |
| 1:58:D0:A1:2C:19:08                 |                              |
```

(*Continued*)

Table 18.26 (*Continued*) SDP Updated Answer (Copyright IETF)

```
| a=setup:active                    | [RFC4145]                      |
| a=dtls-id:1                       | [I-D.ietf-mmusic-dtls-sdp] -   |
|                                   | Bob agrees to use the same     |
|                                   | DTLS association               |
| a=rtcp-mux                        | [RFC5761]                      |
| a=rtcp-mux-only                   | [I-D.ietf-mmusic-mux-exclusiv  |
|                                   | e]                             |
| a=rtcp-rsize                      | [RFC5506]                      |
| a=rtcp-fb:109 nack                | [RFC5104]                      |
| a=extmap:1 urn:ietf:params:rtp-   | [RFC6464]                      |
| hdrext:ssrc-audio-level           |                                |
| a=extmap:2 urn:ietf:params:rtp-   | [I-D.ietf-mmusic-sdp-bundle-n  |
| hdrext:sdes:mid                   | egotiation]                    |
| a=candidate:0 1 UDP 2113667327    | [RFC5245]                      |
| 198.51.100.7 51556 typ host       |                                |
| a=candidate:1 1 UDP 1694302207    | [RFC5245]                      |
| 203.0.113.77 49203 typ srflx      |                                |
| raddr 198.51.100.7 rport 51556    |                                |
| a=end-of-candidates               | [I-D.ietf-mmusic-trickle-ice]  |
| ****** Video m=line *********     | ****************************** |
| m=video 49203 UDP/TLS/RTP/SAVPF   | [RFC4566]                      |
| 120                               |                                |
| c=IN IP4 203.0.113.77             | [RFC4566]                      |
| a=mid:video                       | [RFC5888]                      |
| a=msid:ma tb                      | Identifies RTCMediaStream ID   |
|                                   | (ma) and RTCMediaStreamTrack   |
|                                   | ID (tb)                        |
| a=sendrecv                        | [RFC3264]                      |
| a=rtpmap:120 VP8/90000            | [RFC7741]                      |
| a=rtcp-fb:120 nack                | [RFC5104]                      |
| a=rtcp-fb:120 nack pli            | [RFC5104]                      |
| a=rtcp-fb:120 ccm fir             | [RFC5104]                      |
| a=extmap:2 urn:ietf:params:rtp-   | [I-D.ietf-mmusic-sdp-bundle-n  |
| hdrext:sdes:mid                   | egotiation]                    |
+-----------------------------------+--------------------------------+
```

18.2.5.3 Multi-Resolution, Retransmission, Forward Error Correction Examples

This section deals with scenarios related to multisource, multistream negotiation such as layered coding and simulcast, along with techniques that deal with providing robustness against transmission errors such as Forward Error Correction (FEC) and Retransmission (RTX). Also, note that mechanisms such as FEC and RTX could be envisioned in the above (Figures 1–11) basic scenarios.

18.2.5.3.1 Sendonly Simulcast Session with Two Cameras and Two Encodings per Camera

The SDP below shows an offer/answer exchange with one audio and two video sources. Each of the video sources can be sent at two different resolutions.

One video source corresponds to VP8 encoding, while the other corresponds to H.264 encoding.

[I-D.ietf-mmusic-rid] framework is used to further constrain the media format encodings and map the (PTs) to the "rid" identifiers.

[I-D.ietf-mmusic-sdp-simulcast] framework identifies the simulcast streams via their "rid" identifiers.

Bundle-only attribute is used for the video sources in the offer to ensure enabling video sources in the context of BUNDLE alone.

BUNDLE grouping framework enables multiplexing of all the five streams (one audio stream + four video streams) over a single RTP Session (Tables 18.27 and 18.28).

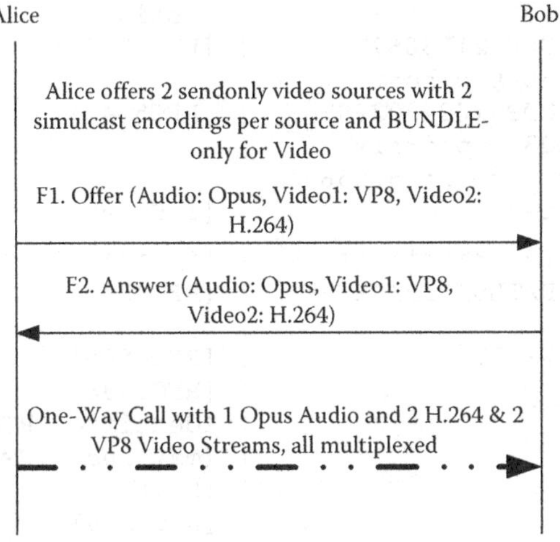

1-way successful simulcast w/BUNDLE

Table 18.27 SDP Offer (Copyright IETF)

Offer SDP Contents	RFC#/Notes
v=0	[RFC4566]
o=- 20519 0 IN IP4 0.0.0.0	[RFC4566]
s=-	[RFC4566]
t=0 0	[RFC4566]
a=group:BUNDLE m0 m1 m2	[I-D.ietf-mmusic-sdp-bundle-negotiation] Alice supports grouping of m=lines under BUNDLE semantics
a=group:LS m0 m1	[RFC5888]
a=ice-options:trickle	[I-D.ietf-mmusic-trickle-ice]
****** Audio m=line *********	****************************

(Continued)

Table 18.27 (*Continued*) SDP Offer (Copyright IETF)

m=audio 54609 UDP/TLS/RTP/SAVPF 109	[RFC4566]
c=IN IP4 203.0.113.141	[RFC4566]
a=mid:m0	[RFC5888]
a=msid:ma ta	Identifies RTCMediaStream ID (ma) and RTCMediaStreamTrack ID (ta)
a=sendonly	[RFC3264]
a=rtpmap:109 opus/48000/2	[RFC7587]
a=maxptime:120	[RFC4566]
a=ice-ufrag:074c6550	[RFC5245]
a=ice-pwd:a28a397a4c3f31747d1ee3474af08a068	[RFC5245]
a=fingerprint:sha-256 19:E2:1C:3B:4B:9F:81:E6:B8:5C:F4:A5:A8:D8:73:04:BB:05:2F:70:9F:04:A9:0E:05:E9:26:33:E8:70:88:A2	[RFC5245]
a=setup:actpass	[RFC4145]
a=dtls-id:1	[I-D.ietf-mmusic-dtls-sdp]
a=rtcp-mux	[RFC5761]
a=rtcp-rsize	[RFC5506]
a=rtcp-fb:109 nack	[RFC5104]
a=extmap:1 urn:ietf:params:rtp-hdrext:ssrc-audio-level	[RFC6464]
a=extmap:2 urn:ietf:params:rtp-hdrext:sdes:mid	[I-D.ietf-mmusic-sdp-bundle-negotiation]
a=candidate:0 1 UDP 2113667327 192.0.2.4 61665 typ host	[RFC5245]
a=candidate:1 1 UDP 694302207 203.0.113.141 54609 typ srflx raddr 192.0.2.4 rport 61665	[RFC5245]
a=end-of-candidates	[I-D.ietf-mmusic-trickle-ice]
****** Video-1 m=line *********	***************************
m=video 0 UDP/TLS/RTP/SAVPF 98 100	bundle-only video line with port number set to zero
c=IN IP4 203.0.113.141	[RFC4566]
a=bundle-only	[I-D.ietf-mmusic-sdp-bundle-negotiation]
a=mid:m1	[RFC5888] Video m=line part of BUNDLE group
a=msid:ma tb	Identifies RTCMediaStream ID (ma) and RTCMediaStreamTrack ID (tb)
a=sendonly	[RFC3264] - Send only video stream
a=rtpmap:98 VP8/90000	[RFC7741]
a=fmtp:98 max-fr=30	[RFC4566]
a=rtpmap:100 VP8/90000	[RFC7741]

(*Continued*)

Table 18.27 (*Continued*) SDP Offer (Copyright IETF)

a=fmtp:100 max-fr=15	[RFC4566]
a=rtcp-fb:* nack	[RFC5104]
a=rtcp-fb:* nack pli	[RFC5104]
a=rtcp-fb:* ccm fir	[RFC5104]
a=extmap:2 urn:ietf:params:rtp-hdrext:sdes:mid	[I-D.ietf-mmusic-sdp-bundle-negotiation]
a=rid:1 send pt=98;max-width=1280;max-height=720;	[I-D.ietf-mmusic-rid] 1:1 rid mapping to payload type and specify resolution constraints
a=rid:2 send pt=100;max-width=640;max-height=480;	[I-D.ietf-mmusic-rid] 1:1 rid mapping to payload type and specify resolution constraints
a=simulcast: send 1;~2	[I-D.ietf-mmusic-sdp-simulcast] Alice can send 2 resolutions identified by the 'rid' identifiers Also, the second stream is initially paused.
****** Video-2 m=line *********	*****************************
m=video 0 UDP/TLS/RTP/SAVPF 101 102	bundle-only video line with port number set to zero
c=IN IP4 203.0.113.141	[RFC4566]
a=bundle-only	[I-D.ietf-mmusic-sdp-bundle-negotiation]
a=mid:m2	[RFC5888] Video m=line part of BUNDLE group
a=msid:ma tc	Identifies RTCMediaStream ID (ma) and RTCMediaStreamTrack ID (tc)
a=sendonly	[RFC3264] - Send only video stream
a=rtpmap:101 H264/90000	[RFC6184]
a=rtpmap:102 H264/90000	[RFC6184]
a=fmtp:101 profile-level-id=42401f;packetization-mode=0;max-fr=30	[RFC6184]Camera-2,Encoding-1
a=fmtp:102 profile-level-id=42401f;packetization-mode=1;max-fr=15	[RFC6184]Camera-2,Encoding-2
a=rtcp-fb:* nack	[RFC5104]
a=rtcp-fb:* nack pli	[RFC5104]
a=rtcp-fb:* ccm fir	[RFC5104]
a=extmap:2 urn:ietf:params:rtp-hdrext:sdes:mid	[I-D.ietf-mmusic-sdp-bundle-negotiation]
a=rid:3 send pt=101;max-width=1280;max-height=720;	[I-D.ietf-mmusic-rid] 1:1 rid mapping to payload type and specify resolution constraints

(*Continued*)

Table 18.27 (*Continued*) SDP Offer (Copyright IETF)

```
| a=rid:4 send pt=102;max-width=640    | [I-D.ietf-mmusic-rid] 1:1    |
| ;max-height=360;                     | rid mapping to payload type  |
|                                      | and specify resolution       |
|                                      | constraints                  |
| a=simulcast: send 3;4                | [I-D.ietf-mmusic-sdp-simulc  |
|                                      | ast] Alice can send 2        |
|                                      | resolutions identified by    |
|                                      | the 'rid' identifiers        |
+--------------------------------------+------------------------------+
```

Table 18.28 SDP Answer (Copyright IETF)

```
+--------------------------------------+------------------------------+
| Answer SDP Contents                  | RFC#/Notes                   |
+--------------------------------------+------------------------------+
| v=0                                  | [RFC4566]                    |
| o=- 20519 0 IN IP4 0.0.0.0           | [RFC4566]                    |
| s=-                                  | [RFC4566]                    |
| t=0 0                                | [RFC4566]                    |
| a=group:BUNDLE m0 m1 m2              | [I-D.ietf-mmusic-sdp-bundle  |
|                                      | -negotiation] Alice          |
|                                      | supports grouping of         |
|                                      | m=lines under BUNDLE         |
|                                      | semantics                    |
| a=group:LS m0 m1                     | [RFC5888]                    |
| a=ice-options:trickle                | [I-D.ietf-mmusic-trickle-ic  |
|                                      | e]                           |
| ****** Audio m=line *********        | **************************** |
|                                      | **                           |
| m=audio 49203 UDP/TLS/RTP/SAVPF 109  | [RFC4566]                    |
| c=IN IP4 203.0.113.77                | [RFC4566]                    |
| a=mid:m0                             | [RFC5888]                    |
| a=msid:ma ta                         | Identifies RTCMediaStream    |
|                                      | ID (ma) and                  |
|                                      | RTCMediaStreamTrack ID (ta)  |
| a=recvonly                           | [RFC3264]                    |
| a=rtpmap:109 opus/48000/2            | [RFC7587]                    |
| a=rtcp-fb:109 nack                   | [RFC5104]                    |
| a=maxptime:120                       | [RFC4566]                    |
| a=ice-ufrag:c300d85b                 | [RFC5245]                    |
| a=ice-pwd:de4e99bd291c325921d5d47ef  | [RFC5245]                    |
| babd9a2                              |                              |
| a=fingerprint:sha-256 6B:8B:F0:65:5  | [RFC5245]                    |
| F:78:E2:51:3B:AC:6F:F3:3F:46:1B:35   |                              |
| :DC:B8:5F:64:1A:24:C2:43:F0:A1:58:D  |                              |
| 0:A1:2C:19:08                        |                              |
| a=setup:active                       | [RFC4145]                    |
| a=dtls-id:1                          | [I-D.ietf-mmusic-dtls-sdp]   |
```

(*Continued*)

Table 18.28 (*Continued*) SDP Answer (Copyright IETF)

a=extmap:2 urn:ietf:params:rtp-hdrext:sdes:mid	[I-D.ietf-mmusic-sdp-bundle-negotiation]
a=candidate:0 1 UDP 2113667327 198.51.100.7 61665 typ host	[RFC5245]
a=candidate:1 1 UDP 694302207 203.0.113.77 49203 typ srflx raddr 198.51.100.7 rport 61665	[RFC5245]
a=end-of-candidates	[I-D.ietf-mmusic-trickle-ice]
****** Video-1 m=line *********	*****************************
m=video 49203 UDP/TLS/RTP/SAVPF 98 100	BUNDLE accepted with port repeated from the audio port
c=IN IP4 203.0.113.77	[RFC4566]
a=mid:m1	[RFC5888] Video m=line part of BUNDLE group
a=msid:ma tb	Identifies RTCMediaStream ID (ma) and RTCMediaStreamTrack ID (tb)
a=recvonly	[RFC3264] - receive only video stream
a=rtpmap:98 VP8/90000	[RFC7741]
a=rtpmap:100 VP8/90000	[RFC7741]
a=fmtp:98 max-fr=30	[RFC4566]
a=fmtp:100 max-fr=15	[RFC4566]
a=rtcp-fb:* nack	[RFC5104]
a=rtcp-fb:* nack pli	[RFC5104]
a=rtcp-fb:* ccm fir	[RFC5104]
a=extmap:2 urn:ietf:params:rtp-hdrext:sdes:mid	[I-D.ietf-mmusic-sdp-bundle-negotiation]
a=rid:1 recv pt=98;max-width=1280;max-height=720;	[I-D.ietf-mmusic-rid] Bob accepts the offered payload format constraints
a=rid:2 recv pt=100;max-width=640;max-height=480;	[I-D.ietf-mmusic-rid] Bob accepts the offered payload format constraints
a=simulcast: recv 1;2	[I-D.ietf-mmusic-sdp-simulcast] Bob accepts the offered simulcast streams and removes the paused state of stream with 'rid' value 2.
****** Video-2 m=line *********	*****************************
m=video 49203 UDP/TLS/RTP/SAVPF 101 102	BUNDLE accepted with port repeated from the audio port
c=IN IP4 203.0.113.77	[RFC4566]
a=mid:m2	[RFC5888] Video m=line part

(*Continued*)

Table 18.28 (*Continued*) SDP Answer (Copyright IETF)

```
|                                    | of BUNDLE group           |
| a=msid:ma tc                       | Identifies RTCMediaStream |
|                                    | ID (ma) and               |
|                                    | RTCMediaStreamTrack ID (tc) |
| a=recvonly                         | [RFC3264]                 |
| a=rtpmap:101 H264/90000            | [RFC6184]                 |
| a=rtpmap:102 H264/90000            | [RFC6184]                 |
| a=fmtp:101 profile-level-id=42401f | [RFC6184]                 |
| ;packetization-mode=1;max-fr=30    |                           |
| a=fmtp:102 profile-level-id=42401f | [RFC6184]                 |
| ;packetization-mode=1;max-fr=15    |                           |
| a=rtcp-fb:* nack                   | [RFC5104]                 |
| a=rtcp-fb:* nack pli               | [RFC5104]                 |
| a=rtcp-fb:*  ccm fir               | [RFC5104]                 |
| a=extmap:2 urn:ietf:params:rtp-    | [I-D.ietf-mmusic-sdp-bundle |
| hdrext:sdes:mid                    | -negotiation]             |
| a=rid:3 recv pt=101;max-width=1280 | [I-D.ietf-mmusic-rid] Bob |
| ;max-height=720;                   | accepts the offered payload |
|                                    | format constraints        |
| a=rid:4 recv pt=102;max-width=640  | [I-D.ietf-mmusic-rid] Bob |
| ;max-height=360;                   | accepts the offered payload |
|                                    | format constraints        |
| a=simulcast: recv 3;4              | [I-D.ietf-mmusic-sdp-simulc |
|                                    | ast] Bob accepts the      |
|                                    | offered simulcast streams. |
+------------------------------------+---------------------------+
```

18.2.5.3.2 Successful Scalable Video Coding Video Session

This section shows an SDP offer/answer for a session with an audio and a single video source. The video source is encoded as layered coding at three different resolutions based on RFC 5583 (see Section 6.1). The video "m=" line shows three streams with the last stream (payload 100) dependent on streams with payload 96 and 97 for decoding (Tables 18.29 and 18.30).

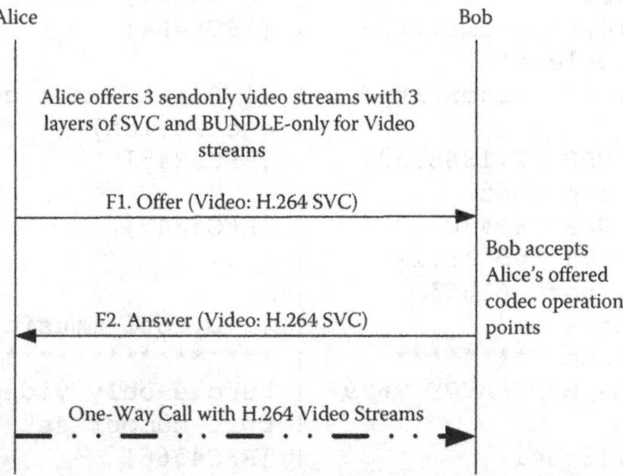

SVC session – 3 layers w/BUNDLE

Table 18.29 SDP Offer with SVC (Copyright IETF)

Offer SDP Contents	RFC#/Notes
v=0	[RFC4566]
o=- 20519 0 IN IP4 0.0.0.0	[RFC4566]
s=-	[RFC4566]
t=0 0	[RFC4566]
a=group:BUNDLE m0 m1	[I-D.ietf-mmusic-sdp-bundle-negotiation] Alice supports grouping of m=lines under BUNDLE semantics
a=group:LS m0 m1	[RFC5888]
a=ice-options:trickle	[I-D.ietf-mmusic-trickle-ice]
****** Audio m=line *********	****************************
m=audio 54609 UDP/TLS/RTP/SAVPF 109	[RFC4566]
c=IN IP4 203.0.113.141	[RFC4566]
a=mid:m0	[RFC5888] Audio m=line part of BUNDLE group with a unique port number
a=msid:ma ta	Identifies RTCMediaStream ID (ma) and RTCMediaStreamTrack ID (ta)
a=sendonly	[RFC3264]
a=rtpmap:109 opus/48000/2	[RFC7587]
a=maxptime:120	[RFC4566]
a=ice-ufrag:074c6550	[RFC5245]
a=ice-pwd:a28a397a4c3f31747d1ee3474af08a068	[RFC5245]
a=fingerprint:sha-256 19:E2:1C:3B:4B:9F:81:E6:B8:5C:F4:A5:A8:D8:73:04 :BB:05:2F:70:9F:04:A9:0E:05:E9:26:33:E8:70:88:A2	[RFC5245]
a=setup:actpass	[RFC4145]
a=dtls-id:1	[I-D.ietf-mmusic-dtls-sdp]
a=rtcp-mux	[RFC5761]
a=rtcp-rsize	[RFC5506]
a=rtcp-fb:109 nack	[RFC5104]
a=extmap:1 urn:ietf:params:rtp-hdrext:ssrc-audio-level	[RFC6464]
a=extmap:2 urn:ietf:params:rtp-hdrext:sdes:mid	[I-D.ietf-mmusic-sdp-bundle-negotiation]
a=candidate:0 1 UDP 2113667327 192.0.2.4 61665 typ host	[RFC5245]
a=candidate:1 1 UDP 694302207 203.0.113.141 54609 typ srflx raddr 192.0.2.4 rport 61665	[RFC5245]
a=end-of-candidates	[I-D.ietf-mmusic-trickle-ice]
****** Video m=line *********	****************************
m=video 0 UDP/TLS/RTP/SAVPF 96 97 100	bundle-only video line with port number set to zero
c=IN IP4 203.0.113.141	[RFC4566]

(*Continued*)

Table 18.29 (*Continued*) SDP Offer with SVC (Copyright IETF)

a=bundle-only	[I-D.ietf-mmusic-sdp-bundle-negotiation]
a=mid:m1	[RFC5888] Video m=line part of BUNDLE group
a=msid:ma tb	Identifies RTCMediaStream ID (ma) and RTCMediaStreamTrack ID (tc)
a=sendonly	[RFC3264] - Send only video stream
a=rtpmap:96 H264/90000	[RFC6184]
a=fmtp:96 profile-level-id=4d0028; packetization-mode=1 ;max-fr=30;max-fs=8040	[RFC6184]H.264 Layer 1
a=rtpmap:97 H264/90000	[RFC6184]
a=fmtp:97 profile-level-id=4d0028 ;packetization-mode=1; max-fr=15 ;max-fs=1200	[RFC6184] H.264 Layer 2
a=rtpmap:100 H264-SVC/90000	[RFC6184]
a=fmtp:100 profile-level-id=4d0028;packetization-mode=1; max-fr=30;max-fs=8040	[RFC6184]
a=depend:100 lay m1:96,97;	[RFC5583]Layer 3 dependent on layers 1 and 2
a=rtcp-fb:* nack	[RFC5104]
a=rtcp-fb:* nack pli	[RFC5104]
a=rtcp-fb:* ccm fir	[RFC5104]
a=extmap:2 urn:ietf:params:rtp-hdrext:sdes:mid	[I-D.ietf-mmusic-sdp-bundle-negotiation]

Table 18.30 SDP Answer with SVC (Copyright IETF)

Answer SDP Contents	RFC#/Notes
v=0	[RFC4566]
o=- 20519 0 IN IP4 0.0.0.0	[RFC4566]
s=-	[RFC4566]
t=0 0	[RFC4566]
a=group:BUNDLE m0 m1	[I-D.ietf-mmusic-sdp-bundle-negotiation]
a=group:LS m0 m1	[RFC5888]
a=ice-options:trickle	[I-D.ietf-mmusic-trickle-ice]
****** Audio m=line *********	****************************
m=audio 49203 UDP/TLS/RTP/SAVPF 109	[RFC4566]
c=IN IP4 203.0.113.77	[RFC4566]
a=mid:m0	[RFC5888]
a=msid:ma ta	Identifies RTCMediaStream ID (ma) and RTCMediaStreamTrack ID (ta)

(*Continued*)

Table 18.30 (*Continued*) SDP Answer with SVC (Copyright IETF)

```
| a=recvonly                         | [RFC3264]                    |
| a=rtpmap:109 opus/48000/2          | [RFC7587]                    |
| a=maxptime:120                     | [RFC4566]                    |
| a=ice-ufrag:074c6550               | [RFC5245]                    |
| a=ice-pwd:a28a397a4c3f31747d1ee34  | [RFC5245]                    |
| 74af08a068                         |                              |
| a=fingerprint:sha-256 6B:8B:F0:65  | [RFC5245]                    |
| :5F:78:E2:51:3B:AC:6F:F3:3F:46:1B  |                              |
| :35 :DC:B8:5F:64:1A:24:C2:43:F0:A  |                              |
| 1:58:D0:A1:2C:19:08                |                              |
| a=setup:active                     | [RFC4145]                    |
| a=dtls-id:1                        | [I-D.ietf-mmusic-dtls-sdp]   |
| a=rtcp-mux                         | [RFC5761]                    |
| a=rtcp-rsize                       | [RFC5506]                    |
| a=rtcp-fb:109 nack                 | [RFC5104]                    |
| a=extmap:1 urn:ietf:params:rtp-    | [RFC6464]                    |
| hdrext:ssrc-audio-level            |                              |
| a=extmap:2 urn:ietf:params:rtp-    | [I-D.ietf-mmusic-sdp-bundle-n |
| hdrext:sdes:mid                    | egotiation]                  |
| a=candidate:0 1 UDP 2113667326     | [RFC5245]                    |
| 198.51.100.7 51556 typ host        |                              |
| a=candidate:1 1 UDP  1694302206    | [RFC5245]                    |
| 203.0.113.77 49203 typ srflx       |                              |
| raddr 198.51.100.7 rport 51556     |                              |
| a=end-of-candidates                | [I-D.ietf-mmusic-trickle-ice] |
| ****** Video m=line *********      | **************************** |
| m=video 49203 UDP/TLS/RTP/SAVPF    | BUNDLE accepted Bundle       |
| 96 100                             | address same as audio m=line. |
| c=IN IP4 203.0.113.77              | [RFC4566]                    |
| a=mid:m1                           | [RFC5888] Video m=line part  |
|                                    | of BUNDLE group              |
| a=msid:ma tb                       | Identifies RTCMediaStream ID |
|                                    | (ma) and RTCMediaStreamTrack |
|                                    | ID (tb)                      |
| a=recvonly                         | [RFC3264] - Receive only     |
|                                    | video stream                 |
| a=rtpmap:96 H264/90000             | [RFC6184]                    |
| a=fmtp:96 profile-level-id=4d0028  | [RFC6184]H.264 Layer 1       |
| ;packetization-mode=1; max-fr=30   |                              |
| ;max-fs=8040                       |                              |
| a=rtpmap:100 H264-SVC/90000        | [RFC6184]                    |
| a=fmtp:100 profile-level-          | [RFC6184]                    |
| id=4d0028;packetization-mode=1;    |                              |
| max-fr=30;max-fs=8040              |                              |
| a=depend:100 lay m1:96;            | [RFC5583] Bob chooses 2 Codec |
|                                    | Operation points             |
| a=rtcp-fb:* nack                   | [RFC5104]                    |
| a=rtcp-fb:* nack pli               | [RFC5104]                    |
| a=rtcp-fb:* ccm fir                | [RFC5104]                    |
| a=extmap:2 urn:ietf:params:rtp-    | [I-D.ietf-mmusic-sdp-bundle-n |
| hdrext:sdes:mid                    | egotiation]                  |
+------------------------------------+------------------------------+
```

18.2.5.3.3 Successful Simulcast Video Session with Retransmission

This section shows an SDP offer/answer exchange for a simulcast scenario with three resolutions and has RFC 4588-style retransmission flows.

[I-D.ietf-mmusic-rid] framework is used to specify all three resolution constraints mapped to a single PT (98).

[I-D.ietf-mmusic-sdp-simulcast] framework identifies the simulcast streams via their "rid" identifiers (Tables 18.31 and 18.32).

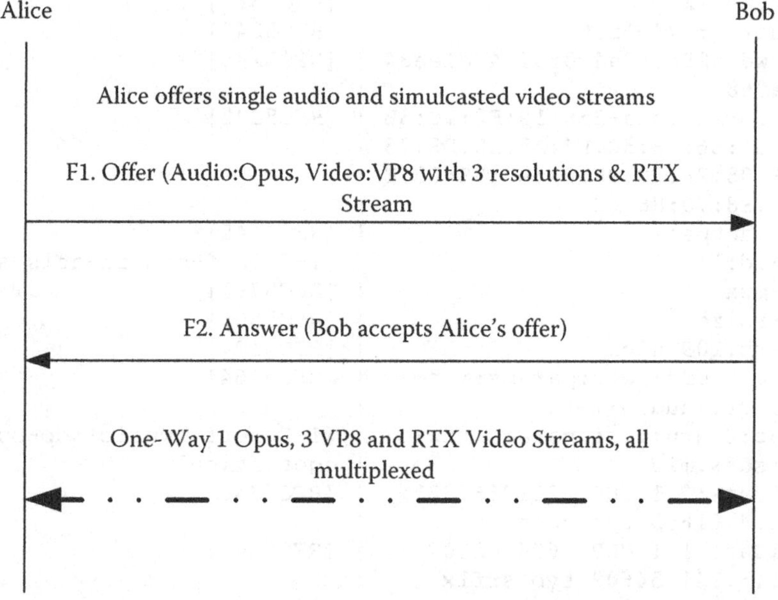

Simulcast streams with retransmission

Table 18.31 SDP Offer w/Simulcast, RTX (Copyright IETF)

Offer SDP Contents	RFC#/Notes
v=0	[RFC4566]
o=- 20519 0 IN IP4 0.0.0.0	[RFC4566]
s=-	[RFC4566]
t=0 0	[RFC4566]
a=group:BUNDLE m0 m1	[I-D.ietf-mmusic-sdp-bundle-negotiation] Alice supports grouping of m=lines under BUNDLE semantics
a=group:LS m0 m1	[RFC5888]
a=ice-options:trickle	[I-D.ietf-mmusic-trickle-ice]
****** Audio m=line *********	****************************
m=audio 54609 UDP/TLS/RTP/SAVPF 109	[RFC4566]

(*Continued*)

Table 18.31 (*Continued*) SDP Offer w/Simulcast, RTX (Copyright IETF)

```
| c=IN IP4 203.0.113.141              | [RFC4566]                     |
| a=mid:m0                            | [RFC5888] Audio m=line part   |
|                                     | of BUNDLE group with a unique |
|                                     | port number                   |
| a=msid:ma ta                        | Identifies RTCMediaStream ID  |
|                                     | (ma) and RTCMediaStreamTrack  |
|                                     | ID (ta)                       |
| a=sendonly                          | [RFC3264]                     |
| a=rtpmap:109 opus/48000/2           | [RFC7587]                     |
| a=maxptime:120                      | [RFC4566]                     |
| a=ice-ufrag:074c6550                | [RFC5245]                     |
| a=ice-pwd:a28a397a4c3f31747d1ee34   | [RFC5245]                     |
| 74af08a068                          |                               |
| a=fingerprint:sha-256 19:E2:1C:3B   | [RFC5245]                     |
| :4B:9F:81:E6:B8:5C:F4:A5:A8:D8:73   |                               |
| :04 :BB:05:2F:70:9F:04:A9:0E:05:E   |                               |
| 9:26:33:E8:70:88:A2                 |                               |
| a=setup:actpass                     | [RFC4145]                     |
| a=dtls-id:1                         | [I-D.ietf-mmusic-dtls-sdp]    |
| a=rtcp-mux                          | [RFC5761]                     |
| a=rtcp-rsize                        | [RFC5506]                     |
| a=rtcp-fb:109 nack                  | [RFC5104]                     |
| a=extmap:1 urn:ietf:params:rtp-     | [RFC6464]                     |
| hdrext:ssrc-audio-level             |                               |
| a=extmap:2 urn:ietf:params:rtp-     | [I-D.ietf-mmusic-sdp-bundle-n |
| hdrext:sdes:mid                     | egotiation]                   |
| a=candidate:0 1 UDP  2113667327     | [RFC5245]                     |
| 192.0.2.4 61665 typ host            |                               |
| a=candidate:1 1 UDP  694302207      | [RFC5245]                     |
| 203.0.113.141 54609 typ srflx       |                               |
| raddr 192.0.2.4 rport 61665         |                               |
| a=end-of-candidates                 | [I-D.ietf-mmusic-trickle-ice] |

| ****** Video m=line *********       | ***************************** |
| m=video 0 UDP/TLS/RTP/SAVPF 98      | bundle-only video line with   |
| 103                                 | port number set to zero       |
| c=IN IP4 203.0.113.141              | [RFC4566]                     |
| a=bundle-only                       | [I-D.ietf-mmusic-sdp-bundle-n |
|                                     | egotiation]                   |
| a=mid:m1                            | [RFC5888]                     |
| a=msid:ma tb                        | Identifies RTCMediaStream ID  |
|                                     | (ma) and RTCMediaStreamTrack  |
|                                     | ID (tb)                       |
| a=sendonly                          | [RFC3264]                     |
| a=rtpmap:98 VP8/90000               | [RFC7741]                     |
| a=fmtp:98 max-fr=30                 | [RFC4566]                     |
| a=rtpmap:103 rtx/90000              | [RFC4588]                     |
| a=fmtp:103 apt=98;rtx-time=200      | [RFC4588]                     |
| a=rtcp-fb:* nack                    | [RFC5104]                     |
| a=rtcp-fb:* nack pli                | [RFC5104]                     |
| a=rtcp-fb:* ccm fir                 | [RFC5104]                     |
```

(*Continued*)

Table 18.31 (*Continued*) SDP Offer w/Simulcast, RTX (Copyright IETF)

```
| a=extmap:2 urn:ietf:params:rtp-      | [I-D.ietf-mmusic-sdp-bundle-n |
| hdrext:sdes:mid                      | egotiation]                   |
| a=rid:1 send pt=98;max-fs=921600     | [I-D.ietf-mmusic-rid]         |
| ;max-fr=30;                          |                               |
| a=rid:2 send pt=98;max-fs=614400     | [I-D.ietf-mmusic-rid]         |
| ;max-fr=15;                          |                               |
| a=rid:3 send pt=98;max-fs=230400     | [I-D.ietf-mmusic-rid]         |
| ;max-fr=30;                          |                               |
| a=simulcast: send 1;2;3              | [I-D.ietf-mmusic-sdp-simulcas |
|                                      | t] Alice can send all the     |
|                                      | simulcast streams             |
+--------------------------------------+-------------------------------+
```

Table 18.32 SDP Answer w/Simulcast, RTX (Copyright IETF)

```
+--------------------------------------+-------------------------------+
| Answer SDP Contents                  | RFC#/Notes                    |
+--------------------------------------+-------------------------------+
| v=0                                  | [RFC4566]                     |
| o=- 20519 0 IN IP4 0.0.0.0           | [RFC4566]                     |
| s=-                                  | [RFC4566]                     |
| t=0 0                                | [RFC4566]                     |
| a=group:BUNDLE m0 m1                 | [I-D.ietf-mmusic-sdp-bundle-n |
|                                      | egotiation] Bob supports      |
|                                      | grouping of m=lines under     |
|                                      | BUNDLE semantics              |
| a=group:LS m0 m1                     | [RFC5888]                     |
| a=ice-options:trickle                | [I-D.ietf-mmusic-trickle-ice] |
| ****** Audio m=line *********        | ***************************** |
| m=audio 49203 UDP/TLS/RTP/SAVPF      | [RFC4566]                     |
| 109                                  |                               |
| c=IN IP4 203.0.113.77                | [RFC4566]                     |
| a=mid:m0                             | [RFC5888]                     |
| a=msid:ma ta                         | Identifies RTCMediaStream ID  |
|                                      | (ma) and RTCMediaStreamTrack  |
|                                      | ID (ta)                       |
| a=recvonly                           | [RFC3264]                     |
| a=rtpmap:109 opus/48000/2            | [RFC7587]                     |
| a=maxptime:120                       | [RFC4566]                     |
| a=ice-ufrag:074c6550                 | [RFC5245]                     |
| a=ice-pwd:a28a397a4c3f31747d1ee34    | [RFC5245]                     |
| 74af08a068                           |                               |
| a=fingerprint:sha-256 6B:8B:F0:65    | [RFC5245]                     |
| :5F:78:E2:51:3B:AC:6F:F3:3F:46:1B    |                               |
| :35 :DC:B8:5F:64:1A:24:C2:43:F0:A    |                               |
| 1:58:D0:A1:2C:19:08                  |                               |
| a=setup:active                       | [RFC4145]                     |
| a=dtls-id:1                          | [I-D.ietf-mmusic-dtls-sdp]    |
```

(*Continued*)

Table 18.32 (*Continued*) SDP Answer w/Simulcast, RTX (Copyright IETF)

SDP	Notes
a=rtcp-mux	[RFC5761]
a=rtcp-rsize	[RFC5506]
a=rtcp-fb:109 nack	[RFC5104]
a=extmap:1 urn:ietf:params:rtp-hdrext:ssrc-audio-level	[RFC6464]
a=extmap:2 urn:ietf:params:rtp-hdrext:sdes:mid	[I-D.ietf-mmusic-sdp-bundle-negotiation]
a=candidate:0 1 UDP 2113667326 198.51.100.7 51556 typ host	[RFC5245]
a=candidate:1 1 UDP 1694302206 203.0.113.77 49203 typ srflx raddr 198.51.100.7 rport 51556	[RFC5245]
a=end-of-candidates	[I-D.ietf-mmusic-trickle-ice]
****** Video m=line *********	****************************
m=video 49203 UDP/TLS/RTP/SAVPF 98 100 101 103	BUNDLE accepted with Bundle address identical to audio m-line
c=IN IP4 203.0.113.77	[RFC4566]
a=mid:m1	[RFC5888] Video m=line part of BUNDLE group
a=msid:ma tb	Identifies RTCMediaStream ID (ma) and RTCMediaStreamTrack ID (tb)
a=recvonly	[RFC3264]
a=rtpmap:98 VP8/90000	[RFC7741]
a=fmtp:98 max-fr=30	[RFC4566]
a=rtpmap:103 rtx/90000	[RFC4588]
a=fmtp:103 apt=98;rtx-time=200	[RFC4588]
a=rtcp-fb:* nack	[RFC5104]
a=rtcp-fb:* nack pli	[RFC5104]
a=rtcp-fb:* ccm fir	[RFC5104]
a=extmap:2 urn:ietf:params:rtp-hdrext:sdes:mid	[I-D.ietf-mmusic-sdp-bundle-negotiation]
a=rid:1 recv pt=98;max-fs=921600;max-fr=30;	[I-D.ietf-mmusic-rid]
a=rid:2 recv pt=98;max-fs=614400;max-fr=15;	[I-D.ietf-mmusic-rid]
a=rid:3 recv pt=98;max-fs=230400;max-fr=30;	[I-D.ietf-mmusic-rid]
a=simulcast: recv 1;2;3	[I-D.ietf-mmusic-sdp-simulcast] Bob accepts the offered simulcast streams

18.2.5.3.4 Successful One-Way Simulcast Session with Two Resolutions and RTX: One Resolution Rejected

This section shows an SDP offer/answer exchange for a simulcast scenario with two resolutions.

It also showcases where Bob rejects one of the simulcast video streams, which results in the rejection of the associated repair stream implicitly (Tables 18.33 and 18.34).

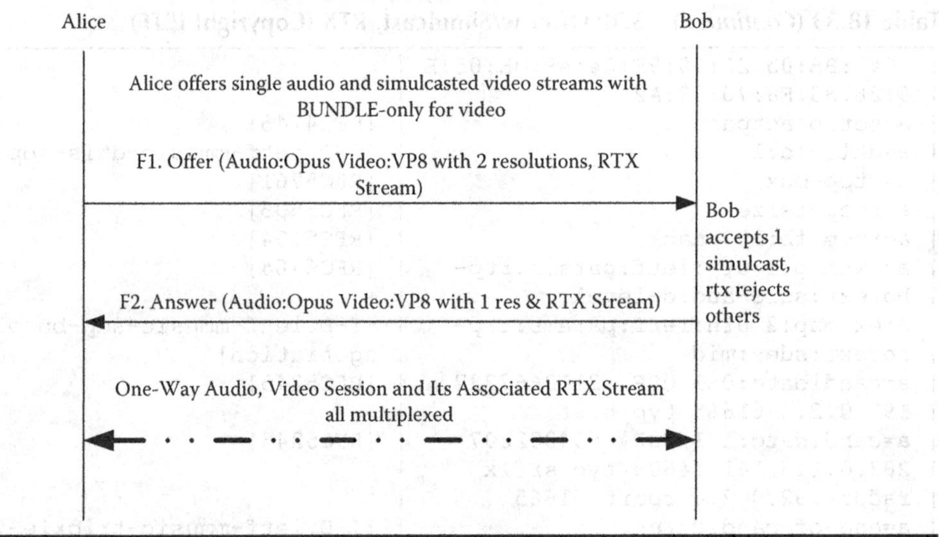

Simulcast streams with retransmission rejected

Table 18.33 SDP Offer w/Simulcast, RTX (Copyright IETF)

Offer SDP Contents	RFC#/Notes
v=0	[RFC4566]
o=- 20519 0 IN IP4 0.0.0.0	[RFC4566]
s=-	[RFC4566]
t=0 0	[RFC4566]
a=group:BUNDLE m0 m1	[I-D.ietf-mmusic-sdp-bundle-negotiation] Alice supports grouping of m=lines under BUNDLE semantics
a=group:LS m0 m1	[RFC5888]
a=ice-options:trickle	[I-D.ietf-mmusic-trickle-ice]
****** Audio m=line *********	****************************
m=audio 54609 UDP/TLS/RTP/SAVPF 109	[RFC4566]
c=IN IP4 203.0.113.141	[RFC4566]
a=mid:m0	[RFC5888]
a=msid:ma ta	Identifies RTCMediaStream ID (ma) and RTCMediaStreamTrack ID (ta)
a=sendonly	[RFC3264]
a=rtpmap:109 opus/48000/2	[RFC7587]
a=maxptime:120	[RFC4566]
a=ice-ufrag:074c6550	[RFC5245]
a=ice-pwd:a28a397a4c3f31747d1ee3474af08a068	[RFC5245]
a=fingerprint:sha-256 19:E2:1C:3B:4B:9F:81:E6:B8:5C:F4:A5:A8:D8:73	[RFC5245]

(*Continued*)

Table 18.33 (*Continued*) SDP Offer w/Simulcast, RTX (Copyright IETF)

```
| :04 :BB:05:2F:70:9F:04:A9:0E:05:E |                                |
| 9:26:33:E8:70:88:A2               |                                |
| a=setup:actpass                   | [RFC4145]                      |
| a=dtls-id:1                       | [I-D.ietf-mmusic-dtls-sdp]     |
| a=rtcp-mux                        | [RFC5761]                      |
| a=rtcp-rsize                      | [RFC5506]                      |
| a=rtcp-fb:109 nack                | [RFC5104]                      |
| a=extmap:1 urn:ietf:params:rtp-   | [RFC6464]                      |
| hdrext:ssrc-audio-level           |                                |
| a=extmap:2 urn:ietf:params:rtp-   | [I-D.ietf-mmusic-sdp-bundle-n  |
| hdrext:sdes:mid                   | egotiation]                    |
| a=candidate:0 1 UDP  2113667327   | [RFC5245]                      |
| 192.0.2.4 61665 typ host          |                                |
| a=candidate:1 1 UDP  694302207    | [RFC5245]                      |
| 203.0.113.141 54609 typ srflx     |                                |
| raddr 192.0.2.4 rport 61665       |                                |
| a=end-of-candidates               | [I-D.ietf-mmusic-trickle-ice]  |
| ****** Video m=line *********     | ****************************** |
| m=video 0 UDP/TLS/RTP/SAVPF 98    | bundle-only video line with    |
| 100 101 103                       | port number set to zero        |
| c=IN IP4 203.0.113.141            | [RFC4566]                      |
| a=bundle-only                     | [I-D.ietf-mmusic-sdp-bundle-n  |
|                                   | egotiation]                    |
| a=mid:m1                          | [RFC5888]                      |
| a=msid:ma tb                      | Identifies RTCMediaStream ID   |
|                                   | (ma) and RTCMediaStreamTrack   |
|                                   | ID (tb                         |
| a=sendonly                        | [RFC3264]                      |
| a=rtpmap:98 VP8/90000             | [RFC7741]                      |
| a=rtpmap:100 VP8/90000            | [RFC7741]                      |
| a=rtpmap:101 rtx/90000            | [RFC4588]                      |
| a=rtpmap:103 rtx/90000            | [RFC4588]                      |
| a=fmtp:98 max-fr=30;max-fs=8040   | [RFC4566]                      |
| a=fmtp:100 max-fr=15;max-fs=1200  | [RFC4566]                      |
| a=fmtp:101 apt=98;rtx-time=200    | [RFC4588]                      |
| a=fmtp:103 apt=100;rtx-time=200   | [RFC4588]                      |
| a=rtcp-fb:* nack                  | [RFC5104]                      |
| a=rtcp-fb:* nack pli              | [RFC5104]                      |
| a=rtcp-fb:* ccm fir               | [RFC5104]                      |
| a=extmap:2 urn:ietf:params:rtp-   | [I-D.ietf-mmusic-sdp-bundle-n  |
| hdrext:sdes:mid                   | egotiation]                    |
| a=rid:1 send pt=98;               | [I-D.ietf-mmusic-rid] 1:1      |
|                                   | mapping between the PT and     |
|                                   | the 'rid' identifier           |
| a=rid:2 send pt=100;              | [I-D.ietf-mmusic-rid] 1:1      |
|                                   | mapping between the PT and     |
|                                   | the 'rid' identifier           |
| a=simulcast: send 1;2             | [I-D.ietf-mmusic-sdp-simulcas  |
|                                   | t]                             |
+-----------------------------------+--------------------------------+
```

Table 18.34 SDP Answer (One Simulcast Rejected) (Copyright IETF)

Answer SDP Contents	RFC#/Notes
v=0	[RFC4566]
o=- 20519 0 IN IP4 0.0.0.0	[RFC4566]
s=-	[RFC4566]
t=0 0	[RFC4566]
a=group:BUNDLE m0 m1	[I-D.ietf-mmusic-sdp-bundle-negotiation] Bob supports grouping of m=lines under BUNDLE semantics
a=group:LS m0 m1	[RFC5888]
a=ice-options:trickle	[I-D.ietf-mmusic-trickle-ice]
****** Audio m=line *********	****************************
m=audio 49203 UDP/TLS/RTP/SAVPF 109	[RFC4566]
c=IN IP4 203.0.113.77	[RFC4566]
a=mid:m0	[RFC5888]
a=msid:ma ta	Identifies RTCMediaStream ID (ma) and RTCMediaStreamTrack ID (ta)
a=recvonly	[RFC3264]
a=rtpmap:109 opus/48000/2	[RFC7587]
a=maxptime:120	[RFC4566]
a=ice-ufrag:074c6550	[RFC5245]
a=ice-pwd:a28a397a4c3f31747d1ee3474af08a068	[RFC5245]
a=fingerprint:sha-256 6B:8B:F0:65:5F:78:E2:51:3B:AC:6F:F3:3F:46:1B:35 :DC:B8:5F:64:1A:24:C2:43:F0:A1:58:D0:A1:2C:19:08	[RFC5245]
a=setup:active	[RFC4145]
a=dtls-id:1	[I-D.ietf-mmusic-dtls-sdp]
a=rtcp-mux	[RFC5761]
a=rtcp-rsize	[RFC5506]
a=extmap:1 urn:ietf:params:rtp-hdrext:ssrc-audio-level	[RFC6464]
a=extmap:2 urn:ietf:params:rtp-hdrext:sdes:mid	[I-D.ietf-mmusic-sdp-bundle-negotiation]
a=candidate:0 1 UDP 2113667326 198.51.100.7 51556 typ host	[RFC5245]
a=candidate:1 1 UDP 1694302206 203.0.113.77 49203 typ srflx raddr 198.51.100.7 rport 51556	[RFC5245]
a=end-of-candidates	[I-D.ietf-mmusic-trickle-ice]
****** Video m=line *********	****************************
m=video 49203 UDP/TLS/RTP/SAVPF 98 101	BUNDLE accepted with Bundle address identical to audio

(Continued)

Table 18.34 (*Continued*) SDP Answer (One Simulcast Rejected) (Copyright IETF)

```
|                                   | m-line                         |
| c=IN IP4 203.0.113.77             | [RFC4566]                      |
| a=mid:m1                          | [RFC5888]                      |
| a=msid:ma tb                      | Identifies RTCMediaStream ID   |
|                                   | (ma) and RTCMediaStreamTrack   |
|                                   | ID (tb)                        |
| a=recvonly                        | [RFC3264]                      |
| a=rtpmap:98 VP8/90000             | [RFC7741]                      |
| a=rtpmap:101 VP8/90000            | [RFC7741]                      |
| a=fmtp:98 max-fr=30;max-fs=8040   | [RFC4566]                      |
| a=fmtp:101 apt=98;rtx-time=200    | [RFC4588]                      |
| a=extmap:2 urn:ietf:params:rtp-   | [I-D.ietf-mmusic-sdp-bundle-n  |
| hdrext:sdes:mid                   | egotiation]                    |
| a=rid:1 recv pt=98;               | [I-D.ietf-mmusic-rid]          |
| a=simulcast: recv 1               | [I-D.ietf-mmusic-sdp-simulcas  |
|                                   | t] Bob rejects the second      |
|                                   | simulcast stream and the       |
|                                   | associated rtx stream.         |
+-----------------------------------+--------------------------------+
```

18.2.5.3.5 Simulcast Video Session with FEC

This section shows an SDP offer/answer exchange for a simulcast video stream at two resolutions and has RFC 5956-style FEC flows.

On completion of the offer/answer exchange mechanism we end up with one audio stream, two simulcast video streams, and two associated FEC streams sent over a single 5-tuple (Tables 18.35 and 18.36).

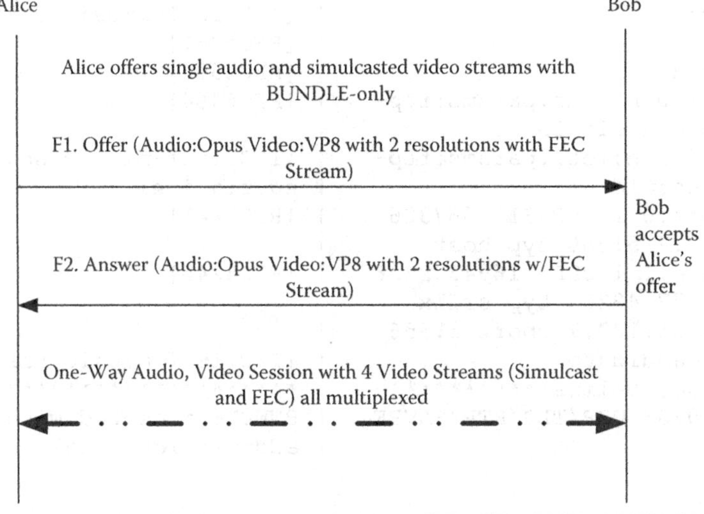

Simulcast streams with forward error correction

Table 18.35 SDP Offer (Copyright IETF)

Offer SDP Contents	RFC#/Notes
v=0	[RFC4566]
o=- 20519 0 IN IP4 0.0.0.0	[RFC4566]
s=-	[RFC4566]
t=0 0	[RFC4566]
a=group:BUNDLE m0 m1	[I-D.ietf-mmusic-sdp-bundle-negotiation] Alice supports grouping of m=lines under BUNDLE semantics
a=group:LS m0 m1	[RFC5888]
a=ice-options:trickle	[I-D.ietf-mmusic-trickle-ice]
****** Audio m=line *********	*****************************
m=audio 54609 UDP/TLS/RTP/SAVPF 109	[RFC4566]
c=IN IP4 203.0.113.141	[RFC4566]
a=mid:m0	[RFC5888]
a=msid:ma ta	Identifies RTCMediaStream ID (ma) and RTCMediaStreamTrack ID (ta)
a=sendonly	[RFC3264]
a=rtpmap:109 opus/48000/2	[RFC7587]
a=maxptime:120	[RFC4566]
a=ice-ufrag:074c6550	[RFC5245]
a=ice-pwd:a28a397a4c3f31747d1ee34 74af08a068	[RFC5245]
a=fingerprint:sha-256 19:E2:1C:3B :4B:9F:81:E6:B8:5C:F4:A5:A8:D8:73 :04 :BB:05:2F:70:9F:04:A9:0E:05:E 9:26:33:E8:70:88:A2	[RFC5245]
a=setup:actpass	[RFC4145]
a=rtcp-mux	[RFC5761]
a=dtls-id:1	[I-D.ietf-mmusic-dtls-sdp]
a=rtcp-rsize	[RFC5506]
a=rtcp-fb:109 nack	[RFC5104]
a=extmap:1 urn:ietf:params:rtp-hdrext:ssrc-audio-level	[RFC6464]
a=extmap:2 urn:ietf:params:rtp-hdrext:sdes:mid	[I-D.ietf-mmusic-sdp-bundle-negotiation]
a=candidate:0 1 UDP 2113667327 192.0.2.4 61665 typ host	[RFC5245]
a=candidate:1 1 UDP 694302207 203.0.113.141 54609 typ srflx raddr 192.0.2.4 rport 61665	[RFC5245]
a=end-of-candidates	[I-D.ietf-mmusic-trickle-ice]
****** Video m=line *********	*****************************
m=video 0 UDP/TLS/RTP/SAVPF 98	bundle-only video line with

(Continued)

Table 18.35 (*Continued*) SDP Offer (Copyright IETF)

100 101 103	port number set to zero
c=IN IP4 203.0.113.141	[RFC4566]
a=bundle-only	[I-D.ietf-mmusic-sdp-bundle-negotiation]
a=mid:m1	[RFC5888] Video m=line part of BUNDLE group
a=msid:ma tb	Identifies RTCMediaStream ID (ma) and RTCMediaStreamTrack ID (tb)
a=sendonly	[RFC3264]
a=rtpmap:98 VP8/90000	[RFC7741]
a=rtpmap:100 VP8/90000	[RFC7741]
a=rtpmap:101 flexfec/90000	[I-D.ietf-payload-flexible-fec-scheme]
a=rtpmap:103 flexfec/90000	[I-D.ietf-payload-flexible-fec-scheme]
a=fmtp:98 max-fr=30;max-fs=8040	[RFC4566]
a=fmtp:100 max-fr=15;max-fs=1200	[RFC4566]
a=fmtp:101 L=5; D=10; ToP=2; repair-window=200000	[I-D.ietf-payload-flexible-fec-scheme]
a=fmtp:103 L=5; D=10; ToP=2; repair-window=200000	[I-D.ietf-payload-flexible-fec-scheme]
a=rtcp-fb:* nack	[RFC5104]
a=rtcp-fb:* nack pli	[RFC5104]
a=rtcp-fb:* ccm fir	[RFC5104]
a=extmap:2 urn:ietf:params:rtp-hdrext:sdes:mid	[I-D.ietf-mmusic-sdp-bundle-negotiation]
a=rid:1 send pt=98;	[I-D.ietf-mmusic-rid] 1:1 mapping between the PT and the 'rid' identifier
a=rid:2 send pt=100;	[I-D.ietf-mmusic-rid] 1:1 mapping between the PT and the 'rid' identifier
a=simulcast: send 1;2	[I-D.ietf-mmusic-sdp-simulcast]

Table 18.36 SDP Answer (Copyright IETF)

Answer SDP Contents	RFC#/Notes
v=0	[RFC4566]
o=- 20519 0 IN IP4 0.0.0.0	[RFC4566]
s=-	[RFC4566]
t=0 0	[RFC4566]
a=group:BUNDLE m0 m1	[I-D.ietf-mmusic-sdp-bundle-negotiation]
a=group:LS m0 m1	[RFC5888]

(*Continued*)

Table 18.36 (*Continued*) SDP Answer (Copyright IETF)

a=ice-options:trickle	[I-D.ietf-mmusic-trickle-ice]
****** Audio m=line *********	****************************
m=audio 49203 UDP/TLS/RTP/SAVPF 109	[RFC4566]
c=IN IP4 203.0.113.77	[RFC4566]
a=mid:m0	[RFC5888] Audio m=line part of BUNDLE group with a unique port number
a=msid:ma ta	Identifies RTCMediaStream ID (ma) and RTCMediaStreamTrack ID (ta)
a=recvonly	[RFC3264]
a=rtpmap:109 opus/48000/2	[RFC7587]
a=maxptime:120	[RFC4566]
a=ice-ufrag:074c6550	[RFC5245]
a=ice-pwd:a28a397a4c3f31747d1ee34 74af08a068	[RFC5245]
a=fingerprint:sha-256 6B:8B:F0:65 :5F:78:E2:51:3B:AC:6F:F3:3F:46:1B :35 :DC:B8:5F:64:1A:24:C2:43:F0:A 1:58:D0:A1:2C:19:08	[RFC5245]
a=setup:active	[RFC4145]
a=dtls-id:1	[I-D.ietf-mmusic-dtls-sdp]
a=rtcp-mux	[RFC5761]
a=rtcp-rsize	[RFC5506]
a=rtcp-fb:109 nack	[RFC5104]
a=extmap:1 urn:ietf:params:rtp- hdrext:ssrc-audio-level	[RFC6464]
a=extmap:2 urn:ietf:params:rtp- hdrext:sdes:mid	[I-D.ietf-mmusic-sdp-bundle-n egotiation]
a=candidate:0 1 UDP 2113667326 198.51.100.7 51556 typ host	[RFC5245]
a=candidate:1 1 UDP 1694302206 203.0.113.77 49203 typ srflx raddr 198.51.100.7 rport 51556	[RFC5245]
a=end-of-candidates	[I-D.ietf-mmusic-trickle-ice]
****** Video m=line *********	****************************
m=video 49203 UDP/TLS/RTP/SAVPF 98 100 101 103	BUNDLE accepted with Bundle Address identical to audio m=line.
c=IN IP4 203.0.113.77	[RFC4566]
a=mid:m1	[RFC5888] Video m=line part of BUNDLE group
a=msid:ma tb	Identifies RTCMediaStream ID (ma) and RTCMediaStreamTrack ID (tb)
a=recvonly	[RFC3264]
a=rtpmap:98 VP8/90000	[RFC7741]
a=rtpmap:100 VP8/90000	[RFC7741]

(*Continued*)

Table 18.36 (*Continued*) SDP Answer (Copyright IETF)

```
| a=rtpmap:101 flexfec/90000          | [I-D.ietf-payload-flexible-fe |
|                                     | c-scheme]                     |
| a=rtpmap:103 flexfec/90000          | [I-D.ietf-payload-flexible-fe |
|                                     | c-scheme]                     |
| a=fmtp:98 max-fr=30;max-fs=8040     | [RFC4566]                     |
| a=fmtp:100 max-fr=15;max-fs=1200    | [RFC4566]                     |
| a=fmtp:101 L=5; D=10; ToP=2;        | [I-D.ietf-payload-flexible-fe |
| repair-window=200000                | c-scheme]                     |
| a=fmtp:103 L=5; D=10; ToP=2;        | [I-D.ietf-payload-flexible-fe |
| repair-window=200000                | c-scheme]                     |
| a=rtcp-fb:* nack                    | [RFC5104]                     |
| a=rtcp-fb:* nack pli                | [RFC5104]                     |
| a=rtcp-fb:* ccm fir                 | [RFC5104]                     |
| a=extmap:2 urn:ietf:params:rtp-     | [I-D.ietf-mmusic-sdp-bundle-n |
| hdrext:sdes:mid                     | egotiation]                   |
| a=rid:1 recv pt=98;                 | [I-D.ietf-mmusic-rid]         |
| a=rid:2 recv pt=100;                | [I-D.ietf-mmusic-rid]         |
| a=simulcast: recv 1;2               | [I-D.ietf-mmusic-sdp-simulcas |
|                                     | t]                            |
+-------------------------------------+-------------------------------+
```

18.2.5.4 Others

The examples in this section provide an SDP offer/answer exchange for a variety of scenarios related to the RTP header extension for conference usages, Legacy Interop scenarios, and more.

18.2.5.4.1 Audio Session: Voice Activity Detection

This example shows Alice indicating the support of the RTP header extension to include the audio level of the audio sample carried in the RTP packet (Tables 18.37 and 18.38).

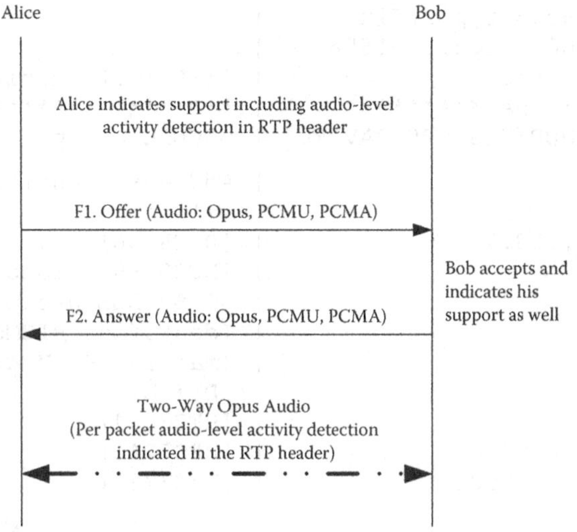

2-way audio with VAD

Table 18.37 SDP Offer (Copyright IETF)

Offer SDP Contents	RFC#/Notes
v=0	[RFC4566]
o=- 20518 0 IN IP4 0.0.0.0	[RFC4566]
s=-	[RFC4566]
t=0 0	[RFC4566]
a=group:BUNDLE audio	[I-D.ietf-mmusic-sdp-bundle-negotiation]
a=ice-options:trickle	[I-D.ietf-mmusic-trickle-ice]
****** Audio m=line *********	*****************************
m=audio 54609 UDP/TLS/RTP/SAVPF 109 0 8	[RFC4566]
c=IN IP4 203.0.113.141	[RFC4566]
a=mid:audio	[RFC5888]
a=msid:ma ta	Identifies RTCMediaStream ID (ma) and RTCMediaStreamTrack ID (ta)
a=sendrecv	[RFC3264]
a=rtpmap:109 opus/48000/2	[RFC7587]
a=rtpmap:0 PCMU/8000	[RFC3551]
a=rtpmap:8 PCMA/8000	[RFC3551]
a=maxptime:120	[RFC4566]
a=ice-ufrag:074c6550	[RFC5245]
a=ice-pwd:a28a397a4c3f31747d1ee34 74af08a068	[RFC5245]
a=fingerprint:sha-256 19:E2:1C:3B:4B:9F:81:E6:B8:5C:F4:A5:A8:D8:73:04 :BB:05:2F:70:9F:04:A9:0E:05:E9:26:33:E8:70:88:A2	[RFC5245]
a=setup:actpass	[RFC4145]
a=dtls-id:1	[I-D.ietf-mmusic-dtls-sdp]
a=rtcp-mux	[RFC5761]
a=rtcp-rsize	[RFC5506]
a=rtcp-fb:* nack	[RFC5104]
a=extmap:1 urn:ietf:params:rtp-hdrext:ssrc-audio-level	[RFC6464]
a=extmap:2 urn:ietf:params:rtp-hdrext:sdes:mid	[I-D.ietf-mmusic-sdp-bundle-negotiation]
a=candidate:0 1 UDP 2113667327 192.0.2.4 61665 typ host	[RFC5245]
a=candidate:1 1 UDP 694302207 203.0.113.141 54609 typ srflx raddr 192.0.2.4 rport 61665	[RFC5245]
a=end-of-candidates	[I-D.ietf-mmusic-trickle-ice]

Table 18.38 SDP Answer (Copyright IETF)

Answer SDP Contents	RFC#/Notes
v=0	[RFC4566]
o=- 16833 0 IN IP4 0.0.0.0	[RFC4566]
s=-	[RFC4566]
t=0 0	[RFC4566]
a=group:BUNDLE audio	[I-D.ietf-mmusic-sdp-bundle-negotiation]
a=ice-options:trickle	[I-D.ietf-mmusic-trickle-ice]
****** Audio m=line *********	*****************************
m=audio 49203 UDP/TLS/RTP/SAVPF 109 0 98	[RFC4566]
c=IN IP4 203.0.113.77	[RFC4566]
a=mid:audio	[RFC5888]
a=msid:ma ta	Identifies RTCMediaStream ID (ma) and RTCMediaStreamTrack ID (ta)
a=sendrecv	[RFC3264] - Bob can send and recv audio
a=rtpmap:109 opus/48000/2	[RFC7587] - Bob accepts only Opus Codec
a=rtpmap:0 PCMU/8000	[RFC3551] PCMU Audio Codec
a=rtpmap:0 PCMA/8000	[RFC3551] PCMA Audio Codec
a=maxptime:120	[RFC4566]
a=ice-ufrag:c300d85b	[RFC5245]
a=ice-pwd:de4e99bd291c325921d5d47efbabd9a2	[RFC5245]
a=fingerprint:sha-256 6B:8B:F0:65:5F:78:E2:51:3B:AC:6F:F3:3F:46:1B:35 :DC:B8:5F:64:1A:24:C2:43:F0:A1:58:D0:A1:2C:19:08	[RFC5245]
a=setup:active	[RFC4145]
a=dtls-id:1	[I-D.ietf-mmusic-dtls-sdp]
a=rtcp-mux	[RFC5761] - Bob can perform RTP/RTCP Muxing on port 49203
a=rtcp-rsize	[RFC5506]
a=rtcp-fb:* nack	[RFC5104]
a=extmap:1 urn:ietf:params:rtp-hdrext:ssrc-audio-level	[RFC6464]
a=extmap:2 urn:ietf:params:rtp-hdrext:sdes:mid	[I-D.ietf-mmusic-sdp-bundle-negotiation]
a=candidate:0 1 UDP 2113667327 198.51.100.7 51556 typ host	[RFC5245]
a=candidate:1 1 UDP 1694302207 203.0.113.77 49203 typ srflx raddr 198.51.100.7 rport 51556	[RFC5245]
a=end-of-candidates	[I-D.ietf-mmusic-trickle-ice]

18.2.5.4.2 Audio Conference: Voice Activity Detection

This example shows the SDP for an RTP header extension that allows RTP-level mixers in audio conferences to deliver information about the audio level of individual participants (Tables 18.39 and 18.40).

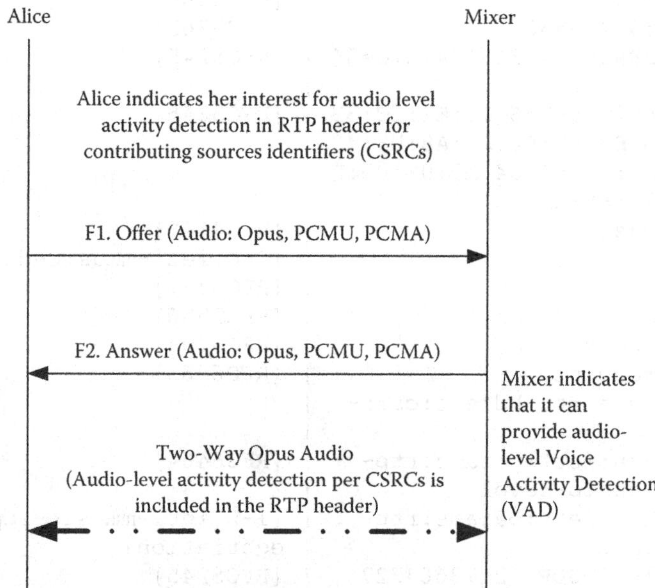

Audio conference with VAD support

Table 18.39 SDP Offer (Copyright IETF)

Offer SDP Contents	RFC#/Notes
v=0	[RFC4566]
o=- 20518 0 IN IP4 0.0.0.0	[RFC4566] - Session Origin Information
s=-	[RFC4566]
t=0 0	[RFC4566]
a=group:BUNDLE audio	[I-D.ietf-mmusic-sdp-bundle-negotiation]
a=ice-options:trickle	[I-D.ietf-mmusic-trickle-ice]
****** Audio m=line *********	*****************************
m=audio 54609 UDP/TLS/RTP/SAVPF 109 0 8	[RFC4566]
c=IN IP4 203.0.113.141	[RFC4566]
a=mid:audio	[RFC5888]
a=msid:ma ta	Identifies RTCMediaStream ID (ma) and RTCMediaStreamTrack ID (ta)

(Continued)

Table 18.39 (*Continued*) SDP Offer (Copyright IETF)

a=sendrecv	[RFC3264] - Alice can send and recv audio
a=rtpmap:109 opus/48000/2	[RFC7587]
a=rtpmap:0 PCMU/8000	[RFC3551] PCMU Audio Codec
a=rtpmap:0 PCMA/8000	[RFC3551] PCMA Audio Codec
a=maxptime:120	[RFC4566]
a=ice-ufrag:074c6550	[RFC5245]
a=ice-pwd:a28a397a4c3f31747d1ee3474af08a068	[RFC5245]
a=fingerprint:sha-256 19:E2:1C:3B:4B:9F:81:E6:B8:5C:F4:A5:A8:D8:73:04 :BB:05:2F:70:9F:04:A9:0E:05:E9:26:33:E8:70:88:A2	[RFC5245]
a=setup:actpass	[RFC4145]
a=dtls-id:1	[I-D.ietf-mmusic-dtls-sdp]
a=rtcp-mux	[RFC5761]
a=rtcp-rsize	[RFC5506]
a=rtcp-fb:* nack	[RFC5104]
a=extmap:1/recvonly urn:ietf:params:rtp-hdrext:csrc-audio-level	[RFC6465]
a=extmap:2 urn:ietf:params:rtp-hdrext:ssrc-audio-level	[RFC6464]
a=extmap:3 urn:ietf:params:rtp-hdrext:sdes:mid	[I-D.ietf-mmusic-sdp-bundle-negotiation]
a=candidate:0 1 UDP 2113667327 192.0.2.4 61665 typ host	[RFC5245]
a=candidate:1 1 UDP 694302207 203.0.113.141 54609 typ srflx raddr 192.0.2.4 rport 61665	[RFC5245]
a=end-of-candidates	[I-D.ietf-mmusic-trickle-ice]

Table 18.40 SDP Answer (Copyright IETF)

Answer SDP Contents	RFC#/Notes
v=0	[RFC4566]
o=- 16833 0 IN IP4 0.0.0.0	[RFC4566] - Session Origin Information
s=-	[RFC4566]
t=0 0	[RFC4566]
a=group:BUNDLE audio	[I-D.ietf-mmusic-sdp-bundle-negotiation]
a=ice-options:trickle	[I-D.ietf-mmusic-trickle-ice]
****** Audio m=line *********	****************************
m=audio 49203 UDP/TLS/RTP/SAVPF	[RFC4566]

(*Continued*)

Table 18.40 (*Continued*) SDP Answer (Copyright IETF)

109 0 98	
c=IN IP4 203.0.113.77	[RFC4566]
a=mid:audio	[RFC5888]
a=msid:ma ta	Identifies RTCMediaStream ID (ma) and RTCMediaStreamTrack ID (ta)
a=sendrecv	[RFC3264]
a=rtpmap:109 opus/48000/2	[RFC7587]
a=rtpmap:0 PCMU/8000	[RFC3551] PCMU Audio Codec
a=rtpmap:0 PCMA/8000	[RFC3551] PCMA Audio Codec
a=maxptime:120	[RFC4566]
a=ice-ufrag:c300d85b	[RFC5245]
a=ice-pwd:de4e99bd291c325921d5d47 efbabd9a2	[RFC5245]
a=fingerprint:sha-256 6B:8B:F0:65 :5F:78:E2:51:3B:AC:6F:F3:3F:46:1B :35 :DC:B8:5F:64:1A:24:C2:43:F0:A 1:58:D0:A1:2C:19:08	[RFC5245]
a=setup:active	[RFC4145]
a=dtls-id:1	[I-D.ietf-mmusic-dtls-sdp]
a=rtcp-mux	[RFC5761]
a=rtcp-rsize	[RFC5506]
a=rtcp-fb:* nack	[RFC5104]
a=extmap:1/sendonly urn:ietf:params:rtp-hdrext:csrc-audio-level	[RFC6465]
a=extmap:2 urn:ietf:params:rtp-hdrext:sdes:mid	[I-D.ietf-mmusic-sdp-bundle-negotiation]
a=candidate:0 1 UDP 2113667327 198.51.100.7 51556 typ host	[RFC5245]
a=candidate:1 1 UDP 1694302207 203.0.113.77 49203 typ srflx raddr 198.51.100.7 rport 51556	[RFC5245]
a=end-of-candidates	[I-D.ietf-mmusic-trickle-ice]

18.2.5.4.3 Successful Legacy Interop Fallback with Bundle Only

In the scenario described below, Alice is a multistream-capable WebRTC endpoint while Bob is a legacy Voice over IP endpoint. The SDP offer/answer exchange demonstrates successful session setup with fallback to audio-only stream negotiated via a bundle-only framework between the endpoints. Specifically,

- The offer from Alice describes two cameras via two video "m=" lines with both marked as bundle only.
- Since Bob does not recognize either the BUNDLE mechanism or the bundle-only attribute, he accepts only the audio stream from Alice.

NOTE: Since Alice is unaware of Bob's support for BUNDLE framework, Alice ensures the inclusion of separate RTP/RTCP ports and candidate information (Tables 18.41 and 18.42).

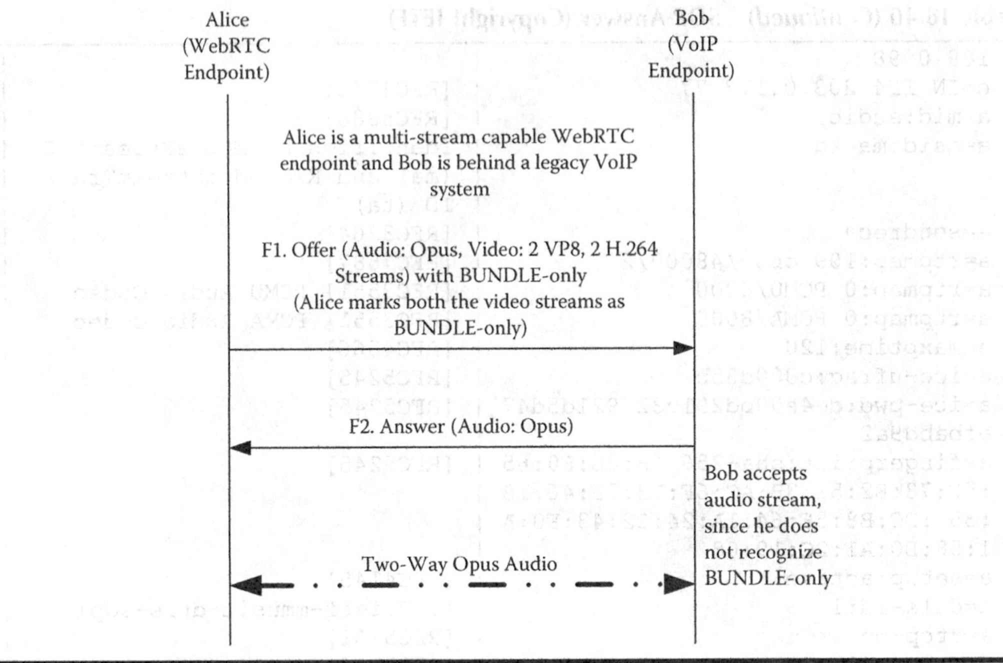

Successful 2-way WebRTC <-> VOIP interop

Table 18.41 SDP Simulcast Bundle-Only (Copyright IETF)

Offer SDP Contents	RFC#/Notes
v=0	[RFC4566]
o=- 20519 0 IN IP4 0.0.0.0	[RFC4566]
s=-	[RFC4566]
t=0 0	[RFC4566]
a=group:BUNDLE m0 m1 m2	[I-D.ietf-mmusic-sdp-bundle -negotiation] Alice supports grouping of m=lines under BUNDLE semantics
a=group:LS m0 m1	[RFC5888]
a=ice-options:trickle	[I-D.ietf-mmusic-trickle-ic e]
****** Audio m=line *********	***************************** **
m=audio 54609 UDP/TLS/RTP/SAVPF 109	[RFC4566]
c=IN IP4 203.0.113.141	[RFC4566]
a=mid:m0	[RFC5888] Audio m=line part of BUNDLE group with a unique port number
a=msid:ma ta	Identifies RTCMediaStream ID (ma) and

(*Continued*)

Table 18.41 (*Continued*) SDP Simulcast Bundle-Only (Copyright IETF)

	RTCMediaStreamTrack ID (ta)
a=sendrecv	[RFC3264]
a=rtpmap:109 opus/48000/2	[RFC7587]
a=rtcp-fb:109 nack	[RFC5104]
a=maxptime:120	[RFC4566]
a=ice-ufrag:074c6550	[RFC5245]
a=ice-pwd:a28a397a4c3f31747d1ee3474 af08a068	[RFC5245]
a=fingerprint:sha-256 19:E2:1C:3B:4 B:9F:81:E6:B8:5C:F4:A5:A8:D8:73:04 :BB:05:2F:70:9F:04:A9:0E:05:E9:26:3 3:E8:70:88:A2	[RFC5245]
a=setup:actpass	[RFC4145]
a=dtls-id:1	[I-D.ietf-mmusic-dtls-sdp]
a=rtcp-mux	[RFC5761]
a=rtcp:64678 IN IP4 203.0.113.141	[RFC3605]
a=rtcp-rsize	[RFC5506]
a=extmap:1 urn:ietf:params:rtp- hdrext:ssrc-audio-level	[RFC6464]
a=extmap:2 urn:ietf:params:rtp- hdrext:sdes:mid	[I-D.ietf-mmusic-sdp-bundle -negotiation]
a=candidate:0 1 UDP 2113667327 192.0.2.4 61665 typ host	[RFC5245]
a=candidate:1 1 UDP 694302207 203.0.113.141 54609 typ srflx raddr 192.0.2.4 rport 61665	[RFC5245]
a=candidate:0 1 UDP 2113667326 192.0.2.4 61667 typ host	[RFC5245]
a=candidate:1 1 UDP 1694302206 203.0.113.141 64678 typ srflx raddr 192.0.2.4 rport 61667	[RFC5245]
****** Video-1 m=line *********	*************************** **
m=video 0 UDP/TLS/RTP/SAVPF 98 100	bundle-only video line with port number set to zero
c=IN IP4 203.0.113.141	[RFC4566]
a=bundle-only	[I-D.ietf-mmusic-sdp-bundle -negotiation]
a=mid:m1	[RFC5888] Video m=line part of BUNDLE group
a=msid:ma tb	Identifies RTCMediaStream ID (ma) and RTCMediaStreamTrack ID (tb)
a=sendrecv	[RFC3264]
a=rtpmap:98 VP8/90000	[RFC7741]
a=imageattr:98 [x=1280,y=720]	[RFC6236]
a=fmtp:98 max-fr=30	[RFC4566]
a=rtcp-fb:* nack	[RFC5104]
a=rtcp-fb:* nack pli	[RFC5104]
a=rtcp-fb:* ccm fir	[RFC5104]

(*Continued*)

Table 18.41 (*Continued*) SDP Simulcast Bundle-Only (Copyright IETF)

a=extmap:2 urn:ietf:params:rtp-	[I-D.ietf-mmusic-sdp-bundle
hdrext:sdes:mid	-negotiation]
****** Video-2 m=line *********	***************************
	**
m=video 0 UDP/TLS/RTP/SAVPF 101 103	bundle-only video line with
	port number set to zero
c=IN IP4 203.0.113.141	[RFC4566]
a=bundle-only	[I-D.ietf-mmusic-sdp-bundle
	-negotiation]
a=mid:m2	[RFC5888] Video m=line part
	of BUNDLE group
a=msid:ma tc	Identifies RTCMediaStream
	ID (ma) and
	RTCMediaStreamTrack ID (tc)
a=sendrecv	[RFC3264]
a=rtpmap:101 H264/90000	[RFC6184]
a=rtpmap:103 H264/90000	[RFC6184]
a=fmtp:101 profile-level-id=4d0028	[RFC6184]Camera-2,Encoding-
;packetization-mode=1;max-fr=30	1 Resolution
a=rtcp-fb:* nack	[RFC5104]
a=rtcp-fb:* nack pli	[RFC5104]
a=rtcp-fb:* ccm fir	[RFC5104]
a=extmap:2 urn:ietf:params:rtp-	[I-D.ietf-mmusic-sdp-bundle
hdrext:sdes:mid	-negotiation]

Table 18.42 SDP Answer (Copyright IETF)

Answer SDP Contents	RFC#/Notes
v=0	[RFC4566]
o=- 20519 0 IN IP4 0.0.0.0	[RFC4566]
s=-	[RFC4566]
t=0 0	[RFC4566]
****** Audio m=line *********	***********************
	*****
m=audio 49203 UDP/TLS/RTP/SAVPF 109	[RFC4566]
c=IN IP4 203.0.113.141	[RFC4566]
a=rtcp:60065 IN IP4 203.0.113.141	[RFC3605]
a=sendrecv	[RFC3264]
a=rtpmap:109 opus/48000/2	[RFC7587]
a=maxptime:120	[RFC4566]
a=ice-ufrag:ufrag:c300d85b	[RFC5245]
a=ice-	[RFC5245]
pwd:de4e99bd291c325921d5d47efbabd9a2	
a=fingerprint:sha-256 6B:8B:F0:65:5F:7	[RFC5245]
8:E2:51:3B:AC:6F:F3:3F:46:1B:35 :DC:B8	
:5F:64:1A:24:C2:43:F0:A1:58:D0:A1:2C:1	

(*Continued*)

Table 18.42 (*Continued*) SDP Answer (Copyright IETF)

9:08	
a=setup:active	[RFC4145]
a=rtcp-rsize	[RFC5506]
a=rtcp-fb:109 nack	[RFC5104]
a=extmap:1 urn:ietf:params:rtp-hdrext :ssrc-audio-level	[RFC6464]
a=candidate:0 1 UDP 2113667327 198.51.100.7 51556 typ host	[RFC5245]
a=candidate:1 1 UDP 694302207 203.0.113.77 49203 typ srflx raddr 198.51.100.7 rport 51556	[RFC5245]
a=candidate:0 2 UDP 2113667326 198.51.100.7 51558 typ host	[RFC5245]
a=candidate:1 2 UDP 1694302206 203.0.113.77 60065 typ srflx raddr 198.51.100.7 rport 51558	[RFC5245]
****** Video m=line *********	****************************
m=video 0 UDP/TLS/RTP/SAVPF 98 100	Bob doesn't recognize bundle-only and hence the m=line is rejected implicitly due to port 0
****** Video m=line *********	****************************
m=video 0 UDP/TLS/RTP/SAVPF 98 100	Bob doesn't recognize bundle-only and hence the m=line is rejected implicitly due to port 0

18.2.5.4.4 Legacy Interop with RTP/Audio Video Profile

In the scenario described below, Alice is a legacy endpoint that sends (RFC 3264, see Section 3.1) an offer with RTP/Audio Video Profile (AVP) based audio and video descriptions along with DTLS fingerprint and RTCP feedback information.

On the other hand, Bob being a WebRTC endpoint follows the procedures in Section 5.1.2 of [I-D.ietf-rtcweb-jsep] and accepts Alice's offer for a DTLS-SRTP-based session with RTCP feedback (Tables 18.43 and 18.44).

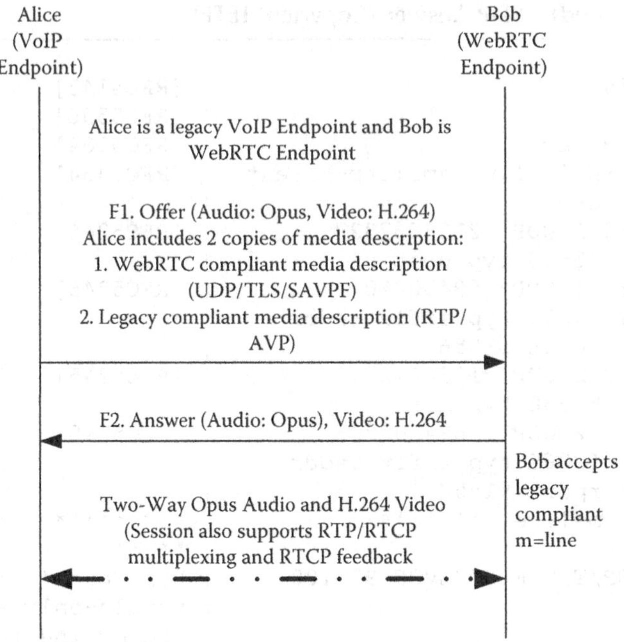

Successful 2-way WebRTC <-> VOIP interop

Table 18.43 SDP Offer (Copyright IETF)

Offer SDP Contents	RFC#/Notes
v=0	[RFC4566]
o=- 20518 0 IN IP4 0.0.0.0	[RFC4566]
s=-	[RFC4566]
t=0 0	[RFC4566]
a=ice-ufrag:074c6550	[RFC5245]
a=ice-pwd:a28a397a4c3f31747d1ee3474af08a068	[RFC5245]
a=rtcp-rsize	[RFC5506]
****** Audio m=line *********	****************************
m=audio 54732 RTP/AVP 109	[RFC4566]Alice includes RTP/AVP audio stream description
c=IN IP4 203.0.113.141	[RFC4566]
a=fingerprint:sha-256 19:E2:1C:3B:4B:9F:81:E6:B8:5C:F4:A5:A8:D8:73:04 :BB:05:2F:70:9F:04:A9:0E:05:E9:26:33:E8:70:88:A2	[RFC5245]
a=rtpmap:109 opus/48000	
a=ptime:20	
a=sendrecv	[RFC3264]

(Continued)

Table 18.43 (*Continued*) SDP Offer (Copyright IETF)

a=rtcp-mux	[RFC5761]Alice still includes RTP/RTCP Mux support
a=rtcp:64678 IN IP4 203.0.113.141	[RFC3605]
a=candidate:0 1 UDP 2113667327 192.0.2.4 54732 typ host	[RFC5245]
a=candidate:1 1 UDP 694302207 203.0.113.141 54732 typ srflx raddr 192.0.2.4 rport 54732	[RFC5245]
a=candidate:0 2 UDP 2113667326 192.0.2.4 64678 typ host	[RFC5245]
a=candidate:1 2 UDP 1694302206 203.0.113.141 64678 typ srflx raddr 192.0.2.4 rport 64678	[RFC5245]
a=rtcp-fb:109 nack	[RFC5104]She adds her intent for NACK RTCP feedback support
****** Video m=line *********	****************************
m=video 62445 RTP/AVP 120	[RFC4566]Alice includes RTP/AVP video stream description
c=IN IP4 203.0.113.141	[RFC4566]
a=fingerprint:sha-256 DC:B8:5F:64:1A:24:C2:43:F0:A1:58:D0:A1:2C:19:08 :6B:8B:F0:65:5F:78:E2:51:3B:AC:6F:F3:3F:46:1B:35	[RFC5245]
a=rtpmap:120 VP8/90000	[RFC7741]
a=sendrecv	[RFC3264]
a=rtcp-mux	[RFC5761]Alice intends to perform RTP/RTCP Mux
a=rtcp:54721 IN IP4 203.0.113.141	[RFC3605]
a=candidate:0 1 UDP 2113667327 192.0.2.4 62445 typ host	[RFC5245]
a=candidate:1 1 UDP 1694302207 203.0.113.141 62537 typ srflx raddr 192.0.2.4 rport 62445	[RFC5245]
a=candidate:0 2 2113667326 192.0.2.4 54721 typ host	[RFC5245]
a=candidate:1 2 UDP 1694302206 203.0.113.141 54721 typ srflx raddr 192.0.2.4 rport 54721	[RFC5245]
a=rtcp-fb:120 nack pli	[RFC5104] Alice indicates support for Picture loss Indication and NACK RTCP feedback
a=rtcp-fb:120 ccm fir	[RFC5104]

Table 18.44 SDP Answer (Copyright IETF)

Answer SDP Contents	RFC#/Notes
v=0	[RFC4566]
o=- 16833 0 IN IP4 0.0.0.0	[RFC4566]
s=-	[RFC4566]
t=0 0	[RFC4566]
****** Audio m=line *********	****************************
m=audio 49203 RTP/AVP 109	[RFC4566] Bob accepts RTP/AVP based audio stream
c=IN IP4 203.0.113.77	[RFC4566]
a=rtpmap:109 opus/48000	
a=ptime:20	
a=sendrecv	[RFC3264]
a=ice-ufrag:c300d85b	[RFC5245]
a=ice-pwd:de4e99bd291c325921d5d47efbabd9a2	[RFC5245]
a=fingerprint:sha-256 BB:05:2F:70:9F:04:A9:0E:05:E9:26:33:E8:70:88:A2 :19:E2:1C:3B:4B:9F:81:E6:B8:5C:F4:A5:A8:D8:73:04	[RFC5245]
a=rtcp-mux	[RFC5761]
a=candidate:0 1 UDP 2113667327 198.51.100.7 49203 typ host	[RFC5245]
a=candidate:1 1 UDP 1694302207 203.0.113.77 49203 typ srflx raddr 198.51.100.7 rport 49203	[RFC5245]
a=rtcp-fb:109 nack	[RFC5104]
****** Video m=line *********	****************************
m=video 63130 RTP/SAVP 120	[RFC4566] Bob accepts RTP/AVP based video stram
c=IN IP4 203.0.113.77	[RFC4566]
a=rtpmap:120 VP8/90000	[RFC7741]
a=sendrecv	[RFC3264]
a=ice-ufrag:e39091na	[RFC5245]
a=ice-pwd:dbc325921d5dd29e4e99147efbabd9a2	[RFC5245]
a=fingerprint:sha-256 BB:0A9:0E:05:E9:26:33:E8:70:88:A25:2F:70:9F:04: :19:E2:1C:3B:4B:9F:81:5:2F:70:9F:04::F4:A5:A8:D8:	[RFC5245]
a=rtcp-mux	[RFC5761]
a=candidate:0 1 UDP 2113667327 198.51.100.7 63130 typ host	[RFC5245]
a=candidate:1 1 UDP 1694302207 203.0.113.77 63130 typ srflx raddr 198.51.100.7 rport 63130	[RFC5245]
a=rtcp-fb:120 nack pli	[RFC5104]
a=rtcp-fb:120 ccm fir	[RFC5104]

18.2.6 Internet Assigned Numbers Authority Considerations

This document requires no actions from Internet Assigned Numbers Authority (IANA).

18.2.7 Security Considerations

The IETF has published separate documents [I-D. Ietf-rtcweb-security-arch] [I-D. Ietf-rtcweb-security] describing the security architecture for WebRTC as a whole. In addition, since the SDP offer and answer messages can contain private information about addresses and sessions to be established between parties, if this information needs to be kept private, some security mechanism (using TLS transport, for example) in the protocol used to carry the offers and answers must be used.

18.2.8 Appendix A of IETF Draft [I-D-ietf-rtcweb-sdp]

18.2.8.1 JSEP SDP Attributes Checklist

This section compiles a high-level checklist of the required SDP attributes to be verified against the examples defined in this specification. The goal here is to ensure that the examples are compliant to the rules defined in Section 5 of the [I-D. Ietf-rtcweb-jsep] specification.

18.2.8.1.1 Common Checklist

This subsection lists SDP attributes that mostly apply at the session level.

- v=0 MUST be the first SDP line.
- o= line MUST follow with values "-" for username, 64-bit value for session ID, and dummy values for "nettype," "addrtype," and "unicast-address" (for example: IN IP4 0.0.0.0).
- o= line MUST have the session version incremented in the cases of subsequent offers.
- s= MUST be the third line with the value of "-."
- t= line MUST follow with the values for "start-time" and "stop-time" set to zeroes.
- a=identity line MUST be included at the session level if the WebRTC identity mechanism is being used.
- a=ice-options:trickle MUST be present at the session level in all offers and answers when supported.

18.2.8.1.2 RTP Media Description Checklist

Following set of checklist items applies to RTP audio and video media descriptions.

- The media description's port value MUST either be set to a dummy value of "9" or use the port from the default candidate, if available.
- The media description's proto value MUST be "UDP/TLS/RTP/SAVPF" for JSEP offers.
- The JSEP answerer MUST support any combination of "RTP/[S]AVP[F]" for interoperability scenarios as defined in Section 5 of [I-D. Ietf-rtcweb-jsep].
- The "c=" line MUST be the first line in a media description. A dummy value of "IN IP 0.0.0.0" is set if there are no candidates gathered, or its value MUST match the default candidate.

- The a=mid attribute MUST be included.
- One of a=sendrecv/a=sendonly/a=recvonly/a=inactive SDP direction attributes MUST be present.
- a=rtpmap and a=fmtp attributes per primary, retransmission, and FEC media format MUST be included.
- a=rtcp-fb lines for each supported feedback mechanism MUST be included when using RTP with feedback.
- a=imageattr may be present for video media descriptions.
- a=msid line MUST be included for all media senders identifying the MediaStreamTrack (i.e., when a=sendonly/a=sendrecv attribute is present) [7].
- a=extmap line identifying the BUNDLE header extension MUST be present.
- a=extmap lines for other supported RTP header extensions MUST be included.
- a=rid line "per encoding" with the direction of "send" MUST be included when further constraining the media format or multiple encodings per media format is needed.
- a=simulcast line MUST be present if there are more than one "a=rid" lines for the media senders.
- a=bundle-only attribute MUST be present for media descriptions that are impacted by various bundle policies (such as max-bundle/balanced)
- For media descriptions that aren't "a=bundle-only" and that have unique addresses, the following attributes MUST be present:
  - a=ice-ufrag and a=ice-pwd
  - a=fingerprint
  - a=setup with value "actpass" in the offers and a value of "active"/"passive" in the answers.
  - a=dtls-id
  - a=rtcp
  - a=rtcp-mux
  - For offerers requiring RTCP to be multiplexed, a "a=rtcp-mux-only" line
  - a=rtcp-rsize
- a=group:BUNDLE line with all the "mid" identifiers part of the BUNDLE group is included at the session level.
- a=group:LS session-level attribute MUST be included with the "mid" identifiers that are part of the lip same synchronization group.

18.2.8.1.3 DataChannel Media Description Checklist

If a datachannel is required, an "application" type media description MUST be included with the following properties:

- Media description's proto value MUST be "DP/DTLS/SCTP" in the JSEP offers.
- A JSEP answerer MUST support reception of "UDP/DTLS/SCTP"/"TCP/DTLS/SCTP"/"DTLS/SCTP" for backwards compatibility reasons.
- A value of "webrtc-datachannel" MUST be used for the media description "fmt" value.
- a=mid line MUST be present.
- a=sctp-port with SCTP port number MUST be included.
- a=max-message-size MAY be included, if appropriate.

18.3 Summary

We have described the use cases of SDP attributes used by the SIP and WebRTC call control protocol for capability negotiations between conferencing parties. A key observation is that the SDP attributes have been used to negotiate call only between two parties. Even for the multipoint multimedia conference call with star-like connectivity topology known as centralized conferencing, the call is negotiated with call-leg individually. This is the shortcoming of the present SDP attributes call control protocol architecture. Moreover, the multipoint multimedia conferencing itself is highly complex, and it is difficult to meet all requirements for the distributed conference architecture.

We have shown the capability negotiations using SDP for SIP in many different scenarios: codec negotiation and selection, hold and resume, the addition and deletion of media streams, and 3PCC. By the same token, capability negations using SDP in WebRTC are explained for a variety of scenarios: audio-only, audio and video, data-only, audio-on-hold, audio with DTMF, one-way audio and video only, audio and video with bundle, and audio and video bundled while data is unbundled. In addition, more complex scenarios, like multisource multiresolution streams with FEC and RTX in a variety of conferencing features and configurations, are described for WebRTC. The audio with voice activity detection is discussed using WebRTC including interoperability with other call control protocols.

18.4 Problems

1. Provide the detailed call flows in SIP for negotiations of codecs between two parties where each party has two audio codecs, but only one of the codecs is common to both of them.
2. Include video codecs in Problem 1 where each party has the same video codec.
3. Include hold and resume features in both Problems 1 and 2.
4. Create a detailed scenario for third-party conferencing using SIP that include all features of Problems 1–3.
5. Add a data stream in Problem 4.
6. Do the call flows, video codec, hold and resume, third party conferencing, and adding a data stream like those of for Problems 1–5 using WebRTC call control protocol.
7. Create a scenario including call flows for both SIP and WebRTC call control protocol where audio and video are bundled.
8. Add a voice activity detection feature to Problem 7.
9. Create a scenario for multiresolution, FEC, and RTX with send-only simulcast session with two cameras and two encodings per camera (e.g., VP8 and H.264) for both SIP and WebRTC and explain in detail the session setup between two parties including call flows.
10. Include the Scalable Video Coding (SVC) video and video with retransmission in Problem 9.

References

1. [I-D-ietf-rtcweb-sdp] Nandakumar, S. and C. Jennings, "Annotated Example SDP for WebRTC," (work in progress), October 12, 2017.
2. [I-D. Ietf-mmusic-sdp-bundle-negotiation] Holmberg, C., Alvestrand, H., and C. Jennings, "Negotiating Media Multiplexing Using the Session Description Protocol (SDP)", draft-ietf-mmusic-sdp-bundle- negotiation-37 (work in progress), March 2017.

3. [I-D. Ietf-mmusic-sdp-simulcast] Burman, B., Westerlund, M., Nandakumar, S., and M. Zanaty, "Using Simulcast in SDP and RTP Sessions", draft-ietf-mmusic-sdp-simulcast-08 (work in progress), March 2017.
4. [I-D. Ietf-mmusic-rid] Thatcher, P., Zanaty, M., Nandakumar, S., Burman, B., Roach, A., and B. Campen, "RTP Payload Format Restrictions", draft-ietf-mmusic-rid-10 (work in progress), March 2017.
5. [I-D. Ietf-rtcweb-jsep] Uberti, J., Jennings, C., and E. Rescorla, "Javascript Session Establishment Protocol", draft-ietf-rtcweb-jsep-20 (work in progress), March 2017.
6. [I-D. Ietf-mmusic-trickle-ice] Ivov, E., Rescorla, E., and J. Uberti, "Trickle ICE: Incremental Provisioning of Candidates for the Interactive Connectivity Establishment (ICE) Protocol", draft-ietf-mmusic-trickle-ice-02 (work in progress), January 2015.
7. [I-D. Ietf-mmusic-msid] Alvestrand, H., "WebRTC MediaStream Identification in the Session Description Protocol", draft-ietf-mmusic-msid-16 (work in progress), February 2017.
8. [I-D.ietf-mmusic-sctp-sdp] Holmberg, C., Shpount, R., Loreto, S., and G. Camarillo, "Session Description Protocol (SDP) Offer/Answer Procedures for Stream Control Transmission Protocol (SCTP) over Datagram Transport Layer Security (DTLS) Transport", draft-ietf-mmusic-sctp-sdp-25 (work in progress), March 2017.
9. [I-D.ietf-rtcweb-data-channel] Jesup, R., Loreto, S., and M. Tuexen, "WebRTC Data Channels", draft-ietf-rtcweb-data-channel-13 (work in progress), January 2015.
10. [I-D.ietf-payload-flexible-fec-scheme] Singh, V., Begen, A., Zanaty, M., and G. Mandyam, "RTP Payload Format for Flexible Forward Error Correction (FEC)", draft-ietf-payload-flexible-fec-scheme-04 (work in progress), March 2017.
11. [I-D.ietf-mmusic-mux-exclusive] Holmberg, C., "Indicating Exclusive Support of RTP/RTCP Multiplexing Using SDP", draft-ietf-mmusic-mux-exclusive-11 (work in progress), February 2017.
12. [I-D.ietf-mmusic-dtls-sdp] Holmberg, C. and R. Shpount, "Using the SDP Offer/Answer Mechanism for DTLS", draft-ietf-mmusic-dtls-sdp-22 (work in progress), March 2017.
13. [I-D.ietf-rtcweb-security-arch] Rescorla, E., "WebRTC Security Architecture", draft-ietf-rtcweb-security-arch-12 (work in progress), June 2016.
14. [I-D.ietf-rtcweb-security] Rescorla, E., "Security Considerations for WebRTC", draft-ietf-rtcweb-security-08 (work in progress), February 2015.
15. [WebRTC] W3C, "WebRTC 1.0: Real-Time Communication between Browsers," http://dev.w3.org/2011/webrtc/editor/webrtc.html.

Appendix A: Augmented Backus-Naur Form

A.1 Overview

SIP uses Augmented Backus-Naur Form (ABNF) syntax for its messages. This section provides formal syntaxes and rules for ABNF specified in RFC 5234. Internet technical specifications often need to define a formal syntax and are free to employ whatever notation their authors deem useful. Over the years, a modified version of Backus-Naur Form (BNF), called ABNF, has become popular among the many Internet specifications. It balances compactness and simplicity with reasonable representational power. In the early days of the Arpanet, each specification contained its own definition of ABNF. This included the email specifications, RFC 0733, and then RFC 0822, which came to be the common citations for defining ABNF. The current RFC 5234 separates those definitions to permit selective reference.

Predictably, it also provides some modifications and enhancements. The differences between standard BNF and ABNF involve naming rules, repetition, alternatives, order-independence, and value ranges. Appendix B supplies rule definitions and encoding for a core lexical analyzer of the type common to several Internet specifications. It is provided as a convenience and is otherwise separate from the meta language defined in the body of this appendix and separate from its formal status.

A.2 Rule Definition

A.2.1 Rule Naming

The name of a rule is simply the name itself, that is, a sequence of characters, beginning with an alphabetical character and followed by a combination of alphabetics, digits, and hyphens (dashes).

NOTE: Rule names are case insensitive.

The names <rulename>, <Rulename>, <RULENAME>, and <rUlENamE> all refer to the same rule. Unlike original BNF, angle brackets ("<," ">") are not required. However, angle brackets may be used around a rule name whenever their presence facilitates discerning the use of a rule name. This is typically restricted to rule name references in freeform prose or to distinguish partial rules that combine to make a string not separated by white space, such as is shown in the discussion about repetition, below.

A.2.2 Rule Form

A rule is defined by the following sequence:

```
name    =   elements crlf
```

where <name> is the name of the rule, <elements> is one or more rule names or terminal specifications, and carriage return followed by line feed <crlf> is the end-of-line indicator. The equal sign separates the name from the definition of the rule. The elements form a sequence of one or more rule names and/or value definitions, combined according to the various operators defined in this document, such as alternative and repetition. For visual ease, rule definitions are left aligned. When a rule requires multiple lines, the continuation lines are indented. The left alignment and indentation are relative to the first lines of the ABNF rules and need not match the left margin of the document.

A.2.3 Terminal Values

Rules resolve into a string of terminal values, sometimes called characters. In ABNF, a character is merely a nonnegative integer. In certain contexts, a specific mapping (encoding) of values into a character set (such as ASCII) will be specified. Terminals are specified by one or more numeric characters, with the base interpretation of those characters indicated explicitly. The following bases are currently defined:

```
b   =   binary
d   =   decimal
x   =   hexadecimal
```

Hence: CR = %d13 and CR = %x0D, respectively, specify the decimal and hexadecimal representation of [US-ASCII] for carriage return. A concatenated string of such values is specified compactly, using a period (".") to indicate a separation of characters within that value. Hence:

```
CRLF = %d13.10
```

ABNF permits the specification of literal text strings directly, enclosed in quotation marks. Hence:

```
command   =    "command string"
```

Literal text strings are interpreted as a concatenated set of printable characters.

NOTE: ABNF strings are case insensitive, and the character set for these strings is US-ASCII. Hence:

```
rulename  = "abc"
```

and:

```
rulename  = "aBc"
```

will match "abc," "Abc," "aBc," "abC," "ABc," "aBC," "AbC," and "ABC."

To specify a rule that is case sensitive, specify the characters individually.

For example:

```
rulename    =    %d97 %d98 %d99
```

or

```
rulename    =    %d97.98.99
```

will match only the string that comprises only the lowercase characters, abc.

A.2.4 External Encodings

External representations of terminal value characters will vary according to constraints in the storage or transmission environment. Hence, the same ABNF-based grammar may have multiple external encodings, such as one for a 7-bit US-ASCII environment, another for a binary octet environment, and still a different one when 16-bit Unicode is used. Encoding details are beyond the scope of ABNF, although Appendix B provides definitions for a 7-bit US-ASCII environment as has been common to much of the Internet. By separating external encoding from the syntax, it is intended that alternate encoding environments can be used for the same syntax.

A.3 Operators

A.3.1 Concatenation: Rule1 Rule2

A rule can define a simple, ordered string of values (i.e., a concatenation of contiguous characters) by listing a sequence of rule names. For example:

```
foo         =    %x61 ; a
bar         =    %x62 ; b
mumble      =    foo bar foo
```

So that the rule <mumble> matches the lowercase string "aba." Linear white space: Concatenation is at the core of the ABNF parsing model. A string of contiguous characters (values) is parsed according to the rules defined in ABNF. For Internet specifications, there is some history of permitting linear white space (space and horizontal tab) to be freely and implicitly interspersed around major constructs, such as delimiting special characters or atomic strings.

NOTE: This specification for ABNF does not provide for implicit specification of linear white space.

Any grammar that wishes to permit linear white space around delimiters or string segments must specify it explicitly. It is often useful to provide for such white space in "core" rules that are then used variously among higher-level rules. The "core" rules might be formed into a lexical analyzer or simply be part of the main ruleset.

A.3.2 Alternatives: Rule1/Rule2

Elements separated by a forward slash ("/") are alternatives. Therefore, foo/bar will accept <foo> or <bar>.

NOTE: A quoted string containing alphabetic characters is a special form for specifying alternative characters and is interpreted as a nonterminal representing the set of combinatorial strings with the contained characters, in the specified order but with any mixture of upper- and lowercase.

A.3.3 Incremental Alternatives: Rule1 =/ Rule2

It is sometimes convenient to specify a list of alternatives in fragments. That is, an initial rule may match one or more alternatives, with later rule definitions adding to the set of alternatives. This is particularly useful for otherwise independent specifications that derive from the same parent ruleset, such as often occurs with parameter lists. ABNF permits this incremental definition through the construct:

```
oldrule  =   / additional-alternatives
```

So that the ruleset

```
ruleset  =  alt1 / alt2
ruleset  =  / alt3
ruleset  =  / alt4 / alt5
```

is the same as specifying

```
ruleset  =  alt1 / alt2 / alt3 / alt4 / alt5
```

A.3.4 Value Range Alternatives: %c##-##

A range of alternative numeric values can be specified compactly, using a dash ("-") to indicate the range of alternative values.

Hence:

```
DIGIT   =   %x30-39
```

is equivalent to:

```
DIGIT      =    "0" / "1" / "2" / "3" / "4" / "5" / "6" / "7" / "8" /
                "9"
```

Concatenated numeric values and numeric value ranges cannot be specified in the same string. A numeric value may use the dotted notation for concatenation, or it may use the dash notation to specify one value range. Hence, to specify one printable character between end-of-line sequences, the specification could be

```
char-line   =    %x0D.0A %x20-7E %x0D.0A
```

A.3.5 Sequence Group: (Rule1 Rule2)

Elements enclosed in parentheses are treated as a single element, whose contents are strictly ordered. Thus,

```
elem (foo / bar) blat
```

matches

```
(elem foo blat) or (elem bar blat),
```

and

```
elem foo / bar blat
```

matches

```
(elem foo) or (bar blat).
```

NOTE: It is strongly advised that grouping notation be used, rather than relying on the proper reading of "bare" alternations, when alternatives consist of multiple rule names or literals. Hence, it is recommended that the following form be used:

```
(elem foo) / (bar blat)
```

It will avoid misinterpretation by casual readers. The sequence group notation is also used within free text to set off an element sequence from the prose.

A.3.6 Variable Repetition: *Rule

The operator "*" preceding an element indicates repetition. The full form is:

```
<a>*<b>element
```

where <a> and <b> are optional decimal values, indicating at least <a> and at most <b> occurrences of the element.

Default values are 0 and infinity so that *<element> allows any number, including zero; 1*<element> requires at least one; 3*3<element> allows exactly 3; and 1*2<element> allows one or two.

A.3.7 Specific Repetition: nRule

A rule of the form:

```
<n>element
```

is equivalent to

```
<n>*<n>element
```

That is, exactly <n> occurrences of <element>. Thus, 2DIGIT is a 2-digit number, and 3ALPHA is a string of three alphabetic characters.

A.3.8 Optional Sequence: [RULE]

Square brackets enclose an optional element sequence:

```
[foo bar]
```

is equivalent to

```
*1(foo bar).
```

A.3.9 Comment: ; Comment

A semicolon starts a comment that continues to the end of line. This is a simple way of including useful notes in parallel with the specifications.

A.3.10 Operator Precedence

The various mechanisms thus far described have the following precedence, from highest (binding tightest) at the top, to lowest (loosest) at the bottom:

Rule name, prose-val, Terminal value

Comment
Value range
Repetition
Grouping, Optional
Concatenation
Alternative

Use of the alternative operator, freely mixed with concatenations, can be confusing. Again, it is recommended that the grouping operator be used to make explicit concatenation groups.

A.4 Definition of ABNF

NOTES:

1. This syntax requires a formatting of rules that is relatively strict. Hence, the version of a ruleset included in a specification might need preprocessing to ensure that it can be interpreted by an ABNF parser.
2. This syntax uses the rules provided in Appendix B.

```
rulelist       =   1*( rule / (*c-wsp c-nl) )

rule           =   rulename defined-as elements c-nl
                        ; continues if next line starts
                        ; with white space

rulename       =   ALPHA *(ALPHA / DIGIT / "-")

defined-as     =   *c-wsp ("=" / "=/") *c-wsp
                        ; basic rules definition and
                        ; incremental alternatives
elements       =   alternation *c-wsp
```

```
c-wsp          =  WSP / (c-nl WSP)

c-nl           =  comment / CRLF
                      ; comment or newline

comment        =  ";" *(WSP / VCHAR) CRLF

alternation    =  concatenation *(*c-wsp "/" *c-wsp concatenation)

concatenation  =  repetition *(1*c-wsp repetition)

repetition     =  [repeat] element

repeat         =  1*DIGIT / (*DIGIT "*" *DIGIT)

element        =  rulename / group / option /
                  char-val / num-val / prose-val

group          =  "(" *c-wsp alternation *c-wsp ")"

option         =  "[" *c-wsp alternation *c-wsp "]"

char-val       =  DQUOTE *(%x20-21 / %x23-7E) DQUOTE
                      ; quoted string of SP and VCHAR
                      ; without DQUOTE

num-val        =  "%" (bin-val / dec-val / hex-val)

bin-val        =  "b" 1*BIT
                  [ 1*("." 1*BIT) / ("-" 1*BIT) ]
                      ; series of concatenated bit values
                      ; or single ONEOF range

dec-val        =  "d" 1*DIGIT
                  [1*("." 1*DIGIT) / ("-" 1*DIGIT)]

hex-val        =  "x" 1*HEXDIG
                  [1*("." 1*HEXDIG) / ("-" 1*HEXDIG)]

prose-val      =  "<" *(%x20-3D / %x3F-7E) ">"
                      ; bracketed string of SP and VCHAR
                      ; without angles
                      ; prose description, to be used as
                      ; last resort
```

A.5 Core ABNF

This appendix contains some basic rules that are in common use. Basic rules are in uppercase. Note that these rules are only valid for ABNF encoded in 7-bit ASCII or in characters sets that are a superset of 7-bit ASCII.

A.5.1 Core Rules

Certain basic rules are in uppercase, such as SP, HTAB, CRLF, DIGIT, ALPHA, etc.

```
ALPHA          =    %x41-5A / %x61-7A ; A-Z / a-z

BIT            =    "0" / "1"

CHAR           =    %x01-7F
                          ; any 7-bit US-ASCII character,
                          ; excluding NUL

CR             =    %x0D
                          ; carriage return

CRLF           =    CR LF
                          ; Internet standard newline

CTL            =    %x00-1F / %x7F
                          ; controls

DIGIT          =    %x30-39
                          ; 0-9

DQUOTE         =    %x22
                          ; " (Double Quote)

HEXDIG         =    DIGIT / "A" / "B" / "C" / "D" / "E" / "F"

HTAB           =    %x09
                          ; horizontal tab
LF             =    %x0A
                          ; linefeed

LWSP           =    *(WSP / CRLF WSP)

                          ; Use of this linear-white-space rule
                          ; permits lines containing only white
                          ; space that are no longer legal in
                          ; mail headers and have caused
                          ; interoperability problems in other
                          ; contexts.
                          ; Do not use when defining mail
                          ; headers and use with caution in
                          ; other contexts.

OCTET          =    %x00-FF
                          ; 8 bits of data

SP             =    %x20
```

```
VCHAR          =    %x21-7E
                         ; visible (printing) characters

WSP            =    SP / HTAB
                          ; white space
```

A.5.2 Common Encoding

Externally, data are represented as "network virtual ASCII" (namely, 7-bit US-ASCII in an 8-bit field), with the high (8th) bit set to zero. A string of values is in "network byte order," in which the higher-valued bytes are represented on the left-hand side and are sent over the network first.

Appendix B: List of Relevant RFCs

RFCs

RFC	*Description*
0733	Standard for the format of ARPA network text messages. D. Crocker, J. Vittal, K.T. Pogran, D.A. Henderson. November 1977. (Obsoletes RFC 0724) (Made obsolete by RFC 0822) (Status: Unknown)
0768	User Datagram Protocol. J. Postel. August 1980. (Also STD 0006) (Status: Internet Standard)
0791	Internet Protocol. J. Postel. September 1981. (Obsoletes RFC0760) (Updated by RFC1349, RFC2474, RFC6864) (Also STD0005) (Status: Internet Standard)
0793	Transmission Control Protocol. J. Postel. September 1981. (Obsoletes RFC 0761) (Updated by RFC 1122, RFC 3168, RFC 6093, RFC 6528) (Also STD 0007) (Status: Internet Standard)
0822	Standard for the Format of ARPA Internet Text Messages. D. Crocker. August 1982. (Obsoletes RFC 0733) (Obsoleted by RFC 2822) (Updated by RFC 1123, RFC 2156, RFC 1327, RFC 1138, RFC 1148) (Also STD0011) (Status: Internet Standard)
1034	Domain Names—Concepts and Facilities. P.V. Mockapetris. November 1987. (Obsoletes RFC 0973, RFC 0882, RFC 0883) (Updated by RFC 1101, RFC 1183, RFC 1348, RFC 1876, RFC 1982, RFC 2065, RFC 2181, RFC 2308, RFC 2535, RFC 4033, RFC 4034, RFC 4035, RFC 4343, RFC 4035, RFC 4592, RFC 5936) (Also STD 0013) (Status: Internet Standard)
1035	Domain Names—Implementation and Specification. P.V. Mockapetris. November 1987. (Obsoletes RFC 0973, RFC 0882, RFC 0883) (Updated by RFC 1101, RFC 1183, RFC 1348, RFC 1876, RFC 1982, RFC 1995, RFC 1996, RFC 2065, RFC 2136, RFC 2181, RFC 2137, RFC 2308, RFC 2535, RFC 2673, RFC 2845, RFC 3425, RFC 3658, RFC 4033, RFC 4034, RFC 4035, RFC 4343, RFC 5936, RFC 5966, RFC 6604) (Also STD 0013) (Status: Internet Standard)
1112	Host extensions for IP multicasting. S.E. Deering. August 1989. (Obsoletes RFC 0988, RFC 1054) (Updated by RFC 2236) (Also STD0005) (Status: Internet Standard)
1305	Network Time Protocol (Version 3) Specification, Implementation and Analysis. D. Mills. March 1992. (Obsoletes RFC0958, RFC 1059, RFC 1119) (Made obsolete by RFC 5905) (Status: Draft Standard)

1319	The MD2 Message-Digest Algorithm. B. Kaliski. April 1992. (Made obsolete by RFC 6149) (Status: Historic)
1321	The MD5 Message-Digest Algorithm. R. Rivest. April 1992. (Updated by RFC 6151) (Status: Informational)
1750	Randomness Recommendations for Security. D. Eastlake 3rd, S. Crocker, J. Schiller. December 1994.(Made obsolete by RFC 4086) (Status: Informational)
1828	IP Authentication using Keyed MD5. P. Metzger, W. Simpson. August 1995. (Status: Historic)
1889	RTP: A Transport Protocol for Real-Time Applications. Audio-Video Transport Working Group, H. Schulzrinne, S. Casner, R. Frederick, V. Jacobson. January 1996. (Made obsolete by RFC 3550) (Status: Proposed Standard)
1890	RTP Profile for Audio and Video Conferences with Minimal Control. Audio-Video Transport Working Group, H. Schulzrinne. January 1996. (Made obsolete by RFC 3551) (Status: Proposed Standard)
1918	Address Allocation for Private Internets. Y. Rekhter, B. Moskowitz, D. Karrenberg, G. J. de Groot, E. Lear. February 1996. (Obsoletes RFC 1627, RFC 1597) (Updated by RFC6761) (Also BCP0005) (Status: Best Current Practice)
2015	MIME Security with Pretty Good Privacy (PGP). M. Elkins. October 1996. (Updated by RFC 3156) (Status: Proposed Standard)
2045	Multipurpose Internet Mail Extensions (MIME) Part One: Format of Internet Message Bodies. N. Freed, N. Borenstein. November 1996. (Obsoletes RFC 1521, RFC 1522, RFC 1590) (Updated by RFC 2184, RFC 2231, RFC 5335, RFC 6532) (Status: Draft Standard)
2046	Multipurpose Internet Mail Extensions (MIME) Part Two: Media Types. N. Freed, N. Borenstein. November 1996. (Obsoletes RFC 1521, RFC 1522, RFC 1590) (Updated by RFC 2646, RFC 3798, RFC 5147, RFC 6657) (Status: Proposed Standard)
2119	Key words for use in RFCs to Indicate Requirement Levels. S. Bradner. March 1997. (Updated by RFC 8174) (Also BCP0014) (Status: Best Current Practice)
2183	Communicating Presentation Information in Internet Messages: The Content-Disposition Header Field. R. Troost, S. Dorner, K. Moore, Ed. August 1997. (Obsoletes RFC 1806) (Updated by RFC 2184, RFC 2231) (Status: Proposed Standard)
2198	RTP Payload for Redundant Audio Data. C. Perkins, I. Kouvelas, O. Hodson, V. Hardman, M. Handley, J.C. Bolot, A. Vega-Garcia, S. Fosse-Parisis. September 1997. (Updated by RFC 6354) (Status: Proposed Standard)
2205	Resource ReSerVation Protocol (RSVP) - Version 1 Functional Specification. R. Braden, Ed., L. Zhang, S. Berson, S. Herzog, S. Jamin. September 1997. (Updated by RFC 2750, RFC 3936, RFC 4495, RFC 5946, RFC 6437, RFC 6780) (Status: Proposed Standard)
2234	Augmented BNF for Syntax Specifications: ABNF. D. Crocker, Ed., P. Overell. November 1997. (Made obsolete by RFC 4234) (Status: Proposed Standard)
2246	The TLS Protocol Version 1.0. T. Dierks, C. Allen. January 1999. (Obsoleted by RFC 4346) (Updated by RFC 3546, RFC 5746, RFC 6176, RFC 7465, RFC 7507, RFC 7919) (Status: Proposed Standard)
2248	Network Services Monitoring MIB. N. Freed, S. Kille. January 1998. (Obsoletes RFC 1565) (Obsoleted by RFC 2788) (Status: Proposed Standard)
2279	UTF-8, a transformation format of ISO 10646. F. Yergeau. January 1998. (Obsoletes RFC 2044) (Made obsolete by RFC 3629) (Status: Draft Standard)

2326	Real Time Streaming Protocol (RTSP). H. Schulzrinne, A. Rao, R. Lanphier. April 1998. (Status: Proposed Standard)
2327	SDP: Session Description Protocol. M. Handley, V. Jacobson. April 1998. (Made obsolete by RFC 4566) (Updated by RFC 3266) (Status: Proposed Standard)
2373	IP Version 6 Addressing Architecture. R. Hinden, S. Deering. July 1998. (Obsoletes RFC 1884) (Made obsolete by RFC 3513) (Status: Proposed Standard)
2387	The MIME Multipart/Related Content-type. E. Levinson. August 1998. (Obsoletes RFC 2112) (Status: Proposed Standard)
2392	Content-ID and Message-ID Uniform Resource Locators. E. Levinson. August 1998. (Obsoletes RFC 2111) (Status: Proposed Standard)
2396	Uniform Resource Identifiers (URI): Generic Syntax. T. Berners-Lee, R. Fielding, L. Masinter. August 1998. (Made obsolete by RFC 3986) (Updates RFC 1808, RFC 1738) (Updated by RFC 2732) (Status: Draft Standard)
2401	Security Architecture for the Internet Protocol. S. Kent, R. Atkinson. November 1998. (Obsoletes RFC 1825) (Made obsolete by RFC 4301) (Updated by RFC 3168) (Status: Proposed Standard)
2421	Voice Profile for Internet Mail - version 2. G. Vaudreuil, G. Parsons. September 1998. (Obsoletes RFC 1911) (Made obsolete by RFC 3801) (Status: Proposed Standard)
2434	Guidelines for Writing an IANA Considerations Section in RFCs. T. Narten, H. Alvestrand. October 1998. (Made obsolete by RFC 5226) (Updated by RFC 3692) (Status: Best Current Practice)
2458	Toward the PSTN/Internet Inter-Networking--Pre-PINT Implementations. H. Lu, M. Krishnaswamy, L. Conroy, S. Bellovin, F. Burg, A. DeSimone, K. Tewani, P. Davidson, H. Schulzrinne, K. Vishwanathan. November 1998. (Status: Informational)
2460	Internet Protocol, Version 6 (IPv6) Specification. S. Deering, R. Hinden. December 1998. (Obsoletes RFC 1883) (Made obsolete by RFC 8200) (Updated by RFC 5095, RFC 5722, RFC 5871, RFC 6437, RFC 6564, RFC 6935, RFC 6946, RFC 7045, RFC 7112) (Status: Draft Standard)
2468	I Remember IANA. V. Cerf. October 1998. (Status: Informational)
2475	An Architecture for Differentiated Services. S. Blake, D. Black, M. Carlson, E. Davies, Z. Wang, W. Weiss. December 1998. (Updated by RFC 3260) (Status: Informational)
2507	IP Header Compression. M. Degermark, B. Nordgren, S. Pink. February 1999. (Status: Proposed Standard)
2542	Terminology and Goals for Internet Fax. L. Masinter. March 1999. (Status: Informational)
2543	SIP: Session Initiation Protocol. M. Handley, H. Schulzrinne, E. Schooler, J. Rosenberg. March 1999. (Made obsolete by RFC 3261, RFC 3262, RFC 3263, RFC 3264, RFC 3265) (Status: Proposed Standard)
2616	Hypertext Transfer Protocol -- HTTP/1.1. R. Fielding, J. Gettys, J. Mogul, H. Frystyk, L. Masinter, P. Leach, T. Berners-Lee. June 1999. (Obsoletes RFC 2068) (Made obsolete by RFC 7230, RFC 7231, RFC 7232, RFC 7233, RFC 7234, RFC 7235) (Updated by RFC 2817, RFC 5785, RFC 6266, RFC 0585) (Status: Draft Standard)
2617	HTTP Authentication: Basic and Digest Access Authentication. J. Franks, P. Hallam-Baker, J. Hostetler, S. Lawrence, P. Leach, A. Luotonen, L. Stewart. June 1999. (Obsoletes RFC 2069) (Made obsolete by RFC 7235, RFC 7615, RFC 7616, RFC 7617) (Status: Draft Standard)

2705	Media Gateway Control Protocol (MGCP) Version 1.0. M. Arango, A. Dugan, I. Elliott, C. Huitema, S. Pickett. October 1999. (Made obsolete by RFC 3435) (Updated by RFC 3660) (Status: Informational)
2732	Format for Literal IPv6 Addresses in URL's. R. Hinden, B. Carpenter, L. Masinter. December 1999. (Made obsolete by RFC 3986) (Updates RFC 2396) (Status: Proposed Standard)
2733	An RTP Payload Format for Generic Forward Error Correction. J. Rosenberg, H. Schulzrinne. December 1999. (Made obsolete by RFC 5109) (Status: Proposed Standard)
2766	Network Address Translation - Protocol Translation (NAT-PT). G. Tsirtsis, P. Srisuresh. February 2000. (Made obsolete by RFC 4966) (Updated by RFC 3152) (Status: Historic)
2818	HTTP Over TLS. E. Rescorla. May 2000. (Updated by RFC 5785, RFC 7230) (Status: Informational)
2822	Internet Message Format. P. Resnick, Ed. April 2001. (Obsoletes RFC 0822) (Made obsolete by RFC 5322) (Updated by RFC 5335, RFC 5336) (Status: Proposed Standard)
2827	Network Ingress Filtering: Defeating Denial of Service Attacks which employ IP Source Address Spoofing. P. Ferguson, D. Senie. May 2000. (Obsoletes RFC 2267) (Updated by RFC3704) (Also BCP0038) (Status: Best Current Practice)
2828	Internet Security Glossary. R. Shirey. May 2000. bytes) (Made obsolete by RFC 4949) (Status: Informational)
2833	RTP Payload for DTMF Digits, Telephony Tones and Telephony Signals. H. Schulzrinne, S. Petrack. May 2000. (Made obsolete by RFC 4733, RFC 4734) (Status: Proposed Standard)
2848	The PINT Service Protocol: Extensions to SIP and SDP for IP Access to Telephone Call Services. S. Petrack, L. Conroy. June 2000. (Status: Proposed Standard)
2871	A Framework for Telephony Routing over IP. J. Rosenberg, H. Schulzrinne. June 2000. (Status: Informational)
2904	AAA Authorization Framework. J. Vollbrecht, P. Calhoun, S. Farrell, L. Gommans, G. Gross, B. de Bruijn, C. de Laat, M. Holdrege, D. Spence. August 2000. (Status: Informational)
2914	Congestion Control Principles. S. Floyd. September 2000. (Updated by RFC 7141) (Also BCP 0041) (Status: Best Current Practice)
2974	Session Announcement Protocol. M. Handley, C. Perkins, E. Whelan. October 2000. (Status: Experimental)
3006	Integrated Services in the Presence of Compressible Flows. B. Davie, C. Iturralde, D. Oran, S. Casner, J. Wroclawski. November 2000. (Status: Proposed Standard)
3015	Megaco Protocol Version 1.0. F. Cuervo, N. Greene, A. Rayhan, C. Huitema, B. Rosen, J. Segers. November 2000. (Obsoletes RFC 2885, RFC 2886) (Made obsolete by RFC 3525) (Status: Proposed Standard)
3016	RTP Payload Format for MPEG-4 Audio/Visual Streams. Y. Kikuchi, T. Nomura, S. Fukunaga, Y. Matsui, H. Kimata. November 2000. Made obsolete by RFC 6416) (Status: Proposed Standard)
3056	Connection of IPv6 Domains via IPv4 Clouds. B. Carpenter, K. Moore. February 2001. (Status: Proposed Standard)

3066	Tags for the Identification of Languages. H. Alvestrand. January 2001. (Obsoletes RFC 1766) (Made obsolete by RFC 4646, RFC 4647) (Status: Best Current Practice)
3102	Realm Specific IP: Framework. M. Borella, J. Lo, D. Grabelsky, G. Montenegro. October 2001. (Status: Experimental)
3103	Realm Specific IP: Protocol Specification. M. Borella, D. Grabelsky, J. Lo, K. Taniguchi. October 2001. (Status: Experimental)
3108	Conventions for the use of the Session Description Protocol (SDP) for ATM Bearer Connections. R. Kumar, M. Mostafa. May 2001. (Status: Proposed Standard)
3174	US Secure Hash Algorithm 1 (SHA1). D. Eastlake 3rd, P. Jones. September 2001. (Updated by RFC 4634, RFC 6234) (Status: Informational)
3235	Network Address Translator (NAT)-Friendly Application Design Guidelines. D. Senie. January 2002. (Status: Informational)
3250	Tag Image File Format Fax eXtended (TIFF-FX) - image/tiff-fx MIME Sub-type Registration. L. McIntyre, G. Parsons, J. Rafferty. September 2002. (Made obsolete by RFC 3950) (Status: Proposed Standard)
3261	SIP: Session Initiation Protocol. J. Rosenberg, H. Schulzrinne, G. Camarillo, A. Johnston, J. Peterson, R. Sparks, M. Handley, E. Schooler. June 2002. (Obsoletes RFC 2543) (Updated by RFC 3265, RFC 3853, RFC 4320, RFC 4916, RFC 5393, RFC 5621, RFC 5626, RFC 5630, RFC 5922, RFC 5954, RFC 6026, RFC 6141) (Status: Proposed Standard)
3262	Reliability of Provisional Responses in Session Initiation Protocol (SIP). J. Rosenberg, H. Schulzrinne. June 2002. (Obsoletes RFC 2543) (Status: Proposed Standard)
3264	An Offer/Answer Model with Session Description Protocol (SDP). J. Rosenberg, H. Schulzrinne. June 2002. (Obsoletes RFC 2543) (Updated by RFC 6157) (Status: Proposed Standard)
3266	Support for IPv6 in Session Description Protocol (SDP). S. Olson, G. Camarillo, A. B. Roach. June 2002. (Made obsolete by RFC 4566) (Updates RFC 2327) (Status: Proposed Standard)
3267	Real-Time Transport Protocol (RTP) Payload Format and File Storage Format for the Adaptive Multi-Rate (AMR) and Adaptive Multi-Rate Wideband (AMR-WB) Audio Codecs. J. Sjoberg, M. Westerlund, A. Lakaniemi, Q. Xie. June 2002. (Made obsolete by RFC 4867) (Status: Proposed Standard)
3279	Algorithms and Identifiers for the Internet X.509 Public Key Infrastructure Certificate and Certificate Revocation List (CRL) Profile. L. Bassham, W. Polk, R. Housley. April 2002. (Updated by RFC 4055, RFC 4491, RFC 5480, RFC 5758) (Status: Proposed Standard)
3280	Internet X.509 Public Key Infrastructure Certificate and Certificate Revocation List (CRL) Profile. R. Housley, W. Polk, W. Ford, D. Solo. April 2002. (Obsoletes RFC 2459) (Made obsolete by RFC 5280) (Updated by RFC 4325, RFC 4630) (Status: Proposed Standard)
3303	Middlebox Communication Architecture and Framework. P. Srisuresh, J. Kuthan, J. Rosenberg, A. Molitor, A. Rayhan. August 2002. (Status: Informational)
3311	The Session Initiation Protocol (SIP) UPDATE Method. J. Rosenberg. October 2002. (Status: Proposed Standard)
3312	Integration of Resource Management and Session Initiation Protocol (SIP). G. Camarillo, Ed., W. Marshall, Ed., J. Rosenberg. October 2002. (Updated by RFC 4032, RFC 5027) (Status: Proposed Standard)

3313	Private Session Initiation Protocol (SIP) Extensions for Media Authorization. W. Marshall, Ed. January 2003. (Status: Informational)
3315	Dynamic Host Configuration Protocol for IPv6 (DHCPv6). R. Droms, Ed., J. Bound, B. Volz, T. Lemon, C. Perkins, M. Carney. July 2003. (Updated by RF C4361, RFC 5494, RFC 6221, RFC 6422, RFC 6644, RFC7083, RFC 7227, RFC 7283, RFC 7550) (Status: Proposed Standard)
3320	Signaling Compression (SigComp). R. Price, C. Bormann, J. Christoffersson, H. Hannu, Z. Liu, J. Rosenberg. January 2003. (Updated by RFC 4896) (Status: Proposed Standard)
3323	A Privacy Mechanism for the Session Initiation Protocol (SIP). J. Peterson. November 2002. (Status: Proposed Standard)
3369	Cryptographic Message Syntax (CMS). R. Housley. August 2002. (Obsoletes RFC 2630, RFC 3211) (Made obsolete by RFC 3852) (Status: Proposed Standard)
3376	Internet Group Management Protocol, Version 3. B. Cain, S. Deering, I. Kouvelas, B. Fenner, A. Thyagarajan. October 2002. (Updates RFC 2236) (Updated by RFC 4604) (Status: Proposed Standard)
3388	Grouping of Media Lines in the Session Description Protocol (SDP). G. Camarillo, G. Eriksson, J. Holler, H. Schulzrinne. December 2002. (Made obsolete by RFC 5888) (Status: Proposed Standard)
3389	Real-time Transport Protocol (RTP) Payload for Comfort Noise (CN). R. Zopf. September 2002. (Status: Proposed Standard
3407	Session Description Protocol (SDP) Simple Capability Declaration. F. Andreasen. October 2002. (Status: Proposed Standard)
3424	IAB Considerations for UNilateral Self-Address Fixing (UNSAF) Across Network Address Translation. L. Daigle, Ed., IAB. November 2002. (Status: Informational)
3428	Session Initiation Protocol (SIP) Extension for Instant Messaging. B. Campbell, Ed., J. Rosenberg, H. Schulzrinne, C. Huitema, D. Gurle. December 2002. (Status: Proposed Standard)
3484	Default Address Selection for Internet Protocol version 6 (IPv6). R. Draves. February 2003. (Made obsolete by RFC 6724) (Status: Proposed Standard)
3485	The Session Initiation Protocol (SIP) and Session Description Protocol (SDP) Static Dictionary for Signaling Compression (SigComp). M. Garcia-Martin, C. Bormann, J. Ott, R. Price, A.B. Roach. February 2003. (Updated by RFC 4896) (Status: Proposed Standard)
3489	STUN - Simple Traversal of User Datagram Protocol (UDP) through Network Address Translators (NATs). J. Rosenberg, J. Weinberger, C. Huitema, R. Mahy. March 2003. (Made obsolete by RFC 5389) (Status: Proposed Standard)
3490	Internationalizing Domain Names in Applications (IDNA). P. Faltstrom, P. Hoffman, A. Costello. March 2003. (Made obsolete by RFC 5890, RFC 5891) (Status: Proposed Standard)
3513	Internet Protocol Version 6 (IPv6) Addressing Architecture. R. Hinden, S. Deering. April 2003. (Obsoletes RFC 2373) (Made obsolete by RFC 4291) (Status: Proposed Standard)
3515	The Session Initiation Protocol (SIP) Refer Method. R. Sparks. April 2003. (Updated by RFC 7647, RFC 8217) (Status: Proposed Standard)

3524	Mapping of Media Streams to Resource Reservation Flows. G. Camarillo, A. Monrad. April 2003. (Status: Proposed Standard)
3525	Gateway Control Protocol Version 1. C. Groves, Ed., M. Pantaleo, Ed., T. Anderson, Ed., T. Taylor, Ed. June 2003. (Obsoletes RFC 3015) (Made obsolete by RFC 5125) (Status: Historic)
3547	The Group Domain of Interpretation. M. Baugher, B. Weis, T. Hardjono, H. Harney. July 2003. (Made obsolete by RFC 6407) (Status: Proposed Standard)
3548	The Base16, Base32, and Base64 Data Encodings. S. Josefsson, Ed. July 2003. (Obsoleted by RFC 4648) (Status: Informational)
3550	RTP: A Transport Protocol for Real-Time Applications. H. Schulzrinne, S. Casner, R. Frederick, V. Jacobson. July 2003. (Obsoletes RFC 1889) (Updated by RFC 5506, RFC 5761, RFC 6051, RFC 6222, RFC 7022, RFC 7160, RFC 7164) (Also STD 0064) (Status: Internet Standard)
3551	RTP Profile for Audio and Video Conferences with Minimal Control. H. Schulzrinne, S. Casner. July 2003. (Obsoletes RFC 1890) (Updated by RFC 5761, RFC 7007) (Also STD 0065) (Status: Internet Standard)
3556	Session Description Protocol (SDP) Bandwidth Modifiers for RTP Control Protocol (RTCP) Bandwidth. S. Casner. July 2003. (Status: Proposed Standard)
3569	3569 An Overview of Source-Specific Multicast (SSM). S. Bhattacharyya, Ed., July 2003. (Status: Informational)
3605	Real Time Control Protocol (RTCP) attribute in Session Description Protocol (SDP). C. Huitema. October 2003. (Status: Proposed Standard)
3611	RTP Control Protocol Extended Reports (RTCP XR). T. Friedman, Ed., R. Caceres, Ed., A. Clark, Ed., November 2003. (Status: Proposed Standard)
3629	UTF-8, a transformation format of ISO 10646. F. Yergeau. November 2003. (Obsoletes RFC 2279) (Also STD0063) (Status: Internet Standard)
3678	Socket Interface Extensions for Multicast Source Filters. D. Thaler, B. Fenner, B. Quinn. January 2004. (Status: Informational)
3711	The Secure Real-time Transport Protocol (SRTP). M. Baugher, D. McGrew, M. Naslund, E. Carrara, K. Norrman. March 2004. (Updated by RFC 5506, RFC 6904) (Status: Proposed Standard)
3725	Best Current Practices for Third Party Call Control (3pcc) in the Session Initiation Protocol (SIP). J. Rosenberg, J. Peterson, H. Schulzrinne, G. Camarillo. April 2004. (Status: Best Current Practice)
3801	Voice Profile for Internet Mail - version 2 (VPIMv2). G. Vaudreuil G. Parsons. June 2004. (Obsoletes RFC 2421, RFC 2423) (Status: Draft Standard)
3830	MIKEY: Multimedia Internet KEYing. J. Arkko, E. Carrara, F. Lindholm, M. Naslund, K. Norrman. August 2004. (Updated by RFC 4738, RFC 6309) (Status: Proposed Standard)
3841	Caller Preferences for the Session Initiation Protocol (SIP). J. Rosenberg, H. Schulzrinne, P. Kyzivat. August 2004. (Status: Proposed Standard)
3850	Secure/Multipurpose Internet Mail Extensions (S/MIME) Version 3.1 Certificate Handling. B. Ramsdell, Ed. July 2004. (Obsoletes RFC 2632) (Made obsolete by RFC 5750) (Status: Proposed Standard)

3851	Secure/Multipurpose Internet Mail Extensions (S/MIME) Version 3.1 Message Specification. B. Ramsdell, Ed. July 2004. (Obsoletes RFC 2633) (Made obsolete by RFC 5751) (Status: Proposed Standard)
3853	S/MIME Advanced Encryption Standard (AES) Requirement for the Session Initiation Protocol (SIP). J. Peterson. July 2004. (Updates RFC 3261)(Status: Proposed Standard)
3862	Common Presence and Instant Messaging (CPIM): Message Format. G. Klyne, D. Atkins. August 2004. (Status: Proposed Standard)
3890	A Transport Independent Bandwidth Modifier for the Session Description Protocol (SDP). M. Westerlund. September 2004. (Status: Proposed Standard)
3925	Vendor-Identifying Vendor Options for Dynamic Host Configuration Protocol version 4 (DHCPv4). J. Littlefield. October 2004. (Status: Proposed Standard)
3935	A Mission Statement for the IETF. H. Alvestrand. October 2004. (Also BCP0095) (Status: Best Current Practice)
3947	Negotiation of NAT-Traversal in the IKE. T. Kivinen, B. Swander, A. Huttunen, V. Volpe. January 2005. (Status: Proposed Standard)
3948	UDP Encapsulation of IPsec ESP Packets. A. Huttunen, B. Swander, V. Volpe, L. DiBurro, M. Stenberg. January 2005. (Status: Proposed Standard)
3950	Tag Image File Format Fax eXtended (TIFF-FX) - image/tiff-fx MIME Sub-type Registration. L. McIntyre, G. Parsons, J. Rafferty. February 2005. (Obsoletes RFC 3250) (Status: Draft Standard)
3952	Real-time Transport Protocol (RTP) Payload Format for internet Low Bit Rate Codec (iLBC) Speech. A. Duric, S. Andersen. December 2004. (Status: Experimental)
3960	Early Media and Ringing Tone Generation in the Session Initiation Protocol (SIP). G. Camarillo, H. Schulzrinne. December 2004. (Status: Informational)
3966	The tel URI for Telephone Numbers. H. Schulzrinne. December 2004. (Status: Proposed Standard) (Obsoletes RFC 2806) (Updated by RFC 5341) (Status: Proposed Standard)
3984	RTP Payload Format for H.264 Video. S. Wenger, M.M. Hannuksela, T. Stockhammer, M. Westerlund, D. Singer. February 2005. (Obsoleted by RFC 6184) (Status: Proposed Standard)
3986	Uniform Resource Identifier (URI): Generic Syntax. T. Berners-Lee, R. Fielding, L. Masinter.January 2005. (Obsoletes RFC 2732, RFC 2396, RFC 1808) (Updates RFC 1738) (Updated by RFC 6874, RFC 7320) (Also STD 0066) (Status: Internet Standard)
4012	Routing Policy Specification Language Next generation (RPSLng). L. Blunk, J. Damas, F. Parent, A. Robachevsky. March 2005. (Updates RFC 2725, RFC 2622) (Updated by RFC 7909) (Status: Proposed Standard)
4028	Session Timers in the Session Initiation Protocol (SIP). S. Donovan, J. Rosenberg. April 2005. (Status: Proposed Standard)
4032	Update to the Session Initiation Protocol (SIP) Preconditions Framework. G. Camarillo, P. Kyzivat. March 2005. (Updates RFC 3312) (Status: Proposed Standard)
4055	Additional Algorithms and Identifiers for RSA Cryptography for use in the Internet X.509 Public Key Infrastructure Certificate and Certificate Revocation List (CRL) Profile. J. Schaad, B. Kaliski, R. Housley. June 2005. (Updates RFC 3279) (Updated by RFC 5756) (Status: Proposed Standard)

4086	Randomness Requirements for Security. D. Eastlake 3rd, J. Schiller, S. Crocker. June 2005. (Obsoletes RFC 1750) (Also BCP0106) (Status: Best Current Practice)
4091	The Alternative Network Address Types (ANAT) Semantics for the Session Description Protocol (SDP) Grouping Framework. G. Camarillo, J. Rosenberg. June 2005. (Made obsolete by RFC 5245) (Status: Proposed Standard)
4092	Usage of the Session Description Protocol (SDP) Alternative Network Address Types (ANAT) Semantics in the Session Initiation Protocol (SIP). G. Camarillo, J. Rosenberg. June 2005. (Made obsolete by RFC 5245) (Status: Proposed Standard)
4103	RTP Payload for Text Conversation. G. Hellstrom, P. Jones. June 2005. (Obsoletes RFC 2793) (Status: Proposed Standard)
4120	The Kerberos Network Authentication Service (V5). C. Neuman, T. Yu, S. Hartman, K. Raeburn. July 2005. (Obsoletes RFC 1510) (Updated by RFC 4537, RFC 5021, RFC 5896, RFC 6111, RFC 6112, RFC 6113, RFC 6649, RFC 6806, RFC 7751, RFC 8062, RFC 8129) (Status: Proposed Standard)
4115	A Differentiated Service Two-Rate, Three-Color Marker with Efficient Handling of in-Profile Traffic. O. Aboul-Magd, S. Rabie. July 2005. (Status: Informational)
4145	TCP-Based Media Transport in the Session Description Protocol (SDP). D. Yon, G. Camarillo. September 2005. (Updated by RFC 4572) (Status: Proposed Standard)
4189	Requirements for End-to-Middle Security for the Session Initiation Protocol (SIP). K. Ono, S. Tachimoto. October 2005. (Status: Informational)
4234	Augmented BNF for Syntax Specifications: ABNF. D. Crocker, Ed., P. Overell. October 2005. (Obsoletes RFC 2234) (Made Obsolete by RFC 5234) (Status: Draft Standard)
4251	The Secure Shell (SSH) Protocol Architecture. T. Ylonen, C. Lonvick, Ed. January 2006. (Status: Proposed Standard)
4291	IP Version 6 Addressing Architecture. R. Hinden, S. Deering. February 2006. (Obsoletes RFC 3513) (Updated by RFC 5952, RFC 6052, RFC 7136, RFC 7346, RFC 7371, RFC 8064) (Status: Draft Standard)
4301	Security Architecture for the Internet Protocol. S. Kent, K. Seo. December 2005. (Obsoletes RFC 2401) (Updates RFC 3168) (Updated by RFC 6040) (Status: Proposed Standard)
4303	IP Encapsulating Security Payload (ESP). S. Kent. December 2005. (Obsoletes RFC 2406) (Status: Proposed Standard)
4306	4306 Internet Key Exchange (IKEv2) Protocol. C. Kaufman, Ed. December 2005. (Obsoletes RFC 2407, RFC 2408, RFC 2409) (Made obsolete by RFC 5996) (Updated by RFC 5282) (Status: Proposed Standard)
4312	The Camellia Cipher Algorithm and Its Use With IPsec. A. Kato, S. Moriai, M. Kanda. December 2005. (Status: Proposed Standard)
4317	Session Description Protocol (SDP) Offer/Answer Examples. A. Johnston, R. Sparks. December 2005. (Status: Informational)
4340	Datagram Congestion Control Protocol (DCCP). E. Kohler, M. Handley, S. Floyd. March 2006. (Updated by RFC 5595, RFC 5596, RFC 6335, RFC 6773) (Status: Proposed Standard)
4346	The Transport Layer Security (TLS) Protocol Version 1.1. T. Dierks, E. Rescorla. April 2006. (Obsoletes RFC 2246) (Made obsolete by RFC 5246) (Updated by RFC 4366, RFC 4680, RFC 4681, RFC 5746, RFC 6176, RFC 7465) (Status: Proposed Standard)

4430	Kerberized Internet Negotiation of Keys (KINK). S. Sakane, K. Kamada, M. Thomas, J. Vilhuber. March 2006. (Status: Proposed Standard)
4434	The AES-XCBC-PRF-128 Algorithm for the Internet Key Exchange Protocol (IKE). P. Hoffman. February 2006. (Obsoletes RFC 3664) (Status: Proposed Standard)
4474	Enhancements for Authenticated Identity Management in the Session Initiation Protocol (SIP). J. Peterson, C. Jennings. August 2006. (Status: Proposed Standard)
4483	A Mechanism for Content Indirection in\ Session Initiation Protocol (SIP) Messages. E. Burger, Ed. May 2006. (Status: Proposed Standard)
4566	Session Description Protocol. M. Handley, V. Jacobson, C. Perkins. July 2006. (Obsoletes RFC 2327, RFC 3266) (Status: Proposed Standard)
4567	Key Management Extensions for Session Description Protocol (SDP) and Real Time Streaming Protocol (RTSP). J. Arkko, F. Lindholm, M. Naslund, K. Norrman, E. Carrara. July 2006. (Status: Proposed Standard)
4568	Session Description Protocol (SDP) Security Descriptions for Media Streams. F. Andreasen, M. Baugher, D. Wing. July 2006. (Status: Proposed Standard)
4570	Session Description Protocol (SDP) Source Filters. B. Quinn, R. Finlayson. July 2006. (Status: Proposed Standard)
4571	Framing Real-time Transport Protocol (RTP) and RTP Control Protocol (RTCP) Packets over Connection-Oriented Transport. J. Lazzaro. July 2006. (Status: Proposed Standard)
4572	Connection-Oriented Media Transport over the Transport Layer Security (TLS) Protocol in the Session Description Protocol (SDP). J. Lennox. July 2006. (Updates RFC 4145) (Status: Proposed Standard)
4574	The Session Description Protocol (SDP) Label Attribute. O. Levin, G. Camarillo. August 2006. (Status: Proposed Standard)
4575	A Session Initiation Protocol (SIP) Event Package for Conference State. J. Rosenberg, H. Schulzrinne, O. Levin, Ed. August 2006. (Status: Proposed Standard)
4579	Session Initiation Protocol (SIP) Call Control - Conferencing for User Agents. A. Johnston, O. Levin. August 2006. (Status: Best Current Practice)
4582	The Binary Floor Control Protocol (BFCP). G. Camarillo, J. Ott, K. Drage. November 2006. (Status: Proposed Standard)
4583	Session Description Protocol (SDP) Format for Binary Floor Control Protocol (BFCP) Streams. G. Camarillo. November 2006. (Status: Proposed Standard)
4585	Extended RTP Profile for Real-time Transport Control Protocol (RTCP)-Based Feedback (RTP/AVPF). J. Ott, S. Wenger, N. Sato, C. Burmeister, J. Rey. July 2006. (Updated by RFC 5506) (Status: Proposed Standard)
4588	RTP Retransmission Payload Format. J. Rey, D. Leon, A. Miyazaki, V. Varsa, R. Hakenberg. July 2006. (Status: Proposed Standard)
4597	Conferencing Scenarios. R. Even, N. Ismail. August 2006. (Status: Informational)
4629	RTP Payload Format for ITU-T Rec. H.263 Video. J. Ott, C. Bormann, G. Sullivan, S. Wenger, R. Even, Ed. January 2007. (Obsoletes RFC 2429) (Updates RFC 3555) (Status: Proposed Standard)
4646	4646 Tags for Identifying Languages. A. Phillips, M. Davis. September 2006. (Obsoletes RFC 3066) (Obsoleted by RFC 5646) (Status: Best Current Practice)

4647	Matching of Language Tags. A. Phillips, M. Davis. September 2006. (Obsoletes RFC 3066) (Also BCP0047) (Status: Best Current Practice)
4648	The Base16, Base32, and Base64 Data Encodings. S. Josefsson. October 2006. (Obsoletes RFC 3548) (Status: Proposed Standard)
4733	RTP Payload for DTMF Digits, Telephony Tones, and Telephony Signals. H. Schulzrinne, T. Taylor. December 2006. (Obsoletes RFC 2833) (Updated by RFC 4734, RFC 5244) (Status: Proposed Standard)
4756	Forward Error Correction Grouping Semantics in Session Description Protocol. A. Li. November 2006. (Obsoleted by RFC 5956) (Status: Proposed Standard)
4787	Network Address Translation (NAT) Behavioral Requirements for Unicast UDP. F. Audet, Ed., C. Jennings. January 2007. (Updated by RFC 6888, RFC 7857) (Also BCP0127) (Status: Best Current Practice)
4796	The Session Description Protocol (SDP) Content Attribute. J. Hautakorpi, G. Camarillo. February 2007. (Status: Proposed Standard)
4855	4855 Media Type Registration of RTP Payload Formats. S. Casner. February 2007. (Obsoletes RFC 3555) (Status: Proposed Standard)
4867	RTP Payload Format and File Storage Format for the Adaptive Multi-Rate (AMR) and Adaptive Multi-Rate Wideband (AMR-WB) Audio Codecs. J. Sjoberg, M. Westerlund, A. Lakaniemi, Q. Xie. April 2007. (Obsoletes RFC 3267) (Status: Proposed Standard)
4916	Connected Identity in the Session Initiation Protocol (SIP). J. Elwell. June 2007. (Updates RFC 3261) (Status: Proposed Standard)
4960	Stream Control Transmission Protocol. R. Stewart, Ed. September 2007. (Obsoletes RFC 2960, RFC 3309) (Updated by RFC 6096, RFC 6335, RFC 7053) (Status: Proposed Standard)
4961	Neighbor Discovery for IP version 6 (IPv6). T. Narten, E. Nordmark, W. Simpson, H. Soliman. September 2007. (Obsoletes RFC 2461) (Updated by RFC 5942, RFC 6980, RFC 7048, RFC 7527, RFC 7559, RFC 8028) (Status: Draft Standard)
4975	The Message Session Relay Protocol (MSRP). B. Campbell, Ed., R. Mahy, Ed., C. Jennings, Ed. September 2007. (Status: Proposed Standard)
4976	Relay Extensions for the Message Sessions Relay Protocol (MSRP). C. Jennings, R. Mahy, A. B. Roach. September 2007. (Status: Proposed Standard)
5027	Security Preconditions for Session Description Protocol (SDP) Media Streams. F. Andreasen, D. Wing. October 2007. (Updates RFC 3312) (Status: Proposed Standard)
5104	Codec Control Messages in the RTP Audio-Visual Profile with Feedback (AVPF). S. Wenger, U. Chandra, M. Westerlund, B. Burman. February 2008. (Updated by RFC 7728, RFC 8082) (Status: Proposed Standard)
5109	RTP Payload Format for Generic Forward Error Correction. A. Li, Ed. December 2007. (Obsoletes RFC 2733, RFC 3009) (Status: Proposed Standard)
5117	RTP Topologies. M. Westerlund, S. Wenger. January 2008. (Obsoleted by RFC 7667) (Status: Informational)
5124	Extended Secure RTP Profile for Real-time Transport Control Protocol (RTCP)-Based Feedback (RTP/SAVPF). J. Ott, E. Carrara. February 2008. (Status: Proposed Standard)
5125	Reclassification of RFC 3525 to Historic. T. Taylor. February 2008. (Obsoletes RFC 3525) (Status: Informational)

5139	Revised Civic Location Format for Presence Information Data Format Location Object (PIDF-LO). M. Thomson, J. Winterbottom. February 2008. (Updates RFC 4119) (Status: Proposed Standard)
5226	Guidelines for Writing an IANA Considerations Section in RFCs. T. Narten, H. Alvestrand. May 2008. (Obsoletes RFC 2434) (Also BCP 0026) (Status: Best Current Practice)
5234	Augmented BNF for Syntax Specifications: ABNF. D. Crocker, Ed., P. Overell. January 2008. (Obsoletes RFC 4234) (Updated by RFC 7405) (Also STD 0068) (Status: Internet Standard)
5239	A Framework for Centralized Conferencing. M. Barnes, C. Boulton, O. Levin. June 2008. (Status: Proposed Standard
5245	Interactive Connectivity Establishment (ICE): A Protocol for Network Address Translator (NAT) Traversal for Offer/Answer Protocols. J. Rosenberg. April 2010. (Obsoletes RFC 4091, RFC 4092) (Updated by RFC 6336) (Status: Proposed Standard)
5246	The Transport Layer Security (TLS) Protocol Version 1.2. T. Dierks, E. Rescorla. August 2008. (Obsoletes RFC 3268, RFC 4346, RFC 4366) (Updates RFC 4492) (Updated by RFC 5746, RFC 5878, RFC 6176, RFC 7465) (Status: Proposed Standard)
5280	Internet X.509 Public Key Infrastructure Certificate and Certificate Revocation List (CRL) Profile. D. Cooper, S. Santesson, S. Farrell, S. Boeyen, R. Housley, W. Polk. May 2008. (Obsoletes RFC 3280, RFC 4325, RFC 4630) (Updated by RFC 6818) (Status: Proposed Standard)
5282	Using Authenticated Encryption Algorithms with the Encrypted Payload of the Internet Key Exchange version 2 (IKEv2) Protocol. D. Black, D. McGrew. August 2008. (Updates RFC 4306) (Status: Proposed Standard)
5285	A General Mechanism for RTP Header Extensions. D. Singer, H. Desineni. July 2008. (Status: Proposed Standard)
5322	Internet Message Format. P. Resnick, Ed.October 2008. (Obsoletes RFC 2822) (Updates RFC 4021) (Updated by RFC 6854) (Status: Draft Standard)
5389	Session Traversal Utilities for NAT (STUN). J. Rosenberg, R. Mahy, P. Matthews, D. Wing. October 2008. (Obsoletes RFC 3489) (Updated by RFC 7350) (Status: Proposed Standard)
5405	Unicast UDP Usage Guidelines for Application Designers. L. Eggert, G. Fairhurst. November 2008. (Also BCP 0145) (Status: Best Current Practice)
5432	Quality of Service (QOS) Mechanism Selection in the Session Description Protocol (SDP). Polk, S. Dhesikan, G. Camarillo. March 2009. (Status: Proposed Standard)
5464	The IMAP METADATA Extension. C. Daboo. February 2009. (Status: Proposed Standard)
5479	Requirements and Analysis of Media Security Management Protocols. D. Wing, Ed., S. Fries, H. Tschofenig, F. Audet. April 2009. (Status: Informational)
5506	Support for Reduced-Size Real-Time Transport Control Protocol (RTCP): Opportunities and Consequences. I. Johansson, M. Westerlund. April 2009. (Updates RFC 3550, RFC 3711, RFC 4585) (Status: Proposed Standard)
5541	Encoding of Objective Functions in the Path Computation Element Communication Protocol (PCEP). JL. Le Roux, JP. Vasseur, Y. Lee. June 2009. (Status: Proposed Standard)

5547	A Session Description Protocol (SDP) Offer/Answer Mechanism to Enable File Transfer. M. Garcia-Martin, M. Isomaki, G. Camarillo, S. Loreto, P. Kyzivat. May 2009. (Status: Proposed Standard)
5576	Source-Specific Media Attributes in the Session Description Protocol (SDP). J. Lennox, J. Ott, T. Schierl. June 2009. (Status: Proposed Standard)
5583	Signaling Media Decoding Dependency in the Session Description Protocol (SDP). T. Schierl, S. Wenger. July 2009. (Status: Proposed Standard)
5626	Managing Client-Initiated Connections in the Session Initiation Protocol (SIP). C. Jennings, Ed., R. Mahy, Ed., F. Audet, Ed. October 2009. (Updates RFC 3261, RFC 3327) (Status: Proposed Standard)
5646	Tags for Identifying Languages. A. Phillips, Ed., M. Davis, Ed. September 2009. (Obsoletes RFC4646) (Also BCP0047) (Status: Best Current Practice)
5750	Secure/Multipurpose Internet Mail Extensions (S/MIME) Version 3.2 Certificate Handling. B. Ramsdell, S. Turner. January 2010. (Obsoletes RFC 3850) (Status: Proposed Standard)
5751	Secure/Multipurpose Internet Mail Extensions (S/MIME) Version 3.2 Message Specification. B. Ramsdell, S. Turner. January 2010. (Obsoletes RFC 3851) (Status: Proposed Standard)
5760	RTP Control Protocol (RTCP) Extensions for Single-Source Multicast Sessions with Unicast Feedback. J. Ott, J. Chesterfield, E. Schooler. February 2010. (Updated by RFC 6128) (Status: Proposed Standard)
5761	Multiplexing RTP Data and Control Packets on a Single Port. C. Perkins, M. Westerlund. April 2010. (Updates RFC 3550, RFC 3551) (Status: Proposed Standard)
5762	RTP and the Datagram Congestion Control Protocol (DCCP). C. Perkins, April 2010. (Updated by RFC 6773) (Status: Proposed Standard)
5763	Framework for Establishing a Secure Real-time Transport Protocol (SRTP) Security Context Using Datagram Transport Layer Security (DTLS). J. Fischl, H. Tschofenig, E. Rescorla. May 2010. (Status: Proposed Standard)
5764	Datagram Transport Layer Security (DTLS) Extension to Establish Keys for the Secure Real-time Transport Protocol (SRTP). D. McGrew, E. Rescorla. May 2010. (Status: Proposed Standard)
5766	Traversal Using Relays around NAT (TURN): Relay Extensions to Session Traversal Utilities for NAT (STUN). R. Mahy, P. Matthews, J. Rosenberg. April 2010. (Status: Proposed Standard)
5853	Requirements from Session Initiation Protocol (SIP) Session Border Control (SBC) Deployments. J. Hautakorpi, Ed., G. Camarillo, R. Penfield, A. Hawrylyshen, M. Bhatia. April 2010. (Status: Informational)
5888	The Session Description Protocol (SDP) Grouping Framework. G. Camarillo, H. Schulzrinne. June 2010. (Obsoletes RFC 3388) (Status: Proposed Standard)
5890	NSIS Protocol Operation in Mobile Environments. T. Sanda, Ed., X. Fu, S. Jeong, J. Manner, H. Tschofenig. March 2011. (Status: Informational)
5891	Authorization for NSIS Signaling Layer Protocols. J. Manner, M. Stiemerling, H. Tschofenig, R. Bless, Ed. February 2011. (Status: Experimental)
5898	Connectivity Preconditions for Session Description Protocol (SDP) Media Streams. F. Andreasen, G. Camarillo, D. Oran, D. Wing. July 2010. (Status: Proposed Standard)

5905	Network Time Protocol Version 4: Protocol and Algorithms Specification. D. Mills, J. Martin, Ed., J. Burbank, W. Kasch. June 2010. (Obsoletes RFC 1305, RFC 4330) (Updated by RFC 7822) (Status: Proposed Standard)
5939	Session Description Protocol (SDP) Capability Negotiation. F. Andreasen. September 2010. (Updated by RFC 6871) (Status: Proposed Standard)
5956	Forward Error Correction Grouping Semantics in the Session Description Protocol. A. Begen. September 2010. (Obsoletes RFC4756) (Status: Proposed Standard)
5974	NSIS Signaling Layer Protocol (NSLP) for Quality-of-Service Signaling. J. Manner, G. Karagiannis, A. McDonald. October 2010. (Status: Experimental)
5996	Internet Key Exchange Protocol Version 2 (IKEv2). C. Kaufman, P. Hoffman, Y. Nir, P. Eronen. September 2010. (Obsoletes RFC 4306, RFC 4718) (Made obsolete by RFC 7296) (Updated by RFC 5998, RFC 6989) (Status: Proposed Standard)
6064	SDP and RTSP Extensions Defined for 3GPP Packet-Switched Streaming Service and Multimedia Broadcast/Multicast Service. M. Westerlund, P. Frojdh. January 2011. (Status: Informational)
6080	A Framework for Session Initiation Protocol User Agent Profile Delivery. D. Petrie, S. Channabasappa, Ed. March 2011. (Status: Proposed Standard)
6120	Extensible Messaging and Presence Protocol (XMPP): Core. P. Saint-Andre. March 2011. (Obsoletes RFC 3920) (Updated by RFC 7590) (Status: Proposed Standard)
6144	Framework for IPv4/IPv6 Translation. F. Baker, X. Li, C. Bao, K. Yin. April 2011. (Status: Informational)
6146	Stateful NAT64: Network Address and Protocol Translation from IPv6 Clients to IPv4 Servers. M. Bagnulo, P. Matthews, I. van Beijnum. April 2011. (Status: Proposed Standard)
6149	MD2 to Historic Status. S. Turner, L. Chen. March 2011. (Obsoletes RFC 1319) (Status: Informational)
6157	IPv6 Transition in the Session Initiation Protocol (SIP). G. Camarillo, K. El Malki, V. Gurbani. April 2011. (Updates RFC 3264) (Status: Proposed Standard)
6184	RTP Payload Format for H.264 Video. Y.-K. Wang, R. Even, T. Kristensen, R. Jesup. May 2011. (Obsoletes RFC3984) (Status Proposed Standard)
6190	RTP Payload Format for Scalable Video Coding. S. Wenger, Y.-K. Wang, T. Schierl, A. Eleftheriadis. May 2011. (Status: Proposed Standard)
6191	Reducing the TIME-WAIT State Using TCP Timestamps. F. Gont. April 2011. (Also BCP0159) (Status: Best Current Practice)
6193	Media Description for the Internet Key Exchange Protocol (IKE) in the Session Description Protocol (SDP). M. Saito, D. Wing, M. Toyama. April 2011. (Status: Informational)
6236	Negotiation of Generic Image Attributes in the Session Description Protocol (SDP). I. Johansson, K. Jung. May 2011. (Status: Proposed Standard)
6238	TOTP: Time-Based One-Time Password Algorithm. D. M'Raihi, S. Machani, M. Pei, J. Rydell. May 2011. (Status: Informational)
6239	6239 Suite B Cryptographic Suites for Secure Shell (SSH). K. Igoe. May 2011. (Status: Informational)
6263	Application Mechanism for Keeping Alive the NAT Mappings Associated with RTP / RTP Control Protocol (RTCP) Flows. X. Marjou, A. Sollaud. June 2011. (Status: Proposed Standard)

6333	Dual-Stack Lite Broadband Deployments Following IPv4 Exhaustion. A. Durand, R. Droms, J. Woodyatt, Y. Lee. August 2011. (Updated by RFC 7335) (Status: Proposed Standard)
6347	Datagram Transport Layer Security Version 1.2. E. Rescorla, N. Modadugu. January 2012. (Obsoletes RFC 4347) (Updated by RFC 7507, RFC 7905) (Status: Proposed Standard)
6406	Session PEERing for Multimedia INTerconnect (SPEERMINT) Architecture. D. Malas, Ed., J. Livingood, Ed. November 2011. (Status: Informational)
6407	The Group Domain of Interpretation. B. Weis, S. Rowles, T. Hardjono. October 2011. (Obsoletes RFC 3547) (Status: Proposed Standard)
6464	A Real-time Transport Protocol (RTP) Header Extension for Client-to-Mixer Audio Level Indication. J. Lennox, Ed., E. Ivov, E. Marocco. December 2011. (Status: Proposed Standard)
6503	Centralized Conferencing Manipulation Protocol. M. Barnes, C. Boulton, S. Romano, H. Schulzrinne. March 2012. (Status: Proposed Standard)
6504	Centralized Conferencing Manipulation Protocol (CCMP) Call Flow Examples. M. Barnes, L. Miniero, R. Presta, S P. Romano. March 2012. (Status: Informational)
6544	6TCP Candidates with Interactive Connectivity Establishment (ICE). J. Rosenberg, A. Keranen, B. B. Lowekamp, A. B. Roach. March 2012. (Status: Proposed Standard)
6724	Locating Services for Calendaring Extensions to WebDAV (CalDAV) and vCard Extensions to WebDAV (CardDAV). C. Daboo. February 2013. (Updates RFC 4791, RFC 6352) (Status: Proposed Standard)
6726	FLUTE - File Delivery over Unidirectional Transport. T. Paila, R. Walsh, M. Luby, V. Roca, R. Lehtonen. November 2012. (Obsoletes RFC 3926) (Status: Proposed Standard)
6849	An Extension to the Session Description Protocol (SDP) and Real-time Transport Protocol (RTP) for Media Loopback. H. Kaplan, Ed., K. Hedayat, N. Venna, P. Jones, N. Stratton. February 2013. (Status: Proposed Standard)
6871	Session Description Protocol (SDP) Media Capabilities Negotiation. R. Gilman, R. Even, F. Andreasen. February 2013. (Updates RFC 5939) (Status: Proposed Standard)
6888	Common Requirements for Carrier-Grade NATs (CGNs). S. Perreault, Ed., I. Yamagata, S. Miyakawa, A. Nakagawa, H. Ashida. April 2013. (Updates RFC4787) (Also BCP0127) (Status: Best Current Practice)
6914	SIMPLE Made Simple: An Overview of the IETF Specifications for Instant Messaging and Presence Using the Session Initiation Protocol (SIP). J. Rosenberg. April 2013. (Status: Informational)
6947	The Session Description Protocol (SDP) Alternate Connectivity (ALTC) Attribute. M. Boucadair, H. Kaplan, R. Gilman, S. Veikkolainen. May 2013. (Status: Informational)
7006	7006 Miscellaneous Capabilities Negotiation in the Session Description Protocol (SDP). M. Garcia-Martin, S. Veikkolainen, R. Gilman. September 2013. (Status: Proposed Standard)
7160	Support for Multiple Clock Rates in an RTP Session. M. Petit-Huguenin, G. Zorn, Ed., April 2014. (Updates RFC 3550) (Status: Proposed Standard)
7195	Session Description Protocol (SDP) Extension for Setting Audio and Video Media Streams over Circuit-Switched Bearers in the Public Switched Telephone Network (PSTN). M. Garcia-Martin, S. Veikkolainen. May 2014. (Status: Proposed Standard)

7225	Discovering NAT64 IPv6 Prefixes Using the Port Control Protocol (PCP). M. Boucadair. May 2014. (Status: Proposed Standard)
7235	Hypertext Transfer Protocol (HTTP/1.1): Authentication. R. Fielding, Ed., J. Reschke, Ed. June 2014. (Obsoletes RFC 2616, RFC 2617) (Status: Proposed Standard)
7258	Pervasive Monitoring Is an Attack. S. Farrell, H. Tschofenig. May 2014. (Also BCP0188) (Status: Best Current Practice)
7296	Internet Key Exchange Protocol Version 2 (IKEv2). C. Kaufman, P. Hoffman, Y. Nir, P. Eronen, T. Kivinen. October 2014. (Obsoletes RFC 5996) (Updated by RFC 7427, RFC 7670, RFC 8247) (Also STD0079) (Status: Internet Standard)
7371	Updates to the IPv6 Multicast Addressing Architecture. M. Boucadair, S. Venaas. September 2014. (Updates RFC 3306, RFC 3956, RF C4291) (Status: Proposed Standard)
7405	Case-Sensitive String Support in ABNF. P. Kyzivat. December 2014. (Updates RFC 5234) (Status: Proposed Standard)
7433	A Mechanism for Transporting User-to-User Call Control Information in SIP. A. Johnston, J. Rafferty. January 2015. (Status: Proposed Standard)
7742	WebRTC Video Processing and Codec Requirements. A.B. Roach. March 2016. (Status: Proposed Standard)
8200	Internet Protocol, Version 6 (IPv6) Specification. S. Deering, R. Hinden. July 2017. (Obsoletes RFC2460) (Also STD0086) (Status: Internet Standard)

Index

A

B

C

D

N

T